T0343172

INVERTEBRATE LEARNING AND MEMORY

INVERTEBRATE LEARNING AND MEMORY

Edited by

RANDOLF MENZEL
Freie Universität Berlin, Berlin, Germany

PAUL R. BENJAMIN
School of Life Sciences, University of Sussex, Brighton, UK

AMSTERDAM • BOSTON • HEIDELBERG • LONDON
NEW YORK • OXFORD • PARIS • SAN DIEGO
SAN FRANCISCO • SINGAPORE • SYDNEY • TOKYO
Academic Press is an imprint of Elsevier

Academic Press is an imprint of Elsevier
32 Jamestown Road, London NW1 7BY, UK
525 B Street, Suite 1800, San Diego, CA 92101-4495, USA

British Library Cataloguing-in-Publication Data
A catalogue record for this book is available from the British Library

Library of Congress Cataloging-in-Publication Data
A catalog record for this book is available from the Library of Congress

ISBN : 978-0-12-415823-8

For information on all Academic Press publications
visit our website at elsevierdirect.com

Typeset by MPS Limited, Chennai, India
www.adi-mps.com

Printed and bound by CPI Group (UK) Ltd, Croydon, CR0 4YY

13 14 15 16 17 10 9 8 7 6 5 4 3 2 1

Contents

8. Issues in Invertebrate Learning Raised by Robot Models

BARBARA WEBB

4 MECHANISMS FROM THE MOST IMPORTANT SYSTEMS

4.1 Nematodes/*Caenorhabditis elegans*

9. Mechanosensory Learning and Memory in *Caenorhabditis elegans*

ANDREA H. McEWAN AND CATHARINE H. RANKIN

10. Molecular and Cellular Circuits Underlying *Caenorhabditis elegans* Olfactory Plasticity

JOY ALCEDO AND YUN ZHANG

11. Thermosensory Learning in *Caenorhabditis elegans*

HIROYUKI SASAKURA AND IKUE MORI

12. Age-Dependent Modulation of Learning and Memory in *Caenorhabditis elegans*

SHIN MURAKAMI

13. Salt Chemotaxis Learning in *Caenorhabditis elegans*

YUICHI IINO

4.2 Mollusks

4.2.1 *Gastropods*

14. A Systems Analysis of Neural Networks Underlying Gastropod Learning and Memory

PAUL R. BENJAMIN

15. Comparison of Operant and Classical Conditioning of Feeding Behavior in *Aplysia*

RICCARDO MOZZACHIODI, DOUGLAS A. BAXTER AND JOHN H. BYRNE

16. Mechanisms of Short-Term and Intermediate-Term Memory in *Aplysia*

ROBERT D. HAWKINS, IGOR ANTONOV AND IKSUNG JIN

17. Synaptic Mechanisms of Induction and Maintenance of Long-Term Sensitization Memory in *Aplysia*

DAVID L. GLANZMAN

18. Roles of Protein Kinase C and Protein Kinase M in *Aplysia* Learning

MARGARET HASTINGS, CAROLE A. FARAH AND WAYNE S. SOSSIN

19. Multisite Cellular and Synaptic Mechanisms in *Hermissenda* Pavlovian Conditioning

TERRY CROW AND NAN GE JIN

20. Molecular and Cellular Mechanisms of Classical Conditioning in the Feeding System of *Lymnaea*

GYÖRGY KEMENES

21. Operant Conditioning of Respiration in *Lymnaea*

KEN LUKOWIAK AND SARAH DALESMAN

22. Associative Memory Mechanisms in Terrestrial Slugs and Snails

ALAN GELPERIN

4.2.2 Cephalopods

23. Observational and Other Types of Learning in *Octopus*

PIERO AMODIO AND GRAZIANO FIORITO

24. The Neurophysiological Basis of Learning and Memory in Advanced Invertebrates

BINYAMIN HOCHNER AND TAL SHOMRAT

25. Learning, Memory, and Brain Plasticity in Cuttlefish (*Sepia officinalis*)

LUDOVIC DICKEL, ANNE-SOPHIE DARMAILLACQ, CHRISTELLE JOZET-ALVES AND CÉCILE BELLANGER

4.3 Crustacea

26. A Multidisciplinary Approach to Learning and Memory in the Crab *Neohelice* (*Chasmagnathus*) *granulata*

DANIEL TOMSIC AND ARTURO ROMANO

4.4 Insects

4.4.1 *Drosophila*

27. *Drosophila* Memory Research through Four Eras

SETH M. TOMCHIK AND RONALD L. DAVIS

28. Visual Learning and Decision Making in *Drosophila melanogaster*

AIKE GUO, HUIMIN LU, KE ZHANG, QINGZHONG REN AND YAH-NUM CHIANG WONG

4.4.2 *Honeybees*

29. In Search of the Engram in the Honeybee Brain

RANDOLF MENZEL

30. Neural Correlates of Olfactory Learning in the Primary Olfactory Center of the Honeybee Brain

JEAN-CHRISTOPHE SANDOZ

31. Memory Phases and Signaling Cascades in Honeybees

ULI MÜLLER

32. Pheromones Acting as Social Signals Modulate Learning in Honeybees

ELODIE URLACHER, JEAN-MARC DEVAUD AND ALISON R. MERCER

33. Extinction Learning and Memory Formation in the Honeybee

DOROTHEA EISENHARDT

34. Glutamate Neurotransmission and Appetitive Olfactory Conditioning in the Honeybee

GÉRARD LEBOULLE

35. Cellular Mechanisms of Neuronal Plasticity in the Honeybee Brain

BERND GRÜNEWALD

36. Behavioral and Neural Analyses of Punishment Learning in Honeybees

STEVANUS RIO TEDJAKUMALA AND MARTIN GIURFA

37. Brain Aging and Performance Plasticity in Honeybees

DANIEL MÜNCH AND GRO V. AMDAM

4.4.3 Ants

38. Learning and Recognition of Identity in Ants

PATRIZIA D'ETTORRE

39. Bounded Plasticity in the Desert Ant's Navigational Tool Kit

RÜDIGER WEHNER AND WOLFGANG RÖSSLER

40. Learning and Decision Making in a Social Context

NIGEL R. FRANKS AND ANA B. SENDOVA-FRANKS

4.4.4 Other Insect Species

41. Olfactory and Visual Learning in Cockroaches and Crickets

MAKOTO MIZUNAMI, YUKIHISA MATSUMOTO, HIDEHIRO WATANABE AND HIROSHI NISHINO

42. Individual Recognition and the Evolution of Learning and Memory in *Polistes* Paper Wasps

ELIZABETH A. TIBBETTS AND MICHAEL J. SHEEHAN

List of Contributors

Joy Alcedo Friedrich Miescher Institute for Biomedical Research, Basel, Switzerland; Wayne State University, Detroit, Michigan

Gro V. Amdam Norwegian University of Life Sciences, Ås, Norway; Arizona State University, Tempe, Arizona

Piero Amodio Associazione Cephalopod Research, CephRes-ONLUS, Napoli, Italy

Igor Antonov Columbia University, New York, New York

Douglas A. Baxter The University of Texas Medical School at Houston, Houston, Texas

Cécile Bellanger Normandie Univ, UCBN, GMPc, EA 4259, F-14032 Caen, France; Centre de Recherches en Environnement Côtier, Station Marine de l'UCBN, 14530 Luc-sur-Mer, France

Paul R. Benjamin University of Sussex, Brighton, United Kingdom

John H. Byrne The University of Texas Medical School at Houston, Houston, Texas

Enrico Cataldo Università di Pisa, Pisa, Italy

Yah-Num Chiang Wong Institute of Neuroscience, Shanghai Institutes for Biological Sciences, Chinese Academy of Sciences, Shanghai, China

Terry Crow University of Texas Medical School, Houston, Texas

Sarah Dalesman University of Calgary, Calgary, Alberta, Canada

Anne-Sophie Darmaillacq Normandie Univ, UCBN, GMPc, EA 4259, F-14032 Caen, France; Centre de Recherches en Environnement Côtier, Station Marine de l'UCBN, 14530 Luc-sur-Mer, France

Ronald L. Davis The Scripps Research Institute Florida, Jupiter, Florida

Patrizia d'Ettorre University of Paris 13, Sorbonne Paris Cité, Villetaneuse, France

Jean-Marc Devaud National Center for Scientific Research, University Paul Sabatier, Toulouse, France

Ludovic Dickel Normandie Univ, UCBN, GMPc, EA 4259, F-14032 Caen, France; Centre de Recherches en Environnement Côtier, Station Marine de l'UCBN, 14530 Luc-sur-Mer, France

Sören Diegelmann Leibniz Institut für Neurobiologie (LIN), Abteilung Genetik von Lernen und Gedächtnis, Magdeburg, Germany

Dorothea Eisenhardt Freie Universität Berlin, Berlin, Germany

Carole A. Farah McGill University, Montreal, Quebec, Canada

André Fiala Georg-August-University of Göttingen, Göttingen, Germany

Graziano Fiorito Associazione Cephalopod Research, CephRes-ONLUS, Napoli, Italy; Stazione Zoologica Anton Dohrn, Napoli, Italy

Nigel R. Franks University of Bristol, Bristol, United Kingdom

Alan Gelperin Princeton University, Princeton, New Jersey; and Monell Chemical Senses Center, Philadelphia, Pennsylvania

Bertram Gerber Otto von Guericke Universität Magdeburg, Institut für Biologie, Magdeburg, Germany; Leibniz Institut für Neurobiologie (LIN), Abteilung Genetik von Lernen und Gedächtnis, Magdeburg, Germany; Center for Behavioral Brain Science, Magdeburg, Germany

Martin Giurfa Université de Toulouse, Centre de Recherches sur la Cognition Animale, Toulouse, France; Centre National de la Recherche Scientifique, Centre de Recherches sur la Cognition Animale, Toulouse, France

David L. Glanzman University of California at Los Angeles, Los Angeles, California

Bernd Grünewald Goethe-Universität Frankfurt am Main, Germany

Aike Guo Institute of Neuroscience, Shanghai Institutes for Biological Sciences, Chinese Academy of Sciences, Shanghai, China; Institute of Biophysics, Chinese Academy of Sciences, Beijing, China

Margaret Hastings McGill University, Montreal, Quebec, Canada

Robert D. Hawkins Columbia University, New York, New York; New York State Psychiatric Institute, New York, New York

M. Heisenberg University of Würzburg, Würzburg, Germany

Binyamin Hochner Hebrew University, Jerusalem, Israel

Yuichi Iino The University of Tokyo, Tokyo, Japan

Iksung Jin Columbia University, New York, New York

Nan Ge Jin University of Texas Medical School, Houston, Texas

Christelle Jozet-Alves Normandie Univ, UCBN, GMPc, EA 4259, F-14032 Caen, France; Centre de Recherches en Environnement Côtier, Station Marine de l'UCBN, 14530 Luc-sur-Mer, France

György Kemenes University of Sussex, Brighton, United Kingdom

Gérard Leboulle Freie Universität Berlin, Berlin, Germany

Huimin Lu Institute of Biophysics, Chinese Academy of Sciences, Beijing, China

Ken Lukowiak University of Calgary, Calgary, Alberta, Canada

Yukihisa Matsumoto Hokkaido University, Sapporo, Japan

Andrea H. McEwan University of British Columbia, Vancouver, British Columbia, Canada

Randolf Menzel Freie Universität Berlin, Berlin, Germany

Alison R. Mercer University of Otago, Dunedin, New Zealand

Birgit Michels Leibniz Institut für Neurobiologie (LIN), Abteilung Genetik von Lernen und Gedächtnis, Magdeburg, Germany

Makoto Mizunami Hokkaido University, Sapporo, Japan

Ikue Mori Nagoya University, Nagoya, Japan

Riccardo Mozzachiodi Texas A&M University–Corpus Christi, Corpus Christi, Texas

Uli Müller Saarland University, Saarbrücken, Germany

Daniel Münch Norwegian University of Life Sciences, Ås, Norway

Shin Murakami Touro University California, Vallejo, California

Hiroshi Nishino Hokkaido University, Sapporo, Japan

Wolfgang Rössler University of Würzburg, Würzburg, Germany

Catharine H. Rankin University of British Columbia, Vancouver, British Columbia, Canada

Qingzhong Ren Institute of Neuroscience, Shanghai Institutes for Biological Sciences, Chinese Academy of Sciences, Shanghai, China

Thomas Riemensperger Georg-August-University of Göttingen, Göttingen, Germany

Arturo Romano Universidad de Buenos Aires, CONICET, Argentina

Jürgen Rybak Max Planck Institute for Chemical Ecology, Jena, Germany

Jean-Christophe Sandoz CNRS, Gif-sur-Yvette, France

Hiroyuki Sasakura Nagoya University, Nagoya, Japan

Timo Saumweber Leibniz Institut für Neurobiologie (LIN), Abteilung Genetik von Lernen und Gedächtnis, Magdeburg, Germany

Michael Schleyer Leibniz Institut für Neurobiologie (LIN), Abteilung Genetik von Lernen und Gedächtnis, Magdeburg, Germany

Ana B. Sendova-Franks University of the West of England, Bristol, United Kingdom

Michael J. Sheehan University of Michigan, Ann Arbor, Michigan

Tal Shomrat Hebrew University, Jerusalem, Israel

Wayne S. Sossin McGill University, Montreal, Quebec, Canada

Stevanus Rio Tedjakumala Université de Toulouse, Centre de Recherches sur la Cognition Animale, Toulouse, France; Centre National de la Recherche Scientifique, Centre de Recherches sur la Cognition Animale, Toulouse, France

Elizabeth A. Tibbetts University of Michigan, Ann Arbor, Michigan

Seth M. Tomchik The Scripps Research Institute Florida, Jupiter, Florida

Daniel Tomsic Universidad de Buenos Aires, CONICET, Argentina

Elodie Urlacher University of Otago, Dunedin, New Zealand; National Center for Scientific Research, University Paul Sabatier, Toulouse, France

Hidehiro Watanabe Fukuoka University, Fukuoka, Japan

Barbara Webb The University of Edinburgh, Edinburgh, United Kingdom

Rüdiger Wehner University of Zürich, Zürich, Switzerland

Ke Zhang Institute of Neuroscience, Shanghai Institutes for Biological Sciences, Chinese Academy of Sciences, Shanghai, China

Yun Zhang Harvard University, Cambridge, Massachusetts

SECTION 1

INTRODUCTION

CHAPTER

1

Beyond the Cellular Alphabet of Learning and Memory in Invertebrates

Randolf Menzel[*] and Paul R. Benjamin[†]

[*]Freie Universität Berlin, Berlin, Germany [†]University of Sussex, Brighton, United Kingdom

INTRODUCTION

In 1984, Hawkins and Kandel published a seminal paper titled "Is There a Cell-Biological Alphabet for Simple Forms of Learning?".[1] Based on their early findings of the cooperative regulation of adenylyl cyclase in sensory neurons of *Aplysia*, an overarching concept was presented which opened our mind to molecular mechanisms of experience-dependent neural plasticity. Several basic forms of nonassociative and associative learning (habituation, sensitization, and classical conditioning) were explained on the level of rather simple molecular reaction cascades in specific neurons. At that time, these were radical ideas, and even today we struggle with the question whether cognitive faculties such as learning and memory formation can be reduced to ubiquitous cellular functions, and what such a reduction might mean. The concepts presented in this paper were also radical in the sense that they broke with the speculation that the information of acquired memories is stored in molecules like RNA. Meanwhile, it is well accepted in neuroscience that neural circuits acquire new information by changing network properties on the level of specified neurons and their synaptic connections. Multiple key elements contribute to these adaptations, and it is the task of today's neuroscience to unravel the complex hierarchies of interactions from the molecular to the systems level in solving the problem of predicting future behavior from experience in the past.

Invertebrate nervous systems, because of their relative simplicity, offer significant advantages for multidisciplinary molecular, cellular, genetic, and behavioral investigations of the mechanisms underlying learning, memory formation, and memory retrieval. Memory formation in invertebrates occurs within small circuits containing a few hundred neurons rather than the millions of the mammalian brain, and it is often possible to study the role of individual neurons that play a key role in these adaptive neural processes. Because of this tractability of invertebrate model systems, many important discoveries have been made in invertebrates (e.g., the role of second messengers such as cAMP, protein kinases, and transcription factors such as CREB (Chapters 15–18, 20, 27, 31, and 35); the role of neural plasticity in addition to and separate from synaptic plasticity (Chapters 14, 15, 19, 20, and 35); the function of identified neurons in the evaluating pathways (examples in many chapters in Section 4); and the differential role of circuits in storing and retrieving memory (Chapters 27, 29, and 31)) that have been found to be generally applicable to higher organisms. The discovery that synaptic plasticity involves both pre- and postsynaptic interactive processes characteristic of Hebbian long-term potentiation in vertebrates again emphasizes the general importance of the results from invertebrate systems (Chapters 17 and 24). Furthermore, a good deal of information is emerging on the molecular and neural mechanisms underlying the different phases of memory consolidation (from short-term to intermediate-term and long-term memory). Training triggers a cascade of molecular events with phase-dependent requirements for protein kinases (Chapters 16–18, 27, 31, and 35) and neuropeptides (Chapter 17). A fascinating recent discovery is that a self-sustaining prion-like protein ApCPEB (*Aplysia* cytoplasmic polyadenylation element binding protein) is involved in memory maintenance promoting persistent facilitation of synaptic transmission (Chapter 17).

Invertebrate Learning and Memory.
DOI: http://dx.doi.org/10.1016/B978-0-12-415823-8.00001-0

BEYOND THE CELLULAR ALPHABET: CIRCUIT AND NETWORK LEVELS OF ANALYSIS, THE NECESSARY STEP

The previous cellular studies emphasized changes at a single locus. However, there is increasing evidence that most forms of learning in invertebrates involve changes at multiple sites in the brain (e.g., Chapters 14, 19, 25, 27, and 29). The identification of these multiple sites of plasticity requires a systems approach to the analysis of learning and memory, which makes it important to first identify the electrical changes that result from training and then attempt to relate these changes to behavioral plasticity. The role of individual neurons and their synaptic connectivity can then be investigated in the context of behavior in a 'top-down' approach. The ability to identify changes in sensory, interneuronal, and motoneuron pathways contributing to the learning process has been one of the successes of this systems approach to invertebrate learning and memory. From this work, it is realized that we cannot hope to understand the nature of the 'engram' without a knowledge of the electrical and cellular changes at all levels of the networks involved in establishing and retrieving the engram (Chapters 14, 19, 29, and 36). This systems analysis is far from complete, especially in the more complex behaviors of social insects and cephalopod mollusks, but considerable progress has been made. Computational modeling of learning networks is an important component of the systems approach (Chapter 7), and links to robotics (Chapter 8) are providing a complementary type of approach to understanding adaptive behavior in invertebrates.

Will it be ever possible to "read" the content of the memory trace, the engram? This question requires a shift from the analysis of mechanisms to that of processes. Finally, we want to understand where and how particular memory contents are stored, and how they are activated for behavioral control. A helpful but rather simple-minded approach is to visualize the changes of neural functions in the course of learning. Such an approach is manifested in the search for structural plasticity as induced during long-term memory formation (Chapter 31) or in the calculation of changes of patterns in neural activity before and after learning (Chapters 14, 20, and 29). These patterns of changes, both in time and in space, should constitute the engram as read by the complete nervous system of the respective animal, and may be even accessible to the human mind for elementary forms of learning that lead to memory traces in restricted parts of the nervous system. However, even in invertebrates the engram will usually involve several to many circuits distributed throughout the nervous system, making it very difficult to relate stored information to neural circuits. The conceptual and experimental problems should not demotivate us to hunt for the engram by shifting our attempts from single neuron analysis to network analyses. Indeed, such a shift appears achievable with molecular-genetic techniques in *Caenorhabditis elegans* (Chapters 9–13) and *Drosophila* (Chapters 5 and 27).

DO INVERTEBRATES HAVE COGNITIVE ABILITIES?

Despite having small brains, invertebrates show a remarkable ability to carry out complex tasks in their natural environment, to learn and to form long-term memory through stepwise consolidation processes. Suggesting that invertebrate animals have cognitive abilities implies that they have sophisticated behavioral capabilities that transcend the elementary forms of adaptive responses to environmental changes. It requires the selection of different options from a repertoire of learned behaviors that allows an animal to respond selectively to novel external stimuli. It is clear that insects, particularly social insects, have this kind of capability (Chapter 3) but also cephalopod mollusks (Chapters 23 and 25) and perhaps terrestrial slugs (Chapter 22). It is interesting that all three groups of animals have special learning 'centers' in the brains (mushroom body in insects, vertical lobes in octopus, and procerebral lobes in terrestrial slugs) with intricate interneuronal organization that would be required for complex information processing at the neural level. Cognitive behavior is suggested by examples of exploration, instrumental and observation learning, expectation, learning in a social context, and planning of future actions (Chapters 3, 23, and 25). In some examples, such as second-order conditioning, it is possible to suggest neural network mechanisms derived from the neural network models of simple forms of first-order conditioning (Chapter 14), but in most examples we have little or no knowledge of the neural mechanisms involved. Computational modeling where explicit mechanisms have to be proposed may be useful here (Chapters 7 and 8). Until we have more detailed information on the mechanisms and processes involved in examples of cognitive behaviors, it will be difficult to know whether cognitive mechanisms have unique information processing compared with 'noncognitive' adaptive behaviors. Could it be that the most complex forms of invertebrate cognitive behavior represent a loop too difficult to close in terms of neural mechanisms? As in other disciplines, cognitive neuroscience advances depend on new methods and new concepts, probably in this order. Invertebrate neuroscience works at the forefront of both aspects, methods and concepts (Chapters 5, 6, and 8). Transgenic or transfected animals will offer opportunities particularly when combined with recording techniques. We are facing an exciting future of invertebrate neuroscience, and we hope this volume will help to prepare for these endeavors.

Reference

1. Hawkins RD, Kandel ER. Is there a cell-biological alphabet for simple forms of learning? *Psychol Rev*. 1984;91:375–391.

SECTION 2

CONCEPTS OF INVERTEBRATE COMPARATIVE COGNITION

CHAPTER

2

Action Selection
The Brain as a Behavioral Organizer

M. Heisenberg
University of Würzburg, Würzburg, Germany

INTRODUCTION

Animals have to generate the right behavior at the right moment and the right place. Action selection is the main task of the brain. Behavioral brain science, particularly with regard to small animals such as insects, has attributed a pivotal role in this process to the sensory stimuli, which are thought to be transformed into motor commands to initiate behavior. Here, I offer an alternative view based on the autonomy and, more precisely, the self of an animal. I argue that most behaviors are active and that animals, like humans, are agents.[1,2] Sensory stimuli trigger behavior only in exceptional cases. Some guide the ongoing behavior, and probably all of them are evaluated in the search process.

In this chapter, the concept of the brain as a behavioral organizer is examined. It is shown that as soon as one takes the organism as an autonomous agent, the behavioral organizer can readily replace the stimulus–response doctrine. In the new concept, there is no need to assume that all behaviors occurring result from sets of sensory stimuli that triggered them. Behavior is initiated from within, spontaneously, actively. This is part of the organism's self-ness. It will also be apparent that there is no alternative to the brain as a behavioral organizer: If an organism has many behavioral options, the brain has to search for the right one. This can be a most demanding task.

The behaviors that matter for action selection come in discrete modules. These are readily discernible in insects. A behavioral module in its simplest form is a precisely orchestrated sequence of muscular contractions and relaxations driven by a central pattern generator (CPG).[3] Most of the time, a CPG is silent. Its motor pattern is inhibited. Most behavioral modules are mutually exclusive. While inhibited, we call a module a behavioral option. Relief from inhibition of the CPG releases the motor pattern.

From birth to death, the overt life of an animal or human is a sequence of activated behavioral modules. Terminating an ongoing module and selecting the module to be activated next is what a brain does. If an animal has many options, this process may well be a search rather than just a choice. The search is shaped by the network of motivational states, dispositions, moods and feelings, sensory information, attention, memories, the current physiological and anatomical state and availability of the options (modules), etc. Because the search evaluates not so much the options per se but, rather, their inferred consequences, these must be represented in the brain and must be continuously updated. This is why the search is so demanding.

The search among the options includes their ranking. Some options may linger in the background for most of a lifetime with low priority, waiting for the right occasion. On the other hand, the search must be organized such that many options can be activated at any moment. A particular combination of sensory stimuli may prompt the immediate release of a certain module (stimulus–response). Once the search process has converged to a choice among few options, we call it a decision.

To compare the inferred consequences of behavioral options, the brain has to quantify their benefits and costs on a common value scale and has to store the probabilities for these to become true. Behaviors as different as, for instance, hiding, courting, and migration are difficult to compare. Immediate and delayed benefits of an option may have to be weighed against

Invertebrate Learning and Memory.
DOI: http://dx.doi.org/10.1016/B978-0-12-415823-8.00002-2

each other, and in some cases the potential value of a behavioral module can only be assessed if one or even multiple follow-up rounds of choice are included in the consideration.

It has long been recognized that action selection must be based on the inferred consequences of these actions.[4] We should be impressed how well this principle works in the brain, but it also requires considerable error-friendliness.

In the following, I focus on a particular brain, that of the fly *Drosophila melanogaster*, assuming that all auto-mobile animals with brains have the same basic organization of behavior. Because the fly brain, based on the number of neurons, is approximately 1 million times smaller than the human brain, its organization may eventually be more readily understood.

BEHAVIORAL MODULES

Let us discuss the fly's known repertoire of behavioral modules: heartbeat, opening and closing of spiracles, swallowing, egg laying, eclosion, feeding, defecating, flight (hover flight or forward flight), various postures, flight start, landing, walking at different gaits, jumping, climbing, different kinds of grooming, courtship with its several components, copulation, and fighting at different levels of escalation. In addition, one observes movements of body parts, such as the head, antennae, maxillary palps, proboscis, single legs, and abdomen. Surely, this list is not complete.

Not all modules are part of the selection process. For instance, heartbeat goes on throughout life, and eclosion occurs only once at the onset. To what degree modules such as opening and closing of the spiracles, swallowing, defecating, and egg laying are part of action selection is not known. Might flies be able to control the excretion of pheromones? In most cases of modules involving legs and wings, only one at a time can be activated, with some obvious exceptions such as landing and flight. 'No behavior' can be any of at least three states: rest, sleep, and hibernation. In other insects, a further state of 'no behavior'—freezing—is observed.

The prevalent and probably oldest module in the context of action selection is locomotion (walking and running). Walking comes in bouts of variable duration and occurs at different speeds. The fly has two walking patterns—a tetrapod gait for lower speeds and a tripod gait for higher speeds. Even with these two motor patterns, the range of speeds is small.[5]

It may be useful and even necessary to differentiate some of the options according to the goals the fly pursues with them. This brings up the peculiar quality of space, which contains near to infinitely many locations. Taking into account that a fly can walk or fly to any of them, one arrives at a large number of behavioral options among which the brain must search. The same module (e.g., walking) may be activated with different goals and in different contexts, with very different inferred consequences.

To some extent, flies can also compose behavioral sequences from simpler modules and improve them by practice.[6] If they have to climb across a wide gap, they place their hind and middle legs strategically in order to reach out as far as possible. This again increases the number of different behavioral options. Finally, flies can even try out new combinations and sequences of modules in new situations. Already for a fly, the repertoire of options is open, and action selection based on the inferred consequences of the actions is a highly involved process.

OUTCOME EXPECTATION

As stated previously, for suitable options to be selected, these must be represented in the brain by their inferred consequences. Such representations have been called 'outcome expectations'.[7] The term *expectation*, which is used in a similar way in learning/memory research,[8] alludes to the unpredictability of the future. It tries to capture the possible deviation of the later realization from what is inferred. Even the effects of the most stereotype sensorimotor reflexes cannot be predicted with certainty.

For some modules, such as feeding or copulation, the outcome is part of the behavior. They are called consummatory acts. Their immediate consequences seem certain: food uptake and transfer of sperm, respectively. However, their longer range outcome is not. Food may be poisonous, the copulation partner may have deleterious genes, etc. Moreover, the consummatory act has costs. At the feeding site or during copulation, the animal may be exposed to predators and other dangers. The same outcome may be reached with lower costs at a later time. This shows that even with consummatory acts, difficult trade-offs may be involved that need to be taken into account during the search process.

There are many examples of the importance of outcome expectations in the search process in *Drosophila*. I mention only a few. In dark, narrow tubes, threatened flies are strongly attracted by near-ultraviolet (UV) light. Arguably, this is because in nature the sky is the only source of bright UV-rich light, and the fly wants to escape from the tube and get into flight. This interpretation is corroborated by the observation that rendering the fly flightless but fully capable of walking—for example, by cutting the wings—abolishes the high attractiveness of UV light. Why? Getting into the open without being able to fly away would put it at higher predation risk than in the tube.[9] The second example is much simpler: *Drosophila* trained to expect food with a certain odor will later search for the food in the odor's vicinity, but this conditioned search does not occur if the animal is satiated.[10]

Walking gets the fly to a new location in space offering new opportunities and risks. Surely, the fly must have an inherited (phylogenetically fixed) 'expectation' of these very general consequences of locomotion. If enclosed without food and water and left to itself, the fly switches between rest and activity periods in a characteristic sequence of ordered randomness.[11] This may be the optimal trade-off between saving energy and not missing a chance for escape.

As mentioned previously, locomotion most often occurs with a goal such as escape, finding food, finding shelter, finding a mate, or just reaching a strategic position (e.g., open space). These different options may have vastly different inferred consequences. How goals influence the search process is a major theme in behavioral research.

It is well-known that while occurring, walking and flight are accompanied by outcome expectations. For instance, if a tethered fly at a torque meter surrounded by a vertical drum with black and white stripes is made to control with its yaw torque the rotations of the drum in a horizontal plane (flight simulator), it can be shown to 'expect' the drum to rotate against the direction of its intended turns.[12] In this situation, the fly can be trained by an infrared laser beam and chromatic changes in the illumination to avoid certain orientations relative to the drum. The fly turns away from the sectors where it would be heated before reaching them. It anticipates the heat as the outcome of a certain turning maneuver.[13]

Flies can learn not only about the consequences of their behaviors but also about conditions of the outside world (classical learning[14]). This kind of adaptation in the search 'space' is not associated with a particular behavioral option but with many or all of them.

Strauss and Pick[6] showed that before crossing a gap, flies estimate the success of their climbing attempt. Because the width of a gap they can cross depends on their body size, they have to individually learn how large a gap they might still be able to cross (B. Kienitz, T. Krause, and R. Strauss, personal communication). In ants, locomotion has been shown to have an outcome expectation regarding step size because these insects estimate the wrong distance traveled if their step size is manipulated.[15]

Motivational states such as hunger, thirst, fear, rage, love, or tiredness are thought to be important organizers of the search process. In *Drosophila*, it has been difficult to study them. Arousal has been defined as a period of increased and faster walking after a stressful experience.[16] There is a strong case for pain.[17] Are flies in love while courting? We still do not know. Courtship has been studied in flies for more than a century. The readiness of females to copulate is increased by male courtship song, and this conditioning effect lasts longer than the sound.[18] Is there a hunger state in flies? This is suggested by the observation that food deprivation has a variety of behavioral effects. For instance, it raises locomotor activity, changes the attractiveness of certain odors, and lowers the threshold for the attractiveness of sugar. Moreover, neuropeptide F (dNPF), which regulates feeding and is released from the respective neurons under starvation, is also involved in the readout of appetitive odor memories.[10]

To conclude this section, let us return to the role of sensory stimuli in the search process. Signals from all sensory modalities arrive in the brain all the time. The brain makes use of them in its search for the right behavior. They need to be interpreted with regard to their behavioral significance. They may modify the outcome expectations and change the hierarchy of the behavioral options for the next selection. They can affect motivational states and cause a reassessment of goals. They provide what I have called 'orientedness,' the disposition to behave with the right changes in orientation and position when necessary. Stimuli triggering a behavior are the rare exception. Such sensorimotor reflexes mostly serve in emergencies. We can safely assume that flies, like humans, avoid settling in environments in which these are frequently needed. Surely, guiding ongoing behavior is a prevalent task of the sensors.

THE ACTIVE BRAIN

Whereas it is evident that the incoming sensory stimuli serve the brain in its search for the right behavior, it may be less obvious why initiating activity is so important in this process. The answer is that behavior deals with an open future that is only partially predictable. For flies, as for humans, most situations are unique and often new, with risks and opportunities. Flies need to be inventive and to take their chances. For many situations, the search does not converge onto a single behavioral option with a clearly superior outcome expectation. Two or more options may score the same low. Often, but not always, the right timing of a behavior is important. Flies, like humans, have to solve problems by trying something new, activating an option never activated in such a situation before.[9] The importance of activity is well expressed by the 'golden rule' of behavior: Do not wait until you are forced by circumstance. It applies not only to flies and humans but also to companies and countries.

Because animals and humans are autonomous agents, their behavior may occur spontaneously, by itself, involving the catalytic element of chance. More than for most other animals, behavioral research on *Drosophila* provides compelling evidence that the search process makes use of chance as an adaptive element. One such example is the decision conflict: If flies in a narrow tube with a light on one end (as discussed previously) are shaken to the other end, their probability of immediately 'running' or 'not running' toward the light is 80 and 20%, respectively. If the 'runners' and 'nonrunners' are separated and the experiment is repeated with either group, one finds in both of them again runners and nonrunners with the same 80/20 ratio as in the first experiment.[19] In each round, the search process arrives at the same ambiguous answer and the flies again take their chance.

A second example is the fly's ability to solve problems by trying out. With little technical sophistication, one can design a setup in which the fly has to generate a certain behavior chosen by the experimenter in order to switch off a threatening heat or electric shock. In most cases, the fly quickly finds out what to do.[20]

In tethered flight, the fly can beat its wings, turn its abdomen, lift its antennae, and move its legs but cannot execute its intended flight maneuvers. It must be trying to escape this disastrous situation. Under these conditions, if one records the fly's yaw torque and forward thrust, one observes that the fly continuously modulates these flight forces. In addition, it probably activates many other behaviors compatible with flight. What one observes is that it immediately takes advantage of any consequences of its behaviors that it can detect. If it measures a coincidence between one of its actions and an incoming sensory signal, it tries to confirm it and, if successful, uses this minute degree of freedom to make the best of it. The feedback signal may hint at how to further improve the situation for an eventual escape.[21] These examples show that the search process to a large extent involves learning about the consequences of one's own behavior. Much of what happens is new. Fortunately, this truism applies not only to the outside world but also to the universe of behavioral options and their consequences.

A tethered fly may be in a state of maximal arousal, and its continuous activity may not be representative. However, the sequence of turning attempts in this situation can be shown to be a well-organized probabilistic behavior pattern.[22] Also, walking flies left alone in a dark chamber organize their time between activity and rest periods.[11] Behavioral science has long adopted methods to account for the continuous stream of activity in observational data (ethograms[23]), even under more natural conditions.

ACTION SELECTION

I have argued that action selection, the search for the right behavior, is the basic task of the fly brain. Many of the behavioral properties mentioned previously are in one way or another part of this process. However, rarely has action selection explicitly been addressed. Kravitz's group[24,25] studied it in male–male interactions and identified three octopaminergic neurons in the subesophageal ganglion that influence the decision between courtship and aggression. Guo and co-workers[26] investigated behavioral choices in stationary flight by setting up a decision conflict. They trained flies to turn toward or away from two kinds of landmarks that differed in two features, color and shape. Once one landmark had become attractive and the other repulsive, they switched the combinations of features for the subsequent retrieval test such that now all landmarks had one attractive and one repulsive feature. The strengths of the memory traces had been balanced to make the landmarks equally attractive for the flies after the switch. Surprisingly, when for the retrieval test the intensity of the colors was increased, the flies opted for this feature and ignored the shapes. Conversely, if the shapes had been made more distinct for the test, flies disregarded the colors.[27] Flies did not take into account that the changed features differed from those presented during acquisition. The search process appeared to be governed more by the salience of the sensory stimuli than by the reliability of the memory traces.

These examples show how little we know about the search process. It is not yet possible to describe its outline, even for a brain as small as that of the fly.

CONCLUSIONS

This chapter does not pretend to reinvent behavioral brain science. In the search process, the brain serves all that brains are known to serve: It stores and integrates experiences from the past to secure the future; it mitigates in advance the squeezes of demands and constraints; and, among others, it establishes and preserves—in humans over many decades—the behavioral uniqueness and continuity of the individual. As far as mental processes have an evolutionary origin and can be ascribed to the brain, they are taken in this approach as properties of the search process. Motivational states such as moods, emotions, and feelings direct the search process to certain regions of the search space and anything like thinking can be understood as a landscaping activity in this space for future searches. With time, brains have evolved increasingly more of these landscaping activities. For instance, an egocentric representation of outside space with its general properties of up and down, right and left, front and rear, and far and close would improve the search in many circumstances. It would improve 'orientedness' to which I alluded to previously. Likewise, the representation of time, as provided by the circadian clock, would be understood as a property of the search process.

As an overarching framework of functional brain science, this perspective tells us to understand sensory integration, motivational control, learning and memory, cognition, intentions, selective attention, decision making, motor programming, and planning as one integrated process: trying to do the right thing. Animals and humans are autonomous, auto-mobile agents. Activity is the most fundamental property of behavior, deserving more recognition than it has received in the past.

Acknowledgments

I thank Bertram Gerber and Randolf Menzel, who contributed ideas and focus.

References

1. Heisenberg M. Initiale Aktivität und Willkürverhalten bei Tieren. *Naturwiss*. 1983;70:70–78.
2. Heisenberg M. Voluntariness (Willkürfähigkeit) and the general organization of behaviour. In: Greenspan RJ, Kyriacou CP, eds. *Flexibility and Constraint in Behavioral System*. Hoboken, NJ: Wiley; 1994:147–156.
3. Bässler U. On the definition of central pattern generator and its sensory control. *Biol Cybern*. 1986;54:65–69.
4. Elsner B, Hommel B. Effect anticipation and action control. *J Exp Psychol*. 2001;27:229–240.
5. Strauss R, Heisenberg M. Coordination of legs during straight walking and turning in *Drosophila melanogaster*. *J Comp Physiol A*. 1990;167:403–412.
6. Pick S, Strauss R. Goal-driven behavioral adaptations in gap-climbing *Drosophila*. *Curr Biol*. 2005;15:1473–1478.
7. Schleyer M, Saumweber T, Nahrendorf W, et al. A behavior-based circuit model of how outcome expectations organize learned behavior in larval *Drosophila*. *Learn Mem*. 2011;18:639–653.
8. Rescorla RA, Wagner AR. A theory of pavlovian conditioning: variations in the effectiveness of reinforcement and non-reinforcement. In: Black AH, Prokasy WF, eds. *Classical Conditioning II: Current Research and Theory*. New York: Appleton; 1972.
9. Heisenberg M, Wolf R. In: Braitenberg V, ed. *Vision in Drosophila. Vol. XII, of Studies of Brain Function*. New York: Springer; 1984.
10. Krashes MJ, DasGupta S, Vreede A, White B, Armstrong JD, Waddell S. A neural circuit mechanism integrating motivational state with memory expression in *Drosophila*. *Cell*. 2009;139:416–427.
11. Martin J-R, Ernst R, Heisenberg M. Temporal pattern of locomotor activity in *Drosophila melanogaster*. *J Comp Physiol A*. 1999;184:73–84.
12. Heisenberg M, Wolf R. Reafferent control of optomotor yaw torque in *Drosophila melanogaster*. *J Comp Physiol A*. 1988;163:373–388.
13. Wolf R, Heisenberg M. Visual space from visual motion: turn integration in tethered flying *Drosophila*. *Learn Mem*. 1997;4:318–327.
14. Tang S, Wolf R, Xu S, Heisenberg M. Visual pattern recognition in *Drosophila* is invariant for retinal position. *Science*. 2004;305:1020–1022.
15. Wittlinger M, Wehner R, Wolf H. The ant odometer: stepping on stilts and stumps. *Science*. 2006;312:1965–1967.
16. Lebestky T, Jung-Sook T, Chang C, et al. Two different forms of arousal in *Drosophila* are oppositely regulated by the dopamine D1 receptor ortholog DopR via distinct neural circuits. *Neuron*. 2009;64:522–536.
17. Al-Anzi B, Tracey Jr WD, Benzer S. Response of *Drosophila* to wasabi is mediated by painless, the fly homolog of mammalian TRPA1/ANKTM1. *Curr Biol*. 2006;16:1–7.
18. Kowalski S, Aubin T, Martin J-R. Courtship song in *Drosophila melanogaster*: a differential effect on male–female locomotor activity. *Can J Zool*. 2004;82:1258–1266.
19. Brown W, Haglund K. The landmark interviews. Bringing behavioral genes to light. *J NIH Res*. 1994;6:66–73.
20. Heisenberg M, Wolf R, Brembs B. Flexibility in a single behavioral variable of *Drosophila*. *Learn Mem*. 2001;8:1–10.
21. Wolf R, Heisenberg M. Basic organization of operant behavior as revealed in *Drosophila* flight orientation. *J Comp Physiol A*. 1991;169:699–705.
22. Maye A, Hsieh CH, Sugihara G, Brembs B. Order in spontaneous behaviour. *PLoS ONE*. 2007;2:e443.
23. Reif M, Linsenmair KE, Heisenberg M. Evolutionary significance of courtship conditioning in *Drosophila melanogaster*. *Anim Behav*. 2002;63:143–155.
24. Certel SJ, Savella MG, Schlegel DC, Kravitz EA. Modulation of *Drosophila* male behavioral choice. *Proc Natl Acad Sci USA*. 2007;104:4706–4711.
25. Certel SJ, Leung A, Lin C-Y, Perez P, Chiang A-S, Kravitz EA. Octopamine neuromodulatory effects on a social behavior decision-making network in *Drosophila* males. *PLoS ONE*. 2010;5:e13248.
26. Tang S, Guo A. Choice behavior of *Drosophila* facing contradictory visual cues. *Science*. 2001;294:1543–1547.
27. Zhang K, Guo JZ, Peng Y, Xi W, Guo A. Dopamine-mushroom body circuit regulates saliency-based decision-making in *Drosophila*. *Science*. 2007;316:1901–1904.

CHAPTER

3

Cognitive Components of Insect Behavior

Martin Giurfa,† and Randolf Menzel‡*

*Université de Toulouse, Centre de Recherches sur la Cognition Animale, Toulouse, France †Centre National de la Recherche Scientifique, Centre de Recherches sur la Cognition Animale, Toulouse, France

‡Freie Universität Berlin, Berlin, Germany

INTRODUCTION

Cognition is the integrating process that utilizes many different forms of memory (innate and acquired), creates internal representations of the experienced world, and provides a reference for expecting the future of the animal's own actions.[1,2] It thus allows the animal to decide between different options in reference to the expected outcome of its potential actions. All these processes occur as intrinsic operations of the nervous system, and they provide an implicit form of knowledge for controlling behavior. None of these processes need to—and certainly will not—become explicit within the nervous systems of many animal species (particularly invertebrates and lower vertebrates), but their existence must be assumed given the animal's specific behavioral output. Here, we focus on cognitive components of insect behavior and analyze behavioral outputs that refer to several forms of internal processing. In doing so, we aim to relate the complexity of the insect nervous system to the level of internal processing, which is a major goal of comparative animal cognition.

ACTING UPON THE ENVIRONMENT: EXPLORATION, INSTRUMENTAL LEARNING, AND OBSERVATIONAL LEARNING

Insects, like all animals, explore the environment and by doing so acquire relevant sensory, motor, and integrative information that facilitates learning about relevant events in such environments.[3–5] Honeybees, for instance, explore the environment before they start foraging,[6,7] and they learn the spatial relations of environmental objects during these exploratory flights.[8–10] Fruit flies (*Drosophila melanogaster*) also respond to their placement within a novel open-field arena with a high level of initial activity,[11–13] followed by a reduced stable level of spontaneous activity. This initial elevated component corresponds to an active exploration because it is independent of handling prior to placement within the arena, and it is proportional to the size of the arena.[14] Furthermore, visually impaired flies are significantly impaired in the attenuation of initial activity, thus suggesting that visual information is required for the rapid decay from elevated initial activity to spontaneous activity within the novel open-field arena.[14]

Exploratory behavior facilitates learning by associating the animal's action to the resulting outcome. For example, a hungry animal searching for food in a particular sensory environment learns upon a successful search the relationship between its own actions, the external conditions signaling the outcome, and the valuating signal of the food reward. This kind of association constitutes the basis of operant (instrumental) learning.[15] Operant learning has been intensively studied in insects. A classic protocol for the study of this learning form is the flight simulator in which a *Drosophila* is suspended from the thorax in the middle of a cylindrical arena that allows the presentation of visual landmarks (Figure 3.1). The tethered fly flies stationary and if some of these landmarks are paired with the aversive reinforcement of an unpleasant heat beam pointed on the thorax, the fly learns to fly toward a safe direction, avoiding the dangerous-landmark directions (Figure 3.1).[17,18] The fly learns to control reinforcement delivery as its flight maneuvers determine the switching-off of the heat beam if the appropriate

Invertebrate Learning and Memory.
DOI: http://dx.doi.org/10.1016/B978-0-12-415823-8.00003-4

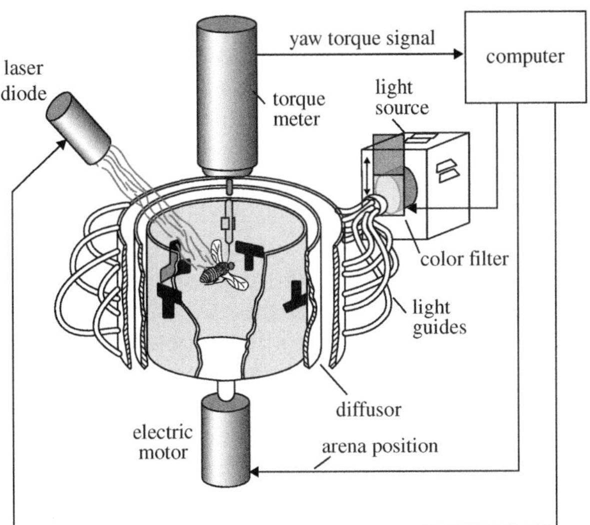

FIGURE 3.1 The flight simulator used for visual conditioning of a tethered fruit fly.[16] (Left) A *Drosophila* is flying stationary in a cylindrical arena. The fly's tendency to perform left or right turns (yaw torque) is measured continuously and fed into a computer, which controls arena rotation. On the screen, four 'landmarks,' two T's and two inverted T's, are displayed in order to provide a referential frame for flight direction choice. A heat beam focused on the fly's thorax is used as an aversive reinforcer. The reinforcer is switched on whenever the fly flies toward a prohibitive direction. Therefore, the fly controls reinforcer delivery by means of its flight direction. (Right) Detail of a tethered fly in suspended flight within the simulator. Source: *Courtesy of B. Brembs.*

flight directions are chosen,[18] thus constituting a case of operant learning (see Chapters 2 and 28).

Furthermore, insects are also endowed with the capacity to learn about the actions produced by others, be they conspecifics or not.[19] Wood crickets (*Nemobius sylvestris*), for instance, learn to hide under leaves by observing experienced conspecifics in the presence of a natural predator, the wolf spider.[20] Observer crickets were placed in a leaf-filled arena accompanied by conspecifics (demonstrators) that were either confronted with a wolf spider and therefore tended to hide under leaves or did not experience this predatory threat. Observers that interacted with spider-experienced conspecifics were more likely to hide under leaves than observers that interacted with conspecifics that had no recent spider experience. This difference persisted 24 hr after demonstrators were removed from the experimental arena, thus showing that perception of danger in observers had been altered by the demonstrators' behavior.[20] Crickets did not hide under leaves when separated from demonstrators by a partition that allowed for pheromone exchange between compartments but not visual or physical contact, nor did they increase their tendency to hide when placed in arenas that had previously contained crickets confronted with spiders. Thus, naive crickets learn from experienced demonstrators how to hide under leaves when facing a potential threat, and this learning requires a direct contact between observers and demonstrators.

An important point raised by this example of observational learning is that it would have to take the form of higher order conditioning because the observer cricket would not actually directly experience the unconditional stimulus of a spider attack, which would result in immediate death, thus making learning superfluous. That insects are capable of such higher order conditioning, specifically second-order conditioning, has been shown in various cases. Honeybees and fruit flies learn such second-order associations. Whereas flies exhibit second-order conditioning in an aversive context, in which they learn to associate an odor (conditioned stimulus 1 (CS1)) with shock (unconditioned stimulus (US)) and then a second odor (conditioned stimulus 2 (CS2)) with the previously conditioned CS1,[21] honeybees learn second-order associations in an appetitive context while searching for food. They learn to connect both two odors (odor 1 + sucrose reward; odor 2 + odor 1[22–24]) and one odor and one color.[25] Although these examples refer to the framework of classical (Pavlovian) conditioning in which animals learn to associate different stimuli,[26] similar explanations could be provided for operant learning situations, thus rendering the higher order conditioning explanation of observational learning plausible.

Observational learning even at a symbolic level is exemplified by dance communication in bees (discussed later).[6]

EXPECTATION

Operant learning means that the animal may develop an expectation about the outcome of its actions. Two forms of expectation can be distinguished: conditioned responding to an experienced stimulus, as in associative learning, and planning of behavior in the absence of the stimuli associated with its outcome. Both of these forms of expectation comprising lower and higher cognitive processes interact in navigation and waggle dance communication in honeybees. Bees navigating toward predictable food sources follow routes and develop visual memories of landmarks seen en route and at the locations of food sources.[8] The locations are qualified in the sense that the insect expects the formerly experienced target signals at specific points of its route. For instance, bees trained to fly along a series of three similar, consecutive compartments in which they have to choose between two patterns, one positive (+) allowing passage to the next compartment and one negative (−) blocking passage, choose between combinations of positive patterns according to their expectation of which should be the positive pattern at a given compartment.[27] If, for instance, bees are trained with a white (+) versus a black disk (−) in the first compartment, a blue (+) versus a yellow disk (−) in the second compartment, and a vertical (+) versus a horizontal black-and-white grating (−) in the third compartment, they prefer the positive white disk over the positive vertical black-and-white grating in the first compartment but they revert this preference if the same stimuli are confronted in the third compartment.[27] Furthermore, bees learn the sequence of four landmarks as cues for turns toward the feeder.[28] Thus, bees exhibit specific expectations along a route about the outcome of landmarks that guide them toward the food source. Similarly, bees trained to fly to different locations in the morning and in the afternoon choose the correct homing direction if released at the wrong time of the day at one of these locations, and they integrate this location-specific information when released halfway between these two locations.[29]

A higher form of expectation can be found after latent learning in navigation and dance communication in honeybees. Bees perform novel shortcuts between two or more locations within a previously explored environment.[30,31] They also fly along shortcuts between a learned location and a location communicated by the waggle dance of a hive mate.[32] They do so without reference to beacons or a structured panorama, excluding the possibility that they somehow rely on snapshot memories established at the respective locations.[33] The fact that they are able in certain circumstances to fly such shortcuts between a communicated location and a location memorized on the basis of their own experience implies that both locations have a common spatial reference framework. Such memory structure could store geometric relations of objects in the explored environment and could be conceptualized as a cognitive (or mental) map because the behavior of bees meets the definition of a cognitive map.[8,30] It would include meaningful objects at their respective locations and on the way toward them, and thus the animal would know at any place where it is relative to potential destinations allowing to plan routes to locations whose signifying signals are not available at the moments decisions are made.

The term *expectation* can be applied at multiple levels of behavioral and neural processes. A low-level process is the efference copy of the neural program initiating the movement that leads to an error signal when compared with the sensory feedback resulting from the movement.[34,35] This error signal is thought to feed into an internal, neural value system leading to associative alterations in the neural circuits initiating the movement. The efference copy can be considered as a neural correlate of expectation because it precedes the conditions of the external world and leads to a correction of neural circuitry. On a formal ground, the neural operation of comparison between the efference copy and sensory feedback is equivalent to the deviation from expectation as derived in computational reinforcement learning by the delta rule.[36] It has been difficult to trace efference copies to neural circuits, but an exciting example exists in insects. A single multisegmental interneuron, the corollary discharge interneuron (CDI), was found in the cricket *Gryllus bimaculatus* that provides presynaptic inhibition to auditory afferents and postsynaptic inhibition to auditory interneurons when the animal produces its own song but not when it hears songs from other animals.[37] When the animal sings without sound (fictive song), the CDI is excited and inhibits the coding of played songs. The authors managed to stimulate CDI intracellularly, resulting in inhibited auditory encoding. They also found that excitation of CDI is specific for self-generated songs and not for other motor patterns, demonstrating that the CDI is both necessary and sufficient for the blocking of sensory input expected to be received from own song production. It will be interesting to determine whether a mismatch between the expected song pattern and the received song produces an error signal that may be used to fine-tune own song production, for example, after some disturbance of the song production by the wings.

The concept of an error signal driving associative learning has a strong impact on neural studies of learning-related plasticity in the nervous system. For example, the dopamine neurons of the ventral tegmentum of the mammalian brain change their response properties to a conditioned stimulus predicting reward (US) according to a modified delta rule.[38] A similar effect was found in an identified neuron in the honeybee brain, which encodes the reinforcing property of the sucrose reward (US) in olfactory learning.[39] This neuron, known as VUMmx1 (ventral unpaired median neuron 1 in the maxillary neuromere), appears to have similar properties as dopaminergic neurons in the mammalian brain (Figure 3.2). During differential conditioning in which a bee is trained to respond to a rewarded odor (CS+) and not to a non-rewarded odor (CS−), intracellular recordings of VUMmx1 activity show that this neuron develops responses to CS+ and stops responding to CS−. If the US is now given after the CS+, one finds no responses to the US anymore, but a US after CS− is well responded to. In other words, the neuron responds to unexpected sucrose presentations but not to an expected one (for further discussion, see Chapter 29).

Octopamine immunoreactive neurons in the *Drosophila* brain correspond to VUMmx1 in structure and function, and they represent the reward function in olfactory learning.[40] Dopaminergic neurons in the *Drosophila* brain act as a value system in the framework of aversive learning.[40] These neurons thus mediate the aversive reinforcing properties of electric shock punishment in odor–shock learning. Signaling from specific subsets of these dopaminergic neurons arborizing at the level of subcompartments of paired, central brain structures called the mushroom bodies, which intervene in the storage and retrieving of olfactory memories,[41–43] are necessary and sufficient to support learning of the odor–shock association.[44] Thus, inhibiting these neurons in genetic mutants impedes aversive learning, whereas artificial activation of these neurons in other types of mutants facilitates odor learning even in the absence of shock.[45–47] Interestingly, dopaminergic neurons in *Drosophila* are weakly activated by odor stimuli before training but respond strongly to electric shocks. However, after one of two odors is paired several times with an electric shock, the neurons acquire the capacity to respond to the odor stimulus.[44] Like VUMmx1, they also respond distinctly to odorants with different outcomes in a differential conditioning experiment with a punished odor (CS+) versus a non-punished odor (CS −): In this case, odor-evoked activity is significantly prolonged only for the CS+. Thus, dopaminergic neurons involved in odor–shock learning not only mediate aversive reinforcing stimulation but also reflect in their activity the training-induced association with the US; in other words, during training they acquire the capability to predict the anticipated punishment (for further discussion, see Chapters 2, 5, 6, 27, and 28). Recently, it was found that a subpopulation of dopamine neurons is involved in coding the reward function in olfactory learning of *Drosophila*. *In vivo* calcium imaging revealed that these neurons are activated by sugar ingestion and the activation is increased on starvation. These dopamine neurons are selectively required for the reinforcing property of, but not a reflexive response to, the sugar stimulus.[48]

These results support the notion that reward is an intrinsic property of structurally and functionally defined neurons in the insect brain. It is controlled by expectation about their own actions that are relative to specific objects in the external world or that are driven internally in order to fulfill expected outcomes that are absent. The latter component—the driving of behavior by expectations of absent outcomes—has been highlighted in experiments in which *Drosophila* larvae are trained with different kinds of appetitive and aversive associations and afterward are tested in retention tests in which the memories induced by this training should be expressed.[49] These experiments show, for instance, that aversive olfactory memories are not expressed if the test situation is performed under extinction conditions—that is, if the previously punished odor is presented without punishment. It is argued that with the expected outcome of punishment being absent, the corresponding avoidance behavior has no reason to be expressed[49] (see Chapter 5). Conversely, after appetitive learning, memories would be expressed only in extinction conditions because the previously rewarded odor would be, in this case, deprived of the expected reward (see Chapter 33 for additional information about context dependence of extinction learning in honeybees). In this case, it makes sense to initiate appetitive search in order to access the reward expected in association with this odor.[49] Thus, conditioned olfactory behavior would reflect specific expectations and would aim at reaching specific goals associated with these expectations.

GENERALIZATION, CATEGORIZATION, AND CONCEPT LEARNING

Extracting information from experienced events and applying it to solve novel situations is a distinctive behavior of 'intelligent' systems. Indeed, experiments showing that animals respond in an adaptive manner to novel stimuli that they have never encountered before and that do not predict a specific outcome based on the animals' past experience are the hallmark of higher forms of flexible behavior. Such a positive transfer of learning (also called stimulus transfer)

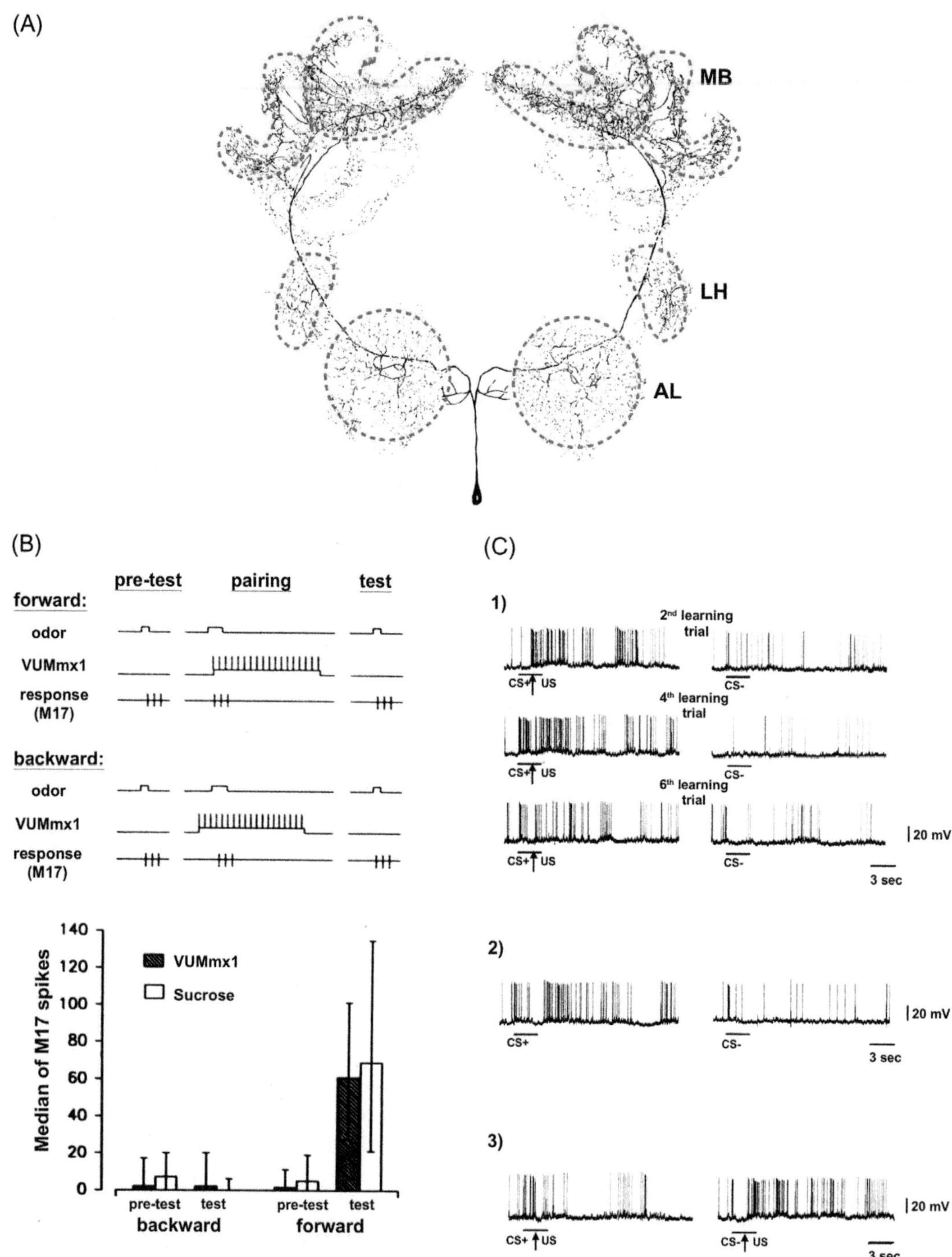

FIGURE 3.2 (A) The VUMmx1 neuron.[39] The soma is located in the maxillary neuromere, and the dendrites arborize symmetrically in the brain and converge with the olfactory pathway at three sites (delimited by a red dashed line), the primary olfactory center, the antennal lobe (AL), the secondary olfactory integration area, the lip region of the mushroom bodies (MB), and the output region of the brain, the lateral horn (LH). VUMmx1 responds to sucrose solution both at the antenna and at the proboscis with long-lasting spike activity and to various visual, olfactory, and mechanosensory stimuli with low-frequency spike activity. (B) Olfactory learning can be induced by substituting the sucrose reward in PER conditioning by an artificial depolarization of VUMmx1 ('sucrose signaling') immediately after odor stimulation. If depolarization precedes olfactory stimulation (backward pairing), no learning is observed. The same forward–backward effect is seen in behavioral PER conditioning. The bees' response is quantified in terms of the number of spikes of M17, a muscle controlling the movement of the proboscis. The results thus show that VUMmx1 constitutes the neural correlate of the US in associative olfactory learning. (C) Intracellular recordings of VUMmx1 during training and tests with a reinforced (CS+; carnation) and a non-reinforced odor (CS−; orange). Such a conditioning leads to an enhanced response of VUMmx1 to CS+ but not to CS−. If the US follows the presentation of the CS+, the response of VUMmx1 to the US is greatly reduced and even inhibited. In contrast, the response of VUMmx1 to the US after the presentation of the CS− remains normal. This indicates that differential conditioning leads to different reward-related responses, depending on whether the reward is expected (after CS+) or not (after CS−).

therefore brings us to a domain that differs from that of elemental forms of learning.[50]

Stimulus transfer admits different levels of complexity that refer to the capacity of transferring specific knowledge to novel events based either on stimulus similarity, and thus on specific physical traits that are recognized in the novel events, or on more abstract relationships that constitute the basis for decisional rules.[51] The first basic process that needs to be mentioned in this context is that of stimulus generalization. Most animals, including insects, have the capacity to record events related with relevant consequences and to signal their reappearance. This requires learning, memorization, and evaluation of perceptual input in relational terms and the capacity of coping with possible distortions of the original stimuli due to noise, extrinsic or intrinsic environmental interferences, positional or developmental changes, etc. Generalization allows for flexible responding when the animal is confronted with these possible interferences because it involves assessing the similarity between the present perceptual input and the previous experience.[52] The evaluation of similarity is performed along one or several dimensions such that stimuli that lie close to each other along a perceptual scale or in a perceptual space are treated as equivalent. As a consequence, generalization processes imply a gradual decrease in responding along a perceptual scale correlated with a progressive decrease in stimulus similarity.[53–55] Stimulus generalization has been shown in insects in perceptual domains as different as the olfactory one,[56–61] the visual one,[62–66] and the gustatory one.[67,68]

The next level of stimulus transfer corresponds to categorization, which is defined as the ability to group distinguishable objects or events on the basis of a common feature or set of features and therefore to respond similarly to them.[51,69] Categorization thus deals with the extraction of these defining features from objects of the animal's environment. Labeling different objects as belonging to the same category implies responding similarly to them; as a consequence, category boundaries are sharper than those corresponding to the gradual decrease of responding along a perceptual scale underlying generalization.[70]

Numerous examples have shown that bees categorize visual stimuli based on unique features or on arrangements of multiple features.[71] For instance, bees categorize visual patterns based on the presence or absence of bilateral symmetry.[72] Bees were trained with triads of patterns in which one pattern was rewarded with sucrose solution and the other two were non-rewarded. For the bees trained for symmetry, the rewarded pattern was symmetric and the non-rewarded patterns were asymmetric. For the bees trained for asymmetry, the rewarded pattern was asymmetric and the two non-rewarded patterns were symmetric. To avoid learning of a specific pattern or triad, bees were confronted with a succession of changing triads during the course of training. Transfer tests presenting stimuli that were unknown to the bees, all non-rewarded, were interspersed during the training with the triads.

Bees trained to discriminate bilaterally symmetric from nonsymmetric patterns learned the task and transferred it appropriately to novel stimuli, thus demonstrating a capacity to detect and categorize symmetry or asymmetry. Interestingly, bees trained for symmetry chose the novel symmetric stimuli more frequently and came closer to and hovered longer in front of them than bees trained for asymmetry did for the novel asymmetric stimuli. It was thus suggested that bees have a predisposition for learning and categorizing symmetry. Such a predisposition can either be innate and could facilitate a better and faster learning about stimuli that are biologically relevant[73] or can be based on the transfer of past experience from predominantly symmetric flowers in the field. A specific ecological advantage would arise from flower categorization in terms of symmetrical versus asymmetrical. The perception of symmetry would be important for pollinators because symmetry of a flower may signal its quality and thus influence mating and reproductive success of plants by affecting the behavior of pollinators.[74] As bees discriminate between symmetry and asymmetry, they should also be capable of performing selective pollination with respect to floral symmetry even within a patch of flowers. This may indicate that plants may have exploited such cognitive capabilities of the pollinators during the evolution of flowers.

A further level of stimulus transfer is termed *concept learning*, which, contrary to categorization based on specific physical features, occurs independently of the physical nature of the stimuli considered (colors, shape, size, etc.)[75,76] and relies on relations between objects.[51,77] Examples of such relations are 'same as,' 'different from,' 'above/below of,' and 'on the left/right of.' Extracting such relations allows transferring a choice to unknown objects that may differ dramatically in terms of their physical features but that may fulfill the learned relation.

Various recent reports have indicated that honeybees learn relational rules of different sorts. These include 'sameness/difference,'[78] 'above/below,'[79] and the mastering of two rules simultaneously—'above/below' (or left/right) and 'different from.'[80]

Learning of the concepts of sameness and difference was demonstrated through the protocols of delayed matching to sample (DMTS) and delayed non-matching to sample (DNMTS), respectively.[78] Honeybees foraging in a Y-maze (Figure 3.3A) were trained in a DMTS experiment in which they were presented with a changing

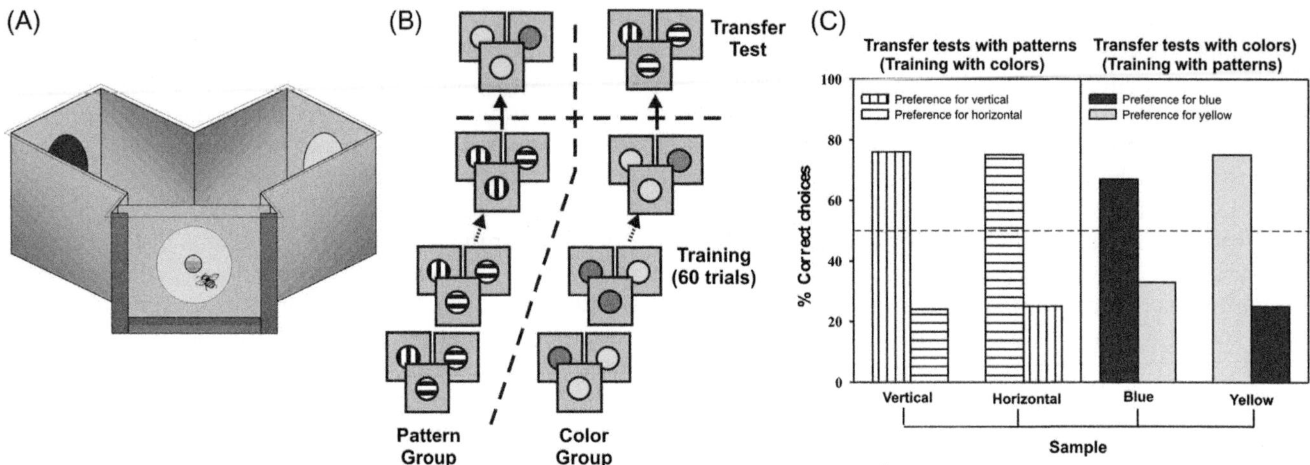

FIGURE 3.3 (A) Y-maze used to train bees in a delayed matching-to-sample task.[78] Bees had to enter into the maze to collect sugar solution on one of the back walls of the maze. A sample was shown at the maze entrance before bees accessed the arms of the maze. (B) Training protocol. A group of bees were trained during 60 trials with black-and-white, vertical and horizontal gratings (pattern group); another group was trained with colors, blue and yellow (color group). After training, both groups were subjected to a transfer test with novel stimuli (patterns for bees trained with colors, and colors for bees trained with patterns). (C) Performance of the pattern group and the color group in the transfer tests with novel stimuli. Both groups chose the novel stimulus corresponding to the sample, although they had no experience with such test stimuli.

non-rewarded sample (i.e., one of two different color disks ('color group') or one of two different black-and-white gratings, vertical or horizontal ('pattern group')) at the entrance of a maze (Figure 3.3B). The bees were rewarded only if they chose the stimulus identical to the sample once within the maze. Bees trained with colors and presented in transfer tests with black-and-white gratings that they had not experienced before solved the problem and chose the grating identical to the sample at the entrance of the maze. Similarly, bees trained with the gratings and tested with colors in transfer tests also solved the problem and chose the novel color corresponding to that of the sample grating at the maze entrance (Figure 3.3C). Transfer was not limited to different types of visual stimuli (pattern vs. color) but could also operate between drastically different sensory modalities such as olfaction and vision.[78] Bees also mastered a DNMTS task, thus showing that they learn a rule of difference between stimuli as well.[78] These results document that bees learn rules relating stimuli in their environment. They were later verified in a study showing that the working memory underlying the solving of the DMTS task lasts for approximately 5 sec,[81] a period that coincides with the duration of other visual and olfactory short-term memories characterized in simpler forms of associative learning in honeybees.[82]

More recently, bees were shown to process two concepts simultaneously, which presupposes an even higher level of cognitive sophistication than dealing with one concept at a time. Following a training in which they had to learn to choose two distinct objects in a specific spatial relationship (above/below or right/left), they mastered two abstract concepts simultaneously, one based on the spatial relationship and another based on the perception of difference.[80] Bees that learned to classify visual targets using this dual concept transferred their choices to unknown stimuli that offered a best match in terms of dual-concept availability: Their components presented the appropriate spatial relationship and differed from one another. These results thus demonstrate that it is possible for a bee to extract at least two different concepts from a set of complex pictures and combine them in a rule for subsequent choices.

MEMORY PROCESSING

Learning does not produce the final memory trace immediately. Time- and event-dependent processes, conceptualized as consolidation processes, form the memory trace. Short-term memory is transformed into midterm and long-term memory, and the molecular, cellular, neural, and systemic processes underlying this transformation are currently intensively studied (see Chapter 27). Stable memory traces need to be moved from a silent into an activated state by retrieval processes. Internal conditions of the animal, external cues, and a neural search process shift a silent memory into an active memory. The expression of the active memory may undergo selection processes before its content is expressed. Animals need to decide between different options as they reside in memory, and the decision process requires access to the expected outcomes. The expected outcomes are also stored in memory, and only when the respective memory contents are retrieved will they be accessible to selection processes. This network of interactions between retrieval, selection, and execution is conceptualized in a particular form of memory—working memory.

Memory systems are also categorized according to their contents. In vertebrates, different contents are related with particular brain structures—for example, procedural memory (cerebellum), episodic memory (hippocampus and prefrontal cortex), and emotional memory (amygdala). Whereas procedural memory certainly exists to a large extent in insects in their ventral ganglia, it is unknown whether memories qualifiable as 'emotional' exist in insects and, if so, whether they reside in modulatory neurons related to reward and punishment and/or in other sets of the widely branching peptidergic neural networks. Higher order forms of memory are usually related to the mushroom bodies (see Chapter 28), but the level of higher order processing mediated by these structures is unknown. Do insects possess a form of episodic memory—the ability to carry out long-term recall of sequences of events or narratives?[83] In humans, birds, and mammals, this property is intimately related to the functions of the hippocampus and cerebral cortex. It is argued that food-storing birds may develop an episodic-like memory about a kind of food stored at a certain place and at a certain time.[84] Pollinating insects certainly control their foraging activities according to the kind of food they collect at a particular place and at a specific time of day, but it is unknown whether they make decisions between options integrating the what, where, and when of potential food sites.

Memory systems are highly dynamic and content sensitive. Any retrieval from the memory store will change its content due to the updating process in working memory. It is precisely this updating process that may lead to extracting rules that underlie implicit forms of abstraction in the visual domain (discussed previously). Furthermore, retrieval from memory store also induces re-learning and consequently consolidation into a new memory form, a process referred to as "reconsolidation."[85] This process has been demonstrated in the honeybee (see Chapter 33).

From an evolutionary standpoint, one may expect that memory dynamics are adapted to choice behavior under natural conditions. Foraging in pollinating insects has a highly regular sequential structure of events ranging from actions within seconds to those separated by months. It thus offers the opportunity to relate memory structure and ecological demands.[82] Choices between flowers within the same patch quickly succeed each other and are performed during early short-term memory. Choices between flowers of different patches occur after the transition to late short-term memory. Successive foraging bouts are interrupted by the return to the hive so that flower choices in a subsequent bout require retrieving information from midterm memory. Finally, interruptions of days, weeks, and months (the latter in the case of overwintering bees) require retrieval from long-term memory.

Internal processing at the level of working memory can be understood as an indication of rudimentary forms of explicit processing and may exist in insects (and cephalopods; see Chapters 23–25) within the context of observatory learning and social communication. Paper wasps recognize each other on an individual basis (see Chapter 42); the ant *Temnothorax albipennis* informs colony members about a new food site by a particular behavior termed tandem running and that has been assimilated to a form of teaching (see Chapter 40); and bees employ a symbolic form of social communication for the transfer of information about spatial food locations. Key components in all these forms of learning and teaching are the retrieval of remote memory and the incorporation of the new information into the existing memory. Working memory provides implicit forms of representation as a substrate for various kinds of neural operations. These may include evaluation of the new information on the background of existing memory, extraction and updating of rules connecting the contents of memory, and decision making in relation to the expected outcome of the animal's actions.

Addressing the properties and functioning of active working memory requires for each paradigm a careful evaluation of whether elemental forms of learning and memory retrieval are sufficient to explain behavioral performances. The tradition of the most parsimonious explanation provides a strong tool in science and is well observed in behavioral studies, particularly in those performed with insects. However, the rigidity of some experimental designs frequently used in laboratory studies of insect behavior might result in the danger that the animal in its restriction can only do what the experimenter allows it to do. The conclusion from such experiments is often that because the animal did what was expected from it, this is the only behavior it possesses. Although scientific progress is bound to search for the most parsimonious explanation, it is not obvious what may be more or less demanding for the small brain of an insect. For example, will it be more difficult to follow a navigation strategy based on route following or on using a cognitive-map? Is it easier to store many sequential images defining a long route or to extract a rule connecting these images? Are neural processes derived from behavioristic learning theory less demanding than those derived from cognitive concepts? The answer at this stage is that we simply do not know, and that the only way to find out is to search for neural mechanisms within a broader conceptual frame. We also need to acknowledge that potential behavioral acts that are not performed by an animal are equally important as expressed behavior. Only by accepting that an attentive brain is constantly producing potential behaviors, most of which are not expressed, will we be able to search for the neural basis of the 'inner doing' as a prerequisite of decision-making processes.

INSECT INTELLIGENCE AND BRAIN STRUCTURE

Thinking about the basic design of a brain subserving the cognitive functions discussed previously, one recognizes a structure of essential modules and their interconnectivity (Figure 3.4).[86] This modular architecture seems to be shared by a broad range of animal species and may even apply to the wormlike creature at the basis of the evolutionary division between protostomes and deuterostomes. These species possess various kinds of perceptual and motor control systems, which constitute the input and output, respectively, of the architecture presented in Figure 3.4. Premotor centers convey information to motor control systems and therefore act as action planning systems. 'Desire' is used here to represent the expected outcomes of behavior, either appetitive or aversive, available to animals via specific signaling pathways. 'Belief,' on the other hand, refers to innate or experience-dependent memories—that is, to the knowledge that the animal has at its disposition and that drives its actions and decisions.

Although the modules depicted in Figure 3.4 may be multiple (multiple perceptual systems, multiple belief-generating systems, multiple desire-generating systems, multiple action-planning systems, and multiple motor control systems), the basic idea of this scheme is that perceptual systems feed onto three downstream systems arranged both serially and in parallel that converge on the action-planning systems,[86] which in turn drive the motor systems. Thus, perceptual systems can reach the action-planning systems directly; in addition, the desire- and belief-generating systems receiving the same perceptual information will act in parallel onto action planning, as well.

When we talk about 'modules' and 'systems,' we mean, in essence, neurons and neural networks. Therefore, if such a scheme should be of any heuristic help in understanding the insect brain, its skeleton needs the flesh of neurons and their functions. The chapters on the insect brain in this volume document

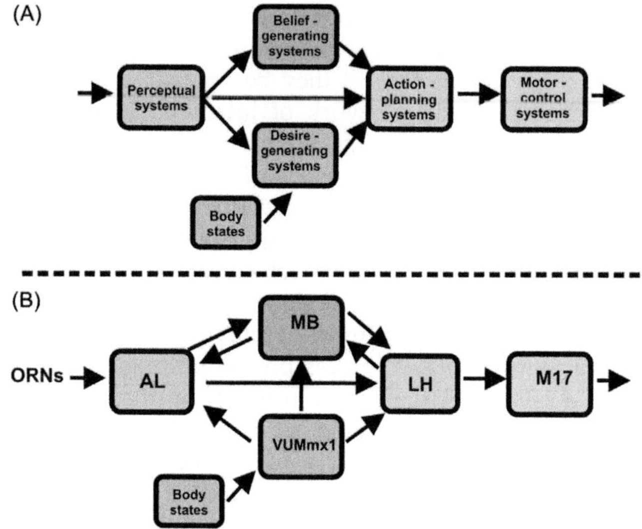

FIGURE 3.4 (A) The cognitive architecture of a generic brain based on interconnected modules for perception, desire- and belief-generating systems, action-planning systems, and motor control systems.[86] Action planning can either be generated by direct inputs from the perceptual systems or result from processes that are generated in parallel pathways weighting the perceptual inputs with respect to learned expectations (belief-generating systems) and signaling of appetitive or aversive outcomes (desire-generating systems). These modules can either be rather basic (as in more simple nervous systems) or highly complex, but their basic structure, particularly with respect to direct and indirect pathways and the necessity of operations between modules, may apply to any nervous system. (B) The cognitive architecture of a honeybee brain illustrated in the case of the olfactory circuit. Olfactory receptor neurons (ORNs) send information about odors to the antennal lobe (AL), which further conveys this information via a direct tract of projection neurons (m-ACT) to the mushroom body (MB), which hosts long-term, olfactory memory traces, and to the lateral horn (LH), a premotor center, via a different tract of projection neurons (l-ACT). The m-ACT tract further projects to the LH and the l-ACT tract to the MB. MBs send feedback neurons to the AL. VUMmx1 is a neuron whose activity mediates the reinforcing properties of appetitive stimuli (e.g., sucrose reward). VUMmx1 contacts the olfactory circuit at the level of the AL, the MB, and the LH, thus favoring the association between odor and sucrose, which is at the basis of olfactory learning. The motor output is represented here by M17, a muscle controlling the appetitive response of proboscis extension, which the bees exhibit to sucrose and/or to an odorant that has been learned to predict sucrose.

such a fleshing out (see Chapters 4, 27, 29, and 41). Sensory systems (vision, olfaction, and mechanosensory) connect to premotor areas via two pathways—a more direct pathway and one via the mushroom body. This is well illustrated in the olfactory circuit of the honeybee, in which olfactory information processed in the first olfactory neuropile, the antennal lobe, is conveyed to the mushroom body and then to the lateral horn, a suspected premotor area, via a medial tract of projection neurons or directly to the lateral horn and then to the mushroom bodies via a lateral tract of projection neurons. The 'desire'-generating systems are multifold and can be retraced to octopamine- and dopamine-containing neurons, signaling appetitive and aversive outcomes, respectively, and to widely branching peptidergic neurons. The VUM neurons are known to feed into three subsystems in parallel—the action-planning systems such as the lateral horn, the belief-generating system (the mushroom body), and the perceptual system (particularly the olfactory antennal lobe). Less is known about the wiring of the dopamine neurons, and it will be interesting to determine whether they also follow this scheme. The inputs from body states onto the desire-generating systems have not yet been identified, but because their function is modulated by body states (e.g., the levels of satiation, sleep, arousal, and attention), we must assume that such input does exist. In brief, such a 'boxicology' of general brain functions, as developed for mammalian brains,[86] applies surprisingly well to the insect brain, and the chapters on insects in this book provide ample evidence for the working of the neurons and networks as components of these boxes.

MINIATURE BRAINS

It is sometimes assumed that 'simple' and 'miniature' nervous systems such as those of arthropods and (most) mollusks implement cognitive faculties by radically different mechanisms compared to vertebrates, relying predominantly or exclusively on innate routines and elemental forms of associative learning. However, as exemplified previously, constructing a great divide between simple and advanced nervous systems will lead us astray because the basic logical structure of the processes underlying spontaneity, decision making, planning, and communication are similar in many respects in large and small brains. Therefore, it seems more productive to envisage differences in quantitative rather than qualitative terms, providing us with a wealth of 'model systems' to elucidate the essence of the basic cognitive processes.

In contrast to studies in mammals and birds,[87] criteria on brain–behavior relations have not been applied systematically to insects. However, it can be tentatively concluded that relatively large insect brains, particularly those with complex mushroom bodies such as those occurring in social Hymenoptera, are equipped with more behavioral flexibility. In the search for neural correlates of behavioral flexibility, features such as 'information processing capacity' (IPC) based on the specific neural features have been invoked.[88–90] These features are, for instance, the number of neurons, dendritic structures, the packing density of synaptic connections, and axonal conduction velocity.[90] However, there is no linear relationship between IPC and these measures because brains are organized to reduce wiring costs and smaller brains require fewer material and less energy for construction and maintenance.[91] Long-distance communication within the brain appears to be a major component of IPC, but it also consumes a high amount of energy for spike propagation, again favoring miniaturization. Splitting information across parallel pathways, reducing feedback neural connections, sparse coding, and synchronous activity are a few of many probable neural processes that keep energy consumption low and information capacity high.[91] Such additional processes could be (1) a globular organization of the neuropil (as found in birds and invertebrates) rather than a sheeted organization (as in the cerebral cortex in mammals); (2) large and widely branching neurons whose dendritic branches may participate in different forms of neural processing, either simultaneously or sequentially; and (3) direct oxygen supply via tracheas, which may make energy consumption more efficient and reduce the size and weight of insect brains. Taken together, these processes may endow small brains with relatively higher IPC compared to large brains.

CONCLUSION

It has often been said that neuroscience lacks a theory (or theories) of the brain.[92,93] Indeed, there appears to be no concept at the level of the neurons, the networks, or the whole brain that is able to provide enough generality for developing such a theory. Potentially, the 'boxicology' of Carruthers[86] (Figure 3.4), together with the hard facts obtained from anatomy, physiology, and behavioral analysis, may provide a path toward developing a theory about how small brains work. Such a theory would need to include the 'inner doing' of the brain—its operations that are not (yet) expressed in behavioral acts and that include operations on representations meaning neural processes at the level of working memory (see Chapter 2). As stated by Carruthers in his book *Architecture of the Mind*,[86]

> To be a believer/desirer ... means possessing distinct content-bearing belief-states and desire-states that are discrete, structured, and causally efficacious in virtue of their structural properties. These are demanding conditions. But not so demanding that the nonhuman animals can be ruled out as candidates immediately. Indeed we propose to argue, on the contrary, that many invertebrates actually satisfy these requirements. (p. 68)

Future studies on insect brains will gain by incorporating these concepts and by relating them to specific neural modularity and connectivity.

References

1. Vauclair J. *Animal Cognition: An Introduction to Modern Comparative Psychology*. Cambridge, MA: Harvard University Press; 1996.
2. Balda RP, Pepperberg IM, Kamil AC. *Animal Cognition in Nature: The Convergence of Psychology and Biology in Laboratory and Field*. San Diego, CA: Academic Press; 1998.
3. Bell WJ. Searching behavior patterns in insects. *Annu Rev Entomol*. 1990;35:447–467.
4. Bell WJ. *Searching Behaviour: The Behavioural Ecology of Finding Resources*. London: Chapman & Hall; 1991.
5. Menzel R, Brembs B, Giurfa M. Cognition in invertebrates. In: Kaas JH, ed. *Evolution of Nervous Systems: Vol. II. Evolution of Nervous Systems in Invertebrates*. Oxford: Academic Press; 2007:403–422.
6. von Frisch K. *The Dance Language and Orientation of Bees*. Cambridge, MA: Harvard University Press; 1967.
7. Capaldi EA, Smith AD, Osborne JL, et al. Ontogeny of orientation flight in the honeybee revealed by harmonic radar. *Nature*. 2000;403:537–540.
8. Menzel R. Navigation and communication in honeybees. In: Menzel R, Fischer J, eds. *Animal Thinking: Contemporary Issues in Comparative Cognition*. Vol 8. Cambridge, MA: MIT Press; 2011:9–22.
9. Zeil J, Kelber A, Voss R. Structure and function of learning flights in bees and wasps. *J Exp Biol*. 1996;199:245–252.
10. Zeil J. Orientation flights of solitary wasps (Cerceris; Specidae; Hymenoptera): I. Description of flight. *J Comp Physiol A*. 1993;172:189–205.
11. Connolly K. Locomotor activity in *Drosophila*: 3. A distinction between activity and reactivity. *Anim Behav*. 1967;15:149–152.
12. Meehan MJ, Wilson R. Locomotor activity in the Tyr-1 mutant of *Drosophila melanogaster*. *Behav Genet*. 1987;17:503–512.
13. Soibam B, Mann M, Liu L, et al. Open-field arena boundary is a primary object of exploration for *Drosophila*. *Brain Behav*. 2012;2:97–108.
14. Liu L, Davis RL, Roman G. Exploratory activity in *Drosophila* requires the kurtz nonvisual arrestin. *Genetics*. 2007;175:1197–1212.
15. Skinner BF. *The Behavior of Organisms*. New York: Appleton-Century-Crofts; 1938.
16. Heisenberg M, Wolf R. *Plasticity of Visuomotor Coordination: Vision in* Drosophila. New York: Springer-Verlag; 1984:168–176
17. Götz KG. Optomotorische Untersuchung des visuellen systems einiger Augenmutanten der Fruchtfliege *Drosophila*. *Kybernetik*. 1964;2:77–92.
18. Heisenberg M, Wolf R, Brembs B. Flexibility in a single behavioral variable of *Drosophila*. *Learn Mem*. 2001;8:1–10.
19. Giurfa M. Social learning in insects: a higher-order capacity? *Front Behav Neurosci*. 2012;6:57.
20. Coolen I, Dangles O, Casas J. Social learning in noncolonial insects? *Curr Biol*. 2005;15:1931–1935.
21. Tabone CJ, de Belle S. Second-order conditioning in *Drosophila*. *Learn Mem*. 2011;18:250–253.
22. Hussaini SA, Komischke B, Menzel R, Lachnit H. Forward and backward second-order Pavlovian conditioning in honeybees. *Learn Mem*. 2007;14:678–683.
23. Bitterman ME, Menzel R, Fietz A, Schäfer S. Classical conditioning of proboscis extension in honeybees (*Apis mellifera*). *J Comp Psychol*. 1983;97:107–119.
24. Takeda K. Classical conditioned response in the honey bee. *J Insect Physiol*. 1961;6:168–179.
25. Grossmann KE. Belohnungsverzögerung beim Erlernen einer Farbe an einer künstlichen Futterstelle durch Honigbienen. *Z Tierpsychol*. 1971;29:28–41.
26. Pavlov I. *Conditioned Reflexes*. New York: Dover; 1927.
27. Collett TS, Fry SN, Wehner R. Sequence learning by honeybees. *J Comp Physiol A*. 1993;172:693–706.
28. Menzel R. Serial position learning in honeybees. *PLoS ONE*. 2009;4:e4694.
29. Menzel R, Geiger K, Müller U, Joerges J, Chittka L. Bees travel novel homeward routes by integrating separately acquired vector memories. *Anim Behav*. 1998;55:139–152.
30. Menzel R, Greggers U, Smith A, et al. Honeybees navigate according to a map-like spatial memory. *Proc Natl Acad Sci USA*. 2005;102:3040–3045.
31. Menzel R, Lehmann K, Manz G, Fuchs J, Koblofsky M, Greggers U. Vector integration and novel shortcutting in honeybee navigation. *Apidologie*. 2012;43:229–243.
32. Menzel R, Kirbach A, Haass WD, et al. A common frame of reference for learned and communicated vectors in honeybee navigation. *Curr Biol*. 2011;21:645–650.
33. Collett TS. Insect navigation en route to the goal: multiple strategies for the use of landmarks. *J Exp Biol*. 1996;199:227–235.
34. von Holst E, Mittelstaedt H. The reafference principle: interaction between the central nervous system and the periphery. *Selected Papers of Erich von Holst: The Behavioural Physiology of Animals and Man*. Vol 1. London: Methuen; 1950:139–173.
35. von Holst E. Relations between the central nervous system and the peripheral organs. *Brit J Anim Behav*. 1954;2:89–86
36. Sutton RS, Barto AG. Toward a modern theory of adaptive networks: expectation and prediction. *Psychol Rev*. 1981;88:135–170.
37. Poulet JF, Hedwig B. The cellular basis of a corollary discharge. *Science*. 2006;311:518–522.
38. Schultz W. Predictive reward signal of dopamine neurons. *J Neurophysiol*. 1998;80:1–27.
39. Hammer M. An identified neuron mediates the unconditioned stimulus in associative olfactory learning in honeybees. *Nature*. 1993;366:59–63.
40. Schwaerzel M, Monastirioti M, Scholz H, Friggi-Grelin F, Birman S, Heisenberg M. Dopamine and octopamine differentiate between aversive and appetitive olfactory memories in *Drosophila*. *J Neurosci*. 2003;23:10495–10502.
41. Busto GU, Cervantes-Sandoval I, Davis RL. Olfactory learning in *Drosophila*. *Physiology*. 2010;25:338–346.
42. Heisenberg M. Mushroom body memoir: from maps to models. *Nat Rev Neurosci*. 2003;4:266–275.
43. Davis RL. Olfactory memory formation in *Drosophila*: from molecular to systems neuroscience. *Annu Rev Neurosci*. 2005;28:275–302.
44. Riemensperger T, Voller T, Stock P, Buchner E, Fiala A. Punishment prediction by dopaminergic neurons in *Drosophila*. *Curr Biol*. 2005;15:1953–1960.
45. Aso Y, Siwanowicz I, Bracker L, Ito K, Kitamoto T, Tanimoto H. Specific dopaminergic neurons for the formation of labile aversive memory. *Curr Biol*. 2010;20:1445–1451.
46. Aso Y, Herb A, Ogueta M, et al. Three dopamine pathways induce aversive odor memories with different stability. *PLOS Genet*. 2012;8:e1002768.
47. Claridge-Chang A, Roorda RD, Vrontou E, et al. Writing memories with light-addressable reinforcement circuitry. *Cell*. 2009;139:405–415.
48. Liu C, Plaçais P-Y, Yamagata N, et al. A subset of dopamine neurons signals reward for odour memory in *Drosophila*. *Nature*. 2012;488:512–516.
49. Gerber B, Hendel T. Outcome expectations drive learned behaviour in larval *Drosophila*. *Proc Biol Sci*. 2006;273:2965–2968.
50. Robertson SI. *Problem Solving*. Hove, UK: Psychology Press; 2001.
51. Zentall TR, Galizio M, Critchfield TS. Categorization, concept learning, and behavior analysis: an introduction. *J Exp Anal Behav*. 2002;78:237–248.

52. Spence KW. The differential response in animal to stimuli varying within a single dimension. *Psychol Rev.* 1937;44:430–444.
53. Estes WK. *Classification and Cognition.* Oxford: Oxford University Press; 1994.
54. Shepard RN. Toward a universal law of generalization for psychological science. *Science.* 1987;237:1317–1323.
55. Ghirlanda S, Enquist M. A century of generalization. *Anim Behav.* 2003;66:15–36.
56. Vareschi E. Duftunterscheidung bei der Honigbiene–Einzelzell-Ableitungen und Verhaltensreaktionen. *Z vergl Physiol.* 1971;75:143–173.
57. Guerrieri F, Schubert M, Sandoz JC, Giurfa M. Perceptual and neural olfactory similarity in honeybees. *PLoS Biol.* 2005;3:e60.
58. Daly KC, Chandra S, Durtschi ML, Smith BH. The generalization of an olfactory-based conditioned response reveals unique but overlapping odour representations in the moth *Manduca sexta. J Exp Biol.* 2001;204:3085–3095.
59. Sandoz JC, Pham-Delègue MH, Renou M, Wadhams LJ. Asymmetrical generalisation between pheromonal and floral odours in appetitive olfactory conditioning of the honey bee (*Apis mellifera* L.). *J Comp Physiol A.* 2001;187:559–568.
60. Eschbach C, Vogt K, Schmuker M, Gerber B. The similarity between odors and their binary mixtures in *Drosophila. Chem Senses.* 2011;36:613–621.
61. Bos N, Dreier S, Jorgensen CG, Nielsen J, Guerrieri FJ, d'Ettorre P. Learning and perceptual similarity among cuticular hydrocarbons in ants. *J Insect Physiol.* 2012;58:138–146.
62. Wehner R. The generalization of directional visual stimuli in the honey bee, *Apis mellifera. J Insect Physiol.* 1971;17:1579–1591.
63. Giurfa M. Colour generalization and choice behaviour of the honeybee *Apis mellifera ligustica. J Insect Physiol.* 1991;37:41–44.
64. Ronacher B, Duft U. An image-matching mechanism describes a generalization task in honeybees. *J Comp Physiol A.* 1996;178:803–812.
65. Gumbert A. Color choices by bumble bees (*Bombus terrestris*): innate preferences and generalization after learning. *Behav Ecol Sociobiol.* 2000;48:36–43.
66. Brembs B, Hempel de Ibarra N. Different parameters support generalization and discrimination learning in *Drosophila* at the flight simulator. *Learn Mem.* 2006;13:629–637.
67. Masek P, Scott K. Limited taste discrimination in *Drosophila. Proc Natl Acad Sci USA.* 2010;107:14833–14838.
68. Akhtar Y, Isman MB. Generalization of a habituated feeding deterrent response to unrelated antifeedants following prolonged exposure in a generalist herbivore, *Trichoplusia ni. J Chem Ecol.* 2004;30:1349–1362.
69. Huber L. Perceptual categorization as the groundwork of animal cognition. In: Taddei-Ferretti C, Musio C, eds. *Downward Processes in the Perception Representation Mechanisms.* Singapore: World Scientific; 1998:287–293.
70. Pastore RE. Categorical perception: some psychophysical models. In: Harnard S, ed. *Categorical Perception. The Groundwork of Cognition.* Cambridge, UK: Cambridge University Press; 1987:29–52.
71. Benard J, Stach S, Giurfa M. Categorization of visual stimuli in the honeybee *Apis mellifera. Anim Cogn.* 2006;9:257–270.
72. Giurfa M, Eichmann B, Menzel R. Symmetry perception in an insect. *Nature.* 1996;382:458–461.
73. Rodriguez I, Gumbert A, Hempel de Ibarra N, Kunze J, Giurfa M. Symmetry is in the eye of the 'beeholder': innate preference for bilateral symmetry in flower-naive bumblebees. *Naturwissenschaften.* 2004;91:374–377.
74. Moller AP, Eriksson M. Pollinator preference for symmetrical flowers and sexual selection in plants. *Oikos.* 1995;73:15–22.
75. Murphy GL. *The Big Book of Concepts.* Cambridge, MA: MIT Press; 2002.
76. Murphy GL. What are categories and concepts? In: Mareschal D, Quinn PC, Lea SEG, eds. *The Making of Human Concepts.* New York: Oxford University Press; 2010:11–28.
77. Zentall TR, Wasserman EA, Lazareva OF, Thompson RKR, Rattermann MJ. Concept learning in animals. *Comp Cogn Behav Rev.* 2008;3:13–45.
78. Giurfa M, Zhang S, Jenett A, Menzel R, Srinivasan MV. The concepts of 'sameness' and 'difference' in an insect. *Nature.* 2001;410:930–933.
79. Avarguès-Weber A, Dyer AG, Giurfa M. Conceptualization of above and below relationships by an insect. *Proc Biol Sci.* 2010;278:898–905.
80. Avarguès-Weber A, Dyer AG, Combe M, Giurfa M. Simultaneous mastering of two abstract concepts by the miniature brain of bees. *Proc Natl Acad Sci USA.* 2012;109:7481–7486.
81. Zhang SW, Bock F, Si A, Tautz J, Srinivasan MV. Visual working memory in decision making by honey bees. *Proc Natl Acad Sci USA.* 2005;102:5250–5255.
82. Menzel R. Memory dynamics in the honeybee. *J Comp Physiol A.* 1999;185:323–340.
83. Clayton NS, Bussey TJ, Dickinson A. Can animals recall the past and plan for the future? *Nat Rev Neurosci.* 2003;4:685–691.
84. Clayton NS, Bussey TJ, Emery NJ, Dickinson A. Prometheus to Proust: the case for behavioural criteria for 'mental time travel'. *Trends Cogn Sci.* 2003;7:436–437.
85. Sara SJ. Retrieval and reconsolidation: toward a neurobiology of remembering. *Learn Mem.* 2000;7:73–84.
86. Carruthers P. *The Architecture of the Mind.* Oxford: Clarendon; 2006.
87. Lefebvre L, Reader SM, Sol D. Brains, innovations and evolution in birds and primates. *Brain Behav Evol.* 2004;63:233–246.
88. Jerison HJ. *Evolution of the Brain and Intelligence.* New York: Academic Press; 1973.
89. Hofman MA. Of brains and mind. A neurobiological treatise on the nature of intelligence. *Evol Cognit.* 2003;9:178–188.
90. Roth G, Dicke U. Evolution of the brain and intelligence. *Trends Cogn Sci.* 2005;9:250–257.
91. Laughlin SB, Sejnowski TJ. Communication in neuronal networks. *Science.* 2003;301:1870–1874.
92. Sejnowski TJ, Koch C, Churchland PS. Computational neuroscience. *Science.* 1988;241:1299–1306.
93. Edelman GM, Tononi G. *Consciousness. How Matter Becomes Imagination.* New York: Penguin Books; 2000.

CHAPTER

4

Exploring Brain Connectivity in Insect Model Systems of Learning and Memory

Neuroanatomy Revisited

Jürgen Rybak

Max Planck Institute for Chemical Ecology, Jena, Germany

INTRODUCTION

The cellular analysis of neuronal networks originated with the advance of microscopy and cell theory in the late 19th century culminating in the neuron doctrine,[1] which states that the elementary unit of the nervous system of any animal is the neuron. The most famous advocate of this doctrine was Ramon y Cajal, who applied this idea to his descriptions of neural networks.[2] Based on the concept of the synapse[3] and the theory of the biophysics of the neuron,[4] intracellular labeling and tract tracing techniques in conjunction with electron microscopy (EM) were and are still most successfully used to decipher neural networks, and in some cases they have led to complete mapping of the nervous system on the synaptic level (e.g., *Caenorhabditis elegans*[5]) or local brain circuits (e.g., stomatogastric ganglion[6] and lamina ganglionaris of the fly (for review, see Meinertzhagen[7]).

A second fundamental concept, dating back to the 19th century, is that brain functions underlying certain behaviors are localized in defined brain regions, as demonstrated by brain injury effects and postmortem inspections of the brain (e.g., the discovery of language areas by Brodmann[8]). Based on these and many other observations, the concept was developed that the brain is composed of functional and corresponding anatomical subregions serving different aspects of sensory, central, and motor behavior (in insects[9]). Depending on the organism, the analysis of anatomical brain circuits follows different strategies: Global analyses of macrocircuits are performed for large vertebrate brains with up to 85 billion neurons (in humans: cerebral cortex, 16×10^9 neurons; cerebellum, 69×10^9 neurons[10]; critically reviewed in Herculano-Houzel[11]), whereas local analyses are applied in imaging techniques (e.g., by indirect brain activity measured by positron emission tomography (PET) and functional magnetic resonance imaging (fMRI)) allowing to relate regions as small as $1\ mm^3$ to mental and behavioral events. Regarding circuit analyses in mammals, one conclusion of these approaches is depicted well by Douglas and Martin[12]:

> For those of us working on the large neocortex of higher mammals, such as cat and monkey, our aim is not to attempt a *C. elegans* circuit diagram, where the origin and destination of every wire is known. Instead our aim is to achieve a solution that is rule-based and probabilistic.

Attempts to study the insect brain may lie somewhere between these two approaches, and they may aim in the longer term for a *C. elegans* full-brain circuitry analysis.

In recent years, spurred by developments in microscopy, genetics, and computer algorithms, the concept of the connectome was coined,[13] which, in analogy to the genome, aims to read the complete brain in terms, but not exclusively, of its synaptic connectivity using light microscopic bulk tracing (e.g., brainbow[14]) and high-throughput electron microscopic analyses.[15] This concept also involves the idea of an integration of data from multiple levels of brain organization,[16] which is gradually being adapted for invertebrate brains.[9,17,18]

Invertebrate Learning and Memory.
DOI: http://dx.doi.org/10.1016/B978-0-12-415823-8.00004-6

INSECT BRAINS ARE SMALL

Horac Barlow[19] claimed that we may understand how the brain works only by studying the activity of individual cells (cited in Douglas and Martin[12]) and by analyzing efficient coding by explaining sensory information with the fewest number of neurons. He argued that knowledge of neural complexity is not necessary in order to develop concepts of how the brain works. As a corollary, invertebrate neuroscience developed the concept of the identified neuron,[20] taking advantage of the ease with which single neurons can be experimentally accessed. Many single neurons were discovered that appeared to be involved in different steps of sensory and motor processing. The advantages of small-brained insects were used to substantiate the identified neuron concept: accessibility, size, easy *in vivo* experimentation (for review, see Menzel[21]) in combination with immunohistochemistry,[22] tract tracing, and EM.[23–25] Thus, it was proposed to use mostly a bottom-up approach for constructing brain circuit models of functions for neurons in major regions of the insect brain (e.g., the antennal lobe[24] and the mushroom bodies[26,27]; Figure 4.1). Neuronal correlates of learning and memory processes are particularly well investigated in two model organisms: *Apis mellifera* and *Drosophila melanogaster*[31–33] (for reviews, see[21,26,27]). The use of transgenic animals with targeted gene expression and experimental manipulation at the neuronal single and population level[34] allows combining of functional studies (two-photon optical imaging) with genetic labeling[35] (see Chapters 6 and 27). These approaches have revolutionized the analysis of brain circuits and established *D. melanogaster* as an important model system for exploring brain circuitry. Although the methodical repertoire in the study of these two species is different, bottom-up approaches are applied in both species to build models of brain circuitry underlying olfactory processing and learning and memory.[36–38] The intricate network of identified circuitry is shown in Figure 4.1A1 for *A. mellifera* and in detail for the microcircuits in the mushroom bodies of both *Apis* and *Drosophila* in Figure 4.1A2 and A3.

METHODS OF ANALYZING INSECT MICROCIRCUITS

The dimensions of invertebrate brains are miniaturized compared to those of vertebrates species, and they have 10^2 to 10^8 neurons compared to vertebrates with 10^7 to 10^{14} neurons. Insect neurons are sparsely connected ($\sim$50 synapses per neuron), having a synaptic density between 0.5 and 2/μm^3 (optic lobes and the larval central brain in *Drosophila*), whereas central neurons of vertebrates have approximately 10 times more synapses per neuron.[7,17] In addition, invertebrate brains differ from their vertebrate counterparts by four characteristic features: (1) The somata are located at the periphery of the neuropil; (2) their neurites are shorter and not myelinated; (3) synapses are frequently polyadic (multiple contacts), with one presynaptic contact as opposed to many postsynaptic partners, leading to higher interconnectedness (in contrast, vertebrate synapses usually contain only one presynaptic and one postsynaptic element); and (4) information is encoded at a higher proportion by graded potentials in insects, leading to lower energy consumption compared with spike neurons of equal information transfer.[39] Taken together, central synapses in insects are packed densely and transmit information particularly efficiently.[40,41]

In the following sections, I review different neuroanatomical approaches and strategies that have been applied to analyze brain structures in insects and other species. These are summarized in Figure 4.2.

Sparse Neuron Reconstruction

Sparse labeling methods (Figure 4.2) such as the Golgi staining often select neurons at random and are not cell type specific.[42] However, because of their sensitive labeling capabilities, distinct cell types can be identified and analyzed by their fine morphology and their plasticity during development investigated. This was done for mushroom body Kenyon cells in honeybee and other species, which are difficult to label with other methods.[43] Electrophysiological studies combined with intracellular staining and immunohistology (e.g., for transmitter identification) play an important role and help in the construction of a bottom-up approach to neural networks with sparse reconstruction methods.[44] These approaches can be combined with light and serial section electron microscopy (for review, see Boeckh and Tolbert[24]).

Filling individual neurons with a tracer or using a whole cell patch electrode randomly located in a soma cluster is an extremely laborious method that cannot easily be adapted to large-scale data acquisition. However, extracellular recordings in conjunction with juxtacellular labeling[45] allow stable and long-time recordings *in vivo* and in deep tissue as well as analyses of parallel processing in neighboring neurons. The problem of neuron identification is resolved to some extent by the use of dye-coated electrodes. Limitations are that postsynaptic potentials cannot be recorded and cell labeling is usually incomplete and less specific.[46,47]

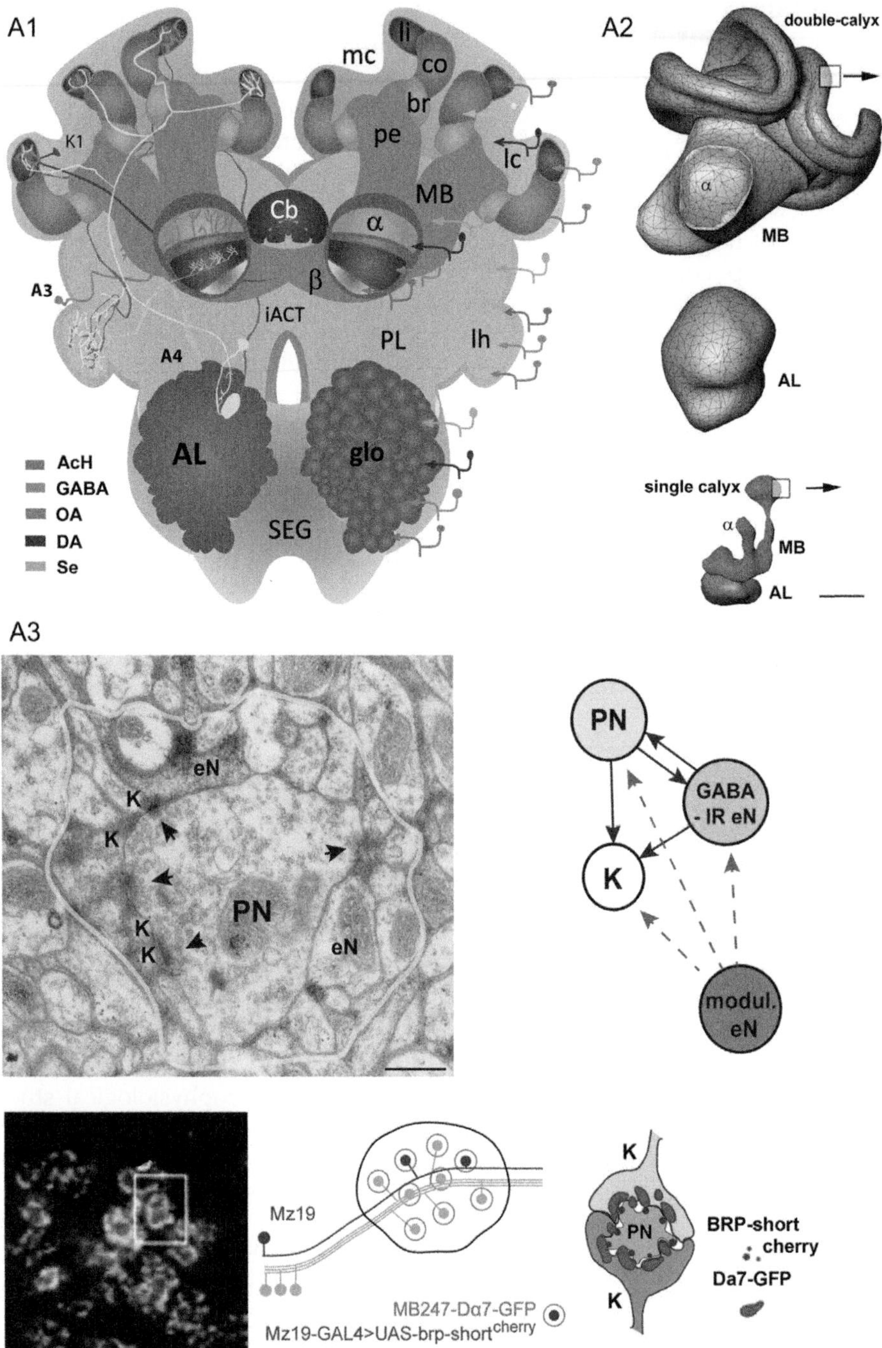

FIGURE 4.1 Neuroanatomy of the olfactory system in two insect species, *Apis mellifera* and *Drosophila melanogaster*. Brain architecture and local microcircuits in *Apis* (top panels in A2 and A3) and *Drosophila* (bottom panels in A2 and A3).(A1) The major brain neuropils, distribution of neurotransmitter, and principal neuron types in the olfactory pathway are shown for *Apis* in more detail. The approximately 160 glomeruli (glo) connect the antennal lobe via the inner (iACT; red) and the outer (yellow) antenna-cerebralis tracts to the mushroom bodies (MB). α, α lobe; β, β lobe; AL, antennal lobe; A3 and A4, extrinsic neurons of the MB; br, basal ring; Cb, central body; co, collar; K1, Kenyon cell, intrinsic neuron of the mb; lc: lateral calyx; lh, lateral horn; li, lip; mc: median calyx; pe, peduncle; PL, protocerebral lobes; SEG, subesophageal ganglion. Neurotransmitter networks: AcH, acetylcholine; DA, dopamine; GABA, γ-aminobutyric acid; OA, octopamine; Se, serotonine. Image courtesy of Alvar Prönneke. (A2) Antennal lobe (AL) and mushroom body (MB) differ in absolute and relative size (relative to brain volume) in the two species. The MB is composed of a double calyx in *Apis* and a single calyx in *Drosophila*. Scale bar $= 50\ \mu m$. Quantitative brain data for *Apis* worker bees: brain volume $\sim 3.77 \times 10^8\ \mu m^3$, with $\sim 8.5 \times 10^5$ neurons. Corresponding brain data for *Drosophila*: brain volume $8.30 \times 10^6\ \mu m^3$, with $\sim 10^5$ neurons. The volume of the paired MB in *Apis* is $8 \times 10^7\ \mu m^3$ (~21% of whole brain volume), with ~300,000 Kenyon cells, whereas the MB in *Drosophila* occupies $2.35 \times 10^5\ \mu m^3$ (2.8% of its brain volume), with ~5000 Kenyon cells. Thus, the relative MB volume in *Apis* is much larger than that in *Drosophila* relative to the whole brain. The AL in *Apis* has a volume of $2.1 \times 10^7\ \mu m^3$ (5.7% of whole brain

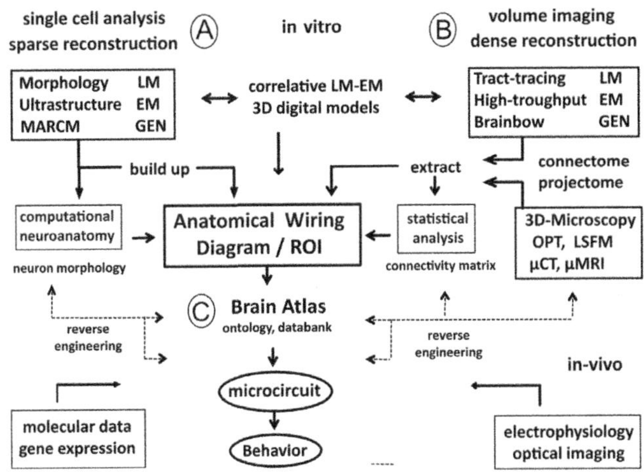

FIGURE 4.2 Flow diagram of neuroanatomical procedures for constructing neural brain circuits in the region of interest (ROI/ anatomical wiring diagram). (A) Sparse reconstruction techniques based on single cell analysis define the anatomical properties as achieved by light microscopy (LM) and electron microscopy (EM) techniques or by genetic (GEN) single cell labeling (MARCM) as applied in transgenic species. Quantitative features of identified neurons can be assessed using digital reconstructions of neurons (computational neuroanatomy). Data sampled from different experiments are 'build up' in anatomical wiring diagrams for the brain region of interest (ROI) (anatomical wiring diagram) or integrated and registered to standard brain atlases (C). (B) Dense reconstruction (or volume imaging) is based on high-throughput techniques permitting scanning of large volumes of neural tissue at different levels of resolution using LM tracer or data from high-throughput EM studies. 'Brainbow' is a genetic multilabeling method at the LM level. 3D microscopy allows data analysis in whole mounts, such as optical tomographic microscopy (OPT) and light sheet-based fluorescence microscopy (LSFM), or scanner techniques: micro computer tomography (µCT) and micro magnet resonance imaging (µMRI). Volume mass data are evaluated statistically, or selected data are extracted by refinement of high-throughput techniques allocating to focus on the ROI. Correlative LM–EM allows to relate data acquired at different levels of resolution. (C) A central role is played by spatial reference systems integrated by image registration techniques—standard brain atlases based on structured neuron nomenclatures (ontology) and composed of structured databases. Brain atlases are also data repositories for data of different modalities (molecular and physiological) from *in vitro* and *in vivo* experiments. The dashed line indicates the approach of reverse engineering, allowing for feedback between different levels of data sampling. Results from functional and behavioral studies can be used to simulate brain function and to create hypothetical brain models at various levels of neural complexity.

Genetic Techniques to Explore Neural Connectivity

The binary Gal4–upstream activating sequence (UAS) system[48] allows ectopic gene expression in a genetically defined subpopulation of cells (see Chapter 6). A transgenic Gal4 animal line determining in which cells a given gene shall be expressed is crossed with a UAS reporter line to determine which gene is being expressed in these cells. Gal4 is a yeast transcriptional activator Gal4 (for which no natural recognition sequence exists in wild-type fly strains) that is under the control of a cell-specific enhancer. The UAS reporter line houses a transgene cloned right behind a Gal4–UAS. Only in cells that express Gal4 and UAS reporter will the gene be expressed. A fluorescent probe tagged with the reporters allows visualization. Specific neuronal populations can be explored by visually screening GAL4 lines with the GAL4–UAS technique.[49–52]

GAL4 lines allow the study of neuron morphology but often have the problem that a dense, spatial clustering of the neuron population occurs. In order to control spatial and temporal expression, and gain sparse reconstruction of single neurons, MARCM (mosaic analysis with a repressible cell marker[53]) is used (Figure 4.2). Expression can be restricted to a neuron's lineage and eventually to single neurons.[9,50,54] Taking advantage of the determined cell lineage scenario in insects, neurons are labeled by heat shock-induced mitotic recombination and site-specific FLP/FRT recombinase during embryogenesis using the repressive marker GAL80.[34] Other binary expression systems that are complementary to the GAL4–UAS system can be used, such as the analogous VP16/LexAop and the MARCM-independent Q system.[55] Here, the DNA-binding and transactivating moieties are independently targeted using distinct promoters to achieve highly restricted, intersectional expression patterns.

The genetic approaches described previously allow the expression of fluorescent calcium probes such as GCaMP, chloemalion, or channelrhodopsin (reviewed in Strutz *et al.*[56]; see also Chapter 6).

◀ volume, and ~26% of the MB volume). The AL volume in *Drosophila* is slightly larger than that of the MB. MB and AL together occupy ~27% of brain volume in *Apis* compared to ~6% in *Drosophila*. Data sources: Chiang et al.,[9] Brandt *et al.*,[28] and the Virtual Fly Brain website (www.virtualflybrain.org). (A3) Microglomeruli in the calyx of *Apis* and *Drosophila*. (Top) *Apis*: On the ultrastructural level, boutons of projection neurons (PN) form the core of microglomeruli, which are presynaptically connected to Kenyon cells (K). Arrows indicate the presynaptic densities in the PN profile. eN, profiles from extrinsic neurons. Scale bar = 0.5 µm. (Right) A scheme of putative synaptic connections of a microglomerulus. Arrows indicate synaptic connections. GABA-ir eN, GABA-like immunoreactivity extrinsic neuron; modul. eN, modulatory interneuron. Modified from Ganeshina and Menzel.[29] (Bottom) *Drosophila*: Model of a microglomerulus based on genetic labeling and fluorescent tags visualized by confocal microscopy. The ectopic expression of synaptic proteins is located presynaptically in a bouton of a projection neuron (inset on the left). Bruchpilot (BRP-short) is a presynaptic protein located in transgenic labeled PN neurons of the *Drosophila* GAL4 line MZ 19 (red dots). Kenyon cells of the GAL4 line MB247 expressing the postsynaptic protein D-alpha-7 tagged with green fluorescent protein (Dα7-GFP) surrounding PN profiles in claw-like endings of Kenyon cell dendrites (K) (see also Figure 4.A3). Source: *After Kremer* et al.[30]

Masuyama *et al.*[57] proposed a novel activity reporter system that can be used to map synaptic circuitry activity dependently. Using this method, the nuclear factor of activated T cells (NFAT) labels active neurons in behaving animals *in vivo*, leading to traces of the active neurons during a particular behavior.

In order to avoid time-consuming EM analyses, a light microscopy (LM)-based connectome method has been proposed in mice,[14] and this was adapted for *Drosophila* in two versions, Brainbow[58] and Flybow[59] (Figure 4.2). It allows multiple-color visualization of a population of neurons, and it helps to estimate putative connectivities through genetic and random expression of up to 100 different hues of fluorescent color tags (XFP). This method is based on a recombinase technique (using Cre-Lox from bacteriophages) that is site specific and can be used for cell lineage tracing studies. In *Drosophila*, the GAL4–UAS system was combined with the Cre-Lox recombinase system to target specific cell populations (lineages). Combining the brainbow technique with promoter-induced synaptic fluorophore provides the possibility of local analysis without using traditional tracer techniques.

A major drawback of the brainbow method is the toxic effect of Cre recombinase. Furthermore, it allows only separation of neuron classes rather than single neurons when combined with the GAL4 method. The brainbow technique is good for sparsely innervating neuron populations because the diffraction limit of the light microscope cannot resolve blending of fluorescent labels in dense labeled tissue, and fine dendritic or axonal terminals cannot be resolved. Mishchenko[60] proposed a 'synaptic brainbow' in which a fluorophore can be expressed at synaptic sites in identified neurons achieved by using the UAS/GAL4 libraries. The connectivity between different classes of neurons may be studied directly by expressing a presynaptic marker in one class of neurons and a postsynaptic marker in another class of neurons localizing synaptic puncta as potential synaptic sites. Circuits involving less than 1000 neurons in a population may be mapped efficiently by multiplexing fluorescent markers.

Targeted genetic expression of pre- and postsynaptic proteins tagged with fluorescent probes in GAL4 lines specific for olfactory projection neurons (MZ19) and Kenyon cells (MB247) of the mushroom bodies (MBs) can be used to define the output and input compartments of identified neurons and reveal microcircuits in the MBs of *Drosophila* (Figure 4.1A3 and 4.3A5). The synaptic structure is composed of bruchpilot, an active zone protein located presynaptically in projection neuron boutons, and the postsynaptic D-alpha-7 subunit of the acetylcholine receptor located on the surface of claw-like spiny Kenyon cells.[30] Moreover, Christiansen *et al.*[61] showed that Kenyon cells are also presynaptic in the MB calyces, thus forming recurrent loops within the glomeruli—a finding that was also reported by Butcher *et al.*[62] in serial section EM.

The transynaptic marker GRASP (GFP reconstitution across synaptic partners) defines cell contacts and synapses *in vitro* and *in vivo*.[63,64] This split GFP genetic technique[65] allows the mapping of putative synaptic partners by labeling synapses based on the proximity of the presynaptic and the postsynaptic plasma membrane. GFP reconstitutes as a fluorescent molecule targeted to specific synaptic sites, such as postsynaptic neurolignins, and attaches to presynaptic neurolexins that act as cell recognition molecules.

Synaptic connections can also be revealed by applying trans-synaptically transported proteins (WGA and plant lectin) or by infecting neurons with neurotropic viruses such as the rabies virus.[66] The rabies virus spreads retrogradely throughout the nervous system via chemical synapses, killing the host by initially infecting muscle cells (via motor endplates) and spreading from the periphery to the central nervous system.[67] Genetic modification of the virus by replacing the glycoprotein by fluorescent tags (EGFP) and pseudotyping its envelop proteins derived from another (noninfectious) virus allows the spread of the virus to be followed, targeting specific cells of interest and their functional synaptic network. This genetically controlled spread allows the tracing of pairwise connected postsynaptic cells and their presynaptic neighbor cells.[68] Dual infections using different strains of the recombinant rabies virus[69] in the olfactory pathway established a long-range connectivity map of the olfactory bulb glomeruli with their cortical areas in transgenic mice.[70]

Improvements can be made by using the genetic approaches. Putative synaptic connections can only be resolved with super-resolution microscopes (STED, PALM, Storm, SIM, and IDLM[71]; see also Table 4.1) or by EM in conjunction with correlative array tomography.[84,88] These methods allow imaging of synaptic structures below the diffraction limit of the light microscope, an essential requirement for identifying synaptic contacts. Mishchenko[89] provides a comparison of two LM methods: two-component fluorescent synaptic markers,[60] as in GRASP,[63] and the recombinase system for stochastic gene expression (brainbow). He derived statistically realistic values for synaptic connectivity in *C. elegans* when measured across many animals by simulating the reconstructions of neural connectivity matrices for measurement of fluorescent puncta, comparing it with ground truth of actually synaptic connectivity derived from EM studies.[5]

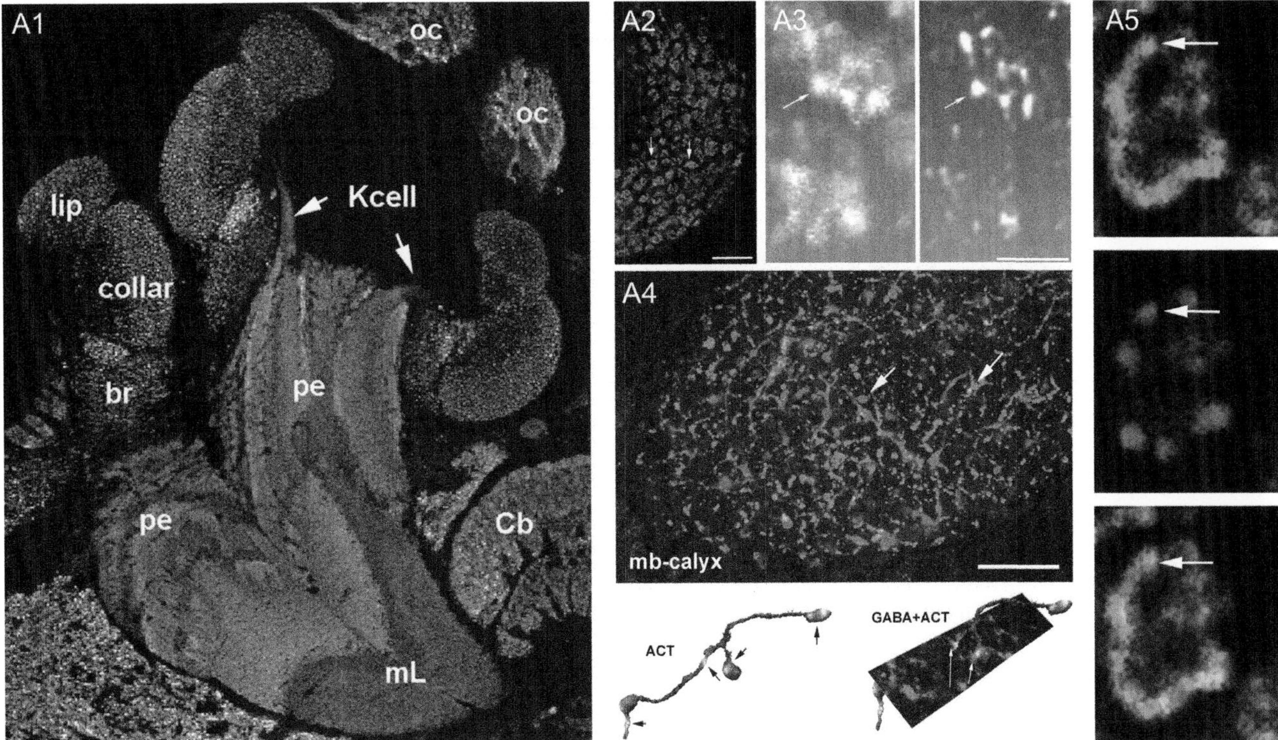

FIGURE 4.3 Confocal microscopy (CSLM) of the mushroom body of *A. mellifera* stained with the synaptic neuropil marker NC82. This procedure allows visualizing brain regions after tagging the secondary antibody with fluorescent fluorophores, here Cy-3. (A1) Nc 82, an presynaptic active zone protein antibody, recognizes synaptic sites. The texture assessed at low resolution reflects the synaptic density in neuropil regions of the mushroom body (mb) and close-by areas of the protocerebral lobe and central body (Cb). br, basal ring; Kcell, axon strands of Kenyon cells; mL, medial lobe (also known as β lobe); oc: ocellus; pe, peduncle. (A2 and A3) High-resolution confocal microscopy allows visualization of microglomeruli in the calyces of the mb. Panel A2 shows punctate structures arranged in bouton-like structures in the collar region (inset in A1), which contain presumably presynaptic densities (arrows). Panel A3 shows a synaptic bouton in the lip region of the calyx before and after deconvolution (20 × oil objective); arrows indicate the presumably presynaptic density. (A4) Dual-channel CSLM analysis allows the mapping of GABA-immunoreactive-like profiles (green) onto axonal terminals of olfactory projection neurons (ACT) (red) in a subregion of the mushroom body calyx (arrows). (Bottom) After reconstruction of single projection neuron arbors (left) and mapping them with appropriate software onto the green channel (right), close attached GABA-immunoreactive profiles can be identified with a high probability as putative synaptic contacts, as indicated by the bright spots (arrows). Scale bars: A2 = 10 μm; A3 = 2 μm; A4 = 20 μm. (A5) Genetic labeling in *D. melanogaster*. (Top) Synaptic sites in a microglomerulus bouton are imaged by dual channels and then merged (arrows). Green indicates postsynaptic sites labeled in a Kenyon cell line expressing the D-alpha-7 protein, a subunit of the acetylcholine receptor. Red dots symbolize the presence of GFP fluorescence-tagged presynaptic protein, bruchpilot, labeled in the GAL4-MZ19 line of olfactory projection neurons. Source: *From Kremer* et al.[30] *See also Figure 4.1A3 (bottom).*

High-Throughput, Dense Reconstructions of Brain Circuitry

Depending on the level of resolution required, the connectome or projectome[13] can be accomplished by mass staining, as in the case of the brainbow, or by establishing a complete synaptic wiring diagram—a synaptome[90]—using the high-throughput EM or whole brain scanning techniques (three-dimensional (3D) optical tomography, high-resolution MRI, and computed tomography)[16] (Figure 4.2B). EM techniques aimed at obtaining maps of synaptic connections in large brain volumes are dense circuit analyses (Figure 4.2). They avoid the major drawbacks of classical EM studies, which are time-consuming, error-prone with regard to serial sectioning, and misalignment of ultrathin sections occurs. Two techniques based on automated data acquisition have been proposed: serial block-face scanning electron microscopy (SBF-SEM)[15,91] and focused ion beam scanning electron microscopy (FIB/SEM).[85,92] Both methods use backscattered electron technology for imaging. In SFB-SEM, resin-embedded blocks, prepared for EM analysis, are used to directly image the region of interest using fully automated diamond knife technology.[15] Images are automatically aligned but lost during sectioning, and only a large field of view can be achieved. The disadvantages are as follows: (1) High acceleration voltage leads to increased hardness of resin in the block, causing cross-linking of resin; (2) image acquisition time is long; (3) due to the

TABLE 4.1 Overview of Microscopic and Genetic Techniques Reviewed[a]

Abbreviation	Technique	Species	Condition	References
Light microscopy, 2D				
CSLM1	Confocal microscopy	A, D	*In vitro/ In vivo*	72
Two-photon	Multiphoton microcopy	A, D	*In vitro/ In vivo*	73, 97
STORM	Stochastic optical reconstruction microscopy		*In vitro/In vivo*	71*
PALM	Photoactivated localization microscopy		*In vitro/In vivo*	71*
STED5	Stimulated emission depletion	D	*In vitro*	74
Light microscopy, 3D				
SIM	Structured illumination microscopy		*In vitro*	75
LSFM	Light sheet fluorescence microscopy		*In vitro*	76
UM	Ultramicroscopy	D	*In vitro*	77,78
OPT	Optical tomography		*In vitro/In vivo*	79,141
SLOTy	Scanning laser optical tomography	L	*In vitro*	80
IDLM	Isotropic diffraction limited microscopy		*In vitro*	60
Genetics				
GRASP	GFP reconstitution across synaptic partners	D	*In vitro/In vivo*	63
MiniSOG	Mini singlet oxygen generator		*In vitro*	81
NFAT	Nuclear factor of activated T cells	D	*In vitro/In vivo*	57
dBrainbow	Brainbow	D	*In vitro*	58
Electron microscopy, 2D				
ss-TEM	Serial section transmission electron microscopy	A, D	*In vitro*	82
ss-SEM	Serial section scanning electron microscopy	A, D	*In vitro*	83
CAT	Correlative array tomography		*In vitro*	84
Electron microscopy, 3D/SEM = scanning electron microscopy imaging				
SBF-SEM	Serial block-face SEM		*In vitro*	15
FIB-SEM	Focused ion beam SEM		*In vitro*	85
μ-CT	Computed tomography	A	*In vitro*	86
μ-NMR	Nuclear magnetic resonance)	A	*In vitro*	87

[a]*These are categorized according to data acquisition (see also Figure 4.2) and species and their availability for* in vivo *approaches, either per se or in conjunction with physiological experiments. Some of the techniques are used in combination, providing the possibility to study neural circuits. References refer to original work or applied usage.*
* *Review: Techniques as applied in* Apis mellifera *(A),* Drosophila melanogaster *(D), and* Locust migratoria *(L).*

high cost of the equipment, only a few specialized users can perform this technique; and (4) sections are lost, so the preparation cannot be processed for multiple labeling (e.g., post-embedding techniques) of synaptic contacts. FIB/SEM:gallium ion milling (i.e., removing thin layers of tissue) is applied combined with block-face image scanning (backscattered electron imaging). No cross-linking of resin is observed as in SFB-SEM. Therefore, deeper tissue penetration and data acquisition at depths up to 900 μm are possible.[85] Both techniques allow for high isotropic resolution ($20 \times 20 \times 25$ nm in SFB-SEM[91] and $4 \times 4 \times 15-40$ nm in FIB[85]), which is particularly important for detection, reliable tracing, and identification of fine neurites below 100 nm. These fine neurites are characteristic features of insect neuropil.[7,17]

The automated image stack acquisition of both techniques requires new methods for automated neuron reconstruction, segmentation, and machine-learning algorithms for automated synapse identification.[93–95] Drawbacks relate to the limits of fully automated segmentation procedures used for profiles, leaving the experimenter with time-consuming manual segmentation and error-proofing. Synapse counts in a small local region of interest are technically possible, but long-distance connections are difficult to analyze.[96] This problem is partially solved by combining EM studies with *in vivo* two photon microscopy.[97] These techniques have yet to be applied to invertebrate brains. A total of 26 TB of data storage would be necessary to store the information of the *Drosophila* brain covering 7 million μm^3 brain volume at an EM magnification of 4000 ×, which would be required to identify synapses.[98] The major drawback of automated high-throughput EM can be partially overcome by using the GAL4–UAS expression system in neuronal populations of interest using genetically induced markers. Horseradish peroxidase (HRP) can be expressed and restricted to the location of the membrane protein CD2.[99,100] In transgenic animals, the HRP::CD2 label allows analyses of the ultrastructure without compromising cytoplasma contents. Using this technique in combination with genetic MARCM and subsequent EM analysis showed that embryonic projection neurons contribute to the adult olfactory circuitry.[101]

Serial section transmission EM (ss-TEM) has been successfully combined with immunofluorescence confocal microscopy in nontransgenic zebra finches.[88] Long strands of ultrathin section were captured on wavers, scanned with backscatter electron technology (as in automated block-face scanning). Different features of the neural circuit can be elucidated by correlated image registration techniques at the LM and EM levels.[84] In addition, super-resolution microscopy (SRM) and genetic techniques (synaptic brainbow[60]) can be applied to obtain ultrastructure resolution. In the SRM nanoscopy, for example, the presence of synaptic proteins or ion channels can be superimposed on EM labeling, thus facilitating the identification of dendritic and axonal compartments.[71] Another method of synapse mapping was proposed by Shu *et al.*[81] for the mouse brain. The mini singlet oxygen generator (MiniSOG), a fluorescence flavoprotein for EM analysis, generates singlet oxygen molecules upon illumination catalyzing diaminobenzidine, an electron dense marker allowing ultrastructural analysis on a large scale. Like GFP, this fluorophore can function as a tag for proteins specific for synaptic sites (e.g., the synaptic cell adhesion molecules SynCAM1 and -2).

Estimates of Synaptic Connectivity Using the Light Microscope

An interesting line of research designed to estimate anatomical and functional connectivity in the neocortex for the thalamocortical pathway of the whisker barrel cortex applies a reverse engineering approach: Reconstructions of single neurons at the light microscopic level using standardized template model and registration techniques are used to match data from several experiments into a common space (3D reference frame). The resulting average neocortical column involves semiautomatic tracing of individual neurons (based on biocytin stainings), including morphometric parameters (somata, bouton, spine, and synaptic density). The estimate of the number and cell type distribution of synapses in this network includes *in vivo* recordings of physiological properties. Together, predictions of functional connectivity and contribution of single neural elements are possible.[102,103] It is interesting in this context that a single neocortical barrel column (~300 × 1000 μm) is equal in size to the MB peduncle and lobes of the honeybee (Figure 4.4). Such reverse engineering approaches including diverse circuit and single neuron parameters are applied in the Blue Brain and Cajal Brain projects.[105] A similar approach for insects has been proposed by Namiki *et al.*[106] Currently, high-throughput LM analyses using mass neuron data in conjunction with image registration into brain templates are only possible in transgenic *Drosophila* (based on GAL4 libraries and single cell analysis MARCM). Connectivity maps based on single overlapping dendritic and axonal endings of single neurons and populations of neurons can be established by matching defined local processing units.[9,50] These approaches provide valuable hints about the relationship between neuropil structures at the mesoscale and information processing units allowing estimates of synaptic density, but they do not provide enough spatial resolution to predict true synaptic contacts. LM contacts do not necessarily verify the presence of true synaptic contacts. In vertebrates, it has been calculated that only 20% of all contacts seen in the LM are in fact synaptic contacts.[96]

A direct LM approach to estimate synaptic distributions is possible by combining immunohistochemistry with high-resolution confocal LM with precise 3D dendritic surface reconstructions. This approach allows for automated co-localization analyses in order to map the distribution of potential synaptic contacts onto dendritic trees or axon terminals.[18,107,108] Hohensee *et al.*[109] claim that this method provides a good estimate of the number of synaptic contacts. Prerequisites for this approach are computer-assisted reconstructions and quantification of morphological parameters as indicated in Figures 4.4A3 and A4.

3D Microscopy Volume Imaging with the Light Microscope

Optical projection tomography (OPT) is based on single projections with full volumetric information. The resulting 3D images do not yet reach cellular

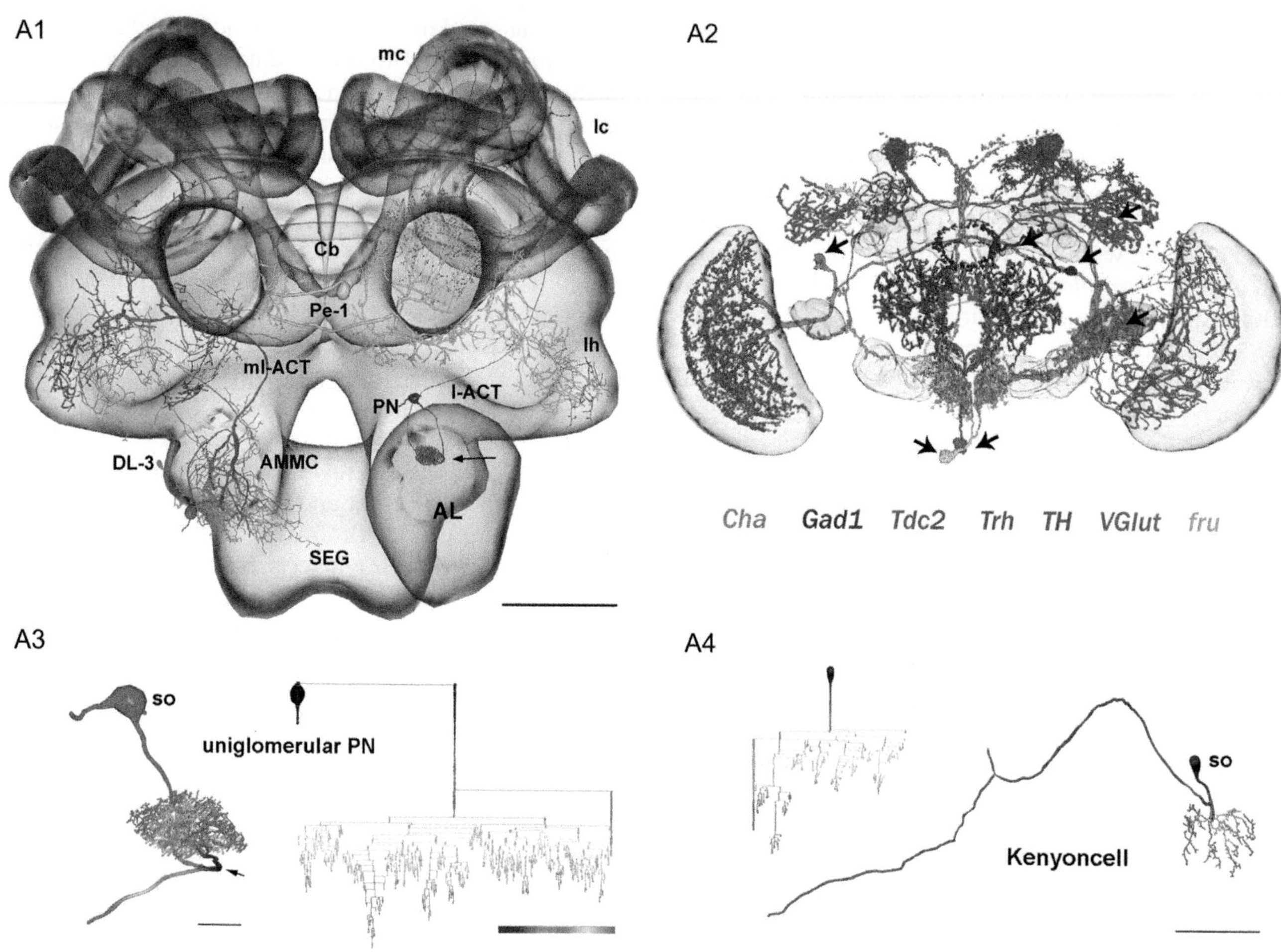

FIGURE 4.4 Standard brain atlases in *Apis* and *Drosophila*. (A1) Single neurons acquired from different experiments are registered in the Honeybee Standard Brain Atlas (major brain neuropils in transparency). Olfactory and mechanosensory pathways in the central bee brain are shown. Left hemisphere: The L3 projection neuron (PN; blue) projects via the lateral antenna-cerebralis tract (l-ACT) to the lateral horn (lh) and mushroom body calyces (mc and lc). The mushroom body output neuron (Pe1; yellow) with dendritic input in the pedunculus and output targets in the protocerebral lobe overlaps with L3 in the lh. Right hemisphere: A mediolateral antenno-cerebralis tract neuron (ml-ACT; red) and a mechanosensory cell (DL-3; green) targeting disperse regions in the central brain, the anteno-mechanomotor center (AMMC) and subesophageal ganglion (SEG). Scale bar = 100 μm. Adapted from Rybak.[104] (A2) Integration of neurons into the standard brain atlas of *Drosophila melanogaster*. Using mosaic analysis with a repressible cell marker (MARCM), single cells can be resolved stemming from different genetic lines. A combined 3D visualization of seven different neuronal populations, each from a separate Gal4 line, in the *Drosophila* standard model brain is shown. Each color represents a Gal4 line and its representative neuron containing the putative neurotransmitter, which in some cases was validated by immunolabeling (*): Gad1-Gal4 (GABA), Cha-Gal4 (acetylcholine), Tdc2-Gal4* (octopamine), Trh-Gal4 (serotonin), TH-Gal4* (dopamine), and VGlut-Gal4 *(glutamate). fru, fruitless (group of Kenyon cell). Arrows indicate the cell body cluster. From Chiang *et al.*[9] (A3 and A4) 3D reconstruction of the glomerular dendritic arborizations of the L3 projection neuron (uniglomerular PN) in the antennal lobe (see arrow in A1) and dendrites of a Kenyon cell (type K1) in the mushroom body calyces, its axonal projections, and their corresponding dendrograms. The neuronal distance measured from the origin (dark blue) of the reconstructed neuron is false color coded (the origin is indicated by an arrow; left, origin for K1 cell; so, soma). Denritic length: uniglomerular PN neuron, ~3200 μm; Kenyon cell, ~720 μm. Scale bars: A3 = 20 μm; A4 = 50 μm. Data for panels A1 and A3 courtesy of Hiro Ai, Daniel Münch, and Gisela Manz. Source: *Adapted from Rybak.*[104]

resolution (best = 20 μm), but they overcome the disadvantages of confocal microscopy (small volumes, optical sectioning requiring reconstruction in 3D, and fluorescence bleaching in out-of-focus plane; for *Drosophila*, see McGurk *et al.*[79]). An improved axial resolution reveals nearly isotropic information. OPT, in combination with good tissue clearing and the use of proper objectives (oil and long working distance), can be applied to fluorescent specimens (e.g., in combination with brainbow techniques) to allow the sampling of structural properties at the cellular level (spines and dendrite morphology) in large volumes. Scanning laser optical tomography (SLOTy) has been proposed for degeneration/regeneration studies in locust olfactory sensory fibers and for studies investigating volumetric plasticity in the central brain of insects.[80] In contrast to conventional OPT, this method allows measurement of several modalities of light, including transmission modulation of the respective structures and fluorescent-tagged proteins such as channels.

Selective plane illumination microscopy and ultramicroscopy[77] are volume imaging techniques of whole specimens spanning the gap between techniques such as fMRI and PET by scanning through 3D space. These methods allow acquisition of morphological and functional data with high resolution in conjunction with specimen clearing agents.[110,111] Next-generation volume imaging methods are expected to avoid the problems of the confocal LM (limited penetration of marker agents and optical beam, minimizing distortion of the tissue[80]). These advances will be of great value with respect to tract tracing (long-range, projectome), for comparative studies of brain neuropils (e.g., glomeruli organization) neuron mass stainings (GAL4 line screening and brainbow) as well as for *in vivo* studies.[141]

Bioinformatics

The development of algorithms has been extremely successful for the evaluation of the localization of single fluorophores in super-resolution microscopy. For example, appropriate statistical procedures have been developed for high-throughput analyses of synaptic connectomes,[112] for the automation of segmentation from 3D image stacks leading to shape models,[18] for machine learning protocols in the automated recognition of image features,[95] and for image registration tools.[113] Open-source imaging processing software (e.g., Image J) is most important for biologists.[114] Image registration techniques allow matching images derived from different preparations, multiple sources, and collected at different resolutions into standard brain models. Knowledge bases are incorporated into ontologies, data repositories, and data banks, greatly enhancing the value of such brain atlases.[115]

A common coordinate system allows the registration of neuronal data by spatial normalization procedures—alignment of images and spatial warping (e.g., see Allen brain mouse and human atlas, http://www.brain-map.org; *Drosophila* atlas, http://www.virtualflybrain.org and http://www.flycircuit.tw[9,116]). Brain images, acquired by mechanical, optical, or tomographic methods, need to be first segmented and then rendered as 3D images. By spanning the gap between grades of resolution (scaling) and image modality such digital models serve many purposes, including as didactic, interactive models during experimentation and as templates in correlative LM–EM studies.[17,104,117,118] Examples are (1) synaptic connectivity within small volumes of the larval *Drosophila* brain based on ss-TEM in conjunction with image registration of sparse labeled confocal images[17] and (2) analysis of the TEM stack at the microcircuit level into a light microscopically derived 3D framework of landmark structures, thus solving the problem of identification of neural elements unlabeled at the ultrastructural level.

To integrate morphological data into a common reference system, two standardization methods have been employed for insect brains (Figure 4.4). First, probabilistic atlases based on comparative volume analyses of brain neuropils allow the quantitization of the variability of the neurons of interest. Such approaches have been successfully applied in developmental studies and studies on neuronal plasticity.[119,120] Second, the iterative shape averaging (ISA) procedure[28,121,122] eliminates specimen shape variability[120,123] and accumulates structural data in the reference space of the standard brain atlas. In comparison to the Virtual Insect Brain protocol, which is based on selecting an individual representative brain (*Drosophila* standard brain atlas) REIN , the ISA averaging procedure is best suited for the registration of neurons collected from different brains http://www.neurobiologie.fu-berlin.de/beebrain/.[104]

Spatially registered data in shared atlases formats are advantageous for comparing data sets among laboratories (data-sharing platforms), providing a platform in which users can register their data and queries. Software tools for warping and spatial register data are an open source of information, including the software package Fiji.[124,125] A controlled vocabulary (ontology[126,127]) and data bank algorithms must be provided.[115] Thus, it is possible to integrate data over a large range of resolution and modalities[9] into a common 3D frame.[128,129] In conjunction with genomic and proteomic data banks, subcellular and cellular networks are connected.

Measures of brain circuit organization achieved by image registration using a brain atlas have been convincingly demonstrated[9] and are reviewed by Kohl and Jefferis.[130] Using GAL4–UAS and single cell MARCM, 16,000 (out of an estimated 100,000) neurons in the *Drosophila* brain were registered to an average brain atlas using the global affine registration algorithm. Neuron classes (defined as originated from clonally related cells and temporally divided by cell birth) could be brought into a new context. A connectivity matrix for the whole *Drosophila* brain was tentatively related to 41 local processing units, six hubs, and 58 tracts as states of information flow in the brain (Figure 4.4A4).

Standard brain atlases are useful in many contexts, particularly in connection with an ontology and data bank. The *Drosophila* standard brain has been successfully connected to an ontology and includes genomic and transcriptomic databases.[127] Further developments will include the incorporation of old data from the literature (camera lucida drawings and histochemical mappings) and bibliographic references (e.g., PubMed, http://www.ncbi.nlm.nih.gov/pubmed; BioInfoBank, http://www.lib.bioinfo.pl)[131] and new data from, for example, cellular expression patterns of synaptic proteins and EM studies.[61,62] Ultimately, the data have to be collected in 3D computer animations. Digital 3D brain atlases can also be used for reverse engineering approaches combining functional and anatomical data.[106,132]

The current situation of standard atlases indicates progress in registration accuracy using either the standard or templates[18,50] or threshold and landmark segmentation.[125] Attempts are being made to develop common platforms for the presentation of 3D atlases, for example, in the case of the digital brain atlasing projects of the ICNF or the virtual fly brain[133] (see http://scalablebrainatlas.incf.org and http://www.virtualflybrain.org).

DISCUSSION AND OUTLOOK

The strength of sparse labeling techniques lies in the identification of individual cells, such as in an electrophysiological experiment combined with anatomical marking. The neuron type, and in some cases even the individual neuron, will be characterized in functional terms based on immunohistological data and those from correlative LM–EM studies. In genetic amenable species such as *Drosophila*, GAL4 techniques in conjunction MARCM allow single cell analysis.

A major question in functional neuroanatomy is how we get from the single neuron level to the network level, which includes the synaptic communication in a population of neurons or certain brain regions. An important step is made by combinations of structural and functional image techniques that combine data from different experiments in a spatial reference frame (see Chapter 6). Atlases integrate our knowledge of anatomical features with molecular and physiological data, and they allow us to reach to a certain degree conclusions about neuronal processing by sampling large amounts of single cells as shown for different insect species on the mesoscopic scale.

Can LM help us to make statements about synaptic connectivity? Genetic labeling techniques such as GRASP allow the identification of synaptic connectivity by expressing fluorescing marker molecules at defined synaptic sites with synaptic partners. In conjunction with high-resolution LM techniques, these data can be sampled on a larger scale. The two-component co-localization synaptic marker[89] at the LM level is compared to other approaches with respect to the reconstruction effort and errors in serial EM and pairwise recordings. However, the gold standard for synaptic connectivity is still the EM because it reveals the essential cellular parameters and distinguishes between chemical and electrical synapses, localizing neurotransmitter vesicles, and so on. The best route to follow here is to use correlative approaches such as array tomography,[88] which can be combined with sparse labeling techniques with great success.

The goal of these efforts is the connectome. The connectivity phenotype on the mesoscale level provides an important missing link between the genotype and behavioral phenotype.[134] In vertebrates, the concept of the connectome applies to the mesoscopic scale simply because of the sheer complexity and number of neurons and synapses that allow only statistical assumptions on true synaptic connections. Therefore, for the analysis of large, complex networks in vertebrates, rule-based and statistical analysis of connectivity has been proposed (e.g., Zamora-Lopez *et al.*[135]; but see also Dercksen *et al.*[103]) in combination with new concepts and experimental approaches, such as network theory.[95,136,137] For insect brains, some of these concepts might be useful considering the complexity of neuropils such as the MB that are unlikely to be analyzed by sparse reconstruction methods only. So far, connectivity matrices in *Drosophila* are based on statistical measures and have been applied successfully on a mesoscopic scale, but this needs to be seen as an intermediate step until more information about both the anatomical and the functional basis is available.

Automated EM such as FIB-SEM and SBF-SEM[85,95] are promising methods for establishing complete connectomes particularly in insect brains, but statistical accounts on classes of synapses make sense only if the network is characterized to a fairly high degree, as in the lamina ganglionaris,[138] or if such automated scans are combined with identified cells as in the retina of vertebrates.[97]

Behavior cannot be predicted from a 'connectome' alone because the brain contains a chemical 'map' of neuromodulation superimposed upon its synaptic connectivity map. Internal states such as hunger or arousal change behavior.[139] Thus, static analyses are insufficient, and *in vivo* studies are needed that allow the tracking of characterized neurons and neural nets in time. Unfortunately, no imaging technique is available to date allowing the direct visualization of microcircuits as a whole with the sufficient temporal or spatial resolution. However, new approaches are on the horizon, such as in the zebrafish.[140] Here, two-photon calcium imaging is used to record from many neurons of larval zebrafish brains involved in sensory feedback and locomotion.

Perhaps the most challenging avenue for circuit-based explanations of brain function and behavioral control lies in the combination of high-throughput technology (e.g., automated EM and genetic labeling such as brainbow) with sparse labeling techniques (e.g., intracellular recording on identified elements) and functional imaging of large sets of neurons.

References

1. Waldeyer-Hartz. Über einige neuere Forschungen im Gebiete der Anatomie des Centralnervensystems. *Deutsch med Wschr.* 1891;17(44):1213–1218.
2. Jones EG. The impossible interview with the man of the neuron doctrine. *J Hist Neurosci.* 2006;15(4):326–340.
3. Sherrington CS. Sherrington, C. S. to Sharpey-Schäfer, E. A. Letter 27, in reference PP/ESS/B21/81897, Wellcome Institute for the History of Medicine: Sharpey-Schäfer papers in the Contemporary Medical Archives Centres.
4. Hodgkin AL, Huxley AF. A quantitative description of membrane current and its application to conduction and excitation in nerve. *J Physiol.* 1952;117:500–544.
5. White JG, Southgate E, Thomson JN. The structure of the nervous system of the nematode *Caenorhabditis elegans*. *Phil Transact R Soc Lond B.* 1986;314:1–340.
6. Nusbaum MP, Beenhakker MP. A small-systems approach to motor pattern generation. *Nature.* 2002;447:343–350.
7. Meinertzhagen IA. The organisation of invertebrate brains: cells, synapses and circuits. *Acta Zool.* 2010;91(1):64–71.
8. Brodmann K. *Vergleichende Lokalisationslehre der Grosshirnrinde in ihren Prinzipien dargestellt auf Grund des Zellenbaues.* Leipzig: Barth; 1909.
9. Chiang AS, Lin CY, Chuang CC. Three-dimensional reconstruction of brain-wide wiring networks in *Drosophila* at single-cell resolution. *Curr Biol.* 2011;21(1):1–11.
10. Williams RW, Herrup K. The control of neuron number. *Annu Rev Neurosci.* 1988;11(1):423–453.
11. Herculano-Houzel S. The human brain in numbers: a linearly scaled-up primate brain. *Front Hum Neurosci.* 2009;3:31.
12. Douglas R, Martin K. What's black and white about the grey matter? *Neuroinformatics.* 2011;9(2):167–179.
13. Sporns O, Tononi G, Kötter R. The human connectome: a structural description of the human brain. *PLoS Comput Biol.* 2005;1(4):e42.
14. Lichtman JW, Livet J, Sanes JR. A technicolour approach to the connectome. *Nat Rev Neurosci.* 2008;9(6):417–422.
15. Denk W. Scanning electron microscopy to reconstruct three-dimensional tissue nanostructure. *PLoS Biol.* 2004;2(11):10.
16. Leergaard TB, Hilgetag CC, Sporns O. Mapping the connectome: multi-level analysis of brain connectivity. *Front Neuroinform.* 2012;6:14.
17. Cardona A, Saalfeld S, Preibisch S. An integrated micro- and macroarchitectural analysis of the *Drosophila* brain by computer-assisted serial section electron microscopy. *PLoS Biol.* 2010;8(10):1–17.
18. Rybak J, Kuss A, Lamecker H. The digital bee brain: integrating and managing neurons in a common 3D reference system. *Front Syst Neurosci.* 2010;4(30):1–14.
19. Barlow H. Single units and sensation: a neuron doctrine for perceptual psychology? *Perception.* 1972;1:371–394.
20. Kupfermann I, Weiss KR. The command neuron concept. *Behav Brain Sci.* 1978;1(01):3–10.
21. Menzel R. The honeybee as a model for understanding the basis of cognition. *Nat Rev Neurosci.* 2012;13(11):758–768.
22. Nässel DR, Homberg U. Neuropeptides in interneurons of the insect brain. *Cell Tissue Res.* 2006;326(1):1–24.
23. Rybak J, Menzel R. Anatomy of the mushroom bodies in the honey bee brain: the neuronal connections of the alpha-lobe. *J Comp Neurol.* 1993;334(3):444–465.
24. Boeckh J, Tolbert LP. Synaptic organization and development of the antennal lobe in insects. *Microsc Res Tech.* 1993;24(3):260–280.
25. Sun XJ, Tolbert LP, Hildebrand JG. Synaptic organization of the uniglomerular projection neurons of the antennal lobe of the moth *Manduca sexta*: a laser scanning confocal and electron microscopic study. *J Comp Neurol.* 1997;379(1):2–20.
26. Menzel R. Searching for the memory trace in a mini-brain, the honeybee. *Learn Memory.* 2001;8(2):53–62.
27. Heisenberg M. Mushroom body memoir: from maps to models. *Nat Rev Neurosci.* 2003;4(4):266–275.
28. Brandt R, Rohlfing T, Rybak J. Three-dimensional average-shape atlas of the honeybee brain and its applications. *J Comp Neurol.* 2005;492(1):1–19.
29. Ganeshina O, Menzel R. GABA-immunoreactive neurons in the mushroom bodies of the honeybee: an electron microscopic study. *J Comp Neurol.* 2001;437(3):335–349.
30. Kremer MC, Christiansen F, Leiss F. Structural long-term changes at mushroom body input synapses. *Curr Biol.* 2010;20(21):1938–1944.
31. Mauelshagen J. Neural correlates of olfactory learning-paradigms in an identified neuron in the honeybee brain. *J Neuro.* 1993;69(2):609–625.
32. Hammer M. An identified neuron mediates the unconditioned stimulus in associative olfactory learning in honeybees. *Nature.* 1993;366(6450):59–63.
33. Grünewald B, Rueckert E, Preibisch S. Physiological properties and response modulations of mushroom body feedback neurons during olfactory learning in the honeybee, *Apis mellifera*. *J Comp Physiol Sens Neural Behav Physiol.* 1999;185(6):565–576.
34. Luo L. Fly MARCM and Mouse MADM: genetic methods of labeling and manipulating single neurons. *Brain Res Rev.* 2007;55(2):220–227.
35. Sachse S, Rueckert E, Keller A. Activity-dependent plasticity in an olfactory circuit. *Neuron.* 2007;56:838–850.
36. Hammer M, Menzel R. Learning and memory in the honeybee. *J Neurosci.* 1995;15(3):1617–1630.
37. Masse NY, Turner GC, Jefferis GS. Olfactory information processing in *Drosophila*. *Curr Biol.* 2009;19(16):R700–R713.
38. Gerber B, Tanimoto H, Heisenberg M. An engram found? Evaluating the evidence from fruit flies. *Curr Opin Neurobiol.* 2004;14(6):737–744.
39. Laughlin SB. The optic lamina of fast flying insects as a guide to neural circuit design. In: Shepherd GM, Grillner S, eds. *Handbook of Brain Microcircuits.* New York: Oxford University Press; 2010:433–440.
40. Prokop A, Meinertzhagen IA. Development and structure of synaptic contacts in *Drosophila*. *Semin Cell Dev Biol.* 2006;17(1):20–30.
41. Niven JE, Farris SM. Miniaturization of nervous systems and neurons. *Curr Biol.* 2012;22(9):R323–R329.
42. Shepherd GM, Greer CA, Mazzarello P. The first images of nerve cells: golgi on the olfactory bulb 1875. *Brain Res Rev.* 2011;:92–105.
43. Farris SM, Robinson GE, Fahrbach SE. Experience- and age-related outgrowth of intrinsic neurons in the mushroom bodies of the adult worker honeybee. *J Neurosci.* 2001;21(16):6395–6404.
44. Lanciego JL, Wouterlood FG. A half century of experimental neuroanatomical tracing. *J Chem Neuroanat.* 2011;42:157–183.
45. Pinault D. A novel single-cell staining procedure performed *in vivo* under electrophysiological control: morpho-functional features of juxtacellularly labeled thalamic cells and other central

neurons with biocytin or neurobiotin. *J Neurosci Methods*. 1996;65 (2):113–136.
46. Duque A, Zaborszky L. Juxtacellular labeling of individual neurons *in vivo*: from electrophysiology to synaptology. In: Zaborszky L, Wouterlood FG, Lanciego JL, eds. *Neuroanatomical Tract-Tracing 3: Molecules, Neurons, and Systems*. New York: Springer; 2006:197–236.
47. Burgalossi A, Herfst L, von Heimendahl M. Microcircuits of functionally identified neurons in the rat medial entorhinal cortex. *Neuron*. 2011;70(4):773–786.
48. Brand A, Perrimon N. Targeted gene expression as a means of altering cell fates and generating dominant phenotypes. *Development*. 1993;118:401–415.
49. Aso Y, Grübel K, Busch S. The mushroom body of adult *Drosophila* characterized by GAL4 drivers. *J Neurogenet*. 2009;23(1-2):156–172.
50. Jefferis GS, Potter CJ, Chan AM. Comprehensive maps of *Drosophila* higher olfactory centers: spatially segregated fruit and pheromone representation. *Cell*. 2007;128(6):1187–1203.
51. Jenett A. A GAL4-driver line resource for *Drosophila* neurobiology. *Cell Rep*. 2012;2(4):991–1001.
52. Elliott DA, Brand AH. The GAL4 system: a versatile system for the expression of genes. *Methods Mol Biol*. 2008;420:79–95.
53. Lee T, Luo LQ. Mosaic analysis with a repressible cell marker for studies of gene function in neuronal morphogenesis. *Neuron*. 1999;22(3):451–461.
54. Lai S-L, Lee T. Genetic mosaic with dual binary transcriptional systems in *Drosophila*. *Nat Neurosci*. 2006;9(5):703–709.
55. Potter CJ, et al. The Q system: a repressible binary system for transgene expression, lineage tracing, and mosaic analysis. *Cell*. 2010;141:536–548.
56. Strutz A, Völler T, Riemensperger T. Calcium imaging of neural activity in the olfactory system of *Drosophila*. In: Martin J-R, ed. *Neuromethods*. New York: Springer Science + Business Media; 2012:43–70.
57. Masuyama K, Zhang Y, Rao Y. Mapping neural circuits with activity-dependent nuclear import of a transcription factor. *J Neurogenet*. 2012;26(1):89–102.
58. Hampel S, Chung P, McKellar CE. *Drosophila* brainbow: a recombinase-based fluorescence labeling technique to subdivide neural expression patterns. *Nat Meth*. 2011;8(3):253–259.
59. Hadjieconomou D, Rotkopf S, Alexandre C. Flybow: genetic multicolor cell labeling for neural circuit analysis in *Drosophila melanogaster*. *Nat Meth*. 2011;8(3):260–266.
60. Mishchenko Y. On optical detection of densely labeled synapses in neuropil and mapping connectivity with combinatorially multiplexed fluorescent synaptic markers. *PLoS ONE*. 2010;5(1): e8853.
61. Christiansen F, Zube C, Andlauer TF. Presynapses in Kenyon cell dendrites in the mushroom body calyx of *Drosophila*. *J Neurosci*. 2011;31(26):9696–9707.
62. Butcher NJ, Friedrich AB, Lu Z. Different classes of input and output neurons reveal new features in microglomeruli of the adult *Drosophila* mushroom body calyx. *J Comp Neurol*. 2012;520 (10):2185–2201.
63. Feinberg EH, VanHoven MK, Bendesky A. GFP reconstitution across synaptic partners (GRASP) defines cell contacts and synapses in living nervous systems. *Neuron*. 2008;57(3):353–363.
64. Gordon MD, Scott K. Motor control in a *Drosophila* taste circuit. *Neuron*. 2009;61(3):373–384.
65. Jones WD. The expanding reach of the GAL4/UAS system into the behavioral neurobiology of *Drosophila*. *BMB Rep*. 2009;42 (11):705–712.
66. Ugolini G. Advances in viral transneuronal tracing. *J Neurosci Methods*. 2010;194(1):2–20.
67. Callaway EM. Transneuronal circuit tracing with neurotropic viruses. *Curr Opin Neurobiol*. 2008;18(6):617–623.
68. Wickersham IR, Sullivan HA, Seung HS. Production of glycoprotein-deleted rabies viruses for monosynaptic tracing and high-level gene expression in neurons. *Nat Protocols*. 2010;5 (3):595–606.
69. Ohara S, Inoue K-i, Witter MP. Untangling neural networks with dual retrograde transsynaptic viral infection. *Front Neurosci*. 2009;3:344–349.
70. Miyamichi K, Amat F, Moussavi F. Cortical representations of olfactory input by trans-synaptic tracing. *Nature*. 2011;472(7342): 191–196.
71. Schermelleh L, Heintzmann R, Leonhardt H. A guide to super-resolution fluorescence microscopy. *J Cell Biol*. 2010;190 (2):165–175.
72. Minsky M. Microscopy apparatus, U.S. Patent US 3013467; 1961.
73. Denk W, Strickler J, Webb W. Two-photon laser scanning fluorescence microscopy. *Science*. 1990;248(4951):73–76.
74. Hell SW, Wichmann J. Breaking the diffraction resolution limit by stimulated emission: stimulated-emission-depletion fluorescence microscopy. *Opt Lett*. 1994;19(11):780–782.
75. Gustafsson MGL. Surpassing the lateral resolution limit by a factor of two using structured illumination microscopy. *J Microsc*. 2000;198(2):82–87.
76. Santi PA. Light sheet fluorescence microscopy. *J Histochem Cytochem*. 2011;59(2):129–138.
77. Dodt H-U, et al. Ultramicroscopy: three-dimensional visualization of neuronal networks in the whole mouse brain. *Nat Meth*. 2007;4(4):331–336.
78. Siedentopf H, Zsigmondy R. Über Sichtbarmachung und Größenbestimmung ultramikroskopischer Teilchen, mit besonderer Anwendung auf Goldrubingläser. *Annalen der Physik*. 1903;10:1–39.
79. McGurk L, Morrison H, Keegan LP. Three-dimensional imaging of *Drosophila melanogaster*. *PLoS ONE*. 2007;2(9):e834.
80. Eickhoff R, Lorbeer R-A, Scheiblich H, et al. Scanning laser optical tomography resolves structural plasticity during regeneration in an insect brain. *PLoS ONE*. 2012;7(7):e41236.
81. Shu X, Lev-Ram V, Deerinck TJ. A genetically encoded tag for correlated light and electron microscopy of intact cells, tissues, and organisms. *PLoS Biol*. 2011;9(4):e1001041.
82. Knoll M, Ruska E. Das elektronenmikroskop. *Zeitschrift für Physik Hadrons Nuclei*. 1932;78(5):318–339.
83. Ardenne M v. Das Elektronerastermikroskop. Praktische Ausführung. *Zeitschrift für technische Physik*. 1938;19:407–416.
84. Micheva KD, O'Rourke SJ, Busse B. Array tomography: a new tool for imaging the molecular architecture and ultrastructure of neural circuits. *Neuron*. 2007;55(1):25–36.
85. Knott G, Marchman H, Lich B. Serial section scanning electron microscopy of adult brain tissue using focused ion beam milling. *J Neurosci*. 2008;28(12):2959–2964.
86. Ribi W, Senden TJ, Sakellariou A. Imaging honey bee brain anatomy with micro-X-ray-computed tomography. *J Neurosci Methods*. 2008;171(1):93–97.
87. Haddad D, Schaupp F, Brandt R. NMR imaging of the honeybee brain. *J Insect Sci*. 2004;4:7.
88. Oberti D, Kirschmann MA, Hahnloser RH. Projection neuron circuits resolved using correlative array tomography. *Front Neurosci*. 2011;5:50.
89. Mishchenko Y. Reconstruction of complete connectivity matrix for connectomics by sampling neural connectivity with fluorescent synaptic markers. *J Neurosci Methods*. 2011;196(2):289–302.
90. DeFelipe J. From the connectome to the synaptome: an epic love story. *Science*. 2010;330:1198–1201.
91. Helmstaedter M, Briggman KL, Denk W. 3D structural imaging of the brain with photons and electrons. *Curr Opin Neurobiol*. 2008;18(6):633–641.

92. Merchan-Perez A, Rodriguez JR, Alonso-Nanclares L. Counting synapses using FIB/SEM microscopy: a true revolution for ultrastructural volume reconstruction. *Front Neuroanat.* 2009;3:18.
93. Morales J, Alonso-Nanclares L, Rodriguez J-R. Espina: a tool for the automated segmentation and counting of synapses in large stacks of electron microscopy images. *Front Neuroanat.* 2011;5:18.
94. Kreshuk A, Straehle CN, Sommer C. Automated detection and segmentation of synaptic contacts in nearly isotropic serial electron microscopy images. *PLoS ONE.* 2011;6(10):e24899.
95. Denk W, Briggman KL, Helmstaedter M. Structural neurobiology: missing link to a mechanistic understanding of neural computation. *Nat Rev Neurosci.* 2012;13(5):351–358.
96. Douglas RJ, Martin KAC. Neuronal circuits of neocortex. *Annu Rev Neurosci.* 2004;27(1):419–451.
97. Bock DD, Lee W-CA, Kerlin AM. Network anatomy and in vivo physiology of visual cortical neurons. *Nature.* 2011;471(7337):177–182.
98. Armstrong JD, van Hemert JI. Towards a virtual fly brain. *Philos Trans Royal Soc Math Phys Eng Sci.* 2009;367(1896):2387–2397.
99. Watts RJ, Schuldiner O, Perrino J. Glia engulf degenerating axons during developmental axon pruning. *Curr Biol.* 2004;14(8):678–684.
100. Rybak J, Talarico G, Ruiz S. Synaptic circuitry of identified neurons in the antennal lobe of *Drosophila melanogaster.* In: *ISOT.* Stockholm; 2012.
101. Marin EC, Watts RJ, Tanaka NK. Developmentally programmed remodeling of the *Drosophila* olfactory circuit. *Development.* 2005;132(4):725–737.
102. Oberlaender M, de Kock CPJ, Bruno RM. Cell type-specific three-dimensional structure of thalamocortical circuits in a column of rat vibrissal cortex. *Cereb Cortex.* 2011;22(10):2375–2391.
103. Dercksen V, Egger R, Hege HC. Synaptic connectivity in anatomically realistic neural networks: modeling and visual analysis. In: Ropinski T, Ynnerman A, Botha C, Roerdink J, eds. *Eurographics Workshop on Visual Computing for Biology and Medicine*; 2012:17–24.
104. Rybak J. The digital honey bee brain atlas. In: Galizia CG, Eisenhardt D, Giurfa M, eds. *Honeybee Neurobiology and Behavior.* Dordrecht, The Netherlands: Springer; 2012:125–140.
105. Markram H. The blue brain project. *Nat Rev Neurosci.* 2006;7(2):153–160.
106. Namiki S, Kanzaki R. Reconstruction of virtual neural circuits in an insect brain. *Front Neurosci.* 2009;3(2):206–213.
107. Evers JF, Schmitt S, Sibila M. Progress in functional neuroanatomy: precise automatic geometric reconstruction of neuronal morphology from confocal image stacks. *J Neurophysiol.* 2005;93(4):2331–2342.
108. Meseke M, Evers JF, Duch C. Developmental changes in dendritic shape and synapse location tune single-neuron computations to changing behavioral functions. *J Neurophysiol.* 2009;102(1):41–58.
109. Hohensee S, Bleiss W, Duch C. Correlative electron and confocal microscopy assessment of synapse localization in the central nervous system of an insect. *J Neurosci Methods.* 2008;168(1):64–70.
110. Hama H, Kurokawa H, Kawano H. Scale: a chemical approach for fluorescence imaging and reconstruction of transparent mouse brain. *Nat Neurosci.* 2011;14(11):1481–1488.
111. Becker K, Jährling N, Saghafi S, Weiler R. Chemical clearing and dehydration of GFP expressing mouse brains. *PLoS ONE.* 2012;7(3):e33916.
112. Helmstaedter M, Briggman KL, Denk W. High-accuracy neurite reconstruction for high-throughput neuroanatomy. *Nat Neurosci.* 2011;14(8):1081–1088.
113. Peng H. Bioimage informatics: a new area of engineering biology. *Bioinformatics.* 2008;24(17):1827–1836.
114. Myers G. Why bioimage informatics matters. *Nat Meth.* 2012;9(7):659–660.
115. Martone ME, Gupta A, Ellisman MH. e-Neuroscience: challenges and triumphs in integrating distributed data from molecules to brains. *Nat Neurosci.* 2004;7(5):467–472.
116. Jones AR, Overly CC, Sunkin SM. The Allen brain atlas: 5 years and beyond. *Nat Rev Neurosci.* 2009;10(11):821–828.
117. Hartenstein V, Cardona A, Pereanu W. Modeling the developing *Drosophila* brain: rationale, technique, and application. *BioScience.* 2008;58(9):823–836.
118. Berry RP, Ibbotson MR. A three-dimensional atlas of the honeybee neck. *PLoS ONE.* 2010;5(5):e10771.
119. Rein K, et al. The *Drosophila* standard brain. *Curr Biol.* 2002;12:227–231.
120. el Jundi B, Heinze S, Lenschow C. The locust standard brain: a 3D standard of the central complex as a platform for neural network analysis. *Front Syst Neurosci.* 2010;3(21):1–15.
121. Rohlfing T, Brandt R, Maurer Jr CR. Bee brains, B-splines and computational democracy: generating an average shape atlas. In: Staib L, eds. *IEEE Workshop on Mathematical Methods in Biomedical Image Analysis: Proceedings: 9-10 December 2001, Kauai, Hawaii.* New York: IEEE, 2001; 187–194.
122. Kvello P, Lofaldli BB, Rybak J. Digital, three-dimensional average shaped atlas of the *Heliothis virescens* brain with integrated gustatory and olfactory neurons. *Front Syst Neurosci.* 2009;3(14):1–14.
123. Kurylas AE, Rohlfing T, Krofczik S. Standardized atlas of the brain of the desert locust, *Schistocerca gregaria. Cell Tissue Res.* 2008;333(1):125–145.
124. Schindelin J, Arganda-Carreras I, Frise E. Fiji: an open-source platform for biological-image analysis. *Nat Meth.* 2012;9(7):676–682.
125. Peng H, Chung P, Long F. BrainAligner: 3D registration atlases of *Drosophila* brains. *Nat Methods.* 2011;8(6):493–500.
126. Kuss A, Prohaska S, Meyer B. Ontology-based visualization of hierarchical neuroanatomical structures. In: Botha CP, Kindlmann G, Niessen WJ, Preim B, eds. *Proceedings of the Eurographics Workshop on Visual Computing for Biomedicine VCBM October 2008*:177–184.
127. Osumi-Sutherland D, Reeve S, Mungall CJ. A strategy for building neuroanatomy ontologies. *Bioinformatics.* 2012;28:1262–1269.
128. Toga AW, Mazziotta JC. *Brain Mapping: The Methods.* 2nd ed. New York: Academic Press; 2002.
129. Mikula S, Trotts I, Stone JM. Internet-enabled high-resolution brain mapping and virtual microscopy. *Neuroimage.* 2007;35(1):9–15.
130. Kohl J, Jefferis GS. Neuroanatomy: decoding the fly brain. *Curr Biol.* 2011;21(1):R19–R20.
131. French L, Pavlidis P. Using text mining to link journal articles to neuroanatomical databases. *J Comp Neurol.* 2012;520(8):1772–1783.
132. Krofczik S, Menzel R, Nawrot MP. Rapid odor processing in the honeybee antennal lobe network. *Front Comput Neurosci.* 2008;2(9):1–13.
133. Bakker R, Potjans J, Wachtler T. Macaque structural connectivity revisited: CoCoMac 2.0. *BMC Neurosci.* 2011;12(suppl 1):P72.
134. Bohland JW, Wu C, Barbas H. *A* proposal for a coordinated effort for the determination of brainwide neuroanatomical connectivity in model organisms at a mesoscopic scale. *PLoS Comput Biol.* 2009;5(3):e1000334.
135. Zamora-Lopez G, Zhou C, Kurths J. Cortical hubs form a module for multisensory integration on top of the hierarchy of cortical networks. *Front Neuroinform.* 2010;4:1.

136. Sporns O. *Networks of the Brain*. Cambridge, MA: MIT Press; 2011.
137. Douglas R, Martin KAC. Canonical cortical circuits. In: Shepherd GM, Grillner S, eds. *Handbook of Brain Microcircuits*. New York: Oxford University Press; 2010.
138. Meinertzhagen IA. Fly photoreceptor synapses: their development, evolution and plasticity. *J Neurobiol*. 1989;20(5):276–294.
139. Inagaki HK, Ben-Tabou de-Leon S, Wong AM. Visualizing neuromodulation *in vivo*: TANGO-mapping of dopamine signaling reveals appetite control of sugar sensing. *Cell*. 2012;148: 583–595.
140. Ahrens MB, Li JM, Orger MB. Brain-wide neuronal dynamics during motor adaptation in zebrafish. *Nature*. 2012;485: 473–477.
141. Rieckher M, Birk UJ, Meyer H, Ripoll J, Tavernarakis N. Microscopic Optical Projection Tomography. *In Vivo. PLoS One*. 2011;6(4):e18963.

CHAPTER

5

'Decision Making' in Larval *Drosophila*

Michael Schleyer[†], *Sören Diegelmann*[†], *Birgit Michels*[†], *Timo Saumweber*[†] *and Bertram Gerber*[†,‡,§]

[†]Leibniz Institut für Neurobiologie (LIN), Abteilung Genetik von Lernen und Gedächtnis, Magdeburg, Germany [‡]Otto von Guericke Universität Magdeburg, Institut für Biologie, Magdeburg, Germany [§]Center for Behavioral Brain Science, Magdeburg, Germany

Should I stay or should I go now?
Should I stay or should I go now?
If I go there will be trouble
And if I stay it will be double.
. . .
This indecision 's bugging me
Exactly whom I'm supposed to be.
—The Clash, 1982

INTRODUCTION

Decision making is a complex psychological process, and as such the neurosciences alone cannot grasp it in all facets. This is particularly obvious for decision making in animals, where verbal behavior cannot be used for the analysis. However, insects allow for experimental approaches with enticing analytical power, and a number of study cases for 'decision making' have been reported: the decision to fight or flight in crickets;[1] the switch of locusts between solitary and gregarious phases,[2] as well as their choice between carbohydrate-rich and protein-rich food sources dependent on physiological state;[3] (for a related follow-up in *Drosophila*, see[4]); choice behavior during foraging in bees;[5,6] the resolution of conflict between contradictory cues in adult *Drosophila*;[7] the organization of gap climbing in adult *Drosophila*[8] or of run–pause–turn behavior in *Drosophila* maggots;[9] and the switch between 'gaits' in leeches,[10,11] included here as insects *honoris causa*. These cases obviously differ in complexity, and indeed the nature of the underlying processes as decisions invites debate. Our approach, however, is to not try to draw a line between processes that are decisions and those that are not but, rather, to characterize aspects of behavioral tasks that we believe bear upon the strength of the decision character of the process. In this context, we argue that the organization of learned behavior in *Drosophila* larvae offers a particularly fruitful study case, holding a balance between sufficiently interesting yet reduced complexity, on the one hand, and experimental tractability, on the other hand.

A decision settles the problem of what is the relatively best thing to do. Here, we view 'decision making' as a process to integrate (1) sensory input, (2) the current status reflecting the evolutionary and individual history of the animal, (3) the available behavioral options, and (4) their expected outcomes. Given that in most cases there are no preconfigured solutions for this integration, decision processes need to negotiate ever-new deals between senses, history, behavioral options, and their expected consequences in a way that remains behaviorally silent until the decision is actually taken. As a corollary, much of the required processing remains offline from immediate sensorimotor loops. Also, because in principle all behavioral options are 'on the table,' of which many are mutually exclusive, we may expect a particularly prominent role of behavioral inhibition. Indeed, a key insight of the early behavioral neurosciences, largely based on the contributions of Erich von Holst, is that sensory processing, rather than triggering motor output, is reconfiguring the inhibitory balance between multiple internally generated behavioral tendencies.[12]

Here, we focus on the decision of behaviorally expressing an associative memory trace—or not.

Invertebrate Learning and Memory.
DOI: http://dx.doi.org/10.1016/B978-0-12-415823-8.00005-8

To this end, we first sketch the architecture of the chemobehavioral system in larval *Drosophila*. Then we present a working hypothesis of memory trace formation regarding the association between odors and taste reinforcement, followed by a discussion of how outcome expectations are implemented to organize conditioned behavior. Finally, we present our thoughts on how these processes bear on decision making and which aspects of the psychological richness of decision making in man are as yet out of scope in maggots.

ARCHITECTURE OF THE CHEMOBEHAVIORAL SYSTEM

Smell and taste serve behavioral organization in different ways. Odors trigger orienting movements and, dependent on the presumed nature of the odor source, organize search or escape behavior and prepare for suitable action in case tracking is successful or should escape fail. In a sense, therefore, olfactory behavior is not about odors but, rather, about odor sources, necessarily involving some 'guesswork' about the relation between the odor and its source. In order for these guesses to be well-informed, olfactory circuits feature stages with an enormous potential to discriminate odors and to attach acquired meaning to them. An interesting set of cases is found for pheromones, which are already endowed with an evolutionarily determined and largely fixed behavioral meaning; as a corollary, pheromone systems are functionally and often structurally separate from the general olfactory system and allow for less mnemonic flexibility.

In contrast, taste operates temporally 'downstream' of smell, in the sense that it operates only upon immediate physical contact with things. Such proximity entails an entanglement with mechanosensation and thus of the what of taste with the where of touch, and allows for relatively direct and local sensorimotor loops to organize eating and drinking (similar arguments apply for contact chemosensation in other contexts, such as predation, defense, social recognition, aggression, and pupariation, as well as courtship, copulation, and oviposition in adults).

The olfactory system of the larva draws on 21 olfactory sensory neurons organized in seven triplets within the dome of the dorsal organ (Figures 5.1D and 5.2).[18–21] They pass olfactory information exclusively ipsilaterally to the larval antennal lobe, each of them targeting one of 21 spherical and anatomically identifiable 'glomerulus' compartments.[22,23] Each olfactory sensory neuron typically expresses one type of olfactory receptor (OR) molecule of the *Or* gene family together with the obligatory olfactory co-receptor ORCO, coded by the *Orco* gene (CG10609, formerly known as *Or83b* gene and OR83b receptor, respectively).[19,24–26] The odor response spectra of these ORs, at the chosen concentrations, are diverse but typically overlapping, ranging from OR94b that responds to a single from the 27 tested odorants to OR42a and OR85c, which each respond to 9 odorants out of this panel.[20] In turn, most odorants activate more than one OR and thus more than one olfactory sensory neuron. Information about odor quality can thus be coded by the combinatorial activation of ligand-specific subsets of the 21 olfactory sensory neurons (for a discussion of the temporal coding aspects of olfaction, see[27]). Antennal lobe interneurons[13] collect from and distribute information to many, if not all, glomeruli. In analogy to adult flies, these connections likely provide inhibitory feedback between glomeruli for gain control across concentrations[28] (note that because these neurons sum up olfactory input across the antennal lobe, they may be involved in determining and/or learning about odor intensity). In turn, in adult flies excitatory projections between antennal lobe glomeruli were found to also shape the pattern of activity across the antennal lobe,[28] and conceivably there are additional functions of these anatomically diverse and interindividually variable neurons (adult;[29] larva[13]).

The information from a given antennal lobe glomerulus is conveyed by typically just one projection neuron[23,30] (but see[31,32]) to two sites: the lateral horn and the mushroom bodies (for a description of up to three projection neurons with apparently broader input regions in the antennal lobe, see[13]). This branched anatomy of projection neurons is characteristic of insects. Compared to the lateral horn pathway, which provides a relatively 'direct' route to the motor system, the loop via the mushroom bodies can be seen as a 'detour.' The mushroom bodies' olfactory input region, the calyx, is organized into approximately 34 anatomically identifiable and roughly circularly arranged glomeruli.[23,30–32] The approximately 21 projection neurons typically signal to one calyx glomerulus each, synapsing onto subsets of the approximately 600 mature mushroom body neurons (also called Kenyon cells[33]); (for a different count based on electron microscopy, see[34]). Interestingly, there is a fairly stereotyped relation between the antennal lobe glomerulus in which a projection neuron receives its input, and the mushroom body calyx glomerulus to which it delivers its output, such that projection neurons can be individually identified just like the olfactory sensory neurons, antennal lobe, and calycal glomeruli. Note that three or four calyx glomeruli are 'orphaned' in the sense that their inputs are not yet characterized. In any event, a given Kenyon cell collects input from an apparently random draw of one to three[23] or up to six[31] calyx glomeruli and thus from up to six

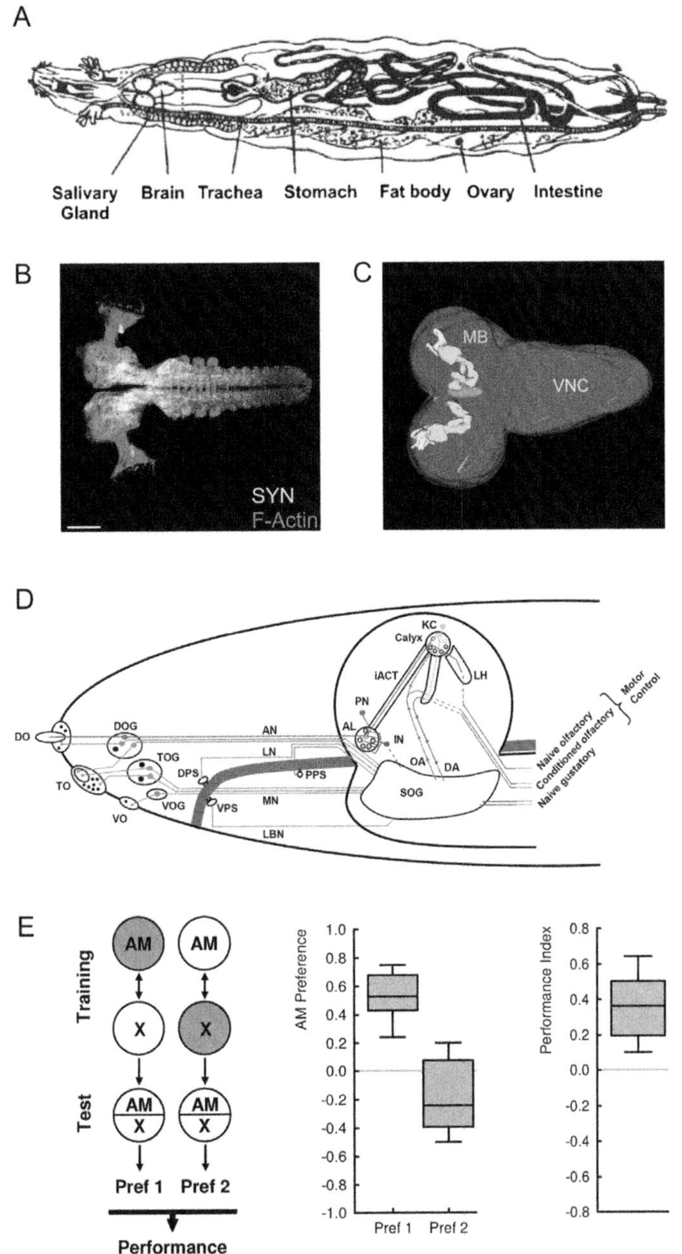

FIGURE 5.1 **Basic organization of the larval chemobehavioral system.** (A) Body plan of larval *Drosophila*. The stippled box approximates the area in panels B and C. (B) Immunoreactivity against Synapsin (white) and F-actin (red) in a third instar larval brain of a wild-type larva. Synapsin immunoreactivity is found throughout the neuropil area of brain and ventral nerve cord. Scale bar = 50 μm. (C) Three-dimensional reconstruction of a third instar larval brain. Mushroom bodies (MB) are displayed in yellow, and Kenyon cell bodies are shown in pale yellow. VNC, ventral nerve cord. (D) Schematic diagram of the chemosensory pathways in the larval head. Olfactory pathways (blue) project into the brain proper and via projection neurons (violet) toward the mushroom body (yellow) and the lateral horn, whereas gustatory afferents (orange) are collected in various regions of the suboesophageal ganglion. The green and red arrows indicate pathways to short-circuit a taste-driven reinforcement signal from the suboesophageal ganglion toward the brain. Abbreviations: AL, antennal lobe; AN, antennal nerve; DA, dopaminergic neurons as participating in reinforcement signaling; DO/DOG, dorsal organ/dorsal organ ganglion; DPS, dorsal pharyngeal sensillae; iACT, inner antennocerebral tract; IN, antennal lobe interneurons[13]; KC, Kenyon cells; LBN, labial nerve; LH, lateral horn; LN, labral nerve; MN, maxillary nerve; OA, octopaminergic neurons as participating in reinforcement signaling; PN, projection neurons; PPS, posterior pharyngeal sensillae; SOG, suboesophageal ganglion; TO/TOG, terminal organ/terminal organ ganglion; VO/VOG, ventral organ/ventral organ ganglion; VPS, ventral pharyngeal sensillae. (E) The larval learning paradigm. A group of 30 larvae is trained such that a sugar reward (green) is paired with *n*-amylacetate (AM) and a pure, tasteless substrate (white) is paired (X) with either a second odor ('two-odor paradigm') or no odor ('one-odor paradigm'). In the subsequent test, the AM preference (Pref 1) is calculated as number of larvae on the AM side minus the number of larvae on the other side divided by the total number. A second group of 30 animals is trained reciprocally (Pref 2). The performance index as a measurement of conditioned behavior is calculated as a the difference of Pref 1 and Pref 2, divided by 2. After training with sugar as reward, appetitive memory results in positive performance index scores (rightmost plot), whereas negative scores would indicate aversive memory. Note that in half of the cases, the sequence of trials is as indicated, whereas in the other half of the cases the sequence of trials is reverse (not shown). For the aversive learning versions of the paradigm, see text. Sources: *Panel A modified from Demerenc and Kaufmann,*[14] *panels B and C based on data in Michels* et al.,[15] *panel D based on Stocker,*[16] *and panel E based on data from Schleyer* et al.[17]

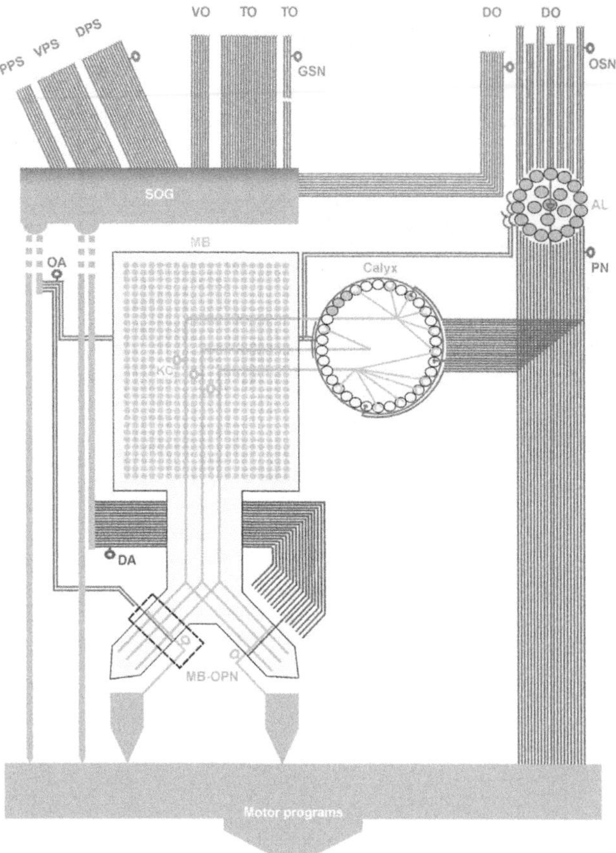

FIGURE 5.2 Simplified wiring diagram of the larval chemobehavioral system. Illustration of the numbers and connectivity architecture of neurons of the larval chemobehavioral system as discussed in this chapter. Note that for clarity, some neurons (e.g., antennal lobe interneurons; see Figure 5.1D or non-Kenyon cell mushroom body intrinsic neurons) are omitted and the topology of the cells, cell bodies, nerves, and brain areas are only partially captured. Second, also for clarity, in many cases only 'example' neurons are drawn in full to illustrate patterns of connectivity—for example, in the antennal lobe; note that the OA neurons likely communicate toward the calyx and the AL in their entirety. Third, the polarity of the aminergic neurons is only partially known. Fourth, the interruption of the pathway of some GSNs originating in the TO indicates that these neurons, together with the OSNs and the GSNs from the DO, travel via the antennal nerve to the SOG. Fifth, the gray fill of some calyx glomeruli indicates that their inputs are not classical PNs. Finally, we draw attention to the lack of knowledge about the connectivity within the SOG as well on the motor side of the system. The area boxed by the stippled line is displayed in Figure 5.3 (see Note Added in Proof). Abbreviations and color code are the same as those for Figure 5.1D plus: GSN, gustatory sensory neurons; MB-OPN, mushroom body output neuron; OSN, olfactory sensory neurons.

projection neurons, covering up to one-third of the projection neuron coding space. In turn, each projection neuron diverges to 30–180 mushroom body neurons (for details of this approximation, see[35]), contributing to up to one-third of the mushroom body coding space. If multiglomerular mushroom body neurons need input from more than one calycal glomerulus to fire, this combined convergent–divergent architecture allows for combinatorial coding and a massive expansion of olfactory coding capability. Importantly, the coding space is immediately and drastically reduced in the very next step toward motor output: The Kenyon cells then connect to relatively few (a reasonable guess may be between one and three dozen[36]) output neurons that via an unknown number of synaptic steps link to the motor system[36]; (for the situation in adults, see[37–39]). These mushroom body output neurons likely receive input from many or all mushroom body cells, thus 'summing up' the activation in their input section of the mushroom body. The mushroom body-to-output neuron synapse can thus be viewed as a 'watershed' in the sense that the combinatorial identity of odor processing is lost at this stage, as a first step to transform descriptive information about "Which odor?" into instructive information regarding "What to do?" In any event, contemplating this sketched circuitry, it is important that olfactory information reaches the motor systems both via the relatively direct lateral horn route and via the mushroom body detour. As argued later, these two routes organize innate versus learned olfactory behavior.

The gustatory sensory system is comparably much less well understood (reviews[35,40–42]). It comprises approximately 90 gustatory sensory neurons[43] (all cell numbers are per body side) located in three external taste organs (terminal organ, ventral organ, and the bulge of the dorsal organ) and three internal taste sensilla (ventral, dorsal, and posterior pharyngeal sensilla).[22,44,45] All these cephalic gustatory sensory neurons bypass the brain hemispheres and instead project to the subesophageal ganglion in a way that depends on the receptor gene(s) expressed and their sense organ of origin.[43] Interestingly, projections from the terminal organ and ventral organ neurons remain ipsilateral, whereas projections from the pharyngeal organs can be either ipsi- or bilateral. From the subesophageal ganglion, taste information is relayed both to modulatory neurons that detour toward the brain and/or the ventral nerve cord[43,46,47] and directly to (pre)motor neurons presumably in the ventral nerve cord (Figure 5.1D), but the details of these connections are unknown. It is known, however, that larvae can detect tastants of various human psychophysical categories, such as sweet, salty, or bitter substances, but may not be able to distinguish different kinds of sweet, for example[48–50]; (with regard to adults, see also[41,51]).

On the molecular level, salt is likely detected by Na^+ channels expressed in the terminal organ,[52] whereas sugar and bitter detection is probably mediated by G protein-coupled receptors of the *Gr* gene family[53,54] (for review, see[41]). Of the 68 known *Gr* genes in

Drosophila, 39 have been shown to be expressed in the taste organs of the larval head, based on *Gr–Gal4* transgene analysis.[43,54] Strikingly, none of the known adult-expressed sugar-sensitive GRs (*Gr5a*, *Gr64a*, and *Gr64f*[51–57]) are among these 39 *Gr* genes expressed in the larvae. Of the known adult-expressed bitter-sensitive GRs (*Gr33a*, *Gr66a*, and *Gr93a*[51,56,58,59]), *Gr33a* and *Gr66a* are also expressed in the larva.[54] Lastly, receptors of the *Ir* family[60] may participate in chemosensation in the larval stages.

It is known from adults that contact chemosensory bristles are innervated by two to four gustatory neurons and one mechanosensory neuron. Based on structural[22] and genetic[61] arguments, a similar taste–touch entanglement can be expected in larvae, at least for the external taste organs.

A WORKING HYPOTHESIS OF MEMORY TRACE FORMATION

Drosophila larvae not only are able to smell and taste but also can associate odors with gustatory reinforcement (Figure 5.1E);[62–65] (for a manual, see[66]). Such taste reinforcement can either be appetitive (sugar and low-concentration salt) or aversive (high-concentration salt and quinine), such that olfactory preference behavior is respectively increased or decreased. However, from the architecture reviewed so far, it appears as if there is no direct connection between olfactory and taste pathways (but see for the anatomical description of neurons innervating both antennal lobe and subesophageal ganglion). How, then, can odors be associated with tastants?

In 1993, Martin Hammer identified the octopaminergic VUM_{mx1} neuron in the honeybee,[98] which receives its gustatory input likely in the subesophageal ganglion and provides output to the antennal lobe, the mushroom body calyx, and the lateral horn. This identified, single neuron is activated by unpredicted sugar as well as by sugar-predicting odors, and it is sufficient to mediate the rewarding effect of sugar in honeybee olfactory learning (for a description of this neuron in adult *Drosophila*, see[67]; this neuron exists in larval *Drosophila* as well[68]). Also in *Drosophila*, there is evidence that octopaminergic neurons can 'short-circuit' taste with smell pathways to mediate appetitive reinforcement signaling:[69] Adult flies lacking octopamine because they lack the required enzyme for its synthesis ($T\beta H^{M18}$ allele of the $T\beta H$ gene, CG1543) are impaired in odor–sugar learning but not in odor–shock learning. In turn, blocking synaptic output from a genetically defined set of dopaminergic neurons (*TH*-Gal4) impaired odor–shock learning but not odor–sugar learning. In *Drosophila* larvae, the net effect of driving relatively large sets of dopaminergic or octopaminergic/tyraminergic neurons (*TH*-Gal4 or *TDC*-Gal4, respectively) can substitute for punishment or reward, respectively, in olfactory learning[70] (see Note added in proof). However, the genetic tools currently available to manipulate dopaminergic or octopaminergic/tyraminergic neurons cover anatomically and functionally heterogeneous sets of neurons. From these sets of neurons, current research focuses on identifying those conferring appetitive or aversive reinforcement signaling for different memory phases and differentiating them from neurons mediating other effects regarding, for example, olfactory processing, gustatory processing, motor control including the regulation of memory retrieval, and mediating satiety states;[47,68,71] (for adult *Drosophila*, see[72–75]). Thus, a reasonable working hypothesis appears to be that if an odor is presented, a particular pattern of olfactory sensory neurons is activated, dependent on the ligand profiles of the respectively expressed receptors, leading to the activation of a particular combination of glomeruli in the antennal lobe, as well as particular sets of projection neurons and the corresponding lateral horn and mushroom body neurons (Figures 5.2 and 5.3). At the same time, a tastant such as sugar activates gustatory sensory neurons that trigger a value signal ('good') via some of the aminergic neurons (in the case of high-concentration salt or quinine, a 'bad' signal is delivered via a different set of aminergic neurons) and sends it to many, if not all, Kenyon cells of the mushroom bodies. Conceivably, only in that subset of Kenyon cells that are activated coincidently by both the odor signal and the value signal a memory trace is formed[77,78] (for more detail, see the following section). This memory trace is thought to consist of an alteration of the connection between the Kenyon cells and their output neurons: If a learned odor is subsequently presented, this altered Kenyon cell output is the basis for conditioned behavior. Therefore, the memory trace-forming convergence in the mushroom body is not between the olfactory and the gustatory pathways directly but, rather, between a side branch of the olfactory pathway and a side branch of the gustatory pathway carrying a valuation signal.

Regarding the following discussion, two additional aspects must be noted. First, the mushroom body loop is largely dispensable, but the projection neurons are required, for basic task-relevant innate olfactory behavior (adult[79] and references therein; larvae[36]). This suggests that such innate olfactory behavior is supported by the direct antennal lobe–lateral horn pathway, whereas conditioned olfactory behavior takes the indirect route via the mushroom bodies. Second, there is no evidence that a given odor would not activate the same subset of Kenyon cells during aversive learning as well as during appetitive learning; this implies that appetitive and aversive memory traces for a given odor may be localized in the same Kenyon cells, but in

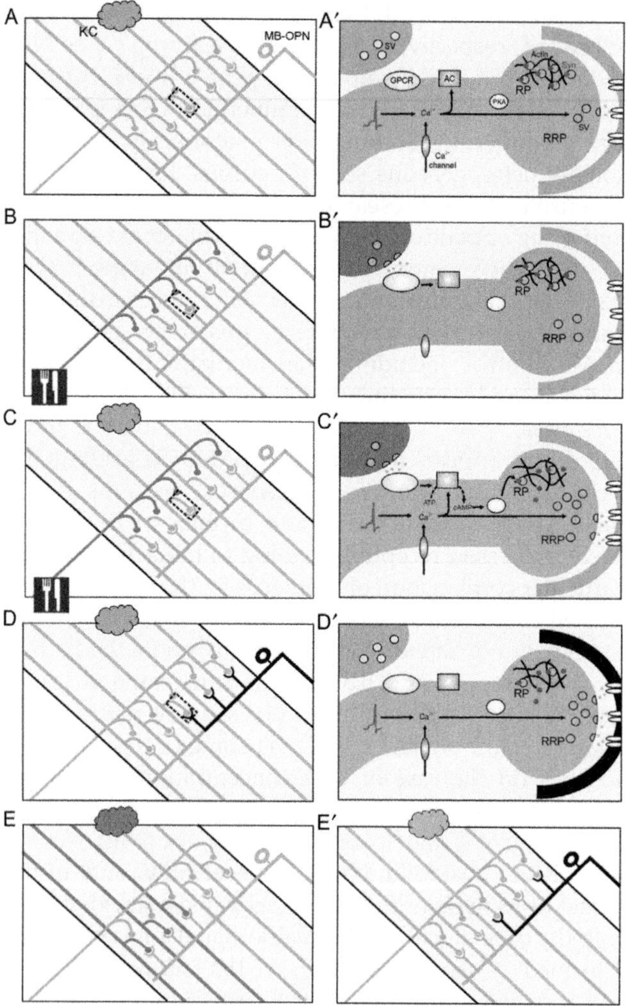

FIGURE 5.3 Working model of associative memory trace formation. The panels on the left (A–D) illustrate the microcircuitry in the mushroom body lobes, and the panels on the right (A′–D′) sketch the molecular processes proposed. (A and A′) Before associative training, the presentation of an odor alone activates a particular subset of Kenyon cells (yellow); the amount of transmitter released from that set of Kenyon cells, however, is not sufficient to activate the output neuron(s). Action-potential-triggered presynaptic calcium influx is sufficient to allow fusion of synaptic vesicles from the readily releasable pool but not for a substantial activation of the adenylate cyclase. (B and B′) Before associative training, an aminergic reward signal alone (green; this signal may be octopaminergic as the color code implies based on the studies discussed in the body text, and/or there may be a dopaminergic reward signal, as the recent study by [76] suggests) conveyed to most, if not all, Kenyon cells does not result in activity in the output neurons, and the activation of the G protein-coupled receptor is not sufficient to substantially activate the adenylate cyclase either. (C and C′) During training, the coincidence of odor-induced activation of the Kenyon cell and of the aminergic reinforcement signal is detected by the adenylate cyclase, such that cAMP is produced, PKA is activated, Synapsin is phosphorylated, and reserve pool vesicles are recruited to the readily releasable pool, where they remain until the Kenyon cell is activated again. That is, during training, the Kenyon cells can draw only on the then-existing readily releasable pool, but not on the reserve-pool of vesicles to be recruited upon an ensuing training trial. Thus, the mushroom body output neuron remains silent during training. (D and D′) After training, the now enlarged pool of readily releasable vesicles in the Kenyon cells is sufficient to allow the trained odor to activate the mushroom body output neuron, which is the basis for conditioned behavior. (E and E′) Presenting previously nontrained odors cannot activate the mushroom body output neuron (E: discrimination) unless the set of Kenyon cells activated by them sufficiently overlaps with that set of Kenyon cells representing the trained odor (E′: generalization). This discrimination–generalization balance could be adaptively adjusted by, for example, changing the excitability of all Kenyon cells and/or of the mushroom body output neuron. Color code and abbreviations are as in previous figures plus AC, adenylate cyclase; ATP, adenosine-triphosphate; Ca^{2+}, calcium ions; cAMP, cyclic adenosine monophosphate; GPCR, G protein-coupled receptor; PKA, protein kinase A; RP, reserve pool; RRP, readily release pool; SV, synaptic vesicle; Syn, Synapsin protein.

functionally and possibly structurally distinct subcellular compartments, connected to separate output neurons (Figure 5.2).

Molecular Coincidence Detection and the AC–cAMP–PKA Cascade

As discussed previously, a memory trace is formed in those Kenyon cells that are activated by both the odor signal and the value signal, but how is this coincidence detected at the molecular level (Figure 5.3)? Based on research in adult flies[77,78] (and references therein), it is likely that the *rutabaga* type I adenylyl cyclase (AC) (*rut* gene, CG9533) acts as molecular coincidence detector between the reinforcer and the olfactory activation of the mushroom body neurons: On the one hand, the odor leads to presynaptic calcium influx, and hence to an activation of calmodulin, in that subset of mushroom body neurons that is activated by this odor. On the other hand, the reinforcer activates aminergic neurons and hence the respective G protein-coupled amine receptors in many, if not all, mushroom body neurons. Critically, it is only by the simultaneous activation of the calmodulin ('odor') and G protein pathways ('reward' or 'punishment,' respectively) that the AC is substantially activated. Thus, it is only in those mushroom body neurons that receive both odor and reinforcement activation that cAMP levels and protein kinase A (PKA) activity are boosted and the respective protein substrates get phosphorylated. Note that reinforcement signals for reward and punishment apparently innervate different domains of the mushroom bodies.[36]

Although this working hypothesis of memory trace formation seems to reasonably integrate most of the available data in adults as well as in larvae, the actual effector proteins that are phosphorylated by PKA to support fly short-term memory remain uncertain (for *Aplysia*, see[80]). To this end, we ventured into an analysis of Synapsin function.[81,15]

Memory Trace and Synapsin Function

Synapsin is an evolutionarily conserved phosphoprotein associated with synaptic vesicles (for a review of *Drosophila* Synapsin, see[82]; for other study cases, see[83]). In flies, Synapsin is coded by a single gene (*syn*, CG3985).[84] It is dispensable for basic synaptic transmission[85] and can bind to both synaptic vesicles and cytoskeletal actin, forming a so-called reserve pool of vesicles. Importantly, phosphorylation of Synapsin allows synaptic vesicles to dissociate from this reserve pool and become part of the readily releasable pool of synaptic vesicles such that they are eligible for release upon a future action potential.

On the behavioral level, larvae carrying the protein-null deletion syn^{97} suffer from a 50% reduction of associative function in the odor–sugar learning paradigm[15,81] (adult odor–shock learning is likewise impaired[85,99]; this defect can be phenocopied by transgenic downregulation of Synapsin using RNAi.[15] The syn^{97} mutant shows intact abilities to recognize gustatory and olfactory stimuli and the respectively needed motor faculties, and it is not differentially sensitive to experimental stress, sensory adaptation, habituation, or satiation compared to its wild-type genetic background.[81] Using a series of rescue experiments in which transgenic Synapsin was expressed in various parts of the larval brain by means of a UAS–*syn*cDNA effector construct, Michels *et al.*[15] analyzed where in the larval brain a Synapsin-dependent memory trace is localized. After showing that an acute rescue is possible (*elav*-Gal4 in combination with *tub*-Gal80[ts]), it was found that Synapsin expression in the mushroom body (*mb*247-Gal4 or *D52H*-Gal4) is sufficient to fully rescue the syn^{97}- mutant defect in associative function (Figure 5.4). Expression of Synapsin in projection neurons (*GH*146-Gal4 or *NP*225-Gal4) is not sufficient for a rescue; notably, this lack of rescue cannot be attributed to adverse effects of the driver constructs used. Furthermore, restoring Synapsin in fairly wide areas of the brain but not in the mushroom bodies (*elav*-Gal4 combined with *mb*247-Gal80) is not sufficient to rescue the learning defect of the syn^{97} mutants, and also in this case the adverse effects of the transgenes cannot be held responsible for such lack of rescue. Thus, it appears that a Synapsin-dependent short-term memory trace is localized to the mushroom bodies, and the mushroom bodies may turn out to be the only site for such a memory trace.

At the molecular level, it was previously argued that the associative coincidence of odor and appetitive reinforcement boosts AC–cAMP–PKA signaling. Interestingly, *Drosophila* Synapsin contains a number of predicted phosphorylation sites, including the evolutionarily conserved PKA/CamK I/IV consensus motif RRFS at Ser6; an evolutionarily nonconserved

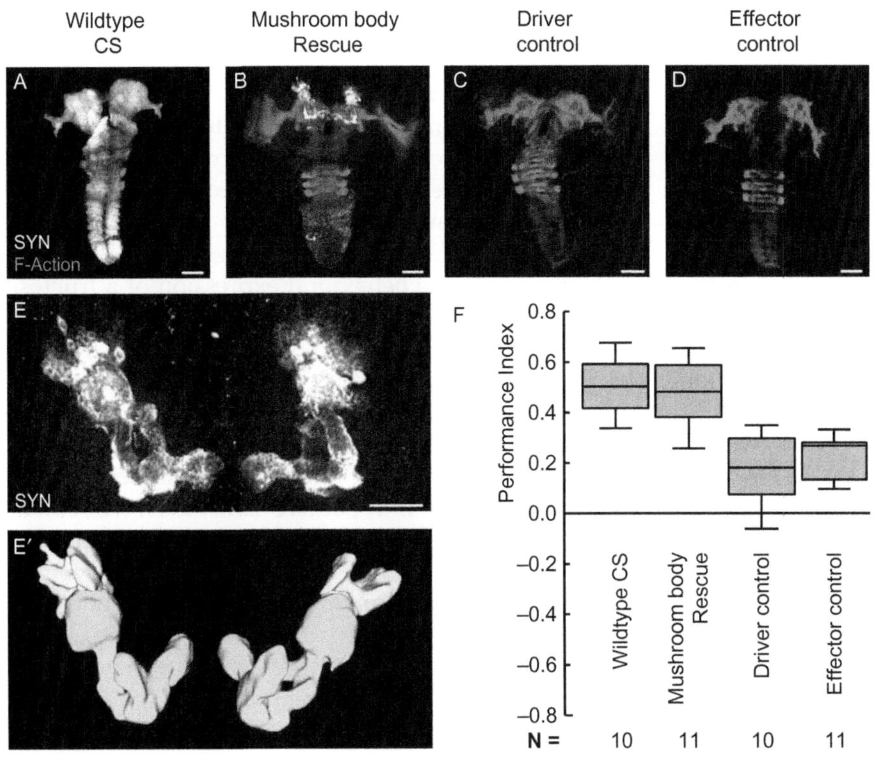

FIGURE 5.4 A local Synapsin-dependent memory trace in the mushroom bodies. (A–D) Anti-Synapsin (white) and anti-F-actin (orange) immunoreactivity of brains of the indicated genotypes. (E) A magnified view of the mushroom bodies from the mushroom body rescue strain using mb247-Gal4 as driver. (E′) Three-dimensional reconstruction of the mushroom bodies. (F) Using the two-odor paradigm, associative function is impaired in both driver and effector control (because these are in the syn^{97}-mutant background) and is fully rescued in the mushroom body rescue strain (which is also in the syn^{97}-mutant background). Scale bar = 50 μm in panels A–D and 25 μm in panel E. The same full rescue is observed upon acute Synapsin expression, as well as for another mushroom-expressing driver strain. No rescue is observed for drivers expressing in projection neurons or when expressing Synapsin in wide areas of the brain but excluding the mushroom bodies (see text). Source: *From Michels* et al.[15]

PKA/CamK I/IV consensus motif RRDS at Ser533; and suspected sites for other kinases, such as CamK II, prolin-dependent kinase, and PKC.[86] Therefore, Michels *et al.*[15] tested whether a Synapsin protein with the two PKA/CamK I/IV sites mutationally disabled (UAS–*syn*cDNA–PKA$^{S6A/S533A}$) can rescue the associative defect of the *syn*[97] mutants. Because no such rescue was observed, it seems likely that Synapsin exerts a function in memory trace formation as a substrate of the AC–cAMP–PKA cascade.

THE DECISION TO BEHAVIORALLY EXPRESS A MEMORY TRACE—OR NOT

The previous discussion suggests that the formation of an odor–sugar associative memory trace involves an interaction between the mushroom body side branch of olfactory processing and a modulatory reward signal branching off of the taste pathway. The experiments reviewed in the following[17,64,87] investigate how the process of 'translating' such a memory trace into conditioned behavior is organized.

When larvae are trained with a medium concentration of fructose, they do not automatically show conditioned behavior during the test (Figure 5.5A, four plots on the left). If tested in the presence of various fructose concentrations, larvae show conditioned behavior when the sugar concentration at the moment of test is lower than the sugar concentration during training, whereas animals tested on a substrate with a sugar concentration equal to or higher than that during training do not. These differences in conditioned behavior are puzzling because the equal training with the medium sugar concentration induces the same memory trace in all these groups. One interpretation of this result could be that medium or higher sugar concentrations during the test prevent conditioned behavior altogether. This is not the case, however: Conditioned behavior actually is possible in the presence of a medium sugar concentration, provided the training-concentration was higher (Figure 5.5A, right plot). Thus, neither the training concentration *per se* nor the testing concentration of sugar *per se* determines conditioned behavior—but their comparison does.

Our interpretation is that conditioned behavior toward food-associated odors is a *search for food*, which is abolished in the presence of the sought-for food. This is a fundamental difference in perspective from regarding conditioned behavior as *response to the odor*. The animals apparently compare the value of the activated memory trace with the value of the testing situation and show appetitive conditioned behavior depending on the result of this comparison. That is, conditioned search is enabled only if the outcome of tracking down the learned odor promises a gain in the sense of yet more-reward than is actually present:

Conditioned search if:
Appetitive memory > observed reward.

This comparison clearly requires the memory trace to be read out. Thus, on the cellular level, the point of comparison between the memory trace and the reward present during testing must be downstream of the memory trace, and according to the previous working hypothesis of memory trace formation, it must be downstream of mushroom body output. Therefore, this is a process different from the effect of satiety, which as suggested by[73] acts on the α/β lobes of the mushroom body to altogether silence mushroom body output at test.

What, then, occurs with respect to conditioned behavior after aversive conditioning? Associative avoidance scores after quinine training are revealed in the presence but not in the absence of quinine (Figure 5.5B). Such behavior can be understood if we regard behavior after aversive conditioning as *escape from quinine*, which is pointless in the absence of a reason for an escape (the same was found for high-salt[17,64] and mechanical disturbance[87] as punishment). Thus, conditioned escape behavior remains disabled unless such conditioned escape offers a gain in the sense of relief from punishment:

No conditioned escape if:
Aversive memory > observed punishment.

Whether there is the same kind of quantitative comparison between memory trace and testing situation as in the appetitive case remains to be directly shown for aversive memories. However, based on experiments using quinine and different high concentrations of salt, this seems likely (see discussion in[17]).

Interestingly, the behavioral expression of a fructose memory trace is independent of the presence of quinine (Figure 5.5C, two leftmost groups), and likewise the presence of fructose does not prompt conditioned escape after quinine training (Figure 5.5C, rightmost group). Thus, the modulation of conditioned behavior by the gustatory testing environment can be exerted independently onto appetitive and aversive conditioned behavior. Once these respective modulations take place, however, the resulting behavioral tendencies can certainly summate: After associating one odor with fructose and the other odor with quinine, scores from a choice situation between these two odors are higher in the quinine testing condition than in the pure testing condition (Figure 5.5D). This, as we argue, is because in the presence of quinine both memory

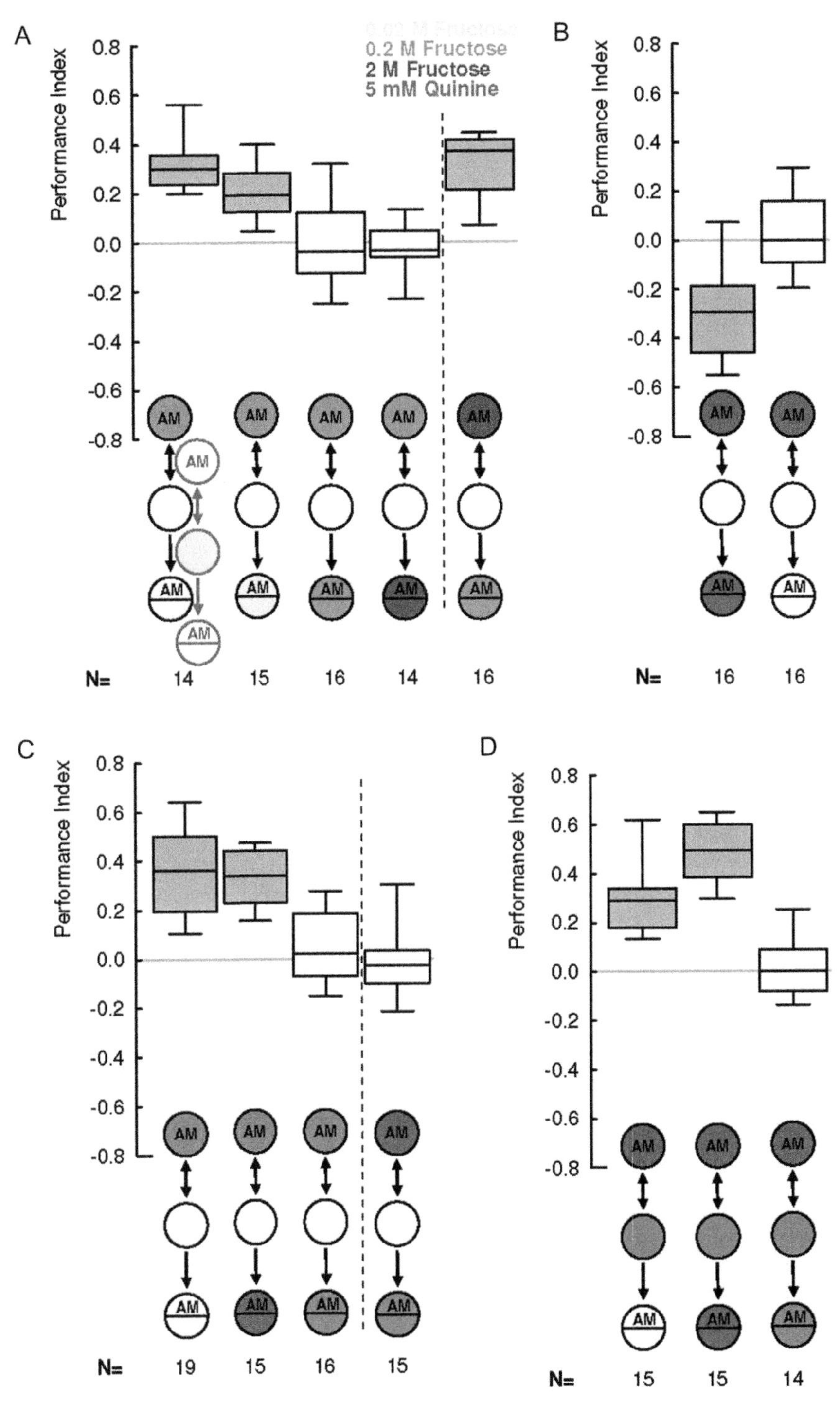

FIGURE 5.5 Expected gain drives learned olfactory behavior. (A) Animals are trained using *n*-amylacetate (AM) and empty odor cups. In the first four plots, a medium fructose concentration (0.2 M) is used as reinforcer during training; the subsequent test is performed in the absence of fructose or in the presence of a lower than trained fructose concentration (0.02 M), the training concentration of fructose (0.2 M), or a higher than trained fructose concentration (2 M). In the right plot, a high fructose concentration (2 M) is used during training, but the test is performed in the presence of the medium (0.2 M) fructose concentration. Memory is behaviorally expressed only if the fructose concentration during training is higher than the fructose concentration at the moment of the test. (B) After aversive-only training, larvae behaviorally express memory only in the presence of quinine. (C) In contrast, after appetitive-only training (three leftmost plots), memory is behaviorally expressed only in the absence of fructose, whereas the presence of quinine has no effect. Conversely, after quinine-only training, the presence of fructose has no effect (right plot). (D) After push–pull training, scores for animals tested on quinine are higher than for those tested on pure, suggesting that only under these conditions both appetitive and aversive memories are behaviorally expressed. Note that the sketches below the boxes show only one possible training regimen; the reciprocally trained group is indicated by a dimmed display in the leftmost plot of panel A, but otherwise omitted. Significant differences from zero (one-sample sign tests; A, $P < 0.05/5$ [Bonferroni correction]; B, $P < 0.05/2$; C, $P < 0.05/4$; and D, $P < 0.05/3$) are indicated by shading of the boxes. Source: *Based on data from Schleyer* et al.[17]

traces can be expressed behaviorally: Conditioned escape from quinine is expressed because quinine is present, and conditioned search for fructose is expressed because fructose is absent.

Thus, conditioned olfactory behavior is organized in a flexible way according to its expected outcome, namely toward finding reward and escaping punishment. Conditioned olfactory behavior is therefore not responsive in nature but, rather, is an action expressed for the sake of those things that are not (yet) there, the sought-for reward and the attempted relief.

Independence of Innate Olfactory and Innate Gustatory Behavior

The experiments detailed previously imply that taste processing affects conditioned olfactory behavior. This

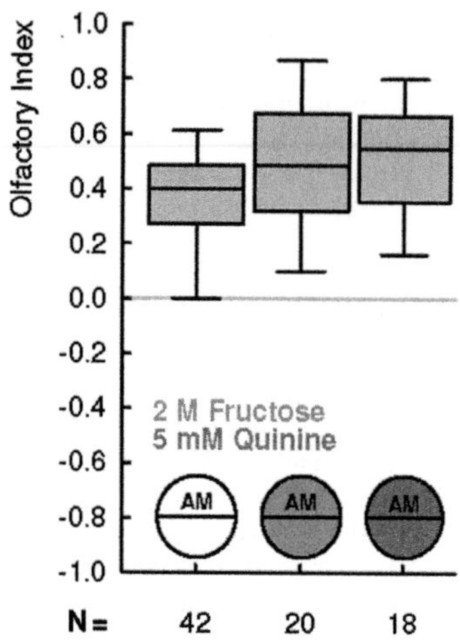

FIGURE 5.6 Innate olfactory behavior is unaltered in the presence of tastants. The olfactory index for *n*-amylacetate (AM) diluted 1:50 in paraffin oil does not differ on the indicated substrates. Pooled data are significantly different from zero ($P < 0.05$, one-sample sign test). Source: *Based on data from Schleyer* et al.[17]

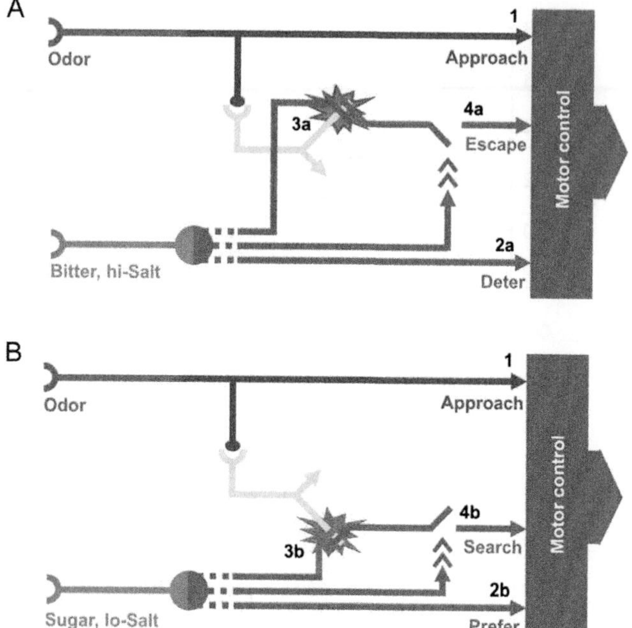

FIGURE 5.7 A minimal, behavior-based circuit for larval decision making. Based on the neuroanatomy presented in Figure 5.2 and the behavioral findings presented in Figures 5.5 and 5.6, we propose a behavior-based circuit of chemosensory behavior and decision making in larval *Drosophila* regarding the decision to behaviorally express an aversive (A) or an appetitive (B) memory trace—or not. (A) (1) Odors innately are usually attractive (Figure 5.6) (for exceptions, see[88,89]). (2a) Larvae innately are deterred by quinine and high concentrations of salt[48,49] and (3a) can associate an odor with such punishment, leading to conditioned aversion to this odor (Figure 5.5B[49,64]). This requires convergence and coincidence detection of olfactory and aversive reinforcement processing (red star symbol). (4a) Olfactory memory traces are behaviorally expressed only if animals expect to improve their situation: Only if at the moment of testing a punishment signal is present that is at least as 'bad' as predicted will conditioned escape behavior be enabled (Figure 5.5B). (B) (2b) Larvae innately prefer sugar and low concentrated salt,[49,50] and (3b) they can associate an odor with such reward, leading to conditioned approach toward this odor (Figure 5.5A[50,62,63]), requiring convergence and coincidence detection of olfactory and appetitive reinforcement processing (green star symbol). (4b) The presence of a reward signal at the moment of the test that is at least as 'good' as predicted disables the expression of conditioned search behavior (Figure 5.5A[17,64]). Source: *Based on Schleyer* et al.[17]

raises the question whether the same is true for innate (in the sense of experimentally naive) olfactory behavior. This is not the case: Larvae show indistinguishable innate olfactory choice behavior on various tastant substrates (Figure 5.6), even when odors are diluted to yield only moderate levels of attraction that arguably are easier to modulate.[17] In turn, innate gustatory behavior is not influenced by odor processing, either.[17]

Thus, innate olfactory and innate gustatory behavior seem to be mutually 'insulated' from each other, and in this sense they seem to be responsive in nature.

A Circuit for Decision Making?

We are thus confronted with two different kinds of olfactory behavior: (1) innate, largely hard-wired and responsive olfactory behavior and (2) flexibly organized learned olfactory behavior that is better captured as an action in pursuit of its outcome. Considering the functional circuitry of the larval chemobehavioral system, these two kinds of olfactory processing can be mapped onto the direct antennal lobe–lateral horn pathway, on the one hand, and the mushroom body loop, on the other hand (Figure 5.7). Regarding the topic of this contribution, namely decision making, we draw attention to the site of the chemobehavioral system labeled 4a and 4b in Figure 5.7. At the moment of testing, an integration takes place at this site between the four aspects of decision making introduced previously: (1) sensory input in terms of the gustatory interneuron(s) reporting on the value of the current situation, (2) the animal's history in terms of the activated memory trace from the mushroom body, (3) the available behavioral options in terms of the connections to motor control, and (4) the predicted consequences of executing these options in terms of the calculated difference between aspects 1 and 2. We therefore argue that the process of behaviorally expressing an odor–taste memory trace, or not doing so, features fundamental properties of decision making. We believe that this process is simple enough that it can eventually be understood at satisfying cellular and molecular detail but is complex enough to warrant experimental effort. Still, this process seems to lack much of the psychological complexity of decision making as we experience it ourselves in our lives. What are the issues?

ASPECTS OF DECISION MAKING

As we have argued, the brain is the organ of behavior organization. At any moment, it structures the solution to possibly the most 'bestial' of all problems: Should I stay or should I go? To this end, the brain integrates (1) sensory input, (2) the subject's individual and evolutionary history as reflected in its internal status, (3) the available behavioral options, and (4) their expected consequences. As a corollary of this integrative function, many events in the brain are related to processing offline from sensorimotor loops. Such offline processing is required because it is usually not clear which behavior will be best: Behaviors that are expressed always if and only if there is a particular input—that is, behaviors that are expressed largely irrespective of other situational or internal conditions—are rare exceptions, in the case of monosynaptic reflexes, or appear pathological in the case of unbreakable habits. Normally, behavior needs to be acutely tailored to suit the configuration of senses, status, options, and expected consequences. These processes, taking place between the moment when uncertainty between behavioral options is registered (the moment when I ask myself whether I should stay or go) and the eventual expression of behavior (me finally staying or going), we experience as us 'taking a decision.'

Obviously, the task at hand when using such a concept of decision making as "experiential corollary of organizing behavior under uncertainty" is to develop behavioral paradigms that confer an experimental handle on such processes. Furthermore, it is obvious that the boundaries of such a concept are fleeting, particularly toward guessing when uncertainty is very high and toward reflexive or habitual behavior when uncertainty is very low. Therefore, we try to 'parametrically' characterize decision making: We characterize five aspects of behavioral tasks that, as we suggest, bear upon the strength of the decision character of the process. These characterizations, based on what we as humans experience when confronted with decisions, will then each be followed by a discussion as to whether these characteristics can be found or can reasonably be sought for in larval *Drosophila*.

Before starting this discussion, note that along sensorimotor loops sensory signals, graded in nature, often need to be dichotomized to command behavior—or not. This process to funnel sensory representations into behavioral categories entails a dilemma: In some situations, it may be warranted to regard recognizably different representations as the same (generalization), whereas in other circumstances it may be appropriate to regard them as different despite recognizable commonalities (discrimination). Clearly, these kinds of processes are fundamental for perception and behavioral organization in any system, and they seem experimentally accessible in *Drosophila*,[90,91] including the adaptive adjustment of the generalization–discrimination balance in larval *Drosophila*.[92]

Options

To take a decision, there have to be options. In the simplest case, the options in the face of a particular situation are to express a given behavior—or not. Only highly specialized monosynaptic reflexes, such as in the context of escape or postural control, may hardly offer at least this one degree of freedom. As more options come into play, the process of behavioral organization increasingly assumes the nature of a decision. Clearly, the more strongly the network is biased toward a given behavior, either because of the way it is wired or because the state of the subject confers an imminent imperative such as after starvation, the less room there is to take a decision. Similarly, if the differences between the predicted consequences of each behavior are too drastic, the process loses its decision character as well, because uncertainty is lost. In turn, the decision character would get lost, too, if the difference in consequences were too small, such that the subject would have to guess.

In the case of *Drosophila* larvae facing the test situation, they may track down the odor, turn their back on the odor, or ignore it. In the latter case, they may orient with respect to other, inadvertently applied stimuli and/or their conspecifics, start feeding on or digging into the substrate, or pupate. This behavioral richness is overlooked when our assay is performed because typically only olfactory behavior with respect to the trained odor is scored. For this scored learned olfactory behavior, we have argued that the larvae have the option to behaviorally express it—or not, and that this process is organized at the site of the circuit labeled by 4a and 4b in Figure 5.7.

Dimensions and Conflict

As the number of dimensions in which two behavioral options differ is increased, the process turns into a decision: All else being equal, you immediately choose the job with the higher salary. Similarly, all else being equal, you choose the job that allows for greater independence. It is only when the two jobs differ in both dimensions and if there is a conflict that a decision is called for—for example, if the better paying job offers less independence. If such a conflict is recognized, you may scale the difference in income between the jobs and the difference in terms of independence according to their priority; if the differential between the jobs remains too small even after such inner bargaining, you may seek a third dimension, such as the duration of the contract, to tip the balance. The more dimensions that are consulted and the more of these iterations that take place, the more the process assumes the nature of a decision.

Regarding the test situation in our assay, we have argued that innate and learned olfactory behavior are organized rather separately. Thus, if the innate and the learned value of an odor are different or even opposite, the larvae may need to prioritize. Likewise, during extinction, the larvae may form an extinction memory, namely that a contingency of odor and reward is no longer valid; it would then need to pit such extinction memory against the 'original' memory that does suggest such a contingency. Similarly, if odor–food memories were specific not only for the quality of the trained odor,[93] but also for its intensity (regarding adults, see[94]), the organization of test behavior along an odor gradient would be a more complicated matter than it appears to be.

Lastly, we note that the two kinds of logical errors one can make—that is, false-positive behavior (tracking down an odor although no reward will be available) and false-negative behavior (not bothering to search for reward although it would be available at the odor source)—represent a fundamental dimensional conflict for larval as well as any sort of decision making. The way this conflict is negotiated is conceivably different for rewards versus punishments (and in both cases is likely subject to state-dependent modulation) and may interact with negotiating the generalization–discrimination balance mentioned previously. Indeed, an adaptive adjustment of the generalization–discrimination balance[92] could be achieved by altering the excitability of mushroom body neurons and/or mushroom body output neurons.

Time and Certainty

Taking decisions takes time: All else being equal, a given behavior loses its credentials as a decision, rather than a guess, as the time allowed to go through the process is shortened. Note that some "flavor' of guessing will always remain in behavioral organization, however, because (1) the subject always has to close the process after finite integration time, (2) predicting the consequences of behavior cannot be fully accurate, and (3) perceptual hypotheses generated from sensory signals are also error-prone. In turn, if for the sake of argument we assume that all perceptual and predictive processes are perfectly accurate and time is not limiting, behavior organization would always converge onto that one optimal solution; there would neither be the initial moment of uncertainty regarding what to do, which is characteristic of the decision process, nor would there be any realistic alternative. Thus, it seems that intermediate levels of perceptual and predictive accuracy as well as intermediate time budgets are characteristic of decision processes: If accuracy is too low or time too short, one cannot decide, but if accuracy is too high and time were endless, there would be no need to decide.

Regarding time, the 3 min allowed for the larvae to distribute seems sufficient for them to 'make up their minds' because scores do not get better if more decision time is allowed.[62] Regarding certainty, one may fancy an experiment in which the larvae can opt out from testing by means of a "don't know" option. As a more sober thought, we note that larvae need more than one odor–reward trial to eventually support conditioned behavior,[63] as is the case for most other classical conditioning paradigms, and likewise associative scores decrease as odor or reward intensities decrease. These three parameters may impact how certain the larvae are about the relation between odor and reward, so one may independently vary these parameters for two odors to pit them against each other. Similarly, it may be possible to introduce variance into the odor–reward relationship. If it turns out that odor–reward memories indeed contain information about these parameters of reliability, it would be a challenge to understand which and how such 'meta-data' are encoded.

As a complication, assays based on the behavior of cohorts of animals, as in our case, or on statistics integrating individual animal behavior into one statistic are ambiguous: A partial associative score may imply that all animals are partially certain, or it may imply that some animals are certain of the contingency between odor and reward whereas other animals are certain that there is no such relation or are agnostic (for a study of this problem in bees, see[95]). Similar questions may be raised on the synaptic level: Do partial scores result from all relevant synapses being partially modified or from fractions of synapses that are while the rest are not?

Pride, Blame, and Person

Once a decision and corresponding action is taken, we take note of the consequences. The clearer the decision character of the process, the more 'personally' we take the consequences—the more readily we take pride in the success or blame ourselves for the failure. This is different in cases in which our behavior had been reflexive or in another way forced, or in cases in which we have guessed and then can merely recognize our good or bad luck. However, both after a successful decision and after a 'lucky punch', we are inclined to repeat the behavior in similar circumstances: If after the first guess we are once again lucky with a similar second guess, we will begin to believe in an action–outcome relation and will incorporate this now 'educated guess' into a future decision process. If such an initial decision indeed yields the intended consequence, we can in the future take less effort in the process, such that eventually the nagging feeling of indecision gets lost. Rather, our behavior will eventually become habitual, forming a reliable aspect of us as persons (see introductory quote).

We believe that the closest behavioral studies can get to a thus-understood 'personal' level is to focus on operant behavior and operant conditioning: A prerequisite for an animal to learn that its actions have particular consequences is that it recognizes itself as 'subject', as author of its actions and decisions. It will be interesting to determine whether *Drosophila* larvae, similar to adults,[96] have such faculties.

Offline Processing

As a last, and indeed fundamental, aspect, note that at the beginning of the decision-making process, more than one but not all behaviors are practically possible: In a fight-or-flight situation, for example, feeding remains suppressed. Such suppression, however, is not limited to those behavioral options that are excluded to begin with. Rather, in a fight-or-flight situation, both fight and flight, despite both being 'primed', both also need to remain disabled until the decision is taken. Obviously, one solution here is lateral inhibition between the two behavioral options (including the use of inhibition to organize the switch between alternative functional configurations of a given network[11]). Thus, decision making will often require heavy inhibition, both to keep behavioral options 'off the table' and to keep inhibition between those options that are being pitted against or between network elements that are part of a combinatorial motor code. This implies that much of neuronal activity, be it excitatory or inhibitory, is related to being able to potentially do things without actually doing them—and not to processing along sensorimotor loops.

CONCLUSION

There seems to be something so fundamental to the uncertainty in the world around us and in our sensory and cognitive faculties that just does not allow for a switchboard-type of brain and that forces us to make decisions and take our chances. Still, the causal texture of the world seems sufficiently obvious and stable for reasoned action and our bodily homeostasis sufficiently stable for personal consistency. It is this sphere of medium uncertainty where decision making and our personal lives are staged.

Acknowledgments

The following bodies funded and/or fund the reviewed research: the Leibniz Institut für Neurobiologie (LIN) Magdeburg; the Wissenschaftsgemeinschaft Gottfried Wilhelm Leibniz (WGL); the universities of Würzburg, Leipzig, and Magdeburg; the Bundesministerium für Bildung und Forschung (Bernstein Focus Insect inspired robotics); the Volkswagen Foundation; and the Deutsche Forschungsgemeinschaft (Heisenberg Programm, IRTG 1156, PP 1392, CRC 554, CRC TR 58-A6, and CRC 779-B11). M.S. is supported by a PhD fellowship of the Studienstiftung des deutschen Volkes. We thank Rupert Glasgow (Zaragoza, Spain), Martin Heisenberg (Würzburg, Germany), Randolf Menzel (Berlin, Germany), Paul Stevenson and Bert Klagges (Leipzig, Germany) for discussions of decision making and Philippe Tobler (Zürich, Switzerland) as well as Ari Berkowitz (Norman, OK) for comments on the manuscript. We appreciate the chance to express our thoughts on decision making despite some disagreement with the editors.

Note Added in Proof

Liu *et al.*[78] showed that the PAM subset of dopaminergic neurons mediates a reward signal, suggesting a heterogeneity of different dopamine neurons mediating reward and punishment. Strikingly, the same conclusion was reached for the mouse by Lammel *et al.*[97]

References

1. Rillich J, Schildberger K, Stevenson PA. Octopamine and occupancy: An aminergic mechanism for intruder-resident aggression in crickets. *Proc Biol Sci.* 2011;278:1873–1880.
2. Ott SR, Verlinden H, Rogers SM, et al. Critical role for protein kinase A in the acquisition of gregarious behavior in the desert locust. *Proc Natl Acad Sci USA.* 2012;109:E381–387.
3. Simpson SJ, James S, Simmonds MS, Blaney WM. Variation in chemosensitivity and the control of dietary selection behaviour in the locust. *Appetite.* 1991;17:141–154.
4. Ribeiro C, Dickson BJ. Sex peptide receptor and neuronal TOR/S6K signaling modulate nutrient balancing in *Drosophila. Curr Biol.* 2010;20:1000–1005.
5. Chittka L, Skorupski P, Raine NE. Speed-accuracy tradeoffs in animal decision making. *Trends Ecol Evol.* 2009;24:400–407.
6. Menzel R, Kirbach A, Haass WD, et al. A common frame of reference for learned and communicated vectors in honeybee navigation. *Curr Biol.* 2011;21:645–650.
7. Tang S, Guo A. Choice behavior of *Drosophila* facing contradictory visual cues. *Science.* 2001;294:1543–1547.
8. Pick S, Strauss R. Goal-driven behavioral adaptations in gap-climbing *Drosophila. Curr Biol.* 2005;15:1473–1478.
9. Gomez-Marin A, Stephens GJ, Louis M. Active sampling and decision making in *Drosophila* chemotaxis. *Nat Comm.* 2011;2:441.
10. Friesen WO, Kristan WB. Leech locomotion: Swimming, crawling, and decisions. *Curr Opin Neurobiol.* 2007;17:704–711.

11. Kristan WB. Neuronal decision-making circuits. *Curr Biol.* 2008;18:R928–R932.
12. Lorenz K. Autobiography. In: Odelberg W, ed. Les Prix Nobel en *1973*. Stockholm: Nobel Foundation; 1973.
13. Thum AS, Leisibach B, Gendre N, Selcho M, Stocker RF. Diversity, variability, and suboesophageal connectivity of antennal lobe neurons in *D. melanogaster* larvae. *J Comp Neurol.* 2011;519:3415–3432.
14. Demerenc M, Kaufmann BP. *Drosophila Guide: Introduction to the Genetics and Cytology of Drosophila Melanogaster*. Washington, DC: Carnegie Institution of Washington; 1972.
15. Michels B, Chen YC, Saumweber T, et al. Cellular site and molecular mode of synapsin action in associative learning. *Learn Mem.* 2011;18:332–344.
16. Stocker RF. Design of the larval chemosensory system. *Adv Exp Med Biol.* 2008;628:69–81.
17. Schleyer M, Saumweber T, Nahrendorf W, et al. A behavior-based circuit model of how outcome expectations organize learned behavior in larval *Drosophila*. *Learn Mem.* 2011;18:639–653.
18. Heimbeck G, Bugnon V, Gendre N, Haberlin C, Stocker RF. Smell and taste perception in *Drosophila melanogaster* larva: Toxin expression studies in chemosensory neurons. *J Neurosci.* 1999;19:6599–6609.
19. Fishilevich E, Domingos AI, Asahina K, Naef F, Vosshall LB, Louis M. Chemotaxis behavior mediated by single larval olfactory neurons in *Drosophila*. *Curr Biol.* 2005;15:2086–2096.
20. Kreher SA, Kwon JY, Carlson JR. The molecular basis of odor coding in the *Drosophila* larva. *Neuron.* 2005;46:445–456.
21. Kreher SA, Mathew D, Kim J, Carlson JR. Translation of sensory input into behavioral output via an olfactory system. *Neuron.* 2008;59:110–124.
22. Python F, Stocker RF. Adult-like complexity of the larval antennal lobe of *D. melanogaster* despite markedly low numbers of odorant receptor neurons. *J Comp Neurol.* 2002;445:374–387.
23. Ramaekers A, Magnenat E, Marin EC, et al. Glomerular maps without cellular redundancy at successive levels of the *Drosophila* larval olfactory circuit. *Curr Biol.* 2005;15:982–992.
24. Benton R, Sachse S, Michnick SW, Vosshall LB. Atypical membrane topology and heteromeric function of *Drosophila* odorant receptors *in vivo*. *PLoS Biol.* 2006;4:e20.
25. Pellegrino M, Steinbach N, Stensmyr MC, Hansson BS, Vosshall LB. A natural polymorphism alters odour and DEET sensitivity in an insect odorant receptor. *Nature.* 2011;478:511–514.
26. Vosshall LB, Hansson BS. A unified nomenclature system for the insect olfactory coreceptor. *Chem Sens.* 2011;36:497–498.
27. Laurent G, Stopfer M, Friedrich RW, Rabinovich MI, Volkovskii A, Abarbanel HD. Odor encoding as an active, dynamical process: Experiments, computation, and theory. *Annu Rev Neurosci.* 2001;24:263–297.
28. Wilson RI. Neural and behavioral mechanisms of olfactory perception. *Curr Opin Neurobiol.* 2008;18:408–412.
29. Chou YH, Spletter ML, Yaksi E, Leong JC, Wilson RI, Luo L. Diversity and wiring variability of olfactory local interneurons in the *Drosophila* antennal lobe. *Nat Neurosci.* 2010;13:439–449.
30. Marin EC, Watts RJ, Tanaka NK, Ito K, Luo L. Developmentally programmed remodeling of the *Drosophila* olfactory circuit. *Development.* 2005;132:725–737.
31. Masuda-Nakagawa LM, Tanaka NK, O'Kane CJ. Stereotypic and random patterns of connectivity in the larval mushroom body calyx of *Drosophila*. *Proc Natl Acad Sci USA.* 2005;102:19027–19032.
32. Masuda-Nakagawa LM, Gendre N, O'Kane CJ, Stocker RF. Localized olfactory representation in mushroom bodies of *Drosophila* larvae. *Proc Natl Acad Sci USA.* 2009;106:10314–10319.
33. Lee T, Lee A, Luo L. Development of the *Drosophila* mushroom bodies: sequential generation of three distinct types of neurons from a neuroblast. *Development.* 1999;126:4065–4076.
34. Technau G, Heisenberg M. Neural reorganization during metamorphosis of the corpora pedunculata in *Drosophila melanogaster*. *Nature.* 1982;295:405–407.
35. Gerber B, Stocker RF. The *Drosophila* larva as a model for studying chemosensation and chemosensory learning: a review. *Chem Sens.* 2007;32:65–89.
36. Pauls D, Selcho M, Gendre N, Stocker RF, Thum AS. *Drosophila* larvae establish appetitive olfactory memories via mushroom body neurons of embryonic origin. *J Neurosci.* 2010;30:10655–10666.
37. Ito K, Suzuki K, Estes P, Ramaswami M, Yamamoto D, Strausfeld NJ. The organization of extrinsic neurons and their implications in the functional roles of the mushroom bodies in *Drosophila melanogaster* Meigen. *Learn Mem.* 1998;5:52–77.
38. Tanaka NK, Tanimoto H, Ito K. Neuronal assemblies of the *Drosophila* mushroom body. *J Comp Neurol.* 2008;508:711–755.
39. Sejourne J, Placais PY, Aso Y, et al. Mushroom body efferent neurons responsible for aversive olfactory memory retrieval in *Drosophila*. *Nat Neurosci.* 2011;14:903–910.
40. Cobb M, Scott K, Pankratz M. Gustation in *Drosophila melanogaster*. *SEB Exp Biol Series.* 2009;63:1–38.
41. Montell C. A taste of the *Drosophila* gustatory receptors. *Curr Opin Neurobiol.* 2009;19:345–353.
42. Tanimura T, Hiroi M, Inoshita T, Marion-Poll F. Neurophysiology of gustatory receptor neurones in *Drosophila*. *SEB Exp Biol series.* 2009;63:59–76.
43. Colomb J, Grillenzoni N, Ramaekers A, Stocker RF. Architecture of the primary taste center of *Drosophila melanogaster* larvae. *J Comp Neurol.* 2007;502:834–847.
44. Singh R, Singh K. Fine structure of the sensory organs of *Drosophila melanogaster* Meigen larva (Diptera: *Drosophilidae*). *Int J Insect Morphol Embryol.* 1984;13:255–273.
45. Gendre N, Luer K, Friche S, et al. Integration of complex larval chemosensory organs into the adult nervous system of *Drosophila*. *Development.* 2004;131:83–92.
46. Melcher C, Pankratz MJ. Candidate gustatory interneurons modulating feeding behavior in the *Drosophila* brain. *PLoS Biol.* 2005;3:e305.
47. Selcho M, Pauls D, Han KA, Stocker RF, Thum AS. The role of dopamine in *Drosophila* larval classical olfactory conditioning. *PLoS One.* 2009;4:e5897.
48. Hendel T, Michels B, Neuser K, et al. The carrot, not the stick: Appetitive rather than aversive gustatory stimuli support associative olfactory learning in individually assayed *Drosophila* larvae. *J Comp Physiol A Neuroethol Sens Neural Behav Physiol.* 2005;191:265–279.
49. Niewalda T, Singhal N, Fiala A, Saumweber T, Wegener S, Gerber B. Salt processing in larval *Drosophila*: choice, feeding, and learning shift from appetitive to aversive in a concentration-dependent way. *Chem Sens.* 2008;33:685–692.
50. Schipanski A, Yarali A, Niewalda T, Gerber B. Behavioral analyses of sugar processing in choice, feeding, and learning in larval *Drosophila*. *Chem Sens.* 2008;33:563–573.
51. Thorne N, Chromey C, Bray S, Amrein H. Taste perception and coding in *Drosophila*. *Curr Biol.* 2004;14:1065–1079.
52. Liu L, Leonard AS, Motto DG, et al. Contribution of *Drosophila* DEG/ENaC genes to salt taste. *Neuron.* 2003;39:133–146.
53. Clyne PJ, Warr CG, Carlson JR. Candidate taste receptors in *Drosophila*. *Science.* 2000;287:1830–1834.
54. Kwon JY, Dahanukar A, Weiss LA, Carlson JR. Molecular and cellular organization of the taste system in the *Drosophila* larva. *J Neurosci.* 2011;31:15300–15309.

55. Dahanukar A, Lei YT, Kwon JY, Carlson JR. Two *Gr* genes underlie sugar reception in *Drosophila*. *Neuron*. 2007;56: 503–516.
56. Jiao Y, Moon SJ, Montell C. A *Drosophila* gustatory receptor required for the responses to sucrose, glucose, and maltose identified by mRNA tagging. *Proc Natl Acad Sci USA*. 2007;104: 14110–14115.
57. Jiao Y, Moon SJ, Wang X, Ren Q, Montell C. Gr64f is required in combination with other gustatory receptors for sugar detection in *Drosophila*. *Curr Biol*. 2008;18:1797–1801.
58. Scott K, Brady Jr R, Cravchik A, et al. A chemosensory gene family encoding candidate gustatory and olfactory receptors in *Drosophila*. *Cell*. 2001;104:661–673.
59. Lee Y, Moon SJ, Montell C. Multiple gustatory receptors required for the caffeine response in *Drosophila*. *Proc Natl Acad Sci USA*. 2009;106:4495–4500.
60. Benton R, Vannice KS, Gomez-Diaz C, Vosshall LB. Variant ionotropic glutamate receptors as chemosensory receptors in *Drosophila*. *Cell*. 2009;136:149–162.
61. Awasaki T, Kimura K. Pox-neuro is required for development of chemosensory bristles in *Drosophila*. *J Neurobiol*. 1997;32:707–721.
62. Scherer S, Stocker RF, Gerber B. Olfactory learning in individually assayed *Drosophila* larvae. *Learn Mem*. 2003;10:217–225.
63. Neuser K, Husse J, Stock P, Gerber B. Appetitive olfactory learning in *Drosophila* larvae: effects of repetition, reward strength, age, gender, assay type and memory span. *Anim Behav*. 2005;69: 891–898.
64. Gerber B, Hendel T. Outcome expectations drive learned behaviour in larval *Drosophila*. *Proc R Soc B*. 2006;273:2965–2968.
65. Saumweber T, Husse J, Gerber B. Innate attractiveness and associative learnability of odors can be dissociated in larval *Drosophila*. *Chem Sens*. 2011;36:223–235.
66. Gerber B, Biernacki R, Thum J. Odor–taste learning in larval *Drosophila*. In: Zhang B, Freeman MR, Waddell S, eds. *Drosophila Neurobiology: A Laboratory Manual*. Cold Spring Harbor, NY: Cold Spring Harbor Laboratory Press; 2010.
67. Busch S, Selcho M, Ito K, Tanimoto H. A map of octopaminergic neurons in the *Drosophila* brain. *J Comp Neurol*. 2009;513:643–667.
68. Selcho M, Pauls D, Jundi BE, Stocker RF, Thum AS. The role of octopamine and tyramine in *Drosophila* larval locomotion. *J Comp Neurol*. 2012;520(16):3764–3785.
69. Schwaerzel M, Monastirioti M, Scholz H, Friggi-Grelin F, Birman S, Heisenberg M. Dopamine and octopamine differentiate between aversive and appetitive olfactory memories in *Drosophila*. *J Neurosci*. 2003;23:10495–10502.
70. Schroll C, Riemensperger T, Bucher D, et al. Light-induced activation of distinct modulatory neurons triggers appetitive or aversive learning in *Drosophila* larvae. *Curr Biol*. 2006;16:1741–1747.
71. Honjo K, Furukubo-Tokunaga K. Distinctive neuronal networks and biochemical pathways for appetitive and aversive memory in *Drosophila* larvae. *J Neurosci*. 2009;29:852–862.
72. Claridge-Chang A, Roorda RD, Vrontou E, et al. Writing memories with light-addressable reinforcement circuitry. *Cell*. 2009;139: 405–415.
73. Krashes MJ, DasGupta S, Vreede A, White B, Armstrong JD, Waddell S. A neural circuit mechanism integrating motivational state with memory expression in *Drosophila*. *Cell*. 2009;139:416–427.
74. Aso Y, Siwanowicz I, Bracker L, Ito K, Kitamoto T, Tanimoto H. Specific dopaminergic neurons for the formation of labile aversive memory. *Curr Biol*. 2010;20:1445–1451.
75. Aso Y, Herb A, Ogueta M, et al. Three dopamine pathways induce aversive odor memories with different stability. *PLoS Genet*. 2012;8(7):e1002768.
76. Liu C, Plaçais PY, Yamagata N, et al. A subset of dopamine neurons signals reward for odour memory in *Drosophila*. *Nature*. 2012;488:512–516.
77. Tomchik SM, Davis RL. Dynamics of learning-related cAMP signaling and stimulus integration in the *Drosophila* olfactory pathway. *Neuron*. 2009;64:510–521.
78. Gervasi N, Tchenio P, Preat T. PKA dynamics in a *Drosophila* learning center: Coincidence detection by rutabaga adenylyl cyclase and spatial regulation by dunce phosphodiesterase. *Neuron*. 2010;65:516–529.
79. Heimbeck G, Bugnon V, Gendre N, Keller A, Stocker RF. A central neural circuit for experience-independent olfactory and courtship behavior in *Drosophila melanogaster*. *Proc Natl Acad Sci USA*. 2001;98:15336–15341.
80. Hawkins RD. A cellular mechanism of classical conditioning in *Aplysia*. *J Exp Biol*. 1984;112:113–128.
81. Michels B, Diegelmann S, Tanimoto H, Schwenkert I, Buchner E, Gerber B. A role for Synapsin in associative learning: the *Drosophila* larva as a study case. *Learn Mem*. 2005;12:224–231.
82. Diegelmann S, Klagges B, Michels B, Schleyer M, Gerber B. Maggot learning and synapsin function. *J Exp Biol*. 2012;:In Press
83. Benfenati F. Synapsins—Molecular function, development and disease. *Sem Cell Dev Biol*. 2011;22:377.
84. Klagges BR, Heimbeck G, Godenschwege TA, et al. Invertebrate synapsins: A single gene codes for several isoforms in *Drosophila*. *J Neurosci*. 1996;16:3154–3165.
85. Godenschwege TA, Reisch D, Diegelmann S, et al. Flies lacking all synapsins are unexpectedly healthy but are impaired in complex behaviour. *Eur J Neurosci*. 2004;20:611–622.
86. Nuwal T, Heo S, Lubec G, Buchner E. Mass spectrometric analysis of synapsins in *Drosophila melanogaster* and identification of novel phosphorylation sites. *J Proteome Res*. 2011;10:541–550.
87. Eschbach C, Cano C, Haberkern H, et al. Associative learning between odorants and mechanosensory punishment in larval *Drosophila*. *J Exp Biol*. 2011;214:3897–3905.
88. Cobb M, Domain I. Olfactory coding in a simple system: Adaptation in *Drosophila* larvae. *Proc Biol Sci*. 2000;267:2119–2125.
89. Colomb J, Grillenzoni N, Stocker RF, Ramaekers A. Complex behavioural changes after odour exposure in *Drosophila* larvae. *Anim Behav*. 2007;73:587–594.
90. Eschbach C, Vogt K, Schmuker M, Gerber B. The similarity between odors and their binary mixtures in *Drosophila*. *Chem Sens*. 2011;36:613–621.
91. Niewalda T, Voller T, Eschbach C, et al. A combined perceptual, physico-chemical, and imaging approach to "odour-distances" suggests a categorizing function of the *Drosophila* antennal lobe. *PLoS One*. 2011;6:e24300.
92. Mishra D, Louis M, Gerber B. Adaptive adjustment of the generalization–discrimination balance in larval *Drosophila*. *J Neurogen*. 2010;24:168–175.
93. Chen YC, Mishra D, Schmitt L, Schmuker M, Gerber B. A behavioral odor similarity "space" in larval *Drosophila*. *Chem Sens*. 2011;36:237–249.
94. Yarali A, Ehser S, Hapil FZ, Huang J, Gerber B. Odour intensity learning in fruit flies. *Proc Biol Sci*. 2009;276:3413–3420.
95. Pamir E, Chakroborty NK, Stollhoff N, et al. Average group behavior does not represent individual behavior in classical conditioning of the honeybee. *Learn Mem*. 2011;18:733–741.
96. Brembs B. The importance of being active. *J Neurogen*. 2009;23: 120–126.
97. Lammel S, Lim BK, Ran C, et al. Input-specific control of reward and aversion in the ventral tegmental area. *Nature*. 2012;491: 212–217.
98. Hammer M. An identified neuron mediates the unconditioned stimulus in associative olfactory learning in honeybees. *Nature*. 1993;366:59–63.
99. Knapek S, Gerber B, Tanimoto H. Synapsin is selectively required for anesthesia-sensitive memory. *Learn Mem*. 2010;17:76–79.

SECTION 3

DEVELOPMENTS IN METHODOLOGY

CHAPTER

6

Optophysiological Approaches to Learning and Memory in *Drosophila melanogaster*

Thomas Riemensperger and André Fiala

Georg-August-University of Göttingen, Göttingen, Germany

INTRODUCTION: STRATEGIES TO DETERMINE NEURONAL SUBSTRATES UNDERLYING LEARNING AND MEMORY

Neuroscientific research on learning and memory ultimately aims to reveal neuronal structures and cellular mechanisms through which experience-dependent information is acquired, stored, and retrieved by neuronal circuitries of the brain.[1] To gain access to the neuronal circuits and biophysical mechanisms mediating changes in behavior, two general strategies are possible: (1) observation of neuronal processes in correlation with learning or memory retrieval and (2) experimental interference with neuronal processes to systematically manipulate learning and memory formation. To facilitate experiments, 'simplified systems' have often been chosen as preferred objects of research. The reasons for using a particular model system, be it an entire animal or a piece of nervous tissue, are usually due to technical advantages. The nervous system of marine snails consists of large neurons that can easily be impaled with recording and stimulation electrodes.[2] Honeybees exhibit a remarkable complexity in their learning capabilities, and electrophysiological approaches, optical imaging techniques, or pharmacological interventions are possible, for which the relative diminutiveness of the brain compared to that of mammals provides clear advantages.[3] Rodent model systems are attractive objects of study because of their relative evolutionary proximity to humans compared to invertebrates. Here, often 'simplified systems' are extracted by using isolated circuits, such as hippocampus slices.[2] For a long time, the fruit fly *Drosophila melanogaster* has been a favorite organism for geneticists, mainly due to the fast reproduction cycle and the large number of offspring that facilitates the screen for mutants.[4] Whereas this has been an excellent model organism for genetic studies, for many decades it was not useful for physiological investigations: Its small neurons and fine neurites are not advantageous for electrophysiological analyses of individual neurons, and electrophysiological techniques have long been restricted mainly to extracellular sensillum recordings[5] and recordings from neuromuscular preparations of the larval body wall.[6] However, two developments have made *Drosophila* amenable for physiological studies. First, germline transformation[7] and the versatility of binary expression systems[8] provide the possibility to target transgenes to distinct and defined populations of neurons.[9] Second, new proteins as molecular tools have been designed that can be transgenically expressed. DNA-encoded fluorescence probes can be targeted to defined cells in order to observe diverse parameters of cellular functions, such as calcium influx, second messenger-dependent signaling, and synaptic transmission,[10] and the discovery of light-sensitive cation channels has made it possible to manipulate the membrane potentials of neurons simply by illumination—a technology that is generally termed 'optogenetics.'[11,12] Because these molecular techniques—optical imaging of cellular processes using DNA-encoded probes and optical activation of neurons using light-gated cation channels—resemble the use of recording and stimulation electrodes in electrophysiology, we refer to these in combination as optophysiological approaches.

Understanding biochemical and physiological mechanisms that mediate learning and memory formation requires some knowledge about where in the brain those changes may happen that are causative for

Invertebrate Learning and Memory.
DOI: http://dx.doi.org/10.1016/B978-0-12-415823-8.00006-X

it. This classical 'localization problem' is not easy to solve, and it cannot be solved with a single experimental approach.[13] To determine whether distinct changes in neuronal activity (here subsuming all possible neuronal processes that can potentially be altered during learning, such as changes in synaptic transmitter release, postsynaptic excitation, or excitability of circuits in general) are indeed the biophysical substrates through which learning and memory observed in behavior are manifested, several experimental tests have been formulated.[13–16] Although experiments to determine whether neuronal substrates are necessary and sufficient to promote a certain type of learning differ slightly among researchers,[13–16] they generally include (1) disruptive alterations of neuronal functions, (2) detectability of changes in correlation with experience-dependent changes in behavior, and (3) artificial mimicry of changes in neuronal function that can substitute for a natural change in behavior. Here, we summarize how the use of optophysiological tools, among others, may contribute to such experiments.

To illustrate the technical approaches, we restrict ourselves to differential associative olfactory learning and the formation of short-term memory in *D. melanogaster*,[17,18] a learning paradigm that is widely used (Figure 6.1A). In this classical conditioning procedure, one odor as conditioned stimulus (CS+) is presented to a group of fruit flies in temporal coincidence with an electric shock as unconditioned stimulus (US). A second odor is presented without any punishment (CS−). In a subsequent test situation, the animals can chose between the two arms of a T-maze that contain either the CS + or the CS−. Learning and short-term memory are assessed by calculating the proportion of animals avoiding the odor that has been associated with the electric shock in comparison to the total number of flies.[18] The neuronal pathways through which olfactory information is processed in the *Drosophila* brain have been characterized to some extent (Figures 6.1B and 6.1C).[20–22] Fruit flies perceive odors by olfactory sensory neurons housed within sensillae of various morphological types that are located on the third antennal segments and the maxillary palps.[23] Olfactory sensory neurons project via the antennal nerves to the antennal lobes, the primary olfactory neuropils of the insect brain, and ramify their terminal aborizations in spherical structures called glomeruli.[24] The basic logic of connectivity is simple: Those sensory neurons that express a given specific olfactory receptor target one or very few identifiable glomeruli.[24–26] Excitatory and inhibitory local interneurons that interconnect glomeruli process the odor information with the antennal lobe,[27–29] and olfactory projection neurons convey the odor information further to the lateral protocerebrum and the mushroom body.[30–32] The logic of odor coding is crucially different between the antennal lobe and the mushroom body. Whereas odors are represented at the level of the antennal lobes in terms of spatiotemporal patterns of overlapping glomerular activity,[33,34] the intrinsic mushroom body neurons (Kenyon cells) show a sparse response to particular odor stimuli—that is, only a small fraction of the approximately 2500 Kenyon cells per hemisphere selectively respond to a particular odor with the generation of very few action potentials.[35,36] This particular coding scheme appears favorable for selectively assigning positive or negative values to a given odor representation through associative learning (Figure 6.1C).[21,37] Modulatory neurons that release biogenic amines as transmitters (e.g., dopamine and octopamine) and that broadly innervate mushroom bodies are assumed to carry the value information evoked by the US.[21,37] The following sections summarize the experimental approaches that have been used to test this idea.

DISRUPTIVE ALTERATIONS: ABLATION, MUTATION, AND BLOCK OF SYNAPTIC TRANSMISSION

The initial and most evident starting point to localize memory traces in a brain relies on disabling learning and memory formation or retrieval, which is typically achieved by physical ablations or functional impairment at various levels of complexity—that is, by impairment of brain structures, defined neurons, or molecules. In honeybees, for example, local cooling using cold needles,[38] pharmacological block of sodium channels using local anesthetics,[39,40] or depleting the brain of biogenic amines using reserpine[41] have been used to impair associative olfactory learning and memory. In *Drosophila*, the first experiments that demonstrated the requirement of mushroom bodies for associative olfactory learning and short-term memory relied—very much in line with the tradition of *Drosophila* as a genetic model organism—on mutants with defective mushroom body development.[42] Of course, if a gene mutation gives rise to a malformation of mushroom bodies, side effects on neurons up- or downstream of the mushroom bodies are difficult to exclude, leaving some uncertainty about the localization of the structure required for proper learning. Improvement has been achieved by a more selected disruption of mushroom body formation through feeding of hydroxyurea that prevents cell division of mushroom body precursor cells.[43] A second line of evidence for the requirement of mushroom bodies for olfactory learning and memory relied on learning mutants.[44,45] Mutant flies are typically called 'learning mutants' if their naive behavioral response to the

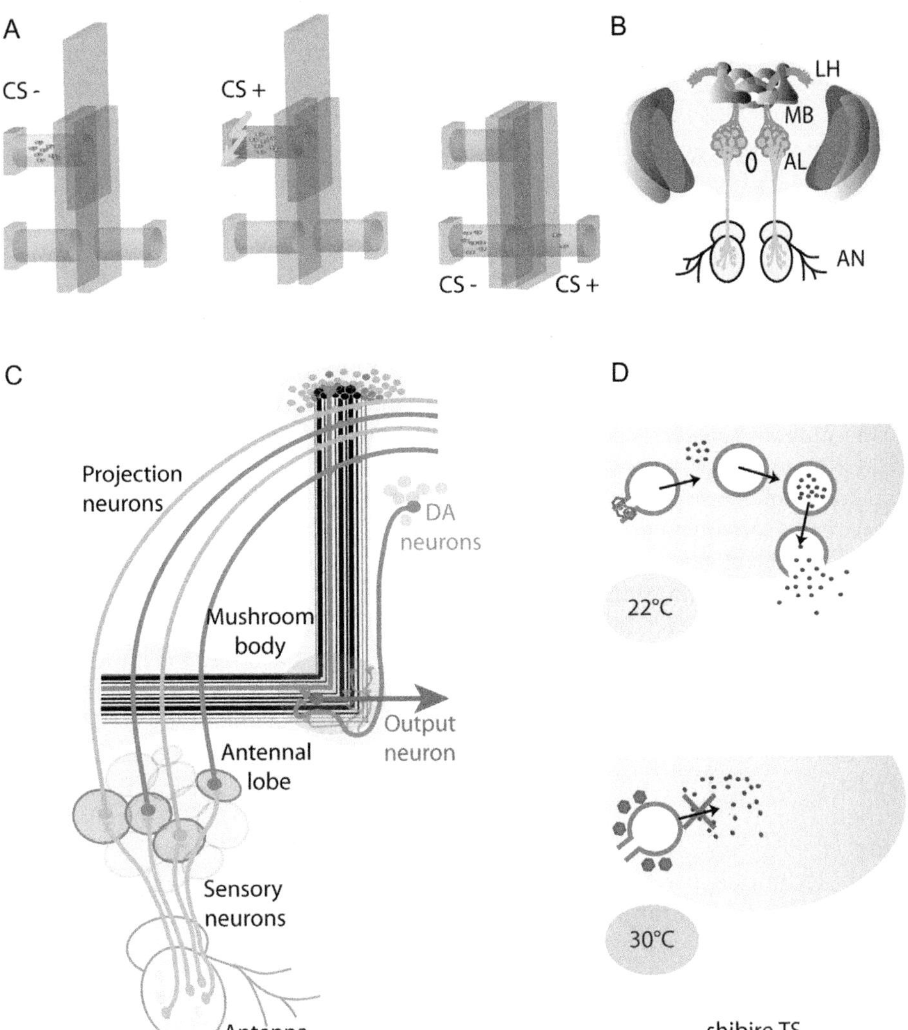

FIGURE 6.1 Classical olfactory conditioning in *Drosophila*. (A) Schematic depiction of a differential conditioning paradigm.[18] During training, flies are sequentially exposed to a non-reinforced odor (CS−) and, with a delay, an odor (CS+) that is temporally paired with an electric shock (unconditioned stimulus). Subsequently, the flies are transferred to a T-maze in which they approach or escape either of the two presented odors. (B) Illustration of the olfactory pathway in the *Drosophila* brain. Odors are perceived by receptors located on the antennae (AN) and conveyed by olfactory sensory neurons (yellow) to the antennal lobes (AL). Olfactory projection neurons (green) convey the information to the mushroom bodies (MB) and the lateral horn (LH). (C) Hypothetical neuronal circuit mediating olfactory associative learning. Odor stimuli are encoded at the level of the antennal lobes as combinatorial glomerular activity patterns and conveyed to the intrinsic mushroom body neurons (Kenyon cells). Here, olfactory information is represented as sparse neuronal activity. Punishment is mediated by dopaminergic neurons, and in coincidence with odor-evoked activity transmission by Kenyon cell output synapses is modified. (D) Temperature-dependent suppression of neurotransmitter release using shibire^{ts}, a temperature-sensitive variant of the protein dynamin.[19] A temperature shift from the permissive (22°C) toward the restrictive (30°C) temperature suppresses synaptic vesicle recycling, ultimately causing a disruption of synaptic transmission.

odors used for training as well as their avoidance of the electric shock is unaltered but learning or short-term memory is impaired.[44,45] Studies on learning mutants provide information on gene products required for particular aspects of olfactory learning, for example, with respect to their requirement for short- or long-term memory formation, aversive or appetitive learning, or consolidation.[44,45] However, two strategies have helped to define the local requirement of the mutated gene products for learning and memory formation. First, several of the genes mutated in learning mutants are preferentially expressed in mushroom bodies—for example, the type I adenylyl cyclase *rutabaga*,[46] the phosphodiesterase *dunce*,[47] and the dominant isoform of the catalytic subunit of cAMP-dependent protein kinase *DC0*.[48] Second, overexpression of the wild-type *rutabaga* gene specifically in the mushroom bodies in the mutant background restored olfactory short-term memory formation.[49] The acute role of the gene product encoded by *rutabaga* could be confirmed by overexpressing the wild-type gene in the mushroom bodies only in the adult stage,[50] thereby ruling out potential developmental defects. The mutations causing deficits in associative olfactory learning and short-term memory often affect genes involved in second-messenger cascades, highlighting a crucial role of metabotropic, modulatory signaling pathways. It is therefore in full accordance that overexpression of a constitutively active form of a G_S protein[51] or impairment of G_O signaling through the expression of pertussis toxin in mushroom bodies[52,53] also impairs associative olfactory learning.

All of the studies mentioned so far used techniques to irreversibly affect learning and memory. However, a technique described by Kitamoto[19] offers the possibility to block synaptic transmission reversibly (Figure 6.1D). A dominant and temperature-sensitive variant of dynamin (*shibire^{ts}*), a protein required for synaptic vesicle recycling, is trapped in an inactive state if the temperature is raised to approximately 30°C. If *shibire^{ts}* is expressed in subsets of neurons and the animals are kept for several minutes at the restrictive temperature, the vesicle pools are depleted and the synaptic transmission is blocked. Because of the reversibility of this effect, training and test phases can be affected separately simply by changing from the permissive to the restrictive temperature or vice versa between training and test situation. This technique has been used by several research groups,[54–56] which found that if the synaptic output from Kenyon cells was blocked only during training, a subsequent memory could be observed in the test phase. If synaptic transmitter release was blocked only during test, a recall of the memory was prevented.[54–56] This finding led to the conclusion that neuronal changes underlying learning rely on plastic processes within or upstream of Kenyon cell presynapses because output from these cells is not required for the acquisition of a short-term memory.[37] However, closer examination by Krashes *et al.*[57] revealed a requirement of synaptic output from a specific subset of Kenyon cells for the retrieval and also for the acquisition of an olfactory short-term memory.

Based on all of the previously mentioned studies, it has been concluded that mushroom bodies are required for associative olfactory learning and the formation of short-term memory in *Drosophila*. The question of whether neurons mediate the reinforcing properties of the US, the punitive electric shock, has likewise been addressed by Schwaerzel *et al.*[58] using *shibire^{ts}*. If synaptic output from a large proportion of neurons releasing dopamine is reversibly blocked only during training, aversive olfactory learning is impaired.[58] The idea that the reinforcing function of the US is mediated by dopaminergic neurons has been substantiated by the fact that appetitive learning using a rewarding sugar stimulus as US, which is also dependent on mushroom body functioning,[58] remains unaffected by the block of these dopaminergic cells. However, note that the large population of dopaminergic cells, which has been targeted in this experiment, does not cover all dopaminergic neurons. This is important because flies lacking a specific dopamine receptor show impairment in both aversive and appetitive learning.[59]

The experiments described so far have provided an enormous body of information about the requirement of Kenyon cells for aversive olfactory learning and short-term memory formation and about the requirement for certain dopaminergic neurons for aversive odor learning. However, these approaches do not indicate whether the cells under investigation are actually modulated in their excitation or their transmitter release due to experience.

DETECTABILITY: OPTICAL IMAGING USING DNA-ENCODED FLUORESCENCE PROBES

The 'conventional' approach to monitoring neuronal activity relies on electrophysiological recordings of changes in membrane potential. These techniques are well-established for insects larger than fruit flies. However, intracellular recordings from neurons in the central brain of *Drosophila* have not been reported. Only recently have patch-clamp recordings from olfactory neurons of the *Drosophila* central brain been described.[27] Although the temporal resolution of electrophysiological recordings is unsurpassable, only one or a small number of neurons can be monitored simultaneously. Optical imaging methods provide a way to record the activity of larger populations of neurons simultaneously but with a much lower spatial and temporal resolution depending on the fluorescence probe and the microscopic setup used. Intracellular calcium represents a parameter that is closely correlated with changes in membrane potential[60] and is widely used to monitor a biophysical correlate of neuronal activity.[61] In insects, the first milestone experiments to optically image neuronal activity in an intact brain were performed in honeybees.[62,63] Here, bath-applied calcium-sensitive dyes have been used to visualize odor-evoked activity in the brain,[62] and with this technique, learning-induced changes in intracellular calcium influx have been determined in both the antennal lobes[64] and the mushroom bodies.[65] A limitation of bath-applied calcium-sensitive dyes is the inability to selectively target distinct neuronal populations—a problem that can be overcome to some extent by targeted injection of dyes.[66–69] Transfusion of defined olfactory neurons with calcium-sensitive dyes has been achieved by injections into axonal tracts[66–69] or groups of somata[68, 69] that could be localized by their position in the brain.

The invention of DNA-encoded fluorescence proteins that have been engineered to report intracellular changes in calcium concentration has provided a step forward because these sensors can be expressed in specific and defined neurons in transgenic organisms.[10] Since the first report of the use of a genetically encoded calcium indicator in the *Drosophila* brain,[33] many studies have exploited this method.[70] In the context of olfactory learning in *Drosophila*, two main questions have been addressed. First, can a change in intracellular calcium be observed in the olfactory pathway in correlation with an olfactory training or memory retrieval? Second, do dopaminergic neurons indeed respond to the US, the electric shock? With respect to the first question, changes in odor-evoked calcium activity dependent on the associative training have been observed in different neuronal structures at different timescales after training.[71–74] The second question has been addressed by Riemensperger *et al.*[75]. In this study, a DNA-encoded calcium sensor was expressed in dopaminergic neurons, and calcium activity was monitored in neuronal arborizations targeting the mushroom body (Figure 6.2). The flies were subjected to a simulated training procedure under the optical imaging microscope with the brain exposed. It was shown that indeed dopaminergic neurons targeting the mushroom body showed strong increases in intracellular calcium in response to the electric shock, confirming that dopaminergic neurons are capable of mediating punishment information during olfactory learning. Moreover, dopaminergic neurons respond slightly to the odors used as CS+ and CS−. Interestingly, the response to the CS+ is prolonged after paring the odor with the US, in contrast to the response evoked by the CS−. This finding shows not only that dopaminergic neurons can act on Kenyon cells but also that, conversely, the training affects the response properties of dopaminergic cells. Note that these studies provide valuable information about whether the neurons under investigation show a correlation between their response properties and the CS–US association. However, it cannot be concluded from these studies that the changes observed are required or causative for any behavioral change.

MIMICRY: OPTOGENETIC AND THERMOGENETIC ACTIVATION OF NEURONS

To define which elements of a neuronal response circuit are sufficient to trigger a simple stimulus–response behavior, one can artificially activate individual neurons within the circuit and observe whether the behavioral response is fully executed. It appeared logical that a similar strategy can be used to define which elements within a neuronal circuit are indeed sufficient to induce a memory. If one bypasses 'natural' stimuli (CS or US) used to train the animal by activating the neurons hypothesized to mediate the respective stimuli, a memory might be artificially created. This experimental concept has been applied to the investigation of associative olfactory learning in honeybees. The activation of an identified octopaminergic neuron in the honeybee brain substituted for an unconditioned stimulus in appetitive associative odor learning,[77] and local injections of octopamine into the bee's brain along with odor stimulation likewise induced an appetitive memory.[78] A similar approach has been introduced by Schroll *et al.*[79] using optogenetic tools. A number of methods attempting to render nerve cells sensitive to light have been reported, all of which have particular advantages and limitations in applicability.[11,12] A breakthrough in optogenetics came with the description of the microbial light-sensitive protein channelrhodopsin-2 (ChR2) by Nagel *et al.*[80] (Figure 6.3A). This light-gated cation channel can be expressed in neurons that can subsequently be depolarized by illumination with blue light, as has been demonstrated in cultured hippocampal neurons.[82] In our studies, we did not use adult fruit flies as described previously. Rather, we studied *Drosophila* larvae because they provide clear advantages for this purpose. The larval cuticle is relatively transparent compared to that of adult fruit flies, and larvae incorporate large amounts of food containing all-*trans*-retinal, a co-factor required for the proper functioning of ChR2 in flies.[79]*Drosophila* larvae are also able to associate an odor with a rewarding or punishing stimulus through learning.[83–85] Here, olfactory learning is, just like in adult flies, dependent on mushroom bodies[86,87] and dopamine,[86,88] Moreover, mutations that affect olfactory learning in adults also do so in larvae.[86] In our experiment (Figures 6.3B–6.3D), ChR2 was expressed in dopaminergic neurons and the larvae were exposed to an odor simultaneously with blue light illumination. Subsequently, the flies were exposed to a second odor but in darkness, and this procedure was repeated three times. When the larvae were subjected to a test situation, they significantly avoided the odor paired with the artificial, light-induced, ChR2-mediated activation of dopaminergic neurons.[79] In a similar experiment, we showed that activating octopaminergic neurons during the presentation of an odor triggers the formation of an appetitive memory.[79] These experiments demonstrated that coactivation of the respective modulatory neurons in coincidence with the odor is sufficient to cause the formation of an olfactory memory. In adult flies, ChR2

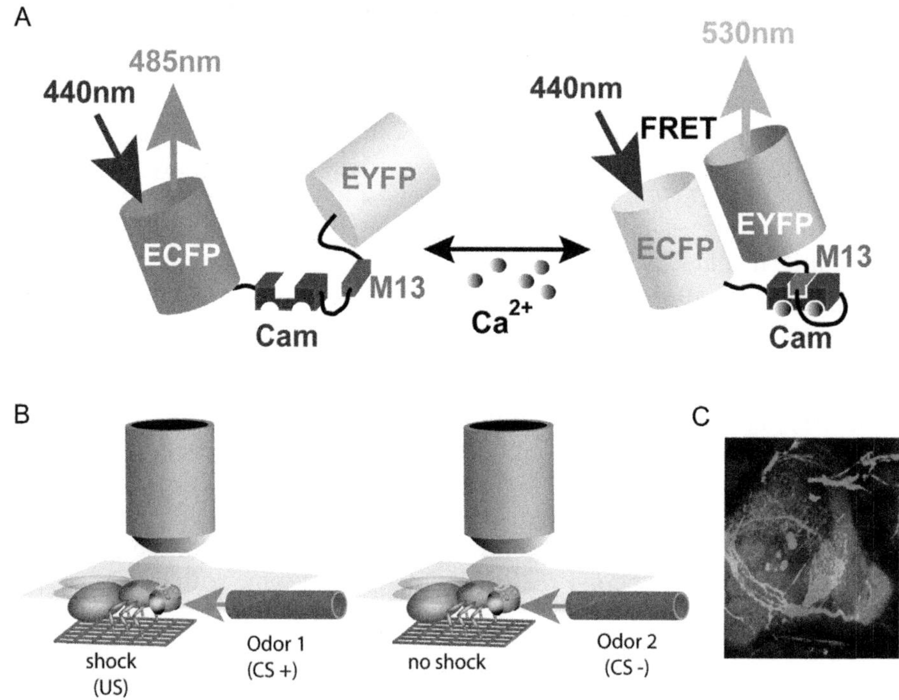

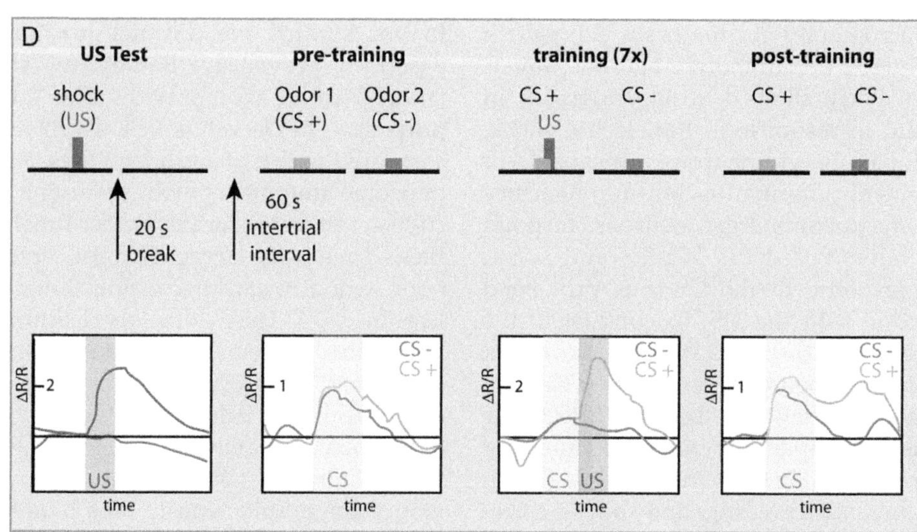

FIGURE 6.2 **Ca^{2+} imaging in dopamine-releasing neurons during associative olfactory learning.** (A) Schematic illustration of the genetically encoded Förster resonance energy transfer (FRET)-based Ca^{2+} indicator Cameleon 2.1.[76] A cyan fluorescence protein (ECFP) and a yellow fluorescence protein (EYFP) are linked to a Ca^{2+} binding domain of calmodulin (Cam) and a calmodulin binding peptide (M13). Ca^{2+} binding to Cam causes a conformational change, bringing ECFP in closer proximity to EYFP, allowing a FRET from ECFP to EYFP. Thereby, changes in intracellular Ca^{2+} concentration can be monitored. (B) Flies are fixed with dental glue to a thin plastic foil and put onto a metal grid used to apply electric shocks. The animals' antennae are oriented toward an airstream to which odors can be applied. For optical access to the brain, a small window is cut into the head capsule. (C) Confocal microscopy image of the mushroom body expressing the red fluorescent protein dsRed and the dopaminergic neurons expressing Cameleon 2.1. (D) Schematic illustration of optical calcium imaging during 10 phases of an experiment. In the first phase, the fly was stimulated with the electric shock (US) and Ca^{2+} activity in dopaminergic neurons was compared to background activity without US. Second, two odors were sequentially presented, both of which evoked slight increases in intracellular calcium. In the third phase, one odor (CS +) was paired with the US and the CS– presented as a control. After seven repetitions of the training trial, CS– and CS + were again presented without any reinforcement. Calcium activity in response to the CS + is prolonged after training. Source: *Data from Riemensperger et al.*[75]

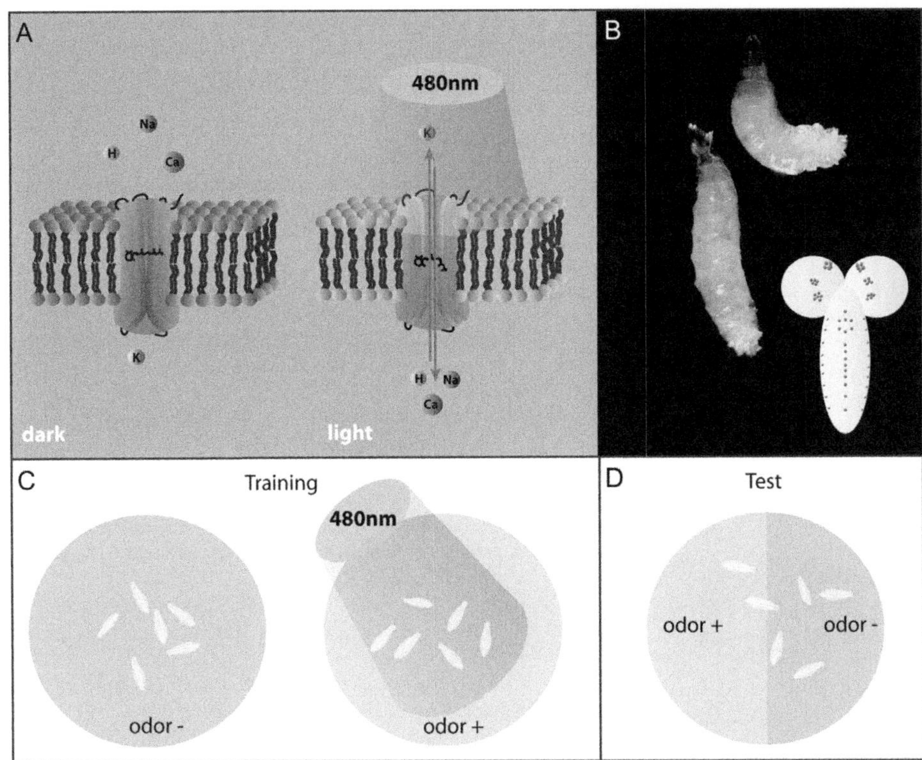

FIGURE 6.3 Optogenetic mimicry of aversive odor learning in *Drosophila* larvae. (A) Depiction of the light-sensitive cation channel Channelrhodopsin-2 (ChR2).[80] Illumination with blue light induces a conformational change, ultimately opening the channel and causing membrane depolarization. (B) Schematic illustration of the expression of ChR2 in dopaminergic neurons in *Drosophila* larvae. (C) Training in an olfactory conditioning paradigm.[79] Larvae expressing ChR2 in dopaminergic neurons were placed in small groups in a dish half-filled with agarose. Larvae were exposed to one odor (CS +) in the presence of light and to a control odor (CS−) in darkness. (D) In the test phase, larvae were placed in the middle of a cell culture dish with the two different odors spotted on the opposite sides of the dish.[79] Source: *Schematic illustration in panel B modified from Monastirioti.*[81]

shows relatively weak efficiency in causing light-induced activation of neurons, perhaps due to the low food uptake of the animals that limits the amount of all-*trans*-retinal in the cells, the nontransparent cuticle, or lower expression levels. For adult flies, an optogenetic approach alternative to ChR2 has been described by Lima and Miesenböck.[89] An ATP-sensitive ionotropic receptor ($P2\times_2$) is expressed in neurons of interest. When ATP coupled to a light-sensitive 'cage' is injected into the animal, firing of the neurons that express $P2\times_2$ is caused by uncaging ATP with a strong flash of UV light (Figure 6.4A). Using this approach, experiments on larvae described above were conceptually recapitulated in adult flies by Claridge-Chang *et al.*[90] If dopaminergic neurons are activated simultaneously with the presentation of an odor, the animals acquire an aversive memory for that odor. Moreover, the neurons that are sufficient for this type of associative aversive learning could be confined to a small cluster of cells innervating the mushroom body.[90] An even more sophisticated experiment has been performed by Aso *et al.*[91] By expressing the temperature-sensitive cation channel dTRPA1[92] (Figure 6.4B), even smaller clusters of dopaminergic neurons were depolarized through temperature increase while the animals were exposed to an odor—a technique that can be called 'thermogenetics.' Interestingly, the activation of very distinct dopaminergic neurons was sufficient for the acquisition of a labile odor memory.[91]

In summary, it can clearly be shown using several approaches that dopaminergic neurons are sufficient to cause the formation of an aversive odor memory if their activity is paired with an odor stimulus. Of course, the described approaches to test whether the excitation of a specific neuron or neuronal class is sufficient to induce a short-term memory do not prove that the neurons under investigation are the only ones causing the learned change in behavior, nor do they show that they actually mediate learning under 'natural' experimental conditions. However, the previously described experiments illustrate how one can test for the sufficiency of particular neurons to induce a behavioral change. Unfortunately, an experimental approach to mimic the CS through artificial stimulation has not been achieved for obvious reasons. If the neuronal substrate underlying associative learning of an odor is manifested by changes in synaptic transmission from Kenyon cells to downstream neurons,[16,21,37] and olfactory information is reflected in the mushroom body in terms of a sparse code comprising only very few Kenyon cells,[35,36] it is obviously difficult to selectively activate those cells that encode a particular odor.

CONCLUSIONS

We have highlighted three conceptual approaches that can be used to determine neuronal substrates

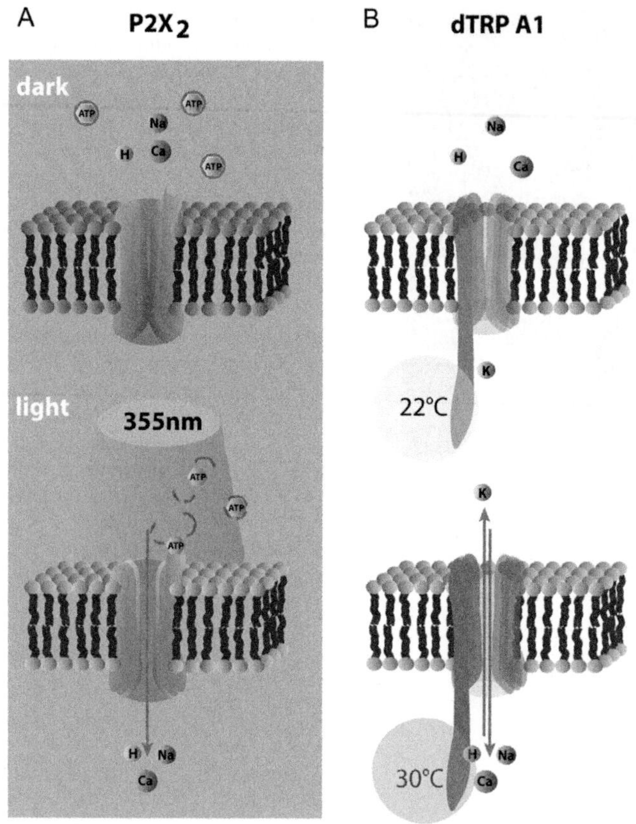

FIGURE 6.4 Optogenetic and thermogenetic approaches used to mimic dopamine signaling in adult *Drosophila*.[90,91] (A) Optogenetic depolarization of neurons using the ATP-dependent cation channel P2 × 2. Caged ATP is applied to the extracellular space and uncaged by a flash of UV light. (B) The expression of the temperature-sensitive TRP channel dTRPA1 can be used for an artificial activation of neuronal subsets through temperature increase.

mediating learning and memory, which necessarily includes a localization of these substrates. Impairing specific neurons or neuronal functions informs about their requirement for learning and/or memory formation. Monitoring neuronal activity or cellular processes involved in neuronal function provides information about biophysical processes correlating with learning or memory. Artificial induction of neuronal activity provides information about a potential sufficiency of the respective neurons for learning and memory formation. A major difficulty in defining an essential memory trace concerns redundancy in complex neuronal circuits: Particular neurons can be sufficient but not necessary for learning or vice versa. The utilization of all three approaches may therefore lead to a more comprehensive picture than the use of any technique alone. Due to the sophisticated genetic expression systems in combination with the steady development of optophysiological or equivalent molecular tools, the fruit fly *D. melanogaster* constitutes an ideal model organism for this task.

Acknowledgment

We are grateful to Prof. Erich Buchner, who has supported our research for many years and contributed to it significantly.

References

1. Albright TD, Jessell TM, Kandel ER, Posner MI. Neural science: a century of progress and the mysteries that remain. *Neuron*. 2000;25(Suppl):S1–55.
2. Hawkins RD, Kandel ER, Siegelbaum SA. Learning to modulate transmitter release: themes and variations in synaptic plasticity. *Annu Rev Neurosci*. 1993;16:625–665.
3. Menzel R, Muller U. Learning and memory in honeybees: from behavior to neural substrates. *Annu Rev Neurosci*. 1996;19:379–404.
4. Rubin GM. *Drosophila melanogaster* as an experimental organism. *Science*. 1988;240(4858):1453–1459.
5. Clyne P, Grant A, O'Connell R, Carlson JR. Odorant response of individual sensilla on the *Drosophila* antenna. *Invert Neurosci*. 1997;3(2-3):127–135.
6. Broadie KS. Synaptogenesis in *Drosophila*: coupling genetics and electrophysiology. *J Physiol Paris*. 1994;88(2):123–139.
7. Rubin GM, Spradling AC. Genetic transformation of *Drosophila* with transposable element vectors. *Science*. 1988;218(4570):348–353.
8. Wimmer EA. Innovations: applications of insect transgenesis. *Nat Rev Genet*. 2003;4(3):225–232.
9. Venken KJ, Simpson JH, Bellen HJ. Genetic manipulation of genes and cells in the nervous system of the fruit fly. *Neuron*. 2011;72(2):202–230.
10. Miyawaki A. Fluorescence imaging of physiological activity in complex systems using GFP-based probes. *Curr Opin Neurobiol*. 2003;13(5):591–596.
11. Deisseroth K, Feng G, Majewska AK, Miesenböck G, Ting A, Schnitzer MJ. Next-generation optical technologies for illuminating genetically targeted brain circuits. *J Neurosci*. 2006;26(41):10380–10386.
12. Fiala A, Suska A, Schlüter OM. Optogenetic approaches in neuroscience. *Curr Biol*. 2010;20(20):R897–R903.
13. Lavond DG, Kim JJ, Thompson RF. Mammalian brain substrates of aversive classical conditioning. *Annu Rev Psychol*. 1993;44:317–342.
14. Martin SJ, Grimwood PD, Morris RG. Synaptic plasticity and memory: an evaluation of the hypothesis. *Annu Rev Neurosci*. 2000;23:649–711.
15. Thompson RF. In search of memory traces. *Annu Rev Psychol*. 2005;56:1–23.
16. Gerber B, Tanimoto H, Heisenberg M. An engram found? Evaluating the evidence from fruit flies. *Curr Opin Neurobiol*. 2004;14(6):737–744.
17. Quinn WG, Harris WA, Benzer S. Conditioned behavior in *Drosophila melanogaster*. *Proc Natl Acad Sci USA*. 1974;71(3):708–712.
18. Tully T, Quinn WG. Classical conditioning and retention in normal and mutant *Drosophila melanogaster*. *J Comp Physiol A*. 1985;157(2):263–277.
19. Kitamoto T. Conditional modification of behavior in *Drosophila* by targeted expression of a temperature-sensitive shibire allele in defined neurons. *J Neurobiol*. 2001;47(2):81–92.
20. Vosshall LB, Stocker RF. Molecular architecture of smell and taste in *Drosophila*. *Annu Rev Neurosci*. 2007;30:505–533.

21. Fiala A. Olfaction and olfactory learning in *Drosophila*: recent progress. *Curr Opin Neurobiol*. 2007;17(6):720–726.
22. Masse NY, Turner GC, Jefferis GS. Olfactory information processing in *Drosophila*. *Curr Biol*. 2009;19(16):R700–R713.
23. de Bruyne M, Warr CG. Molecular and cellular organization of insect chemosensory neurons. *Bioessays*. 2006;28(1):23–34.
24. Vosshall LB, Wong AM, Axel R. An olfactory sensory map in the fly brain. *Cell*. 2000;102(2):147–159.
25. Fishilevich E, Vosshall LB. Genetic and functional subdivision of the *Drosophila* antennal lobe. *Curr Biol*. 2005;15(17):1548–1553.
26. Couto A, Alenius M, Dickson BJ. Molecular, anatomical, and functional organization of the *Drosophila* olfactory system. *Curr Biol*. 2005;15(17):1535–1547.
27. Wilson RI, Laurent G. Role of GABAergic inhibition in shaping odor-evoked spatiotemporal patterns in the *Drosophila* antennal lobe. *J Neurosci*. 2005;25(40):9069–9079.
28. Olsen SR, Bhandawat V, Wilson RI. Excitatory interactions between olfactory processing channels in the *Drosophila* antennal lobe. *Neuron*. 2007;54(1):89–103.
29. Shang Y, Claridge-Chang A, Sjulson L, Pypaert M, Miesenbock G. Excitatory local circuits and their implications for olfactory processing in the fly antennal lobe. *Cell*. 2007;128(3):601–612.
30. Wong AM, Wang JW, Axel R. Spatial representation of the glomerular map in the *Drosophila* protocerebrum. *Cell*. 2002;109 (2):229–241.
31. Marin EC, Jefferis GS, Komiyama T, Zhu H, Luo L. Representation of the glomerular olfactory map in the *Drosophila* brain. *Cell*. 2002;109(2):243–255.
32. Tanaka NK, Awasaki T, Shimada T, Ito K. Integration of chemosensory pathways in the *Drosophila* second-order olfactory centers. *Curr Biol*. 2004;14(6):449–457.
33. Fiala A, Spall T, Diegelmann S, et al. Genetically expressed cameleon in *Drosophila melanogaster* is used to visualize olfactory information in projection neurons. *Curr Biol*. 2002;12(21):1877–1884.
34. Wang JW, Wong AM, Flores J, Vosshall LB, Axel R. Two-photon calcium imaging reveals an odor-evoked map of activity in the fly brain. *Cell*. 2003;112(2):271–282.
35. Luo SX, Axel R, Abbott LF. Generating sparse and selective third-order responses in the olfactory system of the fly. *Proc Natl Acad Sci USA*. 2010;107(23):10713–10718.
36. Honegger KS, Campbell RA, Turner GC. Cellular-resolution population imaging reveals robust sparse coding in the *Drosophila* mushroom body. *J Neurosci*. 2011;31(33):11772–11785.
37. Heisenberg M. Mushroom body memoir: from maps to models. *Nat Rev Neurosci*. 2003;4(4):266–275.
38. Erber J, Masuhr T, Menzel R. Localization of short-term memory in the brain of the bee, *Apis mellifera*. *Physiol Entomol*. 1980;5 (4):343–358.
39. Müller D, Staffelt D, Fiala A, Menzel R. Procaine impairs learning and memory consolidation in the honeybee. *Brain Res*. 2003;977(1):124–127.
40. Devaud JM, Blunk A, Podufall J, Giurfa M, Grünewald B. Using local anaesthetics to block neuronal activity and map specific learning tasks to the mushroom bodies of an insect brain. *Eur J Neurosci*. 2007;26(11):3193–3206.
41. Menzel R, Heyne A, Kinzel C, Gerber B, Fiala A. Pharmacological dissociation between the reinforcing, sensitizing, and response-releasing functions of reward in honeybee classical conditioning. *Behav Neurosci*. 1999;113(4):744–754.
42. Heisenberg M, Borst A, Wagner S, Byers D. *Drosophila* mushroom body mutants are deficient in olfactory learning. *J Neurogenet*. 1985;2(1):1–30.
43. de Belle JS, Heisenberg M. Associative odor learning in *Drosophila* abolished by chemical ablation of mushroom bodies. *Science*. 1994;263(5147):692–695.
44. Davis RL. Physiology and biochemistry of *Drosophila* learning mutants. *Physiol Rev*. 1996;76(2):299–317.
45. Waddell S, Quinn WG. Flies, genes, and learning. *Annu Rev Neurosci*. 2001;24:1283–1309.
46. Han PL, Levin LR, Reed RR, Davis RL. Preferential expression of the *Drosophila* rutabaga gene in mushroom bodies, neural centers for learning in insects. *Neuron*. 1992;9(4):619–627.
47. Nighorn A, Healy MJ, Davis RL. The cyclic AMP phosphodiesterase encoded by the *Drosophila* dunce gene is concentrated in the mushroom body neuropil. *Neuron*. 1991;6(3):455–467.
48. Skoulakis EM, Grammenoudi S. Dunces and da Vincis: the genetics of learning and memory in *Drosophila*. *Cell Mol Life Sci*. 2006;63(9):975–988.
49. Zars T, Fischer M, Schulz R, Heisenberg M. Localization of a short-term memory in *Drosophila*. *Science*. 2000;288(5466):672–675.
50. McGuire SE, Le PT, Osborn AJ, Matsumoto K, Davis RL. Spatiotemporal rescue of memory dysfunction in *Drosophila*. *Science*. 2003;302(5651):1765–1768.
51. Connolly JB, Roberts IJ, Armstrong JD, et al. Associative learning disrupted by impaired Gs signaling in *Drosophila* mushroom bodies. *Science*. 1996;274(5295):2104–2107.
52. Ferris J, Ge H, Liu L, Roman G. G(0) signaling is required for *Drosophila* associative learning. *Nat Neurosci*. 2006;9 (8):1036–1040.
53. Madalan A, Yang X, Ferris J, Zhang S, Roman G. G(0) activation is required for both appetitive and aversive memory acquisition in *Drosophila*. *Learn Mem*. 2011;19(1):26–34.
54. McGuire SE, Le PT, Davis RL. The role of *Drosophila* mushroom body signaling in olfactory memory. *Science*. 2001;293 (5533):1330–1333.
55. Dubnau J, Grady L, Kitamoto T, Tully T. Disruption of neurotransmission in *Drosophila* mushroom body blocks retrieval but not acquisition of memory. *Nature*. 2001;411(6836): 476–480.
56. Schwaerzel M, Heisenberg M, Zars T. Extinction antagonizes olfactory memory at the subcellular level. *Neuron*. 2002;35 (5):951–960.
57. Krashes MJ, Keene AC, Leung B, Armstrong JD, Waddell S. Sequential use of mushroom body neuron subsets during *Drosophila* odor memory processing. *Neuron*. 2007;53(1): 103–115.
58. Schwaerzel M, Monastirioti M, Scholz H, Friggi-Grelin F, Birman S, Heisenberg M. Dopamine and octopamine differentiate between aversive and appetitive olfactory memories in *Drosophila*. *J Neurosci*. 2003;23(33):10495–10502.
59. Kim YC, Lee HG, Han KA. D1 dopamine receptor dDA1 is required in the mushroom body neurons for aversive and appetitive learning in *Drosophila*. *J Neurosci*. 2007;27(29):7640–7647.
60. Berridge MJ. Neuronal calcium signaling. *Neuron*. 1998;21 (1):13–26.
61. Grienberger C, Konnerth A. Imaging calcium in neurons. *Neuron*. 2012;73(5):862–885.
62. Joerges J, Küttner A, Galizia CG, Menzel R. Representations of odours and odour mixtures visualized in the honeybee brain. *Nature*. 1997;387(6630):285–288.
63. Galizia CG, Küttner A, Joerges J, Menzel R. Odour representation in honeybee olfactory glomeruli shows slow temporal dynamics: an optical recording study using a voltage-sensitive dye. *J Insect Physiol*. 2000;46(6):877–886.
64. Faber T, Joerges J, Menzel R. Associative learning modifies neural representations of odors in the insect brain. *Nat Neurosci*. 1999;2(1):74–78.
65. Faber T, Menzel R. Visualizing mushroom body response to a conditioned odor in honeybees. *Naturwissenschaften*. 2001;88 (11):472–476.

66. Sachse S, Galizia CG. Role of inhibition for temporal and spatial odor representation in olfactory output neurons: a calcium imaging study. *J Neurophysiol*. 2002;87(2):1106–1117.
67. Sachse S, Galizia CG. The coding of odour-intensity in the honeybee antennal lobe: local computation optimizes odour representation. *Eur J Neurosci*. 2003;18(8):2119–2132.
68. Szyszka P, Ditzen M, Galkin A, Galizia CG, Menzel R. Sparsening and temporal sharpening of olfactory representations in the honeybee mushroom bodies. *J Neurophysiol*. 2005;94(5):3303–3313.
69. Szyszka P, Galkin A, Menzel R. Associative and non-associative plasticity in Kenyon cells of the honeybee mushroom body. *Front Syst Neurosci*. 2008;2:3.
70. Riemensperger T, Pech U, Dipt S, Fiala A. Optical calcium imaging in the nervous system of *Drosophila melanogaster*. *Biochim Biophys Acta*. 2012;1820(8):1169–1178.
71. Yu D, Akalal DB, Davis RL. *Drosophila* alpha/beta mushroom body neurons form a branch-specific, long-term cellular memory trace after spaced olfactory conditioning. *Neuron*. 2006;52(5):845–855.
72. Akalal DB, Yu D, Davis RL. A late-phase, long-term memory trace forms in the γ neurons of *Drosophila* mushroom bodies after olfactory classical conditioning. *J Neurosci*. 2010;30(49):16699–16708.
73. Akalal DB, Yu D, Davis RL. The long-term memory trace formed in the *Drosophila* α/β mushroom body neurons is abolished in long-term memory mutants. *J Neurosci*. 2011;31(15):5643–5647.
74. Séjourné J, Plaçais PY, Aso Y, et al. Mushroom body efferent neurons responsible for aversive olfactory memory retrieval in *Drosophila*. *Nat Neurosci*. 2011;14(7):903–910.
75. Riemensperger T, Völler T, Stock P, Buchner E, Fiala A. Punishment prediction by dopaminergic neurons in *Drosophila*. *Curr Biol*. 2005;15(21):1953–1960.
76. Miyawaki A, Llopis J, Heim R, et al. Fluorescent indicators for Ca^{2+} based on green fluorescent proteins and calmodulin. *Nature*. 1997;388(6645):882–887.
77. Hammer M. An identified neuron mediates the unconditioned stimulus in associative olfactory learning in honeybees. *Nature*. 1993;366(6450):59–63.
78. Hammer M, Menzel R. Multiple sites of associative odor learning as revealed by local brain microinjections of octopamine in honeybees. *Learn Mem*. 1998;5(1-2):146–156.
79. Schroll C, Riemensperger T, Bucher D, et al. Light-induced activation of distinct modulatory neurons triggers appetitive or aversive learning in *Drosophila* larvae. *Curr Biol*. 2006;16(17):1741–1747.
80. Nagel G, Szellas T, Huhn W, et al. Channelrhodopsin-2, a directly light-gated cation-selective membrane channel. *Proc Natl Acad Sci USA*. 2003;100(24):13940–13945.
81. Monastirioti M. Biogenic amine systems in the fruit fly *Drosophila melanogaster*. *Microsc Res Tech*. 1999;45(2):106–121.
82. Boyden ES, Zhang F, Bamberg E, Nagel G, Deisseroth K. Millisecond-timescale, genetically targeted optical control of neural activity. *Nat Neurosci*. 2005;8(9):1263–1268.
83. Scherer S, Stocker RF, Gerber B. Olfactory learning in individually assayed *Drosophila* larvae. *Learn Mem*. 2003;10(3):217–225.
84. Gerber B, Stocker RF. The *Drosophila* larva as a model for studying chemosensation and chemosensory learning: a review. *Chem Senses*. 2007;32(1):65–89.
85. Khurana S, Abu Baker MB, Siddiqi O. Odour avoidance learning in the larva of *Drosophila melanogaster*. *J Biosci*. 2009;34(4):621–631.
86. Honjo K, Furukubo-Tokunaga K. Distinctive neuronal networks and biochemical pathways for appetitive and aversive memory in *Drosophila* larvae. *J Neurosci*. 2009;29(3):852–862.
87. Pauls D, Selcho M, Gendre N, Stocker RF, Thum AS. *Drosophila* larvae establish appetitive olfactory memories via mushroom body neurons of embryonic origin. *J Neurosci*. 2010;30(32):10655–10666.
88. Selcho M, Pauls D, Han KA, Stocker RF, Thum AS. The role of dopamine in *Drosophila* larval classical olfactory conditioning. *PLoS ONE*. 2009;4(6):e5897.
89. Lima SQ, Miesenböck G. Remote control of behavior through genetically targeted photostimulation of neurons. *Cell*. 2005;121(1):141–152.
90. Claridge-Chang A, Roorda RD, Vrontou E, et al. Writing memories with light-addressable reinforcement circuitry. *Cell*. 2009;139(2):405–415.
91. Aso Y, Siwanowicz I, Bräcker L, Ito K, Kitamoto T, Tanimoto H. Specific dopaminergic neurons for the formation of labile aversive memory. *Curr Biol*. 2010;20(16):1445–1451.
92. Hamada FN, Rosenzweig M, Kang K, et al. An internal thermal sensor controlling temperature preference in *Drosophila*. *Nature*. 2008;454(7201):217–220.

CHAPTER

7

Computational Analyses of Learning Networks

*Douglas A. Baxter**, *Enrico Cataldo*† *and John H. Byrne**

*The University of Texas Medical School at Houston, Houston, Texas †Università di Pisa, Pisa, Italy

INTRODUCTION

As summarized in the present volume, the mechanisms that underlie learning and memory in invertebrates are the subject of intense study. Behavioral studies encompass many learning protocols, ranging from simple forms of nonassociative and associative learning such as habituation, sensitization, and classical and operant conditioning to higher forms of learning, such as observational learning. Moreover, mechanistic analyses span levels of biological organization ranging from gene regulatory networks to signal transduction cascades, single cells, and neural networks, as well as spanning temporal domains ranging from milliseconds to days. The vast amounts of data, their diversity, and their inherent complexity make it difficult to arrive at an intuitive and comprehensive understanding of relationships among molecular, cellular, and network properties and cognitive processes such as learning and memory.

Mathematical models and computer simulations provide quantitative and modifiable frameworks for representing, integrating, and manipulating complex data sets. The development and analyses of computational models of learning and memory mechanisms are helping to elucidate biological principles that are common in diverse species, common to more than one form of learning, and common to learning-induced changes in different types of behaviors. This chapter provides a brief overview of several well-characterized computational models of the biological processes that underlie learning and memory in invertebrates. It focuses on analyses of nonassociative and associative learning.

OLFACTORY LEARNING IN INSECTS

Several olfactory-mediated behaviors in insects can be modified by classical conditioning, which consists of learning the contingency between an odor (the conditioned stimulus (CS)) and an aversive or appetitive stimulus (the unconditioned stimulus (US)). Computational studies provide insights into processes within the olfactory neural network (Figure 7.1), within specific neuronal types (Figure 7.1B), and within molecular pathways (Figure 7.1C) that contribute to classical conditioning.

Neurons located in the mushroom bodies (MBs) are essential to olfactory learning and long-term memory (for reviews, see [1–3,19–26]). Kenyon cells (KCs) are the intrinsic neurons of the MBs, and Hodgkin–Huxley-type models (Figure 7.1B) have been used to study the biophysical properties of KCs and their responses to synaptic inputs.[7–9] Simulations indicate that an A-type K^+ conductance (G_A) and a slow transient K^+ conductance ($G_{K,ST}$) play important roles in determining the spiking characteristics of the model. These two conductances are the primary determinants of the delayed spiking responses in KCs during constant-current stimuli. Moreover, $G_{K,ST}$ reduces the ability of oscillatory stimuli to drive the model. These results suggest that the spiking characteristic of KCs *in vivo* could be profoundly altered by the modulation of G_A and $G_{K,ST}$. Moreover, the voltage-dependent subthreshold properties of KCs underlie supralinear summation of coincident EPSPs, and the the intrinsic properties of KC and circuit properties combine to generate a brief integration window, thereby enhancing coincidence detection in KCs.[7–9,17,21,27,28] Thus, KCs are capable of detecting convergence of an odorant (i.e., a CS) and a reinforcement stimulus (i.e., a US).

Invertebrate Learning and Memory.
DOI: http://dx.doi.org/10.1016/B978-0-12-415823-8.00007-1

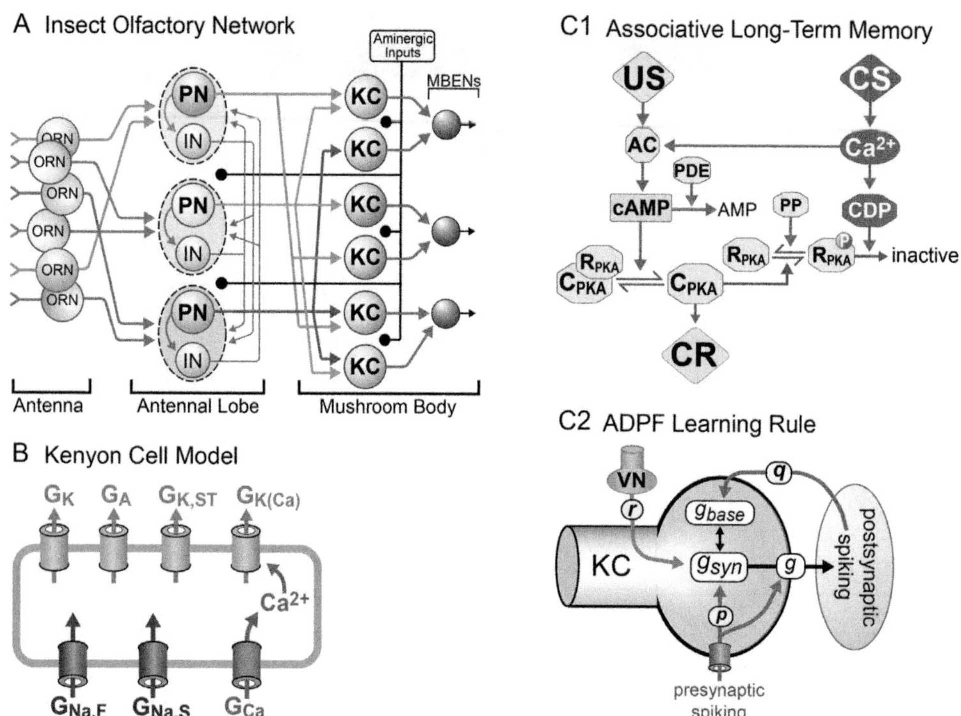

FIGURE 7.1 Molecular, cellular, and network models of insect olfactory learning. (A) Simplified schematic of the neural pathways in the insect olfactory system (for reviews, see[1–7]). In the antenna, each olfactory receptor neuron (ORN) expresses one receptor type. Three types of ORNs are depicted, as indicated by the colors blue, green, and brown. All ORNs with the same receptor type converge onto a single glomerulus (dashed circles) in the antennal lobe (AL) (arrows indicate conventional synaptic projections, and solid circles indicate diffuse modulatory projections). Projection neurons (PNs) and local interneurons (INs) are the two principal types of neurons in the AL. INs receive inputs from ORNs and PNs. Both excitatory and inhibitory INs form extensive interglomerular connections. PNs project to higher brain areas, primarily the mushroom body (MB). In the MB, PNs form diverging connections with Kenyon cells (KCs). Synaptic outputs from KCs converge on mushroom body extrinsic neurons (MBENs), which project to other brain areas. (B) An example of a Hodgkin–Huxley-type model of a KC.[8,9] The KC model includes descriptions of (1) a Na^+ conductance, which is modeled as having fast and slow components ($G_{Na,F}$ and $G_{Na,S}$); (2) a Ca^{2+} conductance (G_{Ca}); (3) a delayed K^+ conductance (G_K); (4) an A-type K^+ conductance (G_A); (5) a Ca^{2+}-dependent K^+ conductance ($G_{K(Ca)}$); and (6) a slow transient K^+ conductance ($G_{K,ST}$) (arrows indicate current flow). (C) Models of molecular processes that underlie associative learning and long-term memory (arrows indicate enzymatic activation and/or synthesis).[10–18] (C1) The unconditioned stimulus (US), which is mediated by aminergic inputs, activates an adenylyl cyclase (AC), which then increases cAMP levels, albeit weakly. The conditioned stimulus (CS), which is mediated by Ca^{2+} influx, primes the AC and increases cAMP levels beyond those levels produced by the US alone. Convergence of Ca^{2+} influx (i.e., CS) and US-mediated activation of AC represents associative aspects of classical conditioning, and the molecular dynamics dictate the temporal relationship between the US and CS that is necessary for associative changes to occur. The CS must precede the US, and the two stimuli must be in close temporal proximity. cAMP binding to the regulatory subunit of PKA (R_{PKA}) stimulates the dissociation and activation of the catalytic subunit (C_{PKA}). Degradation of cAMP is mediated by the activity of a phosphodiesterase (PDE). C_{PKA} covalently modifies target proteins, which modulate neuronal function by altering spike activity and/or synaptic transmission. This modulation leads to the conditioned response (CR) in response to a subsequent CS. Long-term memory is mediated by an autophosphorylation step that induces persistent activation of C_{PKA}. Following dissociation of C_{PKA}/R_{PKA}, R_{PKA} is phosphorylated by C_{PKA}. This autophosphorylation enhances the degradation of R_{PKA} via a Ca^{2+}-dependent proteinase (CDP). A protein phosphatase (PP) mediates the dephosphorylation of R_{PKA}. (C2) Model of activity-dependent presynaptic facilitation (the ADPF learning rule), which simulates a form of value-driven learning. Presynaptic spiking generates an activity trace (p), which is analogous to intracellular Ca^{2+} and has a half-life of 50 msec. The temporal convergence of p and a modulator signal r, which represents activity in a value neuron (VN), increases a variable that represents synaptic strength, (g_{syn}). g_{syn} is analogous to a readily releasable pool of transmitter vesicles. Each presynaptic spike also drives a postsynaptic conductance g, which can elicit postsynaptic activity. Postsynaptic spikes initiate a retrograde signal q, which is a second activity trace with a half-life of 200 msec. q increases the value of g_{base}, which is analogous to a reserve pool of transmitter vesicles. g_{base} decays very slowly and represents long-term memory.

The role(s) of projection neurons (PNs) and interneurons (INs) intrinsic properties in overall network dynamics has also been examined in Hodgkin–Huxley-type models and neural networks. Belmabrouk *et al.*[29] found that I_A and I_{SK} (a small-conductance, Ca^{2+}-dependent K^+ current) in PNs may combine with inhibitory IN inputs (Figure 7.1A) to transform odorant-induced neural activity into a dual latency code for odorant identity and intensity. The mean latency encodes stimulus intensity, whereas the latency-invariant activity encodes odorant identity. Bazhenov *et al*[30]. (see also[27,31]) examined the role of inhibitory INs in the oscillatory output of the antennal lobe (AL). Simulations indicate that an inhibitory subnetwork within the AL helps to synchronize PN activity. Only a population of PNs firing in synchrony provides sufficient input to activate KCs. Thus, network architecture and the intrinsic properties of PNs and INs combine to encode sensory information into an output that is tuned to KC properties.

Analyses of single gene mutations that disrupt olfactory learning and memory in insects indicate that the cAMP/protein kinase A (PKA) cascade is critical for memory formation (for reviews, see[2,32,33]). A type of adenylyl cyclase (AC) appears to be a molecular site of convergence between US and CS, as was initially found in studies of associative plasticity of sensory neurons of *Aplysia*.[34–36] Computational models of the cAMP/PKA cascade (Figure 7.1C1) manifest features analogous to associative learning, such as acquisition and contiguity detection, if elevated cAMP levels and increased PKA activation are assumed to strengthen synaptic strength or neuronal activity.[10,12] By including autophosphorylation and activity-dependent proteolysis, which lead to persistent PKA activation, this model can also simulate long-term memory.[11,14]

Empirical studies are motivating extensions to the cAMP/PKA model. For example, some sites within the olfactory network express a Hebbian form of synaptic plasticity in which coincident pre- and postsynaptic activity is necessary for associative changes in neuronal and/or synaptic plasticity.[13,37,38] Several formulations of Hebbian-like plasticity have been used in olfactory models. Smith *et al.*[17] propose a model of Hebbian-like synaptic plasticity, which they term activity-dependent presynaptic facilitation (the ADPF learning rule; Figure 7.1C2). In ADPF, long-term associative increases in synaptic strength (synaptic strength is determined by the magnitude of g_{syn}) are dependent on the temporal overlap of three variables: (1) p, which represents presynaptic activity; (2) r, which represents activity of a modulatory neuron; and (3) q, which represents postsynaptic activity. In the absence of r and q, synaptic strength, g_{syn}, will not increase. In the presence of r, g_{syn} increases but only for a brief time. Only the convergence of all three variables consolidates learning into long-term memory because changes in g_{base} (g_{base} is analogous to a reserve pool of transmitter vesicles) decay more slowly than changes to g_{syn}. Simulations indicate that networks with the ADPF learning rule acquire and recall CS–US associations and that a sparse representation of the CS enhances learning. The sparse representation results from inhibitory elements in the circuit and divergent connectivity between the AL and MBs. As noted previously, simulations of Hodgkin–Huxley models indicate that KCs function as coincidence detectors. Thus, a sparse representation in combination with the ADPF learning rule helps prevent spurious associative changes in the MBs by incoherent inputs.

Yarali *et al.*[18] proposed an alternative model to the Hebbian-like STDP learning rule. This alternative model simulates bidirectional changes in synaptic strength. Thus, the model simulates two opposing kinds of learning (punishment vs. relief). In the alternative model, bidirectional regulation of AC emerges from the dynamics of Ca^{2+}-dependent and G protein-dependent regulation of AC. If the CS (Ca^{2+} influx) precedes the US (release of a modulatory transmitter such as dopamine), the subsequent US-mediated activation of AC is enhanced (Figure 7.1C1). Alternatively, if the US (binding of a transmitter to a G protein-coupled receptor) precedes the CS, the US stimulates dissociation of the G protein complex and G protein-dependent activation of AC, which is prone to dissociation into an inactive AC. Thus, during punishment learning, in which a CS precedes an aversive US, a Ca^{2+}-sensitive AC detects the convergence of the CS and the US. The resultant increase in cAMP/PKA activation strengthens KC output to avoidance circuitry. During relief learning, in which a CS follows an aversive US, AC is inactive because of the prior US and KC output fails to strengthen.

Other network models have investigated the learning capabilities of networks that incorporate Hebbian-like learning (i.e., spike time-dependent plasticity, STDP). The formulations of STDP, however, are phenomenological rather the biologically realistic.[15,28,30,39,40] These modeling studies compare the consequences of plasticity at different locations in the olfactory system. For example, Finelli *et al.*[28] show that STDP at PN-to-KC connections filters and condenses the AL oscillatory output to the MBs. This sparse signal is more amenable to associative plasticity in KCs. Linster and Smith[15] have shown that a STDP learning rule at the PN-to-modulatory neuron connection (not shown in Figure 7.1A) simulates additional features of associative learning, such as generalization and overshadowing. By including a second locus of plasticity at the modulatory-to-inhibitory IN, the model simulates blocking and unblocking. Similarly, Wessnitzer *et al.*[40] showed that STDP at the PN-to-KC connection simulates discriminations involving compound stimuli. These studies illustrate that the role of a learning rule depends, in part, on network architecture and the location of plasticity within the network.

NONASSOCIATIVE AND ASSOCIATIVE LEARNING IN GASTROPODS

Olfactory Learning *Limax*

Olfactory learning in the terrestrial mollusk *Limax maximus* exhibits first-order classical conditioning as well as a variety of higher-order conditioning phenomena, such as US pre-exposure effects, second-order conditioning, and blocking (for review, see[41]). To examine neuronal and network processes that may underlie this repertoire of learning phenomena, Goel and Gelperin[42] (see also[43,44]] constructed a neural network model (Figure 7.2A). Synaptic plasticity in the network is governed by a learning rule that embodied the concept of activity-dependent heterosynaptic facilitation (see Figure 7.3A and[61,62]). Specifically, activity in the facilitatory neuron (FN) strengthens all co-active synapses, and synaptic plasticity occurs only during FN activity. For example, co-activating the US with the CS1 (i.e., first-order classical conditioning) strengthens the CS1-to-motor neuron (MN) and CS1-to-FN connections. Strengthening these connections allows CS1 to acquire properties similar to US, such as eliciting the CR. Moreover, preconditioning CS1 subsequently allows this stimulus to reinforce CS2 activity during second-order conditioning. This simple network simulates (1) first-order conditioning, in which CS1 elicits the CR after training; (2) second-order conditioning, in which a previously conditioned CS1 reinforces a naive CS2 in the absence of a US; and (3) blocking, in which a preconditioned CS1 subsequently blocks conditioning of CS2 during pairing of a compound CS (CS1/CS2) with the US. This model of *Limax* learning illustrates that a single learning rule (e.g., heterosynaptic facilitation) can simulate multiple forms of learning when incorporated into an appropriate neural network (see also Figure 7.3).

Classical Conditioning of Feeding Behavior in *Lymnaea*

Feeding behavior of the pond snail *Lymnaea stagnalis* can be modified by classical conditioning (for review, see[63]). This behavior is controlled by a central pattern generator (CPG), and the neurons and synaptic connections in the CPG are well-characterized (for review, see[64]). Moreover, neuronal correlates of learning are known. Thus, *Lymnaea* is an excellent candidate for system-level analysis of learning. As a first step toward simulating learning in *Lymnaea*, Vavoulis *et al.*[46] developed a four-cell model of the feeding CPG (Figure 7.2B2). The neural network included cells N1, N2, and N3, which mediate the rhythmic neural activity underlying feeding movements, and cell SO, which is a modulatory neuron. Individual neurons in the neural network were represented by two-compartment models (Figure 7.2B1). The axonal compartment includes a fast, transient Na^+ current (I_{NA}) and a delayed-like K^+ current (I_K), which mediate spike activity. The somatic compartment includes currents (I_{ACh}, I_{NaL}, or I_T), which mediate slowly developing, long-lasting changes of the membrane potential, such as plateau potentials in N1 and N2 and postinhibitory rebound in N3. The model simulates the rhythmic neural activity that mediates feeding behavior.

As a second step in modeling learning in *Lymnaea*, Vavoulis *et al.*[45] modeled the cerebral giant cells (CGCs). The CGCs are modulatory interneurons and are a locus of plasticity following appetitive classical conditioning.[65] The CGC is modeled as a single compartment, which includes (1) transient and persistent Na^+ currents (I_{NaT} and I_{NaP}, respectively), (2) an A-type and a delayed-type K^+ current (I_A and I_D, respectively), and (3) low- and high-voltage-activated Ca^{2+} currents (I_{LVA} and I_{HVA}, respectively). Two of the currents, I_{NaP} and I_D, are increased following conditioning.[45,65] Thus, the effects of conditioning are simulated by increasing the maximal conductances of I_{NaP} and I_D. Simulations reproduce some of the previously identified neuronal changes following conditioning, including a depolarization in CGC without a change in tonic firing in CGC. To maintain the spike waveform in CGC, however, it is necessary to hypothesize an increase in I_{HVA}. The effects of conditioning on I_{HVA} have yet to be examined empirically. Thus, the possible role of I_{HVA} represents an important prediction of the model. An important next step will be to combine the CPG and CGC models and examine the extent to which the currently identified cellular correlates of conditioning can reproduce learning-induced changes in behavior.

Classical Conditioning of Phototaxic Behavior in *Hermissenda*

Naive *Hermissenda crassicornis* (a marine mollusk) move toward light (phototaxis) and contract their foot in response to turbulence. However, pairing light (the CS) with rotation (the US) in a classical conditioning protocol delays phototaxis and light begins to elicit foot contraction (the CR). Photoreceptors are activated by light (the CS), and they receive inputs from hair cells in the statocyst (the US). Thus, photoreceptors are the first site of convergence between the CS and the US, and an associative memory is stored as changes in the intrinsic and synaptic properties of the photoreceptors (for reviews, see[66,67]). To examine the ways in which learning-induced changes in photoreceptors alter phototaxis, Werness *et al.*[54] developed a four-cell model that includes a photoreceptor (B-type photoreceptor), two hair cells, and an interneuron (Figure 7.2C1). The network also includes two learning rules: one that alters the membrane conductance of the photoreceptor and another that alters the hair cell-to-photoreceptor connection during paired presentations of the CS and the US. This combination of learning rules and network

A Neural Network for the logic of *Limax* Learning

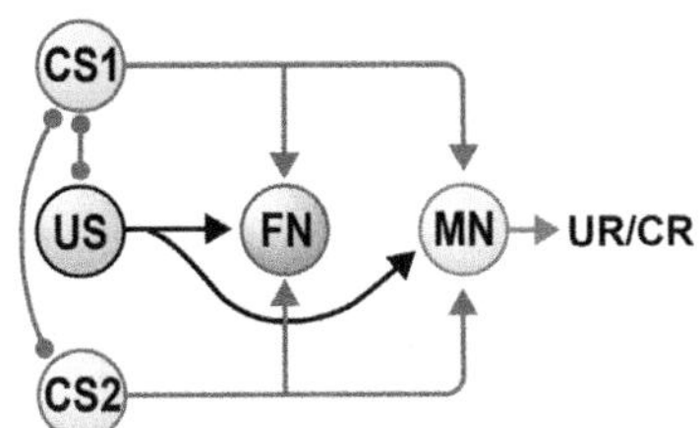

C1 Associative Learning in Network Model of *Hermissenda* Visual-Vestibular System

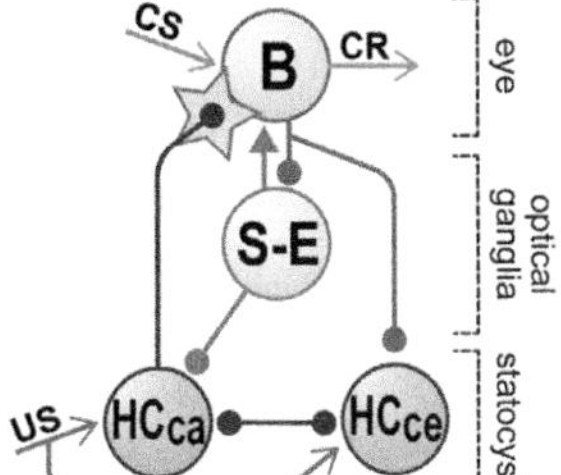

B1 Two-Compartment Model of Neurons in the *Lymnaea* Feeding Network

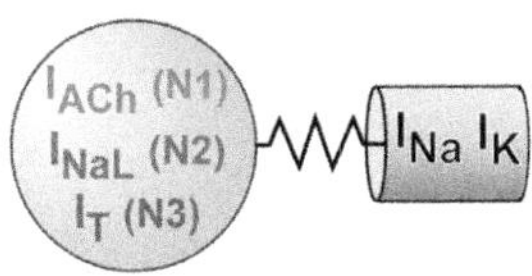

C2 Multi-Compartment Model of *Hermissenda* Type-B Photoreceptor

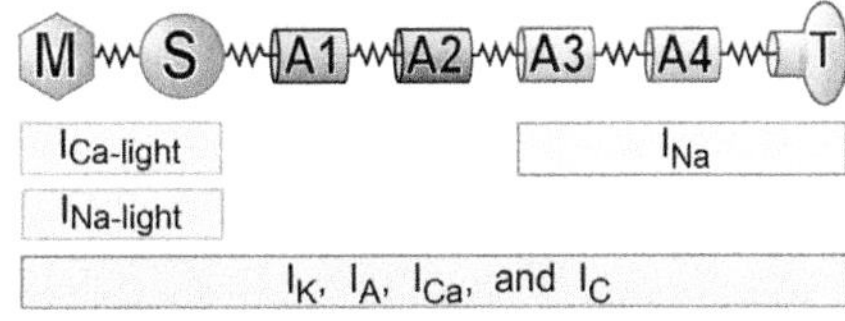

B2 *Lymnaea* Feeding Model Network

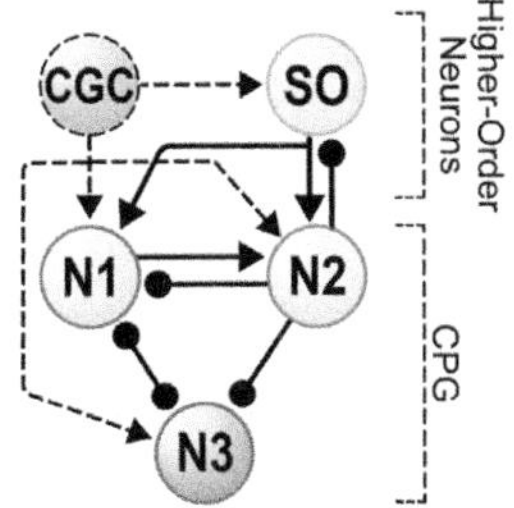

C3 Network Architecture Influences Associative Learning in Model of *Hermissenda* Eye

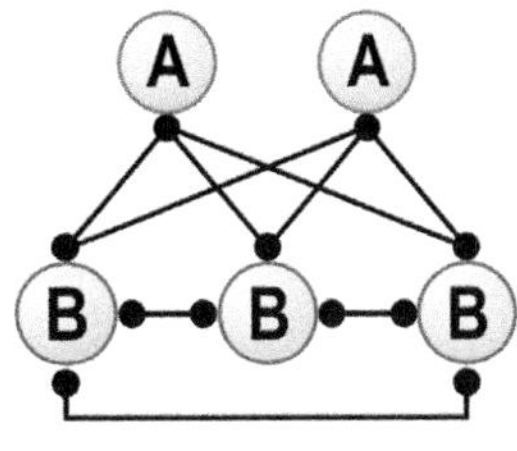

FIGURE 7.2 **Models of associative learning in gastropods.** (A) Neural network architecture for modeling the logic of *Limax* learning.[42,44] Neuronal activity that represents conditioned stimuli (CS1 for first-order classical conditioning and CS1 and CS2 for second-order classical conditioning) excites (arrows) a facilitatory neuron (FN), which also is driven by the neural representation of the unconditioned stimulus (US), and motor neurons (MNs). In addition, the FN feeds back onto its presynaptic inputs (not shown), and this feedback mediates plasticity (see below). MN activity is the output of the system—that is, the unconditioned response (UR) and/or the conditioned response (CR). The CS neurons form reciprocal inhibitory connections (small solid circles), as do the CS1 and US neurons. The learning rule for this network embodies features of activity-dependent heterosynaptic facilitation (see Figure 7.3A). The model assumes that overlapping activity in the FN and its presynaptic elements strengthens the excitatory synapses of the presynaptic elements (both CS-to-FN and CS-to-MN excitatory connections). Moreover, the model assumes that plasticity only occurs during FN activity. (B) Neuronal and network models for simulating classical conditioning of feeding in *Lymnaea*.[45,46] (B1) General features of two-compartment neuronal model (a somatic compartment and an axonal compartment). Spiking properties are restricted to the axonal compartment, whereas slowly-developing, long-lasting processes such as postinhibitory rebound potentials (cell type N3) and plateau potentials (cell types N1 and N2) are restricted to the somatic compartment. (B2) Fictive feeding is simulated using a four-cell network, which contained the modulatory, slow oscillatory neurons (SO) and central pattern generator (CPG) interneurons N1, N2, and N3 (arrows, excitatory connections; small solid circles, inhibitory connections). Each of the four cells is represented by a two-compartment model (panel B1). In a separate study, learning-induced changes in the cerebral giant cell (CGC) were modeled using a single-compartment model. The CGC model includes descriptions of fast, transient Na^+ current (I_{Na}), a delayed K^+ current (I_K), a persistent Na^+ current (I_{NaP}), an A-type K^+ current (I_A), a low-voltage activated Ca^{2+} current (I_{LVA}), and a high-voltage activated Ca^{2+} current (I_{HVA}). Although the CGC model is not included in the network model, the functional connections of the CGC are indicated by dashed lines. (C) Associative learning in neuronal and network models of *Hermissenda*.[47–54] (C1) A four-cell neural network that simulates phototaxis and turbulence-induced inhibition of phototaxis (solid arrows indicate excitatory connections, solid circles represent inhibitory connections, and open arrows represent inputs and outputs to the network). The four cells are (1) a type-B photoreceptor (B), (2) a S-E interneuron (S-E), (3) a caudal hair cell (HC_{ca}), and (4) a cephalic hair cell (HC_{ce}). Light (CS) activates the photoreceptor, whereas rotation (US) activates the hair cells. The membrane potential of the photoreceptor is taken as the output (CR) of the network. Two phenomenological learning rules are included. The first rule strengthens the HC_{ca}-to-B synapse (indicated by star). In this learning rule, the HC_{ca} input to B is inhibitory when B is at rest. When B is depolarized, the synaptic has an early inhibition followed by excitation. Following three pairings of B depolarization and HC_{ca} activation, the HC_{ca}-to-B contact becomes exclusively excitatory. The second learning rule reduces the membrane conductance of B during pairing of the CS and US. The learning-induced decrease in membrane conductance decays very slowly. (C2) More detailed models of photoreceptors incorporating multiple compartments. In this example, the photoreceptor is represented by seven compartments: (1) a microvilli compartment (M), (2) a somatic compartment (S), (3) four axonal compartments (A1–A4), and (4) a synaptic terminal compartment (T). Ionic currents are distributed nonuniformly (as indicated in the boxes below the compartments). Light-activated currents ($I_{Ca\text{-}light}$ and $I_{Na\text{-}light}$) are located in the M and S compartments. A fast, transient Na^+ current (I_{Na}), which mediates spiking, is limited to compartments A3, A4, and T. All compartments include delayed- and A-type K^+ currents, a non-inactivating Ca^{2+} current (I_{Ca}), and a slow, Ca^{2+}-activated K^+ current (I_C). (C3) The multicompartment model of photoreceptors (panel C2) is combined into several networks. As illustrated here, the *Hermissenda* eye includes two A-type and three B-type photoreceptors. All cells form reciprocal inhibitory connections. Fost and Clark[53] tested the relationship between network architecture and features of learning. Several architectures were examined (not shown). In these networks, the number of cells and synaptic connections is systematically altered to examine the ways in which network architecture affects memory.

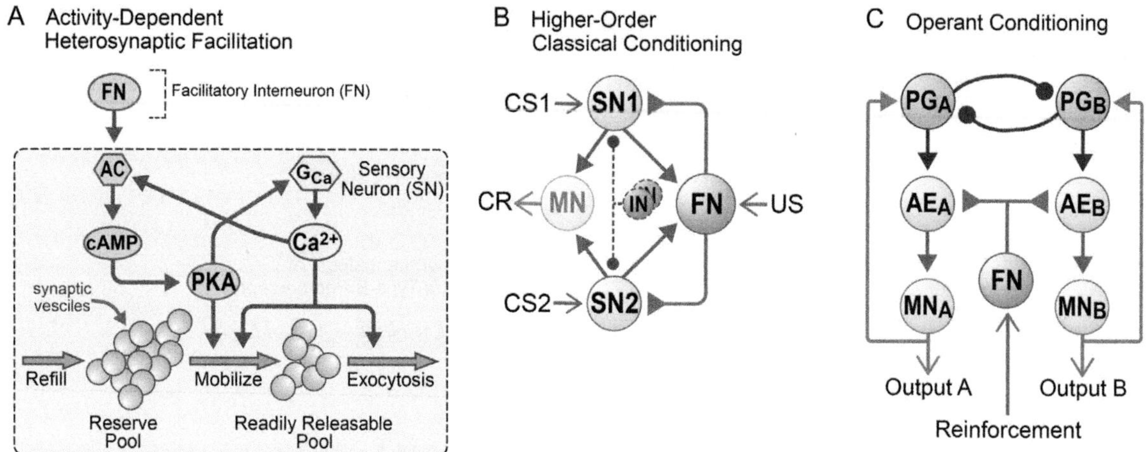

FIGURE 7.3 Activity-dependent heterosynaptic facilitation reproduces features of nonassociative and associative learning.[55–60] (A) Simplified schematic of activity-dependent heterosynaptic facilitation (ADHF). Empirical studies of plasticity in sensory neurons (SNs) of *Aplysia* led to the development of a single-cell model of learning. Exocytosis depletes a readily-releasable pool (RRP) of vesicles. This depletion leads to homosynaptic depression, which contributes to habituation. The RRP is replenished via mobilization of vesicles from a reserve pool (RP). The mobilization process has several components, such as Ca^{2+}-dependent mobilization and PKA-dependent mobilization. Strong stimuli, such as a sensitizing stimulus or a US, cause the release of a modulatory transmitter (e.g., 5-HT) from facilitatory interneurons (FNs). Activity in FN activates an adenylyl cyclase (AC) in the SN. Activation of the cAMP/PKA pathway modulates several membrane conductances in the SN (e.g., a Ca^{2+} conductance, G_{Ca}) as well as mobilization. These targets of the cAMP/PKA pathway work in concert to enhance synaptic strength (i.e., heterosynaptic facilitation), which contributes to sensitization. If prior to FN activity there is activity in the SN, then elevated Ca^{2+} levels in the SN prime the AC, which in turn leads to greater cAMP/PKA activation and greater increases in synaptic strength (i.e., activity-dependent heterosynaptic facilitation). Activity-dependent heterosynaptic facilitation can reproduce several features of first-order classical conditioning. Arrows indicate activation and/or enhancement. (B) Activity-dependent heterosynaptic facilitation also simulates higher-order features of classical conditioning. Buonomano *et al.*[55] simulated two network architectures—a three-cell network and a five-cell network. Both networks contain two sensory neurons (SN1 and SN2), which represent two CS pathways (CS1 and CS2). An additional input to the network (the US) is represented by activity in FN. The five-cell network also included two inhibitory interneurons (INs), which mediate lateral inhibition between the SNs (dashed line). Output from the network (the CR) is represented by activity in a motor neuron (MN), which was not explicitly modeled. Solid arrows represent excitation, solid circles represent inhibition, solid triangles represent modulatory inputs, and open arrows represent inputs (CSs and US) and outputs (CR). Each SN includes a model of activity-dependent heterosynaptic facilitation (the ADHF learning rule; see panel A), and all SN synapses are modified by ADHF. (C) Features of operant conditioning are simulated by incorporating activity-dependent heterosynaptic facilitation into an appropriate neural network architecture. To generated a behavior (i.e., the operand), two spontaneously active, mutually inhibitory pattern-generating neurons (PG_A and PG_B) are simulated. The PGs activate two adaptive elements (AE_A and AE_B). Each AE includes a model of activity-dependent heterosynaptic facilitation. The AEs, in turn, drive activity in two motor neurons (MN_A and MN_B), which represent the network outputs. In addition, the MNs feedback and excite the PG elements. Finally, reinforcement is introduced via FN that modulates the synaptic strengths of the AEs. Solid arrows represent excitation, solid circles represent inhibition, solid triangles represent modulatory inputs, and open arrows represent the input/output of the network.

architecture simulates several features of classical conditioning, including contingency sensitivity, long-term memory, and extinction. Simulation studies, however, did not evaluate the specific contribution(s) that individual network elements or specific features of the learning rules made to the simulated learning. Moreover, the learning rules were phenomenological and did not have any biologically plausible underpinnings. Thus, the simulated network does not address issues related to which molecular, neuronal, and/or network properties are necessary and sufficient to account for conditioning of phototaxis.

More detailed and realistic models of the photoreceptors have been developed.[47–52] These models represent the photoreceptors as multiple compartments (Figure 7.2C2) and include empirically derived and realistic descriptions of the membrane current and/or second-messenger cascades in photoreceptors. Simulations of the multicompartment models indicate that (1) modulation of multiple membrane currents is necessary to increase the excitability of photoreceptors, and (2) synaptic enhancement requires processes beyond simple broadening of the presynaptic spike. Thus, these more realistic models predict that some spike duration-independent mechanisms are necessary for synaptic strengthening to occur. These single-cell models, however, do not address issues related to the specific role that synaptic versus intrinsic neuronal plasticity play in behavioral conditioning (for discussion, see[68–70]).

Finally, the multicompartment models were combined into several alternative network architectures that differed in the number of cells and pattern of synaptic connectivity. Simulations investigated the relationship between network architecture and features of learning.[53] The number of cells (two-cell networks vs. five-cell networks) and pattern of synaptic connections (sparse connectivity patterns vs. fully interconnected networks) were varied. These networks did not include learning rules. Rather, neuronal correlates of learning (i.e., changes in membrane currents and/or synaptic strength following conditioning) were selectively implemented. Simulations indicate that an emergent property of the fully interconnected network (Figure 7.2C3) is an increase in the rate of learning and memory capacity of the network.

Empirical studies indicate that associative memory is distributed among cells downstream of the photoreceptors. Thus, future computational models of associative learning in *Hermissenda* should use expanded networks and multiple sites of plasticity. More comprehensive network models will help to more fully understand the ways in which learning emerges from the interplay of molecular, synaptic, neuronal, and network processes.

Nonassociative and Associative Learning in Defensive Withdrawal Reflexes of *Aplysia*

Possibly the most influential model system for studying the neuronal and synaptic plasticity that underlie learning and memory is the sensorimotor connection that mediates defensive withdrawal reflexes in *Aplysia* (for reviews, see[56,71–75]). Processes that mediate short- and long-term learning (both nonassociative and associative) have been characterized, and computational models have been developed that simulate some of these processes.

Nonassociative Synaptic Plasticity

Gingrich and Byrne[57] developed a model of synaptic plasticity in sensory neurons (Figure 7.3A) that simulates homosynaptic depression and heterosynaptic facilitation, which mediate habituation and sensitization, respectively. In the model, repetitive stimuli deplete the readily releasable pool (RRP) of transmitter vesicles in the sensory neuron (SN). This depletion leads to less transmitter release (i.e., homosynaptic depression) and less activation of the postsynaptic MN (i.e., a diminished response similar to habituation). The RRP is replenished by processes that are termed mobilization. Mobilization represents the movement of vesicles from a reserve pool (RP) to the RRP. Mobilization is regulated by several factors, including (1) a difference between vesicle concentrations in the RP and the RRP, (2) the levels of intracellular Ca^{2+}, and (3) the activation of PKA. Sensitizing stimuli activate a FN, which releases 5-HT. 5-HT, in turn, activates an adenylyl cyclase (AC) and increases cAMP levels, which activate PKA. Thus, the sensitizing stimulus leads to more mobilization, enhanced transmitter release and greater activation of the MN (i.e., sensitization). Because there is no requirement for coincident activity in the SN and FN, this form of synaptic plasticity is nonassociative and is termed heterosynaptic facilitation.

Associative Synaptic Plasticity

Based in part on empirical studies by Walters and Byrne[62] and Ocorr *et al.*,[35] Gingrich and Byrne[58] expanded their model by including Ca^{2+}-dependent regulation of AC activity. The influx of Ca^{2+} during SN activity primes the AC and thereby amplifies cAMP production elicited by the FN. Thus, coincident activity in the SN and the FN produces greater PKA activity than is elicited by the FN alone. This form of associative plasticity is termed activity-dependent heterosynaptic facilitation (ADHF). The ADHF learning rule simulates several features of classical conditioning, including a characteristic interstimulus interval (ISI) relationship and pairing-specific increases in the CS-elicited response.

The models of heterosynaptic facilitation[57] and activity-dependent heterosynaptic facilitation[58] have been adapted to simulate learning in other model systems. For example, the models of Aszodi *et al.*[10,11] (Figure 7.1C1), Smith *et al.*[17] (Figure 7.1C2), and Goel and Gelperin[42] (Figure 7.2A) are derived, in part, from the model of heterosynaptic facilitation and activity-dependent heterosynaptic facilitation (Figure 7.3A). These similarities illustrate the commonality among learning rules.

Incorporating the ADHF Learning Rule into Different Network Architectures

Differences in training protocols are often used to distinguish different forms of learning. For example, pairing a CS with a US is a defining feature of classical conditioning. In contrast, contingent reinforcement of a spontaneously generated behavior (termed an operand) is a defining feature of operant conditioning. Although different forms of associative learning can be distinguished on the basis of training protocols, the extent to which these different forms of learning share a common mechanism is not well understood. To address this issue, several modeling studies incorporated the ADHF learning rule into small neural networks and examined the ability of these networks to simulate higher-order features of classical conditioning (Figure 7.3B) and operant conditioning (Figure 7.3C).[55,59,60,76]

Buonomano *et al.*[55] incorporated ADHF into two small networks—a three-cell network and a five-cell network (Figure 7.3B). Both networks included two sensory neurons (SN1 and SN2), which represent two CSs (CS1 and CS2), and a facilitatory neuron (FN), which represents the US. The outputs of the network are the amounts of transmitter released by SN1 and SN2, which in turn excite a MN. The MN is not explicitly modeled, but the predicted excitation at the sensorimotor connections represents the CR. The five-cell version of the network also included inhibitory interneurons that mediated lateral inhibition between the two SNs. (Here, results from simulating the two networks are not discussed separately.) The original ADHF learning rule did not readily simulate second-order conditioning and blocking. However, by incorporating a few minor adjustments (e.g., increasing Ca^{2+}-dependent production of cAMP, increasing cAMP-dependent mobilization, and incorporating burst-like properties in FN), the neural network simulates second-order conditioning and blocking. An important feature of this network necessary for simulating these higher-order features of classical conditioning is the connections that the SNs make with the FN. Consequently, as the strength of a SN connection increases with repeated trials, the SN becomes capable of activating the FN independent of the US and thereby 'takes control' of the FN.

Similarly, Raymond *et al.*[60] incorporated the ADHF learning rule into a small network (Figure 7.3C) that included (1) a half-center oscillator (PG_A and PG_B), which functions as a CPG; (2) two adaptive elements (AE_A and AE_B), which incorporate the ADHF rule; and (3) two motor neurons (MN_A and MN_B), which mediate the operant. By incorporating a few minor changes in the AEs (e.g., enabling prolonged spiking of the AEs, reducing homosynaptic depression by increasing RRP, reducing the rate of transmitter release, and removing the dynamics of the RP), this network simulates features of operant conditioning, including (1) the effectiveness of contingent versus random reinforcement; (2) extinction; (3) sensitivity to reversed contingencies; (4) the magnitude of reinforcement; and (5) delay of reinforcement and partial reinforcement protocols. These simulations illustrate that a network with a learning rule derived from a form of associative synaptic plasticity involved in classical conditioning can simulate features of operant conditioning.[58] Thus, there need not be fundamentally different cellular mechanisms for these two forms of associative learning.

Predicting Enhanced Training Protocols for the Induction of Long-Term Facilitation and Long-Term Sensitization

The model of heterosynaptic facilitation (Figure 7.3A) simulates short-term facilitation (STF), the duration of which persists for several minutes. The consolidation of learning into long-term memory (LTM) that persists for days or longer, however, requires additional processes. Insights into these additional processes come from studies of long-term facilitation (LTF) at sensorimotor connections.

The transformation of STF into LTF (i.e., consolidation) requires activation of two biochemical cascades: PKA and extracellular signal-regulated kinase (ERK). The requirement for both PKA and ERK suggests that the strength of LTF depends on synergism between these kinases. Following a single 5-HT or sensitization stimulus, however, PKA activates rapidly but transiently, with activity returning to near basal levels within 15 min.[77] In contrast, ERK activates more slowly, with maximal activation occurring approximately 45 min after a single trial.[78] These results indicate that a single stimulus will produce little overlap between PKA and ERK and therefore little or no LTF. However, the results suggest that multiple, properly timed stimuli can increase the overlap and hence the synergism of the two cascades. This increased interaction between the two kinase cascades leads to LTF.

To examine this hypothesis, Zhang et al.[79] developed a simplified mathematical model of the PKA and ERK cascades (Figure 7.4A). In the model, interactions between PKA and ERK cascades are represented by a phenomenological variable denoted *inducer*. The peak levels of *inducer* predict the efficacy of stimulus protocols. As a point of reference, the simulated peak level of *inducer* produced by five 5-min pulses of 5-HT with uniform 20-min ISIs was selected. This 'standard' protocol has been widely used in experimental studies since its introduction by Montarolo *et al.*[81] To identify a protocol that produces the highest peak level of *inducer*, thousands of alternative protocols were simulated. Simulation of these protocols reveal considerable variability in the peak level of *inducer*. The protocol that maximizes the peak level of *inducer* (the 'enhanced' protocol) consisted of nonuniformly spaced 5-HT applications with ISIs of 10, 10, 5, and 30 min. The peak level is approximately 50% higher than the in standard protocol. Subsequent empirical studies confirm that LTF following the enhanced protocol is greater and lasts longer. Moreover, the enhanced protocol increases levels of phosphorylated CREB1 (a transcription factor important in the consolidation of LTM). Finally, behavioral training confirms that LTM is also increased by the enhanced protocol. These results illustrate the feasibility of using computational models to design training protocols that improve memory.

Examining the Role(s) of a Positive Feedback Loop in the Consolidation of Long-Term Memory

The model of Zhang *et al.*[79] illustrates the ways in which interactions between two kinase cascades can

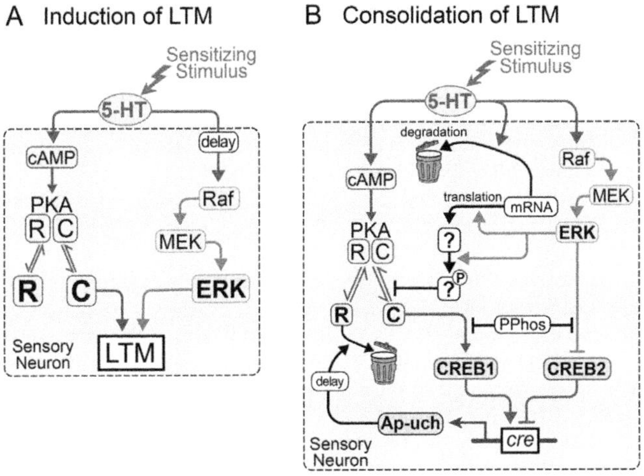

FIGURE 7.4 **Molecular models of the induction and consolidation of long-term memory (LTM) in sensory neurons of *Aplysia*.**[79,80] (A) Schematic of a model for the induction of nonassociative LTM. A sensitizing stimulus causes the release of 5-HT. The 5-HT stimulus leads to increased cAMP levels and activation of PKA (i.e., the dissociation of the regulatory (R) and catalytic (C) subunits). PKA is activated rapidly but transiently. In addition, the 5-HT stimulus activates the Raf/MEK/ERK cascade. The activation of ERK, however, is slow, with maximal activation occurring approximately 45 min after the sensitizing stimulus. An explicit delay is included to represent the slow development of the ERK signal. Because concurrent activation of these two kinase cascades is necessary for the induction of LTM, the model equates the interaction between these kinases cascades to the induction of LTM. (B) Schematic of a model for consolidation of nonassociative LTM. Exposure to 5-HT elevates cAMP, which binds to the PKA holoenzyme, causing it to dissociate and release free catalytic (C) and regulatory (R) subunits. 5-HT also activates ERK (via Raf and MEK) and stimulates translation and phosphorylation of an unidentified protein (question mark), which slows the reassociation of C and R subunits, leading to intermediate-term PKA activation. ERK also phosphorylates the transcription factor CREB2, which relieves repression of *Aplysia* ubiquitin hydrolase (Ap-uch) transcription. In contrast, PKA-mediated phosphorylation of another transcription factor, CREB1, enhances expression of Ap-uch. Phosphorylation of CREB2 by ERK combined with phosphorylation of CREB1 by PKA induces expression of Ap-uch. Ap-uch enhances degradation of R after a delay, leading to long-term PKA activation, which equates to consolidation of LTM. A protein phosphatase (PPhos) counteracts PKA- and ERK-mediated phosphorylation. Solid arrows represent activation, whereas bars represent inhibition.

influence the strength and duration of LTM. A more detailed model by Pettigrew *et al.*[80] examines the dynamical properties of regulatory motifs correlated with different temporal domains of memory (Figure 7.4B). The model incorporates descriptions of the PKA and ERK cascades, interaction between the two kinase systems, downstream targets of the kinases, and positive feedback within the network.

The model simulates several features of LTF, such as short-, intermediate-, and long-term phases of PKA activation, and a threshold for the induction of long- versus short-term plasticity. The model also simulates phosphorylation of CREB1 and consequent induction of the immediate-early gene *Aplysia* ubiquitin hydrolase (*Ap-uch*), which is essential for long-term synaptic facilitation. Simulations investigated mechanisms responsible for different profiles of synaptic facilitation following massed versus spaced exposures to 5-HT, and they suggest that a novel regulatory motif (gated positive feedback) is important for LTF. Moreover, simulations suggested that zero-order ultrasensitivity may underlie a requirement of a threshold number of exposures to 5-HT for LTF induction. Finally, the model predicts a yet to be identified protein that mediates some aspects of the PKA/ERK interactions.

CONCLUSIONS

The computational analyses of learning networks, which are reviewed in this chapter, highlight several common themes. First, processes that underlie the convergence of stimuli during associative learning often involve the dual regulation of enzymatic activity within a single cell (Figures 7.1C1 and 7.3A).[11,12,57,58,66] Second, features of associative learning can be reproduced by simple extensions to processes underlying nonassociative learning. For example, sensitization of withdrawal reflexes in *Aplysia* is mediated, at least in part, by 5-HT-induced AC activity, whereas classical conditioning is mediated, at least in part, by combining 5-HT- and Ca^{2+}-mediated AC activity (Figures 7.1C1 and 7.3A).[34–36,58] Third, different forms of learning do not require unique molecular mechanisms. Rather, learning rules that simulate one form of associative learning (e.g., classical conditioning) can simulate more complex features and/or different forms of learning (e.g., high-order classical conditioning and operant conditioning) when incorporated in an appropriate neural network (Figures 7.2A, 7.3B, and 7.3C).[15,40,42,55,59,60,82] These similarities suggest conservation of processes that mediate learning in different species.

Most computational analyses of learning networks in invertebrates have focused on presynaptic learning rules (Figures 7.1C1 and 7.3A). Empirical studies, however, indicate that both pre- and postsynaptic processes (i.e., Hebbian-like plasticity) contribute to plasticity in some invertebrate systems.[13,37,38,73] Although no complete mechanistic account of Hebbian synaptic plasticity in invertebrates exists, computational analyses of learning networks that include Hebbian-like plasticity (Figure 7.1C2) simulate many features of classical conditioning.[17,28,30,40] Studies have yet to address the issue of what learning and memory processes may emerge from networks that combine presynaptic plasticity and Hebbian-like plasticity.

The majority of computational analyses overlook another important mechanism of learning: changes in the intrinsic excitability of cells.[68–70] Fost and Clark[53] considered the relative contributions of increased excitability and increased synaptic strength in models of the *Hermissenda* eye. They concluded that these two mechanisms work in concert and are not mutually exclusive. However, this issue should be addressed in more complex neural networks (e.g., gastropod feeding systems)[45,46,69] or extended network models of phototaxis in *Hermissenda*.

Finally, the success of Zhang *et al.*[79] in predicting novel training protocols that significantly increase the strength and duration of LTM illustrates the feasibility of using computational methods to assist in the design of training procedures that remediate learning deficits. Developing biologically realistic models with practical predictive power represents a potential new strategy for the rescue of deficits in motor learning and cognition in humans.[83]

Acknowledgment

This work was supported by National Institutes of Health grants R01-MH058321, R01-NS073974, and R01-NS019895.

References

1. Busto GU, Cervantes-Sandoval I, Davis RL. Olfactory learning in *Drosophila. Physiology*. 2010;25:338–346.
2. Davis RL. Traces of *Drosophila* memory. *Neuron*. 2011;14:8–19.
3. Heisenberg M. Mushroom body memoir: from maps to models. *Nature*. 2003;4:266–275.
4. Jortner RA, Farivar SS, Laurent G. A simple connectivity scheme for sparse coding in an olfactory system. *J Neurosci*. 2007;27: 1659–1669.
5. Martin JP, Beyerlein A, Dacks AM, et al. The neurobiology of insect olfaction: sensory processing in a comparative context. *Prog Neurobiol*. 2011;95:427–447.
6. Masse NY, Turner GC, Jefferis GSXE. Olfactory information processing in *Drosophila. Curr Biol*. 2009;19:R700–R713.
7. Perez-Orive J, Bazhenov M, Laurent G. Intrinsic and circuit properties favor coincidence detections for decoding oscillatory input. *J Neurosci*. 2004;24:6037–6047.
8. Pelz C, Jander J, Rosenboom H, Hammer M, Menzel R. IA in Kenyon cells of the mushroom body of honeybees resembles shaker currents: kinetic, modulation by K^+, and simulation. *J Neurophysiol*. 1999;81:1749–1759.
9. Wustenberg DG, Boytcheva M, Grunewald B, Byrne JH, Menzel R, Baxter DA. Current-and voltage-clamp recordings and computer simulations of Kenyon cells in the honeybee. *J Neurophysiol*. 2004;92:2589–2603.
10. Aszodi A, Friedrich P. Molecular kinetic modeling of associative learning. *Neuroscience*. 1987;22:37–48.
11. Aszodi A, Muller U, Friedrich P, Spatz H-C. Signal convergence on protein kinase A as a molecular correlated of learning. *Proc Natl Acad Sci USA*. 1991;88:5832–5836.
12. Buxbaum JD, Dudai Y. A quantitative model for the kinetics of cAMP-dependent protein kinase (type II) activity: long-term activation of the kinase and its possible relevance to learning and memory. *J Biol Chem*. 1989;264:9344–9351.
13. Cassenaer S, Laurent G. Hebbian STDP in mushroom bodies facilitates the synchronous flow of olfactory information in locusts. *Nature*. 2007;448:709–713.
14. Friedrich P. Protein structure: the primary substrate for memory. *Neuroscience*. 1990;35:1–7.
15. Linster C, Smith BH. A computational model of the response of honey bee antennal lobe circuitry to odor mixtures: overshadowing, blocking and unblocking can arise from lateral inhibition. *Behav Brain Res*. 1997;87:1–14.
16. Muller U, Spatz HC. Ca^{2+}-dependent proteolytic modification of the cAMP-dependent protein kinase in *Drosophila* wild-type and dunce memory mutant. *J Neurogenet*. 1989;6:95–114.
17. Smith D, Wessnitzer J, Webb B. A model of associative learning in the mushroom body. *Biol Cybern*. 2008;99:89–103.
18. Yarali A, Nehrkorn J, Tanimotor H, Herz AVM. Event timing in associative learning: from biochemical reaction dynamics to behavioral observation. *PLoS One*. 2012;7:e32885.
19. Berry J, Krause WC, Davis RL. Olfactory memory traces in *Drosophila. Prog Brain Res*. 2008;169:293–304.
20. Davis RL. Physiology and biochemistry of *Drosophila* learning mutants. *Physiol Rev*. 1996;76:299–317.
21. Demmer H, Kloppenburg P. Intrinsic membrane properties and inhibitory synaptic input of Kenyon cells as mechanisms for sparse coding? *J Neurophysiol*. 2009;102:1538–1550.
22. Gerber B, Tanimoto H, Heisenberg M. An engram found? Evaluating the evidence from fruit flies. *Curr Opin Neurobiol*. 2004;280:4017–4020.
23. Giurfa M. Behavioral and neural analysis of associative learning in the honeybee: a taste from magic well. *J Comp Physiol A Neuroethol Sens Neural Behav Physiol*. 2007;193:801–824.
24. Haehnel M, Menzel R. Long-term memory and response generalization in mushroom body extrinsic neurons in the honeybee *Apis melifera*. *J Exp Biol*. 2012;215:559–565.
25. Menzel R. Search for the memory trace in a mini-brain, the honeybee. *Learn Mem*. 2001;8:53–62.
26. Thum AS, Jenett A, Ito K, Heisenberg M, Tanimoto H. Multiple memory traces for olfactory reward learning in *Drosophila. J Neurosci*. 2007;27:11132–11138.
27. Assisi C, Bazhenov M. Synaptic inhibition controls transient oscillatory synchronization in a model of the insect olfactory system. *Front Neuroeng*. 2012;5:7.
28. Finelli LA, Haney S, Bazhenov M, Stopfer M, Sejnowski TJ. Synaptic learning rules and sparse coding in a model sensory system. *PLoS Comput Biol*. 2008;4:e1000062.
29. Belmabrouk H, Nowotny T, Rospars J-P, Martinez D. Interaction of cellular and network mechanisms for efficient pheromone coding in moths. *Proc Natl Acad Sci USA*. 2011;108:19790–19795.
30. Bazhenov M, Stopfer M, Sejnowski TJ, Laruent G. Fast odor learning improves reliability of odor responses in the locust antennal lobe. *Neuron*. 2005;:483–492.
31. Assisi C, Stopfer M, Bazhenov M. Using the structure of inhibitory networks to unravel mechanisms of spatiotemporal patterning. *Neuron*. 2011;69:373–386.
32. Gervasi N, Tchenio P, Preat T. PKA dynamics in a *Drosophila* learning center: coincidence detection by rutabaga adenylyl cyclase and spatial regulation by dunce phosphodiesterase. *Neuron*. 2010;65:516–529.

33. McGuire SE, Deshazer M, Davis RL. Thirty years of olfactory learning and memory research in *Drosophila melanogaster*. *Prog Neurobiol*. 2005;76:328–347.
34. Abrams TW, Yovell Y, Onyike CU, Cohen JE, Jarrard HE. Analysis of sequence-dependent interactions between calcium and transmitter stimuli in activating adenylyl cyclase in *Aplysia*: possible contributions to CS–US sequence requirement during conditioning. *Learn Mem*. 1998;4:496–509.
35. Ocorr KA, Walters ET, Byrne JH. Associative conditioning analog selectively increases cAMP levels of tail sensory neurons in *Aplysia*. *Proc Natl Acad Sci USA*. 1985;82:2548–2552.
36. Yovell Y, Abrams TW. Temporal asymmetry in activation of *Aplysia* adenylyl cyclase may explain properties of conditioning. *Proc Natl Acad Sci USA*. 1992;89:6526–6530.
37. Cassenaer S, Laurent G. Conditional modulation of spike-timing-dependent plasticity for olfactory learning. *Nature*. 2012;482:47–52.
38. Rath L, Giovanni-Galizia C, Szyszka P. Multiple memory traces after associative learning in the honey bee antennal lobe. *Eur J Neurosci*. 2011;34:352–360.
39. Martinez D, Montejo N. A model of stimulus-specific neural assemblies in the insect antennal lobe. *PLoS Comput Biol*. 2008;4: e10000139.
40. Wessnitzer J, Young JM, Armstrong JD, Webb B. A model of non-elemental olfactory learning in *Drosophila*. *J Comput Neurosci*. 2012;32:197–212.
41. Watanabe S, Kirino Y, Gelperin A. Neural and molecular mechanisms of microcognition in *Limax*. *Learn Mem*. 2008;15: 633–642.
42. Goel P, Gelperin A. A neuronal network for the logic of *Limax* learning. *J Comput Neurosci*. 2006;21:259–270.
43. Ermentrout B, Wang JW, Flores J, Gelperin A. Model for olfactory discrimination and learning in *Limax* procerebrum incorporating oscillatory dynamics and wave propagation. *J Neurophysiol*. 2001;85:1444–1452.
44. Sekiguch T, Furudate H, Kimura T. Internal representation and memory formation of odor preference based oscillatory activities in terrestrial slug. *Learn Mem*. 2010;17:372–380.
45. Vavoulis DV, Nikitin E, Kemenes I, et al. Balanced plasticity and stability of the electrical properties of a molluscan modulatory interneuron after classical conditioning: a computational study. *Front Behav Neurosci*. 2010;4:19.
46. Vavoulis DV, Straub VA, Kemenes I, Kemenes G, Feng J, Benjamin PR. Dynamic control a central pattern generator circuit: a computation model of the snail feeding network. *Eur J Neurosci*. 2007;25:2805–2818.
47. Blackwell KT. Paired turbulence and light do not produced a supralinear calcium increased in *Hermissenda*. *J Comput Neurosci*. 2004;17:81–99.
48. Blackwell KT. Ionic currents underlying difference in light response between type A and type B photoreceptors. *J Neurophysiol*. 2006;95:3060–3072.
49. Cai Y, Baxter DA, Crow T. Computational study of enhanced excitability in *Hermissenda*: membrane conductances modulated by 5-HT. *J Comput Neurosci*. 2003;15:105–121.
50. Cai Y, Flynn M, Baxter DA, Crow T. Role of A-type K^+ channels in spike broadening observed in soma and axon of *Hermissenda* type-B photoreceptors: a simulation study. *J Comput Neurosci*. 2006;21:89–99.
51. Flynn M, Cai Y, Baxter DA, Crow T. A computational study of the role of spike broadening in synaptic facilitation of *Hermissenda*. *J Comput Neurosci*. 2003;15:29–41.
52. Fost JW, Clark GA. Modeling *Hermissendia*: I. Differential contributions of I_A and I_C to type-B cell plasticity. *J Comput Neurosci*. 1996;3:137–153.
53. Fost JW, Clark GA. Modeling *Hermissendia*: II. Effects of variations in type-B cell excitability, synaptic strength, and network architecture. *J Comput Neurosci*. 1996;3:155–172.
54. Werness SA, Fay SD, Blackwell KT, Vogl TP, Alkon DL. Associative learning in a network model of *Hermissenda crassicornis*: I. Theory. *Biol Cybern*. 1992;68:125–133.
55. Buonomano DV, Baxter DA, Byrne JH. Small Networks of empirically derived adaptive elements simulate some higher-order features of classical conditioning. *Neural Netw*. 1990;3: 507–523.
56. Byrne JH, Kandel ER. Presynaptic facilitation revisited: state and time dependence. *J Neurosci*. 1996;16:425–435.
57. Gingrich KJ, Byrne JH. Simulation of synaptic depression, post-tetanic potentiation, and presynaptic facilitation of synaptic potentials from sensory neurons mediating gill-withdrawal reflex in *Aplysia*. *J Neurophysiol*. 1985;53:652–669.
58. Gingrich KJ, Byrne JH. Single-cell neuronal model for associative learning. *J Neurophysiol*. 1987;57:1705–1715.
59. Hawkins RD. A simple circuit model for high-order features of classical conditioning. In: Byrne JH, Berry WO, eds. *Neural Models of Plasticity*. San Diego: Academic Press; 1989:74–93.
60. Raymond JL, Baxter DA, Buonomano DV, Byrne JH. A learning rule based on empirically-derived activity-dependent neuromodulation supports operant conditioning in a small network. *Neural Netw*. 1992;5:789–803.
61. Hawkins RD, Abrams TW, Carew TJ, Kandel ER. A cellular mechanism of classical conditioning in *Aplysia*: activity-dependent amplification of presynaptic facilitation. *Science*. 1983;219:400–405.
62. Walters ET, Byrne JH. Associative conditioning of single sensory neurons suggests a cellular mechanism for learning. *Science*. 1983;219:405–408.
63. Benjamin PR, Staras K, Kemenes G. A systems approach to the cellular analysis of associative learning the pond snail *Lymnaea*. *Learn Mem*. 2000;7:124–131.
64. Benjamin PR. Distributed network organization underlying feeding behavior in the mollusk *Lymnaea*. *Neural Syst Circuits*. 2012;2:4.
65. Nikitin ES, Vavoulis DV, Kemenes I, et al. Persistent sodium current is a nonsynaptic substrate for long-term associative memory. *Curr Biol*. 2008;18:1221–1226.
66. Blackwell KT. Subcellular, cellular, and circuit mechanisms underlying classical conditioning in *Hermissenda crassicornis*. *Anat Rec B New Anat*. 2006;289:25–37.
67. Crow T. Pavlovian conditioning of *Hermissenda*: current cellular, molecular, and circuit perspectives. *Learn Mem*. 2004;11:229–238.
68. Benjamin PR, Kemenes G, Kemenes I. Non-synaptic neuronal mechanisms of learning and memory in gastropod molluscs. *Front Biosci*. 2008;13:4051–4057.
69. Mozzachiodi R, Byrne JH. More than synaptic plasticity: role of nonsynaptic plasticity in learning and memory. *Trends Neurosci*. 2010;33:17–26.
70. Zhang W, Linden DJ. The other side of the engram: experience-driven changes in neuronal intrinsic excitability. *Nat Rev Neurosci*. 2003;4:885–900.
71. Antzoulatos EG, Byrne JH. Learning insights transmitted by glutamate. *Trends Neurosci*. 2004;27:555–560.
72. Sharma SK, Carew TJ. The roles of MAPK cascades in synaptic plasticity and memory in *Aplysia*: facilitatory effects and inhibitory constraints. *Learn Mem*. 2004;11:373–378.
73. Glanzman DL. New tricks for an old slug: the critical role of postsynaptic mechanisms in learning and memory in *Aplysia*. *Prog Brain Res*. 2008;169:277–292.
74. Kandel ER. The molecular biology of memory storage: a dialogue between genes and synapses. *Science*. 2001;294:1030–1038.

75. Lee YS, Bailey CH, Kandel ER, Kaang BK. Transcriptional regulation of long-term memory in the marine snail. *Aplysia Mol Brain*. 2008;1:3.
76. Byrne JH, Baxter DA, Buonomano DV, Raymond JL. Neuronal and network determinants of simple and higher-order features of associative learning: experimental and modeling approaches. *Cold Spring Harb Symp Quant Biol*. 1990;55:175–186.
77. Müller U, Carew TJ. Serotonin induces temporally and mechanistically distinct phases of persistent PKA activity in *Aplysia* sensory neurons. *Neuron*. 1998;21:1423–1434.
78. Philips GT, Tzvetkova EI, Carew TJ. Transient mitogen-activated protein kinase activation is confined to a narrow temporal window required for the induction of two-trial long-term memory in *Aplysia*. *J Neurosci*. 2007;27:13701–13705.
79. Zhang Y, Liu R-Y, Heberton GA, et al. Computational design of enhanced learning protocols. *Nat Neurosci*. 2012;15:294–297.
80. Pettigrew D, Smolen P, Baxter DA, Byrne JH. Dynamic properties of regulatory motifs associated with induction of three temporal domains of memory in *Aplysia*. *J Comput Neurosci*. 2005;18: 163–181.
81. Montarolo PG, Goelet P, Castellucci VF, Morgan J, Kandel ER, Schacher S. A critical period for macromolecular synthesis in long-term heterosynaptic facilitation in *Aplysia*. *Science*. 1986;234: 1249–1254.
82. Hawkins RD, Kandel ER. Is there a cell-biological alphabet for simple forms of learning? *Psychol Rev*. 1984;3:375–391.
83. Liu R-Y, Zhang Y, Baxter, DB, Smolen P, Cleary LJ, Byrne JH. Deficit in long-term synaptic plasticity is rescued by a computationally-predicted stimulus protocol. *J Neurosci*. in press.

CHAPTER

8

Issues in Invertebrate Learning Raised by Robot Models

Barbara Webb
The University of Edinburgh, Edinburgh, United Kingdom

INTRODUCTION

One way to understand how a natural system works is to try to replicate it by building a mechanism that works the same way. In the domain of behavioral neurobiology, such a mechanism may take the form of a robot—a machine that, like an organism, can sense and interact autonomously with the real world, controlled by a computational brain. Although much work in this area considers biological systems principally as inspiration for designing more effective robots, some research more explicitly considers the robot to be a scientific model.[1] That is, the robot embodies specific hypotheses about the function in the natural system and allows those hypotheses to be refined and tested by their ability to reproduce the observed phenomena in biology.

Advances in technology have made it increasingly plausible to build robots that approach real animals in complexity, although it is clear they still fall far short in capability. Invertebrate systems have been popular targets because they are often considered to be relatively simple, with smaller brains and more stereotyped behaviors.[2] A wide range of basic sensorimotor capabilities of invertebrates have been explored in robotic implementations to date, including visual stabilization,[3] odor tracking,[4] auditory localization,[5] and six-legged walking.[6] However, mimicking how invertebrates flexibly combine many such capabilities and adapt their behavior to interact flexibly with changing environmental contingencies through learning remains a major challenge.

One of the main lessons learned from robotic modeling is that many 'well-understood' biological systems are not yet understood sufficiently well to replicate. Very few neural circuits have been fully described from sensory inputs to motor outputs. For example, in building robot models of cricket phonotaxis, it became clear that although some key sensory interneurons had been very well characterized, relatively little data were available on brain circuits and descending neurons. The precise dynamics of phonotactic behavior—the final output of this neural processing—have only recently begun to be explored in detail.[7–9] Building a robot also forces consideration of the closed loop—how action affects sensory input. In phonotaxis, this includes very significant effects on the directionality of sound that arise from the walking pattern because the cricket's ears are located on its forelegs.

This book represents an impressive demonstration of how much is currently understood about invertebrate learning. Nevertheless, our attempts to build embodied models of invertebrate learning, which I describe in this chapter, raise a number of open questions. These particularly concern the overall behavioral framework within which specific circuit-level or neurophysiological learning mechanisms are acting to change behavior.

ROBOT MODELS OF INVERTEBRATE LEARNING

The earliest examples of biologically inspired robotics[10] were used to embody learning mechanisms. Ross[11] described an electronic mechanism attached to a robotic device able to learn (by physically rewiring itself in response to particular patterns of input) to traverse a maze—later developed into a three-wheeled 'robot rat'—based on Hull's 'law of effect.' Walter[12] described in detail an electrical 'conditioned Reflex analog' (CORA) that can be combined with his

Invertebrate Learning and Memory.
DOI: http://dx.doi.org/10.1016/B978-0-12-415823-8.00008-3

'tortoise' robot and used to associate a whistle with either innately attractive light stimuli or innately aversive touch stimuli. Interestingly, he reported an unexpected result from this embodiment: The response to touch in his robot involves an internal feedback loop, which means that once some association of sound to touch is established, it tends to reinforce itself in the absence of further touch; the response to light does not have this feedback character. As a consequence, 'attractive reflexes' take longer to learn and are more easily forgotten than 'defensive reflexes,' a difference that is also observed in some biological systems. A later extension to the CORA circuit that enables stimulus generalization and discrimination, as well as extinction and recovery, to be demonstrated was described by Angyan[13] and was implemented on a robot by Zemanek *et al.*[14]

Formal and computational approaches to learning theory have been hugely influential in more recent robotics research.[15,16] Currently, the dominant paradigm is 'reinforcement learning.' This characterizes the problem facing an animal or robot as how, in a given state of the world, to choose an action so as to maximize long-term reward (the collective set of action choices over all states is called the 'policy'). Learning processes allow the system to acquire or improve its knowledge of the value of states and actions and/or to predict the state transitions and rewards resulting from actions, thus enabling an 'optimal' choice of action in each state. A number of algorithms have been devised to solve this problem, such as temporal difference learning[17] (ultimately derived from the Rescorla–Wagner rule[18]), and have been utilized in a wide variety of robotic applications. Moreover, the discovery of several parallels between dopamine signaling in vertebrate brains and the prediction error used in these algorithms has tied such approaches more closely to neurophysiology.[19,20]

Within this framework, it is often assumed that "learning to predict the values of states is classical conditioning; using this information to improve policies or choose actions is instrumental conditioning".[21] However, this is an incomplete account. For example, it cannot fully explain the particular conditioned responses observed in classical conditioning, in which reward is presented independently of action, nor the fact that there are conditioning paradigms in which a particular action will increase even though it decreases the reward obtained.[22]

A wide range of robot models have also been developed to reflect more specific aspects of the neural circuitry involved in learning in vertebrates, such as hypothesized mechanisms in hippocampus[23] or cerebellum,[24] but it is beyond the scope of this chapter to attempt to review such work. By contrast, and perhaps surprisingly, there are only a handful of studies in which robot models of learning have been explicitly based on data from invertebrate systems.

Scutt and Damper[25] and Damper *et al.*[26] introduced a neural circuit for associative learning that uses the mechanism proposed by Hawkins and Kandel[27] to account for conditioning in *Aplysia*. The robot is hardwired to produce a turn away from a bump on either side. Distance sensors are initially connected by weak synapses to both directions of turn, but there are synapse-on-synapse facilitating connections from the bump sensors to the relevant distance–turn synapse. Thus, the connection from the distance sensors to the appropriate turn will strengthen if the distance sensor fires shortly before the bump sensor. As a consequence, the robot begins to turn away from the side where the distance sensor is active before the bump occurs.

A model tested in a simulated agent,[28] and later (albeit unsuccessfully) on a robot,[29] was inspired by the role of the VUMmx1 neuron in bee learning.[30] In effect, this model reduces to use of the Rescorla–Wagner rule to increase the 'value' of the stimulus (weights for active sensors) proportionally to the difference of received reward from predicted value, coupled to a controller that (while reward is not received, and learning is not taking place) will move up the gradient of value of the current sensory input. This model could account for some aspects of experience-dependent choice in foraging bees. In particular, the model produced a similar tradeoff in the expected value of reward and its variability.[31]

Modeling of olfactory learning in the antennal lobe and mushroom body (MB) circuit in flies and bees has been a recent focus of interest, but most of these models do not have a behavioral embodiment (either simulated or as a robot) but, rather, evaluate the ability of the circuit to learn to classify complex input patterns (simulating vision or olfaction). Examples include the model by Linster and Smith,[32] which focuses on the antennal lobe circuitry and assumes Hebbian strengthening of synapses from projection neurons to VUMmx1 due to simultaneous activation; that by Muezzinoglu *et al.*,[33] which uses input data from artificial olfactory sensors and a dynamic model of the antennal lobe to process the data, which is then classified using a standard machine learning method (a support vector machine); and that by Nowotny *et al.*,[34,35] which models antennal lobe and MB processing, with learning occurring at the output of the Kenyon cell (KC) neurons, either with a Hebbian spike timing-dependent plasticity (STDP) rule[34] or through a global reinforcement signal,[35,36] which was tested on classification of hand-written digits (a benchmark test for computer classification algorithms).

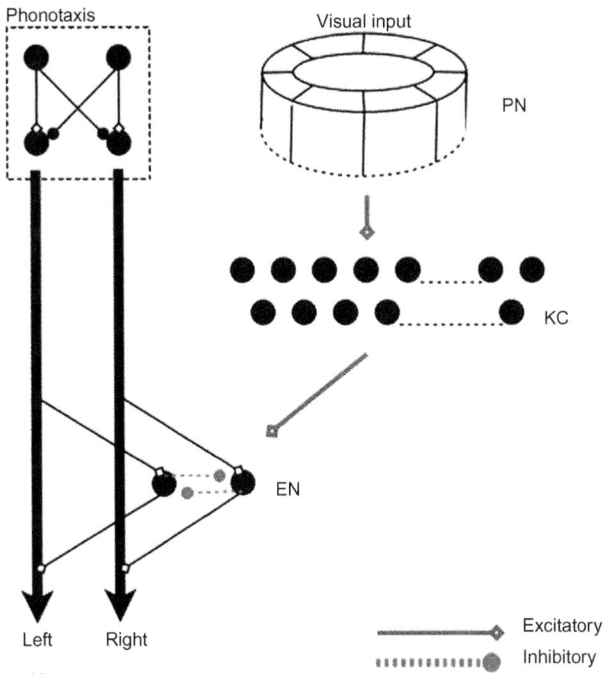

FIGURE 8.1 Model of auditory–visual association in bush crickets. The phonotaxis circuit produces left or right output according to the sound direction and also activates mushroom body extrinsic neurons (EN). Visual input is mapped to a Kenyon cell (KC) layer, and simultaneous activation of KC and EN strengthens the connection between them, such that the system associates the current visual input with the heading direction to sound. After association, the bush cricket can use vision to maintain an appropriate course during gaps in sound.[39] Source: *From Wessnitzer and Webb.*[38]

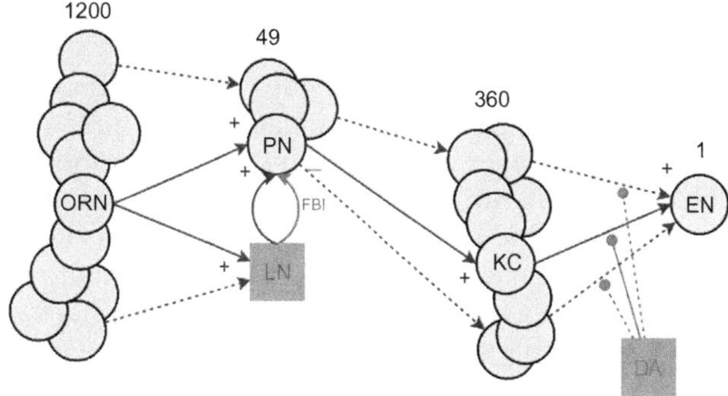

FIGURE 8.2 Model of olfactory learning circuit in *Drosophila*. Olfactory receptor neurons (ORN) activate projection neurons (PN) in the antennal lobe, with local neurons (LN) providing normalization. The PN have a divergent connection onto Kenyon cells (KC), which converge on a single extrinsic neuron (EN). Learning occurs at the synapses of KC onto EN, such that the response to patterns of KC firing evoked by different odors are strengthened by the concurrent presentation of a reinforcing stimulus, such as electric shock, signaled by dopaminergic neurons (DA). It is assumed that activation of the EN leads to avoidance behavior. Source: *From Wessnitzer et al.*[41]

We have used simulated agent models in testing simplified models of this circuitry in the context of associating a visual pattern with a punishment[37] and associating a sound direction with a visual cue[38] (Figure 8.1). In both cases, our assumption was that the conditioned behavior was generated by the conditioned stimulus (CS) coming to activate the same output pathway as the unconditioned response (UR); in other words, a CS–UR association was formed due to UR activation by the unconditioned stimulus (US) being paired with CS activation. In two models more closely linked to neurophysiology,[40,41] we simply assumed that output from the MB, which changes after learning, directly controls a conditioned response; in fact, the models are evaluated in terms of changes in the spiking response of the output neuron rather than actual behavior. The model described by Wessnitzer *et al.*[41] has been reimplemented in two robot instantiations. The key properties of this model (Figure 8.2) are as follows:

- A mapping from simulated odors and odor combinations onto antennal lobe activation, which results in a distributed coding in which odor quality is encoded by a spatial activity pattern across glomeruli, whereas odor concentration is related to intensity and spread of this glomerular pattern.
- A normalization of this activation assumed to arise from lateral inhibition by antennal lobe local neurons.[42]
- A divergence from projection neurons to KCs such that each KC receives approximately 10 inputs[43] and requires more than 2 simultaneous inputs to fire. This results in a sparse recoding of odor patterns that enhances the separation of patterns that overlap at the antennal lobe level.
- A convergence from KCs to a few extrinsic neurons (ENs), with adaptive synapses between the KCs and the ENs.[44]
- A learning rule[45] that uses STDP to 'tag' synapses when KCs and ENs spike close in time,[46] combined with an aminergic reinforcement signal that consolidates changes in conductance for tagged synapses.[47]

Although this model still simplifies many aspects of the circuit (e.g., ignoring the evidence that synaptic modification may also take place at the level of the antennal lobe or input to the MBs, that the role of lateral inhibition in the antennal lobe may be more complex, that there are feedback connections from extrinsic neurons to the KCs, and that there are different temporal stages and likely different neuropils involved in consolidation of memory), it could be shown to learn both simple elemental discriminations and also nonelemental discriminations such as negative patterning. Although this model was based on *Drosophila* circuitry, these learning capabilities have only been clearly demonstrated in bees[48] and were not observed in flies tested using the odor shock paradigm.[49]

A revised version of this model was implemented on parallel programmable hardware in the form of a field programmable gate array and used to control a robot base.[50] Providing a robot with real olfactory sensors that resemble biological olfactory sensors is still a technical challenge, so the robot instead uses light and distance sensing in a setup similar to the olfactory learning paradigm described by Claridge-Chang *et al.*[51] The robot can move between two ends of an enclosed arena, which differ in ceiling height (the CS, representing odor) and light level (the US, representing shock). In this model, we assume that the role of learning in the MB is to reduce the inhibitory influence of MB output on a direct odor-avoidance pathway, assumed to operate via the lateral horn. Before learning, all input patterns excite this pathway, but in parallel the same pattern, via KCs, activates the EN, which inhibits any motor response. The learning parameters are set such that pairing of a CS with a US tends to reduce the synaptic output strength of KCs encoding the particular CS pattern. The EN consequently fires fewer spikes, inhibition is reduced, and the direct pathway drives a conditioned response. A similar mechanism has been described in the honeybee by Okada *et al.*[52]: Decreased response of an MB extrinsic neuron (PE1) after conditioning correlates with an increase in behavioral response, suggesting that learning can be interpreted as selective relaxation of inhibition over sensorimotor connections.

We have used an alternative form of behavioral control of a robot with the same core learning circuit (Jonas Klein, unpublished results). In this case, the ENs are treated directly as motor neurons. Each side of the robot has its own sensory input (using color maps to represent odors and odor combinations), projection neuron layer, KC layer, and EN layer; the firing of the EN determines the wheel speed on that side of the robot. While viewing one color, the robot is 'shocked'; as a consequence, the pattern of KC outputs activated by that color increases in strength to drive the EN. When the robot is then placed in an arena with a choice of colors, it will move faster on the color that has been shocked and tend to turn away from the shocked color at the boundary between the colors because any increase in that color viewed on one side speeds up the corresponding wheel. As a result, the robot spends proportionally more time on the 'unshocked' color area of the arena.

It is not claimed that either of these robot models is a good representation of how classical conditioning controls behavior in insects. However, what they highlight is that understanding learning is not just about understanding changes in synaptic strength. It is also crucial to understand how those synapses fit within a circuit that controls ongoing behavior. Yet very little is known about the output pathways from the MB. Do extrinsic neurons directly drive motor responses? Do they inhibit or excite underlying control circuits? Do they indirectly control behavior by signaling the value of a CS, perhaps by activating aminergic neurons? Do they constitute a circuit that explicitly predicts the occurrence of the US? Can analyzing these different possible mechanisms from a robotic perspective provide any insight?

WHAT IS ASSOCIATED WITH WHAT IN CLASSICAL CONDITIONING?

As discussed in the introduction, a contribution made by the robotic modeling approach to behavioral neurobiology is to enforce consideration of neural mechanisms within the closed loop of behavior. A consequence, with regard to understanding the mechanisms of learning, is that we need to be explicit about what pathways between sensing and acting are modified during associative conditioning, such that a different behavioral response is observed to the same sensory input. Consider the problem in the form of a very simple robot controller (Figure 8.3A).

Let us assume that the robot has several sensors that are connected to sensory units (interneurons) activated by the presence of a particular stimulus. Let us also assume the robot has motor command units that when activated produce a coordinated response, such as approach to or avoidance of the stimulus. We can imagine that some of the sensory units are already connected (perhaps indirectly) to one or another motor command unit such that the presentation of that stimuli produces an innate reflex response. Other sensory units may have weak connections (perhaps insufficient to trigger actions) to multiple motor command units. In classical conditioning, presentation of one of the reflex-producing stimuli (a US that produces a UR) is paired with one of the stimuli that has weak or no effect on action (the CS). The outcome (after repeated pairing) is that subsequent presentation of the CS produces a clear response (the conditioned response (CR)). The obvious question is how has the robot been rewired to produce this new behavior?

Although discussion of this question, in various forms, has a long history, it seems that many alternative views of the 'wiring change' that occurs in classical conditioning are still in play, both in the field of learning in general and in invertebrate learning in particular. For example, there are descriptions and models of classical conditioning that treat it strictly as CS–UR association[53]; that is, the CS comes to activate the same motor unit as the US originally activated, usually through a

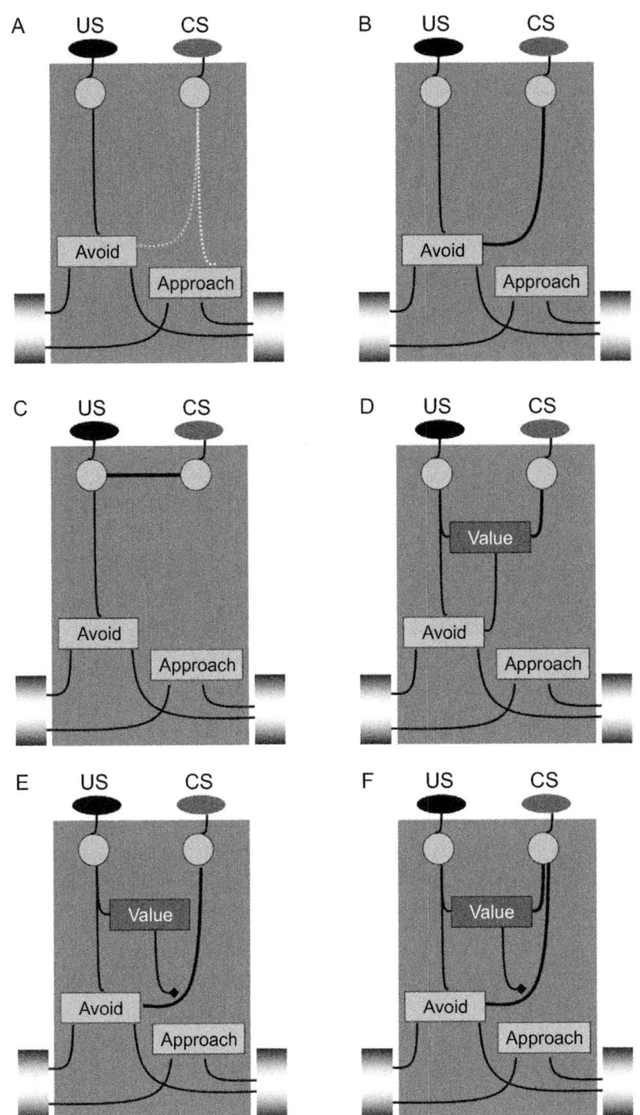

FIGURE 8.3 Alternative ways for a robot to wire a conditioned stimulus to behavior: (A) Before conditioning, the unconditioned stimulus (US) evokes a behavior but the conditioned stimulus (CS) does not. (B) After conditioning, the CS connects to the same behavior that was evoked by the US. (C) After conditioning, the CS connects to the US. (D) After conditioning, the CS has the same value as the US. (E) After conditioning, the value of the US has strengthened the connection of the CS to some behavior (this is not necessarily the same as the behavior evoked by the US). (F). Both D and E occur.

Hebbian strengthening of the CS–UR connection after their simultaneous activation (Figure 8.3B). On the other hand, many accounts assume that the crucial connection is formed between the two sensory units activated by the CS and the US; in other words, classical conditioning is "learning about relationships between stimuli,"[54] such that presentation of the CS creates an activation or expectancy[55] of the US, which causes the response (Figure 8.3C). Of course, it is possible that the connection from the CS could occur at some intermediate point along the US–UR pathway, or indeed at multiple points.[56] Another view[57] is that the robot must have a 'value' unit, activated initially by the US, which drives or gates expression of behavior, and learning involves the strengthening of a connection from the CS to the same 'value' unit (Figure 8.3D). Of course, these are not mutually exclusive mechanisms. To illustrate that this divergence of views exists even for researchers investigating the same system, consider the following accounts of the odor–shock conditioning paradigm in *Drosophila*. A common description is in terms of an association or internal connection being formed between the CS (odor) and the US (shock). For example:

> "In Pavlovian conditioning, animals learn association between a conditioned stimulus (CS) and an unconditioned stimulus (US) through training. Because associative strength between CS and US is thought to determine learning efficacy, elucidation of the neuronal mechanism that underlies CS–US association in the brain is critical to understand the principle of memory formation."[58]

One consequent assumption is that the expressed behavior, avoidance, occurs to the odor because the behavior normally occurring to shock is now elicited via the odor–shock connection formed in the brain (Figures 8.3C and D):

> "Flies naturally avoid electric shock, which serves as the unconditioned stimulus (US). When this shock is paired with delivery of a particular odor, the conditioned stimulus (CS), flies learn this association and subsequently avoid the odor."[59]

In fact, flies' natural response to shock is more often to freeze or jump than to show directed avoidance. The conditioned behavior of odor avoidance is thus sometimes described more explicitly as anticipatory, due to expectation of shock:

> "*Drosophila* exposed to odors paired with electric shock learn this association and display their memory as a selective avoidance of the odor paired with the negative reinforcer."[60]

On the other hand, some descriptions seem more consistent with the view that what has been learned is a CS–UR association (Figure 8.3B):

> "Following association of the two stimuli, the CS alone will elicit a conditioned response characteristic of the US."[61]

Others focus not on association of odor to shock but, rather, to a change in the value or valence of the odor:

> "The temporal pairing of a neutral stimulus with a reinforcer (reward or punishment) can lead to classical conditioning, a simple form of learning in which animal assigns a value (positive or negative) to the formerly neutral stimulus."[62]

> "When a scent is paired repeatedly with electric foot shock, it acquires persistent negative valence—an aversive memory is formed."[63]

The assumption here would seem to be that the avoidance behavior is driven by the negative valence of the odor (Figure 8.3D) rather than by an expectation of shock per se. However, a simpler interpretation of the neural circuit underlying odor–shock learning is also sometimes suggested:

> "The MB model of olfactory short-term memory in *Drosophila* [Heisenberg 2003] proposes that output synapses of the KCs representing the CS+ increase their gain in the course of conditioning to drive a MB output neuron (conditioned response)."[64]

Heisenberg's model[65] assumes that the increase in gain is due to modulation of these synapses by connections from dopaminergic neurons (encoding value?) that are activated by shock (the US). However, note that though this involves a convergence of US and CS signals, it is not the connection between these pathways that is actually strengthened. The 'association' formed is between CS and CR (Figure 8.3E), not between the CS and the US–UR pathway, or CS and value.[1] An interesting consequence of this account is that there is no necessity for the conditioned response to have any particular resemblance to the unconditioned response. A striking example in which the unconditioned response and conditioned response appear opposite in character is provided by Niewalda *et al.*[67] (see also Russell *et al.*[68]). *Drosophila* larvae find low levels of salt attractive, but higher concentrations become aversive. When larvae are trained by pairing an odor with salt at a concentration level slightly higher than that in their diet, which the larvae find aversive in choice tests, they show an increase in attraction to the odor. Such a result would not be possible for a robot that has modified its wiring as shown in Figures 8.3B–D and indeed suggests that the reinforcement value of a US may sometimes be different from its valence as judged by its innate effect on behavior.

CONCLUSION

I described several robot models of insect learning that aim to represent, albeit with many simplifications, our current understanding of the neural circuitry that supports memory formation in mushroom bodies. Recent models share the basic assumption, presented abstractly in Figure 8.3E, that the key association learned in associative learning is between conditioned stimuli and conditioned responses. This is certainly not the full story, but it suggests that a greater focus on the output pathways of the mushroom bodies, to understand how they influence the execution of conditioned responses, is urgently needed if we are to understand and model this circuit correctly.

It is interesting to compare this account of insect learning to the neural circuits identified in learning in some other organisms. The circuit for classical conditioning in *Aplysia*, as described in Baxter and Byrne,[69] involves the alteration of the response of neurons within the buccal central pattern generator, which mediates feeding behavior, to activation from the anterior tentacle nerve (conveying the CS of touch to the lips). This CS–CR pathway is strengthened via a dopaminergic signal conveyed by the esophageal nerve and evoked by food in the foregut (the US). Similarly, the vertebrate cerebellar circuit involved in eyeblink conditioning,[70] although more complex, in essence involves a US signal carried by climbing fibers that modulates the effect of mossy fiber activation, signaling the CS, on output pathways of the cerebellum, to produce a CR (which, Thompson and Steinmetz[70] note, "differ[s] fundamentally in a number of respects" from the UR). In the vertebrate amygdala circuit that underlies auditory fear conditioning, a key mechanism[71] is an increase in the degree to which auditory stimuli (CS) activate neurons in the lateral nucleus of the amygdala (LA), which project to the central nucleus and thus to descending outputs controlling conditioned fear responses. This CS–CR strengthening depends on simultaneous input to the LA from a US pathway via the midbrain periaqueductal gray.[72] In summary, each of these circuits seems to most resemble Figure 8.3E.

It is sometimes useful to step back from neurophysiological and biophysical complexities of networks to consider the overall functional problem. Robot models can be helpful in this regard because they remind us that the function of learning is not, ultimately, to store information in the brain but instead to change behavior. Of course, there can be danger in overabstraction as well: If models are ultimately to be explanatory for biology, it is still necessary to anchor models in real behavioral data and known internal structure.[73] I thus expect and hope that continued close interaction between robotics and biology will be fruitful in further revealing the mechanisms of learning.

[1]Although there are indications[62] that the CS may also come to affect the dopaminergic signal, which would be a CS–value association (Figure 8.3F), it is not necessary to assume this plays a direct role in producing the CR. Alternative roles could be to support second-order conditioning or to enhance or limit the continued strengthening of the CS–CR connection. However, there is evidence that dopaminergic signals may play a role in the motivational gating of CRs.[66]

Acknowledgments

I thank Darren Smith, Jan Wessnitzer, Jo Young, Elias Alevizos, and Jonas Klein.

References

1. Webb B. Can robots make good models of biological behavior? *Behav Brain Sci*. 2001;24:1033–1094.
2. Brooks RA. From earwigs to humans. *Robotics Auton Syst*. 1997;20:291–304.
3. Franceschini N, Pichon JM, Blanes C. From insect vision to robot vision. *Philos Trans Royal Soc B*. 1992;337:283–294.
4. Kanzaki R. Behavioral and neural basis of instinctive behavior in insects: odor-source searching strategies without memory and learning. *Robotics Auton Syst*. 1996;18:33–43.
5. Reeve R, Webb B. New neural circuits for robot phonotaxis. *Philos Trans Royal Soc A*. 2003;361:2245–2266.
6. Duerr V, Krause A, Schmitz J, Cruse H. Neuroethological concepts and their transfer to walking machines. *Int J Robotics Res*. 2003;22:151–167.
7. Poulet JFA, Hedwig B. Auditory orientation in crickets: pattern recognition controls reactive steering. *Proc Nat Acad Sci USA*. 2005;102:15665–15669.
8. Baden T, Hedwig B. Front leg movements and tibial motoneurons underlying auditory steering in the cricket *Gryllus bimaculatus* de Geer. *J Exp Biol*. 2008;211:2123–2133.
9. Witney AG, Hedwig B. Kinematics of phonotactic steering in the walking cricket *Gryllus bimaculatus* (de Geer). *J Exp Biol*. 2011;214:69–79.
10. Cordeschi R. *The Discovery of the Artificial: Behavior, Mind and Machines before and Beyond Cybernetics. Vol. 28 of Studies in Cognitive Systems*. Dordrecht: Kluwer; 2002.
11. Ross T. Machines that think: a further statement. *Psychol Rev*. 1935;42:387–393.
12. Walter WG. *The Living Brain*. Harmondsworth, UK: Penguin; 1961.
13. Angyan AJ. Machina Reproducatrix: an analogue model to demonstrate some aspects of neural adaptation. In: *Mechanisation of Thought Processes Volume II*, National Physical Laboratory Symposium No. 10; 1958.
14. Zemanek H, Hans K, Angyan AJ. A model for neurophysiological functions. In: Colin C, ed. *Fourth London Symposium on Information Theory*; 1961.
15. Kaelbling LP, Littman ML, Moore AW. Reinforcement learning: a survey. *J Artif Intell Res*. 1996;4:237–285.
16. Kober J, Peters J. Reinforcement learning in robotics: a survey. In: Reinforcement Learning: State of the Art (Wiering M., Otterlo M., eds.), Vol. 12 of Adaptation, Learning, and Optimization. Berlin: Springer 2012:579–610.
17. Sutton RS, Barto AG. *Reinforcement Learning: An Introduction. Adaptive Computation and Machine Learning*. Cambridge, MA: MIT Press; 1998.
18. Rescorla Robert A, Wagner AR. A theory of pavlovian conditioning: variations in the effectiveness of reinforcement and nonreinforcement. In: Black AH, Prokasy WF, eds. *Classical Conditioning II: Current Research and Theory*. New York: Appleton-Century-Crofts; 1972:64–99.
19. Montague PR, Dayan P, Sejnowski TJ. A framework for mesencephalic dopamine systems based on predictive hebbian learning. *J Neurosci*. 1996;16:1936–1947.
20. Waelti P, Dickinson A, Schultz W. Dopamine responses comply with basic assumptions of formal learning theory. *Nature*. 2001;412:43–48.
21. Sejnowski TJ, Dayan P, Montague PR. *Predictive hebbian learning. Proceedings of the Eighth Annual Conference on Computational Learning Theory—COLT '95*. New York: ACM Press; 1995:15–18.
22. Dayan P, Balleine BW. Reward, motivation, and reinforcement learning. *Neuron*. 2002;36:285–298.
23. Burgess N, Donnett JG, Jeffery KJ, O'Keefe J. Robotic and neuronal simulation of the hippocampus and rat navigation. *Philos Trans Royal Soc B*. 1997;352:1535–1543.
24. Carrillo RR, Ros E, Boucheny C, Coenen OJ. A real-time spiking cerebellum model for learning robot control. *BioSystems*. 2007;94:18–27.
25. Scutt TW, Damper RI. *Biologically-motivated learning in adaptive mobile robots. SMC '97 IEEE International Conference on Systems, Man and Cybernetics*. Hoboken, NJ: IEEE Press; 1997:475–480
26. Damper RI, French RLB, Scutt TW. ARBIB: an autonomous robot based on inspirations from biology. *Robotics Auton Syst*. 2000;31:247–274.
27. Hawkins RD, Kandel ER. Is there a cell biological alphabet for simple forms of learning? *Psychol Rev*. 1984;91:375–391.
28. Montague PR, Dayan P, Person C, Sejnowski TJ. Bee foraging in uncertain environments using predictive Hebbian learning. *Nature*. 1995;377:725–728.
29. Pérez-Uribe A, Hirsbrunner B. Learning and foraging in robot-bees. In: Meyer J-A, Berthoz A, Floreano D, Roitblat, Wilson SW, eds. *From Animals to Animats: Proceedings of the International Conference on the Simulation of Adaptive Behavior*. Cambridge, MA: MIT Press; 2000:185–194.
30. Hammer M. An identified neuron mediates the unconditioned stimulus in associative olfactory learning in honeybees. *Nature*. 1993;366:59–63.
31. Real LA. Uncertainty and pollinator–plant interactions: the foraging behavior of bees and wasps on artificial flowers. *Ecology*. 1981;62:20–26.
32. Linster C, Smith BH. A computational model of the response of honeybee antennal lobe circuitry to odor mixtures: overshadowing, blocking and unblocking can arise from lateral inhibition. *Behav Brain Res*. 1997;87:1–14.
33. Muezzinoglu M, Vergara A, Huerta R, et al. Artificial olfactory brain for mixture identification. In: NIPS. *Neural Information Processing Systems Foundation*. La Jolla, CA; 2009.
34. Nowotny T, Huerta R, Abarbanel H, Rabinovich M. Self-organization in the olfactory system: one shot odor recognition in insects. *Biol Cybern*. 2005;93:436–446.
35. Nowotny T. Divergence alone cannot guarantee stable sparse activity patterns if connections are dense. *BMC Neurosci*. 2009;10 (suppl 1):P188.
36. Huerta R, Nowotny T. Fast and robust learning by reinforcement signals: explorations in the insect brain. *Neural Comput*. 2009;21:2123–2151.
37. Wessnitzer J, Webb B, Smith D. A model of non-elemental associative learning in the mushroom body neuropil of the insect brain. In: *Proceedings of the International Conference on Adaptive and Natural Computing Algorithms ICANNGA'07*. Warsaw, Poland; 2007:488–497.
38. Wessnitzer J, Webb B. A neural model of cross-modal association in insects. In: *Proceedings of the European Symposium on Artificial Neural Networks ESANN'07*. Belgium: Bruges; 2007:415–421.
39. Helversen D, Wendler G. Coupling of visual to auditory cues during phonotactic approach in the phaneropterine bush cricket *Poecilimon affinis*. *J Comp Physiol A*. 2000;186:729–736.
40. Smith D, Wessnitzer J, Webb B. A model of associative learning in the mushroom body. *Biol Cybern*. 2008;99:89–103.

41. Wessnitzer J, Young J, Armstrong J, Webb B. A model of non-elemental olfactory learning in *Drosophila*. *J Comput Neurosci.* 2012;32:187–212.
42. Olsen SR, Bhandawat V, Wilson RI. Divisive normalization in olfactory population codes. *Neuron.* 2010;66:287–299.
43. Turner G, Bazhenov M, Laurent G. Olfactory representations by *Drosophila* mushroom body neurons. *J Neurophysiol.* 2008;99:734–746.
44. Tanaka N, Tanimoto H, Ito K. Neuronal assemblies of the *Drosophila* mushroom body. *J Comp Neurol.* 2008;508:711–755.
45. Izhikevich E. Solving the distal reward problem through linkage of STDP and dopamine signaling. *Cerebral Cortex.* 2007;17:2443–2452.
46. Cassenaer S, Laurent G. Hebbian STDP in mushroom bodies facilitates the synchronous flow of olfactory information in locusts. *Nature.* 2007;448:709–713.
47. Tomchik SM, Davis RL. Dynamics of learning-related cAMP signaling and stimulus integration in the *Drosophila* olfactory pathway. *Neuron.* 2009;64:510–521.
48. Deisig N, Lachnit H, Giurfa M, Hellstern F. Configural olfactory learning in honeybees: negative and positive patterning discrimination. *Learn Mem.* 2001;8:70–78.
49. Young JM, Wessnitzer J, Armstrong JD, Webb B. Elemental and non-elemental olfactory learning in *Drosophila*. *Neurobiol Learn Mem.* 2011;96:339–352.
50. Alevizos E. Implementation of Neural Plasticity Mechanisms on Reconfigurable Hardware for Robot Learning. Master's thesis, University of Edinburgh, Edinburgh, UK; 2011.
51. Claridge-Chang A, Roorda R, Vrontou E, et al. Writing memories with light-addressable reinforcement circuitry. *Cell.* 2009;139:405–415.
52. Okada R, Rybak J, Manz G, Menzel R. Learning-related plasticity in PE1 and other mushroom body-extrinsic neurons in the honeybee brain. *J Neurosci.* 2007;27:11736–11747.
53. Wörgötter F, Porr B. Temporal sequence learning, prediction, and control: a review of different models and their relation to biological mechanisms. *Neural Comput.* 2005;17:245–319.
54. Brembs B, Plendl W. Double dissociation of protein-kinase C and adenylyl cyclase manipulations on operant and classical learning in *Drosophila*. *Curr Biol.* 2008;18:1168–1171.
55. Lovibond PF, Shanks DR. The role of awareness in pavlovian conditioning: empirical evidence and theoretical implications. *J Exp Psychol Anim Behav Process.* 2002;28:3–26.
56. Holland PC. Event representation in pavlovian conditioning: image and action. *Cognition.* 1990;37:105–131.
57. Dayan P, Niv Y, Seymour B, Daw ND. The misbehavior of value and the discipline of the will. *Neural Networks.* 2006;19:1153–1160.
58. Honjo K, Furukubo-Tokunaga K. Distinctive neuronal networks and biochemical pathways for appetitive and aversive memory in *Drosophila* larvae. *J Neurosci.* 2009;29:852–862.
59. Wu C-L, Shih M-FM, S-Y J, et al. Heterotypic gap junctions between two neurons in the *Drosophila* brain are critical for memory. *Curr Biol.* 2011;21:848–854.
60. Akalal D-BG, Yu D, Davis RL. A late-phase, long-term memory trace forms in the γ neurons of *Drosophila* mushroom bodies after olfactory classical conditioning. *J Neurosci.* 2010;30:16699–16708.
61. Gervasi N, Tchénio P, Preat T. PKA dynamics in a *Drosophila* learning center: coincidence detection by Rutabaga adenylyl cyclase and spatial regulation by Dunce phosphodiesterase. *Neuron.* 2010;65:516–529.
62. Riemensperger T, Voeller T, Stock P, Buchner E, Fiala A. Punishment prediction by dopaminergic neurons in *Drosophila*. *Curr Biol.* 2005;15:1953–1960.
63. Claridge-Chang A, Roorda RD, Vrontou E, et al. Writing memories with light-addressable reinforcement circuitry. *Cell.* 2009;139:405–415.
64. Masek P, Heisenberg M. Distinct memories of odor intensity and quality in *Drosophila*. *Proc Natl Acad Sci USA.* 2008;105:15985–15990.
65. Heisenberg M. Mushroom body memoir: from maps to models. *Nature.* 2003;4:266–275.
66. Krashes M, DasGupta S, Vreede A, White B, Armstrong D, Waddell S. A neural circuit mechanism integrating motivational state with memory expression in *Drosophila*. *Cell.* 2009;139:416–427.
67. Niewalda T, Singhal N, Fiala A, Saumweber T, Wegener S, Gerber B. Salt processing in larval *Drosophila*: choice, feeding, and learning shift from appetitive to aversive in a concentration-dependent way. *Chem Senses.* 2008;33:685–692.
68. Russell C, Wessnitzer J, Young JM, Armstrong JD, Webb B. Dietary salt levels Affect salt preference and learning in larval *Drosophila*. *PLoS ONE.* 2011;6:e20100.
69. Baxter DA, Byrne JH. Feeding behavior of *Aplysia*: a model system for comparing cellular mechanisms of classical and operant conditioning. *Learn Mem.* 2006;13:669–680.
70. Thompson RF, Steinmetz JE. The role of the cerebellum in classical conditioning of discrete behavioral responses. *Neuroscience.* 2009;162:732–755.
71. Maren S, Quirk GJ. Neuronal signalling of fear memory. *Nat Rev Neurosci.* 2004;5:844–852.
72. McNally GP, Johansen JP, Blair HT. Placing prediction into the fear circuit. *Trends Neurosci.* 2011;34:283–292.
73. Webb B. Animals versus animats: or why not model the real iguana? *Adapt Behav.* 2009;17:269–286.

SECTION 4

MECHANISMS FROM THE MOST IMPORTANT SYSTEMS

4.1 Nematodes/*Caenorhabditis elegans*

CHAPTER

9

Mechanosensory Learning and Memory in *Caenorhabditis elegans*

Andrea H. McEwan and Catharine H. Rankin

University of British Columbia, Vancouver, British Columbia, Canada

INTRODUCTION TO *CAENORHABDITIS ELEGANS* LEARNING AND MEMORY

Caenorhabditis elegans is a 1-mm-long soil-dwelling nematode selected by Sydney Brenner in 1965 to study animal behavior and development (particularly neural development).[1] *Caenorhabditis elegans* have become a popular model organism because of their rapid lifecycle, large number of progeny, and ease of cultivation in a laboratory setting. In addition, because *C. elegans* are self-fertilizing hermaphrodites, each progeny represents a genetic clone.

One of the first major studies in *C. elegans* resulted in the publication of the complete cell lineage for every one of the 959 cells in the adult hermaphrodite.[2] With publication of this and several follow-up papers, it became clear that unlike mammals in which neural development involves overgrowth and competitive pruning of cells, the *C. elegans* lineage is invariant. The advantage of the invariant lineage is that it allowed Brenner, Sulston, and Horvitz to identify cell death genes responsible for the death of specific sister cells in identified lineages.[3] For this work, they were awarded the Nobel Prize in medicine in 2002. Although this was a major breakthrough in understanding development, it served to inhibit the study of learning and memory in *C. elegans* because people generalized from a determinate developmental process to the functioning of the nervous system. The common thought during the 1970s and 1980s was that with only 302 neurons and determinate development, there would be no mechanism for learning and memory in *C. elegans* and that all behavior would be "hardwired." This was reinforced by the publication of "The Mind of the Worm" by White *et al.* in 1986, which was an electron microscopy (EM) serial section reconstruction of the nervous system of *C. elegans* including annotations of all chemical synapses and gap junctions as determined by EM structural analysis.[4] White *et al.* examined a very small number of worms (two or three) and concluded that synapses were in the same places in all animals. In addition, *C. elegans* was once criticized for having few morphological and behavioral phenotypes and was described by critics as being a "featureless tube" that was only capable of forward and backward movement.[5] This criticism was stifled after the publication of its first genetic map,[6] which contained more than 100 genetic loci, each with behavioral and/or morphological markers. As a result of the knowledge of determinate development and the fixed synaptic connections, during the 1970s and 1980s no studies were performed that examined learning in *C. elegans*. Interestingly, a paper was published in 1975 by Hedgecock and Russell that suggested that *C. elegans* could change their behavior (change their preferred temperature) as a result of experience; however, the authors were very careful not to use the words "learning" and "plasticity".[7] It was not until 1990 when the first paper showing that *C. elegans* could learn and remember was published[8] that people started to think more about the capabilities of this microscopic worm. Today, a PubMed search with key words "*C. elegans* learning" shows several hundred papers. As the number of chapters on *C. elegans* learning and memory in this volume highlights, *C. elegans* exhibit a great deal of plasticity and a broad range of learning and memory abilities.

By the time *C. elegans* were shown to be capable of learning and remembering, the *C. elegans* research

Invertebrate Learning and Memory.
DOI: http://dx.doi.org/10.1016/B978-0-12-415823-8.00009-5

community had accumulated a great deal of data that greatly facilitated studies of neural circuits underlying behavior and genes that might play a role in specific behaviors. Much of this data was published or stored on publicly accessible databases giving researchers access to a wealth of background knowledge. This data included the cell lineage data, a fully mapped nervous system and a constantly evolving physical map of the genome, and, as of 1998, the sequenced genome of *C. elegans*.[9] During the 1990s and 2000s, these data sets were archived and made publicly available through what is now WormBase (www.wormbase.org), an online database of all published (and a great deal of unpublished) *C. elegans* data.

A number of techniques have been developed or adapted for use in *C. elegans* in order to study neuronal function and behavior on a variety of different levels. These include laser ablation, a type of microsurgery in which individual cells are killed by exposure to a laser beam[10]; channelrhodopsin, a light-gated ion channel that can be used to control neuronal excitability[11]; cameleon, a genetically encoded calcium sensor used for detecting changes in calcium concentration in neurons (an indicator of neural activity)[12]; and worm trackers, a type of automated analysis system that analyzes video data and characterizes locomotion and behavior.[13–19] Another important technological development came from the use of dual worm tracker/calcium imaging instruments.[20–22] These devices use calcium imaging coupled with nematode tracking in order to correlate calcium transients with the behavior of freely moving nematodes. These innovations greatly facilitated behavioral research by allowing researchers to measure behaviors accurately without subjectivity and to alter behavior in a discrete and controlled way.

Caenorhabditis elegans sense the environment, including (but not limited to) tastes, smells, temperature, and physical interactions, through a collection of sensory neurons with specialized functions. This chapter focuses on mechanosensory learning, whereas other chapters more closely describe plasticity in chemosensory and thermosensory systems. *Caenorhabditis elegans* have a number of different types of mechanosensory neurons, including mechanosensory neurons in the nose, pharynx, tail, and along the body. Different mechanosensory neurons may be activated depending on the intensity and the location of the stimulus. As a result, in a laboratory setting, specific techniques are used to deliver, for example, a "harsh touch" versus a "light touch" or nose touch versus head touch. This chapter focuses on the five body touch neurons that project neurites anteriorly or posteriorly. These cells inform the animal about light touch to its body.[23]

CHARACTERISTICS OF SHORT-TERM TAP HABITUATION

The first *C. elegans* behavior in which learning was studied in detail was a mechanosensory response—the "tap withdrawal response".[8] The tap withdrawal response is the worm equivalent of a startle response. This response is elicited by a tap to the side of the petri dish on which worms are cultured. The tap to the petri dish sends vibrations through the agar that are sensed by *C. elegans* through mechanosensory neurons. In response to the tap, the adult worms typically respond by initiating a reversal (crawling backward), sometimes switching direction before moving forward again (Figure 9.1).

In 1990, Rankin *et al.* found that with repeated taps to the petri dish, *C. elegans* would reverse the direction of movement with progressively shorter distances. A total of 30 taps were delivered with a fixed interstimulus interval (ISI) of 10 sec between taps. To rule out the possibility that the worms were responding less because of motor or sensory fatigue, they applied a mild electric shock to the agar and found that worms would reverse with a much larger distance (a process called dishabituation). This demonstrated that the decrement in response to repeated taps was not due to motor or sensory fatigue but, rather, fit the description of the nonassociative form of learning called habituation. This was the first evidence that *C. elegans* were capable of learning. Although this classical protocol for testing short-term habituation in *C. elegans* has evolved over the two decades it has been studied, the basic framework has remained constant. In a typical tap habituation study, populations of age synchronous animals are cultured on a petri dish seeded with a thin lawn of *Escherichia coli*. When animals are 4 days old, they are transferred to a test plate containing a thin *E. coli* lawn and stimulated repeatedly with tap stimuli delivered at a constant ISI, typically 10-sec ISI or 60-sec

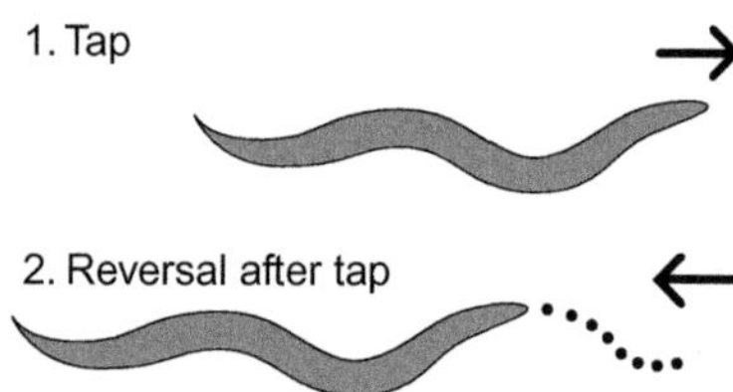

FIGURE 9.1 A schematic of the tap withdrawal response. *Caenorhabditis elegans* are cultured on petri dishes streaked with bacteria. (1) A worm moving forward on agar surface. (2) After the application of a tap stimulus to the side of a petri dish, vibrations propagate through the agar and are sensed by mechanosensory neurons. After a worm senses such a stimulus, it will typically reverse (crawl backwards) before initiating forward movement again. Dots indicate the distance traveled backwards in response to the tap.

ISI. The response to the taps is recorded, and then the magnitude of the tap response is measured. The original tap habituation studies analyzed reversal responses by measuring the size of reversals each worm made in response to each tap to the side of a petri dish. Animals that failed to reverse in response to tap were considered to have reversed a magnitude of 0 units. In this way, both response magnitude and response frequency were incorporated into the "response magnitude" metric. Current studies using machine scoring of data separate these into two discrete measurements: "frequency/probability" habituation and "distance" habituation. In the first measurement, the proportion of worms that respond to the tap is determined (regardless of the distance they respond), whereas in the second measurement the distance analysis is calculated for only those worms that respond to the tap.[19] All results discussed here are presented as the response magnitude (merged distance and probability value) unless stated otherwise.

After it was established that *C. elegans* could learn, researchers focused on characterizing learning and memory to identify the properties of habituation and to define the limits of *C. elegans* capability to learn and remember. To do this, behavior was studied using different metrics and training protocols. These investigations resulted in data detailing (1) the effect of ISI on the rate of habituation, (2) the relationship between the ISI and the rate at which the animals recovered from habituation (ISI-dependent spontaneous recovery), and (3) the effect of habituation on the rate of spontaneous reversals. Furthermore, the training protocols required to generate persistent habituation (intermediate and long-term habituation) were developed, and these forms of memory were investigated. This collection of work has resulted in a better understanding of habituation at the behavioral and circuit level. Each of these levels of organization is discussed in this chapter.

One general characteristic of habituation in all organisms is that animals habituate faster as the frequency of the stimulation is increased. A typical graph plotting response magnitude over number of stimuli follows a negative exponential curve. That is, there is an initial rapid decline in response after the first few stimuli followed by a more gradual decrement until a asymptote is reached. When the stimulus is delivered with a lower frequency, this curve tends to flatten out and become more linear.[24] The effect of ISI on habituation was studied in *C. elegans* (Figure 9.2) to determine whether this pattern was preserved. Investigations into the effect of stimulation frequency on habituation kinetics confirmed that habituation with longer ISIs results in shallower habituation curves.[25b]

After habituation training, the tap response of *C. elegans* shows spontaneous recovery, so the decrement in

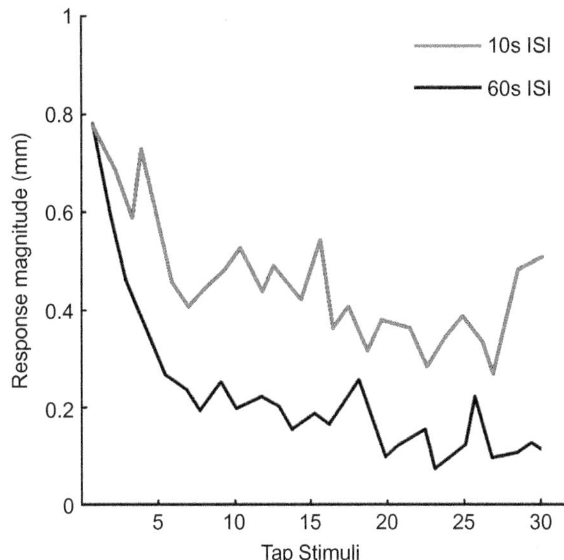

FIGURE 9.2 A typical habituation curve for *C. elegans* tapped at a 10- or 60-sec ISI. The distance an animal reverses (in millimeters) is plotted over the number of taps delivered. Like other organisms (Groves and Thompson 1970), worms given stimuli at a long ISI (60 sec) habituate more slowly compared to those given stimuli at a shorter ISI (10 sec). Source: *Adapted with permission from Broster and Rankin.*[25]

the size of response caused by habituation may recover to naive response levels after the stimulus is withheld for a period of time. Rankin and Broster[25b] examined the relative rates of spontaneous recovery in worms that were habituated at a 2-, 10-, 30-, or 60-sec ISI (Figure 9.3). The degree of recovery was measured at four time points ranging from 30 sec to 30 min after training. They consistently found that regardless of the number of stimuli or level of decrement (after asymptote was reached), animals stimulated with a shorter ISI recovered from habituated levels faster. This is the opposite pattern of what would be expected if the decrement in response were caused by motor fatigue or sensory adaptation when the animals showing the deepest decrement under the faster stimulation paradigm would recover more slowly.[25b] This gives an additional way to distinguish between habituation and fatigue, which is advantageous because the relationship between the cellular mechanisms of habituation and dishabituation are not well understood. ISI-dependent recovery offers an alternate way to demonstrate habituation in mutant strains of worms that do not dishabituate.

In addition to stimulus-induced reversals, *C. elegans* will also reverse spontaneously at a frequency of approximately two reversals per minute.[26] Wild-type worms that have been habituated to tap do not show a change in either the rate or the magnitude of spontaneous reversals.[27] In addition, reversals evoked by a

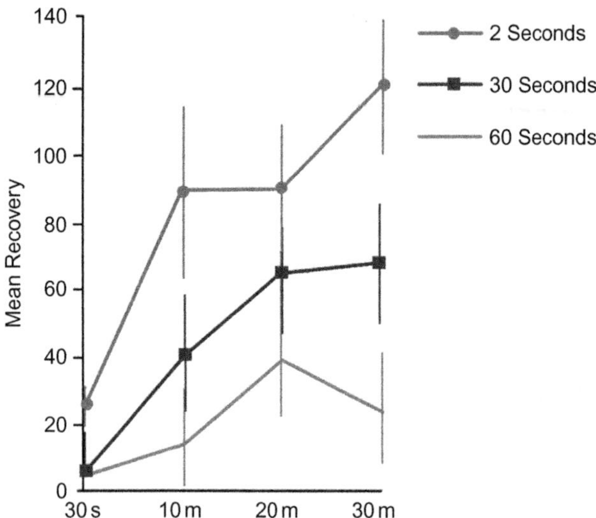

FIGURE 9.3 Spontaneous recovery of the tap response over three time points. After a rest period, the response recovers and worms respond to the tap stimulus with the same magnitude as the initial response. Animals that were habituated at a longer ISI require more time to recover to naive levels compared to animals habituated at a shorter ISI. Source: *Adapted with permission from Rankin and Broster.*[25b]

heated probe, which use the same command interneurons for backward movement as does tap, do not change in size immediately after tap habituation. These data suggest that cellular changes that underlie tap habituation occur before the level of the command interneurons (i.e., in the sensory neurons or the sensory neuron synapses onto the command interneurons).

DEVELOPMENT OF TAP HABITUATION

When *C. elegans* hatch, they have four of the five mechanosensory neurons in the tap circuit—two ALM neurons in the anterior and two PVM neurons in the tail. The fifth cell, an additional head neuron AVM, is added late in larval development. This developmental change prompted investigations into whether tap habituation changed as a function of age of the worms. Chiba and Rankin[28] investigated developmental changes in the tap withdrawal response and quantified the differences between the proportions of animals that accelerated (defined as an increase in forward movement relative to the worm's speed 2 sec before the stimulus) or reversed in response to a tap. The results demonstrated that larval animals were equally likely to accelerate in response to tap as to reverse. Indeed, larval animals accelerated forward approximately half of the time; however, after AVM was incorporated into the tap circuit (late L4), animals reversed the vast majority of time. Chiba and Rankin hypothesized that the appearance of AVM biased the circuit with three head inputs and two tail inputs; this allowed the circuit driving reversals to compete more effectively against the circuit driving forward acceleration.

Several studies investigated the characteristics of habituation in worms during the period of their development.[29–31] Because larval animals respond to tap with both accelerations and reversals, it was necessary to independently assess habituation of each of these responses. By laser ablating either the two PLM tail or the two ALM head mechanosensory neurons, the habituation kinetics of both reversals and accelerations forward were studied.[30] *Caenorhabditis elegans* progress through four larval stages (termed L1–L4) and reach reproductive maturity after approximately 3 days. Neurons in the tap withdrawal circuit, except for AVM and PVD, develop in the embryo. AVM develops postembryonically and likely enters the tap withdrawal circuit between L4 and young adult.[23] Similarly, PVD develops postembryonically beginning at L3. In the laser ablation study, worms of six different ages were studied, with taps delivered at both 10- and 60-sec ISIs: larval stages 1–4, termed "L1," "L2," "L3," and "L4"; "young adults" (3 days old); and "4-day-old" adults. The most robust age effect found was that the initial response of animals became larger as they progressed through developmental stages.[30] This change was consistent for both reversals and accelerations, suggesting that there was a general increase in responsiveness as worms aged/grew larger. Between L1 and adult, 54 new motor neurons also develop,[3] which may account for the age-dependent increased response. Aspects of adult habituation behavior, such as ISI-dependent differences in habituation kinetics and ISI-dependent recovery, were also found in all stages of larval animals.[30] This suggested that the major components driving habituation behavior were already functional by L1. Interestingly, the addition of AVM and PVD to the circuit did not result in dramatic changes in habituation (in both reversals and accelerations). However, it was found that animals habituated slightly more slowly in the period when AVM enters the circuit. There were no discernible behavioral differences when PVD was hypothesized to enter the circuit.[30]

Beck and Rankin[29] examined short-term habituation in adult and aged worms (4, 7, and 12 days old). They investigated both spontaneous and tap-driven reversal behavior. In an initial 10-min baseline recording, they found that 12-day-old animals reversed significantly shorter distances to tap compared to 4- or 7-day-old animals. To test the effects of age on habituation, 4-, 7-, and 12-day-old animals were given habituation training at both 10- and 60-sec ISI. Older animals habituated more quickly for both reversal distance and

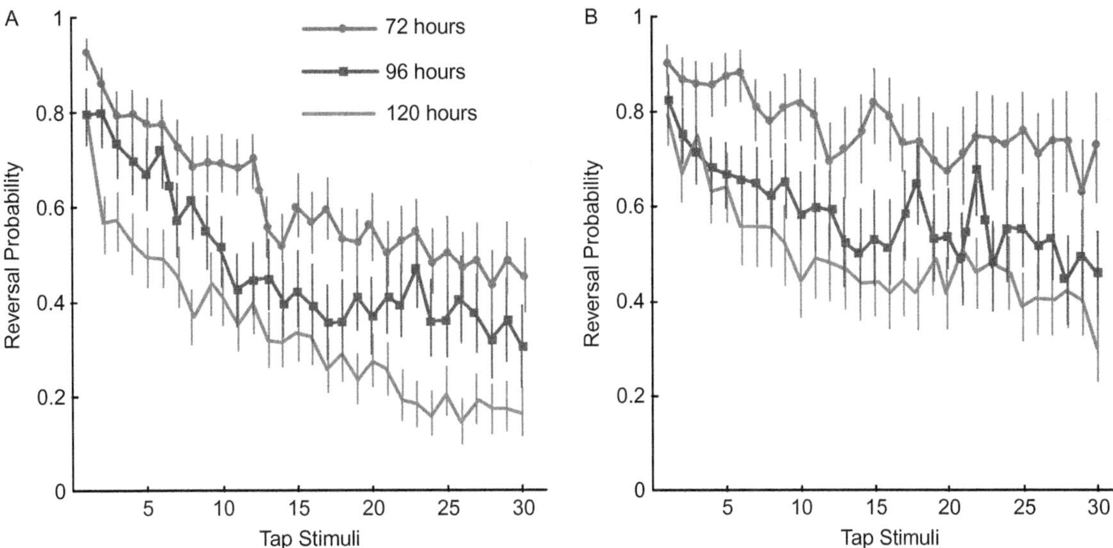

FIGURE 9.4 Age-dependent changes in habituation of response probability. (A) Response probability habituates more slowly for 72-hr-old worms compared to 120-hr-old animals given the same intensity of tap. (B) The rate of habituation of response probability gradually increases as the animal ages. Habituation rate for response distance is not age-dependent.[31]

reversal probability compared to 4- and 7-day-old animals (Figure 9.4).[29] In addition, recovery from habituation took longer in older animals. Together, these data show a profound difference in response and habituation kinetics between young and old animals. Young animals responded more readily to taps and each response was larger compared to those of old animals. Older animals habituated more quickly to stimuli and retained habituation training longer. This pattern may reflect a survival strategy in nematodes: Younger and reproductively active animals need to remain vigilant, whereas older animals, that have already produced progeny, can more readily afford to ignore stimuli and conserve energy.[29]

The Multi-Worm Tracker (MWT)[19] was used to more thoroughly investigate age-dependent differences in habituation during the reproductive phase by examining the habituation kinetics in populations of worms 72, 84, 96, 108, and 120 hr old.[31] Through this analysis, it was found that the proportion of animals responding to repeated taps (regardless of distance) decremented more slowly in young adults (72 hr old) compared to older adults (120 hr old). In other words, the reversal probability habituated more slowly in young adults compared to older adults. As in the original developmental study,[28] this study also found that the initial reversal distance was significantly larger in older adults compared to young adults. However, this relationship broke down at 120 hr of age, possibly due to other effects of aging that hinder locomotion.[31]

To investigate the mechanism of the age-dependent habituation, transgenic worms expressing a light-sensitive ion channel called channelrhodopsin in the mechanosensory neurons were tested.[11] Shining a short pulse of blue light onto the worms causes these channels to open and elicits a reversal response equivalent to the behavior seen in response to a tap. Because channelrhodopsin bypasses mechanotransduction, all animals receive the same level of activation in their sensory cells. If the circuit is modulated downstream of mechanosensory neurons, one would still see the age effect when these cells are artificially activated with light. Conversely, the absence of this phenomenon in light-activated neurons would mean that young animals sense the tap stimuli differently compared to old animals. Indeed, regardless of age, there are no significant differences in habituation curves between 72- and 120-hr animals. This suggested an age-dependent difference in the sensation of stimuli.

Studies in many different model organisms have shown that the intensity of stimuli can modulate the rate of habituation. As the intensity of the stimuli increases, the rate of habituation decreases.[32] It is possible that differences in habituation kinetics observed between old and young animals are due to changes in the perceived intensity of the stimulus. This was tested by habituating groups of 72- and 120-hr-old animals to four different intensities of tap. Repeated tap stimulation at higher intensity resulted in slower habitation of reversal probability in both 72- and 120-hr-old animals. Interestingly, older animals were less sensitive to different intensities: Whereas 72-hr-old animals showed significantly greater decrement as stimulus intensity increased across the four intensities, 120-hr-old

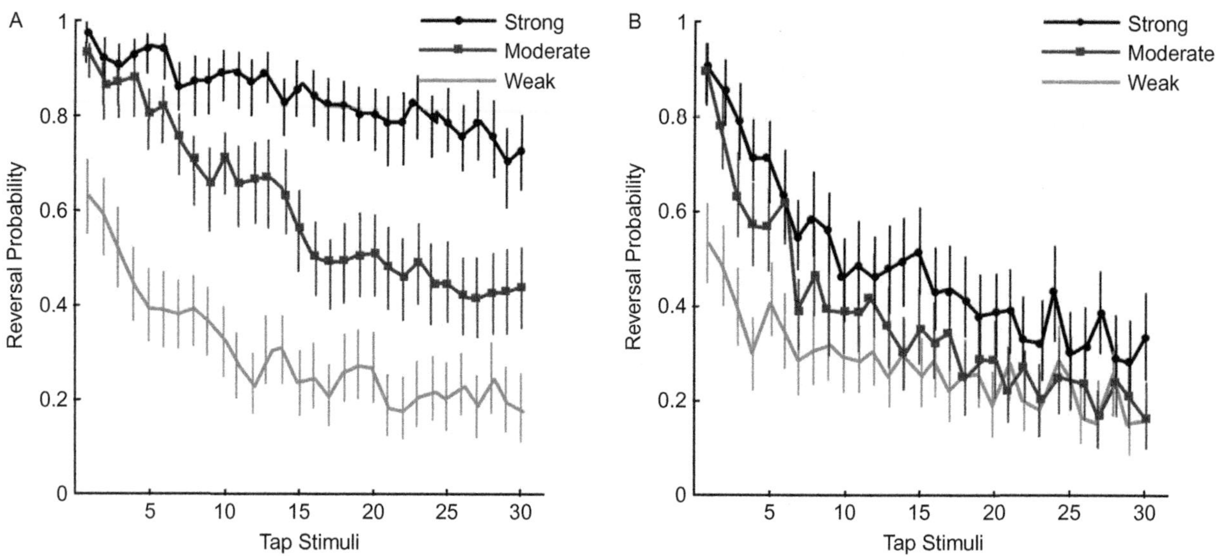

FIGURE 9.5 **Habituation kinetics depend on stimulus intensity.** Habitation curves for response probability of young (72 hr) and old (120 hr) worms undergoing habituation training with different intensities of stimuli. (A) Younger and old animals habituate more quickly to the weakest stimuli. (B) Older animals show less discrimination between different stimulus intensities than do younger animals.[31]

animals only showed significant differences in habituation between the strongest and the weakest tap intensities (Figure 9.5). Taken together, these data confirmed that the rate of habituation is dependent on the perceived intensity of the stimuli and that younger worms differentiate between the intensity of stimuli to a greater extent than older adults.[31]

There are several possible explanations for this age-dependent decrease in sensitivity to stimulus intensity. One explanation may be that there are physiological changes in the worm's flexible exoskeleton, called the cuticle. As the worm ages, hypodermal cells are thought to continue to secrete collagen and thus contribute to increasing cuticle thickness.[33] The increased cuticle thickness may hinder the transmission of mechanical cues in older animals and subsequently change the habituation curve. Other possible explanations for the age-dependent differences include changes in the input resistance of the touch cells or differences in gene expression of components of the transduction machinery across the adult stage.

Interestingly, whereas the difference in reversal probability was dependent on age, the distance worms reversed in response to tap was not (save for the initial tap).[31] Similarly, the rate of habituation for reversal distance was not significantly different except at the most intense stimulus. This demonstrates, in general, that the absolute levels of activity of the sensory neurons ALM, AVM, and PLM do not dictate the magnitude of response at the habituated level. This might be explained by the reciprocally antagonistic nature of the tap withdrawal circuit. That is, although a more intense stimulus may cause an increase in reversal distance; this is dampened by the equally amplified antagonistic response from the posterior sensory cells. Ultimately, the anterior and posterior maintain a balance of activation at each stimulus intensity, which is translated into a fixed reversal distance throughout adulthood.

CIRCUITRY UNDERLYING TAP HABITUATION

Serial EM experiments performed over two decades resulted in a complete map of *C. elegans* neuronal circuitry, including information on the location of synapses.[2] The touch circuit was defined through laser ablation studies[23]; the sensory neurons and interneurons required to produce a behavioral response after either a head touch or a tail touch were identified. The body touch circuit includes sensory neurons ALML/R, PLML/R, and AVM; the interneurons AVAL/R, AVBL/R, and AVDL/R; and possibly PVC and/or RIM. AVA, AVB, and AVD are each bilateral interneurons responsible for transmitting sensory information to motor neurons. Ablation of the command interneurons AVA, AVD, AVE, and RIM resulted in worms that were unable to initiate backward movement.

In their laser ablation analysis of the tap withdrawal circuit, Wicks and Rankin[34] began with the touch circuit and systematically ablated sensory neurons (three types) and interneurons (five types) either singly or in combination to test for alteration in the behavioral responses to tap. They also ablated neurons outside of

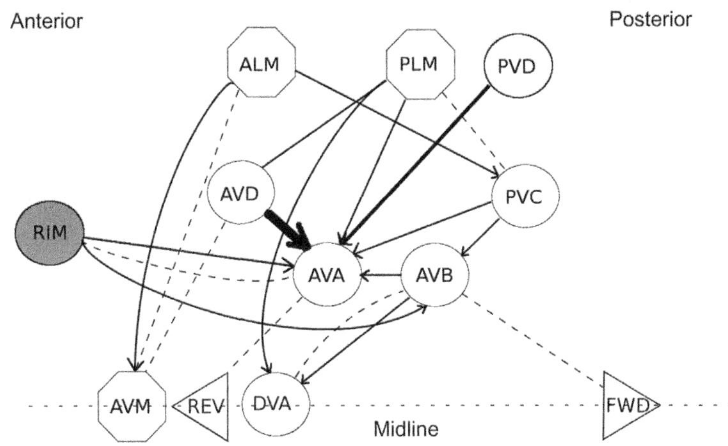

FIGURE 9.6 A schematic summary of connections in the tap circuit. The circuit consists of five sensory neurons (ALML, ALMR, PLML, PLMR, and AVM) shown as octagons, one additional pair of stretch/harsh touch receptors (PVDL and PVDR), a single stretch sensory/interneuron (DVA), four pairs of interneurons (AVA, AVB, AVD, and PVC), and two pools of motor neurons (triangles) that mediate forward and backward movement. Evidence also suggests that the interneuron, RIM, may also contribute to the tap withdrawal response. Dashed lines represent gap junctions, and solid lines represent chemical synapses. The thickness of the lines reflects the number of synapses observed in the electron microscopic serial section reconstruction by White *et al.*[4]

the touch circuit in order to identify additional neurons that might be required specifically in the tap withdrawal response. The results indicated that the tap withdrawal circuit was mediated by five sensory neurons (ALML, ALMR, PLML, PLMR, and AVM), four pairs of command interneurons (AVA, AVB, AVD, and PVC), along with PVDL, PVDR, and DVA (with PVDL and PVDR sensing harsh touch and DVA involved in proprioception) (Figure 9.6). A general feature found in the study was that ablations in the anterior touch cells resulted in a decrease in reversal frequency (and an increase in the number of forward accelerations), whereas ablations in the posterior touch cells increased the frequency of reversals. This suggested that the tap withdrawal response was composed of two antagonistic subcircuits—one that drove accelerations and one that drove reversals. By ablating members of one circuit, tap-induced behavior was biased toward the output of the remaining circuit.

Of the five mechanosensory neurons, only ablation in PLML and/or PLMR demonstrated functional asymmetry. Ablating both PLMs resulted in significantly larger reversal responses compared to those of controls.[4] Although PLM is bilateral, the density of synapses on PLMR is much greater than that on PLML.[4] Supporting this, ablating PLMR resulted in a much stronger reversal phenotype compared to PLML ablation. Double ablations in PLM and either the PVD sensory neurons or the DVA interneurons resulted in a decrease in response magnitude compared to that in PLM ablated animals.[34] Both PVD and DVA neurons synapse onto AVA and PVC (forward and backing interneurons), suggesting that they act to integrate information from the anterior and posterior circuit. Wicks *et al.* speculated that PVD is required to give information on the background vibrations and modulate the level of excitability of the anterior and posterior circuit. This means that DVA and/or PVD would set the sensitivity of the rest of the circuit based on the background vibrational "noise." Taken together, these data indicate that PVD may sense background vibrations and modulate AVA and PVC excitability accordingly. Li *et al.*[35] characterized DVA as a stretch receptor neuron that both positively and negatively regulates locomotion.

The circuit analysis indicated that the tap stimulus excites mechanosensory receptors in both the anterior and the posterior of the worm; therefore, the resultant habituation curve is actually the summation of two competing subcircuits—one driving forward behavior (accelerations) and one driving reversals (Figure 9.6). Wicks and Rankin[36] investigated habituation differences between the reversal- and the acceleration-driving circuits. To gain insight into the shape of the habituation curve for each response, either the anterior mechanosensory neurons or the posterior mechanosensory neurons were laser ablated. Next, ISI differences in habituation kinetics were measured for accelerations (representing the posterior subcircuit) or reversals (representing the anterior subcircuit).[36] At a 10-sec ISI, habituation of accelerations occurred more slowly than reversal habituation and, unlike the 60-sec ISI response, showed facilitation after the first few taps. This suggests that the forward circuit contributes more competition to the reversal circuit in the early stages of habituation at 10-sed ISI. The facilitation of the response seen early in acceleration habituation at 10-sec ISI may explain why the response magnitude drops rapidly at 10-sec ISI compared to 60-sec ISI. Taken together, these studies demonstrate that the rate of habituation of the forward circuit and that of the reverse circuit differ. The amount of competition between each circuit varies over the course of habituation and alters the kinetics of habituation accordingly.[36]

Although the tap withdrawal response was originally defined as containing eight sensory neurons (ALML, ALMR, AVM, PLML, PLMR, PVDL, PVDR, and DVA) and four pairs of command interneurons

(AVA, AVB, AVD, and PVC), there may be other members of the circuit. Evidence from Piggott *et al.*[26] suggests that another subcircuit may also be involved in the tap withdrawal response. They began by ablating command interneurons AVA, AVD, and AVE and found that although the frequency of spontaneous reversals was decreased, animals were still capable of reversing backwards. Through laser ablation studies and imaging techniques, this group identified a second subcircuit, which included the pair of RIM neurons that governed reversal behavior. This introduced the possibility that the tap withdrawal response was governed by three subcircuits—a circuit driving forward behavior and two redundant circuits driving reversal behavior.

The secondary subcircuit was identified by first ablating RIM to search for behavioral defects.[26] Interestingly, ablation of RIM resulted in an increase in spontaneous reversal frequency. Using calcium imaging, Piggott *et al.* demonstrated that RIM-associated reversal behavior corresponded with a decrease in calcium transients, suggesting that it worked to inhibit reversal behavior. In order to identify more members of the circuit, the group examined the neurons that connected to RIM and focused on AIB, an upstream interneuron that synapsed onto RIM. Ablation of AIB resulted in a reduced frequency of spontaneous reversals. Together, this demonstrated that the secondary circuit mediating reversal responses consisted of AIB and RIM. RIM acts to inhibit backward movement while AIB suppresses RIM activity.[26] The role of AIB and RIM in the tap response has not been studied; however, there appears to be a role for RIM in context conditioning for habituation.

After the neurons that mediate the tap withdrawal response were defined, Wicks, Roehrig, and Rankin[37] sought to determine whether the chemical synapses from the mechanosensory neurons onto the command interneurons were excitatory or inhibitory. The *C. elegans* nervous system consists of simple, mostly unbranched processes and, as for the larger relative of *C. elegans*, *Ascaris*, is thought to propagate signals passively (electronic potential) rather than with all-or-none action potentials.[38] Through modeling and based on electrophysiology data from a larger related species (*Ascaris*), Wicks *et al.* were able to make hypotheses about the polarity of neurons in the circuit.

In general, the behavior produced by various cell ablations suggested that the gap junctions within the circuit were excitatory and the synaptic connections were inhibitory.[23,37] This meant that the role of gap junctions between command interneurons and motor neurons is to drive forward or backward movement, and the role of (presumably inhibitory) chemical synapses between sensory neurons and interneurons is to modulate the response. Wicks *et al.*'s[37] model largely supported this view because sensory neurons ALM, AVM, and PLM were hypothesized to be inhibitory.

In EM studies, the interneurons AVD and AVA show a high density of chemical synapses. AVD is thought to be a connector between ALM and AVM to the interneuron that drives backward movement, AVA.[23] Supporting this, Wicks *et al.*'s[37] model hypothesized that the synapses between AVD and AVA are likely to be excitatory.

NEUROTRANSMITTERS INVOLVED IN TAP HABITUATION

Although the directionality of the circuit had been proposed, the neurotransmitter used to modulate the tap withdrawal response had not been defined. *Caenorhabditis elegans* express many different neurotransmitter types, including dopamine,[39] serotonin,[40,41] octopamine,[40] GABA,[42] and glutamate.[43,44] Two major lines of evidence suggested that synaptic transmission from the touch cells to the command interneurons is glutamatergic. First, three classes of glutamate receptor subunits are expressed singly or in combination in the command interneurons AVA, AVB, AVD, and PVC. These are AMPA-type glutamate receptor subunits *glr-1*[43,44] and *glr-2*, expressed in all aforementioned interneurons; NMDA-type glutamate receptor subunits *nmr-1* and *nmr-2*, expressed in interneurons AVA, AVD, AVE, RIM, and PVC[45]; and glutamate gated chloride channel subunits such as *avr-14*, which is expressed on the tap mechanosensory neurons and may serve as an inhibitory autoreceptor.

The second line of evidence came from the analysis of a gene called *eat-4*, which is a *C. elegans* homolog of mammalian VGlut1, a vesicular glutamate transporter. Examination of the expression pattern of *eat-4* showed that it is expressed in a number of neurons, including the mechanosensory neurons.[46] Thus, it is very probable that the synaptic connections between sensory cells and command interneurons are glutamatergic.

LOCUS OF PLASTICITY IN TAP HABITUATION

After defining the tap withdrawal circuit and the parameters that alter behavioral responses, research was directed toward finding the locus of plasticity in habituation. One way of narrowing down potential sites is to examine circuits that govern other behaviors and determine if they overlap with the tap withdrawal circuit. If there is overlap, one would expect that habituation of members of the tap withdrawal circuit would affect the other circuits as well.

One type of behavior that can be used to assess the site of plasticity is the rate of spontaneous reversals. To determine whether the neurons that govern spontaneous reversal overlap with the site of plasticity for tap habituation, Wicks and Rankin[47] examined spontaneous reversals in animals that had undergone habituation training.[47] They found that habituated worms did not show a decrease in either the frequency or the magnitude of spontaneous reversals, suggesting that the site of plasticity in habituation and the neurons governing spontaneous reversal are non-overlapping.[47] Laser ablation of AVA resulted in a significant decrease in the magnitude of reversal in response to tap and the frequency of spontaneous reversals.[10] Conversely, ablation of the sensory neurons ALM, PLM, and AVM and the interneuron AVD did not affect the rate of spontaneous reversals.[34] This suggests that the site of habituation is in the mechanosensory neurons and/or AVD, most likely encoded in alterations in the synaptic connection between the sensory neurons and the command interneurons.

In 2011, Zhen *et al.*[22] investigated the role of the command interneurons and motor neurons in directional movement. They measured the calcium transients in the command interneurons in freely moving worms. Backward movement corresponded with an increase in calcium transients in AVA and AVE; however, there was no change in calcium transients in AVD interneuron, suggesting that it is involved only in stimulus-evoked reversals rather than spontaneous reversals.[22]

Although most mechanosensory studies have examined tap responses, one study thoroughly probed the habituation kinetics of the head touch response. Kitamura *et al.*[48] laser ablated mechanosensory neurons, command interneurons, and combinations thereof and measured the rate at which each animal habituated to head touch. Habituation kinetics approximates a negative exponential curve, and the rate of habituation was measured by finding the half-life of the curve. They found that ablations in AVD or combinations of AVD ablations (except AVD/AVA, which were not tested due to defects in backward movement) resulted in a rapid habituation phenotype. This suggests that the site of plasticity in head touch habituation involved AVD in intact animals. They also found that ablation of ALMR/AVM phenocopied the rapid habituation phenotype of AVD. Because the ALML−AVDR connection is intact in ALMR/AVM ablated animals, these data suggest that ALML function upstream to regulate habituation.[48] ALML synapses onto AVDR and PVC, and although the synaptic connection between ALML and AVD had previously been shown to be important for habituation, the relationship between ALML and PVC was less clear. They therefore ablated AVD, AVM, and ALMR so that only the remaining synaptic connections between ALML and PVC could drive behavior. These lesions also resulted in rapid habituation, demonstrating that synaptic connections between ALML and AVD and between ALML and PVC play a role in habituation kinetics. Interestingly, double ablations in AVD and PVC resulted in a normal habituation curve. Because the only intact chemical synapse in these animals is between AVM and AVB, these neurons must compensate for dysfunction of the rest of the circuit.[48]

Kitamura *et al.*[48] argue that each neuron within the touch circuit is capable of mediating habituation to head touch. Should part of the circuit be compromised, other areas of the circuit compensate. In wild-type animals, the major site of plasticity for habituation to head-touch is likely the ALML-to-AVD chemical synapse (Figure 9.7A). In animals in which AVD interneurons are ablated, habituation occurs through a compensatory mechanism between ALML and PVC. Finally, if ALML is ablated, synaptic connections between AVM and AVB regulate habituation (Figure 9.7B).

Although AVD is a strong candidate for the principal site of plasticity in tap habituation, it is also possible that changes in RIM alter habituation kinetics. RIM forms chemical synapses onto AVA and potentially modulates reversal responses with AVD. Wicks and Rankin[47] demonstrated that the act of habituating a worm to tap did not alter the rate of spontaneous reversals after it had reached the habituated level. Because laser ablation of RIM resulted in an increase in spontaneous reversals, if RIM activity was inhibited in response to tap habituation, one would expect the rate of spontaneous reversals to increase. If RIM is a site of plasticity for tap habituation, it is likely encoded at the synapse between RIM and AVA.

In summary, the tap withdrawal circuit consists of five sensory neurons (ALML, ALMR, PLML, PLMR, and AVM), one additional pair of stretch/harsh touch receptors (PVDL and PVDR), a single stretch receptor/interneuron (DVA), five pairs of interneurons (AVA, AVB, AVD, PVC, and possibly RIM), and two pools of motor neurons that mediate forward and backward movement.[23,48,36] The sensory neurons of the tap withdrawal circuit are connected through gap junctions and chemical synapses whereby the electrical signals are thought to be excitatory and the chemical synapses from the sensory neurons onto the interneurons are thought to be inhibitory.[23,34] The presence of glutamate receptors in the command interneurons AVA, AVB, AVD, and PVC, along with the habituation defects seen with mutations in the glutamate vesicular transporter *eat-4*, suggests that the chemical signal from the touch neurons is mediated through glutamatergic signaling.[43,44,27] Overall, there are two (or possibly three) subcircuits mediating the tap withdrawal response: circuitry

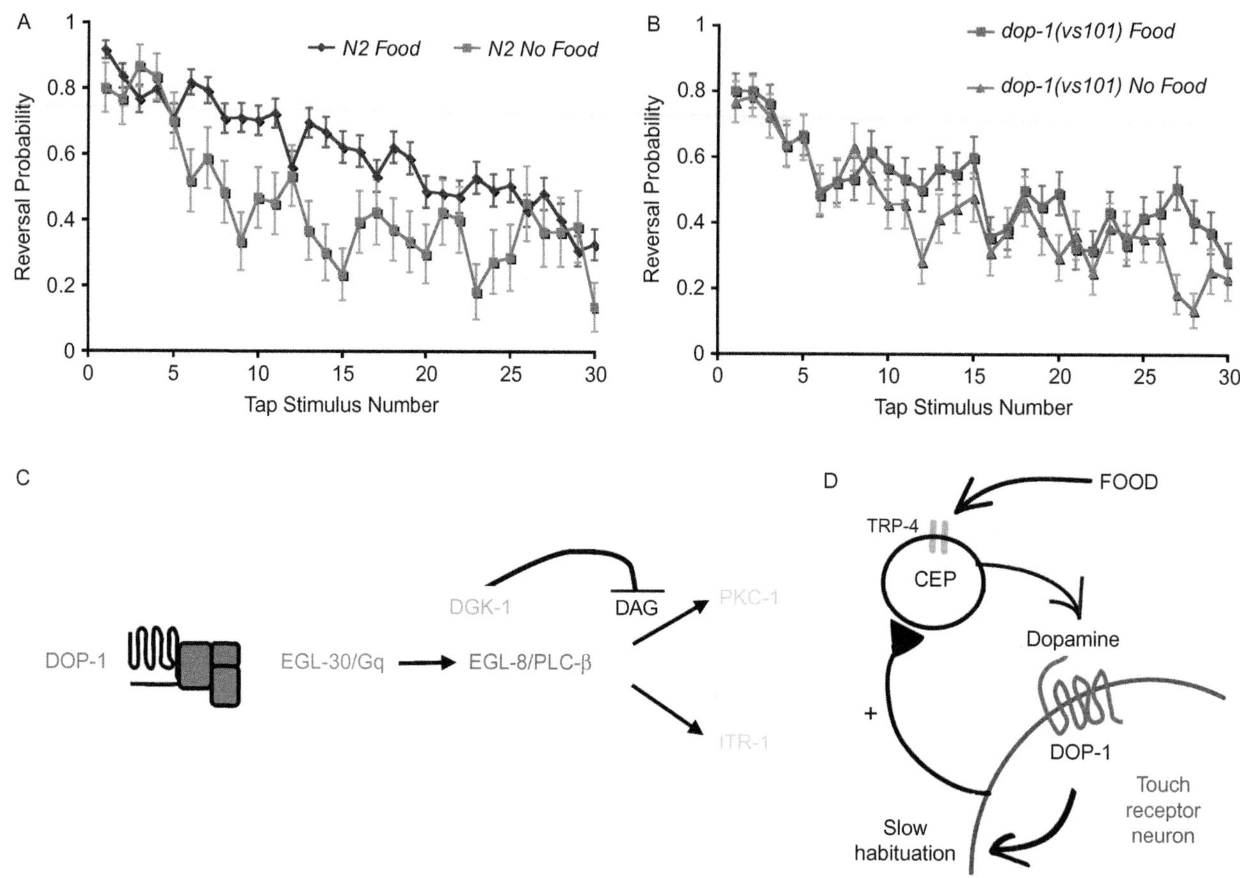

FIGURE 9.7 (A and B) Habituation curves of wild-type and *dop-1* mutants trained either on or off food. Habituation curves of *dop-1* mutants are not significantly different from the habituation curves of wild-type animals off of food. (C) Schematic detailing the molecular pathway in Qq signaling. (D) CEP participates in a feedback loop to slow habituation through increased dopamine signaling. Source: *Adapted with permission from Kindt* et al.[56]

that processes posterior sensory information (PLM along with the interneurons and motor neurons that mediate a forward response), circuitry that processes anterior sensory information (ALM/AVM along with the interneurons and motor neurons that mediate a reversal response), and secondary circuitry that works cooperatively with the command interneurons to mediate a reversal response. The distance that worms move backward in response to tap appears to be controlled by the balance of activation between the two circuits. There is a bias toward reversal movement such that equal activation of the forward and reverse circuit (as seen after a tap stimulus) will typically result in a reversal response.

GENES THAT PLAY A ROLE IN TAP HABITUATION

After characterizing the circuit governing the tap withdrawal response, researchers turned to identifying the genes that regulate habituation. Understanding the genes involved in habituation will give insight into the mechanisms behind synaptic plasticity, an important process in learning and memory. To identify genes that might play a role in tap habituation, Rankin and Wicks[27] used a candidate gene approach and first tested a gene expressed in the tap sensory neurons, *eat-4*. *eat-4* is the *C. elegans* homolog of VGLUT1, a glutamate vesicle transporter whose main function is to transport glutamate into synaptic vesicles.[49] Mutations in *eat-4* result in altered glutamate transmission. The initial response to tap of *eat-4* worms was not significantly different from that of intact wild-type animals; however, *eat-4* worms habituated much more rapidly compared to wild-type worms.[27] This behavioral pattern was found for habituation at both 10- and 60-sec ISIs. This suggested that either *eat-4* is required for sustained glutamate release or glutamate release is not required for the naive response.[27]

To allow for investigation of the "pure" reversal circuit and to rule out the possibility that the electrical connections in the forward circuit were simply canceling out the actions of the reverse circuit (and ostensibly causing rapid "habituation"), the responses of animals were investigated after laser ablating PLM.[27] Indeed, *eat-4* mutants without PLM sensory neurons showed more rapid decrement of responding the PLM ablated wild-type worms, demonstrating that *eat-4* is required for normal habituation.[36]

Interestingly, *eat-4* mutants fail to show dishabituation after being administered a mild electric shock. Although classically, the ability to dishabituate was considered a key feature of habituation,[32] several behavioral tests can be used to demonstrate that the decrement is a result of habituation and not fatigue. One of these tests is ISI-dependent spontaneous recovery.[50] After habituation at a longer ISI, *eat-4* animals spontaneously recovered more slowly than *eat-4* worms habituated at a 10-sec ISI, which fits the characteristics of habituation and not fatigue or adaptation. In addition, neither the rate nor the magnitude of spontaneous reversals in *eat-4* mutants changed after the animals had undergone habituation training. If the decrement in response were caused by fatigue, one would expect the frequency and/or magnitude of spontaneous reversals to be affected as well. Taken together, these data indicate that *eat-4* mutants have a habituation defect that is likely the result of a decrease in glutamate transmission.[27] One possible interpretation of these findings is that *eat-4* mutants experience synapse fatigue resulting from the reduced ability to load glutamate into vesicles. However, there were no differences in the initial response to tap in *eat-4* worms compared to wild-type controls, and *eat-4* worms do recover from habituation, although more slowly than wild-type worms, suggesting there is a second, slower-acting pathway involved in loading glutamate into vesicles. It also suggests that vesicle release from the ready-releasable pool is not dependent on *eat-4*.

Continuing with the candidate gene approach, the next genes to be studied were those involved in the dopamine pathway. A dopamine receptor gene, *dop-1*, was found to be expressed in the tap sensory neurons.[51] In mammalian systems, dopamine signaling is implicated in motivation, reward, and salience.[52] In *C. elegans*, four classes of dopaminergic neurons are thought to sense mechanical stimuli and are required for altering locomotion under different environmental conditions.[53] Dopamine modulates a locomotion behavior that ultimately results in animals staying in food patches for longer periods of time; this is known as the basal slowing response, in which worms slow their basal rate of locomotion after encountering a lawn of bacteria. This is hypothesized to ensure that worms do not stray from their food source.

There are eight dopamine-producing neurons in hermaphroditic *C. elegans*: CEPDL and -R, CEPVL and -R, ADEL and -R, and PDEL and -R. Dopamine signaling is thought to occur extrasynaptically, whereby dopamine binds to dopamine receptors that may be located distally from the site of dopamine release.[54] Dopamine binds four dopamine receptors named *dop-1–dop-4*.[54,51,39] Only *dop-1* is expressed in the tap withdrawal circuit, specifically in the neurons ALM and PLM.[51] Given that both the presence of food and dopaminergic signaling modulate locomotion kinetics, it is possible that the presence of food (*E. coli*) also alters habituation kinetics. To test this, researchers measured both response frequency and response distance separately and found that, indeed, *C. elegans* showed more rapid response probability habituation when tested in the absence of food compared to when they were on food (there were no differences in response distance) (Figure 9.7A).[55] Expanding on this result, Kindt *et al.*[56] sought to identify whether food-dependent modulation of habituation was mediated through dopamine. In this experimental paradigm, *dop-1* and controls were analyzed under "no food" or "food" conditions. If habituation kinetics were modulated by food through dopamine signaling, one would expect *dop-1* mutants and wild-type animals (off food) to habituate at the same rate (because dopamine signaling would be reduced in both cases). Indeed, the habituation curve of wild-type worms off food was not significantly different from that of *dop-1* mutants (Figure 9.7B).[56] To further investigate the relationship between food, dopamine, and habituation, two other mutants were tested for habituation defects: *cat-2* and *dat-1*. *cat-2* encodes tyrosine hydroxylase, which catalyzes the rate-limiting step in dopamine synthesis.[57,58] Animals carrying a mutation in *cat-2* habituated more quickly compared to wild-type (on food) and were not significantly different from wild-type when tested off of food. Next, *dat-1* a plasma membrane dopamine transporter was tested for habituation defects. In *dat-1* mutants, dopamine reuptake is decreased, meaning that dopamine persists in synaptic clefts for longer periods of time, which ultimately leads to strengthened dopamine signaling. Unlike *dop-1* and *cat-2*, *dat-1* mutants habituated more slowly compared to wild-type animals on food. Interestingly, double mutations of *cat-2* and *dop-1* resulted in a fast-habituation phenotype that was more severe than either *dop-1* or *cat-2*. This may be explained by the fact that *cat-2* mutants still exhibit some dopamine neurotransmission.[56] However, if *dop-1* were the only receptor that could mediate dopamine transmission in the tap withdrawal circuit, one would expect the double mutant phenotype to be no more severe than the *dop-1* null mutation, suggesting other dopamine receptors may play a role in tap habituation. Thus, this demonstrates that food modulates habituation rate through dopaminergic signaling and that decreasing dopamine neurotransmission resulted in faster habituation for tap response probability.[56]

To identify the mechanisms downstream of *dop-1*, candidate genes were tested for their influence on habituation kinetics.[56] Because DOP-1 is a G-protein-coupled receptor, strains carrying mutations in G_0, G_q, or G_s α subunits were tested for defects in habituation.[56] In mammalian systems, G_s acts through adenylate kinase to produce the second messenger cAMP, G_0 inhibits G_s, and G_q acts upstream of the phospholipase C (PLC-β) to produce the second messengers inositol triphosphate and diacylglycerol. PLC-β also controls the release of Ca^+ from the endoplasmic reticulum. Candidate *C. elegans* homologs of G_0, G_q, and G_s are *goa-1*, *egl-30*, and gsa-1, respectively. The habituation kinetics of strains lacking functional *egl-30*/G_q phenocopied that of *dop-1* mutants, suggesting that food-dependent habituation modulation acts through the G_q pathway. To test this, putative downstream effectors, *egl*-1 and *dgk-1*, were tested for anomalous habituation kinetics. Indeed, *egl-8* (which encodes PLC-β) habituated more rapidly compared to wild-type animals (on food). Conversely, *dgk-1*, which is thought to antagonize PLC-β signaling, habituated more slowly. Taken together, the genetic interrogation of G-α signaling pathways strongly suggests that food-dependent changes in habituation occur through G_q signaling (Figure 9.7C).[56]

Next, Kindt *et al.*[56] investigated the neuron(s) involved in dopamine-dependent modulations in habituation. To do this, they used *in vivo* calcium imaging, a technique in which calcium levels are visualized using fluorescence of a genetically encoded calcium sensor. Calcium levels are used as a measure of cell activity; an increased calcium signal in response to a stimulus indicates that the cell has been activated. The level of activation can be determined based on the strength of the change in the calcium signal. By monitoring calcium transients of ALM and PLM in *dop-1* and wild-type animals undergoing habituation training (on food), they showed that calcium transients detected in ALM (but not PLM) decremented at a faster rate in *dop-1* mutants compared to wild type. Because ALM cell excitability is reduced in *dop-1* mutants, this suggests that dopamine transmission acts on ALM to increase cell excitability. In addition, calcium transients of ALM were imaged in *egl-30* and *egl-8* mutants (genes in the *dop-1* pathway). Again, these mutants had a more rapid decrement in calcium levels, which is consistent with *dop-1* and supports the behavioral data.[56]

Although behavioral analysis and calcium imaging identified the pathway responsible for altered habituation in response to food, a question that remained was how food regulated the release of dopamine from dopamine-producing cells in the first place. There are three types of dopaminergic neurons: CEP, ADE, and PDE.[39] These neurons are putative mechanosensory neurons that can sense the texture of the ground (or bacteria) beneath them.[59] Although none of these neurons synapse onto ALM, ALM does synapse onto CEP, which suggests a role for CEP in food-dependent habituation effects (see www.wormweb.org). CEP neurons express *trp-4*, a transient receptor potential channel that functions as a stretch receptor in the *C. elegans* nose.[35] Like *dop-1*, mutations in *trp-4* abolish the basal slowing response, suggesting a role in sensing food. A habituation assay also demonstrated that, like *dop-1*, *trp-4* mutants habituate rapidly. Together, these results suggested that the CEP neurons sense mechanical stimulation through the *trp-4* channel. To test this directly, calcium imaging was done in CEP neurons in a *trp-4* mutant background and in wild-type animals undergoing anterior body touch. In this assay, worms were immobilized on agarose pads, and CEP was stimulated using a probe that was pressed against the animal's body (harsh touch) or vibrated (or buzzed) proximal to the nose.[56] This type of stimulation was defined as a nose buzz. In *trp-4* mutants, a calcium response was seen after a harsh nose touch but not a buzz, demonstrating that the *trp-4* channel was required specifically to sense gentle nose touches. This response was seen after approximately 130 msec. Because ALM synapses onto CEP, Kindt *et al.*[56] tested whether activation of ALM could induce calcium transients in CEP. ALM was stimulated by applying the buzz stimulus to the anterior portion of the worm's body within ALM's receptive field (gentle body touch). Calcium transients indicate a delayed CEP activation after ALM stimulation (approximately 200 msec after stimulation), supporting the idea that the activation was indirect. To ensure that the observed CEP activation was a result of signaling from ALM and not (unintended) direct activation CEP from the buzz stimulus, mutants that are insensitive to body touch were tested for activation of CEP. Indeed, CEP remained inactive in mutants that are insensitive to body touch. These data strongly support the idea that the bacterial lawn is sensed through the *trp-4*-expressing CEP neurons and that ALM participates in a positive feedback loop with the dopaminergic CEP neurons to further modulate tap habituation (Figure 9.7D).

During their analysis of dopamine-mediated sensory plasticity, Kindt *et al.*[56] found another modulator of habituation: Ca^{2+}. They used two variants of a genetically encoded calcium reporter: YC2.12 and YC3.12. The former was shown to bind more tightly to calcium, which increases its signal but results in the sequestration of calcium, and when expressed in ALM, the worms habituated more rapidly than wild type.

Conversely, worms carrying YC3.12, which binds more loosely to Ca^{2+}, did not show this defect. To further investigate this phenomenon, two mutants defective in Ca^{2+} homeostasis and release from the ER (*crt-1* and *itr-1*, respectively) were tested in a habituation assay and, like *dop-1*, were found to habituate rapidly. Calcium transients of ALM were measured in *crt-1* and *itr-1* mutants and demonstrated a rapid attenuation of calcium responses compared to control. These data further support the idea that PLC-β-dependent calcium release is involved in modulating habituation.[56]

Another study uncovered a possible link between the decrement in calcium current in the touch cells and the behavioral decrement in response to repeated stimuli.[6] Cai *et al.* identified a potassium channel complex consisting of a pore-forming subunit (*kth-1*) and an accessory subunit (*mps-1*). *mps-1*, expressed in ALM and PLM, encodes a single-pass transmembrane protein that phylogenically belongs to the KCNE family of K^+ channel modulating proteins. *mps-1* phosphorylates *kth-1*, and this phosphorylation results in reduce K^+ conductance. Loss of *mps-1* or *mps-1* kinase activity (using a constitutively inactive *mps-1* mutant) resulted in slow habituation with rapid recovery.[60]

Kindt *et al.*[56] found that repeated stimulation of ALM resulted in decreased calcium current. Similar results were found whereby repeated stimulation resulted in decreased Ca^{2+} influx mediated by the voltage-gated calcium channel *egl-19*.[61] Based on these data, Cai *et al.*[60] proposed that with repeated stimulation, *mps-1* phosphorylates *kth-1* and attenuates K^+ currents. Because K^+ currents are required to repolarize neurons after excitation, reduction in these currents would induce the cumulative inactivation of *egl-19*. The final mechanism put forward is that repeated stimulation results in the autophosphorylation of the *mps-1–kth-1* channel, thereby reducing K^+ conductance, delaying repolarization, and inhibiting Ca^{2+} conductance through the *egl-19* channel. This ultimately results in reduced cell excitability and decreased behavioral response.[60]

Although some of the mechanisms for habituation are starting to be uncovered, there is an ongoing search for genes involved in habituation. Xu *et al.*[62] performed the first forward screen for habituation mutants. In their experimental paradigm, the researchers screened approximately 30,000 mutagenized worms for strains with habituation deficits. From the screen, Xu *et al.* identified three novel genes involved in habituation, which were subsequently named *hab-1*, *hab-2*, and *hab-3*. Unfortunately, these genes were not mapped or sequenced, and so their function is unknown.

Until recently, only a handful of habituation mutants had been studied. This was largely because of the time-consuming nature of collecting behavioral data. In the original tap habituation experiments, reversals from individual worms were captured on videotape. The worm movements were traced onto acetate sheets, which were then scanned into a computer for analysis.[8] Manual data collection and analysis resulted in a bottleneck for behavioral studies.

Recently, behavioral studies have been greatly facilitated by the development of worm trackers.[13–18] There are two general types of worm trackers—trackers that follow and measure the behavior of individual worms and trackers that are capable of measuring behavioral data from a population of worms.

Single worm trackers are adept at detecting precise locomotion behavior; however, one limitation is that it still takes a considerable amount of time to get a large number of subjects. Population trackers can be used to evaluate data sets with large sample sizes and can measure probabilistic behavior. Data capturing and analysis typically occurs in two phases. First, behavioral data are recorded on video; next, video analysis software is used to analyze the behavior. Not surprisingly, storing large volumes of high-quality video on a computer can be impractical, making it a limiting factor in data collection.[19] A newly developed system called the MWT gets around this problem. The MWT comprises a camera, a platform on which *C. elegans* plates are mounted, and a push-solenoid positioned next to the petri dish. To deliver a tap stimulus automatically to the plate, a current is applied to the solenoid in predetermined intervals. Video recordings are not made; however, a skeletonized cartoon of each animal is generated, from which multiple morphological and behavioral properties are collected. Once collected, the behavioral properties of the worm can be extracted and compiled into meaningful data.

In a typical MWT experiment, a plate of age-synchronized worms is mounted on the MWT platform. Naive behavior is recorded for 10 min, after which the solenoid automatically delivers the tap stimulus for the duration of the experiment. The MWT quantifies a plethora of behaviors, including basal speed; the rate and size of spontaneous reversals; reversal probability; and distance, duration, and speed after a tap.[19] Currently, the MWT reports 23 different behaviors; however, many more are possible through the creation of appropriate software plug-in modules. In a preliminary screen of 30 candidate genes, 2 genes not previously associated with habituation were shown to have strong habituation phenotypes: *tom-1*, a gene that encodes a synaptic vesicle protein, tomosyn, was shown to have very rapid response probability habituation, and *adp-1*, an unmapped gene originally identified in a screen for worms deficient in chemosensory adaptation, showed very slow response probability habituation.[19] These mutations support the notion of a major role for synaptic release in habituation of response probability (*tom-1* and *eat-4* have very similar phenotypes) and re-open the question of the relationship between chemosensory adaptation and habituation.

Using the MWT, massive data sets of behavioral data can be collected relatively quickly. Indeed, a behavior screen[63] resulted in the characterization of 504 *C. elegans* strains with mutations in nervous system genes. The analysis of these 504 worm strains resulted in the identification of hundreds of mutations that affected one or more aspect of habituation.

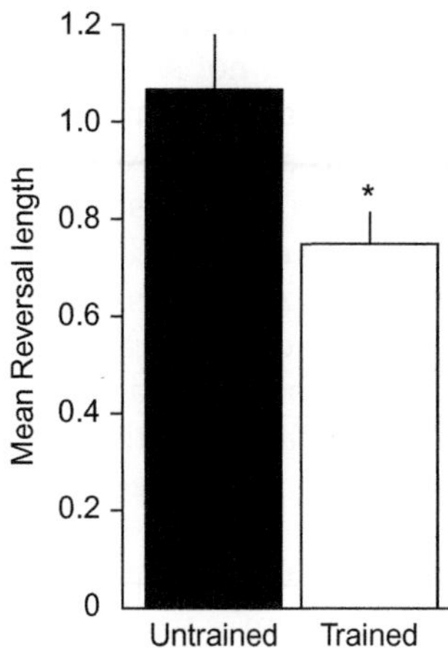

FIGURE 9.8 **Typical responses in nematodes that have undergone long-term habituation training (white) 24 hr prior to testing compared to those that had no training (black).** The average responses to tap (in millimeters) are significantly smaller in *C. elegans* that have undergone long-term memory LTM training compared to naive worms. $*p < 0.05$.

LONG-TERM MEMORY FOR TAP HABITUATION

Caenorhabditis elegans are not only capable of short-term memory of habituation but also exhibit long-term persistence of habituation, termed long-term habituation.[8,55,64] This long-term memory for habituation shows similarities to forms of long-term memory in humans, rodents, *Aplysia*, and *Drosophila*. Specifically, in all these systems, some types of long-term memory have been shown to require spaced or distributed training in which repeated brief bouts of training are followed by rest periods (hours to days). This spaced procedure was first described in 1885 by Ebbinghaus[65] as he learned lists of nonsense syllables. He found that training spaced out over time produced longer-lasting memories than did massed training in which all training was done in a single large block, even if the total time of training did not differ. Since that time, spaced training has been used for long-term memory for long-term habituation and sensitization in *Aplysia*[66] and for classical conditioning in *Drosophila*.[67]

In 1997, Beck and Rankin[68] first tested massed versus spaced training in *C. elegans* in an attempt to produce long-term habituation. They found that as in other systems, spaced training produced memory that could be demonstrated 24 hr after training, whereas massed training did not (Figure 9.8). In addition, parametric analyses of the long-term habituation assay revealed that long-term memory only occurred when tap stimuli within a block were delivered at longer ISIs; that is, training at 10-sec ISI did not result in the retention of habituation, but training at 60-sec ISI did. Based on this research, the paradigm used to generate long-term memory of habituation in *C. elegans* consists of the presentation of three to five training blocks in which each individual block consists of 20 tap stimuli at a 60-sec ISI with 1 hr in between training blocks.

The differences in efficacy in memory formation between animals trained at 10- or 60-sec ISIs support the hypothesis that the mechanisms mediating plasticity at short ISIs are either different or only partially overlap, with long-term memory resulting only from longer stimulation intervals. One possibility is that short ISIs (high-frequency stimulation) might cause the activation of inhibitory signals that prevent the formation of long-lasting synaptic changes. Conversely, longer intervals may cause an accumulation of positive signaling factors (e.g., calcium or calmodulin) that are important for long-term synaptic changes.

A standard definition of long-term memory is that it requires the synthesis of new proteins.[69] This requirement was tested for the formation of long-term habituation in *C. elegans*. Unfortunately, drugs that inhibit transcription or translation that are often used in memory studies in other organisms are quite toxic to *C. elegans* if given for the 6- or 7-hr training period and affect the worm's response to tap regardless of training. Therefore, another means had to be found to test this. A well-known disruptor of ongoing protein synthesis is heat shock.[70,64] When cells experience heat shock, they stop making other proteins and focus on making heat shock proteins to protect themselves from denaturation. Beck and Rankin[70] found that inhibition of protein synthesis by heat shock (32°C for 45 min) during the rest periods of training prevented the formation of long-term habituation. Indeed, heat-shocking worms for only 15 min either immediately or 15 min after training abolished the formation of long-term memory of habituation. In contrast, 15 min of heat shock 30 min after each training block did not significantly disrupt memory, suggesting that the first 30 min after training is a critical period of memory consolidation.[70]

Another characteristic of long-term memory in many organisms is that it requires activation of the transcription factor, cAMP response element binding protein (CREB). A number of behavioral studies in vertebrate[71–74] and invertebrate[75,76] organisms have demonstrated the critical role of CREB in forming new memories. The homolog of CREB in *C. elegans* is called *crh-1*. *crh-1* and mammalian CREB share 80% identity with the kinase-inducible domain and 95% identity with the leucine zipper domain (which is required for CREB dimerization and DNA binding).[77] Behavioral studies show that *crh-1* is not required for short-term habituation at either 60- or 10-sec ISI; in contrast, animals with mutations in *crh-1* fail to show long-term memory of habituation. These results demonstrate that the *C. elegans* homolog of CREB is required specifically for long-term habituation.[78] The locus of activity of *crh-1* was investigated by a series of cell-specific rescue experiments. In these experiments, functional *crh-1* was reintroduced into *crh-1*-deficient worms in subsets of neurons; the results showed that *crh-1* expression was required in AVA, AVD, or RIM (or combinations thereof) and not the sensory neurons to form long-term memory.[78]

Rose *et al.*[55] first reported that glutamate signaling was required for long-term memory for habituation by showing that *eat-4* (vesicular glutamate transporter) mutants were not able to form long-term memory of habituation to tap. Because presynaptic glutamate release is important for the formation of long-term memory, the next step was to identify the postsynaptic glutamate receptor required for its formation. In earlier research, *glr-1* had been identified as encoding a *C. elegans* ionotropic glutamate receptor subunit with high similarity to mammalian AMPA-type glutamate receptors.[43,44] To test the role of *glr-1* in habituation, *glr-1* worms were given both short-term and long-term habituation training. Although the initial responses to tap for *glr-1* mutant worms were typically smaller (as measured by reversal distance), *glr-1* worms showed no deficits in short-term habituation. In contrast, *glr-1* mutants showed no long-term memory of habituation.

Because *glr-1* was found to be essential for the formation of long-term memory, Rose *et al.*[55] next examined whether habituation training altered *glr-1* expression levels. To do this, *glr-1* expression was visualized using a GLR-1::GFP translational reporter (green fluorescent protein is a genetically encoded fluorophore that can be fused to a gene expression to investigate levels of gene expression). Using the GLR-1::GFP construct, Rongo and Kaplan[79] determined that *glr-1* is expressed in clusters along the ventral nerve cord and in the nerve ring. After imaging GLR-1::GFP in the posterior ventral nerve cord, Rose *et al.* found that although the number of clusters of fluorescence did not differ between trained and control worms, the length of the clusters was significantly shorter for worms given spaced training 24 hr before imaging compared to control, untrained worms. This translated to an overall reduction of GLR-1::GFP expression.[55] To determine whether the decreased GLR-1::GFP expression was related to the formation of long-term memory, Rose *et al.* measured expression levels in animals in which long-term memory of habituation was blocked by heat shock. If the formation of long-term memory coincides with a reduction in GLR-1::GFP expression in the ventral nerve cord, then blocking long-term memory should also block the change in *glr-1* expression. Indeed, expression levels in control animals compared to those of animals that had undergone habituation training with heat shock were not different from each other.[55]

Another feature of long-term memory is that it can become labile after it has been recalled.[80] That is, recalling a memory makes it vulnerable to disruption. A recalled memory may need to be reformed, or reconsolidated, in order to be maintained after recall. A series of experiments investigated whether reconsolidation also occurs in *C. elegans*.[81] To test this, first the extent of long-term memory for habituation was tested. Animals were trained with four blocks of 20 taps, and memory was tested 48 hr after training. Trained worms had significantly smaller responses in the testing phase compared to control animals, demonstrating that they could remember for 48 hr. Next, reconsolidation was examined by delivering reminder taps 24 hr after training and testing for memory at 48 hr. To determine whether memory was labile upon reactivation, the researchers applied heat shock after the reminder taps to determine if it would disrupt memory of habituation. Indeed, when animals received heat shock after the reminder taps, they did not show memory during the testing phase 24 hr later. Animals that did not receive heat shock or that received heat shock without the reminder taps showed normal memory at 48 hr. The same effect was seen when the glutamate receptor antagonist DNQX was applied during the reminder taps. This demonstrated that if a memory is reactivated in worms, it will undergo reconsolidation so that it is re-encoded.[81]

Next, Rose *et al.*[55] investigated the effect of reconsolidation on GLR-1::GFP. Recall that GLR-1::GFP expression decreases in animals that show long-term memory of habituation.[55] Researchers compared GLR-1::GFP expression levels in worms that received only reminder taps with those in worms that had undergone heat shock after the reminder taps.[81] The results showed that GLR-1::GFP expression was significantly decreased in animals that only had the reminder taps, whereas in animals that received reminder plus heat shock to produce reconsolidation blockade, GLR-1::GFP levels were similar to those of untrained animals. Together, these results demonstrated that memory in worms requires reconsolidation upon the reactivation of a memory. Reconsolidation blockade with heat shock or DNQX abolished memory of habituation. Similarly, reconsolidation blockade resulted in GLR-1::GFP expression levels that reverted back to levels seen in untrained worms.

CONTEXT: SHORT-TERM AND LONG-TERM MEMORY

Habituation was traditionally classified as a non-associative form of learning because it was assumed that only modifications to the stimulus could alter habituation kinetics. However, investigations into habituation in other organisms demonstrated a role for an associative effect of context (background cues present at both training and testing) in the retention of habituation training.[82–84] Rankin[85] investigated the role of context conditioning of short-term habituation in *C. elegans* by habituating animals with training and testing in either the same or different contexts.

Caenorhabditis elegans are capable of chemotaxis, defined as directional movement toward or away from chemical cues. Worms can sense both volatile odors and the taste of dissolved compounds and are attracted to four classes of molecules: cations Na^+, Li, K^+, and Mg^+.[86] An ideal contextual cue would be one that the worm found salient but that did not alter basal locomotion or habituation kinetics on its own. It was previously determined that 0.4 M sodium acetate acted as an attractive cue for *C. elegans*.[27] Before using it as a context cue, it was first necessary to examine habituation in the presence and absence of this chemical cue to establish that the cue itself did not alter habituation kinetics. To expose animals to the sodium acetate context, petri dishes were washed with sodium acetate and allowed to dry for 1 hr. Next, habituation training was given on either the context plate or a plain agar plate. There was no significant difference in the initial response magnitude for worms on the sodium acetate plate compared to those on the plain plate, nor was there a difference between habituated levels for the two groups. This demonstrated that sodium acetate did not arouse or sedate worms or change habituation.[85]

Next, the effect of context on retention of short-term habituation was tested to determine whether habituation in a context would alter future habituation performance. Worms underwent habituation training in different environmental conditions (either sodium acetate-treated agar or plain agar), were moved to a plain agar rest plate for 1 hr, and then were tested for retention of habituation on either sodium acetate-treated agar or plain agar. The average magnitude of the first two responses was compared between the training and testing blocks. Animals were divided into four training groups: a group that was trained on plain plates and tested on plain plates (the PlPl group), a group trained on plain plates and tested on sodium acetate plates (the PlNa group), a group trained on sodium acetate plates and tested on plain plates (the NaPl group), and a group that was trained on sodium acetate plates and tested on sodium acetate plates (NaNa). Both 10- and 60-sec ISIs were used to determine whether context conditioning was ISI dependent.

The results showed that for both 10- and 60-sec ISI groups, context did enhance the retention of the original training, but only if the context was the same in both conditions. The initial response during the testing of habituation was significantly lower in the NaNa group compared to any other group. This decrement was not present in any other group tested (Figure 9.9).[85]

During associative learning, an animal learns that the presence of one stimulus (the conditioned

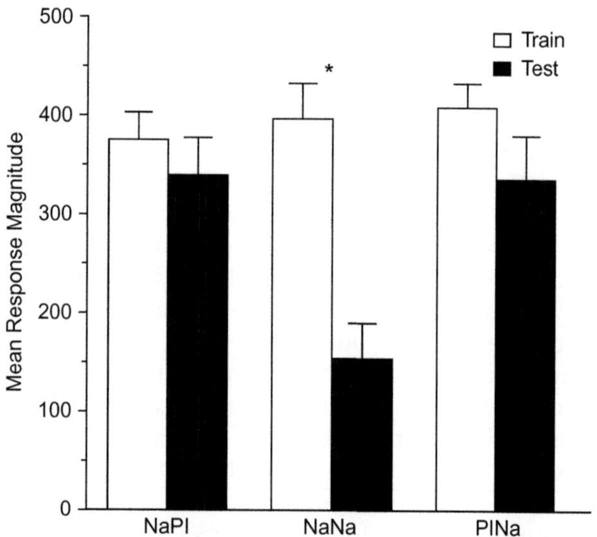

FIGURE 9.9 Animals trained and tested in the same chemosensory context retain short-term memory of habituation. Worms were trained in a sodium acetate (Na) environment and tested in a plain agar environment (Pl) (denoted NaPl), trained and tested in a sodium acetate environment (NaNa), or trained in a plain agar environment and tested in a sodium acetate environment (PlNa). Worms trained and tested in the same environment exhibited significantly smaller responses in the first two taps of the testing block compared to the training block; other groups showed no significant differences.[85] $*p < 0.05$

stimulus) predicts the presence of another stimulus (the unconditioned stimulus).[87] In this experiment, sodium acetate served as the conditioned stimulus, whereas the taps during habituation training acted as the unconditioned stimulus. Extinction occurs when trained animals are exposed to the conditioned stimulus without the unconditioned stimulus (reviewed in Myers and Davis[88]). In this scenario, the conditioned stimulus no longer predicts the unconditioned stimulus, and the responses reset to the time before associative learning occurred. To support the notion that the decrement in response between the training and testing block was the result of context classical conditioning, Rankin[85] tested whether the effect could be extinguished. In this behavioral test, trained animals rested on a plate in the presence of sodium acetate between training and testing. If the decrement in response between training and testing taps is the result of associative learning, spending time in the sodium acetate environment without the mechanical stimuli should cause extinction of the context effect. As predicted, exposure to sodium acetate during the rest period between training and testing completely abolished the effect of context on habituation.[85]

Another feature of associative learning is that for it to occur, the context should be novel. Exposure of a contextual cue before the presentation of the unconditioned stimulus is known to either reduce or eliminate context conditioning. This effect is called latent inhibition.[89] To test whether exposure to the contextual cue before habituation training would abolish the decrement in response in the test block, animals were exposed to sodium acetate 1 hr before habituation training, trained in the presence of sodium acetate, transferred to plain plates for the rest period, and tested in the presence of sodium acetate. As expected, latent inhibition training abolished the formation of associative learning between acetate and tap stimuli.[85]

In this first set of experiments, no context conditioning was seen in the group trained and tested on plain agar. The hypothesis was that this group was experiencing both latent inhibition and extinction when they were placed on a plain plate 1 hr prior to training and moved to a fresh plain plate for 1 hr between training and testing. To test whether plain agar could serve as a context, worms were placed on sodium acetate-treated plates 1 hr before training and then transferred to plain plates for training. Worms were then placed on sodium acetate plates during a 1-hr rest period and transferred to plain plates again for testing. Similar to context conditioning using sodium acetate, worms could form associations with plain plates and tap stimulus as long as they did not undergo latent inhibition or extinction training. The results of these elegant experiments clearly show that tap habituation, a phenomenon that was once considered purely nonassociative learning in *C. elegans*, is sensitive to and can be modulated by context cues.[27]

After establishing that context cues enhanced short-term habituation, the next problems to tackle were to identify the genes involved in context conditioning and to determine whether context conditioning could facilitate long-term memory of habituation. Because context conditioning using taste cues requires the researcher to transfer worms to different plates multiple times (and thus subject the worms to extra stimulation), the protocol established in Rankin and Wicks[27] was adapted to use a volatile odorant. In this system, a droplet of the volatile odorant, diacetyl, is placed on the inside of the lid of a petri dish. This allows the smell of diacetyl to permeate the microenvironment. To remove the olfactory cue, one then replaces the lid rather than transferring the worms. To study the effect of olfactory cues on habituation, animals were habituated with 30 taps at a 10-sec ISI. Next, the plate lid was replaced with a fresh lid to remove the smell of diacetyl and animals were given a 1-hr rest period. Short-term retention of habituation training was tested by replacing the lid with either a fresh diacetyl lid or one that did not contain the odorant. Next, a testing round was performed in which the responses of the first two taps were measured. Under these training conditions, the average of the first two responses was significantly smaller in animals trained and tested in the same context compared to the group of animals tested without the context cue. Taken together, these results show that retention of nonassociative learning is enhanced when using either a taste or a smell as a contextual cue.

A similar paradigm was used to test whether context cues could enhance long-term memory of habituation. In this experiment, animals were trained by giving four blocks of 20 mechanosensory stimuli at a 60-sec ISI. Animals trained and tested using the 60-sec ISI with a context cue present during both training and testing did not show significant enhancement of memory over and above the normal long-term memory produced by this procedure. Surprisingly, worms trained with four blocks of 20 taps at a 10-sec ISI with a context cue could form long-term memory of habituation. Recall that without contextual cues, long-term memory of habituation will not form using a 10-sec ISI.

To further investigate the observation that context cues enhance long-term memory of habituation, mutations in two genes known to be required for long-term memory of habituation, *crh-1* and *glr-1*, were tested for their ability to form context memory. Mutations in either gene caused deficits in context conditioning of long-term memory for habituation and showed that long-term memory for context is dependent on both glutamate neurotransmission and CREB transcription.

Because a context cue led to the formation of a long-term memory under conditions not normally conducive to memory formation, effort was next focused on finding the mechanisms that governed this effect. NMDA-type ionotropic glutamate receptor subunits *nmr-1* and *nmr-2* are known to be involved in salt chemotaxis classical conditioning in *C. elegans*. In salt chemotaxis classical conditioning, animals are incubated on plates with a high salt concentration and that lack food. The normally attractive stimulus, NaCl, becomes associated with starvation and nematodes learn to avoid salt. Kano *et al.*[90] demonstrated through genetic studies that *nmr-1* and *nmr-2* expressed in RIM mediated the learned association between NaCl and starvation. Given the role of *nmr-1* and *nmr-2* in salt chemotaxis, Lau *et al.*[91] tested *nmr-1* mutants to determine whether NMDA-type ionotropic glutamate receptors played a role in context conditioning of tap responses. First, *nmr-1* mutants were tested for short-term habituation defects without contextual cues, and although the initial response was significantly lower compared to wild-type, the rate of habituation and the habituated level were not significantly different. Next, short-term context conditioning in *nmr-1* mutants was tested. *nmr-1* mutants were exposed to diacetyl during short-term habituation training, given a 1-hr rest period without diacetyl, and then tested in the presence of diacetyl. *nmr-1* mutants did not show enhanced retention of short-term habitation. *nmr-1* worms also showed normal long-term memory for habituation when trained at a 6-sec ISI, but they did not show context-dependent long-term memory for habituation when trained at a 10-sec ISI. Thus, *nmr-1* is required in context conditioning to enhance both short- and long-term memory.

Next, *nmr-1* was reintroduced into specific cells in an *nmr-1* mutant background. *nmr-1* is normally expressed in the command interneurons AVA, AVD, and AVE, along with the interneuron RIM. Rescue with the endogenous promoter or cell-specific rescue of *nmr-1* in RIM both rescued the short- and the long-term context-dependent memory of habitation. This indicated that *nmr-1* expression in RIM is critical for retention of habituation in both short-term and long-term training paradigms.[90] RIM receives input from the salt-sensing sensory neurons including AWA, a sensory neuron required for detecting attractive olfactory cues such as diacetyl. RIM forms electrical connections to AVA and AVE and chemical synapses onto AVA and AVB. Because RIM activity inhibits reversal behavior, *nmr-1* may act in RIM to diminish the magnitude of tap responses by suppressing AVA activity or modulating the AIB–RIM disinhibitory circuit.

In summary, tap habituation is subject to the effects of context, and as such, context conditioning can modulate a worm's performance in a subsequent habituation assay. These experiments convincingly demonstrated that context-dependent changes in habituation were due to associative learning by showing that two phenomenon unique to associative learning—extinction and latent inhibition—could occur in this paradigm. These results were extended by the identification of *nmr-1* as a critical gene required for context-dependent changes in habituation and the finding that *nmr-1* acts in RIM to modulate this phenomenon. The results also showed for the first time that nematodes are capable of forming long-term memory of habituation at a 10-sec ISI if they are trained and tested in the presence of salient contextual cues.

CONCLUSIONS

Although *C. elegans* was once thought of as a featureless tube that had little potential for behavioral research, it has proven to be an invaluable tool in the study of learning and memory. *Caenorhabditis elegans* are adept at learning about the different stimuli they experience in their environment and modifying their behavior accordingly. Behavioral studies have shown that *C. elegans* form short-term, intermediate-term, and long-term memories of their previous experience. They also attend to a novel context and show better retention of that stimulus if they encounter it again in the original context. Each type of memory is formed by a diverse set of molecular mechanisms that act at the cellular and synapse level to change cell excitability and synaptic strength between neurons. Considering that the *C. elegans* hermaphrodite has only 302 neurons, the mechanisms regulating the activity of the neurons must be hugely complex in order to be capable of encoding each type of memory.

Caenorhabditis elegans is also a relevant model for how we learn and remember; the process in which a worm encodes memory is not especially different at the molecular level from how other organisms encode memory. For example, *C. elegans* and mammals encode long-term memories through the action of CREB, a transcription factor found in both organisms, and memory is often mediated by changes in the expression levels of AMPA-type glutamate receptors.[92] With this in mind, the next major challenge in studies of habituation is to identify new molecular players in habituation in *C elegans*, fit them into gene pathways, and then test those pathways in mammalian learning and memory. Along this vein, automated worm trackers have been developed in order to facilitate behavioral studies. Using one type of tracker, we have identified more than 100 mutants (out of 522 studied) defective in habituation.[63] More high-throughput studies will undoubtedly add to this list. With the use of automated studies in *C. elegans*, researchers will be able to gain important clues to help understand the mysteries of learning and memory.

References

1. Wood B. *Introduction to* C. elegans *Biology*. Cold Spring Harbor, NY: Cold Spring Harbor Laboratory Press; 1988.
2. Sulston J, Schierenberg E, White J, Thompson J. The embryonic cell lineage of the nematode *Caenorhabditis elegans*. *Dev Biol*. 1983;1000:64–119.
3. Sulston J, Horvits H. Post-embryonic cell lineages of the nematode *Caenorhabditis elegans*. *Dev Biol*. 1977;82:110–156.
4. White J, Southgate E, Brenner S. The structure of the nervous system of the nematode *C. elegans*. *Phil Trans R Soc Lond B*. 1986;:314.
5. Riddle D, Blumenthal T, Meyer B, Priess J, eds. *C. elegans II*. Cold Spring Harbor, NY: Cold Spring Harbor Laboratory Press; 1997.
6. Brenner S. The genetics of *Caenorhabditis elegans*. *Genetics*. 1974;77:71–94.
7. Hedgecock EM, Russell RL. Normal and mutant thermotaxis in the nematode *Caenorhabditis elegans*. *Proc Natl Acad Sci USA*. 1975;72:4061–4065.
8. Rankin C, Chiba C, Beck C. *Caenorhabditis elegans*: a new model system for the study of learning and memory. *Learn Memory*. 1990;37:89–92.
9. Consortium CES. Genome sequence of the nematode *C. elegans*: a platform for investigating biology. *Science*. 1998;282:2012–2020.
10. Bargmann C, Avery L. Laser killing of cells in *Caenorhabditis elegans*. *Methods Cell Biol*. 1995;48:225–250.
11. Nagel G, Brauner M, Liewald J, Adeishvili N, Bamberg E, Gottschalk A. Light activation of channelrhodpsin-2 excitable cells of *Caenorhabditis elegans* triggers rabid behavioural response. *Curr Biol*. 2005;15:2279–2284.
12. Kerr R. Intracellular Ca^{2+} imaging in *C. elegans*. *Methods Cell Biol*. 2006;351:253–264.
13. Fontaine E, Burdick J, Barr A. Automated tracking of multiple *C. elegans*. *Conf Proc IEEE Eng Med Biol Soc*. 2006;1:3716–3719.
14. Geng W, Cosman P, Berry CC, Feng Z, Schafer WR. Automatic tracking, feature extraction and classification of *C. elegans* phenotypes. *IEEE Trans Biomed Eng*. 2004;51:1811–1820.
15. Huang KM, Cosman P, Schafer WR. Automated tracking of multiple *C. elegans* with articulated models. *Proc Biomed Imaging*. 2007;:1240–1243.
16. Powell JR, Kim DH, Ausubel FM. The G protein-coupled receptor FSHR-1 is required for the *Caenorhabditis elegans* innate immune response. *Proc Natl Acad Sci USA*. 2009;106: 2782–2787.
17. Ramot D, Johnson BE, Berry TL, Carnell L, Goodman MB. The parallel worm tracker: a platform for measuring average speed and drug-induced paralysis in nematodes. *PLoS ONE*. 2008;3: e2208.
18. Simonetta SH, Golombek DA. An automated tracking system for *Caenorhabditis elegans* locomotor behavior and circadian studies application. *J Neurosci Methods*. 2007;161:273–280.
19. Swierczek NA, Giles AC, Rankin CH, Kerr RA. High-throughput behavioral analysis in *C. elegans*. *Nat Methods*. 2011;8:592–598.
20. Ben Arous J, Tanizawa Y, Rabinowitch I, Chatenay D, Schafer WR. Automated imaging of neuronal activity in freely behaving *Caenorhabditis elegans*. *J Neurosci Methods*. 2010;187:229–234.
21. Clark DA, Gabel CV, Gabel H, Samuel ADT. Temporal activity patterns in thermosensory neurons of freely moving *Caenorhabditis elegans* encode spatial thermal gradients. *J Neurosci*. 2007;27:6083–6090.
22. Kawano T, Po MD, Gao S, Leung G, Ryu WS, Zhen M. An imbalancing act: gap junctions reduce the backward motor circuit activity to bias *C. elegans* for forward locomotion. *Neuron*. 2011;72:572–586.
23. Chalfie M, Sulston JE, White JG, Southgate E, Thomson JN, Brenner S. The neural circuit for touch sensitivity in *Caenorhabditis elegans*. *J Neurosci*. 1985;5:956–964.
24. Thompson RF, Spencer WA. Habituation: a model phenomenon for the study of neuronal substrates of behavior. *Psychol Rev*. 1966;73:16.
25. Rankin CH, Broster BS. Factors affecting habituation and recovery from habituation in the nematode *Caenorhabditis elegans*. *Behav Neurosci*. 1992;106:239.

25b. Broster BS, Rankin CH. Effects of changing interstimulus interval during habituation in *Caenorhabditis elegans*. *Behav Neurosci*. 1994;108:1019–1029.

26. Piggott BJ, Liu J, Feng Z, Wescott SA, Xu X. The neural circuits and synaptic mechanisms underlying motor initiation in *C. elegans*. *Cell*. 2011;147:922–933.
27. Rankin C, Wicks S. Mutations of the *Caenorhabditis elegans* brain-specific inorganic phosphate transporter eat-4 affect habituation of the tap-withdrawal response without affecting the response itself. *J Neurosci*. 2000;20:4337–4344.
28. Chiba CM, Rankin CH. A developmental analysis of spontaneous and reflexive reversals in the nematode *Caenorhabditis elegans*. *J Neurobiol*. 1990;21:543–554.
29. Beck C, Rankin C. Effects of aging on habituation in the nematode *Caenorhabditis elegans*. *Behav Processes*. 1993;28:145–163.
30. Rankin CH, Gannon T, Wicks SR. Developmental analysis of habituation in the nematode *C. elegans*. *Dev Psychobiol*. 2000;36: 261–270.
31. Timbers TA, Giles AC, Ardiel EL, Kerr RA, Rankin CH. Intensity discrimination deficits cause habituation changes in middle-aged Caenorhabditis elegans. *Neurobiol Aging*. 2013;34 (2):621–631.
32. Groves PM, Thompson RF. Habituation: a dual-process theory. *Psychol Rev*. 1970;77:419.
33. Herndon L, Schmeissner P, Dudaronek J, et al. Stochastic and genetic factors influence tissue-specific decline in ageing *C. elegans*. *Nature*. 2002;419:808–814.
34. Wicks SR, Rankin CH. Integration of mechanosensory stimuli in *Caenorhabditis elegans*. *J Neurosci*. 1995;15:2434–2444.
35. Li W, Feng Z, Sternberg PW, Xu XZA. *C. elegans* stretch receptor neuron revealed by a mechanosensitive TRP channel homologue. *Nature*. 2006;440:684–687.
36. Wicks S, Rankin C. The integration of antagonistic reflexes revealed by laser ablation of identified neurons determines habituation kinetics of the *Caenorhabditis elegans* tap withdrawal response. *J Comp Physiol A*. 1996;179:675–685.
37. Wicks S, Roehrig C, Rankin C. A dynamic network simulation of the nematode tap withdrawal circuit: predictions concerning synaptic function using behavioral criteria. *J Neurosci*. 1996;16: 4017–4031.
38. Davis R, Stretton A. Passive membrane properties of motor neurons and their role in long-distance signaling in the nematode Ascaris. *J Neurosci*. 1989;9:403–414.
39. Sulston J, Dew M, Brenner S. Dopaminergic neurons in the nematode *Caenorhabditis elegans*. *J Comp Neurol*. 1975;163:215–226.

40. Horvitz HR, Chalfie M, Trent C, Sulston JE, Evans PD. Serotonin and octopamine in the nematode *Caenorhabditis elegans. Science.* 1982;216:1012–1014.
41. Loer CM, Kenyon CJ. Serotonin-deficient mutants and male mating behavior in the nematode *Caenorhabditis elegans. J Neurosci.* 1993;13:5407–5417.
42. McIntire SL, Jorgensen E, Horvitz HR. Genes required for GABA function in *Caenorhabditis elegans. Nature.* 1993;364:334–337.
43. Hart A, Sims S, Kaplan J. Synaptic code for sensory modalities revealed by *C. elegans* GLR-1 glutamate receptor. *Nature.* 1995;378:82–85.
44. Maricq AV, Peckol E, Driscoll M, Bargmann CI. Mechanosensory signalling in *C. elegans* mediated by the GLR-1 glutamate receptor. *Nature.* 1995;378:78–81.
45. Brockie PJ, Mellem JE, Hills T, Madsen DM, Maricq AV. The *C. elegans* glutamate receptor subunit NMR-1 is required for slow NMDA-activated currents that regulate reversal frequency during locomotion. *Neuron.* 2001;31:617–630.
46. Lee RYN, Sawin ER, Chalfie M, Horvitz HR, Avery L. EAT-4, a homolog of a mammalian sodium-dependent inorganic phosphate cotransporter, is necessary for glutamatergic neurotransmission in Caenorhabditis elegans. *J Neurosci.* 1999;19:159–167.
47. Wicks SR, Rankin CH. Effects of tap withdrawal response habituation on other withdrawal behaviors: the localization of habituation in the nematode *Caenorhabditis elegans. Behav Neurosci.* 1997;111:342.
48. Kitamura K, Amano S, Hosono R. Contribution of neurons to habituation to mechanical stimulation in *Caenorhabditis elegans. J Neurobiol.* 2001;46:29–40.
49. Bellocchio EE, Reimer RJ, Fremeau Jr RT, Edwards RH. Uptake of glutamate into synaptic vesicles by an inorganic phosphate transporter. *Science.* 2000;289:957–960.
50. Rankin CH, Abrams T, Barry RJ, et al. Habituation revisited: an updated and revised description of the behavioral characteristics of habituation. *Neurobiol Learn Memory.* 2009;92:135–138.
51. Sanyal S, Wintle RF, Kindt KS, et al. Dopamine modulates the plasticity of mechanosensory responses in *Caenorhabditis elegans. EMBO J.* 2004;23:473–482.
52. Berridge KC. Reward learning: reinforcement, incentives, and expectations. *Psychol Learn Motivation.* 2000;40:223–278.
53. Bargmann C. Chemosensation in *C. elegans.* In: WormBook: The Online Review of *C. elegans* Biology, Online; 2006.
54. Chase D, Pepper J, Koelle M. Mechanism of extrasynaptic dopamine signaling in *Caenorhabditis elegans. Nat Neurosci.* 2004;7: 1096–1123.
55. Rose J, Kaun K, Chen S, Rankin C. GLR-1, a non-NMDA glutamate receptor homolog, is critical for long-term memory in *Caenorhabditis elegans. J Neurosci.* 2003;23:9595–9599.
56. Kindt KS, Quast KB, Giles AC, et al. Dopamine mediates context-dependent modulation of sensory plasticity in *C. elegans. Neuron.* 2007;55:662–676.
57. Grima B, Lamouroux A, Blanot F, Biguet NF, Mallet J. Complete coding sequence of rat tyrosine hydroxylase mRNA. *Proc Nat Acad Sci USA.* 1985;82:617.
58. Neckameyer WS, Quinn WG. Isolation and characterization of the gene for *Drosophila* tyrosine hydroxylase. *Neuron.* 1989;2: 1167–1175.
59. Sawin ER, Ranganathan R, Horvitz HR. *C. elegans* locomotory rate is modulated by the environment through a dopaminergic pathway and by experience through a serotonergic pathway. *Neuron.* 2000;26:619–631.
60. Cai SQ, Hernandez L, Wang Y, Park KH, Sesti F. MPS-1 is a K^+ channel β-subunit and a serine/threonine kinase. *Nat Neurosci.* 2005;8:1503–1509.
61. Suzuki H, Kerr R, Bianchi L, et al. *In vivo* imaging of *C. elegans* mechanosensory neurons demonstrates a specific role for the MEC-4 channel in the process of gentle touch sensation. *Neuron.* 2003;39:1005–1017.
62. Xu X, Sassa T, Kunoh K, Hosono R. A mutant exhibiting abnormal habituation behaviour in *Caenorhabditis elegans. J Neurogenetics.* 2002;16:29–44.
63. Giles AC. :Unpublished PhD Dissertation *Candidate Gene and High Throughput Genetic Analysis of Habituation in Caenorhabditis Elegans.* Vancouver, BC: University of British Columbia; 2012
64. Rose JK, Kaun KR, Rankin CH. A new group-training procedure for habituation demonstrates that presynaptic glutamate release contributes to long-term memory in *Caenorhabditis elegans. Learn Mem.* 2002;9:130–137.
65. Ebbinghaus H. *Memory: A Contribution to Experimental Psychology.* New York: Columbia University Teachers College; 1885.
66. Pinsker HM, Hening WA, Carew TJ, Kandel ER. Long-term sensitization of a defensive withdrawal reflex in *Aplysia. Science.* 1973;182:1039–1042.
67. Tully T, Quinn WG. Classical conditioning and retention in normal and mutant *Drosophila melanogaster. J Comp Physiol A Neuroethology Sensory Neural Behav Physiol.* 1985;157:263–277.
68. Beck CDO, Rankin CH. Long-term habituation is produced by distributed training at long ISIs and not by massed training or short ISIs in *Caenorhabditis elegans. Learn Behav.* 1997;25: 446–457.
69. Squire L. Protein synthesis and memory: a review. *Psychol Bull.* 1984;96:518–559.
70. Beck C, Rankin CH. Heat shock disrupts long-term memory consolidation in *Caenorhabditis elegans. Learn Mem.* 1995;2: 161–177.
71. Bernabeu R, Bevilaqua L, Ardenghi P, et al. Involvement of hippocampal cAMP/cAMP-dependent protein kinase signaling pathways in a late memory consolidation phase of aversively motivated learning in rats. *Proc Nat Acad Sci USA.* 1997;94: 7041.
72. Bourtchuladze R, Frenguelli B, Blendy J, Cioffi D, Schutz G, Silva AJ. Deficient long-term memory in mice with a targeted mutation of the cAMP-responsive element-binding protein. *Cell.* 1994;79:59.
73. Guzowski JF, McGaugh JL. Antisense oligodeoxynucleotide-mediated disruption of hippocampal cAMP response element binding protein levels impairs consolidation of memory for water maze training. *Proc Nat Acad Sci USA.* 1997;94:2693.
74. Josselyn S, Kida S, Silva A. Inducible repression of CREB function disrupts amygdala-dependent memory. *Neurobiol Learn Memory.* 2004;82:159–163.
75. Kaang BK, Kandel ER, Grant SCN. Activation of CAMP-responsive genes by stimuli that produce long-term facilitation in Aplysia sensory neurons. *Neuron.* 1993;10:427–435.
76. Yin J, Wallach J, Del Vecchio M, et al. Induction of a dominant negative CREB transgene specifically blocks long-term memory in *Drosophila. Cell.* 1994;79:49.
77. Kimura Y, Corcoran EE, Eto K, et al. A CaMK cascade activates CRE-mediated transcription in neurons of *Caenorhabditis elegans. EMBO Rep.* 2002;3:962–966.
78. Timbers TA, Rankin CH. Tap withdrawal circuit interneurons require CREB for long-term habituation in *Caenorhabditis elegans. Behav Neurosci.* 2011;125:560.
79. Rongo C, Kaplan JM. CaMKII regulates the density of central glutamatergic synapses *in vivo. Nature.* 1999;402:195–199.
80. Nader K, Schafe GE, Le Doux JE. Fear memories require protein synthesis in the amygdala for reconsolidation after retrieval. *Nature.* 2000;406:722–726.
81. Rose JK, Rankin CH. Blocking memory reconsolidation reverses memory-associated changes in glutamate receptor expression. *J Neurosci.* 2006;26:11582–11587.

82. Saraco MG, Maldonado H. Ethanol affects context memory and long-term habituation in the crab *Chasmagnathus*. *Pharmacol Biochem Behav*. 1995;51:223–229.
83. Tomsic D, Massoni V, Maldonado H. Habituation to a danger stimulus in two semiterrestrial crabs: ontogenic, ecological and opioid modulation correlates. *J Comp Physiol A Neuroethology Sensory Neural Behav Physiol*. 1993;173:621–633.
84. Tomsic D, Pedreira ME, Romano A, Hermitte G, Maldonado H. Context-us association as a determinant of long-term habituation in the crab *Chasmagnathus*. *Learn Behav*. 1998;26:196–209.
85. Rankin CH. Context conditioning in habituation in the nematode *Caenorhabditis elegans*. *Behav Neurosci*. 2000;114:496–505.
86. Ward S. Chemotaxis by the nematode *Caenorhabditis elegans*: identification of attractants and analysis of the response by use of mutants. *Proc Nat Acad Sci*. 1973;70:817.
87. Pavlov IP, Anrep GV. *Conditioned Reflexes*. Mineola, NY: Dover; 2003.
88. Myers K, Davis M. Mechanisms of fear extinction. *Mol Psychiatry*. 2006;12:120–150.
89. Lubow R, Moore A. Latent inhibition: the effect of nonreinforced conditioning of the rabbit pinna response. *J Comp Physiol Psychol*. 1959;68:415–419.
90. Kano T, Brockie PJ, Sassa T, et al. Memory in *Caenorhabditis elegans* is mediated by NMDA-type ionotropic glutamate receptors. *Curr Biol*. 2008;18:1010–1015.
91. Lau HL, Timbers TA, Mahmoud R, Rankin CH. Genetic dissection of memory for associative and non-associative learning in *Caenorhabditis elegans*. *Genes brains Behav*. 2013;12(2):210–223.
92. Malinow R, Malenka RC. AMPA receptor trafficking and synaptic plasticity. *Ann Rev Neuro sci*. 2002;25:103–126.

CHAPTER

10

Molecular and Cellular Circuits Underlying *Caenorhabditis elegans* Olfactory Plasticity

Joy Alcedo and Yun Zhang†*

*Friedrich Miescher Institute for Biomedical Research, Basel, Switzerland; Wayne State University, Detroit, Michigan
†Harvard University, Cambridge, Massachusetts

Dwelling in compost-rich soil, the bacteria-feeding nematode *Caenorhabditis elegans* uses its olfactory system as a major means to perceive the outside world and generate optimized behavioral responses. Consistent with this, approximately 5% of the *C. elegans* genome is devoted to putative G protein-coupled receptors that worms can use to detect chemical cues, many of which are odorants. Therefore, it is critical for the survival of the nematodes that they can modulate their olfactory sensorimotor responses to changing environments in an experience-dependent manner at any stage in their life span.

CAENORHABDITIS ELEGANS OLFACTORY SYSTEM

Caenorhabditis elegans can detect hundreds of odors, many of which are bacterially generated.[1–3] The perception of a few of these odors has been mapped to several pairs of sensory neurons, namely AWA, AWB, AWC, ADL, and ASH.[1–4] Together with another seven pairs of sensory neurons, these neurons are contained within the amphid, a sensory organ located within the head of the worm. The sensory cilia of AWA, AWB, and AWC are embedded within the amphid sheath cells, whereas the ciliated endings of ADL and ASH are directly exposed to the outside through the amphid socket[5,6] (Figure 10.1A). These anatomical features allow the perception of olfactory stimuli present within the animals' environment.

Like vertebrates, *C. elegans* uses seven transmembrane domain G protein-coupled receptors (GPCRs) to detect olfactory cues.[8] There are as many as 1000 GPCRs encoded in the *C. elegans* genome, many of which are expressed in the olfactory sensory neurons and localized to the ciliated endings and thus may be used to perceive odors.[9–13] However, unlike the mammalian olfactory system, in which each individual sensory neuron likely expresses one type of olfactory receptors,[14–16] the small number of *C. elegans* sensory neurons combined with a large number of putative odor receptors necessitate the expression of multiple types of olfactory receptors in at least some of the olfactory sensory neurons. Indeed, profiling the expression patterns of approximately 100 candidate odor receptors has shown the presence of many such receptors in individual *C. elegans* sensory neurons.[9–13]

The *C. elegans* genome contains 20 putative α subunits, 2 β subunits, and 2 γ subunits of the heterotrimeric G proteins.[17] Among the 20 G_α subunits, two G_i-like G_α proteins, ODR-3 and GPA-2, play a major role in transducing the signals generated by detecting cues sensed by the olfactory receptors (Figure 10.1B). These G proteins regulate different molecules in different neurons to process distinct sets of olfactory information. For example, in the AWB and AWC neurons, the ODR-3 G protein acts through the ODR-1 and DAF-11 guanylyl cyclases to control the closing or opening of the nucleotide-gated cation channels TAX-2/TAX-4 via cyclic GMP (cGMP; Figure 10.1B).[18,19] In the AWA and ASH neurons, ODR-3 regulates the TRPV-like cation channels OSM-9 and OCR-2, likely via generation of long-chain polyunsaturated fatty acids.[20]

Olfactory sensory signaling can be transmitted to downstream neurons, such as the interneurons AIB and AIY (Figure 10.2C), via synaptic neurotransmitters. Exposure of the AWB or AWC olfactory neurons

Invertebrate Learning and Memory.
DOI: http://dx.doi.org/10.1016/B978-0-12-415823-8.00010-1

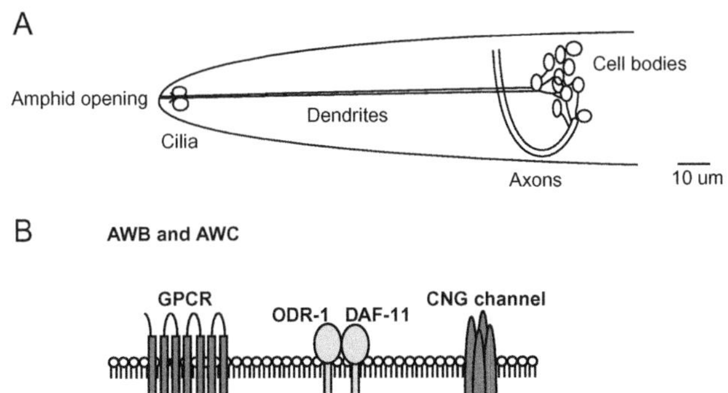

FIGURE 10.1 ***Caenorhabditis elegans*** **olfactory sensory neurons and a mechanism for processing olfactory information.** (A) Anatomical schematic diagram for the amphid organ, which contains ciliated sensory neurons that include the olfactory neurons. (B) The signaling events that regulate olfactory perception by the AWB and AWC neurons. Source: *Panel A modified from Roayaie* et al.[7]

to odors they detect suppresses neuronal activity and is reflected by decreased signals in intracellular calcium recordings, using genetically encoded calcium sensors.[22,23] The neuronal response of AWC has been shown to be transduced to the downstream circuit through glutamatergic neurotransmission, which inhibits or activates the postsynaptic interneurons AIY and AIB, respectively.[23] Bacteria-conditioned medium, which serves as a source of many odors, also regulates the activity of AWB and AWC, [22,23] implicating olfactory sensorimotor responses in food-seeking-related behaviors. Not surprisingly, the availability and quality of food serve as primary modulators of worm olfaction in various ways, and the signaling transduction pathways that process olfactory information in sensory neurons and downstream circuits provide the substrates where these modulations can occur.

CAENORHABDITIS ELEGANS OLFACTORY PLASTICITY

The small and compact *C. elegans* nervous system has exhibited an amazing amount of plasticity, including behaviors that resemble nonassociative and associative learning. Much of these behavioral plasticities result from olfactory modulation. Interestingly, these modulations can occur at different stages, either during larval development or in adulthood, and can also be influenced by the aging process. In this chapter, we review forms of olfactory plasticity displayed by worms of different ages and the corresponding underlying mechanisms. We also discuss the implications of *C. elegans* studies in understanding the function of vertebrate brains.

Larval Stage Olfactory Plasticity: Olfactory Imprinting

Olfactory imprinting refers to forms of olfactory learning that occur during a particular developmental stage or in association with a specific physiological state.[24] This type of olfactory plasticity has been widely found and can take place either during development or in adults, such as the bond between a mother sheep and her lamb as established during parturition on the basis of olfactory cues[25] and the juvenile salmon's olfactory memory of its birth stream.[26] Intriguingly, *C. elegans* juveniles also appear to imprint some chemosensory cues, such as the attractive odors isoamyl alcohol and benzaldehyde, which are sensed by the AWC olfactory neurons.[27] Exposure of *C. elegans* larvae to these odors specifically enhances the attractive response toward these cues in the resulting adults.[27] However, this enhanced olfactory response takes place only if the early exposure to the imprinting odors occurs during a particular time window within the first larval stage,[27] suggesting the existence of a critical period for this form of olfactory plasticity that is reminiscent of the critical period for visual plasticity in mammals.[28]

It was shown that successful olfactory imprinting in *C. elegans* requires that early exposure is accompanied by food or serotonin, a neurotransmitter indicative of food to the worms in many contexts.[27] This suggests that the physiological consequence of the imprinting phenomenon may include the neuronal representation of a signal indicative of a favorable environment. The power of *C. elegans* genetics has allowed the identification of a GPCR, SRA-11, which acts specifically in one pair of the AWC postsynaptic interneurons, AIY, for imprinting to occur.[27] These molecular and cellular characterizations could provide a basis for further exploration of mechanistic insights that are generally relevant to the imprinting behaviors, particularly how information acquired at an early age is maintained throughout development and how such information is represented within the nervous system and accessed during adulthood.

Adult Stage Olfactory Plasticity

Olfactory Adaptation

Adaptation is a common form of sensory plasticity. Similar to other sensory modalities, adaptation in olfaction refers to the decreased olfactory response to an odor after prolonged exposure to the same odor.[29]

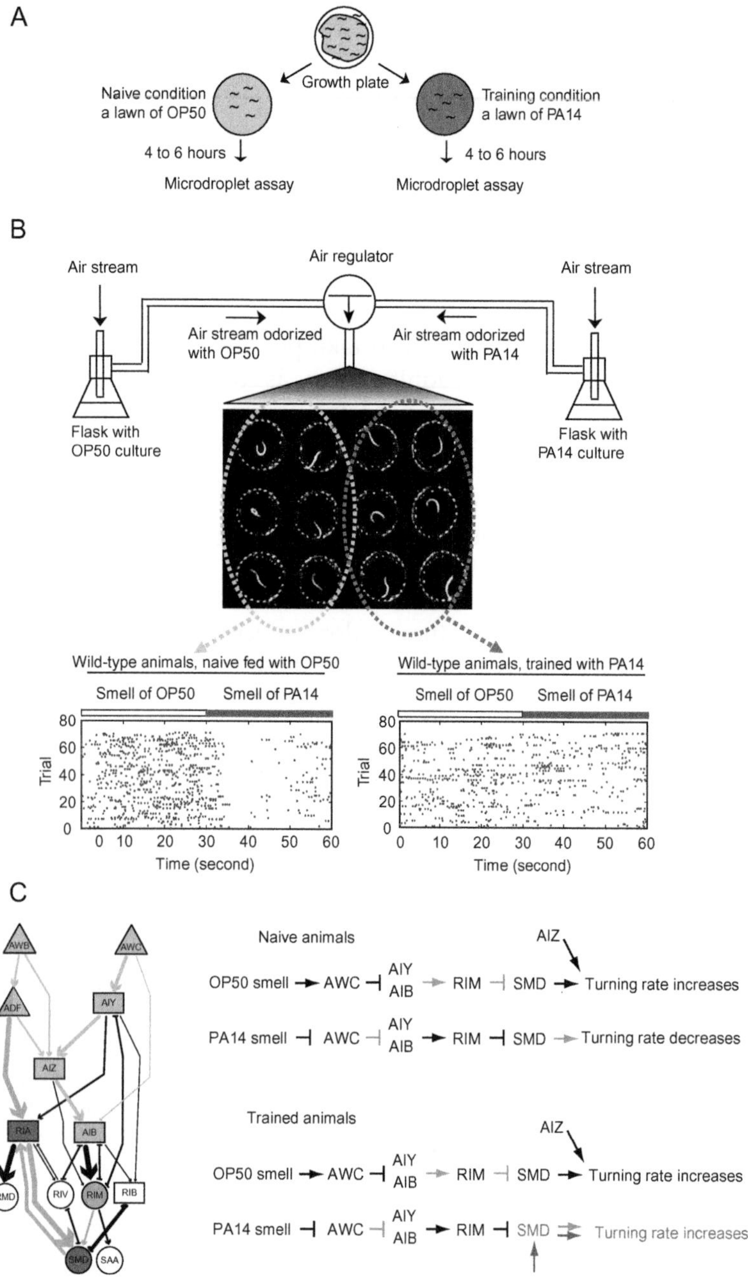

FIGURE 10.2 The aversive olfactory learning of *C. elegans* on the pathogenic bacteria *Pseudomonas aeruginosa* PA14 and the underlying neural circuits. (A) A training protocol for aversive olfactory learning in adult *C. elegans*. (B) The microdroplet assay that quantifies the learning ability of individual animals. One adult animal is placed within each droplet, which is then imaged within a gas-regulated enclosed chamber using a CCD camera at 10 Hz. The animals then receive olfactory inputs via exposure to two alternating airstreams that have been passaged through either a liquid culture of benign *E. coli* OP50 or the pathogenic *P. aeruginosa* PA14. In each experiment, animals are exposed to 12 successive cycles of alternating 30-sec OP50- versus PA14-odorized airstreams. A representative image and the data output of 12 animals in one assay are shown, in which each dot in the vertically arranged raster panels represents one event where the animal bends its body strongly to form an omega (Ω) shape, known as an Ω bend. The frequency of Ω bends correlates with the animal's olfactory preference toward a sensory cue. Attractive odors suppress the rate of Ω bends displayed by swimming animals, whereas repulsive odors increase it.[21] Each row in the raster panels represents one cycle of the olfactory assay on one animal, which is tested for 12 cycles continuously per assay. The result indicates that naive animals prefer the smell of PA14 to the smell of OP50, as demonstrated by the fewer turns they generate toward PA14. In contrast, animals trained with PA14 display a similar amount of turns toward the smell of OP50 and PA14, indicating that training decreases the olfactory preference toward PA14. (C) Two different neural circuits underlie naive versus learned olfactory preferences. The sensorimotor circuit that regulates naive preference is highlighted in blue, and the modulatory circuit that is required for the learned preference is highlighted in pink and red. In naive animals, the differential neuronal responses of AWC sensory neurons to the smell of OP50 versus PA14 propagate through the blue-highlighted downstream circuit to generate preference. In trained animals, RIA interneurons and SMD motor neurons regulate the turning rate toward the smell of PA14 to generate the trained preference. *Source: Modified from Ha* et al.[22]

This odor-specific behavioral alteration is reversible[29] and enhances the ability of the animal to detect novel and more informative environmental cues. In humans, olfactory adaptation is characterized by increased detection threshold and decreased perceived intensity of an odor.[29] In *C. elegans*, prolonged exposure to odors in the absence of food causes adaptation,[30,31] which shares the defining features of odor adaptation in other model organisms, such as fruit flies and humans.[29,32] The degree of adaptation, which is specific to the odor involved, depends on the concentration of the odor and the exposure time.[29–32] These commonalities support the significance and relevance of the *C. elegans* model in dissecting the neuronal substrates of olfactory adaptation.

Although psychophysical studies of olfactory adaptation using human subjects have been aided by language, studies using animals have provided more accessible systems to uncover the neural basis of this behavioral plasticity. Olfactory adaptation can occur with different temporal dynamics, ranging from rapid adaptation that happens in minutes to long-term olfactory regulation over the course of hours or days.[29] In vertebrates, activation of olfactory sensory neurons increases intracellular cyclic AMP levels, which triggers the opening of cyclic nucleotide-gated (CNG) ion channels. This activation leads to calcium influx, which modulates CNG channels in a calcium/calmodulin-dependent manner, generating a calcium-mediated negative feedback onto olfactory signaling.[33–39] However, olfactory CNG channels also have a high sensitivity to cGMP,[38,39] which is consistent with the observation that a cGMP-mediated pathway is also involved in vertebrate olfactory adaptation.[40]

In *C. elegans*, the molecular and cellular mechanisms of olfactory adaptation have been investigated largely via the responses of the AWC olfactory neurons, which involve the modulation of a cGMP-mediated G protein pathway that transduces olfactory responses.[20] Whereas two G_i-like G proteins in AWC activate two guanylyl cyclases in response to odor stimuli, odor adaptation in this neuron has been proposed to lead to downregulation of these cyclases.[19] This in turn would dampen intracellular cGMP levels that regulate the activity of a CNG channel that is composed of the β subunit TAX-2 and the α subunit TAX-4.[19,20]

Similar to vertebrates and humans, the degree of odor adaptation in *C. elegans* also depends on exposure time. An exposure as short as 5 min is sufficient to suppress an attractive response to several odors that are sensed by the AWA or AWC neurons (also known as early adaptation).[1,41] However, adaptation that is induced by a 30-min exposure can recover in 30 min or 1 hr, which is a short-term form of adaptation.[41] In contrast, longer exposure leads to adaptation that persists for 2 hr or more, which is a long-term form of adaptation.[30,31,41,42] The dissection of the mechanistic underpinnings of these different phases of odor adaptation has revealed intriguingly sophisticated mechanisms, as we further elaborate here.

EARLY ADAPTATION

The early adaptation of AWC-sensed odors requires the function of a G_γ protein GPC-1 within the AWC neuron.[43] It is also interesting to note that the function of a RAS-MAPK pathway in the AIY interneurons, the major postsynaptic neuron of AWC, is required for animals to adapt to several AWC-sensed odors,[41] implicating a circuit-level mechanism in odor adaptation.

SHORT-TERM VERSUS LONG-TERM ADAPTATION

On the other hand, both short-term and long-term odor adaptation within the AWC neurons have been shown to require the cGMP-dependent protein kinase (PKG), EGL-4, albeit through different mechanisms.[42] A mutation in the putative EGL-4 phosphorylation site within the C-terminus of the CNG TAX-2 produces a TAX-2 isoform that is sufficient to generate AWC-mediated olfactory responses but defective in the short-term olfactory adaptation triggered by a 30-min or 1-hr exposure to particular odors.[42] However, this mutation does not interfere with the long-term AWC olfactory adaptation that is induced by prolonged exposure,[42] suggesting distinct mechanisms for these two temporally different forms of adaptation. Although the direct phosphorylation of TAX-2 by EGL-4 in response to a short-term odorant exposure remains to be tested, these findings are analogous to the calcium-mediated modulation of CNG channels during olfactory adaptation in vertebrates.[33–39]

The molecular and genetic dissection of EGL-4 function is particularly informative about the mechanisms underlying long-term olfactory adaptation. In humans, the nuclear translocation of the cGMP-dependent kinase (G kinase) leads to changes in gene expression, which involves a physical interaction between the G kinase and the transcriptional regulator TFII-I.[44,45] In *C. elegans*, the intact nuclear localization signal (NLS) of EGL-4, homologous to the NLS of the human G kinase, is also required for the long-term olfactory adaptation in AWC, suggesting that EGL-4 similarly regulates persistent odor adaptation via gene expression changes at the level of transcription.[42] Indeed, the translocation of EGL-4 into the AWC nucleus is not only temporally correlated with, but also sufficient to, generate the decreased responses caused by prolonged exposure to AWC-sensed odors, which mimics the behavioral consequences of odor adaptation.[46] Nuclear translocation of EGL-4 facilitates the long-term change in AWC signaling physiology, likely through stable regulation of some downstream factors.[47] In addition, an unusual nature of the genetic lesion in the *egl-4(ky95)* allele has surprisingly uncovered a role for translational regulation of EGL-4 in generating long-term olfactory adaptation.[48] The mutation in *egl-4 (ky95)* alters a region in the *egl-4* 3′ UTR that contains a conserved Nanos response element (NRE),[48] which binds the Pumilio/Fem-3 (PUF) family of translational regulators.[49,50] The *egl-4* 3′ UTR NRE can bind the *C. elegans* PUF FBF-1 protein; and both FBF-1 and the *egl-4* 3′ UTR NRE are required for long-term adaptation.[48] Mechanistically, this binding enhances the translation of EGL-4 within the cilia and the cell body of AWC, which is induced by prolonged odor exposure, leading to long-term adaptation.[48]

Although EGL-4 functions in the sensory neurons to regulate short-term and long-term adaptation,[46,48] it remains to be determined to what extent the downstream circuit is involved in odor adaptation, especially given the role of the AIY interneuron in early adaptation.[41] However, it has been shown that AWC odor adaptation is also regulated by the antagonistic roles of two G_α proteins, $G_{o\alpha}$ (GOA-1) and $G_{q\alpha}$ (EGL-30), via diacylglycerol and calcium signaling within the AWC neurons.[51] Given the known function of GOA-1 and EGL-30 in regulating neurotransmission,[52–54] these findings further support the idea that regulated neurotransmission from the AWC sensory neurons to a downstream circuit mediates olfactory adaptation.

Because the worms display different phases of adaptation (early, short-term, and long-term adaptation), it is essential to understand how the activity of the key sensory neurons, such as AWC, responds to varying degrees of odorant exposure and how these cellular responses are linked with behavioral changes. It will also be interesting to learn to what degree the mechanisms underlying the different phases of odor adaptation are linked to each other and how they transition between these behavioral phases. Notwithstanding, considering the high degree of conservation across species, the mechanistic dissection of *C. elegans* odor adaptation through the combined use of molecular, cellular, and behavioral methods, which include intracellular calcium imaging[55] and optogenetic tools,[56,57] should yield further insights into olfactory plasticity in other animals, including humans.

CONTEXT MODULATION OF ADAPTATION

Interestingly, olfactory adaptation appears to be affected by the accompanying context, similar to previous observations that context modulates learning and memory recall in various organisms.[58,59] If worms are adapted to an odor, such as benzaldehyde, within a specific context, such as in the presence of ethanol or sodium chloride, worms can display stronger adaptation within the same context during the retest.[60,61] Dopaminergic neurotransmission is required for the enhancing effect of the ethanol contextual cue on benzaldehyde adaptation,[60] which is dissimilar from the dopaminergic modulation of the worm's context-dependent habituation to the mechanical stimuli of tapping.[62,63]

It is also intriguing to note that olfactory adaptation can be altered in that it can lead to behavioral switches—that is, aversion to attractants such as butanone—upon further prolonged exposure of up to 120 min.[64] On the other hand, odor adaptation can also be suppressed by the presence of food.[31,65] These observations have led to the hypothesis that odor adaptation may have features of associative learning, whereby worms in the latter paradigm associate the odor with the food content of the environment.[65] Accordingly, these different behavioral paradigms could serve as models to investigate neural mechanisms that underlie the integration of multiple sensory modalities, which involve not only olfaction but also the different contextual cues.

Olfactory Conditioning

Because worm odor adaptation is subject to context modulation in that it does not preclude associations with other existing cues, it is not surprising that *C. elegans* also displays different forms of learning that result from pairing an odorant with another stimulus.[22,65–71] In the classical conditioning paradigm of associative learning, a stimulus to which animals exhibit a mild or neutral response, known as the conditioned stimulus (CS), is paired with another stimulus that elicits an aversive or appetitive response, known as the unconditioned stimulus (US). Massed training—that is, continued pairing of CS with US without intertrial intervals—leads to short-term or intermediate-term memory, which does not require transcription or translation. Spaced training—that is, repeated training with intertrial intervals—results in long-term memory, which depends on mRNA and protein syntheses.[72–78] Whereas some *C. elegans* olfactory training configurations that meet the defining features of classical conditioning produce short-term and long-term changes in olfaction, the pairing of an olfactory cue with the presence or the absence of food or the physiological consequences of ingesting toxic food also generates robust and specific olfactory plasticity.[22,65–71] The effects of food-related cues in the conditioning of *C. elegans* olfactory responses are consistent with a biological relevance for worm olfactory behavior (i.e., food-seeking-related navigations). Here, we refer to forms of *C. elegans* olfactory plasticity that are generated by the simultaneous presence of an olfactory cue with another conditioning cue, including food cues, as olfactory conditioning.

CLASSICAL CONDITIONING

Using training protocols that resemble classical conditioning, the *C. elegans* olfactory response can be conditioned with either an aversive cue or an appetitive cue. In one case, worms are trained by brief ($<$1 sec) and simultaneous exposure to an attractive odor, 1-propanol (CS), and an aversive tastant, hydrochloric acid (US). The trained animals are then placed on food with no exposure to either chemical during intertrial intervals. This aversive conditioning dramatically decreases the attractive response to 1-propanol specifically.[66] In another case, worms are trained by simultaneously exposing them to an AWC-sensed attractant butanone (CS) and food (US), which are benign bacterial strains, for 30 min[71] or longer.[70] This paired exposure, which may or may not involve spaced training, in which worms are placed on a plate that has no food or odor during the intertrial intervals, further increases the animal's attraction toward butanone.[70,71] In both types of conditioning, aversive and appetitive, spaced training generates learned olfactory responses that last up to 24 hr or even longer, whereas massed training, which lacks the intertrial intervals, generates olfactory changes that are retained for only a few hours.[70,71] Similar to other organisms, in both of these aversive and attractive olfactory learning processes, the short-term memory generated by massed training does not require transcription or translation, but the long-term memory generated by spaced training depends on both.[66,71] Interestingly, long-term olfactory memory in *C. elegans* requires CRH-1, the worm homolog of CREB,[66,71] a transcription factor that regulates long-term memory in *Aplysia*, *Drosophila*, and mammals.[78] Strikingly, overexpression of CRH-1 in the worm nervous system enhances long-term memory,[71] suggesting an instructive role for CRH-1 in the learning process. These mechanistic characterizations validate the *C. elegans* olfactory conditioning paradigms as useful models in which to further dissect the molecular and cellular basis of associative learning. Given the availability of a defined synaptic wiring diagram[79] and of tools for manipulating neuronal and gene activities in specific neurons, it will be particularly interesting to map the functional site of CRH-1 in the *C. elegans* learning behaviors and to explore how the neurons respond to the training conditions and how the regulated activity of CRH-1 leads to changes in behavior.

GARCIA'S EFFECT

In addition to classical conditioning, *C. elegans* displays a form of olfactory learning that is analogous to Garcia's effect,[22,67] which is a common form of associative learning whereby animals learn to avoid the taste or smell of a food source that makes them ill.[80] Because *C. elegans* feeds on bacteria in its natural habitat, pathogenic bacteria that can infect them are a potential danger. Ingestion of infectious bacteria, such as some *Pseudomonas aeruginosa* strains and *Serratia marcescens* strains, results in death over the course of a few days.[81–83] Interestingly, naive worms slightly prefer the odors of these pathogenic bacteria, when the worms have no prior exposure to these pathogens and have only been cultivated on the common laboratory food source, the benign *Escherichia coli* strain OP50; however, animals that have been exposed to and ingested pathogenic bacteria strongly avoid the smell of the pathogen.[22,67] This form of aversive learning can be generated either by exposing adult animals for several hours or exposing them throughout their lives to the pathogen.[22,67] Serotonergic neurotransmission is required for this process because mutants defective in serotonin biosynthesis cannot learn.[67] By manipulating the cell-specific expression of the rate-limiting enzyme for serotonin biosynthesis, tryptophan hydroxylase TPH-1, the serotonin signal from a specific pair of serotonergic neurons, ADF, has been demonstrated to regulate this learning process.[67] This requires the downstream function of a serotonin-gated chloride channel MOD-1 in several interneurons that are important for sensorimotor responses.[67] In addition, both the *tph-1* transcription and the serotonin content of the ADF neurons increase after training, suggesting a reinforcing role for the ADF–serotonin signal in generating a learned olfactory aversion to the pathogenic bacteria.[67]

Although Garcia's effect has been found in many animals, including rodents, fish, and humans,[84,85] the underlying cellular basis is not understood. The identification of the role of the ADF serotonergic signaling pathway and the stereotypic *C. elegans* wiring diagram provide an opportunity to dissect the cellular mechanisms for this phenomenon. Accordingly, using laser ablation, a well-established technique used to eliminate single neurons in live *C. elegans*,[86] and an automated learning assay that quantifies olfactory responses of individual animals,[21] the neural network underlying aversive olfactory learning has been mapped functionally from olfactory sensory neurons to motor neurons[22] (Figure 10.2). This analysis reveals that olfactory learning results from the interplay between two connected neural circuits.[22] One circuit mediates the animals' olfactory preference under naive conditions, and intracellular calcium imaging indicates that this preference is determined by the intrinsic properties of the olfactory sensory neurons present within the circuit (Figure 10.2). The second circuit, which is specifically required to generate the learned preference upon training, consists of the ADF serotonergic neurons, their postsynaptic interneurons, and the corresponding downstream motor neurons.[22] Further behavioral analysis demonstrates that modulations of the downstream interneuron and motor neuron activities in this latter circuit contribute to learning (Figure 10.2).[22] Thus, these findings reveal organizational principles that allow a neural network for learning to encode both naive response and experience-dependent plasticity, which would allow the animal to produce the optimal behavior under specific conditions. Comparison of this neural network with those that regulate the switch between feeding and escape in sea slugs[87] and between fear extinction and renewal in mice[88] suggests a possible general principle: the switch between alternative behavioral states might be generated by the differential usage of anatomically distinct but connected neural circuits.

The mapping of the functional network of learning should provide a platform to further address how a learned behavior is generated by experience. With the advent of optogenetics[56,57] and imaging tools[55] that can be successfully applied to the worm's nervous system, it would be particularly interesting to determine the nature of the neuronal physiological correlates of learning and the causal roles of these learning correlates with regard to behavioral changes. Moreover, it would be intriguing to address how experience is translated into alterations in serotonin signaling and how the downstream circuit interprets and responds to these signal changes to produce the optimal behavioral output.

OTHER FOOD-DEPENDENT EFFECTS ON OLFACTORY PLASTICITY

In addition to classical conditioning and aversive olfactory learning, food-related cues, which modulate many *C. elegans* behaviors,[4,89–92] can acutely and reversibly alter olfactory sensorimotor responses. Octanol repels *C. elegans* and elicits reversals within a few seconds of exposure when animals are on food.[4] At lower octanol concentrations, removal of food increases the animal's response time to the odorant, suggesting that the presence of food increases octanol sensitivity.[4] In contrast, at higher octanol concentrations, worms respond to the odor with a similar speed whether food is present or absent.[4] Subsequent laser ablation analysis reveals that two different sets of sensory neurons are recruited to generate responses to low versus high octanol concentrations and that food availability has differential effects on these two sets of neurons.[4] In *C. elegans*, serotonin levels have been shown to correlate closely with the food content present in the environment.[67,90,93–98] Indeed, a serotonin signal has been demonstrated to act at different functional sites of an aversive olfactory sensorimotor circuit to regulate the food-dependent sensitivity to octanol.[4,92] Using cell-specific manipulation of gene activities, serotonergic signaling has been found to operate through different receptors at different sites of this circuit.[92] These include the mammalian 5-HT_6 homolog SER-5 in the sensory neuron ASH, the primary sensory neuron that detects octanol; the serotonin-gated chloride channel MOD-1 in the first-order interneurons of the olfactory circuits; and the serotonin-gated GPCR SER-1 in interneurons and motor neurons.[92] Together, these studies show that precise serotonin levels mediate the flexible usage of the animal's sensory system,[4,92] revealing an unexpected level of sophistication in the way that serotonin modulates neural circuit function. Because serotonin regulates a wide range of behaviors in many animals, including humans,[99] these studies demonstrate the power of using *C. elegans* in dissecting the mechanistic insights into the neural signaling involved in different behaviors.

Aging-Dependent Changes in Olfactory Plasticity

Similar to vertebrates, including humans,[100] age has a significant effect on *C. elegans* neuronal plasticity.[71,101] Indeed, several studies have shown that the worm's behavioral responses become altered with age, and that these alterations are irrespective of the age-dependent locomotory changes that could potentially influence such responses.[71,101,102] For example, associative learning behavior decreases in older adult *C. elegans*, whether this involves an association between an olfactory attractant and food[71] or between temperature and food.[101] As expected, spaced learning abilities, which yield longer-lasting memories, decay more slowly with age than do massed learning abilities, which involve one-time associations between two stimuli and thus generate shorter-lasting memories.[71]

Many factors influence the aging process, which suggests that there will be several mechanisms that will affect the olfactory learning behavior of aged worms. Two distinct mechanisms that affect aging of *C. elegans*, as well as that of higher animals, involve an insulin/insulin-like signaling pathway and calorie restriction, which is a moderate restriction of food intake levels.[103,104] Accordingly, both mechanisms have been shown to have differential effects on massed and spaced learning behaviors, as well as short-term and long-term memories, in young versus old adult *C. elegans*.[71]

Effects of Insulin Signaling

Although a reduction in insulin signaling initially seems to slightly decrease the massed learning abilities of young adult *C. elegans*, this reduced signaling ultimately prolongs short-term memory.[71,101] Similarly, low insulin signaling extends the long-term memory of young adults, but with little or no effect on their spaced learning behaviors.[71] Together, these observations suggest that the insulin pathway inhibits the maintenance of memory in young worms. The existence of a mechanism whereby memory is actively suppressed in young adults might suggest a need for a high degree of plasticity in the animal's responses to the changing quality of its environment, especially during the early phases of its reproductive period.

Interestingly, in older *C. elegans*, the insulin pathway appears to have no or less of an effect on memory maintenance but is more important in learning, because older mutants with reduced insulin signaling show a slower decay in massed or spaced learning than do age-matched wild-type animals.[71] The positive effect of reduced insulin signaling on attractive learning in older animals might be simply due to a passive mechanism in that animals with low insulin pathway activity are less susceptible to the degenerative effects of aging. However, the intriguing possibility remains that the insulin pathway actively inhibits learning in older worms. Although this pathway has been shown to promote aversive learning, in which worms are trained to avoid sodium chloride after a prior exposure to the salt in the absence of food,[105] this pathway is also known to inhibit temperature and food-content associations in young adults.[101,106] These differing activities of the pathway in regulating different learning behaviors suggest that the pathway acts in a context-dependent manner. This is supported by observations that the regulation of different behavioral and physiological processes by two insulin-like peptides (ILPs) that can act as ligands for this pathway depends on the spatial context of ILP expression.[105–107] Thus, considering the large number of ILPs that can modulate the activity of this pathway,[108,109] it is certainly possible that different ILPs will also be required for the differing temporal requirements for this pathway in learning versus memory maintenance in young, as opposed to older, adult *C. elegans*.

Effects of Restriction of Food Intake Levels

Unlike insulin signaling, restricting food intake levels has no effect on the short-term memory of young adult *C. elegans* and even impairs the long-term memories, without affecting the learning abilities, of these animals.[71] On the other hand, in older adult worms and again unlike reduced insulin signaling, restricting food levels has significant effects on both learning and memory maintenance.[71] Food-level restriction improves the learning behavior of older animals, whether they are exposed to massed or spaced training, and increases both short-term and long-term memories of these animals.[71] Together, these findings suggest that insulin and food-level restriction employ different mechanisms to affect learning and memory, which is consistent with previous observations that insulin pathway components and food restriction can also influence animal life span independently of each other.[110,111] However, despite the differences between these two mechanisms, one should note that the effects of food-level restriction and insulin signaling on *C. elegans* learning and memory appear to converge on regulating the activity of the CREB-like transcription factor CRH-1.[71]

Bearing in mind that many processes are conserved across species, it should not be surprising that food-level restriction and insulin signaling will also affect learning and memory maintenance in other animals. Indeed, as in older *C. elegans*,[71] calorically restricted, old rodents exhibit better memory than their well-fed counterparts.[112] In addition, similar to young *C. elegans*, young rodents with impaired neuronal insulin signaling display improved memory.[113] Nonetheless, in *Drosophila*, a null mutation in a downstream effector of insulin signaling, the insulin receptor substrate-like protein CHICO,[114] leads to decreased olfactory associative learning,[115] suggesting that the insulin pathway in flies promotes learning. Thus, it remains to be seen how conserved the mechanisms will be through which food-level restriction versus insulin signaling affect learning and memory maintenance of animals at different ages. For example, because the impairment in insulin signaling in the previously mentioned rodent versus *Drosophila* studies differs in spatial or temporal context,[113–115] it is likely that the effect of the insulin pathway on the learning process across species will also involve context dependence, as in *C. elegans*.[105–107] Consistent with this hypothesis, both rodents and *Drosophila* also contain large ILP families that are expressed in different subsets of neuronal and non-neuronal cells, which may regulate distinct learning paradigms.[116–119]

Effects of Other Mechanisms

At least one other signaling pathway has been shown to influence age-dependent changes in olfactory behavior.[69] In contrast to younger *C. elegans* adults, older animals show an increase in attraction to high concentrations of attractive odorants but no change in repulsion to repellent odors.[69] This behavioral change requires the neurotransmitter serotonin and the presence of the bacterial food source from which the attractive odorants in this study are derived.[69] The requirement for serotonin and food association for this age-dependent phenotype brings to mind another observation in which altered serotonin signaling is implicated in the *C. elegans* aging process via food-level restriction and at least partly independently of insulin signaling.[120] However, whereas a role for serotonin in food-level restriction is consistent with the food effects on serotonin activity,[97] loss of serotonin signaling by itself appears to have little or no effect on the aging of *C. elegans*, unlike the restriction of food intake levels.[120] This might suggest that the serotonin pathway modulates different dietary influences on physiology—for example, the food-type versus food-level effects on longevity.[121] Similarly, the same could be true for behavior, so that complete loss of serotonin signaling would represent a mix of independent effects on behavior, as well as physiology, from distinct diet-associated mechanisms. In agreement with this view, the serotonergic effect on *C. elegans* life span[120] does not appear to share the same signaling pathway components as the serotonergic effect on the age-dependent increases in odorant attraction.[69]

The finding that odorant attraction increases with age, but only toward high odor concentrations,[69] may indicate that the sensitivity and/or the adaptive responses of the chemoreceptive apparatus decrease with age. However, these older *C. elegans* do not display decreased kinetics in chemoattraction nor decayed adaptive responses, which suggests that the age-dependent increase in chemoattraction is likely due to a prolonged period of association between the attractive odors and the appetitive food source by the animals.[69] This idea is supported by the fact that no similar age-dependent behavioral change is observed against repulsive odors that signify toxic substances, to which the animals were not exposed during the previously discussed study.[69] Accordingly, this age-dependent behavioral change may be reminiscent of use-dependent neuronal plasticity that has been described in vertebrates,[122] which would presumably lead to strengthening of the relevant circuitry in older animals.

SUMMARY

Despite some divergence between the neuronal physiology of *C. elegans* and that of vertebrates,[123,124] intriguing commonalities have emerged from studies described in this chapter at different scales, from behavioral phenomenology to molecular, cellular, and circuit substrates. At the level of behavior, *C. elegans* has demonstrated a plethora of plasticities, including olfactory imprinting during a specific juvenile stage,[27] reminiscent of the critical period in vertebrates,[28] and sensory adaptation to a series of olfactory cues at various concentrations that have been found in many animals.[29–32] *C. elegans* has also shown both appetitive and aversive olfactory conditioning that resemble classical conditioning and aversive olfactory learning against dangerous food sources, which is analogous to taste aversion, a common form of learning in other organisms, including humans.[19,22,29–32,42,65–78] Mechanistically, the molecular and cellular underpinnings of these forms of neural plasticity are conserved. For example, the small neurotransmitter serotonin, which regulates aversive olfactory learning[22,67] and the food-dependent olfactory responses of *C. elegans*,[4,69,90,92,93] also acts as a reinforcing cue in learning in other animals.[99] Moreover, CREB, the key transcription factor that is involved in regulating long-term memory in *Aplysia*, *Drosophila*, and mammals,[78] is required for long-term olfactory memory in worms.[66,71] At the level of neural circuits, the fundamental organizational principle of the neural network underlying *C. elegans* aversive olfactory learning[22] is comparable to the alternative usage of differential circuits in fear extinction versus renewal in mice[88] and the feeding-escape circuits in sea slugs.[87] Finally, the observed age-dependent changes in neuronal plasticity in many adult animals might well be influenced by mechanisms that are also conserved from *C. elegans* to mammals. Thus, the comparisons across species that were discussed in this chapter reveal widely conserved mechanisms that underlie the different forms of neural plasticity and highlight important aspects of *C. elegans* neurobiology that contribute to our general understanding of how the mammalian brain could function or decline.

Acknowledgments

We apologize to the authors whose work we were unable to cite due to space constraints. This work was supported by the Novartis Research Foundation and the Swiss National Science Foundation to J. A. and by The Esther A. and Joseph Klingenstein Fund, March of Dimes Foundation, The Alfred P. Sloan Foundation, The John Merck Fund, and the National Institutes of Health (grant R01 DC009852) to Y. Z.

References

1. Bargmann CI, Hartwieg E, Horvitz HR. Odorant-selective genes and neurons mediate olfaction in *C. elegans*. *Cell*. 1993;74: 515–527.
2. Zechman JM, Labows Jr JN. Volatiles of *Pseudomonas aeruginosa* and related species by automated headspace concentration-gas chromatography. *Can J Microbiol*. 1985;31:232–237.
3. Troemel ER, Kimmel BE, Bargmann CI. Reprogramming chemotaxis responses: Sensory neurons define olfactory preferences in *C. elegans*. *Cell*. 1997;91:161–169.
4. Chao MY, Komatsu H, Fukuto HS, Dionne HM, Hart AC. Feeding status and serotonin rapidly and reversibly modulate a *Caenorhabditis elegans* chemosensory circuit. *Proc Natl Acad Sci USA*. 2004;101:15512–15517.
5. Ward S, Thomson N, White JG, Brenner S. Electron microscopical reconstruction of the anterior sensory anatomy of the nematode *Caenorhabditis elegans*. *J Comp Neurol*. 1975;160:313–337.
6. Perkins LA, Hedgecock EM, Thomson JN, Culotti JG. Mutant sensory cilia in the nematode *Caenorhabditis elegans*. *Dev Biol*. 1986;117:456–487.
7. Roayaie K, Crump JG, Sagasti A, Bargmann CI. The G alpha protein ODR-3 mediates olfactory and nociceptive function and controls cilium morphogenesis in *C. elegans* olfactory neurons. *Neuron*. 1998;20:55–67.
8. Robertson HM, Thomas JH. The putative chemoreceptor families of *C. elegans*. *WormBook*. 2006;1–12.
9. Chen N, Pai S, Zhao Z, et al. Identification of a nematode chemosensory gene family. *Proc Natl Acad Sci USA*. 2005;102: 146–151.
10. Colosimo ME, Brown A, Mukhopadhyay S, et al. Identification of thermosensory and olfactory neuron-specific genes via expression profiling of single neuron types. *Curr Biol*. 2004;14: 2245–2251.
11. McCarroll SA, Li H, Bargmann CI. Identification of transcriptional regulatory elements in chemosensory receptor genes by probabilistic segmentation. *Curr Biol*. 2005;15:347–352.
12. Troemel ER, Chou JH, Dwyer ND, Colbert HA, Bargmann CI. Divergent seven transmembrane receptors are candidate chemosensory receptors in *C. elegans*. *Cell*. 1995;83:207–218.
13. Dwyer ND, Troemel ER, Sengupta P, Bargmann CI. Odorant receptor localization to olfactory cilia is mediated by ODR-4, a novel membrane-associated protein. *Cell*. 1998;93:455–466.
14. Axel R. Scents and sensibility: A molecular logic of olfactory perception (Nobel lecture). *Angew Chem Int Ed Engl*. 2005;44: 6110–6127.
15. Buck L, Axel R. A novel multigene family may encode odorant receptors: A molecular basis for odor recognition. *Cell*. 1991;65: 175–187.

16. Chess A, Simon I, Cedar H, Axel R. Allelic inactivation regulates olfactory receptor gene expression. *Cell*. 1994;78:823–834.
17. Jansen G, Thijssen KL, Werner P, van der Horst M, Hazendonk E, Plasterk RH. The heterotrimeric G protein genes of *Caenorhabditis elegans*. *Ernst Schering Res Found Workshop*. 2000: 13–34.
18. Birnby DA, Link EM, Vowels JJ, Tian H, Colacurcio PL, Thomas JH. A transmembrane guanylyl cyclase (DAF-11) and Hsp90 (DAF-21) regulate a common set of chemosensory behaviors in *Caenorhabditis elegans*. *Genetics*. 2000;155:85–104.
19. L'Etoile ND, Bargmann CI. Olfaction and odor discrimination are mediated by the *C. elegans* guanylyl cyclase ODR-1. *Neuron*. 2000;25:575–586.
20. Bargmann CI. Chemosensation in *C. elegans*. *WormBook*. 2006; 1–29.
21. Luo L, Gabel CV, Ha HI, Zhang Y, Samuel AD. Olfactory behavior of swimming *C. elegans* analyzed by measuring motile responses to temporal variations of odorants. *J Neurophysiol*. 2008;99:2617–2625.
22. Ha HI, Hendricks M, Shen Y, et al. Functional organization of a neural network for aversive olfactory learning in *Caenorhabditis elegans*. *Neuron*. 2010;68:1173–1186.
23. Chalasani SH, Chronis N, Tsunozaki M, et al. Dissecting a circuit for olfactory behaviour in *Caenorhabditis elegans*. *Nature*. 2007;450: 63–70.
24. Hudson R. Olfactory imprinting. *Curr Opin Neurobiol*. 1993;3: 548–552.
25. Poindron P, Levy F, Krehbiel D. Genital, olfactory, and endocrine interactions in the development of maternal behaviour in the parturient ewe. *Psychoneuroendocrinology*. 1988;13:99–125.
26. Hasler A, Scholz AT, Horrall RM. Olfactory imprinting and homing in salmon. *Am Sci*. 1978;66:347–355.
27. Remy JJ, Hobert O. An interneuronal chemoreceptor required for olfactory imprinting in *C. elegans*. *Science*. 2005;309:787–790.
28. Hubel DH, Wiesel TN. The period of susceptibility to the physiological effects of unilateral eye closure in kittens. *J Physiol*. 1970;206:419–436.
29. Dalton P. Psychophysical and behavioral characteristics of olfactory adaptation. *Chem Senses*. 2000;25:487–492.
30. Colbert HA, Bargmann CI. Odorant-specific adaptation pathways generate olfactory plasticity in *C. elegans*. *Neuron*. 1995;14: 803–812.
31. Colbert HA, Bargmann CI. Environmental signals modulate olfactory acuity, discrimination, and memory in *Caenorhabditis elegans*. *Learn Mem*. 1997;4:179–191.
32. Wuttke MS, Tompkins L. Olfactory adaptation in *Drosophila* larvae. *J Neurogenet*. 2000;14:43–62.
33. Zufall F, Leinders-Zufall T. The cellular and molecular basis of odor adaptation. *Chem Senses*. 2000;25:473–481.
34. Chen TY, Yau KW. Direct modulation by Ca(2 +)-calmodulin of cyclic nucleotide-activated channel of rat olfactory receptor neurons. *Nature*. 1994;368:545–548.
35. Liu M, Chen TY, Ahamed B, Li J, Yau KW. Calcium-calmodulin modulation of the olfactory cyclic nucleotide-gated cation channel. *Science*. 1994;266:1348–1354.
36. Kurahashi T, Menini A. Mechanism of odorant adaptation in the olfactory receptor cell. *Nature*. 1997;385:725–729.
37. Varnum MD, Zagotta WN. Interdomain interactions underlying activation of cyclic nucleotide-gated channels. *Science*. 1997;278: 110–113.
38. Nakamura T, Gold GH. A cyclic nucleotide-gated conductance in olfactory receptor cilia. *Nature*. 1987;325:442–444.
39. Zufall F, Firestein S, Shepherd GM. Analysis of single cyclic nucleotide-gated channels in olfactory receptor cells. *J Neurosci*. 1991;11:3573–3580.
40. Zufall F, Leinders-Zufall T. Identification of a long-lasting form of odor adaptation that depends on the carbon monoxide/cGMP second-messenger system. *J Neurosci*. 1997;17:2703–2712.
41. Hirotsu T, Iino Y. Neural circuit-dependent odor adaptation in *C. elegans* is regulated by the Ras-MAPK pathway. *Genes Cells*. 2005;10:517–530.
42. L'Etoile ND, Coburn CM, Eastham J, Kistler A, Gallegos G, Bargmann CI. The cyclic GMP-dependent protein kinase EGL-4 regulates olfactory adaptation in *C. elegans*. *Neuron*. 2002;36: 1079–1089.
43. Yamada K, Hirotsu T, Matsuki M, Kunitomo H, Iino Y. GPC-1, a G protein gamma-subunit, regulates olfactory adaptation in *Caenorhabditis elegans*. *Genetics*. 2009;181:1347–1357.
44. Gudi T, Lohmann SM, Pilz RB. Regulation of gene expression by cyclic GMP-dependent protein kinase requires nuclear translocation of the kinase: Identification of a nuclear localization signal. *Mol Cell Biol*. 1997;17:5244–5254.
45. Casteel DE, Zhuang S, Gudi T, et al. cGMP-dependent protein kinase I beta physically and functionally interacts with the transcriptional regulator TFII-I. *J Biol Chem*. 2002;277: 32003–32014.
46. Lee JI, O'Halloran DM, Eastham-Anderson J, et al. Nuclear entry of a cGMP-dependent kinase converts transient into long-lasting olfactory adaptation. *Proc Natl Acad Sci USA*. 2010;107: 6016–6021.
47. O'Halloran DM, Altshuler-Keylin S, Lee JI, L'Etoile ND. Regulators of AWC-mediated olfactory plasticity in *Caenorhabditis elegans*. *PLoS Genet*. 2009;5:e1000761.
48. Kaye JA, Rose NC, Goldsworthy B, Goga A, L'Etoile ND. A 3'UTR pumilio-binding element directs translational activation in olfactory sensory neurons. *Neuron*. 2009;61:57–70.
49. Murata Y, Wharton RP. Binding of pumilio to maternal hunchback mRNA is required for posterior patterning in *Drosophila* embryos. *Cell*. 1995;80:747–756.
50. Zamore PD, Williamson JR, Lehmann R. The Pumilio protein binds RNA through a conserved domain that defines a new class of RNA-binding proteins. *RNA*. 1997;3:1421–1433.
51. Matsuki M, Kunitomo H, Iino Y. $G_o\alpha$ regulates olfactory adaptation by antagonizing $G_q\alpha$-DAG signaling in *Caenorhabditis elegans*. *Proc Natl Acad Sci USA*. 2006;103:1112–1117.
52. Bastiani CA, Gharib S, Simon MI, Sternberg PW. *Caenorhabditis elegans* $G\alpha_q$ regulates egg-laying behavior via a PLCβ-independent and serotonin-dependent signaling pathway and likely functions both in the nervous system and in muscle. *Genetics*. 2003;165:1805–1822.
53. Brundage L, Avery L, Katz A, et al. Mutations in a *C. elegans* $G_q\alpha$ gene disrupt movement, egg laying, and viability. *Neuron*. 1996;16:999–1009.
54. Lackner MR, Nurrish SJ, Kaplan JM. Facilitation of synaptic transmission by EGL-30 $G_q\alpha$ and EGL-8 PLCβ. *Neuron*. 1999;24: 335–346.
55. Tian L, Hires SA, Mao T, et al. Imaging neural activity in worms, flies and mice with improved GCaMP calcium indicators. *Nat Methods*. 2009;6:875–881.
56. Leifer AM, Fang-Yen C, Gershow M, Alkema MJ, Samuel AD. Optogenetic manipulation of neural activity in freely moving *Caenorhabditis elegans*. *Nat Methods*. 2011;8:147–152.
57. Stirman JN, Crane MM, Husson SJ, et al. Real-time multimodal optical control of neurons and muscles in freely behaving *Caenorhabditis elegans*. *Nat Methods*. 2011;8:153–158.
58. Liu L, Wolf R, Ernst R, Heisenberg M. Context generalization in *Drosophila* visual learning requires the mushroom bodies. *Nature*. 1999;400:753–756.
59. Holland PC, Bouton ME. Hippocampus and context in classical conditioning. *Curr Opin Neurobiol*. 1999;9:195–202.

60. Bettinger JC, McIntire SL. State-dependency in *C. elegans*. *Genes Brain Behav*. 2004;3:266–272.
61. Law E, Nuttley WM, van der Kooy D. Contextual taste cues modulate olfactory learning in *C. elegans* by an occasion-setting mechanism. *Curr Biol*. 2004;14:1303–1308.
62. Rankin CH. Context conditioning in habituation in the nematode *Caenorhabditis elegans*. *Behav Neurosci*. 2000;114:496–505.
63. Kindt KS, Quast KB, Giles AC, et al. Dopamine mediates context-dependent modulation of sensory plasticity in *C. elegans*. *Neuron*. 2007;55:662–676.
64. Tsunozaki M, Chalasani SH, Bargmann CI. A behavioral switch: cGMP and PKC signaling in olfactory neurons reverses odor preference in *C. elegans*. *Neuron*. 2008;59:959–971.
65. Nuttley WM, Atkinson-Leadbeater KP, Van Der Kooy D. Serotonin mediates food-odor associative learning in the nematode *Caenorhabditis elegans*. *Proc Natl Acad Sci USA*. 2002;99: 12449–12454.
66. Amano H, Maruyama IN. Aversive olfactory learning and associative long-term memory in *Caenorhabditis elegans*. *Learn Mem*. 2011;18:654–665.
67. Zhang Y, Lu H, Bargmann CI. Pathogenic bacteria induce aversive olfactory learning in *Caenorhabditis elegans*. *Nature*. 2005;438:179–184.
68. Stetak A, Horndli F, Maricq AV, van den Heuvel S, Hajnal A. Neuron-specific regulation of associative learning and memory by MAGI-1 in *C. elegans*. *PLoS ONE*. 2009;4:e6019.
69. Tsui D, van der Kooy D. Serotonin mediates a learned increase in attraction to high concentrations of benzaldehyde in aged *C. elegans*. *Learn Mem*. 2008;15:844–855.
70. Torayama I, Ishihara T, Katsura I. *Caenorhabditis elegans* integrates the signals of butanone and food to enhance chemotaxis to butanone. *J Neurosci*. 2007;27:741–750.
71. Kauffman AL, Ashraf JM, Corces-Zimmerman MR, Landis JN, Murphy CT. Insulin signaling and dietary restriction differentially influence the decline of learning and memory with age. *PLoS Biol*. 2010;8:e1000372.
72. Tully T, Quinn WG. Classical conditioning and retention in normal and mutant *Drosophila melanogaster*. *J Comp Physiol A*. 1985;157:263–277.
73. DeZazzo J, Tully T. Dissection of memory formation: From behavioral pharmacology to molecular genetics. *Trends Neurosci*. 1995;18: 212–218.
74. Hammer M, Menzel R. Learning and memory in the honeybee. *J Neurosci*. 1995;15:1617–1630.
75. Davis HP, Squire LR. Protein synthesis and memory: A review. *Psychol Bull*. 1984;96:518–559.
76. Epstein HT, Child FM, Kuzirian AM, Alkon DL. Time windows for effects of protein synthesis inhibitors on Pavlovian conditioning in *Hermissenda*: Behavioral aspects. *Neurobiol Learn Mem*. 2003;79:127–131.
77. Goelet P, Castellucci VF, Schacher S, Kandel ER. The long and the short of long-term memory—A molecular framework. *Nature*. 1986;322:419–422.
78. Silva AJ, Kogan JH, Frankland PW, Kida S. CREB and memory. *Annu Rev Neurosci*. 1998;21:127–148.
79. White JG, Southgate E, Thomson JN, Brenner S. The structure of the ventral nerve cord of *Caenorhabditis elegans*. *Philos Trans R Soc Lond B Biol Sci*. 1976;275:327–348.
80. Garcia J, Kimeldorf DJ, Koelling RA. Conditioned aversion to saccharin resulting from exposure to gamma radiation. *Science*. 1955;122:157–158.
81. Hodgkin J, Kuwabara PE, Corneliussen B. A novel bacterial pathogen, *Microbacterium nematophilum*, induces morphological change in the nematode *C. elegans*. *Curr Biol*. 2000;10: 1615–1618.
82. Pujol N, Link EM, Liu LX, et al. A reverse genetic analysis of components of the Toll signaling pathway in *Caenorhabditis elegans*. *Curr Biol*. 2001;11:809–821.
83. Tan MW, Mahajan-Miklos S, Ausubel FM. Killing of *Caenorhabditis elegans* by *Pseudomonas aeruginosa* used to model mammalian bacterial pathogenesis. *Proc Natl Acad Sci USA*. 1999;96:715–720.
84. Bernstein IL. Taste aversion learning: A contemporary perspective. *Nutrition*. 1999;15:229–234.
85. Carew TJ, Sahley CL. Invertebrate learning and memory: From behavior to molecules. *Annu Rev Neurosci*. 1986;9:435–487.
86. Fang-Yen C, Gabel CV, Samuel AD, Bargmann CI, Avery L. Laser microsurgery in *Caenorhabditis elegans*. *Methods Cell Biol*. 2012;107:177–206.
87. Jing J, Gillette R. Escape swim network interneurons have diverse roles in behavioral switching and putative arousal in Pleurobranchaea. *J Neurophysiol*. 2000;83:1346–1355.
88. Herry C, Ciocchi S, Senn V, Demmou L, Muller C, Luthi A. Switching on and off fear by distinct neuronal circuits. *Nature*. 2008;454:600–606.
89. Shtonda BB, Avery L. Dietary choice behavior in *Caenorhabditis elegans*. *J Exp Biol*. 2006;209:89–102.
90. Sawin ER, Ranganathan R, Horvitz HR. *C. elegans* locomotory rate is modulated by the environment through a dopaminergic pathway and by experience through a serotonergic pathway. *Neuron*. 2000;26:619–631.
91. Gray JM, Hill JJ, Bargmann CI. A circuit for navigation in *Caenorhabditis elegans*. *Proc Natl Acad Sci USA*. 2005;102: 3184–3191.
92. Harris GP, Hapiak VM, Wragg RT, et al. Three distinct amine receptors operating at different levels within the locomotory circuit are each essential for the serotonergic modulation of chemosensation in *Caenorhabditis elegans*. *J Neurosci*. 2009;29:1446–1456.
93. Horvitz HR, Chalfie M, Trent C, Sulston JE, Evans PD. Serotonin and octopamine in the nematode *Caenorhabditis elegans*. *Science*. 1982;216:1012–1014.
94. Segalat L, Elkes DA, Kaplan JM. Modulation of serotonin-controlled behaviors by Go in *Caenorhabditis elegans*. *Science*. 1995;267:1648–1651.
95. Waggoner LE, Zhou GT, Schafer RW, Schafer WR. Control of alternative behavioral states by serotonin in *Caenorhabditis elegans*. *Neuron*. 1998;21:203–214.
96. Niacaris T, Avery L. Serotonin regulates repolarization of the *C. elegans* pharyngeal muscle. *J Exp Biol*. 2003;206:223–231.
97. Sze JY, Victor M, Loer C, Shi Y, Ruvkun G. Food and metabolic signalling defects in a *Caenorhabditis elegans* serotonin-synthesis mutant. *Nature*. 2000;403:560–564.
98. Estevez M, Estevez AO, Cowie RH, Gardner KL. The voltage-gated calcium channel UNC-2 is involved in stress-mediated regulation of tryptophan hydroxylase. *J Neurochem*. 2004;88: 102–113.
99. Kandel ER. The molecular biology of memory storage: A dialogue between genes and synapses. *Science*. 2001;294: 1030–1038.
100. Bishop NA, Lu T, Yankner BA. Neural mechanisms of ageing and cognitive decline. *Nature*. 2010;464:529–535.
101. Murakami H, Bessinger K, Hellmann J, Murakami S. Aging-dependent and -independent modulation of associative learning behavior by insulin/insulin-like growth factor-1 signal in *Caenorhabditis elegans*. *J Neurosci*. 2005;25:10894–10904.
102. Herndon LA, Schmeissner PJ, Dudaronek JM, et al. Stochastic and genetic factors influence tissue-specific decline in ageing *C. elegans*. *Nature*. 2002;419:808–814.
103. Kenyon CJ. The genetics of ageing. *Nature*. 2010;464:504–512.

104. Weindruch R, Walford RL. *The Retardation of Aging and Disease by Dietary Restriction*. Springfield, IL: Charles C. Thomas; 1988.
105. Tomioka M, Adachi T, Suzuki H, Kunitomo H, Schafer WR, Iino Y. The insulin/PI 3-kinase pathway regulates salt chemotaxis learning in *Caenorhabditis elegans*. *Neuron*. 2006;51:613–625.
106. Kodama E, Kuhara A, Mohri-Shiomi A, et al. Insulin-like signaling and the neural circuit for integrative behavior in *C. elegans*. *Genes Dev*. 2006;20:2955–2960.
107. Cornils A, Gloeck M, Chen Z, Zhang Y, Alcedo J. Specific insulin-like peptides encode sensory information to regulate distinct developmental processes. *Development*. 2011;138: 1183–1193.
108. Pierce SB, Costa M, Wisotzkey R, et al. Regulation of DAF-2 receptor signaling by human insulin and *ins-1*, a member of the unusually large and diverse *C. elegans* insulin gene family. *Genes Dev*. 2001;15:672–686.
109. Li W, Kennedy SG, Ruvkun G. *daf-28* encodes a *C. elegans* insulin superfamily member that is regulated by environmental cues and acts in the DAF-2 signaling pathway. *Genes Dev*. 2003;17:844–858.
110. Lakowski B, Hekimi S. The genetics of caloric restriction in *Caenorhabditis elegans*. *Proc Natl Acad Sci USA*. 1998;95: 13091–13096.
111. Giannakou ME, Goss M, Partridge L. Role of dFOXO in lifespan extension by dietary restriction in *Drosophila melanogaster*: Not required, but its activity modulates the response. *Aging Cell*. 2008;7:187–198.
112. Pitsikas N, Algeri S. Deterioration of spatial and nonspatial reference and working memory in aged rats: protective effect of life-long calorie restriction. *Neurobiol Aging*. 1992;13:369–373.
113. Irvine EE, Drinkwater L, Radwanska K, et al. Insulin receptor substrate 2 is a negative regulator of memory formation. *Learn Mem*. 2011;18:375–383.
114. Bohni R, Riesgo-Escovar J, Oldham S, et al. Autonomous control of cell and organ size by CHICO, a *Drosophila* homolog of vertebrate IRS1-4. *Cell*. 1999;97:865–875.
115. Naganos S, Horiuchi J, Saitoe M. Mutations in the *Drosophila* insulin receptor substrate, CHICO, impair olfactory associative learning. *Neurosci Res*. 2012;73:49–55.
116. Sherwood OD. Relaxin's physiological roles and other diverse actions. *Endocr Rev*. 2004;25:205–234.
117. Brogiolo W, Stocker H, Ikeya T, Rintelen F, Fernandez R, Hafen E. An evolutionarily conserved function of the *Drosophila* insulin receptor and insulin-like peptides in growth control. *Curr Biol*. 2001;11:213–221.
118. Ikeya T, Galic M, Belawat P, Nairz K, Hafen E. Nutrient-dependent expression of insulin-like peptides from neuroendocrine cells in the CNS contributes to growth regulation in *Drosophila*. *Curr Biol*. 2002;12:1293–1300.
119. Rulifson EJ, Kim SK, Nusse R. Ablation of insulin-producing neurons in flies: Growth and diabetic phenotypes. *Science*. 2002;296:1118–1120.
120. Petrascheck M, Ye X, Buck LB. An antidepressant that extends life span in adult *Caenorhabditis elegans*. *Nature*. 2007;450: 553–556.
121. Maier W, Adilov B, Regenass M, Alcedo J. A neuromedin U receptor acts with the sensory system to modulate food type-dependent effects on *C. elegans* life span. *PLoS Biol*. 2010;8: e1000376.
122. Hebb D. *The Organization of Behavior*. New York: Wiley; 1949.
123. Goodman MB, Hall DH, Avery L, Lockery SR. Active currents regulate sensitivity and dynamic range in *C. elegans* neurons. *Neuron*. 1998;20:763–772.
124. Mellem JE, Brockie PJ, Madsen DM, Maricq AV. Action potentials contribute to neuronal signaling in *C. elegans*. *Nat Neurosci*. 2008;11:865–867.

CHAPTER

11

Thermosensory Learning in *Caenorhabditis elegans*

Hiroyuki Sasakura and Ikue Mori
Nagoya University, Nagoya, Japan

Learning and memory are critical for survival and reproduction of any animal. Invertebrates show rich repertories of sensory behavior, thus facilitating the analysis of learning and memory. Studies on marine molluscas, such as *Aplysia* and *Hermissenda*, revealed the essential mechanism of synaptic plasticity.[1,2] Ecological and physiological studies on honeybee unraveled the amazing behavioral ability of insects and the underlying neural functions.[3–5] Behavioral molecular genetics in *Drosophila melanogaster* and *Caenorhabditis elegans* dissected the mechanism for neural plasticity.[6–9] Studies on other invertebrates with unique properties have yielded additional information on learning and memory. The main message from invertebrate studies is that neural logic is conserved between invertebrates and vertebrates including mammals. Thus, invertebrate neuroscience is a powerful tool for understanding our brain and mind. In this chapter, thermosensory learning of *C. elegans* is viewed as a suitable behavioral system for resolving the mechanism of neural plasticity.

CAENORHABDITIS ELEGANS NEUROSCIENCE

The nematode *C. elegans* hermaphrodite is composed of 959 somatic cells, 302 of which are neurons. The positions and morphologies of neural cell bodies and processes, and all of the chemical synapses and gap junctions, were resolved by serial reconstruction with electron microscopy, and these structures are thought to be almost invariable between individuals.[10,11] Despite the simple nervous system, *C. elegans* exhibits a rich repertoire of sensory and plastic behavior.[9,12–18] Many of the neurotransmitters and neuromodulators used in *C. elegans*, such as acetylcholine, glutamate, dopamine, serotonin, GABA, and neuropeptides, are homologous to those used in mammals.[19–24] The *C. elegans* genome project revealed that genes required for neural development and function are highly conserved in mammalian genes.[25,26]

Nonetheless, the most conspicuous difference between *C. elegans* and mammalian nervous systems is that there seems to be no action potentials in *C. elegans*. No voltage-gated sodium channel is predicted to exist in the *C. elegans* genome, and electrophysiological experiments suggest that *C. elegans* neurons are isopotential and do not generate Na^+ action potentials.[25–27] Ca^{2+} influx and subsequent sequential events are likely to be important for neural excitation in *C. elegans*. Similarly, *C. elegans* utilizes novel classes of ion channels, such as inhibitory glutamate-gated, acetylcholine-gated, and serotonin-gated chloride channels and excitatory GABA-gated cation channels.[20–23,28–31] These novel ion channels have been found only in invertebrates, thereby providing the opportunity to search selective drugs for nematode and insect parasites in vertebrates, including humans.[32]

An essential part of neural properties in *C. elegans* is amazingly similar to those in mammals. High similarity to the mammalian system is an advantage of *C. elegans* neuroscience. In addition, application of rapidly advanced technology such as imaging of neural activity and optogenetics enables worm researchers to study the dynamics of neurons and circuits.[33–38] Classical forward and reverse genetics is still powerful, aided by whole genome sequencing.[39] Recent work on behavioral variation between related species made it possible to analyze multiple gene effects on certain behaviors.[40–44] RNAi knockdown experiments are also

Invertebrate Learning and Memory.
DOI: http://dx.doi.org/10.1016/B978-0-12-415823-8.00011-3

useful for *C. elegans* neuroscience studies.[45] All of these advantages allow comprehensive and high-resolution studies that are potentially useful for understanding human brain and neuronal disorders.

BEHAVIORAL PLASTICITY IN C. *ELEGANS*

Caenorhabditis elegans apparently exhibits behaviors that reflect learning and memory.[9,12–17,46–48] Two forms of learning are generally recognized—nonassociative learning and associative learning. Nonassociative learning is defined as a change of responses to a stimulus without the requirement for positive or negative reinforcement. By contrast, associative learning is defined as the process by which an association occurs between two different stimuli. *Caenorhabditis elegans* possesses both associative and nonassociative learning. Habituation to mechanical stimulus and adaptation to chemicals are regarded as nonassociative learning.[49,50] The enhancement of avoidance behavior from deleterious chemicals after pre-exposure to them is also regarded as nonassociative learning.[51] *Caenorhabditis elegans* associates two different stimuli and shows memory-based plastic behaviors, which are considered as associative learning.

Associative learning between food state (the presence or absence of food) and several environmental stimuli, such as odorants, water-soluble chemicals, and temperature, has also been analyzed. In most cases, food state was used as the unconditioned stimulus (US) and environmental stimulus was used as the conditioned stimulus (CS).[52–61] Some studies avoided using food as the US because of its indispensability for animals and instead used two different chemical cues.[62,63] *Caenorhabditis elegans* naively prefers a certain concentration of oxygen level and shows plasticity to oxygen concentrations.[64–67] Some pathogenic bacteria infect animals, resulting in death after several days. To protect from pathogenic bacterial infection, *C. elegans* has a behavioral strategy to judge food quality and leave pathogenic bacteria by associative learning.[41,42,68,69] Thus, *C. elegans* is likely to express plasticity to any kind of environmental stimuli.

Different strains of *C. elegans*, Bristol and Hawaiian, show completely different behaviors on bacterial food lawn. The Bristol strain shows solitary behavior, in which the animals independently move on the bacterial lawn, whereas the Hawaiian strain shows social behavior by aggregating on the edge of the lawn. Surprisingly, the difference between solitary and social behavior is regulated in single polymorphism in the *npr*-1 gene that encodes a homolog of the mammalian neuropeptide Y receptor.[40]

When animals are placed in the situation in which several alternative choices are available, a decision-making process produces a final choice. Thus, *C. elegans* exhibits the behavioral choice that is regarded as decision making. For example, *C. elegans* avoids harmful Cu^{2+} ions. When attractive odorant is spotted over the Cu^{2+} barrier, *C. elegans* decides whether to migrate to the odor source by crossing over the Cu^{2+} barrier or to withdraw from the Cu^{2+} barrier, depending on the balance between Cu^{2+} and odor concentrations.[70] Bristol and Hawaiian strains vary in their tendency to migrate from or remain on the small lawn of bacterial food. Polymorphisms in a noncoding region of the *tyra*-3 gene, encoding a catecholamine receptor cooperating with polymorphisms in the *npr*-1 gene, regulate whether *C. elegans* leaves or remains on the food.[43,71]

In the behavioral experiments described previously, memory was mostly maintained in the order of minutes or hours. The intriguing question is whether much longer lasting memory, which could be regarded as long-term memory, exists in *C. elegans*. Several lines of evidence suggest that *C. elegans* indeed possesses long-term memory. *Caenorhabditis elegans* shows olfactory imprinting, in which exposure to attractive odors at the first larval stage in the presence of food leads to enhancement of attractive responses to the experienced odors at the adult stage.[72] Upbringing dramatically influences human personality, and the importance of the conditions of upbringing is also observed in *C. elegans*. Animals grown in a group exhibited a much stronger response to the mechanical stimulus than solitary-grown animals at adult stage. This enhancement may be generated by an activity-dependent process during development, in which group-grown animals collided with each other, thereby leading to strengthening of the response to touch.[73]

In summary, *C. elegans* exhibits nonassociative and associative learning as well as short-term and long-term memory. *Caenorhabditis elegans* also shows choice selection behavior that is regarded as decision making. Hence, *C. elegans* is an ideal system to study the principle of learning and memory at the neuronal and neural circuit level.

THERMOTAXIS IN C. *ELEGANS*

Environmental temperature affects many aspects of the physiology of organisms, such as growth, reproduction, development, behavior, and even life span.[74–78] In 1975, Hedgecock and Russell first reported thermotaxis in *C. elegans* (Figure 11.1).[59] They established two types of assay systems to examine the thermotactic behavior in *C. elegans* (Figures 11.1B and 11.1C): population accumulation

(A) *Proc. Nat. Acad. Sci. USA*
Vol. 72, No. 10, pp. 4061–4065, October 1975
Genetics

Normal and mutant thermotaxis in the nematode *Caenorhabditis elegans*

(behavior/mutants/temperature/chemotaxis)

EDWARD M. HEDGECOCK AND RICHARD L. RUSSELL*

Division of Biology, California Institute of Technology, Pasadena, Calif. 91125

Communicated by William B. Wood, May 12, 1975

ABSTRACT When grown at a temperature from 16° to 25° and placed on a thermal gradient, the nematode *Caenorhabditis elegans* migrates to its growth temperature and then moves isothermally. Behavioral adaptation to a new temperature takes several hours. Starved animals, in contrast, disperse from the growth temperature. Several mutants selected for chemotaxis defects have thermotaxis defects as well; these behaviors depend on some common gene products. New mutants selected directly for thermotaxis defects have unusual phenotypes which suggest mechanisms for thermotaxis.

(B)

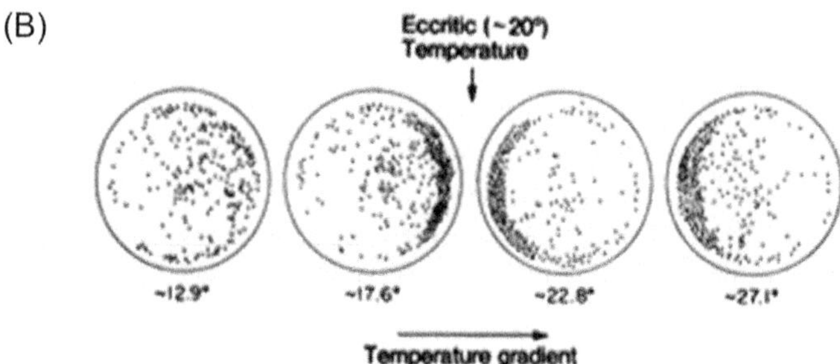

FIG. 1. The distribution of 20°-raised asynchronous adults in the accumulation assay. Animals were applied to the plate centers, run for 1 hr, chloroformed, and their final positions were recorded by dots. Summed results of five separate experiments.

(C)

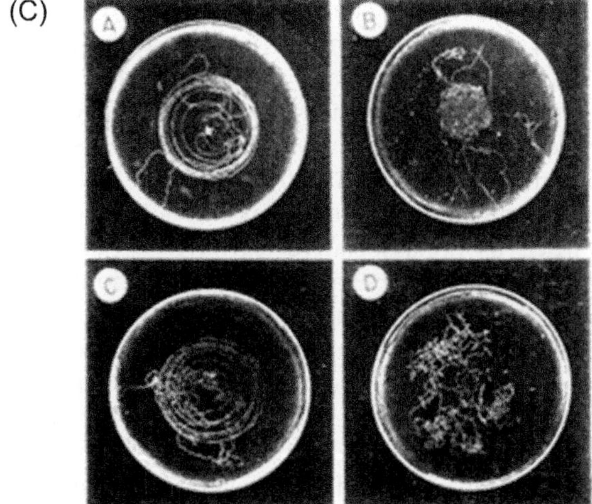

FIG. 8. Tracking of 20°-raised nematodes on radial temperature gradients: A, wild type; B, EH67 (cryophilic); C, DD79; D, DD71 (atactic).

FIGURE 11.1 The first report of thermotaxis in *C. elegans* by Hedgecock and Russell in 1975.[59] (A) Title and abstract. (B) The distribution of animals cultivated at 20°C in the accumulation assay. (C) Track of wild-type and mutant animals cultivated at 20°C on a radial temperature gradient plate. Source: *Reproduced from Hedgecock and Russell.*[59]

on a linear gradient and individual tracking on a radial gradient. In the population accumulation assay, a stable and reproducible linear thermal gradient was established by connecting two thermostatically regulated water baths, 5° and 35°C, by a 61 × 10 × 1.3-cm aluminum slab bridge tightly bolted at each end to a 10-cm aluminum cube immersed in a bath. Four plastic petri plates (9-cm-diameter circular type) containing 35 mL agar culture medium were placed on the aluminum gradient slab at regular intervals (Figure 11.1B). The temperature of the agar surface was monitored with a glass probe thermistor. A uniform stable thermal gradient of 0.5°C/cm was established on the agar surface. The temperature at the center of each plate was 12.9°, 17.6°, 22.8°, and 27.1°C, respectively (Figure 11.1B). The animals cultivated at 16°, 20°, or 25°C with food *Escherichia coli* were collected and washed with buffer and applied to the centers of agar plates. After animals were allowed to move freely for 1 hr, they migrated to the regions that approximately corresponded to the past cultivation temperature. Cultivation temperature-shift experiments indicated that resetting to a new cultivation temperature took 2–4 hr.

Thermotaxis also changed in food-deprived and overcrowded conditions. The animals cultivated at 20°C under high density that leads to the starved condition rather dispersed from 20°C. These observations suggest that thermotaxis is a highly sensitive environmental condition and also highly modifiable based on past experience.

The second type of assay was an individual assay on radial temperature gradients. Individual worms grown at 20°C were placed on 9-cm plastic petri plates containing 10 mL agar. The plates were then moved to

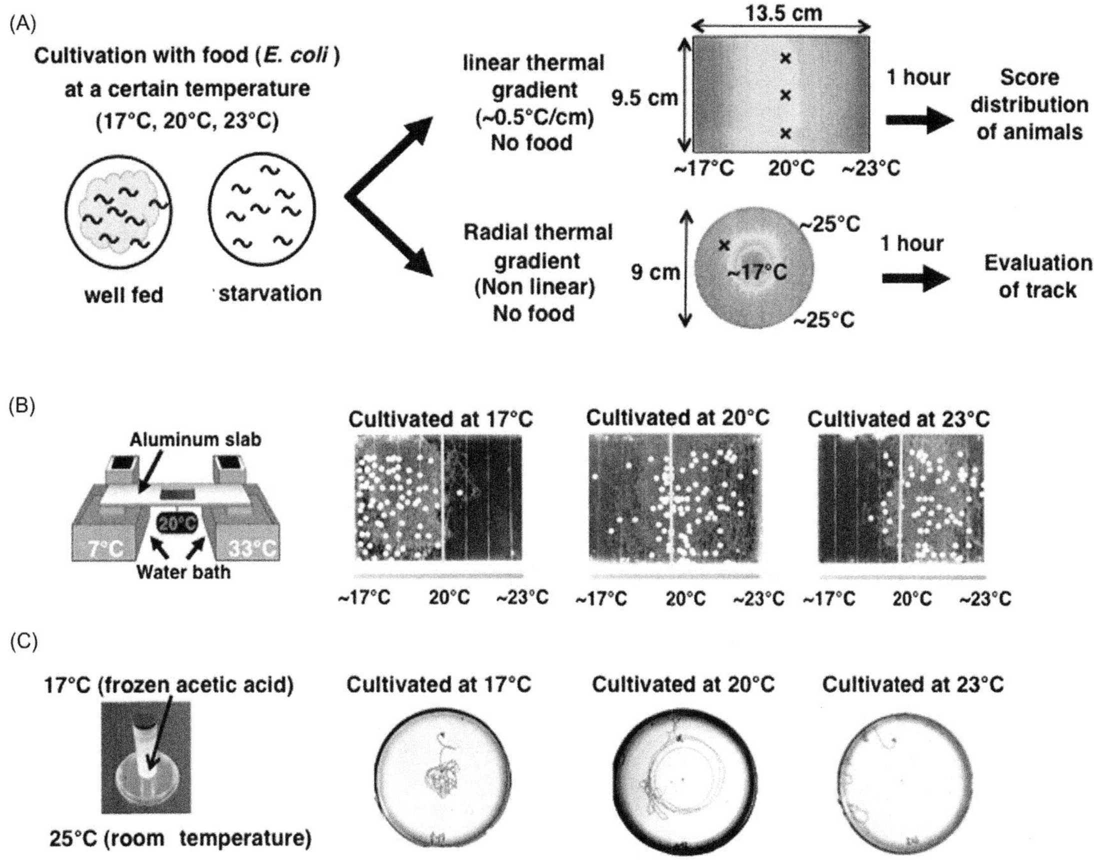

FIGURE 11.2 Thermotaxis assays with modification of Hedgecock and Russell's[59] method. (A) Population assay on a linear thermal gradient and individual assay on a radial thermal gradient. Worms cultivated with food (well-fed condition) or without food (starvation condition) are assayed on linear or radial thermal gradient. X indicates the initial location where animals are placed. (B) Assay apparatus of the population assay on linear thermal gradient and photographs after assays. White dots on the photographs show the final positions of animals. (C) Assay system of the individual assay on radial thermal gradient and tracks of animals after assays.

room temperature (24–26°C) and inverted. A 20-mL glass vial filled with frozen glacial acetic acid (melting point 16.6°C) was placed at the center of each plate on the plastic bottle just above the agar. A stable radial temperature gradient was established on which the worms could move (Figure 11.1C and 11.2C). *Caenorhabditis elegans* cultivated at 20°C with food exhibited isothermal tracks on radial gradients. Isothermal tracking near the past experienced temperature can provide the opportunity to directly observe the memory process.

Hedgecock and Russell[59] reported that several mutants defective in chemotaxis also showed abnormal thermotaxis, predicting that the common mechanism exists for chemo- and thermosensation. They also screened for and isolated thermotaxis defective mutants. Some of the mutants were specifically defective in thermotaxis, suggesting that a specific mechanism exists for thermotaxis. The Hedgecock and Russell paper is a milestone not only for first reporting the thermotaxis, a plastic behavior in *C. elegans*, but also for predicting the underlying conceptual mechanism for thermotaxis.

In 1995, Mori and Ohshima[60] faithfully took over the work by Hedgecock and Russell[59] and reproduced their results on thermotactic behavior, leading to the identification of neural circuits and neurons regulating thermotaxis (Figures 11.2 and 11.3C). Some studies successfully reproduced the themotactic behavior, whereas other studies argued that the thermotactic behaviors they observed were different from those observed by Hedgecock and Russell and by Mori and Ohshima.[46,47,79] They argued that although *C. elegans* migrated to the colder regions, they never migrated to the warmer regions. The essence of the argument is that *C. elegans* possesses the cryophilic drive but never possesses the thermophilic drive. The reason for this discrepancy turned out to be the difference in steepness of thermal gradients in the assays. In 2010, Jurado *et al.*[80] clearly showed that the steepness of the thermal gradient determines whether *C. elegans* migrates to the warmer regions or not. When the steepness is less than 1.0°C/cm, the distribution of animals on the temperature gradient is similar to those reported by Hedgecock and Russell. By contrast, when the steepness is more than 1.0°C/cm, the distribution changes: Whereas animals cultivated at low temperatures accumulated at the cultivation temperature, animals cultivated at high temperature showed athermotactic phenotype. Indeed, the steepness of thermal gradients in studies arguing that *C. elegans* only possesses the cryophilic drive was more than 1.0°C/cm. The importance of steepness was mentioned in the original Hedgecock and Russell paper. They thoughtfully considered natural soil condition for *C. elegans*. They cited the environmental physics paper[81] and noted the following:

> In a moist, organic soil subject to diurnal surface temperature fluctuations, the soil temperature varies as a propagating sine wave (about 2 cm/hr) whose amplitude decays exponentially with depth. A nematode in the upper few centimeters of soil may experience vertical gradients as large as 0.5°C/cm and temporal gradients of 0.5°C/hr.

A thermal gradient as large as 0.5°C/cm is a reasonable experimental condition to study the neural regulation of thermotaxis, and a thermal gradient more than 1.0°C/cm is quite extreme. Although assays using a sharp thermal gradient could be useful for analyzing neural responses on the extreme conditions,[82] such assays should be considered as basically different from Hedgecock and Russell's[59] original assay that aimed to reveal thermosensory behavior in the natural environment.

NEURAL CIRCUIT FOR THERMOTAXIS

Ablations of a specific neuron in live animals by a laser microbeam and the evaluation of the consequent effect on thermotaxis led to the identification of neurons and circuits for thermotaxis.[60] The AFD sensory neurons had been supposed to be thermosensory neurons because the *ttx-1*(*p678*) cryophilic mutant lacked microvillus-like projections in the sensory ending of AFD neurons but was normal for other sensory behaviors (Figures 11.3A and 11.3B).[84,86] Killing both left and right AFD neurons completely abolished isothermal tracking behavior and caused athermotactic or cryophilic phenotype. The abnormalities of AFD-killed animals are independent of cultivation temperature. The reciprocal experiment in which all amphid sensory neurons except AFD were killed showed normal thermotaxis, suggesting that AFD is the site of thermosensation. However, the appearance of cryophilic movement in AFD-killed animals suggests that AFD is not the sole neuron for thermosensation and another neuron(s) may sense the temperature. AWC neurons, designated as olfactory neurons that sense several volatile odorants,[16,87] proved to be another example of thermosensory neurons.[88,89] Killing both AFD and AWC resulted in athermotactic phenotype, and Ca^{2+} imaging showed that AWC indeed responded to the temperature change.

To identify additional neurons required for thermotaxis, the amphid interneurons that connect with AFD by chemical synapses and gap junctions were killed.[60] When AIY interneurons postsynaptic to AFD were killed, almost all of the AIY-killed animals exhibited clear cryophilic phenotype and lost the ability to move isothermally. One of the major postsynaptic partners of AIY is AIZ, another amphid interneuron. The AIZ-killed animals showed thermophilic phenotype; they sought higher temperature than their cultivation temperature, and then they left isothermal tracks. These results suggest that AIY is responsible for thermophilic drive, whereas AIZ is responsible for cryophilic drive, and that the counterbalanced regulation between two interneurons is critical for thermotaxis. Whereas killing AFD thermosensory neurons or AIY interneurons resulted in the loss of the ability to make isothermal tracking, by killing AIZ, the ability to make isothermal tracking was retained. Thus, AFD and AIY are crucial for isothermal tracking, whereas AIZ is not required for isothermal tracking and may be important for maintaining the balance between thermophilic and cryophilic drives (Figure 11.3C).

The RIA interneuron could receive synaptic inputs from both AIY and AIZ interneurons. Based on neural connections revealed by White *et al.*,[10] RIA locates at the site for integrating various sensory inputs. RIA-killed animals showed defective thermotaxis that was qualitatively different from the defects of AFD-, AIY-, or AIZ-killed animals. RIA may integrate signals from both AIY and AIZ and regulate the motor output.

A neural model for thermotaxis circuit was proposed based on laser ablation experiments (Figure 11.3C).[60] Temperature is sensed primarily by AFD and secondarily by AWC. Thermal information is relayed to AIY. AIY regulates AIZ inhibition, and the counterbalanced regulation between AIY and AIZ regulates the direction of movement through locomotion controlling neuron RIA. The thermotaxis neural circuit is as simple and clear as any other circuits known in other organisms. This circuit provides the opportunity to study entire steps from sensory input to neural computation and final motor output control.[46,90]

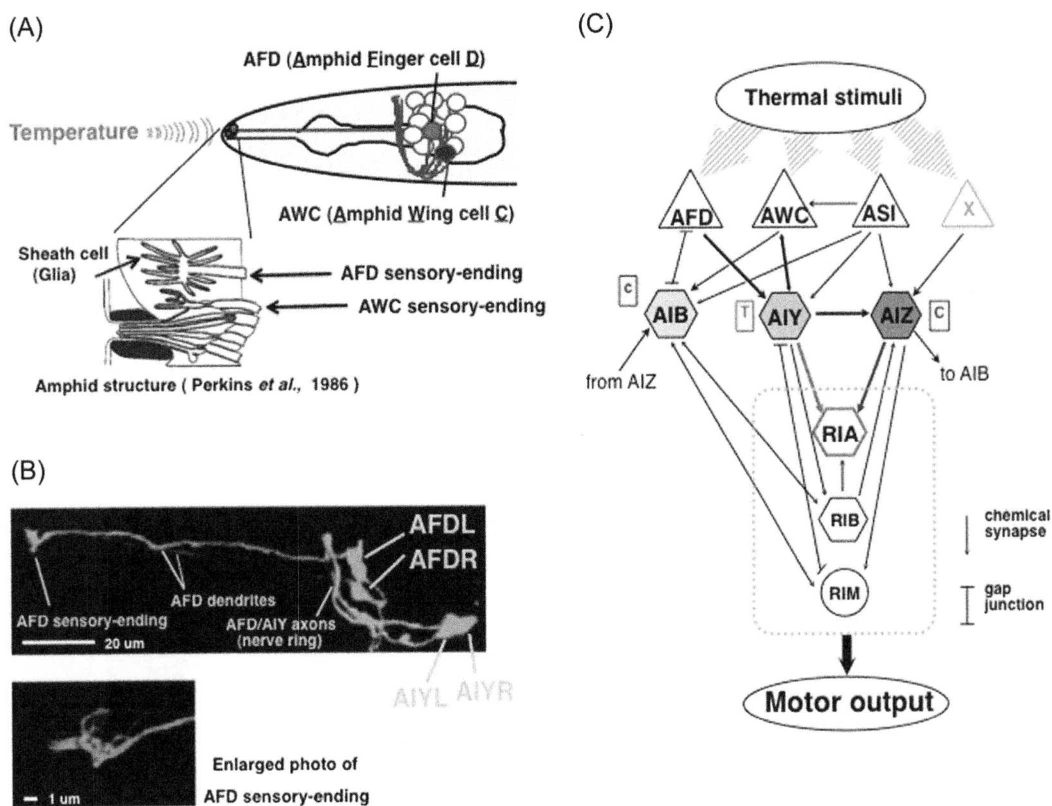

FIGURE 11.3 **Thermosensory neurons and neural circuit for thermotaxis.** (A) *Caenorhabditis elegans* senses environmental temperature through two pairs of sensory neurons, AFD (amphid finger cell D) and AWC (amphid wing cell C). Amphid is the name of the sensory organ.[83] The sensory ending structures of AFD and AWC are specialized. AFD sensory ending has a finger-like structure and is thought to be important for thermosensation. AWC has a large flat cilia-like wing.[84] (B) AFD thermosensory neurons and AIY interneurons are visualized with GFP. (C) Neural circuit for thermotaxis.[47] Studies have shown that the ASI neuron, which is regarded as a pheromone-sensing neuron, also senses temperature.[82,85] ASI in combination with AFD and AWC is required for robust cryophilic movement on a sharp thermal gradient (1.0°C/cm). Source: *Panel A reproduced from Perkins* et al.[84]

THERMOSENSORY SIGNALING

Genes participating in thermosensory function were identified by forward as well as reverse genetics (Figure 11.4).[46,79,90] The cGMP-dependent cation channel composed of TAX-2 and TAX-4 is crucial for activation of AFD.[95,96] The three guanylate cyclases—GCY-8, GCY-18, and GCY-23—specifically expressed in AFD produce cGMP as a second messenger. GCY-8, GCY-18, and GCY-23 redundantly function for thermosensation in AFD,[97] but each cyclase has a different role in the execution of isothermal behavior and the response to temperature.[98] In addition, CaMKI/Ca^{2+}-calmodulin-dependent protein kinase I regulates thermosensation as well as thermosensory signal-dependent gene expression.[99] The sensory ending of AFD is embedded into the sheath cell that is thought to be equivalent to glia.[84] Ablation of sheath glia does not eliminate AFD function but results in thermophilic behavior, suggesting that the interaction between AFD and sheath glia is also essential for thermosensation.[100]

TAX-6/calcineurin functions as a gain controller for the responses to thermal stimulus. The loss-of-function mutants of *tax-6* gene exhibited thermophilic phenotype that is interpreted as hypersensitive to thermal stimulus. By contrast, the introduction of the gain-of-function form of *tax-6* into wild-type AFD neurons caused animals to be cryophilic. These results suggest that TAX-6/calcineurin controls the gain of AFD and that AFD has the ability to sense a wide range of temperatures ranging from low to high.[101]

Loss-of-function mutants of the *ttx-4* gene also show the thermophilic phenotype, which is qualitatively different from that of *tax-6*. *ttx-4* encodes protein kinase C (PKC) epsilon/eta, which belongs to a novel type of PKC family, nPKC, that is activated by diacylglycerol (DAG) alone. Given that the *ttx-4* (*pkc-1*) gene was suggested to play an important role in neuromuscular junction and synaptic transmission in ASE

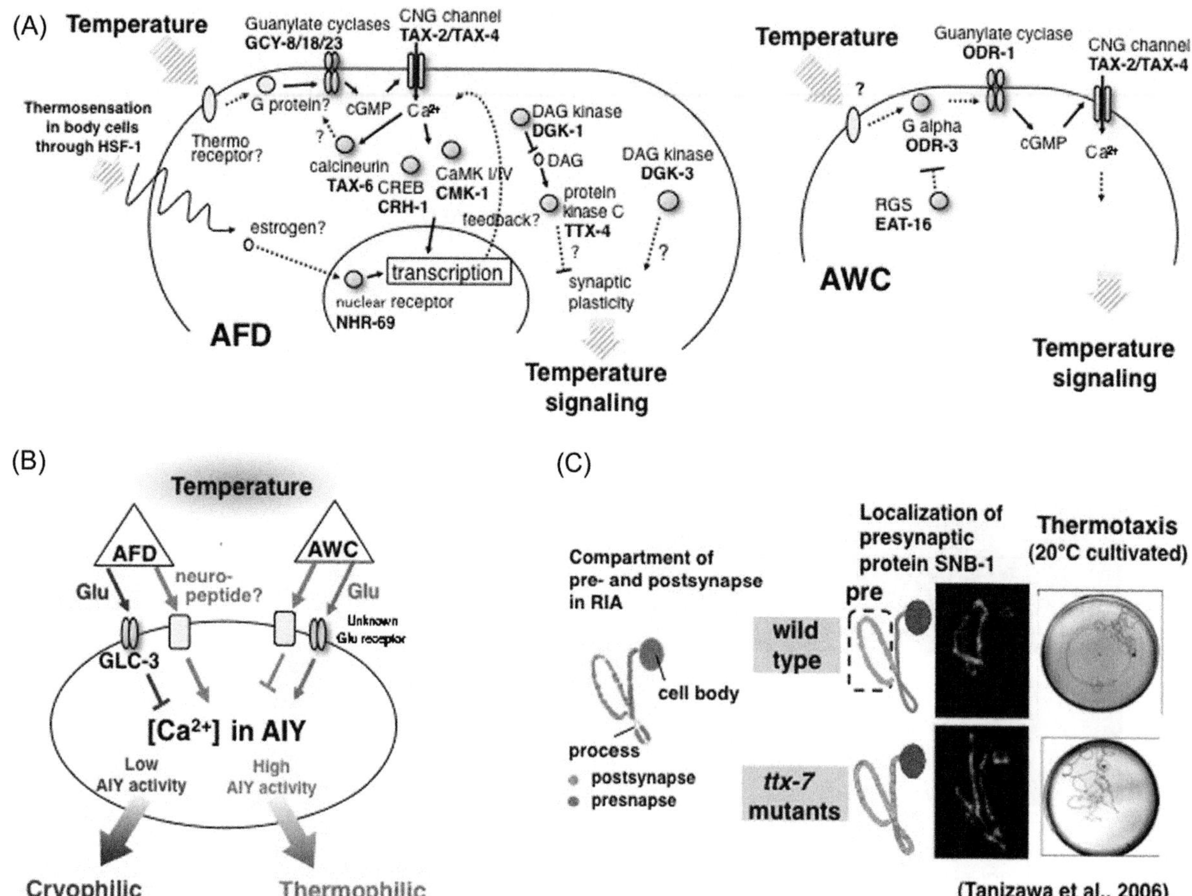

FIGURE 11.4 **Molecular basis of thermosensation and thermal information flow.** (A) Models of the molecular mechanism for thermosensation in AFD and AWC. (B) Models of information flow from thermosensory neurons AFD and AWC to AIY. Vesicular glutamate transporter VGLUT/EAT-4-dependent glutamatergic transmission from AFD (blue) downregulated the activity of AIY through a glutamate-gated chloride channel GLC-3. This inhibition of AIY drives cryophilic movements. Neuropeptide-dependent excitatory signaling from AFD to AIY is proposed by studies of optogenetics. In contrast to the case of AFD, EAT-4-dependent glutamatergic transmission from AWC (red) upregulated the activity of AIY through an unknown glutamate receptor(s), thereby inducing thermophilic drive. The inhibitory signaling from AWC to AIY is also estimated by molecular genetic analysis.[91–93] (C) The compartment of pre- and postsynaptic regions and thermotaxis are regulated by PIP_2 signaling through IMPase. Clear compartmentalization of pre and postsynaptic regions on RIA process was revealed by EM analysis.[10] The photograph shows the localization of presynaptic protein SNB-1::VENUS and thermotaxic behavior of wild-type and *ttx*-7 mutant animals. SNB-1::VENUS is specifically localized to the presynaptic region in wild-type mutants, whereas it is distributed over the postsynaptic region in *ttx*-7 mutants. *ttx*-7 mutants show abnormal thermotaxis, similar to that of RIA-killed animals.[94] Source: *Panel C reproduced from Tanizawa* et al.[94]

taste-sensing neurons, TTX-4 may negatively regulate synaptic transmission from AFD to AIY.[102–104]

The *dgk*-3 gene encodes DAG kinase DGK-3, and *dgk*-3 mutants have shown poorer isothermal tracking.[89] The expression of DGK-3 in AFD rescued the defect, suggesting that DGK-3 is important for isothermal tracking. Because isothermal tracking reflects the animals' ability to remember the past experience of temperature and the AFD neuron is responsible for isothermal tracking, DGK-3 is likely required directly for the memory process in thermotactic behavior.

Temperature is also sensed by non-neuronal body cells such as intestine and body wall muscle cells through heat shock transcription factor HSF-1.[105] The HSF-1-mediated thermosensation cell non-autonomously regulates the activity of AFD through estrogen signaling.[105] Thus, neurohormonal modulation of hard-wired neural circuits also contributes to behavioral plasticity in *C. elegans*.

The thermosensory signaling pathway in AWC is in principle similar to that in AFD (Figure 11.4A), namely cGMP is used as a second messenger and the cGMP-dependent cation channel, composed of TAX-2 and TAX-4, is crucial for activation of AWC. The G protein signaling pathway, which involves G_α/ODR-3, a regulator of G protein signaling RGS/EAT-16, and a guanylate cyclase ODR-1, is important for transmission of temperature information.[88,106] Most of molecular components used in AWC thermosensation are shared by olfactory sensation. How the thermal and olfactory signals are discriminated in AWC is currently an intriguing question.[16]

AFD THEMOSENSORY NEURONS MEMORIZE CULTIVATION TEMPERATURE: THE SENSORY NEURON ACTS AS A MEMORY DEVICE

In isothermal movement, the animals usually move perpendicularly to a thermal gradient, in which a small temperature change has to be accessed without changing the overall body direction (Figures 11.1C, 11.2A, and 11.2C). The observation of isothermal behavior estimated that AFD has the ability to recognize a temperature change less than 0.1°C.[59,60] The importance of AFD as a thermosensory neuron was confirmed by Ca^{2+} imaging (Figures 11.5A and 11.5B),[107,108] in which AFD responded to both warming and cooling and discriminated temperature changes as little as 0.05°C upon application of various patterns of temperature stimuli.[108] Electrophysiological experiments also showed that AFD responded to both warming and cooling by opening and closing ion channels, respectively.[111]

Surprisingly, AFD functions as a temperature memory neuron in addition to its role in thermosensation (Figures 11.5A and 11.5B).[107,108] Ca^{2+} imaging revealed that temperature memory was stored in AFD: The AFD neuron of the animals cultivated at 20°C responded to the warming stimulus above 19°C. Similarly, the AFD of animals cultivated at 15° or 25°C responded to the warning above the threshold temperature that was near the cultivation temperature of 15° and 25°C, respectively. The sensory ending of AFD disconnected from the cell body by a laser beam still retained the ability to memorize the temperature.[108] These results suggest that threshold temperature for Ca^{2+} influx is measured and stored at the AFD sensory ending. Similar to AFD neurons, AWC neurons appear to have the ability to memorize temperature.[88]

CREB (cAMP-response element binding protein) is a transcriptional factor that regulates neural plasticity from invertebrates to mammals.[1] The mutants in the

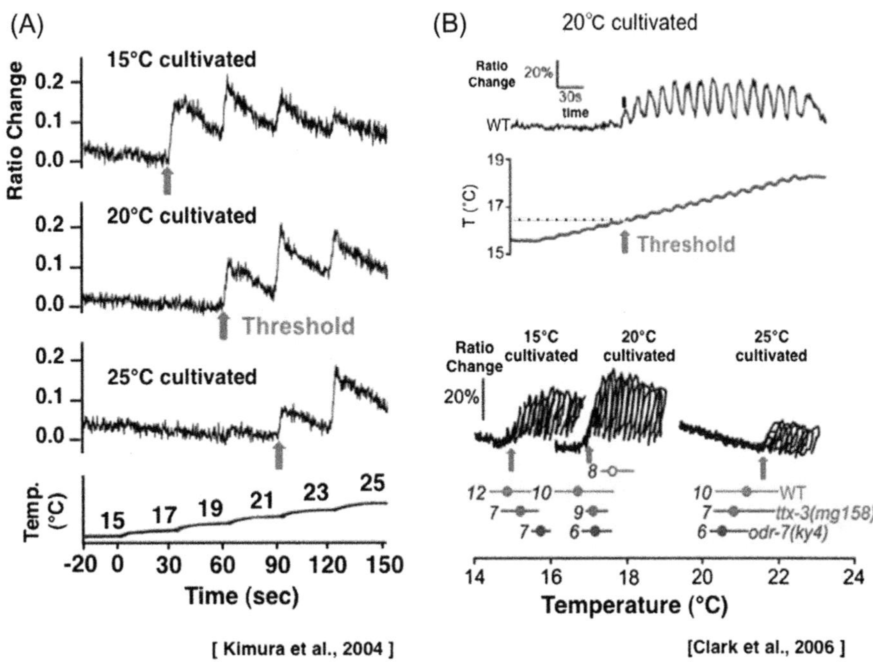

FIGURE 11.5 Memorization of temperature in AFD thermosensory neurons. (A) AFD responded to the stepwise warming with dependency on the past cultivation temperature.[107] When temperature was raised from 15° to 25°C by 2°C increments in a steplike manner, the AFD neuron of wild-type animals cultivated at 20 °C responded to every warming step above 19°C in Ca^{2+} imaging. Similarly, the AFD of animals cultivated at 15° or 25°C responded to warming above the threshold temperature that is near the cultivation temperature of 15° or 25°C, respectively. (B) AFD responded to the sinusoidal temperature oscillation.[108] When temperature was raised with small-amplitude sine wave, the AFD neuron of wild-type animals cultivated at 20°C began to response near 17°C. The AFD of animals cultivated at 15° or 25°C responded to sinus warming above the threshold temperature that is near but lower than the cultivation temperature. The AWA sensory neurons and AIY interneurons are the primary presynaptic and postsynaptic partners of AFD neurons. AWA and AIY are functionary defective in *odr-7*(*ky*4) and *ttx-3*(*mg*158) mutants, in which AWA and AIY are developmentally undifferentiated.[109,110] Operating ranges of AFD Ca^{2+} dynamics were almost normal in *odr-7*(*ky*4) and *ttx-3*(*mg*158) mutants, suggesting that pre- and postsynaptic partner neurons do not affect AFD Ca^{2+} dynamics. Source: *Panel A reproduced from Kimura* et al.[107] *Panel B reproduced from Clark* et al.[108]

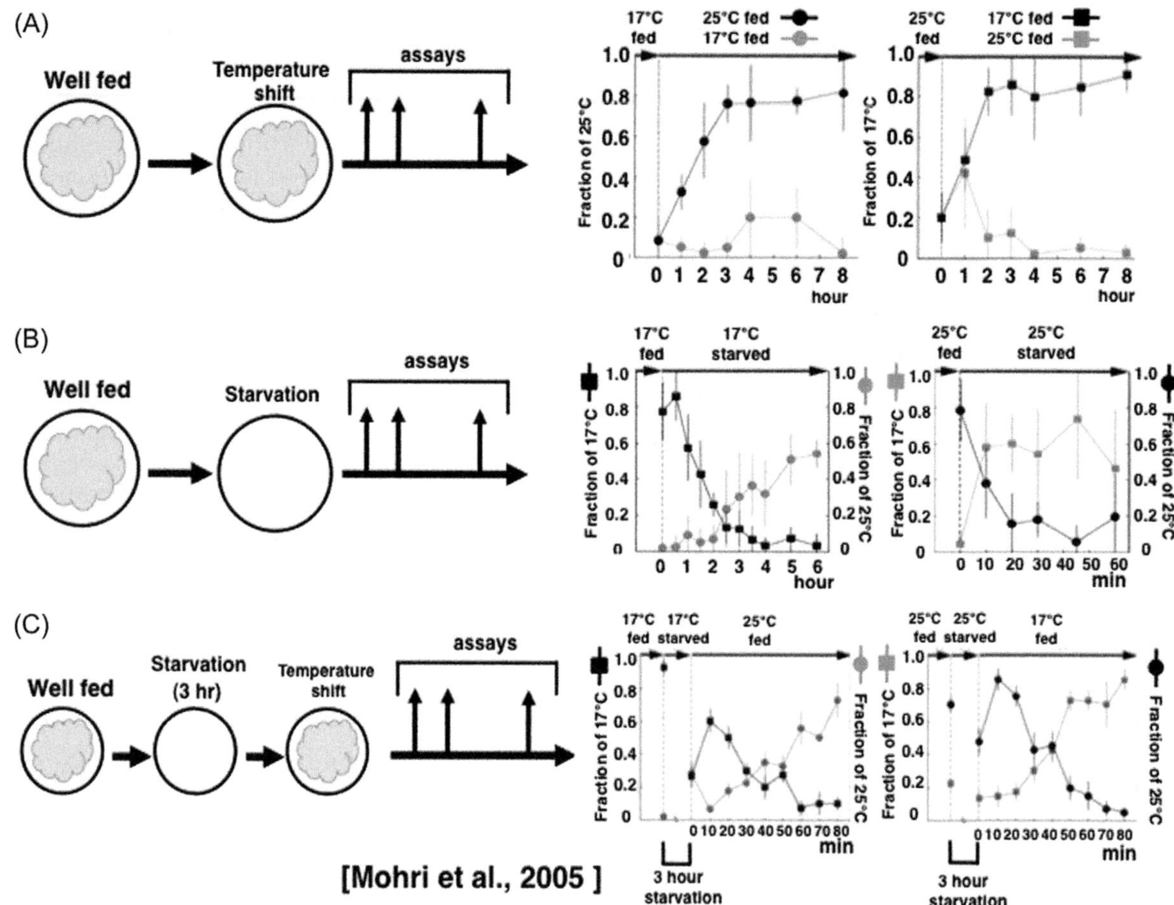

FIGURE 11.6 Associative learning between temperature and food state. (A) The memory formation to a new temperature takes a few hours for well-fed animals.[61] The panel on the left shows the experimental procedure. Graphs on the right show the results of time-course experiments on temperature shift from 17° to 25°C or from 25° to 17°C in fed condition, respectively. The *y* axis indicates the fraction of animals that migrated to 25° and 17°, respectively. (B) Feeding state modulates thermotaxis behavior.[61] The panel on the left shows the procedure of the experiments. The results of the experiments are shown on the right. Well-fed animals cultivated at 17°C (left) or 25°C (right) were transferred to conditioning starvation plate at 17° or 25°C, respectively. (C) The memory formation process for temperature and feeding states are biologically discrete.[61] The panel on the left shows the procedure of the experiments. The results of the experiments are shown on the right. Animals conditioned to be starved at 17°C (left) or 25°C (right) for 3 hr were recultivated with food at 25° or 17°C, respectively. Source: *Reproduced from Mohri* et al.[61]

crh-1 gene encoding a *C. elegans* homolog of CREB showed abnormal thermotaxis, and the expression of *crh*-1*cDNA* only in AFD almost completely rescued the defects.[112] Hence, CREB may be required for memory in AFD or presynaptic plasticity in AFD.

ASSOCIATIVE LEARNING BETWEEN TEMPERATURE AND FOOD

Caenorhabditis elegans eats bacteria as a nutrient.[8] The presence or absence of food is a determinant that largely affects the behavior of *C. elegans*.[9,12–18,48,113,114] Dynamic thermotactic responses can also be observed in association with the presence or absence of food.[46,47] Mohri *et al.*[61] rigorously and comprehensively analyzed the plasticity of thermotaxis using individual assays with radial temperature gradient. After shifting the cultivation temperature from 17° to 25°C for 4 hr, well-fed animals migrated to a new cultivation temperature of 25°C, whereas it took only 2 hr for well-fed animals to migrate to a new temperature after shifting from 25° to 17°C (Figure 11.6A). Then, experiments were conducted in which the cultivation temperature was kept constant and only feeding state was changed (Figure 11.6B). When the animals were cultivated at 17°C in food-deprived condition for 3 hr, they seldom migrated to the region of 17°C. When the animals were cultivated at 25°C in food-deprived condition for 30 min, they seldom migrated to the region of 25°C. Thus, the starvation experience induced avoidance from the cultivation temperature, and it required a much shorter time period to induce cultivation temperature avoidance when cultivated at high temperature in the food-deprived condition.

Temperature higher than 25°C is known to threaten the survival of *C. elegans*. For example, the animals become sterile after being cultivated at temperatures higher than 26°C.[115,116] Dauer formation, an alternative developmental stage resistant to harsh conditions, is generated after 27°C cultivation.[85,117] Perhaps, *C. elegans* possesses the behavioral strategy to keep away from harmful high temperature. Several groups reported that although *C. elegans* migrated to the colder regions, they never migrated to the warmer regions on a sharp thermal gradient.[47,79] In addition, 17°C-cultivated animals started to migrate down the temperature gradient toward the preferred colder regions rather quickly after being placed in the 20°C region of a shallow gradient, whereas 23°C-cultivated animals moved around near 20°C for 10 min before migrating to the preferred high temperature.[118] These observations are consistent with the idea that *C. elegans* has a native tendency to prefer colder temperature to higher temperature, thereby making it difficult to move them to higher temperature regions.[47]

Rapid changes in the food supply to starved animals are extremely influential in thermotaxis.[61] Starved animals previously cultivated either at 17° or at 25°C migrated to the cultivation temperature after recultivation at the same temperature with food for only 10 min. To further investigate the strong and quick behavioral switch from aversion to attraction with respect to the cultivation temperature, two parameters—cultivation temperature and feeding state—were simultaneously changed (Figure 11.6C).[61] When animals starved at 17°C for 3 hr were recultivated with food at 25°C for 10 min, most animals migrated to the former cultivation temperature of 17°C, a noxious temperature associated with food-deprived experience. Likewise, when animals starved at 25°C for 3 hr were recultivated with food at 17°C, most animals initially migrated to 25°C, although they eventually migrated to 17°C—a temperature associated with food availability. These results suggest that food intake quickly causes the transition from aversion to attraction in thermotactic response, and they imply that the formation of memory for temperature and that for feeding state are distinct processes. A current plausible model is that temperature memorized in AFD is transmitted to downstream neurons and integrated with food state (Figure 11.7). Consistent with this hypothesis, food state did not affect the neural activity of AFD.[119]

REGULATION OF ASSOCIATIVE LEARNING BY INSULIN AND MONOAMINE SIGNALING

aho (*abnormal hunger orientation*) mutants defective in forming an association between temperature and food state were screened and characterized.[61] *aho* mutants exhibited the normal thermotaxis in the well-fed state, and many of them exhibited normal sensorimotor responses. They are novel thermotaxis defective mutants in which modulation of thermotaxis by feeding state is specifically impaired.

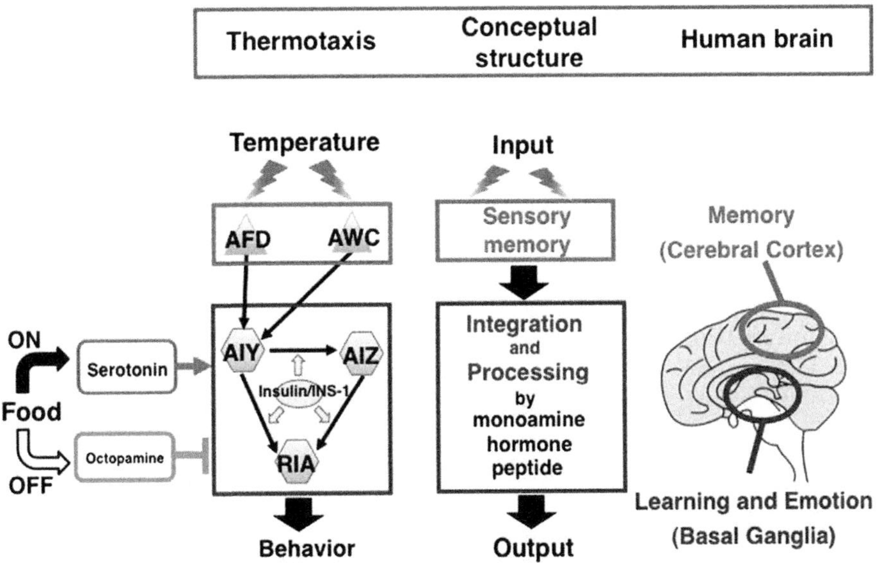

FIGURE 11.7 Proposed analogy between the thermotaxis neural circuit in *C. elegans* and the human brain. In *C. elegans*, temperature is sensed and thermal information is stored in AFD (and probably in AWC). Stored information in AFD is transmitted to the thermotaxis core interneurons AIY, AIZ, and RIA. Thermal information and food state are integrated and processed in those interneurons by monoamines and insulin to generate output behavior. In human brain, working memory is coded in cerebral cortex, and the coded information is conveyed to basal ganglia, in which learning and emotion proceed with modulation through monoamines. We propose here the functional analogy between simple neural circuits in *C. elegans* thermotaxis and the functionally layered structure of human brain.

aho-2 is one of the *aho* mutants severely defective in food-associated thermotactic plasticity. The *aho*-2 gene is identical to *ins*-1 gene, which encodes a homolog of human insulin.[119] INS-1 acts cell non-autonomously, which is consistent with studies that have found that insulin endocrine signaling is essential for neural plasticity in mammalian brain (Figure 11.7).[120] The genetic epistasis analysis showed that INS-1 antagonizes insulin receptor DAF-2 and its downstream PI3-kinase AGE-1 in food-associated thermotactic plasticity. The defects of *age*-1 mutants were rescued by expressing the *age*-1 gene in any of three core interneurons required for thermotaxis—AIY, AIZ, or RIA. These results suggest that INS-1 acts to any interneuron of thermotaxis circuits for associative learning. Consistent with the notion that integration of food and temperature information occurs within interneurons, the absence of food did not affect the response of AFD neurons. *ins*-1 mutants are also defective in associative learning between food and salt, suggesting that insulin signaling is generally important for neural plasticity in *C. elegans*.[57]

HEN-1, another secreted protein with a low-density lipoprotein receptor motif involved in integration processing between Cu^{2+} ion and odorants, is also required for thermotactic plasticity. Genetic analysis suggests that INS-1 and HEN-1 act in parallel, although it remains to be determined how INS-1 and HEN-1 pathways interact with each other.[70]

A novel hydrolase, AHO-3, is required in AWC neurons and other unidentified neurons for food-associated thermotactic plasticity.[121] Expression of a human homolog of AHO-3, FAM108B1, which is expressed in human brain, restored the defect in *aho*-3 mutants, suggesting that this novel type of hydrolase is functionally conserved between *C. elegans* and humans. Although the role of AHO-3 remains to be elucidated, palmitoylation of an N-terminal cysteine cluster was found to be essential for food-associated thermotactic plasticity. Because ABHD12, an AHO-3-related protein in *C. elegans*, was suggested to hydrolyze endocannabinoid 2-arachidonoylglycerolendocan, it is likely that AHO-3 hydrolyzes some type of endocannabinoid, which is important for food-associated thermotactic plasticity.

An ortholog of human calcineurin, α subunit TAX-6, is required in two pairs of interneurons—AIZ and RIA—for food-associated thermotactic plasticity.[122] The thermal response of the AIZ interneuron was downregulated after starvation in wild-type animals, whereas *tax*-6 (sensory +, inter−) mutants, whose abnormal thermosensation in AFD was rescued by expressing *tax-6cDNA* in sensory neurons, are impaired in starvation-induced downregulation in AIZ activity. This result suggests that feeding state-dependent thermotactic plasticity is controlled in thermotactic interneurons through TAX-6/calcineurin. Given that INS-1 acts to core interneurons of the thermotaxis circuit and TAX-6/calcineurin functions in both AIZ and RIA, it is likely that temperature information and food state are associated in the interneurons (Figure 11.7). Consistent with this model, thermal responses of AFD are indistinguishable between the fed and the starved state.[119]

Neuromodulatory monoamines affect many aspects of *C. elegans* behavior.[9,12–18,23] Exogenous serotonin mimicked the presence of food, whereas exogenous octopamine mimicked the absence of food in food-associated thermotactic plasticity.[61] This result suggests that the balanced regulation by two monoamines in interneurons is a key process for thermotactic plasticity. The regulation of plasticity through monoamines and how insulin signaling relates to monoamine signaling are subjects for future study.

INFORMATION FLOW FROM AFD AND AWC TO AIY

Both AFD and AWC neurons synapse onto AIY neurons (Figure 11.3C). A reasonable working model is that temperature information in AFD and AWC is conveyed to AIY interneurons that are responsible for thermophilic movement and isothermal behavior.[60,123] The function of NCS-1/neuronal calcium sensor 1 in AIY is important for isothermal behavior.[124] The AIY neuron is also known to be a key neuron for other kinds of neural plasticity, such as information processing between the Cu^{2+} ion and odorants, and olfactory imprinting.[70,72] It is thus likely that the synaptic regulation from AFD and AWC to AIY may be important for thermotactic plastic behavior. The molecular mechanism for information flow from AFD and AWC to AIY has been partially revealed (Figure 11.4B).[91,125] A study using mutants for the *eat*-4 gene encoding a vesicular glutamate transporter (VGLUT)[126] showed that EAT-4-dependent glutamatergic transmission from AFD downregulated the activity of AIY through the glutamate-gated chloride channel GLC-3. The inhibition of AIY resulted in cryophilic phenotype. In contrast, EAT-4-dependent glutamatergic transmission from AWC upregulated the activity of AIY through an unidentified receptor or channel, thereby inducing thermophilic drive. Hence, bidirectional glutamatergic transmission from AFD and AWC encoding opposite information flows may balance the activity of the single interneuron AIY.[91,125]

AFD-ablated animals exhibited cryophilic or athermotactic phenotypes, whereas animals defective in EAT-4-dependent transmission between AFD and AIY exhibited thermophilic phenotype. These results indicate the existence of excitatory signals from AFD to AIY besides the EAT-4-mediated inhibitory signal. Optogenetics clearly revealed the bidirectional neurotransmission of AFD.[92] The combination of a light-driven cation channel channelrhodopsin-2 (ChR2) and electrophysiology showed the tonic and graded excitatory synaptic transmissions from AFD to AIY. The signal from AFD was linearly scaled down in AIY, and synaptic release showed no obvious facilitation or depression. This excitatory synaptic transmission is probably peptidergic because evoked synaptic current is greatly reduced in *unc*-31 mutants in which a synaptic protein UNC-31 required for peptide release is impaired. The other study using a light-driven chloride channel halorhodopsin (NpHR) in AFD showed that the magnitude of AFD activity is important for regulating AIY activity, thereby controlling behavior. The activation of NpHR in AFD partially reduced the neural activity of AFD as revealed by Ca^{2+} indicator YC3.60 and a membrane voltage indicator Mermaid.[93] The reduced response of AFD caused the hyperactivation of AIY, thereby driving thermophilic movement. The hyperactivation of AIY occurred through EAT-4–dependent glutamatergic transmission because no enhancement of AIY activity was observed in *eat*-4 mutants. The complete inactivation of AFD in *gcy*-23 *gcy*-8 *gcy*-18 triple mutants in which thermosensory signaling is shut off, reduced both excitatory and inhibitory signals, resulting in the loss of AIY response. It is thus plausible that the different activity states of AFD generate the diverse behavioral outputs that are balanced by two opposing types of neurotransmission. Synaptic activity of AFD measured by the fluorescence of pH-sensitive green fluorescence protein showed that AFD synaptic release is more active when the environmental temperature is higher or lower than cultivation temperature and that AFD synaptic release is inactive when the environmental temperature is near cultivation temperature. These results suggest that AFD encodes a direct comparison between environmental and memorized temperature.[127] The elucidation of the whole picture of synaptic transmission from AFD and AIY as well as the elucidation of the underlying regulatory mechanism remain to be achieved.

RIA INTERNEURON AS AN INTEGRATOR AND LOCOMOTION CONTROLLER

AIY is known to be a cholinergic neuron.[128] However, information flows from AIY to the counterbalancing interneuron AIZ and from AIY to the locomotion-controlling neuron RIA remain to be analyzed. The information flow from AIZ to RIA is also unknown. Based on the neural connections reported by White *et al.* (1986), RIA locates at the site for integrating various sensory inputs.[10] RIA may integrate thermotactic signals from both AIY and AIZ to regulate the motor output (Figure 11.3C). RIA has numerous synapses onto SMD and RMD motoneuron types, which control head muscles navigating orientation of movement.[10] The morphological characteristics of RIA are also unique. RIA is a monopolar neuron that extends the axon from cell body to the nerve ring. The nerve ring is the largest axon bundle in *C. elegans* and regarded as a central nervous system. Of the 302 neurons in the adult hermaphrodite, 180 neurons project axons (processes) into the nerve ring where synaptic connections between various neurons are formed.[129] The RIA axon enters the nerve ring and loops round in the ring. RIA has the most numerous synapses in *C. elegans* neurons except for the motoneurons that innervate muscles with which they make multiple synapses.[10] The compartmentalization of the pre- and postsynaptic area is a clear feature of the RIA axon: The majority of the presynaptic specifications are confined to the distal region, whereas most of the postsynaptic specializations are localized in the proximal region.[10,94] Compartmentalized Ca^{2+} dynamics in RIA correlate with the head movement of *C. elegans*: The activation of the dorsal segment of the RIA axon imaged by the Ca^{2+} indicator is correlated with head movement of the dorsal direction, whereas the activation of the ventral segment of the RIA axon is correlated with head movement of the ventral direction.[130] TTX-7, a *C. elegans* homolog of IMPase (inositol monophosphatase), regulates the synaptic compartmentalization and behavior (Figure 11.4C).[94] IMPase is regarded as a bipolar disorder-relevant molecule due to its lithium sensitivity.[131,132] In *ttx*-7 mutants and lithium-treated wild-type animals, presynaptic marker synaptobrevin::VENUS is mislocalized to the entire axonal process, and thermotaxis is severely impaired. Both defects were rescued by the expression of IMPase or inositol application at adult stage. The synaptic and thermotactic behavioral defects of IMPase mutants were suppressed by mutations in two enzymes, phospholipase Cβ and synaptojanin, which presumably reduce the level of membrane phosphatidylinositol 4, 5-bisphosphate (PIP_2).[90] These results suggest that IMPase and the metabolic balance of PIP_2 on plasma membrane are required in the mature nervous system for regulating the correct localization of synaptic components in the central interneuron.

CONCLUSION AND PERSPECTIVE

The studies on thermotaxis of *C. elegans* originated by Hedgecock and Russell have yielded much knowledge about thermosensory transduction, neural information flow, neural computation, and the neural basis of plasticity. However, the knowledge obtained to date is still fragmentary. Future studies on thermotaxis will yield fruitful results that may be useful for understanding the human brain. The most intriguing advantage in the study of thermotaxis in *C. elegans* is our knowledge of the underlying simple and well-understood neural circuit. Based on this circuit, one can study neural function from sensation to motor output as well as the principle of neural plasticity at high resolution, leading to the understanding of human brain operation.

The neural circuit of thermotactic plasticity represents a dramatic conceptual analogy to the human brain (Figure 11.7). The temperature information stored in AFD (and probably AWC) is transmitted to the interneurons AIY, AIZ, and RIA, in which the temperature information is associated with the feeding and/or starvation signals to generate associative learning. We propose that the neural circuit for thermotactic plasticity is amazingly similar to two functional parts of the human brain—the cerebral cortex and basal ganglia. The cerebral cortex encodes working memory that is required for a temporal storage of information. Basal ganglia play an important role in learning, emotion, and motivation. Namely, the sensory neuron AFD (and probably AWC), working as a memory storage device, is equivalent to the cerebral cortex, whereas three interneurons—AIY, AIZ, and RIA—which play a role in learning and likely receive neuromodulatory monoamines in response to feeding state, are equivalent to basal ganglia (Figure 11.7). Given that neural plasticity was obtained at the early stage of evolution for animals to survive in the natural environment, logic of neural function may be conserved between *C. elegans* and human.

Acknowledgments

We thank Hitoshi Okamoto for stimulating discussion; Nana Nishio, Tsubasa Kimata, Paola Jurado, Shunji Nakano, Yuki Tsukada, and Kyogo Kobayashi for materials for the figures; and members of I. Mori's lab for discussion and comments on the manuscript. This work was supported by CREST-JST, Strategic Research Program for Brain Sciences, MEXT, and Grant-in-Aid for Scientific Research on Innovative Area—Neural Diversity and Neocortical Organization from MEXT (to I.M.).

References

1. Kandel ER. The molecular biology of memory storage: A dialogue between genes and synapses. *Science.* 2001;294 (5544):1030–1038.
2. Farley J, Alkon DL. Cellular mechanisms of learning, memory, and information storage. *Annu Rev Psychol.* 1985;36:419–494.
3. Frisch Kv. *The Dance Language and Orientation of Bees.* Cambridge, MA: Harvard University Press; 1993.
4. Heisenberg M. What do the mushroom bodies do for the insect brain? An introduction. *Learn Mem.* 1998;5(1–2):1–10.
5. Menzel R, De Marco RJ, Greggers U. Spatial memory, navigation and dance behaviour in *Apis mellifera. J Comp Physiol A Neuroethol Sens Neural Behav Physiol.* 2006;192 (9):889–903.
6. Quinn WG, Harris WA, Benzer S. Conditioned behavior in *Drosophila melanogaster. Proc Natl Acad Sci USA.* 1974;71 (3):708–712.
7. Waddell S, Quinn WG. Flies, genes, and learning. *Annu Rev Neurosci.* 2001;24:1283–1309.
8. Brenner S. The genetics of *Caenorhabditis elegans. Genetics.* 1974;77 (1):71–94.
9. de Bono M, Maricq AV. Neuronal substrates of complex behaviors in *C. elegans. Annu Rev Neurosci.* 2005;28:451–501.
10. White JG, Southgate E, Thomson JN, Brenner S. The structure of the nervous system of the nematode *Caenorhabditis elegans. Philos Trans R Soc Lond B Biol Sci.* 1986;314(1165):1–340.
11. Varshney LR, Chen BL, Paniagua E, Hall DH, Chklovskii DB. Structural properties of the *Caenorhabditis elegans* neuronal network. *PLoS Comput Biol.* 2011;7(2):e1001066.
12. Chalfie M, Jorgensen EM. *C. elegans* neuroscience: Genetics to genome. *Trends Genet.* 1998;14(12):506–512.
13. Bargmann CI, Kaplan JM. Signal transduction in the *Caenorhabditis elegans* nervous system. *Annu Rev Neurosci.* 1998; 21:279–308.
14. Rankin CH. From gene to identified neuron to behaviour in *Caenorhabditis elegans. Nat Rev Genet.* 2002;3(8):622–630.
15. Hobert O. Behavioral plasticity in *C. elegans*: Paradigms, circuits, genes. *J Neurobiol.* 2003;54(1):203–223.
16. Bargmann CI. Chemosensation in *C. elegans. WormBook.* 2006;1–29.
17. Giles AC, Rose JK, Rankin CH. Investigations of learning and memory in *Caenorhabditis elegans. Int Rev Neurobiol.* 2006;69:37–71.
18. Hart AC, Chao MY. *From odors to behaviors in* Caenorhabditis elegans. *The Neurobiology of Olfaction.* Boca Raton, FL: CRC Press; 2010.
19. Rand JB, Nonet ML. Synaptic transmission. *C. Elegans II.* 1997;611–643.
20. Jorgensen EM. GABA. *WormBook.* 2005;1–13.
21. Brockie PJ, Maricq AV. Ionotropic glutamate receptors: Genetics, behavior and electrophysiology. *WormBook.* 2006;1–16.
22. Rand JB. Acetylcholine. *WormBook.* 2007;1–21.
23. Chase DL, Koelle MR. Biogenic amine neurotransmitters in *C. elegans. WormBook.* 2007;1–15.
24. Li C, Kim K. Neuropeptides. *WormBook.* 2008;1–36.
25. *C. elegans* sequencing Consortium Genome sequence of the nematode *C. elegans*: A platform for investigating biology. *Science*1998;282(5396):2012–2018.
26. Bargmann CI. Neurobiology of the *Caenorhabditis elegans* genome. *Science.* 1998;282(5396):2028–2033.
27. Goodman MB, Hall DH, Avery L, Lockery SR. Active currents regulate sensitivity and dynamic range in *C. elegans* neurons. *Neuron.* 1998;20(4):763–772.
28. Cully DF, Vassilatis DK, Liu KK, et al. Cloning of an avermectin-sensitive glutamate-gated chloride channel from *Caenorhabditis elegans. Nature.* 1994;371(6499):707–711.
29. Mongan NP, Baylis HA, Adcock C, Smith GR, Sansom MS, Sattelle DB. An extensive and diverse gene family of nicotinic acetylcholine receptor alpha subunits in *Caenorhabditis elegans. Recept Channels.* 1998;6(3):213–228.

30. Ranganathan R, Cannon SC, Horvitz HR. MOD-1 is a serotonin-gated chloride channel that modulates locomotory behaviour in *C. elegans*. *Nature*. 2000;408(6811):470–475.
31. Beg AA, Jorgensen EM. EXP-1 is an excitatory GABA-gated cation channel. *Nat Neurosci*. 2003;6(11):1145–1152.
32. Raymond V, Sattelle DB. Novel animal-health drug targets from ligand-gated chloride channels. *Nat Rev Drug Discov*. 2002;1(6): 427–436.
33. Kerr RA. Imaging the activity of neurons and muscles. *WormBook*. 2006;1–13.
34. Schafer WR. Deciphering the neural and molecular mechanisms of *C. elegans* behavior. *Curr Biol*. 2005;15(17):R723–729.
35. Schafer WR. Neurophysiological methods in *C. elegans*: an introduction. *WormBook*. 2006;1–4.
36. Zhang F, Wang LP, Brauner M, et al. Multimodal fast optical interrogation of neural circuitry. *Nature*. 2007;446(7136):633–639.
37. Han X, Boyden ES. Multiple-color optical activation, silencing, and desynchronization of neural activity, with single-spike temporal resolution. *PLoS ONE*. 2007;2(3):e299.
38. Tian L, Hires SA, Mao T, et al. Imaging neural activity in worms, flies and mice with improved GCaMP calcium indicators. *Nat Methods*. 2009;6(12):875–881.
39. Hobert O. The impact of whole genome sequencing on model system genetics: Get ready for the ride. *Genetics*. 2010;184(2): 317–319.
40. de Bono M, Bargmann CI. Natural variation in a neuropeptide Y receptor homolog modifies social behavior and food response in *C. elegans*. *Cell*. 1998;94(5):679–689.
41. Reddy KC, Andersen EC, Kruglyak L, Kim DH. A polymorphism in npr-1 is a behavioral determinant of pathogen susceptibility in *C. elegans*. *Science*. 2009;323(5912):382–384.
42. Chang HC, Paek J, Kim DH. Natural polymorphisms in *C. elegans* HECW-1 E3 ligase affect pathogen avoidance behaviour. *Nature*. 2011;480(7378):525–529.
43. Bendesky A, Tsunozaki M, Rockman MV, Kruglyak L, Bargmann CI. Catecholamine receptor polymorphisms affect decision-making in *C. elegans*. *Nature*. 2011;472(7343):313–318.
44. Bendesky A, Bargmann CI. Genetic contributions to behavioural diversity at the gene–environment interface. *Nat Rev Genet*. 2011;12(12):809–820.
45. Fire A, Xu S, Montgomery MK, Kostas SA, Driver SE, Mello CC. Potent and specific genetic interference by double-stranded RNA in *Caenorhabditis elegans*. *Nature*. 1998;391(6669):806–811.
46. Mori I, Sasakura H, Kuhara A. Worm thermotaxis: A model system for analyzing thermosensation and neural plasticity. *Curr Opin Neurobiol*. 2007;17(6):712–719.
47. Kimata T, Sasakura H, Ohnishi N, Nishio N, Mori I. Thermotaxis of *C. elegans* as a model for temperature perception, neural information processing and neural plasticity. *Worm*. 2012;1(1):31–41.
48. Ardiel EL, Rankin CH. An elegant mind: learning and memory in *Caenorhabditis elegans*. *Learn Mem*. 2012;17(4):191–201.
49. Colbert HA, Bargmann CI. Odorant-specific adaptation pathways generate olfactory plasticity in *C. elegans*. *Neuron*. 1995;14 (4):803–812.
50. Giles AC, Rankin CH. Behavioral and genetic characterization of habituation using *Caenorhabditis elegans*. *Neurobiol Learn Mem*. 2009;92(2):139–146.
51. Kimura KD, Fujita K, Katsura I. Enhancement of odor avoidance regulated by dopamine signaling in *Caenorhabditis elegans*. *J Neurosci*. 2010;30(48):16365–16375.
52. Torayama I, Ishihara T, Katsura I. *Caenorhabditis elegans* integrates the signals of butanone and food to enhance chemotaxis to butanone. *J Neurosci*. 2007;27(4):741–750.
53. Bauer Huang SL, Saheki Y, VanHoven MK, et al. Left–right olfactory asymmetry results from antagonistic functions of voltage-activated calcium channels and the Raw repeat protein OLRN-1 in *C. elegans*. *Neural Dev*. 2007;2:24.
54. Saeki S, Yamamoto M, Iino Y. Plasticity of chemotaxis revealed by paired presentation of a chemoattractant and starvation in the nematode *Caenorhabditis elegans*. *J Exp Biol*. 2001;204(Pt 10): 1757–1764.
55. Hukema RK, Rademakers S, Dekkers MP, Burghoorn J, Jansen G. Antagonistic sensory cues generate gustatory plasticity in *Caenorhabditis elegans*. *EMBO J*. 2006;25(2):312–322.
56. Hukema RK, Rademakers S, Jansen G. Gustatory plasticity in *C. elegans* involves integration of negative cues and NaCl taste mediated by serotonin, dopamine, and glutamate. *Learn Mem*. 2008;15(11):829–836.
57. Tomioka M, Adachi T, Suzuki H, Kunitomo H, Schafer WR, Iino Y. The insulin/PI 3-kinase pathway regulates salt chemotaxis learning in *Caenorhabditis elegans*. *Neuron*. 2006;51(5):613–625.
58. Yamada K, Hirotsu T, Matsuki M, et al. Olfactory plasticity is regulated by pheromonal signaling in *Caenorhabditis elegans*. *Science*. 2010;329(5999):1647–1650.
59. Hedgecock EM, Russell RL. Normal and mutant thermotaxis in the nematode *Caenorhabditis elegans*. *Proc Natl Acad Sci USA*. 1975;72(10):4061–4065.
60. Mori I, Ohshima Y. Neural regulation of thermotaxis in *Caenorhabditis elegans*. *Nature*. 1995;376(6538):344–348.
61. Mohri A, Kodama E, Kimura KD, Koike M, Mizuno T, Mori I. Genetic control of temperature preference in the nematode *Caenorhabditis elegans*. *Genetics*. 2005;169(3):1437–1450.
62. Morrison GE, Wen JY, Runciman S, van der Kooy D. Olfactory associative learning in *Caenorhabditis elegans* is impaired in lrn-1 and lrn-2 mutants. *Behav Neurosci*. 1999;113(2):358–367.
63. Amano H, Maruyama IN. Aversive olfactory learning and associative long-term memory in *Caenorhabditis elegans*. *Learn Mem*. 2011;18(10):654–665.
64. Gray JM, Karow DS, Lu H, et al. Oxygen sensation and social feeding mediated by a *C. elegans* guanylate cyclase homologue. *Nature*. 2004;430(6997):317–322.
65. Cheung BH, Cohen M, Rogers C, Albayram O. de Bono M. Experience-dependent modulation of *C. elegans* behavior by ambient oxygen. *Curr Biol*. 2005;15(10):905–917.
66. Chang AJ, Bargmann CI. Hypoxia and the HIF-1 transcriptional pathway reorganize a neuronal circuit for oxygen-dependent behavior in *Caenorhabditis elegans*. *Proc Natl Acad Sci USA*. 2008;105(20):7321–7326.
67. Ma DK, Vozdek R, Bhatla N, Horvitz HR. CYSL-1 interacts with the O_2-sensing hydroxylase EGL-9 to promote H2S-modulated hypoxia-induced behavioral plasticity in *C. elegans*. *Neuron*. 2012;73(5):925–940.
68. Zhang Y, Lu H, Bargmann CI. Pathogenic bacteria induce aversive olfactory learning in *Caenorhabditis elegans*. *Nature*. 2005;438 (7065):179–184.
69. Shtonda BB, Avery L. Dietary choice behavior in *Caenorhabditis elegans*. *J Exp Biol*. 2006;209(Pt 1):89–102.
70. Ishihara T, Iino Y, Mohri A, et al. HEN-1, a secretory protein with an LDL receptor motif, regulates sensory integration and learning in *Caenorhabditis elegans*. *Cell*. 2002;109(5):639–649.
71. Alkema MJ, Hunter-Ensor M, Ringstad N, Horvitz HR. Tyramine functions independently of octopamine in the *Caenorhabditis elegans* nervous system. *Neuron*. 2005;46(2): 247–260.
72. Remy JJ, Hobert O. An interneuronal chemoreceptor required for olfactory imprinting in *C. elegans*. *Science*. 2005;309(5735): 787–790.

73. Rose JK, Sangha S, Rai S, Norman KR, Rankin CH. Decreased sensory stimulation reduces behavioral responding, retards development, and alters neuronal connectivity in *Caenorhabditis elegans*. *J Neurosci*. 2005;25(31):7159–7168.
74. Hensel H. Neural processes in thermoregulation. *Physiol Rev*. 1973;53(4):948–1017.
75. Eckert R, Randell D, Augustine G, eds. *Animal Physiology*. New York: Freeman; 1988.
76. Conti B, Sanchez-Alavez M, Winsky-Sommerer R, et al. Transgenic mice with a reduced core body temperature have an increased life span. *Science*. 2006;314(5800):825–828.
77. Lee S-J, Kenyon C. Regulation of the longevity response to temperature by thermosensory neurons in *Caenorhabditis elegans*. *Curr Biol*. 2009;19(9):715–722.
78. Mori I, Sasakura H. Aging: Shall we take the high road? *Curr Biol*. 2009;19(9):R363–364.
79. Garrity PA, Goodman MB, Samuel AD, Sengupta P. Running hot and cold: Behavioral strategies, neural circuits, and the molecular machinery for thermotaxis in *C. elegans* and *Drosophila*. *Genes Dev*. 2010;24(21):2365–2382.
80. Jurado P, Kodama E, Tanizawa Y, Mori I. Distinct thermal migration behaviors in response to different thermal gradients in *Caenorhabditis elegans*. *Genes Brain Behav*. 2010;9(1):120–127.
81. Monteith JL. *Principles of Environmental Physics*. New York: Elsevier; 1973:126–133
82. Beverly M, Anbil S, Sengupta P. Degeneracy and neuromodulation among thermosensory neurons contribute to robust thermosensory behaviors in *Caenorhabditis elegans*. *J Neurosci*. 2011;31(32):11718–11727.
83. Sulston JE, Schierenberg E, White JG, Thomson JN. The embryonic cell lineage of the nematode *Caenorhabditis elegans*. *Dev Biol*. 1983;100(1):64–119.
84. Perkins LA, Hedgecock EM, Thomson JN, Culotti JG. Mutant sensory cilia in the nematode *Caenorhabditis elegans*. *Dev Biol*. 1986;117(2):456–487.
85. Hu PJ. Dauer. *WormBook*. 2007;1–19.
86. Satterlee JS, Sasakura H, Kuhara A. Berkeley M, Mori I, Sengupta P. Specification of thermosensory neuron fate in *C. elegans* requires ttx-1, a homolog of otd/Otx. *Neuron*. 2001;31(6):943–956.
87. Bargmann CI, Hartwieg E, Horvitz HR. Odorant-selective genes and neurons mediate olfaction in *C. elegans*. *Cell*. 1993;74(3):515–527.
88. Kuhara A, Okumura M, Kimata T, et al. Temperature sensing by an olfactory neuron in a circuit controlling behavior of *C. elegans*. *Science*. 2008;320(5877):803–807.
89. Biron D, Shibuya M, Gabel C, et al. A diacylglycerol kinase modulates long-term thermotactic behavioral plasticity in *C. elegans*. *Nat Neurosci*. 2006;9(12):1499–1505.
90. Kimata T, Tanizawa Y, Can Y, Ikeda S, Kuhara A, Mori I. Synaptic polarity depends on phosphatidylinositol signaling regulated by myo-inositol monophosphatase in *Caenorhabditis elegans*. *Genetics*. 2012;191(2):509–521.
91. Ohnishi N, Kuhara A, Nakamura F, Okochi Y, Mori I. Bidirectional regulation of thermotaxis by glutamate transmissions in *Caenorhabditis elegans*. *EMBO J*. 2011;30(7):1376–1388.
92. Narayan A, Laurent G, Sternberg PW. Transfer characteristics of a thermosensory synapse in *Caenorhabditis elegans*. *Proc Natl Acad Sci USA*. 2011;108(23):9667–9672.
93. Kuhara A, Ohnishi N, Shimowada T, Mori I. Neural coding in a single sensory neuron controlling opposite seeking behaviours in *Caenorhabditis elegans*. *Nat Commun*. 2011;2:355.
94. Tanizawa Y, Kuhara A, Inada H, Kodama E, Mizuno T, Mori I. Inositol monophosphatase regulates localization of synaptic components and behavior in the mature nervous system of *C. elegans*. *Genes Dev*. 2006;20(23):3296–3310.
95. Coburn CM, Bargmann CI. A putative cyclic nucleotide-gated channel is required for sensory development and function in *C. elegans*. *Neuron*. 1996;17(4):695–706.
96. Komatsu H, Mori I, Rhee JS, Akaike N, Ohshima Y. Mutations in a cyclic nucleotide-gated channel lead to abnormal thermosensation and chemosensation in *C. elegans*. *Neuron*. 1996;17(4):707–718.
97. Inada H, Ito H, Satterlee J, Sengupta P, Matsumoto K, Mori I. Identification of guanylyl cyclases that function in thermosensory neurons of *Caenorhabditis elegans*. *Genetics*. 2006;172(4):2239–2252.
98. Wasserman SM, Beverly M, Bell HW, Sengupta P. Regulation of response properties and operating range of the AFD thermosensory neurons by cGMP signaling. *Curr Biol*. 2011;21(5):353–362.
99. Satterlee JS, Ryu WS, Sengupta P. The CMK-1 CaMKI and the TAX-4 cyclic nucleotide-gated channel regulate thermosensory neuron gene expression and function in *C. elegans*. *Curr Biol*. 2004;14(1):62–68.
100. Bacaj T, Tevlin M, Lu Y, Shaham S. Glia are essential for sensory organ function in *C. elegans*. *Science*. 2008;322(5902):744–747.
101. Kuhara A, Inada H, Katsura I, Mori I. Negative regulation and gain control of sensory neurons by the *C. elegans* calcineurin TAX-6. *Neuron*. 2002;33(5):751–763.
102. Okochi Y, Kimura KD, Ohta A, Mori I. Diverse regulation of sensory signaling by *C. elegans* nPKC-epsilon/eta TTX-4. *EMBO J*. 2005;24(12):2127–2137.
103. Sieburth D, Madison JM, Kaplan JM. PKC-1 regulates secretion of neuropeptides. *Nat Neurosci*. 2007;10(1):49–57.
104. Adachi T, Kunitomo H, Tomioka M, et al. Reversal of salt preference is directed by the insulin/PI3K and Gq/PKC signaling in *Caenorhabditis elegans*. *Genetics*. 2010;186:1309–1319.
105. Sugi T, Nishida Y, Mori I. Regulation of behavioral plasticity by systemic temperature signaling in *Caenorhabditis elegans*. *Nat Neurosci*. 2011;14(8):984–992.
106. Biron D, Wasserman S, Thomas JH, Samuel ADT, Sengupta P. An olfactory neuron responds stochastically to temperature and modulates *Caenorhabditis elegans* thermotactic behavior. *Proc Natl Acad Sci USA*. 2008;105(31):11002–11007.
107. Kimura KD, Miyawaki A, Matsumoto K, Mori I. The *C. elegans* thermosensory neuron AFD responds to warming. *Curr Biol*. 2004;14(14):1291–1295.
108. Clark DA, Biron D, Sengupta P, Samuel AD. The AFD sensory neurons encode multiple functions underlying thermotactic behavior in *Caenorhabditis elegans*. *J Neurosci*. 2006;26(28):7444–7451.
109. Hobert O, D'Alberti T, Liu Y, Ruvkun G. Control of neural development and function in a thermoregulatory network by the LIM homeobox gene lin-11. *J Neurosci*. 1998;18(6):2084–2096.
110. Sengupta P, Colbert HA, Bargmann CI. The *C. elegans* gene odr-7 encodes an olfactory-specific member of the nuclear receptor superfamily. *Cell*. 1994;79(6):971–980.
111. Ramot D, MacInnis BL, Goodman MB. Bidirectional temperature-sensing by a single thermosensory neuron in C. elegans. *Nat Neurosci*. 2008;11(8):908–915.
112. Nishida Y, Sugi T, Nonomura M, Mori I. Identification of the AFD neuron as the site of action of the CREB protein in *Caenorhabditis elegans* thermotaxis. *EMBO Rep*. 2011;12(8):855–862.
113. Horvitz HR, Chalfie M, Trent C, Sulston JE, Evans PD. Serotonin and octopamine in the nematode *Caenorhabditis elegans*. *Science*. 1982;216(4549):1012–1014.

114. Sawin ER, Ranganathan R, Horvitz HR. *C. elegans* locomotory rate is modulated by the environment through a dopaminergic pathway and by experience through a serotonergic pathway. *Neuron*. 2000;26(3):619–631.
115. Petrella LN, Wang W, Spike CA, Rechtsteiner A, Reinke V, Strome S. synMuv B proteins antagonize germline fate in the intestine and ensure *C. elegans* survival. *Development*. 2011;138 (6):1069–1079.
116. McMullen PD, Aprison EZ, Winter PB, Amaral LA, Morimoto RI, Ruvinsky I. Macro-level modeling of the response of *C. elegans* reproduction to chronic heat stress. *PLoS Comput Biol*. 2012;8(1):e1002338.
117. Ailion M, Thomas JH. Dauer formation induced by high temperatures in *Caenorhabditis elegans*. *Genetics*. 2000;156 (3):1047–1067.
118. Ito H, Inada H, Mori I. Quantitative analysis of thermotaxis in the nematode *Caenorhabditis elegans*. *J Neurosci Methods*. 2006;154(1–2):45–52.
119. Kodama E, Kuhara A, Mohri-Shiomi A, et al. Insulin-like signaling and the neural circuit for integrative behavior in *C. elegans*. *Genes Dev*. 2006;20(21):2955–2960.
120. Fernandez AM, Torres-Aleman I. The many faces of insulin-like peptide signalling in the brain. *Nat Rev Neurosci*. 2012;13 (4):225–239.
121. Nishio N, Mohri-Shiomi A, Nishida Y, et al. A novel and conserved protein AHO-3 is required for thermotactic plasticity associated with feeding states in *Caenorhabditis elegans*. *Genes Cells*. 2012;17(5):365–386.
122. Kuhara A, Mori I. Molecular physiology of the neural circuit for calcineurin-dependent associative learning in *Caenorhabditis elegans*. *J Neurosci*. 2006;26(37):9355–9364.
123. Hobert O, Mori I, Yamashita Y, et al. Regulation of interneuron function in the *C. elegans* thermoregulatory pathway by the ttx-3 LIM homeobox gene. *Neuron*. 1997;19(2):345–357.
124. Gomez M, De Castro E, Guarin E, et al. Ca^{2+} signaling via the neuronal calcium sensor-1 regulates associative learning and memory in *C. elegans*. *Neuron*. 2001;30(1):241–248.
125. Ardiel EL, Rankin CH. Some like it hot: Decoding neurotransmission in the worm's thermotaxis circuit. *EMBO J*. 2011;30 (7):1192–1194.
126. Lee RY, Sawin ER, Chalfie M, Horvitz HR, Avery L. EAT-4, a homolog of a mammalian sodium-dependent inorganic phosphate cotransporter, is necessary for glutamatergic neurotransmission in *Caenorhabditis elegans*. *J Neurosci*. 1999;19(1): 159–167.
127. Samuel ADT, Silva RA, Murthy VN. Synaptic activity of the AFD neuron in *Caenorhabditis elegans* correlates with thermotactic memory. *J Neurosci*. 2003;23(2):373–376.
128. Kratsios P, Stolfi A, Levine M, Hobert O. Coordinated regulation of cholinergic motor neuron traits through a conserved terminal selector gene. *Nat Neurosci*. 2012;15(2):205–214.
129. Ware RW, Clark D, Crossland K, Russell RL. The nerve ring of the nematode *Caenorhabditis elegans*: Sensory input and motor output. *J Comp Neurol*. 1975;162(1):71–110.
130. Hendricks M, Ha H, Maffey N, Zhang Y. Compartmentalized calcium dynamics in a *C. elegans* interneuron encode head movement. *Nature*. 2012;487(7405):99–103.
131. Allison JH, Stewart MA. Reduced brain inositol in lithium-treated rats. *Nat New Biol*. 1971;233(43):267–268.
132. Berridge MJ, Downes CP, Hanley MR. Neural and developmental actions of lithium: a unifying hypothesis. *Cell*. 1989;59 (3):411–419.

CHAPTER

12

Age-Dependent Modulation of Learning and Memory in *Caenorhabditis elegans*

Shin Murakami

Touro University California, Vallejo, California

INTRODUCTION

Age-related memory impairment (AMI) is an altered ability to learn new information and to recall previously learned information during normal aging. In *Caenorhabditis elegans*, the onset of AMI has been observed in the early- to mid-phase of reproduction.[1] Similarly, in humans, the earliest signs of AMI can be seen at approximately age 40 years, although the onset of AMI has been under discussion in humans.[2] AMI is considered a normal state, which is the first step toward the transition state, mild cognitive impairment (MCI), followed by the disease state, dementia (see Table 12.1 for terminology and abbreviation); note that the major cause of dementia in humans is Alzheimer's disease. Clinically, memory impairment progresses in the order of AMI (normal state)→MCI (transition state)→dementia (disease state). The term aging-associated memory impairment (AAMI) has been intended to refer to the transition state in humans—that is, memory impairment in humans age 50 years or older.[2] AAMI is similar to MCI in its original meaning, but it also refers to the normal state, which has created confusion. For this reason, the term AMI is used in this chapter to mean normal phase to minimize possible confusion. This chapter describes AMI, which is an emerging field of research.

Classical conditioning (or Pavlovian conditioning) associates two stimuli to predict important events, which is a type of associative learning and memory. Animals use a stimulus associated with food (or no food) to find food (or to avoid no food). They can also learn to associate a stimulus and various environmental conditions (heavy metals, noxious stimuli, infectious bacteria, predators, etc.) to avoid harmful environments. Studies suggest that associative learning and memory are vulnerable to aging in a wide variety of species, including *C. elegans*,[1,4–9] as well as *Aplysia*,[10] honeybees,[11] fruit flies,[12,13] snails,[14] rodents,[15–18] and humans.[19,20] AMI has been observed in selective aspects of associative learning and memory that include short-term memory (STM) and long-term memory (LTM). In mammals, aging affects working memory that uses the multiple buffers provided by STM.[21,22] In addition, implicit memory, including classical conditioning and skill learning, declines with increasing age. The mammalian AMI has been discussed elsewhere.[2,23–25]

CLASSIFICATION OF LEARNING AND MEMORY

This section describes three types of memory classifications. First, psychological classification includes implicit memory (or nondeclarative memory) and explicit memory (or declarative memory). Implicit memory includes memory for skills, habits, and behaviors. Classical conditioning belongs to the implicit memory category. Most *C. elegans* associative learning and memory paradigms involve classical conditioning. Explicit memory is a memory that allows one to recall actual events, including memory of episodes and facts. It can also be defined by the phrase, "You know that you know".[22]

Second, associative and nonassociative learning and memory are included under the category of episodic memory with some exceptions, including classical conditioning, which is implicit memory. Associative learning and memory associate multiple types of information or behavior. Associative learning and memory

Invertebrate Learning and Memory.
DOI: http://dx.doi.org/10.1016/B978-0-12-415823-8.00012-5

TABLE 12.1 Terminology and Abbreviation in Aging and Memory Impairment

Abbreviation	Term	Characteristics
AACI	Age-associated cognitive impairment	Memory impairment in comparison with age-matched controls
AAMI	Age-associated memory impairment	Same as AMI, but originally used to refer to the transition phase similar to MCI
AMI	Age-related memory impairment	Impairment of memory in normal aging
Dementia		Disease state of cognitive impairment
LTM	Long-term memory	Memory that lasts days to decades
MCI	Mild cognitive impairment	Transition state between normal and disease state of cognitive impairment
STM	Short-term memory	Transient memory that lasts for a short time

Source: *Modified from Murakami and Johnson.*[3]

overlap with implicit and explicit memory depending on the types of information used. Other types of learning and memory are called nonassociative memory, including habituation and sensitization.[26–28]

Third, another classification includes STM and LTM using the criteria of memory duration. STM is a transient memory for a short time ranging from seconds to hours; LTM typically lasts days or longer. In *C. elegans*, LTM is sensitive to cold shock and to drugs that inhibit transcription and translation. The term *working memory* has been developed to explain STM more accurately. In the original definition, working memory uses STM as a buffer for the maintenance and manipulation of memory.[29,30] STM and working memory are similar, with some differences; however, in a broad definition, working memory appears to include LTM.[21]

REDUCED PLASTICITY BUT WELL-RETAINED 'OLD MEMORY'

It is a generally consistent finding that aging specifically reduces the learning and memorizing of new information and procedures in a variety of species from *C. elegans* to humans. In contrast, old memory and well-learned procedures (procedural memory for long term) are kept relatively intact (for simplicity, the term *old memory* is used throughout the rest of the chapter).[1,31] In *C. elegans*, old memory has been observed in the memory of the culture conditions in which an association between food and it smell is formed. For example, when worms are cultured in the media with food, they memorize the smell of the food (food–food smell association) by a process of odor imprinting,[32] butanone enhancement,[33] or other mechanisms. This type of memory lasts longer than regular STM. In normal aging, plasticity of learning and memory is reduced. More precisely, the early phase of AMI includes a behavioral shift from responding to a new culture condition to retaining old memories of the existing culture condition.

Early AMI occurs during the reproduction phase, which appears to be relevant to evolution. Specifically, during reproduction, animals appear to minimize behavioral and biological processes that are less relevant to reproduction. This appears to be an example of a tradeoff between reproduction and learning and memory. Old memory, in contrast, is advantageous under conditions that do not change over long periods of time. The early phase of AMI may not only benefit reproduction but also ensure adjustment to consistent environments from the past, although it is disadvantageous under changing environments. However, the reproductive advantage does not apply to the late phase of AMI, during which a wide variety of deteriorations occur (e.g., cessation of reproduction, structural deteriorations of various tissues, and increases in macromolecular damage and mortality).

TWO PHASES OF AMI

The current understanding is that AMI has diverse functions in learning and memory that are age-dependent and can be divided into early and late phases.[1] Early AMI occurs before the cessation of the reproductive phase (early to mid-reproduction); the late phase of AMI occurs at approximately the end of the reproduction phase and includes deteriorations of diverse behaviors. Table 12.2 summarizes age-related changes in associative learning behaviors.

Early Behavioral Markers of AMI

In *C. elegans*, early AMI selectively affects STM and LTM induced by procedures that include starvation. It has been claimed that LTM induced by a spaced training (food–butanol presentation spaced by starvation) declines earlier than STM, which occurs during the

TABLE 12.2 Summary of Age-Related Behavioral Changes in *C. elegans*

Early changes	Change	Function	References
Basal slowing response	Enhanced		51
Starvation–temperature association	Reduced		Murakami, unpublished results
Starvation–benzaldehyde association	Reduced	STM	Murakami, unpublished results
Starvation–butanol association[a]	Reduced	STM, LTM	35
Early–late changes			
Food–benzaldehyde association[b]	Increased	LTM	77
Late changes			
Locomotion	Reduced	Motor activity	51, 55
Chemosensory to benzaldehyde	Reduced	Sensory function	23,77
Chemosensory to octanol	Unchanged?	Sensory function	23,35
Food–temperature association	Reduced	LTM?[c]	55
Not clear			
Habituation	Increased		8

[a]*In the starvation–butanol association paradigm, long-term memory (LTM) decline appears to occur slightly earlier than short-term memory (STM) decline, although it may be due to the number of starvation cycles used in a spaced training (discussed in the text).*
[b]*Food–benzaldehyde association has been assessed during the late reproduction period,[9] and therefore it is unclear whether or not it belongs to early changes. The behavior is dependent on a long-term exposure to food and food smell during early life.*
[c]*Thermotaxis has been claimed as LTM,[34] although it should be tested (see text).*

early to mid-reproduction period.[4] However, the claim remains in doubt because the procedure uses starvation responses, which are altered during reproduction. The early changes of STM include declines in associative learning behaviors with starvation conditioning, including starvation–temperature association (thermotaxis avoidance) and starvation–benzaldehyde association (chemotaxis avoidance). Slowing locomotory responses (called basal and enhanced slowing responses) show a property of associative memory. Basal slowing response is increased, which is suppressed by mutations in the serotonin pathway.[5] Increased attractance to benzaldehyde is observed later (mid- to late reproduction phase) than early AMI,[9] which is also dependent on the serotonin pathway.

Late Markers of AMI

The late-phase change (late post-reproduction period) is associated with declines in chemosensory and locomotory behavior. It has been suggested that temperature–food association is modulated by the insulin/insulin-like growth factor 1 (IGF-1) pathway.[7]

Other Behavioral Markers

Similar to mammals, not all functions of learning and memory decline with increasing age. Some of the early changes are increased (e.g., basal slowing response and an attraction to benzaldehyde). A form of nonassociative learning, habituation (a reduced response after repeated mechanical stimuli), is increased in old animals, whereas recovery from habituation is slower in old compared to young worms.[35] Sensory functions are generally well-preserved during aging, including odor detection; simple thermotaxis[7]; mechanosensation[36]; and chemosensory responses to benzaldehyde, butanol, diacetyl, isoamyl alcohol, Cl^- (NH_4Cl), and octanol.[4,7,9,36]

AGING-RELATED CHANGES IN ASSOCIATIVE LEARNING AND MEMORY

Three types of associative learning behaviors are known in *C. elegans*. This section gives an overview of the effects of aging on each of these types of associative learning behavior: thermotaxis, locomotory associative memory (called basal/enhanced showing response), and chemotaxis (with conditioning).

Thermotaxis was the first associative learning behavior to be studied during aging.[7] In a classic conditioning with food and temperature, worms learn to associate a specific temperature with food, moving toward the temperature after conditioning. This type of thermotaxis when tracking an isotherm is called an isothermal tracking behavior.[37] Thermotactic learning has been claimed to be long-term plasticity (or LTM).[34] The food–temperature association has been shown to decline with increasing age[7]; food–temperature memory can be reconditioned during aging, suggesting that food–temperature memory does not become an old memory. Age-related decline in thermotaxis can be suppressed by the long-lived mutations in the insulin/IGF-1 pathway (including the *daf-2* (insulin/IGF-1 receptor) mutations) and the *age-1* (phosphatidylinositol 3-OH kinase) mutation. The *age-1* mutation causes a threefold extension in the worm's life span that ensures thermotactic associative learning and memory during aging.[6,7] Note that under some experimental conditions, not seen in nature, worms do not show thermotactic associative learning.[38,39] Experimental conditions and behavioral strategies for thermotactic learning and memory have been discussed previously.[39,40]

Locomotory associative memory shows an increase in strength unexpectedly early, which provides evidence for early AMI.[5] When worms are well-fed, they move more slowly on food to avoid moving out of the food area. This is called basal slowing response. When worms are starved, they slow down more, which is called enhanced slowing response. Old worms show enhanced slowing on food even when they are well-fed. Two types of evidence suggest that age-related behavioral phenotype is modulated by the serotonin pathway. First, expression of the serotonin biosynthesis gene *tph-1* (tryptophan hydroxylase-1) is increased. Second, the age-related phenotype is suppressed by reducing the serotonin signal, including the mutation in the *ser-4* (serotonin/octomamine receptor-4) gene and its inhibitor, mianserin, which is known as an antidepressant. Interestingly, another *ser* (serotonin/octomamine receptor) gene, *ser-1*, and its inhibitor, methiothepin, can suppress age-related reduction in locomotory rate. Therefore, the serotonin/octopamine pathways regulate the early phase of behavioral aging in a receptor-dependent manner.

Chemotactic learning and memory have been studied by using the odors (benzaldehyde and butanone) that are sensed by the AWC sensory neurons (starvation–odor association). First, an attraction to the odor benzaldehyde is increased in old worms, which is only observed at a high concentration (10% or higher) of benzaldehyde.[9] The old worms used are 7 days old at 20°C, which is approximately the mid- to late reproduction period. Sensory neurons have high- and low-affinity odor receptors. It has been speculated that specificity to a high concentration may result from alterations in the response of low-affinity receptors for benzaldehyde in old worms. The age-related benzotaxis is dependent on pre-exposure to food and the smell similar to that of benzaldehyde in culture. Similarly, pre-exposure to butanone with food results in an enhanced attraction at a high concentration (0.1% or higher) (butanone enhancement). Butanone enhancement is not limited to butanone but, rather, extends to other odors, including isoamyl alcohol and 2,3-pentanedione.[33]

Second, early AMI is also observed in chemotaxis avoidance to butanone (starvation–butanone association) that declines in the early phase of aging.[4] The chemotactic memory is a type of LTM that lasts 16 hr or longer after conditioning; note that a typical STM lasts up to 2 hr. The procedure uses spaced training in which repeated cycles of food–butanone presentation are spaced by starvation with no butanone. The procedure presumably triggers a complex set of mechanisms, including butanone–food association, butanone enhancement,[33] starvation responses (e.g., sensitization of sensory functions by starvation), and repeated training effects. In addition, it also uses a high concentration (10%) of butanone, which has been claimed to trigger an increase in chemotaxis to butanone in old worms.[9] Thus, interpretation of the results requires caution. Interestingly, the results suggest a decline in 16-hr LTM, starting at the early phase of the reproduction period.[4] The timing is similar to that of the decline observed in aging of locomotory associative memory, which is presumably STM.[5] The mutations in the *crh-1*/CREB (cyclic AMP response element binding protein) gene abolish LTM but not STM, suggesting that the gene is specific to the LTM formed by food–butanone association. However, *crh-1(RNAi)* can abolish STM that associates starvation and benzaldehyde (Murakami, unpublished results). Thus, the specificity of *crh-1* to STM and LTM remains to be clarified. The *daf-2* (insulin/IGF-1 receptor) mutation does suppress reduced STM in old worms,[4] which is consistent with the earlier study of thermotactic associative learning.[7] In contrast, the *daf-2* mutation does not suppress reduced LTM in old worms.[4] Overexpression of the *crh-1*/CREB gene and phosphorylation of CREB correlated with the occurrence of LTM.

ENDOCRINE DISTURBANCE AS A CAUSE OF EARLY AMI

Studies suggest an involvement of endocrine pathways in AMI. During aging, a wide variety of alterations in endocrine systems may affect AMI,[1,41] including sex hormones,[42,43] insulin/IGF-1 and serotonin,[44,45] and inflammation mediators.[46] This section discusses two of these pathways that are known to affect aging in *C. elegans*—memory and AMI.

Serotonin as a Cause of Behavioral Aging in Early Phase

The first evidence for endocrine alterations as a cause of AMI was obtained from the study of the serotonin pathway. It has been shown that expression of the serotonin biosynthesis gene, tryptophan hydroxylase-1 (*tph-1*), is increased during aging.[1,5] Reducing the serotonin signal by serotonergic receptor inhibitors and by blocking the serotonin/octopamine pathway consistently rescued age-related changes in associative learning behaviors, including basal/slowing response and chemotaxis avoidance.[5,9] Importantly, serotonin pathways regulate life span.[47] Thus, it is likely that the serotonin pathways regulate aging and age-related changes in the associative learning behaviors.

The results also provide an important prediction: Age-related increases in serotonin may cause global changes in behaviors that can be affected by serotonin. It remains to be clarified whether or not the observation is limited to serotonin or more general to other neurotransmitters and hormones. Importantly, increased serotonin levels in the mid-phase of aging are consistent with previous studies in rodents.[5] Specifically, serotonin release is increased in the rat brain.[48] Serotonin levels are reduced in the late phase of aging because of the loss of serotonin-releasing neurons.[49]

Insulin/IGF-1 Pathway

The insulin/IGF-1 pathway modulates thermotaxis and chemotaxis avoidance (to sodium chloride and benzaldehyde).[1,6] It also regulates the late phase of age-related changes in learning behavior (temperature–food association).[7] The insulin/IGF-1 pathway includes *daf-2* (insulin-like receptor gene) and *age-1* (phosphatidylinositol-3 OH kinase gene), which negatively regulates the FOXO/Forkhead transcription factor encoded by *daf-16*. Importantly, the *daf-2* mutant shows impaired chemotactic learning with starvation conditioning in young worms, including starvation–benzaldehyde association and starvation–sodium chloride association.[9] Thus, it is not clear how the mutants of the insulin/IGF-1 pathway affect AMI in the learning paradigms, although it seems likely that age-related alterations in the pathway should contribute to AMI. More details have been described previously.[1]

Importantly, the serotonin pathways and the insulin/IGF-1 pathway regulate life span. It has been suggested that the serotonin receptors SER-1 and SER-4 antagonistically regulate life span (inactivation of SER-1 extends life span, whereas inactivation of SER-4 reduces it).[47] In contrast, the insulin/IGF-1 pathway negatively regulates life span (i.e., mutations of the pathway genes increase life span) through the DAF-16/FOXO transcription factor, CREB, and several transcription factors. The transcription factors include HSF-1 (which mediates heat shock response), PHA-1/FOXA (which mediates a nutrient/autophagy signal from LET-363/TOR), SKN-1/Nrf2 (which mediates an oxidative defense signal from Ins/IGF-1, PMK-1/P38, and LET-60/RAS), and EOR-1/PLZF (which mediates protein turnover from LET-60/RAS).[50–53] Thus, the insulin/IGF-1 pathway functions together with a wide variety of signal transduction pathways in response to multiplex stress (e.g., heat, ultraviolet stress, oxidative stress, and heavy metals), nutrients, autophagy, protein turnover, and reproduction (LET-60/RAS and possibly KRI-1—not indicated previously).

Other Factors that Affect AMI

In addition to endocrine changes, age-related biological alterations that influence associative learning and memory include (but are not limited to) reduced mitochondrial functions, increased oxidative stress, physiological alterations, neuroinflammation, and epigenetic changes.[1] In addition, neural regeneration may play a critical role in the prevention of aging in the nervous system.[54] It has become clear that neural loss plays a minor role, and this has been discussed elsewhere. In fact, the nervous system is structurally well conserved during aging in *C. elegans*.[55] Note that the number of principal neurons remains relatively constant in rodents and mammals during aging.[56,57]

Oxidative stress occurs when generation of reactive oxygen species (ROS) overwhelms the biological ability to scavenge ROS. Although the role of oxidative stress in aging has been challenged,[58–60] it is clear that oxidative stress has an impact on AMI. In fact, increased oxidative stress impairs thermotactic learning behavior (food–temperature association).[1,8] The timing of the markers for oxidative stress is similar to the late phase of AMI,[1] although the earliest marker of oxidative damage on the lipid (4-HNE) may be in part overlapped with the later phase of early AMI.[61]

NEURAL REGULATION OF MEMORY AND AMI

Thermotactic and chemotactic memory use the neural circuit, including sensory neurons (e.g., AWC and AFD for thermotactic memory), interneurons (e.g. AIY, AIZ, and RIA for thermotactic memory), and motor neurons that generate the behavioral outputs. Figure 12.1 summarizes the parts of neural circuits that are relevant for thermotactic and chemotactic memory. AWC is a pair of sensory neurons for odors (e.g., benzaldehyde and butanol) and cold temperature; AFD is a pair of sensory neurons for high temperature. In thermosensation, cultivation temperature (T_c) is compared

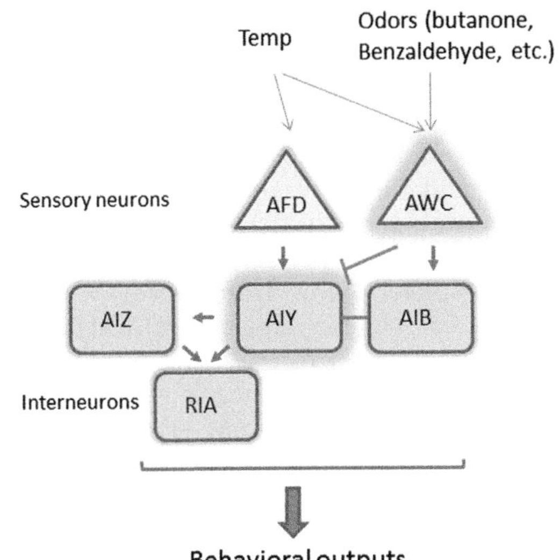

FIGURE 12.1 A simplified neural circuit relevant to AMI. Thermotaxis sensory neurons include AFD and AWC (blue triangles). Interneurons are shown as pink squares. AWC also senses odors for butanone and benzaldehyde. AWC and AIY (highlighted with purple shadows) are likely involved in thermotactic memory and chemotactic memory for benzaldehyde.

with colder or warmer temperatures. Colder temperature ($<T_c$) is mediated by the AWC sensory neurons, whereas warmer temperature ($>T_c$) is mediated by the AFD sensory neurons.[39,40] AWC and AFD coordinate cryophilic locomotion (also called negative thermotaxis) and thermophilic locomotion (also called positive thermotaxis), respectively, based on the cultivation temperature (i.e., threshold). The AWC sensory neurons communicate with the AIY interneurons through the GLC-3 glutamate-activated chloride channel,[62] suggestive of the use of glutamate signaling. Consistent with this interpretation, AWC and AFD sensory neurons appear to use glutamate as a neurotransmitter.[63] The interneurons for thermotaxis (AIY, AIZ, and RIA) have been suggested as a core site for thermotactic learning and memory[7,64]; therefore, they are important for understanding AMI. The long-lived *age-1* PI3K mutation of the insulin/IGF-1 pathway can increase temperature–food association and delay declines in the retention of the associative memory trace.[7] Expression of the *age-1* gene (the PI3K gene in the insulin/IGF-1 pathway) in AIY restores increased thermotactic phenotype of the *age-1* mutation. The results support the role of the insulin/IGF-1 pathway in AMI at the level of the interneurons. In addition, thermotaxis requires the *age-1* gene in the AIY, AIZ, and RIA interneurons.[7,64] Moreover, an insulin-like gene, *ins-1*, is required for thermotaxis. Thus, the insulin/IGF-1 pathway is likely to be a modulator of thermotactic learning and memory that functions in AIY and the other interneurons, AIZ and RIA.[7,64]

In contrast to the interneurons for thermotaxic memory, AWC-dependent chemotactic memory has been previously studied. The AWC sensory neurons sense smell and taste, and they appear to modulate chemotactic memory in young worms (Figure 12.1). The mutants of the insulin/IGF-1 pathway (*ins-1*, *daf-2*, and *age-1*) impair multiple types of chemotactic memory, including starvation–benzaldehyde and starvation–sodium chloride,[65,66] and therefore hamper the interpretation of AMI studies. Of them, the study of starvation–benzaldehyde association has provided an interesting implication for AMI. First, an increased food–benzaldehyde association has been observed during the mid- to late reproduction period (discussed previously). Second, expression of the *age-1* gene in AWC and some other neurons rescues impaired starvation–benzaldehyde association in *age-1(hx546)*.[65] Expression of *age-1* in the non-AWC neurons does not rescue the phenotype. INS-1 (a ligand of DAF-2 insulin/IGF-1 receptor) in ASI (sensory neuron) and AIA (interneuron) is sufficient to mediate starvation–benzaldehyde association. Although INS-1 is diffusible outside of the neurons, it seems plausible that ASI and AIA are candidates for the site that modulates starvation–benzaldehyde association at AWC. Third, the insulin/IGF-1 pathway appears to be partly required for learning (memory acquisition) and to be essential for memory retrieval. In temperature-shift experiments, the *daf-2* mutant shows modest impairment in memory acquisition at the permissive temperature (15°C) and more severe impairment in retrieval at the semirestrictive temperature (23°C).[65] Thus, it seems that *daf-2* has two functions in this type of learning and memory.

CREB transcription factor is known to be involved in memory formation. Similar to *Aplysia*, CREB appears to play a role in sensory neurons in *C. elegans*. CREB is known to play a role in STM and LTM in *Aplysia* and mammals.[67–69] A role in olfactory LTM in *C. elegans* (starvation–butanol association) was discussed previously. In addition to the role of interneurons in thermotaxis, CREB is required in the AFD thermosensory neurons for a normal thermotaxis,[27] suggesting that thermotaxis is in part modulated in the sensory neurons. AWC sensory neurons are essential for olfactory memory for benzaldehyde–starvation, highlighting a role of sensory neurons in associative learning and memory. The results raise an important concept that associative learning and memory use multiple sites in the central nervous system. It is consistent with the observation in fruit flies.[70]

Importantly, CREB also modulates other biological mechanisms, including aging, energy homeostasis in response to starvation, and endoplasmic reticulum stress. CREB and its co-factor, CRTC-1 (CREB-regulated transcriptional coactivator-1), mediate a signal from AMPK and calcineurin that regulate aging.[50] Microarray analysis of *crh-1/CREB* null suggests that genome-wide expression pattern is significant, similar to that in activated *aak-1/AMPK* and *tax-6/calcinurin* null strains.

MIDLIFE CRISIS THEORY AND EPIGENETIC CHANGES

The midlife crisis theory of aging assumes an age-related crisis that includes deteriorations during the middle of the life span.[1] It has been developed to incorporate biological and evolutional aspects of aging. Midlife is defined as approximately one-third of the life span in the middle, which roughly corresponds to the reproduction period. What is the onset of midlife crisis? It can be defined as the timing of earliest age-related phenotype observed (i.e., the onset of aging). The onset of midlife depends on species, reproduction period, life span of the particular species (short-lived species or long-lived species), and other factors. In addition, the early phase of aging does not appear as deteriorations in the structures of neural and other tissues but, rather, as shifts in biological capabilities, including decreases in plasticity, repair, and recovery (e.g., wound healing and regeneration), which limit biological capabilities (or, more precisely, reserves) to recover from harmful events. For example, *C. elegans* is a short-lived species in which aging appears to occur more rapidly than accumulation of damage. It is suggested that drift of developmental pathway drives the transcription circuit to aging through the *elt-5/elt-6* GATA transcription factors[71,72] and through microRNAs.[73–76] An alternative mechanism other than accumulation of damage appears to drive aging in *C. elegans*.

In the midlife, there are a wide variety of age-related phenotypes, ranging from minor to major deficits. They include, but are not limited to, increased mortality, altered learning behaviors, and deterioration of reproductive tissues. Importantly, midlife covers both early and late AMI: Early AMI starts at the early phase of midlife; and late AMI starts roughly at the mid- to late phase of midlife. As discussed previously, oxidative stress is restricted to late AMI, which fits well with the expression of the marker of oxidative protein damage, protein carbonyl.[1,77,78] The late phase of AMI is associated with various age-related deteriorations in behavior and reproductive tissues. As discussed previously, ROS scavenging enzymes are well retained at the timing of early AMI.

The molecular mechanism of midlife crisis remains unclear. However, it seems plausible that endocrine alteration and disturbance, including altered serotonin and insulin/IGF-1 pathways, are causes of the crisis in *C. elegans* for several reasons. First, alterations in the serotonin pathways should impact on associative learning behaviors in the early phase of middle life, presumably leading to early AMI. Second, the insulin/IGF-1 pathway should be involved in food and starvation signals, and therefore alterations in the insulin/IGF-1 pathway should affect classical conditioning that uses food or starvation as an unconditional stimulus. Third, it is consistent with deterioration of reproductive tissue, which should impact hormonal balance at approximately the onset of late AMI. Deterioration of reproductive tissue partly explains altered learning and memory in the midlife. Endocrine disturbance through the serotonin pathways and possibly the insulin/IGF-1 pathway is currently the best-characterized candidate that starts the onset of AMI. Additional details are discussed here and presented in Figure 12.2A.

Epigenetics and AMI

Learning and memory are somatic epigenetic changes in the central nervous system. Somatic epigenetic changes are distinct from reproductive epigenetic changes, some of which are heritable to the next generation (referred to as epigenesis). Why is epigenesis important with regard to AMI? Studies suggest that epigenesis and somatic epigenetic changes may share a common mechanism that affects life span, stress resistance, and AMI. In long-lived dwarf mice, epigenetic changes occur in the cells, leading to increased resistance to multiple stresses (multiplex stress resistance).[3,45] In this paradigm, altered profiles of endocrine in dwarf mice presumably cause epigenetic changes.[45,79,80] Similarly, it is possible that age-related changes in endocrine systems cause epigenetic changes in the cells, leading to changes in age-related phenotypes. In fact, epigenetic changes occur in *C. elegans*, contributing to variable life spans among individual worms. Interestingly, the epigenetic changes in *C. elegans* involve specific modification of chromatin remodeling, including H3 lysine trimethylation (H3K27me3 or H3K27 methylation) complex, at cellular and systemic levels.[81] H3K27 methylation is one of the histone modifications (including methylation at H4K12, H3K27, and H3K9) that can inhibit gene expression through chromatid remodeling. Another type of histone modification, acetylation, has been hypothesized to cause cognitive aging in mammals.[82] Chromatid remodeling is also regulated by phosphorylation.

Figure 12.2 shows a hypothetical model for the 'midlife crisis' theory that is modified to incorporate AMI, endocrine disturbance, and epigenetic changes. In the model (Figure 12.2A), signals from the completion of development trigger the onset of adulthood. It

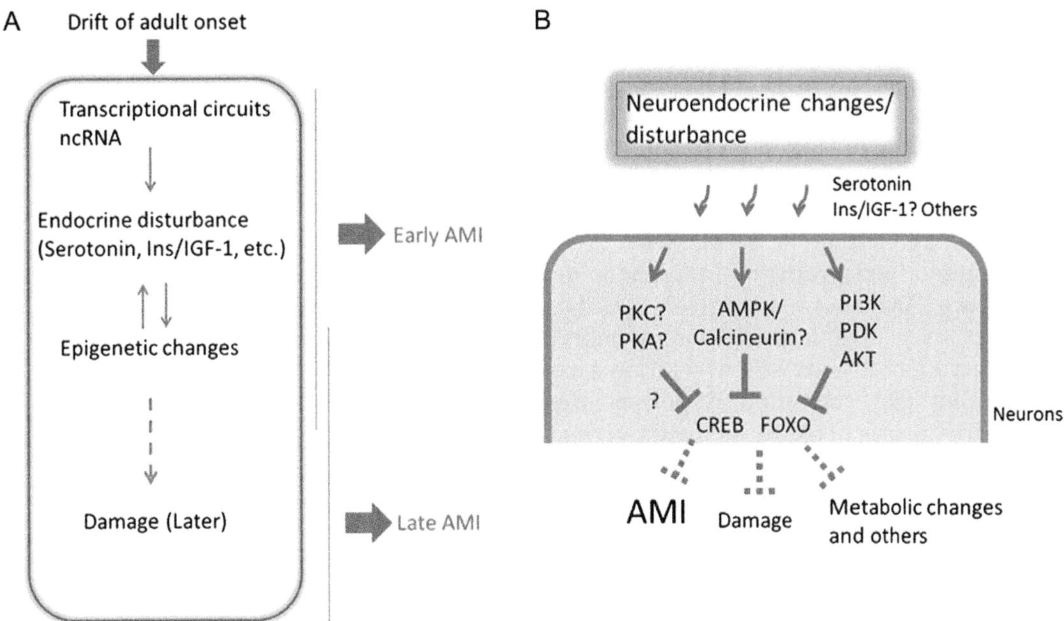

FIGURE 12.2 Models for AMI. (A) Age-related events that may contribute to early and late AMI. The drift of adult onset is mediated by multiple pathways, including the transcription factors and microRNAs. The adult onset triggers age-related endocrine changes, which may in turn causes epigenetic changes in neurons and other cells. This early change does not appear to include macromolecular damage in *C. elegans*. Oxidative and other damage seems to be correlated with late AMI and a wide variety of deteriorations. (B) Effects of age-related neuroendocrine disturbance on neurons. Neuroendocrine changes should affect downstream pathways to cause AMI and other age-related phenotypes. In the model, the changes directly affect AMI through downstream pathways and indirectly affect AMI through macromolecular damage. Serotonin signaling is possibly mediated by PKA and PKC signal transduction pathways. AMPK and calcinurin are shown to modulate CREB transcription factor through its co-factor CRTC-1. Insulin/IGF-1 can activate downstream components including PI3K, PDK, AKT, and FOXO transcription factor. Although CREB acts in AFD thermosensory neurons, CREB is expressed in most, if not all, neurons.

is known that the developmental signals are mediated by microRNA and transcription factors.[71–76] The developmental signals together with intrinsic and environmental signals alter neuroendocrine systems, including serotonin, insulin/IGF-1, and possibly others. Altered neuroendocrine functions can directly alter learning and memory through the secondary signal transduction pathways; candidates for the secondary pathways are indicated in Figure 12.2B. Alternatively, endocrine changes alter the serotonin and insulin/IGF-1 pathways, which in turn lead to reduced stress defense systems and increased damage to macromolecules. Based on observations in dwarf mice and *C. elegans*, age-related epigenetic changes should occur during the phase of endocrine disturbance.[45,79–81] Thus, the role of endocrine disturbance shifts to macromolecular damage, which leads to the late phase of AMI.

Importantly, it is possible that age-related epigenetic changes include altered chromatid structure in the nervous system, contributing to AMI. The possibility does not contradict a recent hypothesis of cognitive aging that states that dysregulation of epigenetic control mechanisms and the accumulation of aberrant epigenetic marks underlie aging-related cognitive dysfunction. The major difference from the current discussion is that the new hypothesis is more specific, assuming dysregulation of epigenetic control by experience- and age-associated chromatid remodeling with an emphasis on histone acetylation, another type of histone modification. In *C. elegans*, deterioration of biological functions is associated with the late phase of AMI, which may be different from the endocrine disturbance described here; nonetheless, the dysregulation of epigenetic control remains to be tested. The new hypothesis may best fit late AMI rather than early AMI. Although speculative, histone modification/chromatin remodeling and age-related transcription factors, including FOXO and CREB,[50] may be closely related, which affects expression of the genes and noncoding RNA, contributing to aging and AMI. It is worth exploring such interactions from the transcription factor networks to epigenetic control (Figure 12.2A).

PERSPECTIVES

AMI is one of the most diverse phenotypes that occurs during aging. In a short-lived species such as *C. elegans*, aging appears to be positively driven by the transcription factor networks. Aging is not only coupled with the end of development but also appears to be affected by epigenetic changes or other factors that alter gene expression. In an evolutionary sense, this can be viewed as an increased fitness and tradeoff at the expense of fitness during reproduction. The midlife crisis theory of aging explains age-related phenotypes from a young period to a transition phase and to a final phase of aging that is irreversible. It is intriguing that despite a wide variety of alterations in learning and memory, AMI shows a relatively specific profile that affects associative learning and memory associated with starvation. It is consistent with an earlier study suggesting that starvation responses are altered.[5] In the early phase of AMI, well-fed worms show locomotory associative memory similar to that of starved worms; more precisely, it appears as if well-fed worms have been starved.[5] In humans, AMI occurs in the prefrontal cortex and hippocampus, affecting working memory and specific types of memory, respectively.[24] Conceptually, it is important that the early phase of AMI is associated only modestly with macromolecular damage and more strongly associated with endocrine alteration and disturbance,[1] including alterations in serotonin, insulin/IGF-1, and possibly other signaling. Serotonin and insulin/IGF-1 are well-known modulators of a wide variety of biological processes, including cognitive functions and behavior (appetite and satiety), nutrition, energy expenditure, metabolism, stress resistance, and aging. In mammals, endocrine alterations occur gradually and earlier than menopause, which occurs at approximately the onset of the late phase of AMI.[1,41–44,46,47] Somatopause may overlap with the early phase because it is a gradual change in hormonal levels often seen in humans in the mid-30s. This chapter also highlighted the role of endocrine signals in epigenetic changes that should broadly affect aging and the nervous functions. An increasing number of analytical techniques are available, including genomics, high-throughput screening analysis, and automated imaging analysis (fluidic and behavioral), and these should help advance the field. Together with the classical and current analyses, the study of *C. elegans* AMI should be exciting and contribute to the understanding of genetic and nongenetic regulation of AMI.

References

1. Murakami S, Cabana K, Anderson D. Current advances in the study of oxidative stress and age-related memory impairment in *C. elegans*. In: Farooqui T, Farooqui A, eds. *Molecular Aspects of Oxidative Stress on Cell Signaling in Vertebrates and Invertebrates*. Hoboken, NJ: Wiley; 2011:347–360.
2. Salthouse TA. *Major Issues in Cognitive Aging*. Oxford: Oxford University Press; 2010.
3. Murakami S, Johnson TE. A genetic pathway conferring life extension and resistance to UV stress in *Caenorhabditis elegans*. *Genetics*. 1996;143:1207–1218.
4. Kauffman AL, Ashraf JM, Corces-Zimmerman MR, Landis JN, Murphy CT. Insulin signaling and dietary restriction differentially influence the decline of learning and memory with age. *PLoS Biol*. 2010;8:e1000372.
5. Murakami H, Bessinger K, Hellmann J, Luerman GC, Murakami S. Manipulation of serotonin signal suppresses early phase of behavioral aging in *Caenorhabditis elegans*. *Neurobiol Aging*. 2008;29:1093–1100.
6. Murakami S. *C. elegans* as a model system to study aging of learning and memory. *Mol Neurobiol*. 2007;35:85–94.
7. Murakami H, Bessinger K, Hellmann J, Murakami S. Aging-dependent and independent regulation of learning by insulin/IGF-1 signal in *C. elegans*. *J Neurosci*. 2005;25:10894–10904.
8. Murakami S, Murakami H. The effects of aging and oxidative stress on learning behavior in *C. elegans*. *Neurobiol Aging*. 2005;26:899–905.
9. Tsui D, van der Kooy D. Serotonin mediates a learned increase in attraction to high concentrations of benzaldehyde in aged *C. elegans*. *Learn Mem*. 2008;15:844–855.
10. Moroz LL, Kohn AB. Do different neurons age differently? Direct genome-wide analysis of aging in single identified cholinergic neurons. *Front Aging Neurosci*. 2010;2:6.
11. Amdam GV, Fennern E, Baker N, Rascón B. Honeybee associative learning performance and metabolic stress resilience are positively associated. *PLoS ONE*. 2010;5:e9740.
12. Tonoki A, Davis RL. Aging impairs intermediate-term behavioral memory by disrupting the dorsal paired medial neuron memory trace. *Proc Natl Acad Sci USA*. 2012; [Epub ahead of print].
13. Yamazaki D, Horiuchi J, Miyashita T, Saitoe M. Acute inhibition of PKA activity at old ages ameliorates age-related memory impairment in *Drosophila*. *J Neurosci*. 2010;30:15573–15577.
14. Hermann PM, Lee A, Hulliger S, Minvielle M, Ma B, Wildering WC. Impairment of long-term associative memory in aging snails (*Lymnaea stagnalis*). *Behav Neurosci*. 2007;121:1400–1414.
15. Kennard JA, Woodruff-Pak DS. Age sensitivity of behavioral tests and brain substrates of normal aging in mice. *Front Aging Neurosci*. 2011;3:9.
16. Sharma S, Haselton J, Rakoczy S, Branshaw S, Brown-Borg HM. Spatial memory is enhanced in long-living Ames dwarf mice and maintained following kainic acid induced neurodegeneration. *Mech Ageing Dev*. 2010;131:422–435.
17. Villarreal JS, Dykes JR, Barea-Rodriguez EJ. Fischer 344 rats display age-related memory deficits in trace fear conditioning. *Behav Neurosci*. 2004;118:1166–1175.
18. Woodruff-Pak DS, Foy MR, Akopian GG, Lee KH, Zach J, Nguyen KP, et al. Differential effects and rates of normal aging in cerebellum and hippocampus. *Proc Natl Acad Sci USA*. 2010;107:1624–1629.
19. Labar KS, Cook CA, Torpey DC, Welsh-Bohmer KA. Impact of healthy aging on awareness and fear conditioning. *Behav Neurosci*. 2004;118:905–915.
20. Shing YL, Werkle-Bergner M, Brehmer Y, Müller V, Li SC, Lindenberger U. Episodic memory across the lifespan: the contributions of associative and strategic components. *Neurosci Biobehav Rev*. 2010;34:1080–1091.
21. Cowan N. What are the differences between long-term, short-term, and working memory? *Prog Brain Res*. 2008;169:323–338.
22. Gluck MA, Mercado E, Myers CE. *Learning and Memory: From Brain to Behavior*. New York: Worth; 2008.
23. Eichenbaum H, Robitsek RJ. Olfactory memory: a bridge between humans and animals in models of cognitive aging. *Ann N Y Acad Sci*. 2009;1170:658–663.

24. Morrison JH, Baxter MG. The ageing cortical synapse: hallmarks and implications for cognitive decline. *Nat Rev Neurosci*. 2012;13: 240–250.
25. Weiler JA, Bellebaum C, Daum I. Aging affects acquisition and reversal of reward-based associative learning. *Learn Mem*. 2008;15:190–197.
26. Ardiel EL, Rankin CH. An elegant mind: learning and memory in *Caenorhabditis elegans*. *Learn Mem*. 2010;17:191–201.
27. Nishida Y, Sugi T, Nonomura M, Mori I. Identification of the AFD neuron as the site of action of the CREB protein in *Caenorhabditis elegans* thermotaxis. *EMBO Rep*. 2011;12:855–862.
28. Rankin CH, Beck CD, Chiba CM. *Caenorhabditis elegans*: a new model system for the study of learning and memory. *Behav Brain Res*. 1990;37:89–92.
29. Baddeley AD, Hitch G. Working memory. In: Bower GH, ed. *The Psychology of Learning and Motivation: Advances in Research and Theory*. 8. New York: Academic Press; 1974:47–89.
30. Miller GA, Galanter E, Pribram KH. *Plans and Structure of Behavior*. New York: Holt, Rinehart & Winson; 1960.
31. Johnson MK, Reeder JA, Raye CL, Mitchell KJ. Second thoughts versus second looks: an age-related deficit in selectively refreshing just-active information. *Psychol Sci*. 2002;13:64–67.
32. Remy JJ, Hobert O. An interneuronal chemoreceptor required for olfactory imprinting in *C. elegans*. *Science*. 2005;309:787–790.
33. Torayama I, Ishihara T, Katsura I. *Caenorhabditis elegans* integrates the signals of butanone and food to enhance chemotaxis to butanone. *J Neurosci*. 2007;27:741–750.
34. Biron D, Shibuya M, Gabel C, Wasserman SM, Clark DA, Brown A, et al. A diacylglycerol kinase modulates long-term thermotactic behavioral plasticity in *C. elegans*. *Nat Neurosci*. 2006;9:1499–1505.
35. Beck CDO, Rankin CH. Effects of aging on habituation in the nematode *Caenorhabditis elegans*. *Behav Processes*. 1993;28: 145–163.
36. Glenn CF, Chow DK, David L, Cooke CA, Gami MS, Iser WB, et al. Behavioral deficits during early stages of aging in *Caenorhabditis elegans* result from locomotory deficits possibly linked to muscle frailty. *J Gerontol A Biol Sci Med Sci*. 2004;59:1251–1260.
37. Hedgecock EM, Russell RL. Normal and mutant thermotaxis in the nematode *Caenorhabditis elegans*. *Proc Natl Acad Sci USA*. 1975;72:4061–4065.
38. Chi CA, Clark DA, Lee S, Biron D, Luo L, Gabel CV, et al. Temperature and food mediate long-term thermotactic behavioral plasticity by association-independent mechanisms in *C. elegans*. *J Exp Biol*. 2007;210:4043–4052.
39. Jurado P, Kodama E, Tanizawa Y, Mori I. Distinct thermal migration behaviors in response to different thermal gradients in *Caenorhabditis elegans*. *Genes Brain Behav*. 2010;9:120–127.
40. Garrity PA, Goodman MB, Samuel AD, Sengupta P. Running hot and cold: behavioral strategies, neural circuits, and the molecular machinery for thermotaxis in *C. elegans* and *Drosophila*. *Genes Dev*. 2010;24:2365–2382.
41. Rehman HU, Masson EA. Neuroendocrinology of female aging. *Gend Med*. 2005;2:41–56.
42. Araujo AB, Wittert GA. Endocrinology of the aging male. *Best Pract Res Clin Endocrinol Metab*. 2011;25:303–319.
43. Kermath BA, Gore AC. Neuroendocrine control of the transition to reproductive senescence: lessons learned from the female rodent model. *Neuroendocrinology*. 2012;96:1–12.
44. Holzenberger M. Igf-I signaling and effects on longevity. *Nestle Nutr Workshop Ser Pediatr Program*. 2011;68:237–245.
45. Murakami S. Stress resistance in long-lived mouse models. *Exp Gerontol*. 2006;41:1014–1019.
46. Barrientos RM, Frank MG, Watkins LR, Maier SF. Aging-related changes in neuroimmune-endocrine function: implications for hippocampal-dependent cognition. *Horm Behav*. 2012;62:219–227.
47. Murakami H, Murakami S. Serotonin receptors antagonistically modulate *C. elegans* longevity. *Aging Cell*. 2007;6:483–488.
48. Cassel JC, Schweizer T, Lazaris A, Knörle R, Birthelmer A, Gödtel-Armbrust U, et al. Cognitive deficits in aged rats correlate with levels of L-arginine, not with nNOS expression or 3,4-DAP-evoked transmitter release in the frontoparietal cortex. *Eur Neuropsychopharmacol*. 2005;15:163–175.
49. Mattson MP. Pathways towards and away from Alzheimer's disease. *Nature*. 2004;430:631–639.
50. Mair W, Morantte I, Rodrigues AP, Manning G, Montminy M, Shaw RJ, et al. Lifespan extension induced by AMPK and calcineurin is mediated by CRTC-1 and CREB. *Nature*. 2011;470:404–408.
51. Martin GM. The biology of aging: 1985–2010 and beyond. *FASEB J*. 2011;25:3756–3762.
52. Panowski SH, Dillin A. Signals of youth: endocrine regulation of aging in *Caenorhabditis elegans*. *Trends Endocrinol Metab*. 2009;20 (6):259–264.
53. Rongo C. Epidermal growth factor and aging: a signaling molecule reveals a new eye opening function. *Aging*. 2011;3:896–905.
54. Chiu H, Alqadah A, Chuang CF, Chang C. *C. elegans* as a genetic model to identify novel cellular and molecular mechanisms underlying nervous system regeneration. *Cell Adh Migr*. 2011;5:387–394.
55. Herndon LA, Schmeissner PJ, Dudaronek JM, Brown PA, Listner KM, Sakano Y, et al. Stochastic and genetic factors influence tissue-specific decline in ageing *C. elegans*. *Nature*. 2002;419:808–814.
56. Duan H, Wearne SL, Rocher AB, Macedo A, Morrison JH, Hof PR. Age-related dendritic and spine changes in corticocortically projecting neurons in macaque monkeys. *Cereb Cortex*. 2003;13:950–961.
57. Merrill DA, Roberts JA, Tuszynski MH. Conservation of neuron number and size in entorhinal cortex layers II, III, and V/VI of aged primates. *J Comp Neurol*. 2000;422:396–401.
58. Doonan R, McElwee JJ, Matthijssens F, Walker GA, Houthoofd K, Back P, et al. Against the oxidative damage theory of aging: superoxide dismutases protect against oxidative stress but have little or no effect on life span in *Caenorhabditis elegans*. *Genes Dev*. 2008;22:3236–3241.
59. Honda Y, Tanaka M, Honda S. Modulation of longevity and diapause by redox regulation mechanisms under the insulin-like signaling control in *Caenorhabditis elegans*. *Exp Gerontol*. 2008;43: 520–529.
60. Yang W, Li J, Hekimi S. A measurable increase in oxidative damage due to reduction in superoxide detoxification fails to shorten the life span of long-lived mitochondrial mutants of *Caenorhabditis elegans*. *Genetics*. 2007;177:2063–2074.
61. Ayyadevara S, Dandapat A, Singh SP, Siegel ER, Shmookler Reis RJ, Zimniak L, et al. Life span and stress resistance of *Caenorhabditis elegans* are differentially affected by glutathione transferases metabolizing 4-hydroxynon-2-enal. *Mech Ageing Dev*. 2007;128:196–205.
62. Chalasani SH, Chronis N, Tsunozaki M, Gray JM, Ramot D, Goodman MB, et al. Dissecting a circuit for olfactory behaviour in *Caenorhabditis elegans*. *Nature*. 2007;450:63–70.
63. Lee RY, Sawin ER, Chalfie M, Horvitz HR, Avery L. EAT-4, a homolog of a mammalian sodium-dependent inorganic phosphate cotransporter, is necessary for glutamatergic neurotransmission in *Caenorhabditis elegans*. *J Neurosci*. 1999;19:159–167.
64. Kodama E, Kuhara A, Mohri-Shiomi A, Kimura KD, Okumura M, Tomioka M, et al. Insulin-like signaling and the neural circuit for integrative behavior in *C. elegans*. *Genes Dev*. 2006;20:2955–2960.
65. Lin CH, Tomioka M, Pereira S, Sellings L, Iino Y, van der Kooy D. Insulin signaling plays a dual role in *Caenorhabditis elegans* memory acquisition and memory retrieval. *J Neurosci*. 2010;30: 8001–8011.

66. Tomioka M, Adachi T, Suzuki H, Kunitomo H, Schafer WR, Iino Y. The insulin/PI 3-kinase pathway regulates salt chemotaxis learning in *Caenorhabditis elegans*. *Neuron*. 2006;51:613–625.
67. Benito E, Barco A. CREB's control of intrinsic and synaptic plasticity: implications for CREB-dependent memory models. *Trends Neurosci*. 2010;33:230–240.
68. Kandel ER. The molecular biology of memory storage: a dialog between genes and synapses. *Biosci Rep*. 2001;21:565–611.
69. Lee YS, Bailey CH, Kandel ER, Kaang BK. Transcriptional regulation of long-term memory in the marine snail. *Aplysia Mol Brain*. 2008;1:3.
70. Chen CC, Wu JK, Lin HW, Pai TP, Fu TF, Wu CL, et al. Visualizing long-term memory formation in two neurons of the *Drosophila* brain. *Science*. 2012;335:678–685.
71. Budovskaya YV, Wu K, Southworth LK, Jiang M, Tedesco P, Johnson TE, et al. An elt-3/elt-5/elt-6 GATA transcription circuit guides aging in *C. elegans*. *Cell*. 2008;134:291–303.
72. Tonsaker T, Pratt RM, McGhee JD. Re-evaluating the role of ELT-3 in a GATA transcription factor circuit proposed to guide aging in *C. elegans*. *Mech Ageing Dev*. 2012;133:50–53.
73. Boehm M, Slack F. A developmental timing microRNA and its target regulate life span in *C. elegans*. *Science*. 2005;310:1954–1957.
74. Ibáñez-Ventoso C, Yang M, Guo S, Robins H, Padgett RW, Driscoll M. Modulated microRNA expression during adult lifespan in *Caenorhabditis elegans*. *Aging Cell*. 2006;5:235–246.
75. Pincus Z, Smith-Vikos T, Slack FJ. MicroRNA predictors of longevity in *Caenorhabditis elegans*. *PLoS Genet*. 2011;7:e1002306.
76. Smith-Vikos T, Slack FJ. MicroRNAs and their roles in aging. *J Cell Sci*. 2012;125:7–17.
77. Adachi H, Ishii N. Effects of tocotrienols on life span and protein carbonylation in *Caenorhabditis elegans*. *J Gerontol A Biol Sci Med Sci*. 2000;55:B280–B285.
78. Levine RL, Stadtman ER. Oxidative modification of proteins during aging. *Exp Gerontol*. 2001;36:1495–1502.
79. Murakami S, Salomon A, Miller RA. Multiplex stress resistance in cells in long-lived Dwarf mice. *FASEB J*. 2003;17:1565–1566.
80. Salmon AB, Murakami S, Bartke A, Kopchick J, Yasumura K, Miller RA. Fibroblast cell lines from young adult mice of long-lived mutant strains are resistant to multiple forms of stress. *Am J Physiol Endocrinol Metab*. 2005;289:E23–E29.
81. Greer EL, Maures TJ, Ucar D, Hauswirth AG, Mancini E, Lim JP, et al. Transgenerational epigenetic inheritance of longevity in *Caenorhabditis elegans*. *Nature*. 2011;479:365–371.
82. Penner MR, Roth TL, Barnes CA, Sweatt JD. An epigenetic hypothesis of aging-related cognitive dysfunction. *Front Aging Neurosci*. 2010;2:9.

CHAPTER

13

Salt Chemotaxis Learning in *Caenorhabditis elegans*

Yuichi Iino

The University of Tokyo, Tokyo, Japan

SALT CHEMOTAXIS IN *CAENORHABDITIS ELEGANS*

As introduced in Chapters 9–12, *Caenorhabditis elegans* senses various chemical, mechanical, and thermal stimuli and responds to them. Of these responses, chemotaxis, in which an animal moves toward or sometimes away from the source of a chemical, is a convenient behavior for the study of sensory perception and behavioral plasticity. *Caenorhabditis elegans* shows chemotaxis to various volatile and water-soluble chemicals,[1,2] which are often categorized as odor and taste substances by analogy with vertebrates. The latter includes inorganic cations such as Na^+, K^+, Li^+, and Mg^{2+} and anions such as Cl^-, Br^-, and I^-, as well as low-molecular-weight polarized organic compounds such as cAMP and cGMP. Also, some amino acids attract *C. elegans*. Counterintuitively, however, *C. elegans* is usually not attracted to sugars. This chapter provides an overview of the chemotaxis to inorganic salts and its plasticity.

Laser ablation experiments performed in 1991 by Cori Bargmann[3] showed that a pair of amphid sensilla at the head are the major organ used for detecting the chemicals and mediating chemotaxis to these chemicals. Of the 12 pairs of amphid sensory neurons, several pairs of neurons were important for chemotaxis to the water-soluble chemicals. Especially when ASE neurons were killed, chemotaxis to cAMP, biotin, NH_4Cl, NaAc, lysine, and serotonin was severely impaired, indicating that ASE neurons are most important and play major roles in sensing the water-soluble chemicals. In addition, ADF, ASG, ASI, and ASK played minor but substantial roles because chemotaxis was further reduced by killing each of these neurons along with ASE, with combinatory killing of three or more neurons showing additive effects in most cases. The contribution of each neuron was somewhat different between different chemicals to be sensed. On the other hand, odorants are sensed by a different set of sensory neurons such as AWA and AWC, as described in Chapter 10.

The amphid sensilla and amphid sensory neurons are bilaterally symmetrical, and most amphid neurons also appear to be functionally symmetrical. However, gene expression analyses showed that the left and right ASE neurons express different sets of genes.[4] This observation prompted researchers to test the left/right difference of ASE neurons by individual laser ablation, and it was found that left and right ASE neurons actually sense different ions. ASER is more important than ASEL for chemotaxis to Cl^-, Br^-, I^-, and K^+, whereas ASEL is more important for Na^+, Li^+, and Mg^{2+}.[5] Receptors for these ions have not been identified, but strong candidates are guanylyl cyclases. There are 27 *gcy* genes encoding receptor-type transmembrane guanylyl cyclases identified in the genome of *C. elegans*.[4] Of these, at least 9 are expressed in either ASER or ASEL in a left/right-biased manner. For example, *gcy-5* is expressed almost exclusively in ASER, whereas *gcy-6* is expressed in ASEL. In addition, knockout of some of these genes causes impaired chemotaxis to a subset of ions, suggesting that some members of the family do confer sensory specificity to the ASE neurons.[5]

In the past, it was difficult to monitor the activities of neurons in *C. elegans*. In pioneering work, Miriam Goodman and Shawn Lockery performed whole-cell patch clamp recording of the ASER neuron[6] and showed that the ASER membrane has a high resistance

Invertebrate Learning and Memory.
DOI: http://dx.doi.org/10.1016/B978-0-12-415823-8.00013-7

due to depolarization-activated K^+ and Ca^{2+} channels and therefore small input currents can elicit a large depolarization, but they could not examine the response of the neuron to sensory stimuli. After GFP-based genetically encoded calcium probes were developed, the sensory responses of sensory neurons and interneurons were measured by calcium imaging. The first surprising results were that ASER and ASEL showed opposite responses. ASER showed calcium response when NaCl was removed, whereas ASEL responded when NaCl was added.[7] The OFF response of ASER and ON response of ASEL were maintained when tested with other ions.[5] Other than ASE neurons, ADF and ASH responded to changes in NaCl concentrations, but no response was detected in ASI or ASG. Of these, the response of ADF was largely diminished in the *unc-13* mutant, which is defective in synaptic transmission, suggesting that ADF is activated by synaptic inputs from other neurons rather than sensing salt by itself.[8]

SALT CHEMOTAXIS LEARNING: THE BEHAVIOR

For all animals, the ability to search and find food is essential for optimizing the chance of survival. Therefore, many animals utilize their own experiences and learn to associate ambient conditions with availability of food, and they later pursue the conditions where food was previously found or avoid the conditions where food was not found. In the case of salt chemotaxis in *C. elegans*, the direction and extent of chemotaxis change depending on past experience with salt and food. This was first demonstrated by placing the worms on a food-free plate containing either sodium acetate or ammonium chloride for several hours and then testing chemotaxis (in the absence of food) to the same chemical (Figure 13.1A).[9] Whereas well-fed animals are attracted to these chemicals as described previously, prestarved animals are no longer attracted to the chemicals. As a control, animals starved in the absence of the salts showed excellent chemotaxis to the salts, indicating that association of salt and starvation causes the learned ignoring of the salt. It was later found that this association actually causes the animals to avoid, rather than ignore, salts.[10,11] Conditioning with one salt affects chemotaxis to other salts, at least to some extent, whereas chemotaxis to odorants is not affected. Therefore, the specificity seems to depend on the sensory neuron that senses the chemicals (described later in more detail). This form of learned salt avoidance lasts for approximately 1 hr[9–11]; therefore, this is likely a form of short-term memory. Similar but slightly different assays have also

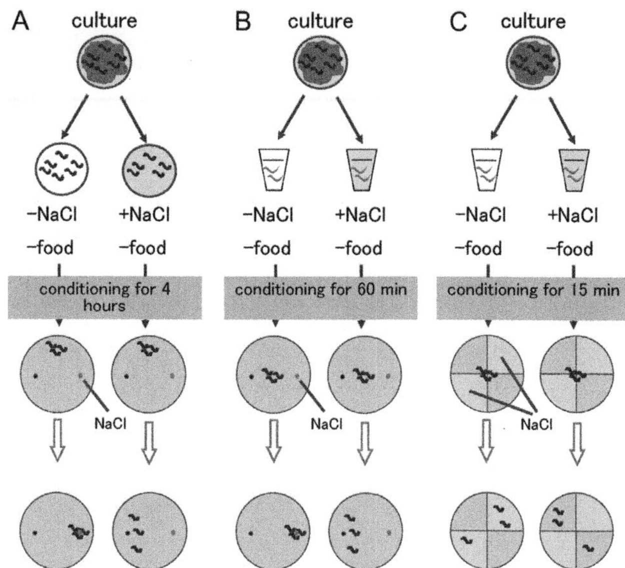

FIGURE 13.1 Different assays for salt chemotaxis plasticity. (A) Worms were kept on plates containing (experimental) or not containing (control) NaCl without food for 4 hr (conditioning step), and then their chemotaxis was tested on a plate on which salt was applied on one spot. The other spot indicates a position symmetrical to the salt spot on which no chemical was applied. (B and C) Similar to panel A, except that worms were conditioned in a salt-containing buffer (experimental) or blank buffer (control) for 1 hr. In panel C, chemotaxis was tested on a quadrant plate with juxtaposed salt-containing or noncontaining agar blocks. Note that behavioral plasticity observed in all these assays is called salt chemotaxis learning in this chapter for simplicity. Source: *Modified from (A) Saeki* et al.,[9] *(B) Tomioka* et al.,[10] *and (C) Hukema* et al.[11]

been reported (Figures 13.1B and C).[10,11] In these assays, worms were conditioned in a salt-containing buffer and then chemotaxis was tested on agar plates with salt gradient or juxtaposed agar blocks with different salt concentrations. Note that behavior in general is sensitive to subtle experimental parameters and therefore what is true in one assay might not be true in other assays. For simplicity, however, I collectively call these forms of behavioral plasticity 'salt chemotaxis learning' in this chapter, in which attraction of *C. elegans* animals to salt is decreased or turned into aversion after exposure to salt without food, although other terminology, such as 'gustatory plasticity' or 'salt chemotaxis plasticity,' is also used in literature.

In this type of learning, conditioning is established by a combination of two stimuli—salt as conditioned stimulus (CS) and starvation as unconditioned stimulus (US). Only when these two stimuli are presented are the animals conditioned (hereafter called salt-conditioned) and the response to the salt changed. The outcome is aversion rather than attraction to salt, and the salt aversion behavior is considered a conditioned response (CR).

THE ROLE OF ASE NEURONS IN SALT CHEMOTAXIS LEARNING

To understand the mechanism of salt chemotaxis learning, as in other types of learning, we need to know the molecules that are involved in the learning. This can be achieved in principle by finding mutants defective in the learning. In addition, we need to know how the CS and US—in this case, salt and lack of food, respectively—are sensed and where in the nervous system these pieces of information are integrated. Also, pertinent questions are the following: Where is the memory stored? and How are the different behavioral responses to salt elicited based on the memory? Unfortunately, we still do not know the exact answers to these questions, but some pieces of the puzzle are emerging.

To identify mutants defective in salt chemotaxis learning, Renate Hukema *et al.*[11] screened 32 mutants that had been isolated in the *C. elegans* research community for other phenotypes or by reverse genetic approaches. Substantial numbers of them were found to be defective in salt chemotaxis learning. The first obvious mutant was *che-1*, which encodes a GLASS-type transcription factor that plays a central role in ASE fate determination.[11] This mutant, which lacks functional ASE, essentially lost chemotaxis to NaCl. In addition, avoidance of NaCl after salt conditioning was also lost. Therefore, this mutant showed essentially no chemotaxis to NaCl before and after salt conditioning. The *che-36* mutant, which is defective in ASE and AWC, behaved similarly.[11] These results are easily understood assuming that ASE neurons sense NaCl during chemotaxis behavior, either attractive or aversive. However, it must be kept in mind that ASE is likely to be required for sensing the CS during conditioning, and lack of ASE may lead to failure of conditioning. Other mutants identified by Hukema *et al.* suggested involvement of other sensory neurons as well, but before describing these, we discuss other molecules that are related to the functions of ASE neurons.

THE INSULIN/PHOSPHATIDYLINOSITOL 3-KINASE PATHWAY

Masahiro Tomioka *et al.*[10] screened existing mutants and found that the insulin/phosphatidylinositol (PI) 3-kinase pathway was essential for salt chemotaxis learning. *Caenorhabditis elegans* has a single gene encoding insulin/insulin-like growth factor (IGF) receptor homolog in the genome. This gene, *daf-2*, is known to be involved in many biological processes. When larval worms are overcrowded and get starved, they change the morphology to what is called dauer larva and become resistant to various stresses such as starvation, heat, and desiccation. When the *daf-2* activity is lacking, larval worms become dauer larva even when they are not crowded and are well fed. *daf-2* is also required for progress through the first larval stage, and therefore *daf-2* null animals arrest development at the early larval stage. *daf-2* is also important for longevity control, and *daf-2* reduction-of-function mutants have a long life span. It was found that *daf-2* temperature-sensitive mutants, when raised at a permissive temperature and tested at a restrictive temperature, show a defect in salt chemotaxis learning. Downstream of DAF-2 insulin/IGF receptor, the PI 3-kinase pathway is known to act for dauer formation, longevity, and other processes. PI 3-kinase phosphorylates the 3-position of inositol in phosphatidylinositol phosphates; for example, phosphatidylinositol 4,5-bisphosphate (PIP_2) is converted to phosphatidylinositol 3,4,5-triphosphate (PIP_3). PIP_3 generated by this reaction recruits phophoinositide-dependent kinase PDK and AKT kinase and activates them. Mutants of *age-1* PI 3-kinase, *pdk-1* PDK and *akt-1* AKT kinase, were all defective in salt chemotaxis learning, suggesting that this pathway was also found to act a role in learning. Naive chemotaxis, namely chemotaxis of unstarved animals, did not change much, suggesting that the insulin/PI 3-kinase pathway is necessary for either conditioning or conditioned response. There are more than 30 insulin-like peptides in the genome of *C. elegans*. A deletion mutant of one of these, *ins-1*, was defective in salt chemotaxis learning, whereas other insulin-like peptides act for other processes. It therefore seems that different insulin-like ligands are involved in different *daf-2*-dependent processes.

To determine where in the nervous system the insulin/PI 3-kinase pathway acts, Tomioka *et al.*[10] performed cell-specific rescue experiments and showed that expression of *age-1* in only the ASER neuron rescues the defect of *age-1* mutants in salt chemotaxis learning, whereas expression in ASEL or other neurons does not rescue the defect. *daf-18* PTEN in the same pathway was also found to act in ASER. Therefore, although NaCl, which consists of Na^+ and Cl^- ions, was used in these experiments, and the Na^+ ion is sensed mainly by ASEL and weakly by ASER, the insulin-PI 3-kinase pathway seems to act solely in ASER. To confirm this, Tomioka *et al.* omitted the Cl^- ion, which is sensed by ASER, and performed salt chemotaxis learning assays using sodium acetate for conditioning and chemotaxis. *age-1* mutants showed a mild defect in salt chemotaxis learning, and this defect was rescued by expression of *age-1* cDNA in ASER but not in ASEL (our unpublished results). Therefore, the PI 3-kinase pathway acts only in ASER, and plasticity of ASEL depends on other molecular pathways.

THE GQ/DIACYLGLYCEROL/PROTEIN KINASE C PATHWAY

To identify genes that genetically interact with the insulin/PI 3-kinase pathway, Takeshi Adachi *et al.*[12] performed suppressor screens. *daf-18* encodes a homolog of mammalian phosphatase, PTEN, which catalyzes dephosphorylation of PIP_3 to PIP_2, namely a reverse reaction of PI 3-kinase. Therefore, the *daf-18* loss-of-function mutants have elevated levels of PIP_3[13] and show phenotypes that are roughly opposite to the loss-of-function mutants of the PI 3-kinase pathway, such as defective dauer formation and short life span. In salt chemotaxis learning, *daf-18* mutants show greatly reduced chemotaxis even before salt conditioning. Therefore, it appears that ectopic activation of the PI 3-kinase pathway mimics conditioning, suggesting a key role for this pathway.

daf-18 animals were mutagenized and suppressor mutants with restored chemotaxis to NaCl were collected. One of the suppressors was a putative gain-of-function mutant of *egl-30*. The *egl-30* gene encodes the sole Gq α subunit in *C. elegans*. The Gq-type G protein is known to regulate phospholipase C, which catalyzes generation of inositol triphosphate (IP_3) and diacylglycerol (DAG) from PIP_2. Loss-of-function mutants of *egl-30* are lethargic, and gain-of-function mutants of the same gene are hyperactive.[14] Extensive studies on this pathway have shown that the Gq/PLC/DAG pathway positively regulates synaptic transmission from ventral cord motor neurons, which are involved in locomotion.[15] This pathway has also been shown to act in sensory neurons and bias the sensory response behavior. For example, in thermosensitive neurons, activation of the DAG pathway causes migration to lower temperatures,[16] and in olfactory neurons activation of the pathway promotes attraction to odorants.[17] In ventral cord motor neurons, two proteins—UNC-13, a protein essential for release of synaptic vesicles, and nPKC PKC-1 ('novel'-type protein kinase C)—have DAG-binding domains and are DAG targets important for synaptic transmission.[15] nPKC PKC-1 also acts in themosensory neurons, olfactory neuron, and the salt-sensing neuron.

In salt chemotaxis learning, expression of *egl-30(gf)* or *pkc-1(gf)* (gf = gain of function) in the ASER sensory neuron suppresses salt avoidance caused by learning so that these transgenic animals migrate to the salt even after salt conditioning.[10,12] Addition of DAG analog, phorbol myristate acetate (PMA), also causes suppression of learned avoidance. Considering the roles for the pathway in other neurons, it is likely that the Gq/DAG/nPKC pathway promotes a subset of synaptic transmission from the ASER neuron and thereby promotes attraction to the salt. The PI 3-kinase pathway may act by negatively regulating the Gq/DAG/nPKC pathway, or the two pathways may act in parallel in opposite directions.

OTHER GENES ACTING IN ASER

The importance of the ASER neuron for salt chemotaxis learning has also been suggested by the discovery of other molecules acting in ASER for this form of learning. CASY-1 is an ortholog of calsyntenin/alcadein, which is a mammalian transmembrane protein with an extracellular cadherin domain that is highly expressed in the brain.[18]*casy-1* mutants show a defect in salt chemotaxis learning, and this defect is rescued by expression of *casy-1* in ASER. The *casy-1* mutants also have defects in olfactory learning, temperature learning, and integration of attractive and aversive signals; therefore, they may be generally important for learning and higher order sensory processing.

PITP-1 is a class IIA phosphatidylinositol transfer protein that is important for intracellular phospholipid cycling. *pipt-1* mutants have defects in salt chemotaxis learning, but they also show reduced chemotaxis in naive conditions.[19] Both of these defects are mostly rescued by expression of *pitp-1* in ASE neurons. Two pathways important for salt chemotaxis and salt chemotaxis learning—the PI 3-kinase pathway and the DAG pathway, the latter of which involves hydrolysis of PIP_2 by PLC—require phosphatidylinositol phosphates and therefore lack of PITP-1 may impair these pathways. Interestingly, PITP-1 was strongly localized to synapses, raising the possibility that local regulation of phospholipid turnover is important for synaptic plasticity affecting salt chemotaxis.

THE EGL-8/DIACYLGLYCEROL/PROTEIN KINASE D PATHWAY ACTING IN ASEL

In the previously discussed studies, NaCl was used mainly as a testing salt, which may be related to the identification of ASER as a major site of action of regulatory molecules. On the other hand, when sodium acetate was used as both conditioning salt and testing salt, the role of a different set of DAG pathway molecules was identified. In this assay, mutants of *egl-8* PLC, *tpa-1* nPKC, and *dkf-2* protein kinase D (PKD) showed defects, whereas *pkc-1* nPKC and *pkc-2* cPKC ('conventional' protein kinase C) did not show defects.[20] PKD has two DAG binding domains and is recruited to the membranes in a DAG-dependent manner. Also, it is phosphorylated and activated by PKC.

Because the *dkf-2* mutant did not show any defect when choline chloride was used as a salt, it is likely that the *egl-8/tpa-1/dkf-2* pathway acts in ASEL and only the plasticity depending on ASEL is affected in these mutants. Two predicted isoforms of DKF-2, DKF-2A and DKF-2B, are generated by alternative splicing of the *dkf-2* gene; DKF-2B is expressed in several neurons including ASE, and DKF-2A is expressed in intestine. Interestingly, rescue experiments suggest that DKF-2 in both tissues is required for salt chemotaxis learning. It still needs to be tested whether *egl-30* Gq/*pkc-1* nPKC acting for ASER functions and *egl-8* PLC/*tpa-1* nPKC/*dkf-2* PKD acting for ASEL functions form totally different pathways or some of the pathway components act for both neurons.

INVOLVEMENT OF OTHER SENSORY NEURONS

Although salt-sensing sensory neurons ASER and ASEL are undoubtedly important, salt chemotaxis learning seems to involve a large number of other neurons. Among the sensory neurons, Hukema *et al.*[11] identified ASI, ASH, ADL, and ADF sensory neurons as important for learned avoidance of salt. For example, animals in which ASI neurons or ASH neurons were killed by cell-specific expression of caspase showed decreased avoidance of salt after salt conditioning. As mentioned previously, many genes important for the learning were identified in this study, and some of them were shown to act in a variety of sensory neurons in cell-specific rescue experiments. For example, *gpc-1*—encoding a γ subunit of G protein—acts in ASI, ASH, and ADL, and *odr-3*—encoding one of the 20 G protein α subunits in *C. elegans*—acts in ADF. These results are interpreted to mean that ASE neurons support attraction to salt, whereas other neurons mediate avoidance of salt. This latter function is usually concealed but activated only after salt conditioning. Alternatively, these sensory neurons may be involved in sensing and transmitting the starvation signal (US). ASI neurons are involved in regulation of dauer formation, a developmental switch regulated by food availability and pheromones, making it likely that they participate in US sensing. On the other hand, ASH and ADL neurons are known to sense aversive chemicals and mediate negative chemotaxis to them, making it likely that these neurons may mediate salt avoidance. As stated previously, ASE-less mutants, which are defective in attraction to salt, do not avoid salt even after conditioning, leading to a model in which ASE sends some signals, such as humoral neuropeptide signals, to activate the aversive sensory neurons (Figure 13.2).

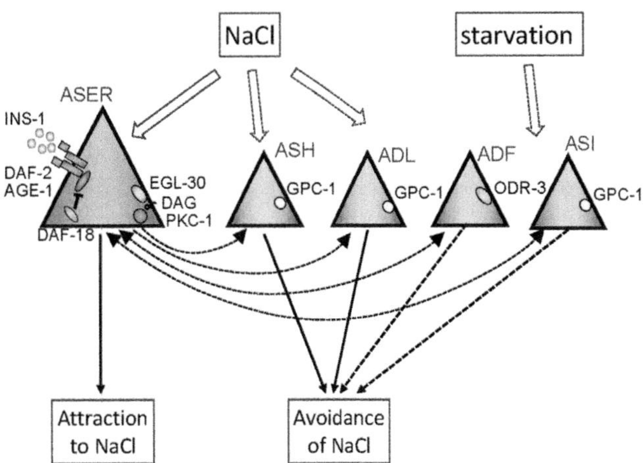

FIGURE 13.2 A model explaining involvement of sensory neurons in salt chemotaxis learning. ASER is a primary sensory neuron that senses NaCl, a CS, and testing salt, but other neurons also sense NaCl under limited conditions. Some of these non-ASER neurons mediate avoidance of NaCl. Molecules determined to act in one or more of these sensory neurons by cell-specific rescue are shown in each neuron. Non-ASER neurons are assumed to be activated by ASER or to send US signal to ASER (dashed arrows). Source: *Results of Hukema* et al.[11] *and Tomioka* et al.[10] *were combined in this figure.*

This view was also supported by the analysis of *daf-18* PTEN mutants with elevated PI 3-kinase pathway activity. Naive *daf-18* mutants avoid NaCl, especially when the ASEL neuron, which mediates attraction to salt, is eliminated.[12] This salt avoidance behavior was abolished when the *daf-18* mutant also had a mutation in *gcy-22*, a guanylyl cyclase essential for salt sensing in the ASER neuron.[5] One interpretation of this observation is that ASER sends the humoral signal to avoidance neurons only when stimulated by salt (CS). The other possibility is that ASER senses salt during chemotaxis and drives avoidance behavior by itself in *daf-18* mutants. When sensory function of sensory neurons other than ASER was eliminated by a mutation in *dyf-11*, which is required for maintenance of sensory cilia, *daf-18* mutants no longer avoided salt, supporting the former possibility.[12]

In addition to the amphid sensory neurons, *gcy-35* encoding soluble guanylyl cyclase is also required for efficient salt chemotaxis learning by acting in AQR, PQR, and URX neurons.[11] These neurons are exposed to body fluid, suggesting that internal body conditions may also affect learning.

ROLES OF INTERNEURONS

It is important to understand how sensory inputs to sensory neurons are processed in the neural circuit to generate switching of attractive behavior to aversive behavior. However, not much is known about the functions of these interneurons. ASE neurons send major synaptic outputs to three interneurons—AIA, AIB, and AIY. Of these, laser ablation of AIA neurons caused severe defect in salt chemotaxis learning.[10] The result was interpreted to be related to insulin secretion function of these interneurons, but other functions are also possible. Regarding the interneurons further downstream in the circuit, the role of RIA neurons was identified through analyses of the *magi-1* gene. *magi-1* encodes an ortholog of mammalian MAGI, a multi-PDZ domain-containing scaffold protein known to interact with a number of neuronal proteins, including the NMDA receptor, the AMPA receptor-regulating protein, and β-catenin.[21] *magi-1* mutants show mild defects in olfactory learning, temperature learning, and salt chemotaxis learning. These defects were rescued by expression of wild-type *magi-1* in RIA neurons. It was further shown that killing RIA by expression of caspase causes defects in salt chemotaxis learning as well as olfactory learning. In the future, it will be interesting to determine how RIA neurons regulate learning by interacting with sensory neurons and other interneurons in the neural circuit.

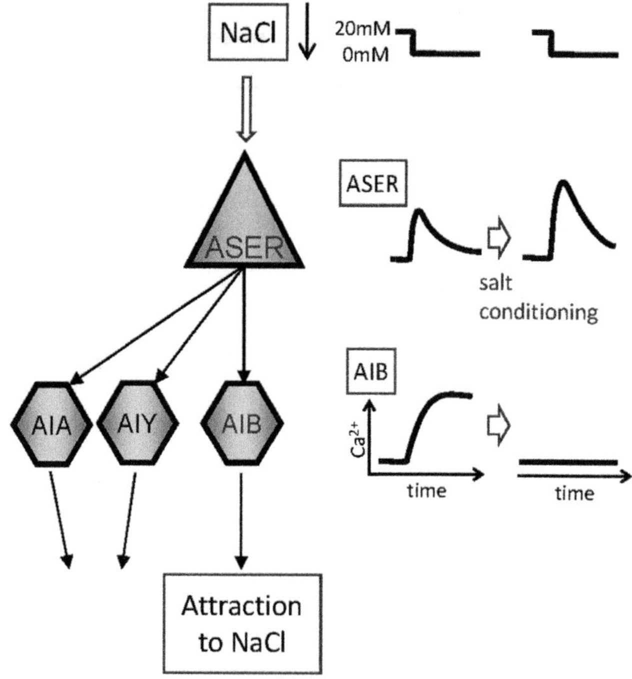

FIGURE 13.3 Change in the response of ASER and AIB neurons. In these experiments, salt concentration was decreased from 20 to 0 mM, and the responses of ASER and AIB neurons were monitored by calcium imaging. Traces for ASER and AIB show the calcium response of each neuron before or after pre-exposure to 20 mM NaCl for 10 min to mimic salt conditioning. After salt pre-exposure, the response of ASER increases, whereas the response of AIB is eliminated. Source: *Adapted from Oda* et al.[22]

CHANGES IN NEURONAL ACTIVITIES CAUSED BY LEARNING

To determine which neurons change the activity after learning, Shigekazu Oda *et al.*[22] performed calcium imaging and examined the response of neurons to NaCl stimulus. They used a poly(dimethylsiloxane)-based microfluidic device called the olfactory chip[23] to immobilize the animals and compared the response of neurons to salt stimulus before and after 10 min of exposure to 20 mM NaCl. After salt pre-exposure, the response of ASER to the removal of NaCl increased, whereas that of the AIB interneuron, which is a direct synaptic target of ASER, was almost lost (Figure 13.3). Visualization of synaptic transmission from ASER using a fluorescent reporter of synaptic vesicle release, synaptobrevin::pHluorin expressed in ASER, indicated that release of synaptic vesicles from ASER decreases after pre-exposure to NaCl. Importantly, all these processes are at least partly dependent on the insulin/PI 3-kinase pathway. Therefore, one of the sites affected by salt chemotaxis learning seems to be the ASER–AIB synapses. However, it is probably not the sole site of change because downregulation of the neuronal response of AIB alone cannot explain negative chemotaxis, and killing AIB neurons alone does not abolish chemotaxis.[10,24] However, these results suggest that AIB is an important component of salt chemotaxis and salt chemotaxis learning and is a good monitor of functional changes of ASER and downstream neural circuits. The response of another ASER-follower neuron, AIA, was also observed in this study, and the response seemed to slightly decrease after salt conditioning, but the difference was not statistically significant. Unfortunately, to date, these are the only neurons monitored before and after salt conditioning, and more extensive studies are required.

HOW IS THE STARVATION SIGNAL TRANSMITTED?

Salt chemotaxis learning occurs by association of salt stimulus with starvation, and it is natural to assume that salt is sensed by ASE neurons. How then is the starvation signal transmitted? Strong candidates are monoamine transmitters. Many behaviors in *C. elegans* are modulated by availability of food. In many, but not all, of these behaviors, serotonin and dopamine mimic the presence of food and octopamine mimics the absence of food.

In the case of salt chemotaxis learning, mutants of *cat-2*, encoding tyrosine hydroxylase necessary for dopamine synthesis, or those of dopamine receptors *dop-1*, *dop-2*, or *dop-3* showed defects.[25] ASIC-1, a degenerin/epithelial sodium channel (DEG/ENaC) family member whose mammalian homolog is enriched in synaptic density of mammalian hippocampus and other brain regions, is specifically expressed in dopaminergic neurons in *C. elegans*. Its deficiency causes reduced synaptic release, as measured using synaptobrevin::pHluorin, from dopaminergic neurons.[26] Consistently, *asic-1* mutants showed defects in salt chemotaxis learning, as well as olfactory learning that also depends on starvation. In the latter paradigm, synaptic release from dopaminergic neurons was increased by conditioning. Thus, dopamine appears to carry the information of the US (starvation) signal, at least in part, but it is unclear why dopamine mimics the presence of food in other behaviors. Dopamine may transmit both reward and punishment signals in different neurons as in *Drosophila*.[27]

Serotonin also seems to have roles in learning. Mutants of the genes involved in serotonin biosynthesis—*tph-1*, *cat-4*, and *bas-1*—showed reduced ability of salt chemotaxis learning. Mutants for serotonin receptors, *ser-1* and *mod-1*, also showed defects.[25] Increased serotonin signaling by exogeneous application of serotonin also compromised the learning. Therefore, excess or shortage of serotonin appears to impair learning, but more detailed studies to identify, for example, the serotonin-sensitive neurons are needed to clarify the exact role of serotonin.

These effects of monoamine neurotransmitters are relatively mild. On the other hand, deficits in insulin signaling severely impair salt chemotaxis learning. In addition, salt chemotaxis behavior of animals with ectopic activation of the insulin/PI 3-kinase signaling due to the *daf-18* mutation mimics starved animals even in the presence of food; therefore, insulin signaling appears to transmit the starvation signal. It is interesting that in other insulin-dependent phenomena such as dauer formation and longevity, insulin appears to transmit food signal, whereas in the case of salt chemotaxis learning, the pathway supports starvation-dependent learning. It will be interesting to determine how secretion of INS-1, the learning-specific insulin, is controlled by the presence or absence of food.

MOLECULAR PATHWAYS FOR MEMORY RETENTION

As mentioned previously, salt chemotaxis learning is probably classified as short-term memory. Then, how is the memory retained? Is there any specific mechanism for that? Takashi Kano *et al*. tracked this problem by observing the distribution of worms on test plates for an extended period.[28] When wild-type animals conditioned on a salt-containing plate for 4 hr were kept on a chemotaxis plate, they initially avoided the salt and kept avoiding for approximately 1 hr, but after that period they were gradually attracted to the salt. These authors found that mutants of *nmr-1*, which encodes a homolog of the NMDA-type glutamate receptor known to be critical for various forms of synaptic plasticity and behavioral learning in mammals, show shorter retention of aversion memory; they switched the behavior to attraction to NaCl faster than wild type. By cell-specific rescue experiments, they found that *nmr-1* acts in RIM neurons. RIM is one of the two classes of neurons that express *tdc-1*, which encodes tyrosine decarboxylase in the biosynthesis pathway of tyramine and octopamine. The *tdc-1* mutant showed a phenotype similar to *nmr-1*. Therefore, the function of *nmr-1* in prolonging the memory may be related to the action of these monoamine transmitters.

Later, mutants of *add-1* were found to show a similar fast recovery phenotype.[29] *add-1* encodes a homolog of α-adducin, a membrane cytoskeleton protein that regulates actin filaments. ADD-1 apparently co-localized with the AMPA-type glutamate receptor GLR-1 and regulated the amount of synaptic GLR-1.[29] Although so far unexamined, ADD-1 may also regulate NMR-1 and may affect memory retention through this function.

LONG-TERM MEMORY

Long-term memory in *C. elegans* has been demonstrated and extensively examined for mechanosensory learning (see Chapter 9). On the other hand, paradigms for long-term memory based on chemotaxis have only recently been developed in two laboratories. Both of these reports used odorants for chemotaxis and observed an association between butanone and food[30] or 1-propanol and hydrochloride.[31] In these assays, retention of the memory for approximately 1 day was demonstrated. Given many similarities between sensation of salt and odor, it is likely that a modified assay of salt chemotaxis learning will also lead to generation of a long-term memory.

EXPERIENCE-DEPENDENT SALT CHEMOTAXIS IN FED ANIMALS

So far, this chapter has described salt chemotaxis learning, in which association of salt and starvation leads to avoidance of salt, and attraction of unstarved animals to salt was considered a default behavior. In fact, since the initial observation by Samuel Ward,[1] salts have been known as chemoattractants. However, why should *C. elegans* be attracted to salt, which worms do not feed on? We have recently found that this is related to the fact that standard laboratory cultivation media for *C. elegans* contains approximately 50 mM of NaCl. When NaCl was omitted from the culture plates, wild-type *C. elegans* was no longer attracted to NaCl. Furthermore, it was found that *C. elegans* was attracted to the salt concentration at which it was previously cultivated with food (our unpublished observations). Therefore, salt chemotaxis is in fact a highly plastic behavior; worms associate not only the presence or the absence of salt but also the concentration of salt with availability of food, and they memorize the association. Based on this memory, they determine the exact behavior in a salt concentration gradient.

CONCLUSION

Salt chemotaxis learning is a very robust learning paradigm and therefore many researchers in *C. elegans* neuroscience have used the assay to find mutants defective in the learning. Based on the knowledge of connectivity of the whole neural circuit, cell-specific rescue has been a convenient method to map the genes in the neural circuit. The studies described in this chapter suggest that much of the learning and memory processes may be carried out in the sensory neurons. This demonstrates a remarkable ability of a small number of neurons to support learning and memory. However, despite intensive study of this form of learning, the exact mode and mechanism of transmission of CS and US and association between them are still obscure. Also, how the rest of the neural circuit is involved in memory formation, retention, and expression remains to be determined. Part of the reason for the limited understanding seems to be the difficulty detecting the change in the amount and localization of relevant molecules due to the small size and relative impenetrability of the animal. However, it is hoped that through the development of imaging techniques, these issues will be addressed in the near future and the dynamics of the whole neural circuit will be resolved soon.

References

1. Ward S. Chemotaxis by the nematode *Caenorhabditis elegans*: identification of attractants and analysis of the response by use of mutants. *Proc Natl Acad Sci USA*. 1973;70(3):817–821.
2. Bargmann CI, Hartwieg E, Horvitz HR. Odorant-selective genes and neurons mediate olfaction in *C. elegans*. *Cell*. 1993;74 (3):515–527.
3. Bargmann CI, Horvitz HR. Chemosensory neurons with overlapping functions direct chemotaxis to multiple chemicals in *C. elegans*. *Neuron*. 1991;7(5):729–742.
4. Yu S, Avery L, Baude E, Garbers DL. Guanylyl cyclase expression in specific sensory neurons: a new family of chemosensory receptors. *Proc Natl Acad Sci USA*. 1997;94(7):3384–3387.
5. Ortiz CO, Faumont S, Takayama J, et al. Lateralized gustatory behavior of *C. elegans* is controlled by specific receptor-type guanylyl cyclases. *Curr Biol*. 2009;19(12):996–1004.
6. Goodman MB, Hall DH, Avery L, Lockery SR. Active currents regulate sensitivity and dynamic range in *C. elegans* neurons. *Neuron*. 1998;20(4):763–772.
7. Suzuki H, Thiele TR, Faumont S, Ezcurra M, Lockery SR, Schafer WR. Functional asymmetry in *Caenorhabditis elegans* taste neurons and its computational role in chemotaxis. *Nature*. 2008;454(7200):114–117.
8. Thiele TR, Faumont S, Lockery SR. The neural network for chemotaxis to tastants in *Caenorhabditis elegans* is specialized for temporal differentiation. *J Neurosci*. 2009;29(38):11904–11911.
9. Saeki S, Yamamoto M, Iino Y. Plasticity of chemotaxis revealed by paired presentation of a chemoattractant and starvation in the nematode *Caenorhabditis elegans*. *J Exp Biol*. 2001;204(Pt 10): 1757–1764.
10. Tomioka M, Adachi T, Suzuki H, Kunitomo H, Schafer WR, Iino Y. The insulin/PI 3-kinase pathway regulates salt chemotaxis learning in *Caenorhabditis elegans*. *Neuron*. 2006;51(5):613–625.
11. Hukema RK, Rademakers S, Dekkers MP, Burghoorn J, Jansen G. Antagonistic sensory cues generate gustatory plasticity in *Caenorhabditis elegans*. *EMBO J*. 2006;25(2):312–322.
12. Adachi T, Kunitomo H, Tomioka M, et al. Reversal of salt preference is directed by the insulin/PI3K and Gq/PKC signaling in *Caenorhabditis elegans*. *Genetics*. 2010;186(4):1309–1319.
13. Solari F, Bourbon-Piffaut A, Masse I, Payrastre B, Chan AM, Billaud M. The human tumour suppressor PTEN regulates longevity and dauer formation in *Caenorhabditis elegans*. *Oncogene*. 2005;24(1):20–27.
14. Brundage L, Avery L, Katz A, et al. Mutations in a *C. elegans* Gqalpha gene disrupt movement, egg laying, and viability. *Neuron*. 1996;16(5):999–1009.
15. Sieburth D, Madison JM, Kaplan JM. PKC-1 regulates secretion of neuropeptides. *Nat Neurosci*. 2007;10(1):49–57.
16. Okochi Y, Kimura KD, Ohta A, Mori I. Diverse regulation of sensory signaling by *C. elegans* nPKC-epsilon/eta TTX-4. *EMBO J*. 2005;24(12):2127–2137.
17. Tsunozaki M, Chalasani SH, Bargmann CI. A behavioral switch: cGMP and PKC signaling in olfactory neurons reverses odor preference in *C. elegans*. *Neuron*. 2008;59(6):959–971.
18. Ikeda DD, Duan Y, Matsuki M, et al. CASY-1, an ortholog of calsyntenins/alcadeins, is essential for learning in *Caenorhabditis elegans*. *Proc Natl Acad Sci USA*. 2008;105(13):5260–5265.
19. Iwata R, Oda S, Kunitomo H, Iino Y. Roles for class IIA phosphatidylinositol transfer protein in neurotransmission and behavioral plasticity at the sensory neuron synapses of *Caenorhabditis elegans*. *Proc Natl Acad Sci USA*. 2011;108(18): 7589–7594.
20. Fu Y, Ren M, Feng H, Chen L, Altun ZF, Rubin CS. Neuronal and intestinal protein kinase D isoforms mediate Na^+ (salt taste)-induced learning. *Sci Signal*. 2009;2(83):ra42.
21. Stetak A, Horndli F, Maricq AV, van den Heuvel S, Hajnal A. Neuron-specific regulation of associative learning and memory by MAGI-1 in *C. elegans*. *PLoS ONE*. 2009;4(6):e6019.

22. Oda S, Tomioka M, Iino Y. Neuronal plasticity regulated by the insulin-like signaling pathway underlies salt chemotaxis learning in *Caenorhabditis elegans*. *J Neurophysiol*. 2011;106(1):301–308.
23. Chronis N, Zimmer M, Bargmann CI. Microfluidics for *in vivo* imaging of neuronal and behavioral activity in *Caenorhabditis elegans*. *Nat Methods*. 2007;4(9):727–731.
24. Iino Y, Yoshida K. Parallel use of two behavioral mechanisms for chemotaxis in *Caenorhabditis elegans*. *J Neurosci*. 2009;29(17):5370–5380.
25. Hukema RK, Rademakers S, Jansen G. Gustatory plasticity in *C. elegans* involves integration of negative cues and NaCl taste mediated by serotonin, dopamine, and glutamate. *Learn Mem*. 2008;15(11):829–836.
26. Voglis G, Tavernarakis N. A synaptic DEG/ENaC ion channel mediates learning in *C. elegans* by facilitating dopamine signalling. *EMBO J*. 2008;27(24):3288–3299.
27. Liu C, Placais PY, Yamagata N, et al. A subset of dopamine neurons signals reward for odour memory in *Drosophila*. *Nature*. 2012;488(7412):512–516.
28. Kano T, Brockie PJ, Sassa T, et al. Memory in *Caenorhabditis elegans* is mediated by NMDA-type ionotropic glutamate receptors. *Curr Biol*. 2008;18(13):1010–1015.
29. Vukojevic V, Gschwind L, Vogler C, et al. A role for alpha-adducin (ADD-1) in nematode and human memory. *EMBO J*. 2012;31(6):1453–1466.
30. Kauffman AL, Ashraf JM, Corces-Zimmerman MR, Landis JN, Murphy CT. Insulin signaling and dietary restriction differentially influence the decline of learning and memory with age. *PLoS Biol*. 2010;8(5):e1000372.
31. Amano H, Maruyama IN. Aversive olfactory learning and associative long-term memory in *Caenorhabditis elegans*. *Learn Mem*. 2011;18(10):654–665.

4.2 Mollusks

4.2.1 Gastropods

CHAPTER

14

A Systems Analysis of Neural Networks Underlying Gastropod Learning and Memory

Paul R. Benjamin

University of Sussex, Brighton, United Kingdom

INTRODUCTION

A systems approach to learning and memory aims to relate changes in the functional properties of neuronal networks to behavioral plasticity. Underlying the changes in network function are different types of synaptic and nonsynaptic modifications[1,2] that are known to occur at a number of locations within the 'learning circuit'.[3–6] Understanding how these multiple changes are integrated to generate network correlates of behavioral learning is the major goal. The advantage of using gastropod mollusks is that behavioral studies can be directly linked to network and cellular levels of analysis by taking advantage of the ability to identify individual neurons with known electrical properties and synaptic connectivity. Simple forms of associative learning, such as classical and operant conditioning, and nonassociative forms, such as habituation and sensitization, can be investigated. The use of a wide variety of learning paradigms in a number of different gastropod species allows comparisons to be made of underlying neural network mechanisms involved in memory formation in different types of learning (e.g., associative vs. nonassociative and classical vs. operant conditioning). Important progress has been made in understanding the mechanisms of synaptic plasticity in gastropod memory formation, but endogenous cellular mechanisms such as changes in neuronal excitability have also been increasingly recognized as contributing to memory formation. How these synaptic and cellular processes are integrated to generate network correlates of behavioral memory is an important focus of this chapter.

BEHAVIOR AND MODEL NETWORKS

A brief review of the neural circuitry underlying three gastropod behaviors is given here because these examples have been most useful in studying the synaptic and cellular changes underlying memory formation.

Aplysia Gill–Siphon Defensive Withdrawal Reflex

A brief tactile stimulus applied to the siphon of *Aplysia* elicits gill and siphon withdrawal into the protective mantle cavity (Figure 14.1A1). The withdrawal responses in these two organs are mediated by identified sensory and motor neurons located in the abdominal ganglia (Figure 14.1A2), and because the strength of these responses is modified by experience, they have been extensively used to study the mechanisms underlying both associative and nonassociative forms of learning.[7] Early studies of habituation focused on gill withdrawal plasticity, but the ease of identifying siphon responses in the live animal and the identification of a distinct set of siphon motor neurons (the LFS cells) has meant that the siphon circuit alone has been used mainly for more recent studies of sensitization and classical conditioning.[8,9] A variety of centrally located siphon sensory neurons and excitatory and inhibitory interneurons provide synaptic input to the LFS motor neurons (Figure 14.1A2),[10] but the direct LE mechanosensory to LFS monosynaptic pathway (Figure 14.1A2) has been a major focus for learning

Invertebrate Learning and Memory.
DOI: http://dx.doi.org/10.1016/B978-0-12-415823-8.00014-9

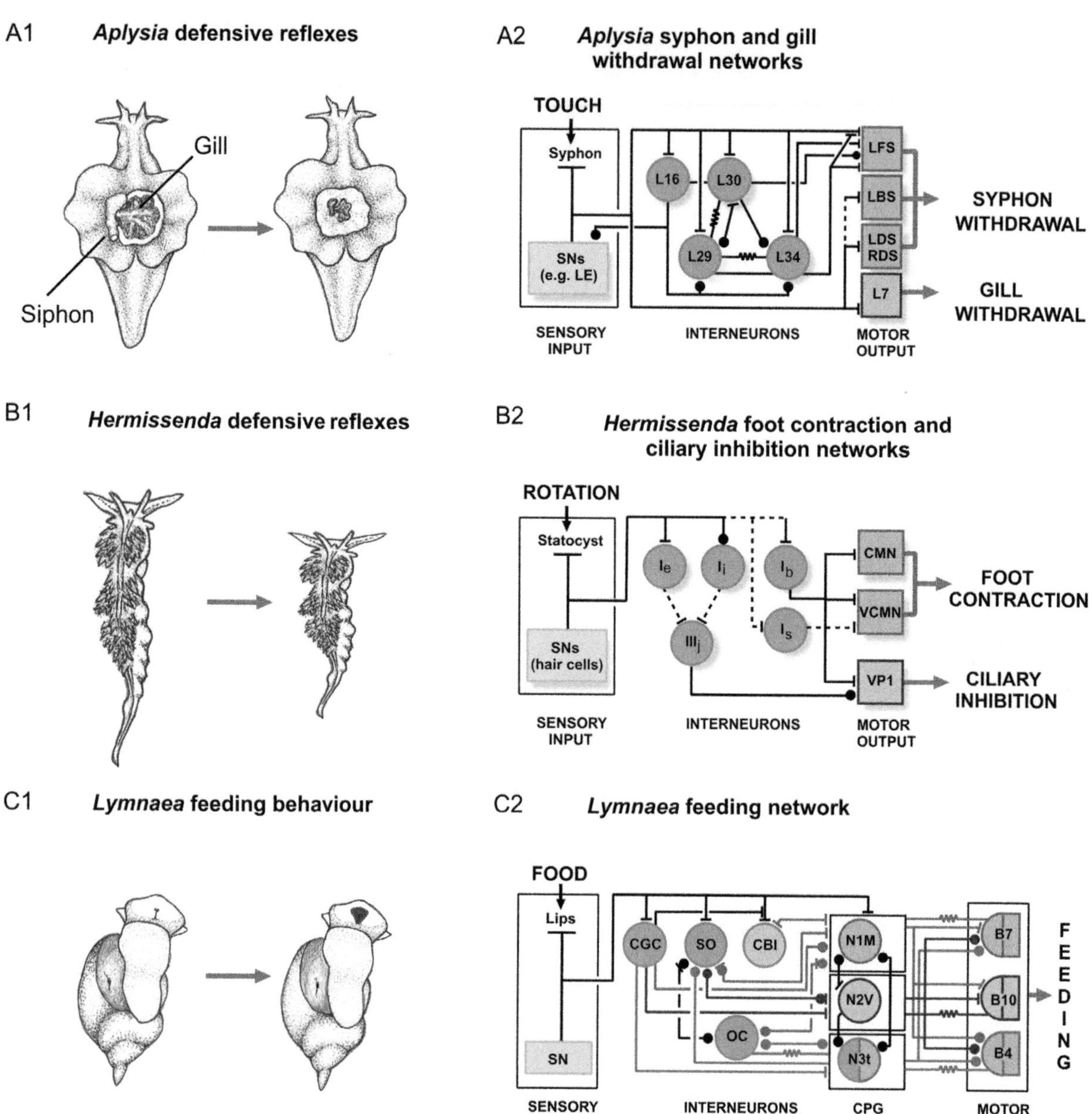

FIGURE 14.1 Molluscan behaviors and underlying neural circuitries most widely used in studies of the cellular mechanisms of learning and memory. (A1) Touch-evoked withdrawal reflex of the gill and syphon in *Aplysia californica.* (A2) Sensory neurons (SNs; yellow square) activated by touch to the syphon provide direct excitatory synaptic inputs to motor neurons innervating the gill and syphon (blue and green squares, respectively). In addition, there are indirect excitatory and inhibitory connections from the sensory to motor neurons, which are mediated by a set of intrinsic interneurons (orange circles) interconnected by both chemical and electrical synapses. Touch can normally only evoke a weak contraction of the syphon and gill, but the reflex becomes stronger after sensitization and classical conditioning. (B1) Foot contraction and ciliary inhibition evoked by rotation in *Hermissenda crassicornis.* (B2) Sensory neurons activated by rotation provide polysynaptic excitatory and inhibitory inputs to motor neurons responsible for foot contraction (green squares) and inhibition of ciliary motor neurons that control forward locomotion (blue square). Dashed lines represent polysynaptic connections from potential interneurons not yet identified. When a visual input is repeatedly paired with rotation, it will become effective in evoking both foot contraction and ciliary inhibition (see Figure 14.3). (C1) Feeding behavior in *Lymnaea stagnalis.* (C2) Chemosensory neurons (SN) located in the lip structures detect the presence of food or chemostimulants, such as sucrose. Excitatory inputs from the sensory pathways are distributed in parallel to modulatory interneurons (dark orange circles) and command-like interneurons (light orange circle) and also to CPG neurons (green, blue, and pink circles). The CPG interneurons drive the motor neurons active during the same phases of the feeding cycle (green indicates the protraction phase, blue the rasp phase, and pink the swallow phase). The CGCs may presynaptically modulate certain types of chemosensory inputs to the CBIs (see Figure 14.4). The complex synaptic connectivity between the CBIs and N1 interneurons leads to activation of the whole CPG, with the protraction (N1), rasp (N2), and swallow (N3) phases of the feeding cycle following in sequence due to the synaptic connectivity within the CPG and their intrinsic properties. Several subtypes of each of the N cells have been characterized, but because of the complexity of their synaptic connections with modulatory interneurons and motor neurons, only one subtype is shown. The rhythmic pattern generated by the CPG drives protraction (B7; green circle), rasp (B10; blue circle), and swallow phase motor neurons (B4; pink circle), leading to sequences of muscular activity and feeding movements executed by the mouth (shown), the radula, and the buccal mass. CBI, cerebrobuccal interneuron; CGC, cerebral giant cell; CPG, central pattern generator; OC, octopamine-containing neuron; SO, slow oscillator. Source: *Dr. Ildiko Kemenes produced the drawings used in this figure.*

studies, simplifying interpretation of the data. Monosynaptic excitatory connections from the LE to LFS motor neurons have been estimated to mediate approximately one-third of the siphon reflex response that corresponds to siphon flaring in the intact animal.[8] The remainder of the response is mediated by peripheral motor neurons,[11] other unidentified sensory neurons,[12] and polysynaptic inputs onto the LFS neurons from excitatory and inhibitory interneurons.[10] For habituation studies, sequences of weak touch stimuli are applied to the siphon, and the strength of the gill–siphon withdrawal reflex is measured either mechanically or by using a photocell to record displacement. Electrical shocks are applied to the tail or neck for sensitization of the siphon reflex, and repeated shocks (unconditioned stimulus (US)) are paired with siphon touch (conditioned stimulus (CS)) for behavioral studies of aversive classical conditioning. The pathway by which the electrical sensitizing stimuli influence the gill–siphon circuitry recruits the extrinsic serotonergic cerebral CB1 neurons that have projections to LE sensory neurons in the abdominal ganglion.[13] Other types of intrinsic interneurons (e.g., L29 and L30; Figure 14.1A2) also play a role in the network plasticity, and most information is available for habituation and sensitization.[14,15]

Hermissenda Statocyst-Mediated Behaviors

High-speed rotation or orbital shaking elicits two types of behavioral response in *Hermissenda*: foot shortening (Figure 14.1B1) and inhibition of the normally positive phototactic response, indicated by a reduced forward locomotion toward light.[16] The type of stimuli used to elicit this response in the laboratory mimics the natural response to mechanical turbulence caused by wave action that results in the animal clinging to the substrate for protection. The sensory response is mediated by a pair of gravity detector organs called the statocysts that have hair cell receptors analogous in function to the vestibular hair cells of vertebrates. In *Hermissenda*, depolarization of the hair cells by rotation generates responses in interneurons in the central ganglia that excite foot motor neurons responsible for foot shortening (CMN and VCMN) and inhibit foot motor neurons (VP1) that control ciliary beating to inhibit forward locomotion (Figure 14.1B2). The interneurons involved in the two behavioral responses are different, giving independent pathways for the control of the two unconditioned responses. Rotation of the body is used as the aversive US in behavioral associative conditioning studies, and a light flash is used as the CS. Repeated pairing of these two stimuli over several days leads to a reduction of the phototactic response to the CS and foot shortening similar to the previously described response to the US. Photoreceptors in the eye respond to the light CS, and these cells have a variety of excitatory and inhibitory synaptic connections with sensory receptors and interneurons of the US statocyst pathway (Figure 14.1B2). However, these synaptic connections are ineffective in eliciting unconditioned responses in naive animals.[16]

Lymnaea Feeding

Feeding in *Lymnaea* consists of a sequence of repetitive movements called rasps. During each rasp, the mouth opens (Figure 14.1C1) and a toothed radula is scraped forward over the food substrate (protraction phase). Food is then lifted into the mouth (retraction phase), which closes while the food is being swallowed (swallow phase), and the sequence is repeated.[17,18] Rhythmic movements of the feeding muscles are driven by a network of motor neurons (B1–B10) that, in turn, are driven by synaptic inputs from a feeding CPG network of interneurons (Figure 14.1C2). Each phase of the feeding rhythm is generated by one of three main types of CPG interneurons—N1 (protraction), N2 (retraction), and N3 (swallow)—providing sequences of excitatory and inhibitory synaptic inputs to motor neurons active in different phases of the feeding rhythm (Figure 14.1C2). CPG-driven rhythmic electrical activity can be recorded in the feeding network even in the absence of feeding muscles, and this is called fictive feeding. Activity in the motor neurons and CPG neurons is controlled by identified higher order interneurons such as the cerebral giant cells (CGCs) and cerebrobuccal interneurons (CBIs) (Figure 14.1C2). These higher order neurons have been the major focus of learning and memory studies.[19,20] The CGCs act as gating neurons in the feeding circuit.[21] Increased CGC spiking activity during feeding facilitates feeding responses to food. The CBI cells are command-like neurons involved in the initiation of feeding.[18] Sucrose is an effective chemical stimulus for feeding and is therefore used as the US for reward conditioning. At the cellular level, sucrose applied to the lips or esophagus in semi-intact preparations induces fictive feeding in motor neurons and interneurons of the feeding network.

The CS used for reward classical conditioning in *Lymnaea* is either a chemical (amyl acetate) or a tactile stimulus (a gentle brushstroke applied to the lips). In both experiments, a strong feeding stimulus, sucrose, is used as the US. The chemical CS was previously thought to have no effect on feeding ('neutral stimulus'), either at the behavioral or the electrophysiological level, but it has recently been shown to have stimulatory or inhibitory effects in naive animals, depending on concentration.[22] Touch to the lips, monitored either at the behavioral or the electrophysiological level, cannot initiate or maintain feeding in naive animals, but nevertheless touch produced a complex sequence of inhibitory and excitatory synaptic inputs on all neurons of the feeding network.[23] Aversive classical conditioning of feeding is also possible in *Lymnaea* using sucrose as the CS and KCl as the US.[24] This requires multiple trials to be successful, whereas another recently developed aversive paradigm using amyl acetate (at low concentrations) as the CS and quinine as the US only requires one trial.[25]

THE COMPLEXITY OF GASTROPOD LEARNING

Molluskan studies are focused on forms of memory such as classical/operant conditioning and sensitization. Initially, simple forms of associative and nonassociative learning behavior were investigated. However, gastropod mollusks are capable of showing more complex types of associative learning behavior with features that are similar to those found in vertebrates. For instance, differential conditioning has been described in a number of gastropods.[26–31] In addition, stimulus generalization, goal tracking, and context dependence (increased learning in a novel environment) were found in *Lymnaea* tactile conditioning.[27,32,33] State-dependent learning has been shown in chemical conditioning in *Lymnaea*. Here, one-trial conditioning is possible, but only if snails are starved 2–5 days prior to training.[34]

The importance of environmental context for learning has been investigated in detail in operant conditioning of the aerial respiratory behavior of *Lymnaea*. When the environment is made hypoxic, the snails float to the water surface and perform rhythmic opening and closing movements of their pulmonary opening, the pneumostome, and this respiratory behavior was used as the operant for behavioral conditioning.[35] Gentle tactile stimulation of the pneumostome area evokes pneumostome closure and stops respiratory behavior. During operant conditioning, a tactile stimulus is applied to the pneumostome repeatedly, each time the aerial respiration is attempted. Conditioning reduces the number of openings, latency to first opening, and total breathing time compared with pretraining or compared with yoked controls. To study the role of environmental 'context' in determining the expression of long-term memory (LTM),[36] one group of snails were trained under standard hypoxic conditions, but the context of training was altered in a second group by training the snails in standard conditions but with the addition of extracts of a food odorant, derived from carrot (the context). The presence of the odor did not prevent learning and formation of LTM, but expression of the memory trace depended on testing the snails in the context in which they were trained. Thus, snails trained with the odor context did not demonstrate LTM unless they were tested in the presence of the odor and vice versa for snails trained under standard conditions. Context also influenced the extinction of the LTM trace.[37] An operant training procedure of two 45-min training sessions followed by a third 18 hr later produced an extended duration of LTM lasting for at least 5 days. If after the last training session the snails were subjected to three 45-min extinction 'training' sessions (breathing with no reinforcing touch stimulus), then LTM was not observed on the next day. Extinction did not occur, however, if the extinction training was carried out in a context that was different from the context of the associative training. Another type of chemical signal that influences operant conditioning is the scent of a crayfish predator. 'Stress' induced by this chemical has been shown to enhance learning as part of a range of changed behavioral responses induced by the presence of a predator.[38] Two types of effects were seen depending on the duration and spacing of the training. In one experiment, snails were trained for 30 min to produce an intermediate-term memory (ITM) lasting 3 hr. Exposure to crayfish effluent (CE) during training produced a memory that persisted for 48 hr consistent with formation of LTM. In a second type of experiment training consists of two periods of 30 min of training separated by a 1-hr interval. Here, the presence of CE during training extends the LTM duration from 1 day to at least 8 days. A context-dependent element is found in these experiments so that memory recall depends on the presence of CE. However, when snails trained in CE are challenged during recall with a background of carrot or standard hypoxic conditions (no added chemical), the LTM memory trace is still expressed, indicating context generalization.

'Higher order' forms of learning, such as second-order conditioning, have been demonstrated in the terrestrial slug *Limax*[39] and in the gill withdrawal reflex of *Aplysia*.[29] Formation of food preferences in *Limax* depends on the consumption of food, and an associative memory trace is formed between the odors and the consumption of the food, depending on the food's nutritional value. Slugs are attracted by odor to locomote toward a food source such as carrot or potato, and when the food is located, they evert their lips to taste the food and then consume the food with rhythmic feeding movements. If the food odors are paired with the bitter taste of quinidine, then the slug learns to avoid the food odor, unlike naive and unpaired groups of slugs that show no change in their food preference.[39] A single pairing of a CS odor (carrot or potato) and the aversive US reduces the time spent near the CS. The association of the food odor (CS) with the bitter taste (US) is an example of first-order classical conditioning, but the paradigm was also extended to demonstrate second-order conditioning.[39] Slugs are first presented with carrot/quinidine (phase 1) and then given the same number of pairings of two odors, potato and carrot (phase 2). To ensure that changes in the slug's preference for potato depended on it receiving both phase 1 and phase 2 pairings, control groups were included. One control group receives paired presentations of odor and quinidine in phase 1, but during phase 2 potato and carrot is unpaired. The other control group received unpaired carrot odor and quinidine in phase 1 but paired carrot and potato odor in phase 2. Slugs that received pairings during both phases of training displayed a reduced preference for potato odor when presented alone compared with slugs from the two control groups, indicating that second-order conditioning has occurred.

Another type of higher order conditioning behavior shown in *Limax* is known as 'blocking.' Sahley *et al.*[39] showed that prior conditioning to one stimulus, carrot odor, reduces or blocks learning to a second stimulus, potato odor, when a compound odor consisting of both carrot and potato odors is subsequently paired with quinidine. In contrast, control groups of slugs without prior training to carrot learn to avoid both carrot odor and potato odor when the compound presentation of both odors is paired with quinidine.

SYNAPTIC MECHANISMS FOR LEARNING

Aplysia Gill–Siphon Withdrawal Reflex: Multiple Types of Learning in the Same Network

Nonassociative Learning: Habituation and Sensitization

The gill–siphonal withdrawal reflex of *Aplysia* shows both habituation and sensitization.[40] When a weak touch (usually by a calibrated water jet) is applied repeatedly to the siphon or adjacent mantle skin at intervals of between 30 sec and 3 min, the gill withdrawal response habituates (decrements) to approximately 30% of control values after 10–15 trials. This short-term habituation lasts for several hours, but if these trials (10 per day) are repeated over 4 days, then a long-term habituation of the gill withdrawal response is induced that lasts for up to 3 weeks. If, prior to touch, a single strong noxious stimulus, such as an electric shock, is applied to the tail or neck, the subsequent touch-evoked response is enhanced or sensitized for a few hours. Long-term sensitization can be induced by the application of multiple shocks. Applying four shocks a day to the head for 4 days significantly increased the duration of withdrawal so that it was greater than that of controls up to 1 week after training. Both of these nonassociative phenomena are central processes that involve changes in synaptic strength at the sensory to motor synapses that mediate the normal gill–siphonal withdrawal reflex.

Behavioral habituation was found to be paralleled by suppression of neurotransmitter release from the presynaptic terminals of the touch-sensitive mechanosensory neurons. A consequent reduction in the size of the motoneuron EPSP (excitatory postsynaptic potential) is followed by a reduction in the activation of the motor neurons, which consequently fire less. This homosynaptic depression (HSD) was thought to be due to a progressive reduction in an inward Ca^{2+} current[41] that is required for transmitter release from secretory vesicles in the sensory neuron terminals. However, doubt about the key role of calcium in HSD in short-term habituation (STH) has emerged from experiments by Armitage and Siegelbaum,[42] who found that presynaptic injection of EGTA (blocks slow changes in internal calcium) into sensory neurons failed to block synaptically induced HSD in motor neurons in culture. Also, calcium influx measured by a calcium-sensitive dye injection into sensory neuron varicosities, thought to be the sites of transmitter release, did not alter during induction of HSD. A non-calcium-dependent mechanism for HSD, perhaps directly coupled to the release process, was suggested by these experiments. There appears to be a consensus that STH involves purely presynaptic mechanisms,[42,43] but recent work by Ezzeddine and Glanzman[43] suggests that long-term habituation (LTH) may involve postsynaptic mechanisms as well. In a behavioral preparation, these authors showed that induction of LTH depended on the activation of specific types of glutamate receptors presumed to be located on the postsynaptic motor neurons. LTH was blocked in the presence of either the NMDA antagonist APV or the AMPA antagonist DNQX. DNQX had no effect on STH, as shown in previous experiments.[42]

Sensitizing stimuli have the opposite effect to habituation, causing an increase in transmitter release.[44] Sensitizing stimuli result in the release of the transmitter 5-HT from extrinsic facilitatory interneurons[13] onto the sensory neuron terminals of the mechanosensory neurons, and this transmitter acts presynaptically to facilitate synaptic transmission at the sensorimotor junction. This causes the motor neurons to fire more after sensitization, thereby increasing the strength of the withdrawal reflex. This type of heterosynaptic facilitation can be mimicked in reduced preparations and in a simplified cell culture system if tail shock is replaced by single applications of 5-HT to the sensory neurons. The mechanisms underlying presynaptic facilitation involve two types of synaptic mechanisms that result in an increase in transmitter release reviewed in Hawkins *et al.*[7] The first synaptic mechanism involves an increase in the duration of the spike in the sensory neuron. Tail shock or a brief application of 5-HT causes an increase in cAMP in the mechanosensory neurons. This increase in cAMP levels activates protein kinase A (PKA) and then phosphorylates a receptor on a potassium channel type, thereby closing it. Spike broadening of the sensory neuron action potential follows from this blockage and results in an increase in calcium influx and a greater release of the sensory transmitter, presumed to be glutamate.[45] Synaptic facilitation due to 5-HT is also mediated by PKC,[46,47] and CaMKII also contributes to spike broadening in behavioral short-term sensitization.[48] A nonsynaptic mechanism, again involving the depression of potassium channels, increases the excitability of the sensory neurons, making it more likely that the sensory neurons will respond to siphonal touch after sensitization. These types of presynaptic mechanisms are produced by single brief applications of 5-HT (1 min) or single shocks to the tail, and they are believed to underlie behavioral short-term sensitization or short-term facilitation (STF) of synaptic transmission seen in reduced preparations and in cell culture.

An intermediate form of behavioral sensitization (ITM) of siphonal withdrawal and correlated synaptic facilitation, from 10 to 85 min in duration, is induced by repeated application of spaced shocks to the tail[49] or repeated application of 5-HT to the ganglion.[50] An intermediate type of synaptic facilitation (ITF) that is presumed to underlie ITM involves persistent activation of PKA and requires protein but not RNA synthesis.[49] The MAPK pathway is also implicated in this type of sensitization.[51] Five shocks to the tail or 5-HT increase the levels of phosphorylated MAPK in siphonal sensory neurons. Blocking of MAPK phosphorylation prevents behavioral ITM and ITF. Further experiments indicate that MAPK is involved in the induction of ITM but not its expression and maintenance.[51] More recent experiments show that both pre- and postsynaptic mechanisms are involved in intermediate-term behavioral sensitization and the underlying synaptic facilitation.[48,52] In these experiments, ITF is shown to require presynaptic covalent modifications by PKA and CaMKII and postsynaptic modifications of Ca^{2+} and CaMKII. Both pre- and postsynaptic protein synthesis are required.[48,53] There is evidence that another type of protein kinase, PKC, also contributes to both pre- and postsynaptic mechanisms of intermediate synaptic facilitation. Most of the evidence for PKC comes from *in vitro* experiments,[54–56] and PKC does not appear to make a major contribution *in vivo* when ITM and ITF are induced in the semi-intact preparation by the usual four or five shocks applied to the tail.[48,56] Another type of intermediate memory, 'site-specific sensitization,' is independent of protein synthesis, where maintenance of the memory is dependent on the persistent activation of PKC.[57]

Work on long-term sensitization (LTS) and its synaptic correlate, long-term facilitation (LTF), has mainly focused on presynaptic mechanisms. LTS/LTF lasting 24 hr or more is triggered by repeated spaced application of 5-HT or sensitizing shock stimuli. It involves gene transcription as well as protein synthesis.[2] Presynaptic LTF requires the long-term activation of both PKA and MAPK by 5-HT. An important step in the expression of LTF is the release from the sensory neuron of the neuropeptide sensorin. Sensorin release is triggered by the 5-HT-mediated activation of presynaptic PKA. Released sensorin binds to presynaptic autoreceptors that activate the MAPK signaling pathway within the sensory neurons.[58] During LTF, persistently activated PKA and MAPK are co-translocated to the sensory neuron nucleus, where they activate a cascade of gene regulatory processes.[2] As in STF, PKA can also phosphorylate proteins that form the K^+ channel in the sensory neurons that both facilitate synaptic transmission and neuronal excitability. The induction of a persistent form of PKA ensures that the effects of PKA on these two processes are prolonged compared with STF. It seems likely that many of the postsynaptic mechanisms that are involved in prolonged synaptic enhancement of intermediate-term sensitization may also be present in LTS.[59] The enhancement of glutamate receptor function in the motor neurons discovered in ITF also occurs in LTF,[60,61] as does an elevation of intracellular Ca^{2+}.[62]

The maintenance of LTS and LTF has been the subject of recent important papers, and it was shown that the *Aplysia* PKC isoform, PKM Apl III, is involved in memory maintenance.[63] The *Aplysia* prion-like protein, ApCPEB,[64] also has been implicated in the maintenance of LTF. The results from this work are consistent with the idea that ApCPEB can act as a self-sustaining prion-like protein and allow persistent facilitation of synaptic transmission observed in LTF.[64–67]

Role of Intrinsic Polysynaptic Pathways in Habituation and Sensitization

As discussed previously, the synapses between central sensory neurons and motor neurons serve as important sites of synaptic plasticity for nonassociative forms of learning. However, interneurons intrinsic to the withdrawal circuit, such as L29 and L30 (Figure 14.1B2), also play a significant role, indicating the presence of parallel processing. In habituation, repetitive stimulation of either the skin or the central sensory neurons leads to a progressive decrement of monosynaptic excitatory sensory connections to both motor neurons and intrinsic facilitatory interneurons such as L29, and this leads to an eventual loss of L29 firing.[10] This is an important part of the mechanism for habituation because L29 normally provides a significant component of the pathway for withdrawal due to its direct synaptic connections with siphon LFS motor neurons and presynaptic facilitation of LE sensory neurons (Figure 14.1B2).[10] Repetitive stimulation of the skin also activates the inhibitory interneurons L30, which by inhibiting the L29s play an important role in reducing their spike activity. It is known that L30 inhibitory synapses undergo activity-dependent potentiation lasting for up to 10 s of seconds, and this form of homosynaptic inhibition of L29 is also likely to have a role in habituation.[15] On the other hand, sensitization of the siphonal withdrawal reflex by tail shock is accompanied by a reduction in network inhibition.[68] This is due to heterosynaptic inhibition of the L30 interneurons.[15] These results suggest a dual mechanism for nonassociative plasticity, with the direction of the response determined by the balance of inhibition and excitation in the withdrawal response network. The balance is toward inhibition for habituation and toward excitation for sensitization.

Associative Learning: Classical Conditioning

Carew and colleagues showed that the gill and siphon withdrawal reflex is subject to aversive classical conditioning as well as habituation and sensitization.[69] In their associative conditioning paradigm, they paired a weak tactile stimulus (the CS) to the siphon with a strong electric shock to the tail (the US). After 15 trials, the CS came to elicit a stronger gill withdrawal than controls; the effect lasted for several days. Later, they showed that differential classical conditioning also worked.[70] A weak tactile stimulus applied to the siphon or mantle shelf was either paired with shock (CS+) or not paired (CS−) in the same animal. Differences between the effects on gill withdrawal at the two sites were enhanced, even after a single trial. Cellular analysis of differential conditioning, using a reduced preparation, showed that the mechanism underlying classical conditioning of the gill withdrawal response was the elaboration of presynaptic facilitation that was previously shown to underlie sensitization, except that the effect of the pairing of CS and US produced an even greater facilitation of the withdrawal response. This effect was called activity-dependent presynaptic facilitation (ADPF).[26] Evidence for this mechanism was obtained in a reduced preparation in which the effects of touch were mimicked by stimulation of two different sensory neurons that made excitatory synaptic connections with siphonal motor neurons. Differential conditioning produced a significantly greater enhancement of the sensorimotor synapse if weak sensory neuron spike activation (by current injection) was followed by the tail shock (CS+) than if the sensory neuron stimulation was unpaired (CS−) or if shock occurred alone (sensitization). This result supported the notion of ADPF because of its dependence on temporal pairing. Like sensitization, this mechanism was shown to be presynaptic and could also be mimicked by the application of serotonin. A pairing-specific enhancement of Ca^{2+} influx into the presynaptic terminal of the sensory neurons was seen that was due to a reduction in the opening of the same 5-HT-sensitive K^+ channel that was shown to be involved in sensitization. Again, the activation of the cAMP-dependent PKA pathway is involved in phosphorylation of potassium channels and their subsequent closing, but it is enhanced compared with sensitization. With a larger number of pairings, a longer term form of synaptic plasticity is induced which is caused by persistent increases in the levels of cAMP, and a cascade of molecular events involving PKA and MAPK occur that lead to gene regulation and synaptic remodeling.[7] As was case with nonassociative LTF, associative conditioning depends on an increase in the level of the peptide sensorin, but different combinations of signaling pathways are involved.[71] When a tetanus (20 Hz for 2 sec) of the sensory neurons was paired with one application of 5-HT in cell culture, there was a rapid increase in PKC- and protein synthesis-dependent sensorin synthesis that did not require the sensory neuron cell body but did require the motor neuron. The secretion of the newly synthesized sensorin by 2 hr after pairing required both PKA and PKC activities to produce associative LTF. The secreted sensorin led to the phosphorylation and translocation of MAPK into the nuclei of the sensory neurons (cf. long-term sensitization).

The early model for aversive classical conditioning was entirely presynaptic, but the discovery that sensorimotor synapses in culture or in reduced preparations exhibit NMDA-dependent long-term potentiation (LTP)[72,73] led to the now generally accepted hypothesis that associative learning involves postsynaptic processes utilizing Hebbian-type LTP as well as presynaptic ADPF (Figure 14.2). Use of the more intact preparation by Antonov *et al.*[74] allowed classical conditioning of the behavioral siphonal withdrawal reflex to be correlated with synaptic plasticity. Pairing siphonal touch with tail shock induced a parallel associative

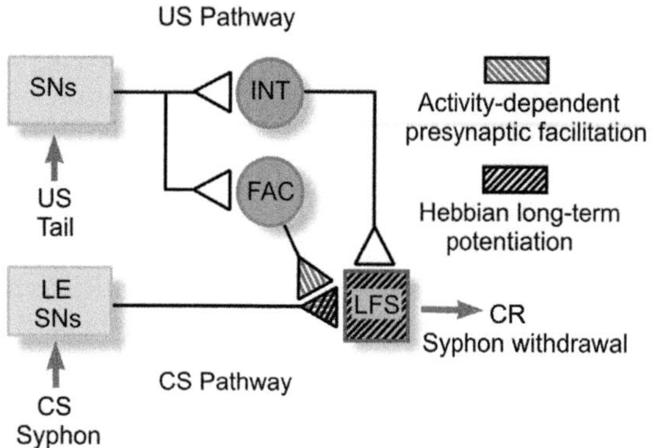

FIGURE 14.2 Aversive classical conditioning of the syphon withdrawal reflex in *Aplysia*. During aversive classical conditioning, the conditioned stimulus (CS; touch to the syphon) weakly activates LE sensory neurons (SNs), which make monosynaptic connections onto LFS motor neurons. The unconditioned stimulus (US; electric shock to the tail) strongly activates tail sensory neurons. These sensory neurons excite both facilitatory interneurons (FAC) that produce presynaptic facilitation at the LE-to-LFS synapse and other types of interneurons (INT) that postsynaptically excite the LFS motor neurons. Thus, both the sensory neurons and the motor neurons are sites of CS and US pathway convergence and can act as coincidence detectors. The shading indicates neuronal elements that must be activated conjointly for the induction of either activity-dependent presynaptic facilitation (red) or Hebbian LTP (black) at the LE–LFS synapses.[9] After aversive classical conditioning, the synaptic connections of the LE sensory neurons with the LFS motor neurons are strengthened and the excitability of the LE sensory neurons is increased. These changes lead to an increased syphon withdrawal response (conditioned response (CR)) to the touch CS. Source: *Reproduced from Antonov* et al.[9] *with permission from Elsevier.*

enhancement of siphonal withdrawal and synaptic strength. Application of the NMDA receptor blocker APV to the ganglion or injection of the calcium chelator BAPTA into the siphonal motor neurons blocked conditioning, providing direct evidence that Hebbian LTP is part of the mechanism for behavioral classical conditioning.[9] To compare the role of pre- and postsynaptic mechanisms in synaptic plasticity, the peptide inhibitor of PKA (PKAi) was injected into LE sensory neurons and BAPTA into LFS siphonal motor neurons.[9] Both procedures reduced the pairing-specific facilitation of PSPs in the motor neurons during siphonal conditioning. Thus, pre- and postsynaptic mechanisms appear to be contributing to the plastic changes in synaptic strength between the sensory neurons and the motoneurons (Figure 14.2). There also appears to be a retrograde signal involved, as well as orthograde, because injecting BAPTA into the motoneurons blocks the changes in the cellular membrane properties of the sensory neurons (increase in membrane resistance), so some of the changes occurring in the two sides of synapse following conditioning are coordinated. The nature of the retrograde signal is unknown, but it was suggested to be the diffusible gas nitric oxide (NO) because injecting the NO scavenger myoglobin into the sensory neuron blocks the facilitation of the PSP during conditioning.[7] However, more recent experiments suggest that NO is not the retrograde transmitter because injecting an NO scavenger into the motoneuron reduced synaptic facilitation but had no effect on the sensory neuron's membrane properties.[75] However, NO is still thought to be important in classical conditioning by acting directly on both sensory and motor neurons to effect different processes of synaptic facilitation, but the origin of the gas is probably the intrinsic interneuron L29 rather than the siphon motor neurons.[75]

On the basis of experiments in which 5-HT was used to mimic classical conditioning, Roberts and Glanzman[73] speculated that the rapid-onset ADPF mechanism is responsible for short-term synaptic facilitation. Hebbian LTP has a longer onset and leads to more persistent synaptic plasticity for medium-term and perhaps long-term synaptic facilitation.

Hermissenda Phototaxic Behavior: Associative Learning with Two Reflexive Responses

Behavioral training in *Hermissenda* involves the pairing of the CS, a light flash, with the US, a mechanical perturbance such as rotation. Pairing induces a short-term memory that can last for a few minutes (e.g., single trial) or LTM lasting for days or weeks depending on the number of trials (e.g., 150 trials per day over 3 days gives a week-long memory). Neural correlates can be studied in semi-isolated central nervous system (CNS) preparations made from behaviorally trained animals, or single-trial *in vitro* conditioning can be induced by pairing a light flash with CNS application of 5-HT, a key transmitter in the US pathway. Conditioning in the *Hermissenda* system involves the development of two different behavioral responses to the CS— foot contraction and inhibition of ciliary locomotion.[16] Both these conditioned responses are thought to develop independently due to the involvement of different components of the central circuit involved in the two behaviors (Figure 14.1B2).[76] Functionally, there is a learning-induced 'transfer' of the ability to activate these circuits from the US to the CS. Here, the focus is on the part of the circuit that underlies the conditioned inhibition of foot ciliary motor neurons that control forward locomotion because most is known about how conditioning changes this circuit. After conditioning, the CS is able to activate the same interneuronal pathway that normally mediates the inhibition of ciliary motor neurons by the mechanical US in naive animals. In earlier papers, it was proposed that changes in the cellular properties of photoreceptor cells and the strength of synaptic connections between the two classes of photoreceptors synaptic connections solely could account for the changes underlying behavioral conditioning,[77] but more recent identification of elements of the interneuronal circuit has identified an intermediate level of processing between sensory and ciliary motoneurons (Figure 14.1B2). Within this more complete circuit, Crow and Tian[78] identified further sites of synaptic and nonsynaptic plasticity that are also involved in the conditioned response so that a more complex multisite model for associative conditioning in *Hermissenda* has emerged.[16,79]

The electrical changes following conditioning are thought to be due to interactions between convergent CS (photic) and US (mechanical) synaptically mediated pathways that exist at various levels in the ciliary control network (Figure 14.3A). There are two types of synaptic connections between the CS and US pathways at the level of the sensory receptors. Reciprocal inhibitory monosynaptic connections occur between the statocyst hair cells and the B-type photoreceptors in the eye, and another unidirectional excitatory polysynaptic pathway, mediated by 5-HT, exists between the hair cells and the photoreceptors (Figure 14.3A). The presence of this 5-HT-mediated US excitatory pathway is of particular significance because this chemical is thought to underlie the ability of the US to induce changes in the CS photoreceptor pathway following conditioning. Conditioning induces cellular changes in both type A and type B photoreceptors, which increase the firing of the cells in response to the CS. The role of CS and US convergent inputs in conditioning is best understood at the level of the primary sensory neurons, but convergent interactions are also important at the level of the cerebropleural interneurons known as the I_e and I_i cells. Both cell types receive conjoint synaptic input from the photoreceptors and statocyst hair cells (Figure 14.3A). There are no synaptic connections between the two types of I cells.[80] Photoreceptors and statocysts form monosynaptic excitatory connections with the I_e interneurons and monosynaptic inhibitory connections with the I_i interneurons. Both I cell types receive larger compound synaptic inputs in response to the CS in conditioned animals compared with controls.[78] Larger EPSPs in the I_e cells increase spike activity, and larger inhibitory postsynaptic potentials (IPSPs) in the I_i cells reduce ongoing spike activity. This increase in the amplitude of the PSPs in the I cells is due to increased CS-induced spike activity in the photoreceptors following conditioning but also to an increase in the strength of the B photoreceptor monosynaptic connections to both I cell types (Figure 14.3B). In addition, the intrinsic excitability of the I_e

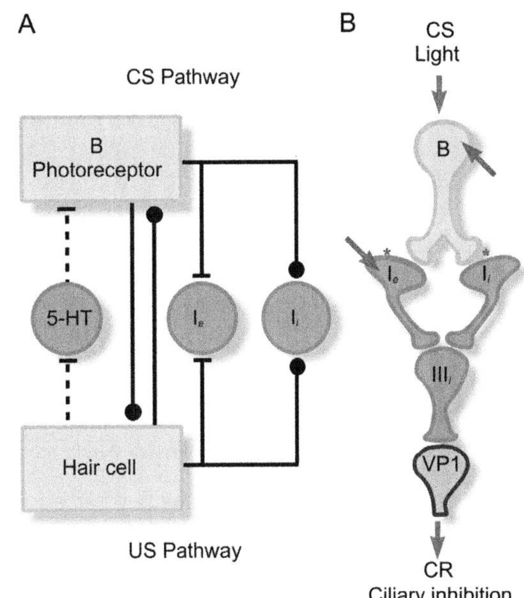

FIGURE 14.3 Aversive classical conditioning of the phototactic response in *Hermissenda*. (A) Sites of convergence between identified components of the CS and US pathways. Statocyst hair cells have indirect excitatory synaptic connections with B-type photoreceptors through a proposed 5-HT-mediated interneuronal pathway. Hair cells and photoreceptors also have reciprocal inhibitory monosynaptic connections. Both hair cells and B-type photoreceptors form monosynaptic connections with both type I_e and type I_i interneurons (see Figure 14.1B2), which are therefore further sites of synaptic interactions between the CS and the US pathway.[16] Dots indicate inhibitory chemical synaptic connections, and bars indicate excitatory synaptic connections. (B) Components of the CS pathway involved in ciliary inhibition after classical conditioning. Changes in both cellular excitability (green arrows) and synaptic efficacy (green asterisks) contribute to CS-elicited inhibition of ciliary locomotion. The net effect of cellular and synaptic plasticity is to increase the spike activity of type III_i inhibitory interneurons during light, which produces an inhibition of spiking activity in the VP1 ciliary activating motor neurons in conditioned animals. Source: *From Benjamin PR, Kemenes G. Behavioral and circuit analysis of learning and memory in mollusks. In: R Menzel ed.* Learning Theory and Behavior. *Vol. 1. Oxford: Elsevier; 2008:587–604. Used with permission from Elsevier.*

interneurons also appears to be increased after conditioning, a third mechanism leading to the increased spike response of the I_e cells to the CS (Figure 14.3B). More recent experiments suggest that 5-HT may be mediating some of the effects of conditioning on the interneurons, particularly in the I_e cells. 5-HT produces an increase in spontaneous and light-induced spike activity and enhanced intrinsic excitability similar to those seen after conditioning.[81]

Recording the VP1 ciliary motor neurons shows that light inhibits the tonic spike activity in conditioned animals compared with controls. Reduction of VP1 spiking reduces foot ciliary activity and inhibits forward locomotion. Facilitation of the synaptic connection between the B-type photoreceptors and type I_e

interneurons in conjunction with intrinsic enhanced excitability in type B photoreceptors and type I_e interneurons would result in an increase in spike activity in the type III_i inhibitory interneurons and inhibition of the VP1 ciliary motor neurons via the III_i-to-VP1 monosynaptic connection (Figure 14.3B). Thus, this combined set of synaptic and cellular changes located within the ciliary control circuit can elegantly account for the light-elicited inhibition of locomotion following associative conditioning.

Lymnaea and *Aplysia* Feeding: Associative Learning in CPG-Driven Rhythmic Networks

The feeding system of *Lymnaea* has been used extensively to investigate reward-based classical conditioning. Tactile, chemical, and visual cues have all been used in classical conditioning experiments.[82] Using the single-trial chemical conditioning paradigm, an electrophysiological correlate of the conditioned response was recorded as a sequence of CS-driven bursts ('fictive' feeding activity) in motor neurons recorded in semi-intact preparation made from behaviorally conditioned snails.[83] This CPG-driven feeding activity in the motor neurons depends on the activity of neurons at all levels of the feeding network, so the conditioned fictive feeding activity recorded in motor neurons is a systems 'readout' of the memory trace in the whole feeding system. More detailed experiments found that conditioning affects central but not peripheral processing of chemosensory information, with the cerebral ganglia being an important site of plasticity.[84] Cerebral plasticity originates from the feeding interneuron type called the CBI (Figure 14.1C2). These CBIs act as initiating neurons for the feeding network[18] so that when they fire the feeding CPG is activated, followed by rhythmic ingestion movements. Neural firing in the CBIs is enhanced in response to the CS after conditioning.[84] This increase in CBI firing is due to an increase in strength of the excitatory synapse between the CS chemosensory neurons and the CBIs. How does this occur? Considerable experimental evidence suggests that it is due to presynaptic facilitation of the sensory neuron to CBI synapse by the modulatory neuron type known as the CGC,[85] the details of which are considered in the next section. The CGCs are known to have a basic role in normal activation of feeding by food, but independent of this function, they also play a role in learning. Cellular changes that occur in the CGCs after conditioning are sufficient to explain the enhancement of the CS effects on feeding following conditioning, but they cannot be the whole story because the onset of CGC depolarization is at 16–24 hr after training, whereas a behavioral memory trace is

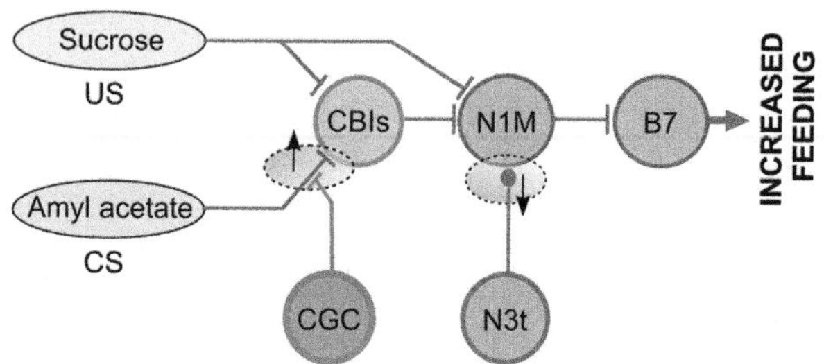

FIGURE 14.4 Synaptic and nonsynaptic mechanisms of memory after single-trial chemical conditioning of *Lymnaea*. After conditioning, the modulatory CGC is persistently depolarized (bold orange outline), which facilitates (upward arrow) the CS-to-CBIs synaptic connection and thereby activates the N1M, a key CPG interneuron of the feeding system (see Figure 14.1C2). The spike frequency of the N3t is reduced after conditioning, reducing the suppressive inhibitory synaptic control of the N1M (downward arrow) and thus allowing it to be more readily activated by the CS. Both mechanisms allow the N1M to drive motor neurons such as the B7 that generate the increased feeding response following conditioning.

present as early as 10 min after training, so an alternative mechanism must also be present to explain the early memory trace. Significantly, there is a second type of electrical change in the feeding circuit that occurs as early as 1 hr after one-trial *in vitro* conditioning. This is a conditioning-induced reduction in tonic (continuous) inhibitory synaptic input to the feeding CPG.[86] By reducing the frequency of this 'background' inhibitory input, the threshold for the CS to activate the feeding network is reduced, making it more likely that the feeding network (CPG) will respond to the CS.[86] This inhibitory input originates from one of the CPG interneuron types known as the N3t (N3tonic). How these two types of mechanisms (presynaptic facilitation and reduction in tonic synaptic inhibition) act in the feeding network following conditioning is shown in Figure 14.4.

Electrical correlates of *Lymnaea* tactile reward conditioning were also recorded at different sites in the feeding network. After 15 behavioral training trails over 3 days, touching the lips of the intact snail induces a pattern of feeding movements significantly greater than that in controls (Figure 14.5A). Similar significant changes were recorded at the level of motor neurons in semi-intact preparations made from the same snails (Figure 14.5B1).[87] Electrophysiological correlates of behavioral differential classical conditioning were also obtained.[30] In these experiments, the lips and the tentacle were used as CS+ (reinforced conditioned stimulus) or CS− (non-reinforced conditioned stimulus) sites for behavioral tactile conditioning. In a second experimental group, the CS+/CS− negative sites were reversed. Following successful behavioral conditioning, the touch stimulus evoked CPG-driven

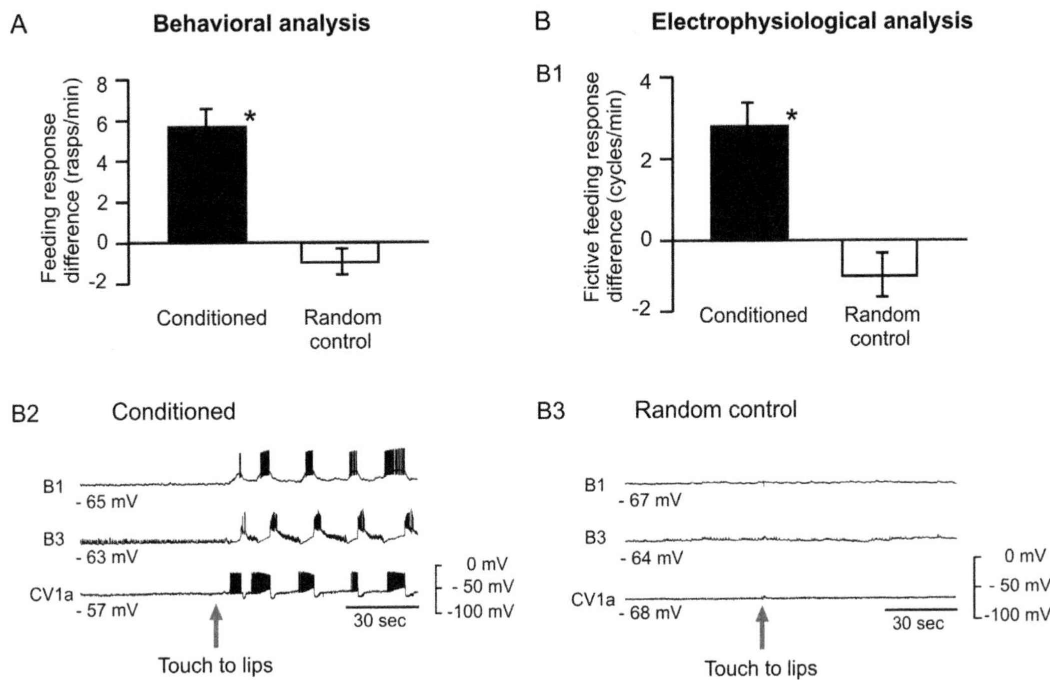

FIGURE 14.5 Contribution of nonsynaptic long-term changes in membrane potential to tactile classical conditioning in *Lymnaea*. (A) The behavioral response to the touch CS is significantly (asterisk) stronger in the conditioned versus the control group, indicating successful conditioning. (B1) Behavioral classical conditioning produces a CS-induced systems-level electrical activity in motor neurons (fictive feeding response) recorded in semi-intact preparations. The electrophysiological response to the touch CS, which was calculated as the post-minus the prestimulus response to touch, is significantly stronger in the conditioned versus the control group. (B2) Behavioral classical conditioning produces a CS-induced increase in the fictive feeding response in B1 and B3 feeding motor neurons and the CV1a (cerebral-ventral 1a), a type of CBI command-like interneuron (see Figure 14.1C2). (B3) The lip touch CS evokes no rhythmic fictive feeding activity in a preparation from a control animal. Note that CV1a is more depolarized in the preparation from the conditioned animal (B2) than in the preparation from the control animal (B3), which is a consistent observation in this type of experiment. Depolarization of CV1 contributes significantly to the enhanced feeding response following this type of conditioning (see text). Source: *From Benjamin PR, Kemenes G. Invertebrate models to study learning and memory:* Lymnaea. *In: Squire L ed.* Encyclopedia of Neuroscience. *Vol. 5. Oxford: Academic Press, 2009:197-204. Used with permission from Elsevier.*

fictive feeding in CS+ but not CS− sites in both experimental groups. More detailed changes were also recorded in other parts of the feeding network. For instance, conditioning-induced increases in spike activity were recorded in the touch-sensitive CS pathway between cerebral and buccal ganglia.[88] CPG interneurons, such as N1M (N1 medial), were also affected by learning. A long-lasting sequence of inhibitory synaptic inputs that occur in the N1M in response to lip touch changes to a strong depolarizing synaptic input after *in vitro* conditioning, and this drives a sustained plateauing pattern in the N1M cell.[82] This is an example of synaptic plasticity affecting an important CPG element of the feeding network that is also involved in feeding initiation.[18] At the motor neuronal level, there is a conditioning-induced increase in the amplitude of an early touch-evoked EPSP and spike activity recorded from the B3 feeding motoneuron.[89] A nonsynaptic neuronal change also occurs after tactile conditioning, and this has been shown to also play an important role in tactile conditioning.[19] This occurs in one of the CBI cell types known as the CV1 (see the following section). This neuron is capable of driving a feeding rhythm via its excitatory synaptic connection with the N1M cell, and spiking activity in this cell normally accompanies unconditioned feeding responses stimulated by sucrose. After tactile conditioning, the CV1 cells are considerably more active following touch in conditioned snails compared with controls and show the typical rhythmic activity (Figure 14.5B2) seen with sucrose.[19] These experiments using tactile conditioning reveal synaptic and nonsynaptic changes in central sensory pathways, command-like and CPG interneurons, and motor neurons, thus strongly supporting a multisite hypothesis for associative conditioning.

Experiments aimed at investigating the neural basis of aversive classical conditioning of the feeding systems in *Lymnaea* (conditioned taste aversion (CTA)) were carried out in isolated brains dissected from conditioned and control animals. The synaptic connection between the CGCs and the N1M was examined, and a significant change in the size of IPSPs recorded in the N1M was recorded following spike induced in the CGCs by artificial depolarization.[24] Because the CGCs are known to play a critical gating role in feeding behavior and the N1M (a pivotal member of the CPG) is also involved in feeding initiation,[18] this enhanced IPSP may be an important cellular correlate of the reduced response to sugar observed in behavioral CTA. The N1M IPSPs examined in these experiments probably originate from the N3t cells of the feeding CPG, indicating that the N1M cells are important in both reward and aversive conditioning of the *Lymnaea* feeding system.

Like *Lymnaea*, the *Aplysia* feeding system has also been subjected to behavioral and electrophysiological analysis of tactile reward conditioning using lip touch as the CS and food reinforcement (seaweed) as the US. An *in vitro* electrophysiological analog of classical conditioning employed electrical stimulation of nerves to activate the CS and US pathways. Lesions of the esophageal nerve (En) blocked classical conditioning, suggesting that this nerve mediates the effects of the US.[88] This nerve also contains dopamine (DA) immunofluorescent fibers, so it is interesting that the DA antagonist, methylergonovine, blocked the acquisition of the *in vitro* analog of reward conditioning.[90] Thus, the behavioral, immunostaining, and pharmacological data are consistent with the hypothesis that DA located within nerve fibers in the En nerve mediates the actions of the US during conditioning. Individual neurons of the *Aplysia* feeding circuit have been analyzed for changes occurring during conditioning. Buccal neurons B31/32 play a key role in initiating buccal motor programs, and these neurons were significantly more depolarized by the CS after conditioning, although their intrinsic properties were unaffected. This suggests that there is greater excitation of the B31/B32 neurons after conditioning due to an enhancement of the CS pathway. At least part of the increased depolarization arises from the monosynaptic connections that the feeding command-like neuron CBI-2 has with the B31/B32 neurons because these cells increase their spiking activity in response to CS stimulation.[91] No change in the strength of the CBI-2-to-B31/B32 synaptic connection was induced by conditioning to suggest that this might play a role in conditioning. It appears that changes in the CS sensory pathways might play a role in tactile sensory conditioning as was found in *Lymnaea*.[89] However, changes in the endogenous plateauing properties of the buccal feeding neuron B51 also play a key role in tactile classical conditioning, and this is considered in the following section.

NONSYNAPTIC CELLULAR MECHANISMS FOR LEARNING

It has long been recognized that changes in synaptic strength play a major role in molluskan learning and memory,[44,73] but an increasing number of examples of nonsynaptic plasticity have been discovered that form part of multisite memory mechanisms in neural circuits underlying behavioral learning.[92,93] These nonsynaptic mechanisms include changes in input resistance, membrane potential, threshold for plateau initiation, and bursting coherence. Changes in these nonsynaptic properties usually alter the excitability of the neurons, with learning-induced changes in endogenous ionic currents providing the underlying mechanism. This contrasts with synaptic plasticity, where the strength of synaptic connections between neurons is changed after learning.

Changes in Input Resistance

Examples of input resistance increases induced by conditioning occur in the photoreceptor cells of *Hermissenda*.[16] Conditioning of phototactic behavior induces cellular changes in both type A and type B photoreceptors, which increase the firing of the cells in response to light (the CS) compared with controls. The CS evokes a larger receptor potential and a persistent enhanced excitability in the B cells as tested by the response to a standard applied depolarizing pulse. An increase in membrane resistance contributes to this increase in spike activity, and a decrease in spike accommodation is also important in producing a sustained response to receptor depolarization in response to the CS. Conditioning reduces the peak amplitude of several types of conductances in the B-type photoreceptors, including a Ca^{2+}-dependent ($I_{K,Ca}$) and voltage-dependent currents (I_A and I_{Ca}).[16]

Similar input resistance increases have been described in the LE mechanoreceptor neurons of the gill–siphon reflex of *Aplysia* following sensitization and classical conditioning.[7,75] This nonsynaptic mechanism increases the excitability of the sensory neurons, tested directly by current injection, making it more likely that the sensory neurons will respond to siphonal touch after conditioning. The increase in excitability is due to a decrease in a specific type of intrinsic potassium current that was identified when serotonin was applied to sensory neurons in the intact ganglion.[94]

Changes in Membrane Potential

Persistent changes in membrane potential occur in whole body withdrawal interneurons in *Helix*[95] and in feeding command-like CBI (e.g., CV1) interneurons in *Lymnaea* involved in feeding initiation.[19] In both snails, the cells are depolarized following conditioning, and this lowers the threshold for firing in response to the CS, thus allowing command cells to directly activate the motor circuits. In *Helix*, there is an additional decrease in threshold for spike initiation in withdrawal interneurons[95] that also promotes the conditioned withdrawal reflex. In *Lymnaea*, a long-lasting membrane depolarization of 11 mV on average was recorded in the CV1 neurons from conditioned (Figure 14.5B2) compared with control snails (Figure 14.5B3)[19] that persisted as long as the electrophysiological and behavioral memory trace (up to 4 days). The depolarization makes the cells more responsive to the CS following tactile conditioning (Figure 14.5B2) and can account for the activation of the feeding response via the CV1 cell's strong excitatory synaptic connection with interneurons of the feeding CPG. The importance of this result is emphasized by experiments in which the membrane potential of the CV1 cells is manipulated to either reverse the effect of behavioral conditioning or mimic the effects of conditioning in naive snails. These experiments showed that the persistent depolarization of the CV1 cells was both sufficient and necessary for the conditioned tactile response in the feeding network.[19] The *Aplysia* homologs of the *Lymnaea* CV1 cells, the CBI-2 cell type, have been examined after *in vitro* classical conditioning of feeding; unlike *Lymnaea*, no changes in the membrane potential or other nonsynaptic cellular properties of the CBIs were reported.[92]

The CGCs in *Lymnaea* also show persistent depolarization of cell body membrane potential of approximately 10 mV compared with controls after one-trial chemical conditioning.[20] This depolarization indirectly increases the strength of postsynaptic responses to stimulation by the chemical CS by a process that involves an increase in intracellular calcium concentration of the CGC proximal axonal processes.[20] The local targets for CGC depolarization are the CBI command-like cells for feeding, and artificial depolarization of the CGCs in naive snails increases the response of the CBI cells to the CS, mimicking the effects of behavioral conditioning. The CGCs are extrinsic to the feeding circuit, and so the change of membrane potential activates feeding responses indirectly by affecting command interneurons intrinsic to the circuit. It appears that the CGCs are increasing the strength of the CS-to-CBI synapse by presynaptic facilitation. This learning-related mechanism is independent of the normal 'gating' function of the CGCs that depends on their tonic firing.[21] Recently, it has been shown that the conditioning-induced depolarization of the CGCs is due to an increase in the size of a persistent sodium current.[96] The measured increase in the size of this current is sufficient to depolarize the CGC by the required amount. Surprisingly, the depolarization of the CGCs does not cause a change in the firing rate of the CGCs or their spike shape. In order to understand the ionic mechanisms of this novel combination of plasticity and stability, a Hodgkin–Huxley-type computer model of the CGCs was made. The model was used to elucidate how learning-induced changes in a measured somal persistent sodium current and a delayed rectifier potassium current[97] led to a persistent depolarization of the CGCs while maintaining firing rate. Included in the model was an additional increase in the conductance of a high-voltage activated Ca^{2+} current that allowed the spike amplitude and spike duration to be maintained. A balanced increase in these three types of conductances was sufficient to explain the novel electrophysiological changes observed after conditioning.[97]

Interestingly, the CGC's changes occur in chemically conditioned snails but not those subjected to tactile conditioning. For tactile conditioning, the CV1 cells, but not the CGCs, are depolarized.[19] The reason for this difference in the two types of classical reward conditioning is unclear, but it is probably linked to differences in the neural pathways activated by amyl acetate versus lip touch.

Cleary *et al.*[98] showed that the membrane potential of tail motor neurons in *Aplysia* was changed after behavioral long-term sensitization. Behavioral long-term sensitization resulted in a hyperpolarization of the resting membrane potential and a concomitant decrease in spike threshold. These two changes would tend to have the opposite effects on motor neuron excitability, and why they happen is unclear.

Changes in the Threshold for Plateau Formation

Changes in the threshold for plateau formation in motor pattern generation occur in the same *Aplysia* interneuron (B51) in both classical and operant conditioning of feeding, allowing comparisons to be made of the nonsynaptic endogenous changes occurring in B51 in the two different types of learning.[99] Changes in the plateauing properties of B51 are important because the neuron plays an important role in decision making in the *Aplysia* feeding network.[100]

During tactile classical conditioning in preparations made from behaviorally trained animals, B51 showed a greater number of CS-evoked plateau potentials compared with controls, and it was synaptically depolarized more by the CS.[91] Neither the input resistance nor the resting potential of the B51 was affected by conditioning, but another intrinsic property, the threshold for plateau initiation, was increased. Paradoxically, this would make the cell less responsive to excitatory synaptic input, but nevertheless the cell still showed more plateau potentials after conditioning, so the synaptic inputs must be sufficiently strong to overcome the effects of reduced excitability.[91] The function of the reduced excitability in classical conditioning is unclear. Pharmacological evidence from a more recent paper suggests that acetylcholine (ACh) mediates the effects of the CS used in ganglionic classical conditioning and the effects of the CS in a single-cell analog of classical conditioning.[101] Paired application of ACh and dopamine (the US) to isolated B51 neurons increased the threshold for plateauing, and this persisted for up to 24 hr. No change in the input resistance was observed after conditioning. Depolarization of B51 by ACh also increased after conditioning. These changes to the endogenous membrane properties and the response to the ACh in the isolated neuron are similar to those observed following behavioral and ganglionic *in vitro* conditioning.[101]

Brembs *et al.*[102] developed a behavioral paradigm for reward operant conditioning using the consummatory (ingestive) phase of the feeding cycle as the operant. A cellular correlate of operant conditioning was monitored by intracellularly recording the B51 feeding interneuron in isolated buccal ganglia made from behaviorally trained animals.[103] Cells from the contingent group show a significant decrease in threshold for plateau formation and a significant increase in input resistance compared with cells from yoked controls but no change in the resting membrane potential. To test whether this is due to an intrinsic change in the B51 cell rather than from synaptic input originating from outside the cell, an analog of operant conditioning was developed in which B51 was grown in culture and electrically triggered bursts of spikes paired with 6-sec puffs of dopamine were applied close to the isolated cell during a 10-min training period. Contingent application of these two stimuli produced a significant reduction in plateau threshold and a significant increase in input resistance compared with unpaired controls, similar to that occurring in the previous intact buccal ganglion preparation. These results indicate that endogenous changes are induced in B51 by operant conditioning. This increase in endogenous excitability would be expected to increase the plateauing activity in B51 as a result of synaptic inputs from other elements of CPG underlying rhythmic feeding activity, and it was shown that the number of spontaneous plateaus was increased 24 hr after *in vitro* conditioning.[103] The molecular mechanisms underlying the changes in the B51 intrinsic properties recently have been investigated,[103,104] and it appears that the second-messenger cAMP is involved together with the protein kinases PKA and PKC. Inhibiting PKA with bath application of Rp-cAMP blocked conditioning; conversely, injecting cAMP into B51 mimicked the effects of conditioning. Blocking PKC with bath application of bisindolymaleimide also blocked conditioning, as did the expression of a dominant-negative isoform of Ca^{2+}-dependent PKC. Activation of PKC also mimicked conditioning but was dependent on both cAMP and PKA, suggesting that PKC acted upstream of PKA activation. Thus, using B51 as a neuronal analog has provided access to mechanisms mediating operant conditioning in which cellular electrical and molecular changes can directly be related to behavior.

It is interesting that the two types of conditioning (classical and operant) in *Aplysia* have the opposite types of effect—increasing the plateau membrane potential threshold in classical conditioning and decreasing it in operant conditioning.[99] This is despite the increase in overall plateauing frequency in both types of conditioning in the intact network.

Changes in Endogenous Bursting Properties

In an alternative operant conditioning paradigm in *Aplysia*, application of an external rewarding stimulus that was contingent with spontaneous bites induced spontaneous feeding cycles that were elevated in frequency and regularity compared with naive or noncontingent animals.[105] An analysis of the neuronal basis of this learning-induced compulsive behavior revealed that three endogenously bursting protraction phase neurons (B30/B31/B65), which are involved in initiating the feeding rhythm, regularize and synchronize their bursting activity through the promotion of stereotyped burst-generating oscillations and an increase in the strength of their electrotonic synaptic connections.[106] The regularity of bursting appears to be due to changes in the endogenous properties of the neurons so that in otherwise inactive networks, depolarization of individual neurons with similar current magnitudes induces recurring bursts of spikes with durations that remain similar from cycle to cycle, unlike neurons from noncontingent preparations in which the cells show highly variable durations. Other changes in endogenous properties, such as an increase in input resistance and a reduced threshold for bursting, contribute to an increase in bursting frequency seen in contingent preparations. The increase in input resistance also contributes to the increase in the strength of the electrical coupling between B30/B31/B65 that plays an important role in the increased synchronicity of bursting in the three neurons. In this example, changes in the endogenous properties and consequent firing patterns of feeding interneurons cause irregular and infrequent feeding behavior to switch to inflexible reward-driven compulsive-like repetitive feeding behavior.

DISCUSSION AND CONCLUSIONS

A Multisite Mechanism for Gastropod Learning

A systems or network approach to learning and memory in gastropods has been successful in relating network changes to behavioral plasticity and identifying the types of changes (synaptic and nonsynaptic) and their locations within the 'learning circuits' that underlie memory. Ideally, the knowledge of these various changes in the neural circuit should be used to predict the behavioral outcomes of training in a quantitative manner by use of computational modeling, and this has been successful in a few cases.[107,108] The neuronal components of gastropod motor circuits used for learning studies can be conveniently divided into sensory neurons, interneurons, and motor neurons, and the synaptic wiring diagram including these cell types is well described in the model systems that form the main content of this chapter (Figure 14.1). Changes in firing rates, synaptic connectivity, and cellular properties (usually endogenous excitability) underlying memory formation have been found to occur in all three classes of neurons to varying degrees. The main aim has been to understand how sensory neurons and their synaptic pathways are modified by learning to produce a changed motor response. This almost always involves sensory synaptic connections to interneurons as well as motor neurons. Interneurons are more important in learning in networks in which there are intermediate layers of organization between sensory and motor neurons, such as in the *Hermissenda* phototactic system[16] and the feeding systems of *Lymnaea*[18] and *Aplysia*.[107] The most complex interneuronal organization occurs in terrestrial slugs and snails, in which there is a separate lobe of a CNS ganglion, the procerebrum (PC), devoted to the processing of odor discrimination. The PC has a complex intrinsic synaptic connectivity and appears to be the primary center for associative discrimination learning.[109] It receives sensory input from a sensory ganglion in the periphery and connects to downstream motor centers that mediate the conditioned response.

Both associative and nonassociative forms of learning involve changes at multiple sites in the network. In the siphonal withdrawal reflex of *Aplysia*, habituation, sensitization, and classical conditioning all involve changes in the strength of synapses between sensory neurons and motoneurons. In addition, changes occur between sensory neurons and modulatory interneurons and between interneurons that form an intrinsic network. Increased firing following short-term sensitization also causes long-term increases in motor neuron firing lasting several minutes.[110] This increase in background firing rate is capable of amplifying siphon neuromuscular contractions. Changes in the intrinsic excitability of sensory neurons also occur in siphonal learning as well as cellular and Ca^{2+}-dependent events on receptors at the motoneuronal level. In *Hermissenda*, the electrical changes that underlie the responses to associative conditioning are due to interactions between the convergent CS (photic) and US (mechanical) synaptically mediated pathways at both sensory and interneuronal levels. In the sensory photoreceptor cells, this induces changes in the endogenous biophysical properties that induce a larger receptor potential and increased firing of the receptor cells. At the interneuronal level, there are increases in response to the CS due to this increase in photoreceptor firing but also because of an increase in the strength of the synapse between the photoreceptor cells and interneurons. There also is an increase in the endogenous excitability of one of these intermediate interneurons. In gastropod feeding systems, there is most evidence for sensory-to-interneuronal rather than sensory-to-motor neuron synaptic plasticity. In *Lymnaea*, there is one example in which a direct monosynaptic pathway from lip mechanoreceptors to a motor neuron is enhanced in amplitude following tactile classical conditioning.[89] However, this is not a major pathway for conditioned response activation. One of the main sites of synaptic plasticity in feeding is the sensory pathways between the sensory neurons and the CBI command-like neurons. Following chemical conditioning, the monosynaptic synaptic pathway between CS pathways and the CBIs is increased in strength sufficiently to activate the CBIs. This is due to presynaptic facilitation of the sensory neurons by the modulatory CGC interneurons.[20] In *Lymnaea* tactile conditioning, there is a persistent depolarization of the CBIs that reduces the threshold for spike initiation and makes it more likely that the CBIs will respond to sensory input. In addition to monosynaptic connections to CBI interneurons, there is a direct pathway from the sensory neurons to the CPG interneurons. This pathway also seems to be enhanced after *Lymnaea* tactile conditioning.[83] At the CPG/rhythm-generating level of the network, there are a number of examples in which changes in the endogenous properties of the interneurons play an important part in both classical and operant conditioning. In the *Aplysia* feeding network, these include changes in the threshold for plateauing in CPG/decision-making interneurons such as the B51 and production of coherent bursting in other protraction phase interneurons. These results from a number of gastropod systems strongly indicate that sensitization and conditioning (both classical and operant) is due to multiple synaptic and endogenous changes occurring at various sites in the network.

Interactions between Inhibitory and Excitatory Synaptic Mechanisms are Important for Memory Formation

Changes in the strength of excitatory synaptic connections have been a major focus of attention in learning studies, but there is increasing evidence that modification of inhibitory synaptic mechanisms is also involved in memory.[6,24] Indeed, interactions between inhibitory and excitatory processes in the determination of behavioral outcomes have been a major source of discussion in both the psychological/ethological and the neurobiological literature for many years.[111–113] In gastropod mollusks, interactions between these two types of synaptic changes have been shown to be important in habituation, sensitization (*Aplysia* gill and siphon withdrawal reflex[64]), and classical conditioning (*Pleurobranchaea* feeding,[114] *Hermissenda* phototaxis,[80] and *Lymnaea* feeding[87]). Fitzgerald *et al.*[115] emphasized that the balance between inhibition and facilitation is largely responsible for determining the net amplitude and direction of change in the siphon withdrawal reflex when comparing different forms of nonassociative plasticity (habituation and sensitization).

The results from aversive classical conditioning of feeding in *Pleurobranchaea* are particularly interesting because they involve learning-induced modulation of 'spontaneous' or background excitatory and inhibitory synaptic inputs whose balance in response to the CS is changed by conditioning.[114] This occurs in the paracerebral (phasic type) command cells that are involved in the initiation of feeding. Conditioned animals show a decreased level of excitatory inputs and an increased level of inhibitory inputs on application of the CS (food or touch), reducing the ability of the paracerebral cells to activate feeding—that is, the balance of inhibition/excitation was changed in favor of inhibition. In *Lymnaea* reward classical conditioning, there is a conditioning-induced reduction in the frequency of spontaneous tonic inhibitory synaptic input to the feeding CPG.[87] By reducing the frequency of this 'background' inhibitory input, the threshold for the CS to activate the feeding network is reduced, making it more likely that the feeding network (CPG) will respond to the CS.[86] Note that the change in inhibition in *Lymnaea* is not induced by the application of the CS,[87] unlike in the *Pleurobranchaea* example.[114] Tonic inhibitory input in *Lymnaea* originates from one of the CPG interneuron types known as the N3t (N3tonic) that inhibits the feeding CPG.[116] There is another source of inhibition that is extrinsic to the feeding circuit, but this is not altered by conditioning.[22] Conditioning in *Lymnaea* also increases the strength of excitatory synaptic input from the CS pathways to the CBIs, thus exciting them to activate the CPG-driven feeding motor program. In naive animals the CS has no ability to activate the CBIs, so this is a new functional pathway.[22]

Issues Concerning the Role of Nonsynaptic Mechanisms in Gastropod Learning

It can be concluded from the evidence reviewed in this chapter that nonsynaptic mechanisms are an important component of molluskan memory. They mainly act to increase the excitability of sensory cells and interneurons (1) by increasing the input resistance of the somal membrane so that synaptic or receptor currents cause greater depolarization and a consequent increase in firing rates, (2) by depolarizing the membrane potential so that the cell is closer to threshold for spike initiation, and (3) by reducing the threshold for plateauing so that triggering synaptic inputs increases the occurrence of plateauing potentials. An example in which persistent depolarization of membrane potential has no effect on firing rate occurs in the modulatory CGCs of *Lymnaea* after chemical conditioning. Here, the persistent depolarization acts to cause presynaptic facilitation of the CS sensory pathway by increasing transmitter release from local CGC proximal neuritic axonal branches rather than by changing CGC firing rate.[20]

It is usually assumed that these changes in excitability arise from persistent changes in the strength of intrinsic ionic currents (e.g., B cell photoreceptors of *Hermissenda*), particularly those mediated by potassium channels. An exception may be the CV1 cells of *Lymnaea*.[19] The input resistance of these cells is unaffected, despite a persistent depolarization following conditioning. This may indicate that the endogenous ion channel properties of CV1 are not involved, although there are a number of examples in which long-lasting depolarization occurs with no net increase in input resistance.[117,118] The persistent change in CV1 membrane potential could have its origins in enhanced extrinsic synaptic input, although this seems unlikely because synaptic input would be expected to change input resistance.

How can changes in general electrical excitability lead to a defined conditioned response to a specific CS? This is not a problem when the excitability change occurs in sensory neurons that form the CS pathway, such as the B-type photoreceptors that mediate the phototactic conditioned response in *Hermissenda*[16] or the touch-sensitive neurons involved in classical conditioning of the siphon–gill withdrawal reflex of *Aplysia*.[75] When excitability changes occur in command-like neurons such as the CV1s,[19] a change in excitability might generalize to a wider range of sensory inputs other than those used for the original conditioning experiments. In the example of the CV1 cells of the *Lymnaea* feeding system, the selectivity of the CS response to lip touch after conditioning is ensured by CV1 being embedded in a small network with a highly specialized function that responds to only a narrow range of food-related sensory inputs. These sensory inputs arise mainly from the lips, which normally come into contact with food during rhythmic feeding, unlike other sensory structures such as the tentacles. Indeed, in behavioral conditioning experiments,[27] no generalization of the classically conditioned response occurred from lips to tentacles. In further electrophysiological experiments, artificial depolarization of the CV1 cells in naive snails selectively increased the response to lip touch with no effect on tentacle touch.[30] This specificity of the conditioned response appears to be due to the selective ability to reinforce synaptic inputs from one site on the body versus another, even within the same sensory modality.

Another question that arises is whether changes in nonsynaptic properties of neurons in mollusks are important for specific phases of memory formation, as has been suggested in vertebrate systems.[119] Evidence for this derives from work on the CGCs in the *Lymnaea* feeding system by Kemenes *et al.*[20] The use of a single-trial training protocol allowed them to follow both the onset and persistence of neuronal changes that paralleled the time course of long-term memory formation measured behaviorally. The delayed onset of the depolarization of the CGC shows that it is not involved in the early expression of the memory trace or any consolidation process, suggesting that the CGCs are involved in LTM alone. Whereas LTM may be supported by the depolarized state of the CGCs, early memory and memory consolidation must involve other mechanisms. This could involve synaptic plasticity, as has been demonstrated in *Lymnaea* reward and aversive conditioning.[24,89] If these synaptic changes persisted in parallel with nonsynaptic mechanisms, then the role of the learning-induced membrane potential depolarization of the CGCs would add a type of extrinsic nonsynaptic plasticity to the feeding circuit to support or provide a backup for learning-induced synaptic plasticity.

Why is the Circuit-Level Organization Underlying Memory so Complex in Gastropods?

It could be argued that the complex circuit-level organization underlying memory in gastropods may be necessary for the more complex forms of learning of which gastropod mollusks are capable. For instance, in context-specific learning, the context modulates the expression of the CS and US association, providing an additional requirement for learning.[36,120] It has been speculated that this would require an 'extra' synaptic connection between the context and CS pathways that would enable the CS to stimulate a greater response in the motor neurons as was suggested for the *Aplysia* gill withdrawal reflex.[119] Complexity at the interneuronal level could explain some aspects of second-order conditioning.[2,121] In the gill withdrawal reflex circuit, there are intrinsic facilitatory interneurons. Following conditioning, these neurons are excited not only by the US but also by the CS. Also, the facilitatory neurons produce facilitation not only at the sensory-to-motoneuron synapse but also at the synapse from the sensory neurons to the facilitatory interneurons. As a result, during stage 1 of secondary conditioning, CS1 acquires a greater ability to excite the facilitatory interneurons, thus allowing them potentially to act as the US in stage 2. This suggested mechanism has been successfully incorporated into a quantitative model for second-order conditioning.[122]

Acknowledgments

I thank Dr. Ildiko Kemenes for her help in producing the figures and the BBSRC for financial support.

References

1. Byrne JH. Cellular analysis of associative learning. *Physiol Rev.* 1987;67:329–439.
2. Kandel ER. The molecular biology of memory storage: a dialogue between genes and synapses. *Science.* 2001;294:1030–1038.
3. Wolpaw JR. The complex structure of a simple memory. *Trends Neurosci.* 1997;20:588–594.
4. Lisberger SG. Cerebellar LTD: a molecular mechanism of behavioural learning? *Cell.* 1998;92:701–704.
5. Benjamin PR, Staras K, Kemenes G. A systems approach to the cellular analysis of associative learning in the pond snail. *Lymnaea Learn Mem.* 2001;7:124–131.
6. Hansel C, Linden DJ, D'Angelo E. Beyond parallel fiber LTD: The diversity of synaptic and non-synaptic plasticity in the cerebellum. *Nat Neurosci.* 2001;4:467–475.
7. Hawkins RD, Kandel ER, Bailey CH. Molecular mechanisms of memory storage in *Aplysia*. *Biol Bull.* 2006;210:174–191.
8. Antonov I, Kandel ER, Hawkins RD. The contribution of facilitate and monosynaptic PSPs to dishabituation and sensitization of the *Aplysia* siphon withdrawal reflex. *J Neurosci.* 1999;19:10438–10450.
9. Antonov I, Antonova I, Kandel ER, Hawkins RD. Activity-dependent presynaptic facilitation and Hebbian LTP are both required and interact during classical conditioning in *Aplysia*. *Neuron.* 2003;37:135–147.
10. Frost WN, Kandel ER. Structure of the network mediating siphon-elicited siphon withdrawal in *Aplysia*. *J Neurophysiol.* 1995;73:2413–2427.
11. Perlman AJ. Central and peripheral control of siphon withdrawal reflex in *Aplysia californica*. *J Neurophysiol.* 1979;42: 510–529.
12. Frost WN, Kaplan TE, Cohen V, Henzi V, Kandel ER, Hawkins RD. A simplified preparation for relating cellular events to

behaviour: contribution of LE and unidentified siphon sensory neurons to mediation and habituation of the *Aplysia* gill- and siphonal-withdrawal reflex. *J Neurosci.* 1997;17:2900–2913.

13. Mackey S, Kandel ER, Hawkins RD. Identified serotonergic neurons LCB1 and RCB1 in the cerebral ganglia of *Aplysia* produce presynaptic facilitation of siphon sensory neurons. *J Neurosci.* 1989;9:4227–4235.
14. Cleary LJ, Byrne JH, Frost WN. Role of interneurons in defensive withdrawal reflexes in *Aplysia*. *Learn Mem.* 1995;2:133–151.
15. Fischer TM, Carew TJ. Activity-dependent regulation of neural networks: the role of inhibitory synaptic plasticity in adaptive gain control in the siphonal withdrawal reflex of *Aplysia*. *Biol Bull.* 1997;192:164–166.
16. Crow T. Pavlovian conditioning in *Hermissenda*; current cellular, molecular, and circuit perspectives. *Learn Mem.* 2004;11:229–238.
17. Benjamin PR, Elliott CJH. Snail feeding oscillator: the central pattern generator and its control by modulatory interneurons. In: Jacklet J, ed. *Neuronal and Cellular Oscillators*. New York: Dekker; 1989:173–214.
18. Benjamin PR. Distributed network organization underlying feeding behaviour in the mollusc *Lymnaea*. *Neural Syst Networks.* 2012;2:4.
19. Jones NG, Kemenes I, Kemenes G, Benjamin PR. A persistent cellular change in a single modulatory neuron contributes to associative long-term memory. *Curr Biol.* 2003;13:1064–1069.
20. Kemenes I, Straub VA, Nikitin ES, et al. Role of delayed nonsynaptic neuronal plasticity in long-term associative memory. *Curr Biol.* 2006;16:1269–1279.
21. Yeoman MS, Pieneman AW, Ferguson GP, Ter Maat A, Benjamin PR. Modulatory role for the serotonergic cerebral giant cells in the feeding system of the snail *Lymnaea*: 1. Fine wire recording in the intact animal and pharmacology. *J Neurophysiol.* 1994;72:1357–1371.
22. Straub VA, Kemenes I, O'Shea M, Benjamin PR. Associative memory stored by functional novel pathway rather than modifications of pre-existing neuronal pathways. *J Neurosci.* 2006;26:4139–4146.
23. Staras K, Kemenes G, Benjamin PR. Electrophysiological and behavioural analysis of lip touch as a component of the food stimulus in the snail *Lymnaea*. *J Neurophysiol.* 1999;81:1261–1273.
24. Kojima H, Nagayama S, Fujita Y, Ito E. Enhancement of an inhibitory input to the feeding central pattern generator in *Lymnaea stagnalis* during conditioned taste-aversion learning. *Neurosci Lett.* 1997;230:179–182.
25. Kemenes I, O'Shea M, Benjamin PR. Different circuit and monoamine mechanisms consolidate long-term memory in aversive and reward classical conditioning. *Eur J Neurosci.* 2011;33: 143–152.
26. Hawkins RD, Abrams TJ, Carew TJ, Kandel ER. A cellular mechanism of classical conditioning in *Aplysia*: activity-dependent amplification of presynaptic facilitation. *Science.* 1983;219: 400–405.
27. Kemenes G, Benjamin PR. Appetitive learning in snails shows characteristics of conditioning in vertebrates. *Brain Res.* 1989;489: 163–166.
28. Sahley CL, Martin KA, Gelperin A. An analysis of associative learning in the terrestrial mollusc *Limax maximus*: II. Appetitive learning. *J Comp Physiol A.* 1990;167:339–345.
29. Hawkins RD, Greene W, Kandel ER. Classical conditioning, differential conditioning, and second order conditioning of the *Aplysia* gill-withdrawal reflex in a simplified mantle organ preparation. *Behav Neurosci.* 1998;112:636–645.
30. Jones NG, Kemenes G, Benjamin PR. Selective expression of electrical correlates of differential appetitive conditioning in a feeding network. *J Neurophysiol.* 2001;85:89–97.
31. Inoue T, Murakami M, Watanabe S, Inokuma Y, Kurino Y. *In vitro* odor-aversion conditioning in a terrestrial mollusc. *J Neurophysiol.* 2006;95:3898–3903.
32. Kemenes G, Benjamin PR. Goal-tracking behaviour in the pond snail *Lymnaea*. *Behav Neural Biol.* 1989;52:260–270.
33. Kemenes G, Benjamin PR. Training in a novel environment improves the appetitive learning performance of the snail, *Lymnaea stagnalis*. *Behav Neural Biol.* 1994;61:139–149.
34. Alexander J, Audesirk TE, Audesirk GJ. One-trial reward learning in the snail *Lymnaea stagnalis*. *J Neurobiol.* 1984;15:67–72.
35. Lukowiak K, Ringseis E, Spencer G, Wildering W, Syed N. Operant conditioning of aerial respiratory behaviour in *Lymnaea stagnalis*. *J Exp Biol.* 1996;199:683–691.
36. Hanley J, Lukowiak K. Context learning and the effect of context on memory retrieval in *Lymnaea*. *Learn Mem.* 2001;8:35–43.
37. McComb C, Sangha S, Quadry S, Yue J, Scheibenstock A, Lukowiak K. Context extinction and associative learning in *Lymnaea*. *Neurobiol Learn Mem.* 2002;78:23–34.
38. Orr MV, Lukowiak K. Electrophysiological and behavioral evidence demonstrating that a predator detection alters adaptive behaviours in the snail *Lymnaea*. *J Neurosci.* 2008;28: 2726–2734.
39. Sahley CL, Rudy JW, Gelperin A. An analysis of associative learning in a terrestrial mollusc: higher order conditioning, blocking and a transient US exposure effect. *J Comp Physiol.* 1981;144:1–8.
40. Kandel ER. *Cellular Basis of Behavior. An Introduction to Behavioural Biology*. San Francisco: Freeman; 1976.
41. Shapiro E, Castellucci VF, Kandel ER. Presynaptic inhibition in *Aplysia* involves a decrease in the Ca^{2+} current of the presynaptic neuron. *Proc Nat Acad Sci USA.* 1980;77:1185–1189.
42. Armitage BA, Siegelbaum SA. Presynaptic induction and expression of homosynaptic depression at *Aplysia* sensorimotor neuron synapses. *J Neurosci.* 1998;18:8770–8779.
43. Ezzeddine Y, Glanzman DL. Prolonged habituation of the gill-withdrawal reflex in *Aplysia* depends on protein synthesis, protein phosphatase activity, and postsynaptic glutamate receptors. *J Neurosci.* 2003;23:9585–9594.
44. Carew TJ, Sahley CL. Invertebrate learning and memory: from behaviour to molecules. *Annu Rev Neurosci.* 1986;9:435–487.
45. Dale N, Kandel ER. L-Glutamate may be the fast excitatory transmitter of *Aplysia* sensory neurons. *Proc Nat Acad Sci USA.* 1993;90:282–285.
46. Braha O, Dale N, Hochner B, Klein M, Abrams TW, Kandel ER. Second messengers involved in two processes of presynaptic facilitation that contribute to sensitization and disinhibition in *Aplysia* sensory neurons. *Proc Nat Acad Sci USA.* 1990;87: 2040–2044.
47. Sossin WS, Schwartz JH. Selective activation of Ca^{2+}-activated PKCs in *Aplysia* neurons by 5-HT. *J Neurosci.* 1992;12:1160–1168.
48. Antonov I, Kandel ER, Hawkins RD. Presynaptic and postsynaptic mechanisms of synaptic plasticity and metaplasticity during intermediate-term memory formation in *Aplysia*. *J Neurosci.* 2010;30:5781–5791.
49. Sutton MA, Masters SE, Bagnall MW, Carew TJ. Molecular mechanisms underlying a unique intermediate phase of memory in *Aplysia*. *Neuron.* 2001;31:143–154.
50. Ghirardie M, Montarolo PG, Kandel ER. A novel intermediate stage in the transition between short- and long-term facilitation in the sensory to motor neuron synapse of *Aplysia*. *Neuron.* 1995;14:413–420.
51. Sharma SK, Sherff CM, Shobe J, Bagnall MW, Sutton MA, Carew TJ. Differential role of mitogen-activated protein kinase in three distinct phases of memory for sensitization in *Aplysia*. *J Neurosci.* 2003;23:3899–3907.

52. Jin I, Kandel ER, Hawkins RD. Whereas short-term facilitation is presynaptic, intermediate-term facilitation involves both presynaptic and postsynaptic protein kinases and protein synthesis. *Learn Mem.* 2012;:96–102.
53. Villareal G, Li Q, Cai D, Glanzman DL. The role of rapid, local, postsynaptic protein synthesis in learning-related synaptic facilitation in *Aplysia*. *Curr Biol.* 2007;17:2073–2080.
54. Chitwood RA, Li Q, Glanzman DL. Serotonin facilitates AMPA-type responses in isolated siphon motor neurons of *Aplysia* in culture. *J Physiol.* 2001;534:501–510.
55. Villareal G, Li Q, Cai D, et al. Role of protein kinase C in the induction and maintenance of serotonin-dependent enhancement of the glutamate response in isolated siphon motor neurons of *Aplysia californica*. *J Neurosci.* 2009;29:5100–5107.
56. Sutton MA, Carew TJ. Parallel molecular pathways mediate expression of distinct forms of intermediate-term facilitation at tail sensory-motor synapses in *Aplysia*. *Neuron.* 2000;26: 219–231.
57. Sutton MA, Bagnall MW, Sharma SK, Shobe J, Carew TJ. Intermediate-term memory for site-specific sensitization in *Aplysia* is maintained by persistent activation of protein kinase C. *J Neurosci.* 2004;24:3600–3609.
58. Hu J-Y, Glickman L, Wu F, Schacher S. Serotonin regulates the secretion and autocrine action of a neuropeptide to activate MAPK required for long-term facilitation in *Aplysia*. *Neuron.* 2004;43:375–385.
59. Glanzman DL. Simple minds: the neurobiology of invertebrate learning and memory. In: North G, Greenspan RJ, eds. *Invertebrate Neurobiology*. Cold Spring Harbor: Cold Spring Harbor Laboratory Press; 2007:347–380.
60. Trudeau LE, Castellucci VF. Postsynaptic modifications in long-term facilitation in *Aplysia*: upregulation of excitatory amino acid receptors. *J Neurosci.* 1995;15:1275–1284.
61. Zhu H, Wu F, Schacher S. Site specific and sensory neuron-dependent increases in post-synaptic glutamate sensitivity accompany serotonin-induced long-term facilitation at *Aplysia* sensorimotor synapses. *J Neurosci.* 1997;17:4976–4986.
62. Cai D, Chen S, Glanzman DL. Postsynaptic regulation of long-term facilitation in *Aplysia*. *Curr Biol.* 2008;18:920–925.
63. Cai D, Pearce K, Chen S, Glanzman DL. Protein kinase M maintains long-term sensitization and long-term facilitation in *Aplysia*. *J Neurosci.* 2011;31:6421–6431.
64. Si K, Lunquist S, Kandel ER. A neuronal isoform of the *Aplysia* CPEB has prion-like properties. *Cell.* 2003;115:879–891.
65. Si K, Giustetto M, Etkin A, et al. A neuronal isoform of CEPB regulates local protein synthesis and stabilizes synapse-specific long-term facilitation in *Aplysia*. *Cell.* 2003;115:893–904.
66. Miniaci MC, Kim JH, Puthenveettil SV, et al. Sustained CPEB-dependent local protein synthesis is required to stabilize synaptic growth for persistence of long-term facilitation. *Neuron.* 2008;59:1024–1036.
67. Si K, White-Grindley M, Kandel ER. *Aplysia* CPEB can form prion-like multimers in sensory neurons that contribute to long-term facilitation. *Cell.* 2010;140:421–435.
68. Trudeau L-E, Castellucci VR. Functional uncoupling of inhibitory interneurons plays an important role in short-term sensitization of *Aplysia* gill and siphon withdrawal reflex. *J Neurosci.* 1993;13:2126–2135.
69. Carew TJ, Walters ET, Kandel ER. Classical conditioning in a simple withdrawal reflex in *Aplysia californica*. *J Neurosci.* 1981;1:1426–1437.
70. Carew TJ, Hawkins RD, Kandel ER. Differential classical conditioning of a defensive withdrawal reflex in *Aplysia californica*. *Science.* 1983;219:397–400.
71. Hu J-Y, Chen Y, Schacher S. Protein kinase C regulates local synthesis and secretion of a neuropeptide required for activity-dependent long-term synaptic plasticity. *J Neurosci.* 2007;27: 8927–8939.
72. Glanzman DL. The cellular basis of classical conditioning in *Aplysia californica*: It's less simple than you think. *Trends Neurosci.* 1995;18:30–36.
73. Roberts AC, Glanzman DL. Learning in *Aplysia*: looking at synaptic plasticity from both sides. *Trends Neurosci.* 2003;26: 662–670.
74. Antonov I, Antonova I, Kandel ER, Hawkins RD. The contribution of activity-dependent synaptic plasticity to classical conditioning in *Aplysia*. *J Neurosci.* 2001;21:6413–6422.
75. Antonov I, Ha T, Antonova I, Moroz LL, Hawkins RD. Role of nitric oxide in classical conditioning of siphonal withdrawal in *Aplysia*. *J Neurosci.* 2007;27:10993–11002.
76. Crow T, Tian L-M. Polysensory interneuronal projections to foot contractile pedal neurons in *Hermissenda*. *J Neurophysiol.* 2009;101:824–833.
77. Crow T, Alkon DL. Associative behavioral modifications in *Hermissenda*: cellular correlates. *Science.* 1980;209:412–414.
78. Crow T, Tian L-M. Facilitation of monosynaptic and complex EPSPs in type I interneurons of conditioned. *Hermissenda J Neurosci.* 2002;22:7818–7824.
79. Crow T, Tian L-M. Pavlovian conditioning in *Hermissenda*: a circuit analysis. *Biol Bull.* 2006;210:289–297.
80. Crow T, Tian L-M. Sensory regulation of network components underlying ciliary locomotion in *Hermissenda*. *J Neurophysiol.* 2008;100:2496–2506.
81. Jin NG, Tian L-M, Crow T. 5-HT and GABA modulate intrinsic excitability of type I interneurons in *Hermissenda*. *J Neurophysiol.* 2009;102:2825–2833.
82. Benjamin PR, Kemenes G. Invertebrate models to study learning and memory: *Lymnaea*. In: Squire L, ed. *Encyclopedia of Neuroscience*. Vol. 5. Oxford: Academic Press; 2009:197–204.
83. Kemenes I, Kemenes G, Andrew RJ, Benjamin PR, O'Shea M. Critical time-window for NO-cGMP-dependent long-term memory formation after one-trial appetitive conditioning. *J Neurosci.* 2002;:1414–1425.
84. Straub VA, Styles BJ, Ireland JS, O'Shea M, Benjamin PR. Central localization of plasticity involved in appetitive conditioning in *Lymnaea*. *Learn Mem.* 2004;11:787–793.
85. Kemenes I, Straub VA, Nikitin ES, et al. Role of delayed nonsynaptic neuronal plasticity on long-term associative memory. *Curr Biol.* 2006;16:1269–1279.
86. Marra V, Kemenes I, Vavoulis D, Feng J-F, O'Shea M, Benjamin PR. Role of tonic inhibition in associative reward conditioning in *Lymnaea*. *Front Behav Neurosci.* 2010;:4. doi:10.3389/fnbeh.2010.00161.
87. Staras K, Kemenes G, Benjamin PR. Neurophysiological correlates of unconditioned and conditioned feeding behaviour in the pond snail *Lymnaea stagnalis*. *J Neurophysiol.* 1998;79: 3030–3040.
88. Lechner HA, Baxter DA, Byrne JH. Classical conditioning of feeding in *Aplysia*: II. Neurophysiological correlates. *J Neurosci.* 2000;20:3377–3386.
89. Staras K, Kemenes G, Benjamin PR. Cellular traces of behavioural classical conditioning can be recorded at several specific sites in a simple nervous system. *J Neurosci.* 1999;19: 347–357.
90. Reyes FD, Mozzachiodi R, Baxter DA, Byrne JH. Reinforcement in an *in vitro* analog of appetitive conditioning of feeding behaviour in *Aplysia*: blockade by dopamine antagonist. *Learn Mem.* 2005;12:216–2020.
91. Mozzachiodi R, Lechner HA, Baxter DA, Byrne JH. *In vitro* analog of classical conditioning of feeding behaviour in *Aplysia*. *Learn Mem.* 2003;10:478–494.

92. Benjamin PR, Kemenes G, Kemenes I. Non-synaptic neuronal mechanisms of learning and memory in gastropod molluscs. *Front Biosci*. 2008;13:4051–4057.
93. Mozzachiodi R, Byrne JH. More than synaptic plasticity: role of non-synaptic plasticity in learning and memory. *Trends Neurosci*. 2009;33:17–26.
94. Klein M, Camardo J, Kandel ER. Serotonin modulates a specific potassium current in the sensory neurons that show presynaptic facilitation in *Aplysia*. *Proc Nat Acad Sci USA*. 1982;79:5713–5717.
95. Gainutdinov KL, Chekmarev LJ, Gainutdinova TH. Excitability increase in withdrawal interneuron after conditioning in the snail. *Neuroreport*. 1988;16:517–520.
96. Nikitin ES, Vavoulis DV, Kemenes I, et al. Persistent sodium current is a non-synaptic substrate for long-term associative memory. *Curr Biol*. 2008;18:1221–1226.
97. Vavoulis DV, Nikitin ES, Kemenes I, et al. Balanced plasticity and stability of the electrical properties of a molluscan modulatory interneuron after classical conditioning: a computational study. *Front Behav Neurosci*. 2010;4:19.
98. Cleary WL, Byrne JH. Cellular correlates of long-term sensitization in *Aplysia*. *J Neurosci*. 1998;18:5988–5998.
99. Lorenzetti FD, Mazzachiodi R, Baxter DA, Byrne JH. Classical and operant conditioning differentially modify the intrinsic properties of an identified neuron. *Nat Neurosci*. 2006;9:17–19.
100. Nargeot R, Simmers J. Neural mechanisms of operant conditioning and learning-induce behavioural plasticity. *Cell Mol Life Sci*. 2011;68:803–816.
101. Lorenzetti DA, Byrne JH. Classical conditioning analog enhanced acetylcholine responses but reduced excitability of an identified neuron. *J Neurosci*. 2011;31:14789–14793.
102. Brembs B, Lorenzetti FD, Reyes FD, Baxter DA, Byrne JH. Operant reward learning in *Aplysia*: neuronal correlates and mechanisms. *Science*. 2002;296:1706–1709.
103. Mozzachiodi R, Lorenzetti FD, Baxter DA, Byrne JH. Changes in neuronal excitability serve as a mechanism of long-term memory for operant conditioning. *Nat Neurosci*. 2008;11:1146–1148.
104. Lorenzetti FD, Baxter DA, Byrne JH. Molecular mechanisms underlying a cellular analog of operant reward learning. *Neuron*. 2008;59:815–828.
105. Nargeot R, Petrissans C, Simmers J. Behavioral and *in vitro* correlates of compulsive-like food-seeking induced by operant conditioning. *J Neurosci*. 2007;27:8059–8070.
106. Nargeot R, Le Bon-Jego M, Simmers J. Cellular and network mechanisms of operant learning-induced compulsive behaviour in *Aplysia*. *Curr Biol*. 2009;19:975–984.
107. Goh F, Gelperin A. A neuronal network for the logic of *Limax* learning. *J Comput Neurosc*. 2006;21:259–270.
108. Zhang Y, Liu R-Y, Heberton GA, et al. Computational design of enhanced learning protocols. *Nat Neurosci*. 2012;15:294–297.
109. Watanabe S, Kirino Y, Gelperin A. Neural and molecular mechanisms of microcognition. *Learn Mem*. 2008;15:633–642.
110. Frost WN, Clark GA, Kandel ER. Parallel processing of short-term memory for sensitization in *Aplysia*. *J Neurobiol*. 1988;19:297–334.
111. Hinde RA. Behavioural habituation. In: Horn G, Hinde RA, eds. *Short-Term Changes in Neural Activity and Behaviour*. Cambridge, UK: Cambridge University Press; 1970:3–40.
112. Groves PM, Thompson RF. Habituation: a dual-process theory. *Psychol Rev*. 1970;77:419–450.
113. Prescott SA. Interactions between depression and facilitation within neural networks: updating the dual process theory. *Learn Mem*. 1998;5:446–466.
114. Davis WJ, Gillette R, Kovac MP, Croll RP, Matera EM. Organization of synaptic inputs to the paracerebral feeding command interneurons of *Pleurobranchaea californica*. *J Neurophysiol*. 1983;49:1557–1572.
115. Fitzgerald K, Wright EA, Marcus E, Carew TJ. Multiple forms of non-associative plasticity in *Aplysia:* a behavioural cellular and pharmacological analysis. *Philos Trans R Soc Lond B*. 1990;329:171–178.
116. Staras K, Kemenes I, Benjamin PR, Kemenes G. Loss of self-inhibition is a cellular mechanism for episodic rhythmic behaviour. *Curr Biol*. 2003;13:116–124.
117. Swandulla D, Lux HD. Changes in ionic conductances induced by cAMP in *Helix* neurons. *Brain Res*. 1984;305:115–122.
118. Kemenes G, S-Rozsa K, Carpenter DO. Cyclic-AMP mediated excitatory responses to leucine enkephalin in *Aplysia* neurons. *J Exp Biol*. 1993;181:321–328.
119. Zhang W, Linden DJ. The other side of the engram: experience-driven changes in neuronal intrinsic properties. *Nat Rev Neurosci*. 2003;4:885–900.
120. Colwill RM, Absher RA, Roberts MV. Conditional discrimination learning in *Aplysia californica*. *J Neurosci*. 1988;8: 4440–4444.
121. Hawkins RD, Kandel ER. Is there a cell biological alphabet for simple forms of learning? *Psychol Rev*. 1984;91:375–391.
122. Hawkins RD. A biologically based computational model for several different forms of learning. *Psychol Learn Mem*. 1989;23:65–108.

CHAPTER

15

Comparison of Operant and Classical Conditioning of Feeding Behavior in *Aplysia*

Riccardo Mozzachiodi[*], *Douglas A. Baxter*[†] *and John H. Byrne*[†]

[*]Texas A&M University–Corpus Christi, Corpus Christi, Texas [†]The University of Texas Medical School at Houston, Houston, Texas

INTRODUCTION

Classical conditioning[1] and operant conditioning[2,3] are two important forms of associative learning that allow animals, including humans, to survive in a changing environment. Despite the individual analysis of the neuronal processes underlying classical and operant conditioning,[4–12] it has been difficult to test experimentally whether, at some fundamental level, they are mechanistically distinct or similar.[13–15] This difficulty is largely due to the lack of a suitable model system that exhibits both operant and classical conditioning of the same behavior and, at the same time, is tractable at the cellular and molecular levels.[16–19]

Feeding behavior in the marine mollusk *Aplysia californica*[20] is a useful model system to overcome the previously mentioned limitations and conduct a comparative analysis of the mechanisms of classical and operant conditioning.[21] Feeding behavior is controlled by a well-characterized neural circuit[22–24] and, importantly, is modulated by both forms of associative learning.[25–28]

This chapter reviews studies that analyzed and compared the mechanisms underlying classical and operant conditioning. These analyses examine the neural substrates of learning at the levels of changes in neural circuit activity, the properties of individual neurons, and molecular processes that mediate the cellular modifications. Intriguing similarities and differences between these two forms of learning are being elucidated. For example, both forms of learning lead to increased biting *in vivo* and increased fictive ingestion in the neural circuit *in vitro*. A reinforcement pathway, which uses dopamine (DA) as the reinforcement transmitter, is common to both classical and operant conditioning. Finally, a cellular locus of plasticity that is common to both forms of learning, neuron B51, is altered in opposite ways by the two conditioning paradigms. The excitability of B51 is increased by operant conditioning, whereas it is decreased by classical conditioning.

FEEDING BEHAVIOR IN *APLYSIA* AND ITS UNDERLYING NEURAL CIRCUIT

The consummatory phase of feeding in *Aplysia* (i.e., biting) consists of a series of rhythmic movements of the mouth (lips and jaws) and buccal mass (odontophore and radula) by which food is transported into the mouth and ultimately into the esophagus.[20,22,23] When a bite occurs, the odontophore, with its two radula halves (toothed grasping surfaces), first rotates forward toward the mouth (i.e., protraction) and the jaws open to accommodate the protracting odontophore. The two halves of the radula, initially separated during protraction, begin to close and grasp the food before the peak of protraction. The radula remains closed as the odontophore retracts (backward rotation), which brings the food into the mouth and buccal cavity, and the jaws close as the odontophore retracts.[29] In addition to bites, rejection movements can also be generated in response to inedible objects taken into the buccal cavity.[29] During rejection, the two halves of the radula are closed as the odontophore protracts toward the mouth, and they open as the odontophore retracts, which ejects the inedible object from the buccal cavity. Thus, both bite and rejection are composed of two

Invertebrate Learning and Memory.
DOI: http://dx.doi.org/10.1016/B978-0-12-415823-8.00015-0

phases: a protraction phase followed by a retraction phase. During a bite, the radula is closed during the retraction phase, whereas during rejection, the radula is open during the retraction phase.

The neural circuitry that mediates consummatory feeding behavior is located primarily in the cerebral and buccal ganglia and has been extensively characterized during the past few decades.[22,23] In particular, the motor activity that controls the rhythmic movements of the odontophore and radula is generated by a central pattern generator (CPG) in the buccal ganglion.[22,23,30,31] Using a variety of intact, semi-intact, and reduced preparations, the patterns of CPG activity responsible for the elicitation of feeding movements (buccal motor patterns (BMPs)) have been examined in great detail (Figure 15.1). During the occurrence of a BMP, large-unit activity in the intrinsic buccal muscle nerve I2 (I2 n.), which controls radula protraction, precedes large-unit activity in buccal muscle nerve n.2,1, which controls radula retraction (Figure 15.1).[29,33] Two major types of BMPs have been identified, one recorded during rejection (rBMP; Figure 15.1A)[29,33] and the other recorded during ingestion (iBMP; Figure 15.1C).[29,33] The two types of BMPs can be distinguished by examining the timing of large-unit activity in the radula nerve (Rn), which represents activity in radula closer motor neuron B8, relative to the onset of activity in n.2,1 and I2 n. During a rBMP, activity in B8 overlaps with large-unit activity in I2 n and precedes that in n.2,1 (Figure 15.1A); whereas during an iBMP, large-unit activity in n.2,1 and activity in B8 primarily overlaps and follows that in I2 n (Figure 15.1C).[28,29,32–40]

The use of isolated ganglia, which retain the ability to express *in vitro* motor patterns analogous to those observed *in vivo*, facilitated the analysis of the role of specific cellular components of the buccal CPG in the generation of BMPs.[31–35,41–45] Neurons with distinct characteristics, including pattern initiation (e.g., B63, B31/32, and B30), pattern switch (e.g., B4/5, B34, and B65), pattern termination (B52), and decision making (B51), have been identified (Figure 15.1B).[22,23] This knowledge of the circuit has enabled the identification of neurons whose properties are modified by classical and operant conditioning.

FEEDING BEHAVIOR IS MODIFIED BY ASSOCIATIVE LEARNING

Associative Paradigm for Learning that Food is Inedible

The first example of conditioning of feeding behavior in *Aplysia* was described by Susswein and colleagues.[46,47] A training procedure was developed in which *Aplysia* were presented with food (i.e., seaweed), which was made inedible by wrapping it in a plastic net.[47] Wrapped, food still elicits bites and is initially brought into the mouth and buccal cavity. However, because netted food cannot be swallowed, it

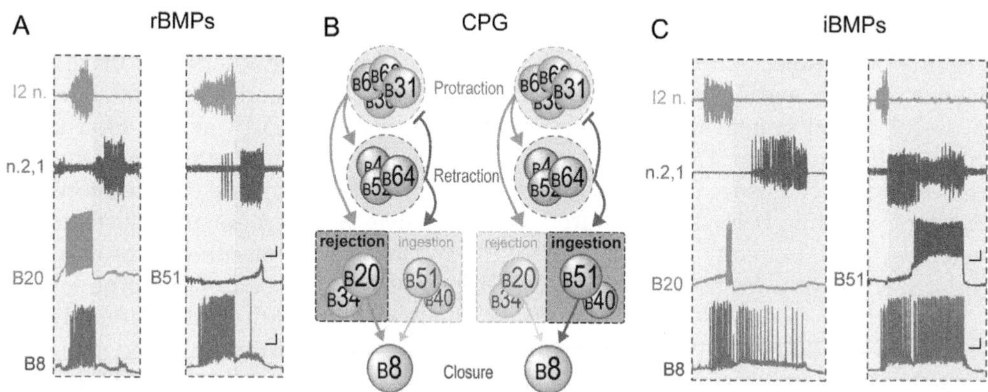

FIGURE 15.1 **Patterns of neural activity expressed by the feeding CPG.** Extracellular recording of activity in the I2 nerve (I2 n.) corresponds to the protraction phase of a BMP, whereas activity in nerve 2,1 corresponds to the retraction phase (see Nargeot *et al.*[32] for detailed descriptions of extracellular recordings and naming convention for nerves). Intracellular recordings from B8 monitored closure activity during the retraction versus protraction and thereby distinguished between fictive ingestion versus rejection. Intracellular recording are also illustrated from B20 and B51, two cells that play important roles in selecting between fictive ingestion versus rejecting. (A) Examples of rBMPs. During fictive rejection, B8 activity occurs primarily during the protraction phase (yellow shading). B20 excites B8 and is preferentially active during the protraction phase of rBMPs. Conversely, B51, which also excites B8, generally is not active during rBMPs. (B) Simplified representation of the feeding CPG. Activity in protraction (yellow shading) and retraction (blue shading) cells occurs during all BMPs. In contrast, subsets of cells are selectively active during fictive rejections (e.g., B20 and B34) or during fictive ingestion (e.g., B40 and B51). (C) Examples of iBMPs. Note the differential activity in B20 and B51 during fictive rejection (panel A) versus fictive ingestion (panel C). Calibration bars indicate a 5-sec span of time for all traces and a 10-mV deflection for the intracellular recordings.

triggers repetitive failed swallowing responses and is eventually rejected.[46,47] The netted food continues to stimulate the lips, producing further biting responses, followed by failed swallows. As training proceeds, most responses fail to lead to entry of netted food into the buccal cavity. In the cases in which netted food enters the buccal cavity, it remains for progressively shorter periods, eliciting fewer attempted swallowing responses. Finally, animals stop responding to the netted food.[46,47] Based on the amount of training, memory can be measured in the short-, intermediate-, and long-term domains.[48,49] Progress has been made in elucidating other features of this conditioning,[48,50,51] as well as the underlying biochemical and molecular mechanisms.[9,52–54]

Appetitive Classical Conditioning

The previously discussed learning protocol produces a suppression of the feeding response. However, feeding behavior can also be modified by appetitive associative paradigms (i.e., classical and operant conditioning), which induce an increase of its expression.[25,27] Colwill *et al.*[26] and Lechner *et al.*[27] developed a training protocol for classical conditioning of biting that produced both short- and long-term memory. Tactile stimulation of the lips with a soft paintbrush served as the conditioned stimulus (CS), whereas seaweed presentation served as the unconditioned stimulus (US). The effects of conditioning were assessed by counting the number of bites elicited by the CS, delivered prior to (pre-test) and after training (post-test; Figure 15.2A). Paired training produced a significant increase in the number of CS-evoked bites, compared to unpaired training, both 60 min and 24 hr after training.[27,28,39]

Further analysis led to the characterization of the reinforcement pathway involved in classical conditioning. Initial findings indicated that food must be ingested for the CS–US association to occur, suggesting that the US pathway originates in the foregut.[27] To identify the anatomical and physiological substrates of the US, extracellular nerve recordings were made in freely behaving *Aplysia* via chronically implanted electrodes hooked on the anterior branch of esophageal nerve (En2),[25] which projects to the foregut. Whereas little nerve activity was observed during biting in the absence of food, brief (~3 sec) bursts of high-frequency nerve activity (~30 Hz) were recorded in En2 when bites were followed by the ingestion of food,[25] suggesting a role for En2 in conveying afferent information related to the US. To further determine the role of foregut afferent pathways in classical conditioning, En2 was bilaterally severed.[27] Lesioned animals did not exhibit classical conditioning compared to similarly trained sham operated animals.[27]

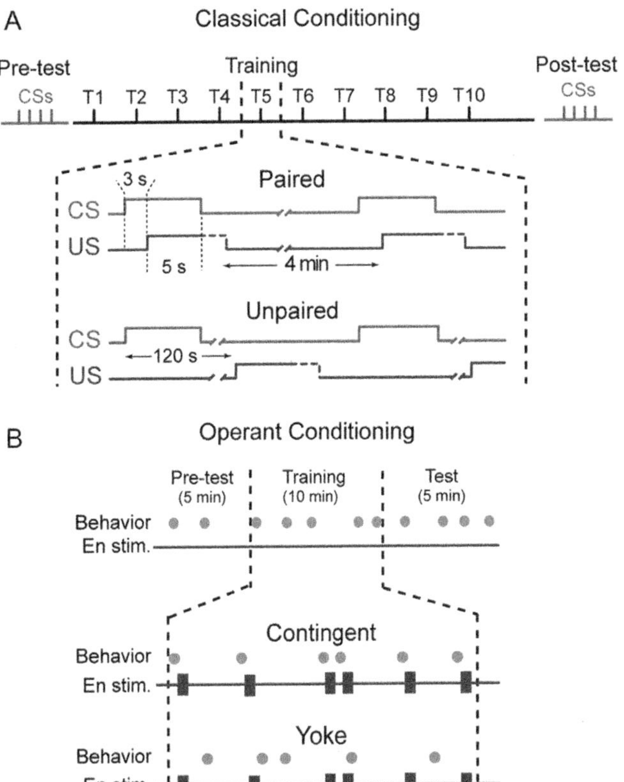

FIGURE 15.2 Training protocols for classical (A) and operant (B) conditioning. Similar protocols were used during *in vivo* behavioral training[25,27] and *in vitro* analogs of conditioning.[32,55] (A) In the *in vivo* classical conditioning protocol, pre-tests and post-tests were identical and consisted of four CSs spaced 1 min apart. Post-tests were conducted either 60 min or 24 hr after training. Training consisted of 10 trials of CS–US presentations. In the paired protocol, the CS preceded the US of 3 sec. Each CS–US paired presentation was delivered with a 4-min intertrial interval (ITI). In the unpaired protocol, the CS and US did not overlap but were spaced 2 min apart. In the experiments analyzing the neurophysiological correlates of conditioning, ganglia were removed within 6 hr after training. In the *in vitro* analog of classical conditioning, the testing procedures, the number of trials, and the ITI were identical to those used *in vivo*. (B) In the *in vivo* operant conditioning protocol, pre-test and post-test were identical and consisted of a 5-min period in which the number of bites was measured. Training consisted of a 10-min period during which En2 stimulations were made contingent or noncontingent (yoke) to the expression of bites. Post-test was conducted either immediately or 24 h after training. In the experiments analyzing the neurophysiological correlates of operant conditioning, ganglia were removed within 100 min after training. In the *in vitro* analog of operant conditioning, the testing procedure, the duration of training, and En2 stimulations were identical to those used *in vivo*.

Appetitive Operant Conditioning

The finding that En2 was a reinforcement pathway facilitated the development of an appetitive operant conditioning behavioral protocol in which biting served as the operant and electrical stimulation of En2 served as reinforcement.[25] During contingent training, bites were immediately followed by reinforcements (Figure 15.2B), which were delivered via stimulating electrodes implanted on the En2. During yoke training, animals received the same number of reinforcements as in the contingent group, but these reinforcements were not made contingent on the expression of bites (Figure 15.2B). Conditioning was measured as the number of bites that occurred during a 5-min post-test observation period. Operant conditioning produced a significantly greater number of bites, compared to that in the yoke group, both immediately and 24 hr after training.[25] Therefore, both classical and operant conditioning led to an increase in feeding, which was achieved through the use of a common reinforcement pathway—the En2.

A contingent-dependent increase in the number of bites was also produced by an operant paradigm in which the reinforcement consisted of a seaweed extract injected into the mouth at the end of a biting cycle instead of En2 electrical stimulation.[56] In naive animals, protraction/retraction cycles of the radula were expressed sporadically with variable interbite intervals. In contrast, conditioned animals expressed movement cycles with an elevated frequency and a stereotyped rhythmic organization. Both the rate increase and regularization, which were retained for several hours after training, depended on the contingency because they were not observed in animals subjected to nonassociative training.[56]

TABLE 15.1 Learning-Induced Changes in Synapses and Electrical Coupling among Cells

Cell	Operant Conditioning	Classical Conditioning
B4[1,2]		— CS input
B30[3]	↑ Coupling to B63	
B31[1,2]		↑ CS input
B51[4]		↑ CS-evoked cholinergic EPSP
B63[3]	↑ Coupling to B30 and B65	
B65[3]	↑ Coupling to B63	
CBI-2[2]		↑ CS-evoked EPSP; — CBI-2 to B31 EPSP

[1]*Lechner* et al.[39]
[2]*Mozzachiodi* et al.[55]
[3]*Nargeot* et al.[57]
[4]*Lorenzetti* et al.[28]
CS, conditioned stimulus; ↑, increase; —, no change.

CELLULAR ANALYSIS OF APPETITIVE CLASSICAL AND APPETITIVE OPERANT CONDITIONING OF FEEDING

Neuronal Correlates of Classical and Operant Conditioning

The persistence of the changes in feeding behavior, produced by both classical and operant conditioning, facilitated the identification of the neurophysiological substrates of these two forms of associative learning. For classical conditioning, the cerebral and buccal ganglia were removed within 6 hr after training and electrical stimulation of nerve AT4 was used to activate the CS pathway.[27,28,39] AT4 conveys mechanosensory information from the lip region, which is brushed to deliver the CS during *in vivo* classical conditioning.[27,39,55] A greater number of CS-evoked iBMPs were measured in the paired group compared to unpaired animals.[39] This modification was not accompanied by any change in the rate of spontaneous BMPs and was therefore consistent with the pairing-specific increase in CS-evoked bites observed *in vivo*.[39] At the cellular level, classical conditioning produced an increase in the CS-evoked excitatory synaptic drive to pattern-initiating neuron B31/32 (Figure 15.3A), without any change in its membrane properties (i.e., resting membrane potential, input resistance, and firing threshold; Tables 15.1 and 15.2), suggesting that modification at the level of neurons presynaptic to B31/32 might be responsible for the enhanced synaptic input to B31/32.[39] In addition to the pairing-specific strengthening of the CS-evoked excitatory drive to B31/32, a change in the burst threshold of decision-making neuron B51 was measured following *in vivo* training.[28] This change is discussed in detail later.

For operant conditioning, the contingent-dependent increase in bites was associated with the regularization of the bursting activity of a cluster of pattern-initiating neurons, consisting of B63, B30, and B65.[57] This synchronization of the pattern-initiating neurons appears to be produced by two distinct contingent-dependent mechanisms: (1) decreased burst threshold in B63, B30, and B65, and (2) enhanced electrical coupling between pairs of pattern-initiating neurons (Figure 15.3B; Tables 15.1 and 15.2).[57] In addition to the changes in the pattern-initiating neurons, operant conditioning is associated with modifications of the input resistance and burst threshold of B51.[25] These changes are described in detail later.

Classical and Operant Conditioning of Feeding can be Expressed *in Vitro*

The ability of the isolated ganglia to express BMPs analogous to those recorded *in vivo*[29,32,33] enabled the development of *in vitro* analogs of classical and

TABLE 15.2 Learning-Induced Changes in the Intrinsic Properties of Cells

Cell	Operant Conditioning	Classical Conditioning
B4[1,2]		— R_{in}, — Thresh., — RMP
B30[3]	↑ R_{in}, ↓ Thresh., — RMP	
B31[1,2]		— R_{in}, — Thresh., — RMP
B51[2,4–10]	**↑ R_{in}, ↓ Thresh.**, — RMP	**— R_{in}, ↑ Thresh.**, — RMP, ↑ ACh responses
B63[3]	↑ R_{in}, ↓ Thresh., — RMP	
B65[3]	↑ R_{in}, ↓ Thresh., — RMP	
CBI-2[2]		↑ CS-evoked firing, — R_{in}, — Thresh., — RMP

[1]*Lechner* et al.[39]
[2]*Mozzachiodi* et al.[55]
[3]*Nargeot* et al.[57]
[4]*Brembs* et al.[25]
[5]*Lorenzetti* et al.[28]
[6]*Lorenzetti* et al.[58]
[7]*Lorenzetti* et al.[59]
[8]*Mozzachiodi* et al.[40]
[9]*Nargeot* et al.[44]
[10]*Nargeot* et al.[45]
CS, conditioned stimulus; R_{in}, input resistance; RMP, resting membrane potential; Thresh., threshold; ↑, increase; ↓, decrease; —, no change; boldface type indicates sites that have been examined following both operant and classical conditioning.

operant conditioning. In the *in vitro* analog of classical conditioning, electrical stimulation of AT4 was used as a CS.[27,39,55] As in the protocols of *in vivo* and *in vitro* operant conditioning, electrical stimulation of En2 served as reinforcement.[55,60,61] The testing procedures, the number of training trials, and the intertrial interval (ITI) of the *in vitro* classical conditioning training protocol were identical to those used during *in vivo* training.[27] Preparations that received paired training exhibited a significantly greater number of CS-evoked iBMPs than did the unpaired control group.[28,55] Similar to *in vivo* training, the *in vitro* analog of classical conditioning expresses the increase in the CS-evoked excitatory synaptic drive to B31/32 (Figure 15.3A; Table 15.1).[55] The use of this analog also revealed another cellular change occurring upstream of B31/32 at the level of the cerebral ganglion: the pairing-specific increase of the CS-evoked synaptic drive to command-like interneuron CBI-2 (Tables 15.1 and 15.2).[55] CBI-2 activates the feeding CPG via monosynaptic connections with CPG elements, including B31/32 and B63 (Figure 15.3A).[30,34,35,42,62]

An *in vitro* analog of operant conditioning was also developed.[32] The occurrence of iBMPs (i.e., the *in vitro* operant) was contingently reinforced by brief electrical stimulation of En2 during the training period. A yoked control group was used, in which En2 stimulations were presented in a noncontingent manner.[32] Contingently reinforced preparations produced significantly more iBMPs than did the yoked control group.[32,60] This increase persisted for at least 24 hr.[40]

These *in vitro* analogs of conditioning are validated, in part, by similarities between *in vivo* and *in vitro* learning-induced changes in feeding and fictive feeding, as well as similarities between cellular loci of plasticity. Overall, the previous findings indicate that both *in vitro* analogs led to an in increase in the expression of iBMPs, which is consistent with the increase in bites observed following both *in vivo* classical and operant conditioning.[25,27,39] Moreover, they indicate that both forms of associative learning use the same reinforcement pathway. Histofluorescence analysis indicated that processes within En2 contain dopamine (DA),[43] suggesting that DA may mediate the reinforcement signals in classical and operant conditioning. To test this hypothesis, the effect of the DA antagonist methylergonovine[63,64] was examined in the *in vitro* analogs. Methylergonovine, applied at a concentration that did not impair the activity of the feeding CPG[61,63] and the CS pathway,[61] prevented the acquisition of both classical and operant conditioning *in vitro*.[61,63]

Decision-Making Neuron B51

Among the different types of neurons in the feeding CPG, one neuron (B51) has characteristics that reflect the all-or-nothing nature of bites.[40,45,65] B51 expresses an intrinsically generated all-or-nothing prolonged burst of action potentials known as plateau potentials.[25,40,45,58,65] The occurrence of plateau potentials in B51 is highly correlated with the expression of iBMPs (Figure 15.1C).[45] In contrast, during rBMPs, B51 only exhibits a slow depolarization, which does not reach burst threshold (Figure 15.1A).[45] Although B51 does not initiate BMPs (a function attributed to other cells, including pattern-initiating neurons such as B31/32 and B63[22,23,57]), it appears to contribute importantly to the neural output of the biting circuit. Indeed, when B51 is artificially depolarized to elicit a burst of spikes during the retraction phase of a BMP, the probability of eliciting an iBMP is enhanced,[40,45] whereas when B51 is artificially hyperpolarized during the retraction phase of a BMP, the probability of eliciting an iBMP is reduced.[45] Therefore, neuron B51 appears to operate as if it 'decides' whether or not the feeding CPG generates a bite.

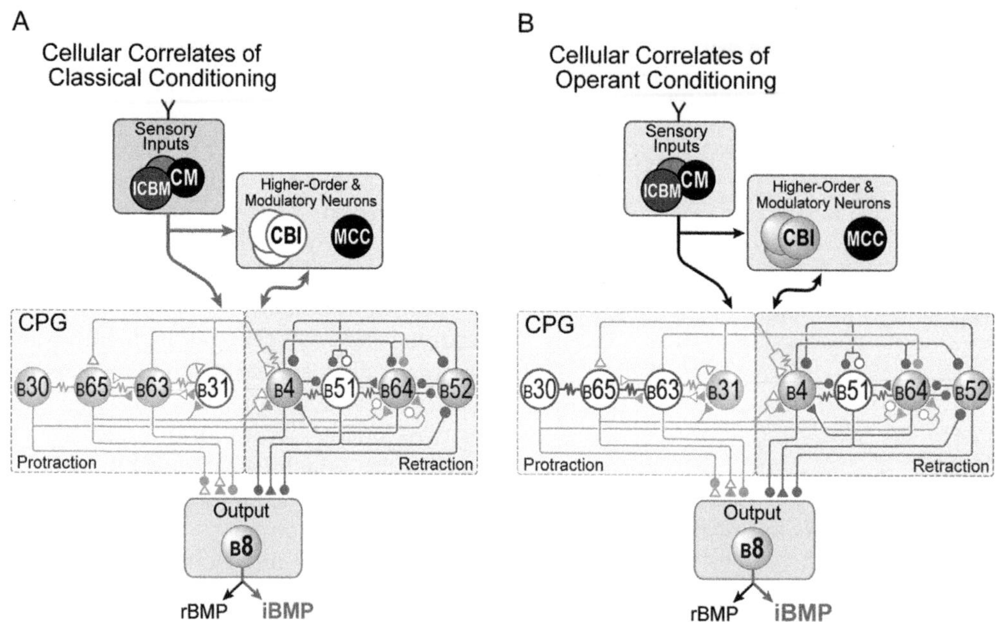

FIGURE 15.3 **Neuronal correlates of classical and operant conditioning.** Simplified schematic of the feeding neural circuit. Sensory neurons in the cerebral ganglia (mechanoafferents ICBM and CM) convey information to higher-order cells in the feeding circuit, such as the command-like neurons CBIs, and modulatory cells, such as the metacerebral cell (MCC), also located in the cerebral ganglia. Sensory information is also conveyed directly to the CPG in the buccal ganglia. In addition, the CPG receives inputs from the command-like cells and modulatory cells. Cells in the CBG can be classified, in part, by their activity during a BMP. Some cells are active during the protraction phase (yellow shading), whereas others are active during the retraction phase (blue shading). Activity in radula closure motor neuron B8 occurs during the protraction phase in rBMPs and during the retraction phase in iBMPs. Identified loci of plasticity are indicated by red and white shading. (A) Following classical conditioning, the CS-evoked excitatory inputs to CBI-2 and to B31 are strengthened, the CS-evoked firing in CBI-2 is enhanced, and the threshold of B51 is increased. (B) Following operant conditioning, the electrical coupling among B30, B63, and B65 is strengthened, and the excitability of B51, B30, B63, and B65 is increased. Several additional sites have been examined and not found to be modified by learning (not shown).

B51 Neuronal Plasticity in Operant and Classical Conditioning: Common Site but Different Changes

The parallel development of *in vivo* and *in vitro* paradigms for operant and classical conditioning, which both similarly modify the same behavior/fictive behavior and share the same reinforcement pathway, offered the unprecedented opportunity to compare and contrast the means by which the activity of a behaviorally relevant neuron is altered by two distinct forms of associative learning. Because of its decision-making features, attention focused on B51.

In vivo and *in vitro* contingent reinforcement produced identical changes in B51 membrane properties. In particular, a greater input resistance and a lower burst threshold were measured in the contingent group compared to the yoke group (Table 15.2).[25,40,44] These changes are indicative of an increase in B51 intrinsic excitability and are consistent with the increased number of BMPs that were associated with a plateau potential in B51 produced by contingent reinforcement *in vitro*[40,44,45] and, in turn, with the increased number of bites observed following *in vivo* training.[25] Because of the all-or-nothing nature of B51 plateau potential,[44,45,65] the increased excitability can translate into an increased likelihood for a plateau potential to occur and, consequently, into a greater bias for the feeding CPG to generate iBMPs (Figure 15.3B). Nargeot *et al.*[45] demonstrated that increased excitability of B51 could be induced by contingently reinforcing the occurrence of artificially induced plateau potentials as a proxy for an endogenous plateau potential generated during an iBMPs. These results indicated that (1) B51 plateau potential can be considered a cellular representation of the operant and (2) B51 may function as a coincidence detector for the operant and the reinforcement. Interestingly, iontophoretic injection of cAMP into B51 from naive ganglia produced two effects 24 hr after treatment: (1) changes in input resistance and burst threshold analogous to those induced by contingent reinforcement[40] and (2) a reconfiguration of the feeding CPG with increased expression of plateau potentials.[40] These findings provide compelling evidence that altering intrinsic excitability in an individual neuron might *per se* sustain robust neurophysiological modifications of a neural circuit and, ultimately, changes in behavior.[66]

Both *in vivo* and *in vitro* classical conditioning altered the strength of the CS-evoked synaptic drive to B51. In particular, the excitatory synaptic input was enhanced following training, whereas the inhibitory synaptic input was not altered (Table 15.1).[28] The enhanced excitatory synaptic drive to B51 is consistent with the CS-evoked increases observed in both the number of plateau potentials and the number of iBMPs following paired training (Figure 15.3A).[28] The effects of classical conditioning on B51 membrane properties were also examined. Unexpectedly, classical conditioning decreased B51 excitability through an increase in B51 burst threshold (Table 15.2). This finding is in contrast to the increase in excitability that has been previously reported following classical conditioning in several invertebrate and vertebrate model systems[66] and would tend to make B51 less likely to reach the firing threshold and elicit an iBMP. However, because both B51 plateau potentials and iBMPs were increased in response to the CS, it appears that the factors that enhance the recruitment of B51 overpower the diminished excitability and bias B51 toward producing more plateau potentials, resulting in a greater number of iBMPs and, in turn, an increased number of bites following classical conditioning (Figure 15.3A).[28] What, then, is the role of decreasing the excitability of B51 following appetitive classical conditioning? One hypothesis is that when classical conditioning occurs, the animal learns not only when to bite but also when not to bite. Specifically, the decreased excitability would bias B51 toward being less likely to be active and thus decrease the probability of iBMPs when the CS is not present. This pairing-specific decrease in the excitability of B51 might be an adaptive mechanism to help shape the CS specificity produced by classical conditioning. The combination of the strengthening of the CS-evoked drive to CBI-2 and B31/32 and the enhanced CS-evoked excitatory input to B51 might explain, at least in part, the increase in bites observed following *in vivo* classical conditioning. Strengthening the CS pathway to CBI-2 and B32/31 would likely increase the number of CS-evoked BMPs. In addition, the increase in the recruitment of B51 would, in turn, bias the CPG toward the expression of iBMPs. Therefore, the effects produced by classical conditioning, which are distributed among elements of the feeding CPG, both induce an increase in the number of BMPs and, through the increased recruitment of B51, bias the nature of the BMPs toward ingestion.

In summary, both operant and classical conditioning modify the threshold for eliciting plateau potentials in neuron B51, but in opposite directions, revealing a fundamental difference in the cellular mechanisms underlying these two main forms of associative learning (Tables 15.1 and 15.2).

COMPARISON OF THE MOLECULAR MECHANISMS OF OPERANT AND CLASSICAL CONDITIONING

Although the *in vitro* analogs helped to characterize the circuit and cellular mechanisms of classical and operant conditioning, a further reduced level of analysis was required to explore the fundamental molecular and biophysical underpinnings of associative learning. For example, because cells such as B51 are embedded in the feeding CPG, it is difficult to examine the biophysical, biochemical, and molecular mechanisms occurring at the single-cell level. Consequently, cellular analogs of operant and classical conditioning, consisting of single isolated B51 neurons in culture, were developed and validated.[25,40,58,59]

In the single-cell analog of operant conditioning, a plateau potential in B51 served as the cellular representation of the operant, and a brief iontophoretic application of DA onto the neuron served as the representation of reinforcement.[25] During a 10-min training period, seven plateau potentials were induced in B51 by intracellular current injection. Each plateau potential was followed immediately by a 'puff' of DA in the contingent training, whereas the puff was delayed by 40 sec during noncontingent training. Contingent reinforcement produced an increase in the input resistance and a decrease in the burst threshold, compared to noncontingent training.[25] Notably, these changes were similar to those produced by *in vivo* and *in vitro* operant conditioning, and the changes were maintained for up to 24 hr.[40,58]

This single-cell analog was used to dissect the molecular mechanisms of operant conditioning. During a plateau potential (i.e., the cellular representation of the operant), Ca^{2+} entry through voltage-gated channels activates a Ca^{2+}-dependent isoform of protein kinase C (PKC), apl-I, which translocates from the cytoplasm to the plasma membrane, where it becomes active (Figure 15.4A).[58] DA (i.e., the cellular representation of the reinforcement) binds to a D1-like receptor expressed primarily at the B51 axon hillock (Figure 15.4A).[58] The D1-like receptor is presumably coupled to an *Aplysia* isoform of type II adenylyl cyclase, AC*apl*B,[67] which displays a high degree of PKC sensitivity.[68–70] These findings indicate that the cellular representations of the operant and the reinforcement activate distinct signaling cascades in B51 (Figure 15.4A). Importantly, both PKA and PKC were found to be necessary for the single-cell analog of operant conditioning.[58] The PKA pathway appears to function downstream of the PKC pathway, as demonstrated by the observation that injection of cAMP into B51 fully mimics the effects

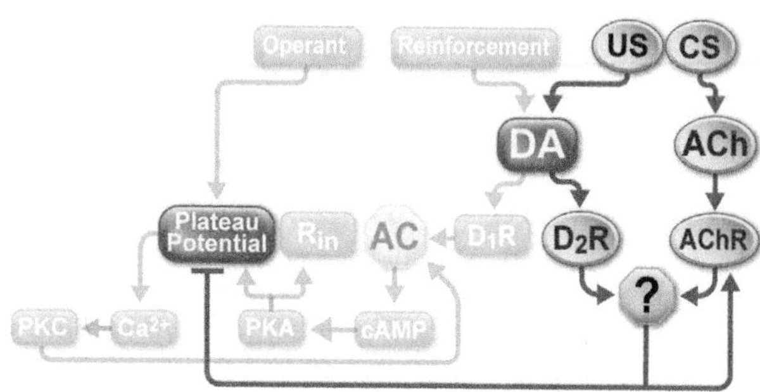

FIGURE 15.4 Models of molecular networks mediating operant and classical conditioning in B51. (A) Operant conditioning. The operant is represented by a plateau potential, which leads to an influx of Ca^{2+} and activation of PKC. PKC then phosphorylates adenylyl cyclase (AC) and primes AC for enhanced synthesis of cAMP. The reinforcement is mediated by dopamine (DA), which binds to D_1R and likely acts through AC to increase the production of cAMP. If a plateau potential immediately precedes the reinforcement and AC is phosphorylated by PKC, then the production of cAMP is greater than what would occur after either behavior alone or DA alone. After a sufficient number of contingent reinforcements, the increased levels of cAMP activate PKA sufficiently to increase the excitability of B51. (B) Classical conditioning. The conditioned stimulus (CS) is mediated by ACh, and the unconditioned stimulus (US) is mediated by DA, which presumably binds to a D_2-like receptor. The site of convergence between the US and the CS is not known. Similarly, the effector molecules downstream from the CS and the US are unknown. Nevertheless, when pairing of the CS and the US occurs repeatedly, a decrease in excitability is produced as well as an enhanced response to ACh. Red components of the model are associated with operant conditioning, whereas green components are associated with classical conditioning. Yellow components represent sites of convergence between either the operant and reinforcement or the CS and the US. Blue components represent sites that overlap between operant and classical conditioning. Arrows represent an enhancement or positive interaction, whereas T bars represent an inhibition or negative interaction.

of contingent reinforcement in a PKC-independent manner.[58] Conversely, although PKC activation by phorbol diacetate could also mimic the changes produced by contingent reinforcement, its effect was PKA dependent.[58] These results support a model of operant conditioning in which adenylyl cyclase serves as coincident detector for B51 plateau potential and reinforcement-mediated release of DA onto B51 (Figure 15.4A). During contingent reinforcement, the Ca^{2+}-dependent PKC activates the D1-dependent adenylyl cyclase, switched on by DA, in a manner stronger than DA alone (Figure 15.4A). The subsequent increase in cytosolic cAMP activates the downstream PKA, which in turn leads to the observed changes in excitability, presumably through phosphorylation of specific membrane channels. Interestingly, the single-cell analog of operant conditioning also increased the levels of phospho-CREB1 in B51, which may represent the pathway for the induction of the long-term changes in B51 excitability produced by contingent reinforcement.[40,58]

A single-cell analog of classical conditioning has also been developed.[59] Iontophoretic applications of DA onto B51 were used as cellular representation of the US.[59] For the CS pathway, the neurotransmitter that mediates the CS-evoked excitatory synaptic input to B51 was characterized. AT4 electrical stimulation produces a biphasic synaptic input to B51, consisting of an initial inhibitory synaptic potential (IPSP), followed by a delayed excitatory synaptic potential (EPSP).[28] Classical conditioning selectively enhances the EPSP, without altering the IPSP, following both *in vivo* and *in vitro* paired training.[28] The CS-mediated EPSP to B51 appears to be cholinergic because it is blocked by the acetylcholine (ACh) receptor antagonist hexamethonium (Figure 15.4B).[59] In addition, iontophoretic applications of ACh to B51 produce a depolarization, similar to the EPSP evoked by AT4 stimulation.[59] Therefore, a single-cell analog of classical conditioning was designed by presenting 10 paired iontophoretic applications of ACh and DA to B51, with a 2-min ITI.[59] This single-cell analog of classical conditioning produced two pairing-specific changes in B51 compared to the unpaired protocol in which the CS and US were spaced 1 min apart: (1) an increase in the ACh-evoked depolarization and (2) an increase in the burst threshold, without effects on the input resistance.[59] Notably, both changes persisted for at least 24 hr.[59] DA successfully served as cellular representation of the US in the analog of classical conditioning, as it did for the reinforcement signal for the single-cell analog of operant conditioning. However, whereas the D1 receptor agonist chloro-APB could substitute for the reinforcement in the single-cell analog of operant conditioning, it failed to induce pairing-specific changes in the ACh-evoked depolarization or the burst threshold in the single-cell analog of classical conditioning. These results suggest that the US is mediated by either a D2-like receptor (D2–D4) or a combination of D1-like (D1 and D5) and D2-like receptors (Figure 15.4B).[59] These findings support a model in which the coincidence detector for classical conditioning involves a temporal contingency between the release and the action of ACh and DA in B51 (Figure 15.4B).

CONCLUSIONS

Feeding behavior in *Aplysia* has proven to be an excellent model system for comparing and contrasting the cellular and molecular mechanisms of associative learning. This chapter focused on the changes produced by appetitive operant and classical conditioning on B51 activity. However, modifications have been identified in other neurons of the feeding neural circuit following operant (B30, B63, and B65)[56,57] and classical conditioning (CBI-2 and B31/32).[39,55] Synergism among different neuronal elements modified by conditioning has emerged as a general principle for the associative storage of information in both vertebrate[7,71] and invertebrate animals.[72–74]

Because of the distributed cellular substrates underlying associative learning, it is important to determine the contribution of each locus of plasticity to the expression of the learned behavioral changes. The presence of multiple sites underlying associative plasticity also raises the issue regarding to what extent the cellular mechanisms for plasticity at each site are conserved. For example, is the convergence of Ca^{2+} entry and activation of DA receptors observed in B51 following operant conditioning also used to bring about the contingent-dependent changes in B63? Continued analysis of the cellular and molecular mechanisms underlying appetitive classical and appetitive operant conditioning of feeding in *Aplysia* will provide important insights into the similarities and differences between these two main forms of associative learning.

Acknowledgments

This work was supported by National Institutes of Health grant MH58321 to J.H.B. and National Science Foundation grant IOS1120304 to R.M. We thank Curtis Neveu for providing the recordings illustrated in Figure 15.1.

References

1. Pavlov I. *Conditioned Reflexes: An Investigation of the Physiological Activity of the Cerebral Cortex.* London: Oxford University Press; 1927.
2. Skinner B. *The Behavior of Organisms: An Experimental Analysis.* New York: Appleton-Century-Crofts; 1938.
3. Thorndike E. *Animal Intelligence: Experimental Studies.* New York: Macmillian; 1911.
4. Brembs B. Operant conditioning in invertebrates. *Curr Opin Neurobiol.* 2003;13:710–717.
5. Carew TJ, Sahley CL. Invertebrate learning and memory: from behavior to molecules. *Annu Rev Neurosci.* 1986;9:435–487.
6. Kelley AE. Memory and addiction; shared neural circuitry and molecular mechanisms. *Neuron.* 2004;44:161–179.
7. Kim JJ, Thompson RF. Cerebellar circuits and synaptic mechanisms involved in classical eyeblink conditioning. *Trends Neurosci.* 1997;20:177–181.
8. Menzel R, Müller U. Learning and memory in honeybees: from behavior to neural substrates. *Annu Rev Neurosci.* 1996;19:379–404.
9. Michel M, Green CL, Gardner JS, Organ CL, Lyons LC. Massed training-induced intermediate-term operant memory in *Aplysia* requires protein synthesis and multiple persistent kinase cascades. *J Neurosci.* 2012;32:4581–4591.
10. Saar D, Barkai E. Long-term modifications in intrinsic neuronal properties and rule learning in rats. *Eur J Neurosci.* 2003;17: 2727–2734.
11. Schafe GE, Nader K, Blair HT, LeDoux JE. Memory consolidation of Pavlovian fear conditioning: a cellular and molecular perspective. *Trends Neurosci.* 2001;24:540–546.
12. Schultz W. Getting formal with dopamine and reward. *Neuron.* 2002;36:241–263.
13. Dayan P, Balleine BW. Reward, motivation, and reinforcement learning. *Neuron.* 2002;36:285–298.
14. Gormezano I, Tait RW. The Pavlovian analysis of instrumental conditioning. *Pavlov J Biol Sci.* 1976;11:37–55.
15. Rescorla RA, Solomon RL. Two-process learning theory: relationships between pavlovian conditioning and instrumental learning. *Psychol Rev.* 1967;74:151–182.
16. Abramson CI, Feinman RD. Operant punishment of eye elevation in the green crab, *Carcinus maenas. Behav Neural Biol.* 1987;48:259–277.
17. Abramson CI, Feinman RD. Classical conditioning of the eye withdrawal reflex in the green crab. *J Neurosci.* 1988;8: 2907–2912.
18. Gong Z, Xia S, Liu L, Feng C, Guo A. Operant visual learning and memory in *Drosophila* mutants *dunce, amnesiac* and *radish. J Insect Physiol.* 1998;44:1149–1158.
19. Putz G, Bertolucci F, Raabe T, Zars T, Heisenberg M. The *S6KII* (rsk) gene of *Drosophila melanogaster* differentially affects an operant and a classical learning task. *J Neurosci.* 2004;24: 9745–9751.
20. Kupfermann I. Feeding in *Aplysia*: a simple system for the study of motivation. *Behav Biol.* 1974;10:1–26.
21. Baxter DA, Byrne JH. Feeding behavior of *Aplysia*: a model system for comparing cellular mechanisms of classical and operant conditioning. *Learn Mem.* 2006;13:669–680.
22. Cropper EC, Evans CG, Hurwitz I, et al. Feeding neural networks in the mollusc *Aplysia. Neurosignals.* 2004;13:70–86.
23. Elliott C, Susswein A. Comparative neuroethology of feeding control in molluscs. *J Exp Biol.* 2002;205:877–896.
24. Nargeot R, Simmers J. Neural mechanisms of operant conditioning and learning-induced behavioral plasticity in *Aplysia. Cell Mol Life Sci.* 2011;68:803–816.
25. Brembs B, Lorenzetti F, Reyes F, Baxter DA, Byrne JH. Operant reward learning in *Aplysia*: neuronal correlates and mechanisms. *Science.* 2002;296:1706–1709.
26. Colwill R, Goodrum K, Martin A. Pavlovian appetitive discriminative conditioning in *Aplysia californica. Anim Learn Behav.* 1997;25:268–276.
27. Lechner HA, Baxter DA, Byrne JH. Classical conditioning of feeding in *Aplysia*: I. Behavioral analysis. *J Neurosci.* 2000;20: 3369–3376.
28. Lorenzetti FD, Mozzachiodi R, Baxter DA, Byrne JH. Classical and operant conditioning differentially modify the intrinsic

properties of an identified neuron. *Nat Neurosci.* 2006; 9:17–19.

29. Morton D, Chiel H. *In vivo* buccal nerve activity that distinguishes ingestion from rejection can be used to predict behavior transitions in *Aplysia. J Comp Physiol A.* 1993;172:17–32.
30. Church P, Lloyd P. Activity of multiple identified motor neurons recorded intracellularly during evoked feeding like motor programs in *Aplysia. J Neurophysiol.* 1994;72:1794–1809.
31. Susswein AJ, Byrne JH. Identification and characterization of neurons initiating patterned neural activity in the buccal ganglia of *Aplysia. J Neurosci.* 1988;8:2049–2061.
32. Nargeot R, Baxter DA, Byrne JH. Contingent-dependent enhancement of rhythmic motor patterns: an *in vitro* analog of operant conditioning. *J Neurosci.* 1997;17:8093–8105.
33. Morton D, Chiel H. The timing of activity in motor neurons that produce radula movements distinguishes ingestion from rejection in *Aplysia. J Comp Physiol A.* 1993;173:519–536.
34. Jing J, Weiss KR. Neural mechanisms of motor program switching in *Aplysia. J Neurosci.* 2001;15:7349–7362.
35. Jing J, Weiss KR. Interneuronal basis of the generations of related but distinct motor programs in *Aplysia*: implications for current neuronal models of vertebrate intralimb coordination. *J Neurosci.* 2002;22:6228–6238.
36. Jing J, Weiss KR. Generation of variants of a motor act in a modular and hierarchical motor network. *Curr Biol.* 2005;15: 1712–1721.
37. Jing J, Cropper EC, Hurwitz I, Weiss KR. The construction of movement with behavior-specific and behavior-independent modules. *J Neurosci.* 2004;24:6315–6325.
38. Jing J, Vilim FS, Cropper EC, Weiss KR. Neural analog of arousal: persistent conditional activation of a feeding modulator by serotonergic initiators of locomotion. *J Neurosci.* 2008;28: 12349–12361.
39. Lechner HA, Baxter DA, Byrne JH. Classical conditioning of feeding in *Aplysia*: II. Neurophysiological correlates. *J Neurosci.* 2000;20:3377–3386.
40. Mozzachiodi R, Lorenzetti FD, Baxter DA, Byrne JH. Changes in neuronal excitability serve as mechanism of long-term memory for operant conditioning. *Nat Neurosci.* 2008;10:1146–1148.
41. Hurwitz I, Kupfermann I, Susswein AJ. Different roles of neurons B63 and B34 that are active during the protraction phase of buccal motor programs in *Aplysia californica. J Neurophysiol.* 1997;78:1305–1319.
42. Hurwitz I, Kupfermann I, Weiss KR. Fast synaptic connections from CBIs to pattern-generating neurons in *Aplysia*: initiation and modification of motor programs. *J Neurophysiol.* 2003;89: 2120–2136.
43. Kabotyanski E, Baxter DA, Byrne JH. Identification and characterization of catecholaminergic neuron B65, which initiates and modifies patterned activity in the buccal ganglia of *Aplysia. J Neurophysiol.* 1998;79:605–621.
44. Nargeot R, Baxter DA, Byrne JH. *In vitro* analog of operant conditioning in *Aplysia*: I. Contingent reinforcement modifies the functional dynamics of an identified neuron. *J Neurosci.* 1999;15:2247–2260.
45. Nargeot R, Baxter DA, Byrne JH. *In vitro* analog of operant conditioning in *Aplysia*: II. Modifications of the functional dynamics of an identified neuron contributes to motor pattern selection. *J Neurosci.* 1999;19:2261–2272.
46. Susswein A, Schwartz M. A learned change of response to inedible food in *Aplysia. Behav Neural Biol.* 1983;39:1–6.
47. Susswein A, Schwartz M, Feldman E. Learned changes of feeding behavior in *Aplysia* in response to edible and inedible foods. *J Neurosci.* 1986;6:1513–1527.
48. Botzer D, Markovich S, Susswein A. Multiple memory processes following training that a food is inedible in *Aplysia. Learn Mem.* 1998;5:204–219.
49. Schwarz M, Feldman E, Susswein AJ. Variables affecting long-term memory of learning that a food is inedible in *Aplysia. Behav Neurosci.* 1991;105:193–201.
50. Katzoff A, Ben-Gedalya T, Susswein AJ. Nitric oxide is necessary for multiple memory processes after learning that food is inedible in *Aplysia. J Neurosci.* 2002;22:9581–9594.
51. Katzoff A, Miller N, Susswein AJ. Nitric oxide and histamine signal attempts to swallow: a component of learning that food is inedible in *Aplysia. Learn Mem.* 2010;17:50–62.
52. Cohen-Armon M, Visochek L, Katzoff A, et al. Long-term memory requires polyADP-ribosylation. *Science.* 2004;304: 1820–1822.
53. Levitan D, Lyons LC, Perelman A, et al. Training with inedible food in *Aplysia* causes expression of C/EBP in the buccal but not cerebral ganglion. *Learn Mem.* 2008;15:412–416.
54. Michel M, Green CL, Lyons LC. PKA and PKC are required for long-term but not short-term *in vivo* operant memory in *Aplysia. Learn Mem.* 2011;18:19–23.
55. Mozzachiodi R, Lechner HA, Baxter DA, Byrne JH. *In vitro* analog of classical conditioning of feed behavior in *Aplysia. Learn Mem.* 2003;10:478–494.
56. Nargeot R, Petrissans C, Simmers J. Behavioral and *in vitro* correlates of compulsive-like food seeking induced by operant conditioning in *Aplysia. J Neurosci.* 2007;27:8059–8070.
57. Nargeot R, Le Bon-Jego M, Simmers J. Cellular and network mechanisms of operant learning-induced compulsive behavior in *Aplysia. Curr Biol.* 2009;19:975–998.
58. Lorenzetti FD, Baxter DA, Byrne JH. Molecular mechanisms underlying a cellular analog of operant reward learning. *Neuron.* 2008;59:815–828.
59. Lorenzetti FD, Baxter DA, Byrne JH. Classical conditioning analog enhanced acetylcholine responses but reduced excitability of an identified neuron. *J Neurosci.* 2011;31: 14789–14793.
60. Brembs B, Baxter DA, Byrne JH. Extending *in vitro* conditioning in *Aplysia* to analyze operant and classical processes in the same preparation. *Learn Mem.* 2004;11:412–420.
61. Reyes FD, Mozzachiodi R, Baxter DA, Byrne JH. Reinforcement in an *in vitro* analogue of appetitive classical conditioning of feeding behavior in *Aplysia:* blockade by a dopamine antagonist. *Learn Mem.* 2005;12:216–220.
62. Rosen S, Teyke T, Miller M, Weiss KR, Kupfermann I. Identification and characterization of cerebral-to-buccal interneurons implicated in the control of motor programs associated with feeding in *Aplysia. J Neurosci.* 1991;11:3630–3655.
63. Nargeot R, Baxter DA, Patterson GW, Byrne JH. Dopaminergic synapses mediate neuronal changes in an analogue of operant conditioning. *J Neurophysiol.* 1999;81:1983–1987.
64. Teyke T, Rosen SC, Weiss KR, Kupfermann I. Dopaminergic neuron B20 generates rhythmic neuronal activity in the feeding motor circuitry of *Aplysia. Brain Res.* 1993;630:226–237.
65. Plummer M, Kirk M. Premotor neurons B51 and B52 in the buccal ganglia of *Aplysia californica. J Neurophysiol.* 1990;63: 539–557.
66. Mozzachiodi R, Byrne JH. More than synaptic plasticity: role of nonsynaptic plasticity in learning and memory. *Trends Neurosci.* 2010;33:17–26.
67. Lin AH, Cohen JE, Wan Q, et al. Serotonin stimulation of cAMP-dependent plasticity in *Aplysia* sensory neurons is mediated by calmodulin-sensitive adenylyl cyclase. *Proc Natl Acad Sci USA.* 2010;107:15607–15612.

68. Bol GF, Hulster A, Pfeuffer T. Adenylyl cyclase type II is stimulated by PKC via C-terminal phosphorylation. *Biochim Biophys Acta*. 1997;1358:307–313.
69. Jacobowitz O, Chen J, Premont RT, Iyengar R. Stimulation of specific types of Gs-stimulated adenylyl cyclases by phorbol ester treatment. *J Biol Chem*. 1993;268:3829–3832.
70. Zimmermann G, Taussig R. Protein kinase C alters the responsiveness of adenylyl cyclases to G protein alpha and beta gamma subunits. *J Biol Chem*. 1996;271:27161–27166.
71. Raymond JL, Lisberger SG, Mauk MD. The cerebellum: a neuronal learning machine? *Science*. 1996;272:1126–1131.
72. Jones NG, Kemenes I, Kemenes G, Benjamin PR. A persistent cellular change in a single modulatory neuron contributes to associative long-term memory. *Curr Biol*. 2003;17: 1064–1069.
73. Nikitin ES, Vavoulis DV, Kemenes I, et al. Persistent sodium current is a nonsynaptic substrate for long-term associative memory. *Curr Biol*. 2008;18:1221–1226.
74. Staras K, Kemenes G, Benjamin PR. Cellular traces of behavioral classical conditioning can be recorded at several specific sites in a simple nervous system. *J Neurosci*. 1999;19: 347–357.

CHAPTER

16

Mechanisms of Short-Term and Intermediate-Term Memory in *Aplysia*

Robert D. Hawkins†, Igor Antonov* and Iksung Jin**

*Columbia University, New York, New York †New York State Psychiatric Institute, New York, New York

INTRODUCTION

Memory and synaptic plasticity in many systems, including *Aplysia* and hippocampus, have different stages, which are defined primarily by their time courses but also involve different training protocols and different cellular and molecular mechanisms. Thus, relatively weak stimulation produces short-term plasticity, which lasts minutes and involves covalent modifications in either the presynaptic (*Aplysia*) or the postsynaptic (hippocampus) compartment of existing synapses. By contrast, strong stimulation produces long-term plasticity that lasts hours to days and involves protein and RNA synthesis and synaptic remodeling or the growth of new synapses, which require both presynaptic and postsynaptic changes. In addition, an intermediate-term stage has been identified that lasts tens of minutes to hours and often involves covalent modifications and protein but not RNA synthesis, as well as the recruitment of synaptic proteins but not growth, and thus combines elements of the short-term and long-term stages.

These findings have raised several questions. How are the different stages related? Are they induced in series, or are they induced independently in parallel? How does plasticity spread from one side of the synapse during the short-term stage to both sides during the long-term stage, and when does that spread first occur? In this chapter, we address these questions in the context of a review of short-term and intermediate-term memory and synaptic plasticity in *Aplysia*. Long-term plasticity has been reviewed elsewhere.[1,2]

SIMPLE FORMS OF LEARNING IN *APLYSIA*

The mechanisms of several simple forms of learning have been studied extensively in the marine mollusk *Aplysia*, which has a number of experimental advantages (for references, see Hawkins *et al.*[2]). The nervous system of *Aplysia* consists of approximately 10,000 neurons, many of which are uniquely identifiable individuals. In addition, studies of learning have focused primarily on defensive withdrawal reflexes, which have simple circuits consisting of only tens or perhaps hundreds of neurons. Most studies have examined the gill and siphon withdrawal reflex, in which a light touch to the siphon (an exhalant funnel for the gill) produces contraction of the gill and siphon. Other studies have examined the tail withdrawal reflex or the tail-elicited siphon withdrawal reflex, but the results of all of these studies have generally been similar.

Despite its simplicity, the gill and siphon withdrawal reflex undergoes a variety of different forms of learning. Repeated low-frequency stimulation of the siphon produces a gradual decrease in the amplitude of the withdrawal reflex, or habituation. Following habituation, a noxious stimulus such as a shock to the tail produces an enhancement of subsequent responses to siphon stimulation, or dishabituation. Tail shock also produces enhancement of nonhabituated siphon responses, or sensitization. Furthermore, the duration of the sensitization depends on the strength of the tail shock: A single shock typically produces short-term sensitization lasting minutes, whereas repeated shocks can produce long-term sensitization lasting days.

Invertebrate Learning and Memory.
DOI: http://dx.doi.org/10.1016/B978-0-12-415823-8.00016-2

In addition to these nonassociative forms of learning, the reflex also undergoes two associative forms of learning—classical and operant conditioning. In classical conditioning, stimulation of the siphon just before the tail shock (paired training) produces a larger and longer lasting enhancement of the withdrawal reflex than tail shock alone (sensitization) or explicitly unpaired training with the two stimuli. In operant conditioning, a mild shock to the siphon whenever the gill relaxes produces a subsequent increase in both tonic and spontaneous phasic contractions of the gill compared with yoked controls.[3]

A basic question in psychology has been how these different forms of learning are related. Are dishabituation and sensitization manifestations of the same underlying process, or are they fundamentally different? What about short-term and long-term learning, nonassociative and associative learning, and classical and operant conditioning? Because the withdrawal reflex exhibits all of these forms of learning under similar conditions, it is advantageous for addressing these questions. However, answering them with behavioral methods alone has proven to be difficult. For example, it has been difficult to determine whether dishabituation and sensitization are similar or different behaviorally, in part because the onset of sensitization (but not dishabituation) is opposed by transient inhibition of the reflex following tail shock.[4,5] However, these questions can be addressed more successfully at the cellular and molecular levels. To date, all of the different forms of learning of the withdrawal reflex have been analyzed at these levels except for operant conditioning. Operant conditioning has also been analyzed for another behavior in *Aplysia*,[6,7] but this review is limited to studies of the withdrawal reflex.

SHORT-TERM PLASTICITY

The neural circuit for the reflex consists in part of monosynaptic connections from siphon sensory neurons (SNs) to gill and siphon motor neurons (MNs), as well as polysynaptic connections involving several excitatory and inhibitory interneurons. It is possible to record the activity of these identified neurons and their synaptic connections during learning in a semi-intact preparation of the siphon withdrawal reflex and thus to examine the contributions of plasticity at different sites in the circuit to behavioral learning. Such experiments have shown that several forms of plasticity at the SN–MN synapses (including homosynaptic depression, heterosynaptic facilitation, and activity-dependent facilitation) contribute to habituation, dishabituation, sensitization, and classical conditioning of the reflex, and that plasticity at other sites also contributes.[8,9]

The mechanisms of plasticity at the SN–MN synapses have been examined more extensively in neural analogs of learning in isolated ganglia or in cell culture, in which tail shock is replaced by either nerve shock or application of serotonin (5-HT), an endogenous facilitatory transmitter that is released following tail shock. The initial studies of short-term facilitation at the SN–MN synapses provided some of the earliest evidence that the same type of plasticity can involve different molecular mechanisms depending on the experimental protocol.[10] Facilitation by 1-min application of 5-HT to rested synapses (an analog of sensitization) involves cAMP; protein kinase A (PKA); decreased K^+ current; and increased spike width, Ca^{2+} influx, and transmitter release from the SNs (Figure 16.1, ST). Unlike facilitation by 1-min 5-HT, facilitation by 10-min 5-HT involves PKC, which also produces spike broadening in the SNs. Furthermore, facilitation by 1-min 5-HT at depressed synapses (an analog of dishabituation) involves PKC as well, but in this case it acts by a spike broadening-independent mechanism, perhaps vesicle mobilization (Figure 16.1, DIS). Like mammalian PKC, PKC in *Aplysia* has three isoforms: conventional (Apl I), which depends on diacylglycerol and Ca^{2+}; novel (Apl II), which depends on diacylglycerol but not Ca^{2+}; and atypical (Apl III), which depends on neither.[11,12] Subsequent studies have shown that facilitation at depressed synapses involves presynaptic PKC Apl II,[13,14] and that facilitation at nondepressed synapses also involves CamKII.[15,16]

These results suggest that although dishabituation and sensitization both involve facilitation at the SN–MN synapses, they may involve fundamentally different mechanisms at the molecular level. Similar experiments have also shed light on the relationship between sensitization and classical conditioning. Spike activity in the SN just before the 5-HT application (an analog of classical conditioning) produces activity-dependent enhancement of facilitation, which involves Ca^{2+} priming of the adenylyl cyclase, increased production of cAMP and activation of PKA, etc. (Figure 16.1, CC). This result suggests that conditioning is mechanistically similar to sensitization. In addition, spike activity in the SN paired with depolarization of the MN (another analog of conditioning) produces a fundamentally different type of mechanism, Hebbian potentiation, which involves activation of NMDA-type receptors and postsynaptic Ca^{2+}. Subsequent studies of behavioral conditioning in the semi-intact siphon withdrawal preparation have shown that both of these mechanisms contribute, and that they interact through retrograde signaling.[17]

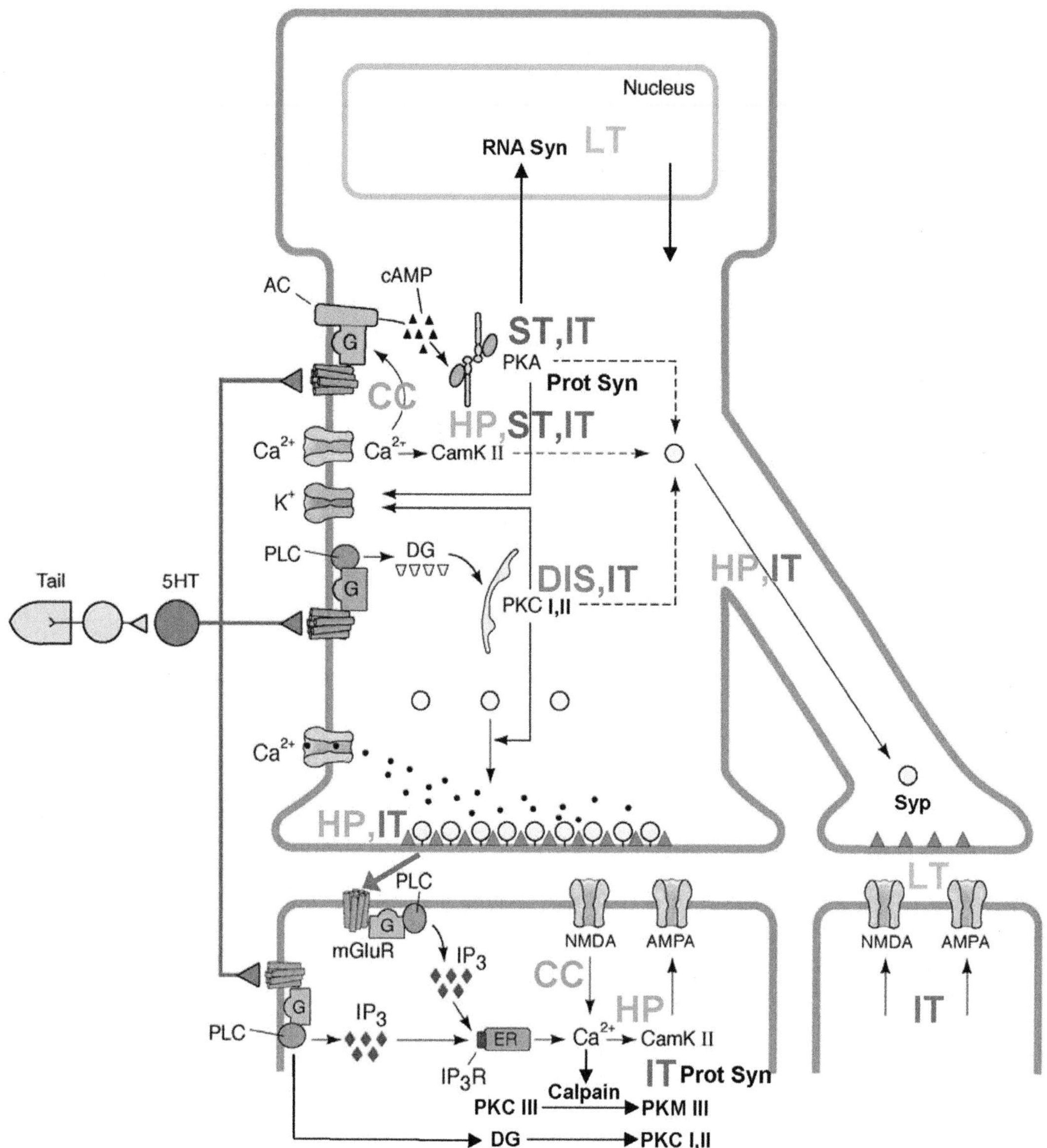

FIGURE 16.1 Cellular and molecular mechanisms of plasticity at sensorimotor neuron synapses that contribute to simple forms of learning in *Aplysia*. Dishabituation (DIS) involves presynaptic PKC. Short-term (ST) sensitization involves presynaptic PKA and CamKII. Intermediate-term (IT) sensitization involves presynaptic PKA and CamKII or PKC, protein synthesis, and spontaneous transmitter release. In addition, it involves postsynaptic mGluRs, CamKII or PKC, protein synthesis, and membrane insertion of AMPA-like receptors, as well as recruitment of pre- and postsynaptic proteins to new synaptic sites. Long-term (LT) sensitization involves gene regulation and growth of new synapses. Classical conditioning (CC) involves priming of presynaptic adenylyl cyclase by Ca^{2+}, and postsynaptic NMDA receptors. Homosynaptic potentiation (HP) involves presynaptic CamKII and recruitment of presynaptic proteins to new sites. It also involves postsynaptic mGluRs, CamKII, and possible insertion of AMPA-like receptors.

THE RELATIONSHIP BETWEEN SHORT- AND LONG-TERM PLASTICITY, AND THE DISCOVERY OF INTERMEDIATE-TERM PLASTICITY

In all of these cases, however, short-term plasticity involves covalent modifications of proteins in existing synapses. By contrast, five tail shocks or five applications of 5-HT separated by 15 min produce long-term (24-hr) facilitation, which involves protein and RNA synthesis and the growth of new synapses and is thus fundamentally different from short-term facilitation (Figure 16.1, LT). What is the relation between these different stages of plasticity? Are they independent and induced in parallel, or does one induce the other in series?

Carew and colleagues first addressed this question in studies of facilitation by 5-HT in isolated ganglia.

They found that a low concentration of 5-HT or application of 5-HT directly to the SN cell body could produce long-term facilitation without short-term facilitation.[18] Furthermore, when they followed the time course of facilitation, short-term facilitation decayed to baseline before the onset of long-term facilitation.[19] These results suggest that short-term and long-term facilitation are independent and induced in parallel. However, similar experiments in culture produced somewhat different results: Short-term facilitation could be induced by a lower concentration of 5-HT than long-term facilitation, and there was no 'dip' in the time course of facilitation between short-term and the onset of long-term facilitation.[20] These results suggested that in some circumstances, short-term and long-term facilitation might be induced at least partly in series.

During the course of these parametric studies of short-term and long-term facilitation, Ghirardi *et al.*[20] obtained evidence for a third stage that they called intermediate-term facilitation, which is typically induced by an intermediate level of 5-HT (four or five pulses of a low concentration), lasts hours, and involves PKA and protein synthesis but not RNA synthesis (Figure 16.1, IT). They also found that a higher concentration of 5-HT induces both intermediate-term and long-term facilitation, and that the mechanisms of the facilitation depend not only on the time after 5-HT but also on the concentration of 5-HT. For example, facilitation 30 min after the 5-HT, which is in the intermediate-term range, can depend on PKA only (with 1 or 10 nM 5-HT); PKA and protein synthesis (with 50 nM 5-HT); or PKA, protein synthesis, and RNA synthesis (with 100 nM or 10 μM 5-HT). These results illustrate that intermediate-term facilitation (like the other stages) is not a unitary entity but, rather, can involve a number of different mechanisms depending on the protocol.

If the stages are not single entities, then asking if they are in parallel or series is not really a meaningful question. However, asking about the relationship between particular mechanisms is more meaningful. For example, these data suggest that PKA, protein synthesis, and RNA synthesis could be in series because they are always recruited in that order with increasing levels of either 5-HT or time.

MECHANISMS OF INDUCTION, MAINTENANCE, AND EXPRESSION OF INTERMEDIATE-TERM FACILITATION

The experiments of Ghirardi *et al.*[20] did not distinguish between mechanisms of induction, maintenance, or expression of the facilitation, nor did they examine whether those mechanisms are pre- or postsynaptic. Carew and colleagues studied mechanisms of induction and maintenance of intermediate-term sensitization and intermediate-term facilitation with a repeated pulses protocol (five spaced tail shocks or five pulses of 5-HT) similar to that used by Ghirardi *et al.*[20] They found that induction requires protein but not RNA synthesis, and maintenance involves persistent activation of PKA but not PKC.[21,22] In addition, MAP kinase is required for induction but not maintenance.[23]

Site-Specific Sensitization

Carew and colleagues also studied mechanisms of an activity-dependent form of plasticity, site-specific sensitization (one tail shock at the same site as testing), and a cellular analog of that protocol (one pulse of 5-HT coincident with spike activity in the SN). Unlike sensitization with the repeated pulses protocol, induction of intermediate-term activity-dependent sensitization or facilitation does not require protein or RNA synthesis but, rather, requires calpain-dependent proteolysis of PKC to form PKM, and maintenance involves persistent activation of PKM rather than PKA.[21,24] Subsequent studies have shown that induction also involves presynaptic MAP kinase and PKC Apl I.[14,25] Thus, with both protocols, induction involves multiple mechanisms including MAP kinase and maintenance involves persistent kinase activity but of different kinases (PKA or PKM) with the different protocols.

10-Min 5-HT

Glanzman and colleagues have been studying intermediate-term facilitation with another protocol, 10-min application of 5-HT, and have focused on postsynaptic mechanisms. Facilitation of SN–MN PSPs in culture by 10-min 5-HT lasts more than 30 min and is blocked by postsynaptic injection of a Ca^{2+} chelator (BAPTA), inhibitors of Ca^{2+} release from intracellular stores, or an inhibitor of exocytosis (botulinum toxin B).[26] Facilitation in the ganglion by nerve shock has similar properties, and expression of the facilitation depends on AMPA-like receptors. Furthermore, dishabituation in a semi-intact preparation is also reduced by postsynaptic injection of botulinum toxin B. These results suggest that intermediate-term memory and synaptic facilitation involve postsynaptic Ca^{2+} and AMPA receptor insertion.

To examine possible postsynaptic mechanisms independent of presynaptic mechanisms, Glanzman and colleagues studied facilitation of the response to focal application of glutamate (the glutamate excitatory potential (Glu-EP)) in isolated MNs. Similar to the SN–MN EPSP, the Glu-EP undergoes intermediate-term facilitation following 10-min 5-HT. Furthermore, induction of that facilitation is blocked by injection of BAPTA or botulinum toxin B into the MN, and expression depends on AMPA receptors.[27] Likewise, intermediate-term facilitation of both the SN–MN EPSP and the Glu-EP involve protein synthesis in the MN.[28] They also both involve phospholipase C (PLC), which produces inositol triphosphate (IP_3) and diacylglycerol (which in turn stimulate Ca^{2+} release and activate PKC Apl I and II).[29] Consistent with this result, induction of facilitation of the Glu-EP involves PKC Apl I or II as well as calpain-dependent proteolysis of PKC Apl III to form PKM, and maintenance involves persistent activation of PKM.[11,30] Subsequent studies have shown that intermediate-term facilitation of the SN–MN EPSP also involves postsynaptic PKC Apl III.[31] Collectively, these results suggest that induction, maintenance, and expression of intermediate-term facilitation by 10-min 5-HT could be entirely postsynaptic.

PRE- AND POSTSYNAPTIC MECHANISMS OF INTERMEDIATE-TERM PLASTICITY

The previously discussed studies suggest that whereas short-term facilitation and intermediate-term activity-dependent facilitation involve presynaptic mechanisms, intermediate-term facilitation by 10-min 5-HT involves postsynaptic mechanisms. However, in each case, only one side of the synapse was examined, and it was not known whether the same protocol might involve mechanisms on both sides. We and our colleagues have addressed that question and have found that a number of different types of intermediate-term plasticity involve both pre- and postsynaptic mechanisms.

Homosynaptic Potentiation

We first examined an activity-dependent form of plasticity that was originally referred to as post-tetanic potentiation but later as the more generic homosynaptic potentiation, in which brief tetanic stimulation (20 Hz for 2 sec) of the SN produces intermediate-term potentiation of the SN–MN PSP in culture (Figure 16.1, HP).[32] This form of plasticity, which currently has no known behavioral function, involves both pre- and postsynaptic Ca^{2+} and postsynaptic depolarization.[33] It also involves presynaptic ryanodine receptors, CamKII, actin, and a rapid increase in clusters or puncta of the vesicle-associated protein synaptophysin.[34,35] In addition, it involves postsynaptic type I metabotropic glutamate receptors, IP_3, CamKII, and possibly AMPA receptor insertion, and it also involves pre- or postsynaptic MAP kinase but not PKA or PKC Apl I or II.

Classical Conditioning

We next examined an analog of classical conditioning, referred to as pairing-specific facilitation, in which tetanic stimulation of the SN immediately precedes brief application of 5-HT in culture.[36] Like homosynaptic potentiation, pairing-specific facilitation involves both pre- and postsynaptic Ca^{2+} and postsynaptic depolarization, as well as presynaptic PKA.[37] Similarly, the pairing-specific facilitation at SN–MN synapses during classical conditioning in the semi-intact siphon withdrawal preparation also involves presynaptic PKA and postsynaptic Ca^{2+}, as well as NMDA receptor activation.[17] These results suggest that the conditioning involves both activity-dependent enhancement of presynaptic facilitation (induced by presynaptic activity paired with a modulatory transmitter such as 5-HT) and Hebbian potentiation (induced by presynaptic activity paired with postsynaptic depolarization). Furthermore, some of the PKA-dependent presynaptic changes are blocked by injecting a Ca^{2+} chelator into the postsynaptic neuron, suggesting that these two types of associative mechanisms interact through retrograde signaling. In addition to the unknown retrograde signal, a paracrine signal—nitric oxide—is released by identified facilitatory interneurons and acts directly in both the SNs and MNs to contribute to different components of the facilitation.[38]

Sensitization and Dishabituation

Antonov *et al.*[15] also examined mechanisms of facilitation at the SN–MN synapses during short-term and intermediate-term sensitization and dishabituation in the semi-intact siphon withdrawal preparation. The facilitation during short-term sensitization involves PKA, CamKII, and transient spike broadening in the SN, but it does not involve Ca^{2+} or CamKII in the MN and thus appears to be entirely presynaptic. The facilitation during intermediate-term sensitization also involves PKA, CamKII, and transient spike broadening in the SN. In addition, it involves Ca^{2+} and CamKII in the MN and protein synthesis in both neurons, and it is thus both pre- and postsynaptic. Sensitization is not affected by the broad-spectrum PKC inhibitor chelerythrine and thus does not involve any isoform of PKC. By contrast, intermediate-term dishabituation is blocked by chelerythrine but not by inhibitors of PKA, CamKII, or protein synthesis. Furthermore, the facilitation during dishabituation involves PKC in the SN, but it does not involve Ca^{2+} in the MN. These results suggest that unlike intermediate-term sensitization, intermediate-term dishabituation is entirely presynaptic. Moreover, the presynaptic mechanisms are different than those during intermediate-term sensitization, and they are more similar to those during site-specific sensitization.[24,25] These results are also generally similar to those for neural analogs of sensitization and dishabituation,[10] and they show that not only the molecular mechanisms but also the site of plasticity depends on the stage (short term vs. intermediate term) and the type (sensitization vs. dishabituation) of learning.

1- and 10-Min 5-HT

Jin *et al.*[39] performed similar experiments in cell culture, in which a mechanistic analysis would be more feasible. Facilitation by 1-min application of 5-HT to rested synapses (an analog of short-term sensitization) involves PKA and CamKII in the SN, but it does not involve Ca^{2+} in the MN or PKC in either neuron. By contrast, facilitation by 1-min application of 5-HT to depressed synapses (an analog of short-term dishabituation) involves PKC in the SN but not in the MN. Thus, the two types of short-term plasticity involve different kinases, but in both cases the plasticity appears to be entirely presynaptic. Facilitation by 10-min application of 5-HT to rested synapses (an analog of intermediate-term sensitization) also involves PKC (but not PKA or CamKII) in the SN. In addition, it involves Ca^{2+} and CamKII (but not PKC Apl I or II) in the MN and protein synthesis in both neurons, and it is thus both pre- and postsynaptic.

These results are similar to those for facilitation during short-term and intermediate-term sensitization, except that intermediate-term facilitation by 10-min 5-HT involves presynaptic PKC rather than PKA and CamKII, in agreement with previous experiments in culture.[10] The postsynaptic mechanisms are similar to those in previous studies of facilitation by 10-min 5-HT,[26,28] as well as those during intermediate-term sensitization[15] and also homosynaptic potentiation.[34] In particular, in each case, intermediate-term plasticity at SN–MN synapses involves postsynaptic CamKII and not PKC Apl I or II. By contrast, facilitation of the Glu-EP in isolated motor neurons by 10-min 5-HT involves PKC Apl I or II.[29,30] Thus, the postsynaptic mechanisms of intermediate-term plasticity at SN–MN synapses may be somewhat different from those in isolated motor neurons.

Recruitment of Synaptic Proteins

Collectively, these results suggest that intermediate-term facilitation is the first stage to involve both pre- and postsynaptic molecular mechanisms. It also involves recruitment of synaptic proteins. Intermediate-term facilitation by four pulses of a low concentration of 5-HT is accompanied by filling of presynaptic varicosities with the vesicle-associated protein synaptophysin within 3 hr but not by the formation of new varicosities.[40] Like facilitation of the PSP with this protocol,[20] the increase in clusters of synaptophysin does not last 24 hr and does not require protein or RNA synthesis. By contrast, intermediate- and long-term facilitation by five pulses of a higher concentration of 5-HT are accompanied by both filling of varicosities and the formation of new varicosities within 12–18 hr. Again like facilitation of the PSP and the increase in varicosities,[41] the increase in clusters of synaptophysin following five pulses of 5-HT lasts 24 hr and requires protein synthesis. Intermediate- and long-term facilitation by five pulses of 5-HT are also accompanied by increases in clusters of the postsynaptic proteins ApGluR1 and ApNR1 within 12 hr, whereas short-term facilitation by a single pulse of 5-HT is not.[42] These results suggest that the intermediate-term stage is the first to involve recruitment of both pre- and postsynaptic proteins, which could be initial steps in the formation of new synapses during long-term facilitation.

SPONTANEOUS TRANSMITTER RELEASE IS CRITICAL FOR THE INDUCTION OF INTERMEDITATE- AND LONG-TERM FACILITATION

If short-term facilitation is presynaptic but intermediate- and long-term facilitation involve both pre- and postsynaptic mechanisms, how are the postsynaptic mechanisms first recruited? There are at least two basic possibilities, which are not mutually exclusive: The pre- and postsynaptic mechanisms might be induced by activation of pre- and postsynaptic 5-HT receptors in parallel, or activation of presynaptic 5-HT receptors might increase spontaneous release of glutamate, which then activates postsynaptic glutamate receptors to induce the postsynaptic mechanisms in series (Figure 16.1). In part because the postsynaptic mechanisms of intermediate-term facilitation are similar to those induced by glutamate release during homosynaptic potentiation,[34] Jin *et al.*[43] decided to investigate the possible role of spontaneous transmitter release from the presynaptic neuron as an anterograde signal for recruiting postsynaptic mechanisms of intermediate- and long-term facilitation.

We first recorded spontaneous miniature excitatory postsynaptic currents (mEPSCs) or miniature excitatory postsynaptic potentials (mEPSPs) in cell culture interleaved with either intermediate-term facilitation of the evoked EPSP induced by 10-min 5-HT or long-term facilitation induced by five pulses of 5-HT. In both cases, there was a substantial increase in the frequency of mEPSCs and a more modest increase in their amplitude during or soon after the 5-HT. Furthermore, these increases correlated with subsequent facilitation of the evoked EPSP, consistent with the idea that spontaneous release may contribute to the induction of the facilitation.

To provide a more direct test of this idea, we examined whether manipulations that reduce spontaneous release also reduce the facilitation. As a first step, we injected the SN with the light chain of botulinum toxin D (BoTx D), which interferes with transmitter release by cleaving the vesicle-associated protein VAMP/synaptobrevin.[44] Presynaptic injection of a low concentration of BoTx D, which did not reduce the pretest EPSP, reduced the frequency of mEPSPs before and during the 5-HT. It also reduced both intermediate- and long-term facilitation of the evoked EPSP, with no effect on test-alone controls. We obtained similar results with presynaptic injection of the slow Ca^{2+} chelator EGTA.

We also examined whether manipulations that increase spontaneous release more selectively than 5-HT enhance the facilitation. First, to increase spontaneous release without also directly affecting the postsynaptic neuron, we expressed in the SN an *Aplysia* octopamine receptor (OAR) that is coupled to adenylyl cyclase and is not normally expressed in SNs.[45] Application of octopamine to co-cultures with OAR-expressing SNs produced a substantial increase in the frequency of mEPSCs and a more modest increase in their amplitude. A 10-min application of octopamine also produced intermediate-term facilitation of the evoked EPSP, and five pulses of octopamine produced long-term facilitation.

As another way to increase spontaneous release, we used α-latrotoxin (LaTx), which stimulates the release of docked vesicles from presynaptic terminals.[46] LaTx produced a substantial increase in the frequency of mEPSCs with no increase in their amplitude. A 10-min application of LaTx by itself did not produce facilitation of the evoked EPSP. However, when we applied LaTx together with a low concentration of 5-HT, which by itself produced modest facilitation, the combination produced greater facilitation than the 5-HT alone. Together with the results described previously, these findings suggest that mechanisms recruited by spontaneous release are necessary for the induction of intermediate- and long-term facilitation and act synergistically with additional mechanisms (activated, for example, by presynaptic cAMP) to produce the facilitation.

SPONTANEOUS TRANSMITTER RELEASE FROM THE PRESYNAPTIC NEURON RECRUITS POSTSYNAPTIC MECHANISMS OF INTERMEDIATE- AND LONG-TERM FACILITATION

Jin *et al.*[47] next investigated whether spontaneous transmitter release recruits postsynaptic mechanisms during the induction of intermediate- and long-term facilitation, focusing primarily on the intermediate-term stage. Fluorescent *in situ* hybridization revealed that the MNs express an *Aplysia* homolog of the group I metabotropic glutamate receptor mGluR5. An inhibitor of these receptors, MPEP, reduced both intermediate-term facilitation by 10-min 5-HT and long-term facilitation by five pulses of 5-HT. Likewise, postsynaptic injection of an antisense oligonucleotide for mGluR5 reduced both intermediate- and long-term facilitation.

mGluR5 is often linked to IP_3 production and Ca^{2+} release from intracellular stores. Using a fluorescent reporter,[48] we found that 10-min application of 5-HT produced an increase in IP_3 production in the synaptic region (initial segment) of the MN, and that increase was reduced by injection of BoTx D (which decreases spontaneous release) into the SN. In addition, we found that, like facilitation by 10-min application of 5-HT, facilitation by 10-min application of octopamine to cultures with OAR-expressing SNs was blocked by MPEP or injection of the Ca^{2+} chelator BAPTA into the MN. Furthermore, 10-min octopamine or LaTx also produced a gradual increase in Ca^{2+} in the synaptic region of the MN, and that increase was reduced by MPEP as well. These results suggest that increased spontaneous transmitter release from the presynaptic neuron during 10-min 5-HT or octopamine activates metabotropic glutamate receptors, which stimulate IP_3 production and Ca^{2+} release in the postsynaptic neuron.

We also examined postsynaptic mechanisms of expression of intermediate-term facilitation by 10-min 5-HT. An inhibitor of AMPA receptors reduced facilitation after washout of the 5-HT, suggesting that expression of the later part of facilitation may involve upregulation of AMPA-like receptors (see also[26]). To examine this mechanism more directly, we expressed in the MN the *Aplysia* homolog of the AMPA receptor subunit GluR1 tagged with a pH-dependent variant of GFP, which increases its fluorescence intensity following fusion of transport vesicles with the surface membrane. In addition, we expressed in the SN the presynaptic protein synaptophysin tagged with mCherry. A 10-min application of 5-HT produced an increase in the intensity of puncta of ApGluR1-pHluorin in the synaptic region of the MN, and this increase was reduced by MPEP or injection of BoTx D into the SN. A 10-min application of 5-HT also produced increases in the number of puncta of ApGluR1 and overlap of puncta of ApGluR1 with synaptophysin, and these increases were reduced by injection of BoTx D into the SN as well. However, there was no increase in puncta of synaptophysin during 10-min 5-HT, although there are increases at later times during intermediate- and long-term facilitation.[40]

These results suggest that an increase in spontaneous transmitter release from the presynaptic neuron during short-term facilitation induces membrane insertion of postsynaptic ApGluR1 at existing puncta and the formation of new puncta during intermediate-term facilitation. The increase in ApGluR1 puncta precedes an increase in presynaptic synaptophysin puncta and therefore may be a first step in a sequence that can lead to new synapse assembly during long-term facilitation.

CONCLUSIONS

At the beginning of this chapter, we posed two questions: How are the different stages of memory and synaptic plasticity related, and when and how does plasticity spread from one side of the synapse to both sides? Studies of the mechanisms of short-term and intermediate-term plasticity in *Aplysia* have begun to answer these questions. Whereas short-term plasticity is presynaptic, intermediate-term plasticity involves both pre- and postsynaptic mechanisms. Furthermore, studies of the postsynaptic mechanisms of intermediate-term facilitation suggest that they can be induced either in parallel or in series with those of short-term facilitation (Figure 16.1). On the one hand, 5-HT can act in parallel on presynaptic receptors to induce mechanisms of short-term facilitation and on postsynaptic receptors to induce mechanisms of intermediate-term facilitation. On the other hand, one of the early effects of activating presynaptic 5-HT receptors is to enhance spontaneous release of glutamate, which can then act on postsynaptic metabotropic glutamate receptors to induce mechanisms of intermediate-term facilitation in series with those of short-term facilitation. The postsynaptic 5-HT and metabotropic glutamate receptors are both linked to PLC, which can stimulate production of diacylglycerol and IP_3.

Several lines of evidence suggest that the series (metabotropic glutamate receptor) pathway plays a more important role than the parallel (postsynaptic 5-HT receptor) pathway. Intermediate-term facilitation and postsynaptic mechanisms including increases in IP_3, Ca^{2+}, and membrane insertion and clustering of AMPA-like receptors are all largely blocked by presynaptic BoTx D (which reduces spontaneous release) or an antagonist of metabotropic glutamate receptors. Conversely, stimulation of presynaptic octopamine receptors alone is sufficient to induce the facilitation and an increase in postsynaptic Ca^{2+}. Furthermore, the parallel and series pathways may also at least in part activate different postsynaptic kinases, with their contributions depending on the experimental preparation. Thus, intermediate-term facilitation of the Glu-EP in isolated motor neurons (which is induced by 5-HT receptors on the MN) involves both PKC Apl I or II and PKC Apl III.[30] Intermediate-term facilitation at SN–MN synapses (where postsynaptic mGluRs can contribute) involves postsynaptic PKC Apl III but not PKC Apl I or II, and it also involves postsynaptic CamKII.[15,31,34,39] Furthermore, facilitation at SN–MN synapses during intermediate-term sensitization in a semi-intact preparation does not involve any of the PKC isoforms, but does involve postsynaptic CamKII.[15] Thus, one possible explanation for these results is that the parallel (5-HT receptor) pathway preferentially stimulates diacylglycerol and thus PKC Apl I and II, whereas the series (mGluR) pathway preferentially stimulates IP_3 and Ca^{2+} and thus CamKII, which plays a larger role at synapses. Either pathway may also stimulate proteolysis of postsynaptic PKC Apl III to form PKM, at least in cultured neurons (Figure 16.1).

More generally, these results suggest that, like synapse formation,[49] the different stages of synaptic plasticity may involve different pre- and postsynaptic mechanisms coordinated by back-and-forth signaling in a chain or cascade that can culminate in growth (Figure 16.2). Consistent with this idea, we have found that spontaneous transmitter release from the presynaptic neuron during short-term facilitation recruits postsynaptic molecular mechanisms of intermediate-term facilitation, including IP_3, Ca^{2+}, and the formation of clusters of AMPA-like glutamate receptors. Postsynaptic Ca^{2+} is in turn necessary for long-term facilitation, perhaps through retrograde signaling to presynaptic neurexin, ApCAM, or Trk receptors.[51–54] The new postsynaptic clusters of AMPA-like receptors may also participate in retrograde signaling and recruit presynaptic clusters of synaptophysin during a later stage of intermediate-term facilitation and growth of varicosities during long-term facilitation.[40,55,56]

These results are similar to those of theoretical 'cascade' models of memory storage.[50] Such models progress through a series of increasingly stable states and are thus able to exhibit both plasticity and long-term stability, which are essential features of memory but tend to be mutually exclusive in simpler models. This advantage of a cascade of mechanisms might also explain why plasticity involves so many different

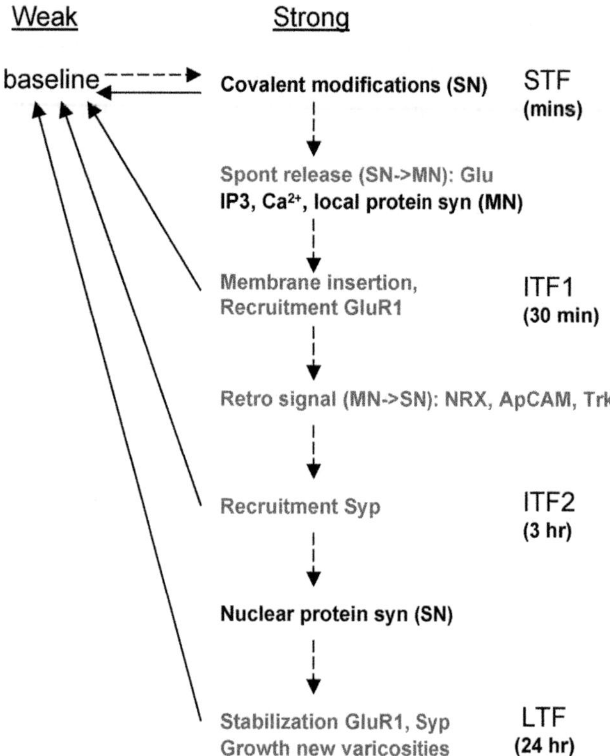

FIGURE 16.2 **Cascade model of synaptic plasticity in *Aplysia*.** In cascade models,[50] synapses have two levels of strength (weak and strong) and several increasingly long-lasting states. In *Aplysia*, relatively weak stimulation produces short-term facilitation (STF) that lasts minutes, stronger stimulation produces intermediate-term facilitation (ITF) that lasts minutes to hours, and even stronger stimulation produces long-term facilitation (LTF) that lasts days. The different stages of facilitation may involve a series or cascade of pre- and postsynaptic mechanisms that is initiated by spontaneous transmitter release during STF, progresses through two stages of ITF, and can culminate in synaptic growth during LTF. The mechanisms in this growth cascade are a subset of all mechanisms involved in facilitation, and some other mechanisms (not shown) may act in parallel and contribute only to specific stages. Thus, the stages *per se* do not necessarily form a cascade. Dashed lines, transitions that are initiated by different durations or patterns of 5-HT; solid lines, spontaneous transitions; red, extracellular signaling molecules; blue, structural modifications; NRX, neurexin; Syp, synaptophysin. (Please refer to color plate section.)

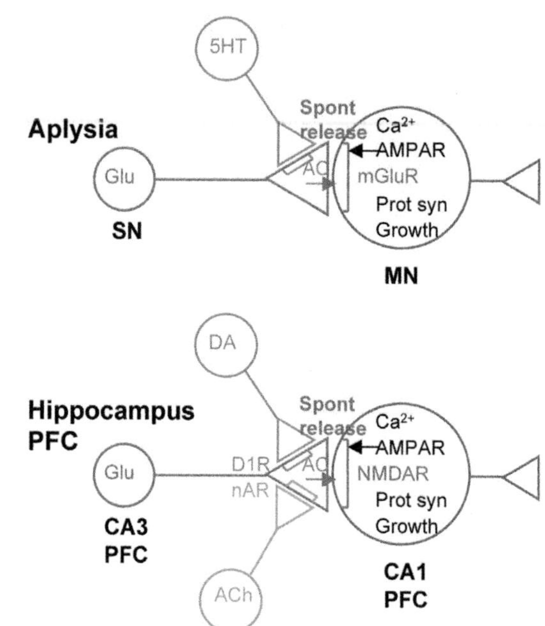

FIGURE 16.3 The role of spontaneous release in cellular and molecular mechanisms of intermediate- and long-term facilitation in *Aplysia* (top) compared with its hypothesized role in learning-related synaptic plasticity in hippocampus and prefrontal cortex (PFC) (bottom). 5-HT, serotonin (green); AC, adenylyl cyclase; ACh, acetylcholine (yellow); D1R, D1 dopamine receptor; DA, dopamine (green); Glu, glutamate (red); nAR, nicotinic acetylcholine receptor.

mechanisms and why they make different contributions under different experimental conditions. Furthermore, recent evidence suggests that long-term potentiation in hippocampus involves a cascade of mechanisms roughly similar to the one we have proposed for facilitation in *Aplysia*.[57,58] Thus, long-term plasticity may involve such cascades more generally.

Our results also suggest that spontaneous transmitter release acts as an anterograde signal for the spread of plasticity during the intermediate-term stage. Although spontaneous release is enhanced by activation of presynaptic 5-HT receptors in *Aplysia*, different types of presynaptic receptors could play an analogous role in mammals (Figure 16.3). For example, presynaptic D1 receptors are present on glutamatergic afferents to hippocampus and prefrontal cortex (PFC).[59,60] As in *Aplysia*, these receptors act through adenylyl cyclase and PKA to enhance spontaneous release of glutamate.[61,62] Similarly, activation of presynaptic nicotinic acetylcholine receptors enhances spontaneous release of glutamate in hippocampus and PFC.[63,64] Glutamate can then activate postsynaptic NMDA receptors (in addition to or instead of metabotropic glutamate receptors, as in *Aplysia*) to elevate Ca^{2+} and engage mechanisms of early phase long-term potentiation (LTP), including AMPA receptor insertion,[65–67] and of late-phase LTP, including protein synthesis and growth.[1,68,69] All these mechanisms of plasticity are similar to those we have described in *Aplysia*.

Synaptic plasticity in hippocampus is thought to be involved in reference memory, and plasticity in PFC is thought to be involved in working memory.[70,71] Thus, in addition to acting as an anterograde signal during synaptic plasticity related to memory in *Aplysia*, spontaneous release may play a similar role in synaptic plasticity in brain regions involved in memory in mammals, and it could contribute to disorders that affect plasticity in these regions, including Alzheimer's disease,[72–74] schizophrenia,[75–77] and attention deficit hyperactivity disorder.[78,79]

Acknowledgments

We thank Tom Abrams for his comments. Preparation of this manuscript was supported by NIH grant GM097502.

References

1. Bailey CH, Barco A, Hawkins RD, Kandel ER. Molecular studies of learning and memory in *Aplysia* and hippocampus: a comparative analysis of implicit and explicit memory storage. In: Byrne J, ed. *Learning and Memory: A Comprehensive Reference*. Oxford: Elsevier; 2008.
2. Hawkins RD, Kandel ER, Bailey CH. Molecular mechanisms of memory storage in *Aplysia*. *Biol Bull*. 2006;210:174–191.
3. Hawkins RD, Clark GA, Kandel ER. Operant conditioning of gill withdrawal in *Aplysia*. *J Neurosci*. 2006;26:2443–2448.
4. Hawkins RD, Cohen TE, Kandel ER. Dishabituation in *Aplysia* can involve either reversal of habituation or superimposed sensitization. *Learn Mem*. 2006;13:397–403.
5. Marcus EA, Nolen TG, Rankin CH, Carew TJ. Behavioral dissociation of dishabituation, sensitization, and inhibition in *Aplysia*. *Science*. 1988;241:210–213.
6. Michel M, Green CL, Gardner JS, Organ CL, Lyons LC. Massed training-induced intermediate-term operant memory in *Aplysia* requires protein synthesis and multiple persistent kinase cascades. *J Neurosci*. 2012;32:4581–4591.
7. Mozzachiodi R, Lorenzetti FD, Baxter DA, Byrne JH. Changes in neuronal excitability serve as a mechanism of long-term memory for operant conditioning. *Nat Neurosci*. 2008;11:1146–1148.
8. Antonov I, Kandel ER, Hawkins RD. The contribution of facilitation of monosynaptic PSPs to dishabituation and sensitization of the *Aplysia* siphon withdrawal reflex. *J Neurosci*. 1999;19:10438–10450.
9. Antonov I, Antonova I, Kandel ER, Hawkins RD. The contribution of activity-dependent synaptic plasticity to classical conditioning in *Aplysia*. *J Neurosci*. 2001;21:6413–6422.
10. Byrne JH, Kandel ER. Presynaptic facilitation revisited: state and time dependence. *J Neurosci*. 1996;16:435-435
11. Bougie JK, Lim T, Farah CA, et al. The atypical protein kinase C in *Aplysia* can form a protein kinase M by cleavage. *J Neurochem*. 2009;109:1129–1143.
12. Kruger KE, Sossin WS, Sacktor TC, Bergold PJ, Beushausen S, Schwartz JH. Cloning and characterization of Ca(2 +)-dependent and Ca(2 +)-independent PKCs expressed in *Aplysia* sensory cells. *J Neurosci*. 1991;11:2302–2313.
13. Manseau F, Fan X, Hueftlein T, Sossin W, Castellucci VF. Ca^{2+}-independent protein kinase C Apl II mediates the serotonin-induced facilitation at depressed *Aplysia* sensorimotor synapses. *J Neurosci*. 2001;21:1247–1256.
14. Zhao Y, Leal K, Abi-Farah C, Martin KC, Sossin WS, Klein M. Isoform specificity of PKC translocation in living *Aplysia* sensory neurons and a role for Ca^{2+}-dependent PKC APL I in the induction of intermediate-term facilitation. *J Neurosci*. 2006;26: 8847–8856.
15. Antonov I, Kandel ER, Hawkins RD. Presynaptic and postsynaptic mechanisms of synaptic plasticity and metaplasticity during intermediate-term memory formation in *Aplysia*. *J Neurosci*. 2010;30:5781–5791.
16. Nakanishi K, Zhang F, Baxter DA, Eskin A, Byrne JH. Role of calcium–calmodulin-dependent protein kinase II in modulation of sensorimotor synapses in *Aplysia*. *J Neurophysiol*. 1997;78: 409–416.
17. Antonov I, Antonova I, Kandel ER, Hawkins RD. Activity-dependent presynaptic facilitation and Hebbian LTP are both required and interact during classical conditioning in *Aplysia*. *Neuron*. 2003;37:135–147.
18. Emptage NJ, Carew TJ. Long-term synaptic facilitation in the absence of short-term facilitation in *Aplysia* neurons. *Science*. 1993;262:253–256.
19. Mauelshagen J, Parker GR, Carew TJ. Dynamics of induction and expression of long-term synaptic facilitation in *Aplysia*. *J Neurosci*. 1996;16:7099–7108.
20. Ghirardi M, Montarolo PG, Kandel ER. A novel intermediate stage in the transition between short- and long-term facilitation in the sensory to motor neuron synapses of *Aplysia*. *Neuron*. 1995;14:413–420.
21. Sutton MA, Carew TJ. Parallel molecular pathways mediate expression of distinct forms of intermediate-term facilitation at tail sensory-motor synapses in *Aplysia*. *Neuron*. 2000;26:219–231.
22. Sutton MA, Masters SE, Bagnall MW, Carew TJ. Molecular mechanisms underlying a unique intermediate phase of memory in *Aplysia*. *Neuron*. 2001;31:143–154.
23. Sharma SK, Sherff CM, Shobe J, Bagnall MW, Sutton MA, Carew TJ. Differential role of mitogen-activated protein kinase in three distinct phases of memory for sensitization in *Aplysia*. *J Neurosci*. 2003;23:3899–3907.
24. Sutton MA, Bagnall MW, Sharma SK, Shobe J, Carew TJ. Intermediate-term memory for site-specific sensitization in *Aplysia* is maintained by persistent activation of protein kinase C. *J Neurosci*. 2004;24:3600–3609.
25. Shobe JL, Zhao Y, Stough S, et al. Temporal phases of activity-dependent plasticity and memory are mediated by compartmentalized routing of MAPK signaling in *Aplysia* sensory neurons. *Neuron*. 2009;61:113–125.
26. Li Q, Roberts AC, Glanzman DL. Synaptic facilitation and behavioral dishabituation in *Aplysia*: dependence on release of Ca^{2+} from postsynaptic intracellular stores, postsynaptic exocytosis, and modulation of postsynaptic AMPA receptor efficacy. *J Neurosci*. 2005;25:5623–5637.
27. Chitwood RA, Li Q, Glanzman DL. Serotonin facilitates AMPA-type responses in isolated siphon motor neurons of *Aplysia* in culture. *J. Physiol*. 2001;534:501–510.
28. Villareal G, Li Q, Cai D, Glanzman DL. The role of rapid, local, postsynaptic protein synthesis in learning-related synaptic facilitation in *Aplysia*. *Curr Biol*. 2007;17:2073–2080.
29. Fulton D, Condro MC, Pearce K, Glanzman DL. The potential role of postsynaptic phospholipase C activity in synaptic facilitation and behavioral sensitization in *Aplysia*. *J Neurophysiol*. 2008;100:108–116.
30. Villareal G, Li Q, Cai D, et al. Role of protein kinase C in the induction and maintenance of serotonin-dependent enhancement of the glutamate response in isolated siphon motor neurons of *Aplysia californica*. *J Neurosci*. 2009;29:5100–5107.
31. Bougie JK, Cai D, Hastings M, et al. Serotonin-induced cleavage of the atypical protein kinase C Apl III in *Aplysia*. *J Neurosci*. 2012;32:14630–14640.
32. Eliot LS, Kandel ER, Hawkins RD. Modulation of spontaneous transmitter release during depression and posttetanic potentiation of *Aplysia* sensory-motor neuron synapses isolated in culture. *J Neurosci*. 1994;14:3280–3292.
33. Bao J-X, Kandel ER, Hawkins RD. Involvement of pre- and postsynaptic mechanisms in posttetanic potentiation at *Aplysia* synapses. *Science*. 1997;275:969–973.

34. Jin I, Hawkins RD. Presynaptic and postsynaptic mechanisms of a novel form of homosynaptic potentiation at *Aplysia* sensory-motor neuron synapses. *J Neurosci*. 2003;23:7288–7297.
35. Jin I, Udo H, Hawkins RD. Rapid increase in clusters of synaptophysin at onset of homosynaptic potentiation in *Aplysia*. *Proc Natl Acad Sci USA*. 2011;108:11656–11661.
36. Eliot LS, Hawkins RD, Kandel ER, Schacher S. Pairing-specific, activity-dependent presynaptic facilitation at *Aplysia* sensory-motor neuron synapses in isolated cell cultures. *J Neurosci*. 1994;14:368–383.
37. Bao JX, Kandel ER, Hawkins RD. Involvement of presynaptic and postsynaptic mechanisms in a cellular analog of classical conditioning at *Aplysia* sensory-motor neuron synapses in isolated cell culture. *J Neurosci*. 1998;18:458–466.
38. Antonov I, Ha T, Antonova I, Moroz L, Hawkins RD. Role of nitric oxide in classical conditioning of siphon withdrawal in *Aplysia*. *J Neurosci*. 2007;27:10993–11002.
39. Jin I, Kandel ER, Hawkins RD. Whereas short-term facilitation is presynaptic, intermediate-term facilitation involves both presynaptic and postsynaptic protein kinases and protein synthesis. *Learn Mem*. 2011;18:96–102.
40. Kim J-H, Udo H, Li H-L, et al. Presynaptic activation of silent synapses and growth of new synapses contribute to intermediate and long-term facilitation in *Aplysia*. *Neuron*. 2003;40:151–165.
41. Bailey CH, Montarolo P, Chen M, Kandel ER, Schacher S. Inhibitors of protein and RNA synthesis block structural changes that accompany long-term heterosynaptic plasticity in *Aplysia*. *Neuron*. 1992;9:749–758.
42. Li HL, Huang BS, Vishwasrao H, et al. Dscam mediates remodeling of glutamate receptors in *Aplysia* during *de novo* and learning-related synapse formation. *Neuron*. 2009;61:527–540.
43. Jin I, Puthenveettil S, Udo H, Karl K, Kandel ER, Hawkins RD. Spontaneous transmitter release is critical for the induction of long-term and intermediate-term facilitation in *Aplysia*. *Proc Natl Acad Sci USA*. 2012;109:9131–9136.
44. Yamasaki S, Hu Y, Binz T, et al. Synabtobrevin/vesicle-associated membrane protein (VAMP) of *Aplysia californica*: structure and proteolysis by tetanus toxin and botulinal neurotoxins type D and F. *Proc Natl Acad Sci USA*. 1994;91:4688–4692.
45. Chang DJ, Li XC, Lee YS, et al. Activation of a heterologously expressed octopamine receptor coupled to adenylyl cyclase produces all the features of presynaptic facilitation in *Aplysia* sensory neurons. *Proc Natl Acad Sci USA*. 2000;97:1829–1834.
46. Ceccarelli B, Hurlbut WP. Ca^{2+}-dependent recycling of synaptic vesicles at the frog neuromuscular junction. *J Cell Biol*. 1980;87: 297–303.
47. Jin I, Udo H, Rayman JB, et al. Postsynaptic mechanisms recruited by spontaneous transmitter release during long-term and intermediate-term facilitation in *Aplysia*. *Proc Natl Acad Sci USA*. 2012;109:9137–9142.
48. Violin JD, Zhang J, Tsien RY, Newton AC. A genetically encoded fluorescent reporter reveals oscillatory phosphorylation by protein kinase C. *J Cell Biol*. 2003;161:899–909.
49. McAllister AK. Dynamic aspects of CNS synapse formation. *Annu Rev Neurosci*. 2007;30:425–450.
50. Fusi S, Drew PJ, Abbott LF. Cascade models of synaptically stored memories. *Neuron*. 2005;45:599–611.
51. Cai D, Chen S, Glanzman DL. Postsynaptic regulation of long-term facilitation in *Aplysia*. *Curr Biol*. 2008;18:920–925.
52. Choi YB, Li HL, Kassabov SR, et al. Neurexin–neuroligin trans-synaptic interaction mediates learning-related synaptic remodeling and long-term facilitation in *Aplysia*. *Neuron*. 2011;70: 468–481.
53. Hu JY, Chen Y, Bougie JK, Sossin WS, Schacher S. Aplysia cell adhesion molecule and a novel protein kinase C activity in the postsynaptic neuron are required for presynaptic growth and initial formation of specific synapses. *J Neurosci*. 2010;30: 8353–8366.
54. Kassabov SR, Monje FJ, Fiumara F, Kandel ER. Identification of ApTrk a Trk receptor homolog expressed in *Aplysia* sensory and motor neurons critical for the induction and maintenance of serotonin dependent long term facilitation. *Soc Neurosci Abstr*. 2007;473:6.
55. Lee SJ, Uemura T, Yoshida T, Mishina M. GluRδ2 assembles four neurexins into trans-synaptic triad to trigger synapse formation. *J Neurosci*. 2012;32:4688–4701.
56. Ripley B, Otto S, Tiglio K, Williams ME, Ghosh A. Regulation of synaptic stability by AMPA receptor reverse signaling. *Proc Natl Acad Sci USA*. 2011;108:367–372.
57. Antonova I, Lu F-M, Zablow L, Udo H, Hawkins RD. Rapid and long-lasting increase in sites for synapse assembly during late-phase potentiation in rat hippocampal neurons. *PLoS ONE*. 2009;4(11):e7690.
58. Johnstone VP, Raymond CR. A protein synthesis and nitric oxide-dependent presynaptic enhancement in persistent forms of long-term potentiation. *Learn Mem*. 2011;18:625–633.
59. Bergson C, Mrzljak L, Smiley JF, Pappy M, Levenson R, Goldman-Rakic PS. Regional, cellular, and subcellular variations in the distribution of D1 and D5 dopamine receptors in primate brain. *J Neurosci*. 1995;15:7821–7836.
60. Paspalas CD, Goldman-Rakic PS. Presynaptic D1 dopamine receptors in primate prefrontal cortex: target-specific expression in the glutamatergic synapse. *J Neurosci*. 2005;25:1260–1267.
61. Bouron A, Reuter H. The D1 dopamine receptor agonist SKF-38393 stimulates the release of glutamate in the hippocampus. *Neuroscience*. 1999;94:1063–1070.
62. Wang Z, Feng XQ, Zheng P. Activation of presynaptic D1 dopamine receptors by dopamine increases the frequency of spontaneous excitatory postsynaptic currents through protein kinase A and protein kinase C in pyramidal cells of rat prelimbic cortex. *Neuroscience*. 2002;112:499–508.
63. Sharma G, Vijayaraghavan S. Modulation of presynaptic store calcium induces release of glutamate and postsynaptic firing. *Neuron*. 2003;38:929–939.
64. Wang BW, Liao WN, Chang CT, Wang SJ. Facilitation of glutamate release by nicotine involves the activation of a Ca^{2+}/calmodulin signaling pathway in rat prefrontal cortex nerve terminals. *Synapse*. 2006;59:491–501.
65. Gurden H, Takita M, Jay TM. Essential role of D1 but not D2 receptors in the NMDA receptor-dependent long-term potentiation at hippocampal-prefrontal cortex synapses *in vivo*. *J Neurosci*. 2000;20:RC106.
66. Huang YY, Simpson E, Kellendonk C, Kandel ER. Genetic evidence for the bidirectional modulation of synaptic plasticity in the prefrontal cortex by D1 receptors. *Proc Natl Acad Sci USA*. 2004;101:3236–3241.
67. Otmakhova NA, Lisman JE. D1/D5 dopamine receptor activation increases the magnitude of early long-term potentiation at CA1 hippocampal synapses. *J Neurosci*. 1996;16: 7478–7486.
68. Huang YY, Kandel ER. D1/D5 receptor agonists induce a protein synthesis-dependent late potentiation in the CA1 region of the hippocampus. *Proc Natl Acad Sci USA*. 1995;92: 2446–2450.
69. Navakkode S, Sajikumar S, Frey JU. Synergistic requirements for the induction of dopaminergic D1/D5-receptor-mediated LTP in hippocampal slices of rat CA1 *in vitro*. *Neuropharmacology*. 2007;52:1547–1554.
70. Goldman-Rakic PS, Castner SA, Svensson TH, Siever LJ, Williams GV. Targeting the dopamine D1 receptor in

schizophrenia: insights for cognitive dysfunction. *Psychopharmacology*. 2004;174:3–16.

71. Morris RG, Moser EI, Riedel G, et al. Elements of a neurobiological theory of the hippocampus: the role of activity-dependent synaptic plasticity in memory. *Philos Trans R Soc Lond B Biol Sci*. 2003;358:773–786.
72. Khan GM, Tong M, Jhun M, Arora K, Nichols RA. Beta-amyloid activates presynaptic alpha7 nicotinic acetylcholine receptors reconstituted into a model nerve cell system: involvement of lipid rafts. *Eur J Neurosci*. 2010;31:788–796.
73. Nimmrich V, Grimm C, Draguhn A, et al. Amyloid beta oligomers (A beta (1–42) globulomer) suppress spontaneous synaptic activity by inhibition of P/Q-type calcium currents. *J Neurosci*. 2008;28:788–797.
74. Puzzo D, Privitera L, Leznik E, et al. Picomolar amyloid-beta positively modulates synaptic plasticity and memory in hippocampus. *J Neurosci*. 2008;28:14537–14545.
75. Chen L, Yang CR. Interaction of dopamine D1 and NMDA receptors mediates acute clozapine potentiation of glutamate EPSPs in rat prefrontal cortex. *J Neurophysiol*. 2002;87: 2324–2336.
76. Murphy BL, Arnsten AF, Goldman-Rakic PS, Roth RH. Increased dopamine turnover in the prefrontal cortex impairs spatial working memory performance in rats and monkeys. *Proc Natl Acad Sci USA*. 1996;93:1325–1329.
77. Radek RJ, Kohlhass KL, Rueter LE, Mohler EG. Treating the cognitive deficits of schizophrenia with alpha4beta2 neuronal nicotinic receptor agonists. *Curr Pharm Des*. 2010;16:309–322.
78. Gamo NJ, Wang M, Arnsten AF. Methylphenidate and atomoxetine enhance prefrontal function through α_2-adrenergic and dopamine D1 receptors. *J Am Acad Child Adolesc Psychiatry*. 2010;49:1011–1023.
79. Young JW, Crawford N, Kelly JS, et al. Impaired attention is central to the cognitive deficits observed in alpha 7 deficient mice. *Eur Neuropsychopharmacol*. 2007;17:145–155.

CHAPTER

17

Synaptic Mechanisms of Induction and Maintenance of Long-Term Sensitization Memory in *Aplysia*

David L. Glanzman

University of California at Los Angeles, Los Angeles, California

INTRODUCTION

One of the more remarkable features of the brain is its ability to transform experience into memories that persist for a lifetime. That memories can remain more or less stable over the course of many decades despite the significant physical changes that occur in the brain with age is one of the grand mysteries of neuroscience. The past 30 years have witnessed rapid progress in our understanding of how long-term memories are induced in the brain; by comparison, we understand far less about how memories are maintained.

Much of what we know about the cell biology of long-term memory has come not from studies of the mammalian brain but, rather, from studies of the nervous systems of invertebrate organisms, the most influential of which, arguably, involve the marine snail *Aplysia californica*. Why this relatively humble mollusk should have played such a key role in mechanistic studies of long-term memory requires some explanation. *Aplysia* has several important features that facilitate relating cellular and molecular changes to learned behavioral changes. First, its central nervous system (CNS) possesses only approximately 20,000 neurons,[1] compared to, for example, approximately 21 million in the brain of a rat.[2] This greatly simplifies the task of identifying learning-related sites of change in the nervous system. The logic here is similar to that employed by the legendary Spanish neuroanatomist, Ramon y Cajal, who chose the embryos of animals such as dogs and cats for his initial investigations of the fine structure of the brain using the Golgi stain. In justifying this approach, Cajal[3] stated, "Since the full grown forest [i.e., the adult brain] turns out to be impenetrable and indefinable, why not revert to the study of the young wood, in the nursery stage, as we might say?" Similarly, the simplicity of the *Aplysia* nervous system renders the specific pattern of cellular changes produced by a learning experience more readily visible in *Aplysia* than in the relatively impenetrable neural thicket of the mammalian brain. Another major advantage of *Aplysia* is that neurons that mediate specific behaviors can be identified and then individually dissociated and placed into cell culture. This means that behaviorally relevant neural circuits can be re-created *in vitro*. Importantly, these *in vitro* circuits have been shown to exhibit most, if not all, of the same forms of synaptic plasticity that they exhibit *in vivo*. This feature has been extensively exploited, most prominently by Eric Kandel and colleagues, with the result that arguably more is known about the cell biology of learning and memory in *Aplysia* than in any other organism.[4] The monosynaptic connection between the sensory and motor neurons that mediate the defensive withdrawal reflex of the gill and siphon (the sensorimotor synapse) has been a particularly fruitful system for mechanistic analyses of learning. Among the major contributions of research on the sensorimotor synapse is an understanding of the critical role in long-term, learning-related synaptic plasticity of protein synthesis and gene transcription,[5] particularly that stimulated by the transcription factor cyclic adenosine monophosphate (cAMP) response element binding protein (CREB).[6–9] Importantly, these processes have each

Invertebrate Learning and Memory.
DOI: http://dx.doi.org/10.1016/B978-0-12-415823-8.00017-4

been found to play a critical role in mammalian learning-related long-term synaptic plasticity as well,[10–13] which indicates that there has been broad conservation of the cellular and molecular mechanisms of learning and memory over the course of evolution.

This review focuses on one specific form of long-term memory in *Aplysia*, the memory for long-term sensitization (LTS) of the defensive withdrawal reflex.[14] Knowledge regarding underlying mechanisms has advanced further for LTS than for any other form of long-term memory in *Aplysia*. This review begins with a description of the cellular and molecular mechanisms that mediate induction of LTS. It then describes what is known about the mechanisms that underlie the maintenance of LTS. Finally, future directions toward an understanding of this form of nonassociative learning and memory are discussed. Although sensitization in *Aplysia* involves nonsynaptic mechanisms, particularly changes in neuronal excitability, this chapter focuses on the synaptic mechanisms of LTS. (For a review on nonsynaptic mechanisms of learning in *Aplysia* and other organisms, see Mozzachiodi and Byrne.[15])

LONG-TERM SENSITIZATION IN *APLYSIA*: MECHANISMS OF INDUCTION

When an *Aplysia* receives a noxious stimulus, such as an electrical shock to its tail, the endogenous monoaminergic transmitter serotonin (5-HT) is released into the animal's CNS from a network of central serotonergic neurons.[16–19] This release of 5-HT triggers a number of cellular changes that lead to enhancement of the defensive withdrawal reflex.[20] Much of the analysis of these sensitization-related cellular changes has focused on the monosynaptic connections between the central sensory and motor neurons that mediate the reflex. Two such connections that have been particularly well studied with respect to sensitization have been that between the sensory and motor neurons that mediate contraction of the gill and siphon[21–23] and the synaptic connection between the pleural sensory neurons and the pedal motor neurons, which mediates contraction of the tail.[24,25] The central sensory and motor neurons that mediate the gill- and siphon-withdrawal reflex are located in the abdominal ganglion.[26] Tail withdrawal is mediated by sensory neurons in the paired pleural ganglia, whereas the motor neurons that mediate this reflex are located in the paired pedal ganglia.[27]

Tail (or tail nerve) shock and 5-HT facilitate transmission at the sensorimotor synapse.[25,28–32] (Other endogenous transmitters can produce facilitation of the sensorimotor synapse, particularly the small cardioactive peptides[30,33] and nitric oxide, released by the L29 facilitatory interneurons[34]; however, thus far, only 5-HT has been shown to support long-term facilitation of the sensorimotor synapse.) The facilitation of the sensorimotor synapse plays a prominent role in both short-term[31,35] and long-term[36] sensitization.

A single tail shock, or a single, brief (≤5 min) pulse of 5-HT, will produce only short-term (<30 min) behavioral and synaptic enhancement. However, several spaced tail shocks, or repeated application of 5-HT, can result in long-term (≥24 hr) behavioral and synaptic changes in *Aplysia*.[5,14,36,37] The long-term memory for sensitization differs in several fundamental respects from the short-term form. First, the long-term memory, unlike the short-term form, depends on protein synthesis and gene transcription.[5,7–9,38–48] Second, LTS, unlike short-term sensitization (STS), typically, although not invariably,[49] involves significant structural reorganization, particularly the growth of new presynaptic varicosities.[50–56]

Comment on Cellular Locus of Inductive Processes

A complication in describing the cellular and molecular mechanisms that underlie the induction—as well as the maintenance—of LTS in *Aplysia* is identifying the cellular site where the mechanisms operate. For decades it was assumed that the changes mediating the long-term behavioral and synaptic changes that characterized LTS were confined to the presynaptic sensory neuron.[4,57] Although, as summarized later, there are indeed important presynaptic changes that contribute to the induction of the long-term memory for sensitization, it is now apparent that such changes are insufficient to account for the memory, and that postsynaptic changes are also critical (for review, see Glanzman[58]). Furthermore, in at least some cases, the changes, whether presynaptic or postsynaptic, are not cell autonomous but require as yet unidentified trans-synaptic signals; these signals may be retrograde[59–61] or anterograde.[62] The role of postsynaptic mechanisms and trans-synaptic signals in long-term memory in *Aplysia* had been neglected (but see refs.[61,63–65]) prior to the past decade. There have been far more studies of presynaptic changes that underlie long-term memory than of postsynaptic changes; biochemical and molecular studies of LTS have been particularly weighted toward presynaptic mechanisms. A cursory survey of the literature on LTS in *Aplysia* might therefore lead the naive reviewer to conclude that presynaptic mechanisms play a larger role than postsynaptic mechanisms in this form of learning. However, the present prominence of presynaptic mechanisms in studies of long-term memory in *Aplysia* is unlikely to reflect any underlying biological reality,[58] and the role of postsynaptic biochemical and molecular changes in LTS represents a rich, mostly untapped, vein of discovery for future investigation.

Biochemical and Molecular Mechanisms of Induction

5-HT, released into the CNS of *Aplysia* by sensitizing stimuli such as tail shock,[17] activates the second messenger cAMP within sensory neurons.[28,66–71] Repeated or prolonged application of sensitizing stimuli, or of 5-HT, produces prolonged elevation of cAMP within the sensory neurons,[70] resulting in persistent activity of protein kinase A (PKA).[72] The persistent activity of PKA, a heterodimer with two regulatory and two catalytic subunits, results, at least in part, from a selective downregulation of the regulatory subunits of the protein kinase.[45] The decrease in the ratio of regulatory to catalytic subunits requires protein synthesis[45] and is caused by selective proteolysis of the regulatory subunits.[73]

The loss of PKA's regulatory subunits and its consequent persistent activation, produced by repeated stimulation with 5-HT, can lead to a translocation of the catalytic subunits from the cytoplasm of the sensory neuron to its nucleus.[67] The nuclear translocation of the PKA catalytic subunits, in turn, activates cyclic AMP response element binding (CREB) protein, a transcriptional activator[7,8,74]; the activation of CREB is a key step in the induction of long-term memory in *Aplysia*. CREB represents a family of transcription factors.[75] In *Aplysia*, the activator isoform is termed CREB1. The activity of CREB1 is normally inhibited by a repressor CREB isoform, known as CREB2[41]; stimuli that induce LTS or long-term facilitation (LTF) cause the derepression of CREB1 by CREB2, thereby leading to CRE-dependent gene expression.[6] It is believed that this derepression is mediated by extracellular signal-regulated kinase (ERK)/mitogen-activated kinase (MAPK); cytoplasmic MAPK is activated within sensory neurons by prolonged 5-HT treatment and translocates to the nucleus, where it phosphorylates CREB2.[76–80] Among the genes stimulated by CREB1 are the immediate early gene ubiquitin-C hydrolase,[81] which encodes an enzyme that enhances proteasome activity,[82] as well as genes that encode two other transcription factors, *Aplysia* Activating Factor (ApAF)[83] and the CCAAT box/enhanced binding protein (C/EBP).[84] ApAF and C/EBP together, possibly with other as yet unidentified transcription factors, trigger a second wave of gene activation; one of the consequences of this later wave of transcription is believed to be the growth of new synaptic connections (discussed later).[53,85]

Presynaptic Mechanisms of LTS and LTF

LTS is associated with facilitation of the monosynaptic connection between the central sensory and motor neurons that mediate the siphon- and tail-withdrawal reflex.[36,37,86] Because short-term facilitation (STF) of the sensorimotor synapse is largely, perhaps exclusively, the result of presynaptic changes,[29,87–89] it was seductive to extend the presynaptic model to LTF. Unquestionably, LTF results partly from enhanced presynaptic release of glutamate, the neurotransmitter used by *Aplysia* sensory neurons.[90–93] Nonetheless, the idea that LTF results exclusively,[4,57,87] or even predominantly, from presynaptic mechanisms is no longer tenable.[58,59,62,94]

A major advance in the mechanistic analysis of LTF in *Aplysia* was the demonstration of LTF in sensorimotor co-cultures by Montarolo *et al.* in 1986.[5] The methodology of Montarolo *et al.* has been used to make many of the most important discoveries regarding long-term memory in *Aplysia*. For *in vitro* studies of long-term synaptic plasticity, identified sensory and motor neurons are individually dissociated from central ganglia of *Aplysia* and placed into cell culture together (Figure 17.1A_1). Under the right conditions, the neurons will form chemical synaptic connections (Figure 17.1A_2); the strength of these connections gradually increases over the first several days *in vitro* and then reaches a stable value typically after 3–5 days. (The precise time at which the synapses stabilize in strength depends on factors such as the specific identities of the pre- and postsynaptic neurons—for example, whether the giant gill and mantle motor neuron L7[1,5] or one of the small siphon motor neurons[23,96] is used for the postsynaptic neuron—the condition of the animals from which the neurons were dissociated, and the cell culture medium used.) After the synaptic strength has stabilized, as indicated by measurements of the excitatory postsynaptic potential (EPSP), the experiment is begun. LTF is induced by five spaced pulses of 5-HT; each pulse is typically 5 min long and separated from the other pulses by a 15-min period during which the drug is washed out of the culture dish (Figure 17.1B_1). This training protocol, referred to as the 5 × 5-HT protocol, was originally designed to mimic the spaced delivery of tail shocks used to induce behavioral LTS. (If activation of the sensory neuron is paired with the 5-HT pulses, the result is an associative form of LTF.[97,98] Activity-dependent LTF, an associative form of synaptic plasticity that mediates so-called 'site-specific' sensitization[99,100] and also plays a role in classical conditioning of the withdrawal reflex,[101–104] differs somewhat mechanistically from activity-independent LTF.[105] Furthermore, the mechanism of associativity for activity-dependent LTF in

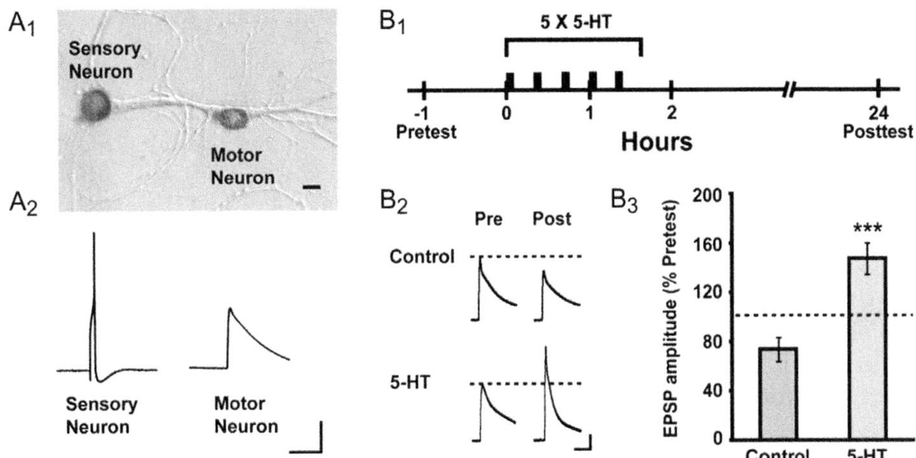

FIGURE 17.1 Long-term facilitation of the sensorimotor synapse *Aplysia* in dissociated cell culture. (A_1) Sensorimotor co-culture. Here, the presynaptic neuron is a pleural sensory neuron,[24] and the postsynaptic neuron is a small siphon (LFS) motor neuron.[23] Scale bar = 20 μm. (A_2) Sample electrophysiological records from a pretest of a synapse. Scale bars = 20 mV and 200 msec. (B_1) Experimental protocol for the demonstration of 24-hr LTF. The co-culture was treated with five spaced 5-min applications of 5-HT (100 μM); the drug was washed out of the cell culture for 15 min between 5-HT pulses. (B_2) Sample EPSPs. Each pair of traces shows EPSPs recorded from the same sensorimotor synapse on Day 1 (Pre) and approximately 24 hr later (Post). Training with 5-HT produced LTF (lower traces). Scale bars = 10 mV and 80 msec. (B_3) Summary of the group data. The mean normalized EPSP in the synapses trained with 5-HT (5-HT group, $n = 10$) at 24 hr after training was significantly greater than that in the control (untrained) group ($n = 10$; $^{***}p < 0.001$). Source: *Reproduced with permission from Cai* et al.[95]

site-specific sensitization differs from that in classical conditioning. See Chapter 18.) Training with the 5 × 5-HT protocol can induce LTF that persists for 24[5] to 72 hr[106,107] (Figure 17.1$B_{2,3}$). Commonly, the 5-HT is simply added to the cell culture dish, but in some studies the 5-HT has been locally perfused in spaced bouts over a region of sensorimotor synaptic contact in order to induce synapse-specific LTF.[106,108] (Although the 5 × 5-HT delivery method, which uses a uniform interval between the pulses of 5-HT, is certainly effective, it has recently been shown that a protocol involving nonuniform interstimulus intervals results in enhanced LTF.[109])

LTF, like LTS, requires protein synthesis and CREB-dependent gene transcription.[5,7,8,41,110] Both presynaptic[45,106,108,111] and postsynaptic[59,62,112] protein synthesis are involved, and a critical component of this LTF-related protein synthesis is local.[42,106,108,111–114] Among the local mRNAs that are translated by prolonged 5-HT stimulation are those that encode the cytoskeletal proteins, α_1-tubulin and β-thymosin, which may play roles in synaptic growth, and ribosomal proteins.[113] The latter may increase the number of translationally competent ribosomes at the stimulated synapses. By thereby producing a localized site for translation, this mechanism could serve to restrict the products of learning-related transcription to stimulated synapses; such a mechanism may help to account for synapse specificity during learning.[42,108,115]

One important function of local protein synthesis during LTF is to stimulate a retrograde signal that travels to the nucleus and activates gene transcription. This phenomenon was first described by Martin *et al.*[108] in their study of synapse-specific LTF. In this study, two sensory neurons were co-cultured with, and synaptically connected to, a single L7 motor neuron. When one of the two regions of sensorimotor contact, but not the sensory neuron cell body, was given 5 × 5-HT training via local perfusion, the trained sensorimotor synapse exhibited LTF, whereas the second, untrained, synapse did not. By contrast, local 5 × 5-HT treatment failed to induce LTF of the stimulated synapse when the protein synthesis inhibitor emetine was included in the 5-HT-containing solution. Martin and colleagues further showed that when a brief pulse of 5-HT—which, alone, produces only STF—was applied to one of the two regions of sensorimotor contact, and the other region of sensorimotor contact was given the 5 × 5-HT training, the result was that both synapses underwent LTF. This result implies that the 1 × 5-HT-trained synapse was able to 'capture' the somal products required for LTF that are produced by the 5 × 5-HT treatment of the other synapse; in other words, the 1 × 5-HT treatment induces a 'tag' of the synapse that is necessary, albeit not sufficient, for the synapse to undergo LTF. Induction of this tag does not depend on protein synthesis because it is not prevented if the synaptic region is exposed to emetine (a long-lasting protein synthesis inhibitor) together with the single pulse of 5-HT. Taken together, Martin *et al.*'s results indicate that local 5 × 5-HT training induces two processes required for LTF—a protein synthesis-dependent retrograde signal that activates gene transcription and a protein synthesis-independent tag that permits the synapse to capture the somal proteins/mRNAs whose production results from the 5×5-HT stimulated transcription. (Note that similar results have been reported for long-term potentiation (LTP) in the mammalian hippocampus.[116,117]) In a study of sensorimotor synapses in pleural–pedal ganglia, Sherff and Carew[118] also found evidence for a protein synthesis-dependent retrograde signal that can be induced by relatively brief 5-HT treatment and that supports the induction of LTF.

One presynaptic protein that is locally synthesized in response to prolonged 5-HT stimulation is the neuropeptide sensorin.[119] First identified in *Aplysia*, sensorin is found exclusively in sensory neurons.[120] When a sensorimotor co-culture is stimulated with multiple pulses of 5-HT, there is a rapid, protein synthesis-dependent increase in sensorin expression in the presynaptic varicosities of sensory neurons, as assessed with immunohistochemistry.[121–123] Interestingly, this increase in local translation of sensorin requires a postsynaptic signal; the increase does not occur if the motor neuron is injected with the rapid Ca^{2+} chelator 1,2-bis (*o*-aminophenoxy)ethane-*N*,*N*,*N*′,*N*′-tetraacetic acid (BAPTA) or in the varicosities of isolated sensory neurons stimulated with 5×5-HT.[59,60] The enhanced local expression of sensorin is followed by release of the neuropeptide from the varicosities[123,124]; released sensorin binds to autoreceptors on the sensory neuron, thereby activating presynaptic MAPK, which then translocates to the sensory neuron's nucleus.[124] Release of sensorin is necessary for LTF: Treatment of sensorimotor co-cultures with an anti-sensorin antibody blocks the induction of LTF.[124] Furthermore, release of sensorin requires PKA activity. (Note that it also requires protein kinase C activity. See Hu *et al.*[105]) This finding led Schacher and colleagues[123,124] to propose that sensorin release and binding to autoreceptors is a necessary intermediate step between the activation of PKA and the activation and nuclear translocation of MAPK during LTF.

In addition to inducing an increase in local synthesis of sensorin, multiple pulses of 5-HT cause an increase in transcription of sensorin[119]; the sensorin mRNAs are exported to synapses in sensorimotor co-cultures, and this synaptic localization of the mRNAs requires an appropriate target motor neuron.[119,125,126] This result addresses a crucial general question for molecular models of long-term memory, namely whether the products of nuclear transcription can support synapse specificity during learning-related long-term synaptic plasticity.[127] Evidence from studies of LTP in the mammalian hippocampus and hippocampal-dependent learning suggest that mRNAs can indeed be targeted to active synaptic regions.[128–130] Martin and colleagues[126] have shown that the synaptic localization of sensorin reporter mRNAs in *Aplysia* sensorimotor co-cultures is accomplished through *cis*-acting localization elements (LEs). LEs within the 3′ UTR are sufficient for localization of the reporter mRNA to distal sensory neurites, whereas LEs within the 5′ UTR are required for synaptic localization of the reporter mRNAs.[126] Recently, this group identified a 66-nucleotide-long element in the 5′ UTR of sensorin that is both necessary and sufficient for synaptic localization of the mRNA.[131]

Structural Changes Underlying Long-Term Sensitization Memory

Early ultrastructural work in intact animals showed that LTS is associated with presynaptic morphological changes. In particular, the induction of LTS in *Aplysia* results in an increase in the number of presynaptic varicosities, the number of active zones associated with sensory neuron terminals, and the number of vesicles per active zone.[50,51,132] Furthermore, generally, there is enhanced outgrowth of the neurites of sensory neurons during LTS,[37,51,133] although some training protocols that induce LTS do not induce outgrowth of sensory neurites.[49]

Glanzman *et al.*[61] extended these results to the *in vitro* system; they demonstrated that LTF results in an increase in both the number of presynaptic varicosities and the number of sensory neurites in sensorimotor co-cultures. Since then, much of the work on structural changes associated with long-term, sensitization-related memory in *Aplysia* has used the *in vitro* system. The increase in presynaptic growth in *Aplysia* sensory neurons due to LTF requires protein and RNA synthesis,[134] can be induced by injecting cAMP into the sensory neuron, and is mediated by CREB1.[108] Just as is the case for LTS, however, there are training protocols that can induce LTF that do not produce the presynaptic outgrowth,[42] although the facilitation induced by these protocols does not persist for significantly longer than 24 hr.

One of the requirements for the structural changes that typically accompany LTS is the downregulation of an *Aplysia* neural cell adhesion molecule (ApCAM). 5×5-HT stimulation causes a downregulation of the ApCAM protein in sensory neurons; it also leads to a decrease in the number of ApCAM molecules on the surface of sensory neurites.[135] The decrease in the amount of surface ApCAM molecules results from their endocytosis.[54] Furthermore, applying an antibody to ApCAM to isolated sensory neurons causes their neurites to defasciculate[135]; such defasciculation may well be a critical step in the 5-HT-induced outgrowth of sensory neurites. The downregulation of presynaptic ApCAM expression and the internalization of surface ApCAM molecules on sensory neurites and subsequent neuritic defasciculation are likely to be initial steps in the process of structural change that accompanies LTS and LTF.[136,137]

In addition to playing a critical role in the restructuring of sensory neurons during long-term learning in *Aplysia*, internalization of ApCAM also subserves retrograde synapse-to-nucleus signaling. Lee *et al.*[74] found that ApCAM is associated with a transcriptional activator, CAM-associated protein (CAMAP). In its nonactivated state, CAMAP is bound to the cytoplasmic tail of ApCAM molecules in the cell membrane of the sensory processes. Repeated pulses of 5-HT stimulate PKA, which phosphorylates CAMAP; phosphorylation of CAMAP, in turn, causes its dissociation from ApCAM, an essential step in the internalization of ApCAM.[54,135] The freed CAMAP then translocates to the nucleus, where it serves as a coactivator, together with CREB1, of the downstream transcription factor C/EBP, an immediate gene whose activation is critical for the consolidation of LTF.[84]

Whereas the majority of the work on structural changes related to long-term memory in *Aplysia* has focused on presynaptic changes, LTS also involves postsynaptic structural changes. In an electron microscopic study, Bailey and Chen[63] quantified changes in the number of spine-like filapodia on the neurites of the identified motor neuron L7 that occurred in intact animals following LTS-inducing training. They observed an increase in the number of small ($<0.5\ \mu m$), spine-like L7 processes. Furthermore, they also saw an increase in the number of presynaptic contacts per square micrometer of L7's dendritic surface. These results suggest that LTS causes a coordinated growth of pre- and postsynaptic structures, resulting in increased synaptic contact between the sensory and motor neurons.

Postsynaptic and Trans-Synaptic Mechanisms

As noted previously, persistent memory in *Aplysia* involves postsynaptic as well as presynaptic changes. One such change is an increase in postsynaptic Ca^{2+}. Cai *et al.*[59] reported that LTF of the *in vitro* sensorimotor synapse is blocked by an injection of the rapid chelator BAPTA into the motor neuron prior to $5 \times$ 5-HT stimulation. This result implies that multiple pulses of 5-HT cause a rise in postsynaptic intracellular Ca^{2+}. Similar results have been obtained if a postsynaptic injection of BAPTA is made prior to a 10-min exposure to 5-HT,[88,89] which causes intermediate-term facilitation (ITF)—that is, facilitation that persists from 30 min to approximately 3 hr after the exposure to 5-HT.[138] The 5-HT-induced rise in postsynaptic Ca^{2+} results from release from intracellular stores—both inositol-1,4,5-trisphosphate (IP_3)-mediated and ryanodine receptor-mediated stores.[88,139] The increase in intracellular Ca^{2+} activates postsynaptic calcium/calmodulin II (CaMKII).[89] The downstream effects of this pathway's activity remain to be determined, but they are likely to involve at least some of those described below.

Prolonged 5-HT treatment in both the intact nervous system[65] and sensorimotor co-cultures[64] produces functional upregulation of α-amino-3-hydroxy-5-methyl-4-isoxazole propionic acid-type glutamate receptors (AMPARs). The early evidence for this effect came from studies in which the effect of brief pulses of AMPAR agonists delivered to *Aplysia* motor neurons was quantified electrophysiologically before and after prolonged exposure to 5-HT. It was found that this treatment resulted in a greater evoked potential in the motor neurons 24 hr later.[64,65] A 10-min-long, ITF-inducing exposure to 5-HT has a similar effect.[88,139] Together, these results imply that prolonged 5-HT stimulation triggers an increase in the number of postsynaptic AMPARs. Investigations of mammalian LTP and learning have shown that long-term synaptic plasticity and learning are mediated, in part, by modulation of AMPAR trafficking.[140,141] In the case of the mammalian CNS, modulation of AMPAR receptor trafficking can be initiated by activation of postsynaptic *N*-methyl-D-aspartate receptors (NMDARs). The evidence from the work on *Aplysia* suggests that a monoamine can similarly initiate modulation of AMPAR trafficking. In mammals, activation of NMDARs causes additional AMPARs to be inserted into dendrites, either at extrasynaptic sites or directly into the postsynaptic membrane.[141] The extrasynaptically inserted AMPARs then diffuse into the synapse.[142] Direct insertion of the AMPARs is believed to be mediated by exocytosis from a recycling pool of receptors.[143,144]

The evidence from studies of ITF and LTF in *Aplysia* supports the following scheme: 5-HT exposure causes in increase in postsynaptic Ca^{2+} due to release from intracellular stores; the increased intracellular Ca^{2+} then drives insertion of additional AMPARs into postsynaptic sites via an exocytotic mechanism.[59,62,88,139] This could involve direct synaptic insertion from a recycling pool of vesicles or lateral diffusion following exocytotic insertion into extrasynaptic sites. Unfortunately, currently, no antibodies to *Aplysia* AMPARs are available; therefore, one cannot directly monitor possible movement of endogenous AMPARs into and out of postsynaptic sites. On the other hand, several *Aplysia* AMPA-type receptors have been cloned,[94,145] which has permitted experimenters to fabricate exogenous constructs of green fluorescent protein (GFP)-tagged *Aplysia* AMPA-type receptors and express these constructs in *Aplysia* motor neurons. Using this method, together with fluorescence confocal microscopy, the experimenters monitored the effect of 5-HT treatment on the exogenous AMPA-type receptors. The resulting data are consistent with the idea that LTF[94] and ITF[62] in *Aplysia* involve modulation of postsynaptic AMPAR trafficking. In the case of LTF, the 5×5-HT training produced enhanced clustering of AMPA-type receptors in the motor neurons of sensorimotor co-cultures,[94] whereas in the case of ITF, a 10-min exposure to 5-HT caused a rapid increase in the number of GFP puncta in the postsynaptic cell membrane, reflecting receptor insertion.

Dopamine causes the insertion of new AMPARs into the dendrites of hippocampal neurons, and this insertion depends on local protein synthesis.[146] Postsynaptic protein synthesis may also play a role in the functional upregulation of AMPA-type receptors caused by prolonged 5-HT stimulation in *Aplysia*. Although it was originally concluded that postsynaptic protein synthesis was not mechanistically involved in LTF,[65,108] recent work has established that proteins synthesized by the motor neuron are, in fact, critical for this form of synaptic plasticity.[59,62] Furthermore, at least some of these postsynaptic proteins result from local synthesis.[112] In addition to mediating the increased number of postsynaptic AMPA-type receptors,[65] postsynaptic protein synthesis probably subserves other important mechanistic roles in LTF. For example, it is likely to mediate the growth of new postsynaptic spines.[63] Another potential role for protein synthesis is in the regulation of postsynaptic protein kinase C (PKC). Recent work has implicated the constitutively active protein kinase M (PKM) fragment of the atypical *Aplysia* PKC isoform, named PKM Apl III,[147] in the maintenance of the memory for ITF,[148,149] as well as LTF and LTS.[95] In *Aplysia*, PKM Apl III is formed by Ca^{2+}-dependent proteolytic cleavage of PKC Apl III, the atypical isoform.[147] Bougie *et al.*[149] used a Förster resonance energy transfer reporter construct to measure 5-HT-induced cleavage of PKC Apl III in *Aplysia* motor neurons. (Note that 5-HT does not induce cleavage of exogenous PKC Apl III in *Aplysia* sensory neurons.[149]) They observed that a 10-min exposure to 5-HT caused cleavage of the atypical PKC, and this cleavage depended on calpain and protein synthesis. (Interestingly, the mammalian homolog of PKM Apl III, PKMζ, is generated by local translation of the mRNA, which is formed by transcription from an alternate start site within the PKCζ gene.[150])

A fascinating theme to emerge from recent studies of learning-related persistent synaptic plasticity in *Aplysia* is the importance of trans-synaptic mechanisms. The first evidence for a critical role for retrograde signaling in LTF came from the study of Glanzman *et al.*[61] of *in vitro* changes in presynaptic structure that mediate LTF. These investigators examined the effect of multiple pulses of 5-HT on the structure of isolated sensory neurons in culture and sensory neurons of sensorimotor co-cultures. Whereas the 5-HT treatment triggered significant outgrowth of sensory processes and varicosities in the co-cultures, it had no effect on the structure of isolated sensory neurons. These results suggest that a signal from the motor neuron is required for the 5-HT-stimulated growth of the sensory neuron. Later work established that LTF depended on postsynaptic Ca^{2+} because it could be blocked by injecting BAPTA into the target motor neuron prior to the 5×5-HT treatment.[59,62] (Postsynaptic BAPTA also blocks the induction of ITF[88,89]) This result implies that at least some of the presynaptic changes associated with LTF do not result exclusively from sensory neuron-autonomous processes but, rather, require a retrograde signal for their expression. In support of this idea, one critical presynaptic mechanism that mediates LTF, the synaptic translation of the sensory neuron peptide sensorin, has been shown to require Ca^{2+} signaling in the motor neuron.[59,60] A retrograde signal implicated in LTF is an interaction between the synaptic adhesion molecules neurexin and neuroligin.[151] Choi *et al.*[152] identified *Aplysia* homologs of neurexin and neuroligin, ApNRX and ApNLG; ApNRX was found to be present in sensory neurons, whereas ApNLG was present in motor neurons. Recombinant versions of the two molecules were found to bind to each other. Furthermore, overexpression of ApNRX in sensory neurons and overexpression of ApNLG in motor neurons of sensorimotor co-cultures produced a long-term increase in the strength of the sensorimotor synapse, whereas injection of antisense ApNRX in the presynaptic sensory neuron or antisense ApNLG in the postsynaptic motor neuron blocked the induction of LTF as well as the presynaptic structural changes associated with LTF. These results represent strong evidence for a role in LTF for trans-synaptic signaling via a neurexin–neuroligin interaction; whether other trans-synaptic pathways mediate retrograde signaling during LTF remains to be determined.

Recently, anterograde signaling has also been implicated in LTF in *Aplysia*. Jin *et al.*[153] reported that spontaneous release of transmitter from the sensory neuron was required for the induction of ITF and LTF. In addition, this group found that spontaneous presynaptic release activates the group I metabotropic receptor mGluR5 in the motoneuron; activation of mGluR5, in turn, elevates postsynaptic Ca^{2+} through release from IP_3-mediated stores.[62] This finding is consistent with previous evidence implicating release of Ca^{2+} from IP_3-mediated postsynaptic stores in ITF in *Aplysia*.[88,139] Intriguingly, Jin *et al.*[62] found that injecting botulinum toxin, an exocytotic inhibitor, into presynaptic sensory neurons blocked the increase in AMPA-type receptors in the postsynaptic neuron induced by a 10-min application of 5-HT. These results suggest that modulation of postsynaptic AMPA-type receptor trafficking is regulated by an anterograde signal linked to enhanced presynaptic release.

The mechanistic model for LTS in *Aplysia* that has emerged within the previous decade is strikingly complex (Figure 17.2). Both pre- and postsynaptic changes are involved in the induction and stabilization of long-term memory; moreover, these changes are coordinated by retrograde and anterograde signaling. It is perhaps surprising that a simple, nonassociative form of learning in a relatively simple invertebrate organism has proved to involve such a rich repertoire of cellular and molecular processes. No doubt, there remain mediatory processes to be identified.

MAINTENANCE OF LTS MEMORY IN APLYSIA

Although significant progress has been made toward a mechanistic understanding of the induction and early stabilization of LTS (Figure 17.2), far less is understood about how the long-term memory persists within the nervous system of *Aplysia*. However, the previous decade has witnessed accelerated interest in memory maintenance in *Aplysia*, and we now have some insights into this fascinating problem.

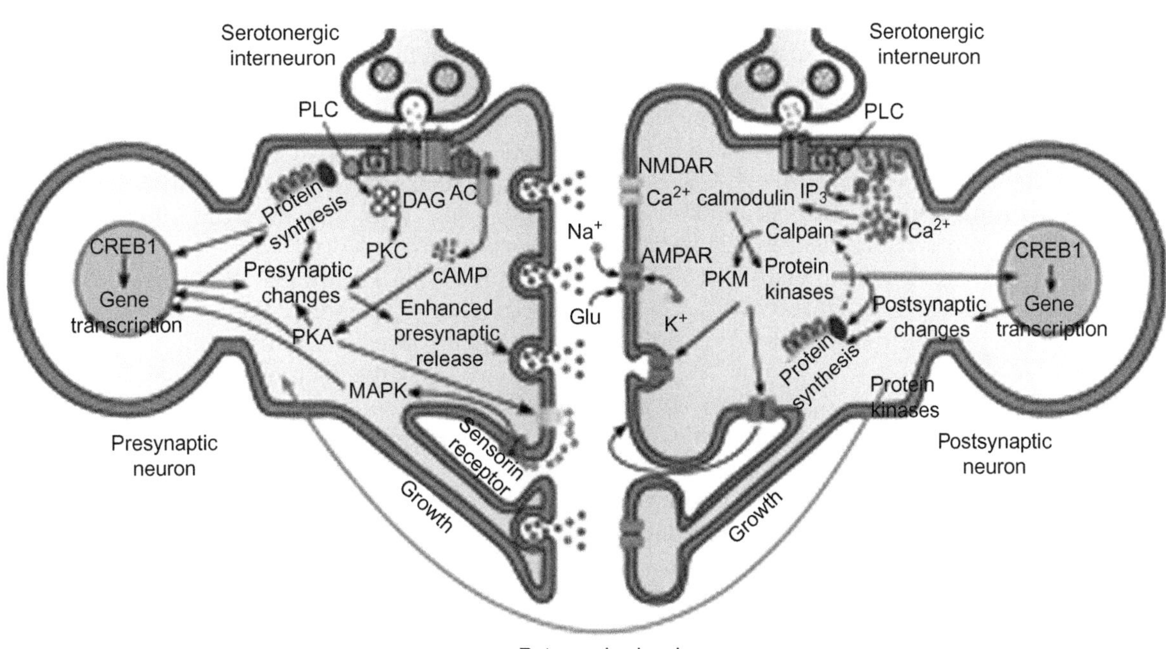

FIGURE 17.2 Cellular model for sensitization-related long-term facilitation. Prominent features of the model are postsynaptic modulatory input from monoaminergic interneurons and retrograde signaling. Prolonged stimulation with 5-HT causes both pre- and postsynaptic long-term changes, including enhanced release of glutamate, the presynaptic transmitter, and modulation of postsynaptic AMPA-type receptor trafficking. The latter process involves exocytotic insertion of AMPA receptors into the cell membrane of the motor neuron, either at extrasynaptic sites, after which the receptors are transported to the synapse, or directly into the postsynaptic membrane. This process is mediated by G protein-stimulated release of Ca^{2+} from intracellular stores,[88,139] as well as by PKM[154,155] and possibly other kinases, and it requires enhanced spontaneous presynaptic release (not depicted).[62,153] The elevated postsynaptic intracellular Ca^{2+} is also responsible for activating PKM Apl III through calpain-dependent proteolytic cleavage[149]—which itself depends on protein synthesis—as well as for triggering, either directly or indirectly (possibly through local protein synthesis[112]), the activation of one or more retrograde signals. The retrograde signals, in turn, contribute critically to presynaptic changes and enhanced presynaptic release. Another process that is critical for the long-term presynaptic changes is the local synthesis and release of the peptide sensorin, which binds to autoreceptors on the sensory neuron, thereby activating presynaptic MAPK.[105,124,156] Not shown are potential anterograde signals[62] nor processes that mediate the early maintenance of LTF.[111,157] Also, both LTS and LTF involve long-term structural modifications, including the growth of presynaptic processes and varicosities[50,51,61] as well as of postsynaptic spines.[63]

CPEB and the Prion Hypothesis

One idea proposed to explain the persistence of the memory for LTF, and by extension for LTS, centers on the translational activator cytoplasmic polyadenylation element binding protein (CPEB). CPEB mediates polyadenylation-induced translation and plays a key role in cell division during development.[158] A CPEB isoform has been identified in the nervous system of *Aplysia* (ApCPEB)[111]; this isoform has several properties that make it attractive as a memory maintenance molecule in *Aplysia*. mRNAs of ApCPEB are present in the neurites of sensory neurons, and translation of ApCPEB mRNA is stimulated locally within the neurites by a prolonged application of 5-HT.[111] Another suggestive property of ApCPEB is that it can exist in two states—a monomeric state in which the protein is inactive, or acts as a repressor, and an active, multimeric state. In the multimeric state, ApCPEB can recruit monomeric proteins to the multimer and thereby become self-sustaining.[157] Thus, multimeric ApCPEB can act as a self-sustaining, local translational hub within sensory neurites and thereby promote ongoing protein synthesis at the synapse. In this way, ApCPEB could serve as both a synaptic marker and a mechanism for sustaining synapse-specific, long-term memories.[108] Consistent with this idea, 5 × 5-HT stimulation can induce multimerization of ApCPEB in *Aplysia* sensory neurons, and injecting an antibody to multimeric ApCPEB into sensory neurons 24 hr after 5 × 5-HT treatment disrupts the maintenance of LTF.[157] Studies of long-term memory in *Drosophila*[159,160] provide further support for the involvement of CPEB in memory persistence.

The amyloid-like properties of ApCPEB in *Aplysia* have led to the suggestion that this molecule is a prion.[157,161] This is an intriguing idea because it suggests a functional role for prions in the nondiseased brain, in contrast to its pathogenic role in devastating diseases such as Creutzfeldt–Jacob disease.[162] Regardless of whether or not ApCPEB is a true prion, however, it is unlikely to mediate the maintenance of long-term memory in *Aplysia* beyond the first 2 days. Evidence for this assertion comes from a study by Miniaci *et al.*,[106] who examined the effect on the memory for LTF of inhibiting protein synthesis at various times after treating sensorimotor co-cultures with 5 × 5-HT. Consistent with a role for ApCPEB in the early maintenance of long-term synaptic memory, treatment with the protein synthesis inhibitor emetine at 24 or 48 hr after 5-HT training disrupted LTF, as indicated by tests at later times; by contrast, emetine treatment at 72 hr after 5-HT training did not affect LTF. Similar results were obtained from local application of an antisense ApCPEB oligonucleotide to sensorimotor synapses. Thus, by 72 hr after training, local translation by ApCPEB would appear not to be critical for the maintenance of LTF.

PKM and the Maintenance of Long-Term Memory in *Aplysia*

Work in mammals has implicated the constitutively active fragment of atypical PKCζ, PKMζ, in the persistence of long-term synaptic plasticity and memory.[154,163–166] As mentioned previously, the nervous system of *Aplysia* contains an isoform of PKMζ, PKM Apl III. This kinase can be inhibited by the zeta inhibitory peptide (ZIP), which encodes the amino acid sequence of the pseudosubstrate region of the inhibitory domain of the atypical PKC isoform, PKC Apl III.,[147] as well as—at micromolar concentrations—by the PKC inhibitor chelerythrine.[148] We have used these two compounds to test for a role for PKM Apl III in LTS and LTF in *Aplysia*. We have found that treatment with either ZIP or chelerythrine disrupts the maintenance of LTS.[95] (In our experiments, the drugs were injected directly into the hemolymph of the animals at various times after sensitization training with tail shocks.) ZIP or chelerythrine eliminated the memory for LTS, even when applied as late as 7 days after training; furthermore, the memory did not reappear for at least 2 days after the drug injection, nor could it be reinstated by a brief bout of sensitization training. Injection of the scrambled ZIP peptide did not affect the long-term memory. Both ZIP and chelerythrine were equally effective at disrupting the maintenance of long-term synaptic memory in *Aplysia*. In our experiments, sensorimotor co-cultures, after an initial test of synaptic strength, received 5 × 5-HT stimulation; 24 hr later, some of the co-cultures were incubated for 1 hr in either ZIP or chelerythrine. The synapses were then retested at 48 hr after training. LTF was reversed at 48 hr in co-cultures treated with either ZIP or chelerythrine, whereas co-cultures incubated at 24 hr in either scrambled ZIP or control perfusion medium exhibited significant facilitation. Similar results have been reported by Hu *et al.*,[156] who trained co-cultures with 5 × 5-HT for two consecutive days; such training yields LTF that persists for 1 week. These investigators found that treatment with chelerythrine on the second day of 5-HT treatment eliminated all LTF, causing a return of the sensorimotor EPSP to its baseline amplitude by Day 3 of the experiment; moreover, no facilitation was evident on Day 7. Treatment with the PKM inhibitor on the second day of 5-HT training also blocked the increase in presynaptic sensorin expression normally observed 24 hr later. Taken together, the results of these two studies provide strong support for the notion that PKC activity, specifically that of PKM Apl III, supports the long-term maintenance of memory in *Aplysia*.

Memory Reconsolidation in *Aplysia*

One of the most intriguing of memory phenomena is so-called memory reconsolidation. Until relatively recently, the predominant idea regarding memory persistence was that memories destined for long-term storage underwent a single, protein synthesis-dependent process of consolidation, after which they were stable, barring injury or disease.[167] Opposed to this idea have been data indicating that memories could become labile and subject to disruption upon reactivation.[168–171] Recent years have witnessed increasing evidence, from studies of both vertebrates and invertebrates, of the lability of reactivated long-term memories..[172–181] Two studies have now demonstrated reconsolidation of long-term synaptic and behavioral memory in *Aplysia*; specifically, both Cai *et al.*[107] and Lee *et al.*[182] showed that the memory for LTS and LTF can become disrupted by inhibition of protein synthesis following reactivation of the behavioral/synaptic memory. (A methodological difference between the two studies is that Cai and colleagues used brief sensitization training/5-HT stimulation to reactivate the memory for LTS/LTF, whereas Lee *et al.* used test-type stimulation—weak tactile stimulation of the siphon in the case of LTS and brief homosynaptic activation of the synapse in the case of LTF—for memory reactivation.) The study by Hu *et al.*[156] provides additional support for the idea that the memory for LTF can undergo reconsolidation; they found that if a protein synthesis inhibitor was applied to sensorimotor co-cultures during the second day of 5-HT training, both the new facilitation, resulting from training on the second day, and the persistent facilitation, resulting from the first day of 5-HT training, were disrupted.

Lee *et al.*'s[182] results implicate the ubiquitin/proteosome pathway in memory reconsolidation in *Aplysia*. These investigators found that the ubiquitin/proteasome inhibitor β-lactone blocked the disruption of long-term behavioral and synaptic memory when β-lactone was applied together with the protein synthesis inhibitor immediately following memory reactivation. These results suggest that when the long-term memory is reactivated, it is destabilized by protein degradation via the ubiquitin/proteasome pathway; subsequently—barring inhibition of protein synthesis—the memory becomes restabilized via new translation.[183,184]

The data from the studies of memory 'erasure' through inhibition of PKM and of memory reconsolidation in *Aplysia* have important implications. It has been contentious in the field of memory research whether, when long-term memories are apparently erased by inhibition of PKM or through disruption of reconsolidation, the memories are permanently eliminated or whether they are, instead, temporarily unable to be retrieved (e.g., see Lattal and Abel[185]). Another issue of contention is the extent to which reconsolidation of memory resembles original consolidation.[174,176,186] These issues are difficult to resolve in the absence of a realistic synaptic model of memory erasure and reconsolidation. The recent studies in *Aplysia* provide such a model. The cellular and molecular consequences of inhibiting PKM Apl III, or of disrupting memory reconsolidation, can be examined over several days at a single synapse in cell culture, one amenable to rigorous mechanistic analyses. Furthermore, the results from studies of the *Aplysia in vitro* sensorimotor synapse are likely to have significant ecological validity. The extensive body of prior work on the mechanisms underlying behavioral plasticity of the defensive withdrawal reflex, some of which has been reviewed here, has established that not only can the findings from experiments on sensorimotor co-cultures be extrapolated to the *in vivo* synapse but they will also prove relevant to actual learning in the intact animal. One can therefore anticipate that future studies in *Aplysia* will provide valuable insights into the processes that underlie memory erasure and reconsolidation.

SUMMARY

During the past four decades, sensitization of the defensive withdrawal reflex of *Aplysia* has served as a valuable model system for understanding the cellular and molecular mechanisms that mediate long-term memory. Importantly, discoveries from work on LTS in *Aplysia*, such as the importance of CREB-dependent transcription, have generalized to other systems, particularly mammalian systems. Moreover, I believe that the mine of mechanistic insights into long-term memory opened by work on learning and memory in *Aplysia* is far from being exhausted. As suggested by the recent work on PKM Apl III and on memory reconsolidation, in particular, one can anticipate that studies of LTS and LTF in *Aplysia* will be a source of fundamental knowledge about how the brain stores and maintains memories for many years to come.

Acknowledgments

I thank Wayne Sossin for critically reading the manuscript and Diancai Cai for assistance with the figures.

References

1. Kandel ER. *Cellular Basis of Behavior: An Introduction to Behavioral Biology*. San Francisco: Freeman; 1976.
2. Korbo L, Pakkenberg B, Ladefoged O, Gundersen HJG, Arlien-Søborg P, Pakkenberg H. An efficient method for estimating the total number of neurons in rat brain cortex. *J Neurosci Methods.* 1990;31(2):93–100.
3. Ramon y, Cajal S. *Recollections of My Life*. Cambridge, MA: MIT Press; 1989.
4. Kandel ER. The molecular biology of memory storage: a dialogue between genes and synapses. *Science*. 2001;294:1030–1038.
5. Montarolo PG, Goelet P, Castellucci VF, Morgan J, Kandel ER, Schacher S. A critical period for macromolecular synthesis in long-term heterosynaptic facilitation in *Aplysia*. *Science*. 1986;234:1249–1254.
6. Bartsch D, Ghirardi M, Skehel PA, et al. Aplysia CREB2 represses long-term facilitation: relief of repression converts transient facilitation into long-term functional and structural change. *Cell*. 1995;83:979–992.
7. Dash PK, Hochner B, Kandel ER. Injection of the cAMP-responsive element into the nucleus of *Aplysia* sensory neurons blocks long-term facilitation. *Nature*. 1990;345(6277):718–721.
8. Kaang BK, Kandel ER, Grant SG. Activation of cAMP-responsive genes by stimuli that produce long-term facilitation in *Aplysia* sensory neurons. *Neuron*. 1993;10(3):427–435.
9. Goelet P, Castellucci VF, Schacher S, Kandel ER. The long and the short of long-term memory—A molecular framework. *Nature*. 1986;322:419–422.
10. Frey U, Krug M, Reymann KG, Matthies H. Anisomycin, an inhibitor of protein synthesis, blocks late phases of LTP phenomena in the hippocampal CA1 region *in vitro*. *Brain Res*. 1988;452 (1–2):57–65.
11. Impey S, Mark M, Villacres EC, Poser S, Chavkin C, Storm DR. Induction of CRE-mediated gene expression by stimuli that generate long-lasting LTP in area CA1 of the hippocampus. *Neuron*. 1996;16(5):973–982.
12. Frey U, Frey S, Schollmeier F, Krug M. Influence of actinomycin D, a RNA synthesis inhibitor, on long-term potentiation in rat hippocampal neurons in vivo and *in vitro*. *J Physiol*. 1996;490(Pt 3):703–711.
13. Nguyen PV, Abel T, Kandel ER. Requirement of a critical period of transcription for induction of a late phase of LTP. *Science*. 1994;265(5175):1104–1107.
14. Pinsker HM, Hening WA, Carew TJ, Kandel ER. Long-term sensitization of a defensive withdrawal reflex in *Aplysia*. *Science*. 1973;182:1039–1042.
15. Mozzachiodi R, Byrne JH. More than synaptic plasticity: role of nonsynaptic plasticity in learning and memory. *Trends Neurosci*. 2010;33(1):17–26.
16. Mackey SL, Kandel ER, Hawkins RD. Identified serotonergic neurons LCB1 and RCB1 in the cerebral ganglia of *Aplysia* produce presynaptic facilitation of siphon sensory neurons. *J Neurosci*. 1989;9(12):4227–4235.
17. Marinesco S, Carew TJ. Serotonin release evoked by tail nerve stimulation in the CNS of *Aplysia*: characterization and relationship to heterosynaptic plasticity. *J Neurosci*. 2002;22(6):2299–2312.
18. Marinesco S, Kolkman KE, Carew TJ. Serotonergic modulation in *Aplysia*: I. Distributed serotonergic network persistently activated by sensitizing stimuli. *J Neurophysiol*. 2004;92(4): 2468–2486.
19. Hawkins RD. Localization of potential serotonergic facilitator neurons in *Aplysia* by glyoxylic acid histofluorescence combined with retrograde fluorescent labeling. *J Neurosci*. 1989;9(12): 4214–4226.
20. Glanzman DL, Mackey SL, Hawkins RD, Dyke AM, Lloyd PE, Kandel ER. Depletion of serotonin in the nervous system of *Aplysia* reduces the behavioral enhancement of gill withdrawal as well as the heterosynaptic facilitation produced by tail shock. *J Neurosci*. 1989;9(12):4200–4213.
21. Byrne JH, Castellucci VF, Carew TJ, Kandel ER. Stimulus-response relations and stability of mechanoreceptor and motor neurons mediating defensive gill-withdrawal reflex in *Aplysia*. *J Neurophysiol*. 1978;41(2):402–417.
22. Carew TJ, Castellucci VF, Byrne JH, Kandel ER. Quantitative analysis of relative contribution of central and peripheral neurons to gill-withdrawal reflex in *Aplysia californica*. *J Neurophysiol*. 1979;42(2):497–509.
23. Frost WN, Clark GA, Kandel ER. Parallel processing of short-term memory for sensitization in *Aplysia*. *J Neurobiol*. 1988; 19:297–334.
24. Walters ET, Byrne JH, Carew TJ, Kandel ER. Mechanoafferent neurons innervating tail of *Aplysia*: I. Response properties and synaptic connections. *J Neurophysiol*. 1983;50:1522–1542.
25. Walters ET, Byrne JH, Carew TJ, Kandel ER. Mechanoafferent neurons innervating tail of *Aplysia*: II. Modulation by sensitizing stimulation. *J Neurophysiol*. 1983;50(6):1543–1559.
26. Kandel ER. *Behavioral Biology of Aplysia: A Contribution to the Comparative Study of Opisthobranch Molluscs*. San Francisco: Freeman; 1979.
27. Cleary LJ, Byrne JH, Frost WN. Role of interneurons in defensive withdrawal reflexes in *Aplysia*. *Learn Mem*. 1995;2(3–4):133–151.
28. Brunelli M, Castellucci V, Kandel ER. Synaptic facilitation and behavioral sensitization in *Aplysia*: possible role of serotonin and cyclic AMP. *Science*. 1976;194:1178–1181.
29. Castellucci V, Kandel ER. Presynaptic facilitation as a mechanism for behavioral sensitization in *Aplysia*. *Science*. 1976;194 (4270):1176–1178.
30. Pieroni JP, Byrne JH. Differential effects of serotonin, FMRFamide, and small cardioactive peptide on multiple, distributed processes modulating sensorimotor synaptic transmission in *Aplysia*. *J Neurosci*. 1992;12(7):2633–2647.
31. Antonov I, Kandel ER, Hawkins RD. The contribution of facilitation of monosynaptic PSPs to dishabituation and sensitization of the *Aplysia* siphon withdrawal reflex. *J Neurosci*. 1999;19(23): 10438–10450.
32. Hochner B, Klein M, Schacher S, Kandel ER. Additional component in the cellular mechanism of presynaptic facilitation contributes to behavioral dishabituation in *Aplysia*. *Proc Natl Acad Sci USA*. 1986;83(22):8794–8798.
33. Abrams TW, Castellucci VF, Camardo JS, Kandel ER, Lloyd PE. Two endogenous neuropeptides modulate the gill and siphon withdrawal reflex in *Aplysia* by presynaptic facilitation involving cAMP-dependent closure of a serotonin-sensitive potassium channel. *Proc Natl Acad Sci USA*. 1984;81(24):7956–7960.
34. Antonov I, Ha T, Antonova I, Moroz LL, Hawkins RD. Role of nitric oxide in classical conditioning of siphon withdrawal in *Aplysia*. *J Neurosci*. 2007;27(41):10993–11002.
35. Cohen TE, Kaplan SW, Kandel ER, Hawkins RD. A simplified preparation for relating cellular events to behavior: mechanisms contributing to habituation, dishabituation, and sensitization of the Aplysia gill-withdrawal reflex. *J Neurosci*. 1997;17(8): 2886–2899.
36. Frost WN, Castellucci VF, Hawkins RD, Kandel ER. Monosynaptic connections made by the sensory neurons of the gill- and siphon-withdrawal reflex in *Aplysia* participate in the storage of long-term memory for sensitization. *Proc Natl Acad Sci USA*. 1985;82:8266–8269.
37. Cleary LJ, Lee WL, Byrne JH. Cellular correlates of long-term sensitization in *Aplysia*. *J Neurosci*. 1998;18(15):5988–5998.

38. Zwartjes RE, West H, Hattar S, et al. Identification of specific mRNAs affected by treatments producing long-term facilitation in *Aplysia*. *Learn Mem*. 1998;4(6):478–495.
39. Castellucci VF, Blumenfeld H, Goelet P, Kandel ER. Inhibitor of protein synthesis blocks long-term behavioral sensitization in the isolated gill-withdrawal reflex of *Aplysia*. *J Neurobiol*. 1989;20 (1):1–9.
40. Cai D, Chen S, Fan X, Sossin WS, Glanzman DL. Postsynaptic inhibition of protein kinase C Apl III disrupts intermediate-term facilitation in *Aplysia* Paper presented at Neuroscience 2011, Washington, DC.
41. Bartsch D, Casadio A, Karl KA, Serodio P, Kandel ER. CREB1 encodes a nuclear activator, a repressor, and a cytoplasmic modulator that form a regulatory unit critical for long-term facilitation. *Cell*. 1998;95(2):211–223.
42. Casadio A, Martin KC, Giustetto M, et al. A transient, neuron-wide form of CREB-mediated long-term facilitation can be stabilized at specific synapses by local protein synthesis. *Cell*. 1999;99 (2):221–237.
43. Rajasethupathy P, Fiumara F, Sheridan R, et al. Characterization of small RNAs in *Aplysia* reveals a role for miR-124 in constraining synaptic plasticity through CREB. *Neuron*. 2009;63(6): 803–817.
44. Liu R-Y, Shah S, Cleary LJ, Byrne JH. Serotonin- and training-induced dynamic regulation of CREB2 in *Aplysia*. *Learn Mem*. 2011;18(4):245–249.
45. Bergold PJ, Sweatt JD, Winicov I, Weiss KR, Kandel ER, Schwartz JH. Protein synthesis during acquisition of long-term facilitation is needed for the persistent loss of regulatory subunits of the *Aplysia* cAMP-dependent protein kinase. *Proc Natl Acad Sci USA*. 1990;87(10):3788–3791.
46. Eskin A, Garcia KS, Byrne JH. Information storage in the nervous system of *Aplysia*: specific proteins affected by serotonin and cAMP. *Proc Natl Acad Sci USA*. 1989;86(7):2458–2462.
47. Noel F, Scholz KP, Eskin A, Byrne JH. Common set of proteins in *Aplysia* sensory neurons affected by an *in vitro* analogue of long-term sensitization training, 5-HT and cAMP. *Brain Res*. 1991;568(1–2):67–75.
48. Liu Q-R, Hattar S, Endo S, et al. A developmental gene (Tolloid/BMP-1) is regulated in *Aplysia* neurons by treatments that induce long-term sensitization. *J Neurosci*. 1997;17(2): 755–764.
49. Wainwright ML, Zhang H, Byrne JH, Cleary LJ. Localized neuronal outgrowth induced by long-term sensitization training in *Aplysia*. *J Neurosci*. 2002;22(10):4132–4141.
50. Bailey CH, Chen M. Morphological basis of long-term habituation and sensitization in *Aplysia*. *Science*. 1983;220(4592): 91–93.
51. Bailey CH, Chen M. Long–term memory in *Aplysia* modulates the total number of varicosities of single identified sensory neurons. *Proc Natl Acad Sci USA*. 1988;85:2373–2377.
52. Bailey CH, Kandel ER. Structural changes accompanying memory storage. *Annu Rev Physiol*. 1993;55:397–426.
53. Bailey CH, Kandel ER. Synaptic remodeling, synaptic growth and the storage of long-term memory in *Aplysia*. *Prog Brain Res*. 2008;169:179–198.
54. Bailey CH, Chen M, Keller F, Kandel ER. Serotonin-mediated endocytosis of apCAM: an early step of learning-related synaptic growth in *Aplysia*. *Science*. 1992;256(5057):645–649.
55. Kim JH, Udo H, Li HL, et al. Presynaptic activation of silent synapses and growth of new synapses contribute to intermediate and long-term facilitation in *Aplysia*. *Neuron*. 2003;40(1):151–165.
56. Udo H, Jin I, Kim JH, et al. Serotonin-induced regulation of the actin network for learning-related synaptic growth requires Cdc42, N-WASP, and PAK in *Aplysia* sensory neurons. *Neuron*. 2005;45(6):887–901.
57. Dale N, Schacher S, Kandel ER. Long-term facilitation in *Aplysia* involves increase in transmitter release. *Science*. 1988;239: 282–285.
58. Glanzman DL. Common mechanisms of synaptic plasticity in vertebrates and invertebrates. *Curr Biol*. 2010;20(1):R31–R36.
59. Cai D, Chen S, Glanzman DL. Postsynaptic regulation of long-term facilitation in *Aplysia*. *Curr Biol*. 2008;18(12):920–925.
60. Wang DO, Kim SM, Zhao Y, et al. Synapse- and stimulus-specific local translation during long-term neuronal plasticity. *Science*. 2009;324(5934):1536–1540.
61. Glanzman DL, Kandel ER, Schacher S. Target-dependent structural changes accompanying long-term synaptic facilitation in *Aplysia* neurons. *Science*. 1990;249(4970):799–802.
62. Jin I, Udo H, Rayman JB, Puthanveettil S, Kandel ER, Hawkins RD. Spontaneous transmitter release recruits postsynaptic mechanisms of long-term and intermediate-term facilitation in *Aplysia*. *Proc Natl Acad Sci USA*. 2012;109(23):9137–9142.
63. Bailey CH, Chen M. Long-term sensitization in *Aplysia* increases the number of presynaptic contacts onto the identified gill motor neuron L7. *Proc Natl Acad Sci USA*. 1988;85(23):9356–9359.
64. Zhu H, Wu F, Schacher S. Site-specific and sensory neuron-dependent increases in postsynaptic glutamate sensitivity accompany serotonin-induced long-term facilitation at *Aplysia* sensorimotor synapses. *J Neurosci*. 1997;17(13):4976–4986.
65. Trudeau LE, Castellucci VF. Postsynaptic modifications in long-term facilitation in *Aplysia*: upregulation of excitatory amino acid receptors. *J Neurosci*. 1995;15(2):1275–1284.
66. Ocorr KA, Byrne JH. Membrane responses and changes in cAMP levels in *Aplysia* sensory neurons produced by serotonin, tryptamine, FMRFamide and small cardioactive peptideB (SCPB). *Neurosci Lett*. 1985;55(2):113–118.
67. Bacskai BJ, Hochner B, Mahaut SM, et al. Spatially resolved dynamics of cAMP and protein kinase A subunits in *Aplysia* sensory neurons. *Science*. 1993;260(5105):222–226.
68. Braha O, Edmonds B, Sacktor T, Kandel ER, Klein M. The contributions of protein kinase A and protein kinase C to the actions of 5-HT on the L-type Ca^{2+} current of the sensory neurons in *Aplysia*. *J Neurosci*. 1993;13(5):1839–1851.
69. Kandel ER, Schwartz JH. Molecular biology of learning: modulation of transmitter release. *Science*. 1982;218:433–443.
70. Sweatt JD, Kandel ER. Persistent and transcriptionally-dependent increase in protein phosphorylation in long-term facilitation of *Aplysia* sensory neurons. *Nature*. 1989;339(6219): 51–54.
71. Bernier L, Castellucci VF, Kandel ER, Schwartz JH. Facilitatory transmitter causes a selective and prolonged increase in adenosine 3′:5′-monophosphate in sensory neurons mediating the gill and siphon withdrawal reflex in *Aplysia*. *J Neurosci*. 1982;2 (12):1682–1691.
72. Greenberg SM, Castellucci VF, Bayley H, Schwartz JH. A molecular mechanism for long-term sensitization in *Aplysia*. *Nature*. 1987;329:62–65.
73. Bergold PJ, Beushausen SA, Sacktor TC, Cheley S, Bayley H, Schwartz JH. A regulatory subunit of the cAMP-dependent protein kinase down-regulated in *Aplysia* sensory neurons during long-term sensitization. *Neuron*. 1992;8(2): 387–397.
74. Lee SH, Lim CS, Park H, et al. Nuclear translocation of CAM-associated protein activates transcription for long-term facilitation in *Aplysia*. *Cell*. 2007;129(4):801–812.
75. Alberini CM. Transcription factors in long-term memory and synaptic plasticity. *Physiol Rev*. 2009;89(1):121–145.

76. Ormond J, Hislop J, Zhao Y, et al. ApTrkl, a Trk-like receptor, mediates serotonin- dependent ERK activation and long-term facilitation in *Aplysia* sensory neurons. *Neuron*. 2004;44(4): 715–728.
77. Sharma SK, Carew TJ. The roles of MAPK cascades in synaptic plasticity and memory in Aplysia: facilitatory effects and inhibitory constraints. *Learn Mem*. 2004;11(4):373–378.
78. Sharma SK, Sherff CM, Shobe J, Bagnall MW, Sutton MA, Carew TJ. Differential role of mitogen-activated protein kinase in three distinct phases of memory for sensitization in *Aplysia*. *J Neurosci*. 2003;23(9):3899–3907.
79. Guan Z, Giustetto M, Lomvardas S, et al. Integration of long-term-memory-related synaptic plasticity involves bidirectional regulation of gene expression and chromatin structure. *Cell*. 2002;111(4):483–493.
80. Martin KC, Michael D, Rose JC, et al. MAP kinase translocates into the nucleus of the presynaptic cell and is required for long-term facilitation in *Aplysia*. *Neuron*. 1997;18(6): 899–912.
81. Hegde AN, Inokuchi K, Pei W, et al. Ubiquitin C-terminal hydrolase is an immediate-early gene essential for long-term facilitation in *Aplysia*. *Cell*. 1997;89(1):115–126.
82. Zhao Y, Hegde AN, Martin KC. The ubiquitin proteasome system functions as an inhibitory constraint on synaptic strengthening. *Curr Biol*. 2003;13(11):887–898.
83. Bartsch D, Ghirardi M, Casadio A, et al. Enhancement of memory-related long-term facilitation by ApAF, a novel transcription factor that acts downstream from both CREB1 and CREB2. *Cell*. 2000;103(4):595–608.
84. Alberini CM, Ghirardl M, Metz R, Kandel ER. C/EBP is an immediate-early gene required for the consolidation of long-term facilitation in *Aplysia*. *Cell*. 1994;76(6):1099–1114.
85. Bailey CH, Kandel ER, Si K. The persistence of long-term memory: a molecular approach to self-sustaining changes in learning-induced synaptic growth. *Neuron*. 2004;44(1):49–57.
86. Zhang F, Goldsmith JR, Byrne JH. Neural analogue of long-term sensitization training produces long-term (24 and 48 h) facilitation of the sensory-to-motor neuron connection in *Aplysia*. *J Neurophysiol*. 1994;72(2):778–784.
87. Byrne JH, Kandel ER. Presynaptic facilitation revisited: state and time dependence. *J Neurosci*. 1996;16:425–435.
88. Li Q, Roberts AC, Glanzman DL. Synaptic facilitation and behavioral dishabituation in *Aplysia*: dependence upon release of Ca^{2+} from postsynaptic intracellular stores, postsynaptic exocytosis and modulation of postsynaptic AMPA receptor efficacy. *J Neurosci*. 2005;25:5623–5637.
89. Jin I, Kandel ER, Hawkins RD. Whereas short-term facilitation is presynaptic, intermediate-term facilitation involves both pre-synaptic and postsynaptic protein kinases and protein synthesis. *Learn Mem*. 2011;18(2):96–102.
90. Dale N, Kandel ER. L-Glutamate may be the fast excitatory transmitter of *Aplysia* sensory neurons. *Proc Natl Acad Sci USA*. 1993;90(15):7163–7167.
91. Antzoulatos EG, Byrne JH. Learning insights transmitted by glutamate. *Trends Neurosci*. 2004;27(9):555–560.
92. Levenson J, Endo S, Kategaya LS, et al. Long-term regulation of neuronal high-affinity glutamate and glutamine uptake in *Aplysia*. *Proc Natl Acad Sci USA*. 2000;97(23):12858–12863.
93. Levenson J, Sherry DM, Dryer L, Chin J, Byrne JH, Eskin A. Localization of glutamate and glutamate transporters in the sensory neurons of *Aplysia*. *J Comp Neurol*. 2000;423(1):121–131.
94. Li HL, Huang BS, Vishwasrao H, et al. Dscam mediates remodeling of glutamate receptors in *Aplysia* during *de novo* and learning-related synapse formation. *Neuron*. 2009;61(4):527–540.
95. Cai D, Pearce K, Chen S, Glanzman DL. Protein kinase M maintains long-term sensitization and long-term facilitation in *Aplysia*. *J Neurosci*. 2011;31(17):6421–6431.
96. Lin XY, Glanzman DL. Long-term potentiation of *Aplysia* sensorimotor synapses in cell culture: regulation by postsynaptic voltage. *Proc Biol Sci*. 1994;255:113–118.
97. Schacher S, Wu F, Sun ZY. Pathway-specific synaptic plasticity: activity-dependent enhancement and suppression of long-term heterosynaptic facilitation at converging inputs on a single target. *J Neurosci*. 1997;17(2):597–606.
98. Eliot LS, Hawkins RD, Kandel ER, Schacher S. Pairing-specific, activity-dependent presynaptic facilitation at *Aplysia* sensory-motor neuron synapses in isolated cell culture. *J Neurosci*. 1994;14:368–383.
99. Walters ET. Multiple sensory neuronal correlates of site-specific sensitization in *Aplysia*. *J Neurosci*. 1987;7(2):408–417.
100. Walters ET. Site-specific sensitization of defensive reflexes in *Aplysia*: a simple model of long-term hyperalgesia. *J Neurosci*. 1987;7(2):400–407.
101. Antonov I, Antonova I, Kandel ER, Hawkins RD. Activity-dependent presynaptic facilitation and Hebbian LTP are both required and interact during classical conditioning in *Aplysia*. *Neuron*. 2003;37(1):135–147.
102. Carew TJ, Hawkins RD, Kandel ER. Differential classical conditioning of a defensive withdrawal reflex in *Aplysia californica*. *Science*. 1983;219:397–400.
103. Hawkins RD, Abrams TW, Carew TJ, Kandel ER. A cellular mechanism of classical conditioning in *Aplysia*: activity-dependent amplification of presynaptic facilitation. *Science*. 1983;219:400–405.
104. Walters ET, Byrne JH. Associative conditioning of single sensory neurons suggests a cellular mechanism for learning. *Science*. 1983;219:405–408.
105. Hu JY, Chen Y, Schacher S. Protein kinase C regulates local synthesis and secretion of a neuropeptide required for activity-dependent long-term synaptic plasticity. *J Neurosci*. 2007;27 (33):8927–8939.
106. Miniaci MC, Kim JH, Puthanveettil SV, et al. Sustained CPEB-dependent local protein synthesis is required to stabilize synaptic growth for persistence of long-term facilitation in *Aplysia*. *Neuron*. 2008;59(6):1024–1036.
107. Cai D, Pearce K, Chen S, Glanzman DL. Reconsolidation of long-term memory in *Aplysia*. *Curr Biol*. 2012;22:1783–1788.
108. Martin KC, Casadio A, Zhu H, et al. Synapse-specific, long-term facilitation of *Aplysia* sensory to motor synapses: a function for local protein synthesis in memory storage. *Cell*. 1997;91 (7):927–938.
109. Zhang Y, Liu R-Y, Heberton GA, et al. Computational design of enhanced learning protocols. *Nat Neurosci*. 2012;15:294–297.
110. Michael D, Martin KC, Seger R, Ning MM, Baston R, Kandel ER. Repeated pulses of serotonin required for long-term facilitation activate mitogen-activated protein kinase in sensory neurons of *Aplysia*. *Proc Natl Acad Sci USA*. 1998;95(4):1864–1869.
111. Si K, Giustetto M, Etkin A, et al. A neuronal isoform of CPEB regulates local protein synthesis and stabilizes synapse-specific long-term facilitation in *Aplysia*. *Cell*. 2003;115(7):893–904.
112. Villareal G, Li Q, Cai D, Glanzman DL. The role of rapid, local postsynaptic protein synthesis in learning-related synaptic facilitation in *Aplysia*. *Curr Biol*. 2007;17:2073–2080.
113. Moccia R, Chen D, Lyles V, et al. An unbiased cDNA library prepared from isolated *Aplysia* sensory neuron processes is enriched for cytoskeletal and translational mRNAs. *J Neurosci*. 2003;23(28):9409–9417.

114. Martin KC, Barad M, Kandel ER. Local protein synthesis and its role in synapse-specific plasticity. *Curr Opin Neurobiol.* 2000;10(5):587–592.
115. Martin KC, Kosik KS. Synaptic tagging: who's it? *Nat Rev Neurosci.* 2002;3(10):813–820.
116. Frey U, Morris RGM. Synaptic tagging and long-term potentiation. *Nature.* 1997;385:533–536.
117. Frey U, Morris RGM. Synaptic tagging: implications for late maintenance of hippocampal long-term potentiation. *Trends Neurosci.* 1998;21:181–188.
118. Sherff CM, Carew TJ. Coincident induction of long-term facilitation in *Aplysia*: cooperativity between cell bodies and remote synapses. *Science.* 1999;285(5435):1911–1914.
119. Sun ZY, Wu F, Schacher S. Rapid bidirectional modulation of mRNA expression and export accompany long-term facilitation and depression of *Aplysia* synapses. *J Neurobiol.* 2001;46(1): 41–47.
120. Brunet JF, Shapiro E, Foster SA, Kandel ER, Iino Y. Identification of a peptide specific for *Aplysia* sensory neurons by PCR-based differential screening. *Science.* 1991;252(5007): 856–859.
121. Liu K, Hu JY, Wang D, Schacher S. Protein synthesis at synapse versus cell body: enhanced but transient expression of long-term facilitation at isolated synapses. *J Neurobiol.* 2003;56(3): 275–286.
122. Santarelli L, Montarolo P, Schacher S. Neuropeptide localization in varicosities of *Aplysia* sensory neurons is regulated by target and neuromodulators evoking long-term synaptic plasticity. *J Neurobiol.* 1996;31(3):297–308.
123. Hu JY, Wu F, Schacher S. Two signaling pathways regulate the expression and secretion of a neuropeptide required for long-term facilitation in *Aplysia*. *J Neurosci.* 2006;26(3):1026–1035.
124. Hu JY, Glickman L, Wu F, Schacher S. Serotonin regulates the secretion and autocrine action of a neuropeptide to activate MAPK required for long-term facilitation in *Aplysia*. *Neuron.* 2004;43(3):373–385.
125. Hu JY, Meng X, Schacher S. Target interaction regulates distribution and stability of specific mRNAs. *J Neurosci.* 2002;22 (7):2669–2678.
126. Lyles V, Zhao Y, Martin KC. Synapse formation and mRNA localization in cultured *Aplysia* neurons. *Neuron.* 2006;49(3): 349–356.
127. Lisman J. What does the nucleus know about memories? *J NIH Res.* 1995;7:43–46.
128. Steward O, Wallace CS, Lyford GL, Worley PF. Synaptic activation causes the mRNA for the IEG Arc to localize selectively near activated postsynaptic sites on dendrites. *Neuron.* 1998;21 (4):741–751.
129. Steward O, Worley PF. Selective targeting of newly synthesized Arc mRNA to active synapses requires NMDA receptor activation. *Neuron.* 2001;30(1):227–240.
130. Miller S, Yasuda M, Coats JK, Jones Y, Martone ME, Mayford M. Disruption of dendritic translation of CaMKIIα simpairs stabilization of synaptic plasticity and memory consolidation. *Neuron.* 2002;36(3):507–519.
131. Meer EJ, Wang DO, Kim S, Barr I, Guo F, Martin KC. Identification of a *cis*-acting element that localizes mRNA to synapses. *Proc Natl Acad Sci USA.* 2012;109(12):4639–4644.
132. Bailey CH, Chen M. Time course of structural changes at identified sensory neuron synapses during long-term sensitization in *Aplysia*. *J Neurosci.* 1989;9(5):1774–1780.
133. Wainwright ML, Byrne JH, Cleary LJ. Dissociation of morphological and physiological changes associated with long-term memory in *Aplysia*. *J Neurophysiol.* 2004;92(4):2628–2632.
134. Bailey CH, Montarolo P, Chen M, Kandel ER, Schacher S. Inhibitors of protein and RNA synthesis block structural changes that accompany long-term heterosynaptic plasticity in *Aplysia*. *Neuron.* 1992;9(4):749–758.
135. Mayford M, Barzilai A, Keller F, Schacher S, Kandel ER. Modulation of an NCAM-related adhesion molecule with long-term synaptic plasticity in *Aplysia*. *Science.* 1992;256(5057): 638–644.
136. Zhu H, Wu F, Schacher S. *Aplysia* cell adhesion molecules and serotonin regulate sensory cell–motor cell interactions during early stages of synapse formation *in vitro*. *J Neurosci.* 1994;14 (11):6886–6900.
137. Han J-H, Lim C-S, Lee Y-S, Kandel ER, Kaang B-K. Role of *Aplysia* cell adhesion molecules during 5-HT-induced long-term functional and structural changes. *Learn Mem.* 2004;11 (4):421–435.
138. Sutton MA, Carew TJ. Behavioral, cellular, and molecular analysis of memory in *Aplysia* I: intermediate-term memory. *Integ Comp Biol.* 2002;42:725–735.
139. Chitwood RA, Li Q, Glanzman DL. Serotonin facilitates AMPA-type responses in isolated siphon motor neurons of *Aplysia* in culture. *J Physiol.* 2001;534:501–510.
140. Kessels HW, Malinow R. Synaptic AMPA receptor plasticity and behavior. *Neuron.* 2009;61(3):340–350.
141. Shepherd JD, Huganir RL. The cell biology of synaptic plasticity: AMPA receptor trafficking. *Annu Rev Cell Dev Biol.* 2007;23 (1):613–643.
142. Makino H, Malinow R. AMPA receptor incorporation into synapses during LTP: the role of lateral movement and exocytosis. *Neuron.* 2009;64(3):381–390.
143. Lüscher C, Xia H, Beattie EC, et al. Role of AMPA receptor cycling in synaptic transmission and plasticity. *Neuron.* 1999;24:649–658.
144. Tao-Cheng J-H, Crocker VT, Winters CA, Azzam R, Chludzinski J, Reese TS. Trafficking of AMPA receptors at plasma membranes of hippocampal neurons. *J Neurosci.* 2011;31 (13):4834–4843.
145. Yung I, Chun S, Kapya E, et al. Cloning of glutamate receptors from the central nervous system of *Aplysia*. *Soc Neurosci Abstr.* 2002;28(376):377.
146. Smith WB, Starck SR, Roberts RW, Schuman EM. Dopaminergic stimulation of local protein synthesis enhances surface expression of GluR1 and synaptic transmission in hippocampal neurons. *Neuron.* 2005;45(5):765–779.
147. Bougie JK, Lim T, Farah CA, et al. The atypical protein kinase C in *Aplysia* can form a protein kinase M by cleavage. *J Neurochem.* 2009;109(4):1129–1143.
148. Villareal G, Li Q, Cai D, et al. Role of protein kinase C in the induction and maintenance of serotonin-dependent enhancement of the glutamate response in isolated siphon motor neurons of *Aplysia californica*. *J Neurosci.* 2009;29(16):5100–5107.
149. Bougie J, Cai D, Hastings M, et al. Serotonin-induced cleavage of the atypical protein kinase C Apl III in *Aplysia*. *J Neurosci.* 2012;32:14630–14640.
150. Hernandez AI, Blace N, Crary JF, et al. Protein kinase Mζ synthesis from a brain mRNA encoding an independent protein kinase Cζ catalytic domain: implications for the molecular mechanism of memory. *J Biol Chem.* 2003;278(41): 40305–40316.
151. Lisé MF, El-Husseini A. The neuroligin and neurexin families: from structure to function at the synapse. *Cell Mol Life Sci.* 2006;63(16):1833–1849.
152. Choi Y-B, Li H-L, Kassabov SR, et al. Neurexin-neuroligin transsynaptic interaction mediates learning-related synaptic

remodeling and long-term facilitation in *Aplysia*. *Neuron*. 2011;70(3):468–481.
153. Jin I, Puthanveettil S, Udo H, Karl K, Kandel ER, Hawkins RD. Spontaneous transmitter release is critical for the induction of long-term and intermediate-term facilitation in *Aplysia*. *Proc Natl Acad Sci USA*. 2012;109(23):9131–9136.
154. Sacktor TC. How does PKMζ maintain long-term memory? *Nat Rev Neurosci*. 2011;12(1):9–15.
155. Ling DS, Benardo LS, Sacktor TC. Protein kinase Mζ enhances excitatory synaptic transmission by increasing the number of active postsynaptic AMPA receptors. *Hippocampus*. 2006;16:443–452.
156. Hu JY, Baussi O, Levine A, Chen Y, Schacher S. Persistent long-term synaptic plasticity requires activation of a new signaling pathway by additional stimuli. *J Neurosci*. 2011;31(24): 8841–8850.
157. Si K, Choi YB, White-Grindley E, Majumdar A, Kandel ER. *Aplysia* CPEB can form prion-like multimers in sensory neurons that contribute to long-term facilitation. *Cell*. 2010;140(3): 421–435.
158. Richter JD. CPEB: a life in translation. *Trends Biochem Sci*. 2007;32(6):279–285.
159. Keleman K, Kruttner S, Alenius M, Dickson BJ. Function of the *Drosophila* CPEB protein Orb2 in long-term courtship memory. *Nat Neurosci*. 2007;10(12):1587–1593.
160. Majumdar A, Cesario Wanda C, White-Grindley E, et al. Critical role of amyloid-like oligomers of *Drosophila* Orb2 in the persistence of memory. *Cell*. 2012;148(3):515–529.
161. Si K, Lindquist S, Kandel ER. A neuronal isoform of the *Aplysia* CPEB has prion-like properties. *Cell*. 2003;115(7):879–891.
162. Aguzzi A, Zhu C. Five questions on prion diseases. *PLoS Pathog*. 2012;8(5):e1002651.
163. Pastalkova E, Serrano P, Pinkhasova D, Wallace E, Fenton AA, Sacktor TC. Storage of spatial information by the maintenance mechanism of LTP. *Science*. 2006;313(5790):1141–1144.
164. Shema R, Haramati S, Ron S, et al. Enhancement of consolidated long-term memory by overexpression of protein kinase Mζ in the neocortex. *Science*. 2011;331(6021):1207–1210.
165. Shema R, Sacktor TC, Dudai Y. Rapid erasure of long-term memory associations in the cortex by an inhibitor of PKMζ. *Science*. 2007;317(5840):951–953.
166. Ling DS, Benardo LS, Serrano PA, et al. Protein kinase Mζ is necessary and sufficient for LTP maintenance. *Nat. Neurosci*. 2002;5(4):295–296.
167. McGaugh JL. Memory—A century of consolidation. *Science*. 2000;287(5451):248–251.
168. Bregman N, Nicholas T, Lewis DJ. Cue-dependent amnesia: permanence and memory return. *Physiol Behav*. 1976;17(2): 267–270.
169. Misanin JR, Miller RR, Lewis DJ. Retrograde amnesia produced by electroconvulsive shock after reactivation of a consolidated memory trace. *Science*. 1968;160(827):554–555.
170. Przybyslawski J, Sara SJ. Reconsolidation of memory after its reactivation. *Behav Brain Res*. 1997;84(1–2):241–246.
171. Sara SJ, Roullet P, Przybyslawski J. Consolidation of memory for odor-reward association: beta-adrenergic receptor involvement in the late phase. *Learn Mem*. 1999;6(2):88–96.
172. Debiec J, LeDoux JE, Nader K. Cellular and systems reconsolidation in the hippocampus. *Neuron*. 2002;36(3):527–538.
173. Nader K, Schafe GE, Le Doux JE. Fear memories require protein synthesis in the amygdala for reconsolidation after retrieval. *Nature*. 2000;406(6797):722–726.
174. Dudai Y. The restless engram: consolidations never end. *Annu Rev Neurosci*. 2012;35:227–247.
175. Eisenberg M, Kobilo T, Berman DE, Dudai Y. Stability of retrieved memory: inverse correlation with trace dominance. *Science*. 2003;301(5636):1102–1104.
176. Taubenfeld SM, Milekic MH, Monti B, Alberini CM. The consolidation of new but not reactivated memory requires hippocampal C/EBPβ. *Nat Neurosci*. 2001;4(8):813–818.
177. Alberini CM. The role of reconsolidation and the dynamic process of long-term memory formation and storage. *Front Behav Neurosci*. 2011;5:12.
178. Sangha S, Scheibenstock A, Lukowiak K. Reconsolidation of a long-term memory in *Lymnaea* requires new protein and RNA synthesis and the soma of right pedal dorsal 1. *J Neurosci*. 2003;23(22):8034–8040.
179. Lagasse F, Devaud J-M, Mery F. A switch from cycloheximide-resistant consolidated memory to cycloheximide-sensitive reconsolidation and extinction in *Drosophila*. *J Neurosci*. 2009;29 (7):2225–2230.
180. Nader K, Hardt O. A single standard for memory: the case for reconsolidation. *Nat Rev Neurosci*. 2009;10(3):224–234.
181. Rose JK, Rankin CH. Blocking memory reconsolidation reverses memory-associated changes in glutamate receptor expression. *J Neurosci*. 2006;26(45):11582–11587.
182. Lee S-H, Kwak C, Shim J, et al. A cellular model of memory reconsolidation involves reactivation-induced destabilization and restabilization at the sensorimotor synapse in *Aplysia*. *Proc Natl Acad Sci USA*. 2012;109(35):14200–14205.
183. Kaang BK, Choi JH. Protein degradation during reconsolidation as a mechanism for memory reorganization. *Front Behav Neurosci*. 2011;5:2.
184. Lee SH, Choi JH, Lee N, et al. Synaptic protein degradation underlies destabilization of retrieved fear memory. *Science*. 2008;319(5867):1253–1256.
185. Lattal KM, Abel T. Behavioral impairments caused by injections of the protein synthesis inhibitor anisomycin after contextual retrieval reverse with time. *Proc Natl Acad Sci USA*. 2004;101 (13):4667–4672.
186. Alberini CM. Mechanisms of memory stabilization: are consolidation and reconsolidation similar or distinct processes? *Trends Neurosci*. 2005;28(1):51–56.

CHAPTER

18

Roles of Protein Kinase C and Protein Kinase M in *Aplysia* Learning

Margaret Hastings, Carole A. Farah and Wayne S. Sossin

McGill University, Montreal, Quebec, Canada

INTRODUCTION

Memory formation results from the activation of cellular and molecular cascades that are highly conserved throughout evolution.[1] In particular, protein kinase Cs (PKCs) play important roles in memory formation in multiple organisms ranging from invertebrates to primates.[2–8] Different PKC isoforms exist with different requirements for co-factors for activation and different downstream effectors.[4,9] Furthermore, activation of PKC isoforms is sensitive to the pattern of stimulation, and different PKC isoforms induce distinct types of synaptic plasticity. Moreover, a persistently active catalytic fragment of PKC known as a protein kinase M (PKM) can play a role in memory maintenance by forming a molecular memory trace.[3,5,6] In this chapter, we review the role of PKCs and PKMs in learning and memory formation in the marine mollusk *Aplysia californica*, whose simple and accessible nervous system has proved to be invaluable for the investigation of cellular and molecular mechanisms underlying learning and memory formation.[4,10] First, we define the PKC family in *Aplysia*, identifying the domains important for kinase regulation. Next, we discuss the rules of isoform-specific PKC activation and isoform-specific requirements for PKC during learning paradigms in *Aplysia*. In this system, PKMs are thought to be formed by calpain-dependent cleavage, and we describe the families of calpains present in *Aplysia*. Finally, we discuss how PKCs interact with other signaling pathways involved in synaptic plasticity. Overall, the focus of this review is to define how distinct PKC isoforms are tuned to participate in distinct memory traces.

THE PKC FAMILY

Isoforms of PKC in *Aplysia*

There are four known families of PKCs in metazoans: the conventional or classical PKCs (cPKCs), which are activated by calcium, including the vertebrate PKCs α, β, and γ; the novel calcium-independent PKCs (nPKCs) of the epsilon family, also referred to as novel type I, including the vertebrate PKCs ε and η; the nPKCs of the delta family, also referred to as novel type II, including the vertebrate PKCs δ and θ; and the atypical PKCs (aPKCs), which include the vertebrate PKCs ζ and ι (Figure 18.1).[9,11] These families diverged soon after the origin of multicellular organisms.[12] The novel PKCs retain the original organization of PKCs present in unicellular organisms such as yeast, whereas the introduction of the new calcium-sensitive C2 domain in conventional PKCs and the PB1 domain in atypical PKCs defined the divergence of the PKC families in early metazoans.[12] In *Aplysia californica*, there are three known nervous system-enriched PKCs: the conventional PKC Apl I, the novel type I PKC Apl II of the epsilon family, and the atypical PKC Apl III.[4,13–15] Although there is clearly a novel type II nPKC of the delta family in the genome of *Aplysia*, this isoform is not highly expressed in the nervous system of *Aplysia*[4,15] and is not discussed further in this chapter. The PKC Apl I and PKC Apl III mRNAs are both subject to alternative splicing with one splice isoform enriched in the nervous system. For PKC Apl I, the nervous system-enriched splice isoform shares a conserved carboxy (C)-terminal with a similarly spliced vertebrate PKC, PKCβ1,[14,16] and this conserved C-terminal may

Invertebrate Learning and Memory.
DOI: http://dx.doi.org/10.1016/B978-0-12-415823-8.00018-6

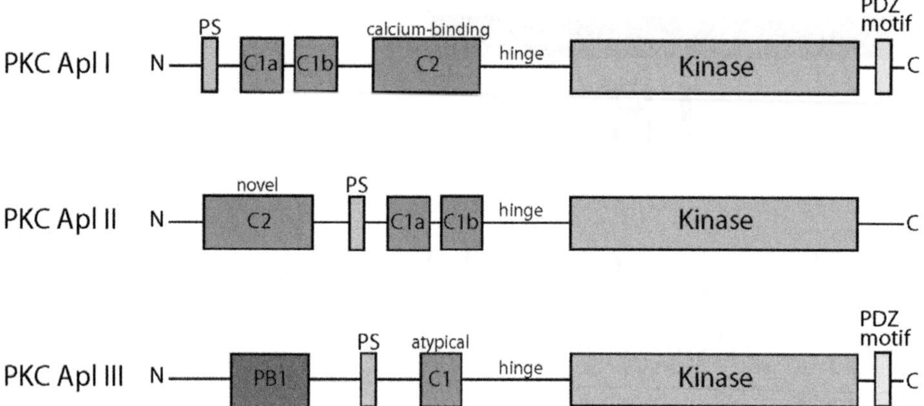

FIGURE 18.1 Schematic representation of the nervous system-enriched PKC isoforms in *Aplysia*. The C2 domain of PKC Apl II does not bind calcium, and the C1 domain of PKC Apl III does not bind DAG. A putative conserved PDZ motif is present at the C-terminal of PKC Apl I and PKC Apl III. PS, pseudosubstrate.

provide a site for interaction with PDZ-containing proteins. For PKC Apl III, the nervous system-enriched splice isoform contains a calpain cleavage site.[13]

Differences in the Regulatory Domain Define Distinct Activators for PKC Isoforms

All PKCs have a catalytic domain located at the C-terminal and a regulatory domain located at the N-terminal (Figure 18.1). The regulatory domain contains a pseudosubstrate sequence that in the inactive form of the kinase is lodged in the active site of the catalytic domain, blocking kinase activation. In order for PKCs to become active, a conformational change is required to move the pseudosubstrate away from the active site and allow substrate binding.[17,18] The major difference between the PKC families is in the domains present in their regulatory domain that regulate this conformational change. Conventional PKCs have two tandem C1 domains and a calcium-sensitive C2 domain (Figure 18.1). The C1 domains bind to diacylglycerol (DAG), whereas the C2 domain mediates calcium-dependent binding to phosphatidylserine and phosphoinositide-4,5-bisphosphate in the membrane.[19–21] The C1 domains of the conventional PKCs have a lower affinity for DAG than those of the novel PKCs[22] and thus require the additional lipid interactions with the C2 domain for translocation to membranes.[12,23]

Novel PKCs of the delta and epsilon family also have two C1 domains and one C2 domain, but their C2 domain is located N-terminal to the pseudosubstrate and to the C1 domains, whereas it is located after the pseudosubstrate and C1 domains in conventional PKCs (Figure 18.1). Although C2 domains are often used as calcium sensors,[24] the protein domain existed before it gained the ability to bind calcium[25] and the C2 domain of the novel PKCs is orthologous to the C2 domain present in unicellular organisms that does not bind calcium.[12] In the delta family, the C2 domain acts as a phosphotyrosine binding molecule,[26] whereas the role in the epsilon family is more controversial. The C2 domain of PKC Apl II, an ortholog of PKCε, is a negative regulator of the kinase and acts to inhibit binding of DAG to the C1 domains and, thus, enzyme activation.[27–30] Indeed, removal of the C2 domain lowers the amount of lipid required to activate the enzyme.[29,30] Binding of phosphatidic acid (PA) to the C1b domain removes C2 domain-mediated inhibition and allows binding of DAG to the C1 domain and translocation of the kinase.[27] Arginine 273 in the C1b domain was shown to be critical for PKC Apl II binding to PA, and mutating this residue to alanine removes C2 domain-mediated inhibition.[27] Thus, both PA and DAG are required to translocate PKC Apl II to membranes. Arginine 273 is conserved in vertebrate novel PKCs, and phosphatidic acid is also critical for the activation of PKCε.[31,32] The C2 domain of PKCε was also reported to assist activation of the kinase through binding to both lipids and receptors for activated C kinase (RACKs).[33,34] It remains possible that once C2 domain-mediated inhibition is removed, the C2 domain of PKC Apl II binds to phospholipids in the membrane and/or RACKs to contribute to protein translocation,[27,35] although some evidence argues against this.[27,36,37] Finally, although we initially reported that phosphorylation of the C2 domain increased binding to lipids,[38] our recent evidence suggests that this phosphorylation is important in removing PKC Apl II from membranes after activation, not in increasing lipid binding.[39]

Atypical PKCs are named after their atypical C1 domain, which cannot bind to DAG. Atypical PKCs do not have a C2 domain but, rather, a PB1 domain at their N-terminal, which mediates protein–protein interactions that can lead to kinase activation.[40,41] PKC Apl III shuttles between the nuclear and cytoplasmic compartments due to an unidentified sequence in the regulatory domain of PKC Apl III.[13] Many important

PB1 domain binding proteins are conserved and are involved in regulating cellular polarity,[42] but none of these have been characterized in *Aplysia.*

The Catalytic and C-Terminal Domains are Sites of Regulation of PKCs by Phosphorylation

All PKCs contain a conserved catalytic domain followed by a conserved C-terminal domain. Although the kinase domains show some differences in substrate specificity,[43] it is not clear if these are sufficient to define distinct substrates or whether protein–protein interactions that determine the location and closely associated proteins are more important in defining specific substrates of the kinases.[4] PKCs, like most other AGC kinases, contain three phosphorylation sites in the kinase and associated C-terminal domain that regulate kinase activity.[44–46] These sites are known as the phosphoinositide kinase 1 (PDK-1) site, the turn site, and the hydrophobic or phosphoinositide-dependent kinase 2 site (PDK-2).

PDK-1 has been cloned from *Aplysia*,[47] and phosphospecific antibodies have been raised to the PDK-1 site for all three *Aplysia* PKCs.[13,48] Although the name of the kinase stems from the PI-3 kinase dependence of its phosphorylation of PKB or AKT, the dependence for phosphoinositides does not emanate from a requirement for phosphoinositides to directly activate PDK-1 but instead from controlling the interaction between PDK-1 and its substrates.[45] Thus, for PKC Apl I and PKC Apl II, PDK-1 phosphorylation was insensitive to PI-3 kinase inhibitors.[48] In contrast, PDK-1 phosphorylation of the atypical PKC Apl III was sensitive to PI-3 kinase inhibitors, similar to that of vertebrate atypical PKCs.[13,45] The turn site and hydrophobic site have been postulated to be either autophosphorylated or phosphorylated by TORC2.[44] The hydrophobic site has been investigated using a phosphospecific antibody to PKC Apl II. Phosphorylation of the site was seen in a kinase dead PKC and was not blocked by PKC inhibitors, suggesting phosphorylation by a distinct kinase.[49] There appears to be a role for TORC2 in this phosphorylation because decreasing the levels of the TORC2-specific adaptor Rictor lowers phosphorylation of PKC Apl II at the hydrophobic site.[50]

The Hinge Domain and PKM Formation

The first PKC identified was a proteolytic product termed PKM based on its dependence on magnesium for activity.[51] This cleavage occurred in the hinge domain separating the catalytic and regulatory domains. These kinases lack the pseudosubstrate and thus do not require the presence of lipids to induce conformational changes to remove the pseudosubstrate from the catalytic domain. Although PKMs still require phosphorylation for activation, there is little evidence that PKMs undergo regulated phosphorylation.[13,52] In rodents, the atypical PKMζ has been found to play a unique and crucial role in memory maintenance.[5] PKMζ is proposed to be necessary and sufficient for the maintenance of late long-term potentiation (L-LTP), a cellular model for consolidated long-term memory.[5] Furthermore, local infusion of two distinct inhibitors that target PKMζ into specific brain regions, including hippocampus, insular cortex, or amygdala, resulted in erasure of consolidated long-term memory in a wide range of behavioral tasks in rodents.[5] These findings point to a widespread and fundamental role for PKMζ in the persistence of memory.

In vertebrates, PKMζ is transcribed from an independent promoter located in an intron in the PKCζ gene, and the translation of this mRNA is increased by stimuli that induce LTP.[53,54] The *Aplysia* ortholog of PKCζ, PKC Apl III, shows no evidence of an internal promoter in its gene.[13] Although the absence of evidence is not evidence of absence, bioinformatic studies on the evolution of the alternative start site in PKCζ are consistent with its origin at the time of the duplication of the two vertebrate atypical isoforms, zeta and iota.[13] First, PKCζ, but not PKCι, contains a highly conserved methionine that serves as the initiation codon for PKM. Not only is this methionine not present in PKCι but also it is not conserved in atypical PKCs outside of vertebrates.[13] Moreover, homology surrounding the promoter, located in an intron of the PKCζ gene, is absent in genes encoding atypical PKCs outside of vertebrates. Together with the absence of evidence for an alternative mRNA for PKC Apl III, it is highly unlikely that *Aplysia* contains an alternative mRNA encoding a PKM form of PKC Apl III. Instead, a PKM Apl III seems to be formed by cleavage.

Overexpression of PKC Apl III at high levels in isolated *Aplysia* sensorimotor neurons resulted in proteolysis, evidenced by redistribution of the N- and C-terminal portions of the kinase within the cell, and this cleavage was decreased by treatment with calpain inhibitor V.[13] Cleavage could also be induced in cultured sensory neurons expressing low levels of PKC Apl III by treating the cells with the calcium ionophore ionomycin, providing further support for calpain-mediated activation being a rate-limiting factor for PKM Apl III formation in *Aplysia*.[13] *In vitro* biochemical assays confirmed calpain-mediated cleavage of PKC Apl III, and immunoblots with antibodies targeting different regions of the PKC Apl III molecule confirmed that the cleavage occurred in the hinge region located between the regulatory and the kinase domains (Figure 18.1), consistent with PKM formation.[13] Interestingly, the nervous system-enriched spliced form of PKC Apl III contains a site for calpain-mediated cleavage in the region of alternative splicing, and lower levels of the PKC with the splice inserts were required to induce cleavage of PKC Apl III into PKM Apl III.[13] Calpains can also cleave PKC Apl I and PKC Apl II in the hinge domain.[55] It is not clear whether calpain-mediated cleavage of PKCs plays a role in the formation of PKMs in vertebrates, in addition to the known transcriptional and translational regulation of PKMζ formation. It is interesting to note that LTP is blocked by calpain inhibitors,[56,57] although there are many additional calpain substrates that could explain this blockade.[58]

Calpain Families in *Aplysia*

There are multiple conserved calpain subfamilies with different features reflecting differences in regulation and substrates. Preliminary work has revealed that several of these calpain family members are expressed in the *Aplysia* nervous system. Thus, a compelling next step is to determine the calpain(s) responsible for PKM formation in *Aplysia*.

Calpains can be broadly classified as classical or nonclassical based on their structure (Figure 18.2). In addition to the catalytic protease domain (dII) common to all calpain isoforms, classical calpains contain a C2-like domain (dIII) that mediates calcium-regulated binding to phospholipids,[59] a calcium-binding penta-EF-hand domain (dIV), and an N-terminal domain (dI) that is weakly inhibitory and undergoes autolysis as part of the activation process (Figure 18.2).[60,61] The *Aplysia* genome contains a classical calpain. In vertebrates, several of the classical forms are regulated by a small subunit,[62] but the small subunit did not exist before chordates and given the high homology of its penta-EF-hand domain to classical calpains, it is likely the result of gene duplication of a classical isoform followed by truncation. The small subunit both acts as a chaperone (some isoforms of calpains are degraded in the absence of this subunit) and is involved in activation through calcium-dependent disassociation.[63] This mechanism of regulation may still be present in the absence of the small subunit and would reflect calcium-dependent dissociation of calpain dimers.[64] Nonclassical calpains include three prominent subfamilies—Tra, PalB, and SOL[63,65]—of which we have detected PalB and Sol in the *Aplysia* genome (Figure 18.2; M. Hastings and W. S. Sossin, unpublished results). Tra members are found in the genomes of other mollusks, and the absence of a Tra ortholog may represent the incompleteness of the *Aplysia* genome at this time. Tra calpain subfamily members possess a structure similar to that of the classical calpains, but instead of the classical EF-hand domain, they have a highly diverged C2 domain.[63,65] Calcium-dependent proteolysis has been observed for Tra calpain homologs in several species.[66,67] The ancient PalB subfamily of calpains[65] originally identified in fungus[68] also lacks the classical C-terminal EF-hand domain, possessing duplicate C2-like domains instead,[63] the first of which is greatly diverged. PalB calpains also contain one or more N-terminal microtubule interacting and trafficking (MIT) domains, involved in protein–protein interactions.[69] PalB homologs appear to be catalytically active, but whether calcium is required for activation is unknown. Several PalB interacting proteins do have calcium binding domains, suggesting that calcium-dependent activation could be indirect in this case.[69] Finally, small optic

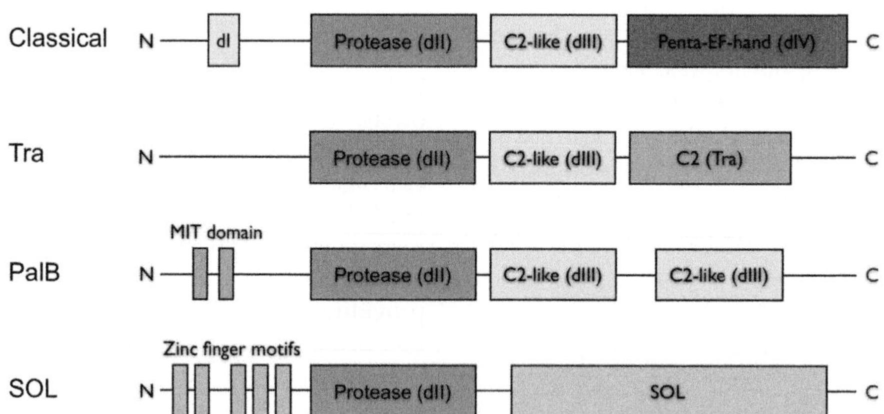

FIGURE 18.2 Schematic representation of the calpain families. All families share a conserved catalytic domain (dII), and all except Sol have a following C2-like domain (dIII). Each family is distinguished by a distinct domain. The classical family has a penta-EF-hand domain (dIV) that is largely responsible for its calcium sensitivity. The Tra family has a distinct C2 (TRA) domain. The PalB family has MIT domains and a duplication of the C2-like domain (dIII). Finally, Sol has a zinc finger region at its N-terminus and a conserved domain unique to this family called the Sol domain. The conserved function of these additional domains in the atypical calpains is not known. MIT, microtubule interacting and trafficking domain.

lobes (SOL) calpain homologs, originally identified in *Drosophila*, lack the C2-like and EF-hand domains of classical calpains, suggesting reduced regulation by calcium.[70,71] Instead, SOL calpains have a unique domain of unknown function at the C-terminal end, known as the SOL homology domain, and a zinc finger motif is present at the N terminal.[70,71] Although all three catalytic residues are present, suggesting proteolytic potential, proteolytic activity of SOL calpains has yet to be observed.

In view of the diverse structures of the different calpain isoforms and the implications of these structures for their localization and activation, identifying the calpain isoform(s) responsible for PKM formation will help identify upstream events that regulate PKM formation in *Aplysia* and give insight into whether cleavage of PKCs into PKMs is conserved in vertebrates as well.

ISOFORM-SPECIFIC ROLES OF PKCS DURING DISTINCT LEARNING PARADIGMS

Different learning experiences produce distinct memory traces as a result of the unique pattern of neuronal activity they produce. Several forms of PKC-dependent plasticity occur at the sensorimotor synapse of *Aplysia*, varying in their duration and the stimulation used to evoke them (Table 18.1). PKCs have also been implicated in plasticity occurring in other neurons of *Aplysia* as well (Table 18.1). Due to the small number of PKC isoforms expressed in the *Aplysia* nervous system and the amenability of this system to cell culture and microinjection techniques, it has been possible to begin to work out in many cases the specific PKC isoform(s) activated during memory formation and the PKCs required for the form of plasticity that underlies memory formation (Table 18.1). This work has provided insight into the mechanisms by which particular patterns of neuronal stimulation can selectively activate specific PKCs.

The main reagents used in these studies were fluorescent protein–PKC fusion proteins that allow the measurement of PKC translocation from the cytosol to the membrane using live imaging of cultured neurons. The advantages and technical points of this technique have been reviewed and will not be detailed here.[72] Also of importance are kinase dead fluorescent protein–PKC fusion proteins that act as isoform-specific dominant negatives.[37] These kinases have been well characterized and have activities similar to those of their nontagged versions. Also, the critical phosphorylation sites have been converted to aspartic acids to improve their ability to attain a correctly folded status.[37] These are important tools because there are no pharmacological PKC inhibitors that distinguish between classical and novel PKCs in *Aplysia*.[4]

Activation and Roles of PKC Apl I

PKC Apl I as a Coincidence Detector

In heterologous cells, PKC Apl I was translocated to the membrane, a sign of activation, in response to low levels of the cell-permeable DAG analog 1,2-dioctanoyl-*sn*-glycerol (DOG), combined with the calcium ionophore ionomycin, whereas neither DOG nor ionomycin alone translocated the kinase at subthreshold concentrations, suggesting a synergism between DAG and calcium in activating PKC Apl I in cells.[73] Due to its requirement for both Ca^{2+} binding to the C2 domain and DAG binding to the C1 domain, the *Aplysia* classical PKC Apl I has the potential to act as a

TABLE 18.1 Summary of the Distinct Aspects and Distinct Roles of PKCs and PKMs in *Aplysia*

	Isoform			
	PKC Apl I	**PKC Apl II**	**PKC Apl III**	**PKM**
Activators	Calcium/DAG	PA/DAG phosphorylation	PI-3 kinase phosphorylation	Calpain
Plasticity/paradigm, where kinase is activated or required	Activity-dependent facilitation (sensory neuron) Plateau potentials (B51 neuron) Bursts of action potentials (sensory neuron)	Reversal of depression (sensory neuron) Massed applications of 5-HT (sensory neuron) Synapse formation (motor neuron)	Massed applications of 5-HT (sensory neuron)	Activity-dependent facilitation (neuron not identified) Massed applications of 5-HT (motor neuron) Spaced applications of 5-HT (neuron not identified)
Behavior/memory	Site-specific sensitization Operant conditioning	Dishabituation	Unknown	Long-term sensitization Intermediate-term aversive operant conditioning

coincidence detector. Serotonin (5-hydroxytryptamine (5-HT)) is the facilitating transmitter in sensory neurons and is sufficient to induce synaptic facilitation.[10] In cultured *Aplysia* sensory neurons overexpressing fluorescently tagged PKC Apl I, translocation occurred when the neuron was stimulated to fire in the presence of 5-HT but not when stimulation or 5-HT was administered alone.[73] These findings are consistent with PKC Apl I serving as a coincidence detector, being activated in neurons specifically by Ca^{2+} influx due to neuronal activity combined with DAG production downstream of a G protein-coupled 5-HT receptor.

Role of MAP Kinases in PKC Apl I Activation

MAP kinases (MAPKs) are required for the translocation of PKC Apl I by the combination of 5-HT and activity.[74] The activation of MAPKs is downstream of tyrosine kinase signaling activated by 5-HT.[74] How MAPKs play a role in activation of PKC Apl I is still unclear. Because MAPKs are not required for 5-HT-mediated activation of PKC Apl II,[74] it is unlikely that production of DAG is regulated by MAPK. MAPKs may directly regulate PKC Apl I by phosphorylation, but there is no evidence for this. Because PKC Apl I activation is highly dependent on calcium entry, a likely role for MAPKs is to enhance the amount of calcium entering the neuron. It will be important in the future to determine if the amount of calcium entry during the pairing of 5-HT and activity is reduced by inhibition of MAPKs.

Activation of PKC Apl I by Calcium Entry Alone

There is also evidence that PKC Apl I may be activated by calcium alone. Operant conditioning in *Aplysia* is driven by the conjunction of dopamine release and plateau potentials in B51 neurons.[75] In these neurons, PKC Apl I can be translocated to membranes by plateau potentials.[76] DAG is produced mainly by the large family of phospholipase C (PLC) isoforms.[16] The large amounts of calcium entering during the plateau potential could activate PLCδ, a calcium-activated PLC isoform, and an ortholog of PLCδ is present in the *Aplysia* genome.[16] Burst-dependent protection (the inhibition of depression by firing sensory neurons in bursts) is also dependent on PKC Apl I.[77] However, in sensory neurons, action potentials did not translocate PKC Apl I in the absence of 5-HT.[73] One caveat of the translocation assay is that it may not detect activation of prelocalized kinases and thus some kinase activation may be missed. Indeed, burst-dependent protection is dependent on localization of PKC Apl I through the PDZ protein PICK1 binding to the C-terminus of the nervous system-enriched splice form of PKC Apl I.[77] Another example of prelocalization of PKC is in bag cell neurons of *Aplysia*, in which PKC is critically important in activating a nonselective cation channel important for the burst discharge that causes egg-laying hormone release, but PKC is localized to the channel prior to its activation.[78]

Protein Kinase C Apl I Roles in Synaptic Plasticity

ACTIVITY-DEPENDENT FACILITATION

Previously, we described how PKC Apl I was translocated to membranes by the combination of activity and 5-HT in sensory neurons. The physiological relevance of this coordinated regulation of PKC Apl I is strongly supported by experiments showing that overexpression of a dominant-negative mutant of PKC Apl I, but not PKC Apl II, in the sensory neuron disrupts a form of intermediate-term facilitation (ITF) that occurs at *Aplysia* sensorimotor synapses following simultaneous 5-HT treatment and sensory neuron firing.[73] These stimuli also lead to a long-term facilitation that requires synthesis and release of sensorin.[79] Both the increase in local synthesis and the secretion of sensorin require PKC after this stimulation, although the isoform specificity has not been determined in this case.[79] This type of facilitation is thought to underlie a form of memory in the behaving animal known as site-specific sensitization, whereby the animal's siphon withdrawal reflex in response to tail touch is enhanced specifically at the site of tail shock.[55]

SITE-SPECIFIC SENSITIZATION VERSUS FORWARD CONDITIONING

It is interesting to contrast site-specific sensitization (which requires PKC as the associative molecule coupling action potential firing with serotonin) with forward associative conditioning (in which adenylyl cyclase (AC) is the putative associative molecule coupling action potential and firing).[80] During site-specific sensitization, one would suspect that calcium entry to the sensory neuron and DAG production occur at the same time. Indeed, because PKC Apl I binds directly to both calcium and DAG, the appearance of both at the same time seems to be a prerequisite for activation. In contrast, AC appears to be sensitive to the order of the synergistic activators.[81] Calcium and 5-HT do not activate AC synergistically in the steady state.[81] Peak AC stimulation is greater when the calcium pulse immediately precedes the 5-HT pulse than when the 5-HT pulse occurs first.[81] This is represented in the animal by forward conditioning, in which the conditioned stimulus (a light siphon touch leading to calcium release) is administered immediately before the unconditioned stimulus (a tail shock leading to 5-HT release) to converge on AC activation.[82–84] One mechanism that may contribute to this sequence preference is that Ca^{2+}/calmodulin binding to AC accelerates the rate of AC activation by receptor G_s.[81] Furthermore, activation of *Aplysia* AC by a Ca^{2+} pulse rises with a delay compared with activation by a 5-HT pulse, and a late pulse of Ca^{2+}, which arrives after 5-HT, acts via calmodulin to accelerate the decay of AC activation by receptor G_s.[80] Onyike and coworkers[85] identified the calmodulin/AC interaction as the site of cellular memory for the Ca^{2+} transient that ends even before AC is fully activated.

PKC APL I AND OPERANT CONDITIONING IN B51 CELLS

Another form of plasticity in *Aplysia* for which PKC Apl I has been implicated is in the cellular model of operant conditioning of appetitive behavior. Appetitive operant conditioning involves a long-lasting increase in feeding behavior that is produced by 'reinforcing' the animal's spontaneous biting behavior with electrical stimulation of the esophageal nerve, meant to simulate the reinforcing effects of feeding.[75] The resulting increase in appetitive behavior correlates with an increase in excitability of buccal ganglia neuron B51, and these changes can be replicated in the isolated B51 cell in culture by contingently pairing plateau potentials, elicited by electrical stimulation, with the subsequent administration of exogenous dopamine.[75] In this single-cell paradigm, changes in membrane excitability were blocked in the presence of a PKC inhibitor and could be mimicked by activating PKC with a phorbol ester (a pharmacological analog of DAG).[76] As discussed previously, PKC Apl I can be translocated during the *in vitro* analog of operant conditioning.[76] Moreover, overexpression of a dominant-negative mutant form of PKC Apl I blocked the changes in burst threshold and input resistance induced by the single-cell conditioning protocol or phorbol esters, whereas a dominant-negative PKC Apl II mutant had no effect, establishing PKC Apl I as the specific isoform required for this form of plasticity.[76]

Activation and Roles of PKC Apl II

Activation of PKC Apl II by the Conjunction of DAG and PA

Like PKC Apl I, PKC Apl II requires DAG binding to the C1 domain for activation. However, unlike PKC Apl I, the PKC Apl II C2 domain cannot bind calcium. PKC Apl II can be translocated to the membrane in heterologous cells at low levels of DOG, without the requirement for additional Ca^{2+} that constrained PKC Apl I.[73] Similarly, whereas PKC Apl I requires concurrent 5-HT and neuronal activity for translocation in isolated *Aplysia* sensory neurons, PKC Apl II translocates to the plasma membrane upon 5-HT treatment alone and in the absence of neuronal activity.[73] Although this can be partly explained by the higher affinity of the C1 domain of PKC Apl II for DAG compared to the C1 domain of PKC Apl I,[27] there are other factors contributing to PKC Apl II isoform-specific regulation. As discussed previously, PA synergizes with DAG to translocate PKC Apl II both by removing C2 domain-mediated inhibition and through direct recruitment through C1 domain interactions.[27,28] This synergism is important for 5-HT-induced PKC Apl II translocation in *Aplysia* sensory neurons because inhibition of either PLC, which produced DAG, or phospholipase D (PLD), the enzyme that produces PA, equally blocked the translocation of PKC Apl II in response to 5-HT.[27,28] Thus, PKC Apl II may act as a coincident sensor for the production of DAG and PA in sensory neurons.

Is DAG Production the Rate-Limiting Step for PKC Activation?

Because PKC Apl I and PKC Apl II both require an additional second messenger in conjunction with DAG for activation, one must consider whether production of DAG is really the rate-limiting step in PKC activation. Although PLC inhibitors do block PKC Apl II translocation, they may be inhibiting basal DAG production as opposed to 5-HT-induced PLCs. There is no direct biochemical evidence for DAG production downstream of 5-HT. Moreover, PLC activation usually also leads to the production of the second messenger IP3 that activates the IP3 receptor in the endoplasmic reticulum, allowing for calcium release from internal stores. However, 5-HT does not lead to internal store calcium release in sensory neurons, despite the fact that injection of IP3 does.[86] Moreover, both PLC activation and release of calcium from internal stores downstream of 5-HT have been suggested to be important for plasticity in motor neurons,[87,88] but 5-HT does not translocate PKC Apl II in motor neurons.[73] PLC activation is usually downstream of G_q-coupled receptors. However, attempts to link the putatively G_q-coupled ortholog of 5-HT2 receptors in *Aplysia*, 5-HT_{2Apl}, to the activation of PKC have so far been unsuccessful.[89] Another surprising finding is that activation of PKC Apl II by 5-HT in sensory neurons is reduced by the general tyrosine kinase inhibitor genestein.[89] Whether this is due to a requirement for an intracellular tyrosine kinase, such as an Src family member, or transactivation of a tyrosine kinase by 5-HT, such as *Aplysia* Trk-like receptor,[90,91] is not clear. It is also not known if the tyrosine kinase activity is required upstream of PLC or PLD activation or if it plays some other role. A full understanding of how PKC Apl II is activated is an important area for future research.

Activation of PKC Apl II by Phosphorylation

The previously mentioned studies examined PKC activation by translocation, but PKC activity can also be modulated by phosphorylation, independently of translocation. Continuous application of 5-HT, an *in vitro* analog of massed training, leads to an increase in PKC Apl II activity.[92] This increase is not due to PKM formation because the increased activity is still sensitive to a regulatory domain directed inhibitor, calphostin C.[92] The increased activity of PKC Apl II is correlated with an increase in phosphorylation of PKC Apl II at both the PDK-1 and the hydrophobic sites.[48,49] Interestingly, increases in phosphorylation of PKC Apl I at the PDK-1 site were not observed.[48] The increase in phosphorylation at the hydrophobic site is mediated by the TOR complex (probably TORC2) because it is blocked by an inhibitor of the TOR kinase (Torin 1).[50] Phosphorylation at the hydrophobic site directly affects PKC Apl II activity but does not regulate translocation,[49] suggesting that the increased phosphorylation could account for the increased activity of PKC Apl II after continuous application of 5-HT.

PKC Apl II Responses Can Differentiate between Spaced and Massed Applications of 5-HT in Sensory Neurons

Memory retention is highly sensitive to the pattern of trials used during training. Training distributed over time (spaced training) is superior to training presented with little or no rest intervals (massed training) at generating long-term memories.[93–97] In culture, spaced applications of 5-HT are superior to massed applications at generating long-term facilitation of cultured sensorimotor neuron synapses.[98] PKC Apl II activation is highly sensitive to the pattern of 5-HT application.[99] Spaced applications lead to desensitization of PKC Apl II translocation by a PKA-dependent mechanism, whereas massed application leads to persistent translocation of PKC Apl II.[99] Furthermore, regulation of PKC translocation is mediated by competing feedback mechanisms that act through protein synthesis.[99] Mathematical modeling of the PKC Apl II response to the different patterns of stimulation allowed the following to be determined: (1) increased desensitization due to PKA-mediated heterologous desensitization was coupled to a faster recovery than the homologous desensitization that occurs in the absence of PKA activity, and (2) the major determinant of how the system responds to spacing is the production and degradation rate of proteins.[100] Indeed, our results suggested that increased desensitization during spaced applications of 5-HT was due to the short half-life of a hypothetical protein, which prevented homologous desensitization. One pulse of 5-HT would synthesize this protein, which would protect against desensitization for the short period before it was degraded.[100] Massed application of 5-HT would constantly replenish this protein, whereas spaced applications of 5-HT, with an interapplication interval longer than the half-life of the protein, would overshoot the protective period and cause increased desensitization.[100]

Protein Kinase C Apl II and Reversal of Depression

As described previously, PKC Apl II can be activated by 5-HT alone. Surprisingly, this activation apparently plays no role in short-term facilitation at sensorimotor neuron synapses. Short-term facilitation is not blocked by either PKC inhibitors[101] or dominant negatives of PKC Apl I or PKC Apl II.[37,73] However, PKC is critical for the actions of 5-HT if the synapse is first depressed.[101] This form of plasticity, which is analogous to behavioral dishabituation of the defensive withdrawal reflex, is blocked by overexpression of a dominant-negative form of PKC Apl II but not PKC Apl I in the sensory neuron.[37] Presumably, PKC Apl II can phosphorylate targets important for reversing cellular depression but cannot phosphorylate targets important for facilitation. Interestingly, phorbol esters, pharmacological analogs of DAG that activate both PKC Apl I and PKC Apl II, increase synaptic strength at sensorimotor neuron synapses through phosphorylation of SNAP-25.[102] The prediction would be that SNAP-25 phosphorylation would be mediated by PKC Apl I, but not PKC Apl II, or that phorbol esters target PKCs to phosphorylate targets distinct from those targeted by 5-HT.

Protein Kinase C Apl II and Intermediate-Term Facilitation after Massed Application of 5-HT

Ninety-minute bath application of 5-HT to isolated pleural pedal ganglia produces intermediate-[103] and long-term[104] facilitation at sensorimotor synapses. As discussed previously, this is associated with an increase in PKC Apl II phosphorylation and an increase in PKC Apl II activity. However, there is no direct evidence that the memory trace is mediated by PKC Apl II. This application, unlike spaced training, does not lead to persistent activation of PKA.[105]

When 5-HT stimulation in sensorimotor neuron cultures is prolonged from 1 to 10 min, facilitation becomes quite complicated. A 10-min application of 5-HT is sufficient to induce facilitation that is dependent on PKC but not PKA.[106] At this time, facilitation also depends on postsynaptic calcium entry and both presynaptic and postsynaptic protein synthesis.[87,106,107] Although it is appealing to believe that this is due to presynaptic PKC Apl II because PKC Apl II is strongly activated at this time,[27] this has not been directly shown. Moreover, as discussed later, this stimulation also appears sufficient to induce PKM activation in the motor neuron.[108] Thus, the specific isoforms required for the induction or the maintenance of this form of ITF are not currently known. Although facilitation was partly blocked by a pseudosubstrate inhibitor based on the classical PKC sequence injected into the presynaptic cell, these inhibitors do not distinguish between PKC isoforms.[109]

PKC Apl II and Synapse Formation

PKC has been shown to play a key role in regulating synapse formation between the sensory and motor neurons.[110] Dominant-negative PKC Apl II expressed in the motor neuron, but not in the sensory neuron, blocked initial synapse formation in culture, and PKC Apl II was highly enriched at sites of synapse formation.[111] Whether PKC Apl II also plays a role in the synapse formation after learning has not been determined. The role of PKC Apl II is downstream of the adhesion protein ApCAM.[111] Interestingly, PKCs have been implicated in actions downstream of the vertebrate ortholog of Apcam, NCAM, through activation of FGF receptors by NCAM.[112] This would be consistent with the important role of tyrosine kinases in the activation of PKC Apl II.[89,113]

Activation and Roles of PKC Apl III

Activation of PKC Apl III by Phosphorylation

A 10-min treatment of cultured *Aplysia* sensory neurons with 5-HT increased phosphorylation of overexpressed PKC Apl III, as measured by immunoreactivity with a phospho-specific antibody targeting the PDK phosphorylation site. This increase was blocked by a PI3 kinase inhibitor,[13] suggesting that PKC Apl III may mediate some of the actions of 5-HT downstream of PI3 kinase activation, such as the remodeling of the cytoskeleton.[114] Indeed, atypical PKCs play conserved roles in cell polarity and regulation of the cytoskeleton.[115] To date, most of the research on PKC Apl III has focused on its possible role in formation of PKM, and this is detailed later in a general discussion of PKMs.

A Possible Role for PKC Apl III in the Nucleus

The presence of PKC APl III, but not the other isoforms of PKC, in the nucleus[13] would be consistent with roles for PKC Apl III in activating transcription. Interestingly, degradation of the CREB repressor in *Aplysia* is stimulated by PKC phosphorylation,[116] suggesting a role for PKC Apl III in assisting the transcriptional activation underlying long-term facilitation.

Activation and Roles of PKMs

As discussed previously, there is no evidence in *Aplysia* for mRNAs directly encoding PKMs, analogous to the PKMζ mRNA transcript in vertebrates. Thus, we focus on the activation of PKMs by calpain-dependent cleavage.

Activation of PKCs by Cleavage

Calpains have been shown to be required for two distinct forms of ITF at the sensorimotor neuron synapse that require PKC (Table 18.1). Neither the mechanisms underlying the activation of calpains nor the calpain isoform activated are known. Although both of these forms of plasticity are associated with calcium entry,[87,117] and increased calcium can stimulate cleavage of overexpressed PKC Apl III,[13] it cannot yet be concluded that a classical calpain is important for these events. Additional work is required to identify the factors regulating PKM formation in *Aplysia*. We have generated fluorescence resonance energy transfer (FRET) constructs from PKCs containing CFP on the N-terminus and YFP on the C-terminus to measure calpain activation using a loss of FRET assay, similar to other live protease assays.[118] Using this construct, we have shown that 5-HT can induce calpain-dependent cleavage of PKC Apl III in motor neuron processes.[119]

The Role of PKMs in Maintaining Long-Term Facilitation and Long-Term Memory

Studies in *Aplysia* have demonstrated that the role of PKMs in the persistence of memory shown in vertebrate systems is conserved. Both zeta inhibitory peptide (ZIP), the same myristoylated pseudosubstrate peptide used to disrupt PKMζ activity in rodents, and chelerythrine, a PKC inhibitor that has a much higher affinity for PKMs than for PKCs,[108,120] disrupted 5-HT-induced long-term facilitation in cultured *Aplysia* sensorimotor neurons when administered 24 hr after induction of facilitation and also completely abolished long-term sensitization of the siphon withdrawal reflex in the animal, even when applied as late as 7 days after training.[6] ZIP is based on a pseudosubstrate sequence of the atypical PKCs, and this sequence is highly conserved. However, neither the ZIP peptide nor chelerythrine can distinguish between PKMs derived from distinct PKCs in *Aplysia*.[109] Thus, although this data implicates PKMs in mediating the maintenance of facilitation and memory in this system, it is not clear which isoform(s) the PKM is generated from to mediate long-term facilitation.

The Role of PKMs in Maintaining an Increase in AMPA Receptors in Motor Neurons

Although the mechanism by which PKMζ maintains memory in rodents is unclear, it is thought to involve regulation of the trafficking of AMPA-type glutamate receptors in the postsynaptic cell,[121,122] and there is evidence that an *Aplysia* PKM similarly maintains a 5-HT-mediated increase in the postsynaptic response to glutamate.[108] That this form of facilitation is maintained by ongoing PKM Apl III activity is suggested by the observation that enhancement of glutamate-evoked responses in isolated motor neurons can be reversed by low concentrations of chelerythrine but not by bisindolylmaleimide-1—a pharmacological profile consistent with an atypical PKM.[108] Furthermore, induction of facilitation was blocked in the presence of the calpain inhibitor, consistent with a requirement for cleavage of PKC to form a PKM.[108] These findings suggest that the regulation of postsynaptic glutamate receptor trafficking by atypical PKMs is a conserved mechanism for maintaining memory in vertebrates and invertebrates.

The Role of PKMs in Intermediate Memory of Aversive Operant Conditioning

Aplysia can learn that a food is inedible, which is an aversive form of operant conditioning.[123] This form of memory, similar to sensitization, leads to multiple memory traces depending on the type of training and the time the memory persists.[124] When the training is given in a massed manner, there is an intermediate form of memory that depends on PKM formation, based on the inhibition of memory retention by chelerythrine.[125] The maintenance is likely to be due to PKM Apl III because it is not blocked by bisindolylmaleimide-1, which blocks the other PKMs. Unlike sensitization, the long-term retention of this memory is not sensitive to inhibitors of PKM.[125] The cellular locus for this memory has not been established.

PKMs and Activity-Dependent Plasticity

Another form of plasticity has also been found to rely on a PKM in *Aplysia*. As discussed previously, whereas a single application of 5-HT to the isolated *Aplysia* pleural–pedal ganglion produces only short-term facilitation of the tail sensorimotor synapse, the combination of a single application of 5-HT with activity in the sensory neuron produces a more prolonged, activity-dependent facilitation whose induction depends on PKC Apl I.[73] The maintenance of this form of plasticity depends on persistent PKC activity, as does the behavioral correlate of this facilitation—site-specific sensitization.[55,117] This is likely mediated by a PKM. First, the behavior is disrupted both by inhibitors targeting the catalytic but not the regulatory domain of PKC and by calpain inhibitors.[55] However, the maintenance of this plasticity was blocked by bisindolylmaleimide-1,which is a much better inhibitor of PKM Apl I than PKM Apl III.[108,109] The high dose of bisindolylmaleimide-1 used in these experiments (10 μM) does not allow one to rule out the involvement of PKM Apl III. Thus, whereas it is likely that maintenance of this form of plasticity requires PKM Apl I, it remains possible that although induction of this form of plasticity requires PKC Apl I, its maintenance could be mediated by PKM Apl III.

INTERACTION OF PKCS WITH OTHER SIGNAL TRANSDUCTION PATHWAYS

So far, we have discussed the mechanisms by which PKCs are activated and underlie distinct forms of synaptic plasticity. However, many stimuli that induce memory also activate other pathways as well as PKC. There is also a great deal of information on how these pathways interact with PKC.

PKC versus PKA Signaling: Synergy or Opposition?

In addition to PKC, PKA has been shown to be a key signaling molecule underlying synaptic plasticity in *Aplysia*.[1] Furthermore, several reports have provided evidence of cross talk between PKC and PKA pathways during plasticity. These reports point to a complex relationship between these signaling pathways, at times antagonistic and at other times cooperative.

Activation of PKA by PKC through AC

The initial evidence of positive modulation of AC by PKC came from the observation that phorbol ester-induced PKC activation in isolated *Aplysia* pleural sensory neuron clusters produces an increase in cAMP compared to control conditions.[126] This finding is consistent with observed increases in excitability following exposure of *Aplysia* sensory neurons to phorbol esters.[127,128] These results imply the existence of a PKC-regulated AC in *Aplysia*. Indeed, PKC-mediated positive regulation of cAMP production appears to contribute to at least one form of plasticity in *Aplysia*—the single-cell analog of appetitive operant conditioning discussed previously. In this system, evidence suggests PKC is upstream of PKA. When the plasticity was induced by injection of cAMP, it was unaltered by the presence of PKC inhibitors; in contrast, when plasticity was induced by phorbol ester-mediated activation of PKC, it was blocked by cAMP antagonists.[76] The most attractive possibility is that the synergy takes place at the level of a PKC-regulated AC,[76] similar to analogous regulation in vertebrate ACs.[16] It is thus curious why in sensory neurons, in which cAMP can also be increased by phorbol esters, PKC is not upstream of PKA as well. This may be due to differences in the isoform of PKC: PKC Apl I underlies plasticity in B51, whereas only PKC Apl II is activated by 5-HT in sensory neurons.

Inhibition of PKA by PKC

One study suggested that in sensory neurons, activation of PKC prevented activation of PKA. Serotonin-induced spike broadening and enhancement of excitability, both PKA-dependent processes, were reduced following prolonged incubation with phorbol esters and could be rescued by the cell-permeable cAMP analog 8-bromo-cAMP.[126] These observations indicate that PKC can antagonize PKA activation in a manner that could impact 5-HT-dependent synaptic plasticity. This may be important in distinguishing spaced versus massed training in *Aplysia*.[100,106]

Inhibition of PKC by PKA

As discussed previously, PKA is important for desensitization of the PKC response in sensory neurons; spaced training leads to strong desensitization of the ability of 5-HT to translocate PKC Apl II. This would be consistent with an inhibitory role for PKC Apl II in the induction of long-term facilitation, which is much stronger after spaced training than after massed training. Arguing against this is the finding that two spaced trials separated by 45 min are sufficient to induce LTM,[129] whereas this spacing does not lead to desensitization of the PKC response in isolated sensory neurons.[99]

Activation of PKC by PKA

There is no evidence in *Aplysia* for activation of PKC by either PKA or the PKA-activator cAMP. In sensory neurons, PKC activation (assayed by the ability to reverse synaptic depression) is not blocked when the G_s-coupled 5-HT_{7Apl} receptor is removed.[130] However, in vertebrates, an ortholog of PKC Apl II, PKCε, is activated downstream of G_s-coupled receptors through cAMP activation of the guanine exchange factor EPAC, followed by activation of PLCε.[131] When the 5-HT_{7Apl} receptor is expressed in a heterologous system, Sf9 cells, 5-HT translocates PKC Apl II.[89] It is possible that in some neurons of *Aplysia*, PKC activation will be downstream of cAMP.

Inhibition of PKC Isoforms May Also Play a Role in Synaptic Plasticity

Although the activation of PKC isoforms has been shown to underlie several forms of synaptic facilitation, the inhibition of PKCs may also play a role in synaptic plasticity. Inhibitory G protein-coupled receptors in *Aplysia*, including serotonin and dopamine receptors, as well as FMRFamide receptors block PKC Apl II translocation in sensory neurons.[132] Pleural sensory neurons are heterogeneous for their inhibitory response to endogenous transmitters, with approximately 60% of cultured *Aplysia* pleural sensory neurons exhibiting an inhibitory response to dopamine, 40% to FMRFamide, and 5–7% to 5-HT.[132] Interestingly, dopamine can reverse PKC Apl II translocation in sensory neurons during the continued application of 5-HT or after DOG-mediated translocation of PKC Apl II, suggesting that inhibition is not through blockade of DAG production.[132] Because the dopamine-mediated effects could be rescued by PA, they may work by blocking PLD activation.[132] However, PA could not recover the dopamine blockade of PKC-dependent synaptic facilitation due to the additional dopamine-mediated inhibition of non-L-type calcium channels required for transmitter release.[132] These findings indicate that dopamine and, to a lesser extent, FMRFamide and 5-HT negatively modulate synaptic strength at a subset of *Aplysia* sensorimotor synapses, in part through modulation of PKC.

CONCLUSION

This chapter reviewed the roles of PKC and PKM in inducing and maintaining the synaptic changes that underlie learning and memory formation in *A. californica*. PKCs can act as coincidence detectors to integrate multiple signals, and in some cases the pattern of application of the stimulation dictates the activation of specific PKC isoforms. Many important challenges remain, including determining how tyrosine kinase activation regulates PKC Apl II, identifying the mechanisms underlying calpain activation of PKMs, and identifying the substrates that are phosphorylated downstream of the activation of the different PKC isoforms and how they contribute to generating distinct forms of synaptic plasticity and different forms of memory.

Acknowledgments

W.S.S. is a James McGill Professor, and M.H is supported by a Canadian Institute of Health Research (CIHR) Studentship. This work was supported by CIHR grant MOP 12046.

References

1. Kandel ER. The molecular biology of memory storage: a dialogue between genes and synapses. *Science*. 2001;294(5544): 1030–1038.
2. Serrano P, Friedman EL, Kenney J, et al. PKMzeta maintains spatial, instrumental, and classically conditioned long-term memories. *PLoS Biol*. 2008;6(12):2698–2706.
3. Drier EA, Tello MK, Cowan M, et al. Memory enhancement and formation by atypical PKM activity in *Drosophila melanogaster*. *Nat Neurosci*. 2002;5(4):316–324.
4. Sossin WS. Isoform specificity of protein kinase Cs in synaptic plasticity. *Learn Mem*. 2007;14(4):236–246.
5. Sacktor TC. How does PKMzeta maintain long-term memory?. *Nat Rev Neurosci*. 2011;12(1):9–15.
6. Cai D, Pearce K, Chen S, Glanzman DL. Protein kinase M maintains long-term sensitization and long-term facilitation in *Aplysia*. *J Neurosci*. 2011;31(17):6421–6431.
7. Rosenegger D, Lukowiak K. The participation of NMDA receptors, PKC, and MAPK in the formation of memory following operant conditioning in *Lymnaea*. *Mol Brain*. 2010;3:24.
8. Brembs B, Plendl W. Double dissociation of PKC and AC manipulations on operant and classical learning in *Drosophila*. *Curr Biol*. 2008;18(15):1168–1171.
9. Newton AC. Protein kinase C: structural and spatial regulation by phosphorylation, cofactors, and macromolecular interactions. *Chem Rev*. 2001;101(8):2353–2364.
10. Kandel E. The molecular biology of memory storage: a dialogue between genes and synapses. *Science*. 2001;294(5544):1030–1038.
11. Corbalan-Garcia S, Gomez-Fernandez JC. Protein kinase C regulatory domains: the art of decoding many different signals in membranes. *Biochim Biophys Acta*. 2006;1761(7):633–654.
12. Farah CA, Sossin WS. The role of C2 domains in PKC signaling. *Adv Exp Med Biol*. 2012;740:663–684.
13. Bougie JK, Lim T, Farah CA, et al. The atypical protein kinase C in *Aplysia* can form a protein kinase M by cleavage. *J Neurochem*. 2009;109(4):1129–1143.
14. Kruger KE, Sossin WS, Sacktor TC, Bergold PJ, Beushausen S, Schwartz JH. Cloning and characterization of Ca(2 +)-dependent and Ca(2 +)-independent PKCs expressed in *Aplysia* sensory cells. *J Neurosci*. 1991;11(8):2303–2313.
15. Sossin WS, Diaz AR, Schwartz JH. Characterization of two isoforms of protein kinase C in the nervous system of *Aplysia californica*. *J Biol Chem*. 1993;268(8):5763–5768.
16. Sossin WS, Abrams TW. Evolutionary conservation of the signaling proteins upstream of cyclic AMP-dependent kinase and protein kinase C in gastropod mollusks. *Brain Behav Evol*. 2009;74 (3):191–205.
17. House C, Kemp BE. Protein kinase C contains a pseudosubstrate prototope in its regulatory domain. *Science*. 1987;238(4834): 1726–1728.
18. Makowske M, Rosen OM. Complete activation of protein kinase C by an antipeptide antibody directed against the pseudosubstrate prototope. *J Biol Chem*. 1989;264(27):16155–16159.
19. Evans JH, Murray D, Leslie CC, Falke JJ. Specific translocation of protein kinase Calpha to the plasma membrane requires both Ca^{2+} and PIP2 recognition by its C2 domain. *Mol Biol Cell*. 2006; 17(1):56–66.
20. Gallegos LL, Newton AC. Spatiotemporal dynamics of lipid signaling: protein kinase C as a paradigm. *IUBMB Life*. 2008;60 (12):782–789.
21. Sanchez-Bautista S, Marin-Vicente C, Gomez-Fernandez JC, Corbalan-Garcia S. The C2 domain of PKCalpha is a Ca^{2+}-dependent PtdIns(4,5)P2 sensing domain: a new insight into an old pathway. *J Mol Biol*. 2006;362(5):901–914.
22. Dries DR, Gallegos LL, Newton AC. A single residue in the C1 domain sensitizes novel protein kinase C isoforms to cellular diacylglycerol production. *J Biol Chem*. 2007;282(2):826–830.
23. Johnson JE, Giorgione J, Newton AC. The C1 and C2 domains of protein kinase C are independent membrane targeting modules, with specificity for phosphatidylserine conferred by the C1 domain. *Biochemistry*. 2000;39(37):11360–11369.
24. Rizo J, Sudhof TC. C2-domains, structure and function of a universal Ca^{2+}-binding domain. *J Biol Chem*. 1998;273(26): 15879–15882.
25. Zhang D, Aravind L. Identification of novel families and classification of the C2 domain superfamily elucidate the origin and evolution of membrane targeting activities in eukaryotes. *Gene*. 2011;469(1–2):18–30.
26. Benes CH, Wu N, Elia AE, Dharia T, Cantley LC, Soltoff SP. The C2 domain of PKCdelta is a phosphotyrosine binding domain. *Cell*. 2005;121(2):271–280.
27. Farah CA, Nagakura I, Weatherill D, Fan X, Sossin WS. Physiological role for phosphatidic acid in the translocation of the novel protein kinase C Apl II in *Aplysia* neurons. *Mol Cell Biol*. 2008;28(15):4719–4733.
28. Pepio AM, Sossin WS. The C2 domain of the Ca(2 +)-independent protein kinase C Apl II inhibits phorbol ester binding to the C1 domain in a phosphatidic acid-sensitive manner. *Biochemistry*. 1998;37(5):1256–1263.
29. Pepio AM, Fan X, Sossin WS. The role of C2 domains in Ca^{2+}-activated and Ca^{2+}-independent protein kinase Cs in *Aplysia* [published erratum appears in *J Biol Chem* 1998 Aug 28;273 (35):22856]. *J Biol Chem*. 1998;273(30):19040–19048.
30. Sossin WS, Fan X, Saberi F. Expression and characterization of *Aplysia* protein kinase C: a negative regulatory role for the E region. *J Neurosci*. 1996;16(1):10–18.
31. Hucho TB, Dina OA, Levine JD. Epac mediates a cAMP-to-PKC signaling in inflammatory pain: an isolectin B4(+) neuron-specific mechanism. *J Neurosci*. 2005;25(26):6119–6126.
32. Jose Lopez-Andreo M, Gomez-Fernandez JC, Corbalan-Garcia S. The simultaneous production of phosphatidic acid and diacylglycerol is essential for the translocation of protein kinase Cε to the plasma membrane in RBL-2H3 cells. *Mol Biol Cell*. 2003;14 (12):4885–4895.
33. Corbalan-Garcia S, Sanchez-Carrillo S, Garcia-Garcia J, Gomez-Fernandez JC. Characterization of the membrane binding mode of the C2 domain of PKC epsilon. *Biochemistry*. 2003;42(40): 11661–11668.
34. Yao L, Fan P, Jiang Z, Gordon A, Mochly-Rosen D, Diamond I. Dopamine and ethanol cause translocation of εPKC associated with εRACK: cross-talk between cAMP-dependent protein kinase A and protein kinase C signaling pathways. *Mol Pharmacol*. 2008;73(4):1105–1112.
35. Farah CA, Sossin WS. The role of C2 domains in PKC signaling. *Adv Exp Med Biol*. 2012;740:663–683.
36. Farah CA, Sossin WS. A new mechanism of action of a C2 domain-derived novel PKC inhibitor peptide. *Neurosci Lett*. 2011; 504(3):306–310.
37. Manseau F, Fan X, Hueftlein T, Sossin W, Castellucci VF. Ca^{2+}-independent protein kinase C Apl II mediates the serotonin-induced facilitation at depressed *Aplysia* sensorimotor synapses. *J Neurosci*. 2001;21(4):1247–1256.
38. Pepio AM, Sossin WS. Membrane translocation of novel protein kinase Cs is regulated by phosphorylation of the C2 domain. *J Biol Chem*. 2001;276(6):3846–3855.
39. Farah CA, Lindeman AA, Siu V, Gupta MD, Sossin WS. Autophosphorylation of the C2 domain inhibits translocation of the novel protein kinase C (nPKC) Apl II. *J Neurochem*. 2012;123 (3):360–372.

40. Ohno S. Intercellular junctions and cellular polarity: the PAR–aPKC complex, a conserved core cassette playing fundamental roles in cell polarity. *Curr Opin Cell Biol.* 2001;13(5): 641–648.
41. Moscat J, Diaz-Meco MT. The atypical protein kinase Cs: functional specificity mediated by specific protein adapters. *EMBO Rep.* 2000;1(5):399–403.
42. Moscat J, Diaz-Meco MT, Albert A, Campuzano S. Cell signaling and function organized by PB1 domain interactions. *Mol Cell.* 2006;23(5):631–640.
43. Nishikawa K, Toker A, Johannes FJ, Songyang Z, Cantley LC. Determination of the specific substrate sequence motifs of protein kinase C isozymes. *J Biol Chem.* 1997;272(2):952–960.
44. Jacinto E, Lorberg A. TOR regulation of AGC kinases in yeast and mammals. *Biochem J.* 2008;410(1):19–37.
45. Mora A, Komander D, van Aalten DM, Alessi DR. PDK1, the master regulator of AGC kinase signal transduction. *Semin Cell Dev Biol.* 2004;15(2):161–170.
46. Freeley M, Kelleher D, Long A. Regulation of protein kinase C function by phosphorylation on conserved and non-conserved sites. *Cell Signal.* 2011;23(5):753–762.
47. Khan A, Pepio AM, Sossin WS. Serotonin activates S6 kinase in a rapamycin-sensitive manner in *Aplysia* synaptosomes. *J Neurosci.* 2001;21(2):382–391.
48. Pepio AM, Thibault GL, Sossin WS. Phosphoinositide-dependent kinase phosphorylation of protein kinase C Apl II increases during intermediate facilitation in *Aplysia*. *J Biol Chem.* 2002;277(40): 37116–37123.
49. Lim T, Sossin WS. Phosphorylation at the hydrophobic site of protein kinase C Apl II is increased during intermediate term facilitation. *Neuroscience.* 2006;141(1):277–285.
50. Labban M, Dyer JR, Sossin WS. Rictor regulates phosphorylation of the novel protein kinase C Apl II in *Aplysia* sensory neurons. *J Neurochem.* 2012;122:1108–1117.
51. Inoue M, Kishimoto A, Takai Y, Nishizuka Y. Studies on a cyclic nucleotide-independent protein kinase and its proenzyme in mammalian tissues: II. Proenzyme and its activation by calcium-dependent protease from rat brain. *J Biol Chem.* 1977;252(21): 7610–7616.
52. Kelly MT, Crary JF, Sacktor TC. Regulation of protein kinase Mzeta synthesis by multiple kinases in long-term potentiation. *J Neurosci.* 2007;27(13):3439–3444.
53. Hernandez AI, Blace N, Crary JF, et al. Protein kinase M zeta synthesis from a brain mRNA encoding an independent protein kinase C zeta catalytic domain: implications for the molecular mechanism of memory. *J Biol Chem.* 2003;278(41):40305–40316.
54. Hrabetova S, Sacktor TC. Bidirectional regulation of protein kinase M zeta in the maintenance of long-term potentiation and long-term depression. *J Neurosci.* 1996;16(17):5324–5333.
55. Sutton MA, Bagnalll MW, Sharma SK, Shobe J, Carew TJ. Intermediate-term memory for site-specific sensitization in *Aplysia* is maintained by persistent activation of protein kinase C. *J Neurosci.* 2004;24:3600–3609.
56. Denny JB, Polan-Curtain J, Ghuman A, Wayner MJ, Armstrong DL. Calpain inhibitors block long-term potentiation. *Brain Res.* 1990;534(1–2):317–320.
57. Oliver MW, Baudry M, Lynch G. The protease inhibitor leupeptin interferes with the development of LTP in hippocampal slices. *Brain Res.* 1989;505(2):233–238.
58. Lynch G, Rex CS, Gall CM. LTP consolidation: substrates, explanatory power, and functional significance. *Neuropharmacology.* 2007;52(1):12–23.
59. Tompa P, Emori Y, Sorimachi H, Suzuki K, Friedrich P. Domain III of calpain is a Ca^{2+}-regulated phospholipid-binding domain. *Biochem Biophys Res Commun.* 2001;280(5):1333–1339.
60. Elce JS, Hegadorn C, Arthur JS. Autolysis, Ca^{2+} requirement, and heterodimer stability in m-calpain. *J Biol Chem.* 1997;272 (17):11268–11275.
61. Ohno S, Emori Y, Imajoh S, Kawasaki H, Kisaragi M, Suzuki K. Evolutionary origin of a calcium-dependent protease by fusion of genes for a thiol protease and a calcium-binding protein?. *Nature.* 1984;312(5994):566–570.
62. Suzuki K, Hata S, Kawabata Y, Sorimachi H. Structure, activation, and biology of calpain. *Diabetes.* 2004;53(Suppl 1):S12–S18.
63. Ono Y, Sorimachi H. Calpains—An elaborate proteolytic system. *Biochim Biophys Acta.* 2012;1824(1):224–236.
64. Ravulapalli R, Diaz BG, Campbell RL, Davies PL. Homodimerization of calpain 3 penta-EF-hand domain. *Biochem J.* 2005;388(Pt 2):585–591.
65. Croall DE, Ersfeld K. The calpains: modular designs and functional diversity. *Genome Biol.* 2007;8(6):218.
66. Sokol SB, Kuwabara PE. Proteolysis in *Caenorhabditis elegans* sex determination: cleavage of TRA-2A by TRA-3. *Genes Dev.* 2000; 14(8):901–906.
67. Waghray A, Wang DS, McKinsey D, Hayes RL, Wang KK. Molecular cloning and characterization of rat and human calpain-5. *Biochem Biophys Res Commun.* 2004;324(1):46–51.
68. Denison SH, Orejas M, Arst Jr. HN. Signaling of ambient pH in *Aspergillus* involves a cysteine protease. *J Biol Chem.* 1995;270 (48):28519–28522.
69. Maki M, Maemoto Y, Osako Y, Shibata H. Evolutionary and physical linkage between calpains and penta-EF-hand Ca(2+) -binding proteins. *FEBS J.* 2012;279(8):1414–1421.
70. Delaney SJ, Hayward DC, Barleben F, Fischbach KF, Miklos GL. Molecular cloning and analysis of small optic lobes, a structural brain gene of *Drosophila melanogaster*. *Proc Natl Acad Sci USA.* 1991;88(16):7214–7218.
71. Kamei M, Webb GC, Heydon K, Hendry IA, Young IG, Campbell HD. Solh, the mouse homologue of the *Drosophila melanogaster* small optic lobes gene: organization, chromosomal mapping, and localization of gene product to the olfactory bulb. *Genomics.* 2000;64(1):82–89.
72. Farah CA, Sossin WS. Live-imaging of PKC translocation in Sf9 cells and in *Aplysia* sensory neurons. *J Vis Exp.* 2011;(50):e2516.
73. Zhao Y, Leal K, Abi-Farah C, Martin KC, Sossin WS, Klein M. Isoform specificity of PKC translocation in living *Aplysia* sensory neurons and a role for Ca^{2+}-dependent PKC APL I in the induction of intermediate-term facilitation. *J Neurosci.* 2006;26(34):8847–8856.
74. Shobe JL, Zhao Y, Stough S, et al. Temporal phases of activity-dependent plasticity and memory are mediated by compartmentalized routing of MAPK signaling in *Aplysia* sensory neurons. *Neuron.* 2009;61(1):113–125.
75. Brembs B, Lorenzetti FD, Reyes FD, Baxter DA, Byrne JH. Operant reward learning in *Aplysia*: neuronal correlates and mechanisms. *Science.* 2002;296(5573):1706–1709.
76. Lorenzetti FD, Baxter DA, Byrne JH. Molecular mechanisms underlying a cellular analog of operant reward learning. *Neuron.* 2008;59(5):815–828.
77. Wan Q, Jiang XY, Negroiu AM, Lu SG, McKay KS, Abrams TW. Protein kinase C acts as a molecular detector of firing patterns to mediate sensory gating in *Aplysia*. *Nat Neurosci.* 2012;15: 1144–1152.
78. Wilson GF, Magoski NS, Kaczmarek LK. Modulation of a calcium-sensitive nonspecific cation channel by closely associated protein kinase and phosphatase activities. *Proc Natl Acad Sci USA.* 1998;95(18):10938–10943.
79. Hu JY, Chen Y, Schacher S. Protein kinase C regulates local synthesis and secretion of a neuropeptide required for activity-dependent long-term synaptic plasticity. *J Neurosci.* 2007;27(33): 8927–8939.

80. Abrams TW, Yovell Y, Onyike CU, Cohen JE, Jarrard HE. Analysis of sequence-dependent interactions between transient calcium and transmitter stimuli in activating adenylyl cyclase in *Aplysia*: possible contribution to CS–US sequence requirement during conditioning. *Learn Mem*. 1998;4(6):496–509.
81. Yovell Y, Abrams TW. Temporal asymmetry in activation of *Aplysia* adenylyl cyclase by calcium and transmitter may explain temporal requirements of conditioning. *Proc Natl Acad Sci USA*. 1992;89(14):6526.
82. Hawkins RD, Abrams TW, Carew TJ, Kandel ER. A cellular mechanism of classical conditioning in *Aplysia*: activity-dependent amplification of presynaptic facilitation. *Science*. 1983; 219(4583):400–405.
83. Ocorr KA, Walters ET, Byrne JH. Associative conditioning analog selectively increases cAMP levels of tail sensory neurons in *Aplysia*. *Proc Natl Acad Sci USA*. 1985;82(8):2548–2552.
84. Hawkins RD, Carew TJ, Kandel ER. Effects of interstimulus interval and contingency on classical conditioning of the *Aplysia* siphon withdrawal reflex. *J Neurosci*. 1986;6(6): 1695–1701.
85. Onyike CU, Lin AH, Abrams TW. Persistence of the interaction of calmodulin with adenylyl cyclase: implications for integration of transient calcium stimuli. *J Neurochem*. 1998;71(3): 1298–1306.
86. Blumenfeld H, Spira ME, Kandel ER, Siegelbaum SA. Facilitatory and inhibitory transmitters modulate calcium influx during action potentials in *Aplysia* sensory neurons. *Neuron*. 1990;5(4):487–499.
87. Li Q, Roberts AC, Glanzman DL. Synaptic facilitation and behavioral dishabituation in *Aplysia*: dependence on release of Ca^{2+} from postsynaptic intracellular stores, postsynaptic exocytosis, and modulation of postsynaptic AMPA receptor efficacy. *J Neurosci*. 2005;25(23):5623–5637.
88. Fulton D, Condro MC, Pearce K, Glanzman DL. The potential role of postsynaptic phospholipase C activity in synaptic facilitation and behavioral sensitization in *Aplysia*. *J Neurophysiol*. 2008;100(1):108–116.
89. Nagakura I, Dunn TW, Farah CA, Heppner A, Li FF, Sossin WS. Regulation of protein kinase C Apl II by serotonin receptors in *Aplysia*. *J Neurochem*. 2010;115(4):994–1006.
90. Nagakura I, Ormond J, Sossin WS. Mechanisms regulating ApTrkl, a Trk-like receptor in *Aplysia* sensory neurons. *J Neurosci Res*. 2008;86(13):2876–2883.
91. Ormond J, Hislop J, Zhao Y, et al. ApTrkl, a Trk-like receptor, mediates serotonin-dependent ERK activation and long-term facilitation in *Aplysia* sensory neurons. *Neuron*. 2004;44(4): 715–728.
92. Sossin WS. An autonomous kinase generated during long-term facilitation in *Aplysia* is related to the Ca(2 +)-independent protein kinase C Apl II. *Learn Mem*. 1997;3(5):389–401.
93. Commins S, Cunningham L, Harvey D, Walsh D. Massed but not spaced training impairs spatial memory. *Behav Brain Res*. 2003;139(1–2):215–223.
94. Isabel G, Pascual A, Preat T. Exclusive consolidated memory phases in *Drosophila*. *Science*. 2004;304(5673):1024–1027.
95. Josselyn SA, Shi C, Carlezon Jr. WA, Neve RL, Nestler EJ, Davis M. Long-term memory is facilitated by cAMP response element-binding protein overexpression in the amygdala. *J Neurosci*. 2001;21(7):2404–2412.
96. Sutton MA, Ide J, Masters SE, Carew TJ. Interaction between amount and pattern of training in the induction of intermediate- and long-term memory for sensitization in *Aplysia*. *Learn Mem*. 2002;9(1):29–40.
97. Naqib F, Sossin WS, Farah CA. Molecular determinants of the spacing effect. *Neural Plast*. 2012;2012:581291.
98. Mauelshagen J, Sherff CM, Carew TJ. Differential induction of long-term synaptic facilitation by spaced and massed applications of serotonin at sensory neuron synapses of *Aplysia californica*. *Learn Mem*. 1998;5(3):246–256.
99. Farah C, Weatherill D, Dunn T, Sossin W. PKC differentially translocates during spaced and massed training in *Aplysia*. *J Neurosci*. 2009;29(33):10281.
100. Naqib F, Farah CA, Pack CC, Sossin WS. The rates of protein synthesis and degradation account for the differential response of neurons to spaced and massed training protocols. *PLoS Comput Biol*. 2011;7(12):e1002324.
101. Ghirardi M, Braha O, Hochner B, Montarolo PG, Kandel ER, Dale N. Roles of PKA and PKC in facilitation of evoked and spontaneous transmitter release at depressed and nondepressed synapses in *Aplysia* sensory neurons. *Neuron*. 1992;9 (3):479–489.
102. Houeland G, Nakhost A, Sossin WS, Castellucci VF. PKC modulation of transmitter release by SNAP-25 at sensory-to-motor synapses in *Aplysia*. *J Neurophysiol*. 2007;97(1):134–143.
103. Yanow SK, Manseau F, Hislop J, Castellucci VF, Sossin WS. Biochemical pathways by which serotonin regulates translation in the nervous system of *Aplysia*. *J Neurochem*. 1998;70(2): 572–583.
104. Zhang F, Endo S, Cleary LJ, Eskin A, Byrne JH. Role of transforming growth factor-beta in long-term synaptic facilitation in *Aplysia*. *Science*. 1997;275(5304):1318–1320.
105. Muller U, Carew TJ. Serotonin induces temporally and mechanistically distinct phases of persistent PKA activity in *Aplysia* sensory neurons. *Neuron*. 1998;21(6):1423–1434.
106. Jin I, Kandel ER, Hawkins RD. Whereas short-term facilitation is presynaptic, intermediate-term facilitation involves both presynaptic and postsynaptic protein kinases and protein synthesis. *Learn Mem*. 2011;18(2):96–102.
107. Villareal G, Li Q, Cai D, Glanzman DL. The role of rapid, local, postsynaptic protein synthesis in learning-related synaptic facilitation in *Aplysia*. *Curr Biol*. 2007;17(23):2073–2080.
108. Villareal G, Li Q, Cai D, et al. Role of protein kinase C in the induction and maintenance of serotonin-dependent enhancement of the glutamate response in isolated siphon motor neurons of *Aplysia californica*. *J Neurosci*. 2009;29(16):5100–5107.
109. Sossin WS, Hastings M, Bougie JK. Neither pseudosubstrate based inhibitors, nor chelerythrine, selectively inhibit PKMs made from different isoforms of PKC. *Soc Neurosci Abstr*. 2011; [238.15]
110. Hu JY, Chen Y, Schacher S. Multifunctional role of protein kinase C in regulating the formation and maturation of specific synapses. *J Neurosci*. 2007;27(43):11712–11724.
111. Hu JY, Chen Y, Bougie JK, Sossin WS, Schacher S. *Aplysia* cell adhesion molecule and a novel protein kinase C activity in the postsynaptic neuron are required for presynaptic growth and initial formation of specific synapses. *J Neurosci*. 2010;30(25): 8353–8366.
112. Kolkova K, Novitskaya V, Pedersen N, Berezin V, Bock E. Neural cell adhesion molecule-stimulated neurite outgrowth depends on activation of protein kinase C and the Ras-mitogen-activated protein kinase pathway. *J Neurosci*. 2000;20 (6):2238–2246.
113. Sossin WS, Chen CS, Toker A. Stimulation of an insulin receptor activates and down-regulates the Ca^{2+}-independent protein kinase C, Apl II, through a Wortmannin-sensitive signaling pathway in *Aplysia*. *J Neurochem*. 1996;67(1):220–228.
114. Udo H, Jin I, Kim JH, et al. Serotonin-induced regulation of the actin network for learning-related synaptic growth requires Cdc42, N-WASP, and PAK in *Aplysia* sensory neurons. *Neuron*. 2005;45(6):887–901.

115. Etienne-Manneville S, Hall A. Cell polarity: Par6, aPKC and cytoskeletal crosstalk. *Curr Opin Cell Biol*. 2003;15(1):67–72.
116. Upadhya SC, Smith TK, Hegde AN. Ubiquitin-proteasome-mediated CREB repressor degradation during induction of long-term facilitation. *J Neurochem*. 2004;91(1):210–219.
117. Sutton MA, Carew TJ. Parallel molecular pathways mediate expression of distinct forms of intermediate-term facilitation at tail sensory-motor synapses in *Aplysia*. *Neuron*. 2000;26(1): 219–231.
118. Kelly JC, Cuerrier D, Graham LA, Campbell RL, Davies PL. Profiling of calpain activity with a series of FRET-based substrates. *Biochim Biophys Acta*. 2009;1794(10):1505–1509.
119. Bougie JK, Cai D, Hastings M, et al. Serotonin-induced cleavage of the atypical protein kinase C Apl III in *Aplysia*. *J Neurosci*. 2012;32(42):14630–14640.
120. Ling DS, Benardo LS, Serrano PA, et al. Protein kinase Mzeta is necessary and sufficient for LTP maintenance. *Nat Neurosci*. 2002;5(4):295–296.
121. Migues PV, Hardt O, Wu DC, et al. PKMzeta maintains memories by regulating GluR2-dependent AMPA receptor trafficking. *Nat Neurosci*. 2010;13(5):630–634.
122. Yao Y, Kelly MT, Sajikumar S, et al. PKM zeta maintains late long-term potentiation by *N*-ethylmaleimide-sensitive factor/GluR2-dependent trafficking of postsynaptic AMPA receptors. *J Neurosci*. 2008;28(31):7820–7827.
123. Chiel HJ, Susswein AJ. Learning that food is inedible in freely behaving *Aplysia californica*. *Behav Neurosci*. 1993;107(2):327–338.
124. Botzer D, Markovich S, Susswein AJ. Multiple memory processes following training that a food is inedible in *Aplysia*. *Learn Mem*. 1998;5(3):204–219.
125. Michel M, Green CL, Gardner JS, Organ CL, Lyons LC. Massed training-induced intermediate-term operant memory in *Aplysia* requires protein synthesis and multiple persistent kinase cascades. *J Neurosci*. 2012;32(13):4581–4591.
126. Sugita S, Baxter DA, Byrne JH. Modulation of a cAMP/protein kinase A cascade by protein kinase C in sensory neurons of *Aplysia*. *J Neurosci*. 1997;17(19):7237–7244.
127. Manseau F, Sossin WS, Castellucci VF. Long-term changes in excitability induced by protein kinase C activation in *Aplysia* sensory neurons. *J Neurophysiol*. 1998;79(3):1210–1218.
128. Sugita S, Goldsmith JR, Baxter DA, Byrne JH. Involvement of protein kinase C in serotonin-induced spike broadening and synaptic facilitation in sensorimotor connections of *Aplysia*. *J Neurophysiol*. 1992;68(2):643–651.
129. Philips GT, Tzvetkova EI, Carew TJ. Transient mitogen-activated protein kinase activation is confined to a narrow temporal window required for the induction of two-trial long-term memory in *Aplysia*. *J Neurosci*. 2007;27(50):13701–13705.
130. Lee YS, Choi SL, Lee SH, et al. Identification of a serotonin receptor coupled to adenylyl cyclase involved in learning-related heterosynaptic facilitation in *Aplysia*. *Proc Natl Acad Sci USA*. 2009;106(34):14634–14639.
131. Oestreich EA, Malik S, Goonasekera SA, et al. Epac and phospholipase Cepsilon regulate Ca^{2+} release in the heart by activation of protein kinase Cepsilon and calcium-calmodulin kinase II. *J Biol Chem*. 2009;284(3):1514–1522.
132. Dunn TW, Farah CA, Sossin WS. Inhibitory responses in *Aplysia* pleural sensory neurons act to block excitability, transmitter release, and PKC Apl II activation. *J Neurophysiol*;107 (1):292–305.

CHAPTER

19

Multisite Cellular and Synaptic Mechanisms in *Hermissenda* Pavlovian Conditioning

Terry Crow and Nan Ge Jin

University of Texas Medical School, Houston, Texas

INTRODUCTION

Historically, invertebrates were selected for studies that examined the relationship between the nervous system and behavior based on the relative simplicity of their nervous systems. The experimental approaches of these investigations were derived from the tradition in biology of pursuing questions regarding the nervous system and behavior using simplified representations to determine the fundamental or common rules of neural organization. Therefore, experimental questions of nervous system organization and behavior focused on studies of multiple patterns of synaptic convergence, divergence, and feedback among neurons with an emphasis on networks. The use of less complex representations was necessary because most nervous systems are so complex, containing many neurons with variable properties and elaborate patterns of synaptic connectivity, that the goal of elucidating organizational principles with current technology was unattainable. To achieve these goals required a nervous system with a relatively small number of identifiable cells amenable to electrophysiological and biophysical approaches to establish the pattern of synaptic connections and investigate the dynamic properties of the neurons. Therefore, determining physiological mechanisms and strategies of neural connectivity required the use of simpler networks. Most early approaches to an analysis of learning in gastropod mollusks were derived from the simpler network tradition to determine fundamental principles of neural organization.

Using Pavlovian conditioning as a model of associative learning in conjunction with a reductionist strategy has generated several mechanisms of contiguity between the conditioned stimulus (CS) and unconditioned stimulus (US). Because the primary focus has been on contiguity, other variables important for associative learning, such as contingency or the predictive relationship between the CS and the US, have not been adequately addressed. Moreover, the primary emphasis placed on contiguity has often led to a narrowly focused and overly simplistic view of the associative process. A comprehensive analysis of the cellular basis of associative learning will have to account for complex features and broader behavioral questions that have long been central to the study of learning in vertebrates.[1] An understanding of associative processes underlying Pavlovian conditioning requires an explanation of not only the mechanism of temporal contiguity between the CS and the US but also how conditioning-dependent changes in the relevant circuitry of the nervous system are expressed in the generation of multiple conditioned responses. There is now an appreciation that an adequate explanation of the learning process in the relatively simple nervous systems of many invertebrates is not all that different from the proposed goals of studying associative learning in the more complex nervous systems of vertebrates.[2] The goals can be summarized by the following experimental questions: (1) What are the conditions that produce associative learning? (2) What is learned or what is the content of learning? (3) How is learning expressed in the generation of behavior? Mechanisms based on more than simple coincidence detection must operate in the formation of associations. For example, both excitatory and inhibitory associations can be formed. The engagement of multiple mechanisms at both cellular and subcellular levels within networks is consistent with the richness and complexity of learned behavior. This broader approach requires the identification and

Invertebrate Learning and Memory.
DOI: http://dx.doi.org/10.1016/B978-0-12-415823-8.00019-8

characterization of mechanisms operating within interacting neural networks. The study of cellular/molecular changes within networks opens new possibilities for identifying mechanisms that support fundamental properties of the associative process.

This chapter reviews the research on the Pavlovian CR complex conducted at different levels of analysis involving the identification of proteins and post-translational modifications regulated by conditioning, changes in synaptic efficacy, and intrinsic excitability in sensory neurons and interneurons expressed within neural networks in *Hermissenda*.

The nudibranch mollusk *Hermissenda crassicornis* is one preparation that has contributed to an understanding of Pavlovian conditioning at the cellular, molecular, and network level. Pavlovian conditioning in *Hermissenda* involves changes in both intrinsic enhanced excitability and synaptic efficacy at multiple sites within the neural circuit supporting the generation of the CR. The modifications produced by Pavlovian conditioning involve the actions of several second messengers and the engagement of multiple cellular mechanisms within identified sensory neurons (photoreceptors) and interneurons that are expressed by alterations in the properties of channels in excitable membranes. Initial acquisition and long-term retention involve both presynaptic and postsynaptic mechanisms. Consistent with research conducted with numerous diverse species, the development of memory over time in *Hermissenda* occurs in distinct stages with different underlying mechanisms. These mechanistically distinct stages support most examples of long-term memory formation. An early stage dependent on post-translational modifications of existing proteins involving several signaling pathways is expressed by changes in both synaptic connections and intrinsic excitability. An intermediate stage that is dependent on translation, post-translational modifications, but not transcription, and long-term memory that requires post-translational modifications, new mRNA and protein synthesis, and the expression of both long-term changes in intrinsic cellular excitability and synaptic strength.

PAVLOVIAN CONDITIONING AND THE CR COMPLEX

Pavlovian conditioning studies of diverse species have shown that conditioning involves the acquisition of multiple conditioned responses produced by the same conditioning procedure.[3–8] Classical conditioning of *Hermissenda* also involves multiple CRs and follows the Pavlovian tradition in which the CS and US elicit different responses prior to training. Stimulation of the CS and US pathways conveys different sensory information to the central nervous system. Before conditioning, light, the CS, does not elicit either of the unconditioned responses that have been studied in *Hermissenda*—foot shortening or inhibition of forward ciliary locomotion. Stimulation of statocyst hair cells of the graviceptive system using rotation or orbital shaking, the US, elicits foot shortening and a reduced rate of forward ciliary locomotion.[9–13] Pavlovian conditioning produces both light-elicited inhibition of ciliary locomotion, which results in a suppression of *Hermissenda*'s normal positive phototaxis,[10,14–16] and CS-elicited foot shortening[12] (Figure 19.1). Both conditioned foot contraction and inhibition of ciliary locomotion involve the development or emergence of a new response to the CS, not the potentiation, through US presentations of an already existing response to the CS referred to as alpha conditioning or reflex potentiation.[1,17] In both of the CRs, there is a transfer of functional aspects of the response-evoking properties of the US to the CS.[10,12,13] The two CRs are proposed to develop independently,[13] which is consistent with results showing that the neural networks supporting foot contraction and ciliary locomotion involve different neuronal components with few common interneurons.[18–23]

Pavlovian conditioning in *Hermissenda* exhibits many of the characteristics of Pavlovian conditioning in vertebrates, such as the acquisition of a CR complex, extinction, CS specificity, and conditioned inhibition.[10,12,13,16,24–26] The temporal association of the CS and US involves both contiguity[10] and contingency.[27,28] Conditioning in the two different behavioral response systems supporting the two CRs is sensitive to both CS–US contiguity and forward interstimulus interval manipulations.[29] Retention of conditioned behavior persists for several days to weeks depending on the number of conditioning trials presented in initial acquisition.[10,30,31]

NEURAL NETWORK

The anatomy and synaptic organization of the two sensory structures (visual and graviceptive) mediating the CS and the US have been described in detail.[32–36] In addition, the sites of convergence providing for synaptic interactions between the CS and US pathways have been identified and are summarized in Figure 19.2.[18,20–23,32,33,37–41] Anatomical, ultrastructural, and electrophysiological analyses indicate that synaptic interactions between sensory neurons and interneurons are in the neuropil of the cerebropleural ganglion.[42,43] The CS produces a depolarizing generator potential and an increase in spike activity of the

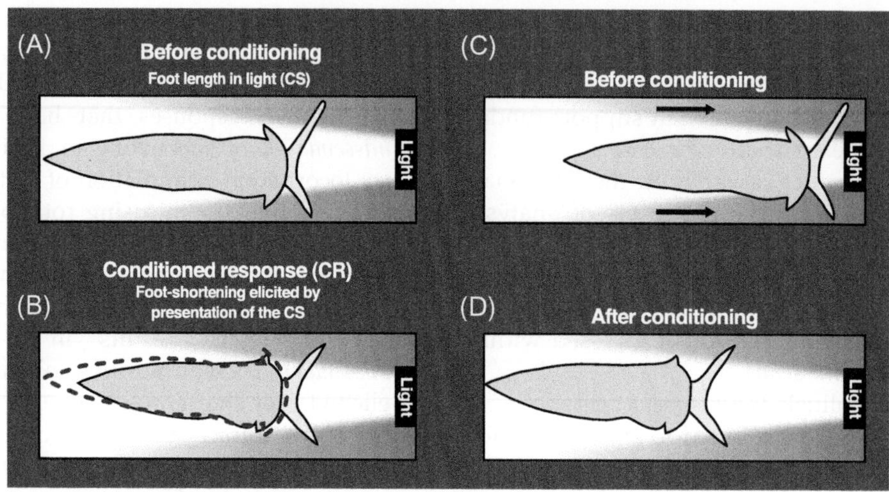

FIGURE 19.1 Pavlovian conditioned CR complex in *Hermissenda*: foot-shortening and inhibition of locomotion. (A) Foot length measured during the presentation of the CS (light) before conditioning. (B) Foot-shortening CR elicited by the CS after conditioning. The black dashed line indicates foot length measured during the presentation of the CS before conditioning. (C) Light-elicited ciliary locomotion assessed before conditioning. (D) Inhibition of light-elicited ciliary locomotion after Pavlovian conditioning. Random or pseudorandom presentations of the CS and US do not produce either inhibition of ciliary locomotion or CS-elicited foot shortening. Source: *Adapted with permission from Crow T. Pavlovian conditioning of* Hermissenda*: Current cellular, molecular and circuit perspectives.* Learning and Memory *2004;11:229–238.*

five photoreceptors in each eye.[35,44] The sensory neurons of the pathway mediating the US consist of 13 hair cells in each gravity detecting statocyst. Rotation or graviceptive stimulation produces a depolarizing generator potential and an increase in spike frequency of the stimulated hair cells.[45] The first site of convergence between the CS and US pathways is between the photoreceptors and hair cells. Statocyst hair cells project monosynaptically to photoreceptors and receive monosynaptic input from photoreceptors (Figure 19.2). Stimulation of statocyst hair cells elicits a monosynaptic GABAergic inhibitory postsynaptic potential (IPSP) in type B photoreceptors.[46–48] Based on behavioral, physiological, and immunohistochemical studies, it is proposed that hair cells also project polysynaptically to photoreceptors through a serotonergic modulatory pathway.[49–59] Statocyst hair cells and photoreceptors also form monosynaptic connections with type I_e and I_i interneurons[37,39,40] and polysynaptic connections with type I_b and I_s interneurons[18,20,23] (Figure 19.2). Type I_b and I_s interneurons project to identified efferent neurons innervating foot contraction and shortening along the rostrocaudal axis of the foot.[23]

The CS modulates ciliary locomotion through monosynaptic connections between sensory neurons (photoreceptors) and type I_e and I_i interneurons and monosynaptic connections between I_e and III_i inhibitory interneurons.[18,39] Polysynaptic connections between I_i and III_i inhibitory interneurons also support the generation of light-elicited ciliary locomotion.[18] Type III_i inhibitory interneurons form monosynaptic connections with identified ciliary efferent neurons located in the pedal ganglia. Activation of ciliary efferent neurons is produced by a light-dependent reduction in the spike activity of type III_i inhibitory interneurons. Ciliary locomotion is reduced or inhibited by the US due to hair cell excitation of type I_e interneurons that in turn excite type III_i inhibitory interneurons, resulting in inhibition of ciliary efferent neurons. An additional pathway that may modulate ciliary locomotion consists of monosynaptic excitatory projections from type I_b interneurons to ciliary efferent neurons.[20] The identification of interneurons projecting to efferent neurons that innervate the foot has provided for the analysis of reflex movements of the foot modified by Pavlovian conditioning.[20,23,60,61]

LONG-TERM MEMORY FOLLOWING MULTITRIAL CONDITIONING

Sensory Neurons

The analysis of electrophysiological and biophysical modifications detected following multitrial Pavlovian conditioning has focused on two sites of convergence between the CS and US pathways. The first site is in the sensory neurons (photoreceptors) of the pathway mediating the CS.[15] Neural modifications detected in the sensory neurons of conditioned animals involve

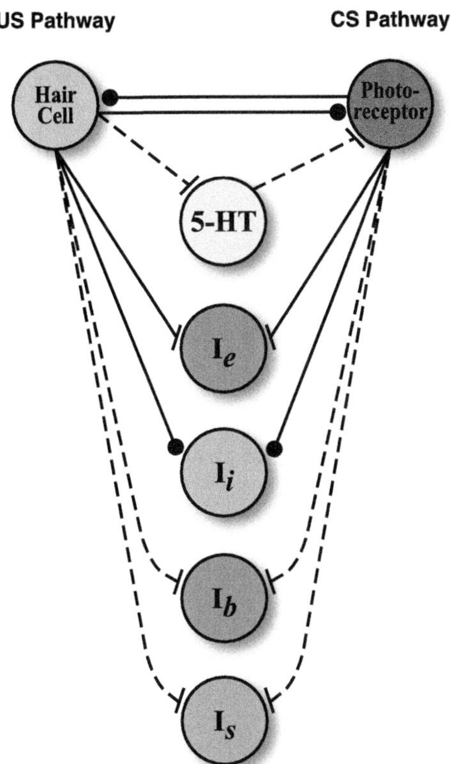

FIGURE 19.2 Diagram showing sites of synaptic convergence between identified neurons of the CS and US pathways. The formation of intrinsic enhanced cellular excitability and changes in the strength of synaptic connections between sensory neurons (photoreceptors) and type I interneurons support the generation of CS-evoked inhibition of locomotion. Proposed modifications in type I_b and I_s interneurons may contribute to the development of the foot-shortening CR. Hair cells of the graviceptive system and photoreceptors of the visual system form monosynaptic connections with identified type I interneurons and polysynaptic connections with I_b and I_s interneurons. Solid circles indicate inhibitory synaptic connections, and bars indicate excitatory synaptic connections. Solid lines represent established monosynaptic connections, and dashed lines indicate polysynaptic connections with potential interneurons not yet identified. Source: *Adapted with permission from Crow* et al. *Mechanism of memory formation and storage in* Hermissenda. *In:* Encyclopedia of Behavioral Neuroscience. *Oxford: Elsevier, 2010, pp 192–199.*

both enhanced excitability that is intrinsic to identified type A and type B photoreceptors[11,15,62–67] and facilitation of synaptic connections between identified sensory neurons[66,68,69] and interneurons.[41] The intrinsic modifications in sensory neurons are expressed by enhancement of the amplitude of CS-elicited generator potentials, increased spike frequency elicited by the CS and extrinsic current, decreased spike frequency accommodation, and a reduction in the peak amplitude of voltage-dependent (I_A, I_{Ca}) and Ca^{2+}-dependent ($I_{K(Ca)}$) currents.[11,13,15,46,47,61–66,68,70–76] The increase in the amplitude of CS-elicited generator potentials is in part the result of a reduction in I_A and $I_{K(Ca)}$. In sensory neurons of conditioned animals, the peak amplitude of I_A is significantly reduced and exhibits more rapid inactivation compared to controls.[63] However, both the delayed rectifier (I_K) and the inward rectifier (I_h) may play a role in conditioning-dependent enhanced excitability. The application of 5-HT to the isolated nervous system enhances the peak amplitude of I_h and decreases the peak amplitude of I_K and I_A in type B-photoreceptors.[49] In addition, 5-HT reduces the amplitude of $I_{K(Ca)}$ and decreases I_{Ca} in type B photoreceptors.[58] The reduction in $I_{K(Ca)}$ produced by 5-HT is a consequence of the decrease in I_{Ca} by 5-HT rather than a direct effect of 5-HT on $I_{K(Ca)}$. In conditioned animals, type A photoreceptors exhibit a decrease in the amplitude of light-elicited generator potentials, enhanced excitability to extrinsic current, increases in CS-elicited spike activity, and a significant increase in the magnitude of I_K.[65,66,75,77] Multitrial conditioning does not result in changes in either I_A or $I_{K(Ca)}$ in type A photoreceptors, in contrast to the modifications in type B photoreceptors.[77] In addition, the amplitude of the monosynaptic IPSP between the medial type B photoreceptor and the medial type A photoreceptor is significantly enhanced in conditioned animals.[68,69]

Interneurons

The second site of synaptic facilitation has been identified at the monosynaptic connection between sensory neurons and type I_e and I_i interneurons.[41] Conditioning-dependent changes in interneurons involve both synaptic efficacy and intrinsic enhanced excitability. As shown in Figure 19.3, the CS elicits a significant increase in the spike activity of type I_e interneurons in conditioned animals compared to pseudorandom controls. In conditioned animals, the complex excitatory postsynaptic potential (EPSP) elicited by the CS also exhibits a significant increase in amplitude. Moreover, the amplitude of the monosynaptic EPSP recorded in type I_e interneurons evoked by a single spike in the sensory neuron is significantly enhanced (Figure 19.4). In addition to conditioning-dependent synaptic facilitation, type I_e interneurons also express intrinsic enhanced excitability with conditioning.[19] Extrinsic current pulses elicit significantly more spikes in type I_e interneurons of conditioned animals compared to pseudorandom controls (Figure 19.5). Therefore, multitrial conditioning in *Hermissenda* results in both presynaptic and postsynaptic modifications. The enhanced excitability of sensory neurons (type B photoreceptors), expressed by an increase in both the amplitude of CS-elicited generator potentials and the number of action potentials elicited by the CS, may be a major contributor to changes in the duration

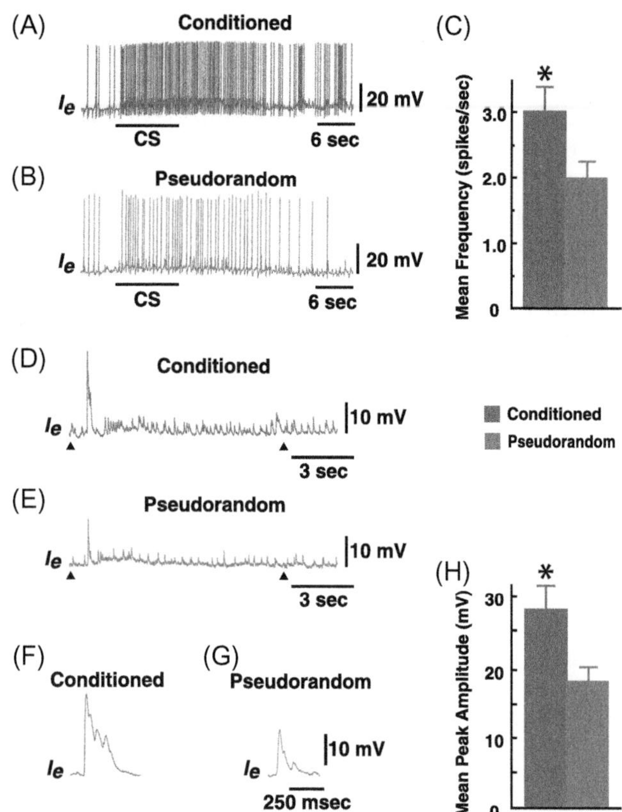

FIGURE 19.3 Example of the CS-elicited increase in spike activity for conditioned and pseudorandom controls. CS-elicited depolarization and increased spike activity recorded in a type I_e interneuron from a conditioned animal (A) and pseudorandom control (B). The bar beneath the recordings indicates the presentation of the 10-sec CS. (C) Group data depicting the mean increase, relative to baseline activity, in spike frequency elicited by the CS recorded from type I_e interneurons from conditioned and pseudorandom controls. $^*p < 0.025$. (D and E) Conditioning enhances CS-elicited complex EPSP amplitude. CS-elicited complex EPSP recorded in a type I_e interneuron from a conditioned animal (D) and a pseudorandom control (E). The arrowhead beneath each recording indicates the onset and offset of the 10-sec CS. The I_e interneurons were briefly hyperpolarized to approximately −80 mV to block spike activity during the presentation of the CS. The initial component of the complex EPSP is shown on a faster time base in panels F and G. (H) Group data depicting the mean ± SEM peak amplitude of the complex EPSPs recorded from I_e interneurons in the conditioned group and pseudorandom controls. $^*p < 0.01$. Source: *Adapted from Crow T, Tian LM. Facilitation of monosynaptic and complex PSPs in type I interneurons of conditioned* Hermissenda. J Neurosci *2002b;22:7818–7824. Copyright 2002 by the Society for Neuroscience; used with permission.*

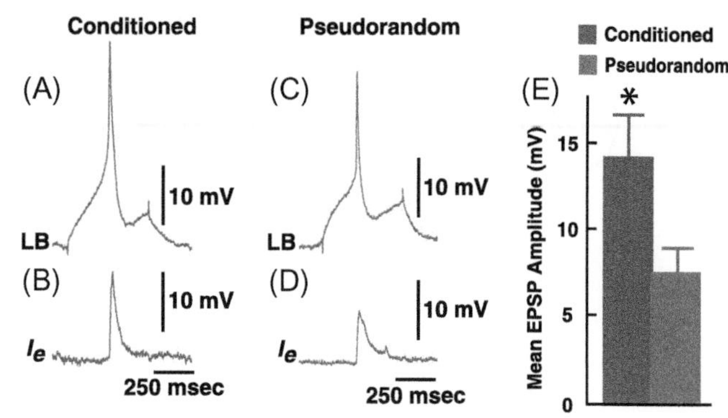

FIGURE 19.4 Conditioning results in facilitation of I_e monosynaptic EPSPs. A single lateral B spike generated by an extrinsic current pulse (A) elicited a monosynaptic EPSP in a type I_e interneuron (B) from a conditioned preparation. (C) A single lateral B spike elicited a smaller EPSP in a type I_e interneuron (D) from a pseudorandom control animal. (E) Group data (mean ± SEM) for the amplitude of I_e interneuron monosynaptic EPSPs recorded from conditioned animals ($n = 6$) and pseudorandom controls ($n = 6$). $^*p < 0.05$. Source: *Adapted from Crow T, Tian LM. Facilitation of monosynaptic and complex PSPs in type I interneurons of conditioned* Hermissenda. J Neurosci *2002b;22:7818–7824. Copyright 2002 by the Society for Neuroscience; used with permission.*

and amplitude of CS-elicited complex PSPs and increased CS-elicited spike activity in type I_e interneurons of conditioned animals.[41] However, facilitation of the amplitude of the monosynaptic EPSP between the sensory neuron and type I_e interneurons of conditioned animals may involve both pre- and postsynaptic mechanisms.

An example of light-elicited spike activity recorded from an identified ciliary efferent neuron in a conditioned animal and a pseudorandom control is shown in Figure 19.6. The presentation of the CS in the conditioned animal evoked an initial decrease in the spike activity of the ciliary efferent neuron, followed by inhibition of firing (Figure 19.6A). In contrast, the pseudorandom control exhibited an increase in the spike activity recorded from the ciliary efferent neuron during light (Figure 19.6B). The analysis of the group data revealed a dramatic inhibition of spike activity elicited by the CS in conditioned animals compared to pre-CS baseline activity and pseudorandom controls. Collectively, the facilitation of the monosynaptic EPSP in I_e interneurons combined with intrinsic enhanced excitability of I_e interneurons produces synaptic excitation of type III_i inhibitory interneurons. As summarized in the diagram of circuit components of the CS pathway in Figure 19.7, the increased activity of interneurons elicited by the CS produces inhibition of ciliary neuron activity and inhibition of forward locomotion (CR).[18]

CELLULAR AND MOLECULAR MECHANISMS UNDERLYING SHORT-, INTERMEDIATE-, AND LONG-TERM MEMORY FORMATION

Contemporary views of memory and its formation over time indicate that both declarative and

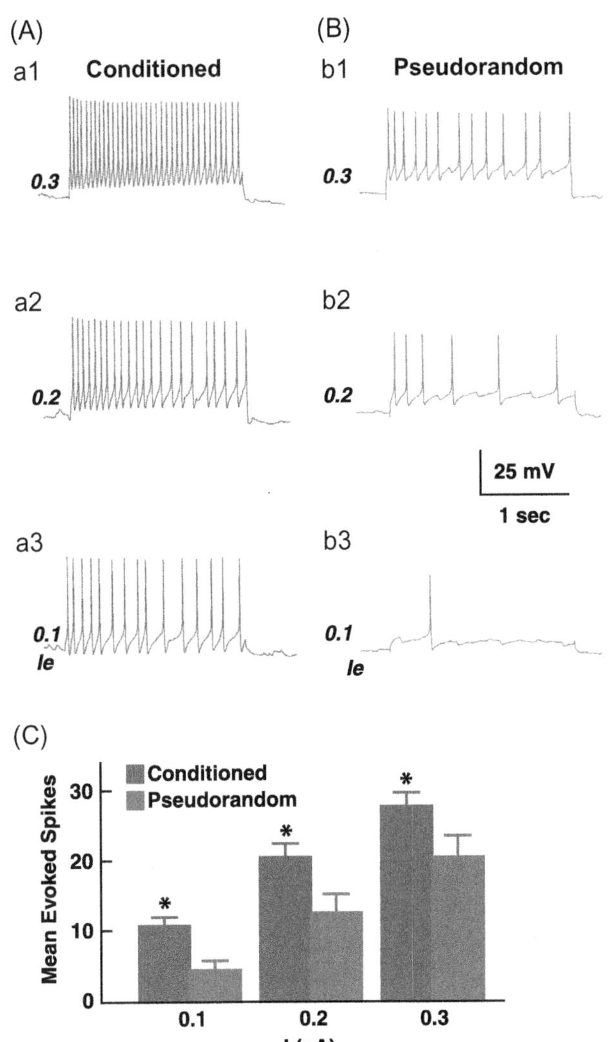

FIGURE 19.5 Conditioning produces intrinsic enhanced excitability of identified I_e interneurons as compared to pseudorandom controls. (A) Excitability of a type I_e interneuron from a conditioned animal with 2-sec depolarizing extrinsic current pulses of increasing intensity (0.1–0.3 nA). (B) Excitability of the type I_e interneuron from a pseudorandom control assessed with 2-sec depolarizing current pulses of increasing intensity (0.1–0.3 nA). (C) Group data showing the mean evoked spikes (±SEM) recorded in type I_e interneurons from conditioned animals ($n = 10$) and pseudorandom controls ($n = 12$) for the three depolarizing current levels. $*p < 0.05$. Source: *Adapted from Crow T, Tian LM. Neural correlates of Pavlovian conditioning in components of the neural network supporting ciliary locomotion in* Hermissenda. Learning and Memory *2003b;10:209–216. Copyright 2003 by Cold Spring Harbor Laboratory Press; used with permission.*

nondeclarative forms of memory involve multiple stages with different underlying mechanistic requirements. A number of *in vivo* and *in vitro* procedures involving one or several training trials have been employed in Pavlovian conditioning studies of *Hermissenda* to examine the early events supporting the formation of short-, intermediate-, and long-term memory underlying both CRs. The different protocols involving one or several conditioning trials produce behavioral changes and physiological modifications that can be detected within minutes following training.[78–84] Because the two sensory pathways mediating the CS and the US are totally intact in the isolated nervous system, *in vitro* Pavlovian conditioning procedures can be applied to the isolated circumesophageal nervous system. Pairing the CS with mechanical perturbations of the statocyst produced by piezoelectric stimulation (US) sufficient to depolarize hair cells, or rotation of the isolated nervous system (US), produces electrophysiological correlates in sensory neurons that are similar to correlates produced by multitrial *in vivo* procedures.[69,83,84] A multitrial *in vitro* procedure involving pairing the CS with extrinsic current depolarization of identified statocyst hair cells (nominal US) also produces conditioning correlates in sensory neurons.[81] The results of these investigations depend on the various conditioning protocols, the efficacy of the US, and the duration of CS–US stimulation.

ONE-TRIAL CONDITIONING

To more precisely control when the initial learning occurs and to not confound time after conditioning, when memory is tested, with degree of learning produced by varying numbers of conditioning trials, a one-trial *in vivo* conditioning procedure was developed that produces a pairing-dependent long-term inhibition of normal light-elicited ciliary locomotion.[85] Pairing the CS with the direct application of one of the proposed transmitters of the US pathway (5-HT, nominal US) to the exposed nervous system of otherwise intact *Hermissenda* produces inhibition of light-elicited ciliary locomotion when the animals are tested 24 hr following the conditioning trial. An *in vitro* analog of the one-trial procedure involving pairing the CS with 5-HT application to the isolated circumesophageal nervous system has been used to examine mechanisms underlying the development of short-, intermediate-, and long-term memory. In addition, a one-trial *in vitro* procedure consisting of pairing the CS with mechanical perturbation of the statocyst produces a significant Ca^{2+}-dependent increase in input resistance of sensory neurons (a correlate of enhanced excitability) that is detected within minutes postconditioning.[83] Moreover, a one-trial *in vitro* procedure involving GABA application to the region of the sensory neuron terminal branches (nominal US) paired with a 10-sec depolarization of the sensory neuron (nominal CS) resulted in an increase in the input resistance of the sensory neuron that persisted for at least 10 min.[86] These studies indicate that the application of a neurotransmitter

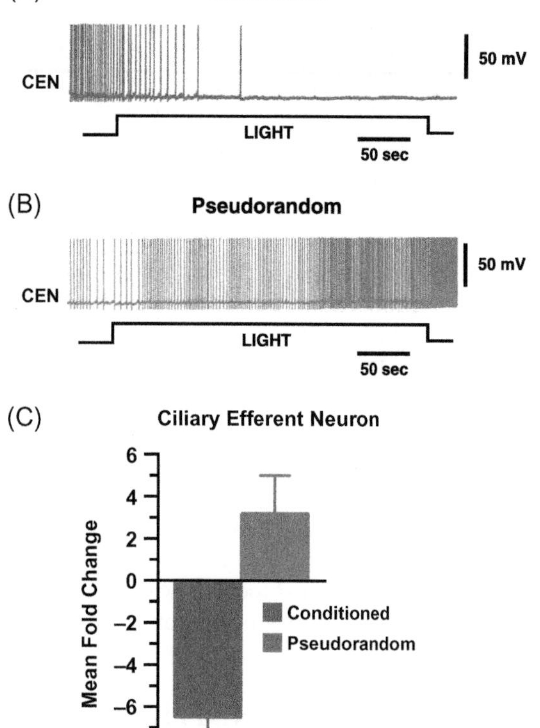

FIGURE 19.6 Examples of light-elicited changes in spike activity recorded from ciliary efferent neurons from conditioned animals and pseudorandom controls. (A) Light-elicited decrease in the spike activity recorded from a ciliary efferent neuron in a conditioned animal. (B) Light-elicited increase in the spike activity recorded from a ciliary efferent neuron in a pseudorandom control. (C) Group data depicting the mean fold change in spike activity (±SEM) recorded in 5 min of light relative to a 5-min period in the dark immediately preceding light onset collected from conditioned animals ($n = 15$) and pseudorandom controls ($n = 15$). $*p < 0.01$. Source: *Panels A and B adapted from Crow T, Tian LM. Neural correlates of Pavlovian conditioning in components of the neural network supporting ciliary locomotion in* Hermissenda. Learning and Memory *2003b;10:209–216. Copyright 2003 by Cold Spring Harbor Laboratory Press; used with permission.*

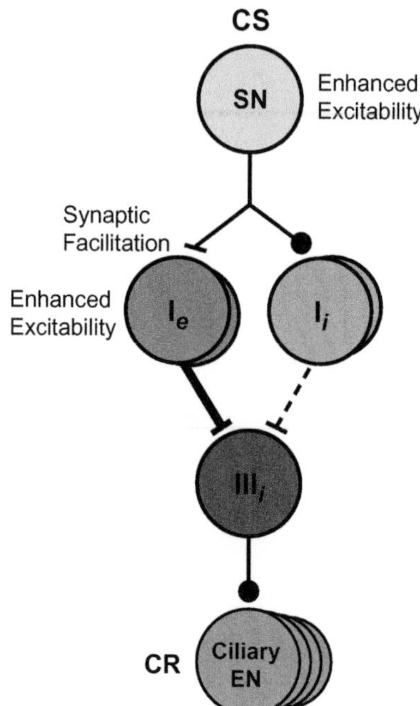

FIGURE 19.7 Components of the neural network supporting visually mediated ciliary locomotion in *Hermissenda*. Only synaptic connections with a single sensory neuron are shown. However, each sensory neuron projects to a different aggregate of type I_e and I_i interneurons. Monosynaptic connections are depicted by solid lines, and polysynaptic connections are depicted by dashed lines. Solid circles denote inhibitory synapses, and bars denote excitatory synapses. Conditioning results in the induction of intrinsic enhanced excitability in the sensory neurons and type I_e interneurons. Synaptic facilitation occurs at the monosynaptic connection between sensory neurons and type I interneurons. The CS presentation generates inhibition of ciliary efferent neuron activity and inhibition of ciliary locomotion.

when paired with depolarization resulting in a brief period of Ca^{2+} elevation is sufficient to produce enhanced excitability, and the procedures may engage the essential components for the formation of associations underlying conditioning.

SECOND MESSENGERS

Protein Kinase C

Protein kinase C (PKC) activation contributes to enhanced excitability and synaptic facilitation underlying the formation of short- and long-term memory in *Hermissenda*.[84,87–90] PKC is translocated from the cytosol to membrane in the nervous system of *Hermissenda* by treatment with a phorbol ester (TPA). The bath application of a phorbol ester (PDB) or injection of PKC into type B photoreceptors results in a reduction in the peak amplitude of two K^+ currents, I_A and $I_{K,(Ca)}$, that resemble changes in conductances detected following multitrial Pavlovian conditioning.[88]

In intact animals, nine conditioning trials produce a CS-elicited foot-shortening CR that is detected within minutes after the last conditioning trial.[84] An *in vitro* conditioning procedure consisting of nine training trials of the CS paired with rotation of the isolated circumesophageal nervous system (US) enhances sensory neuron excitability and increases the amplitude of the plateau phase of the CS-elicited generator potential. The conditioning-dependent change in excitability of sensory neurons is blocked by the broad-spectrum kinase inhibitor H-7 applied in artificial seawater during *in vitro* conditioning. A second *in vitro* procedure involving pairing the CS with extrinsic current depolarization of the sensory neurons (nominal US) produced enhanced excitability that is also blocked by preconditioning application of H-7 or sphingosine.[84]

One-trial *in vivo* conditioning consisting of pairing the CS with the application of 5-HT to the exposed but otherwise intact circumesophageal nervous system produces short-term, intermediate-term, and long-term enhanced excitability of sensory neurons.[52,87,91] The induction of short-term enhanced excitability following one-trial conditioning is blocked by the protein kinase inhibitors H-7 and sphingosine and by downregulation of PKC produced by pretreatment with TPA.[87,92] However, whereas H-7, sphingosine, or downregulation of PKC by TPA block short-term enhanced excitability, the same treatments do not block long-term enhanced excitability produced by one-trial conditioning.[92] Therefore, short- and long-term enhanced excitability produced by one-trial *in vivo* conditioning involve independent or parallel processes and different contributions of second messengers. Thus, the expression of long-term memory produced by one-trial conditioning does not depend on the induction of short-term memory. Consistent with previous studies, the induction of enhanced excitability in sensory neurons produced by five *in vitro* conditioning trials involving the CS paired with depolarizing current stimulation of an identified statocyst hair cell is blocked by pretreatment with PKC inhibitors.[89] However, the contribution of PKC to the expression of long-term enhanced excitability depends on the conditioning protocol and the number of conditioning trials. Previously established long-term enhancement produced by one-trial *in vivo* conditioning is not reversed by the application of the broad-spectrum kinase inhibitor H-7 or the PKC inhibitors sphingosine and staurosporine.[91] In contrast, long-term enhanced excitability in sensory neurons produced by multitrial Pavlovian conditioning is attenuated by H-7, or sphingosine, suggesting that long-term enhanced excitability is dependent on persistent kinase activity.[89]

Lateral type A photoreceptors exhibit an increase in the number of spikes elicited by the CS and extrinsic current (enhanced excitability) following multitrial Pavlovian conditioning.[65] Injection of the PKC inhibitor peptide PKC(19–36) into lateral type A photoreceptors 24–48 hr following multitrial conditioning reverses enhanced excitability within 16 min postinjection, suggesting that either a long-lived activator or a constitutively active kinase contributes to the expression of enhanced excitability in lateral A photoreceptors.[66] Injection of the control noninhibitory peptide [glu[27]] PKC(19–36) does not reverse enhanced excitability in lateral A photoreceptors of conditioned animals. PKC activation also contributes to the induction of 5-HT-dependent synaptic facilitation, but persistent PKC activity is not required for long-term synaptic facilitation. Short-term synaptic facilitation of the connection between type B and type A photoreceptors is produced by bath application of 5-HT.[66,93] Injection of the PKC inhibitor peptide PKC(19–36) into medial type B photoreceptors blocks 5-HT-induced synaptic facilitation of the IPSP recorded in the medial type A photoreceptor.[66] However, injection of PKC(19–36) into medial type B photoreceptors following multitrial Pavlovian conditioning does not reduce or reverse established synaptic facilitation of the IPSP recorded in medial type A photoreceptors. Thus, PKC contributes to the induction of short-term synaptic facilitation of the monosynaptic connection between type B and type A photoreceptors but not to the expression of long-term synaptic facilitation of the same monosynaptic connection between type B and type A photoreceptors.

Extracellular Signal-Regulated Protein Kinase

One-trial *in vitro* conditioning of the isolated nervous system involving the CS paired with 5-HT results in the increased ^{32}P labeling of a protein with an apparent molecular weight consistent with extracellular signal-regulated kinase (ERK). The increased phosphorylation of the protein following one-trial conditioning is blocked by pretreatment with the MEK1 inhibitor PD098059.[79] Assays of ERK activity with brain myelin basic protein as a substrate show greater ERK activity for nervous systems from one-trial *in vitro* conditioned animals compared to controls that received the CS and 5-HT unpaired. In addition, Western blot analysis of phosphorylated ERK with a phosphoERK antibody shows a significant increase in ERK phosphorylation after one-trial conditioning compared with unpaired controls. The increased phosphorylation is blocked by pretreatment with a MEK1 inhibitor (PD098059). Following a multitrial conditioning procedure consisting of 10–15 trials, circumesophageal nervous systems from conditioned animals exhibit significantly greater ERK phosphorylation compared with pseudorandom controls.[79] Evidence indicates that PKC contributes to the 5-HT-dependent activation of the ERK pathway. The phorbol ester TPA increases ERK phosphorylation that is blocked by pretreatment with PKC inhibitors. TPA-dependent ERK phosphorylation is also blocked by the MEK1 inhibitors PD0988059 or U0126. The increased phosphorylation of ERK by 5-HT is attenuated, but not blocked, by pretreatment with the Ca^{2+} chelator BAPTA-AM or pretreatment with PKC inhibitors Gö6976 or GF109203X.[94] This suggests that Ca^{2+}-dependent PKC activation contributes to ERK phosphorylation, although a PKC-independent pathway also contributes to 5-HT-dependent ERK phosphorylation and activation.

LONG-TERM MEMORY DEPENDS ON TRANSLATION AND TRANSCRIPTION

The existence of mechanistic differences between short- and long-term enhanced excitability is illustrated by studies showing that inhibition of protein synthesis during one-trial *in vivo* conditioning blocks long-term enhanced excitability without affecting the induction or expression of short-term enhanced excitability.[95] Moreover, long-term enhanced excitability produced by one-trial conditioning is blocked by inhibition of mRNA synthesis, which does not affect the induction of short-term enhanced excitability.[96] This result indicates that long-term memory following one-trial *in vivo* conditioning is dependent on both translation and transcription.

The time-dependent development of enhanced excitability following one-trial *in vivo* conditioning is biphasic; enhancement reaches a peak at 3 hr, decreases toward baseline control levels at 5 or 6 hr, and increases to a plateau at 16–24 hr postconditioning.[97] Enhanced excitability following one-trial conditioning also involves an intermediate phase of memory consolidation that requires protein synthesis but not mRNA synthesis.[98] The phosphorylation of cytoskeletal-related protein Csp24 is associated with the intermediate phase but not the short-term phase. Reducing the concentrations of 5-HT used in one-trial conditioning produces a short-term (<1 hr) associative enhancement of excitability that does not involve the post-translational modification of Csp24.[99]

The conditioned foot contraction CR is expressed at a retention interval of 5 min following two or nine conditioning trials.[100] However, nine conditioning trials are required for 90-min retention. *In vivo* incubation of animals with the protein synthesis inhibitor anisomycin during conditioning does not affect the expression of the CR at the 5-min retention interval, but it does attenuate conditioning at the 90-min interval for the group receiving nine conditioning trials. A protocol involving *in vitro* conditioning of the isolated nervous system produces similar results as the effects of anisomycin on conditioned behavior. Two conditioning trials produces a short-term protein synthesis-independent increase in excitability that decrements within 45 min, and nine conditioning trials produces a persistent protein synthesis-dependent increase in sensory neuron excitability detected at 90 min.[100] Applying anisomycin 5 min after the ninth conditioning trial does not affect the retention of enhanced excitability. However, a study has challenged the view that protein synthesis occurring after the learning event is necessary and sufficient for the formation of long-term memory. PKC activation produced by bryostatin application on days before conditioning leads to the expression of proteins that can support long-term memory produced by later Pavlovian conditioning. Two conditioning trials typically results in a short-term (~7 min) foot-shortening CR. A 4-hr exposure to bryostatin on 2 days preceding conditioning results in a long-term (>1 week) CR produced by two conditioning trials that is not blocked by anisomycin.[101]

PROTEINS REGULATED BY PAVLOVIAN CONDITIONING: PROTEOMIC ANALYSES

Different *in vivo* and *in vitro* conditioning protocols have been used to study proteins regulated by conditioning. Multitrial and one-trial *in vitro* conditioning procedures produce changes in protein abundance and post-translational modifications. Proteomic analyses of short- and intermediate-term memory following *in vitro* conditioning and post-translational modifications following multitrial conditioning have resulted in the identification of regulated proteins.[102,103] Changes in protein abundance at two different times following one-trial conditioning (30 min and 3 hr) were quantified using two-dimensional difference gel electrophoresis (2D-DIGE) (Figure 19.8). Protein identification was determined by tandem mass spectrometry. Proteins were identified that exhibited statistically significant increased abundance at both 30 min and 3 hr postconditioning and significant increased abundance only at 30 min or only at 3 hr postconditioning. Proteins were also identified that expressed a significant decrease in abundance detected at both 30 min and 3 hr postconditioning or a decrease in abundance only at 3 hr postconditioning. The proteomic analysis indicated that proteins involved in diverse cellular functions such as translational regulation, cell signaling, cytoskeletal regulation, metabolic activity, and protein degradation contribute to the initial formation of memory produced by one-trial *in vitro* conditioning.

Modification of protein phosphorylation regulated by multitrial Pavlovian conditioning was assessed by densitometric analysis of ^{32}P-labeled proteins or Pro-Q phosphoprotein dye resolved by 2DE. Mass spectrometric analysis of protein digests from analytical gels or preparative gels provided for the identification of phosphoproteins regulated by conditioning. A number of phosphoproteins exhibited statistically significant increased phosphorylation in conditioned groups compared to pseudorandom controls. However, only a few of the full-length cDNAs have been cloned and the phosphoproteins fully characterized.

Calexcitin (CE) is a GTP- and Ca^{2+}-binding protein found in *Hermissenda* photoreceptors.[104–106] CE is

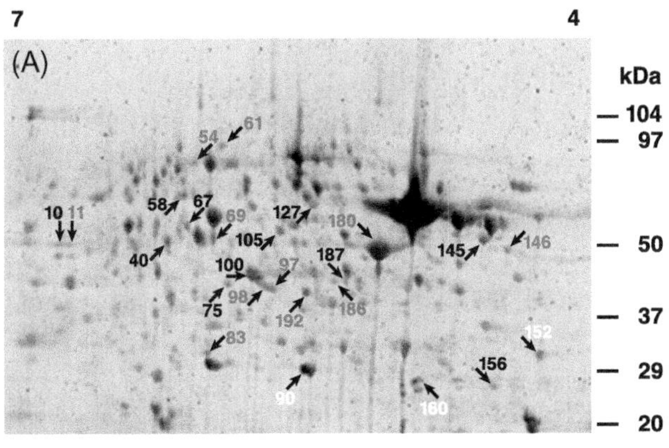

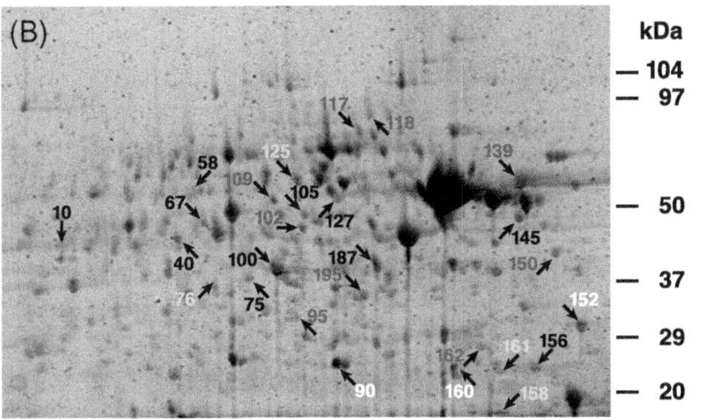

FIGURE 19.8 One-trial *in vitro* conditioning regulates protein abundance in circumesophageal nervous systems at different times postconditioning. (A) 2D-DIGE image showing statistically significant differentially regulated protein spots (arrows) in the 30-min postconditioning group. (B) 2D-DIGE image showing statistically significant differentially regulated spots (arrows) in the 3-hr postconditioning group. Protein spots designated by black numbers showed significantly increased abundance at both 30 min and 3 hr, those designated by blue numbers showed significantly increased abundance only at 30 min, and those designated by red numbers showed significantly increased abundance only at 3 hr. Protein spots designated by white numbers exhibited significantly decreased abundance at both 30 min and 3 hr, and spots designated by yellow numbers showed significantly decreased abundance only at 3 hr. Source: *From Crow T, Xue-Bian JJ. Proteomic analysis of short- and intermediate-term memory in* Hermissenda. Neuroscience *2011;192:102–111. Copyright by Elsevier; used with permission.*

activated by elevated Ca^{2+} and binds to the ryanodine receptor (RyR) to increase cytosolic Ca^{2+} concentration.[107–109] CE is phosphorylated by PKC, which results in translocation of CE to membrane compartments, where it decreases K^+ currents. In addition to CE, one-trial and multitrial conditioning regulates the abundance and phosphorylation of proteins in the CS pathway and circumesophageal nervous system.[96–98,102,103,110,111] Condition stimulus pathway protein 24 (Csp24) is a cytoskeleton-related protein that is homologous to members of the family of multidomain β-thymosin repeat proteins.[99,112,113] Actin co-immunoprecipitates with Csp24 and is co-localized with Csp24 in the cytosol of B photoreceptor cell bodies.[112] The abundance and phosphorylation of Csp24 is regulated by Pavlovian conditioning and is involved in both intermediate-term and long-term memory. In addition, recombinant Csp24 binds to and sequesters G-actin *in vitro*, and phosphorylation of Csp24 by one-trial *in vitro* conditioning increases the co-immunoprecipitation of actin with anti-Csp24.[114]

MECHANISMS OF CS–US ASSOCIATIONS IN SENSORY NEURONS

The photoreceptors in the eyes of *Hermissenda* exhibit a spatial segregation of function. Phototransduction takes place in the apical region where the rhabdomere abuts the lens, and spike generation occurs near the distal end of the axon close to the location of synapses on the terminal processes. Therefore, the CS and US have spatially separate physiological consequences in type B photoreceptors. Both light and depolarization increase cytosolic Ca^{2+} levels in photoreceptors.[47,48,72,76,115] Light activates phospholipase C to produce an increase in inositol trisphosphate (IP_3) and diacylglycerol (DAG).[116,117] IP_3 opens rhabdomeric Na^+ and Ca^{2+} channels, which results in a depolarizing generator potential and Ca^{2+} influx.[72] IP_3 also binds to its receptor (IP_3R), which triggers Ca^{2+} release from the endoplasmic reticulum.[118] The Ca^{2+} influx from the rhabdomere and the IP_3R-gated storage compartment can cause Ca^{2+} release from the RyR-gated compartment.[118] In addition, the induction of 5-HT-dependent enhanced excitability in type B photoreceptors is Ca^{2+} dependent because BAPTA loading of photoreceptors before 5-HT application blocks the induction of enhanced excitability.[119]

Rotation (US) produces a depolarizing generator potential in identified statocyst hair cells and elicits a monosynaptic GABAergic IPSP in the photoreceptors.[46–48,120] The US is also proposed to activate a polysynaptic serotonergic pathway that projects to type B photoreceptors.[52,55,85] Both 5-HT[56,59] and GABA[59] are linked to a pertussis toxin-sensitive G protein. These proteins can activate multiple second messenger systems, several of which have been implicated in one-trial and multitrial classical conditioning. The primary focus of 5-HT effects has been the modulation of membrane conductances in type B photoreceptors[49,53,59] and the regulation of intrinsic excitability by changes in voltage-dependent currents in I_e and I_i interneurons.[121,122] It is proposed that GABA binding to G protein-coupled receptors on photoreceptors activates phospholipase A_2 to liberate arachidonic acid (AA), which interacts with Ca^{2+} to synergistically stimulate PKC,[76] and create a back-propagating wave of Ca^{2+} released from intracellular stores.[47,123] When the CS and US are repeatedly paired, the Ca^{2+} influx due to light, IP_3R stores, RyR stores, and voltage-gated Ca^{2+} channels sums together.[118] The large increase in cytosolic Ca^{2+} combined with DAG and AA act to synergistically activate PKC by translocation of PKC to the membrane.[124] Each pairing of the CS and US has been proposed to incrementally increase the proportion of PKC translocated to the membrane that would contribute to the phosphorylation of K^+ channels.[76,104,125] Phosphorylation of CE also results in binding to the Ca^{2+}-ATPase transporter to increase the rate of Ca^{2+} removal from the cytosol.[104] Multitrial conditioning increases the phosphorylation of CE[106] and increases CE in B photoreceptors, specifically in Ca^{2+} sequestering organelles such as endoplasmic reticulum (ER) and within mitochondria and photopigments.[105] The increased CE levels in B photoreceptors of conditioned animals result in increased excitability via K^+ channel inactivation and internal Ca^{2+} release from the ER due to increased CE binding to ryanodine receptors (Figure 19.9).

Csp24 also contributes to enhanced excitability in sensory neurons by its effects on K^+ currents. Csp24 is phosphorylated by procedures that produce intermediate-term and long-term enhanced excitability but not after *in vitro* procedures that result in only short-term enhanced excitability of photoreceptors.[99] Several signaling pathways regulate Csp24 phosphorylation; thus, it can integrate a number of signals that result in cytoskeletal remodeling. Inhibitors of PKC and MEK1 reduce Csp24 phosphorylation produced by *in vitro* conditioning. In addition to PKC and ERK regulation of Csp24, Rho GTPase activity and its downstream target ROCK contribute to the post-translational regulation of Csp24 through an inhibitory pathway.[126] The ROCK inhibitor Y-27632 significantly increases Csp24 phosphorylation, and the Rho activator lysophosphatidic acid decreases Csp24 phosphorylation.[126] In addition, the application of 5-HT to the isolated nervous system decreases Rho activity and increases the phosphorylation of Csp24. Inhibition of cyclin-dependent kinase 5 by butyrolactone also reduces Csp24 phosphorylation. Incubation of isolated *Hermissenda* nervous systems with *Csp* antisense oligonucleotides decreases Csp24 expression, and treatment with antisense oligonucleotides before one-trial *in vitro* conditioning blocks intermediate-term enhanced excitability without affecting the induction of short-term immediate enhanced excitability.[113] Because Csp24 is associated with the actin cytoskeleton, its regulation by

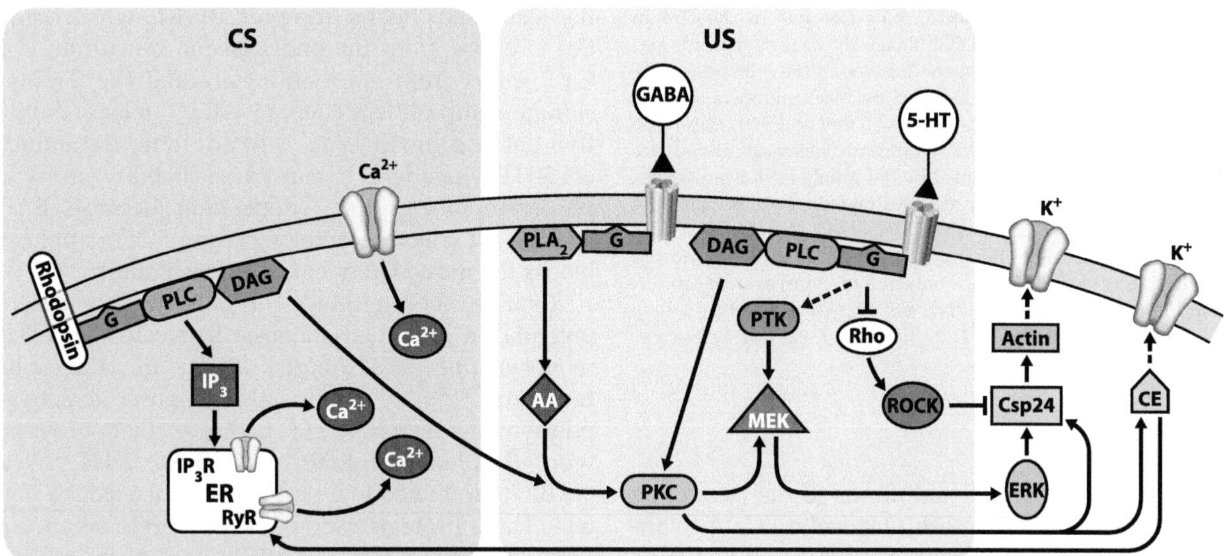

FIGURE 19.9 Proposed model of the mechanisms generating the CS–US association in sensory neurons produced by Pavlovian conditioning in *Hermissenda*. Acquisition involves the interaction of Ca^{2+} with the second messenger pathways regulated by neurotransmitter release in the US pathway. Light (CS) activates phospholipase C (PLC) to produce an increase in inositol trisphosphate (IP_3) and diacylglycerol (DAG). The depolarizing generator potential and IP_3 effects on endoplasmic reticulum (ER) result in an increase in intracellular Ca^{2+}. Transmitters in the US pathway bind to G protein-coupled receptors (G) to activate phospholipase A_2 (PLA_2); increase arachidonic acid (AA); and activate protein kinase C (PKC), nonreceptor protein tyrosine kinase (PTK), extracellular signal-regulated kinase (ERK), and the Rho/ROCK pathway. Enhanced excitability is a consequence of short-term and long-term modification of K^+ channels by CE and Csp24. Proteomic analyses have identified cytoskeletal-related proteins that are regulated by conditioning. Source: *Adapted with permission from Crow* et al. *Molecular mechanisms of associative learning in* Hermissenda. *In: David Sweatt J (ed.)* Molecular Mechanisms of Memory, *Vol. 4. Oxford: Elsevier, 2008, pp. 119–132. Copyright 2008 by Elsevier.*

conditioning may influence K^+ channel activity by the spatial and temporal control of actin dynamics. One-trial *in vitro* conditioning of isolated type B photoreceptors produces a significant reduction in the amplitude of I_A and a depolarized shift in the steady-state activation curve of I_A without altering the inactivation curve.[127] The conditioning-dependent changes in I_A are blocked by incubation of the isolated photoreceptors with *Csp* antisense oligonucleotide. Therefore, Csp24 contributes to the regulation of voltage-gated K^+ channels associated with intrinsic enhanced excitability underlying Pavlovian conditioning. The distribution of Csp24-like immunoreactivity in lateral type B photoreceptors is also changed by one-trial *in vitro* conditioning. Conditioning results in a significant decrease in immunoreactivity in the soma and a significant increase in immunoreactivity in the terminal arborizations of identified lateral type B photoreceptors.[128]

SUMMARY

Pavlovian conditioning in *Hermissenda* results in both intrinsic enhanced cellular excitability and modifications in synaptic efficacy at multiple loci within the neural network responsible for the generation of the CR complex. The first site of storage for the memory of the associative experience is in the primary sensory neurons of the CS pathway. The modifications in the sensory neurons are spatially segregated. There are alterations in the properties of K^+ channels in the soma that result in an enhancement of the amplitude of the CS-elicited generator potential and a concomitant change in channels in the spike-generating zone that results in a decrease in spike frequency accommodation. In addition, changes in synaptic efficacy result in facilitation of the monosynaptic connections between identified type B and type A photoreceptors and between identified photoreceptors and interneurons. Because the second site of memory storage is in the type I interneurons, the memory for Pavlovian conditioning involves both presynaptic and postsynaptic mechanisms.

Acquisition of short-term, intermediate-term, and long-term memory for Pavlovian conditioning involves the activation of several second messenger cascades, post-translational modification of proteins, and the synthesis of mRNA and proteins. Acquisition engages the interaction of elevated intracellular Ca^{2+} and arachidonic acid to activate PKC and ERK that is dependent on CS–US pairings. The mechanism for intrinsic enhanced excitability is different from the mechanisms supporting modifications in synaptic efficacy because long-term synaptic facilitation detected following multitrial conditioning is not PKC dependent.

Two proteins that have been fully characterized and that are regulated by Pavlovian conditioning are CE and Csp24. The binding of CE to the plasma membrane decreases K^+ conductances and releases Ca^{2+} from internal stores. Csp24 abundance and phosphorylation is regulated by one-trial and multitrial conditioning, associated with actin, and contributes to long-term intrinsic enhanced excitability produced by the depolarized shift in the steady-state activation of I_A and the concomitant reduction in peak I_A. Therefore, the expression of Csp24 is important in both intermediate-term and long-term memory involving intrinsic enhanced excitability.

The analysis of mechanisms of memory in *Hermissenda* raises a number of questions that are important for an understanding of memory produced by Pavlovian conditioning. How are post-translational modifications in proteins supporting short-term memory transformed into long-term memory involving both intrinsic enhanced excitability and changes in synaptic efficacy? What are the contributions of presynaptic and postsynaptic modifications to short-term, intermediate-term, and long-term memory? How does the regulation of CE and Csp24 by conditioning result in an alteration in the properties of K^+ channels in excitable membranes? Finally, how are modifications in intrinsic excitability and synaptic strength at multiple sites integrated within a neural network to result in a reconfiguration that supports the defining characteristics of Pavlovian conditioning and generation of the conditioned responses?

References

1. Sahley CL, Crow T. Invertebrate learning: current perspectives. In: Martinez JL, Kesner RP, eds. *Learning and Memory*. New York: Academic Press; 1998:197–209.
2. Rescorla RA. *Pavlovian Second-Order Conditioning: Studies in Associative Learning*. New York: Wiley; 1980.
3. Ayers ED, Powell DA. Multiple response measures during classical conditioning. *J Neurosci Methods*. 2002;114:33–38.
4. Black AH, de Toledo K. The relationship among classically conditioned responses: heart rate and skeletal behavior. In: Black AH, Prokasy EF, eds. *Classical Conditioning II: Current Research and Theory*. New York: Appleton-Century-Crofts; 1972:290–311.
5. Gantt WH. Cardiovascular component of the conditioned reflex to pain, foods and other stimuli. *Physiol Rev*. 1960;40:266–291.
6. Prokasy WF. Acquisition of skeletal conditioned responses in pavlovian conditioning. *Psychophysiol*. 1984;21:1–13.
7. Schneiderman N. Response system divergencies in aversive classical conditioning. In: Black AL, Prokasy WF, eds. *Classical Conditioning II: Current Research and Theory*. New York: Appleton-Century-Crofts; 1972:341–376.
8. Weinberger NM, Diamond DM. Physiological plasticity in auditory cortex: rapid induction by learning. *Prog Neurobiol*. 1987;29:1–55.
9. Alkon DL. Associative training of *Hermissenda*. *J Gen Physiol*. 1974;64:70–84.

10. Crow TJ, Alkon DL. Retention of an associative behavioral change in *Hermissenda*. *Science*. 1978;201:1239–1241.
11. Farley J, Alkon DL. Associative neural and behavioral change in *Hermissenda*: consequences of nervous system orientation for light- and pairing-specificity. *J Neurophysiol*. 1982;48: 785–807.
12. Lederhendler II, Gart S, Alkon DL. Classical conditioning of *Hermissenda*: origin of a new response. *J Neurosci*. 1986;6: 1325–1331.
13. Matzel LD, Schreurs BG, Alkon DL. Pavlovian conditioning of distinct components of *Hermissenda's* responses to rotation. *Behav Neural Biol*. 1990;54:131–145.
14. Crow T. Conditioned modification of phototactic behavior in *Hermissenda*: I. Analysis of light intensity. *J Neurosci*. 1985;5: 209–214.
15. Crow T, Alkon DL. Associative behavioral modification in *Hermissenda*: cellular correlates. *Science*. 1980;209:412–414.
16. Crow T, Offenbach N. Modification of the initiation of locomotion in *Hermissenda*: behavioral analysis. *Brain Res*. 1983;271: 301–310.
17. Schreurs BG. Classical conditioning of model systems: a behavioral review. *Psychobiology*. 1989;17:145–155.
18. Crow T, Tian LM. Interneuronal projections to identified cilia-activating pedal neurons in *Hermissenda*. *J Neurophysiol*. 2003;89: 2420–2429.
19. Crow T, Tian LM. Neural correlates of Pavlovian conditioning in components of the neural network supporting ciliary locomotion in *Hermissenda*. *Learn Mem*. 2003;10:209–216.
20. Crow T, Tian LM. Statocyst hair cell activation of identified interneurons and foot contraction motor neurons in *Hermissenda*. *J Neurophysiol*. 2004;92:2874–2883.
21. Crow T, Tian LM. Pavlovian conditioning in *Hermissenda*: a circuit analysis. *Biol Bull*. 2006;210:289–297.
22. Crow T, Tian LM. Sensory regulation of network components underlying ciliary locomotion in *Hermissenda*. *J Neurophysiol*. 2008;100:2496–2506.
23. Crow T, Tian LM. Polysensory interneuronal projections to foot contractile pedal neurons in *Hermissenda*. *J Neurophysiol*. 2009;101:824–833.
24. Britton G, Farley J. Behavioral and neural bases of noncoincidence learning in *Hermissenda*. *J Neurosci*. 1999;19: 9126–9132.
25. Richards WG, Farley J, Alkon DL. Extinction of associative learning in *Hermissenda*: behavior and neural correlates. *Behav Brain Res*. 1984;14:161–170.
26. Walker TL, Campodonico JL, Cavallo JS, Farley J. AA/12-lipoxygenase signaling contributes to inhibitory learning in *Hermissenda* type B photoreceptors. *Front Behav Neurosci*. 2010;4:1–13.
27. Farley J. Contingency learning and causal detection in *Hermissenda*: I Behavior. *Behav Neurosci*. 1987;101:13–27.
28. Farley J. Contingency learning and causal detection in *Hermissenda*: II. Cellular mechanisms. *Behav Neurosci*. 1987;101: 28–56.
29. Matzel LD, Schreurs BG, Lederhendler I, Alkon DL. Acquisition of conditioned associations in *Hermissenda*: additive effects of contiguity and the forward interstimulus interval. *Behav Neurosci*. 1990;104:597–606.
30. Alkon DL. Learning in a marine snail. *Sci Am*. 1983;249:70–83.
31. Harrigan JF, Alkon DL. Individual variation in associative learning of the nudibranch mollusc *Hermissenda crassicornis*. *Biol Bull*. 1985;168:222–238.
32. Alkon DL. Neural organization of a molluscan visual system. *J Gen Physiol*. 1973;61:444–461.
33. Alkon DL. Intersensory interactions in *Hermissenda*. *J Gen Physiol*. 1973;62:185–202.
34. Alkon DL, Bak A. Hair cell generator potentials. *J Gen Physiol*. 1973;61:619–637.
35. Alkon DL, Fuortes MG. Responses of photoreceptors in *Hermissenda*. *J Gen Physiol*. 1972;60:631–649.
36. Detwiler PB, Alkon DL. Hair cell interactions in the statocyst of *Hermissenda*. *J Gen Physiol*. 1973;62:618–642.
37. Akaike T, Alkon DL. Sensory convergence on central visual neurons in *Hermissenda*. *J Neurophysiol*. 1980;44:501–513.
38. Alkon DL, Akaike T, Harrigan J. Interaction of chemosensory, visual, and statocyst pathways in *Hermissenda crassicornis*. *J Gen Physiol*. 1978;71:177–194.
39. Crow T, Tian LM. Monosynaptic connections between identified A and B photoreceptors and interneurons in *Hermissenda*: evidence for labeled lines. *J Neurophysiol*. 2000;84: 367–375.
40. Crow T, Tian LM. Morphological characteristics and central projections of two types of interneurons in the visual pathway of *Hermissenda*. *J Neurophysiol*. 2002;87:322–332.
41. Crow T, Tian LM. Facilitation of monosynaptic and complex PSPs in type I interneurons of conditioned *Hermissenda*. *J Neurosci*. 2002;22:7818–7824.
42. Crow T, Heldman E, Hacopian V, Enos R, Alkon DL. Ultrastructure of photoreceptors in the eye of *Hermissenda* labelled with intracellular injections of horseradish peroxidase. *J Neurocytol*. 1979;8:181–195.
43. Kawai R, Crow T. Fluorescent labeling of terminal processes of identified photoreceptors in *Hermissenda*. *Soc Neurosci Abstr*. 2004;50:2004.
44. Dennis MJ. Electrophysiology of the visual system in a nudibranch mollusc. *J Neurophysiol*. 1967;30:1439–1465.
45. Alkon DL. Responses of hair cells to statocyst rotation. *J Gen Physiol*. 1975;66:507–530.
46. Alkon DL, Anderson MJ, Kuzirian AJ, et al. GABA-mediated synaptic interaction between the visual and vestibular pathways of *Hermissenda*. *J Neurochem*. 1993;61:556–566.
47. Blackwell KT. Calcium waves and closure of potassium channels in response to GABA stimulation in *Hermissenda* type B photoreceptors. *J Neurophysiol*. 2002;87:776–792.
48. Sakakibara M, Takagi H, Yoshioka T, Alkon DL. Propagated calcium modulates the calcium-dependent potassium current by the activation of GA_BA_B receptor at the axonal branch in the type B photoreceptor of *Hermissenda*. *Ann N Y Acad Sci*. 1993;707:492–495.
49. Acosta-Urquidi J, Crow T. Differential modulation of voltage-dependent currents in *Hermissenda* type B photoreceptors by serotonin. *J Neurophysiol*. 1993;70:541–548.
50. Auerbach SB, Grover LM, Farley J. Neurochemical and immunocytochemical studies of serotonin in the *Hermissenda* central nervous system. *Brain Res Bull*. 1989;22:353–361.
51. Crow T, Bridge MS. Serotonin modulates photoresponses in *Hermissenda* type-B photoreceptors. *Neurosci Lett*. 1985;60:83–88.
52. Crow T, Forrester J. Light paired with serotonin *in vivo* produces both short- and long-term enhancement of generator potentials of identified B-photoreceptors in *Hermissenda*. *J Neurosci*. 1991;11:608–617.
53. Farley J, Wu R. Serotonin modulation of *Hermissenda* type B photoreceptor light responses and ionic currents: implications for mechanisms underlying associative learning. *Brain Res Bull*. 1989;22:335–351.
54. Grover LM, Farley J, Auerbach SB. Serotonin involvement during *in vitro* conditioning of *Hermissenda*. *Brain Res Bull*. 1989;22:363–372.
55. Land PW, Crow T. Serotonin immunoreactivity in the circumesophageal nervous system of *Hermissenda crassicornis*. *Neurosci Lett*. 1985;62:199–205.

56. Rogers RF, Matzel LD. G-protein mediated responses to localized serotonin application in an invertebrate photoreceptor. *Neuroreport*. 1995;6:2161–2165.
57. Tian LM, Kawai R, Crow T. Serotonin-immunoreactive CPT interneurons in *Hermissenda*: identification of sensory input and motor projections. *J Neurophysiol*. 2006;96:327–335.
58. Yamoah EN, Crow T. Evidence for a contribution of I_{Ca} to serotonergic modulation of $I_{K,Ca}$ in *Hermissenda* photoreceptors. *J Neurophysiol*. 1995;74:1349–1354.
59. Yamoah EN, Crow T. Protein kinase and G-protein regulation of Ca^{2+} currents in *Hermissenda* photoreceptors by 5-HT and GABA. *J Neurosci*. 1996;16:4799–4809.
60. Goh Y, Alkon DL. Sensory, interneuronal, and motor interactions within *Hermissenda* visual pathway. *J Neurophysiol*. 1984;52:156–169.
61. Goh Y, Lederhendler I, Alkon DL. Input and output changes of an identified neural pathway are correlated with associative learning in *Hermissenda*. *J Neurosci*. 1985;5:536–543.
62. Alkon DL, Lederhendler I, Shoukimas JJ. Primary changes of membrane currents during retention of associative learning. *Science*. 1982;215:693–695.
63. Alkon DL, Sakakibara M, Forman R, Harrigan J, Lederhendler I, Farley J. Reduction of two voltage-dependent K^+ currents mediates retention of a learned association. *Behav Neural Biol*. 1985;44:278–300.
64. Crow T. Conditioned modification of phototactic behavior in *Hermissenda*: II. Differential adaptation of B-photoreceptors. *J Neurosci*. 1985;5:215–223.
65. Frysztak RJ, Crow T. Differential expression of correlates of classical conditioning in identified medial and lateral type A photoreceptors of *Hermissenda*. *J Neurosci*. 1993;13:2889–2897.
66. Frysztak RJ, Crow T. Synaptic enhancement and enhanced excitability in presynaptic and postsynaptic neurons in the conditioned stimulus pathway of *Hermissenda*. *J Neurosci*. 1997;17: 4426–4433.
67. West A, Barnes ES, Alkon DL. Primary changes of voltage responses during retention of associative learning. *J Neurophysiol*. 1982;48:1243–1255.
68. Frysztak RJ, Crow T. Enhancement of type B and A photoreceptor inhibitory synaptic connections in conditioned *Hermissenda*. *J Neurosci*. 1994;14:1245–1250.
69. Gandhi CC, Matzel LD. Modulation of presynaptic action potential kinetics underlies synaptic facilitation of type B photoreceptors after associative conditioning in *Hermissenda*. *J Neurosci*. 2000;20:2022–2035.
70. Alkon DL. Calcium-mediated reduction of ionic currents: a biophysical memory trace. *Science*. 1984;30:1037–1045.
71. Alkon DL, Sanchez-Andrés JV, Ito E, Oka K, Yoshioka T, Collin C. Long-term transformation of an inhibitory into an excitatory GABAergic synaptic response. *Proc Natl Acad Sci USA*. 1992;89:11862–11866.
72. Blackwell KT. Evidence for a distinct light-induced calcium-dependent potassium current in *Hermissenda crassicornis*. *J Comput Neurosci*. 2000;9:149–170.
73. Blackwell KT. The effect of intensity and duration on the light-induced sodium and potassium currents in *Hermissenda* type B photoreceptor. *J Neurosci*. 2002;22:4317–4228
74. Collin C, Ikeno H, Harrigan JF, Lederhendler I, Alkon DL. Sequential modification of membrane currents with classical conditioning. *Biophys J*. 1988;54:955–960.
75. Farley J, Richards WG, Grover LM. Associative learning changes intrinsic to *Hermissenda* type A photoreceptors. *Behav Neurosci*. 1990;104:135–152.
76. Muzzio IA, Gandhi CC, Manyam U, Pesnell A, Matzel LD. Receptor-stimulated phospholipase A(2) liberates arachidonic acid and regulates neuronal excitability through protein kinase C. *J Neurophysiol*. 2001;85:1639–1647.
77. Farley J, Han Y. Ionic basis of learning-correlated excitability changes in *Hermissenda* type A photoreceptors. *J Neurophysiol*. 1997;77:1861–1888.
78. Crow T. Conditioned modification of locomotion in *Hermissenda crassicornis*: analysis of time-dependent associative and nonassociative components. *J Neurosci*. 1983;3:2621–2628.
79. Crow T, Xue-Bian JJ, Siddiqi V, Kang Y, Neary JT. Phosphorylation of mitogen-activated protein kinase by one-trial and multi-trial classical conditioning. *J Neurosci*. 1998;18: 3480–3487.
80. Epstein HT, Child FM, Kuzirian AM, Alkon DL. Time windows for effects of protein synthesis inhibitors on Pavlovian conditioning in *Hermissenda*: behavioral aspects. *Neurobiol Learn Mem*. 2003;79:127–131.
81. Farley J, Alkon DL. *In vitro* associative conditioning of *Hermissenda*: cumulative depolarization of type B photoreceptors and short-term associative behavioral changes. *J Neurophysiol*. 1987;57:1639–1668.
82. Kuzirian AM, Epstein HT, Gagliardi CJ, et al. Bryostatin enhancement of memory in *Hermissenda*. *Biol Bull*. 2006;210: 201–214.
83. Matzel LD, Rogers RF. Postsynaptic calcium, but not cumulative depolarization, is necessary for the induction of associative plasticity in *Hermissenda*. *J Neurosci*. 1993;13:5029–5040.
84. Matzel LD, Lederhendler II, Alkon DL. Regulation of short-term associative memory by calcium-dependent protein kinase. *J Neurosci*. 1990;10:2300–2307.
85. Crow T, Forrester J. Light paired with serotonin mimics the effect of conditioning on phototactic behavior of *Hermissenda*. *Proc Natl Acad Sci USA*. 1986;83:7975–7978.
86. Matzel LD, Alkon DL. GABA-induced potentiation of neuronal excitability occurs during contiguous pairing with intracellular calcium elevation. *Brain Res*. 1991;554:77–84.
87. Crow T, Forrester J, Williams M, Waxham MN, Neary JT. Down-regulation of protein kinase C blocks 5-HT-induced enhancement in *Hermissenda* B photoreceptors. *Neurosci Lett*. 1991;121:107–110.
88. Farley J, Auerbach S. Protein kinase C activation induces conductance changes in *Hermissenda* photoreceptors like those seen in associative learning. *Nature*. 1986;319:220–223.
89. Farley J, Schuman E. Protein kinase C inhibitors prevent induction and continued expression of cell memory in *Hermissenda* type B photoreceptors. *Proc Natl Acad Sci USA*. 1991;88: 2016–2020.
90. Neary JT, Naito S, Weer De, Alkon DL. Ca^{2+}/diacylglycerol-activated, phospholipid-dependent protein kinase in the *Hermissenda* CNS. *J Neurochem*. 1986;47:1405–1411.
91. Crow T, Forrester J. Protein kinase inhibitors do not block the expression of established enhancement in identified *Hermissenda* B-photoreceptors. *Brain Res*. 1993;613:61–66.
92. Crow T, Forrester J. Down-regulation of protein kinase C and kinase inhibitors dissociate short- and long-term enhancement produced by one-trial conditioning of *Hermissenda*. *J Neurophysiol*. 1993;69:636–641.
93. Schuman EM, Clark GA. Synaptic facilitation at connections of *Hermissenda* type B photoreceptors. *J Neurosci*. 1994;14: 1613–1622.
94. Crow T, Xue-Bian JJ, Siddiqi V, Neary JT. Serotonin activation of the ERK pathway in *Hermissenda*: contribution of calcium-dependent protein kinase C. *J Neurochem*. 2001;78: 358–364.
95. Crow T, Forrester J. Inhibition of protein synthesis blocks long-term enhancement of generator potentials produced by one-trial *in vivo* conditioning in *Hermissenda*. *Proc Natl Acad Sci USA*. 1990;87:4490–4494.

96. Crow T, Siddiqi V, Dash PK. Long-term enhancement but not short-term in *Hermissenda* is dependent upon mRNA synthesis. *Neurobiol Learn Mem.* 1997;68:343–350.
97. Crow T, Siddiqi V. Time-dependent changes in excitability after one-trial conditioning of *Hermissenda*. *J Neurophysiol.* 1997;78: 460–3464.
98. Crow T, Xue-Bian JJ, Siddiqi V. Protein synthesis-dependent and mRNA synthesis-independent intermediate phase of memory in *Hermissenda*. *J Neurophysiol.* 1999;82:495–500.
99. Crow T, Xue-Bian JJ. Identification of a 24 kDa phosphoprotein associated with an intermediate stage of memory in *Hermissenda*. *J Neurosci.* 2000;20:RC74.
100. Ramirez RR, Gandhi CC, Muzzio IA, Matzel LD. Protein synthesis-dependent memory and neuronal enhancement in *Hermissenda* are contingent on parameters of training and retention. *Learn Mem.* 1998;4:462–477.
101. Alkon DL, Epstein H, Kuzirian A, Bennett MC, Nelson TJ. Protein synthesis required for long-term memory is induced by PKC activation in days before associative learning. *Proc Natl Acad Sci. USA.* 2005;102:16432–16437.
102. Crow T, Xue-Bian JJ. Proteomic analysis of post-translational modifications in conditioned *Hermissenda*. *Neuroscience.* 2010;165:1182–1190.
103. Crow T, Xue-Bian JJ. Proteomic analysis of short- and intermediate-term memory in *Hermissenda*. *Neuroscience.* 2011;192:102–111.
104. Alkon DL, Nelson TJ, Zhao W, Calvallaro S. Time domains of neuronal Ca^{2+} signaling and associative memory: steps through a calexcitin, ryanodine receptor K^+ channels cascade. *Trends Neurosci.* 1998;21:529–537.
105. Kuzirian AM, Epstein HT, Buck D, Child FM, Nelson T, Alkon DL. Pavlovian conditioning-specific increases of the Ca^{2+}- and GTP-binding protein, calexcitin in identified *Hermissenda* visual cells. *J Neurocytol.* 2001;12:993–1008.
106. Neary JT, Crow T, Alkon DL. Change in a specific phosphoprotein band following associative learning in *Hermissenda*. *Nature.* 1981;293:658–660.
107. Ascoli G, Liu KX, Olds JL, et al. Secondary structure of Ca^{2+} induced conformational changes of calexcitin, a learning associated protein. *J Biol Chem.* 1997;272:29771–29779.
108. Nelson TJ, Cavallaro S, Yi CL, et al. Calexcitin: a signaling protein that binds calcium and GTP, inhibits potassium channels, and enhances membrane excitability. *Proc Natl Acad Sci USA.* 1996;93:13808–13813.
109. Nelson TJ, Zhao WQ, Yuan S, Favit A, Pozzo-Miller L, Alkon DL. Calexcitin interaction with neuronal ryanodine receptors. *Biochem J.* 1999;341:423–433.
110. Crow T. Pavlovian conditioning of *Hermissenda*: current cellular, molecular, and circuit perspectives. *Learn Mem.* 2004;11:229–238.
111. Crow T, Siddiqi V, Zhu Q, Neary JT. Time-dependent increase in protein phosphorylation following one-trial enhancement in *Hermissenda*. *J Neurochem.* 1996;66:1736–1741.
112. Crow T, Xue-Bian JJ. One-trial *in vitro* conditioning regulates a cytoskeletal-related protein (CSP24) in the conditioned stimulus pathway of *Hermissenda*. *J Neurosci.* 2002;22:10514–10518.
113. Crow T, Redell JB, Tian LM, Xue-Bian J, Dash PK. Inhibition of conditioned stimulus pathway phosphoprotein 24 expression blocks the development of intermediate-term memory in *Hermissenda*. *J Neurosci.* 2003;23:3415–3422.
114. Redell JB, Xue-Bian JJ, Bubb MR, Crow T. One-trial *in vitro* conditioning regulates an association between the β-thymosin repeat protein Csp24 and actin. *Neuroscience.* 2007;148: 413–420.
115. Connor J, Alkon DL. Light- and voltage-dependent increases of calcium ion concentration in molluscan photoreceptors. *J Neurophysiol.* 1984;51:745–752.
116. Sakakibara M, Alkon DL, Kouchi T, Inoue H, Yoshioka T. Induction of photoresponse by the hydrolysis of polyphosphoinositides in the *Hermissenda* type B photoreceptor. *Biochem Biophys Res Commun.* 1994;202:299–306.
117. Sakakibara M, Alkon DL, Neary JT, Heldman E, Gould R. Inositol trisphosphate regulation of photoreceptor membrane currents. *Biophys J.* 1986;50:797–803.
118. Blackwell KT, Alkon DL. Ryanodine receptor modulation of *in vitro* associative learning in *Hermissenda crassicornis*. *Brain Res.* 1999;20:114–125.
119. Falk-Vairant J, Crow T. Intracellular injections of BAPTA block induction enhancement in *Hermissenda* type B-photoreceptors. *Neurosci Lett.* 1992;14:745–748.
120. Rogers RF, Fass DM, Matzel LD. Current, voltage, and pharmacological substrates of a novel GABA receptor in the visual-vestibular system of *Hermissenda*. *Brain Res.* 1994;650: 93–106.
121. Jin NG, Tian LM, Crow T. 5-HT and GABA modulate intrinsic excitability of type I interneurons in *Hermissenda*. *J Neurophysiol.* 2009;102:2825–2833.
122. Jin NG, Crow T. Serotonin regulates voltage-dependent currents in type $I_{e(A)}$ and I_i interneurons of *Hermissenda*. *J Neurophysiol.* 2011;106:2557–2569.
123. Ito E, Oka K, Collin C, Schreurs BG, Sakakibara M, Alkon DL. Intracellular calcium signals are enhanced for days after pavlovian conditioning. *J Neurochem.* 1994;62: 1337–1344.
124. Lester DS, Collin C, Etcheberrigaray R, Alkon DL. Arachidonic acid and diacylglycerol act synergistically to activate protein kinase C *in vitro* and *in vivo*. *Biochem Biophys Commun.* 1991;179:1522–1528.
125. Muzzio IA, Talk AC, Matzel LD. Incremental redistribution of protein kinase C underlies the acquisition curve during *in vitro* associative conditioning in *Hermissenda*. *Behav Neurosci.* 1997;111:739–753.
126. Crow T, Xue-Bian JJ, Dash PK, Tian LM. Rho/ROCK and Cdk5 effects on phosphorylation of a β-thymosin repeat protein in *Hermissenda*. *Biochem Biophys Commun.* 2004;323: 395–401.
127. Yamoah EN, Levic S, Redell JB, Crow T. Inhibition of conditioned stimulus pathway phosphoprotein 24 expression blocks the reduction in A-type transient K^+ current produced by one-trial *in vitro* conditioning of *Hermissenda*. *J Neurosci.* 2005;25: 4793–4800.
128. Kawai R, Crow T. Paired presentations of light and serotonin produces a change in the distribution of Csp24 in *Hermissenda* lateral type B photoreceptors. *Soc Neurosci Abstr.* 2005;31:2005.

CHAPTER

20

Molecular and Cellular Mechanisms of Classical Conditioning in the Feeding System of *Lymnaea*

György Kemenes

University of Sussex, Brighton, United Kingdom

INTRODUCTION

Soon after the publication of the first papers describing central pattern generator (CPG), motoneuronal, and modulatory components of the feeding network in *Lymnaea*,[1–4] seminal behavioral work established the ability of *Lymnaea* to form long-term associative memory after classical conditioning with a few pairings or even just a single pairing of a nonfood chemical stimulus, amyl acetate, and a food stimulus, sucrose.[5–7] It was also soon shown that *Lymnaea* is capable of forming a positive association between tactile stimuli and food, but only after multiple trials.[8] Both the chemical and the tactile conditioning paradigms revealed a number of characteristics of associative learning in *Lymnaea* that were shared by learning in vertebrates, such as dependence on age and motivational states, stimulus generalization, discriminative learning, and classical–operant interactions.[7–10] Later, it was also demonstrated that *Lymnaea* is able to form negative associations between a food conditioned stimulus (CS) and an aversive chemical unconditioned stimulus (US).[11] The wealth of knowledge about the behavioral features of both aversive and reward classical conditioning of feeding, together with a detailed understanding of the neuronal network underlying the feeding behavior, has made *Lymnaea* a very attractive and highly productive experimental model for top-down analyses of associative learning and memory. A detailed description of the *Lymnaea* feeding behavior and underlying neuronal circuitry has been presented in several review articles.[12–15]

The two seemingly disparate fields of research, physiological analysis of the feeding CPG and behavioral analysis of associative conditioning, started to converge in the mid- to late 1990s when neuronal correlates of behavioral and *in vitro* classical conditioning were first described.[11,16–19] Detailed analyses of molecular mechanisms of associative memory started in the early 2000s after sufficient information had been obtained on both the behavioral and the cellular aspects of associative learning after classical conditioning. In parallel to the molecular analysis, more detailed electrophysiological investigations identified synaptic and nonsynaptic (intrinsic) forms of plasticity induced by conditioned taste aversion (CTA) learning and single-trial food-reward conditioning, respectively.[11,20]

By the mid-2000s, *Lymnaea* had emerged as the most widely used molluskan model system for the top-down analysis of both the molecular and the cellular mechanisms of associative learning, including both classical and operant conditioning. This chapter reviews the molecular- and cellular-level findings obtained using classical conditioning in *Lymnaea*. A review of the results based on operant conditioning in this mollusk is provided elsewhere in this book.

MOLECULAR MECHANISMS OF CLASSICAL CONDITIONING IN THE FEEDING SYSTEM OF *LYMNAEA*

Reward Classical Conditioning of Feeding

Classical (or Pavlovian) food-reward conditioning in *Lymnaea* is based on forming an association between a nonfood tactile or chemical conditioned stimulus (CS) and an unconditioned food stimulus (US). Associations are formed in the feeding network and lead to a learned change in the behavior, with trained animals producing

Invertebrate Learning and Memory.
DOI: http://dx.doi.org/10.1016/B978-0-12-415823-8.00020-4

stronger feeding responses to the tactile or chemical CS compared to control animals.[5,6,8,9]

Single-Trial Reward Conditioning

To date, the most detailed information on the molecular mechanisms of associative long-term memory (LTM) in *Lymnaea* has been gained from experiments using single-trial food-reward (also known as appetitive) classical conditioning. In this paradigm, snails are subjected to a conditioning protocol using a single pairing of amyl acetate (pear drops) as the CS with sucrose as the US. As early as 30 min after training, the explicitly paired (CS + US) experimental group shows significantly greater feeding responses to amyl acetate compared to its own naive responses and those of all the standard control groups (random, explicitly unpaired, CS alone, and US alone).[5,21]

The most important finding from the original behavioral studies was that even a single pairing of CS and US resulted in LTM, which lasted for several weeks.[5] This is a remarkable example of single-trial learning, which is now very effectively used for analyses of the time course of the molecular mechanisms underlying memory processes. The use of this single-trial paradigm for the analysis of the molecular mechanisms of memory consolidation and reconsolidation has two main advantages. First, after single-trial conditioning, translation- and transcription-dependent memory emerges in a matter of hours,[21] allowing this type of memory to be studied on a timescale of a few hours to several weeks. This early emergence of LTM was utilized in *in vitro* conditioning experiments investigating the cellular and molecular mechanisms of memory formation in semi-intact preparations, which are only viable for up to 6 hr.[22] Second, unlike multitrial paradigms, single-trial conditioning allows the analysis of the amnesic effects of sharply timed manipulations of key molecular pathways, both during memory consolidation[21,23–26] and during reconsolidation.[27,28]

At the molecular level, the most important question to be addressed was whether or not LTM forming after single-trial conditioning (similar to "flashbulb" memory known in other systems, including humans[29]) is based on the same conserved pathways that were originally described using multitrial paradigms in other systems. An important clue to suggest that this might be the case is that the broad-spectrum protein synthesis inhibitor anisomycin (ANI) injected in an early time window (10 min to 1 hr) after conditioning blocks the 24-hr memory trace (conventionally regarded as LTM).[21] An earlier (5-hr) memory trace is blocked by both ANI and actinomycin D, an RNA synthesis blocker, confirming that LTM, defined as memory dependent on early post-training translation and transcription, is present as early as 5 hr after single-trial conditioning.[21] This observation indicates that single-trial reward conditioning triggers molecular cascades that are involved in the rapid consolidation of long-lasting memory traces. Interestingly, there is only a single early time window of sensitivity to transcription- and translation blockers,[21] unlike in a number of other studies, which described a temporally distinct second window of protein synthesis-dependent LTM in both vertebrates and invertebrates.[30–33] It has been suggested that "strong" training protocols lead to rapid memory consolidation involving only a single wave of protein synthesis-dependent events, whereas "weak" training protocols are followed by a more prolonged consolidation phase containing two or more windows of requirement for new protein synthesis.[30] According to this categorization, single-trial reward conditioning in *Lymnaea* certainly qualifies as a strong training protocol, explaining both the rapid emergence and the persistence of the associative memory trace, characteristic of flashbulb memory.

The possibility that cooling-induced blockade of LTM stems from the disruption of protein synthesis was also investigated after single-trial food-reward conditioning in *Lymnaea*.[34] By varying the timing of post-training hypothermia, the critical period during which cooling disrupts the consolidation of appetitive LTM was determined. Post-training hypothermia was found to disrupt LTM only when applied immediately after conditioning, whereas delaying the treatment by 10 min left the 24-hr memory trace intact. This brief (<10 min) window of sensitivity differs from the time window previously described for the protein synthesis inhibitor anisomycin, which was effective during at least the first 30 min after conditioning.[21] Therefore, hypothermia and protein synthesis inhibition exhibit distinct time windows of effectiveness in *Lymnaea*, a fact that is inconsistent with the hypothesis that cooling-induced amnesia occurs through the direct disruption of macromolecular synthesis. It was therefore suggested that the very early blocking of the 24-hr memory trace by cooling is interfering with enzymic cascades, including kinases that are known to be activated by chemical conditioning.

CREB AND LTM AFTER SINGLE-TRIAL REWARD CONDITIONING

Detailed molecular analyses identified highly conserved cyclic AMP-responsive element binding protein (CREB) genes (LymCREB1 and LymCREB2) and CREB-like proteins in *Lymnaea*.[35,36] These were important findings because regulation of gene expression during memory consolidation is known to involve a variety of transcription factors, with CREB playing a particularly important role in the switch from short-term to long-term memory storage in a variety of different species and paradigms.[29,37–49] Consistent with a role for CREB in *Lymnaea* LTM is the observation that levels of phosphorylated CREB1 are increased in neurons of the feeding network following single-trial food-reward conditioning.[35]

NITRIC OXIDE AND LTM AFTER SINGLE-TRIAL REWARD CONDITIONING

The first study to investigate the molecular mechanisms of LTM in *Lymnaea* established that consolidation of LTM in *Lymnaea* is dependent on the nitric oxide (NO)-cGMP signaling pathway.[23] There is a critical period of sensitivity up to 5 hr after conditioning when blocking this pathway by drug injection prevents LTM formation, which was established at the level of both whole animals and the neuronal network responsible for generating the feeding behavior.[23]

Further evidence for a role of NO in LTM came from experiments on single CGCs.[50,51] These neurons express mRNA transcripts from two related neuronal nitric oxide synthase (nNOS) genes (*Lym-nNOS1* and *Lym-nNOS2*), and they also express NO-sensitive soluble guanylyl cyclase (sGC). Six hours after one-trial conditioning, *Lym-nNOS1* is upregulated compared with controls. This upregulation of the NOS coding transcript may be due to an earlier downregulation of the *Lymnaea anti-nNOS* transcript at 4 hr that is known to be inhibitory on NOS transcript production. Electrophysiological experiments on isolated CGCs also showed that the CGCs responded to NO by generating a prolonged depolarization of the membrane potential, similar to what was found earlier after single-trial classical conditioning.[20] The NO-induced depolarization was blocked by ODQ,[51] supporting the hypothesis that it is mediated by sGC. This is a rare example of analysis in which the role of a signaling molecule in memory formation could be traced from the behavioral to the network and single neuronal level.

PROTEIN KINASE A, MITOGEN-ACTIVATED PROTEIN KINASE, AND LTM AFTER SINGLE-TRIAL REWARD CONDITIONING

Other highly conserved molecular pathways that have been implicated in LTM after single-trial reward conditioning in *Lymnaea* are the protein kinase A (PKA) and mitogen-activated protein kinase (MAPK)-dependent signaling pathways. Inhibition of PKA catalytic subunit activity or MAPK phosphorylation blocked 24-hr LTM without blocking sensory or motor pathways.[25,28] When measured 30 min after conditioning, increased levels of both PKA activity and MAPK phosphorylation were found,[25] with increased PKA activation also detected when measured in an earlier (5- and 10-min) and a later (1-hr) time window,[28] indicating a more prolonged dependence of 24-hr LTM on PKA compared to protein synthesis. However, a detailed analysis of the time windows of requirement for PKA revealed an important difference between memory at 6 and 24 hr after training. For memory to be expressed at 6 hr, there is a very early (within 5 min) postacquisition requirement for PKA with a narrow time window (~10 min). If a PKA inhibitor is injected more than 10 min after training, the 6-hr memory trace remains unaffected. By contrast, the 24-hr memory trace is impaired by injection of the PKA inhibitor for up to 3 hr post-training. This latter finding shows that the same very early PKA-mediated events that are necessary for 6-hr memory are not sufficient for 24-hr memory trace to form, which is dependent on much more prolonged PKA activity compared to that required for the 6-hr memory trace.

NMDA RECEPTORS AND CaMKII AFTER SINGLE-TRIAL REWARD CONDITIONING

The involvement of NMDA receptors and CaM kinase II (CaMKII) in different phases of associative LTM induced by single-trial reward classical conditioning has also been investigated in *Lymnaea*.[26] First, by using a general CaM kinase inhibitor, KN-62, it was established that CaM kinase activation was necessary for acquisition and late consolidation but not early or intermediate consolidation or retrieval of LTM. Then, Western blot assays of the buccal and cerebral ganglia from classically conditioned animals and treatment of intact animals with CaMKIINtide identified CaMKII as the main CaM kinase required for learning and also for late memory consolidation in a critical time window (~24 hr after learning). Acquisition is dependent on both NMDA receptor and CaMKII activation, but CaMKII-dependent late memory consolidation does not require the activation of NMDA receptors. These findings support the notion that even apparently stable memory traces may undergo further molecular changes and identify NMDA-independent intrinsic activation of CaMKII as a mechanism underlying this "lingering consolidation." This process may facilitate the preservation of LTM in the face of protein turnover or active molecular processes that underlie forgetting.

A summary of the observations concerning the roles of PKA, MAPK, CREB, NOS/NO, NMDA receptors, and CaMKII in 24-hr LTM forming after reward conditioning is shown in Table 20.1. This direct comparison shows an important difference between MAPK and the other signaling molecules investigated in reward conditioning. Unlike the other factors, MAPK is activated not only in response to contingent CS + US application but also when the CS or US are applied alone. This observation, together with the fact that preventing MAPK phosphorylation after training blocks LTM,

TABLE 20.1 Summary of the Role of Various Highly Conserved Molecular Pathways in the Consolidation of Associative Memory after Single-Trial Food-Reward Classical Conditioning in *Lymnaea*

Molecule/Pathway	CS Alone	US Alone	CS/US Unpaired	CS/US Paired	Inhibition Blocks Associative LTM
PKA	–	–	–	↑	Yes
MAPK	↑	↑	↑	↑	Yes
CREB	–	–	–	↑	?
NOS/NO	–	–	–	↑	Yes
CaMKII	–	–	–	↑	Yes

–, no change after training; ↑, activation/upregulation after training; ?, no data.
Data sources: *I. Kemenes* et al.,[23] *Ribeiro* et al.,[25,35] *Michel* et al.,[28] *G. Kemenes* et al.,[27] *Korneev* et al.,[50] *and Wan* et al.[26]

shows that MAPK is necessary but not sufficient for the consolidation of associative LTM after single-trial reward conditioning. PKA, NMDA receptors, CaMKII, CREB, and NOS/NO are selectively activated/upregulated by the associative training protocol, so potentially each of these factors could be sufficient for memory consolidation. However, it is more likely the case that these and other signaling molecules make a synergistic contribution to the memory consolidation process, with none of them alone being sufficient for LTM.

Treatment of isolated cerebral ganglia with the adenylate cyclase activator forskolin resulted in massively increased CREB1 phosphorylation,[35] indicating a potential link between training-induced PKA activation and CREB phosphorylation in conditioned animals. Whether there is any linkage between the NO, PKA, and MAPK signaling pathways has yet to be determined.

A summary of the known time windows of sensitivity of the 24- and 42-hr LTM to inhibition of the various different molecular pathways reviewed previously is shown in Figure 20.1. For all signaling molecules except CaMKII, time windows of requirement for 24-hr LTM were only investigated in the first 6 hr after training, so we cannot rule out later time windows for requirement for new proteins, PKA, and NO for the 24-hr memory trace. The requirement at 24 hr for the 42-hr memory trace was investigated for all the signaling molecules shown in the diagram.

The related observation that *Lym-nNOS1* mRNA levels increase at 6 hr post-training[50] strongly suggests a later role for NO, perhaps in memory maintenance. The level of phosphorylated CREB1 was also found to be high at 6 hr post-training,[35] indicating that new transcription and translation take place after this period. The time windows of requirement for NO and PKA are both wider[23,28] than the time window of requirement for early protein synthesis,[21] indicating that their function goes beyond being upstream

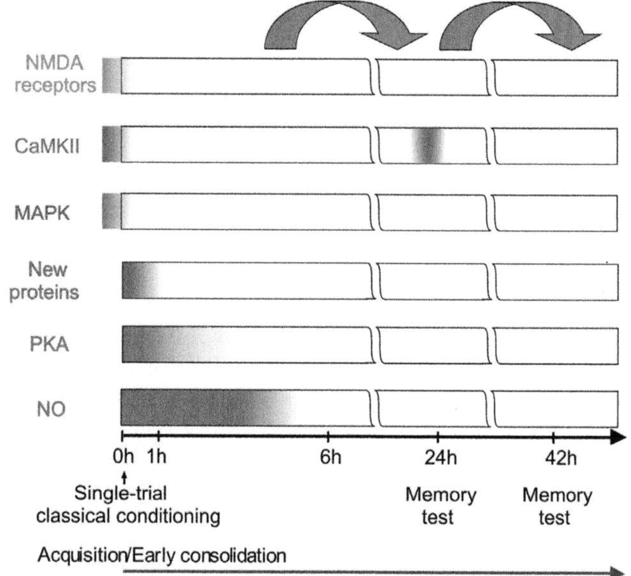

FIGURE 20.1 **Time windows of known molecular requirements for LTM after single-trial food-reward classical conditioning in *Lymnaea*.** The large curved arrows indicate the requirement for the activation or synthesis of molecules in the 0- to 6-hr (acquisition/early and intermediate consolidation) and 24-hr time windows (late consolidation) for memory expressed at 24 and 42 hr, respectively. In the late memory consolidation phase (24 hr post-training), of the molecules investigated in our studies, only the intrinsic activation of CaMKII is required for memory expression at 42 hr post-training. Data sources: *Based on data from I. Kemenes* et al.,[23] *Fulton* et al.,[21] *G. Kemenes* et al.,[27] *Michel* et al.,[28] *Wan* et al.,[26] *and Pirger* et al.[52].

components of molecular pathways leading to new protein synthesis in the first hour after training.

PACAP AND LTM AFTER SINGLE-TRIAL REWARD CONDITIONING

Observations concerning the role of PKA in memory consolidation in *Lymnaea*[27,28] indicate that similar to other systems, activation of adenylate cyclase (AC) is a key step in LTM formation in *Lymnaea*. However, there was no information available on the molecules involved in the learning-induced activation of AC. A 2010 study demonstrated both the presence and biochemical activity of a protein homologous to the vertebrate pituitary adenylate cyclase-activating polypeptide (PACAP) and its receptors in the *Lymnaea* nervous system.[53] This was followed by another study in 2010,[52] which showed that application of a PACAP receptor antagonist at approximately the time of single-trial food-reward training with amyl acetate and sucrose blocked associative LTM. This finding suggested that in this strong food-reward conditioning paradigm, the activation of AC by PACAP was necessary for LTM to form. Interestingly, in the weak multitrial food-reward conditioning paradigm, lip touch paired with sucrose,[8] memory formation was also dependent on PACAP.[52] Significantly, systemic application of PACAP at the beginning of multitrial tactile conditioning accelerated the formation of transcription-dependent memory.[52] In PACAP-treated animals, robust LTM formed after just three trials, whereas control animals required more than six trials to form LTM. This memory-boosting effect of exogenously applied PACAP was blocked by the PACAP receptor antagonist PACAP6-38. These findings therefore provided the first evidence that in the same nervous system PACAP is both necessary and instructive for fast and robust memory formation after reward classical conditioning.

Pretraining application of the PACAP receptor antagonist resulted in a complete abolition of memory after both single-trial chemical and multitrial tactile conditioning but not in a loss of the unconditioned feeding response. Based on this finding, it is tempting to speculate that PACAP is released in response to the chemical and tactile conditioned stimuli, whereas similar to what was found in *Aplysia*, the effect of the unconditioned stimulus on AC may be mediated by different peptide or nonpeptide transmitters, such as SCPs or 5-HT[54] or DA.[55,56] A likely scenario is that in *Lymnaea*, the PACAP-mediated effect of the chemical or tactile CS and the non-PACAP-mediated effect of sucrose US converge on AC, and this convergence provides the molecular basis for coincidence detection, a fundamental requirement for associative learning. *Lymnaea* is known to differentiate learning with amyl acetate[20] from learning with touch[57] at the neuronal level within the same network (the feeding circuitry in this case), but there is no evidence for a similar differentiation at the molecular level within the same neuron. Thus, the same molecules (e.g., PACAP) can fulfill the same role (e.g., activation of AC) in different neurons, leading to learning-induced changes in different pathways (e.g., activated by touch vs. activated by amyl acetate).

cAMP-DEPENDENT MOLECULAR CASCADES AND NEURONAL PLASTICITY CONTRIBUTING TO LTM AFTER SINGLE-TRIAL REWARD CONDITIONING

An important link between cAMP-dependent molecular cascades and neuronal plasticity contributing to LTM was found in voltage-clamp experiments, which demonstrated a long-lasting cAMP-induced increase in a low-threshold persistent sodium current of the CGCs.[58] This current makes an important contribution to the CGC somal membrane potential,[24,59] which in turn becomes persistently depolarized after single-trial reward conditioning.[20] The CGC's synaptic output also became similarly enhanced by somal injection of cAMP[58] and artificial depolarization of the soma membrane,[20] lending further support to the notion that cAMP-dependent cascades might support long-lasting plastic changes in ion channel number and/or function contributing to learning-induced maintained depolarization. The mechanism by which persistent somal depolarization affects synaptic output appears to be based on a maintained increase in background calcium levels in axon terminals of the CGCs presynaptic to their target neurons,[20] similar to what was previously demonstrated in mammalian neurons.[60]

PKA, PROTEIN SYNTHESIS, AND MEMORY RECONSOLIDATION AFTER SINGLE-TRIAL REWARD CONDITIONING

An important finding is that memory reconsolidation after retrieval at 6 hr post-training is both PKA- and protein synthesis-dependent, whereas reconsolidation after retrieval at 24 hr depends on protein synthesis but not on PKA activity.[27] This finding indicates that depending on how recent or remote consolidated memory is relative to the time of training, different molecular pathways are activated by memory retrieval and contribute differentially to memory reconsolidation. At a more general level, this phase-dependent differential molecular requirement for reconsolidation supports the notion that even seemingly fully consolidated memories undergo further selective molecular maturation processes (the 'lingering consolidation' hypothesis[61]), which may only be revealed by analyzing the role of a variety of different specific pathways in memory reconsolidation after retrieval.

A key question concerning the molecular mechanisms of memory reconsolidation is whether or not they are a recapitulation of the processes active during memory consolidation.[61–66] If they are, memory retrieval should reactivate the same molecular cascades that were activated by training, and therefore the same amnestic treatments should impede both processes. A direct comparison of the PKA and protein synthesis dependence of the consolidation of the 6- and 24-hr

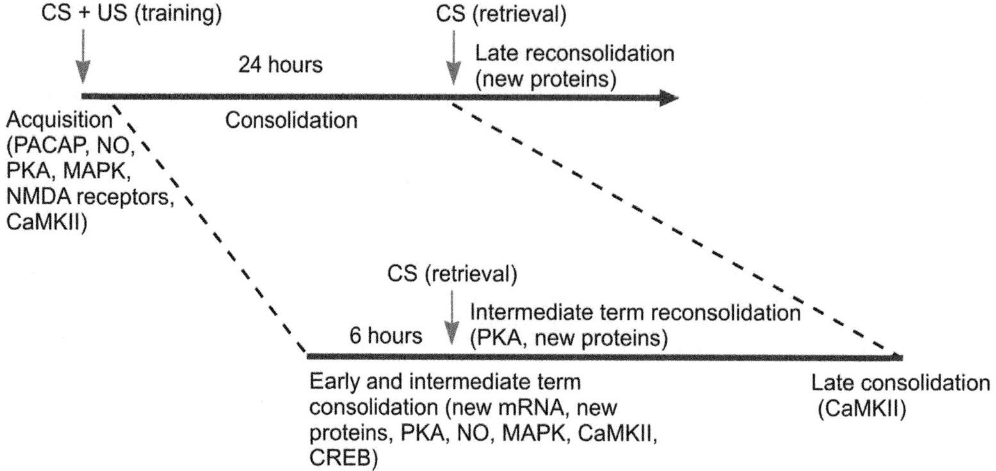

FIGURE 20.2 **The currently known molecular players in acquisition, memory consolidation, and reconsolidation after single-trial food-reward classical conditioning in *Lymnaea*.** Data sources: *Based on data from I. Kemenes* et al.,[23] *Ribeiro* et al.,[25,35] *Fulton* et al.,[21] *G. Kemenes* et al.,[27] *Michel* et al.,[28] *Wan* et al.,[26] *and Pirger* et al.[52]

memory trace and the reconsolidation of memory after retrieval at these two time points supports the notion that reconsolidation is a recapitulation of consolidation only at the general molecular level (protein synthesis) but not at the level of a specific signaling molecule (PKA). It remains to be elucidated if other specific signaling molecules also known to contribute to the consolidation of the 24-hr memory trace (e.g., MAPK, NO, and CaMKII) are activated by memory retrieval.

A summary of the known molecular players in acquisition, memory consolidation, and reconsolidation after single-trial food-reward classical conditioning is shown in Figure 20.2.

Aversive Classical Conditioning of Feeding

Aversive classical conditioning of feeding in *Lymnaea* is based on pairing sucrose as a CS with an aversive chemical US such as KCl, which inhibits feeding and evokes a withdrawal response (CTA). After eight or more trials, trained animals showed a significantly weaker feeding response to sucrose than did controls, and this associative memory lasted for more than 1 month.[11,67]

A neural analysis of CTA was carried out on isolated brains dissected from conditioned and control animals.[11,68] Like reward chemical conditioning, aversive chemical conditioning leads to specific changes in the feeding network, described in previous reviews[67,69,70] and later in this chapter.

PKA, CREB, and LTM after Aversive Conditioning

The transcription and translation dependence of aversive behavioral LTM or the necessity of the activation of specific molecular cascades for its consolidation have not been tested in intact *Lymnaea*, but *in vitro* experiments were performed to investigate the role of the transcription factor CREB and its upstream activators in synaptic plasticity that may be linked to CTA. The injection of cAMP or PKA into the soma of the CGCs in isolated nervous system preparations led to a long-term enhancement of the synapse between the CGC and a follower motoneuron, B1.[71] The injection of a CRE oligonucleotide into the CGCs prevented this cAMP-induced long-lasting synaptic plasticity.[36] These *in vitro* experiments showed that one of the mechanisms of the previously described enhancements in the CGC's synaptic output induced by CTA learning could be the activation of the cAMP/PKA/CREB cascade.

In a more recent study, the expression level of activator and repressor isoforms of *Lym*-CREB1 mRNAs was investigated by a real-time quantitative reverse-transcription polymerase chain reaction method.[72] CTA learning increased *Lym*-CREB1 gene expression, but interestingly, it did not change the activator:repressor mRNA ratio. The repressor isoforms, as well as the activator ones, were expressed in increased amounts in the central nervous system (CNS) after CTA learning, giving rise to the hypothesis that the activator:repressor ratio only becomes higher either during translation or post-translationally, for example, by proteolytic degradation of the repressor isoform of CREB1.[72]

C/EBP and LTM after Aversive Conditioning

A link has also been found between CTA in intact *Lymnaea* and another conserved transcription factor, CCAAT/enhancer binding protein (C/EBP), and this was shown in the buccal ganglia and specifically the B2 gut motoneuron.[73] Both the phosphorylated and total levels of the *Lym*-C/EBP protein increased in the buccal ganglia and B2 during CTA consolidation, although *Lym*-C/EBP mRNA levels were reduced in the same time window (1 hr post-training) when the increases in protein levels were measured. One explanation for this paradoxical observation could be that the existing pool of *Lym*-C/EBP mRNA is rapidly translated and degraded early after CTA learning, resulting in fast turnover of newly transcribed mRNA, which may be required for a prolonged *de novo* synthesis of large amounts of *Lym*-C/EBP necessary for long-term memory after CTA learning. A more detailed analysis of the time course of changes in *Lym*-C/EBP mRNA and protein levels will be necessary to elucidate the relationship between *Lym*-C/EBP gene transcription, mRNA turnover, and protein synthesis in this system.

Although the CRE element upstream of *Lym*-C/EBP has not been investigated, based on data from *Aplysia*, *Helix*, and mammals,[37,74,75] the expression of *Lym*-C/EBP is likely to be regulated by *Lymnaea* CREB. It is also not known which genes are targeted by C/EBP in *Lymnaea*. However, likely candidate downstream targets of C/EBP are the *Lym*-nNOS genes.[50] *Lym*-C/EBP and *Lym*-nNOS are co-localized in the B2 motoneurons, and *Lym*-nNOS genes have three putative *Lym*-C/EBP binding sites,[50] providing the necessary structural conditions for the interaction of C/EBP with NOS genes in the *Lymnaea* feeding network.

Comparison of the Molecular Mechanisms Underlying Reward and Aversive Conditioning of Feeding

An interesting but largely unanswered question concerns the differences in the molecular mechanisms underlying reward and aversive conditioning of feeding. Plastic changes after both types of conditioning were found in the CGCs,[11,20,24] which are therefore regarded as key neurons for both reward and aversive conditioning. In the case of reward conditioning, the plastic change (a persistent somal depolarization) is nonsynaptic in nature and increases the probability of a feeding response to the CS by recruiting feeding command-like neurons into the network of cells activated by the CS.[20] In the case of aversive conditioning, the plastic change increases the amplitude of a CGC-driven inhibitory synaptic input to the N1M type CPG neurons,[11] whose inhibition suppresses feeding.[76] It will be important to determine what molecular differences underlie the different cellular changes induced in the same neuron by reward versus appetitive conditioning and contributing to either the activation or the suppression of the same behavior after classical conditioning.

One study has made direct comparisons at the behavioral pharmacological level regarding the different roles that dopamine and octopamine play in the consolidation of appetitive versus aversive classical conditioning of feeding in *Lymnaea*.[77] Snails were classically conditioned in a single trial to associate the same CS (amyl acetate) with either a punishment (quinine) or a reward (sucrose), showing either a reduced or an elevated feeding response, respectively, to the CS. The study found that consolidation of LTM in reward conditioning depends on dopamine but not octopamine. In contrast, aversive LTM depends on octopamine but not dopamine. Octopamine is the invertebrate equivalent of noradrenalin, so these results on the monoamine dependence of reward and aversive conditioning in *Lymnaea* resemble, at the transmitter receptor level, those in mammals but are the opposite of those in another invertebrate group, the insects.

A general model of how the molecular mechanisms implicated in both reward and aversive conditioning of feeding operate at the neuronal level (e.g., in the CGCs that are targets for plastic changes after both reward and aversive conditioning of feeding) is presented in Figure 20.3.

CELLULAR MECHANISMS OF CLASSICAL CONDITIONING IN THE FEEDING SYSTEM OF *LYMNAEA*

Cellular Mechanisms of Single-Trial Chemical Classical Conditioning of Feeding

An electrophysiological correlate of the conditioned response to amyl acetate was recorded as changes in the fictive feeding responses in motor neurons,[23] but electrical activity following conditioning has also been recorded in other parts of the feeding system in attempts to localize sites of plasticity. The cell bodies of chemosensory neurons are located in lip epithelial tissue and project to the cerebral ganglia via the lip nerves, where they synapse with cerebral ganglion neurons such as the CBIs.[79] Extracellularly recorded spike responses to both the CS and the US can be recorded in the lip nerves from naive animals, and these responses do not change after conditioning. In contrast, neuronal output from the cerebral ganglia is significantly enhanced in response to the CS after conditioning.[79] This indicates that chemical conditioning affects central but not peripheral processing of chemosensory information, with the cerebral ganglia being an important site of plasticity. The fibers that were recorded extracellularly to indicate cerebral plasticity were originating from the CBI interneurons, so their activation is particularly significant. Confirmation that the CBIs do increase their activity after conditioning

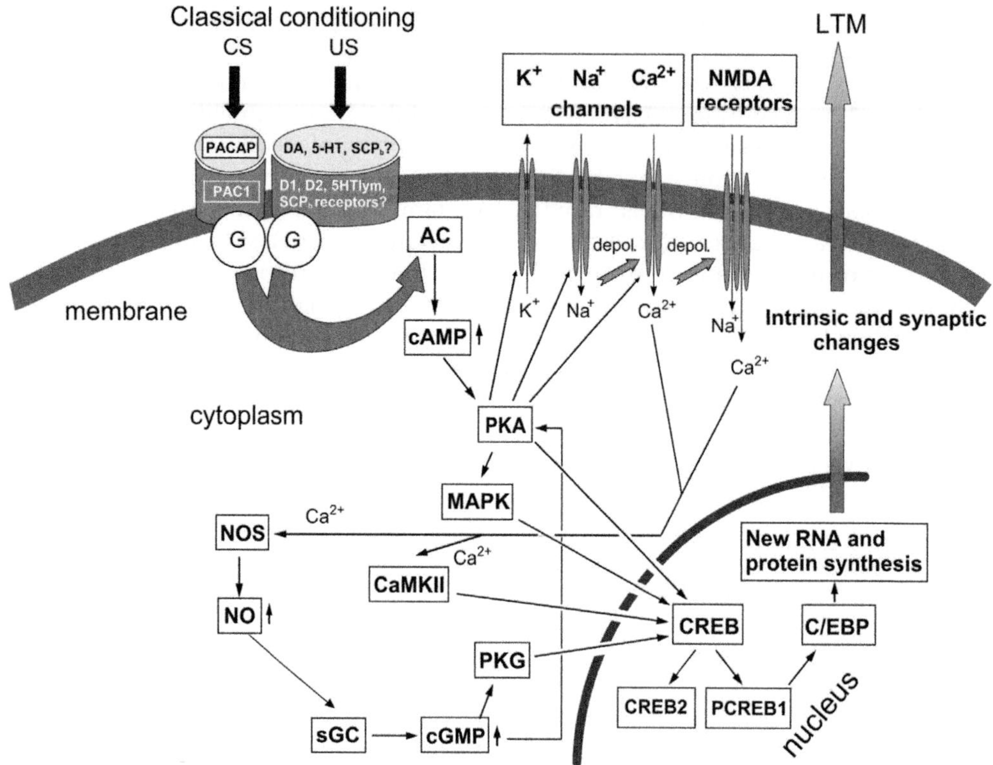

FIGURE 20.3 A general model of how the molecular mechanisms implicated in both reward and aversive conditioning of feeding might operate at the neuronal level. The transmitters and receptors likely involved in the detection of the CS and US, respectively, are only shown for reward conditioning. However, the CS pathway for CTA learning is the same as the US pathway for food-reward learning. The same molecular cascades underlie both the intrinsic changes after food-reward conditioning and synaptic changes after CTA learning. Which type of plasticity will occur is likely defined by the combination of CS and US, which in turn might activate the same cascades or ion channels differentially. Identified molecular components (shown boxed) and data sources: *PACAP and PAC1, Pirger* et al.[53]; *NOS and NO, I. Kemenes* et al.[23] *and Korneev* et al.[50]; *Na^+, Ca^{2+}, and K^+ channels, Staras* et al.[78] *and Nikitin* et al.[24]; *CREB, Ribeiro* et al.[25] *and Sadamoto* et al.[36]; *C/EBP, Hatakeyama* et al.[73]; *RNA and protein synthesis, Fulton* et al.[21]; *MAPK, Ribeiro* et al.[25]; *PKA, G. Kemenes* et al.[27] *and Michel* et al.[28]; *cAMP, Nikitin* et al.[58]; *sGC, Ribeiro* et al.[51]; *AC, Pirger* et al.[52]; *CaMKII and NMDA, Wan et al.*[26].

was obtained by showing increases in feeding patterns to the CS in CV1 neurons, a specific CBI cell type.[20] From this work, it is suggested that the synapses between the primary chemosensory neurons and the CBIs are increased in strength. Two other CS pathways are present in naive animals, but these are not affected by conditioning.[80]

The current network model for chemical conditioning also includes nonsynaptic plasticity. The CGCs are persistently depolarized by approximately 10 mV after behavioral conditioning. This depolarization indirectly increases the strength of postsynaptic responses to CGC stimulation by a process that involves an increase in intracellular calcium concentration.[20] The local target for CGC depolarization is the CBI cells, and artificial depolarization of the CGCs in naive snails increases the response of the CBI cells to the CS, mimicking the effects of behavioral conditioning. It appears that the CGCs are increasing the strength of the CS-to-CBI synapse by presynaptic facilitation. Importantly, the effect of the somal depolarization of the CGC does not spread onto the more distal modulatory connections of the CGC with the CPG or motoneurons,[20] leaving the normal modulatory function of the CGC unaffected by learning.

The onset of the CGC depolarization is between 16 and 24 hr after training, and it persists for at least 14 days, as long as the behavioral memory trace is present. Therefore, it is likely that the CGCs are involved in the maintenance of the late phase of LTM and they encode information that is important for memory recall. Interestingly, the CV1 cells that show a persistent change in membrane potential after tactile conditioning (see later discussion) show no change after chemical conditioning.

It has been shown that the conditioning-induced depolarization of the CGCs is due to an increase in the size of a persistent sodium current.[24] The measured increase in the size of this current is sufficient to depolarize the CGC by the required amount, and this was confirmed by computer modeling. Surprisingly, the depolarization of the CGCs does not cause a change in the firing rate of the CGCs or their spike shape, and this is due to balancing increases in two other currents—a delayed rectifier potassium current (I_D) and a high-voltage-activated calcium current (I_{HVA}).[59] These results indicate that the cellular changes that occur in the CGCs after conditioning are sufficient to explain the enhancement of the CS effects on feeding following conditioning, but they cannot be the whole story because the onset of CGC depolarization is 16–24 hr after training, whereas a behavioral memory trace is present 2 hr after training; thus, an alternative mechanism must also be present to explain the early memory trace. Significantly, a second type of electrical change in the feeding circuit occurs as early as 1 hr after one-trial *in vitro* conditioning. This is a conditioning-induced reduction in tonic inhibitory synaptic input to the feeding CPG.[22] By reducing the frequency of this "background" inhibitory input, the threshold for the CS to activate the feeding network is reduced, making it more likely that the feeding network (CPG) will respond to the CS.[81] This inhibitory input originates from one of the CPG interneuron types known as the N3t (N3 tonic). Computer simulation of N3t–N1M interactions suggests that changes in N3t firing are sufficient to explain the increase in the fictive feeding activity produced by *in vitro* conditioning.[22] It remains to be determined if this mechanism can also fully account for memory expression in intact animals before the onset of CGC depolarization.

Cellular Mechanisms of Multitrial Tactile Classical Conditioning of Feeding

As previously mentioned, *Lymnaea* can also be classically conditioned to a lip touch CS by repeatedly pairing a touch to the lip with food (5–15 trials over 3 days). This type of reward learning shares important characteristics with associative conditioning in vertebrates, such as stimulus generalization and discriminative learning, classical–operant interactions, and strong dependence on both external and internal background variables.[8–10]

Two approaches have been used to investigate the neural basis of tactile reward classical conditioning in *Lymnaea*. One approach was based on the development of an *in vitro* preparation in which electrophysiological manipulation of neuronal pathways aims to mimic the behavioral conditioning paradigm.[16] In this study, a lip touch stimulus was paired with intracellular activation of the modulatory slow oscillator neuron, which can drive fictive feeding. After 6–10 pairings, presentation of the touch stimulus could activate a robust fictive feeding rhythm in feeding motor neurons. A second approach used behavioral conditioning followed by electrophysiological analysis to record changes in electrical activity that follows LTM formation. Using the lip touch behavioral training protocol, snails were subjected to 15 training trials over 3 days, and then these and control animals were dissected for electrophysiological analysis, starting on the day after the last training trial. Touching the lips of the intact snails from the experimental group after training induced a pattern of feeding movements significantly greater than in controls. Similar significant differences were seen between experimental and control animals at the level of the electrophysiologically recorded fictive feeding pattern in motor neurons made from the same snails.[17,18]

The CPG-driven activity in the motor neurons ultimately depends on the activity of neurons at all levels of the feeding network, so the conditioned fictive feeding recorded in the motor neurons is a systems-level "readout" of the memory trace in the whole feeding system. However, more detailed changes can also be recorded in different parts of the network.[18] One of these is the early excitatory postsynaptic potential (EPSP) that occurs in the B3 motor neuron before the onset of the fictive feeding pattern. The amplitude, but not the latency and duration, of the EPSP was significantly enhanced after conditioning. In sated snails, the conditioned fictive feeding response to touch was lost, but the increase in the EPSP amplitude persisted. This suggests that there is unlikely to be a causal link between increases in amplitude in B3 and generation of the fictive feeding pattern.

Electrical correlates of tactile conditioning were also recorded at other levels within the feeding circuit, and these could all be potential sites of plasticity. That sites quite early in the CS pathway could be involved in conditioning was revealed by extracellularly recording mechanosensory fibers located in the connective between the cerebral and the buccal ganglia. Tactile responses could be recorded in these fibers, and following conditioning, the number of spikes occurring early in this response increased compared with that of controls.[18]

Interestingly, a correlate of tactile conditioning could also be recorded in the CPG network. A long-lasting sequence of inhibitory synaptic inputs that occurs in the N1 CPG interneurons in response to lip touch in naive animals changes to a strong sustained depolarizing synaptic input after *in vitro* conditioning, and this drives a sustained plateauing pattern in the N1 cell.[16] This is an example of synaptic plasticity affecting an important CPG component of the feeding network.

One candidate for initiating CPG activity following conditioning is the CBI cell type known as CV1. This neuron is capable of driving a fictive feeding pattern via its connections with the N1M cells of the CPG network, and activity in these cells normally accompanies unconditioned feeding patterns stimulated by sucrose.[76] After conditioning, the CV1 cells are significantly more active following touch in conditioned snails compared with controls, and they show the typical patterned activity seen with sucrose (US) application.[57] More detailed experiments on the role of CV1 cells in tactile conditioning have revealed that nonsynaptic electrical changes play a role in memory.[57] A long-lasting membrane depolarization of 11 mV on average was recorded in CV1s from conditioned compared with control snails that persisted for as long as the electrophysiological and behavioral memory trace. The depolarization makes the cells more responsive to the CS and can account for the activation of the feeding response after conditioning via the CV1 cell's strong excitatory synaptic connection with the CPG. The importance of this result is emphasized by experiments in which the membrane potential of the CV1 cells is manipulated to either reverse the effect of behavioral conditioning or mimic the effects of conditioning in naive snails. These experiments showed that the persistent depolarization of the CV1 cells was both sufficient and necessary for the conditioned tactile response in the feeding network.

Electrical correlates of differential appetitive classical conditioning were also recorded in the *Lymnaea* feeding network.[82] In spaced training (15 trials over 3 days), the lips and the tentacle were used as CS+ (reinforced conditioned stimulus) or CS− (non-reinforced conditioned stimulus) sites for behavioral tactile conditioning. In one group of experimental animals, touch to the lips (the CS+ site) was followed by the sucrose US, but touch to the tentacle (the CS− site) was not reinforced. In a second experimental group, the CS+/CS− sites were reversed. Semi-intact lip−tentacle−CNS preparations were made from both experimental groups and a naive control group. Intracellular recordings were made from the B3 motor neuron of the feeding network, which allowed the monitoring of activity in the CPG interneurons as well as early synaptic inputs evoked by the touch stimulus. Following successful behavioral conditioning, the touch stimulus evoked CPG-driven fictive feeding activity at the CS+ but not the CS− sites in both experimental groups. Naive snails/preparations showed no touch responses. A weak asymmetrical stimulus generalization of conditioned feeding was not retained at the electrophysiological level. An early EPSP response to touch was only enhanced following conditioning in the lip CS+/tentacle CS− group but not in the tentacle CS+/lip CS− group. These results showed that the main features of differential appetitive classical conditioning of feeding (first described by Kemenes and Benjamin[8]) can be recorded at the electrophysiological level, but some characteristics of the conditioned response are selectively expressed in the reduced preparation.

Cellular Mechanisms of Aversive Classical Conditioning of Feeding to a Chemical CS

A cellular analysis of the conditioned taste aversion response was carried out on isolated brains dissected from conditioned and control animals. In particular, the synaptic connection between the modulatory CGCs and the CPG interneuron, N1M, was examined. In conditioned animals compared to controls, a significant increase in the size of inhibitory postsynaptic potentials (IPSPs) recorded in the N1M was observed following an artificial depolarization of the CGCs.[11] Because the CGCs are known to play a critical gating role in feeding behavior[83,84] and the N1M is a pivotal member of the feeding CPG,[85] this enhanced IPSP may be an important cellular correlate of the conditioned taste-aversion learning. The N1M IPSPs examined in these experiments probably originate from the N3t cells of the feeding CPG, indicating that these cells are important in both reward and aversive conditioning of the *Lymnaea* feeding system.

Multiple-site optical recording was also used to analyze the neural activity changes caused by CTA training in *Lymnaea*.[68] Electrical stimulation of the median lip nerve, which transmits food chemosensory signals to the CNS, evoked a large number of spikes in several parts of the buccal ganglion. The effects of CTA training on the spike responses were examined in two areas of the ganglion where the most active neural responses occurred. In one area, which included the location of the N1M CPG interneurons, the number of spikes evoked by median lip nerve stimulation was significantly reduced in conditioned animals compared to controls. In another area positioned between the B3 and B4Cl buccal motoneurons, the evoked spike responses were unaffected by CTA training. These results, taken together with the results indicating an enhancement of an inhibitory input to the N1M cells during CTA, suggest that the food chemosensory signal transmitted to the N1M cells through the median lip nerves is suppressed during CTA, resulting in a decrease of the feeding response to sucrose.

Figure 20.4 summarizes the main differences between the cellular changes resulting from single-trial food-reward conditioning and CTA learning. In both paradigms, sucrose was used as a salient feeding stimulus. However, in food-reward conditioning, sucrose acts as the US paired with a neutral chemical stimulus as the CS, whereas in the CTA paradigm it acts as the CS paired with an aversive chemical stimulus as the US. A key target for plastic

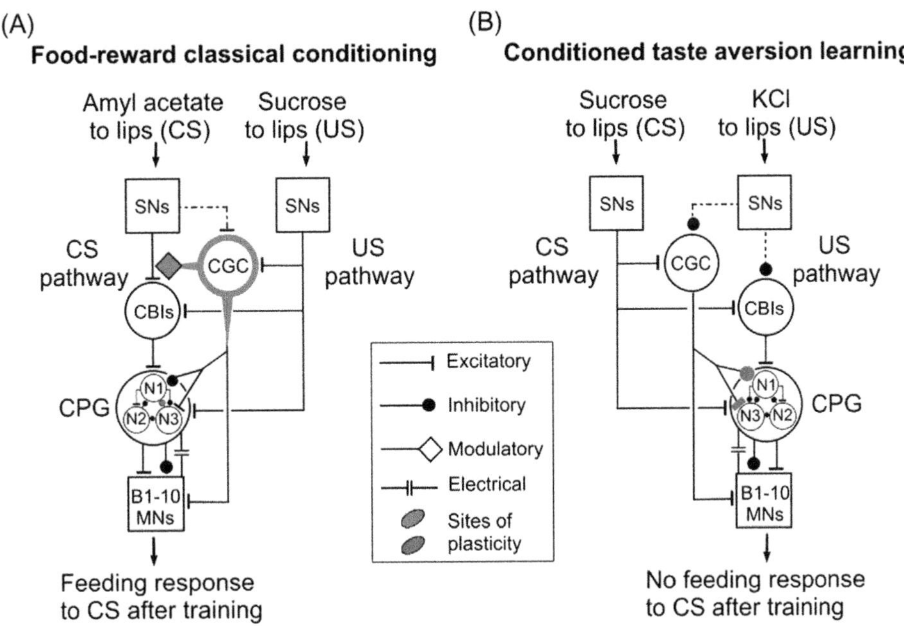

FIGURE 20.4 A comparison of sites of plasticity in the *Lymnaea* feeding system after food-reward versus conditioned taste aversion learning. (A) Single-trial chemical classical conditioning of feeding. In naive animals, the excitatory connections between the CS chemosensory neurons (SNs) and the command-like cerebral–buccal interneurons (CBIs) are weak, the CGCs are at their normal membrane potential (~−65 mV), and the presynaptic modulatory input from the CGCs to the SNs (mediated by an axonal side branch in the cerebral ganglia) is inactive or weak. During training, the food unconditioned stimulus (US; sucrose to lips) activates the feeding CPG via direct and indirect excitatory inputs (via the CBIs) to produce an unconditioned feeding response. In conditioned animals, more than 20 hr after training, the CGC soma and proximal axon segments are depolarized compared to naive and unpaired control animals. This leads to an enhancement of the CGC presynaptic modulatory inputs to SNs and a consequent strengthening of the SN-to-CBI excitatory synapse, which enables a conditioned feeding response to the CS. The bold green outlines of the CGC soma and axonal branches indicate learning-induced nonsynaptic plasticity (membrane depolarization), which increases the strength of CGC-to-SN and SN-to-CBI synaptic connections in a "remote-controlled" manner. The fully tapered bold green line over the main axonal branch indicates that the effect of somal depolarization does not spread onto the more distal connections of the CGC with the CPG or motoneurons, leaving the normal modulatory function of the CGC unaffected by learning. In *in vitro* single-trial food-reward conditioning experiments, a weakening of the N3 tonic to N1M inhibitory synaptic input was also observed after training. (B) Conditioned taste aversion learning. In naive animals, sucrose to the lips activates the feeding CPG via direct and indirect excitatory inputs (via the CBIs) to produce a feeding response. During training, sucrose to the lips (used here as the CS) is paired with the application of an aversive chemical US, KCl. In conditioned animals, the CGC-to-N1M synaptic inhibitory inputs are enhanced both directly and by the mediation of the N3t interneuron, which blocks the feeding response to the CS at the level of the CPG. The putative CS-to-CGC pathways are not yet fully elucidated for either amyl acetate or KCl. The possible KCl-to-CBI inhibitory pathway also remains to be investigated (dashed lines). Therefore, the pathways linking the aversive US to the feeding system are currently unknown. *Data sources: Panel A, I. Kemenes et al.,*[23] *and Marra et al.*[22]*; panel B, Kojima et al.*[11]

changes in both types of learning is the CGC.[11,20] However, LTM after single-trial food-reward conditioning primarily involves nonsynaptic plasticity in the CGCs affecting the CS sensory neuron-to-CBI pathway, whereas CTA learning predominantly leads to synaptic plasticity affecting the CGC-to-CPG pathway.

A Direct Comparison of the Circuit Mechanisms Underlying Reward and Aversive Conditioning of Feeding

The same study that compared the molecular mechanisms underlying reward and aversive classical conditioning in the snail *Lymnaea*[77] showed that there were also differences in the underlying mechanisms at the circuit level. Conditioning-dependent changes in excitatory and inhibitory pathways were revealed in electrophysiological experiments using semi-intact preparations. The experimenters took advantage of the distinct anatomical locations of the inhibitory and excitatory pathways to analyze their role in aversive and reward conditioning.[79,80] Reducing the CNS to the buccal and cerebral ganglia in naive snails eliminates inhibitory pathways that mediate aversive responses while leaving the excitatory chemosensory pathways intact.[80] After aversive conditioning, removal of the same ganglia eliminated the aversive memory trace. At the concentration of 0.004% used in the study by Kemenes *et al.*,[77] amyl acetate has no inhibitory effects on feeding in naive snails,[80] so it seems likely that aversive conditioning is due to an increase in the strength of the inhibitory pathways, which are weak or absent in the naive snail. Removal of the inhibitory circuits had no effect on the ability of the CS to elicit increased feeding responses after reward conditioning, and statistical comparisons between the feeding responses with and without the rest of the brain showed no differences in the size of the response. This suggests that an increase in the excitatory pathways, rather than a reduction in the strength of inhibitory pathways, underlies behavioral reward conditioning.

CONCLUSIONS

Studies based on the classical conditioning paradigms described in this chapter have yielded valuable information on a variety of general and specific molecular and cellular mechanisms contributing to memory consolidation, maintenance, retrieval, and reconsolidation. Memory consolidation after food-reward classical conditioning and reconsolidation after retrieval share the same general molecular requirement (synthesis of new proteins), but the requirement for specific signaling molecules (e.g., PKA) for reconsolidation depends on how recent or remote the consolidated memory is relative to the time of training. Studies of LTM in classical conditioning have emphasized the importance of regulation of gene expression by transcription factors such as CREB and C/EBP and the role of the PKA and MAPK signaling pathways as well as NMDA receptors, CaMKII and PACAP, with the most detailed information gained from experiments using the single-trial food-reward classical conditioning paradigm. The importance of these molecular pathways in *Lymnaea* provides further evidence for the generality of these highly conserved mechanisms in learning, both across phylogenetic groups and across learning paradigms (nonassociative or associative, single- or multitrial, aversive or reward, and operant or classical). NO has been shown to be important for memory consolidation in classical reward conditioning, and this appears to involve regulation of nNOS gene expression in a specific modulatory cell type, the CGCs. The CGCs are key sites of plasticity after both food-reward and conditioned taste aversion learning using a chemical conditioned stimulus, but another type of modulatory neuron, CV1, is key to plastic network-level changes after food-reward conditioning using a tactile conditioned stimulus. Work using *Lymnaea* as an experimental model has thus both provided important insights into the molecular mechanisms of associative learning and memory at the behavioral level and linked these mechanisms to learning-induced cellular and molecular changes in single identified neurons.

References

1. Benjamin PR, Rose RM. Central generation of bursting in the feeding system of the snail, *Lymnaea stagnalis*. *J Exp Biol*. 1979;80:93–118.
2. McCrohan CR, Benjamin PR. Patterns of activity and axonal projections of the cerebral giant cells of the snail, *Lymnaea stagnalis*. *J Exp Biol*. 1980;85:149–168.
3. McCrohan CR, Benjamin PR. Synaptic relationships of the cerebral giant cells with motoneurones in the feeding system of *Lymnaea stagnalis*. *J Exp Biol*. 1980;85:169–186.
4. Rose RM, Benjamin PR. The relationship of the central motor pattern to the feeding cycle of *Lymnaea stagnalis*. *J Exp Biol*. 1979;80:137–163.
5. Alexander JE, Audesirk TE, Audesirk GJ. One-trial reward learning in the snail *Lymnea stagnalis*. *J Neurobiol*. 1984;15:67–72.
6. Alexander JE, Audesirk TE, Audesirk GJ. Rapid, nonaversive conditioning in a freshwater gastropod: II. Effects of temporal relationships on learning. *Behav Neural Biol*. 1982;36:391–402.
7. Audesirk TE, Alexander JE, Audesirk GJ, Moyer CM. Rapid, nonaversive conditioning in a freshwater gastropod: I. Effects of age and motivation. *Behav Neural Biol*. 1982;36:379–390.
8. Kemenes G, Benjamin PR. Appetitive learning in snails shows characteristics of conditioning in vertebrates. *Brain Res*. 1989;489: 163–166.
9. Kemenes G, Benjamin PR. Goal-tracking behavior in the pond snail, *Lymnaea stagnalis*. *Behav Neural Biol*. 1989;52:260–270.
10. Kemenes G, Benjamin PR. Training in a novel environment improves the appetitive learning performance of the snail, *Lymnaea stagnalis*. *Behav Neural Biol*. 1994;61:139–149.
11. Kojima S, Nanakamura H, Nagayama S, Fujito Y, Ito E. Enhancement of an inhibitory input to the feeding central pattern generator in *Lymnaea stagnalis* during conditioned taste-aversion learning. *Neurosci Lett*. 1997;230:179–182.
12. Benjamin PR. Distributed network organization underlying feeding behaviour in the mollusc *Lymnaea*. *Neural Syst Circuits*. 2012;2:4.
13. Benjamin PR, Elliott CJH. Snail feeding oscillator: the central pattern generator and its control by modulatory interneurons. In: Jacklet J, ed. *Neuronal and Cellular Oscillators*. New York: Dekker; 1989:173–214.
14. Benjamin PR, Kemenes G, Staras K. *Molluscan nervous systems*. *Encyclopedia of Life Sciences*. London: Wiley; 2005.
15. Elliott CJ, Susswein AJ. Comparative neuroethology of feeding control in molluscs. *J Exp Biol*. 2002;205:877–896.
16. Kemenes G, Staras K, Benjamin PR. *In vitro* appetitive classical conditioning of the feeding response in the pond snail *Lymnaea stagnalis*. *J Neurophysiol*. 1997;78:2351–2362.
17. Staras K, Kemenes G, Benjamin PR. Neurophysiological correlates of unconditioned and conditioned feeding behavior in the pond snail *Lymnaea stagnalis*. *J Neurophysiol*. 1998;79: 3030–3040.
18. Staras K, Kemenes G, Benjamin PR. Cellular traces of behavioral classical conditioning can be recorded at several specific sites in a simple nervous system. *J Neurosci*. 1999;19:347–357.
19. Whelan HA, McCrohan CR. Food-related conditioning and neuronal correlates in the freshwater snail *Lymnaea stagnalis*. *J Mollusc Stud*. 1996;62:483–494.
20. Kemenes I, Straub VA, Nikitin ES, et al. Role of delayed nonsynaptic neuronal plasticity in long-term associative memory. *Curr Biol*. 2006;16:1269–1279.
21. Fulton D, Kemenes I, Andrew RJ, Benjamin PR. A single time-window for protein synthesis-dependent long-term memory formation after one-trial appetitive conditioning. *Eur J Neurosci*. 2005;21:1347–1358.

22. Marra V, Kemenes I, Vavoulis D, Feng J, O'Shea M, Benjamin PR. Role of tonic inhibition in associative reward conditioning in *Lymnaea*. *Front Behav Neurosci*. 2010;4:161.
23. Kemenes I, Kemenes G, Andrew RJ, Benjamin PR, O'Shea M. Critical time-window for NO-cGMP-dependent long-term memory formation after one-trial appetitive conditioning. *J Neurosci*. 2002;22:1414–1425.
24. Nikitin ES, Vavoulis DV, Kemenes I, et al. Persistent sodium current is a nonsynaptic substrate for long-term associative memory. *Curr Biol*. 2008;18:1221–1226.
25. Ribeiro MJ, Schofield MG, Kemenes I, O'Shea M, Kemenes G, Benjamin PR. Activation of MAPK is necessary for long-term memory consolidation following food-reward conditioning. *Learn Mem*. 2005;12:538–545.
26. Wan H, Mackay B, Iqbal H, Naskar S, Kemenes G. Delayed intrinsic activation of an NMDA-independent CaM-kinase II in a critical time window is necessary for late consolidation of an associative memory. *J Neurosci*. 2010;30(1):56–63.
27. Kemenes G, Kemenes I, Michel M, Papp A, Muller U. Phase-dependent molecular requirements for memory reconsolidation: differential roles for protein synthesis and protein kinase A activity. *J Neurosci*. 2006;26:6298–6302.
28. Michel M, Kemenes I, Müller U, Kemenes G. Different phases of long-term memory require distinct temporal patterns of PKA activity after single-trial classical conditioning. *Learn Mem*. 2008;15:694–702.
29. Carew TJ. Molecular enhancement of memory formation. *Neuron*. 1996;16:5–8.
30. Bourtchouladze R, Abel T, Berman N, Gordon R, Lapidus K, Kandel ER. Different training procedures recruit either one or two critical periods for contextual memory consolidation, each of which requires protein synthesis and PKA. *Learn Mem*. 1998;5:365–374.
31. Epstein HT, Child FM, Kuzirian AM, Alkon DL. Time windows for effects of protein synthesis inhibitors on Pavlovian conditioning in *Hermissenda*: behavioral aspects. *Neurobiol Learn Mem*. 2003;79:127–131.
32. Freeman FM, Rose SP, Scholey AB. Two time windows of anisomycin-induced amnesia for passive avoidance training in the day-old chick. *Neurobiol Learn Mem*. 1995;63:291–295.
33. Grecksch G, Matthies H. Two sensitive periods for the amnesic effect of anisomycin. *Pharmacol Biochem Behav*. 1980;12:663–665.
34. Fulton D, Kemenes I, Andrew RJ, Benjamin PR. Time-window for sensitivity to cooling distinguishes the effects of hypothermia and protein synthesis inhibition on the consolidation of long-term memory. *Neurobiol Learn Mem*. 2008;90:651–654.
35. Ribeiro MJ, Serfozo Z, Papp A, et al. Cyclic AMP response element-binding (CREB)-like proteins in a molluscan brain: cellular localization and learning-induced phosphorylation. *Eur J Neurosci*. 2003;18:1223–1234.
36. Sadamoto H, Sato H, Kobayashi S, et al. CREB in the pond snail *Lymnaea stagnalis*: cloning, gene expression, and function in identifiable neurons of the central nervous system. *J Neurobiol*. 2004;58:455–466.
37. Alberini CM. Genes to remember. *J Exp Biol*. 1997;202: 2887–2891.
38. Fletcher L. Memories are made of this: the genetic basis of memory. *Mol Med Today*. 1997;3:429–434.
39. Frank D, Greenberg ME. CREB: a mediator of long-term memory from mollusks to mammals. *Cell*. 1994;79:5–8.
40. Goda Y. Memory mechanisms: a common cascade for long-term memory. *Curr Biol*. 1995;5:136–138.
41. Josselyn SA, Nguyen PV. CREB, synapses and memory disorders: past progress and future challenges. *Curr Drug Targets CNS Neurol Disord*. 2005;4:481–497.
42. Kandel ER. The molecular biology of memory storage: a dialogue between genes and synapses. *Science*. 2001;294: 1030–1038.
43. Lamprecht R. CREB: a message to remember. *Cell Mol Life Sci*. 1999;55:554–563.
44. Pittenger C, Kandel E. A genetic switch for long-term memory. *C R Acad Sci III*. 1998;321:91–96.
45. Scott R, Bourtchuladze R, Gossweiler S, Dubnau J, Tully T. CREB and the discovery of cognitive enhancers. *J Mol Neurosci*. 2002;19:171–177.
46. Silva AJ, Kogan JH, Frankland PW, Kida S. CREB and memory. *Annu Rev Neurosci*. 1998;21:127–148.
47. Stevens CF. CREB and memory consolidation. *Neuron*. 1994;13: 769–770.
48. Tully T, Bourtchouladze R, Scott R, Tallman J. Targeting the CREB pathway for memory enhancers. *Nat Rev Drug Discov*. 2003;2:267–277.
49. Yin JC, Tully T. CREB and the formation of long-term memory. *Curr Opin Neurobiol*. 1996;6:264–268.
50. Korneev SA, Straub V, Kemenes I, et al. Timed and targeted differential regulation of nitric oxide synthase (NOS) and anti-NOS genes by reward conditioning leading to long-term memory formation. *J Neurosci*. 2005;25:1188–1192.
51. Ribeiro M, Straub VA, Schofield M, et al. Characterization of NO-sensitive guanylyl cyclase: expression in an identified interneuron involved in NO-cGMP-dependent memory formation. *Eur J Neurosci*. 2008;28:1157–1165.
52. Pirger Z, László Z, Kemenes I, Tóth G, Reglodi D, Kemenes G. A homolog of the vertebrate pituitary adenylate cyclase-activating polypeptide is both necessary and instructive for the rapid formation of associative memory in an invertebrate. *J Neurosci*. 2010;30:13766–13773.
53. Pirger Z, Laszlo Z, Hiripi L, et al. Pituitary adenylate cyclase activating polypeptide (PACAP) and its receptors are present and biochemically active in the central nervous system of the pond snail *Lymnaea stagnalis*. *J Mol Neurosci*. 2010;42:464–471.
54. Trudeau LE, Castellucci VF. Contribution of polysynaptic pathways in the mediation and plasticity of *Aplysia* gill and siphon withdrawal reflex: evidence for differential modulation. *J Neurosci*. 1992;12:3838–3848.
55. Brembs B, Lorenzetti FD, Reyes FD, Baxter DA, Byrne JH. Operant reward learning in *Aplysia*: neuronal correlates and mechanisms. *Science*. 2002;296:1706–1709.
56. Nargeot R, Baxter DA, Patterson GW, Byrne JH. Dopaminergic synapses mediate neuronal changes in an analogue of operant conditioning. *J Neurophysiol*. 1999;81:1983–1987.
57. Jones NG, Kemenes I, Kemenes G, Benjamin PR. A persistent cellular change in a single modulatory neuron contributes to associative long-term memory. *Curr Biol*. 2003;13(12): 1064–1069.
58. Nikitin ES, Kiss T, Staras K, O'Shea M, Benjamin PR, Kemenes G. Persistent sodium current is a target for cAMP-induced neuronal plasticity in a state-setting modulatory interneuron. *J Neurophysiol*. 2006;95:453–463.
59. Vavoulis DV, Nikitin ES, Kemenes I, et al. Balanced plasticity and stability of the electrical properties of a molluscan modulatory interneuron after classical conditioning: a computational study. *Front Behav Neurosci*. 2010;4:19.
60. Awatramani GB, Price GD, Trussell LO. Modulation of transmitter release by presynaptic resting potential and background calcium levels. *Neuron*. 2005;48:109–121.
61. Dudai Y, Eisenberg M. Rites of passage of the engram: reconsolidation and the lingering consolidation hypothesis. *Neuron*. 2004;44:93–100.

62. Dudai Y. Molecular bases of long-term memories: a question of persistence. *Curr Opin Neurobiol*. 2002;12:211–216.
63. Morris RG, Inglis J, Ainge JA, et al. Memory reconsolidation: sensitivity of spatial memory to inhibition of protein synthesis in dorsal hippocampus during encoding and retrieval. *Neuron*. 2006;50:479–489.
64. Nader K. Memory traces unbound. *Trends Neurosci*. 2003;26: 65–72.
65. Nader K, Schafe GE, Ledoux JE. The labile nature of consolidation theory. *Nat Rev Neurosci*. 2000;1:216–219.
66. Sara SJ. Retrieval and reconsolidation: toward a neurobiology of remembering. *Learn Mem*. 2000;7:73–84.
67. Ito E, Kobayashi S, Sadamoto H, Hatakeyama D. Associative learning in the pond snail, *Lymnaea stagnalis*. *Zoolog Sci*. 1999;16: 711–723.
68. Kojima S, Hosono T, Fujito Y, Ito E. Optical detection of neuromodulatory effects of conditioned taste aversion in the pond snail *Lymnaea stagnalis*. *J Neurobiol*. 2001;49:118–128.
69. Benjamin PR, Staras K, Kemenes G. A systems approach to the cellular analysis of associative learning in the pond snail *Lymnaea*. *Learn Mem*. 2000;7:124–231.
70. Lukowiak K, Sangha S, McComb C, et al. Associative learning and memory in *Lymnaea stagnalis*: how well do they remember?. *J Exp Biol*. 2003;206:2097–2103.
71. Nakamura H, Kobayashi S, Kojima S, Urano A, Ito E. PKA-dependent regulation of synaptic enhancement between buccal motor neurons and its regulatory interneuron in *Lymnaea stagnalis*. *Zoolog Sci*. 1999;16:387–394.
72. Sadamoto H, Kitahashi T, Fujito Y, Ito E. Learning-dependent gene expression of CREB1 isoforms in the molluscan brain. *Front Behav Neurosci*. 2010;4:25.
73. Hatakeyama D, Sadamoto H, Watanabe T, et al. Requirement of new protein synthesis of a transcription factor for memory consolidation: paradoxical changes in mRNA and protein levels of C/EBP. *J Mol Biol*. 2006;356:569–577.
74. Grinkevich LN. Formation of C/EBP transcription factors and possible pathways for controlling their activity during learning in *Helix*. *Neurosci Behav Physiol*. 2002;32:33–39.
75. Niehof M, Manns MP, Trautwein C. CREB controls LAP/C/EBP beta transcription. *Mol Cell Biol*. 1997;17:3600–3613.
76. Kemenes G, Staras K, Benjamin PR. Multiple types of control by identified interneurons in a sensory-activated rhythmic motor pattern. *J Neurosci*. 2001;21:2903–2911.
77. Kemenes I, O'Shea M, Benjamin PR. Different circuit and monoamine mechanisms consolidate long-term memory in aversive and reward classical conditioning. *Eur J Neurosci*. 2011;1: 143–152.
78. Staras K, Győri J, Kemenes G. Voltage-gated ionic currents in an identified modulatory cell type controlling molluscan feeding. *Eur J Neurosci*. 2002;15:109–119.
79. Straub VA, Styles BJ, Ireland JS, O'Shea M, Benjamin PR. Central localization of plasticity involved in appetitive conditioning in *Lymnaea*. *Learn Mem*. 2004;11:787–793.
80. Straub VA, Kemenes I, O'Shea M, Benjamin PR. Associative memory stored by functional novel pathway rather than modifications of preexisting neuronal pathways. *J Neurosci*. 2006;26: 4139–4146.
81. Benjamin PR, Staras K, Kemenes G. What roles do tonic inhibition and disinhibition play in the control of motor programs? *Front Behav Neurosci*. 2010;4:30.
82. Jones NG, Kemenes G, Benjamin PR. Selective expression of electrical correlates of differential appetitive classical conditioning in a feeding network. *J Neurophysiol*. 2001;85:89–97.
83. Yeoman MS, Kemenes G, Benjamin PR, Elliott CJ. Modulatory role for the serotonergic cerebral giant cells in the feeding system of the snail, *Lymnaea*: II. Photoinactivation. *J Neurophysiol*. 1994;72:1372–1382.
84. Yeoman MS, Pieneman AW, Ferguson GP, Ter Maat A, Benjamin PR. Modulatory role for the serotonergic cerebral giant cells in the feeding system of the snail, *Lymnaea*: I. Fine wire recording in the intact animal and pharmacology. *J Neurophysiol*. 1994;72:1357–1371.
85. Kemenes G, Elliott CJ. Analysis of the feeding motor pattern in the pond snail, *Lymnaea stagnalis*: photoinactivation of axonally stained pattern-generating interneurons. *J Neurosci*. 1994;14: 153–166.

CHAPTER

21

Operant Conditioning of Respiration in *Lymnaea*

The Environmental Context

Ken Lukowiak and Sarah Dalesman

University of Calgary, Calgary, Alberta, Canada

INTRODUCTION

The ability of animals to learn and remember during the course of their lifetime enables them to adapt to various environmental changes, which will directly affect their fitness. Stress modulates both the ability to learn and the ability to form memory, and it also affects the ability to recall memory. The 'fact' that stress alters memory has been a part of the scientific literature since the time of Bacon.[1] How stress alters memory can be seen most easily if one examines the empirical data gathered from studies on rodents and presented in what is now called the Yerkes–Dodson law[2,3] (Figure 21.1). This 'law' shows that stress is an important element in determining both whether information becomes stored as long-term memory (LTM) and for how long. Too much or too little stress impedes LTM formation, whereas a certain amount of 'good stress' enhances LTM formation. Organisms might decide to only expend the 'neuronal cost' (e.g., altered gene activity and new protein synthesis) necessary to form LTM to 'relevant' events. An important factor that helps determine whether a specific 'event' will be encoded into memory is the level of stress perceived by the organism at the time of the occurrence and how the stress relates to the event. Because memory is dynamic, stress and traumatic events have substantial modulatory effects on memory, including false memory and post-traumatic stress syndrome.[4–6]

Here, stress is defined as any significant state that requires a physiological, psychological, or behavioral readjustment or modification necessary to maintain the well-being of the organism.[7–10] Some of the stressors that we encounter in our lives may be physical (e.g., heat or cold shock) or psychological (e.g., public speaking), and these sorts of stressors have the capability to alter memory formation and/or its recall. The effects of stress on memory have been studied in a number of different model organisms, with sometimes contradictory results.[2] That is, in some instances, memory may be enhanced, whereas in others its formation or its recall is blocked. In vertebrates, this is often dependent on the sex of the model system[11] and can be altered by other factors, such as whether the animal has given birth.[12] Given the complexities of the vertebrate brain and the different behaviors tested, in addition to the diverse ways stressors act on memory formation, disagreement in the literature regarding the role that stress plays in LTM formation is not too surprising. We have sought to overcome these difficulties by using a simpler model system, the hermaphroditic pond snail *Lymnaea stagnalis,* and a relatively simple behavior, aerial respiration, which undergoes associative learning and LTM formation.[13–16] We have also utilized ecologically and behaviorally relevant stressors in our experiments to obtain a biologically relevant understanding of how stressors that snails encounter in their natural environment alter memory formation, its strength/persistence, and its recall.[7,8]

Although related, learning and memory are two separate and distinct processes.[17] Behaviorally, learning is the acquisition of a new or altered behavior, whereas memory refers to the retention of what is learned. There are fundamental molecular differences

Invertebrate Learning and Memory.
DOI: http://dx.doi.org/10.1016/B978-0-12-415823-8.00021-6

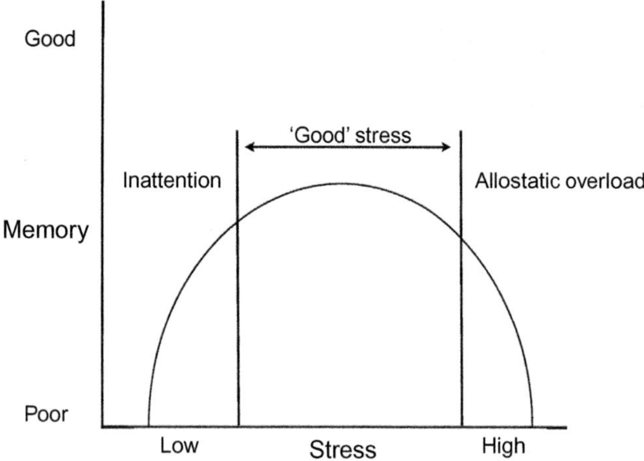

FIGURE 21.1 An idealized version of the so-called Yerkes–Dodson law. Plotted on the x-axis is the level of perceived stress by the subject at some given time. The perceived level of stress runs from 'low' to 'high.' On the y-axis is plotted memory, ranging from 'poor' to 'good.' At low levels of stress, there is 'inattention,' and this results in a 'poorer' memory. At high levels of stress, there is 'allostatic overload' resulting in poor memory. Allostatic overload simply means that the subject is finding it difficult to cope with the stressor and maintain homeostasis. In between the high and low stress levels is the region of the curve, which we have labeled 'good' stress. We define this 'good' stress level as that which we can cope with but which is sufficient to keep our attention. Thus, some levels of stress memory will be better than other levels. This curve can change depending on the age of the individual, the individual's previous history, and the difficulty of the task.

between learning and memory. The process that leads to the formation of memory following learning has come to be known as consolidation. The consolidation hypothesis postulated that after learning occurred, memory existed in a fragile state and was vulnerable to interference. With time, the memory was stabilized and became insusceptible to interference by amnesiac treatments. We now know, however, that a recalled memory can sometimes be modified or blocked entirely—that is, during the process of reconsolidation.[18]

In 1996, we published our first paper[15] describing the ability of aerial respiration in *Lymnaea* to be operantly conditioned (a form of associative learning) and to then form LTM. We initiated these studies because we knew a great deal about the underlying neuronal circuitry, a three-neuron central pattern generator (CPG) that drives this behavior (Figure 21.2). In fact, this may be the only neuronal circuit that drives a tractable behavior where both the sufficiency and the necessity of the circuit have been experimentally worked out.[19,20] We believe that we have come a long way since 1996 in achieving many of our goals. For example, we have demonstrated that one of the CPG neurons, Right Pedal Dorsal 1 (RPeD1)—the neuron that initiates rhythmogenesis—is a necessary site for LTM formation, reconsolidation, extinction, and forgetting.[18,21–23] These topics have been described both in original publications and in a number of review papers,[24,25] and we do not discuss these data in any detail here.

We extended these findings to show that the 'state' of RPeD1 activity is very well correlated with whether or not the snail has received some training resulting in memory formation.[26] In addition, in a strain of *Lymnaea* that possesses enhanced abilities to form LTM—that is, smart snails[27]—RPeD1 activity is much reduced in naive preparations made from these snails compared to RPeD1 activity in naive preparations from strains that do not exhibit enhanced LTM-forming capabilities.[28] Moreover, RPeD1 activity in the naive smart snails appears to resemble RPeD1 activity in the typical snails that have already received some training that eventually may lead to LTM formation.[26,28]

In the natural environment, suboptimal conditions may act as a stressor on the snails and alter their cognitive abilities. Our overall basic finding (i.e., the subject of this chapter) is that the ability of *Lymnaea* to form LTM following associative learning is pliable because memory formation, its persistence, and its recall are all significantly altered by ecologically relevant environmental stressors. Whether memory formation and its maintenance is enhanced or blocked depends on (1) the nature of the stressor; (2) the timing of the application of the stressor relative to both the learning procedure and the consolidation process; (3) how the stressor is perceived by the snail; (4) when the stressor is encountered; and (5) whether the individual stressors are applied separately or in combination, which is rarely tested.

The bulk of learning and memory studies involving *Lymnaea*[7,8,25,29–31] have utilized laboratory-bred specimens derived from snails originally collected (in the 1950s; >250 generations grown in the lab) from ditches in polders near Utrecht in The Netherlands. These are referred to here as 'lab-bred' snails. However, the Lukowiak lab also uses freshly collected as well as offspring of these freshly collected *Lymnaea* stagnalis (i.e., F1's and F2's bred and grown under laboratory conditions) obtained from ponds in Alberta, Canada, and in southwest England (the Somerset Levels), which provide insight into how different populations (i.e., strains) have differing capacities to form memory and deal with stress. In this chapter, we (1) discuss on how ecologically relevant stressors alter LTM formation, its strength, and its persistence; (2) examine how snails respond when faced with more than one stressful stimulus; and (3) describe the strain differences in LTM-forming

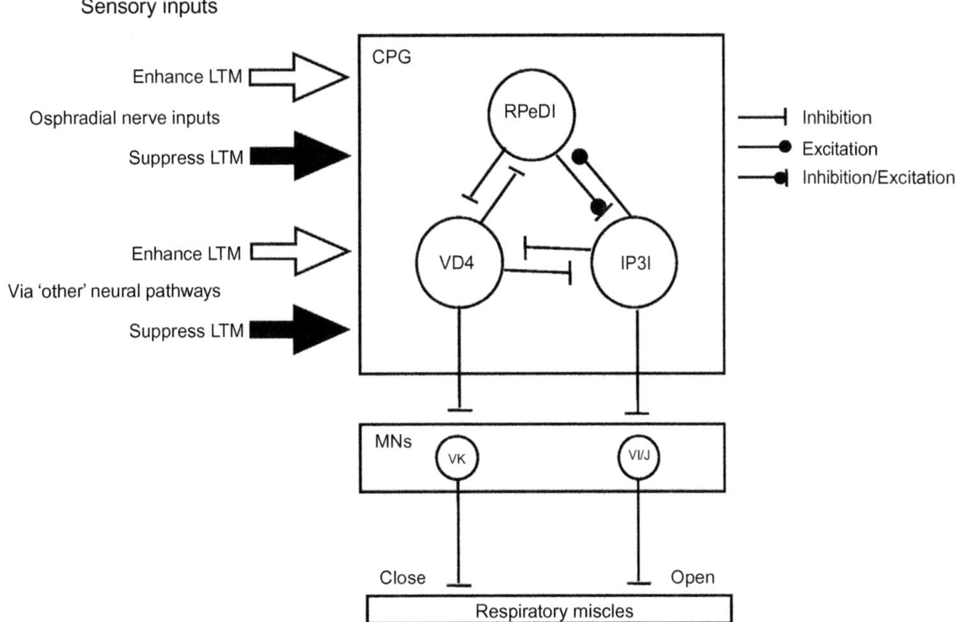

FIGURE 21.2 The central pattern generator (CPG) that drives aerial respiration in *Lymnaea* and the sensory influences that impinge on it. A schematic diagram of the three-neuron respiratory CPG and its follower cells. Depolarization of RPeD1 excites IP3 (inhibition followed by excitation) and inhibits VD4 (inhibition). Once activated, IP_3 excites RPeD1 and a group of motor neurons (VI/J cells) involved in pneumostome opening. IP_3 also inhibits VD4, the interneuron involved in pneumostome closing. The combined inhibitory input from both IP_3 and RPeD1 causes VD4 to fire a burst of action potentials. This bust of activity in VD4 inhibits both RPeD1 and IP_3 and excites a group of motor neurons (VK cells) involved in pneumostome closing. This begins the cycle of alternate bursting activity in VD4 and IP_3 that underlies the generation of the respiratory rhythm. On the left-hand side of the figure are sensory inputs that we have experimentally determined to alter memory formation. Sensory input from the osphradium, a peripheral sensory structure, can either enhance (open arrows) or suppress LTM formation. Thus, predator detection (i.e., the 'smell' associated with crayfish), which enhances LTM formation, is mediated via osphradial input. Detection of low environmental levels of Ca^{2+} in pond water suppresses LTM formation, and this again is mediated via osphradial inputs (solid arrows). We have also found that other sensory inputs not carried by the osphradial nerve alter LTM formation. Again, the same scheme is used; open arrows enhance and solid arrows suppress LTM formation. For example, crowding, which blocks LTM formation, is not mediated by osphradial inputs.

capabilities and our new research into how different strains respond to stress.

In 1929, August Krogh[32] stated, "For such a large number of problems there will be some animal of choice, or a few such animals, on which it can be most conveniently studied." As this 'Krogh principle'[33] has been more widely applied, it has resulted in numerous successes, including the use of mollusks (e.g., *Aplysia californica*), in the field of learning and memory.[34] With regard to how stress alters memory formation and the within-strain variation in memory-forming abilities, a different mollusk, *L. stagnalis*, may be Krogh's "animal of choice." *Lymnaea* has frequently been used to study learning and memory due to a relatively simple neuronal network that can be linked to tractable behaviors [for reviews, see 25,35,36]. We have made use of aerial respiratory behavior in the *Lymnaea* model system because (1) the neural circuit that mediates this tractable behavior is well understood; (2) this behavior can be operantly conditioned, and long-lasting memory can be demonstrated; and (3) it is possible to demonstrate that a single neuron in the snail is a necessary site of LTM formation.

AERIAL RESPIRATORY BEHAVIOR

Lymnaea are bimodal breathers, which means that they can exchange O_2 and CO_2 cutaneously (through their skin) or aerially through an orifice (i.e., the pneumostome).[15] To aerially respire, the snail approaches the water surface, opens its pneumostome, and begins gas exchange by contracting and relaxing mantle muscles. The CPG that drives this behavior has been thoroughly characterized and proven both necessary and sufficient for aerial respiratory behavior in *Lymnaea*. In a hypoxic environment, the frequency of aerial respiration increases significantly.[37,38] The respiratory movements are often repeated several times before the snail submerges. Aerial respiration in *Lymnaea* is thus a simple, rhythmical, and spontaneously occurring behavior, whose breathing frequency can be modulated by simple alterations of its environment.

OPERANT CONDITIONING OF AERIAL RESPIRATORY BEHAVIOR

We have studied LTM formation in *Lymnaea* following operant conditioning of aerial respiration (referred to here as the 'typical training procedure')[15] or a one-trail training (1-TT) procedure.[39] In the typical training procedure, we place snails into a beaker filled with hypoxic pond water (PW). The PW is made hypoxic by bubbling N_2 gas through it for 20 min. In hypoxic conditions, snails crawl to the air–water interface to open their pneumostome and perform gas exchange with the external environment. As they begin to open the pneumostome, we apply a negative reinforcement in the form of a gentle tactile stimulus. We call this a 'poke.' The poke is of sufficient strength to cause pneumostome closure but does not elicit the whole-animal withdrawal response. With repeated pokes (over 30 or 45 min), snails reduce the number of attempted pneumostome openings when placed in the hypoxic context. Memory is operationally defined depending on the specific training procedure used, but it always results in the depression of breathing behavior relative to the naive state. In some of the studies reviewed in this chapter, snails received two operant training sessions separated by a 1-hr interval before being tested for memory some time later, or they received only a single 0.5-hr training session with memory tested at specific times thereafter.

In the 1-TT procedure, we assess memory in a different manner. The total breathing time in the 30-min pre-observation session (Pre-Obs) is compared to that in the 30-min post-observation session (Post-Obs) to assess memory formation. The 1-TT procedure consists of the following: snails are again placed in the hypoxic training beaker; however, the first time each snail initiates pneumostome opening, it is removed and immediately placed in a stressor (e.g., 25 mM KCl, quinine, or garlic) and left there for 30–35 sec. Snails are then returned to their home eumoxic aquarium for 24 hr to test for LTM and then tested for memory by another 30-min observation session in the hypoxic training beaker (Post-Obs). Snails are said to have memory after 1-TT if the total breathing time in the post-training observation session (Post-Obs) is significantly less than that in the pre-training observation session (Pre-Obs).

A number of important controls are utilized in all of the memory studies discussed here in order to ensure that the behavioral changes observed are a result of associative learning. The first of these is the yoked control group. A yoked control snail is given a noncontingent stimulus (i.e., irrespective of pneumostome activity) every time a snail from the experimental cohort receives a contingent stimulus (i.e., a stimulus when pneumostome opening is initiated). As a result, the experimental snail associatively learns and forms memory, but the yoked control snail cannot because there is no association present. The yoked control procedure serves to control for phenomena that may alter behavior or the internal state of the snail, such as (1) stress associated with stimulus application, (2) handling, and (3) the training environment. In addition to the yoked control, a 'change of context' procedure is often used to ensure the behavior of snails is not altered by experimental conditions. When trained snails are placed in a novel context following training, they respond as if they were naive. Typically, we train snails in hypoxic PW that lacks the presence of an odorant (i.e., the standard context). To create a 'novel context,' N_2 is first bubbled through a flask containing, for example, a slurry of carrot and water. The resulting carrot-odorant N_2 is bubbled into the training beaker prior to training or testing.

We are able to easily manipulate the conditions in which snails are maintained and trained to assess how different environmental conditions (e.g., predators, differences in [Ca^{2+}] PW, and crowding) alter the ability of snails to form LTM and alter the activity of the RPeD1 neuron, which is necessary for LTM formation.

ECOLOGICALLY RELEVANT STRESSORS AND LTM FORMATION

Serendipity is a wonderful thing in science. In the summer of 2006, we experienced a major rupture of pipes above our snail tanks in the culture room, and chlorinated tap water flooded into these aquaria. Many snails died, and the rest were not healthy. We needed to perform experiments, and it so happened that a graduate student at the time (Mike Orr) was visiting his father's horse ranch in southern Alberta, where he came across a number of ponds alongside the Belly River that contained *Lymnaea*. The freshly collected *Lymnaea* (Belly snails) were brought back to the lab and we used them. However, we found that these *Lymnaea* possessed superior memory-forming capabilities compared to the snails we had been using (i.e., Dutch 'lab-bred' snails).[40]

Our first hypothesis was that the Belly snails were reared in an enriched environment compared to that of the lab-reared snails. That is, rearing and living in a laboratory setting resulted in a 'dumbing down' of the ability of snails to form LTM. The obvious experiment was to enrich the environment in the lab. We chose to do this by placing a predator (a crayfish) in the aquarium along with the snails. We found that the lab-reared snails trained in crayfish effluent (CE) had enhanced memory-forming capabilities that, although not quite as good as those of the Belly snails, were much better than anything we had ever seen from our lab-reared snails.[41] Thus, it appeared that our hypothesis was proven, except for two additional findings: (1) Eggs collected from Belly snails and reared in the lab continued to have superior memory-forming capabilities compared to Dutch lab-reared snails,[42] and (2) freshly collected snails from canals in polders in Utrecht possessed similar memory-forming capabilities as those of the lab-reared snails.[40] What these data did show was that environmentally relevant stressors could alter LTM formation and that there are inherent differences in different strains of *Lymnaea* to form memory.[43,44] We therefore spent the next few years examining (1) the effects of various 'natural' stressors on the snail's ability to form LTM and (2) studying the differences in innate ability to form LTM in different strains of *Lymnaea*.

WHAT IS STRESSFUL FOR A SNAIL?

Presumably, the imminent threat of death is the most stressful situation in life. In the face of predation, many species rely on experience to adjust their behavior relevant to current risk,[45] and *Lymnaea* is no different, showing that it is able to learn about predation risk.[46] Therefore, the enhancement in memory in the presence of chemicals from a predator is perhaps unsurprising.[41] However, we found that whether a predator enhances memory or not depends on the snail's place of origin; only kairomones from sympatric predators cause memory enhancement.[43]

Early work in our assessment of the effects of predator cues on *Lymnaea* indicated that exposure to CE caused changes in the activity of RPeD1 in the Dutch lab-bred strain, depressing firing and burst rate, such that activity in this neuron became more like that seen in a trained snail.[47] We have begun to assess the mechanism by which predator cues modulate activity in RPeD1. Our first target was to find the sensory system used in memory modulation. An obvious candidate to test was the osphradium, an external chemosensory organ that responds to a wide range of chemical stimuli.[48,49] We severed the osphradial nerve proximal to the osphradium, preventing input from this organ to the central nervous system (CNS). In the absence of osphradial input, we no longer saw depression in the activity of RPeD1 or memory enhancement in response to CE,[50,51] although the snails were still able to form LTM. We also hypothesized that serotonin may be involved in the pathway signaling predator stress in *Lymnaea*. Injecting mianserin, which blocks serotonin receptors, was also effective in blocking the memory-enhancing effects of CE but again did not prevent LTM formation.[50] Further work is required to elucidate whether blocking serotonin receptors blocks signaling pathways from the osphradium or affects the way these signals are processed in the CNS.

RESOURCE RESTRICTION

Memory formation is thought to be a costly process[52,53]; hence, when energy intake is restricted, we might predict that LTM formation would be blocked if the animal were attempting to conserve energetic resources. Contrary to this prediction, short periods of food restriction from 1 to 5 days can enhance memory associated with appetitive[35] or aversive[54] food conditioning. Indeed, a food-restriction protocol is deemed necessary to produce consistent results in these studies. However, we proposed that restricting food intake, and therefore reducing the energy available to the snail, may be adverse to non-food-related memory formation. *Lymnaea* were maintained for 5 days prior to training in the absence of food availability, and their response to operant conditioning was compared to that of snails fed *ad libitum*. Unexpectedly, food-deprived snails were equally capable of forming LTM when trained and tested in PW as satiated snails.[55] However, when food-deprived animals were trained in the presence of a food smell, created by bubbling nitrogen through a carrot juice flask before it was bubbled through the training beaker, snails no longer demonstrated learning or memory.

In addition to food requirements, *Lymnaea* requires a good supply of calcium, both to grow its own shell and also during reproduction to provision embryos.[56,57] *Lymnaea* is a calciphile, meaning it absorbs most of its calcium requirements directly from water. Populations of this species are generally found where environmental calcium is 20 mg/L or greater;[58–60] below this level, growth and survival are reduced.[61] Low-calcium environments also result in shell thinning in mollusks, potentially making them easier prey.[58,60,62–64] Considering this, we predicted that despite populations surviving at 20 mg/L [Ca^{2+}], the animals may be experiencing stress. We decided to compare how animals held at this low concentration (20 mg/L) would fair relative to those held in our standard artificial PW containing 80 mg/L calcium.

We initially assessed basic traits, metabolic rate, and activity, comparing animals in low (20 mg/L) versus standard (80 mg/L) calcium PW. We found that basic metabolic rate increased and activity levels decreased in the low-calcium environment.[65] This was not surprising because calcium uptake requires energy when external [Ca^{2+}] is below approximately 50 mg/L.[66] Next, we assessed the effect of calcium concentration on LTM formation using 1-TT.[39] Following even short (~1 hr) periods of exposure to a low-calcium environment, *Lymnaea* was no longer able to form LTM, although it still demonstrated intermediate-term memory (ITM).[67] There was no alteration in the electrophysiological activity of RPeD1 in naive animals relative to calcium availability, but significant differences between calcium environments were apparent in RPeD1 24 hr following training. Whereas in the standard calcium environment RPeD1 activity was significantly depressed relative to the naive state, in low calcium only a partial change in activity occurred, insufficient to result in a significant alteration in breathing behavior.[68]

We previously hypothesized that forgetting operant conditioning in *Lymnaea* is due to the conflicting association that occurs as a result of 'new learning and memory formation' as the snail is able to breathe freely in the aquaria (i.e., retrograde interference). This theory is supported by our data demonstrating that if we prevent aerial respiration, we effectively block forgetting.[21] We considered that we may be able to block forgetting using the same stressful stimuli that blocks memory formation. We first operantly trained animals in a standard calcium environment and then moved *Lymnaea* immediately into a low-calcium environment.[69] We tested memory 72 hr following training, by which point the Dutch laboratory *Lymnaea* in a standard calcium environment have forgotten.[41,70] Following training in standard conditions, *Lymnaea* subsequently held in low-calcium conditions demonstrated LTM 72 hr later. Therefore, low environmental calcium blocked forgetting of a previously learned behavior.[69]

Our recent work assessed the mechanism by which LTM in Dutch *Lymnaea* is blocked by a low-calcium environment. Previous work[71] demonstrated that *Lymnaea* may orientate toward a calcium-rich environment, suggesting that the snail is able to externally sense calcium concentration in its surroundings. Therefore, we assessed the role of the osphradium in sensing external calcium by severing the osphradial ganglion connecting the osphradium to the CNS and comparing them to those of sham-operated controls.[72] In low-calcium conditions, the osphradially cut group formed LTM, equivalent to animals maintained in our standard calcium PW. However, LTM was blocked in the sham-operated animals. Therefore, we conclude that osphradial input is required to mediate the memory-blocking effect of low environmental calcium in *Lymnaea*. Previous work has shown that *Lymnaea* maintains calcium levels in the hemolymph in low-calcium environments.[66,73] Hence, our results demonstrating external modulation of this stress response are well supported by other studies. Because no significant change occurs in RPeD1 activity in naive animals following exposure to low calcium,[67] we conclude that the effect that low calcium has on the CNS is modulated outside the CPG controlling respiratory behavior.

SOCIAL STRESS

Under natural conditions, *Lymnaea* population density fluctuates widely within and between years. For example, sampling within a single site shows hundredfold changes in the number of individuals within the same month in consecutive years.[74] Indeed, in our experience, we have found broad fluctuations in density both between sampling periods and within a site on a single day. Factors causing this patchiness are proposed to be changes in food availability or temperature gradients.[75] Consequently, the social situation of an individual may change rapidly from extremely crowded conditions to physical isolation. Previous work has shown that crowded conditions reduce reproduction and survival in *Lymnaea*,[74,76–78] indicating that this species may well be experiencing stress at high population density. In addition, isolation of individuals significantly alters their copulatory behavior to preferentially mating in the male role.[79] We were interested in whether either of these social situations, crowding or social isolation, acts as a stress that alters learning and memory.

To assess the effects of crowded conditions on memory formation in the Dutch laboratory strain, we subjected them to crowding (defined here as 20 snails per 100 mL) either immediately prior to or following training. We found that as long as snails were crowded immediately before or after the training procedure, whether for 1 or 24 hr, LTM was blocked.[80] However, *Lymnaea* experiencing crowding stress were still able to form ITM. Therefore, the effect that crowding has on memory formation does not block protein synthesis required for ITM, but crowding stress is incompatible with gene transcription required for LTM formation.[81]

Next, we wanted to assess what it is about crowded conditions that blocks LTM formation. Previous studies on how high population density alters growth and embryonic development postulated that the main effect occurs through waterborne chemicals.[82,83] We placed *Lymnaea* in water in which conspecifics had recently been crowded prior to training, but this had no effect on LTM formation.[80] Next, we hypothesized that effects on LTM might be due to physical restrictions on the space available to each animal. We tested this theory by crowding snails in PW containing empty

Lymnaea shells, equal to the number of live animals we have used previously. Again, this did not block LTM formation.[80] In addition, we severed chemosensory input from osphradium but found that *Lymnaea* still responded to crowding as a memory-blocking stress.[72] We are still searching for the sensory system or systems controlling the response to crowding. However, these data clearly demonstrate that two stressors (low calcium and crowding) that produce the same phenotype in the Dutch lab-bred strain, blocking LTM formation, can be modulated via disparate sensory systems.[72]

In addition to overcrowding acting as a stress for *Lymnaea*, we also chose to examine the other end of the social spectrum, specifically social isolation. Social isolation is known to alter learning and memory in a wide range of vertebrates;[84,85] however, the effect in invertebrates is less well studied. Isolating *Drosophila melanogaster* results in a decline in fiber density in mushroom bodies.[86] Similarly, isolation of *Apis mellifera ligustica* results in a decline in mushroom body volume.[87] Although learning and memory were not directly assessed in either of these studies, the mushroom bodies play an important role in learning and memory in insects.[88] In addition, isolating the sea hare, *Aplysia fasciata*, either during or soon after training to avoid inedible food blocks LTM formation.[89] Prolonged isolation of *Lymnaea* during development leads to delayed reproduction and increased size at the onset of reproduction,[79,90] and acute isolation alters reproductive behavior.[79] Therefore, we decided to assess whether an acute period of isolation acts as a stressor that alters learning and memory formation.

Snails were isolated in perforated containers; several such containers were maintained in the same aquaria using similar protocol as that known to alter reproductive behavior.[79] Therefore, there was potential for waterborne chemical cues to pass between the animals, but they were unable to make physical contact. We considered this to be closer to the situation that they might experience in natural populations at very low density. Contrary to our predictions, we found that social isolation did not alter learning and memory formation in *Lymnaea* from the Dutch lab-bred strain.[91] Based on these results, it seems that social isolation does not act as a stressor altering memory formation.

THERMAL STRESS

As a temperate zone species,[92] the thermal environment experienced by *Lymnaea* will vary considerably during the course of a year. Living in shallow, slow-flowing or stagnant freshwater habitats, these animals receive little buffering from climatic conditions.[74] These populations may experience wide temperature fluctuations during the course of a day, experiencing rapid heating up to approximately 28°C and dropping to close to freezing overnight. In the laboratory, the maximum growth rate of *Lymnaea* is achieved between 11 and 28°C,[93] but the lowest mortality occurs between 15.7 and 20.1°C.[77] Laboratory reared *Lymnaea* are typically maintained at a constant temperature, with most work on learning and memory carried out in this optimal 15–20°C range.

It is clear from the natural habitat colonized by *Lymnaea* that this species is able to tolerate a wide range of temperatures; however, outside the 'optimum' temperature range, we might predict that *Lymnaea* is experiencing thermal stress. Cooling *Aplysia* immediately prior to training did not alter LTM formation following food aversion training; however, cooling immediately following training attenuated LTM.[94] We were curious about whether we would see similar effects on LTM formation in *Lymnaea*. Following both operant training and 1-TT, we found that rapidly cooling the animals immediately following training to 4°C for 1 hr blocked both ITM and LTM,[70,95] similar to effects seen following taste aversion learning.[54] However, a 10- to 15-min break between training and cooling was sufficient to allow memory formation. Interestingly, we also found that a very short bout of cooling (10 min at 4°C) immediately following training enhanced memory formation.[95] Our working theory is that longer periods of cooling immediately following training prevent the molecular mechanisms necessary for memory formation in the CNS, whereas a very short bout may act as a thermal shock to the snail, altering memory formation via stress.[95] Exposing animals to a prolonged period of cooling, after a short break following training, blocks forgetting for at least 8 days.[70] However, in addition to blocking forgetting, cooling can also interfere with a residual memory trace formed following training that results in ITM.[68] Following our work on the effects of cooling, we are now starting to assess how a similarly acute exposure to heat (~30°C) affects memory formation and to assess the mechanisms through which temperature alters learning and memory.

ANTHROPOGENIC STRESS

Although they can hardly be classified as 'natural' environmental stressors, anthropogenic-derived pollutants are now common in the freshwater environment. Most work on pollutants uses relatively high or chronic exposures, measuring LC_{50}. Recently, the focus in pollution assessment has moved to measuring behavioral changes, focusing on locomotion and/or respiration.[96] We were interested in whether pollutants would also alter learning and memory. Thus far, we have focused on two classes of pollutant. One is H_2S, considered both an industrial and an environmental pollutant,[97] and the other is toxic 'heavy' metals, specifically Zn and Cd, which are common pollutants around areas of resource extraction.[98–100] In both cases, we chose to assess the effects of acute exposure to very low concentrations, testing for measurable changes in memory formation.

Acute exposure to H_2S in human subjects results in a wide variety of effects, including dizziness, lack of coordination, headache, loss of concentration, and detrimental effects on cognitive function.[101–103] These studies indicate that H_2S may have important effects on learning and memory formation. However, the picture is still unclear. Whereas H_2S exposure may result in memory loss in rodents, for example, at physiological concentrations, H_2S actually facilitates long-term potentiation (LTP) in neuronal structures thought to be necessary for memory formation in an activity- and dose-dependent manner.[104,105] We therefore assessed the effects of H_2S on memory formation following operant conditioning in *Lymnaea* using a range of doses from 50 to 100 μmol/L during training. At lower concentrations (50 and 75 μmol/L), LTM was impaired relative to that of animals trained under controlled conditions.[106] These H_2S-exposed animals were still able to form LTM, but when the 'grade' distribution was analyzed (i.e., the extent of change in breathing behavior), they did not show the same magnitude of behavioral change compared to control animals. However, when *Lymnaea* were exposed to 100 μmol/L H_2S, both learning and memory were blocked. Whereas H_2S is thought to be necessary for LTP in mammals,[105] higher levels of H_2S have been hypothesized to cause the 'metabolic intoxication' seen in the mammalian brain.[107] We propose that this intoxication may explain the dose-dependent response seen on LTM formation in *Lymnaea*.[106]

Aquatic gastropods are sensitive to sublethal levels of heavy metals, demonstrating alteration in behavioral traits, including changes in feeding behavior,[108] locomotion,[109] and failure to exhibit antipredator behavior.[110,111] In addition, aquatic gastropods may be able to sense and avoid low levels of heavy metals in their environment.[112] Effects of a heavy metal on neurological function had been shown following ingestion of very low levels of Pb in rats [reviewed in 113] and may therefore prove a more sensitive measure of toxicity than measures of locomotion and breathing behavior. We chose to assess how two heavy metals, Cd and Zn singularly and in combination, at levels below those allowable in municipal drinking water (Zn, 1100 μg/L; Cd, 3 μg/L) affect learning and memory in *Lymnaea* following an acute exposure (~48 hr) prior to and during training/testing.

We found no effect of Zn or Cd or a combination of the two on locomotory or aerial respiratory behavior.[114] When *Lymnaea* were exposed to either metal in isolation, they were able to learn and form memory equally well as the control animals. However, when exposed to these metals in combination, both learning and long-term memories were blocked.[114] Severing input from the osphradium prevented the memory-blocking effect of combined metals, indicating that the snail is responding to chemosensory information signaling stress in the CNS rather than direct effects of metal toxicity on the nervous system. Preliminary data indicate that this combined effect appears to be synergistic. This work highlights that neuronal stress effects in *Lymnaea* may be far more sensitive to very low levels of heavy metal pollution than other behavioral traits, such as locomotion and respiratory behavior.

INTERACTION BETWEEN STRESSORS

Typically, the effects of stress, particularly in studies of learning and memory, are assessed using single stressors (including much of our work to date). However, in 'real-life,' animals including humans will be faced with multiple sources of stress, both sequentially and in combination. We were very interested in how *Lymnaea* would respond to multiple sources of stress. For example, can we predict how different stressors will interact to alter learning and memory based on their individual effects? To address this question, we have started to combine exposure to different stressors, where we already know their individual effects on learning and memory in *Lymnaea*. The answer appears to be that the interaction of stressors on memory can demonstrate emergent properties.

One of our first experiments assessed the combined effects of CE and crowding. These stressors have opposing effects: When experienced as a single stressor, CE enhances LTM formation,[41] whereas crowding blocks LTM.[80] This experiment aimed to assess whether the enhancing effects on memory seen in CE would still be present following crowding. Operant conditioning of the Dutch laboratory population for 0.5 hr in control conditions results in ITM but not LTM; however, 0.5-hr training in CE results in LTM lasting at least 24 hr.[41] When we crowded *Lymnaea* immediately prior to 0.5-hr training in CE and tested for LTM 24 hr later, we found that the snails did not show memory.[115] The conclusion from this work was that crowding effectively 'trumped' the effects of CE, blocking the memory-enhancing effects of this stressor. Recent (unpublished) work indicates that the interaction between CE and crowding may not be this simple and may be dependent on other environmental factors.

A second memory-blocking stressor that we used in combination is low environmental calcium (20 mg/L). Similar to crowding, low calcium availability blocks the ability of the Dutch laboratory strain *Lymnaea* to form LTM, but they are still able to learn and form ITM.[67,69] When we operantly trained animals that were held in a low-calcium environment in the presence of CE, they were subsequently able to form LTM lasting 24 hr,[116] suggesting that training in CE may prevent the memory-blocking effects of a low-calcium environment. However, in control conditions, the Dutch strain normally forms memory lasting 24–48 hr following two 0.5-hr training sessions, but if exposed to CE during this training regime, LTM then lasts 8 days.[41] To confirm whether CE was preventing the effects of low calcium on memory, we also tested for LTM 72 hr following training in CE. Our results demonstrated that the snails no longer showed LTM 72 hr following training, when we would expect them to still show strong memory retention following training in CE in a standard calcium environment.[116] In fact, in the presence of both stressors, the snails now show an identical memory phenotype to the one they show under control conditions, so the effects of each stressor appear to have effectively canceled each other out. Sensory input for modulation of memory phenotype occurs via the osphradium for both low-calcium conditions[72] and CE.[50,51] However, we doubt that the interactive effect occurs at the level of sensation because *Lymnaea* is clearly still able to sense CE in a low-calcium environment. We consider that it is more likely that it is the manner in which these inputs are processed in the CNS that results in an identical phenotype to control trained animals.

A third environmental factor we decided to consider in combination with other environmental stressors is that of social isolation. This was particularly interesting to us because we were somewhat surprised that this 'stressor' did not alter learning and memory formation.[91] We decided to test how socially isolated *Lymnaea* formed LTM following operant conditioning in the presence of two additional environmental stressors—low calcium, which typically blocks memory formation,[67,69] and CE, which typically enhances memory formation[41,50]—as well as in the presence of a combination of these two stressors, which results in a memory phenotype identical to that of snails trained under control conditions.[116] Socially isolated snails trained in the presence of CE did not demonstrate any difference relative to grouped animals. However, when we isolated *Lymnaea* in low calcium, which normally blocks LTM, they now formed LTM lasting 24 hr (Figure 21.3).[91]

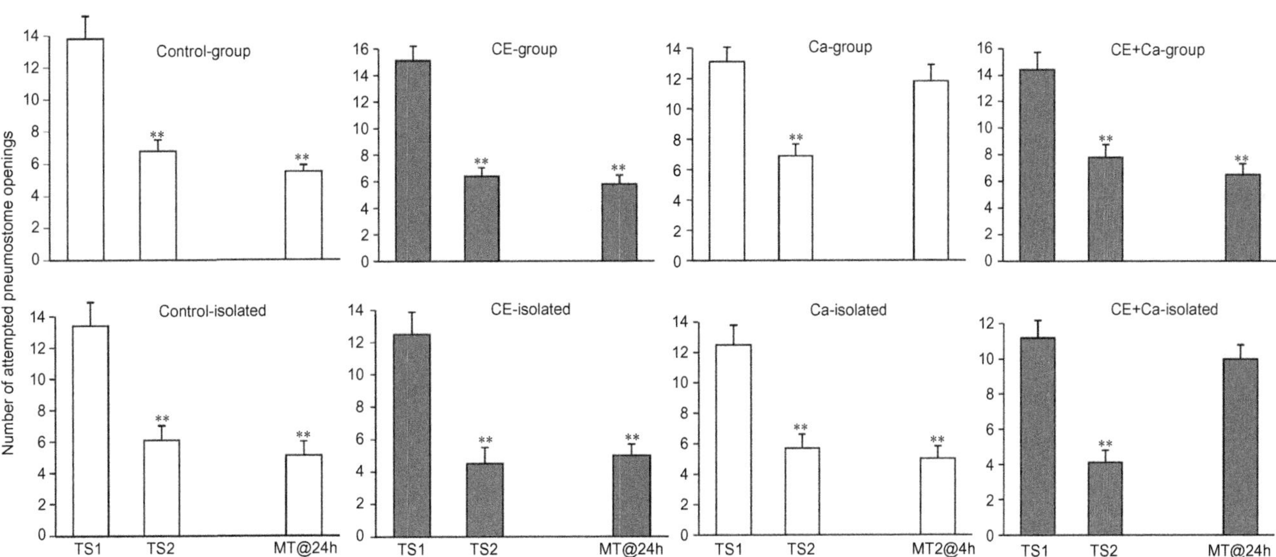

FIGURE 21.3 Social stress in combination with other stressors has differing effects on LTM formation. In the top panels, snails have not been socially isolated, whereas in the bottom panels snails have been socially isolated. Plotted are the mean number of attempted pneumostome openings and the standard error of the mean (SEM). In the left-most panel, it can be seen that social isolation has no effect on memory formation. In the next panel, it can be seen that social isolation also has no effect on the enhancing effect of crayfish effluent (CE) on LTM formation. In the next panel to the right, social isolation has an effect on how low environmental calcium (Ca^{2+}) affects LTM formation. Low Ca^{2+} blocks LTM formation when snails are in a group, but this blocking of LTM is not seen when snails are socially isolated. We could not have predicted this outcome before doing the experiment. Finally, the panel on the far right demonstrates the effect of low Ca^{2+} and CE exposure to socially isolated snails. As can be seen, when snails are not isolated, CE 'overcomes' the effect of low Ca^{2+} on LTM formation. However, when socially isolated, the effects of low Ca + CE result in suppression of memory formation. $^{**}p < 0.01$.. Source: *Data replotted with permission from Dalesman S, Lukowiak K. Social snails: The effect of social isolation on cognition is dependent on environmental context.* J. Exp. Biol. *2011;214:4179–4185*

Although the result that social isolation can enhance memory in low-calcium conditions initially seems surprising, we think this may be due to their change in reproductive behavior. Adult *Lymnaea* normally produce 100–300 eggs per week,[74,78,117] each of which is provisioned with its own small amount of calcium.[56,118] However, switching energy investment primarily toward male reproductive output in isolation[79] may significantly reduce calcium requirements of *Lymnaea* and reduce the stress associated with being held in low-calcium conditions, allowing these animals to form LTM. An even greater surprise was that when we trained isolated snails that had been held in low-calcium conditions in the presence of CE (which normally enhances memory formation), although the animals were able to demonstrate learning, LTM was blocked (Figure 21.3). When conditions are so stressful, it may be that *Lymnaea* is simply too stressed to pay attention to training.[2,3]

Our data present evidence to support negative, neutral, and enhancing effects of social isolation on memory formation, demonstrating that the effects of a stressor may be highly dependent on the context in which it is experienced.[91] These data support our theory that inconsistencies in the literature on the effects of isolation on cognition[85] may be due to changes in other environmental variables, such as differences in animal husbandry. This also shows that we cannot predict the effects that each type of stressor will have in combination on *Lymnaea* memory formation based on the effects of each stressor experienced alone; that is, stressors in combination may have emergent properties. This work provides an intriguing avenue for future exploration, and we already have a wealth of environmental stress combinations to test. Importantly, this raises our awareness and, it is hoped, that of others reading this work in detailing other environmental factors experienced by the animal when testing for stress effects.

POPULATION DIFFERENCES

As mentioned previously, our discovery of population differences in memory retention was purely by chance, but it has provided an intriguing new avenue for research. Whereas several of the populations we sampled exhibit similar LTM retention as that of the Dutch laboratory strain, a few have enhanced LTM formation.[28,42,43,119] For example, following a 0.5-hr operant training session, the Dutch strain and other populations exhibit ITM lasting only approximately 3 hr;[40,119] however, following the same training procedure, our 'smart' snails exhibit LTM lasting approximately 5 days.[27] Whereas heritable differences in learning and memory are found among individuals within a population in other invertebrate species,[120,121] *Lymnaea* is the first invertebrate species in which distinct differences in LTM-forming abilities have been found among natural populations. Although we doubt very much that it will be the only invertebrate in which such population differences are found, we intend to take advantage of what is so far a unique opportunity to develop an understanding of the mechanisms that control these differences and how they affect a population's ability to respond to memory-altering stressors. Thus far, we have focused on proximate causes of differentiation among populations. In addition to differences in memory retention, do these animals differ in other behavioral or physiological traits?

Where we see enhanced LTM to reduce aerial respiration, one possible explanation is that these 'smart' individuals simply breathe more often and therefore receive a greater number of physical stimuli, resulting in stronger memory formation. Our results show that both the total breathing time in hypoxia or eumoxia in the absence of physical stimuli and the average number of stimuli they receive during a typical training session (i.e., the number of pokes in 30 min) do not differ significantly between animals exhibiting good or poor LTM-forming abilities.[28,43,119] Therefore, we cannot explain differences in memory retention simply by the differences in breathing rate among populations.

An alternative way to assess differences between populations is by assessing the underlying neurophysiology. Following operant training to reduce aerial respiration, there are significant changes in some electrophysiological properties of RPeD1, a neuron that forms part of the CPG that controls aerial respiration. In semi-intact preparations, electrophysiological changes correlate well with changes in breathing behavior[122,123] and also differ depending on whether ITM or LTM is formed.[26] We compared electrophysiological properties in RPeD1 in two Canadian populations, one exhibiting enhanced LTM formation (TC1) and one exhibiting poorer LTM-forming capabilities (TC2) similar to the Dutch lab-bred strain. Our findings indicated that the 'smart' TC1 population showed very similar electrophysiological properties in RPeD1 in naive animals as those in the populations without enhanced memory, including TC2 and the Dutch laboratory strain, following 0.5-hr training.[26,28] Therefore, it seems that these 'smart' individuals form LTM more readily because they are already primed to do so. Currently, we do not know why or how RPeD1 is 'primed' in the 'smart' snails.

Having identified populations in both the United Kingdom and Canada that differ in memory-forming ability, we were interested in how these 'smart' snails would respond to environmental stressors relative to the Dutch population. Although we have previously demonstrated that 'smart' snails form LTM in the presence of predator kairomones, we have not previously published data demonstrating that they can extend memory retention beyond that seen in control conditions in the same way that we see in the Dutch strain and other wild populations.[43,119] It is possible that these 'smart' *Lymnaea* demonstrate enhance LTM because they are already at their maximum ability to form LTM following operant conditioning due to their primed state, whereas the Dutch snails are more plastic, only exhibiting maximum LTM formation in the presence of a stressor such as CE. Here, we present previously unpublished data from work with snails from a 'smart' population in the United Kingdom (Chilton Moor). These data demonstrate that whereas 'smart' snails form LTM lasting 5 but not 8 days in control conditions,[27] when we train them in predator cues (CE), we can extend this memory to at least 8 days (Figure 21.4). Therefore, the 'smart' snails are still able to show plasticity in LTM formation, despite starting in a primed state.[28]

In addition to assessing whether all populations respond to predator stress with enhanced LTM formation, which we now believe to be the case, we also wanted to assess whether they all respond to stressors that block memory in the Dutch strain. We therefore chose to assess the effects of crowding and low environmental calcium on LTM formation in two of our 'smart' populations, one from Canada (Trans-Canada 1 (TC1)) and one from the United Kingdom (Chilton Moor (CM)). We found that crowding the snails immediately prior to training blocked LTM in both our 'smart' populations, demonstrating the same phenotypic response as seen in the Dutch strain.[27] However, a low-calcium environment did not block LTM, as seen in the Dutch strain.[67,69,72] but did significantly shorten memory retention from 5 days to less than 72 hr. Therefore, although the directional effect on memory retention (i.e., reducing duration) was similar in the 'smart' snails to what is seen in the Dutch laboratory strain, the snails with enhanced memory appear to be more resistant to low calcium stress. Connections between stress resistance and cognitive abilities have been found elsewhere[120,124,125] and may prove to be highly conserved across species.

Based on our current work, it seems that all snail populations tested, whether they have enhanced memory-forming abilities or not, respond to ecologically relevant stressors in a similar directional pattern. We have been able to identify that LTM formation under control conditions, and also in stressed conditions, appears to vary consistently with memory phenotype (i.e., whether the population is classified as 'smart' or 'dumb'). This is the case even when we compared populations situated on separate continents, in North America and Europe. To us, this indicates that whatever mechanism(s) is controlling LTM capabilities (e.g., causing changes in the properties of RPeD1), and also how memory is altered by stress, is highly conserved across the *L. stagnalis* species, showing similar patterns on both broad and narrow geographic scales.

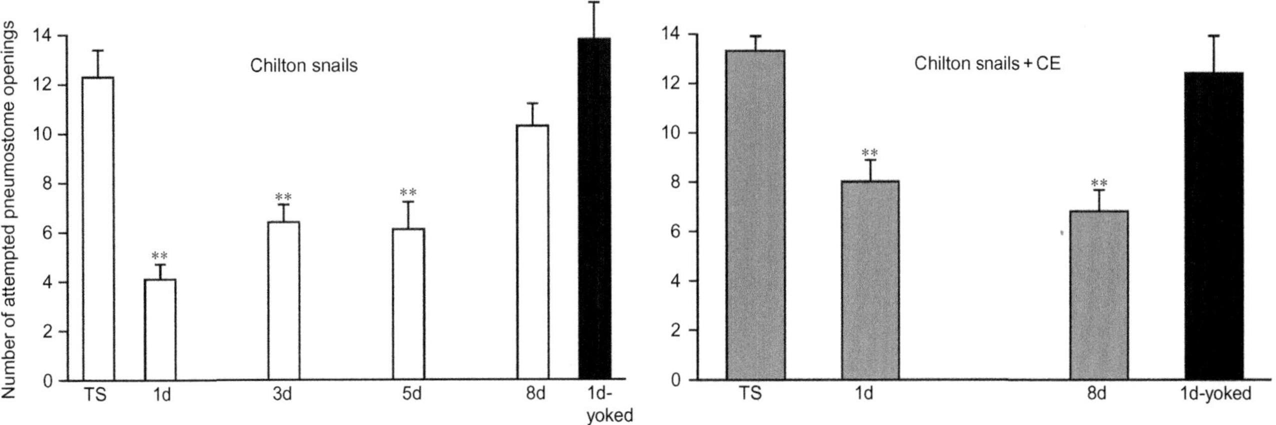

FIGURE 21.4 Smart snails and CE. Plotted are the mean number of attempted pneumostome openings and the standard error of the mean (SEM). (Left) In smart snails obtained from a pond in the Somerset Levels in the United Kingdom, a single 0.5-hr training session results in a memory that persists for up to 5 days. In lab-bred snails, a single 0.5-hr training session results in a memory that only persists for up to 3 hr. However, as shown on the right, predator detection even in these smart snails results in enhanced LTM formation because now memory persists for more than 8 days. In both panels, we present data showing that the yoked control procedure does not result in LTM when tested at the 24-hr mark. $^{**}p < 0.01$.. Source: *Data in the left-hand panel replotted with permission from Dalesman S, Lukowiak K. How stress alters memory in 'smart' snails.* PLoS ONE *2012;7:e32334*

CONCLUSIONS

Memory formation in our model system can be significantly altered by stressors that a snail is likely to encounter in naturally occurring situations in a pond. However, it is almost impossible to accurately predict before performing the experiment whether the stressor will enhance or diminish memory formation. The 'state' of the snail—for example, whether it is from a 'smart' strain—complicates the situation further with regard to memory formation. The specific strain of the snail will determine how it responds to a specific stressor. The situation becomes even more complicated when attempting to predict in advance the outcome on memory formation when stressors occur in combination with each other. It appears that how stressors interact with each other to produce their effect on memory formation is an emergent property of the neuronal networks activated by the stressors. All these data indicate that memory formation is dynamic, that it is subject to how stressors alter the activity of neurons necessary for memory formation, and that the 'state' of the organism must be taken into consideration when attempting to determine the causal mechanisms of memory formation. That is, understanding how memory is formed, how it is altered, and how it is maintained is a very complicated undertaking, even in a relatively simple model system such as Lymnaea.

References

1. Bacon F. *The New Organon*. New York: Bobbs-Merrill; 1620.
2. Shors TJ. Learning during stressful times. *Learn Mem.* 2004;11:137–144.
3. Yerkes RM, Dodson JD. The relation of strength of stimulus to rapidity of habit-formation. *J Comp Neurol Psychol.* 1908;18: 459–482.
4. Kim J, Diamond D. The stressed hippocampus, synaptic plasticity and lost memories. *Nat Rev Neurosci.* 2002;3:453–462.
5. Lukowiak K, Fras M, Smyth K, Wong C, Hittel K. Reconsolidation and memory infidelity in *Lymnaea. Neurobiol Learn Mem.* 2007;87:547–560.
6. Yehuda R, Ledoux J. Response variation following trauma: a translational neuroscience approach to understanding PTSD. *Neuron.* 2007;56:19–32.
7. Lukowiak K, Martens K, Rosenegger D, Browning K, de Caigny P, Orr M. The perception of stress alters adaptive behaviours in *Lymnaea stagnalis. J Exp Biol.* 2008;211:1747–1756.
8. Lukowiak K, Orr M, de Caigny P, et al. Ecologically relevant stressors modify long-term memory formation in a model system. *Behav Brain Res.* 2010;214:18–24.
9. McEwen BS, Sapolsky RM. Stress and cognitive function. *Curr Opin Neurobiol.* 1995;5:205–216.
10. Selye H. The evolution of the stress concept. *Am Sci.* 1973;61:692–699.
11. Andreano JM, Cahill L. Sex influences on the neurobiology of learning and memory. *Learn Mem.* 2009;16:248–266.
12. Maeng LY, Shors TJ. Once a mother, always a mother: maternal experience protects females from the negative effects of stress on learning. *Behav Neurosci.* 2012;126:137–141.
13. Lukowiak K, Adatia N, Krygier D, Syed N. Operant conditioning in *Lymnaea*: evidence for intermediate- and long-term memory. *Learn Mem.* 2000;7:140–150.
14. Lukowiak K, Cotter R, Westly J, Ringseis E, Spencer G, Syed N. Long-term memory of an operantly conditioned respiratory behaviour pattern in *Lymnaea stagnalis. J Exp Biol.* 1998;201: 877–882.
15. Lukowiak K, Ringseis E, Spencer G, Wildering W, Syed N. Operant conditioning of aerial respiratory behaviour in *Lymnaea stagnalis. J Exp Biol.* 1996;199:683–691.
16. Lukowiak K, Sangha S, Scheibenstock A, et al. A molluscan model system in the search for the engram. *J Physiol Paris.* 2003;97:69–76.
17. Milner B, Squire LR, Kandel ER. Cognitive neuroscience and the study of memory. *Neuron.* 1998;20:445–468.
18. Sangha S, Scheibenstock A, Lukowiak K. Reconsolidation of a long-term memory in *Lymnaea* requires new protein and RNA synthesis and the soma of right pedal dorsal 1. *J Neurosci.* 2003;23:8034–8040.
19. Syed NI, Bulloch AGM, Lukowiak K. *In vitro* reconstruction of the respiratory central pattern generator of the mollusk *Lymnaea. Science.* 1990;250:282–285.
20. Syed NI, Ridgway RL, Lukowiak K, Bulloch AGM. Transplantation and functional-integration of an identified respiratory interneuron in *Lymnaea stagnalis. Neuron.* 1992;8:767–774.
21. Sangha S, Scheibenstock A, Martens K, Varshney N, Cooke R, Lukowiak K. Impairing forgetting by preventing new learning and memory. *Behav Neurosci.* 2005;119:787–796.
22. Sangha S, Scheibenstock A, Morrow R, Lukowiak K. Extinction requires new RNA and protein synthesis and the soma of the cell right pedal dorsal 1 in *Lymnaea stagnalis. J Neurosci.* 2003;23:9842–9851.
23. Scheibenstock A, Krygier D, Haque Z, Syed N, Lukowiak K. The soma of RPeD1 must be present for long-term memory formation of associative learning in *Lymnaea. J Neurophys.* 2002;88: 1584–1591.
24. Lattal KM, Radulovic J, Lukowiak K. Extinction: does it or doesn't it? The requirement of altered gene activity and new protein synthesis. *Biol Psychiatry.* 2006;60:344–351.
25. Parvez K, Rosenegger D, Orr M, Martens K, Lukowiak K. Canadian Association of neurosciences review: learning at a snail's pace. *Can J Neurol Sci.* 2006;33:347–356.
26. Braun MH, Lukowiak K. Intermediate and long-term memory are different at the neuronal level in *Lymnaea stagnalis* (L.). *Neurobiol Learn Mem.* 2011;96:403–416.
27. Dalesman S, Lukowiak K. How stress alters memory in "smart" snails. *PLoS ONE.* 2012;7:e32334.
28. Braun MH, Lukowiak K, Karnik V, Lukowiak K. Differences in neuronal activity explain differences in memory forming abilities of different populations of *Lymnaea stagnalis. Neurobiol Learn Mem.* 2012;97:173–182.
29. Fulton D, Kemenes I, Andrew RJ, Benjamin PR. A single time-window for protein synthesis-dependent long-term memory formation after one-trial appetitive conditioning. *Eur J Neurosci.* 2005;21:1347–1358.
30. Ito E, Kobayashi S, Kojima S, Sadamoto H, Hatakeyama D. Associative learning in the pond snail, *Lymnaea stagnalis. Zool Sci.* 1999;16:711–723.

31. Sunada H, Sakaguchi T, Horikoshi T, Lukowiak K, Sakakibara M. The shadow-induced withdrawal response, dermal photoreceptors, and their input to the higher-order interneuron RPeD11 in the pond snail *Lymnaea stagnalis*. *J Exp Biol*. 2010;213: 3409–3415.
32. Krogh A. The progress of physiology. *Am J Physiol*. 1929;90:243–251.
33. Krebs HA. The August Krogh principle: "For many problems there is an animal on which it can be most conveniently studied.". *J Exp Zool*. 1975;194:221–226.
34. Kandel ER. The molecular biology of memory storage: a dialogue between genes and synapses. *Science*. 2001;294:1030–1038.
35. Benjamin PR, Staras K, Kemenes G. A systems approach to the cellular analysis of associative learning in the pond snail *Lymnaea*. *Learn Mem*. 2000;7:124–131.
36. Brembs B. Operant conditioning in invertebrates. *Curr Opin Neurobiol*. 2003;13:710–717.
37. Karnik V, Dalesman S, Lukowiak K. Input from a chemosensory organ, the osphradium, does not mediate aerial respiration in *Lymnaea stagnalis*. *Aquat Biol*. 2012;15:167–173.
38. Lukowiak K, Martens K, Orr M, Parvez K, Rosenegger D, Sangha S. Modulation of aerial respiratory behaviour in a pond snail. *Respir Physiol Neurobiol*. 2006;154:61–72.
39. Martens K, Amarell M, Parvez K, et al. One-trial conditioning of aerial respiratory behaviour in *Lymnaea stagnalis*. *Neurobiol Learn Mem*. 2007;88:232–242.
40. Orr MV, Hittel K, Lukowiak K. Comparing memory-forming capabilities between laboratory-reared and wild *Lymnaea*: learning in the wild, a heritable component of snail memory. *J Exp Biol*. 2008;211:2807–2816.
41. Orr MV, Lukowiak K. Electrophysiological and behavioral evidence demonstrating that predator detection alters adaptive behaviors in the snail *Lymnaea*. *J Neurosci*. 2008;28:2726–2734.
42. Orr M, Hittel K, Lukowiak KS, Han J, Lukowiak K. Differences in LTM-forming capability between geographically different strains of Alberta *Lymnaea stagnalis* are maintained whether they are trained in the lab or in the wild. *J Exp Biol*. 2009;212: 3911–3918.
43. Orr M, Hittel K, Lukowiak K. "Different strokes for different folks": geographically isolated strains of *Lymnaea stagnalis* only respond to sympatric predators and have different memory forming capabilities. *J Exp Biol*. 2009;212:2237–2247.
44. Orr M, Lukowiak K. Sympatric predator detection alters cutaneous respiration in *Lymnaea*. *Comm Integr Biol*. 2010;4:1–4.
45. Ferrari MCO, Wisenden BD, Chivers DP. Chemical ecology of predator–prey interactions in aquatic ecosystems: a review and prospectus. *Can J Zool-Rev Can Zool*. 2010;88:698–724.
46. Dalesman S, Rundle SD, Coleman RA, Cotton PA. Cue association and antipredator behaviour in a pulmonate snail, *Lymnaea stagnalis*. *Anim Behav*. 2006;71:789–797.
47. Orr MV, El-Bekai M, Lui M, Watson K, Lukowiak K. Predator detection in *Lymnaea stagnalis*. *J Exp Biol*. 2007;210:4150–4158.
48. Kamardin NN, Shalanki Y, Sh.-Rozha K, Nozdrachev AD. Studies of chemoreceptor perception in mollusks. *Neurosci Behav Phys*. 2001;31:227–235.
49. Wedemeyer H, Schild D. Chemosensitivity of the osphradium of the pond snail *Lymnaea stagnalis*. *J Exp Biol*. 1995;198:1743–1754.
50. Il-Han J, Janes T, Lukowiak K. The role of serotonin in the enhancement of long-term memory resulting from predator detection in *Lymnaea*. *J Exp Biol*. 2010;213:3603–3614.
51. Karnik V, Braun MH, Dalesman S, Lukowiak K. Sensory input from the osphradium modulates the response to memory enhancing stressors in *Lymnaea stagnalis*. *J Exp Biol*. 2012;215: 536–542.
52. Barnard CJ, Collins SA, Daisley JN, Behnke JM. Odour learning and immunity costs in mice. *Behav Process*. 2006;72: 74–83.
53. Burns JG, Foucaud J, Mery F. Costs of memory: lessons from "mini" brains. *Proc R Soc B-Biol Sci*. 2011;278:923–929.
54. Sugai R, Azami S, Shiga H, et al. One-trial conditioned taste aversion in *Lymnaea*: good and poor performers in long-term memory acquisition. *J Exp Biol*. 2007;210:1225–1237.
55. Haney J, Lukowiak K. Context learning and the effect of context on memory retrieval in *Lymnaea*. *Learn Mem*. 2001;8: 35–43.
56. Ebanks SC, O'Donnell MJ, Grosell M. Acquisition of Ca^{2+} and HCO_3^-/CO_3^{2-} for shell formation in embryos of the common pond snail *Lymnaea stagnalis*. *J Comp Physiol B Biochem Syst Environ Physiol*. 2010;180:953–965.
57. Ebanks SC, O'Donnell MJ, Grosell M. Characterization of mechanisms for Ca^{2+} and HCO_3^-/CO_3^{2-} acquisition for shell formation in embryos of the freshwater common pond snail *Lymnaea stagnalis*. *J Exp Biol*. 2010;213:4092–4098.
58. Boycott AE. The habitats of fresh-water Mollusca in Britain. *J Anim Ecol*. 1936;5:116–186.
59. Madsen H. Effect of calcium concentration on growth and egg laying of *Helisoma duryi*, *Bulinus africanus* and *Bulinus truncatus* (Gastropoda: Planorbidae). *J Appl Ecol*. 1987;24:823–836.
60. Young JO. Preliminary field and laboratory studies on the survival and spawning of several species of Gastropoda in calcium-poor and calcium-rich waters. *Proc Malac Soc Lond*. 1975;41:429–437.
61. McKillop W, Harrison A. Distribution of aquatic gastropods across the interface between the Canadian Shield and limestone formations. *Can J Zool*. 1972;50:1433–1445.
62. Lewis DB, Magnuson JJ. Intraspecific gastropod shell strength variation among north temperate lakes. *Can J Fish Aquat Sci*. 1999;56:1687–1695.
63. Rundle SD, Spicer JI, Coleman RA, Vosper J, Soane J. Environmental calcium modifies induced defences in snails. *Proc R Soc B Biol Sci*. 2004;271:S67–S70.
64. Zalizniak L, Kefford BJ, Nugegoda D. Effects of different ionic compositions on survival and growth of *Physa acuta*. *Aquat Ecol*. 2009;43:145–156.
65. Dalesman S, Lukowiak K. Effect of acute exposure to low environmental calcium alters respiration and locomotion of *Lymnaea stagnalis* (L.). *J Exp Biol*. 2010;213:1471–1476.
66. Greenaway P. Calcium regulation in the freshwater mollusc, *Limnaea stagnalis* (L.) (Gastropoda: Pulmonata): 1. The effect of internal and external calcium concentration. *J Exp Biol*. 1971;54:199–214.
67. Dalesman S, Braun MH, Lukowiak K. Low environmental calcium blocks long term memory formation in a pulmonate snail. *Neurobiol Learn Mem*. 2011;95:393–403.
68. Parvez K, Stewart O, Sangha S, Lukowiak K. Boosting intermediate-term into long-term memory. *J Exp Biol*. 2005;208: 1525–1536.
69. Knezevic B, Dalesman S, Karnik V, Byzitter J, Lukowiak K. Low external environmental calcium levels prevent forgetting in *Lymnaea*. *J Exp Biol*. 2011;214:2118–2124.
70. Sangha S, Morrow R, Smyth K, Cooke R, Lukowiak K. Cooling blocks ITM and LTM formation and preserves memory. *Neurobiol Learn Mem*. 2003;80:130–139.
71. Piggott H, Dussart G. Egg-laying and associated behavioural responses of *Lymnaea peregra* (Müller) and *Lymnaea stagnalis* (L.) to calcium in their environment. *Malacologia*. 1995;37: 13–21.

72. Dalesman S, Karnik V, Lukowiak K. Sensory mediation of memory blocking stressors in the pond snail, *Lymnaea stagnalis*. *J Exp Biol*. 2011;214:2528–2533.
73. Grosell M, Brix KV. High net calcium uptake explains the hypersensitivity of the freshwater pulmonate snail, *Lymnaea stagnalis*, to chronic lead exposure. *Aquat Toxicol*. 2009;91: 302–311.
74. Brown KM. The adaptive demography of four freshwater pulmonate snails. *Evolution*. 1979;33:417–432.
75. Bovbjerg RV. Dispersal and dispersion of pond snails in an experimental environment varying to three factors, singly and in combination. *Physiol Zool*. 1975;48:203–215.
76. Janse C, Slob W, Popelier CM, Vogelaar JW. Survival characteristics of the mollusc *Lymnaea stagnalis* under constant culture conditions: effects of aging and disease. *Mech Ageing Dev*. 1988;42:263–274.
77. McDonald SLC. The biology of *Lymnaea stagnalis* (L.) (Gastropoda: Pulmonata). *Sterkiana*. 1969;36:1–17.
78. Noland LE, Carriker MR. Observations on the biology of the snail *Lymnaea stagnalis appressa* during twenty generations in laboratory culture. *Am Mid Nat*. 1946;36:467–493.
79. P.A.C.M. De Boer, Jansen RF, Koene JM, Ter Maat A. Nervous control of male sexual drive in the hermaphroditic snail *Lymnaea stagnalis*. *J Exp Biol*. 1997;200:941–951.
80. De Caigny P, Lukowiak K. Crowding, an environmental stressor, blocks long-term memory formation in *Lymnaea*. *J Exp Biol*. 2008;211:2678–2688.
81. Sangha S, Scheibenstock A, McComb C, Lukowiak K. Intermediate and long-term memories of associative learning are differentially affected by transcription versus translation blockers in *Lymnaea*. *J Exp Biol*. 2003;206:1605–1613.
82. Crabb E. Growth of a pond snail, *Lymnaea stagnalis appressa*, as indicated by increase in shell size. *Biol Bull*. 1929;56: 41–63.
83. Voronezhskaya EE, Khabarova MY, Nezlin LP. Apical sensory neurones mediate developmental retardation induced by conspecific environmental stimuli in freshwater pulmonate snails. *Development*. 2004;131:3671–3680.
84. Cacioppo JT, Hawkey LC. Perceived social isolation and cognition. *Trends Cogn Sci*. 2009;13:447–454.
85. Fone KCF, Porkess MV. Behavioural and neurochemical effects of post-weaning social isolation in rodents: relevance to developmental neuropsychiatric disorders. *Neurosci Biobehav Rev*. 2008;32:1087–1102.
86. Technau GM. Fiber number in the mushroom bodies of adult *Drosophila melanogaster* depends on age, sex and experience. *J Neurogenet*. 2007;21:183–196.
87. Maleszka J, Barron AB, Helliwell PG, Maleszka R. Effect of age, behaviour and social environment on honey bee brain plasticity. *J Comp Physiol A Neuroethol Sens Neural Behav Physiol*. 2009;195:733–740.
88. Fahrbach SE. Structure of the mushroom bodies of the insect brain. *Annu Rev Entomol*. 2006;51:209–232.
89. Schwarz M, Blumberg S, Susswein AJ. Social isolation blocks the expression of memory after training that a food is inedible in *Aplysia fasciata*. *Behav Neurosci*. 1998;112:942–951.
90. Koene JM, Ter Maat A. Energy budgets in the simultaneously hermaphroditic pond snail, *Lymnaea stagnalis*: a trade-off between growth and reproduction during development. *Belg J Zool*. 2004;134:41–45.
91. Dalesman S, Lukowiak K. Social snails: the effect of social isolation on cognition is dependent on environmental context. *J Exp Biol*. 2011;214:4179–4185.
92. Dillon RT. *The Ecology of Freshwater Molluscs*. Cambridge, UK: Cambridge University Press; 2000.
93. Vaughn CM. Effects of temperature on hatching and growth of *Lymnaea stagnalis*. *Am Mid Nat*. 1953;49:214–228.
94. Botzer D, Markovich S, Susswein AJ. Multiple memory processes following training that a food is inedible in *Aplysia*. *Learn Mem*. 1998;5:204–219.
95. Martens KR, De Caigny P, Parvez K, Amarell M, Wong C, Lukowiak K. Stressful stimuli modulate memory formation in *Lymnaea stagnalis*. *Neurobiol Learn Mem*. 2007;87:391–403.
96. Boyd RS. Heavy metal pollutants and chemical ecology: exploring new frontiers. *J Chem Ecol*. 2010;36:46–58.
97. Roth SH. Hydrogen sulphide. In: Corn M, ed. *Handbook of Hazardous Materials*. New York: Academic Press; 1993.
98. Han FXX, Banin A, Su Y, et al. Industrial age anthropogenic inputs of heavy metals into the pedosphere. *Naturwissenschaften*. 2002;89:497–504.
99. Kelly EN, Schindler DW, Hodson PV, Short JW, Radmanovich R, Nielsen CC. Oil sands development contributes elements toxic at low concentrations to the Athabasca River and its tributaries. *Proc Natl Acad Sci USA*. 2010;107:16178–16183.
100. Schwarzenbach RP, Egli T, Hofstetter TB, von Gunten U, Wehrli B. Global water pollution and human health. *Annu Rev Environ Resour*. 2010;35:109–136.
101. Farahat SA, Kishk NA. Cognitive functions changes among Egyptian sewage network workers. *Toxicol Ind Health*. 2010;26:229–238.
102. Hessel PA, Melenka LS. Health effects of acute hydrogen sulfide exposures in oil and gas workers. *Environ Epidemiol Toxicol*. 1999;1:201–206.
103. Kilburn KH. Exposure to reduced sulfur gases impairs neurobehavioral function. *South Med J*. 1997;90:997–1006.
104. Abe K, Kimura H. The possible role of hydrogen sulfide as an endogenous neuromodulator. *J Neurosci*. 1996;16:1066–1071.
105. Hu LF, Lu M, Wong PTH, Bian JS. Hydrogen sulfide: neurophysiology and neuropathology. *Antioxid Redox Signal*. 2011;15:405–419.
106. Rosenegger D, Roth S, Lukowiak K. Learning and memory in *Lymnaea* are negatively altered by acute low-level concentrations of hydrogen sulphide. *J Exp Biol*. 2004;207: 2621–2630.
107. Wang R. Two's company, three's a crowd: can H_2S be the third endogenous gaseous transmitter? *FASEB J*. 2002;16:1792–1798.
108. Das S, Khangarot BS. Bioaccumulation and toxic effects of cadmium on feeding and growth of an Indian pond snail *Lymnaea luteola* L. under laboratory conditions. *J Hazard Mater*. 2010;182:763–770.
109. Das S, Khangarot BS. Bioaccumulation of copper and toxic effects on feeding, growth, fecundity and development of pond snail *Lymnaea luteola* L. *J Hazard Mater*. 2011;185: 295–305.
110. Lefcort H, Ammann E, Eiger SM. Antipredatory behavior as an index of heavy-metal pollution? A test using snails and caddisflies. *Arch Environ Contam Toxicol*. 2000;38:311–316.
111. Lefcort H, Thomson SM, Cowles EE, et al. Ramifications of predator avoidance: predator and heavy-metal-mediated competition between tadpoles and snails. *Ecol Appl*. 1999;9: 1477–1489.
112. Lefcort H, Abbott DP, Cleary DA, Howell E, Keller NC, Smith MM. Aquatic snails from mining sites have evolved to detect and avoid heavy metals. *Arch Environ Contam Toxicol*. 2004;46:478–484.
113. Toscano CD, Guilarte TR. Lead neurotoxicity: from exposure to molecular effects. *Brain Res Rev*. 2005;49:529–554.
114. Byzitter J, Lukowiak K, Karnik V, Dalesman S. Acute combined exposure to heavy metals (Zn, Cd) blocks memory formation in a freshwater snail. *Ecotoxicology*. 2012;21:860–868.

115. De Caigny P, Lukowiak K. A clash of stressors and LTM formation. *Commun Intergr Biol*. 2008;1:125–127.
116. Dalesman S, Lukowiak K. Interaction between environmental stressors mediated via the same sensory pathway. *Comm Integr Biol*. 2011;4:717–719.
117. Nichols D, Cooke J, Whiteley D. *The Oxford Book of Invertebrates*. Oxford: Oxford University Press; 1971.
118. Taylor HH. The ionic properties of the capsular fluid bathing embryos of *Lymnaea stagnalis* and *Biomphalaria sudanica* (Mollusca: Pulmonata). *J Exp Biol*. 1973;59:543–564.
119. Dalesman S, Rundle SD, Lukowiak K. Microgeographic variability in long-term memory formation in the pond snail, *Lymnaea stagnalis*. *Anim Behav*. 2011;82:311–319.
120. Mery F, Belay AT, So AKC, Sokolowski MB, Kawecki TJ. Natural polymorphism affecting learning and memory in *Drosophila*. *Proc Nat Acad Sci USA*. 2007;104:13051–13055.
121. van den Berg M, Duivenvoorde L, Wang G, et al. Natural variation in learning and memory dynamics studied by artificial selection on learning rate in parasitic wasps. *Anim Behav*. 2011;81:325–333.
122. McComb C, Rosenegger D, Varshney N, Kwok HY, Lukowiak K. Operant conditioning of an in vitro CNS-pneumostome preparation of *Lymnaea*. *Neurobiol Learn Mem*. 2005;84:9–24.
123. Spencer GE, Syed NI, Lukowiak K. Neural changes after operant conditioning of the aerial respiratory behavior in *Lymnaea stagnalis*. *J Neurosci*. 1999;19:1836–1843.
124. Katsnelson E, Motro U, Feldman MW, Lotem A. Individual-learning ability predicts social-foraging strategy in house sparrows. *Proc R Soc B Biol Sci*. 2011;278:582–589.
125. Sokolowski MB, Pereira HS, Hughes K. Evolution of foraging behavior in *Drosophila* by density-dependent selection. *Proc Natl Acad Sci USA*. 1997;94:7373–7377.

CHAPTER

22

Associative Memory Mechanisms in Terrestrial Slugs and Snails

Alan Gelperin

Princeton University, Princeton, New Jersey; and Monell Chemical Senses Center, Philadelphia, Pennsylvania

INTRODUCTION

Animal cognition connotes, among other things, the study of learning and memory, in addition to an array of other neural computations such as selective attention, category learning, spatial learning, numerosity, tool use, and language.[1] Increasingly, the term *cognition* is applied to descriptions of the use of behavioral plasticity in a variety of invertebrates, including mollusks,[2–4] particularly cephalopods.[5,6] As the study of animal cognition matures, it is increasingly clear that the complexity of learning and memory mechanisms in invertebrates is usefully viewed in the broad context of cognitive neuroscience. The following brief survey attempts to contribute to the comparative study of animal cognition by highlighting the diversity, complexity, mechanistic underpinnings, and adaptive value of learning and memory functions demonstrated in terrestrial mollusks.

Our unfolding understanding of comparative cognition in mollusks, including terrestrial slugs and snails, receives impetus from several distinct domains of scientific inquiry. The traditional reductionist approach seeks to reduce a neural computation observed at the behavioral level to its constituent cellular and biophysical mechanisms, with the ultimate goal of identifying the causative cellular changes that mediate the behavioral plasticity. Terrestrial slugs and snails provide favorable material for the reductionist approach due to the technical advantages inherent in the large size and wide distribution of voltage-sensitive ion channels in the somata and processes of many of their central neurons.[7–10] The formalisms of Pavlovian and operant conditioning provide a distinctly different intellectual scaffold for probing the logical operations carried out by neural circuits during learning and memory retrieval, as revealed by conditioning experiments.[11,12] The set of ostensibly mammalian conditioning phenomena, such as first-order conditioning, second-order conditioning, and blocking, is a fertile field within which to prospect for learned logic operations resident in the brains of terrestrial mollusks, among other groups of vertebrate and invertebrate species. A third domain of activity in comparative cognition arises from ecological and evolutionary biology, in which attempts to understand the adaptive behavior of animals in an ecological and evolutionary context yield insights into the roles of learning and memory mechanisms in a wide range of species, including terrestrial mollusks. A novel evolutionary theme emphasized by Niven and Laughlin[13] points out that nervous systems are metabolically expensive cognitive machines and thus selective pressures will tend to optimize their function, which by extension includes the partitioning of behavioral control systems between built-in innate mechanisms and learned responses requiring sites of synaptic plasticity.[14,15] Interestingly, evidence from *Drosophila* indicates that learning functions are not simply maximized over evolutionary time because artificially applied directed selection for learning ability can produce flies that learn faster and remember better than normal flies but at the cost of reduced fitness in other traits, such as larval fitness and life span.[16]

The comparative study of learning and memory mechanisms also reveals the inherent tension between asking a biologically meaningful question about an animal's learning ability and asking a question that is readily instrumented in the laboratory. All animals are smart, if you ask the right question, based on the adaptive challenges the animal has solved using learning.

Invertebrate Learning and Memory.
DOI: http://dx.doi.org/10.1016/B978-0-12-415823-8.00022-8

As James Gould[17] stated, "Natural selection creates niche-specific minds designed to solve particular intellectual challenges." Students of comparative cognition, particularly with regard to invertebrate species, must discern the set of neural computations for which a particular species has evolved a learning mechanism. Then frame experimental questions to dissect that learning mechanism using as input the appropriate set of sensory stimuli and indexing the occurrence of learning and memory storage by measuring the appropriate motor responses. This is part of the research program pursued by neuroethologists, who combine interests in the neural mechanisms of behavior with the study of behavior in its natural setting.[18,19]

LEARNING SOLUTIONS TO LIFESTYLE CHALLENGES BY TERRESTRIAL GASTROPODS

Water Homeostasis

Within the terrestrial gastropods, the challenge of maintaining adequate water balance is particularly acute for terrestrial slugs because they must maintain a moist body surface for respiratory exchange[20] and continually produce pedal mucous for locomotion,[21,22] without recourse to withdrawal into an external shell for mitigation of water loss, as can terrestrial snails. The importance of locating water sources using remote cues such as odors suggested that dehydrated slugs might show rapid and reliable learning of odor cues paired with access to a water source. This turned out to be true. Studies have shown that after rapid dehydration, pairing odor cues with a moist surface available for contact rehydration[23,24] results in rapid development of a preference for the odor paired with the opportunity for contact rehydration, as opposed to an odor paired with the availability of aversive water containing both salt and quinine.[2] Following the dehydration and training procedure shown in Figure 22.1, slugs can rapidly and repeatedly be motivated to seek water rewards and experience odor cues associated with the safe water available for contact rehydration, during which slugs absorb water through an epithelial paracellular pathway in the integument of the foot.[24]

Two peptides have been shown to be involved in this water regulatory mechanism. Injection of angiotensin II, an oligopeptide containing eight amino acids acting as a dipsogenic peptide in mammals, can trigger both water-seeking behavior and contact rehydration by integumental uptake of water via the foot in *Limax maximus*.[25] Another small peptide, arginine vasotocin, is also part of the regulatory mechanism that couples dehydration to water-seeking behavior and contact

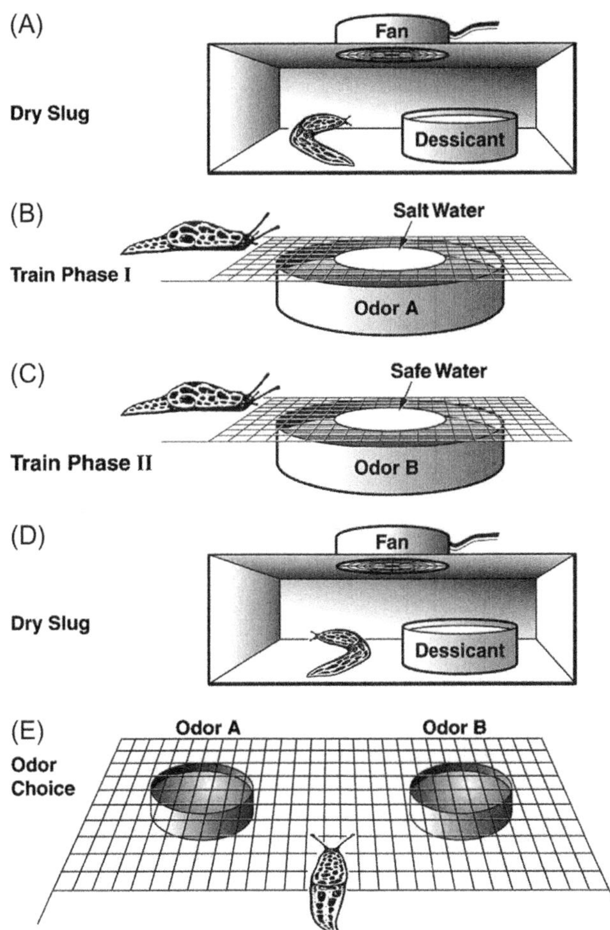

FIGURE 22.1 Diagrammatic presentation of a method for appetitive conditioning of odor preferences using rapid dehydration followed by water reward. (A) The slug is dehydrated to 60–65% of its fully hydrated weight within 30 min. (B) The slug is offered dangerous water with odor A. (C) The slug is offered safe water with odor B. (D) The slug is again dehydrated. (E) The dehydrated slug is offered a choice between spending time over odor A or odor B. Source: *Reproduced with permission from Watanabe S, Kirino Y, Gelperin A. Neural and molecular mechanisms of microcognition in* Limax. Learning Memory *2008; 15(9): 633–642.*

rehydration.[26] These laboratory observations are reinforced by the fact that changes in water balance due to local rainfall are thought to influence dietary choices, particularly the selection of fresh plant material with high water content, in the terrestrial snail *Cepaea nemoralis*.[27] The use of a learning mechanism to make food choices based on hydration state is yet to be demonstrated.

The resistance to dehydration by terrestrial snails is remarkable, allowing them to live in desert environments[28] in which the surface temperatures may reach 70°C and rains may occur no more than once per year.[29] Terrestrial snails from temperate environments, such as the cosmopolitan species *Helix pomatia*, would also be likely candidates to learn odor or surface texture associations with water rewards, delivered as access to bulk water sources or as access to food materials with high water content such as fresh plant material. Conversely, odors or surface textures associated with desiccating conditions might become aversive, as indexed by directed locomotion away from a localized source of the odor associated with desiccation. Water stimuli applied in combination with odor exposure to aestivating *H. pomatia* may well elicit a conditioned odor attraction, particularly because activation of aestivating snails results in heightened activity in both serotonergic and dopaminergic central neurons[30] as well as neural mobilization of a number of peptides.[31] Both serotonin and dopamine, prime candidates for involvement in reinforcement pathways during learning, are significantly altered by the availability of environmental moisture,[32] encouraging the view that learned associations of stimuli associated with either a lack or surfeit of water may occur. Results on the variability of resistance to dehydration in *L. maximus* suggest that this trait may be the subject of selection pressure.[33]

The role of peptides in molluskan information processing networks, including networks with significant synaptic plasticity, has been reviewed.[34]

Temperature

As with all poikilotherms, terrestrial snails and slugs have a preferred temperature and a temperature range bracketing the preferred temperature over which neural and behavioral function can be maintained. Working within the range of tolerated temperatures, application of thermal stresses and the removal of thermal stresses are likely to be potent reinforcing stimuli when coupled with visual, tactile, or chemosensory cues. A training paradigm exploiting temperature preferences as reinforcing stimuli has been demonstrated using crickets[35] and fruit flies.[36] These insects learned to associate distal visual cues with the location of a cool platform within a heated experimental enclosure, a topological analogy with the widely used Morris water maze for studying place learning in rodents. This place learning task could readily be adapted to study the place preferences of terrestrial slugs and snails by creating a substrate for locomotion with a distinctive surface texture or neutral odor co-localized with a subsection of the substrate that is cool within a heated ambient environment. Digital video systems with automated feature recognition to track subject movements within the arena are readily available.[37,38]

Hunting Prey and Mates Using Slime Trails

Carnivorous terrestrial mollusks such as the rosy wolf snail *Euglandina rosea* actively track their prey using directional signals contained in the slime trails of prey species.[39–41] Wolf snails can discriminate slime trails laid by prey species such as *H. pomatia* from slime trails laid by conspecifics and can learn to follow trails containing novel chemical cues that have been associated with exposure to conspecifics.[42] Defense strategies for the terrestrial mollusks involve mobilizing copious amounts of viscous slime that can interfere with predator movements, unlike marine mollusks that employ predator defensive strategies using toxic secretions.[43]

Slime trails are also used to signal reproductive status,[44,45] presumably by incorporation of pheromones[46] into the hydrated matrix of the slime trail. Sexually mature slugs and snails orient to pheromone-containing slime trails and, upon intersecting the trail, efficiently determine the movement direction of the conspecific generating the trail, which is then followed with sufficient speed to overtake the trail producer and initiate further chemical and tactile signaling as a prelude to mating.[47] Perhaps learning is involved in optimizing the ability to decipher the small gradients in chemical cues that signal the direction of movement of the prey or conspecifics that produced the intersected slime trail.

Avoiding Toxic Plant Compounds

Learning about the relative palatability of alternative food sources is a rich source of adaptive challenges potentially solved by a learning mechanism. Plants make a variety of toxic secondary metabolites, such as alkaloids, which are both bitter tasting and toxic to slugs at sufficiently high doses.[48] Slugs will feed on plants or artificial diets containing alkaloids in laboratory tests if no other choice is available, and they will mobilize a variety of postingestive biochemical detoxification mechanisms, most notably cytochrome P450 in the digestive gland.[48] If alternate food sources are available, slugs will shift their food preferences guided by postingestive feedback cues.

Another form of plant toxicity that clearly mobilizes a robust learning mechanism is due to the absence in some plant tissues of essential amino acids needed for normal growth. The plant tissue may trigger ingestion initially but reveal its nutritional inadequacy by an as yet little understood postingestive mechanism. For example, slugs given only maize, known to be deficient in the essential amino acid lysine, as their sole food source grow more slowly, produce fewer fertile eggs, and survive for a shorter period after capture than slugs fed nutritionally complete plant tissue, such as that of dandelion (*Taraxacum officinale*).[49] Slugs have been shown to exhibit learned modifications of feeding preferences after feeding on artificial diets lacking an essential amino acid such as methionine.[50] *Limax* can learn to alter their feeding preferences after a single meal on a methionine-deficient diet. It would be interesting to determine if secondary cues associated with the deficient diet, particularly odor cues, would be rendered aversive by association with the negative postingestive cues produced by the essential amino acid-deficient diet. Sensitive food plant selection experiments would likely reveal that reduced digestive efficiency due to alkaloid content in plant tissues would also promote a learned shift of plant preference away from the plants with high alkaloid content. Cates and Orians[51] documented the differential palatability of 100 plant species to two slug species (*Ariolimax columbianus* and *Arion ater*). They ascribed the differential palatability they measured to the differential investment of the tested plant species in anti-herbivore chemical defenses. A learned component of slug food plant choices is likely to play a role in guiding responses such as those studied by Cates and Orians and in subsequent studies of this type.[52,53] Chevalier *et al.*[54] suggest that aversive ingestive conditioning may explain part of the food selection behavior of *H. aspersa* when given access to different chemotypes of *Lupinus albus* differing in their content of quinolizidine alkaloids as their sole food. Olfactory cues from food plants are clearly critical components of food plant selection by *Helix*, as assessed using a Y-tube olfactometer.[55]

Homing Behavior

Another homeostatic challenge is to find a suitable home site for refuge from predators, sunlight, and drought, preferably a home site with ready access to suitable food plants and potential mates. Both terrestrial snails and slugs can establish favorable home sites from which bouts of foraging are initiated at the onset of the circadian activity cycle[56] and to which individuals can return over new paths not determined by the outbound mucus trail.[57] Capture and release experiments have shown that the garden snail, *H. pomatia*, can return to sites favorable for overwintering with an angular error of less than 30° over distances up to 40 m.[58] These observations suggest that providing artificial home sites with initially neutral but distinctive odor cues may provide another robust method for appetitive odor conditioning in snails and slugs.

The importance of understanding the diversity of physiological parameters outlined here and the tolerated ranges of variation is shown by the demonstration that the tolerance limits of a small set of these physiological parameters determine the distribution and abundance of a terrestrial slug, as studied on an island in the Southern Ocean.[59] Deciphering the role of learning and memory storage mechanisms in the optimization of responses of terrestrial slugs and snails to these environmental stressors provides myriad opportunities for asking experimental questions that have the potential to reveal highly developed learning abilities.

COMPLEXITY OF ODOR CONDITIONING

One approach to exploring cognitive function in terrestrial mollusks is to assess the complexity of the learned logic operations that can be carried out by representatives of this group of animals. This approach has been particularly fruitful in studies of aversive and appetitive odor conditioning using *L. maximus*, so a brief summary of these results is presented, supplemented with data on other species. This discussion is facilitated by a summary of some of the odor conditioning experiments carried out with *Limax* as shown in Figure 22.2.[60]

The Kamin blocking effect[61] was an early striking example of higher order conditioning shown robustly in *Limax* odorant conditioning. We have revisited the question as to how the blocking phenomenon, which requires that conditioned stimuli acquire predictive relationships with unconditioned stimuli,[12] can be implemented in a cellular model with plausible neurophysiological mechanisms—a subject of continuing interest to students of the mammalian[62] and insect[63] central nervous system (CNS). The blocking phenomenon involves two phases of conditioning. In the first phase, a conditioned stimulus (CS1) is paired with the

1st ORDER CONDITIONING : $A^+ + Q^- \rightarrow A^-, B^+$

2nd ORDER CONDITIONING : $A^+ + Q^-; B^+ + A^- \rightarrow A^-, B^-, C^+$

$A^+ + Q^-; (A^- B^+) \rightarrow A^-, B^-, C^+$

COMPOUND CONDITIONING : $(A^+ B^+) + Q^- \rightarrow A^-, B^-, C^+$

BLOCK OF CONDITIONING : $A^+ + Q^-; (A^- B^+) + Q^- \rightarrow A^-, B^+$

EXTINCTION AFTER CONDITIONING : $A^+ + Q^-; (A^- B^+) ; A^-, A^- ... \rightarrow A^+, B^+$

$A^+ + Q^-; B^+ + A^- ; A^-, A^- ... \rightarrow A^+, B^-$

APPETITIVE CONDITIONING : $X^- + F^+ \rightarrow X^+, Y^-$

FIGURE 22.2 A subset of the types of learning demonstrated by behavioral experiments with *Limax*. To naive slugs, A–C are innately attractive (+) odors, X and Y are innately repellent (−) odors, Q is an innately repellent taste, and F is an innately attractive taste. Stimuli in parentheses are presented simultaneously. Source: *Reprinted from* Trends in Neurosciences, *9(7), A. Gelperin, Complex associative learning in small neural networks, 323–328, Copyright 1986, with permission from Elsevier.*

unconditioned stimulus (US). In the second phase, a compound stimulus consisting of CS1 and a second conditioned stimulus (CS2) is presented, followed by presentation of the US. Blocking is evidenced by lack of association between CS2 and the US, as found in other invertebrates and in mammals.

Our circuit solution to the blocking problem in *Limax* requires a facilitator neuron (FN) interposed between sensory inputs from CS1, CS2, and US, on the one hand, and the motor neurons executing the conditioned response, on the other hand. It also requires reciprocal inhibition between the CS1, CS2, and US inputs, which can be amplified by learning via activation of the FN. These interactions were explored in simulations based on use of integrate-and-fire model neurons that had been used previously for modeling spiking cortical neurons and synaptic connections that could show potentiation.[64] There are several cellular loci in the *Limax* CNS that potentially can provide tests of the predictions of the proposed model mechanism for blocking, including the serotonergic metacerebral giant cell,[65] the *Limax* feeding command neurons,[66–68] and the odor-responsive parietal neuron v-PN.[69] The demonstration of odor conditioning by the isolated *Limax* CNS greatly augments the feasibility of these experiments,[70] building on prior work demonstrating taste-aversion learning by the isolated *Limax* CNS.[71]

Serotonergic modulation of sensory neurons is a central element in plasticity of the *Aplysia* gill and siphon withdrawal circuit,[72] whereas depletion of serotonin impairs short-term but not long-term memory in *Limax*.[73] Although slugs will quickly learn to avoid a diet devoid of the essential amino acid tryptophan, if a tryptophan-deficient diet is the only food available, slugs will maintain normal levels of brain serotonin and function of the serotonergic metacerebral cell despite this dietary challenge.[74] Multiple subtypes of serotonin receptors are likely involved in responding to the diverse synaptic outputs of the metacerebral cell, as shown in the feeding circuit in *Lymnaea*.[75] In mammals, cells in the anterior piriform cortex are sensitive to deficiencies in essential amino acids, based on uncharged transfer RNA inducing phosphorylation of eukaryotic initiation factor 2 via a non-derepressing type 2 kinase.[76] These findings in the mammalian cortex provide guidance to the search for neurons in the *Limax* CNS responding to diets devoid of essential amino acids and directing selective food avoidance.

The issue of how compound odor stimuli are stored in the CNS can be explored by training slugs to associate an aversive US with a compound CS under conditions in which they experience the compound of CS1 and CS2 as a spatially and temporally uniform mixture or as a spatially and temporally heterogeneous mixture. The heterogeneous mixture of CS1 and CS2 provides information on the components of the mixture as well as the mixture itself. The uniform mixture is more likely to be treated as a single odor object. A difference in storage mechanism for compound odor stimuli is revealed by measuring the aversion to CS1 and CS2 presented alone after aversive compound conditioning to the mixture. *Limax* appears to store a uniform mixture of CS1 and CS2 as a unique odor object such that strong aversive responses are shown to the CS1 + CS2 compound but clear attraction to CS1 alone and CS2 alone remains intact.[77] Conversely, pairing the US with a spatially and temporally heterogeneous mixture of CS1 and CS2 leads to a different form of sensory storage because after conditioning, slugs show strong aversion to CS1 presented alone and CS2 presented alone as well as strong aversion to the compound stimulus. Similar results were obtained using cooling-induced retrograde amnesia to probe the nature of CS storage after compound conditioning. Using cooling to alter stimulus representations, slugs appeared to learn a binary odor mixture as a unitary object when they had no exposure to the separate elements of the mixture.[78] These experiments emphasize the synthetic role of olfactory processing and the critical role of learning in stimulus processing in both mollusks and mammals.[79]

NEUROGENESIS MAY CONTRIBUTE TO OLFACTORY LEARNING

Another design principle of olfactory information processing shared between mammals and mollusks is the addition of new circuit elements both to the sensory periphery and to the central processing structures—the procerebral (PC) lobe and the olfactory bulb (OB)—in adulthood. The *Helix* PC lobe supports significant adult neurogenesis,[80,81] as does the *Limax* PC lobe.[2] Many neurons in the slug CNS undergo DNA endoreplication[82]; however, neurons in the PC lobe are strictly diploid.[83] The relationship between the diploid phenotype of PC lobe neurons and the maintenance of neurogenesis into adulthood remains to be elucidated. From a comparative perspective, it is interesting that neurogenesis also occurs in the olfactory pathways of decapod crustaceans[84] and mammals. Results using optogenetic activation of newborn neurons in the mouse OB provide direct evidence that activity in adult-born neurons may facilitate odor learning.[85]

PROCEREBRUM AS AN OLFACTORY LEARNING CENTER

The procerebrum or PC lobe is a division of the cerebral ganglion unique to terrestrial slugs and snails that is specialized for the processing of olfactory information. The PC lobe contains several tens of thousands of small (7- to 9-mm soma diameter) local interneurons composed of at least two subpopulations, bursting (B) and nonbursting (NB) neurons.[86] B neurons are a few percent of the total neuronal population and are coupled with each other by chemical and electrical synapses.[87] B neurons have extensive synaptic connections with NB neurons. Traveling waves of activity have been recorded from the *Limax* PC lobe *in vitro*, but evidence for traveling waves in the PC lobe of intact behaving slugs is suggestive but not definitive, based on *in vivo* recordings of neuronal dynamics in the PC lobes of intact, awake, behaving slugs.[88] More direct evidence for oscillations in the local field potential of the PC lobe *in vivo* and its responsiveness to odor stimulation has been obtained in *Helix*.[89] There are direct interactions between the oscillatory circuitry in the PC lobe and motor neurons controlling nose positioning and between PC lobe circuitry and remote neurons in the pedal and buccal ganglia.[90] These anatomical results are consistent with a behavioral interaction between olfactory inputs and feeding motor responses that has been documented in *Limax*.[91] Interactions between the oscillatory activity of the PC lobe and motor control of the superior tentacles for olfactory scanning have also been demonstrated in *Helix*.[92] Subsequent work showed that both serotonin and dopamine are involved in the coordination of olfaction and superior tentacle movements determining the scanning range and frequency of the olfactory epithelium borne at the tips of the superior tentacles.[93] The PC lobe receives a rich supply of serotonergic fibers in both *Limax* and *Helix*.[94] Serotonin is a neuromodulator known to increase the frequency of the oscillation of the local field potential in the PC lobe of *Limax*[95] and excite cultured PC neurons.[96]

Increasing evidence indicates that the PC lobe is critical for odor learning in *Limax*[2,97] and, by extension, critical for odor learning in the array of terrestrial gastropod species possessing a PC lobe. Optical studies of odor-elicited activity in the PC lobe showed changes due to prior conditioning of odorant responses.[97,98] Wavelet-based analysis may help to characterize more completely the complex dynamics of local field potential activity as modified by odor input in the PC lobe.[99] Further evidence for the role of the PC lobe in odor learning is seen in the demonstration that a small set of genes is specifically activated in PC lobe neurons due to odor learning. Antibodies to the gene product for one of the learning-activated genes have been made and used to show enhanced levels of expression of the gene after odor learning.[100] A direct test of the role of the PC lobe in odor learning in *Limax* is based on studies of the effects of lesions of the PC lobe on subsequent odor learning.[101] Ablation of the PC lobe 7 days prior to odor conditioning produced clear deficits in odor conditioning not evident in slugs given control operations. PC lobe ablation after odor conditioning also reduced avoidance of the aversively conditioned odor. Control experiments showed that PC lesioned slugs were not anosmic. These results strongly support evidence from optical, molecular, and activity-dependent dye uptake studies also indicating a selective role of the PC lobe in odor learning.

PROGRESS WITH *LIMAX* ODOR LEARNING

Early experiments by Tatsuhiko Sekiguchi[78,102–104] and Tetsuya Kimura[98,105,106] on the characteristics of odor learning in *Limax*, including learning-dependent update of the fluorescent dye Lucifer yellow, have been extended to relate the nature of the odor learning, appetitive or aversive, to the position of the Lucifer yellow-labeled neurons in the PC lobe.[97] These authors also present a two-layered computational model highlighting the importance of the phase difference between oscillations in the two layers of the model. This model is an extension of previous modeling efforts incorporating oscillations and waves in the local field potential of the *Limax* procerebral lobe.[107–109]

A novel experimental finding in the *Limax* odor memory system is that the odor memory appears to be stored in either the right or the left PC lobe,[110] based on effects of unilateral lesion of either the right or the left PC lobe on odor memory retention after odor learning. Unilateral ablation of one PC lobe after learning results in a memory deficit in 50% of the lesioned slugs. This result was predicted by previous findings that memory storage-dependent labeling of PC neurons by Lucifer yellow occurs on only one side of the brain and that *in vitro* studies of a *Limax* nose–brain preparation yielded strong evidence for crossed inhibition between right and left olfactory processing circuits.[111] Anatomical work in the olfactory processing system of *Helix* demonstrates that inputs to the PC lobe originate from interneurons in the digitate ganglion located immediately internal to the olfactory sensory epithelium in the superior tentacles.[112] Learning-induced changes in the MAPK/ERK cascade are also lateralized in the CNS of *Helix*.[113] The general topic of left–right asymmetries in the nervous systems of invertebrates has been reviewed by Frasnelli *et al.*[114]

A peptide neuromodulator has been added to the growing list of classical, neuromodulatory, and gaseous neurotransmitters shown to be present in the PC lobe of terrestrial mollusks, some with dramatic effects on both the oscillatory dynamics and the synaptic interactions in the PC lobes (cf. Table I in Gelperin[115]). The tetrapeptide FMRFamide, first identified in central ganglia of a bivalve mollusk[116] and known to be widely distributed in the *Limax*[117] and *Helix*[118] CNS, has been shown to decrease the oscillatory frequency of the PC lobe local field potential via a G protein-mediated effect on membrane activity.[119] The behavioral choices of *Helix* are reflected in changes in procerebral oscillations.[89]

Ample evidence indicates that the PC lobe is a critical site for odor learning in terrestrial mollusks,[101,120] in a way analogous to the importance of the mushroom bodies for insect odor learning. The PC lobe is also remarkable in two other respects. First, it is the site of constant neurogenesis due to cell divisions in an apical region of proliferative cells, as revealed by labeling newborn neurons with a brief exposure of the intact slug to bromodeoxyuridine (cf. Figure 3 in Watanabe *et al.*[2]). Second, both of the olfactory inputs to the PC lobe originating in the superior and inferior tentacles can regenerate after extirpation,[121] and even more remarkable, the PC lobe can regenerate.[122,123] The physiological and computational implications of active adult neurogenesis for odor learning and memory storage are a subject of intensive investigation in both molluskan and mammalian systems.[85,124]

The gaseous neurotransmitter and neuromodulator nitric oxide plays a prominent role in the olfactory information processing pathways of terrestrial mollusks[125–127] and mammals.[128,129] The modulation of network activity in the PC lobe of *Limax* by nitric oxide[130,131] is now known to depend on synthetic activity of a novel form of nitric oxide synthase found in the PC lobe[132] with a distinctive and highly conserved genomic structure.[133] Nitric oxide in the PC lobe may have a selective role in appetitive rather than aversive odor learning in *Limax*,[134] whereas in *Helix* nitric oxide may be involved in both memory storage and memory loss mechanisms.[135] The role of nitric oxide in hippocampal memory mechanisms is similarly complex.[136]

Further discoveries of mammalian neurotransmitters playing a role in molluskan olfactory information processing will undoubtedly be forthcoming. For example, a molluskan homolog of the vertebrate pituitary adenylate cyclase-activating polypeptide (PACAP) is involved in food-reward conditioning in *Lymnaea*[31] and is widely distributed in the CNS of *Helix*.[30]

LEARNING OF TENTACLE POSITION

Learned modifications of tentacle positioning have been used in *H. aspersa* to demonstrate blocking of conditioned tentacle lowering[137] and conditioned inhibition, verified using both retardation and summation tests.[138] Latent inhibition, second-order conditioning, and sensory preconditioning have also been demonstrated.[139] The tentacle positioning system has also been used to demonstrate the rewarding properties of direct brain stimulation in *Helix*.[140] Lowering of the tentacles is also used as an index of appetitive conditioning to food[141] and can be dissociated from food finding after conditioning.[142] Some elements of the motor system for tentacle positioning have been elucidated,[143,144] including modulatory giant neurons in *Achatina fulica*[145] and motoneurons in *Ariolimax columbianus*[146,147] and *Helix*.[148] *Tritonia* peptide is widely distributed in central neurons of *H. aspersa*, including motoneurons sending processes to the tentacle retractor muscles,[149] probably representing another example of a peptide co-transmitter co-localized in neurons liberating a classical neurotransmitter such as glutamate,[150,151] acetylcholine,[152,153] or serotonin.[154]

CONCLUSIONS

The evolution of studies in comparative cognition can be viewed as the confluence of two complimentary intellectual streams. On the one hand, an increasing number of experimentalists and theorists interested in learning and memory are finding that an amazing variety of invertebrates can display sophisticated feats of cognitive computation, thought until recently to be the exclusive province of vertebrates, mammals, and humans.[155–159] On the other hand, as the wealth of experimental data detailing the cognitive feats performed by invertebrates increases, the outmoded idea of the Scala naturae, which posits evolution as a linear process with *Homo sapiens* as its ultimate accomplishment, is waning. Charles Darwin was very careful to avoid this linear, progressive view of evolution, stating that "it is absurd to talk of one animal being higher than another," a comment found in his unpublished transmutation notebook "B" of 1837.[160] At the confluence of these two streams emerges the idea that "simple" animals often reveal their possession of surprisingly sophisticated cognitive machinery when experimental questions are asked in the proper context.

References

1. Shettleworth SJ. Clever animals and killjoy explanations in comparative psychology. *Trends Cogn Sci*. 2010;14(11):477–481.
2. Watanabe S, Kirino Y, Gelperin A. Neural and molecular mechanisms of microcognition in *Limax*. *Learn Mem*. 2008;15(9):633–642.
3. Dalesman S, Lukowiak K. Social snails: the effect of social isolation on cognition is dependent on environmental context. *J Exp Biol*. 2011;214(24):4179–4185.

4. Arshavsky YI. Cellular and network properties in the functioning of the nervous system: from central pattern generators to cognition. *Brain Res Rev*. 2003;41(2-3):229–267.
5. Ikeda Y. A perspective on the study of cognition and sociality of cephalopod mollusks, a group of intelligent marine invertebrates. *Jpn Psychol Res*. 2009;51(3):146–153.
6. Mather JA. Cephalopod consciousness: behavioural evidence. *Conscious Cogn*. 2008;17(1):37–48.
7. Kandel ER, Tauc L. Anomalous rectification in the metacerebral giant cells and its consequences for synaptic transmission. *J Physiol*. 1966;183(2):287–304.
8. Kandel ER, Tauc L. Input organization of two symmetrical giant cells in the snail brain. *J Physiol*. 1966;183(2):269–286.
9. Chang JJ, Gelperin A, Johnson FH. Intracellularly injected aequorin detects trans-membrane calcium flux during action potentials in an identified neuron from the terrestrial slug, *Limax maximus*. *Brain Res*. 1974;77:431–442.
10. Antic S, Zecevic D. Optical signals from neurons with internally applied voltage-sensitive dyes. *J Neurosci*. 1995;15:1392–1405.
11. Gelperin A, Hopfield JJ, Tank DW. The logic of *Limax* learning. In: Selverston AI, ed. *Model Neural Networks and Behavior*. New York: Plenum; 1986:237–261.
12. Sahley CL, Martin KA, Gelperin A. Analysis of associative learning in the terrestrial mollusc *Limax maximus*: II Appetitive learning. *J Comp Physiol A*. 1990;167:339–345.
13. Niven JE, Laughlin SB. Energy limitation as a selective pressure on the evolution of sensory systems. *J Exp Biol*. 2008;211(11):1792–1804.
14. Burger JMS, Kolss M, Pont J, Kawecki TJ. Learning ability and longevity: a symmetrical evolutionary trade-off in *Drosophila*. *Evolution*. 2008;62(6):1294–1304.
15. Sznajder B, Sabelis MW, Egas M. The interplay between genetic and learned components of behavioral traits. *J Plant Interact*. 2011;6(2-3):77–80.
16. Kawecki TJ. Evolutionary ecology of learning: insights from fruit flies. *Popul Ecol*. 2010;52(1):15–25.
17. Gould JL. Animal cognition. *Curr Biol*. 2004;14(10):R372–R375.
18. Roeder K. *Nerve Cells and Insect Behavior*. Cambridge, MA: Harvard University Press; 1998.
19. Elliott CJH, Susswein AJ. Comparative neuroethology of feeding control in molluscs. *J Exp Biol*. 2002;205(7):877–896.
20. Prior DJ. Neuronal control of osmoregulatory responses in gastropods. *Adv Comp Environ Physiol*. 1989;5:1–24.
21. Denny MW. Mechanical properties of pedal mucus and their consequences for gastropod structure and performance. *Amer Zool*. 1984;24(1):23–36.
22. Lai JH, del Alamo JC, Rodriguez-Rodriguez J, Lasheras JC. The mechanics of the adhesive locomotion of terrestrial gastropods. *J Exp Biol*. 2010;213(22):3920–3933.
23. Prior DJ. Analysis of contact-rehydration in terrestrial gastropods: osmotic control of drinking behaviour. *J Exp Biol*. 1984;111:63–73.
24. Prior DJ, Uglem GL. Analysis of contact-rehydration in terrestrial gastropods: absorption of C-14 inulin through the epithelium of the foot. *J Exp Biol*. 1984;111:75–80.
25. Makra ME, Prior DJ. Angiotensin-II can initiate contact-rehydration in terrestrial slugs. *J Exp Biol*. 1985;119:385–388.
26. Banta PA, Welsford IG, Prior DJ. Water-orientation behavior in the terrestrial gastropod *Limax maximus*: the effects of dehydration and arginine vasotocin. *Physiol Zool*. 1990;63:683–696.
27. Mensink PJ, Henry HAL. Rain events influence short-term feeding preferences in the snail *Cepaea nemoralis*. *J Molluscan Stud*. 2011;77:241–247.
28. Mizrahi T, Heller J, Goldenberg S, Arad Z. Heat shock proteins and resistance to desiccation in congeneric land snails. *Cell Stress Chaperones*. 2010;15(4):351–363.
29. Schmidt-Nielsen K, Taylor CR, Shkolnik A. Desert snails: problems of heat, water and food. *J Exp Biol*. 1971;55:385–398.
30. Hernadi L, Pirger Z, Kiss T, et al. The presence and distribution of pituitary adenylate cyclase activating polypeptide and its receptor in the snail *Helix pomatia*. *Neuroscience*. 2008;155(2):387–402.
31. Pirger Z, Laszlo Z, Kemenes I, Toth G, Reglodi D, Kemenes GA. Homolog of the vertebrate pituitary adenylate cyclase-activating polypeptide is both necessary and instructive for the rapid formation of associative memory in an invertebrate. *J Neurosci*. 2010;30(41):13766–13773.
32. Hernadi L, Karpati L, Gyori J, Vehovszky A, Hiripi L. Humoral serotonin and dopamine modulate the feeding in the snail, *Helix pomatia*. *Acta Biol Hung*. 2008;59(Suppl):39–46.
33. Gaitan-Espitia JD, Franco M, Bartheld JL, Nespolo RF. Repeatability of energy metabolism and resistance to dehydration in the invasive slug *Limax maximus*. *Invertebr Biol*. 2012;131(1):11–18.
34. Kiss T. Diversity and abundance: the basic properties of neuropeptide action in molluscs. *Gen Comp Endocrinol*. 2011;172(1):10–14.
35. Wessnitzer J, Mangan M, Webb B. Place memory in crickets. *Proc Biol Sci*. 2008;275(1637):915–921.
36. Foucaud J, Burns JG, Mery F. Use of spatial information and search strategies in a water maze analog in *Drosophila melanogaster*. *PLoS ONE*. 2010;5(12):e15231.
37. Grimm B, Schaumberger K. Daily activity of the pest slug *Arion lusitanicus* under laboratory conditions. *Ann Appl Biol*. 2002;141(1):35–44.
38. Schuder I, Port G, Bennison J. The behavioural response of slugs and snails to novel molluscicides, irritants and repellents. *Pest Manag Sci*. 2004;60(12):1171–1177.
39. Clifford KT, Gross L, Johnson K, Martin KJ, Shaheen N, Harrington MA. Slime-trail tracking in the predatory snail, *Euglandina rosea*. *Behav Neurosci*. 2003;117:1086–1095.
40. Davis-Berg EC. The predatory snail *Euglandina rosea* successfully follows mucous trails of both native and non-native prey snails. *Invertebr Biol*. 2012;131(1):1–10.
41. Holland BS, Chock T, Lee A, Sugiura S. Tracking behavior in the snail *Euglandina rosea*: first evidence of preference for endemic vs. biocontrol target pest species in Hawaii. *Am Malacol Bull*. 2012;30(1):153–157.
42. Shaheen N, Patel K, Patel P, Moore M, Harrington MA. A predatory snail distinguishes between conspecific and heterospecific snails and trails based on chemical cues in slime. *Anim Behav*. 2005;70:1067–1077.
43. Derby CD. Escape by inking and secreting: marine molluscs avoid predators through a rich array of chemicals and mechanisms. *Biol Bull*. 2007;213(3):274–289.
44. Takeichi M, Hirai Y, Yusa Y. A water-borne sex pheromone and trail following in the apple snail, *Pomacea canaliculata*. *J Molluscan Stud*. 2007;73:275–278.
45. Ng TPT, Davies MS, Stafford R, Williams GA. Mucus trail following as a mate-searching strategy in mangrove littorinid snails. *Anim Behav*. 2011;82(3):459–465.
46. Gelperin A. Neural computations with mammalian infochemicals. *J Chem Ecol*. 2008;34(7):928–942.
47. Chase R, Darbyson E, Horn KE, Samarova E. A mechanism aiding simultaneously reciprocal mating in snails. *Can J Zool*. 2010;88(1):99–107.
48. Agular R, Wink M. How do slugs cope with toxic alkaloids? *Chemoecology*. 2005;15:167–177.

49. Honemann L, Nentwig W. Does feeding on Bt-maize affect the slug *Arion vulgaris* (Mollusca: Arionidae)? *Biocontrol Sci Technol.* 2010;20:13–18.
50. Delaney K, Gelperin A. Post-ingestive food-aversion learning to amino acid deficient diets by the terrestrial slug *Limax maximus*. *J Comp Physiol A.* 1986;159:281–295.
51. Cates RG, Orians GH. Successional status and the palatability of plants to generalized herbivores. *Ecology.* 1975;56:410–418.
52. Kozlowski J, Kozlowska M. Palatability and consumption of 95 species of herbaceous plants and oilseed rape for *Arion lusitanicus* Mabille 1868. *J Conchology.* 2009;40:79–90.
53. Hahn PG, Dornbush ME. Exotic consumers interact with exotic plants to mediate native plant survival in a Midwestern forest herb layer. *Biol Invasions.* 2012;14:449–460.
54. Chevalier L, Desbuquois C, Papineau J, Charrier M. Influence of the quinolizidine alkaloid content of *Lupinus albus* (Fabaceae) on the feeding choice of *Helix aspersa* (Gastropoda : Pulmonata). *J Molluscan Stud.* 2000;66:61–68.
55. Hanley ME, Collins SA, Swann C. Advertising acceptability: is mollusk olfaction important in seedling selection? *Plant Ecology.* 2011;212(4):727–731.
56. Hommay G, Lorvelec O, Jacky F. Daily activity rhythm and use of shelter in the slugs *Deroceras reticulatum* and *Arion distinctus* under laboratory conditions. *Ann Appl Biol.* 1998;132(1): 167–185.
57. Gelperin A. Olfactory basis of homing behavior in the giant garden slug, *Limax maximus*. *Proc Natl Acad Sci USA.* 1974;71: 966–970.
58. Edelstam C, Palmer C. Homing behaviour in gastropodes. *Oikos.* 1950;2(2):258–270.
59. Lee JE, Janion C, Marais E, van Vuuren BJ, Chown SL. Physiological tolerances account for range limits and abundance structure in an invasive slug. *Proc R Soc B Biol Sci.* 2009;276 (1661):1459–1468.
60. Gelperin A. Complex associative learning in small neural networks. *Trends Neurosci.* 1986;9(7):323–328.
61. Kamin LJ. Predictability, surprise, attention and conditioning. In: Campbell BA, Church RM, eds. *Punishment and Aversive Behavior.* New York: Appleton-Century-Crofts; 1969:279–296.
62. Padlubnaya DB, Parekh NH, Brown TH. Neurophysiological theory of Kamin blocking in fear conditioning. *Behav Neurosci.* 2006;120(2):337–352.
63. Giurfa M, Sandoz JC. Invertebrate learning and memory: fifty years of olfactory conditioning of the proboscis extension response in honeybees. *Learn Mem.* 2012;19(2):54–66.
64. Goel P, Gelperin A. A neuronal network for the logic of *Limax* learning. *J Comput Neurosci.* 2006;21:259–270.
65. Gelperin A. Synaptic modulation by identified serotonergic neurons. In: Jacobs B, Gelperin A, eds. *Serotonin Neurotransmission and Behavior.* Cambridge, MA: MIT Press; 1981:288–301.
66. Delaney K, Gelperin A. Cerebral interneurons controlling fictive feeding in *Limax maximus*: I. Anatomy and criteria for re-identification. *J Comp Physiol A.* 1990;166:297–310.
67. Delaney K, Gelperin A. Cerebral interneurons controlling fictive feeding in *Limax maximus*: II. Initiation and modulation of fictive feeding. *J Comp Physiol A.* 1990;166:311–326.
68. Delaney K, Gelperin A. Cerebral interneurons controlling fictive feeding in *Limax maximus*: III. Integration of sensory inputs. *J Comp Physiol A.* 1990;166:327–343.
69. Inoue T, Inokuma Y, Watanabe S, Kirino Y. *In vitro* study of odor-evoked behavior in a terrestrial mollusk. *J Neurophysiol.* 2004;91:372–381.
70. Inoue T, Murakami M, Watanabe S, Inokuma Y, Kirino Y. *In vitro* odor-aversion conditioning in a terrestrial mollusk. *J Neurophysiol.* 2006;95:3898–3903.
71. Chang JJ, Gelperin A. Rapid taste-aversion learning by an isolated molluscan CNS. *Proc Natl Acad Sci USA.* 1980;77: 6204–6206.
72. Dunn TW, Farah CA, Sossin WS. Inhibitory responses in *Aplysia* pleural sensory neurons act to block excitability, transmitter release, and PKC Apl II activation. *J Neurophysiol.* 2012;107(1): 292–305.
73. Shirahata T, Tsunoda M, Santa T, Kirino Y, Watanabe S. Depletion of serotonin selectively impairs short-term memory without affecting long-term memory in odor learning in the terrestrial slug *Limax valentianus*. *Learn Mem.* 2006;13(3):267–270.
74. Gietzen DW, Harris AS, Carlson S, Gelperin A. Amino acids and serotonin in *Limax maximus* after a tryptophan devoid diet. *Comp Biochem Physiol A.* 1992;101(1):143–149.
75. Kawai R, Kobayashi S, Fujito Y, Ito E. Multiple subtypes of serotonin receptors in the feeding circuit of a pond snail. *Zool Sci.* 2011;28(7):517–525.
76. Rudell JB, Rechs AJ, Kelman TJ, Ross-Inta CM, Hao SZ, Gietzen DW. The anterior piriform cortex is sufficient for detecting depletion of an indispensable amino acid, showing independent cortical sensory function. *J Neurosci.* 2011;31(5):1583–1590.
77. Hopfield JF, Gelperin A. Differential conditioning to a compound stimulus and its components in the terrestrial mollusc *Limax maximus*. *Behav Neurosci.* 1989;103:274–293.
78. Sekiguchi T, Suzuki H, Yamada A, Kimura T. Aversive conditioning to a compound odor stimulus and its components in a terrestrial mollusc. *Zool Sci.* 1999;16:879–883.
79. Wilson DA, Stevenson RJ. Olfactory perceptual learning: the critical role of memory in odor discrimination. *Neurosci Biobehav Rev.* 2003;27:307–328.
80. Zakharov IS, Hayes NL, Ierusalimsky VN, Nowakowski RS, Balaban PM. Postembryonic neuronogenesis in the procerebrum of the terrestrial snail, *Helix lucorum*. *J Neurobiol.* 1998;35: 271–276.
81. Longley RD. Neurogenesis in the procerebrum of the snail *Helix aspersa*: a quantitative analysis. *Biol Bull.* 2011;221(2):215–226.
82. Yamagishi M, Ito E, Matsuo R. DNA endoreplication in the brain neurons during body growth of an adult slug. *J Neurosci.* 2011;31 (15):5596–5604.
83. Chase R, Tolloczko B. Evidence for differential DNA endoreplication during the development of a molluscan brain. *J Neurobiol.* 1987;18:395–406.
84. Sintoni S, Benton JL, Beltz BS, Hansson BS, Harzsch S. Neurogenesis in the central olfactory pathway of adult decapod crustaceans: development of the neurogenic niche in the brains of procambarid crayfish. *Neural Develop.* 2012;7:1.
85. Alonso M, Lepousez G, Wagner S, et al. Activation of adult-born neurons facilitates learning and memory. *Nat Neurosci.* 2012;15 (6):897–904.
86. Wang JW, Denk W, Flores J, Gelperin A. Initiation and propagation of calcium-dependent action potentials in a coupled network of olfactory interneurons. *J Neurophysiol.* 2001;85(2): 977–985.
87. Kazanci FG, Ermentrout B. Pattern formation in an array of oscillators with electrical and chemical coupling. *Siam J Appl Math.* 2007;67(2):512–529.
88. Cooke IRC, Gelperin A. *In vivo* recordings of spontaneous and odor-modulated dynamics in the *Limax* olfactory lobe. *J Neurobiol.* 2001;46:126–141.
89. Samarova E, Balaban P. Changes in frequency of spontaneous oscillations in procerebrum correlate to behavioural choice in terrestrial snails. *Front Cell Neurosci.* 2009;3:8.
90. Gelperin A, Flores J. Vital staining from dye-coated microprobes identifies new olfactory interneurons for optical and electrical recording. *J Neurosci Methods.* 1997;72:97–108.

91. Sahley CL, Martin KA, Gelperin A. Odors can induce feeding motor responses in the terrestrial mollusc *Limax maximus*. *Behav Neurosci*. 1992;106:563–568.
92. Nikitin ES, Zakharov IS, Samarova EI, Kemenes G, Balaban PM. Fine tuning of olfactory orientation behaviour by the interaction of oscillatory and single neuronal activity. *Eur J Neurosci*. 2005;22(11):2833–2844.
93. Roshchin M, Balaban PM. Neural control of olfaction and tentacle movements by serotonin and dopamine in terrestrial snail. *J Comp Physiol A Neuroethol Sens Neural Behav Physiol*. 2012;198 (2):145–158.
94. Battonyai I, Elekes K. The 5-HT immunoreactive innervation of the *Helix* procerebrum. *Acta Biol Hung*. 2012;63:96–103.
95. Gelperin A, Rhines LD, Flores J, Tank DW. Coherent network oscillations by olfactory interneurons: modulation by endogenous amines. *J Neurophysiol*. 1993;69(6):1930–1939.
96. Rhines LD, Sokolove PG, Flores J, Tank DW, Gelperin A. Cultured olfactory interneurons from *Limax maximus*: optical and electrophysiological studies of transmitter-evoked responses. *J Neurophysiol*. 1993;69(6):1940–1947.
97. Sekiguchi T, Furudate H, Kimura T. Internal representation and memory formation of odor preference based on oscillatory activities in a terrestrial slug. *Learn Mem*. 2010;17(8):372–380.
98. Kimura T, Toda S, Sekiguchi T, Kawahara S, Kirino Y. Optical recording analysis of olfactory response of the procerebral lobe in the slug brain. *Learn Mem*. 1998;4:389–400.
99. Schutt A, Ito I, Rosso OA, Figliola A. Wavelet analysis can sensitively describe dynamics of ethanol evoked local field potentials of the slug (*Limax marginatus*) brain. *J Neurosci Methods*. 2003;129(2):135–150.
100. Fukunaga S, Matsuo R, Hoshino S, Kirino Y. Novel kruppel-like factor is induced by neuronal activity and by sensory input in the central nervous system of the terrestrial slug *Limax valentianus*. *J Neurobiol*. 2006;66(2):169–181.
101. Kasai Y, Watanabe S, Kirino Y, Matsuo R. The procerebrum is necessary for odor-aversion learning in the terrestrial slug *Limax valentianus*. *Learn Mem*. 2006;13(4):482–488.
102. Sekiguchi T, Suzuki H, Yamada A, Mizukami A. Cooling-induced retrograde amnesia reflexes pavlovian conditioning associations in *Limax flavus*. *Neurosci Res*. 1994;18:267–275.
103. Sekiguchi T, Yamada A, Suzuki H. Reactivation dependent changes in memory states in the terrestrial slug *Limax flavus*. *Learn Mem*. 1997;4:356–364.
104. Sekiguchi T, Yamada A, Suzuki H, Mizukami A. Temporal analysis of the retention of a food-aversion conditioning in *Limax flavus*. *Zool Sci (Tokyo)*. 1991;8:103–111.
105. Kimura T, Suzuki H, Kono E, Sekiguchi T. Mapping of interneurons that contribute to food aversion conditioning in the slug brain. *Learn Mem*. 1998;4:376–388.
106. Kimura T, Toda S, Sekiguchi T, Kirino Y. Behavioral modulation induced by food odor aversive conditioning and its influence on the olfactory responses of an oscillatory brain network in the slug *Limax marginatus*. *Learn Mem*. 1998;4:365–375.
107. Ermentrout B, Flores J, Gelperin A. Minimal model of oscillations and waves in the *Limax* olfactory lobe with tests of the model's predictive power. *J Neurophysiol*. 1998;79:2677–2689.
108. Ermentrout B, Wang JW, Flores J, Gelperin A. Model for olfactory discrimination and learning in *Limax* procerebrum incorporating oscillatory dynamics and wave propagation. *J Neurophysiol*. 2001;85:1444–1452.
109. Ermentrout GB, Wang JW, Flores J, Gelperin A. Model for transition from waves to synchrony in the olfactory lobe of *Limax*. *J Comput Neurosci*. 2004;17:365–383.
110. Matsuo R, Kobayashi S, Yamagishi M, Ito E. Two pairs of tentacles and a pair of procerebra: optimized functions and redundant structures in the sensory and central organs involved in olfactory learning of terrestrial pulmonates. *J Exp Biol*. 2011;214(6):879–886.
111. Teyke T, Wang JW, Gelperin A. Lateralized memory storage and crossed inhibition during odor processing by *Limax*. *J Comp Physiol A*. 2000;186:269–278.
112. Ierusalimsky VN, Balaban PM. Two morphological sub-systems within the olfactory organs of a terrestrial snail. *Brain Res*. 2010;1326:68–74.
113. Kharchenko OA, Grinkevich VV, Vorobiova OV, Grinkevich LN. Learning-induced lateralized activation of the MAPK/ERK cascade in identified neurons of the food-aversion network in the mollusk *Helix lucorum*. *Neurobiol Learn Mem*. 2010;94 (2):158–166.
114. Frasnelli E, Vallortigara G, Rogers LJ. Left–right asymmetries of behaviour and nervous system in invertebrates. *Neurosci Biobehav Rev*. 2012;36(4):1273–1291.
115. Gelperin A. Oscillatory dynamics and information processing in olfactory systems. *J Exp Biol*. 1999;202:1855–1864.
116. Price DA, Greenberg MJ. Purification and characterization of a cardioexcitatory neuropeptide from central ganglia of a bivalve mollusk. *Prep Biochem*. 1977;7(3-4):261–281.
117. Cooke I, Gelperin A. Distribution of FMRFamide-like immunoreactivity in the nervous system of the slug *Limax maximus*. *Cell Tissue Res*. 1988;253:69–76.
118. Elekes K, Nassel DR. Distribution of FMRFamide-like immunoreactive neurons in the central nervous system of the snail *Helix pomatia*. *Cell Tissue Res*. 1990;262:177–190.
119. Kobayashi S, Hattori M, Karoly E, Ito E, Matsuo R. FMRFamide regulates oscillatory activity of the olfactory center in the slug. *Eur J Neurosci*. 2010;32(7):1180–1192.
120. Matsuo R, Ito E. Recovery of learning ability after the ablation of the procerebrum in the terrestrial slug, *Limax valentianus*. *Acta Biol Hung*. 2008;59:73–76.
121. Matsuo R, Kobayashi S, Tanaka Y, Ito E. Effects of tentacle amputation and regeneration on the morphology and activity of the olfactory center of the terrestrial slug *Limax valentianus*. *J Exp Biol*. 2010;213(18):3144–3149.
122. Matsuo R, Kobayashi S, Murakami J, Ito E. Spontaneous recovery of the injured higher olfactory center in the terrestrial slug *Limax*. *PLoS ONE*. 2010;5(2):e9054.
123. Matsuo R, Ito E. Spontaneous regeneration of the central nervous system in gastropods. *Biol Bull*. 2011;221(1):35–42.
124. Aimone JB, Wiles J, Gage FH. Computational influence of adult neurogenesis on memory encoding. *Neuron*. 2009;61(2): 187–202.
125. Fujie S, Aonuma H, Ito I, Gelperin A, Ito E. The nitric oxide/cyclic GMP pathway in the olfactory processing system of the terrestrial slug *Limax marginatus*. *Zool Sci*. 2002;19: 15–26.
126. Fujie S, Yamamoto T, Murakami J, et al. Nitric oxide synthase and soluble guanylyl cyclase underlying the modulation of electrical oscillations in a central olfactory organ. *J Neurobiol*. 2005;62:14–30.
127. Nacsa K, Elekes K, Serfozo Z. Immunodetection and localization of nitric oxide synthase in the olfactory center of the terrestrial snail, *Helix pomatia*. *Acta Biol Hung*. 2012;63:104–112.
128. Lowe G, Buerk DG, Ma J, Gelperin A. Tonic and stimulus-evoked nitric oxide production in the mouse olfactory bulb. *Neuroscience*. 2008;153(3):842–850.
129. McQuade LE, Ma J, Lowe G, Ghatpande A, Gelperin A, Lippard SJ. Visualization of nitric oxide production in the mouse main olfactory bulb by a cell-trappable copper(II) fluorescent probe. *Proc Natl Acad Sci USA*. 2010;107(19): 8525–8530.

130. Gelperin A. Nitric oxide mediates network oscillations of olfactory interneurons in a terrestrial mollusc. *Nature*. 1994;369: 61–63.
131. Gelperin A, Flores J, Raccuia-Behling F, Cooke IRC. Nitric oxide and carbon monoxide modulate oscillations of olfactory interneurons in a terrestrial mollusc. *J Neurophysiol*. 2000;83: 116–127.
132. Matsuo R, Ito E. A novel nitric oxide synthase expressed specifically in the olfactory center. *Biochem Biophys Res Commun*. 2009;386(4):724–728.
133. Matsuo R, Misawa K, Ito E. Genomic structure of nitric oxide synthase in the terrestrial slug is highly conserved. *Gene*. 2008;415(1-2):74–81.
134. Yabumoto T, Takanashi F, Kirino Y, Watanabe S. Nitric oxide is involved in appetitive but not aversive olfactory learning in the land mollusk *Limax valentianus*. *Learn Mem*. 2008;15(4): 229–232.
135. Balaban PM, Roshchin MV, Korshunova TA. Two-faced nitric oxide is necessary for both erasure and consolidation of memory. *Zh Vyssh Nerv Deiat Im I P Pavlova*. 2011;61(3):274–280.
136. Bon CLM, Garthwaite J. On the role of nitric oxide in hippocampal long-term potentiation. *J Neurosci*. 2003;23:1941–1948.
137. Acebes F, Solar P, Carnero S, Loy I. Blocking of conditioning of tentacle lowering in the snail (*Helix aspersa*). *Q J Exp Psychol*. 2009;62(7):1315–1327.
138. Acebes F, Solar P, Moris J, Loy I. Associative learning phenomena in the snail (*Helix aspersa*): conditioned inhibition. *Learn Behav*. 2012;40(1):34–41.
139. Loy I, Fernandez V, Acebes F. Conditioning of tentacle lowering in the snail (*Helix aspersa*): acquisition, latent inhibition, overshadowing, second-order conditioning, and sensory preconditioning. *Learn Behav*. 2006;34(3):305–314.
140. Balaban PM, Chase R. Self-stimulation in snails. *Neurosci Res Comm*. 1989;4(3):139–146.
141. Ungless MAA. Pavlovian analysis of food-attraction conditioning in the snail *Helix aspersa*. *Anim Learn Behav*. 1998;26:15–19.
142. Ungless MA. Dissociation of food-finding and tentacle-lowering, following food-attraction conditioning in the snail, *Helix aspersa*. *Behav Processes*. 2001;53(1-2):97–101.
143. Hernadi L, Teyke T. Novel triplet of flexor muscles in the posterior tentacles of the snail, *Helix pomatia*. *Acta Biol Hung*. 2012;63:123–128.
144. Krajcs N, Mark L, Elekes K, Kiss T. Morphology, ultrastructure and contractile properties of muscles responsible for superior tentacle movements of the snail. *Acta Biol Hung*. 2012;63: 129–140.
145. Bugai VV, Zhuravlev VL, Safonova TA. Neuroeffector connections of giant multimodal neurons in the African snail *Achatina fulica*. *Neurosci Behav Physiol*. 2005;35(6):605–613.
146. Chan CY, Moffett S. Cerebral moto-neurons mediating tentacle retraction in the land slug *Ariolimax columbianus*. *J Neurobiol*. 1982;13(2):163–172.
147. Chan CY, Moffett S. Neurophysiological mechanisms underlying habituation of the tentacle retraction reflex in the land slug *Ariolimax*. *J Neurobiol*. 1982;13(2):173–183.
148. Nikitin ES, Zakharov IS, Balaban PM. Regulation of tentacle length in snails by odor concentration. *Neurosci Behav Physiol*. 2006;36(1):63–72.
149. Pavlova GA, Willows AOD. Immunological localization of *Tritonia* peptide in the central and peripheral nervous system of the terrestrial snail *Helix aspersa*. *J Comp Neurol*. 2005;491 (1):15–26.
150. Fox LE, Lloyd PE. Glutamate is a fast excitatory transmitter at some buccal neuromuscular synapses in *Aplysia*. *J Neurophysiol*. 1999;82(3):1477–1488.
151. Matsuo R, Kobayashi S, Watanabe S, et al. Glutamatergic neurotransmission in the procerebrum (olfactory center) of a terrestrial mollusk. *J Neurosci Res*. 2009;87(13):3011–3023.
152. D'este L, Casini A, Kimura S, et al. Immunohistochemical demonstration of cholinergic structures in central ganglia of the slug (*Limax maximus, Limax valentianus*). *Neurochem Int*. 2011;58 (5):605–611.
153. Bellier JP, Kimura H. Peripheral type of choline acetyltransferase: biological and evolutionary implications for novel mechanisms in cholinergic system. *J Chem Neuroanat*. 2011;42(4): 225–235.
154. Kiss T, Hernadi L, Laszlo Z, Fekete ZN, Elekes K. Peptidergic modulation of serotonin and nerve elicited responses of the salivary duct muscle in the snail, *Helix pomatia*. *Peptides*. 2010;31 (6):1007–1018.
155. Avargues-Weber A, Dyer AG, Combe M, Giurfa M. Simultaneous mastering of two abstract concepts by the miniature brain of bees. *Proc Natl Acad Sci USA*. 2012;109(19): 7481–7486.
156. Greenspan RJ, van Swinderen B. Cognitive consonance: complex brain functions in the fruit fly and its relatives. *TINS*. 2004;27:707–711.
157. Brembs B. Towards a scientific concept of free will as a biological trait: spontaneous actions and decision-making in invertebrates. *Proc R Soc B Biol Sci*. 2011;278(1707):930–939.
158. Stamps JA, Briffa M, Biro PA. Unpredictable animals: individual differences in intraindividual variability (IIV). *Anim Behav*. 2012;83(6):1325–1334.
159. Giurfa M, Avargues A, Menzel R. Non-elemental learning in invertebrates. In: Breed MD, Moore J, eds. *Encyclopedia of Animal Behavior*. Waltham, MA: Academic Press; 2010:566–572.
160. Gross CG. Reply to: the ladder of progress in neuroscience. *TINS*. 1994;17:227–228.

4.2.2 Cephalopods

CHAPTER

23

Observational and Other Types of Learning in *Octopus*

Piero Amodio and Graziano Fiorito*,†*

*Associazione Cephalopod Research, CephRes-ONLUS, Napoli, Italy †Stazione Zoologica Anton Dohrn, Napoli, Italy

INTRODUCTION

Popular and scientific interests in octopuses and their allies, the cuttlefishes and squids, are interwoven together and find their roots in antiquity. The extensive use of these animals as a source of food and their artistic representations during the Middle Paleolithic or Minoan civilizations are popular examples. This heritage continues to provide an important place for octopuses in the food market for the Mediterranean and other civilizations throughout the world (e.g., Spain, Morocco, and Japan).[1,2] What probably also attracts people and science is the natural curiosity of these creatures, which is interpreted as a sign of their intelligence.[3]

A glance at the videos posted on YouTube and at the scientific literature that has been published during the past 3 years provides an incredible variety of examples of the richness of behavioral adaptations of cephalopod mollusks and octopods in particular.

In the wild, octopuses collect and transport, by 'stilt-walking,' coconut shell halves on soft sandy bottoms and use them to build dens.[4] They are also found to move bipedally along bottoms using a rolling gait.[5] On the other hand, to prevent intense predation pressure, cephalopods evolved an effective and impressive camouflaging ability that exploits features of their surroundings to enable them to 'blend in'[6,7] (for review see[8,9]). Camouflaging does not represent a limitation for communication in cephalopods. The richness and diversity of body patterns allow octopuses and their allies to both disguise themselves and provide the means to signal to others.[8–11] Communication may be public or covert, as occurs also in fish.[12] In fact, cephalopods are capable of using signals that are not 'visible' to human senses, being based on polarized light[13,14]; using this 'special channel,' animals communicate with conspecifics without changing the body colors that provide camouflage.

Finally, cephalopods are known to be learning invertebrates (for reviews, see[15–18]). Learning, camouflage to achieve background matching and signaling to others, bipedalism, and the more complex stilt-walking are key examples of the complexity and richness of the behavioral repertoire of cephalopods. They recently inspired, for example, biorobotics using the octopus as a model of embodied intelligence and for soft robotics,[19–21] and they continue to be a source of characters in advertising.

COMPLEXITY VERSUS SIMPLICITY: EXAMPLES FROM OCTOPUSES

Locomotion

Bipedal locomotion requires muscles to act against a rigid skeleton. In the octopus, this is achieved by the action of transverse, longitudinal, and oblique bundles of muscles in the absence of any skeletal support.[22–24] During walking, a given segment of an octopus arm is stiffened, while the next one is relaxed. Stance and swing[5] are achieved by contraction of local bundles of transverse and longitudinal muscles in the arm (stance), inducing a stiffened segment. A 'relaxed' arm swings due to the contraction of another portion of longitudinal muscles and the resistance of the corresponding transverse ones. In the case of the more complex stilt-walking,[4] the simultaneous and independent neuromuscular coordination of a given segment of

Invertebrate Learning and Memory.
DOI: http://dx.doi.org/10.1016/B978-0-12-415823-8.00023-X

each of the eight arms results in the assemblage of locomotory and postural patterns that allow the animal to tip-tap-toe while carrying the coconut shell. In addition, holding the coconut is due to the contraction of a short part of the longitudinal muscles around the rim of the shell, assisted by the suction capabilities of individual suckers of that arm segment.

Locomotory patterns are performed through refined neural control of different muscolar bundles distributed with similar arrangements along each of the eight arms. These are individually coordinated in each ideal part of a given arm, thus producing the large observable variety of movements, postures, and actions.[9,25] Complex postural and locomotory patterns do not seem to require sophisticated central neural control. The vast majority of patterns of an arm are produced by neural circuitry restricted to the arm with little feedback by higher order motor centers.[26–29] It is noteworthy that the arms contain approximately 60% of the cells constituting the nervous system of an octopus.[30,31]

Locomotory patterns are not limited to ambulatory movements; they are also performed during manipulatory actions involved, for example, in reaching and dealing with prey[28,32–34] or in play-like behavior.[35,36] These require various forms of acquisition of the textural characteristics of the prey through learning and tactile processing (for review, see[15,37]).

Mimicry

In recent years, the mimic octopus, *Thaumoctopus mimicus*, has become a classic example of the dramatic continuous changes of the body form and appearance that occur in cephalopods. The animal is capable of mimicking other, noncephalopod, species. As mentioned previously, octopuses are known to exhibit excellent crypsis and polyphenism achieved by neurally controlled skin patterning.[9,10,38,39] However, the abilities of some sand-dwelling octopuses, such as *T. mimicus* and *Macrotritopus defilippi*, stand out: these animals perform dramatic changes in their body mimicking, such as the swimming behavior and coloration of flatfishes.[40,41] This acts as a primary defense mechanism against visual predators. Predation and competition for resources with vertebrates (i.e., fishes and mammals) provided cephalopods with the evolutionary pressure to acquire special behavioral and neuromuscular adaptive traits.[16,42]

Mimicry, camouflage, crypsis, and polyphenism are achieved by a peculiar 'organ'—the skin.[39] It acts as a barrier and interface between the organism and the environment, thus acquiring a behavioral role.[39,42,43] Changes in the skin occur by a constantly regulated and refined neuromuscular mechanism including more than 200 chromatophores/mm[238] to produce, in *Octopus vulgaris*, more than 60 light and dark chromatic and body patterns. In addition, intricate muscle bundles in the skin are responsible for a number of fine textural components that allow the animal to assume a large variety of appearances resulting from the ability to produce smooth or grained skin and/or the erection of primary papillae on the mantle or head (for review, see[9,39,43,44]). It is also surprising that the richness of body patterns is achieved on the basis of a single, mid-wavelength visual pigment, which makes octopuses essentially colorblind.[45–47]

Body patterns may be exhibited in acute or chronic states. Notably, chronic patterns allow human observers to recognize individual subjects by the unique body color patterns they present on the skin. Thus, the configurations of fixed white markings on the dorsal mantle of an octopus, due to contracted chromatophores in given patches of the skin,[38,39,43] appear different in different individuals. This allows photo identification and gives the human observer the ability to recognize and track octopuses,[48] just as is the case in vertebrates.[49]

SMART VERSUS STUPID: LEARNING AND OTHER FORMS OF BEHAVIORAL PLASTICITY

Is an octopus a stupid or a smart spineless animal? From both popular and scientific knowledge, a contrasting view on octopuses and their allies has emerged.

In the current Japanese language, *Tako* is the term utilized to identify an octopus. Interestingly, a jargonized use of the term *Tako* is applied to identify a 'stupid' person.

Thinking of octopuses as stupid creatures is not novel. Aristotle[50] reported that these animals are very easy to catch because a hand waved at the water surface may be enough to allow humans to attract and quite easily grab an octopus. The very simple fishing techniques utilized by fishermen in almost all waters throughout the world[51] are excellent examples of their natural curiosity and the intrinsic exploratory drive toward natural (or artificial) objects.

On the other hand, they are also considered to be clever animals. An example is given in the anecdotal note by Pliny the Elder,[52] who referred to an octopus using a pebble to keep the valves of *Pinna nobilis*, a large bivalve mussel, open and then happily ripping the animal's flesh for a tasty meal. It is noteworthy that Jeannette Villepreux, Lady Power, witnessed an instance of tool use documenting the interaction between *Pinna* and *Octopus* kept in floating cages in the harbor of Messina.[53] This resembles the behavioral flexibility more recently observed by Finn and coworkers[4] for octopuses manipulating coconut shells to build their own dens.

Octopuses and cephalopods represent good examples of learning in invertebrates.[15,16] The study of their learning capabilities has been approached in different ways and reviewed in several instances. During the past 60 years, a dedicated research effort allowed octopuses to be recognized as capable of various forms of learning, ranging from simple sensitization to associative learning and from classical and instrumental conditioning to spatial learning and problem solving.[8,15,16,37,54–56] Notable examples for *O. vulgaris* are exemplified by contextual and associative learning, problem solving, and include learning from conspecifics.

First, daily presentation of food increases the chance of the animal attacking so that its predatory performance, measured as the time to attack the prey-item from its appearance in the tank, improves with time. This is considered to be an example of sensitization, occurring in tasks involving vision and/or chemotactic behavior.[57,58] Evidence that food enhances the probability to attack a given stimulus in a captive situation is also interpreted as contextual learning,[58] similar to what occurs in other invertebrates.[59–63] The performance of subjects in the 'novel context' also appears to be affected by individual dietary preferences derived from previous feeding habits or by early experiences in the life of the animals.[64–66] Second, 'conditional' learning has been reported for octopuses in both laboratory[67–70] and natural settings.[71]

Finally, problem solving in *O. vulgaris*[72] was first described by Piéron[73] and is one of the best known examples in which the knowledge of the skills of octopuses was transferred to the public domain.[1,74] To open a glass jar containing a reward (e.g., a living crab), the animal has to switch between visual and tactile modalities. This switch is not 'automatic', as suggested by variations in the capability (and learning) to solve the problem among different individuals. Learning the task requires that the animal sort out the species-specific motor programs necessary to solve the problem,[72] as occurs when animals prey upon bivalves.[33,75,76] Problem solving also requires a detour (i.e., a maze?), a task resembling the speculative pounce[77]—a common foraging strategy animals use mostly during exploration of the sea bottom in search of edible items hidden under rocks or in crevices (for review, see[8,9]).

The behavioral flexibility of *O. vulgaris* corresponds to its ecological plasticity, possibly due to changes in the lifestyle during ontogeny,[78] and to the known horizontal and vertical migrations by animals of different ages/stages.[79]

The requirement to deal with different environments, and to be exposed to potentially different degrees of complexity, increases their behavioral flexibility. A comparative analysis of cephalopods' habits reveals that octopus has less feeding specialization and a higher versatility in foraging than other cephalopods.[58]

The few examples cited previously and mentioned in several reviews[8,15,56,80] suggest that octopus is a species capable of rapid learning under conditions similar to those that occur in mammals, thus providing support for an evolutionary convergence of this taxon with higher vertebrates.[42,81]

LEARNING FROM OTHERS IN OCTOPUSES: EXPERIMENTAL EVIDENCE

Visual Discrimination

Fiorito and Scotto[82] reported that social learning occurred in the common octopus (*O. vulgaris*), a solitary living invertebrate. In brief, naive animals (observers) were allowed to witness a simultaneous visual discrimination task performed by previously trained conspecifics (demonstrators). The task consisted of a choice between white and red plastic balls as discriminanda. Demonstrators' individual training required at least 17 trials to achieve the correct performance. Observer octopuses solved the same discrimination task after only 4 observational trials, and they performed in accordance with their demonstrators (Figure 23.1A). The vicariously acquired learning appeared to be stable with time.[82]

The conclusions from Fiorito and Scotto's[82] study were subjected to a number of criticisms, particularly (1) the possible involvement of more general species-specific responses instead of true 'imitative' processes, (2) the absence of any contingent reinforcement that would underlie the observers copying of the demonstrators behavior (but see[8,83]), and (3) a general skepticism of the appropriateness of learning protocols utilized in the task.[84–86] Observer octopuses matched the choice of demonstrators in both conditions (white/red and red/white), and their performance was well above chance (70–86% correct) after only 4 trials and contrary to what is required for individual learning.[82,87]

In addition, Fiorito and Scotto[82] noted that observing octopuses tracked the actions of demonstrators with slight movements of their head and eyes (head bobbing).[88] In fact, observers increased their 'attention' during each of the four (demonstration) trials. In

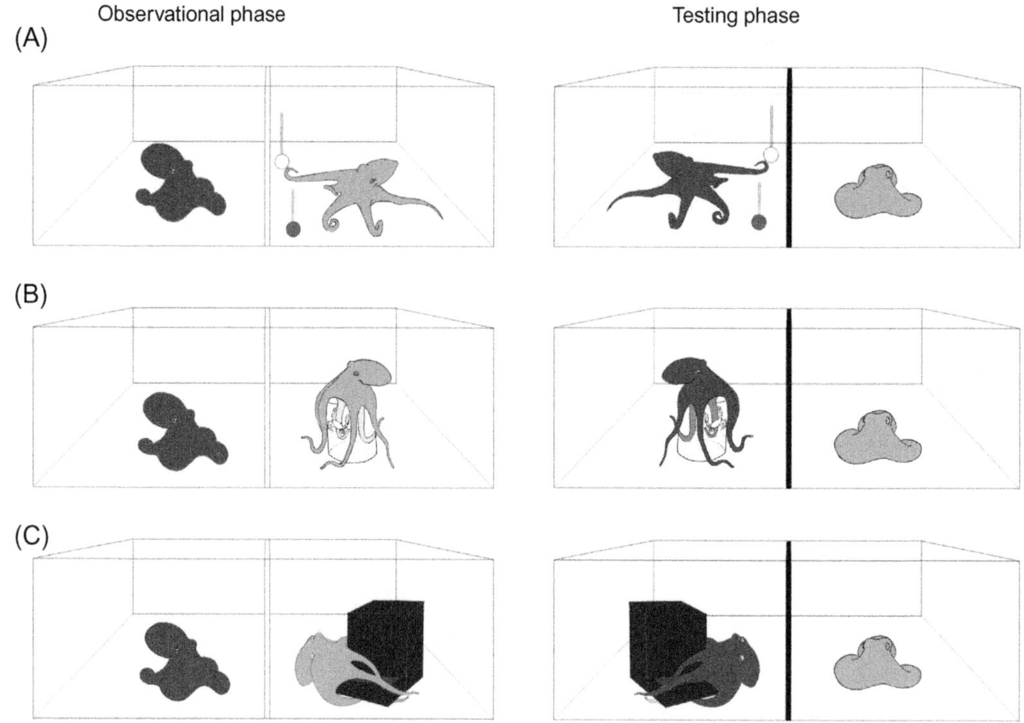

FIGURE 23.1 Sketch of the three different experimental conditions in which learning by observation has been tested in *Octopus vulgaris*: (A) Simultaneous visual discrimination task[82]; (B) problem solving: a glass jar containing a live crab (Borrelli and Fiorito, unpublished data); and (C) problem solving: a black box.[58] During the observational phase, the observers (left tank; dark blue) are exposed to trained demonstrators (right tank; light blue). At the end of this phase (four trials for visual discrimination and two trials for problem solving), a panel is lowered to isolate the two individuals (testing phase), and the task is presented to the observers. Source: *Sketches by Marino Amodio.*

particular, head bobs were more frequent when demonstrators performed the task than during the intertrial period. Although the stimuli utilized in the study were not deceptive (they were different colored balls), it remains an open question whether octopuses have the capacity to track the action patterns of conspecifics.

Finally, observers 'followed' the behavior of their models both toward the preferred stimulus (red ball) and toward the nonpreferred one (white ball), suggesting that a form of matched behavior (stimulus/local enhancement or social facilitation) must have been at least acquired during the observational phase.[87] In addition, while in social context, a behavioral action may be 'imitated' even without contingent reinforcement.[89]

The ability of octopuses to 'copy' the behavior of their conspecifics was replicated in some experiments. Dawes and colleagues[90] reported similar results as those of the original work. Fiorito and Chichery[91] tested the hypothesis that 'observational' learning could be related to the neural circuit known to modulate learning in the octopus. Fiorito and colleagues[92] studied the possible effects of pharmacological interference with the cholinergic system on attentional and/or memory retrieval deficits of the observers.

Problem Solving

Using a different experimental context and as part of a pilot study, Borrelli[58] tested the possible outcome of a 'social learning' experiment with *O. vulgaris* based on a problem box. The problem-solving task appeared to be an alternative type of experiment to test for social learning because previous studies had shown that octopuses are initially unable to open a glass jar containing a crab but subsequently become capable of solving the problem after repeated exposure.[72,73,93] The 'exposure' of octopuses unable to solve the jar problem to a demonstration by trained conspecifics (two trials) was enough for observers to pick up the right cues to become successful in solving the problem (Borrelli and Fiorito, unpublished data; Figure 23.1B). More than six trials were required for inexperienced (i.e., 'unable') octopuses to solve the task by individual trial-and-error learning.[72]

Borrelli[58] found similar results in a short replica of this experiment. Naive animals were selected for their inability to open the transparent jar to reach the prey[72] and were able to solve the problem immediately after being exposed to the demonstrators. Their success, measured as the capability to open the jar and prey upon the crab, increased from 0 to 90% after only two observational trials ($n = 20$). Control octopuses ($n = 15$) were exposed to a jar in the adjacent tank in the absence of any demonstration; they obtained a success rate of only 30% after two pseudo-observational trials. This difference was significant when the performances of control versus observer animals were compared (Borrelli and Fiorito, unpublished data; Figure 23.1B).

In a subsequent study, Borrelli[58] utilized a black cube to test whether octopuses are able to utilize motor information acquired vicariously in a problem-solving task. Experiments were designed with appropriate controls in order to study (1) the spontaneous capability to open the box, (2) the influence of the exposure to a conspecific trained to solve the problem, and (3) the influence of the exposure to the task alone in an adjacent tank (only visual interaction without physical contact) and in the absence of a conspecific. In brief, Borrelli exposed naive individuals to trained octopuses that attacked and solved the black box problem by opening the drawer and capturing the crab hidden inside the box. After two observational trials, the box was presented to the observers. Only half of the octopuses (29 out of 55) were successful after social demonstration (Figure 23.1C). Among the unsuccessful animals, 40% did not even approach the black box, remaining motionless in the tank.[58]

These data suggest that learning by observation of a black box is a difficult task for *O. vulgaris*. The surprisingly low performance observed in these experiments is remarkable compared to the accuracy reached by *O. vulgaris* after observation in other experimental contexts. In analyzing her experiments, Borrelli[58] excluded the possibility that the 'quality' of the performance observed by naive individuals could have influenced the outcome, as has been reported to affect observer efficiency in other animal models.[94,95] In fact, (1) the box was in the observers' visual field almost all the time, (2) the opening side (i.e., the drawer) was visible, and (3) the observers appeared to be alert in the vast majority of cases. However, solving the black box problem depends on the integration of visual and (more importantly) tactile cues and the coordination of motor actions activated by different sensory systems. Therefore, it is not possible to exclude that from the observers' point of view, tactile cues were very difficult to deduce simply by visual observation of the task. In addition, the short amount of time of the demonstration (<2 min) and the fact that demonstrators usually covered the drawer with their arms and interbrachial web interfered with the potential cues necessary to solve the task (even in the most alert octopus).

On the basis of the available knowledge, *O. vulgaris* lacks cross-modality integration when two sensory systems (i.e., visual and chemotactile) are considered. In fact, both neuroanatomical and experimental evidence show that visual and tactile inputs are classified, processed, and stored separately over a series of intersecting neural matrices.[80,96] Although some components share common pathways and sites in the brain (e.g., frontal and vertical lobes), there is no evidence of interactive associations between the two modalities[80,97] as occurs in higher vertebrates such as mammals.[98–101] In octopus, integration between visual and tactile sensorimotor processing appears to occur only at the level of effectors—that is, the peripheral nervous system such as the neural cord in the arm (but see also[27,34,96]). Therefore, it is not surprising that the limit imposed by separate neural processing of sensory information results in the lack of integration of cues from different modalities and thus the low ability to recall the motor patterns required to solve this complex task.

Although future experiments are required to test for 'cross-modality' processing in the octopus, the 'black box' approach could be utilized to test for the spontaneous exploratory capabilities of animals and the potential effect of individual and/or social experience on their capacity to deal with the problem. In addition, a two-action test method and/or experiments based on deceptive stimuli may shed light on social learning in this solitary invertebrate and, more generally, test for their higher order cognitive capabilities.[81]

WHY SHOULD OCTOPUS POSSESS SOCIAL LEARNING SKILLS?

Social learning refers to learning about some aspect of behavior that is influenced by observation of, or interaction with, another individual or its products.[102–104] In 1992, a 'taxonomy', of social influence/social learning phenomena was attempted in vertebrates by Whiten and Ham.[105] This provided a reference for classifying the known range of cases that reported on animals' capability to acquire biologically important information simply from observing the actions of others. Based on this 'taxonomy', Webster and Fiorito[106] categorized known examples of 'social learning' reported to occur in invertebrates and recognized different types ranging from social guidance to social learning *sensu stricto*. Observational conditioning was attributed to the work on octopuses by Fiorito and Scotto.[82]

Questions remain about the biological significance of social learning for *O. vulgaris*. Why should an invertebrate species, notoriously classified as cryptic and solitary, resort to social learning?

After Fiorito and Scotto pioneering study[82] nominated octopus as social learner, an intense scientific debate began. Apart from 'technical' criticisms related to the experimental design (discussed previously), the core of the criticism was based on the difficulty in identifying adaptive functions of social learning mechanisms in nongregarious and short-living species such as the octopus.

Social learning is classically considered an adaptive trait restricted to group-living species, or at least species characterized by relevant bonds among offspring and parents.[107] Moreover, social learning is an essential requirement for the origin and maintenance of traditions in a population. Therefore, it is considered a sophisticated skill requiring high cognitive 'substrate' and long life span. Classical examples are tool use in apes and song learning in birds and cetaceans. Nevertheless, in the past few years, a large number of studies have increased our knowledge about alternative learning processes to trial-and-error learning, refining our classical assumptions about social learning. Here, we focus on the criticisms that have been leveled against observational learning in *Octopus*, trying to challenge them on the basis of the most recent knowledge about social learning.

Are Gregarious Lifestyle and Long Life Span the Requirements for Social Learning to Occur?

As mentioned previously, it is the opinion of some researchers that gregarious lifestyle is an essential requirement for social learning both in vertebrates,[105] (for review, see[108,109]) and in invertebrates.[110] This view is contradicted by other cases in which it does not seem to be the prerequisite in either group of animals.[111,112]

Recent studies support the hypothesis that social learning 'skills' may occur in noncolonial invertebrates (insects[113–115] and crustaceans[116]), as well as in solitary vertebrates (reptiles[117]). Furthermore, species conducting a gregarious lifestyle may also be poor social learners (cattle[118] and horses[119]).

Does Octopus Have a Strictly Solitary Lifestyle?

The answer is simple: We still do not know.

Evidence suggests that, at least at a certain age, individuals of *O. vulgaris* tend to overlap their home range[120] and that encounters are not a rare event[121] (V. Petrella and L. Fabbricatore, personal communication). In addition, there is emerging evidence that social interaction in the common octopus plays a role in several instances other than reproductive contexts. For example, Tricarico and colleagues[122] showed that this species is able to recognize individual conspecifics. Moreover, the large variety of different body patterns performed by octopuses is probably more than just a sophisticated defensive tool. It is clear that communication involving body patterns can also play a relevant role in social interactions.[123]

Finally, octopus social attitude and lifestyle could be more 'plastic' than is usually assumed. For instance, *Octopus*' predatory performances are affected by the presence of conspecifics, making the animals less prone to attack a living crab. However, with time, this situation returns to normal, probably due to habituation in the presence of other conspecifics.[122]

In other species of octopuses, population densities are known to depend on ecological factors such as den availability and predatory pressure.[124] For example, dense assemblage in the wild is characteristic in *Octopus bimaculoides*, which consequently shows a high tolerance of crowding in captivity.[125]

Among mammals, the wolverine (*Gulo gulo*) offers an interesting parallelism.[126] This solitary carnivore has been shown to have latent abilities to aggregate socially because it has adapted to an ecological niche in which resource abundance and distribution inhibit group living.

Possible Adaptive Meaning of Social Learning in Octopus and in Other Solitary and Fast Life History Species

The selective pressures that gave rise to observational learning capability in the common octopus may be a side effect of complex neural and perceptual systems selected during evolution for other functions.

Heyes[103] suggested that the classical dichotomy between 'social' and 'individual' learning should be rejected because it relies on the false supposition that there are two 'parallel' or distinct learning mechanisms. According to Heyes, recent evidence supports the following: (1) Social and asocial learning abilities covary, across and within species; (2) social learning occurs even in solitary animals; and (3) social learning has the same characteristics in different species, including humans. Consequently, social and individual learning depend on the same associative mechanisms encoding information for long-term storage and retrieval by shaping facilitatory and inhibitory links between events and their 'representations'; these mechanisms are conserved in animal phylogeny and mediate social learning.[103] Therefore, social learning occurs on the basis of 'inputs', which Heyes termed perceptual, attentional, and motivational processes that are tuned to a particular channel of social information. In these circumstances, 'weak' levels of perceptual, attentional, and motivational attitudes toward others may explain why individuals of one species (or a given population) may simultaneously be poor social learners and smart at individual learning.[103] When considering social learning, one should assume that this may serve a variety of purposes, such as acquiring useful information about local foraging hot spots or other functions useful in the short term.[127,128]

Social learning is a recent addition to animal behavior studies, and recent advances indicate that it occurs under less rigid constraints. Therefore, it is not surprising to discover that social learning may occur in a nonsocial reptile (a spatial task is acquired 'socially' in a turtle[117]) or that hermit crabs may dynamically use local enhancement for finding food and shelters,[116] even in the absence of social structure of any form, showing resemblances to shoaling in fishes.

CONCLUSIONS

As mentioned previously, observational learning in the cephalopod mollusk *O. vulgaris* has been criticized ever since Fiorito and Scotto[82] published their first paper on this topic. The most conservative criticisms concerned the possible involvement of more general species-specific responses instead of true associative processes.[8,83,129] Criticisms were also centered on the correctness and/or appropriateness of the learning procedures applied in *O. vulgaris*[84,85] (see also general criticisms of learning procedures applied to octopuses in[130,131]).

In light of this general skepticism, which provoked a great debate essentially within the community of cephalopod scholars, a proper definition of *O. vulgaris* social learning capability within the currently accepted 'taxonomy' of social influence and social learning phenomena has still to be agreed upon.

Until recently, experiments showed some intrinsic limits (possibly constraints) of observational learning in the octopus—limits that have been only marginally explored. Further studies are required to unravel the many questions that remain open on this peculiar capability in cephalopods, possibly integrating field and laboratory work and taking into account different factors (if any) that may constrain the ability to learn in the social context.

Acknowledgments

We are indebted to Dr. L. Borrelli for valuable discussions and comments and to our colleagues for their patience.

References

1. Cousteau J-Y, Diolé P. *Octopus and Squid: The Soft Intelligence*. In: Bernard JF, ed. and Trans. Garden City, NY: Doubleday; 1973.
2. Lane FW. *Kingdom of the Octopus: The Life History of the Cephalopoda*. New York: Sheridan House; 1960.
3. Mather JA, Anderson RC, Wood JB. *Octopus: The Ocean's Intelligent Invertebrate*. Portland, OR: Timber Press; 2010.
4. Finn JK, Tregenza T, Norman MD. Defensive tool use in a coconut-carrying octopus. *Curr Biol*. 2009;19:R1069–R1070.
5. Huffard CL, Boneka F, Full RJ. Underwater bipedal locomotion by octopuses in disguise. *Science*. 2005;307:1927.
6. Kelman EJ, Baddeley RJ, Shohet AJ, Osorio D. Perception of visual texture and the expression of disruptive camouflage by the cuttlefish, *Sepia officinalis*. *Proc R Soc B Biol Sci*. 2007;274: 1369–1375.
7. Josef N, Amodio P, Fiorito G, Shashar N. Camouflaging in a complex environment: octopuses use specific features of their surroundings for background matching. *PLoS ONE*. 2012;7: e37579.
8. Hanlon RT, Messenger JB. *Cephalopod Behaviour*. Cambridge, UK: Cambridge University Press; 1996.
9. Borrelli L, Gherardi F, Fiorito G. *A Catalogue of Body Patterning in Cephalopoda*. Napoli, Italy: Stazione Zoologica Anton Dohrn and Firenze University Press; 2006.
10. Barbato M, Bernard M, Borrelli L, Fiorito G. Body patterns in cephalopods: "Polyphenism" as a way of information exchange. *Pattern Recognit Lett*. 2007;28:1854–1864.
11. Zylinski S, How MJ, Osorio D, Hanlon RT, Marshall NJ. To be seen or to hide: visual characteristics of body patterns for camouflage and communication in the Australian giant cuttlefish *Sepia apama*. *Am Nat*. 2011;177:681–690.
12. Partridge JC, Cuthill IC. Animal behaviour: ultraviolet fish faces. *Curr Biol*. 2010;20:R318–R320.
13. Mathger LM, Shashar N, Hanlon RT. Do cephalopods communicate using polarized light reflections from their skin? *J Exp Biol*. 2009;212:2133–2140.
14. Wardill TJ, Gonzalez-Bellido PT, Crook RJ, Hanlon RT. Neural control of tuneable skin iridescence in squid. *Proc R Soc B Biol Sci*. 2012;doi: 10.1098/rspb.2012.1374.
15. Borrelli L, Fiorito G. Behavioral analysis of learning and memory in cephalopods. In: Byrne JJ, ed. *Learning and Memory: A Comprehensive Reference*. Oxford: Academic Press; 2008:605–627.
16. Grasso FW, Basil JA. The evolution of flexible behavioral repertoires in cephalopod molluscs. *Brain Behav Evol*. 2009;74: 231–245.
17. Hochner B, Shomrat T, Fiorito G. The octopus: a model for a comparative analysis of the evolution of learning and memory mechanisms. *Biol Bull*. 2006;210:308–317.
18. Hochner B. Octopuses. *Curr Biol*. 2008;18:R897–R898.
19. Laschi C, Cianchetti M, Mazzolai B, Margheri L, Follador M, Dario P. Soft robot arm inspired by the octopus. *Adv Robot*. 2012;26:709–727.
20. Margheri L, Laschi C, Mazzolai B. Soft-robotic arm inspired by the octopus: I. From biological functions to artificial requirements. *Bioinspir Biomim*. 2012;7:025004.
21. Mazzolai B, Margheri L, Cianchetti M, Dario P, Laschi C. Soft-robotic arm inspired by the octopus: II. From artificial requirements to innovative technological solutions. *Bioinspir Biomim*. 2012;7:025005.
22. Kier WM. The functional morphology of the musculature of squid (*Loliginidae*) arms and tentacles. *J Morphol*. 1982;172: 179–192.
23. Kier WM, Tongues KK. Tentacles and trunks: the biomechanics of movement in muscular hydrostats. *Zool J Linn Soc*. 1985;83: 307–324.

24. Margheri L, Ponte G, Mazzolai B, Laschi C, Fiorito G. Non-invasive study of *Octopus vulgaris* arm morphology using ultrasound. *J Exp Biol.* 2011;214:3727–3731.
25. Mather JA. How do octopuses use their arms? *J Comp Psychol.* 1998;112:306–316.
26. Matzner H, Gutfreund Y, Hochner B. Neuromuscular system of the flexible arm of the octopus: physiological characterization. *J Neurophysiol.* 2000;83:1315–1328.
27. Sumbre G, Gutfreund Y, Fiorito G, Flash T, Hochner B. Control of octopus arm extension by a peripheral motor program. *Science.* 2001;293:1845–1848.
28. Sumbre G, Fiorito G, Flash T, Hochner B. Motor control of flexible octopus arms. *Nature.* 2005;433:595–596.
29. Young JZ. The diameters of the fibres of the peripheral nerves of octopus. *Proc R Soc Lond Ser B Biol Sci.* 1965;162:47–79.
30. Grimaldi AM, Fiorito G, Herculano-Houzel S. The *Octopus vulgaris* brain in numbers: Young (1963) revisited. *J Shellfish Res.* 2011;30:1005.
31. Young JZ. The number and sizes of nerve cells in *Octopus. Proc Zool Soc Lond.* 1963;140:229–254.
32. Chichery MP, Chichery R. Manipulative motor activity of the cuttlefish *Sepia officinalis* during prey capture. *Behav Process.* 1988;17:45–56.
33. Fiorito G, Gherardi F. Prey-handling behaviour of *Octopus vulgaris* (Mollusca, Cephalopoda) on bivalve preys. *Behav Process.* 1999;46:75–88.
34. Sumbre G, Fiorito G, Flash T, Hochner B. Octopuses use a human-like strategy to control precise point-to-point arm movements. *Curr Biol.* 2006;16:767–772.
35. Kuba M, Meisel DV, Byrne RA, Griebel U, Mather JA. Looking at play in *Octopus vulgaris. Berliner Paläobiol Abh.* 2003;3:163–169.
36. Kuba MJ, Byrne RA, Meisel DV, Mather JA. When do octopuses play? Effects of repeated testing, object type, age, and food deprivation on object play in *Octopus vulgaris. J Comp Psychol.* 2006;120:184–190.
37. Sanders GD. The cephalopods. In: Corning WC, Dyal JA, Willows AOD, eds. *Invertebrate Learning. Cephalopods and Echinoderms.* New York: Plenum; 1975:1–101.
38. Messenger JB. Cephalopod chromatophores: neurobiology and natural history. *Biol Rev.* 2001;76:473–528.
39. Packard A. The skin of cephalopods (Coleoids): general and special adaptations. In: Trueman ER, Clarke MR, eds. *Form and Function.* San Diego: Academic Press; 1988:37–67.
40. Hanlon RT, Watson AC, Barbosa A. A "mimic octopus" in the Atlantic: flatfish mimicry and camouflage by *Macrotritopus defilippi. Biol Bull.* 2010;218:15–24.
41. Huffard CL, Saarman N, Hamilton H, Simison W. The evolution of conspicuous facultative mimicry in octopuses: an example of secondary adaptation? *Biol J Linn Soc.* 2010;101:68–77.
42. Packard A. Cephalopods and fish: the limits of convergence. *Biol Rev.* 1972;47:241–307.
43. Packard A. Organization of cephalopod chromatophore systems: a neuromuscular image-generator. In: Abbott NJ, Williamson R, Maddock L, eds. *Cephalopod Neurobiology.* Oxford: Oxford University Press; 1995:331–367.
44. Allen J, Mathger LM, Barbosa A, Hanlon RT. Cuttlefish use visual cues to control three-dimensional skin papillae for camouflage. *J Comp Physiol A-Neuroethol Sens Neural Behav Physiol.* 2009;195:547–555.
45. Brown PK, Brown PS. Visual pigments of the octopus and cuttlefish. *Nature.* 1958;182:1288–1290.
46. Marshall NJ, Messenger JB. Colour-blind camouflage. *Nature.* 1996;382:408–409.
47. Messenger JB. Evidence that *Octopus* is color blind. *J Exp Biol.* 1977;70:49–55.
48. Huffard CL, Caldwell RL, DeLoach N, et al. Individually unique body color patterns in octopus (*Wunderpus photogenicus*) allow for photo identification. *PLoS ONE.* 2008;3:e3732.
49. Hammond PS, Mizroch SA, Donovan GP. Individual recognition of cetaceans. *Use of Photo-Identification and Other Techniques to Estimate Population Parameters: Incorporating the Proceedings of the Symposium and Workshop on Individual Recognition and the Estimation of Cetacean Population Parameters.* 12th ed. Cambridge, UK: International Whaling Commission; 1990.
50. Aristotle. Historia Animalium, English translation by D'Arcy Wenthworth Thompson. In: *The Works of Aristotle Translated into English under the Editorship of J.A. Smith and W.D. Ross.* Vol IV. Oxford: Clarendon; 1910.
51. Boyle PR, Rodhouse P. *Cephalopods. Ecology and Fisheries.* Oxford: Blackwell; 2005.
52. Pliny the Elder. *Naturalis Historia.* Torino: Giulio Einaudi; 1983.
53. Power J. Observations on the habits of various marine animals: observations upon *Octopus vulgaris* and *Pinna nobilis. Ann Mag N Hist.* 1857;20:336.
54. Wells MJ. Learning in the octopus. *Symp Soc Exp Biol.* 1965;20:477–507.
55. Wells MJ. *Octopus: Physiology and Behaviour of an Advanced Invertebrate.* London: Chapman & Hall; 1978.
56. Young JZ. Learning and discrimination in the octopus. *Biol Rev.* 1961;36:32–96.
57. Chase R, Wells MJ. Chemotactic behavior in octopus. *J Comp Physiol A: Sens Neural Behav Physiol.* 1986;158:375–381.
58. Borrelli L. *Testing the Contribution of Relative Brain Size and Learning Capabilities on the Evolution of Octopus vulgaris and Other Cephalopods.* London: Open University; 2007.
59. Haney J, Lukowiak K. Context learning and the effect of context on memory retrieval in *Lymnaea. Learn Mem.* 2001;8:35–43.
60. Law E, Nuttley WM, van der Kooy D. Contextual taste cues modulate olfactory learning in *C. elegans* by an occasion-setting mechanism. *Curr Biol.* 2004;14:1303–1308.
61. Liu L, Wolf R, Ernst R, Heisenberg M. Context generalization in *Drosophila* visual learning requires the mushroom bodies. *Nature.* 1999;400:753–756.
62. Skow CD, Jakob EM. Jumping spiders attend to context during learned avoidance of aposematic prey. *Behav Ecol.* 2006;17:34–40.
63. Tomsic D, Pedreira ME, Romano A, Hermitte G, Maldonado H. Context-US association as a determinant of long-term habituation in the crab *Chasmagnathus. Anim Learn Behav.* 1998;26:196–209.
64. Messenger JB, Sanders GD. Visual preference and two-cue discrimination learning in octopus. *Anim Behav.* 1972;20:580–585.
65. Bradley EA, Messenger JB. Brightness preference in octopus as a function of background brightness. *Mar Behav Physiol.* 1977;4:243–251.
66. Guibe M, Poirel N, Houde O, Dickel L. Food imprinting and visual generalization in embryos and newly hatched cuttlefish, *Sepia officinalis. Anim Behav.* 2012;84:213–217.
67. Papini MR, Bitterman ME. Appetitive conditioning in *Octopus cyanea. J Comp Psychol.* 1991;105:107–114.
68. Hvorecny LM, Grudowski JL, Blakeslee CJ, et al. Octopuses (*Octopus bimaculoides*) and cuttlefishes (*Sepia pharaonis, S. officinalis*) can conditionally discriminate. *Anim Cogn.* 2007;10:449–459.
69. Shomrat T, Zarrella I, Fiorito G, Hochner B. The octopus vertical lobe modulates short-term learning rate and uses LTP to acquire long-term memory. *Curr Biol.* 2008;18:337–342.
70. De Lisa E, De Maio A, Moroz LL, Moccia F, Mennella Faraone MR, Di Cosmo A. Characterization of novel cytoplasmic PARP in the brain of *Octopus vulgaris. Biol Bull.* 2012;222:176–181.

71. Huffard CL, Boneka F, Caldwell RL. Male–male and male–female aggression may influence mating associations in wild octopuses (*Abdopus aculeatus*). *J Comp Psychol*. 2010;124:38–46.
72. Fiorito G, von Planta C, Scotto P. Problem-solving ability of *Octopus vulgaris* Lamarck (Mollusca, Cephalopoda). *Behav Neural Biol*. 1990;53:217–230.
73. Piéron H. Contribution a la psychologie du poulpe: L'acquisition d'habitudes. *Bull Inst psych internat Paris*. 1911;11: 111–119.
74. Wikipedia. Paul the Octopus. Available at <http://en.wikipedia.org/wiki/Paul_the_Octopus> [serial online]; 2012.
75. McQuaid CD. Feeding behaviour and selection of bivalve prey by *Octopus vulgaris* Cuvier. *J Exp Mar Biol Ecol*. 1994;177: 187–202.
76. Steer MA, Semmens JM. Pulling or drilling, does size or species matter? An experimental study of prey handling in *Octopus dierythraeus* (Norman, 1992). *J Exp Mar Biol Ecol*. 2003;290: 165–178.
77. Yarnall JL. Aspects of behaviour of *Octopus cyanea* Gray. *Anim Behav*. 1969;17:747–754.
78. Nixon M, Mangold K. The early life of *Sepia officinalis*, and the contrast with that of *Octopus vulgaris* (Cephalopoda). *J Zool*. 1998;245:407–421.
79. Oosthuizen A, Smale MJ. Population biology of *Octopus vulgaris* on the temperate southeastern coast of South Africa. *J Mar Biol Assoc UK*. 2003;83:535–541.
80. Young JZ. Computation in the learning system of cephalopods. *Biol Bull*. 1991;180:200–208.
81. Edelman DB, Seth AK. Animal consciousness: a synthetic approach. *Trends Neurosci*. 2009;32:476–484.
82. Fiorito G, Scotto P. Observational Learning in *Octopus vulgaris*. *Science*. 1992;256:545–547.
83. Biederman GB, Davey VA. Social learning in invertebrates. *Science*. 1993;259:1627–1628.
84. Boal JG. A review of simultaneous visual discrimination as a method of training octopuses. *Biol Rev*. 1996;71:157–190.
85. Mather JA. Cognition in cephalopods. *Adv Stud Behav*. 1995;24:317–353.
86. Wood JB, Day CL. CephBase. The Cephalopod Page. Available at <http://www.cephbase.utmb.edu> [serial online]; 2003.
87. Fiorito G. Social learning in invertebrates: response. *Science*. 1993;259:1629.
88. Packard A, Sanders GD. Body patterns of *Octopus vulgaris* and maturation of the response to disturbance. *Anim Behav*. 1971;19:780–790.
89. Bonnie KE, de Waal FB. Copying without rewards: socially influenced foraging decisions among brown capuchin monkeys. *Anim Cogn*. 2007;10:283–292.
90. Dawes J, Fernandes J, Roberston JD. *Octopus vulgaris* can learn by visual observation. *Anat Rec*. 1993;(Suppl. 1):46.
91. Fiorito G, Chichery R. Lesions of the vertical lobe impair visual discrimination learning by observation in *Octopus vulgaris*. *Neurosci Lett*. 1995;192:117–120.
92. Fiorito G, Agnisola C, d'Addio M, Valanzano A, Calamandrei G. Scopolamine impairs memory recall in *Octopus vulgaris*. *Neurosci Lett*. 1998;253:87–90.
93. Fiorito G, Biederman GB, Davey VA, Gherardi F. The role of stimulus preexposure in problem solving by *Octopus vulgaris*. *Anim Cogn*. 1998;1:107–112.
94. Biederman GB, Vanayan M. Observational learning in pigeons: the function of quality of observed performance in simultaneous discrimination. *Learn Motiv*. 1988;19:31–43.
95. Robertson HA, Vanayan M, Biederman GB. Observational learning and the role of confinement in pigeons: suppression of learning as a function of observing the performance of a conspecific. *J Gen Psychol*. 1985;112:375–382.
96. Young JZ. Multiple matrices in the memory system of *Octopus*. In: Abbott JN, Williamson R, Maddock L, eds. *Cephalopod Neurobiology*. Oxford: Oxford University Press; 1995:431–443.
97. Allen A, Michels J, Young JZ. Possible interactions between visual and tactile memories in octopus. *Mar Behav Physiol*. 1986;12:81–97.
98. Gottfried JA, Dolan RJ. Response to small: crossmodal integration—Insights from the chemical senses. *Trends Neurosci*. 2004;27:123–124.
99. Gottfried JA, Smith AP, Rugg MD, Dolan RJ. Remembrance of odors past: human olfactory cortex in cross-modal recognition memory. *Neuron*. 2004;42:526–527.
100. Ohara S, Lenz FA, Zhou YD. Modulation of somatosensory event-related potential components in a tactile-visual cross-modal task. *Neuroscience*. 2006;138:1387–1395.
101. Suchan B, Linnewerth B, Koster O, Daum I, Schmid G. Cross-modal processing in auditory and visual working memory. *NeuroImage*. 2006;29:853–858.
102. Galef BG. Social learning and traditions in animals: evidence, definitions, and relationship to human culture. *WIREs Cogn Sci*. 2012;3:581–592.
103. Heyes C. What's social about social learning? *J Comp Psychol*. 2012;126:193–202.
104. Hoppitt W, Laland KN. Social processes influencing learning in animals: a review of the evidence. *Adv Stud Behav*. 2008;:105–165.
105. Whiten A, Ham R. On the nature and evolution of imitation in the animal kingdom: reappraisal of a century of research. *Adv Stud Behav*. 1992;21:239–283.
106. Webster SJ, Fiorito G. Socially guided behaviour in non-insect invertebrates. *Anim Cogn*. 2001;4:69–79.
107. Galef BG, Laland KN. Social learning in animals: empirical studies and theoretical models. *Bioscience*. 2005;55:489–499.
108. Brown C, Laland KN. Social learning in fishes: a review. *Fish Fish*. 2003;4:280–288.
109. Zentall TR. Imitation in animals: evidence, function, and mechanisms. *Cybern Syst*. 2001;32:53–96.
110. Thorne BL, Traniello JFA. Comparative social biology of basal taxa of ants and termites. *Annu Rev Entomol*. 2003;48:283–306.
111. Hosey GR, Wood M, Thompson RJ, Druck PL. Social facilitation in a "non-social" animal, the centipede *Lithobius forficatus*. *Behav Process*. 1985;10:123–130.
112. Lefebvre L, Palameta B, Hatch KK. Is group-living associated with social learning? A comparative test of a gregarious and a territorial columbid. *Behaviour*. 1996;133:241–261.
113. Coolen I, Dangles O, Casas J. Social learning in noncolonial insects? *Curr Biol*. 2005;15:1931–1935.
114. Leadbeater E, Raine NE, Chittka L. Social learning: ants and the meaning of teaching. *Curr Biol*. 2006;16:R323–R325.
115. Leadbeater E, Chittka L. Social learning in insects: from miniature brains to consensus building. *Curr Biol*. 2007;17: R703–R713.
116. Laidre ME. How rugged individualists enable one another to find food and shelter: field experiments with tropical hermit crabs. *Proc R Soc B Biol Sci*. 2010;277:1361–1369.
117. Wilkinson A, Kuenstner K, Mueller J, Huber L. Social learning in a non-social reptile (*Geochelone carbonaria*). *Biol Lett*. 2010;6:614–616.
118. Pfister JA, Price KW. Lack of maternal influence on lamb consumption of locoweed. *J Anim Sci*. 1996;74:340–344.
119. Lindberg AC, Kelland A, Nicol CJ. Effects of observational learning on acquisition of an operant response in horses. *Appl Anim Behav Sci*. 1999;61:187–199.

120. Mather JA, O'Dor RK. Foraging strategies and predation risk shape the natural history of juvenile *Octopus vulgaris*. *Bull Mar Sci*. 1991;49:256–269.
121. Woods J. Octopus-watching off Capri. *Animals (Lond)*. 1965;7:324–327.
122. Tricarico E, Borrelli L, Gherardi F, Fiorito G. I know my neighbour: individual recognition in *Octopus vulgaris*. *PLoS ONE*. 2011;6:e18710.
123. Moynihan M. *Communication and Noncommunication by Cephalopods*. Bloomington: Indiana University Press; 1985.
124. Mather J. Social organization and use of space by *Octopus joubini* in a semi-natural situation. *Bull Mar Sci*. 1980;30:848–857.
125. Forsythe JW, Hanlon RT. Effect of temperature on laboratory growth, reproduction and life span of *Octopus bimaculoides*. *Mar Biol*. 1988;98:369–379.
126. Dalerum F. *Sociality in a Solitary Carnivore, the Wolverine* [PhD thesis]. Stockholm: Stockholm University; 2005.
127. Danchin E, Giraldeau LA, Valone TJ, Wagner RH. Public information: from nosy neighbors to cultural evolution. *Science*. 2004;305:487–491.
128. Whiten A, van Schaik CP. The evolution of animal 'cultures' and social intelligence. *Philos Trans R Soc B Biol Sci*. 2007;362: 603–620.
129. Suboski MD, Muir D, Hall D. Social learning in invertebrates. *Science*. 1993;259:1628–1629.
130. Bitterman ME. Learning in the lower animals. *Am Psychol*. 1966;21:1073.
131. Bitterman ME. Critical commentary. In: Corning WC, Dyal JA, eds. *Invertebrate Learning*. New York: Plenum; 1975: 139–145.

CHAPTER

24

The Neurophysiological Basis of Learning and Memory in Advanced Invertebrates
The Octopus and the Cuttlefish

Binyamin Hochner and Tal Shomrat
Hebrew University, Jerusalem, Israel

INTRODUCTION

It is commonly believed that invertebrates should be used for studying general questions in neurobiology only when their specific features, such as "a small number of large identifiable neurons," generation of simple stereotypical behaviors, or their amenity to genetic manipulation, provide a special experimental advantage. These characteristics aid in unraveling cellular and synaptic processing, as well as the neuroethology of relatively simple behaviors. In contrast, invertebrates are thought not to be very useful for examining the neuroethological bases of complex behavior.

However, this approach—ignoring the potential of invertebrates for the analysis of complex behavior—has neglected three facts. First, some invertebrates exhibit behaviors as complex as those of vertebrates and, correspondingly, possess large nervous systems. Second, the nervous systems of invertebrates showing complex behavior (some insects and the modern cephalopods (Coleoidea) octopus, squid, and cuttlefish) still show the typical "simple" invertebrate organization and thus are more amenable to exploration of the neural processes involved in complex brain functions. Third, unlike vertebrates, certain invertebrates may have highly distinct brain structures—like the mushroom bodies of insects and the vertical lobe of cephalopods. The great anatomical differences from the rest of the nervous system indicate special functions and, indeed, both structures subserve the learning component of these animals' complex behaviors. Thus, comparative functional analysis of these structures may reveal general principles of the neuronal organization of complex behavior and its evolution.

Research on modern cephalopods has shown that this is indeed the case. These invertebrates show behavioral repertoires comparable to those of higher vertebrates, including behaviors associated with the intelligent hunting and defense behaviors of solitary animals.[1–6] Their learning and memory behaviors are also similar to those of vertebrates.[2–4,6–9] However, their brains maintain the much simpler invertebrate organization.[10] This unique combination of a simple nervous system and complex behavior is especially advantageous for tackling the central question of how a nervous system controls complex behaviors. Comparing the cellular processes and neuronal circuitry of their learning and memory systems with those of other invertebrates and vertebrates may advance our understanding of their evolution and function. This comparative evolutionary approach can determine whether mechanisms subserving cognitive functions evolved convergently across widely diverse phyla or, alternatively, whether evolutionarily primitive mechanisms mediating simple forms of behaviors are conserved in more advanced animals and integrated in mediating complex behaviors.

Another important feature of the modern cephalopods has been revealed through fruitful collaborations between roboticists and biologists. In the approach commonly termed "bioinspired robotics," engineers are trying to use biological principles to design more efficient robotic hardware (morphology, material, and actuation) and software (processing of

Invertebrate Learning and Memory.
DOI: http://dx.doi.org/10.1016/B978-0-12-415823-8.00024-1

sensory information, decision making, and motor control).* Analysis of the motor control strategies used to drive the cephalopods' long, flexible, and highly skilled arms is helping roboticists design robots with flexible arms for use in confined spaces.[11]

THE CEPHALOPOD NERVOUS SYSTEM

The nervous system of modern cephalopods (Coleoidea) is uniquely divided into three parts: the central brain (50 million cells in *Octopus vulgaris*); the two optic lobes (each 80 million cells in octopus); and the peripheral nervous system that, uniquely for the octopus due to its eight long flexible arms, is numerically the largest, containing 320 million nerve cells. The central brain comprises tens of lobes in which the cell bodies of the monopolar neurons lie in the outer region of each lobe, whereas their processes form the central neuropil, as is usual in invertebrate ganglia (Figure 24.1).[13,14]

Early experiments indicated that the vertical lobe (VL) is not involved in simple motor functions. Stimulating it or the superior frontal lobes in cuttlefish or octopus evoked no obvious effects, whereas stimulating other parts of the brain caused movements of various body parts.[15,16] Removing the VL also did not appear to affect the animal's general behavior.[9,17,18] Deficiencies were revealed only when animals were required to learn new tasks. After removal or lesions of the VL, octopuses continued to attack crabs despite receiving electrical shocks, unless the intertrial interval was less than approximately 5 min.[17] The VL appears to be specifically involved in long-term and more complex forms of memory.

Lesions in the ventral part of the VL in the cuttlefish (*Sepia officinalis*) led to a marked impairment in the acquisition of spatial learning, whereas lesions in the dorsal part of the VL impaired long-term retention of spatial learning.[19] An interesting result obtained by Sanders and Young in 1940,[20] but with only two animals, was that removal of the cuttlefish VL did not affect the ability of the animals to attack a prawn as long as it remained in the cuttlefish's field of vision. Today, we interpret this result as suggesting that the cuttlefish VL is involved in working memory.

A complex form of learning demonstrated by octopuses is observational learning. A naive octopus learns to attack a previously positively rewarded target after only four observations of a trained octopus attacking the same target. This is much faster than it takes to train the demonstrator octopus.[7] A lesion study has demonstrated that the VL is important for this advanced form of learning.[21]

Finally, the morphological structure of the VL appears relatively simple, but its unique matrix-like organization led Boycott and Young[17] to postulate that the median superior frontal lobe (MSFL)−VL and inferior frontal networks (Figure 24.1) form associative networks for learning and memory. Wells[2] and Young[22] further suggested that the VL matrix is analogous to the mammalian hippocampus memory system and the insect mushroom bodies.[22–24] The VL system is thus an exciting site for exploring the neuronal circuitry and physiological mechanisms involved in various forms and phases of learning and memory.

ANATOMY OF THE VERTICAL LOBE SYSTEM

The octopus VL contains only two types of neuron, amacrine cells and large efferent neurons, both of which are morphologically typical invertebrate monopolar neurons (Figure 24.2).[13,25,26] Twenty-five million amacrine cells, whose approximately 5-μm diameter makes them the smallest neurons in the octopus brain, converge and synapse onto only approximately 65,000 large neurons (~15-μm diameter). The amacrine cells are intrinsic interneurons because their processes remain within the VL (Figures 24.1 and 24.2). In contrast, the large neurons are efferent neurons whose axons form the only output of the VL, leaving organized axon bundles or roots that are easy to identify and record from in brain slice preparations (Figure 24.1B).

The octopus VL receives inputs from the MSFL, which is thought to integrate visual and taste information.[13] Note that although we use the acronym MSFL collectively, the superior frontal lobe (SFL) in the cuttlefish (Decapoda) is not divided into lateral and median parts as in the octopus. The octopus MSFL contains only one morphological type of neuron, whose 1.8 million axons project to the VL in a distinct tract running between the VL neuropil and its outer cell body layer (Figure 24.1). This arrangement allows each MSFL axon to make *en passant* synapses with many amacrine neurons along the VL (Figure 24.2). Young[13] postulated a direct connection to the large

*Note the difference between bioinspiration and biomimetics. Currently, bioinspiration seems a more reasonable approach because we still cannot produce artificial materials even close to the properties of the active materials evolved by natural selection. For example, roboticists are far from producing an active synthetic polymer yielding a strain and speed like myofilaments.

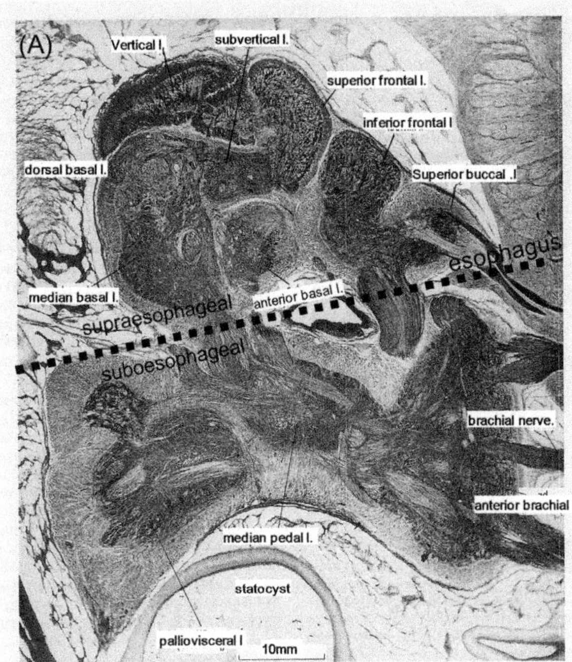

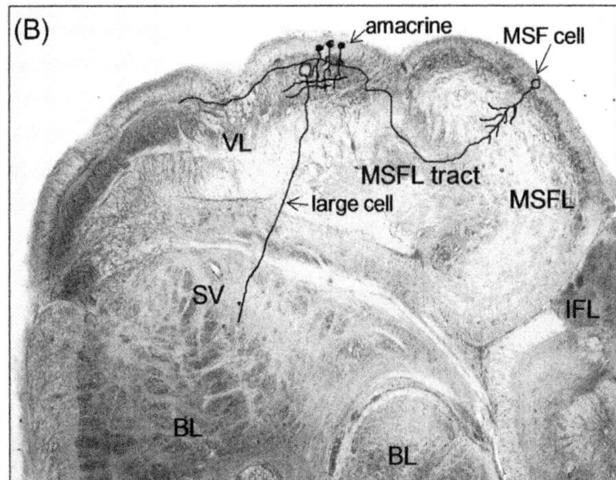

FIGURE 24.1 The morphological organization of the octopus central brain. (A) Sagittal section of the sub- and supraesophageal lobes of *Octopus vulgaris* showing the dorsally located superior frontal (MSFL)–vertical lobes (VL) system and the organization of the lobes discussed in the text. (B) Unstained median sagittal section through the dorsal part of the supraesophageal brain mass with superimposed schematic drawing of the three types of neurons in this system and their connections. This area in the supraesophageal lobes (see panel A) forms the main part of the VL slice preparation. BL, basal lobes; IFL, inferior frontal lobe; SV, subvertical lobe. Source: *Panel A modified from Nixon and Young.*[12]

neurons, but this was not supported by Gray's[25] electron microscopic study nor by recent physiological results. The unusual matrix-like organization of the MSFL–VL complex is shared by the inferior frontal lobe (Figure 24.1), which is involved in chemotactile memory.[2,22]

Although little is known about the targets of the axons of the large neurons, Young[13] suggested that some of the large VL neurons send their axons back into the lateral SFL, creating a recurrent loop between the VL and the SFL. Modern tracing techniques have clearly revealed such connections in cuttlefish (N. Graindorge, thesis). Such recurrent excitatory connections are computationally attractive because they may create a reverberatory circuit, which may subserve working memory by maintaining ongoing electrical activity.

Testing the hypothesis that the VL–SFL system forms associative networks for learning and memory clearly requires physiological characterization of these brain circuits and their plastic properties. New experimental preparations, such as the isolated brain, brain slice preparations, and a preparation for stimulating–recording in freely moving animals,[16,27,28] now allow neurophysiological exploration of this anatomically unique system.

NEUROPHYSIOLOGY OF SFL INPUT TO THE OCTOPUS VERTICAL LOBE

The VL and the inferior frontal lobe are unique structures in the cephalopod brain. Their large numbers of neurons are organized in layers with their processes aligned more or less in parallel, resembling vertebrae brain organization. The insect mushroom body is similarly organized; relatively few projecting neurons originating from the antennal lobe innervate a large number of small amacrine interneurons *en passant*.[28] This organization pattern, typical for "fan-out fan-in" networks, is quite unusual in the invertebrate nervous system. It is highly suggestive that it occurs in invertebrate brain structures involved in higher cognitive functions associated with learning and memory.

Unlike the more random organization of typical invertebrate ganglia, this layered organization generates a measurable external electric field near the active neurons. This field potential shows local field potentials (LFPs; spontaneous, ongoing background activity), evoked potentials and event-related potentials, and a compound field potential similar to an electroencephalogram.[29] Bullock[30] suggested that the existence of such a compound field potential indicates the high level of complexity of the octopus brain.

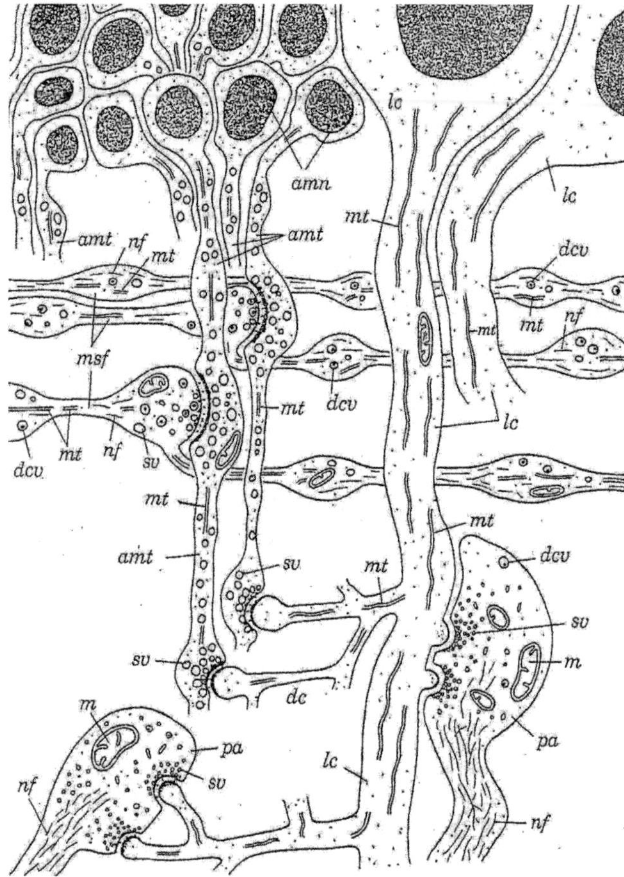

FIGURE 24.2 Diagram showing what is thought to be the basic circuitry of the vertical lobe. amn, amacrine cells; amt, amacrine trunk; dc, dendritic collaterals of large cell; dcv, dense-core vesicle; lc, body or trunk of large cell; m, mitochondrion; msf, median superior frontal axon; mt, microtubule; nf, neurofilaments; pa, possible "pain" axon input to the large cell; sv, synaptic vesicles. Source: *Reproduced from Gray[25] with permission.*

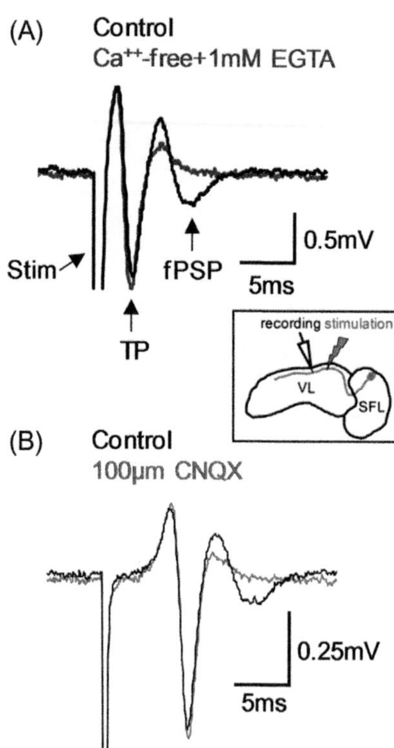

FIGURE 24.3 (A) Superimposed LFPs in control and after blocking postsynaptic field potentials (fPSP) in Ca^{2+}-free EGTA in artificial seawater. Stim, stimulus artifact; TP, tract potential. (B) Similar to panel A, showing blockade of fPSP by AMPA-like glutamatergic receptor antagonists CNQX. Inset: Schematic slice with a typical placement of the recording and stimulating electrodes.

The amount of current flowing in the extracellular space due to neuron activity, such as an action potential, is usually tiny. Because the extracellular impedance is very low (approximately three or four orders of magnitude lower than the neurons' input resistance), only summed field potentials generated by many neurons become large enough to be reliably detected against the background electrical noise. Such summation occurs only if the neurons are synchronously active and their currents flow in the same direction; otherwise, they would cancel each other out. This condition is achieved only when the neurons have long axons running parallel. Such an arrangement, which is common in vertebrae brains, occurs in the MSFL–VL system, in which the MSFL axonal input to the VL is organized in tracts. The cell bodies of the millions of amacrine interneurons lie in the outer zone of the VL lobe, whereas their axons project in parallel into the lobe, perpendicular to the MSFL axonal tract (Figures 24.1 and 24.2). Such an architecture would be expected to generate a significant LFP, like that generated close to the Schaffer collaterals in the hippocampus.

Stimulating the SFL tract with short current pulse evokes a typical LFP waveform (Figure 24.3), a triphasic (positive–negative–positive) tract potential (TP) generated by the volley of action potential propagating along the stimulated axons in the SFL tract. The delay after the stimulus artifact depends on the distance between stimulus and recording electrodes. A mainly negative LFP follows immediately after the second positive wave of the TP. This potential is most likely a glutamatergic postsynaptic field potential (fPSP) because it disappears in zero-calcium physiological solution (Figure 24.3A) and is blocked by AMPA-like antagonists such as CNQX (Figure 24.3B), DNQX, or kynurenate.[31]

The short latency between the peak negativity of the TP and the fPSP onset (~3 msec; Figures 24.3A and 24.3B) suggests a monosynaptic delay. The physiological results thus agree well with Gray's anatomical scheme (Figure 24.2), in which the terminals of the SFL axons synapse directly on the amacrine dendrites. That is, the first synaptic layer of the VL is likely a glutamatergic synaptic input from the SFL neurons onto the amacrine interneurons.

Recording LFPs from slices prepared from the cuttlefish VL show similar LFP characteristics to those found in octopus.[32] Thus, cuttlefish appear to have similar membrane properties and connectivity patterns despite significant differences in the overall anatomy of the two VL systems. The octopus VL is composed of five lobuli, whereas in the cuttlefish, the VL is a dome-shaped structure with no distinct enfoldings as in the octopus. However, the amacrine cells and the large neurons of the two systems are quite similar.

The first layer of the VL is similarly organized to the stratum radiatum of the hippocampus. In both cases, the collaterals/tract terminals synapse *en passant* with the dendrites of the pyramidal/amacrine cells, respectively. It is therefore not surprising that the LFP in the VL (presynaptic TP followed by fPSP; Figure 24.3) is similar to that evoked at the stratum radiatum by stimulating the Schaffer fiber collaterals in CA1. However, because the amacrine cells are inexcitable (i.e., they do not generate regenerative action potentials),[8] the fPSP exhibits an amplitude-independent waveform with no population spikes,[27] unlike the fPSP of the hippocampal CA1 pyramidal neurons.[33]

The amacrine cells innervate the large efferent neurons. In the octopus, this synaptic connection is achieved via specialized "serial" synapses (Figure 24.2),[25] in which the MSFL synaptic terminals contact the amacrine dendrite, which in turn is the presynaptic input to the large cells' (neurons) spines. To investigate the nature of this input to the large neurons, infrared differential interference contrast microscopy aided intracellular recording from their cell bodies, as well as extracellular recording of their spiking activity in their axon bundles. Figure 24.4A shows excitatory postsynaptic potentials (EPSPs) evoked by stimulating the MSFL tract of an octopus and spontaneous EPSPs in a large neuron. Although the fine anatomical details in the cuttlefish are not as well-known as in the octopus, intracellular recording from the large neurons in the cuttlefish VL revealed similar neurophysiological properties (Figure 24.5C) as those in the octopus (Figure 24.4A). Like most invertebrate neurons, the cell bodies of the large neurons are inexcitable, as demonstrated by the small non-overshooting spikelets (arrowheads in Figures 24.4A and 24.5C). These decrease in size passively as they pass along the single neurite to the cell body from a distant spike initiation zone at the junction between the dendritic tree and the axon.

In both cuttlefish and octopus, the synaptic input to the large neurons is cholinergic, with both evoked and spontaneous EPSPs being blocked by hexamethonium (Figures 24.4A and 24.4B), a muscarinic receptor antagonist that also blocks the synaptic potential at the neuromuscular junctions of the octopus arm.[34] Hexamethonium also blocked both spontaneous and evoked spiking activity recorded from the large neuron axonal bundles (Figures 24.4B and 24.4D). As would be expected for glutamatergic synapses, the fPSP of the first synaptic layer in octopus and cuttlefish was unaffected by cholinergic antagonists (Figure 24.4C). Thus, the cholinergic synapse must be the amacrine-to-large neuron synapse. Both cholinergic and glutamatergic antagonists blocked the large neuron output as shown by recording from their axon bundles. These findings suggest that there is no strong direct connection from MSFL axons to the large neurons and that the main connections within the VL are the MSFL inputs onto the amacrine cells, which in turn innervate the large efferent neurons.[32]

The VL system of cuttlefish and octopus thus appears to be organized as a simple feed-forward fan-out fan-in type of network. This type of network architecture is frequently found among biological and artificial networks that learn to classify inputs when endowed with the synaptic plasticity that creates learning ability.[35] The first fan-out synaptic layer may create high dimensionality neural representations of the incoming sensory information in a form suitable for further processing at the fan-in layer.[32] If the VL system of octopus and cuttlefish possesses this architecture and it functions as a learning and memory network, then it is essential that there be synaptic plasticity at one or more of its synaptic sites.

NEURONAL OUTPUT FROM THE VERTICAL LOBES OF OCTOPUS AND CUTTLEFISH DEMONSTRATES ACTIVITY-DEPENDENT LONG-TERM POTENTIATION

The organization of the cephalopod lobes in the central brain offers a relatively simple means of measuring the input/output relationship of the VL stimulating the MSFL tract and of measuring the VL output by either recording intracellularly from the large neuron cell bodies or recording their spiking activity in the axonal bundles leaving the VL. Applying four high-frequency (HF) trains to the MSFL tract (20 pulses at 50 Hz) induces a robust activity-dependent LTP in the output of the VL in both octopus and cuttlefish (Figures 24.5A1 and 24.5B1 and Figures 24.5A2 and 24.5B2 respectively). Thus, the region in the coleoids brains associated with learning and memory is endowed with a property universally believed to be essential for networks mediating

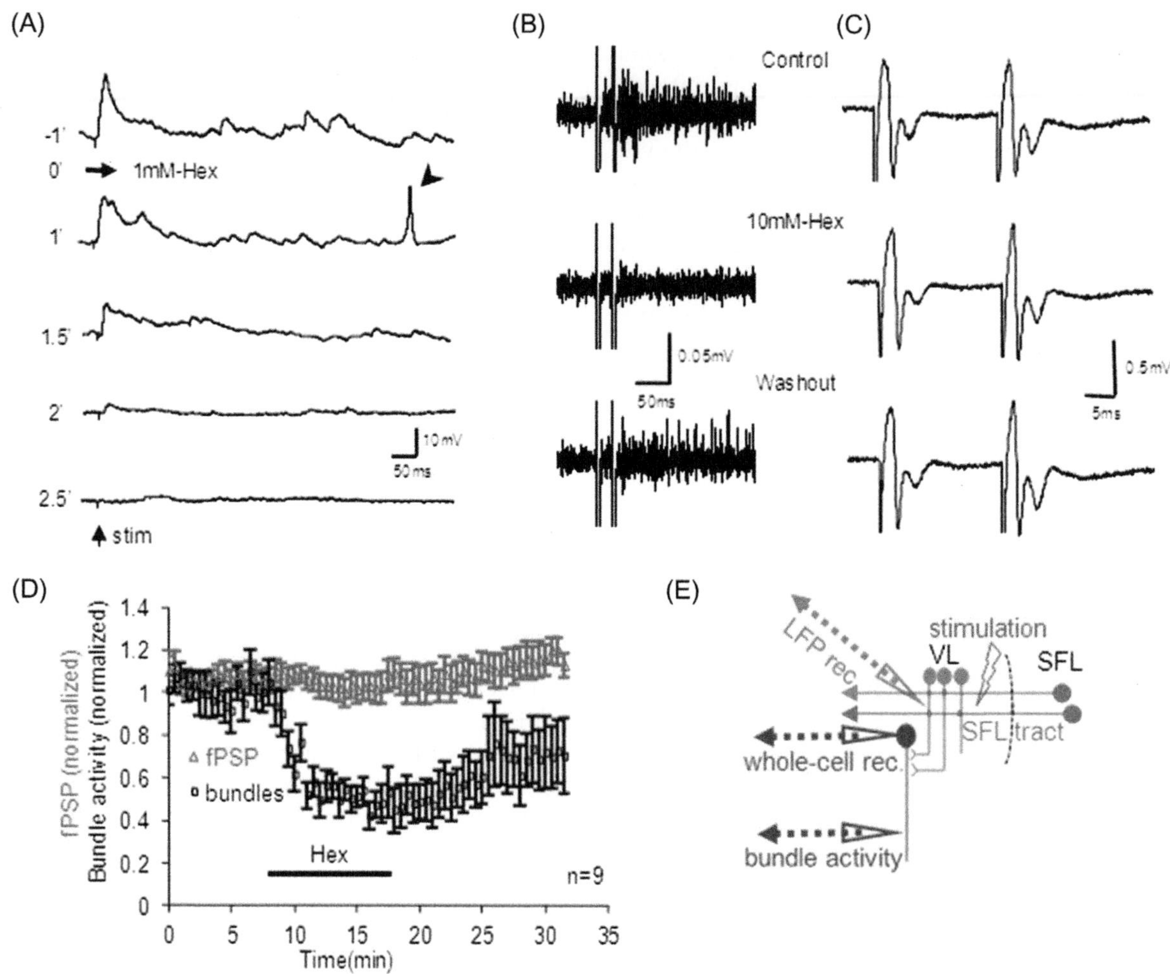

FIGURE 24.4 The synaptic inputs to the large cells are most likely cholinergic. (See panel E for the different recording modes). (A) Hexamethonium blocked the spontaneous and SFL tract-evoked EPSPs recorded intracellularly from a large neuron in the octopus (arrowheads indicate spikelet). (B) Hexamethonium blocked the burst of action potentials recorded extracellularly from large neuron axonal bundles. Twin stimuli were used to obtain a clearly measurable bundle response (B and C). (C) Hexamethonium had no effect on the TP and fPSPs. Records were obtained simultaneously with bundle activity shown in panel B. (D) Summary of nine experiments in octopus as exemplified in panels B and C. Black curve, normalized integrated bundle activity; red curve, fPSP amplitude. Here and in subsequent figures, responses were normalized to the average of 3–10 test responses at the beginning of the experiments. Source: *Modified from Shomrat* et al.[32]

behavioral learning and memory. What are the mechanisms of this neural plasticity? Are they important for cephalopod behavior?

SYNAPTIC PLASTICITY IN THE VERTICAL LOBES OF OCTOPUS AND CUTTLEFISH

Examining the neurophysiological properties of synaptic transmission in the VL network revealed activity-dependent synaptic plasticity. Surprisingly, the synaptic plasticity is located at different sites in the VL of the octopus and cuttlefish.

The local field potential recorded at the MSFL tract of the octopus demonstrates a very robust activity-dependent plasticity. Tetanization led to a long-term potentiation (LTP) with an approximately fourfold increase in fPSP amplitude without affecting the amplitude of the presynaptic TP (Figure 24.6A, inset). The potentiation of the glutamatergic synaptic input to the amacrine neurons is long term. Tetanizing a second time showed that this LTP is a saturated phenomenon because no further long-term enhancement was obtained (Figure 24.6A, second HF). Surprisingly, the exact same experiments in the cuttlefish VL reveal no activity-dependent plasticity (Figure 24.6B).

Does activity-dependent plasticity occur only at the fan-out glutamatergic connections of the octopus VL (Figures 24.1 and 24.5)? As described previously, in the octopus, HF tetanization, which induced LTP of the fPSP, also caused a long-term increase in the extracellular

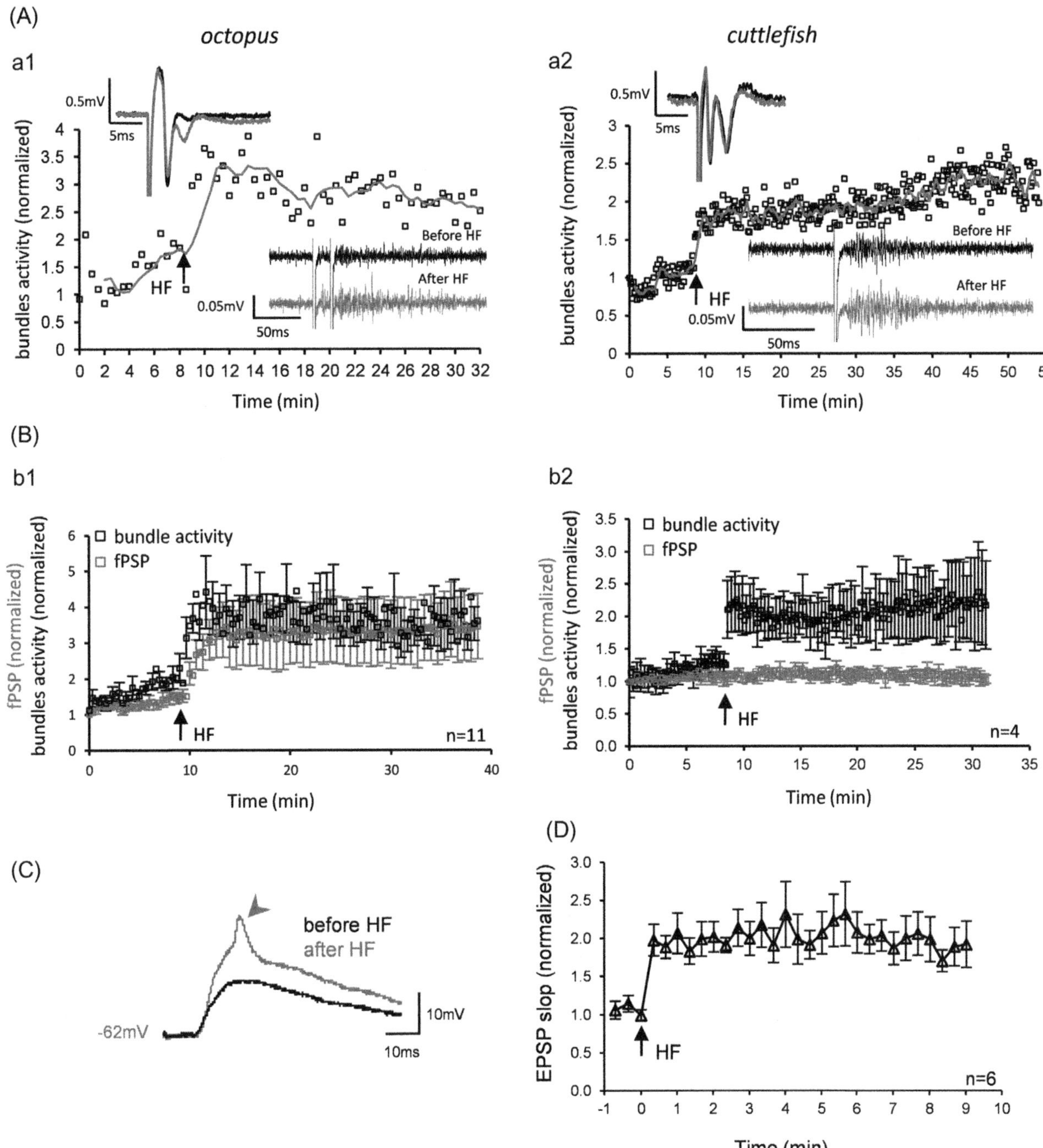

FIGURE 24.5 Input/output relationships of the VL show LTP both in octopus (A1 and B1) and in cuttlefish (A2 and B2). (A1 and A2) Extracellular recordings from the axon bundles of the large cells. The development and maintenance of LTP as measured by the integrated activity in the axon bundles evoked by MSFL tract stimulation. Top insets: Superimposed LFPs before (black) and after (red) high-frequency (HF) stimulation. Bottom insets: activity of the large cells axon bundles before (black) and after (red) HF stimulation. (B1 and B2) Summary of experiments of the type shown in panels A1and A2, respectively. The red curves show the fPSP amplitude. Note the absence of LTP of cuttlefish fPSP. The black curves show the integrated spiking activity recorded from the large cell axonal bundles. Both animals show LTP of the bundle activity. (C) Activity-induced LTP of the EPSP in the large cells of cuttlefish. Whole-cell intracellular recordings showing facilitation of the EPSP and a spikelet (red arrowhead) after HF stimulation of the MSFL tract. (D) Summary of six intracellular recording experiments as in panel C. LTP is expressed as an increase in slope of EPSP onset normalized to control. Source: *Modified from Shomrat* et al.[32]

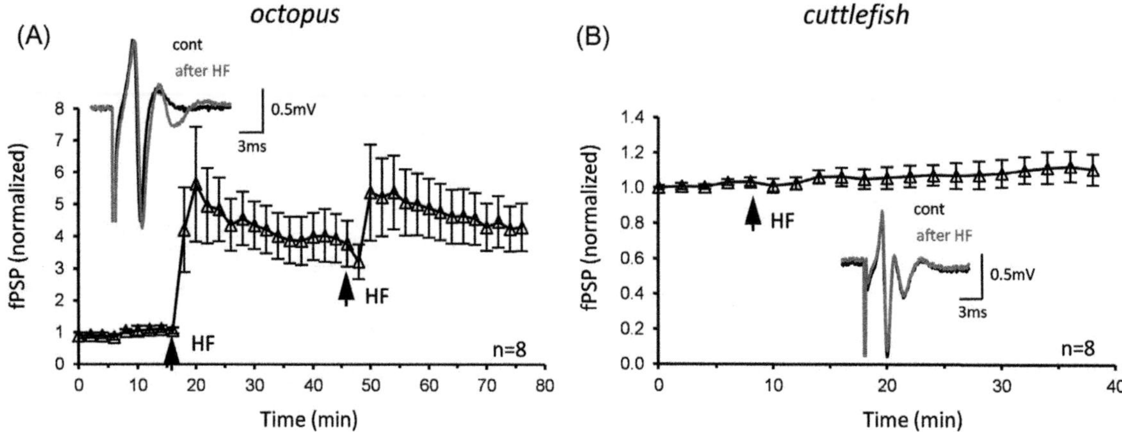

FIGURE 24.6 **Long-term potentiation at the MSF-to-amacrine connections differs dramatically in octopus (left) and cuttlefish (right).** (A) Summary of eight experiments showing the development, maintenance, and saturability of LTP in octopus. (B) Summary of eight experiments showing no significant change in the cuttlefish. LTP was induced by 4 HF trains (20 pulses, 50 Hz, 10-sec intertrain interval). fPSPs were normalized to the average of 10 test fPSPs at the beginning of each experiment. Inset: LFP traces before (black) superimposed on those obtained after HF stimulation (red). Source: *Modified from Shomrat* et al.[32]

spiking activity of the large neuron axon bundles (Figures 24.5A1 and 24.5B1). This result was not surprising: facilitation of the synaptic input to the amacrine cells (the fPSP) should increase their cholinergic input to the large neurons and thus enhance their output (i.e., increase the bundle activity induced by tract stimulation (Figure 24.5B1)). Therefore, this result does not reveal whether there is also synaptic plasticity at the input to the large neurons.

Analyzing the relationship between the amplitude of the fPSP and bundle activity showed this to be linear and unaffected by LTP induction.[32] Because the same fPSP amplitude gave rise to the same level of bundle activity irrespective of LTP, LTP in the octopus VL must occur only at the first fan-out synaptic layer. Indirect support for this interpretation is also apparent in Figure 24.5B1, which shows that the average relative increase in the fPSP amplitude was similar to the relative increase of the average bundle activity of the large neurons. Importantly, this result shows (1) that in the octopus only the glutamatergic input to the amacrine cells undergoes LTP and (2) that the output of the VL (the bundle activity) is linearly related to the input to the VL (the fPSP). This finding has important computational consequences (discussed later).[32]

In the cuttlefish, the site of synaptic plasticity mediating the activity-dependent increase in the VL output (Figures 24.5A2 and 24.5B2) is the cholinergic synaptic input to the large neurons. Intracellular recordings from the large neurons clearly show that HF stimulation of the SFL tract leads to a robust enhancement of the EPSP recorded from the large neuron (Figures 24.5C and 24.5D). Thus, whereas in the octopus the increase in VL output involves LTP of the glutamatergic connection onto the amacrine interneurons (the fan-out connections), in the cuttlefish the LTP occurs at the converging or fan-in cholinergic connections of the amacrine neurons onto the large neurons (see connectivity scheme shown in Figure 24.12).

WHAT DO THE VERTICAL LOBES OF OCTOPUS AND CUTTLEFISH COMPUTE?

It is possible that these differences between octopuses and cuttlefish mediate different behaviors yet to be systematically examined. However, computational considerations suggest a quite different possibility. The VL of both octopus and cuttlefish shows a linear relation between the fPSP amplitude and the integrated level of spiking activity in the axonal bundle of the large neurons (Figure 24.5B1)—that is, a linear input/output relationship of the VL (for a detailed explanation, see Shomrat *et al.*[32]). This linear input/output relationship has important computational consequences. In a linearly operating fan-out fan-in network, similar computation capacity is obtained regardless of whether the plasticity is localized at the fan-out or the fan-in layer.[32]

This computational consideration may answer the puzzling question of why these differences in network organization are found in phylogenetically close relatives possessing similar unusual characteristics (flexible body, similar sensory modalities, chromatophore system, and subcutaneous muscular system) and sharing comparable evolutionary adaptation to the same ecological niche and way of life (e.g., both are mainly solitary predators). If the two networks have indeed evolved to the same computational capacity, either via evolutionary selection or via "self-organizational" mechanisms, then computational constraints rather than specific neural properties may determine the physical network properties.[32]

MECHANISM OF LTP INDUCTION IN THE OCTOPUS VERTICAL LOBE

To understand the functional role of the VL network in learning and memory, we need to determine whether activity-dependent synaptic plasticity is mediated by associative or nonassociative mechanisms. That is, does the synaptic plasticity in the VL fulfill Hebb's rule for associative plasticity? This rule states that the synaptic connection between pre- and postsynaptic cells strengthens only when both are simultaneously and sufficiently active.[36] The NMDA channel, whose discovery was one of the most exciting breakthroughs in modern neuroscience, shows such a coincidence-detecting gating property. We therefore explored the possible involvement of such Hebbian mechanism in the fan-out stage in the octopus. Direct tests for the involvement of an NMDA-like receptor in the postsynaptic current (fPSP) or in LTP induction gave negative results; neither APV, which blocks NMDA-like current in cephalopod chromatophore muscles,[37] nor MK-801 blocked these phenomena[27] (Shomrat and Hochner, unpublished data).

The crucial test for whether the octopus VL has evolved an NMDA-independent Hebbian plasticity is to give the tetanization after completely blocking the postsynaptic response and then, after washing out the antagonists, to check whether LTP was induced. Such experiments show that the LTP in the octopus VL uses both associative and nonassociative induction mechanisms.[27] In slightly less than half the experiments, LTP induction was largely blocked by glutamatergic antagonists (Figure 24.7, left); in the other experiments, there was hardly any blocking effect and LTP appears to have occurred in the absence of a postsynaptic response. If the two processes thus revealed were evenly distributed among the different synaptic connections in the VL, we would expect a normal distribution of the results. The bimodal distribution in Figure 24.7 can only be explained by the two types of plasticity being segregated in anatomically different regions, as in the hippocampus (CA3 vs. CA1).[38] Such differentiation has not been demonstrated morphologically in the octopus VL.

Characterizing associative/Hebbian-type activity-dependent synaptic plasticity requires determining whether LTP results from an increase in the amount of transmitter released or an increase in postsynaptic response or both—an issue that is still not completely resolved in the various hippocampal synapses. Detailed analysis of the changes in the properties of synaptic transmission accompanying LTP induction in the VL suggests that the expression of LTP is mainly, if not exclusively, presynaptic. That is, it occurs in the MSFL neuron terminals at their synapses with the amacrine neurons.[31] It is conceivable that this synapse, with its Hebbian type of LTP induction (Figure 24.7), involves some sort of retrograde messenger, such as nitric oxide (NO). This would transfer the association signal from the postsynaptic cell back to the presynaptic terminals to induce the LTP.[39] Supporting this possibility, drugs blocking NO production do interfere with tactile and visual learning in the octopus.[40,41] Furthermore, NO has been shown to be an important neuromodulator in invertebrates, a well-studied example being the digestive system of gastropods.[42]

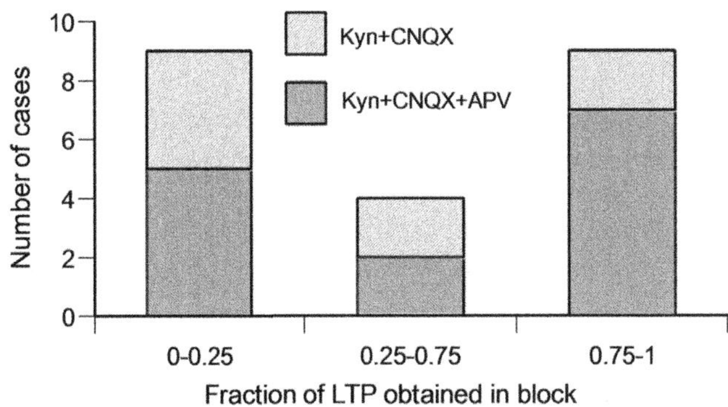

FIGURE 24.7 Dependence of LTP induction on postsynaptic response. Summary of 22 experiments in which the response during HF tetanization was completely blocked by a mixture of 20 mM kynurenate plus 200 μM CNQX and, in some experiments, also 200 μM APV. Three ranges of LTP (expressed as the percentage of total LTP) were induced in the presence of the blocking mixture. The distribution shows a higher proportion of either largely blocked (0–25%) or hardly blocked (75–100%) LTP. Source: *Reproduced from Hochner* et al.[27]

NEUROMODULATION IN THE VERTICAL LOBE

The highly dynamic properties of neural networks mediating learning and memory are achieved not only by various types of activity-dependent plasticity but also by neuromodulators. Neuromodulators may feed in negative or positive heterosynaptic reward signals that modulate the activity-dependent (homosynaptic) processes.[43] Such organization is well documented in both invertebrates and vertebrates and has most likely evolved to support supervised learning mechanisms.[44–46]

The first neuromodulator tested in the octopus VL was serotonin. Serotonin is a well-known neuromodulator in mollusks, involved in both short- and long-term facilitation of the sensorimotor synapse in the defensive reflex of *Aplysia californica*.[47,48] Immunohistochemistry showed serotonin reactivity in

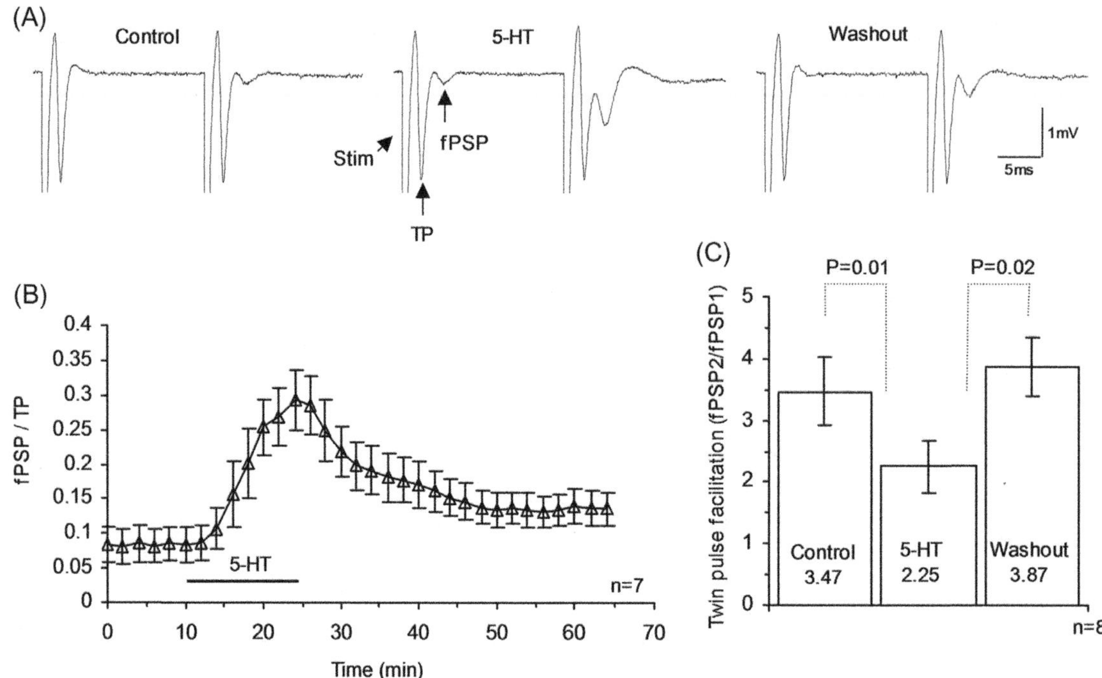

FIGURE 24.8 **5-HT induces short-term synaptic facilitation in the octopus.** (A) Records from one experiment. Responses to stimulation by twin pulses. Stim, stimulation artifact; TP, tract potential. (B) Summary of seven experiments demonstrating the facilitatory effect of 100–200 μM 5-HT and its reversal on washout. Test stimuli were applied every 10 or 20 sec. (C) 5-HT reversibly reduced the level of paired-pulse facilitation measured as the ratio of the second to the first fPSP amplitudes. Source: *Modified from Shomrat* et al.[49]

nerve terminals in the VL neuropil but not in the cell bodies, indicating that serotonin may convey signals to the VL from other brain or body areas.[49] In cuttlefish VL, Boyer *et al.*[50] described distributed serotonin reactivity mainly in fibers in the neuropile, but in contrast to the octopus VL, some cell bodies were also labeled.

The short-term modulatory effects of serotonin on octopus VL function are shown in Figure 24.8. Serotonin (5-HT) at 100–200 μM caused an average of approximately 3.5-fold facilitation of the fPSP. This synaptic connection in the cuttlefish VL, which lacks activity-dependent synaptic plasticity (Figure 24.6B), did not show any modulation by 5-HT.[32] In the octopus, the 5-HT facilitation involves presynaptic modulation of transmitter release because it was accompanied by a reversible reduction in twin-pulse facilitation (i.e., a reduction in the ratio of the second fPSP amplitude to that of the first in twin-pulse stimulation; Figures 24.8A and 24.8C). Repeated exposure to 5-HT did not lead to long-term modulation in the octopus VL, unlike its long-term effects on the *Aplysia* sensorimotor synapse. Also in contrast to *Aplysia*, c-AMP does not appear to be a major second messenger in 5-HT-induced fPSP facilitation in the octopus VL.[49]

It is intriguing that 5-HT shows only a robust short-term facilitatory effect on a synaptic connection undergoing long-term activity-dependent facilitation. The serotonergic system in the octopus VL may have adapted to provide a modulatory signal to the VL rather than inducing long-term plasticity changes as in *Aplysia*. Based on the experiment shown in Figure 24.9, Shomrat *et al.*[49] suggested that this modulatory signal may serve as a reinforcing effect on LTP induction. In this experiment, tetanization trains of only 3 rather than 20 pulses (at 50 Hz) were applied. This reduced intensity induced only a partial and very modest LTP (19.5% of the final LTP). Similar triplet stimulation in the presence of 5-HT, which caused a robust short-term fPSP facilitation, induced a much higher LTP (60.7%; Figure 24.9B). The simple explanation is that 5-HT augments LTP indirectly by enhancing synaptic activity. That is, the serotonergic system of the VL appears to reinforce the induction of activity-dependent plasticity, and 5-HT may thus convey reward signals to the learning and memory network in the VL, as does dopamine in mammals[46] and mollusks[45,51] and octopamine in insects.[52]

ARE THE OCTOPUS VERTICAL LOBE AND ITS LTP INVOLVED IN BEHAVIORAL LEARNING AND MEMORY?

The octopus VL, with its simple circuitry, may be the preparation of choice for investigating how neurons participate in storing memories and how

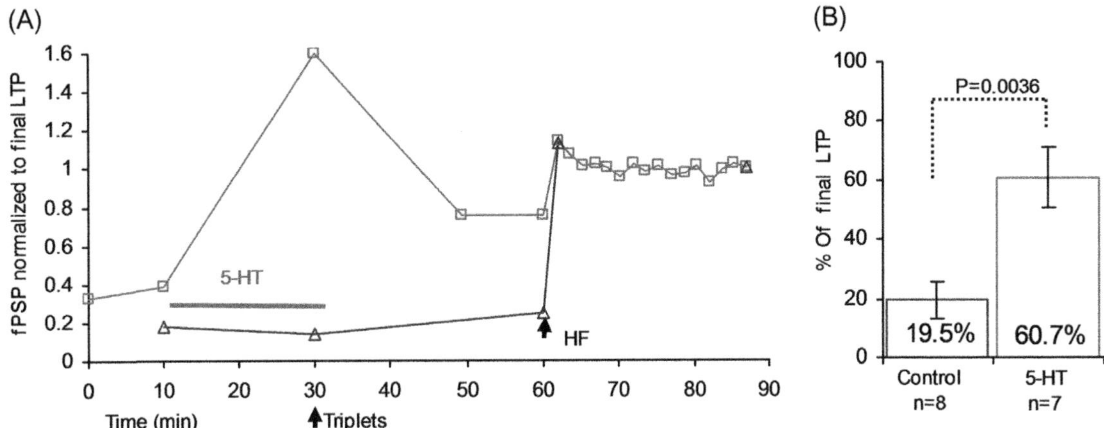

FIGURE 24.9 **5-HT reinforced activity-dependent LTP induction.** (A) Partial LTP was induced by a triplet pulse stimulation protocol. 5-HT (red trace, open squares) caused large short-term facilitation. After 30-min washout (60 min), HF stimulation revealed the residual LTP. HF stimulation gave rise to greater facilitation of the fPSPs in the control experiments (blue trace, open triangles), indicating less recruitment of LTP than in the presence of 5-HT. (B) Summary histogram showing the percentage of the final LTP measured at the end of the experiments exemplified in panel A. LTP was induced by the triplet stimulation protocol with and without 5-HT. Source: *Modified from Shomrat* et al.[49]

memories are retrieved or forgotten. Pioneering steps in this direction were made by Young, Boycott, Wells, and colleagues during the second half of the 20th century. Their behavioral, morphological, and lesion experiments showed that the octopus VL is important for learning and memory but is not the sole brain lobe involved in these functions.[2] The discovery of activity-dependent long-term plasticity and its neuromodulation in the VL provides physiological support for a VL function in learning and memory. However, a direct experimental test was required to confirm the involvement of these physiological processes in learning behavior.

To directly test the involvement of the LTP in the VL in learning and memory, Shomrat *et al.*[28] followed the experiments of Moser *et al.*[53] These showed that artificial saturation of LTP, induced in the rat hippocampus by tetanization, impaired spatial learning when the tetanization was applied prior to learning. This technically challenging experiment in the rat is much simpler in the octopus because the VL lies most dorsally in the brain and is relatively easily accessible to a large electrode for global tetanization. Such tetanization induced on average 56% of the available LTP.

Shomrat *et al.*[28] then tested whether this reduction in the available synaptic plasticity affected a learning task given 75 min after recovery from anesthesia and tetanization. The learning task was a passive avoidance task in which a mild electric shock "taught" the octopus to stop attacking a red ball. The results of LTP saturation by the tetanization were contrasted with the effects of transecting the medial SFL tract to the VL, which disconnects the VL system from its sensory input. In contrast, inducing strong LTP of the glutamatergic synaptic connection of the SFL axons onto the amacrine cells may be viewed as "short-circuiting" the VL (see VL circuitry in Figures 24.4E and 24.12). Neither tetanization nor transection eliminated the ability of the octopuses to learn the task (Figure 24.10). Transection slowed the learning rate relative to that of sham-operated animals (Figure 24.10A). Unexpectedly, saturating the LTP had the opposite effect, enhancing the learning process relative to that of nontetanized animals (Figure 24.10B). Thus, LTP of specific synaptic connections in the VL does not appear to be involved in short-term learning. The VL output most likely controls the process mediating short-term learning that occurs outside the VL, with transection reducing and LTP enhancing the rate of learning by modulating the VL output.

Testing for long-term memory 24 hr after training produced more easily understandable results. Both treatments severely impaired long-term memory (Figure 24.11). Sham-operated and control animals did not demonstrate perfect memory of the task, with approximately 70% of the animals attacking the ball on their first trial. However, they demonstrated a robust saving or recollection as they stopped attacking the ball in the following test trials (Figure 24.11, open symbols). The transected animals (Figure 24.11A) remembered almost nothing of the avoidance task they had learned the day before, whereas the tetanized animals showed severe impairment (Figure 24.11B). This difference was probably due to the tetanization not completely saturating the LTP.[28] These results suggest that memories acquired soon after tetanization or transaction are not consolidated. In contrast, a memory acquired before tetanization or transection, such as to attack a crab or a white ball (associated with positive reward in pretraining), was not impaired.[28] The results

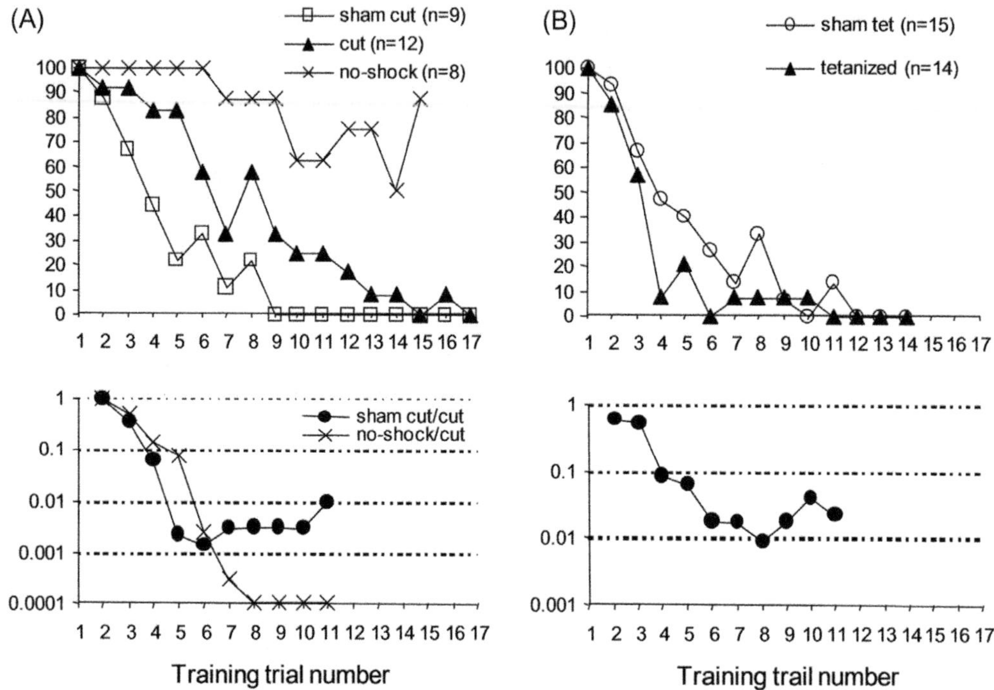

FIGURE 24.10 **Transection slowed and tetanization enhanced short-term learning of an avoidance task.** (A) The MSLF-transected animals showed significantly slower learning curves than the sham controls, with a significant difference from the fourth testing trial on (bottom panel shows cumulative Fisher's test between the different experiments; see Shomrat *et al.*[28]). Nevertheless, the transected animals stopped touching the red ball significantly faster than no-shock controls. (B) Tetanized animals learned faster than the sham-operated animals, and by the eighth trial the level of cumulative Fisher's test fell below 0.01 (0.0094; bottom panel). Source: *Modified from Shomrat* et al.[28]

confirm earlier lesion experiments showing that previous memories were not affected by lesions of the VL,[17] similar to results obtained in mammals and even humans with a severed hippocampus.[54,55]

These experiments clearly demonstrate that the VL and its LTP are not important for the actual storage of long-term memory. Short- and long-term memory traces appear to be stored outside the VL, possibly in the circuitry mediating the attack behavior. Instead, the VL plays an important role in controlling the consolidation of short-term memory into long-term memory occurring elsewhere.

A SYSTEM MODEL FOR OCTOPUS LEARNING AND MEMORY

We conclude by proposing a tentative model for the octopus learning and memory system incorporating the physiological and behavioral findings that we have presented in this chapter (Figure 24.12). The rationale behind this model is summarized in Table 24.1.

To explain how the model functions, we consider the passive avoidance task used by Shomrat *et al.*[56] The octopus learns to refrain from attacking a red ball (actually dark because octopuses are color-blind) by receiving an electric shock on its arms when it attacks. The information on shape and brightness is fed into the attack behavior circuitry that activates the natural attack behavior. This information feeds in parallel to the MSFL. Each quality (brightness and shape) is then transferred by a different set of MSFL neurons to the VL, most likely creating a sparse representation of each sensory quality in the matrix-like connections of the MSFL neurons with the amacrine interneurons. Those amacrine cells receiving inputs from both qualities are more likely to undergo LTP due to their higher level of activity. The "dark versus round" association is reinforced if they are conjugated with the pain signal conveyed to the VL by the serotonergic system. The strengthening of this association during training, in turn, creates a long-term enhancement of the amacrine cell input to the set of large neurons driven by this mutual sensory representation. The VL output generally inhibits the tendency to attack and can be regarded as an inhibitory supervising signal. The enhancement of the output generated by the red ball representation in the VL now specifically inhibits attacking the dark ball. It is not clear whether similar mechanisms mediate positive reward learning, viz., by decreasing the inhibitory drive of the VL.

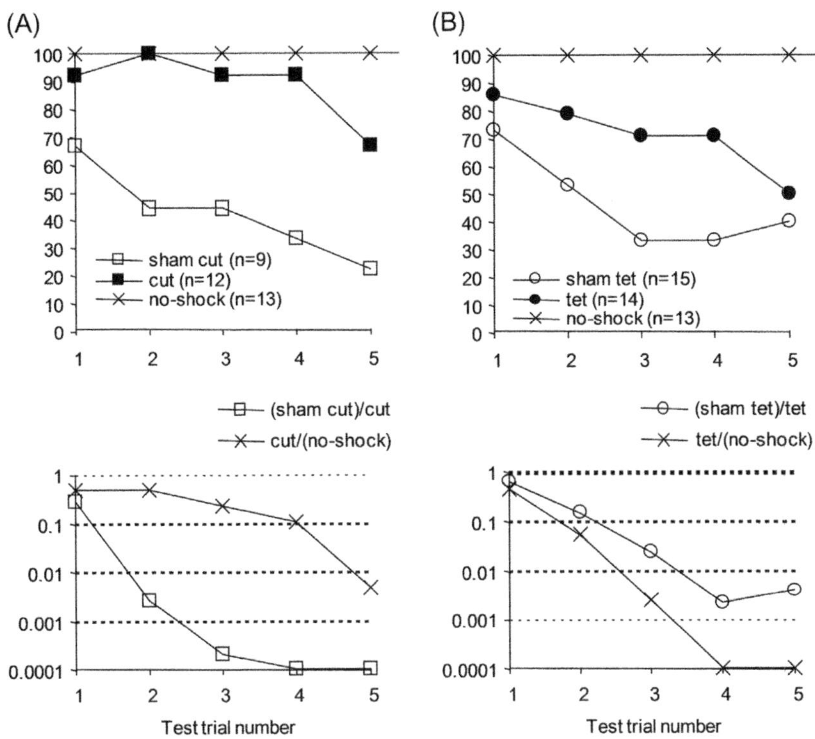

FIGURE 24.11 Tetanization and transection impair long-term recall tested 24 hr after training. The animals were given five test trials without electric shock. Testing revealed no significant difference in long-term memory but impairment in recall in consecutive tests both in transected (A) and in tetanized (B) animals. By the fifth test trial, the experimental animals showed some retention. Cumulative Fisher's exact test between the treated and the sham groups and the treated groups and the noncontingent controls are shown in the bottom panels (see explanation in the legend to Figure 24.10). Source: *Modified from Shomrat* et al.[28]

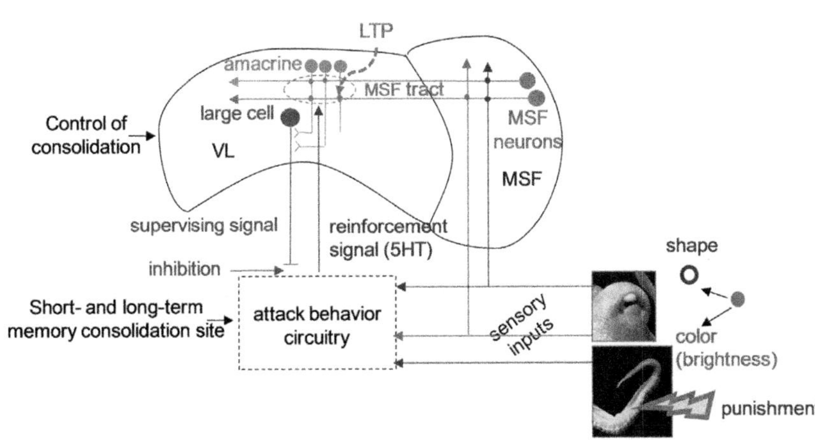

FIGURE 24.12 A tentative model for the octopus learning and memory system. For explanation, see text and Table 24.1.

TABLE 24.1 The Five Basic Findings and Their Interpretation on which the Octopus Learning and Memory Model Is Based

Experimental Finding	Conclusion
1. Transection and tetanization did not affect behavior; octopuses still showed their stereotypical attack behavior.	Sensory inputs feed in parallel to the VL system and to the circuits controlling behavior.
2. Bothe tetanization and transection did not erase old memories.	Long-term memory is stored outside the VL.
3. Tetanization accelerated learning (decreased the tendency to attack) and transection slowed learning (increased the tendency to attack). However, both treatments did not prevent the attack behavior.	The output of the VL modulates the rate of short-term learning that takes place outside the VL system, possibly by inhibiting the attack-mediating circuit.
4. Both treatments prevented the consolidation of short-term into long-term memory.	The LTP in the VL system is crucial for the consolidation of long-term memory outside the VL system.
5. Serotonin reinforced LTP induction.	Serotonin conveys the reward/punishment signal to the VL, which reinforces LTP of specific synapses active during the signal.

CONCLUSION

The results presented in this chapter show that cephalopods are valuable animals for exploring the neural mechanisms subserving complex behaviors. Cephalopod behaviors are mediated by a unique embodiment[57] that includes a unique segregation of the nervous system between three large components: the arms, the eyes, and a central brain. However, the cephalopod nervous system still maintains the basic features of invertebrate morphology and neurophysiology. Thus, as we have summarized here, the neural complexity of cephalopods appears to have been achieved by building complex networks from simpler invertebrate elements.[58] Most likely, evolution and self-organization have shaped simpler neural networks to control complex behavior, and these simpler networks provide a more accessible opportunity for assessing how neural networks are embedded in the organization of complex behaviors. The large but simply organized fan-out fan-in network of cephalopods' vertical lobe is a vivid demonstration of this idea.

Acknowledgments

Our research is supported by the Smith Family Laboratory at the Hebrew University, the United States–Israel Binational Science Foundation, the Israel Science Foundation, and European Commission EP7 projects OCTOPUS and STIFF-FLOP. We thank Jenny Kien for editorial assistance and suggestions.

References

1. Packard A. Cephalopods and fish: the limits of convergence. *Biol Rev.* 1972;47:241–307.
2. Wells MJ. *Octopus.* London: Chapman & Hall; 1978.
3. Grasso FW, Basil JA. The evolution of flexible behavioral repertoires in cephalopod molluscs. *Brain Behav Evol.* 2009; 74(3):231–245.
4. Hochner B, Shomrat T, Fiorito G. The octopus: a model for a comparative analysis of the evolution of learning and memory mechanisms. *Biol Bull.* 2006;210(3):308–317.
5. Hochner B. Octopuses. *Curr Biol.* 2008;18(19):R897.
6. Hanlon RT, Messenger JB. *Cephalopod Behaviour.* Cambridge, UK: Cambridge University Press; 1996.
7. Fiorito G, Scotto P. Observational learning in *Octopus vulgaris. Science.* 1992;256(5056):545–547.
8. Hochner B, Shomrat T, Fiorito G. The octopus: a model for a comparative analysis of the evolution of learning and memory mechanisms. *Biol Bull.* 2006;210(3):308–317.
9. Sanders GD. The cephalopods. In: Corning WC, Dyal JA, Willows AOD, eds. *Invertebrate Learning.* New York: Plenum; 1975:139–145.
10. Budelmann BU. The cephalopods nervous system: what evolution has made of the molluscan design. In: Breidbach O, Kutsuch W, eds. *The Nervous System of Invertebrates: An Evolutionary and Comparative Approach.* Basel: Birkhauser; 1995:115–138.
11. Calisti M, et al. An octopus-bioinspired solution to movement and manipulation for soft robots. *Bioinspir Biomim.* 2011;6 (3):036002.
12. Nixon M, Young JZ. *The Brain and Lives of Cephalopods.* Oxford: Oxford University Press; 2003.
13. Young JZ. *The Anatomy of the Nervous System of Octopus vulgaris.* Oxford: Clarendon Press; 1971 [xxxi, 690].
14. Bullock TH, Horridge GA. *Structure and Function in the Nervous Systems of Invertebrates.* San Francisco: Freeman; 1965.
15. Boycott BB. The functional organization of the brain of the cuttlefish *Sepia officinalis. Proc R Soc Lond B Biol Sci.* 1961;153:503.
16. Zullo L, et al. Nonsomatotopic organization of the higher motor centers in octopus. *Curr Biol.* 2009;19(19):1632–1636.
17. Boycott BB, Young JZ. A memory system in *Octopus vulgaris* Lamarck. *Proc R Soc Lond B Biol Sci.* 1955;143 (913):449–480.
18. Maldonado H. The positive and negative learning process in *Octopus vulgaris* Lamarck: influence of the vertical and median superior frontal lobes. *Z Vgl Physiol.* 1965;51:185–203.
19. Graindorge N, et al. Effects of dorsal and ventral vertical lobe electrolytic lesions on spatial learning and locomotor activity in *Sepia officinalis. Behav Neurosci.* 2006;120(5):1151–1158.
20. Sanders FK, Young JZ. Learning and other functions of the higher nervous centres of *Sepia. J Neurophysiol.* 1940;3:501.
21. Fiorito G, Chichery R. Lesions of the vertical lobe impair visual discrimination learning by observation in *Octopus vulgaris. Neurosci Lett.* 1995;192(2):117.
22. Young JZ. Computation in the learning system of cephalopods. *Biol Bull.* 1991;180(2):200–208.
23. Young JZ. Multiple matrices in the memory system of octopus. In: Abbott JN, Williamson R, Maddock L, eds. *Cephalopod Neurobiology.* Oxford: Oxford University Press; 1995:431–443.
24. Hochner B. Functional and comparative assessments of the octopus learning and memory system. *Front Biosci.* 2010;2:764–771.
25. Gray EG. The fine structure of the vertical lobe of octopus brain. Phil. *Trans R Soc Lond B.* 1970;258:379–394.
26. Gray EG, Young JZ. Electron microscopy of synaptic structure of octopus brain. *J Cell Biol.* 1964;21:87–103.
27. Hochner B, et al. A learning and memory area in the octopus brain manifests a vertebrate-like long-term potentiation. *J Neurophysiol.* 2003;90(5):3547–3554.
28. Shomrat T, et al. The octopus vertical lobe modulates short-term learning rate and uses LTP to acquire long-term memory. *Curr Biol.* 2008;18(5):337–342.
29. Bullock TH, Basar E. Comparison of ongoing compound field potentials in the brains of invertebrates and vertebrates. *Brain Res.* 1988;472(1):57–75.
30. Bullock TH. Ongoing compound field potentials from octopus brain are labile and vertebrate-like. *Electroencephalogr Clin Neurophysiol.* 1984;57(5):473–483.
31. Hochner B, et al. A learning and memory area in the octopus brain manifests a vertebrate-like long-term potentiation. *J Neurophysiol.* 2003;90(5):3547–3554.
32. Shomrat T, et al. Alternative sites of synaptic plasticity in two homologous fan-out fan-in learning and memory networks. *Curr Biol.* 2011;21(21):1773–1782.
33. Miyakawa H, Kato H. Active properties of dendritic membrane examined by current source density analysis in hippocampal CA1 pyramidal neurons. *Brain Res.* 1986;399(2):303–309.
34. Matzner H, Gutfreund Y, Hochner B. Neuromuscular system of the flexible arm of the octopus: physiological characterization. *J Neurophysiol.* 2000;83(3):1315–1328.
35. Vapnik VN. *Statistical Learning Theory.* New York: Wiley; 1998.
36. Hebb DO. *The Organization of Behavior; A Neuropsychological Theory.* New York: Wiley; 1949.

37. Lima PA, Nardi G, Brown ER. AMPA/kainate and NMDA-like glutamate receptors at the chromatophore neuromuscular junction of the squid: role in synaptic transmission and skin patterning. *Eur J Neurosci.* 2003;17(3):507–516.
38. Kandel ER, Schwartz JH, Jessell TM. *Principles of Neural Science.* 4th ed. New York: McGraw-Hill; 2000.
39. Garthwaite J. Concepts of neural nitric oxide-mediated transmission. *Eur J Neurosci.* 2008;27(11):2783–2802.
40. Robertson JD, et al. Nitric oxide is necessary for visual learning in *Octopus vulgaris. Proc Biol Sci.* 1996;263(1377):1739–1743.
41. Robertson JD, Bonaventura J, Kohm AP. Nitric oxide is required for tactile learning in *Octopus vulgaris. Proc Biol Sci.* 1994;256 (1347):269–273.
42. Susswein AJ, Chiel HJ. Nitric oxide as a regulator of behavior: new ideas from aplysia feeding. *Prog Neurobiol.* 2012; 97(3):304–317.
43. Bailey CH, et al. Is heterosynaptic modulation essential for stabilizing Hebbian plasticity and memory? *Nat Rev Neurosci.* 2000; 1(1):11–20.
44. Keene AC, Waddell S. *Drosophila* olfactory memory: single genes to complex neural circuits. *Nat Rev Neurosci.* 2007;8(5):341–354.
45. Kemenes I, O'Shea M, Benjamin PR. Different circuit and monoamine mechanisms consolidate long-term memory in aversive and reward classical conditioning. *Eur J Neurosci.* 2011;33 (1):143–152.
46. Schultz W. Behavioral dopamine signals. *Trends Neurosci.* 2007;30(5):203.
47. Glanzman DL. Common mechanisms of synaptic plasticity minireview in vertebrates and invertebrates. *Curr Biol.* 2010;20(1): R31–R36.
48. Kandel ER. The molecular biology of memory storage: a dialogue between genes and synapses. *Science.* 2001;294(5544): 1030–1038.
49. Shomrat T, et al. Serotonin is a facilitatory neuromodulator of synaptic transmission and "reinforces" long-term potentiation induction in the vertical lobe of *Octopus vulgaris. Neuroscience.* 2010;169(1):52–64.
50. Boyer C, et al. Distribution of neurokinin A-like and serotonin immunoreactivities within the vertical lobe complex in *Sepia officinalis. Brain Res.* 2007;1133(1):53–66.
51. Reyes FD, et al. Reinforcement in an *in vitro* analog of appetitive classical conditioning of feeding behavior in *Aplysia*: blockade by a dopamine antagonist. *Learn Mem.* 2005;12(3):216–220.
52. Cassenaer S, Laurent G. Conditional modulation of spike-timing-dependent plasticity for olfactory learning. *Nature.* 2012;482(7383):47–52.
53. Moser EI, et al. Impaired spatial learning after saturation of long-term potentiation. *Science.* 1998;281(5385):2038–2042.
54. Stellar E. Physiological psychology. *Annu Rev Psychol.* 1957;8:415–436.
55. Corkin S. What's new with the amnesic patient H.M.? *Nat Rev Neurosci.* 2002;3(2):153–160.
56. Shomrat T, et al. The octopus vertical lobe modulates short-term learning rate and uses LTP to acquire long-term memory. *Curr Biol.* 2008;18(5):337–342.
57. Hochner B. An embodied view of octopus neurobiology. *Curr Biol.* 2012;22(20):R887–R892.
58. Emes RD, et al. Evolutionary expansion and anatomical specialization of synapse proteome complexity. *Nat Neurosci.* 2008;11 (7):799–806.

CHAPTER

25

Learning, Memory, and Brain Plasticity in Cuttlefish (*Sepia officinalis*)

Ludovic Dickel[*,†], *Anne-Sophie Darmaillacq*[*,†], *Christelle Jozet-Alves*[*,†] and *Cécile Bellanger*[*,†]

[*]Normandie Univ, UCBN, GMPc, EA 4259, F-14032 Caen, France [†]Centre de Recherches en Environnement Côtier, Station Marine de l'UCBN, 14530 Luc-sur-Mer, France

INTRODUCTION

It is important to remind behavioral neurobiologists and, more widely, experimental psychologists that some animals that are phylogenetically remote from mammals show particularly impressive learning and cognitive capabilities. Along with some other invertebrates, cephalopods are excellent representatives to support this assessment. They are well-known for their fascinating learning capabilities and their well-developed and centralized nervous system, which is arranged around the esophagus like an actual 'brain.' Cephalopod brains are by far the largest of all invertebrate nervous systems in their number of cells (520 million neurons in *Octopus*[1]). This molluskan brain can rival those of many vertebrates, including some bird species, in their brain-to-body weight ratio.[2] This may explain why cephalopods are so fascinating for those interested in the evolution of brain and behavior. They also provide original models for studying neural network bases of learning and memory. Of the 700 cephalopod species and despite the pioneering work on cuttlefish by Sanders and Young,[3] the common octopus (*Octopus vulgaris*) has been the most extensively studied species in this field, beginning in the second half of the 20th century (for review, see Wells[4]). There are various reasons for this. Octopuses are easy to keep under laboratory conditions; they live in a den and promptly catch any object moving near their home, which makes operant learning procedures easier to design. Moreover, octopuses successfully survive brain lesions, which can be carried out with a minimum of general tissue damage and blood loss.

The extraordinary amount of work by Young and collaborators on octopuses was undertaken at the Stazione Zoologica di Napoli (Italy), in the vicinity of which octopuses are abundant. Moreover, this marine station has extensive experience, facilities, and skills in maintaining octopuses under laboratory conditions. As a consequence, brain, learning, and memory capabilities in cuttlefish remained less known during most of the 20th century, despite the remarkable research of Wells[5,6] and Messenger[7–9] on both juveniles and adult animals. In the 1990s, a resurgence of interest for learning in cuttlefish occurred at the Marine Station of Luc-sur-Mer in Chichery's laboratory. It is located on the French coast of the English Channel, where the common cuttlefish (*Sepia officinalis*) comes to reproduce and lay eggs in spring and summer. Originally, the studies conducted in this laboratory followed the pioneering work of Sanders and Young[3] and Messenger.[7,8,10] The former investigated in detail the prey-catching behavior of cuttlefish. *Sepia officinalis* is an active predator, capturing large and very mobile prey such as crabs, fish, and shrimp. The attack is visually guided and consists of a precise sequence of events: (1) prey detection, (2) orientation of the head toward the prey, (3) pursuit of the prey when necessary, (4) positioning with ocular convergence, and (5) prey seizure.[7,11–13] The prey seizure stage can occur in two different ways—by rapidly shooting out the two long tentacles on fish, small crabs, and shrimp (Figure 25.1A)

Invertebrate Learning and Memory.
DOI: http://dx.doi.org/10.1016/B978-0-12-415823-8.00025-3

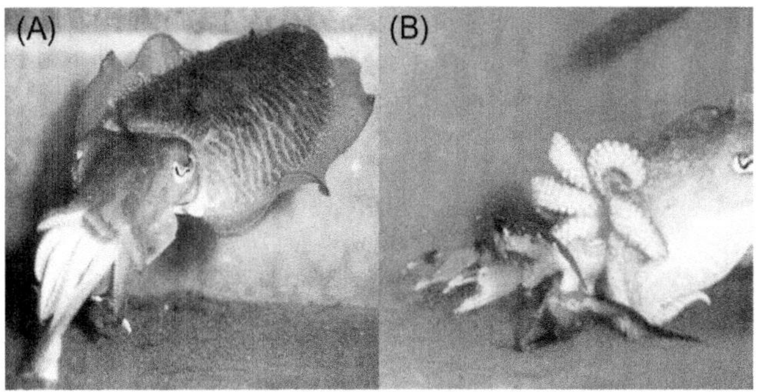

FIGURE 25.1 Prey-catching strategies in cuttlefish: (A) tentacle striking on a prawn and (B) jumping on a crab.

or by jumping on the prey and seizing it with the eight arms (Figure 25.1B).[14] After the prey has been caught, it is manipulated and ingested. These prey-catching behaviors are 'all-or-none' events. They are easily identifiable and quantifiable so that they were advantageously used in assessing learning capabilities in cuttlefish.

Although cuttlefish are very efficient swimmers (with the use of both funnel and fins), they are nectobenthic and spend most of their time on the bottom of their tanks in captivity, hiding in dark shelters when available, digging into the sand, or concealing in their environment. They possess spectacular capabilities in displaying a great variety of neurally controlled body and color patterns that are adapted to the color and the texture of the background (for a review, see Hanlon and Messenger[15]).

In this chapter, we summarize the main data available concerning the anatomical and functional organization of the cuttlefish brain. Then, we review some of different learning procedures that are used in adult and juvenile cuttlefish and, when possible, their neural correlates. Finally, we highlight some of the most promising avenues to follow in the future to better understand the extraordinary behavioral plasticity of these sophisticated invertebrates.

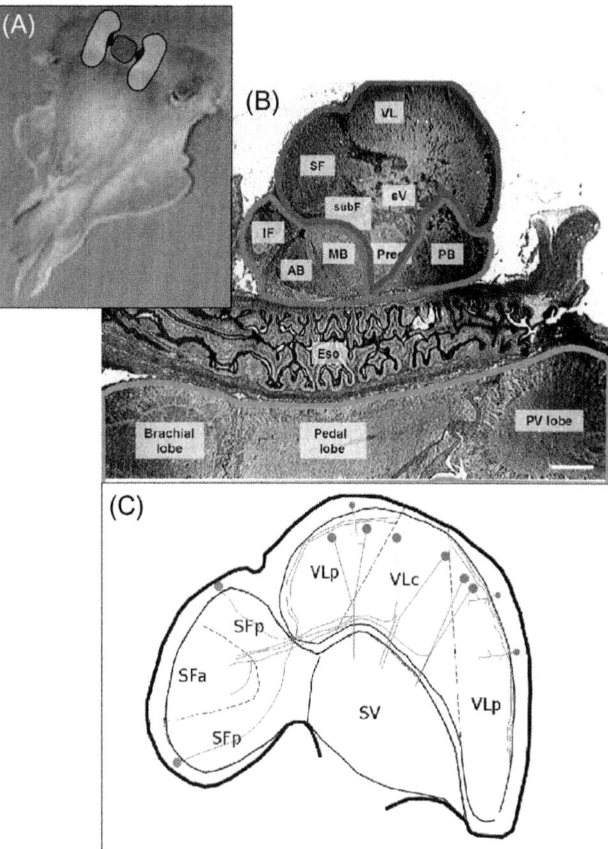

FIGURE 25.2 (A) Front view of a cuttlefish showing the central nervous system: optic lobes (yellow) and central mass (blue). (B) Sagittal section of an adult cuttlefish brain showing the supra- and subesophageal masses. The lower motor centers are surrounded in orange, the intermediate and higher motor centers in red, and the sensory association centers (electrically silent areas) in blue. AB, anterior basal lobe; Eso, esophagus; IF, inferior frontal lobe; MB, median basal lobe; PB, posterior basal lobe; Prec, precommissural lobe; PV lobe, palleovisceral lobe; SF, superior frontal lobe; subF, subfrontal lobe; sV, subvertical lobe; VL, vertical lobe. Scale bar = 450 μm. (C) Diagram of the vertical lobe connections. SFa, anterior part of the superior frontal lobe; SFp, posterior part of the superior frontal lobe; SV, subvertical lobe; VLc, central vertical lobe; VLp, peripheral vertical lobe.

THE CUTTLEFISH BRAIN

The cephalopod central nervous system, lying between the eyes, consists of a central circumesophageal brain connected to two large optic lobes by short optic tracts (Figure 25.2A). The anatomy of the brain of *Sepia* has been described in great detail (for review, see Nixon and Young[16]). Each brain structure, called a lobe, consists of cell layers (cortex) surrounding a neuropil (networks of fibers). Based on the behavioral responses induced by electrical stimulation of different lobes,[17–19] the cuttlefish brain was functionally subdivided into lower, intermediate, and higher motor centers and silent areas, hence considered as sensory association centers involved in the integration of multisensory inputs.

The optic lobes integrate visual inputs from the retina and are involved in the control of motor programs (higher motor centers).[17–19] Moreover, the optic lobes have been suggested to store memory in *Octopus*,[20] particularly in the case of associative learning.

The central mass of the brain (Figure 25.2B) comprises a supra- and a subesophageal mass connected by periesophageal lobes. The sub- and periesophageal lobes, classified as lower and intermediate motor centers, control swimming, escape behavior, ink ejection, and arms and tentacle movements. The ventral part of the supraesophageal mass comprises some higher centers implicated in the control of motor programs; their electrical stimulation induces combined movements of different groups of effectors.[17]

The vertical lobe (VL) system, also called the VL complex by Young,[21] corresponds to the dorsal part of the central mass. It contains associative structures (electrically 'silent areas'[17]) receiving inputs from different sensory systems and is known to be involved in learning and memory (for reviews, see[22,23]). The VL system includes the precommissural, subfrontal, superior frontal, subvertical, and vertical lobes (Figure 25.2B). Young[21] has described in detail this system in squids, but the general organization is strikingly similar in the cuttlefish. At the front of the VL system, the superior frontal lobe shows two distinct parts: the anterior superior frontal lobe, which receives inputs from the VL, and the posterior superior frontal lobe, which sends all its axons to the VL (Figure 25.2C). Each part is recruited at specific stages of learning.[24] The subvertical lobe is positioned behind the superior frontal lobe and below the VL. This structure receives axons from and sends axons to the VL and the superior frontal lobe. Both the superior frontal lobe and the subvertical lobe are centers of sensory multiconvergence, sending pretreated visual and tactile information to the VL. The VL is located in the dorsal part of the supraesophageal mass. Young[21] described two parts in the VL: the central VL corresponding to the whole middle part of the dome and, surrounding it, the peripheral VL (Figure 25.2C). Ablation or partial lesion of the VL induces learning and long-term memory impairments.[3,25] In the dorsal part of the VL, a large tract of fibers coming from the posterior superior frontal lobe is situated just below the cortex (Figure 25.2C). This VL–superior frontal lobe tract may be involved in the regulation of locomotor activity level and in memory.[25] Small cells scattered throughout the neuropil of the VL send axons ending in the subvertical lobe (Figure 25.2C). These VL–subvertical lobe tracts are the sites of long-term potentiation,[26] and their appearance during development is concomitant with the maturation of short-term memory capabilities in early juveniles.[27] Young[28] described a series of intersected matrices within the VL system in *Octopus* and emphasized the similarity of their organization to that of the hippocampus of vertebrates. He distinguished distinct memory matrices for visual and tactile learning.[29] Despite similar anatomical organization, the VL system of cuttlefish seems to show different functional and neural network properties compared to the VL system of octopuses.[26] Young suggested that if a system of matrices for tactile memory exists in the cuttlefish, it would differ from that of octopuses.[16] Finally, as in *Octopus*, VL damage or removal impairs learning and memory but never completely; this suggests a synergistic involvement of other structures (e.g., optic lobes).

BRAIN AND BEHAVIORAL PLASTICITY IN ADULTS

The 'Prawn in a Tube' Procedure

In behavioral neuroscience, one of the most used learning paradigms in cuttlefish is the 'prawn in a tube' (PT) procedure. In this procedure, prawns are enclosed in a transparent tube (made of clear Perspex or glass) in the middle of the cuttlefish tank (Figure 25.3). Under these conditions, a cuttlefish attempts to catch the unreachable prey by rapidly shooting out its two long tentacles, subsequently referred to as a 'strike.'

The number of strikes decreases with time during a continuous presentation (Figure 25.4) or in the course of eight successive 3-min presentations of the apparatus (with an interval of 30 min between presentations).[8] This learning is apparently very simple and

FIGURE 25.3 **Cuttlefish pointing at prawns enclosed in a glass tube.**

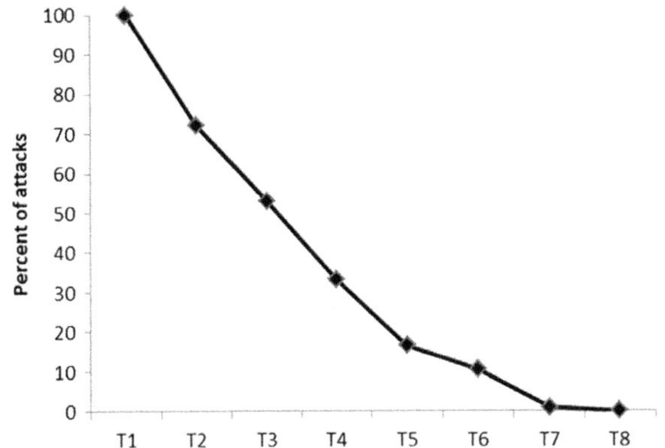

FIGURE 25.4 Percentage of attacks during the presentation of the glass tube containing shrimp. The number of attacks was counted by 3-min blocks and then expressed as a percentage of the number of attacks at T1. The training session was stopped when the cuttlefish reached the learning criterion of less than two attacks during a 3-min block.

practical, but to be efficiently used in functional neurobiology, it was of crucial importance to better characterize the nature of this learning and of the reinforcements that are possibly involved as well as the different cues cuttlefish can rely on to solve and memorize the task. As a consequence, the PT procedure has been extensively investigated in several studies, all based on the remarkable set of observations made by Messenger.[8]

PT Procedure: Habituation or Associative Learning?

When cuttlefish's tentacles were cut, the waning of 'pseudo-strikes' during the tube presentation was significantly lower than the waning of tentacle strikes in the non-operated animals placed in the same conditions. Furthermore, when a crab was offered after the learning, it was rapidly captured by the jumping or striking strategy.[8] This suggests that the waning of attempted captures in this learning was not the result of motor fatigue or of temporary incapacity to make tentacle strikes. Agin *et al.*[30] and Purdy *et al.*[31] assessed whether the decrease in the PT procedure was better interpreted as habituation or associative learning by testing whether the response could be reinstated through dishabituation. Agin *et al.* used a flashing light or a free-moving crab as 'dishabituating' stimulus, whereas Purdy *et al.* used a free-swimming shrimp (these authors used fish in the tube during training). In neither of these works did the authors demonstrate any increase in attempted captures (i.e., any 'dishabituation') on the glass during the PT presentation following the dishabituating stimulus. These works confirmed the original hypothesis of Sanders and Young[3] and Messenger,[8] according to which this

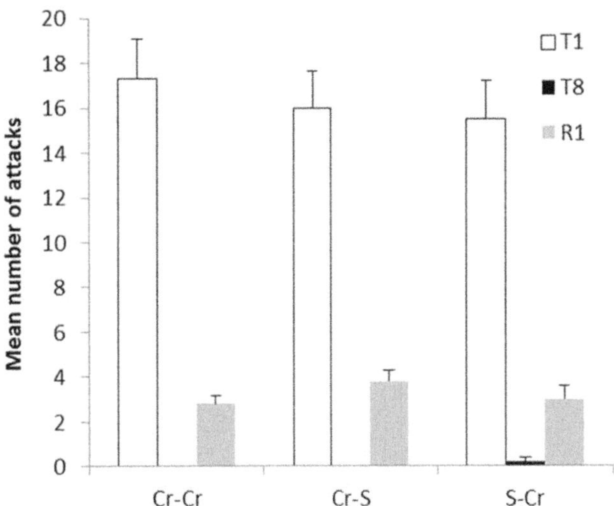

FIGURE 25.5 Mean number of attacks ± SEM during the first 3 min (T1) and the last 3 min of the training session (T8) and the first 3 min of the retention test (R1) in the three conditions: Cr–Cr, crabs were used in both training and the retention test; Cr–S, crabs were used during training and shrimp during the retention test; and S–Cr, shrimp were used during training and crabs during the retention test.

learning was considered to be associative, with each rapid shot of the tentacles (occurring in less than 32 msec at 25°C[8]) being associated with the 'pain' inflicted to the cuttlefish when tentacle clubs hit the transparent wall.

Nature of the Reinforcement of PT Learning

Several authors[3,8,10,25,32–35] noted that the lack of food reward at each attempted capture could also potentially play a role in the inhibition of predatory behavior. To better balance these two types of potential negative reinforcements, Cartron *et al.*[36] demonstrated that a rapid waning of attempted captures also occurs when large crabs (instead of shrimp) are placed in the glass tube (Figure 25.5). As mentioned previously, large crabs are generally caught with cuttlefish's eight arms (jumping strategy) instead of tentacle tips. The distance of attack (distance between the cuttlefish arms and the prey at the end of the positioning phase) is comparable in the striking and jumping strategies, but the attack duration is much longer for the latter (mean duration of jumping, 248 msec).[37] In the 'crab in the tube' procedure, jumps are even reduced in strength by the opening of the arm crown when cuttlefish reach the transparent tube. As a consequence, the 'pain,' if any, inflicted to the cuttlefish when it jumps on a transparent tube is probably largely attenuated at each attempted capture. As a consequence, pain is probably not strongly involved in the learning not to attack a crab in a transparent tube. Interestingly,

Cartron *et al.* observed a small number of tentacle strikes on the glass during training in the 'crab in the tube' procedure (~10% of the total number of attempted captures at the beginning of training), but their numbers did not decrease with time except at the very end of training, when the predatory behavior was totally inhibited. Thus, strikes on the glass may play a role in learning but are not necessary to explain the waning of attempted captures. It is thus unlikely that 'pain' is mainly responsible of the inhibition of the predatory behavior in the PT procedure.

This idea is supported by Dickel *et al.*,[34] who have shown that the higher the number of tentacle strikes at the beginning of the PT learning, the longer the time needed for the animal to inhibit its predatory behavior. Thus, a high number of tentacle impacts on the glass (i.e., the number of 'strike-contingent pain') does not make the animal learn faster. Because it is unlikely that 'pain' is mainly responsible for the inhibition of the predatory behavior in the PT procedure, one can hypothesize that the lack of food reward at each attempted capture serves as a negative reinforcement in this learning in cuttlefish. This is also in agreement with the waning of the number of 'pseudo-strikes' observed in the PT procedure (or on free-swimming prawns) when the cuttlefish's tentacles are removed.[7,8] Finally, note that before the inhibition of their predatory behavior in the PT procedure, cuttlefish alternately switch their prey-capture strategy from striking to jumping. Unfortunately, although this has been mentioned in the literature,[32] the phenomenon has never been further investigated.

The Transparent Tube as a Learning Cue in the PT Procedure

The glass tube is not 'invisible' to cuttlefish, as was originally thought. Cuttlefish are sensitive to the linear polarization of light; this sensitivity has been shown to help cuttlefish detect transparent prey.[38] Therefore, it seems likely that cuttlefish can perceive other transparent objects, such as glass or Perspex tubes, based on this sensitivity. Thus, cuttlefish can use the sight of the transparent apparatus in front of the prawns as an important cue to learn and memorize the PT situation. In agreement with this hypothesis, even just after the learning session, cuttlefish rapidly capture a shrimp[34,35] or another prey type[8,31] that freely swims outside the glass tube in the testing tank.

Does PT Learning Rely on Tactile Information?

Tactile and visual learning systems are considered as 'separated' in cephalopod brain (at least in *Octopus*)[28,29]; thus, it is an important question to ask about the role of tactile inputs in PT learning in cuttlefish. This question was first tackled by Sanders and Young.[3] Although the decision to attack or not a prey is based on visual inputs coming from the prawns, the cuttlefish 'feels' the transparent tube with its tentacles and/or arms at each strike/jump. Does this mean that learning mainly relies on visual information or does it involve the integration of both visual and tactile inputs? Two sets of information clarified this question. First, Messenger[8] has shown that learning occurs even when the cuttlefish has no tentacle—that is, when it has no tactile cues from the tube. As a consequence, it is clear that a cuttlefish can learn even when only visual cues are available. On the other hand, when the tube is made 'invisible' to the cuttlefish (by covering the glass tube with a depolarizing filter), learning is still possible, at a rate comparable to that observed in the PT procedure without a depolarizing filter.[36] With an 'invisible' tube, learning is probably largely based on tactile cues, so in principle both types of cues can be used in the learning task. It is interesting to note that the suppression of tactile cues from the 'transparent' tube, using cuttlefish with no tentacles,[8] is more deleterious for the learning rate than the reduction of visual cues from the tube.[36] One can therefore hypothesize that cuttlefish also use tactile information from the glass tube in the priming of the initial training. Note that during a retention test, cuttlefish that previously had totally inhibited their predatory behavior during the training still attempt to catch the prawns, even if less often than at the beginning of the training.[8,30,32,34,35,39,40] Thus, it is unlikely that cuttlefish only rely on visual information to remember the task, with tactile memory processes probably largely involved in the priming of the retention task. Different authors have also mentioned that in the presence of the PT apparatus, cuttlefish often display slight touches of the glass tube with their arm tips or grasp on it after attempted jumping. By increasing the amount of tactile information from the tube, these behaviors could play a role in facilitating learning in some individuals. This purely observational study has yet to be precisely investigated by quantitative experiments.

Stimulus Specificity of PT Learning

The last important point concerns the stimulus-specificity of this learning. In the PT procedure, does a cuttlefish learn not to capture 'shrimp when enclosed in a tube' (in this case, the cues are both prey type and the transparent tube) or 'any prey enclosed in the tube' (in this case, the learning cue is the transparent tube)? The number of attempted captures during a retention test is not significantly affected by a change of prey type in the transparent tube between the learning and a 1-hr retention test (shrimp vs. crabs or crabs vs. shrimp; Figure 25.5).[36] These observations highlight the crucial importance of the glass tube as a learning cue in this paradigm. However, Purdy *et al.*[31] suggested more selectivity in the learning responses to prey species in another cuttlefish species (*S. pharaonis*). By using fish or shrimp as prey in the tube and daily sessions as training; cuttlefish that switched from fish to shrimp seemed to slightly, but significantly, increase their attack rates, whereas cuttlefish that switched from shrimp to fish did not. The authors interpreted this asymmetrical stimulus specificity as effects of salience or appetence of the two prey types (cuttlefish rely on linear polarization of light to detect silvery fish[41]), which could suggest that fish and shrimp placed in a Perspex apparatus induce a different motivational state in cuttlefish. These interesting data in *S. pharaonis* probably have to be further investigated by using larger samples and a standardized learning protocol in *S. officinalis*.

In summary, the PT learning procedure is probably largely associative, with the presence of the tube (perceived by both visual and tactile cues) being largely associated with the absence of food reward at each attempted capture. It is of crucial importance to highlight that PT learning in cuttlefish is likely to involve both tactile and visual memory systems of the brain.

Memory Trace of PT Learning

Because of its relative convenience, this learning paradigm has been principally used to investigate the formation of memory trace in cuttlefish. One of the most famous pieces of research in this field was published by Messenger,[39] who determined a rate of recovery of predatory behavior as a function of different rest intervals after initial 20-min training. Using a large sample of individuals, the mean number of tentacle strikes observed during a 5-min presentation of the PT during the retention test was compared to the one observed during the first 5 min of the learning. The results are remarkable because the recovery curve was not smooth. There was no recovery of the response (i.e., good retention) in retention tests performed within the 15 min following the training session. Then the response began to recover and then declined until it almost disappeared at 60 min. It then recovered again and, by 90 min, reached an asymptote until 24 hr. Note that the retention of such learning is still considerable at 24 hr. This biphasic retention curve may reflect the presence of more than one memory system with different time courses (short-term and long-term memory systems in cuttlefish) that also has been observed in *Octopus*.[42]

The long-term retention of PT learning is, at least partially, protein synthesis-dependent.[43] Cycloheximide (CMX) injections (translation blocker and inhibitor of protein synthesis) impair recall after a 24-hr retention delay. Amnesia was observed when CMX was injected 1–4 hr after the initial training but not when it was administrated immediately or 6 hr after training. This suggests that *de novo* protein synthesis involved in long-term consolidation occurs during a sensitive window opening within the first hour post-learning and closing during the fifth hour post-learning. However, CXM did not completely abolish the long-term retention of PT learning. This may be due to a moderate degree of cerebral protein synthesis inhibition. A second hypothesis is that consolidation could occur as a multiple 'waves' process starting during the training session.[44] Two main interdependent parameters may be relevant in the process of long-term storage—the time course of the training session and the amount of sensory information received (especially for tactile stimuli)—but their respective importance remains to be determined experimentally.

Brain Correlates of PT Learning

The neural correlates of PT learning have been investigated using two main approaches, lesion and pharmacological, but most studies have focused on the VL system. Brain lesion approaches were used first; the removal of the vertical and superior frontal lobes (Figure 25.2B) had no apparent effect on feeding behavior toward a prawn.[3] Ablation of the superior frontal lobe after the training had no effect on the retention performances, whereas ablation of the VL induced long-term retention deficits (18 hr). Nevertheless, VL-less animals were still capable of relearning. This study suggested the involvement of alternative learning centers such as the optic lobes. A few animals with restricted brain damage (excitotoxic lesions) were tested with the PT procedure in our laboratory (unpublished data). An animal with a lesion in the posterior part of the VL inhibits its tentacle strikes more slowly than a sham animal and shows no 24-hr retention of the task. A lesion in the anterior part of the superior frontal lobe has no effect on the acquisition performance but induces 24-hr retention impairments. These results slightly differ from those of ablation experiments. In both brain lesion experiments (ablation and excitotoxic lesion), local neurons, inputs and outputs, are deleted but ablation disrupts the fibers that only run across the lesioned region to end in another part of the brain, whereas excitotoxic lesion does not. Although these data are preliminary, they suggest that specific attention has to be paid to very precise neural pathways within the VL system. Surprisingly, cytochrome oxidase (CO) activity, commonly used as a marker of neuronal activity, remains unchanged in the VL during consolidation or long-term retention of PT learning, whereas modulations in frontal superior lobe activity have been evidenced in its posterior part (corresponding to outputs to the VL) during consolidation and in its anterior part (corresponding to both inputs from VL, inferior frontal, subvertical, and optic lobes and outputs to the subvertical lobe) after a 24-hr retention test.[24] There is no direct evidence for the involvement of a 'non-VL system' lobe in the PT procedure. However, only a subset of the structures of the VL system (VL, subvertical, and superior frontal lobes; Figure 25.2C) have been investigated as possible structures of interest as neural substrates of learning and memory in cuttlefish. Because there is evidence that tactile information is involved in PT learning, the inferior frontal lobe (Figure 25.2B) should also be considered as a structure of interest. Indeed, this lobe receives tactile proprioceptive inputs and sends axons to the subvertical and superior frontal (probably the anterior part) lobes. One can consider those lobes receiving inputs from the optic lobes as a network of multimodal integration that may integrate visual and tactile information in PT learning. Common neurotransmitters and neuromodulators (acetylcholine, serotonin, GABA, glutamate, octopamine, and oxytocin/vasopressin[45–47]) have been described in the cuttlefish brain, but very little is known about their functional involvement, especially in memory processes. However, it is known that cholinergic neurons are the sites of synaptic long-term potentiation in the VL in cuttlefish.[26] Furthermore, acetylcholine catabolism is increased in the optic lobes after a 24-hr retention delay but not after 2- and 60-min retention delays.[45] This is a further argument for involvement of the optic lobes in long-term retention of PT learning. Two vasopressin/oxytocin superfamily peptides, octopressin and cephalotocin, are also implicated in long-term retention (24 hr) of PT learning.[46] These peptides are widely expressed in cuttlefish brain, particularly in the vertical, optic, and inferior frontal lobes. Finally, different sets of evidence are in agreement with the 'visuo-tactile' matrices suggested in the literature, but numerous gaps remain to be filled (spatiotemporal implication of the actors). In complement with the approaches cited previously, electrophysiological and molecular markers of neuronal activity should be further developed.

Spatial Learning

Evidence for Spatial Skills in Cuttlefish

Aitken *et al.*[48] found high site fidelity and quite small home ranges in the giant Australian cuttlefish (*S. apama*; between 5300 and 27,300 m^2). Another interesting observation is that after a few days of monitoring, cuttlefish seem to have settled into a den (rock crevice, no signal), and thereafter only occasional foraging excursions were observed (<5% of the day).[48] We can hypothesize that cuttlefish may have spatial knowledge of their surroundings at least to optimize their hunting time, and to quickly return to safe places. Karson *et al.*[49] undertook preliminary observations to determine whether *S. officinalis* would explore a completely novel environment—a large artificial pond. After several hours of observation, the authors closed a hole previously allowing cuttlefish to travel from one side to the other side of the pond. This elicited investigation of the closed hole; it suggests that cuttlefish are able to learn where they can find openings in vertical barriers. Cuttlefish orientation has the added complexity of navigating in a three-dimensional environment. Karson *et al.* then designed a wall maze with two holes allowing cuttlefish (*S. officinalis*) to access their home tank—one on the left side (10 cm above the bottom of the tank) and one on the right side (60 cm above the bottom of the tank)—with both holes remaining open throughout the experiment. The testing arena had no horizontal surface on which the cuttlefish could settle, providing high motivation to escape. Cuttlefish demonstrated a significant decrease in exit time within 10 trials.

Karson *et al.*[49] designed another spatial task apparatus: They trained cuttlefish to exit a circular testing arena with two exit holes cut on opposite sides. The exit holes were surrounded by visual cues (striped or spotted panels of fabric), with only one exit hole opened to a resting tank. Cuttlefish demonstrated their ability to solve the task and showed a marked performance improvement over serial reversals (once the cuttlefish reached the acquisition criterion, the open exit hole was closed and the opposite hole was opened). Hvorecny *et al.*[50] tested *S. officinalis* and *S. pharaonis* in a modified version of Karson et al.'s spatial learning apparatus. Two distinct testing arenas were used—one with a brick and the other with algae inside. The open exit hole was indicated by different visual cues in the two mazes (i.e., a striped panel of fabric indicating the open hole in the arena with a brick and a spotted panel of fabric indicating the open hole in the arena with algae). The authors showed that some cuttlefish learned to select the correct exit hole in the two different mazes when trials were intermixed. This experiment demonstrates that cuttlefish are capable of conditional discrimination. All these experiments showed that spatial learning is possible in cuttlefish, but none have studied which sensory senses and spatial mechanisms cuttlefish use to orient.

The complexity of their visual system (for review, see Nixon and Young[16]) and their high performance in visual discrimination tasks[51] suggest that cuttlefish are able to rely on visual landmarks to orient.

Strategy Used to Solve a Spatial Task

The circular testing arena (described previously) has been used successfully to test cuttlefish spatial capabilities.[49,50] Nevertheless, other learning tasks have been designed to ask very precise questions about information cuttlefish attend to and learn to solve a spatial problem. For example, do cuttlefish learn to use visual cues within a maze and/or in the laboratory environment? Alternatively, do cuttlefish use a motor sequence to solve a spatial task? In the studies described previously, the authors trained animals to solve a spatial task and then modified their environment (e.g., by displacing landmarks) to determine which strategy the animals were using to solve the task (e.g., using landmarks or learning a motor sequence).

Alves *et al.*[52] designed a cross-maze apparatus to explore cuttlefish (*S. officinalis*) spatial strategies, a task extensively used in a wide range of models (rats[53] and fish[54]). In this experiment, cuttlefish were rewarded for solving the task with time in the dark on the sandy bottom of a goal compartment located at the end of one arm of the maze. Locating the goal compartment required the cuttlefish to either learn a motor sequence (e.g., the need to turn left to find the entrance of the goal compartment) or orient using visual cues. In a first experiment, only distal visual cues were provided around the maze (water-pipes and sets of shelves of the laboratory room).[52] When the animals reached an acquisition criterion, cuttlefish were again placed in the cross-maze but in the start arm opposite to the one used during training. Nine of 10 cuttlefish swam in the opposite direction of the rewarded goal compartment, indicating that cuttlefish used the previously rewarded motor sequence. In a second experiment,[52] the cross-maze was surrounded by black curtains, and two visual cues were provided just above the water (striped and spotted PVC panels). By changing the right/left location of the two visual cues at the end of acquisition, the authors showed that half of the cuttlefish consistently swam in the opposite direction of the rewarded goal compartment; this result indicates that cuttlefish were using visual cues to solve the task. The other cuttlefish consistently swam in the correct direction of the rewarded goal compartment; this result means that these cuttlefish were using a motor sequence to solve the task. These two experiments successfully demonstrated that cuttlefish can use either visual cues or motor sequence to solve the same spatial task. Moreover, the availability and salience of visual cues seemed to determine whether the cuttlefish used the visual cues or a motor sequence as shown in vertebrates.[53,55] In this experiment, the choice made by a cuttlefish during the probe tests does not exclude the possibility that both strategies occurred in parallel during training (as in Cartron *et al.*'s experiment[56]).

Cartron *et al.*[56] examined the ability of cuttlefish (*S. officinalis*) to solve a Y-maze with two kinds of visual cues: the *e*-vector of a polarized light and two PVC panels (one striped and one spotted PVC panel) placed just above the water surface. During training, both visual cues were available. At the end of training, one kind of cue was randomly eliminated by the experimenters (i.e., either the PVC panels or the filter linearly polarizing the light). All cuttlefish tested were still able to orient when one of the visual cues became unavailable. Cuttlefish were also given one probe trial with conflicting information: The *e*-vector of the polarized light indicated one arm as rewarded, whereas the PVC panels indicated the other arm as rewarded. The latency to choose an arm was significantly greater when cuttlefish had to deal with conflicting information than when the two types of cue (*e*-vector and PVC panels) were congruent. This study showed for the first time the ability of cuttlefish to orient either parallel or perpendicular to the *e*-vector of a polarized light to find a goal compartment. Moreover, it clearly indicates that redundant spatial information is acquired simultaneously in cuttlefish. Such simultaneous learning has been demonstrated in mammals[57] and insects.[58]

Spatial Cognition and Sexual Maturation

Jozet-Alves *et al.*[59] assessed spatial learning performances of male and female cuttlefish (*S. officinalis*), either before or after sexual maturation, in a T-maze (procedure described previously[52]). Sexually mature males were more likely to attend to the visual cues provided above the apparatus to solve the maze compared to sexually mature females and immature cuttlefish. In contrast, sexually mature females have been shown to preferentially rely on a motor sequence (right vs. left turn). However, this difference in strategy did not lead to a sex difference in overall performance: Males and females did not differ in the time they took to learn the spatial task. This study demonstrated for the first time a cognitive dimorphism between sexes in an invertebrate. Several evolutionary hypotheses have been proposed to explain such differences. One of the best supported hypotheses suggests that the sex differences in spatial capabilities will evolve only in species in which range expansion is significantly different between males and females.[60,61] Jozet-Alves *et al.*[59] also showed that sexually mature males traveled a longer distance when placed in an open field compared with the other tested cuttlefish. These results are consistent with the range-size hypothesis.

Brain Correlates of Spatial Learning

Graindorge *et al.*[25] have made electrolytic lesions restricted either to the ventral or to the dorsal part of the VL in *S. officinalis*. Sham-operated and VL lesioned cuttlefish were trained in a modified version of Alves *et al.*'s[52] spatial learning procedure (the cuttlefish were given only three sessions of trials), and their locomotor activity was assessed in an open field. The results showed that ventral lesions of the VL led to impairment in the acquisition of spatial tasks, whereas dorsal lesions increased locomotor activity in an open field. These data highlight direct functional analogies between the VL of cuttlefish and the vertebrate's hippocampus.[25]

Other Learning

Classical Conditioning

Following Thomas's[62] classification of learning, some authors tried to undertake a systematic and comparative analysis of learning in cuttlefish. They began with the demonstration of signal learning. In classical conditioning, a neutral stimulus (the conditioned stimulus (CS)) is presented to an animal just before the presentation of an unconditioned stimulus (US) that elicits an unconditioned response. After repeated presentations of the CS followed by the US, some animals begin to behave toward the CS as if it were the US. In appetitive conditioning, in which the US is food, animals either try to 'eat' the CS, which is called autoshaping, or just approach it using a behavior called sign tracking.[63] Purdy *et al.*[64] showed evidence of sign tracking in *S. officinalis*. In the paired condition, they presented a light for 30 sec followed by food delivery to the side of the tank; this occurred four times a day for 30 days, with an intertrial interval of 20 min. Control cuttlefish (unpaired condition) were presented the light and food with a delay of 2 min. The cuttlefish in the paired condition oriented and positioned themselves toward the light significantly more often than did the control cuttlefish. The cuttlefish performed an anticipatory response, which suggests that they are capable of sign tracking. From an ecological perspective, such learning makes sense. Cuttlefish predate upon prey that often hide in crevices or within algae. One way for a cuttlefish to find food is to orient toward places or stimuli where it has previously successfully caught prey. Unfortunately, only a few cuttlefish were used in this study (six and three in the paired and unpaired conditions, respectively). As a consequence, even if the results appear clear-cut, one should interpret these conclusions with caution. However, they are supported by another study that used a classical conditioning procedure.[51] In this study, the CS was plastic spheres of different brightness paired or not with food (US). Even if cuttlefish were not required to make any particular response, in the paired condition they tended to catch the sphere instead of orienting toward it (a phenomenon known as autoshaping). The difference observed between the experiments by Purdy *et al.*[64] and Cole and Adamo[51] may be due to (1) the nature of the US (the spheres having more features in common with prey than a light) and (2) the shorter delay between CS and US in presentation.[64] The authors claim that this learning could be adaptive because it allows speculative hunting response to recurring environmental features.[65] In terms of costs and benefits, speculative hunting is unlikely to occur in cuttlefish, particularly because cuttlefish are visual hunters.[7] Taken together, these studies show that cuttlefish are capable of associative learning in classical conditioning procedures.

Operant Conditioning

Operant conditioning is another type of associative learning in which an animal learns to associate a behavior with its consequences. Reinforcement and punishment are the core tools of operant conditioning. The former causes a behavior to occur with greater frequency, whereas the latter causes a behavior to occur with lower frequency, sometimes even until its complete inhibition (cf. the PT procedure). Darmaillacq *et al.*[66] designed a paradigm of aversion learning. In this experiment, cuttlefish were trained not to attack their preferred prey. This prey was rendered distasteful with a coating made of quinine (a substance that tastes bitter to humans) and presented to the cuttlefish every 15 min until they stopped attacking the prey twice in a row. In such condition, cuttlefish inhibit their predatory behavior within a few trials (<10), whereas control cuttlefish that tried to catch the prey but did not get it (removed before it is seized) kept on trying to get it. Indeed, after the cuttlefish inhibited their predatory behavior, a choice between the preferred prey and another prey was presented 24 or 72 hr later. For each retention delay, significantly more cuttlefish chose the other prey than their preferred one, although control cuttlefish still chose their preferred food. Beyond the demonstration that cuttlefish are capable of instrumental learning, these results showed for the first time 72-hr retention capabilities in cuttlefish. In an ecological perspective, this learning appears adaptive. Prey choice may be affected by prey defense mechanisms, such as the presence of toxins, escape strategies, and the availability of prey.[65] Quinine is a natural and widespread chemical. Thus, it makes sense that cuttlefish avoid prey that is distasteful, particularly when they experience it several times. Furthermore, cuttlefish are able to store for a long time information that is crucial to their survival and welfare. Unpublished data obtained using CO histochemistry[24] showed that taste aversion learning induced changes in CO staining with a pattern in the cuttlefish brain that depended on the delay after learning. Immediately after training, CO staining was decreased in the posterior superior frontal lobe; after 24 hr, it was increased in the inferior frontal lobe; and after 72 hr, significant changes in CO activity were observed in the VL and the superior frontal lobe (unpublished data). Like previous studies in cuttlefish and vertebrates,[24,67] these results suggest a differential temporal evolution of post-training changes in regional brain activity supporting an evolution of the neural substrate of memory. They also confirm the involvement of the VL system in learning and memory processes. Unlike in the study by Agin *et al.*,[24] the superior frontal lobe is not the only brain region involved in these processes. This difference can be explained because the learning procedures in the two studies are different in terms of learning length, reinforcements, and control groups. However, the results presented here were obtained with a small number of animals and need further investigation.

DEVELOPMENTAL PERSPECTIVES

Behavior and Brain Development in Juveniles

Hatchlings appear to have similar behaviors as those of adults; a few days after hatching, they are active predators able to catch shrimp and shrimp-like prey, fish, and crabs that are large relative to their own size. They have the same prey-capture strategies as adults, but the tentacle strike strategy seems preferentially used, even to catch small crabs.[33,68] They assume the same nectobenthic life and the same defensive strategies (sand digging, changing of body patterns to conceal, inking, etc.) as adults. They do not benefit from parental care, so they have to cope on their own to avoid predators and find prey; thus, one can expect high plasticity of adaptive behavior in early juveniles. As a consequence, cuttlefish is a remarkable model to investigate early development of learning and memory, with the same learning paradigm being applicable from hatching to adulthood.

Hatchlings of cuttlefish probably possess sensory and motor skills equivalent to those of adults.[5,27,33,68,69] These behavioral assessments are in agreement with some preliminary observations of brain development in cuttlefish.[33] A developmental study of the cuttlefish brain from hatching to 30 days of age was performed using immunohistological labeling of 'heavy neurofilaments' (NF-H). The presence of these elements of neuronal cytoskeletons indicates that neuronal networks are structurally stabilized.[70] Low, intermediate, and high motor centers of the brain are well labeled from hatching. In contrast, NF-H progressively appear in the neuropil of associative lobes of the brain ('silent area')—that is, the VL, subvertical lobe, and superior frontal lobe. At 1 month of age, intensity and number of labeled fibers in the VL system are comparable to those observed in the adult VL system.[33] In the optic lobes, whereas the most central part of the medulla and the deep retina contain numerous stabilized fibers from hatching, NF-H labeling fibers in the peripheral medulla of the optic lobes appear later, beginning at the end of the first week of life (Figure 25.6).

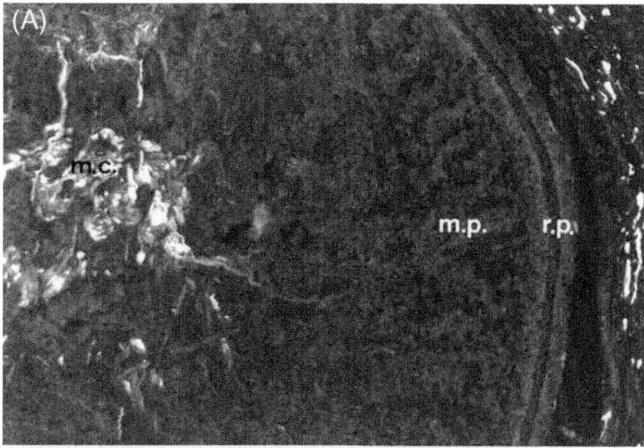

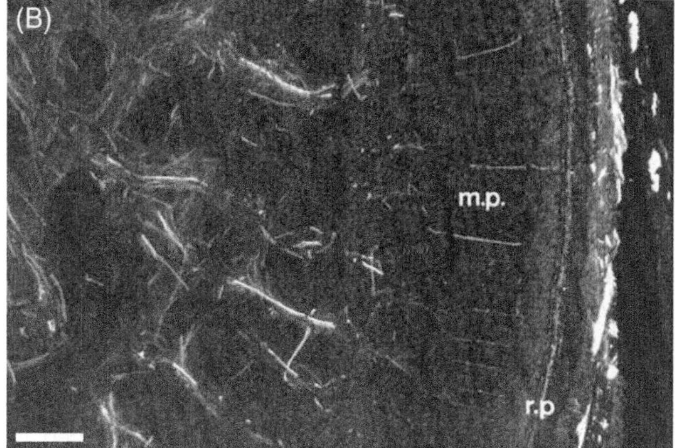

FIGURE 25.6 **Heavy neurofilaments immunoreactivity in optic lobes at hatching (A) and 9 days of age (B).** The microphotographs were taken from sagittal sections. Immunoreactive fibers are visible in the peripheral medulla of the optic lobe at 9 days, whereas they are abundant only in the central medulla and deep retina (well visible at higher magnification; not shown) from hatching. m.c., central medulla; m.p., peripheral medulla; r.p., deep retina. Scale bar = 100 μm.

Development of Long-Term Memory and Brain Correlates

The PT procedure was used for the first time in early juveniles by Wells[6] and Messenger.[9] Using this paradigm with apparatuses of suitable size, Wells was the first to mention the poor learning capabilities of very young cuttlefish. In more detailed studies, Messenger[9] and Dickel *et al.*[35] showed a gradual improvement of training performance in the course of the first 3 months of life. Specifically, 8-day-old animals displayed memory capacities sufficient to inhibit their predatory behavior within a continuous presentation of PT, showing the existence of short-term memory capabilities from this age.[35,40] However, retention of the task at 60 min (corresponding to long-term memory storage[39]) increased progressively between 15 and 60 days of age. In pioneering work, Messenger[9] used eight 3-min presentations of the PT (hereafter referred to as 'trials') with intertrial intervals of 30 min during which the juveniles were left undisturbed. All juveniles were tested 24 hr later. The author showed an increase in both learning (i.e., waning in number of strikes between the first and the eighth trial) and 24-hr retention performance (i.e., difference between the number of strikes during the first trial and the retention test) between 7 and 112 days of age. Messenger hypothesized that the poor memory capabilities of juveniles were related to the immaturity of their vertical and superior frontal lobes. This assessment was further investigated by Dickel *et al.*.[35] By using morphometric measurements of different lobes of the brain and PT learning experiments in juveniles aged 8–90 days, these authors showed that only the growth of the superior frontal lobe and the VL was significantly correlated with the improvement of learning and long-term retention performances with age. This developmental approach confirms Messenger's hypothesis and appears to be in agreement with results obtained from adults in lesion experiments: The VL system seems to be involved in training of an associative learning[71–73] as well as in retention processes in *Octopus*.[20,73] In this species, Young[74] emphasized the likely role played by these structures in building associations among various stimuli. This strong hypothesis was recently supported by both physiological[24] and electrophysiological studies (see Chapter 24).

Influence of Early Experience on Memory and Brain

There is a substantial body of literature on the influence of rearing environment on behavioral and brain maturation in mammalian and avian species (for review, see Renner and Rosenzweig[75]). This factor is crucial on a fundamental level to further investigate the mechanisms of 'phenotypic plasticity'[76] and from an animal husbandry standpoint. The European regulation about the use of animals in laboratories now integrates cephalopods. It requires environmental enrichment in the animal enclosure. As such, one can ask whether enrichment of rearing tanks can have consequences on both brain and behavioral plasticity in cuttlefish.

In a series of experiments, Dickel *et al.*[34] tested three groups of cuttlefish reared in different conditions: standard conditions, SC (large and bare tanks, reared in groups); impoverished conditions, IC (small and bare tanks, solitary condition); and enriched conditions, EC (large tanks with sand, rock shelters, shells, and artificial seaweed). Some animals were maintained in impoverished conditions until 1 month of age and then transferred in enriched conditions (I/EC), and others were kept in enriched conditions until 1 month of age and then transferred to impoverished conditions (E/IC). Using the PT paradigm, retention performance of the juveniles from the SC, IC, and EC groups was assessed at 1 month, and that of juveniles from all groups (SC, IC, EC, I/EC, and EC/IC) was assessed at 3 months.[34] For an easier comparison of 24-hr retention performance between the different groups, the level of learning acquisition was standardized for all animals. Learning and 24-hr retention performance developed faster in EC than in IC cuttlefish at the same age, with learning and retention performance of SC cuttlefish intermediate between EC and IC. Interestingly, at 3 months, cuttlefish from the E/IC group showed the same learning and retention performance as the IC group at the same age, whereas the I/EC group did not differ in its performance from that of the EC group. These results indicate that an enriched environment facilitates development of learning and memory capabilities in young cuttlefish.

Preliminary morphometric studies of brain structures have shown that the VL and medulla of the optics lobes develop faster in cuttlefish reared for 3 months in EC than in cuttlefish from IC.[33,77] There is no significant difference between IC and EC cuttlefish at 2 months. In cuttlefish from the IC and EC groups, cell proliferation was quantified by bromodeoxyuridine (BrdU) labeling of dividing cells of the brain in hatchlings and at 1, 15, 30, and 60 days of age. Brains were removed 4 hr after injection of BrdU, and labeled cells were analyzed by immunohistochemistry.[78] This study focused on the VL system (vertical, subvertical, and superior frontal lobes) and the optic lobes. This experiment showed that cell proliferation, which is intense in all brain structures on day 1, strongly decreases during the first 2 months of life. However, the rate of decrease is lower in the EC group than in the IC group in both neuropil and deep retina of the optic lobes. We did not find any significant differences in the lobes of the VL system. Although preliminary, these data indicate that the maturation of the most associative structures of the brain is sensitive to early experience of juveniles. Enriched environment of rearing would facilitate neurite growth in the VL and in the superior frontal lobe, whereas at least cell proliferation intensity (i.e., the density of labeled cells) is affected by an impoverished environment in structures that are involved in the primary integration of visual information (i.e., deep retina of the optic lobes). However, the latter result has to be considered with caution because the sequence of neuronal development is still unknown. Further investigations also have to be conducted to study the effect of enriched conditions of rearing on the maturation of motor and sensory structures. Taken together, however, these studies indicate that exposure to an enriched environment induces both structural and functional changes in the cuttlefish brain. There is further evidence that early experience also affects the maturation of defensive behavior such as body patterning[78,79] and sand digging.[80] Enrichment is a complex combination of physical activities, learning, and inanimate and social stimulation. It is still unknown which of these factors is critical for brain and behavioral plasticity in cuttlefish juveniles.

Imprinting and Early Cognition

Imprinting, originally described by Lorenz,[81] refers to a specific form of persistent learning that occurs during a sensitive period without any obvious conventional reinforcement.[82] In chicks, the social preference induced in an individual by imprinting is then generalized to all the individuals of the same species.[81] Other imprinting-like behaviors have since been reported, including habitat imprinting[83] and food imprinting.[84] Cuttlefish hatch with internal nutritive reserves. Even if active predation can begin before the inner yolk is entirely used up,[27,68] these reserves allow young cuttlefish not to feed for a number of days. This short period is of critical importance in the life of young cuttlefish because ambushed hatchlings can collect information about their environment (available prey, predation risk, etc.) before they start foraging.[85] When naive cuttlefish start feeding, they spontaneously prefer shrimp-like prey to any other type of prey, such as as crustacean or fish.[6,86] However, it has been shown that a preference for crabs (a nonpreferred prey) could be induced after a mere visual familiarization with crabs during a sensitive period within the first hours of life after hatching.[87,88] Interestingly, the efficiency of this familiarization depends on the length of the exposure as well as the density of prey exposed—that is, the flow of information perceived during the sensitive period. Finally, it appears that the primacy of the early familiarization outweighs untrained preference for shrimp.[88] It has also been shown that if naive juvenile cuttlefish are presented a choice between black and white crabs, they prefer the dark phenotype.[89] However, if they are exposed to white crabs at hatching during the sensitive period, they subsequently prefer white rather than black crabs (two-way choice: white vs. black crabs) and black crabs over shrimp (two-way choice: black crabs vs. shrimp). This result suggests that cuttlefish can make intracategorical discrimination (white vs. black phenotype) and that they can generalize the learning of the characteristics of the prey to which they were familiarized to a novel prey that shares the same morphological features.[89] Early diet selection may utilize food-item generalization to reduce energy and time devoted to information processing.[90] Generalizing among initial prey preferences would allow a young cuttlefish to diversify its diet while dispersing away from the laying site without the increased cost associated with trial-and-error learning.[85]

Taken together, these studies show that the learning of the visual characteristics of a potential prey meets all the criteria of imprinting as described previously: no reinforcement, existence of a sensitive period, persistence,[88] and generalization of the prey preference.[89] It was proposed that food imprinting could account for the prey preference observed in week-old cuttlefish. Food imprinting could be a good compromise between a certain degree of flexibility in response, useful for learning information for which the timing is likely to be predictable—food seen in the first few hours of life—but in which specifying the exact details of the experience is not useful. Such early learning capabilities would allow juveniles to take advantage of a changing environment and to deal with a world in which shrimp (i.e., their 'innate' food preference) would be unavailable. Contrary to filial or sexual imprinting, food imprinting would seem costly and even disadvantageous in the long term,[91] particularly for long-lived individuals living in somewhat changeable environments. As such, it may be helpful for juveniles to become highly attracted by available prey in the weeks following hatching before they starts foraging around. This preference for a prey that was seen for a few hours soon after hatching and expressed 7 days later suggests a rapid acquisition and long-term memory capabilities. This appears to contradict previously mentioned results obtained from studies using the PT procedure in juveniles, in which the authors showed that 8-day-old juveniles had a poor acquisition of the associative task[35,40] and that long-term memory only appeared from the age of 1 month.[9,34,35,40] One possible explanation is that the acquisition and retention processes involved in the establishment of prey preference in juvenile cuttlefish may depend on different rules than those in avoidance learning.[85] In the chick, it has been shown that memories supporting imprinting preferences and those consecutive to the acquisition of a heat-reinforced discrimination using the imprinted objects are functionally different and are located in different areas of the brain.[92] In the cuttlefish, one could argue that the two kinds of memories may involve separate brain structures. Because the VL is very immature in the first weeks of life, good candidates would be the optic lobes. These brain structures process visual information coming directly from the eyes,[93] and they may also be involved in learning and memory.[94] Dickel[33] showed that neural maturation occurs earlier in the optic lobes than in the vertical system (Figure 25.6). This finding is also supported by a recent study in which crabs were exposed to the cuttlefish's right or left eye (unpublished data). Only cuttlefish exposed to crabs with their right eye were significantly more likely to choose crabs than were control cuttlefish that did not see crabs. This study indicates that the right eye pathway involving the optic lobes is sufficient for food imprinting and that a kind of hemispheric specialization may exist in the cuttlefish. Thus, the right visual field would be more generally specialized in foraging and feeding behavior, as in many vertebrate species[95] and in honeybees.[96]

One of the characteristics of cuttlefish is that they are semelparous, which means that adult cuttlefish only mate once in their life and females die after having laid hundreds of eggs. A direct consequence is that embryos and juveniles develop without parental care. Unlike other cuttlefish species, in *S. officinalis* the egg capsule is stained in black by the ink of the female. Immediately after spawning, the capsule is completely opaque; as the embryo grows and the osmotic pressure of the perivitelline fluid increases,[69] it is dilated and peels off. Consequently, the capsule becomes more transparent and, hence, it is likely that several types of information can reach the embryo. It is known that in birds and mammals, the onset of the sensory systems follows a fixed sequence in which the tactile system appears first and then the vestibular, chemical, auditory, and visual systems.[97] In the cuttlefish, the tactile, chemical, and visual systems are functional before hatching, although the precise onset for the tactile and chemical systems is still not known (unpublished data). However, there is some behavioral evidence that the visual system is functional long before hatching. Indeed, Darmaillacq *et al.*[98] showed that embryos exposed to crabs (unpreferred prey) between 1 and 2 weeks before hatching subsequently preferred crabs to shrimp 7 days after hatching. This first demonstration of embryonic visual learning also highlights the extraordinary long-term memory capabilities of this marine invertebrate. In addition, the learning of the characteristics of prey is not limited to the general shape and motion but also includes the phenotype of the prey. Indeed, when offered a choice between white and black crabs, young cuttlefish visually familiarized *in ovo* to white crabs prefer white crabs for their first meal compared to the control cuttlefish that prefer black crabs; they also prefer black crabs to shrimp.[89] This result confirms that juvenile cuttlefish have the capability to categorize and generalize prey, but interestingly, it also shows that embryos are capable of fine perception of the features of prey, notably its brightness. The ability of cuttlefish embryos to perceive visually and to process and learn particular characteristics of prey present in the vicinity of the eggs may confer important adaptive advantages. In mammals, prenatal olfactory learning of the maternal diet may provide the opportunity to learn about a safe and natural diet.[99] In the absence of parental care, neonates have to avoid harmful food on their own. Female cuttlefish generally lay their eggs in shallow water[69] and may choose places where newly hatched juveniles can easily find potential prey. The ability to learn the visual characteristics of the prey *in ovo* would facilitate postnatal imprinting on this kind of prey. Finally, these results imply that it is also very likely that the sensitive window that is active soon after hatching[88] may be open in the few weeks before hatching. Another series of studies showed that chemical information can also reach the embryo. In embryos, the chemical system is functional before the visual system (unpublished data). In a study addressing the effect of a chemical exposure on visual preference after hatching, embryos were exposed to odors from shrimp (*Crangon crangon*; preferred prey), crabs (*Carcinus maenas*; unpreferred prey), mollusks (*Mytilusedulis*; nonprey), or a seawater control (no prey).[100] They were then tested for their visual preference between crabs and shrimp. Cuttlefish that had previous experience with shrimp odor had a visual preference for crabs, whereas cuttlefish that had previously smelled crabs preferred shrimp and cuttlefish that had previously smelled bivalves had no preference. To explain these puzzling results, the authors noted a cross-modal effect between the chemical and the visual systems. Moreover, they hypothesized that an overstimulation of the chemical system during embryonic development could have disturbed the onset of the visual system and hence the visual perception in juveniles.

CONCLUSION

The studies reported in this chapter demonstrate the increasing interest in examining the biological bases of behavioral plasticity in *S. officinalis*. There is a significant effort at an international level to develop efficient and well-characterized behavioral tools to investigate the neural substrates of learning and memory in this species. Based on these tools and on the extraordinary amount of knowledge already collected in *Octopus* (for reviews, see[4,101]), studies on cuttlefish could allow a better understanding of the neural mechanisms of complex learning (imprinting, classical and instrumental conditioning, spatial learning, etc.). In this species, promising perspectives are still open to investigate high-order cognitive processes such as concept formation, representation building, and executive functions.

From a neurobiological approach, cuttlefish brain anatomy is well-known, and ablation or pharmacological lesions as well as more modern methods (neuropharmacology, pathway tracing, *ex vivo* electrophysiology, etc.) have provided precious knowledge on the circuitry and functioning of relevant brain regions (i.e., VL system[26] and optic lobes[93]). However, it is still difficult to apply techniques that would allow the determination of neural activity *in vivo* (calcium imaging, positron emission tomography, and magnetic resonance imaging) or during the different steps of memorization of complex tasks (markers of neuronal plasticity as immediate-early gene expression). Another promising perspective would be to apply neuropharmacological blockers or facilitators of selective neural pathways during learning, based on the pioneering works of Chichery,[102] Chrachri *et al.*,[103] and Messenger.[47] Some of these tracks would have to be conducted on a comparative basis between different species of cephalopods that show considerable differences (morphology, ecology, behavior, etc.), but all share impressive behavioral plasticity and an overall identical brain organization.

In parallel, the cuttlefish provides a unique opportunity to explore the development of brain and cognition in an invertebrate species. Data about learning and memory skills in embryos,[89,98] in hatchlings and juveniles (for review, see Dickel *et al.*[85]), during sexual maturation,[59] and during senescence[32,104,105] show several analogies with higher vertebrates such as birds and mammals. This is of major interest from a fundamental standpoint to better understand mechanisms of developmental plasticity of brain and behavior. It may also produce interesting outcomes for applied studies—to investigate, for example, some developmental disorders as long-lasting effects of early stress on brain and behavior or mechanisms of brain and cognitive aging.

However, behavioral biology research of *S. officinalis* suffers from a lack of knowledge about the behavior of cuttlefish in the field. The behavioral rules and constraints of cuttlefish in its natural environment remain largely unknown. Research in behavioral ecology in this species will allow the refinement and diversification of learning paradigms available for use in the laboratory. As already reported in the literature,[106] these kinds of investigations will be a challenging key to answering questions about the neurobiology of cognition in cuttlefish.

References

1. Young JZ. The number and sizes of nerve cells in *Octopus*. *Proc R Soc Lond*. 1963;140:229–254.
2. Packard A. Cephalopods and fish: the limits of convergence. *Biol Rev*. 1972;47:241–307.
3. Sanders FK, Young JZ. Learning and other functions of the higher nervous centres of Sepia. *J Neurophysiol*. 1940;3:501–526.
4. Wells MJ. *Octopus: Physiology and Behaviour of an Advanced Invertebrate*. London: Chapman & Hall; 1978.
5. Wells MJ. Factors affecting reactions to *Mysids* by newly-hatched *Sepia*. *Behaviour*. 1958;13:96–111.
6. Wells MJ. Early learning in *Sepia*. *Symp Zool Soc Lond*. 1962;8: 149–159.
7. Messenger JB. The visual attack of the cuttlefish *Sepia officinalis*. *Anim Behav*. 1968;16:342–357.
8. Messenger JB. Learning in the cuttlefish *Sepia*. *Anim Behav*. 1973;21:801–826.
9. Messenger JB. Learning performance and brain structure: a study in development. *Brain Res*. 1973;58:519–523.
10. Messenger JB. Prey-capture and learning in the cuttlefish *Sepia*. *Symp Zool Soc Lond*. 1977;38:347–376.
11. Chichery MP, Chichery R. The anterior basal lobe and control of prey-capture in the cuttlefish (*Sepia officinalis*). *Physiol Behav*. 1987;40:329–336.
12. Chichery MP, Chichery R. Manipulative motor activity of the cuttlefish, *Sepia officinalis*, during prey capture. *Behav Process*. 1988;17:45–56.
13. Chichery MP, Chichery R. The predatory behavior of *Sepia officinalis*: ethological and neurophysiological studies. In: Boucaud-Camou E, ed. La Seiche, the Cuttlefish. *First International Symposium on the Cuttlefish* Sepia. Caen: Centre Publ Univ Caen; 1991:141–151.
14. Duval P, Chichery MP, Chichery R. Prey capture by the cuttlefish (*Sepia officinalis* L.): an experimental study of two strategies. *Behav Process*. 1984;9:13–21.
15. Hanlon RT, Messenger JB, eds. *Cephalopod Behaviour*. Cambridge, UK: Cambridge University Press; 1996.
16. Nixon M, Young JZ. *The Brains and Lives of Cephalopods*. Oxford: Oxford University Press; 2003.
17. Boycott B. The functional organization of the brain of the cuttlefish *Sepia officinalis*. *Proc R Soc Lond B*. 1961;153:503–534.
18. Chichery R, Chanelet J. Motor and behavioural responses obtained by stimulation with chronic electrodes of the optic lobe of *Sepia officinalis*. *Brain Res*. 1976;105:525–532.
19. Chichery R, Chanelet J. Motor responses obtained by stimulation of the peduncle lobe of *Sepia officinalis* in chronic experiments. *Brain Res*. 1978;150:188–193.
20. Young JZ. Rates of establishment of representations in the memory of octopuses with and without vertical lobes. *J Exp Biol*. 1961;38:43–60.
21. Young JZ. The nervous system of *Loligo*: V. The vertical complex. *Philos Trans R Soc Lond*. 1979;285:311–354.
22. Sanders GD. The cephalopods. In: Corning WC, Dyal JA, Willows AOD, eds. *Invertebrate Learning*. New York: Plenum; 1975:1–101.
23. Agin V, Chichery R, Dickel L, Darmaillacq AS, Bellanger C. Behavioural plasticity and neural correlates in adult cuttlefish. *Vie Milieu*. 2006;56:81–87.
24. Agin V, Chichery R, Chichery MP. Effects of learning on cytochrome oxidase activity in cuttlefish brain. *Neuroreport*. 2001;12: 113–116.
25. Graindorge N, Alves C, Darmaillacq AS, Chichery R, Dickel L, Bellanger C. Effects of dorsal and ventral vertical lobe electrolytic lesions on spatial learning and locomotor activity in *Sepia officinalis*. *Behav Neurosci*. 2006;120(5):1151–1158.
26. Shomrat T, Graindorge N, Bellanger C, Fiorito G, Loewenstein Y, Hochner B. Alternative sites of synaptic plasticity in two

homologous "fan-out fan-in" learning and memory networks. *Curr Biol.* 2011;21:1773–1782.

27. Dickel L, Chichery MP, Chichery R. Postembryonic maturation of the vertical lobe complex and early development of predatory behavior in the cuttlefish (*Sepia officinalis*). *Neurobiol Learn Mem.* 1997;67:150–160.
28. Young JZ. Computation in the learning system of cephalopods. *Biol Bull.* 1991;180:200–208.
29. Young JZ. Multiple matrices in the memory system of *Octopus.* In: Abbott NJ, Williamson R, Maddock L, eds. *Cephalopod Neurobiology: Neuroscience Studies in Squid, Octopus and Cuttlefish.* London: Oxford University Press; 1995:431–443.
30. Agin V, Chichery R, Dickel L, Chichery MP. The "prawn-in-the-tube" procedure in the cuttlefish: habituation or passive avoidance learning? *Learn Mem.* 2006;13:97–101.
31. Purdy JE, Dixon D, Estrada A, Riedlinger E, Suarez R. Prawn-in-a-tube procedure: habituation or associative learning in cuttlefish? *J Gen Psychol.* 2006;133(2):121–152.
32. Chichery R, Chichery MP. Learning performances and aging in cuttlefish (*Sepia officinalis*). *Exp Gerontol.* 1992;27:234–239.
33. Dickel L. *Comportement Prédateur et Mémoire Chez La Seiche (Sepia officinalis), Approches Développementale et Neuroéthologique.* Caen, France: Thèse de l'Université de Caen Basse-Normandie; 1997.
34. Dickel L, Boal JG, Budelmann BU. The effect of early experience on learning and memory in cuttlefish. *Dev Psychobiol.* 2000;36 (2):101–110.
35. Dickel L, Chichery MP, Chichery R. Increase of learning abilities and maturation of the vertical lobe complex during postembryonic development in the cuttlefish, *Sepia. Dev Psychobiol.* 2001;39:92–98.
36. Cartron L, Darmaillacq A-S, Dickel L. The prawn-in-the-tube procedure: what does cuttlefish learn and memorize? *Behav Brain Res.* 2012; [accepted for publication]
37. Chichery MP. *Approche Neuroéthologique du Comportement Prédateur de La Seiche Sepia Officinalis.* Caen, France: Doctorat d'Etat de l'Université de Caen Basse-Normandie; 1992.
38. Shashar N, Hanlon RT, Petz A. Polarization vision helps detect transparent prey. *Nature.* 1998;393:222–223.
39. Messenger JB. Two stage recovery of a response in *Sepia. Nature.* 1971;232:202–203.
40. Dickel L, Chichery MP, Chichery R. Time differences in the emergence of short- and long-term memory during postembryonic development in the cuttlefish, *Sepia. Behav Process.* 1998;44: 81–86.
41. Shashar N, Hagan R, Boal JG, Hanlon RT. Cuttlefish use polarization sensitivity in predation on silvery fish. *Vis Res.* 2000;40 (1):71–75.
42. Sanders GD, Barlow JJ. Variations in the retention performances during long-term memory formations. *Nature.* 1971;232:203–204.
43. Agin V, Chichery R, Maubert E, Chichery MP. Time-dependent effects of cycloheximide on long term memory in the cuttlefish. *Pharmacol Biochem Behav.* 2003;75:141–146.
44. Hernandez PJ, Abel T. The role of protein synthesis in memory consolidation: progress amid decades of debate. *Neurobiol Learn Mem.* 2008;89(3):293–311.
45. Bellanger B, Dauphin F, Chichery MP, Chichery R. Changes in cholinergic enzyme activities in the cuttlefish brain during memory formation. *Physiol Behav.* 2003;79(3–4):749–756.
46. Bardou I, Leprince J, Chichery R, Vaudry H, Agin V. Vasopressin/oxytocin-related peptides influence long-term memory of a passive avoidance task in the cuttlefish, *Sepia officinalis. Neurobiol Learn Mem.* 2010;93(2):240–247.
47. Messenger JB. Neurotransmitters of cephalopods. *Invertebr Neurosci.* 1996;2:95–114.
48. Aitken JP, O'Dor RK, Jackson GD. The secret life of the giant Australian cuttlefish *Sepia apama* (Cephalopoda): behaviour and energetics in nature revealed through radio acoustic positioning and telemetry (RAPT). *J Exp Mar Biol Ecol.* 2005;320:77–91.
49. Karson MA, Boal JG, Hanlon RT. Experimental evidence for spatial learning in cuttlefish (*Sepia officinalis*). *J Comp Psychol.* 2003;117(2):149–155.
50. Hvorecny LM, Grudowski JL, Blakeslee CJ, et al. Octopuses (*Octopus bimaculoides*) and cuttlefishes (*Sepia pharaonis, S. officinalis*) can conditionally discriminate. *Anim Cog.* 2007;10 (4):449–459.
51. Cole PD, Adamo SA. Cuttlefish (*Sepia officinalis*: Cephalopoda) hunting behavior and associative learning. *Anim Cog.* 2005;8: 27–30.
52. Alves C, Chichery R, Boal JG, Dickel L. Orientation in the cuttlefish *Sepia officinalis*: response versus place learning. *Anim Cog.* 2007;10:29–36.
53. Restle F. Discrimination of cues in mazes: a resolution of the "place-vs-response" question. *Psychol Rev.* 1957;64:217–228.
54. Odling-Smee L, Braithwaite VA. The influence of habitat stability on landmark use during spatial learning in the three-spined stickleback. *Anim Behav.* 2003;65:701–707.
55. Carman HM, Mactutus CF. Proximal versus distal cue utilization in spatial navigation: the role of visual acuity? *Neurobiol Learn Mem.* 2001;78:332–346.
56. Cartron L, Darmaillacq AS, Jozet-Alves C, Shashar N, Dickel L. Cuttlefish rely on both polarized light and landmarks for orientation. *Anim Cog.* 2012;15:591–596.
57. Gibson BM, Shettleworth SJ. Place versus response learning revisited: tests of blocking on the radial maze. *Behav Neurosci.* 2005;119(2):567–586.
58. Steck K, Hansson BS, Knaden M. Desert ants benefit from combining visual and olfactory landmarks. *J Exp Biol.* 2011;214:1307–1312.
59. Jozet-Alves C, Modéran J, Dickel L. Sex differences in spatial cognition in an invertebrate: the cuttlefish. *Proc R Soc Lond B.* 2008;275:2049–2054.
60. Gaulin SJC, FitzGerald RW. Sex differences in spatial ability: an evolutionary hypothesis and test. *Am Nat.* 1986;127:74–88.
61. Gaulin SJC, FitzGerald RW. Sexual selection for spatial-learning ability. *Anim Behav.* 1989;37:322–331.
62. Thomas RK. Evolution of intelligence: an approach to its assessment. *Brain Behav Evol.* 1980;17:454–472.
63. Moore BR. The evolution of learning. *Biol Rev.* 2004;79:301–335.
64. Purdy J, Roberts A, Garcia C. Sign tracking in cuttlefish (*Sepia officinalis*). *J Comp Psychol.* 1999;113:443–449.
65. Davies NB, Krebs JR. Predators and prey. In: Krebs JR, Davies NB, eds. *Behavioural Ecology: An Evolutionary Approach.* Oxford: Blackwell; 1979:21–151.
66. Darmaillacq AS, Dickel L, Chichery MP, Agin V, Chichery R. Rapid taste aversion learning in adult cuttlefish, *Sepia officinalis. Anim Behav.* 2004;68:1291–1298.
67. Bontempi B, Jaffard R, Destrade C. Differential temporal evolution of post-training changes in regional brain glucose metabolism induced by repeated spatial discrimination training in mice: visualization of the memory consolidation process? *Eur J Neurosci.* 1996;8:2348–2360.
68. Boletzky Sv. Juvenile behaviour. In: Boyle PR, ed. *Cephalopod Life Cycles, Comparative Reviews.* Vol 2. New York: Academic Press; 1987:45–84.
69. Boletzky Sv. Sepia officinalis. In: Boyle PR, ed. *Cephalopod Life Cycles, Species Accounts.* Vol 1. New York: Academic Press; 1983:31–52.
70. Carden MJ, Trojanowski JQ, Schaepfer WW, Lee VMY. Two-stage expression of neurofilament polypeptides during rat

neurogenesis with early establishment of adult phosphorylation patterns. *J Neurosci.* 1987;7(11):3489–3504.
71. Boycott BB, Young JZ. A memory system in *Octopus vulgaris. Proc R Soc Lond.* 1955;143:449–480.
72. Maldonado H. The general amplification function of the vertical lobe in *Octopus vulgaris. Zeits Verg Phys.* 1963;47: 215–229.
73. Young JZ. The organization of a memory system. *Proc R Soc Lond.* 1965;163:285–320.
74. Young JZ. Unit processes in the formation of representations in the memory of *Octopus. Proc R Soc Lond.* 1960;153:1–17.
75. Renner MJ, Rosenzweig MR. *Enriched and Impoverished Environments: Effects on Brain and Behavior.* New-York: Springer-Verlag; 1987.
76. West-Eberhard MJ. Phenotypic plasticity and the origins of diversity. *Annu Rev Ecol Syst.* 1989;20:249–278.
77. Jozet-Alves C. *Neuroéthologie de La Cognition Spatiale Chez La Seiche Commune Sepia Officinalis.* Vol 13. Thèse de l'Université de Paris; 2008.
78. Poirier R. *Expérience Précoce et Ontogenèse des Comportements Défensifs Chez La Seiche (Sepia officinalis) : Approches Comportementale et Neurobiologique.* Caen, France: Thèse de l'Université de Caen Basse-Normandie; 2004.
79. Poirier R, Chichery R, Dickel L. Early experience and postembryonic maturation in body patterns in cuttlefish (*Sepia officinalis*). *J Comp Psychol.* 2005;119:230–237.
80. Poirier R, Chichery R, Dickel L. Effect of early experience on sand-digging efficiency in juvenile cuttlefish (*Sepia officinalis*). *Behav Process.* 2004;67(2):273–279.
81. Lorenz K. Der Kumpan in der Umwelt des Vögels. *J Ornithol.* 1935;83:137–213.
82. Bolhuis JJ. Mechanisms of avian imprinting: a review. *Biol Rev.* 1991;66:303–345.
83. Morton ML, Wakamatsu MW, Pereyra ME, Morton GA. Postfledging dispersal, habitat imprinting, and phylopatry in a montane, migratory sparrow. *Ornis Scand.* 1991;22: 98–106.
84. Burghardt GM, Hess EH. Food imprinting in the snapping turtle, *Chelydra serpentina. Science.* 1966;151:108–109.
85. Dickel L, Darmaillacq AS, Poirier R, Agin V, Bellanger C, Chichery R. Behavioural and neural maturation in the cuttlefish *Sepia officinalis. Vie Milieu.* 2006;56(2):89–95.
86. Darmaillacq AS, Chichery R, Poirier R, Dickel L. Effect of early feeding experience on subsequent prey preference by cuttlefish, *Sepia officinalis. Dev Psychobiol.* 2004;45:239–244.
87. Darmaillacq AS, Chichery R, Shashar N, Dickel L. Early familiarization overrides innate prey preference in newly hatched *Sepia officinalis* cuttlefish. *Anim Behav.* 2006;71:511–514.
88. Darmaillacq AS, Chichery R, Dickel L. Food imprinting, new evidence from the cuttlefish *Sepia officinalis. Biol Lett.* 2006;2:345–347.
89. Guibé M, Poirel N, Houdé O, Dickel L. Food imprinting and visual generalization in embryos and newly hatched cuttlefish (*Sepia officinalis*). *Anim Behav.* 2012;84:213–217.
90. Ginane C, Dumont B. Generalization of conditioned food aversions in grazing sheep and its implications for food categorization. *Behav Process.* 2006;73(2):178–186.
91. Healy SD. Imprinting: seeing food and eating it. *Curr Biol.* 2006;16(13):R501–R502.
92. Honey RC, Horn G, Bateson P, Walpole W. Functionally distinct memories for imprinting stimuli: behavioral and neural dissociations. *Behav Neurosci.* 1995;109:689–698.
93. Williamson R, Chrachri A. Cephalopod neural networks. *Neurosignals.* 2004;13:87–98.
94. Young JZ. Influence of previous preferences on the memory of *Octopus vulgaris* after removal of the vertical lobe. *J Exp Biol.* 1965;43:595–603.
95. Rogers LJ. Lateralization in its many forms, and its evolution and development. In: Hopkins WD, ed. The Evolution of Hemispheric Specialization in Primates, *Special Topics in Primatology.* Vol 5. Amsterdam: Elsevier; 2007:23–56.
96. Letzkus P, Boeddeker N, Wood JT, Zhang SW, Srinivasan MV. Lateralization of visual learning in the honeybee. *Biol Lett.* 2008;4:16–19.
97. Gottlieb G. Ontogenesis of sensory function in birds and mammals. In: Tobach E, Aronson LR, Shaw E, eds. *The Biopsychology of Development.* New York: Academic Press; 1971:67–12
98. Darmaillacq AS, Lesimple C, Dickel L. Embryonic visual learning in the cuttlefish, *Sepia officinalis. Anim Behav.* 2008;76(1):131–134.
99. Hudson R, Distel H. The flavor of life: perinatal development of odor and taste preferences. *Schweiz Med Wochenschr.* 1999;129 (5):176–181.
100. Guibé M, Boal JG, Dickel L. Early exposure to odors changes later visual prey preferences in cuttlefish. *Dev Psychobiol.* 2010;52(8):833–837.
101. Borrelli L, Fiorito G. Behavioral analysis of learning and memory in cephalopods. In: Menzel R, Byrne J, eds. *Learning and Memory: A Comprehensive Reference.* Oxford: Elsevier; 2008:605–628.
102. Chichery R. *Etude du Comportement Moteur de La Seiche, Sepia Officinalis : Approches Neurophysiologique et Neuropharmacologique.* Caen, France: Doctorat d'Etat de l'Université de Caen Basse-Normandie; 1980.
103. Chrachri A, Williamson R. Dopamine modulates synaptic activity in the optic lobes of cuttlefish, *Sepia officinalis. Neurosci Lett.* 2005;377:152–157.
104. Chichery MP, Chichery R. Behavioural and neurohistological changes in aging *Sepia. Brain Res.* 1992;574:77–84.
105. Bellanger C, Dauphin F, Belzunces LP, Cancian C, Chichery R. Central acetylcholine synthesis and catabolism activities in the cuttlefish during aging. *Brain Res.* 1997;762(1–2):219–222.
106. Messenger JB. Current issues of cephalopod behavior. *J Mar Biol Assoc.* 1995;75:507–514.

4.3 Crustacea

CHAPTER

26

A Multidisciplinary Approach to Learning and Memory in the Crab *Neohelice* (*Chasmagnathus*) *granulata*

Daniel Tomsic and Arturo Romano

Universidad de Buenos Aires, Conicet, Argentina

INTRODUCTION: CRUSTACEANS AS MODEL SYSTEMS IN NEUROBIOLOGY

Crustaceans have been traditional models for neurobiological studies. Investigations of these animals have provided insights that are of general neurobiological interest concerning matters as diverse as the role of GABA as inhibitory neurotransmitter,[1,2] the identity of the neural substrates involved in behavioral decision making,[3,4] coincidence detection as a mechanism of behavioral responses,[5] and the effects of social hierarchy on the properties of individually identified neurons.[6,7] Some of these achievements were possible because of the particular advantages offered by decapod crustaceans for the neurophysiological approach, namely the presence of giant neurons, which are easily accessible with microelectrodes, in combination with the animal's cooperation in experimental manipulations.[8] Additional advantages described later in this chapter make some crustaceans suitable models for investigating the neurobiology of learning and memory.

LEARNING AND MEMORY IN CRUSTACEANS

Malacostracan crustaceans display a great diversity of body forms, and they include crayfish, crabs, lobsters, shrimp, mantis shrimp, and many other less familiar animals. Most of them are highly active organisms that live in rather complex and changing environments. It is not surprising that these animals make use of information acquired through experience to constantly adapt their behavior. For example, they possess rather sophisticated learning and memory abilities. Indeed, such abilities are well documented by studies performed both in the field and in the laboratory. These studies range from field observations[9] to carefully planned and controlled experiments in the field[10] and in the laboratory.[11,12] The studies encompass different learning and memory skills, such as simple forms of working memories,[13] nonassociative memories (e.g., habituation,[14,15] and sensitization[16,17]), classical,[11] and operant conditioning.[18,19] Using food as positive reinforcement, Abramson and Feinman[20] demonstrated that crabs can easily learn a lever-press paradigm similar to the typical instrumental conditioning used with mammals. Numerous studies performed mainly in crayfish and crabs have used shuttle boxes or T-maze chambers to investigate their learning to avoid electrical shocks[21–23] or to obtain food reward.[24,25] Some of these studies revealed acquisition of long-lasting memories. In addition, enduring memories were shown to be context dependent—that is, to be determined by an association between the training environment and the training stimulus.[26,27]

Many other well-accounted learning and memory studies have been performed in crustaceans and have revealed the capacity of these animals to achieve complex learning, but whose classification in terms of type of learning is not so obvious. For example, shore crabs learn, memorize, and transfer learned handling skills to novel prey,[28] learn to improve prey selectivity,[29]

Invertebrate Learning and Memory.
DOI: http://dx.doi.org/10.1016/B978-0-12-415823-8.00026-5

and learn to locate hidden food.[30] Hermit crabs and crayfish perform individual recognition by learning the opponent's identity.[31,32] Studies on hermit crabs have also addressed their ability to memorize information regarding protective shells, which is a major issue for these animals.[33] Another group of crustaceans that has attracted attention for learning studies is the stomatopods. Some stomatopods learn to recognize and use novel, artificial burrows, whereas others learn to identify novel prey species and handle them for effective predation. Stomatopods learn the identities of individual competitors and mates using both chemical and visual cues (reviewed in Cronin[34]).

Although not always acknowledged, navigating the environment represents a challenge that often requires the use of memory. Fiddler crabs, which feed in loops up to 1 m long, rely on path integration, sometimes called the 'Ariadne's thread mechanism,' based on idiothetic information to integrate their foraging routes and to return to their dens. They walk on the ground and, like ants and spiders, store internal information from their limbs to 'count' the number of steps they take, an indication of short-term memory of acquired information. A direct view of the surrounding landmarks does not help them return home when they are passively displaced or translated.[35] Hence, fiddler crabs rely on an egocentric system of references (idiothetic memory). Conversely, the swimming crab *Thalamita crenata* has good spatial knowledge, based on the storage of memories about landmarks. This orienting mechanism is much more flexible and complex than that of fiddler crabs and may be comparable to the route-based memory of honeybees.[36] Finally, although the ability to form spatial memory representations in the form of a cognitive map relying on a system of orienting cues has not been formally tested in crustaceans, there are indications that they possess such capacity.[37,38]

The brief description provided here shows that behavioral studies on learning and memory in crustaceans are indeed abundant. However, sustained research programs aimed at understanding the neurobiology of these processes are scant. One such line of investigation conducted by Frank Krasne and colleagues was aimed at elucidating the neural basis of learning-induced changes of the tail flip response in the crayfish. These studies uncovered important features of the neurophysiological processes underlying the change of response.[39,40] However, the most compelling investigation on the neurobiology of memory using a crustacean as experimental model is being performed in the semiterrestrial grapsoid crab *Neohelice granulata* (until recently *Chasmagnathus granulatus*).[41] Studies on this crab encompass laboratory and field experiments and a multidisciplinary approach that includes behavioral analyses, pharmacology, molecular biology, electrophysiology, neuroanatomy, and calcium imaging techniques. In this chapter, we present some important contributions to the neurobiology of learning and memory that emerged from investigations conducted during approximately the past 20 years in the crab *Neohelice*.

THE CRAB *NEOHELICE*: HABITAT AND HABITS

Chasmagnathus granulatus has recently been renamed *Neohelice granulata*.[41] Here, we adopt the new name *Neohelice*. However, note that most articles quoted throughout this chapter used the animal's original name, *Chasmagnathus*.

Neohelice is a burrowing semiterrestrial grapsoid crab that inhabits the intertidal zone (mudflat areas) and salt marshes (areas densely vegetated with cord grasses) on the Atlantic coast of South America. In certain regions, their distribution overlaps with the fiddler crab *Uca uruguayensis*. *Neohelice* can constitute dense populations, reaching as high as 100 individuals per square meter or more. Because they are confined geographically to estuaries and bays, which are separated by several hundred kilometers of oceanic coast, populations in different regions are thought to be isolated from one another. Along its wide, but discontinuous, geographic distribution, the species experiences different tidal regimes (from a few centimeters to 9 m), water salinities (from near 0 to 60‰), environmental structures (mud flat vs. vegetated areas), and predation risks (areas where aerial predators are abundant or scarce).[42] Consequently, *N. granulata* has been an excellent model for studying intra- and interpopulation variability (reviewed in Spivak[43]).

Neohelice is a robust midsized running crab, reaching up to 36 mm across the carapace, which digs semipermanent burrows and is active in both air and water. It is a highly visual animal that displays conspicuous visually guided behaviors similar to those described in fiddler crabs.[44] Each of its two eyes, containing approximately 8000 ommatidia, is mounted on a 5-mm-long eyestalk. As in fiddler crabs, the eyes of *Neohelice* possess a zone of higher acuity around the equator.[45] The neuroanatomy and physiology of the visual nervous system of *Neohelice* is probably the best known in any crustacean. *Neohelice* strongly relies on visual information. Among other things, the visual system allows the animal to detect and organize anticipatory responses to the attacks of its aerial predators. These responses, however, need to be adaptive; that is, they must be sensitive to modification by information acquired through new experiences.

CRAB LEARNING IN THE LABORATORY

Our studies in the laboratory have shown the ability of *Neohelice* to acquire and retain different types of memory. Paradigms that proved to induce long-term memory (≥24 hr) successfully include (1) habituation of the escape response to a visual danger stimulus[46–48], (2) habituation of exploratory activity to the contextual environment,[24] (3) sensitization to electrical shocks or to visual moving stimuli,[16,49] (4) passive avoidance learning,[22,50] (5) appetitive conditioning,[24,51,52] (6) contextual learning,[27,53,54] and (7) cued associative learning of autonomic response.[55] Behavioral studies using some of these paradigms were occasionally followed up by pharmacological studies.[16,56–60] However, the paradigm that has been used the most in *Neohelice* as a model for studying the neurobiology of learning and memory with a multidisciplinary approach is 'context-signal memory' (CSM).

CONTEXT-SIGNAL MEMORY

Neohelice is preyed upon by gulls and other sea birds; hence, objects moving over the animal elicit an escape response. In our laboratory, a black rectangular screen moving horizontally overhead represents a visual danger stimulus (VDS) that evokes the crab's escape. A typical trial with a VDS consists of two back-and-forth circular movements of the screen over 9 sec (Figures 26.1A and 26.1B). The escape response is recorded with the animal located inside an actometer, which consists of a bowl-shaped container with a circular flat floor covered to a depth of 0.5 cm with water. The response is captured by a transducer device and recorded with a PC. Our experimental room has 40 actometers, which are isolated from each other by partitions. The basic experimental design used for assessing long-term memory in the crab includes a training and a testing session typically separated by 24 hr and at least two groups of animals, one trained and one untrained. Pharmacological or other intervening experiments include the corresponding additional control and trained pairs of groups (e.g., control and trained, both injected with a certain drug). Each group usually includes 30–40 crabs. In the first session, the trained group receives the repetitive presentation of the VDS, whereas the control group remains in the actometers without any phasic stimulation. At the end of this session, the animals are put in individual containers, where they remain until the testing session. In the second session, all animals are again put in the actometers to be tested with the VDS. Retention is said to indicate acquired memory when the trained group shows a level of responsiveness to the VDS that is statistically lower than that of the control group. A single training session of 45 min, with 15 presentations of the VDS separated by 3 min (spaced training), invariably causes a reduction of the escape response that is retained for at least 5 days.[12,62] Although at the beginning we termed this phenomenon habituation,[47,52,54,58,59] later investigations demonstrated that it is a more complex form of memory. In fact, the long-term modification of the escape response is exhibited only if the animal is tested in the same visual environment in which it was trained. In other words, a change in the visual context between training and testing prevents the memory from being evoked (Figures 26.1A and 26.1B).[27,61,62] Moreover, exposure to the context alone, prior to or following the training—the typical procedure that causes respectively latent inhibition or extinction—impairs the formation or the expression of the crab's memory.[27] These results led to the conclusion that the memory produced by spaced training is determined by an association between the VDS and the context.[27,61] For this reason, during the past decade, we have called this associative memory 'context-signal memory.' CSM thus entails an association between two visual memories, a memory of the context (CM) and a memory of the signal (SM), each of which can be acquired independently.[27,54,62] Further investigations described later in this chapter continue to support these conclusions.[12,61,63,64]

Important Attributes of CSM

In addition to being context dependent, CSM exhibits some principal visual memory attributes, such as the capacity for generalization and stimulus recognition. Like humans, arthropods easily recognize learned objects over large changes in retinal position, a property commonly referred to as 'position invariance'.[65] When a crab attempts to escape from the VDS inside the actometer, the stimulus position continuously changes over its retina. Even in these circumstances, animals are able to acquire a strong memory, hinting at a rather flexible type of learning. But are crabs able to recognize the learned stimulus when it appears in a new spatial position 24 hr after training? To test this possibility, we used two VDSs located in different positions above the crab (Figure 26.2A). Animals were trained with one of the two VDSs, and 24 hr later they were tested with either the same or the different VDS. An equal number of crabs were trained with each of the VDSs. Results in Figure 26.2B show that the responses of animals tested with the same or with the

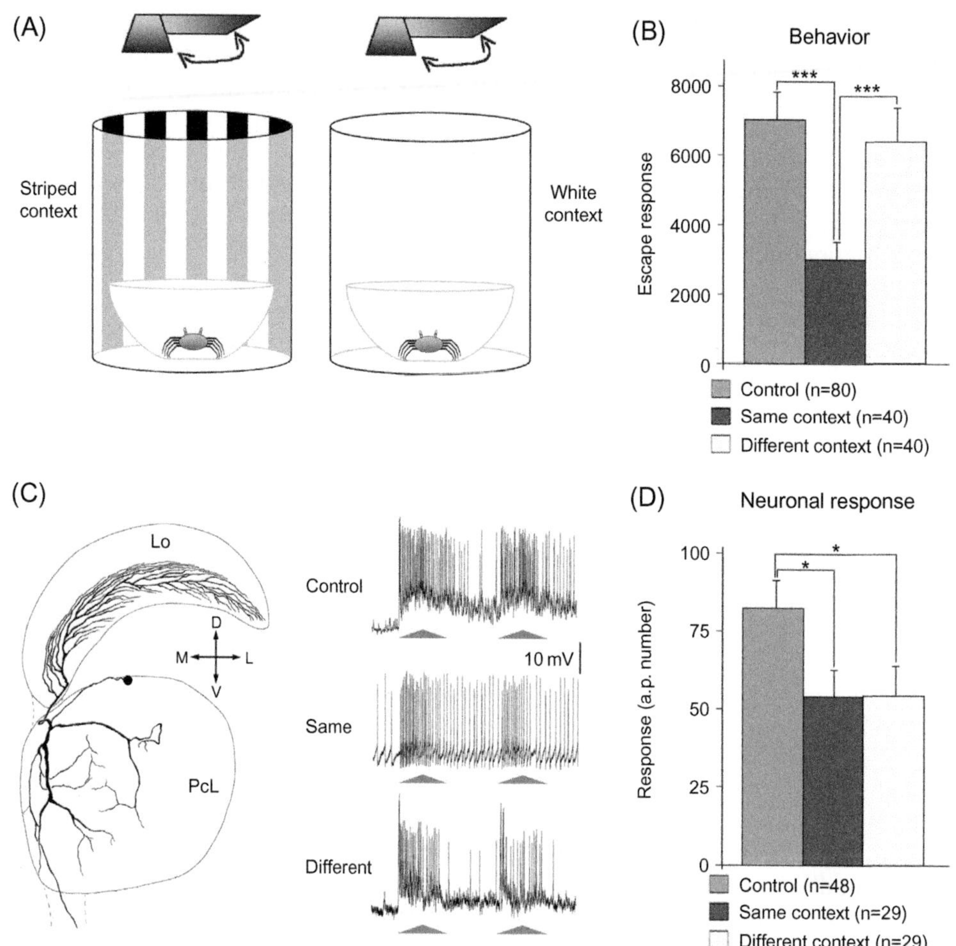

FIGURE 26.1 **Behavioral expression of CSM and neuronal correlates.** (A) The two visual contexts used for training and testing the crab in behavioral and electrophysiological experiments. The intensity of escape response elicited by a VDS (the screen moved above the crab) was recorded in an actometer. Training consisted of 15 trials of VDS presentations, with fixed ITIs of 3 min. Experiments included a control and two trained groups, one tested in the same context used during training and the other in a different context. Both behavioral and electrophysiological tests were performed 24 hr following training. (B) Mean escape responses of the three groups at the testing session. Crabs that were tested in the same context of training expressed CSM, whereas those tested in a different context did not. (C) Reconstruction of an intracellularly stained lobula giant (LG) neuron. The truncated axon projects to the midbrain. Lo, lobula; PcL, lateral protocerebrum; D, dorsal; L, lateral; M, medial; V, ventral. Right side. Representative examples of LG responses to the VDS during testing from animals belonging to the three groups. The triangles below the traces represent the two cycles of stimulus movement comprising a trial (triangle base, 2.2 sec). (D) Mean neuronal responses of the three groups as exemplified in panel C. The learning-induced response modification of LG neurons observed 24 hr after training is not affected by contextual changes. Bars show mean ± SE. $^{*}p < 0.05$; $^{***}p < 0.005$. In panel B, *n* represents the number of crabs in each group. In panel D, *n* represents the number of neurons (one per animal) in each group. Source: *Reproduced from Sztarker and Tomsic*[61]

different VDSs were similar, and the responses of both groups were significantly lower than those of the control group. Therefore, 1 day after training, crabs were able to transfer the learned response to a VDS located in a different position.[61]

Being able to recognize the learned stimulus when it appears in different spatial positions is adaptive. However, animals also need to keep reacting to unknown danger stimuli, requiring that the learning-induced reduction of the escape response should be stimulus specific. These two opposite arguments make the actual behavior a tradeoff between stimulus generalization and stimulus specificity. The final decision is based on the similarity between the new stimulus and the trained one. This similarity is evaluated along one or multiple perceptual dimensions that are relevant for each animal species. The previously described experiment showed that crabs are able to generalize when the new stimulus is presented in a different part of the visual space. We then investigated whether they are

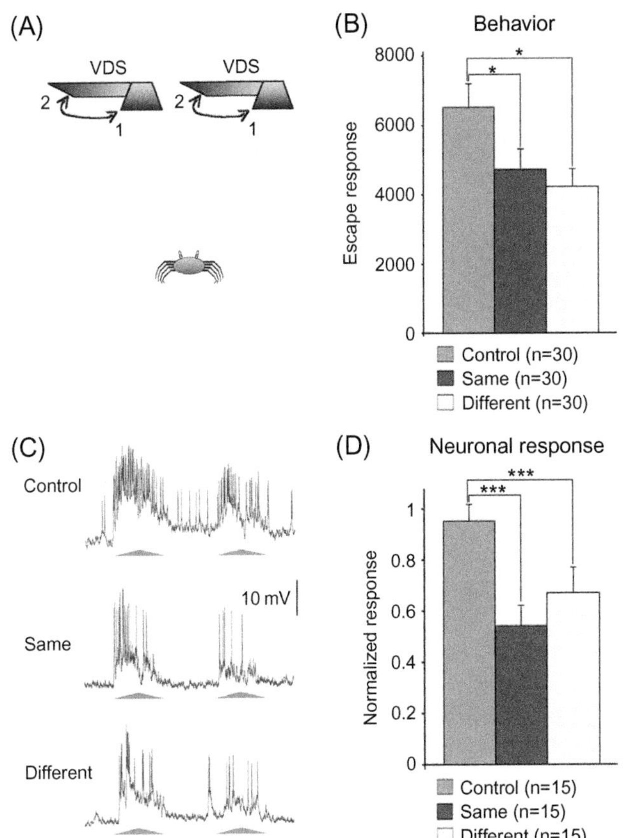

FIGURE 26.2 The crab's ability to generalize stimulus position correlates with response properties of LG neurons. (A) Two VDSs located 25 cm above the crab with similar motion cycles (moved from 1 to 2 and back) were used independently to stimulate animals. The two VDSs were separated by 37 cm, which represents an angular separation of 73° for the crab. Training consisted of 15 trials with fixed ITI of 3 min. Behavioral experiments included a control and two trained groups, one tested with the same VDS as in the training and the other with a different one. (B) Mean escape responses of the three groups at a testing session 24 hr after training. Crabs recognized the learned stimulus regardless of its position in space. (C) Representative example of the responses of one LG to the VDS at the control trial (first training trial) and at test trials performed at the end of training with the same or the different VDS. (D) Mean neuronal responses of experiments as exemplified in panel C. The behavioral ability of spatial generalization is reflected by the response of the LGs. Bars show mean ± SE. $^{*}p < 0.05$; $^{***}p < 0.005$. In panel B, n represents the number of crabs in each group. In panel D, n represents the number of neurons (one per animal) evaluated before training (control) and at the two test trials. Source: *Reproduced from Sztarker and Tomsic.*[61]

still able to generalize when an additional stimulus dimension, such as the sequence of motion direction, is altered. To test this, we used the two VDSs of the previous experiment but inverted the motion cycle of one of them: Starting from the center, each stimulus moved to one or to the other side of the crab (Figure 26.3A). Results in Figure 26.3B show that animals tested with the same VDS used during training responded significantly less than control animals. On the contrary, animals tested with the different VDS responded like controls. This indicates that crabs did not generalize between VDSs that differ in their sequence of motion. In this case, they showed a stimulus-specific memory.[47,61]

Adaptive Value of CSM

Suppressing an escape response to a potentially dangerous stimulus for a long period of time can be highly risky. Thus, why is *Neohelice* so prone to give up escaping to a VDS for such a long time? Before attempting to answer this question, some information about the characteristics of the learned responses to danger stimuli is warranted. First, regarding a stimulus that signals danger in the wild, the greater the ambiguity of the signal, the greater the likelihood to give up responding to such stimulus. Thus, a high degree of behavioral change is expected when the stimulus almost invariably proves to be innocuous, in contrast to the lack of change expected when an unequivocal relationship links the stimulus and a subsequent damage. Second, as we have shown, the learning-induced change of response to VDS in *Neohelice* is stimulus specific. In fact, animals tested with VDSs that were different (or moved differently) from VDSs in the training escaped like untrained animals. In other words, crabs are recognizing a visual motion stimulus that proved to be innocuous from a slightly different one.[47,61] But are there innocuous visual motion stimuli in the natural environment of *Neohelice* that need to be ignored? The answer is yes. This species inhabits an upper intertidal zone densely vegetated by cord grass *Spartina alterniflora* or *Spartina densiflora*—both erect, tough, long-leaved grasses ranging from 0.3 to 2 m tall. Therefore, the high propensity of *Neohelice* to reduce its escape response to an iterated object moving overhead may be explained by the great ambiguity of such a signal because this crab is immersed in an environment featuring wind-induced oscillations of the upper portion of the cord grasses, which may elicit the escape response.[54] This interpretation is supported by the fact that *Pachygrapsus marmoratus*, another grapsoid crab that inhabits a barren biotope in comparison with that of *Neohelice*, is by far less inclined to reduce its response to the VDS.[54] But in such circumstances, how does *Neohelice* cope with the attack of a real predator? Unlike *Pachygrapsus*, which inhabits rock crevices, *Neohelice* invests time and effort into digging its own burrow. The burrow-centered habits may have contributed to *Neohelice*'s ability to learn and recognize the contextual environment. Thus, the crab may recognize the familiar layout of cord grasses that surrounds its burrow. Hence, it may learn to disregard the overhead movements normally occurring

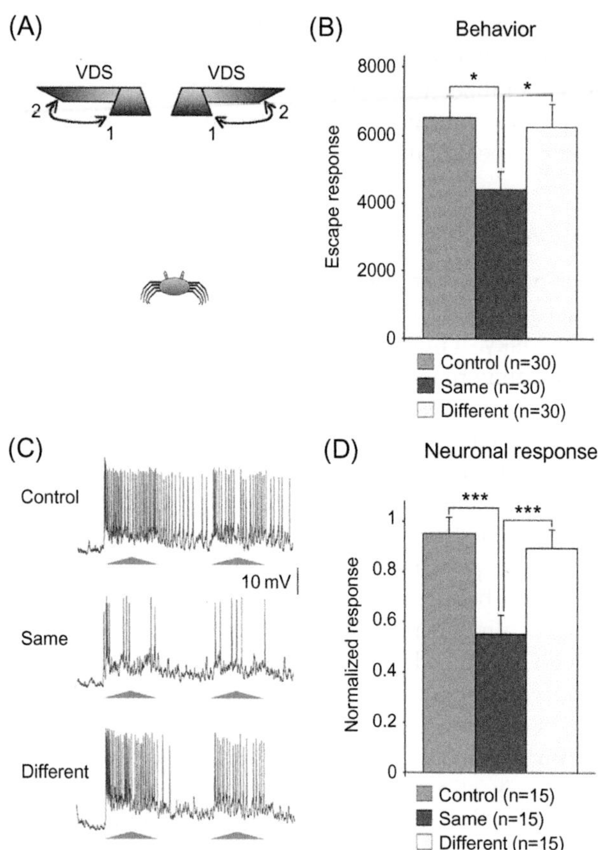

FIGURE 26.3 Stimulus-specific learning is parallel in LGs. (A) Two VDSs located above the crab with inverted motion cycles (moved from 1 to 2 and back) were used independently to stimulate crabs. Behavioral experiments included a control and two trained groups, one tested with the same VDS as in training and the other with a different VDS. (B) Mean escape responses of the three groups at the testing session 24 hr after training. Crabs recognized the learned stimulus and distinguished it from a similar but unlearned stimulus. (C) Representative example of responses of an LG to the VDS at a control trial and at test trials performed with the same or the different VDS. (D) Mean neuronal responses of experiments as exemplified in panel C. The behavioral ability for stimulus specificity is reflected in the responses of the LGs. Source: *Reproduced from Sztarker and Tomsic.*[61]

in this particular context. However, if for any reason the animal travels beyond its home range, it would regain its ability to escape from similar stimuli. Thus, in *Neohelice*, the risk that involves the long-term waning of the escape response is counteracted by the fact that the learned inhibition is stimulus and context specific.

The previous interpretation is based on results obtained in experiments performed in the laboratory. A characteristic of these experiments is that the animal is trained inside a container from which it cannot escape—a condition in which, in addition to learning that the stimulus is not dangerous, the animal can learn that to escape is actually impossible (i.e., an instance of learned helplessness). On the other hand, a replacement of the escape response by a freezing response following training has been described and interpreted as an instance of fear conditioning.[66,67] Further field studies, such as those performed by Fathala *et al.*,[68,69] are necessary to be certain about the adaptive value of CSM of *Neohelice*.

MASSED AND SPACED TRAINING RENDER TWO DIFFERENT KINDS OF MEMORY

In addition to the number of training trials, memory formation is known to critically depend on the frequency of trial presentations. In different animals and using various learning tasks, short intertrial intervals (ITIs), which are known as massed training, result in short or intermediate memories, whereas long ITIs (spaced training) result in long-lasting memories.[70,71] This topic has been extensively investigated in *Neohelice*.[62,72] Fifteen trials of spaced training (ITI: 3 min) result in a long-lasting memory (>5 days), whereas a similar number of massed training trials results in a rapid reduction of the escape response that is only short lasting (a few minutes). Increasing the number of trials with massed training can render an intermediate memory (<2 days). However, when the ITI is extremely short (near 0 sec), long-term memory can never be acquired, even if the training comprises hundreds of trials.[72] The memory induced by spaced training is expressed from the very beginning of the testing session, indicating that the reinstallation of the animal in the training context acts as a reminder of the memory to be evoked.[27,54,64] On the other hand, the intermediate memory induced by extensive massed training is evoked only after a number of test trials have occurred, indicating that some trials at the beginning of the test are required as a reminder for the animal to recall the learned stimulus.[72,73] In fact, at variance with the memory produced by spaced training, which is determined by a context-signal association,[27] the memory induced by massed training is context independent.[62] For this reason, these memories have been respectively termed CSM and SM. Pharmacological, molecular, and electrophysiological experiments demonstrated that CSM and SM are indeed different types of memory. For instance, CSM, but not SM, depends on protein and RNA synthesis,[62] is mediated by the cAMP signal pathway,[74,75] and involves the Rel/NF-κB transcription factor.[76] *In vivo* intracellular recordings identified neurons that reflect the learning and memory of the VDS and, hence, appear to house the SM trace but not all the components of CSM.[61]

In addition, factors other than the visual stimulus and the environmental context critically affect the CSM.

For instance, CSM is frequency specific,[72] depends on the time of day,[77] and is affected by environmental water shortage[78–81] and by the age of the animals.[52] Furthermore, acquisition of CSM may entail the adoption of freezing as an alternative defensive response.[66]

Because the acquisition of SM depends solely on stimulus repetition (i.e., associations are not involved) and at test it requires some recalling trials before being expressed, this memory represents a typical case of habituation. Thus, SM is an elementary component of the more complex CSM. Behavioral and mechanistic evidence for this point and a model of interpretation have been reviewed elsewhere[73,82] and will be further discussed later in this chapter.

ANATOMICAL DESCRIPTION OF BRAIN REGIONS INVOLVED IN CRAB'S VISUAL MEMORY

The CSM of *Neohelice* is based on visual information. Before describing the neurons involved in the visual learning and memory of the crab, a brief description of the organization of the visual nervous system is needed. Each eye of *N. granulata* consists of approximately 8000 ommatidia, which are distributed around the tip of the eyestalk except for a narrow area of cuticle located toward the medial side of the animal. Thus, the visual field of each eye subtends almost the entire panorama surrounding the animal. As is typical of decapods, each of the two optic lobes consists of three nested retinotopic neuropils plus an additional fourth retinotopic neuropil more recently discovered that, with a number of circumscribed protocerebral neuropils, are contained in the eyestalk. These are connected to medial neuropils of the supraesophageal ganglion by discrete protocerebral tracts. Based on modern anatomical and developmental studies, the optic neuropils were renamed as lamina, medulla, and lobula so as to conform to homologous neuropils in insects.[83–86] The principal architectural elements are retinotopic columns intersected by layers. Each column or cartridge represents a visual sampling unit of the retina: Photoreceptors from each ommatidium send their axons into a lamina optic cartridge, each of which is delineated by a rectilinear organization of tangentially directed processes. Efferent neurons extending from optic cartridges project the distal representation of visual sampling units proximally into the medulla. Deeper retinotopic organization is preserved through the medulla by its columnar neurons as they map this geometry into the lobula. In malacostracans as in insects, axons connecting the lamina with the medulla and the medulla with the lobula form chiasmata.

The analysis of the optical apparatus indicated that this crab possesses greater visual acuity around the equator and at the lateral side of the eye. These specializations are thought to be related to the particular features of the animal's ecological environment.[45] Advances in our understanding of the neural architecture, the anatomical organization, and the way in which visual space maps in the optic neuropils of the crab[45,86–88] have make it possible to relate events occurring in a particular position of visual space with physiological responses taking place in specific regions of a visual neuropil. Here, we focus on the lobula because of its central role in visual memory.

Like the medulla, the lobula of *Neohelice* is a dome-shaped structure, slightly elongated in the lateromedial axis. As in the lamina and medulla, the fibroarchitectural appearance of the lobula depends on the section orientation because different cell types have their tangential processes exclusively oriented either along the anteroposterior or the lateromedial axis. Transverse sections show four strata of tangential processes oriented lateromedially. From the periphery to the center, these strata are referred to as the first through fourth lateromedial tangential layers, LMT1–LMT4.[86] These strata are separated by regions containing arborization profiles belonging to columnar elements and local interneurons and the profiles of tangential processes running anteroposteriorly. LMT1 and LMT4 comprise relatively thin tangential processes, whereas LMT2 and LMT3 contain long, wide-diameter tangential fibers that increase in girth toward the medial side of the neuropil. Intracellular staining confirmed that LMT2 and LMT3 layers comprise the bistratified dendritic tree of wide-field motion-sensitive neurons termed bistratified lobula giant (BLG) neurons to distinguish them from the monostratified type described later. The dendritic tree of each of these neurons consists of several branches that run parallel to each other all along the lateromedial axis of the lobula. The dendrites converge toward the medial side of the neuropil into a thicker single axonal trunk that can be followed toward the midbrain.[86,89,90]

Longitudinal sections of the lobula demonstrate five strata composed of tangential processes oriented anteroposteriorly. From the periphery to the center, these are called the anteroposterior tangential levels APT1–APT5.[86] Level APT4 deserves special attention because it is composed of 14 neurons, each of which possesses an exceptionally wide-diameter primary branch (8–10 μm in diameter) oriented along the anteroposterior axis. Primary branches from these 14 cells are arranged in parallel, separated from one another by approximately 35 μm. Each primary branch provides several secondary processes that arise at right angles and thus extend lateromedially within the

LMT3 stratum in close proximity to the proximal tangential processes of the BLG neurons. Each primary branch is connected to a prominent axon that descends through the lateral protocerebrum to reach the optic tract. The descending axons of the 14 neurons converge to form a discrete bundle. These cells, hereafter termed monostratified lobula giants (MLGs), integrate the group of motion-sensitive LG neurons described later.

IN VIVO PHYSIOLOGICAL CHARACTERIZATION OF BRAIN INTERNEURONS

Ideally, neuronal events related to behavioral performance should be investigated while they occur *in vivo*, but physical accessibility and stability limit intracellular studies. Toward this aim, the crab offers unique advantages: (1) The visual neuropils are easily accessed with minimum damage to the animal, (2) the hard carapace provides the mechanical stability required for intracellular recordings, and (3) *Neohelice* has a great resistance to manipulation. Remarkably, following several hours of electrophysiological recording, the crab remains healthy, and days after the experiment no behavioral differences are observed with respect to nontreated animals.[8]

We characterized the electrophysiological properties of different types of neurons by their responses to a pulse of light or to moving stimuli delivered to the intact animal. We also identified the location and morphology of these neurons by intracellularly staining them with neurobiotin. Neurons from the lamina and medulla react more vigorously to a pulse of light than to a moving stimulus,[89,91,92] whereas the recorded neurons from the lobula are clearly tuned to respond to visual motion.[8,89,90,93] Because we are interested in learning and memory processes induced by a motion stimulus, the VDS, the LG neurons are the main focus of our attention.

CHARACTERIZATION OF THE LG NEURONS AND THEIR ROLE IN THE CRAB ESCAPE RESPONSE

In *Neohelice*, the sudden movement of a VDS above the animal elicits a strong escape reaction. We have morphologically identified and physiologically characterized four classes of LGs that responded to the same stimulus that elicits the behavioral reaction. Two of the classes present arborizations in the lobula that are monostratified (MLG types 1 and 2), whereas the other two classes are bistratified (BLG types 1 and 2). Figure 26.1C illustrates a MLG2 type of neuron. Detailed characteristics of the four types of LG neurons have been described elsewhere.[89,90,93–95] Briefly, the response to a moving stimulus consists of a strong discharge of action potentials frequently superimposed on noisy graded potentials (Figures 26.1C, 26.2C, and 26.3C). The response to a single moving object is more intense than that to the movement of the whole visual field, indicating that these neurons are tuned to object detection rather than to flow field processing. As is characteristic in all movement-sensitive neurons, including those of vertebrates, the response of the LGs is relatively independent of the background intensity and the contrast between moving target and background. The response of each neuron is highly consistent upon repeated stimulation, but such consistency can be obtained only when the stimuli are separated by long intervals. Repeated stimulation at intervals shorter than 10 min produces a reduction of the response. The extent and location of the receptive field, as well as the directional sensitivity, vary among the different LG classes. Intracellular recordings also showed that the LGs respond to visual stimuli presented either to the ipsilateral or to the contralateral eye, thus demonstrating that processing of binocular visual information occurs at the level of the lobula.[94] In addition, three of the four classes respond to both visual and mechanical stimuli applied to areas of the body, demonstrating that the integration of multimodal information also occurs in the lobula. The general characteristics of the LGs of *Neohelice* coincide with those of the morphologically unidentified movement fibers largely studied by Wiersma and co-workers in different decapod species (reviewed in Wiersma *et al.*[96]).

A variety of experimental conditions that affected the level of escape, such as seasonal variations, changes in stimulus features, and whether the crab perceived stimuli monocularly or binocularly, also consistently affected the response of LG neurons in a way that closely matched the effects observed at the behavioral level.[61,95] The analysis showed that the firing profile of the LGs corresponds well with that of the behavioral reaction, whereas a comparison of the times taken to reach half of the maximum responses revealed that, on average, the neuronal reaction anticipates the behavioral response by approximately 120 msec.[8,93] The sum of evidence suggests that the LGs play a central role in the circuit that controls the escape response elicited by VDS.

LG NEURONS AND THEIR ROLE IN VISUAL LEARNING AND MEMORY

We found that the rates of reduction and recovery of the response of LGs in both massed and spaced training remarkably coincide with the changes occurring at the behavioral level. Fifteen trials of massed training yield a fast and profound reduction of the response, which recovers completely after 15 min at both behavioral and neuronal levels. In contrast, an equal number of spaced training trials results in a slower and shallower reduction of the response, but the changes last longer. The change of response is specific for the moving stimulus because the response of the neuron to a pulse of light measured before and after repeated presentation of the training stimulus is not affected.[8]

In the experiments described previously, changes in the response of the LGs were assessed up to 15 min following training. Because spaced training induces a memory that lasts for several days, we investigated whether the neuronal changes observed 15 min after training persisted for at least 24 hr after training. To determine this, we trained crabs with 30 spaced trials, and 24 hr later we compared their behavioral and neuronal responses with those of control animals. The comparisons showed that training similarly affected the behavioral and neuronal responses (black bars in Figures 26.1B and 26.1D). Several lines of evidence indicate that the changes observed in LGs do not result from changes occurring in presynaptic neurons. For example, neurons from the medulla were found not to change upon repeated VDS presentations.[8] In addition, unpublished results indicate that the long-term neuronal reduction is not accounted for by changes in the input signal because no changes in the postsynaptic potential (PSP) response were observed. The evidence suggests that learning induces a change in the transfer of information from the input to the output neuronal signals (PSP to elicited spikes). The results demonstrate that long-term memory of the VDS, which is expressed as a reduction in the intensity of the escape reaction, is supported by long-term changes that take place in the giant neurons of the lobula.[8,61,97]

By recording the responses of LG neurons after training, we found that the ability of crabs to generalize the learned stimulus into new space positions (Figure 26.2B) and to distinguish it from a similar but unlearned stimulus (Figure 26.3B)—that is, two of the main attributes of stimulus memory—can be completely explained by the performance of LGs (Figures 26.2D and 26.3D, respectively).[61]

As previously mentioned, the CSM can be evoked only in the training context, implying an association between CM and SM. It has been shown that these two memory components can be acquired independently—CM by allowing the animal to simply explore a new context[27,54] and context-independent SM by performing a massed training (ITI: 2 sec) with a large number of VDS presentations.[62] Therefore, animals seem to store two visual components of the learned experience, one related to the stimulus (SM) and one related to the context (CM), that eventually can be associated to form the CSM. In the previously described experiments, the response of LGs from trained crabs was assessed while keeping the contextual environment constant between training and testing. Hence, an important question that remained was whether the effect of changing the context in the testing session would be reflected by the performance of the LGs. In other words, do these neurons store the entire CSM trace or only the SM component? To test this, we evaluated the response of LGs to the VDS in animals located in the same or in a different visual context compared with where the memory had been acquired. We found that whereas the learning-induced behavioral changes are context dependent (Figure 26.1B, white bar), the neuronal changes induced in the LGs are independent of the visual context (Figure 26.1D, white bar). In other words, the memory trace identified in the LGs only stands for the SM component of the CSM. Consequently, the traces corresponding to the contextual visual information and the context-signal association (the CM and the associative components of the CSM) must be stored in different places. Our results provided the first neurophysiological evidence that memory traces regarding 'what' and 'where' are stored separately in the arthropod brain.[61]

Experiments in insects point to the mushroom bodies as one candidate site for the processing of contextual information.[98–100] There are many functional similarities between the hemiellipsoid bodies of crustaceans and the mushroom bodies of insects.[101] There is thus a possibility that the hemiellipsoid bodies could be involved in processing contextual cues in crabs. If this is correct, the association of CM with SM can still occur somewhere else. The LG neurons are located in the optic lobe and project the axon along the protocerebral tract toward the midbrain. The hemiellipsoid body is located in the lateral protocerebrum and is also connected to the midbrain through the protocerebral tract. We do not know whether these two neural pathways are actually connected in the midbrain to support the SM–CM association. However, experiments described in the following sections—using combinations of behavioral, pharmacological, and molecular approaches—show that enduring synaptic changes related to CSM indeed occur in the midbrain of the crab.

PHARMACOLOGICAL AND MOLECULAR CHARACTERIZATION OF CSM FORMATION AND PROCESSING

A number of studies using a pharmacological approach demonstrated the participation of neurotransmitters such as glutamate,[64,102] acetylcholine,[103] and the neuropeptide angiotensin II[80,81] in the crab memory. In addition to the role of neurotransmitters, a universal feature of long-lasting forms of memory is their dependence on mRNA and protein synthesis.[104] Accordingly, CSM proved to be sensitive to the protein synthesis inhibitor cycloheximide (CHX) during the first hours after training (up to 6 hr) and was also sensitive to the mRNA synthesis inhibitor actinomycin D.[105]

ROLE OF PROTEIN KINASE A IN MEMORY CONSOLIDATION

The research has focused on signal transduction mechanisms that are activated by learning and participate in memory storage. Initially, we studied the participation of the cAMP-dependent protein kinase (PKA). A body of evidence has implicated the cAMP pathway in neural plasticity related to memory formation (e.g., see[106,107]; see also Chapter 31). Although previous studies in the *Aplysia* sensorimotor synapse demonstrated the key role of PKA in neural plasticity, we obtained the first data showing that the manipulation of PKA activity by means of cAMP analogs affects memory formation. Administration of a PKA activator, CPT-cAMP, together with a phosphodiesterase inhibitor, IBMX, before or after a weak training facilitated long-term memory in a dose-dependent manner.[75] Although this result supports the view that cAMP augmentation and PKA activation are important steps in memory consolidation, activation of the cGMP–PKG pathway could not be ruled out because CPT-cAMP also activates cGMP-dependent protein kinase and both IBMX and CPT-cAMP increase cGMP levels. Therefore, we used two membrane-permeable cAMP analogs, Sp-5,6-DCl-cBIMPS and Rp-8-Cl-cAMPS, that are highly specific for PKA. Sp-5,6-DCl-cBIMPS, a PKA activator, facilitated memory, whereas Rp-8-Cl-cAMPS, a PKA inhibitor, induced amnesia.[74]

Further analysis using the PKA inhibitor revealed two critical periods during which PKA activity was necessary for CSM consolidation: one during training and the other between 4 and 8 hr after training.[108] To evaluate whether PKA activation is in fact part of the cellular mechanisms triggered by training, we next measured PKA activity in the supraesophageal ganglion during the critical periods for CSM formation. In agreement with our previous findings, PKA was activated in the brain immediately and 6 hr after training. In contrast, PKA activity increased immediately but not 6 hr after exposure to a novel context.[109]

Two major families of PKA isoforms (PKA I and PKA II) that differ in their respective R subunits (R I and R II) have been described in different animals and tissues. PKA isoforms may show different subcellular distributions, so different activation kinetics and cAMP sensitivity[110] are considered to provide specificity to PKA action.[111] We demonstrated that PKA I and PKA II are present in *Neohelice* neural tissues, showing characteristics similar to those described in other species. Indeed, *Neohelice* PKA I is 10 times more sensitive than PKA II to cAMP.[112] Studying the activity profile of both PKA isoforms after training, we found different activation patterns. PKA II is activated immediately after exposure to the training context without stimulus presentation. For animals trained in the CSM paradigm, the total level of PKA I 6 hr after training significantly increases. Considering the higher sensitivity of PKA I to cAMP, its increase can account for the PKA activation found 6 hr after training and is proposed as a novel mechanism providing the prolonged PKA activation during memory consolidation.

MITOGEN-ACTIVATED PROTEIN KINASES IN CSM

The participation of other protein kinases that belong to the family of mitogen-activated protein kinases (MAPKs), namely extracellular-signal-regulated kinase (ERK) and c-Jun N-terminal kinase (JNK), has been studied. This is a highly conserved family of protein kinases, present in species as different as yeast, worms, mollusks, insects, and mammals.[113,114] These Ser/Thr protein kinases are found in a variety of tissues, regulating development, mitogenesis, and the stress response. Since the initial work of English and Sweatt[115] in long-term potentiation and the work of Martin *et al.*[116] in long-term facilitation in *Aplysia* neurons, substantial evidence has been obtained supporting the involvement of ERK in synaptic plasticity. Subsequent studies have shown the involvement of this kinase in memory processes in rodents,[117,118] *Hermissenda*,[119] and *Aplysia*.[120] Although MAPKs are highly conserved, no evidence has been found for their participation in memory formation in crustaceans. Translocation of ERK to the nucleus and indirect activation of transcription factor CREB was classically proposed as the pathway by which this MAPK acts in memory consolidation. In the crab, we found high immunoreactivity with antibodies that recognize specifically the phosphorylated form of ERK and JNK in the central brain. Using these antibodies, we found that ERK, but not JNK, showed memory-specific activation in the brain 1 hr after training. However, this activation was detected in cytosolic but not in nuclear extracts. This is a striking finding considering the importance of nuclear targets of these

MAPKs in different processes, including learning, and their relevance in mechanisms of gene expression regulation in other species. Coincident with the temporal course of activation, an inhibitor of the ERK pathway, PD098059, showed an amnesic effect when administered 45 min after training but not immediately pre-training or post-training. In addition, the same drug did not affect short-term memory.[121]

All these findings suggest that an extranuclear ERK pool is necessary for long-term memory consolidation. Further research is required to elucidate its specific subcellular localization and potential molecular effectors.

REL/NF-κB, A KEY TRANSCRIPTION FACTOR IN CONSOLIDATION OF CSM

During approximately the past 10 years, experimental data have increasingly supported the participation of the nuclear transcription factor κB (NF-κB) in memory processes. The first evidence indicating that learning activates NF-κB was found in *Neohelice*.[76,122] We detected specific κB DNA binding activity using a double-strand DNA oligonucleotide containing the NF-κB consensus sequence in a gel shift assay. With this technique, we detected one specific complex in the crab central brain. One of the complex components is a 61-kDa protein that could be detected with an antibody against p65/Rel A, a member of the NF-κB family in mammals. The gel shift assay allows an estimation of NF-κB activity, and a high correlation between memory formation and NF-κB activation was demonstrated. Spaced training yielding CSM correlates with NF-κB activation in the cell nucleus of crab central brain. In contrast, NF-κB is not activated after massed training. Crabs that received spaced training showed two phases of Rel/NF-κB activation. The first phase was seen immediately after 10 or more trials, coinciding with the number of trials required for memory formation. Such activation decayed to basal level 3 hr after training. The second phase occurred 6 hr after training, waning to basal level at 12 and 24 hr.[76] Moreover, activated NF-κB was found in synaptic terminals after long-term memory induction. The fact that activation of a transcription factor (TF) was found so far from its normal site of action in the nucleus gave physiological support for the hypothesis that NF-κB has a dual role in synapse-to-nucleus signaling, initially as a synaptic activity detector and then as transcriptional regulator after its retrograde transport to the nucleus.[76,123–125] We next evaluated the effect of the inhibition of this TF pathway by means of the drug sulfasalazine. This drug is a specific inhibitor of IκB kinase (IKK), the protein kinase that activates NF-κB by phosphorylation of IκB that, after this phosphorylation, is degraded by proteasome, allowing the nuclear translocation of the TF. Sulfasalazine administration induced amnesia during the two periods in which NF-κB was active but not before or after the activation periods.[126] This activity profile is similar to that found for PKA activity after training and suggests that these two pathways may be functionally related.

To study the effect of NF-κB inhibition in another step of its activation pathway, we used a cell-permeable 26S proteasome inhibitor, MG132. This drug, among other effects, impedes NF-κB activation by reducing the degradation of its inhibitor IκB. In fact, MG132 impaired training-induced NF-κB activation in crab brain when administered *in vivo*, whereas the same dose impaired long-term memory.[127]

The requirement of this TF in CSM formation led us to question which neurotransmitters or neuromodulators are involved in NF-κB activation during learning. The neuropeptide angiotensin II plays an important role in the crab long-term memory,[79] and we found that NF-κB in the crab brain is activated by angiotensin II administration at doses that facilitated CSM. The activation of the TF was reversed by saralasin, an antagonist of angiotensin II receptors. Accordingly, NF-κB activation was also found after water shortage, a treatment that increased angiotensin II levels in the brain and induced memory facilitation.[81]

EPIGENETIC MECHANISMS IN CSM FORMATION

In order to persist, memory storage requires changes in the gene expression pattern. Involvement of epigenetic mechanisms was postulated in order to subserve a more persistent regulation of specific gene expression needed to maintain long-term changes in neural plasticity. This could be achieved by means of providing relatively stable marks in the chromatin.[128] Chromatin structure and function can be affected by various post-translational modifications of the amino-terminal tails of the nucleosomal histones.[129,130] Lysine acetylation is one of the best characterized histone modifications. Chromatin-modifying enzymes, which carry out protein acetylation and deacetylation, have already been described[131]—histone acetyl transferases, such as CREB binding protein (CBP/p300), and histone deacetylases (HDACs), respectively. Histone acetylation induces transcription activation by increasing DNA accessibility to the transcription machinery.[132,133] We examined whether histone acetylation is involved in consolidation of CSM, whether it depends on the training strength, and whether it is a permanent

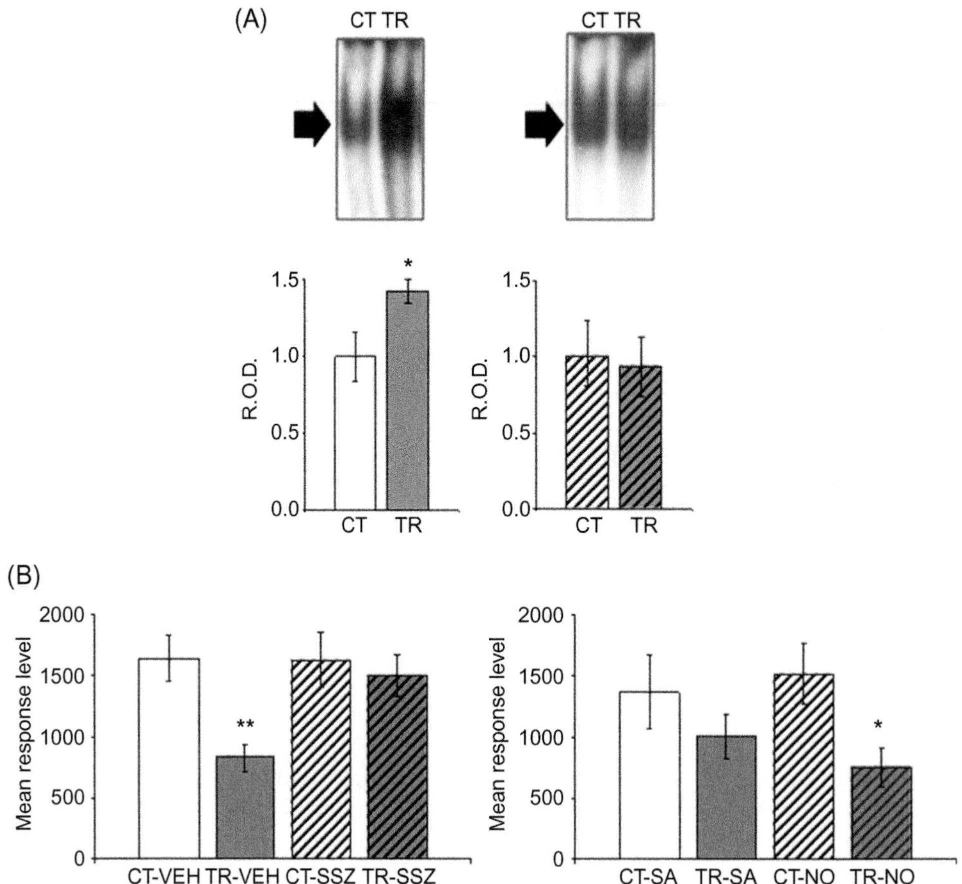

FIGURE 26.4 The role of NF-κB in memory reconsolidation. (A) NF-κB activation after re-exposure to the training context. NF-κB DNA binding activity measured immediately after re-exposure to the training context (left) or after exposure to a different context (right). DNA binding activity was measured by electrophoretic mobility shift assay performed with nuclear extracts obtained from passive control (CT) and spaced trained (TR) groups. The specific band is indicated by arrows. (Bottom histograms) NF-κB activity relative to CT. Mean ± SEM of relative optical density (R.O.D.) values of the specific NF-κB retarded band normalized to CT group media, obtained in four independent experiments. $*p < 0.05$ in t test. (B) Effect of sulfasalazine injection before context re-exposure on retention at testing performed 24 hr after re-exposure. Mean response level ± SEM during testing session showing comparisons between CT groups injected with either vehicle or sulfasalazine. (Left) Animals tested in the same context used at training. CT-VEH, vehicle-injected passive control group; TR-VEH, vehicle-injected spaced trained group; CT-SSZ, sulfasalazine-injected passive control group; TR-SSZ, sulfasalazine-injected spaced training group. $**p < 0.01$. (Right) All animals were injected with sulfasalazine before re-exposure to the same context used at training (SA) or to a different context (NO) and testing 24 hr after re-exposure. $*p < 0.05$. Source: *Data redrawn from Merlo et al.*[142]

or transient mechanism. Our results showed no changes in histone 3 (H3) acetylation during consolidation of a standard training protocol of 15 trials. However, strong training of 30 trials induced a significant increase in the general levels of H3 acetylation 1 hr post-training, returning to basal levels afterward. Accordingly, systemic delivery of the histone deacetylase inhibitors sodium butyrate and trichostatin A allowed a weak training to induce long-term memory in two phases during consolidation. These findings indicate that H3 acetylation (1) is involved in consolidation, (2) occurs only after strong training, and (3) is a transient process. They also indicate that the inhibition of HDAC after a weak training increases H3 acetylation and enhances memory in the two phases.[134]

CSM RECONSOLIDATION AND EXTINCTION: A CORNERSTONE IN THE STUDY OF MEMORY REPROCESSING

In the classical postulation, memory is initially labile and becomes resistant to different disruption treatments during a consolidation period. However, this assumption has been challenged by studies in which

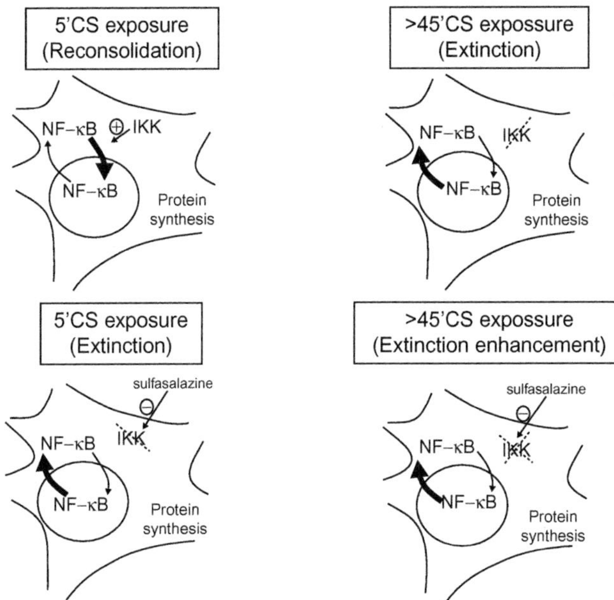

FIGURE 26.5 A model of the role of NF-κB in memory extinction and reconsolidation and the effect of its inhibition. (Top left) Five-minute context re-exposure induces NF-κB-dependent and protein synthesis-dependent memory reconsolidation. (Top right) Prolonged re-exposure induces NF-κB inhibition and protein synthesis-dependent memory extinction. (Bottom left) NF-κB inhibition by sulfasalazine plus 5-min re-exposure mimics memory extinction. (Bottom right) NF-κB inhibition by sulfasalazine plus prolonged re-exposure induces extinction facilitation. Source: *Reproduced from Merlo and Romano.*[143]

experimental data indicate that a consolidated memory may become labile again when retrieved and reactivated by a reminder, opening a new consolidation-like period called reconsolidation[135–137] (see Chapter 33). Seminal work in the crab CSM revealed that this process is not restricted to rodents and showed important parametric characteristics of reconsolidation and its relation with memory extinction. In CSM reconsolidation studies, a brief re-exposure (5 min) to the training context reactivates the memory. This treatment opens a new period of lability in which CHX is again amnesic. However, protein synthesis inhibition was not amnesic when the animals were exposed to a different context than that used in training.[64]

Conversely, the prolonged re-exposure (typically for 2 hr) of the crabs to the training container 1 day after training induced memory extinction. A day after, memory extinction was revealed in a testing session in which the animals showed high levels of escape response. Extinction consolidation, similar to the consolidation of the original memory, is sensitive to CHX when administered soon after re-exposure but not 6 hr later.[12] Thus, these studies using CHX revealed that the memory course after retrieval, either to reconsolidation or to extinction, depends on the remainder duration.[12] This was a key finding for the understanding of memory dynamic and reprocessing that reconciled apparently contradictory experimental data in both vertebrates and invertebrates.

NF-κB and Memory Reconsolidation

Whether reconsolidation is mechanistically similar to the initial consolidation is a matter of debate.[138–141] We tested whether NF-κB is required for reconsolidation and found that NF-κB was specifically reactivated in animals re-exposed 24 hr after training to the same training context but not to a different context (Figure 26.4A). Furthermore, NF-κB was not activated in animals re-exposed to the context after a weak training protocol insufficient to induce long-term memory. In addition, sulfasalazine impaired reconsolidation when administered 20 min before re-exposure to the training context, but it was not effective when a different context was used (Figure 26.4B).[142] These findings are in keeping with the view that basic molecular mechanisms of consolidation are necessary to restore reactivated memory. At the same time, these findings reveal for the first time that NF-κB is activated specifically by retrieval and that this activation is required for memory reconsolidation, supporting the view that this molecular mechanism plays a key role in both consolidation and reconsolidation.

NF-κB and Memory Extinction

As mentioned previously, NF-κB activation plays a critical role in consolidation and reconsolidation. We therefore investigated the possible participation of NF-κB in memory extinction. Under the hypothesis that extinction is a new memory with similar characteristics as the original memory consolidation, we could expect this TF to be required for consolidation of extinction. However, we found that the administration of the NF-κB inhibitor sulfasalazine prior to the extinction session did not impair but, rather, enhanced extinction. Moreover, reinstatement experiments showed that the original memory was not affected and that NF-κB inhibition by sulfasalazine impaired or delayed spontaneous recovery, thus strengthening the ongoing memory extinction process. Interestingly, in animals with a fully consolidated memory, brief re-exposure to the training context induced NF-κB activation in the brain, whereas prolonged re-exposure induced NF-κB inhibition in correlation with memory extinction. Such inhibition was found 45 min after the beginning of re-exposure, and NF-κB activity recovered to basal levels at 2 hr, when

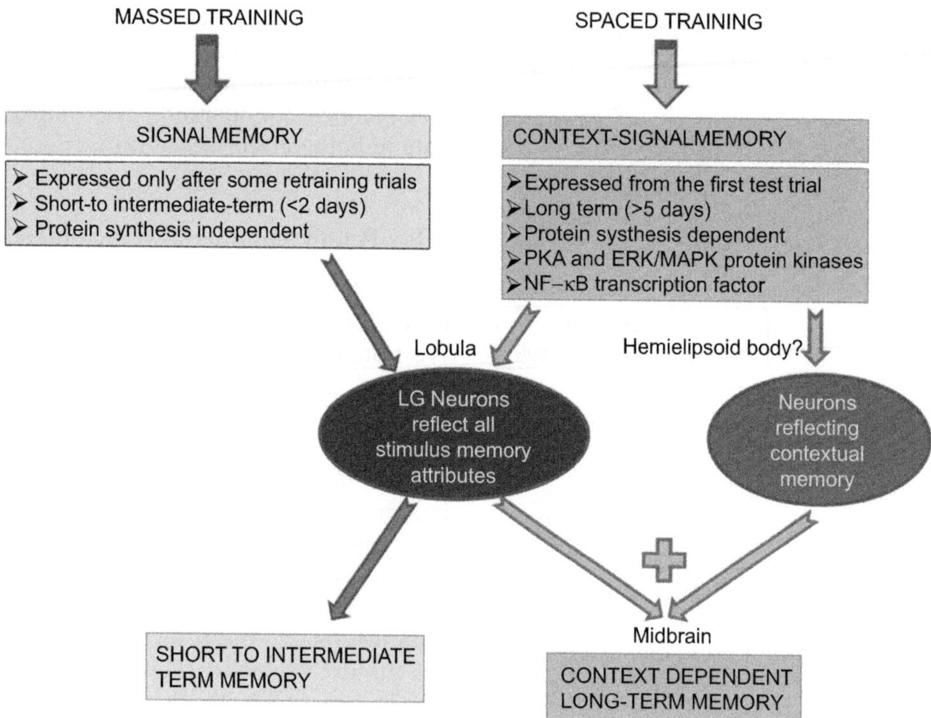

FIGURE 26.6 Schematic model of the components and attributes of signal memory and context-signal memory. Signal memory is acquired by massed training, lasts short to intermediate term, and is context independent. On the other hand, context-signal memory is acquired by spaced training, lasts a long time, and is determined by an association of signal memory (stimulus memory) with a contextual memory. Attributes of the stimulus visual memory such as the ability for spatial generalization and for stimulus recognition can be entirely accounted for by the performance of the lobula giant neurons. Context memory may take place in the hemiellipsoid bodies. Molecular studies indicate that signal memory and contextual memories are likely associated in the midbrain of the crab.

a clear extinction level was already acquired.[143] Together, the data on the NF-κB dynamics in reconsolidation and extinction indicate that during re-exposure to the training context, NF-κB is initially activated. However, after some time an active NF-κB inhibition mechanism takes place, which ceases after prolonged re-exposure at the end of the extinction session. Few data are available in the field of molecular mechanisms for memory reprocessing. The study of NF-κB dynamics in the crab's brain provides new information about the molecular mechanisms involved in the switch that determines if a memory will be reconsolidated or extinguished.

On the basis of these findings, we propose a working model for the role of NF-κB in memory after retrieval (Figure 26.5). The initial process of transcriptional activation induced by retrieval would be mediated by protein kinases. In particular, the activation of IKK and PKA induces NF-κB activation and its translocation to the nucleus. The prolonged presence of the training context would induce activation of other mediators such as protein phosphatases, which may increase the level of NF-κB inhibitor IκB and thus induce its nuclear exportation.[144] Under this interpretation, the administration of NF-κB inhibitors during memory reactivation would reinforce the effect of prolonged exposure to the context, inducing extinction strengthening. Our interpretation is in line with the view that extinction formation recruits some mechanisms different from the ones of the original memory consolidation, and that weakening of the original consolidated circuits is part of the neural correlate of memory extinction. However, we cannot exclude the requirement of reinforcement mechanisms in other circuits, independent of the NF-κB pathway, mediating the same extinction process. In summary, the evidence revised here supports the view that extinction does not require the activity of NF-κB, a key TF involved in consolidation and reconsolidation, but actually requires its inhibition.

CONCLUSION

The study of several species of invertebrates and vertebrates has allowed considerable progress to be made in understanding the mechanisms underlying learning and memory formation.[119,145,146,147,148] Each species offers special advantages for studying certain aspect of learning and memory. The availability of different species in diverse animal groups is also critical for comparative analysis, which ultimately tells us about the communalities and particularities of the memory processes.

For more than 20 years, a research effort has been focused on studying learning and memory in the crab, *N. granulata*. The most compelling studies have been performed in the CSM paradigm. Studies at the behavioral, anatomical, and cellular levels—using pharmacological, electrophysiological, and molecular approaches—have provided an integrated description of the different aspects of learning and memory in this species (Figure 26.6). The different phases of memory processing of CSM are well characterized, which constitutes an advantage for the study of molecular mechanisms of memory. Appropriate controls are available to evaluate the effects of factors such as stress, sensorial stimulation, and motor activity and to dissect them from the specific mechanisms of memory storage. The strong resistance of crabs to experimental manipulations, and the accessibility to an important part of the brain, allows using intracellular recording in the intact animal to investigate the changes occurring in identified neurons during learning. Furthermore, the ecology of the crab makes feasible the investigation of neural changes in animals freely behaving in their natural environment.[149] These and other experimental advantages offered by crabs have led to a number of fundamental discoveries of general interest in the field of learning and memory, including the following: (1) Studies on memory reconsolidation in the crab have led to new interpretations of the way in which memories are stored and upgraded,[12,26,63,64,150] (2) electrophysiological studies have provided the most compelling evidence to date on the identification of individual neurons supporting long-term visual memory in an arthropod,[8,61] and (3) molecular results have provided the first experimental evidence of the role of the transcription factor NF-κB in memory consolidation.[76,122]

Acknowledgment

We dedicate this chapter to the memory of our mentor in science, Professor Héctor Maldonado, whose contributions to the field of learning and memory in invertebrates have been enormous. His earlier extensive studies on the memory of octopus and praying mantis, and later studies on the memory of crab, reflect his convictions about the relevance of these animals for investigating the neurobiology of memory. He was a remarkable scientist as well as a highly dedicated teacher. We had the great fortune of having received his advices and enthusiasm every day for more than 25 years. We will forever be indebted to our friend.

References

1. Florey E, Biderman MA. Studies on the distribution of factor I and acetylcholine in crustacean peripheral nerve. *J Gen Physiol.* 1960;43:509–522.
2. Kravitz EA, Kuffler SW, Potter DD. Gamma-aminobutyric acid and other blocking compounds in crustacea: III. Their relative concentrations in separated motor and inhibitory axons. *J Neurophysiol.* 1963;26:739–751.
3. Wiersma CA, Ikeda K. Interneurons commanding swimmeret movements in the crayfish, *Procambarus clarki* (girard). *Comp Biochem Physiol.* 1964;12:509–525.
4. Wiersma CA. Giant nerve fiber system of the crayfish; A contribution to comparative physiology of synapse. *J Neurophysiol.* 1947;10:23–38.
5. Edwards DH, Yeh SR, Krasne FB. Neuronal coincidence detection by voltage-sensitive electrical synapses. *Proc Natl Acad Sci USA.* 1998;95:7145–7150.
6. Herberholz J, Issa FA, Edwards DH. Patterns of neural circuit activation and behavior during dominance hierarchy formation in freely behaving crayfish. *J Neurosci.* 2001;21:2759–2767.
7. Yeh SR, Fricke RA, Edwards DH. The effect of social experience on serotonergic modulation of the escape circuit of crayfish. *Science.* 1996;271:366–369.
8. Tomsic D, Beron de Astrada M, Sztarker J. Identification of individual neurons reflecting short- and long-term visual memory in an arthropod. *J Neurosci.* 2003;23:8539–8546.
9. Hiatt RW. The biology of the lined shore crab, *Pachygrapsus crassipes* Randall. *Pac Sci.* 1948;2:135–213.
10. Hemmi JM, Merkle T. High stimulus specificity characterizes anti-predator habituation under natural conditions. *Proc R Soc B Biol Sci.* 2009;276:4381–4388.
11. Abramson CI, Feinman RD. Classical conditioning of the eye withdrawal reflex in the green crab. *J Neurosci.* 1998;8: 2907–2912.
12. Pedreira ME, Maldonado H. Protein synthesis subserves reconsolidation or extinction depending on reminder duration. *Neuron.* 2003;38:863–869.
13. Hirsh R, Wiersma CA. The effect of the spacing of background elements upon optomotor memory responses in the crab: the influence of adding or deleting features during darkness. *J Exp Biol.* 1977;66:33–46.
14. Glantz RM. The visually evoked defense reflex of the crayfish: habituation, facilitation, and the influence of picrotoxin. *J Neurobiol.* 1974;5:263–280.
15. Marchand AR, Barnes WJP. Correlates of habituation of a polysynaptic reflex in crayfish *in vivo* and *in vitro*. *Eur J Neurosci.* 1992;4:521–532.
16. Aggio J, Rakitin A, Maldonado H. Serotonin-induced short- and long-term sensitization in the crab *Chasmagnathus*. *Pharmacol Biochem Behav.* 1996;53:441–448.
17. Appleton T, Wilkens JL. Habituation, sensitization and the effect of serotonin on the eyestalk withdrawal reflex of *Cancer magister*. *Comp Biochem Physiol A.* 1990;97:159–163.
18. Abramson CI, Feinman RD. Operant punishment of eye elevation in the green crab, *Carcinus maenas*. *Behav Neur Biol.* 1987;48:259–277.
19. Hoyle G. Learning of leg position by the ghost crab *Ocypode ceratophthalma*. *Behav Biol.* 1976;18:147–163.
20. Abramson CI, Feinman RD. Lever-press conditioning in the crab. *Physiol Behav.* 1990;48:267–272.
21. Cuadras J, Vila E, Balasch J. T-maze shock avoidance in the hermit crab *Dardanus arrosor*. *Rev Esp Fisiol.* 1978;34:273–276.
22. Fernandez-Duque E, Valeggia C, Maldonado H. Multitrial inhibitory avoidance learning in the crab *Chasmagnathus*. *Behav Neur Biol.* 1992;57:189–197.

23. Kawai N, Kono R, Sugimoto S. Avoidance learning in the crayfish (*Procambarus clarkii*) depends on the predatory imminence of the unconditioned stimulus: a behavior systems approach to learning in invertebrates. *Behav Brain Res.* 2004;150:229–237.
24. Dimant B, Maldonado H. Habituation and associative learning during exploratory behavior of the crab *Chasmagnathus*. *J Comp Physiol A*. 1992;170:749–759.
25. Eisenstein EM, Mill PJ. Role of the optic ganglia in learning in the crayfish *Procambarus clarkii* (Girard). *Anim Behav.* 1965;13: 561–565.
26. Perez-Cuesta LM, Maldonado H. Memory reconsolidation and extinction in the crab: mutual exclusion or coexistence? *Learn Mem.* 2009;16:714–721.
27. Tomsic D, Pedreira ME, Hermitte G, Romano A, Maldonado H. Context-US association as a determinant of long-term habituation in the crab *Chasmagnathus*. *Anim Learn Behav.* 1998;26: 196–209.
28. Hughes RN, O'Brien N. Shore crabs are able to transfer learned handling skills to novel prey. *Anim Behav.* 2001;61: 711–714.
29. Micheli F. Behavioural plasticity in prey-size selectivity of the blue crab *Callinectes sapidus* feeding on bivalve prey. *J Anim Ecol.* 1995;64:63–74.
30. Roudez RJ, Glover T, Weis JS. Learning in an invasive and a native predatory crab. *Biol Invasions.* 2008;10:1191–1196.
31. Gherardi F, Tricarico E, Atema J. Unraveling the nature of individual recognition by odor in hermit crabs. *J Chem Ecol.* 2005;31:2877–2896.
32. Hazlett BA. 'Individual' recognition and agonistic behaviour in *Pagurus bernhardus*. *Nature.* 1969;222:268–269.
33. Jackson NW, Elwood RW. Memory of information gained during shell investigation by the hermit crab, *Pagurus bernhardus*. *Anim Behav.* 1989;37:529–534.
34. Cronin TW. Stomatopods. *Curr Biol.* 2006;16:235–236.
35. Cannicci S, Fratini S, Vannini M. Short range homing in fiddler crabs (Ocypodidae, genus *Uca*): a homing mechanism not based on local visual landmarks. *Ethology.* 1999;105:867–880.
36. Cannicci S, Barelli C, Vannini M. Homing in the swimming crab *Thalamita crenata*: a mechanism based on underwater landmark memory. *Anim Behav.* 2000;60:203–210.
37. Cannicci S, Ruwa RK, Vannini M. Homing experiment in the tree-climbing crab *Sesarma leptosoma* (Decapoda, Grapsidae). *Ethology.* 1997;103:935–944.
38. Vannini M, Cannicci S. Homing behaviour and possible cognitive maps in crustacean decapods. *J Exp Mar Biol Ecol.* 1995;193: 67–91.
39. Krasne FB, Bryan JS. Habituation: regulation through presynaptic inhibition. *Science.* 1973;182:590–592.
40. Shirinyan D, Teshiba T, Taylor K, O'Neill P, Lee SC, Krasne FB. Rostral ganglia are required for induction but not expression of crayfish escape reflex habituation: role of higher centers in reprogramming low-level circuits. *J Neurophysiol.* 2006;95: 2721–2724.
41. Sakai K, Turkay M, Yang SL. Revision of the *Helice/ Chasmagnathus* complex (Crustacea: Decapoda: Brachyura). *Abh Senckenbergischen Naturforschenden Gesellshaft.* 2006;565:1–76.
42. Luppi T, Bas C, Méndez Casariego A, et al. Variations in activity patterns in the estuarine crab *Neohelice* (=*Chasmagnathus*) *granulata* in different habitats, seasons and tidal regimes. *Helgoland Mar Res.* 2012;10.1007/s10152-012-0300-9.
43. Spivak E. The crab *Neohelice* (=*Chasmagnathus*) *granulata*: an emergent animal model from emergent countries. *Helgoland Mar Res.* 2010;3:149–154.
44. Zeil J, Hemmi JM. The visual ecology of fiddler crabs. *J Comp Physiol A.* 2006;192:1–25.
45. Berón de Astrada M, Bengochea M, Medan V, Tomsic D. Regionalization in the eye of the grapsid crab *Neohelice granulata* (=*Chasmagnathus granulatus*): variation of resolution and facet diameters. *J Comp Physiol A.* 2012;198:173–180.
46. Brunner D, Maldonado H. Habituation in the crab *Chasmagnathus granulatus*: effect of morphine and naloxone. *J Comp Physiol A.* 1988;162:687–694.
47. Lozada M, Romano A, Maldonado H. Long-term habituation to a danger stimulus in the crab *Chasmagnathus granulatus*. *Physiol Behav.* 1990;47:35–41.
48. Tomsic D, Maldonado H. Central effect of morphine pretreatment on short- and long-term habituation to a danger stimulus in the crab *Chasmagnathus*. *Pharmacol Biochem Behav.* 1990;36: 787–793.
49. Rakitin A, Tomsic D, Maldonado H. Habituation and sensitization to an electrical shock in the crab *Chasmagnathus*: effect of background illumination. *Physiol Behav.* 1991;50:477–487.
50. Denti A, Dimant B, Maldonado H. Passive avoidance learning in the crab *Chasmagnathus granulatus*. *Physiol Behav.* 1988;43: 317–320.
51. Dimant B, Rossen A, Hermitte G. CS-US delay does not impair appetitive conditioning in *Chasmagnathus*. *Behav Processes.* 2002;60:1–14.
52. Tomsic D, Dimant D, Maldonado H. Age related deficits of long-term memory in *Chasmagnathus*. *J Comp Physiol A.* 1996;178: 139–146.
53. Pedreira ME, Dimant B, Maldonado H. Inhibitors of protein and RNA synthesis block context memory and long-term habituation in the crab *Chasmagnathus*. *Pharmacol Biochem Behav.* 1996;54:611–617.
54. Tomsic D, Massoni V, Maldonado H. Habituation to a danger stimulus in two semiterrestrial crabs: ontogenic, ecological and opioid correlates. *J Comp Physiol A.* 1993;173:621–633.
55. Burnovicz A, Hermitte G. Conditioning of an autonomic response in Crustacea. *Physiol Behav.* 2010;101:168–175.
56. Maldonado H, Romano A, Lozada M. Opioid action on response level to a danger stimulus in the crab (*Chasmagnathus granulatus*). *Behav Neurosci.* 1989;103:1139–1143.
57. Romano A, Lozada M, Maldonado H. Effect of naloxone pretreatment on habituation in the crab *Chasmagnathus granulatus*. *Behav Neur Biol.* 1990;53:113–122.
58. Romano A, Lozada M, Maldonado H. Nonhabituation processes affect stimulus specificity of response habituation in the crab *Chasmagnathus granulatus*. *Behav Neurosci.* 1991;105:542–552.
59. Tomsic D, Maldonado H, Rakitin A. Morphine and GABA: effects on perception, escape response and long-term habituation to a danger stimulus in the crab *Chasmagnathus*. *Brain Res Bull.* 1991;26:699–706.
60. Valeggia C, Fernandez-Duque E, Maldonado H. Danger stimulus-induced analgesia in the crab *Chasmagnathus granulatus*. *Brain Res.* 1989;481:304–308.
61. Sztarker J, Tomsic D. Brain modularity in arthropods: individual neurons that support 'what' but not 'where' memories. *J Neurosci.* 2011;31:8175–8180.
62. Hermitte G, Pedreira ME, Tomsic D, Maldonado H. Context shift and protein synthesis inhibition disrupt long-term habituation after spaced, but not massed, training in the crab *Chasmagnathus*. *Neurobiol Learn Mem.* 1999;71:34–49.
63. Pedreira ME, Perez-Cuesta LM, Maldonado H. Mismatch between what is expected and what actually occurs triggers memory reconsolidation or extinction. *Learn Mem.* 2004;11: 579–585.

64. Pedreira ME, Perez-Cuesta LM, Maldonado H. Reactivation and reconsolidation of long-term memory in the crab *Chasmagnathus*: protein synthesis requirement and mediation by NMDA-type glutamatergic receptors. *J Neurosci.* 2002;22: 8305–8311.
65. Tang S, Wolf R, Xu S, Heisenberg M. Visual pattern recognition in *Drosophila* is invariant for retinal position. *Science.* 2004;305:1020–1022.
66. Pereyra P, Gonzalez Portino E, Maldonado H. Long-lasting and context-specific freezing preference is acquired after spaced repeated presentations of a danger stimulus in the crab *Chasmagnathus*. *Neurobiol Learn Mem.* 2000;74:119–134.
67. Pereyra P, Saraco M, Maldonado H. Decreased response or alternative defensive strategy in escape: two novel types of long-term memory in the crab *Chasmagnathus*. *J Comp Physiol A.* 1999;184:301–310.
68. Fathala MV, Iribarren L, Kunert MC, Maldonado H. A field model of learning: 1. Short-term memory in the crab *Chasmagnathus granulatus*. *J Comp Physiol A.* 2010;196:61–75.
69. Fathala MV, Kunert MC, Maldonado H. A field model of learning: 2. Long-term memory in the crab *Chasmagnathus granulatus*. *J Comp Physiol A.* 2010;196:77–84.
70. Menzel R, Manz G, Greggers U. Massed and spaced learning in honeybees: the role of CS, US, the intertrial interval, and the test interval. *Learn Mem.* 2001;8:198–208.
71. Tully T, Preat T, Boynton SC, Del Vecchio M. Genetic dissection of consolidated memory in *Drosophila*. *Cell.* 1994;79:35–47.
72. Pedreira ME, Romano A, Tomsic D, Lozada M, Maldonado H. Massed and spaced training build up different components of long-term habituation in the crab *Chasmagnathus*. *Anim Learn Behav.* 1998;26:34–45.
73. Maldonado H, Romano A, Tomsic D. Long-term habituation (LTH) in the crab *Chasmagnathus*: a model for behavioral and mechanistic studies of memory. *Braz J Med Biol Res.* 1997;30: 813–826.
74. Romano A, Delorenzi A, Pedreira ME, Tomsic D, Maldonado H. Acute administration of a cAMP analog and a phosphodiesterase inhibitor improve long-term habituation of the crab *Chasmagnathus*. *Behav Brain Res.* 1996;75:119–125.
75. Romano A, Locatelli F, Pedreira ME, Delorenzi. A, Maldonado H. Effect of activation and inhibition of cAMP-dependent protein kinase on long-term habituation of the crab *Chasmagnathus*. *Brain Res.* 1996;735:131–140.
76. Freudenthal R, Romano A. Participation of NF-κB transcription factors in long-term memory in the crab *Chasmagnathus*. *Brain Res.* 2000;855:274–281.
77. Pereyra P, De La Iglesia HO, Maldonado H. Training-to-testing intervals different from 24 h impair habituation in the crab *Chasmagnathus*. *Physiol Behav.* 1996;59:19–25.
78. Delorenzi A, Dimant B, Frenkel L, Nahmod VE, Nässel DR, Maldonado H. High environmental salinity induces memory enhancement and increases levels of brain angiotensin-like peptides in the crab *Chasmagnathus granulatus*. *J Exp Biol.* 2000;203: 3369–3379.
79. Delorenzi A, Maldonado H. Memory enhancement by the angiotensinergic system in the crab *Chasmagnathus* is mediated by endogenous angiotensin II. *Neurosci Lett.* 1999;266:1–4.
80. Delorenzi A, Pedreira ME, Romano A, et al. Angiotensin II enhances long-term memory in the crab *Chasmagnathus*. *Brain Res Bull.* 1996;41:211–220.
81. Frenkel L, Freudenthal R, Romano A, Nahmod VE, Maldonado H, Delorenzi A. Angiotensin II and the transcription factor Rel/NF-kappaB link environmental water shortage with memory improvement. *Neurosci.* 2002;115:1079–1087.
82. Maldonado H. Crustacean as model to investigate memory illustrated by extensive behavioral and physiological studies in *Chasmagnathus*. In: Wiese K, ed. *Studies in Crustacean Neuroscience: Comparative Approaches to the Investigation of the Nervous System*. New York: Springer-Verlag; 2002:314–327.
83. Krieger J, Sandeman RE, Sandeman DC, Hansson BS, Harzsch S. Brain architecture of the largest living land arthropod, the giant robber crab *Birgus latro* (Crustacea, Anomura, Coenobitidae): evidence for a prominent central olfactory pathway? *Fron Zool.* 2010;7:25.
84. Sinakevitch I, Douglass JK, Scholtz G, Loesel T, Strausfeld NJ. Conserved and convergent organization in the optic lobes of insects and isopods, with reference to other crustacean taxa. *J Comp Neurol.* 2003;467:150–172.
85. Strausfeld NJ. Brain organization and the origin of insects: an assessment. *Proc R Soc B Biol Sci.* 2009;276:1929–1937.
86. Sztarker J, Strausfeld NJ, Tomsic D. Organization of optic lobes that support motion detection in a semiterrestrial crab. *J Comp Neurol.* 2005;493:396–411.
87. Berón de Astrada M, Medan V, Tomsic D. How visual space maps in the optic neuropils of a crab. *J Comp Neurol.* 2011;519:1631–1639.
88. Sztarker J, Strausfeld NJ, Andrew D, Tomsic D. Neural organization of first optic neuropils in the littoral crab *Hemigrapsus oregonensis* and the semiterrestrial species *Chasmagnathus granulatus*. *J Comp Neurol.* 2009;513:129–150.
89. Berón de Astrada M, Tomsic D. Physiology and morphology of visual movement detector neurons in a crab (Decapoda: Brachyura). *J Comp Physiol A.* 2002;188:539–551.
90. Medan V, Oliva D, Tomsic D. Characterization of lobula giant neurons responsive to visual stimuli that elicit escape behaviors in the crab *Chasmagnathus*. *J Neurophysiol.* 2007;98:2414–2428.
91. Berón de Astrada M, Sztarker J, Tomsic D. Visual interneurons of the crab *Chasmagnathus* studied by intracellular recordings in vivo. *J Comp Physiol A.* 2001;187:37–44.
92. Berón de Astrada M, Tuthill JC, Tomsic D. Physiology and morphology of sustaining and dimming neurons of the crab *Chasmagnathus granulatus* (Brachyura: Grapsidae). *J Comp Physiol A.* 2009;195:791–798.
93. Oliva D, Medan V, Tomsic D. Escape behavior and neuronal responses to looming stimuli in the crab *Chasmagnathus granulatus* (Decapoda: Grapsidae). *J Exp Biol.* 2007;210:865–880.
94. Sztarker J, Tomsic D. Binocular visual integration in the crustacean nervous system. *J Comp Physiol A.* 2004;190:951–962.
95. Sztarker J, Tomsic D. Neuronal correlates of the visually elicited escape response of the crab *Chasmagnathus* upon seasonal variations, stimuli changes and perceptual alterations. *J Comp Physiol A.* 2008;194:587–596.
96. Wiersma CAG, Roach JLM, Glantz RM. Neural integration in the optic system. In: Sandeman DC, Atwood HL, eds. The Biology of the Crustacea, *Vol. 4. Neural Integration and Behavior*. New York: Academic Press; 1982:1–31.
97. Tomsic D, Berón de Astrada M, Sztarker J, Maldonado H. Behavioral and neuronal attributes of short- and long-term habituation in the crab *Chasmagnathus*. *Neurobiol Learn Mem.* 2009;92:176–182.
98. Liu G, Seiler H, Wen A, et al. Distinct memory traces for two visual features in the *Drosophila* brain. *Nature.* 2006;439: 551–556.
99. Liu L, Wolf R, Ernst R, Heisenberg M. Context generalization in *Drosophila* visual learning requires the mushroom bodies. *Nature.* 1999;400:753–756.
100. Menzel R, Giurfa M. Cognitive architecture of a mini-brain: the honeybee. *Trends Cogn Sci.* 2001;5:62–71.

101. McKinzie ME, Benton JL, Beltz BS, Mellon D. Parasol cells of the hemiellipsoid body in the crayfish *Procambarus clarkii*: dendritic branching patterns and functional implications. *J Comp Neurol*. 2003;462:168–179.
102. Troncoso J, Maldonado H. Two related forms of memory in the crab *Chasmagnathus* are differentially affected by NMDA receptor antagonists. *Pharmacol Biochem Behav*. 2002;72:251–265.
103. Berón de Astrada M, Maldonado H. Two related forms of long-term habituation in the crab *Chasmagnathus* are differentially affected by scopolamine. *Pharmacol Biochem Behav*. 1999;63: 109–118.
104. Squire LR, Davis HP. The pharmacology of memory: a neurobiological perspective. *Annu Rev Pharmacol Toxicol*. 1981;21: 323–356.
105. Pedreira ME, Dimant B, Tomsic D, Quesada-Allue LA, Maldonado H. Cycloheximide inhibits long-term habituation and context memory in the crab *Chasmagnathus*. *Pharmacol Biochem Behav*. 1995;52:385–395.
106. Castellucci VF, Nairn A, Greengard P, Schwartz JH, Kandel ER. Inhibitor of adenosine 3″,5′-monophosphate-dependent protein kinase blocks presynaptic facilitation in *Aplysia*. *J Neurosci*. 1982;12:1673–1681.
107. Frey U, Huang YY, Kandel ER. Effects of cAMP simulate a late stage of LTP in hippocampal CA1 neurons. *Science*. 1993;260: 1661–1664.
108. Locatelli F, Maldonado H, Romano A. Two critical periods for cAMP-dependent protein kinase activity during long-term memory consolidation in the crab *Chasmagnathus*. *Neurobiol Learn Mem*. 2002;77:234–249.
109. Locatelli F, Romano A. Differential role of cAMP-dependent protein kinase isoforms during long-term memory consolidation in the crab *Chasmagnathus*. *Neurobiol Learn Mem*. 2005;83: 232–242.
110. Cadd GG, Uhler MD, McKnight G. Holoenzymes of cAMP-dependent protein kinase containing the neural form of type I regulatory subunit have an increased sensitivity to cyclic nucleotides. *J Biol Chem*. 1990;265:19502–19506.
111. Skålhegg BS, Tasken K. Specificity in the cAMP/PKA signalling pathway: differential expression, regulation, and subcellular localization of subunits of PKA. *Front Biosci*. 2000;5:678–693.
112. Locatelli F, Lafourcade C, Maldonado H, Romano A. Characterisation of cAMP-dependent protein kinase isoforms in the brain of the crab *Chasmagnathus*. *J Comp Physiol B*. 2001;171:33–40.
113. Noselli S. JNK signaling and morphogenesis in *Drosophila*. *Trends Genet*. 1998;14:33–38.
114. Villanueva A, Lozano J, Morales A, et al. JKK-1 and MEK-1 regulate body movement coordination and response to heavy metals through JNK-1 in *Caenorhabditis elegans*. *EMBO J*. 2001;20:5114–5128.
115. English JD, Sweatt JD. Activation of p42 mitogen-activated protein kinase in hippocampal long term potentiation. *J Biol Chem*. 1996;271:24329–24332.
116. Martin KC, Michael D, Rose JC, et al. MAP kinase translocates into the nucleus of the presynaptic cell and is required for long-term facilitation in *Aplysia*. *Neuron*. 1997;18:899–912.
117. Alonso M, Viola H, Izquierdo I, Medina JH. Aversive experiences are associated with a rapid and transient activation of ERKs in the rat hippocampus. *Neurobiol Learn Mem*. 2002;77: 119–124.
118. Atkins CM, Selcher JC, Petraitis JJ, Trzaskos JM, Sweatt JD. The MAPK cascade is required for mammalian associative learning. *Nature Neurosci*. 1998;1:602–609.
119. Crow T, Xue-Bian J-J, Siddiqi V, Kang Y, Neary JT. Phosphorylation of mitogen-activated protein kinase by one-trial and multi-trial classical conditioning. *J Neurosci*. 1998;18: 3480–3487.
120. Sharma SK, Sherff CM, Shobe J, Bagnall MW, Sutton MA, Carew TJ. Differential role of mitogen-activated protein kinase in three distinct phases of memory for sensitization in *Aplysia*. *J Neurosci*. 2003;23:899–907.
121. Feld M, Dimant B, Delorenzi A, Coso O, Romano A. Extranuclear activation of ERK/MAPK is required for long-term memory consolidation in the crab *Chasmagnathus*. *Behav Brain Res*. 2005;158:251–261.
122. Freudenthal R, Locatelli F, Hermitte G, et al. κ-B like DNA-binding activity is enhanced after a spaced training that induces long-term memory in the crab *Chasmagnathus*. *Neurosci Lett*. 1998;242:143–146.
123. Meberg PJ, Kinney WR, Valcourt EG, Routtenberg A. Gene expression of the transcription factor NF-κB in hippocampus: regulation by synaptic activity. *Mol Brain Res*. 1996;38: 179–190.
124. Meffert MK, Chang JM, Wiltgen BJ, Fanselow MS, Baltimore D. NF-κB function in synaptic signaling and behavior. *Nature Neurosci*. 2003;6:1072–1078.
125. Wellmann H, Kaltschmidt B, Kaltschmidt C. Retrograde transport of transcription factor NF-κB in living neurons. *J Biol Chem*. 2001;276:11821–11829.
126. Merlo E, Freudenthal R, Romano A. The IκB kinase inhibitor sulfasalazine impairs long-term memory in the crab *Chasmagnathus*. *Neuroscience*. 2002;:161–172.
127. Merlo E, Romano A. The proteasome inhibitor MG132 impairs long-term memory in the crab *Chasmagnathus*. *Neuroscience*. 2007;147:46–52.
128. Levenson JM, Sweatt JD. Epigenetic mechanisms: a common theme in vertebrate and invertebrate memory formation. *Cell Mol Life Sci*. 2006;63(9):1009–1016.
129. Kouzarides T. Chromatin modifications and their functions. *Cell*. 2007;128:693–705.
130. Strahl BD, Allis CD. The language of covalent histone modifications. *Nature*. 2000;403:41–45.
131. Sterner DE, Berger SL. Acetylation of histones and transcription-related factors. *Microbiol Mol Biol Rev*. 2000;64: 435–459.
132. Norton VG, Imai BS, Yau P, Bradbury EM. Histone acetylation reduces nucleosome core particle linking number change. *Cell*. 1989;108:449–457.
133. Vettese-Dadey M, Grant PA, Hebbes TR, Crane-Robinson C, Allis CD, Workman JL. Acetylation of histone H4 plays a primary role in enhancing transcription factor binding to nucleosomal DNA *in vitro*. *EMBO J*. 1996;15: 2508–2518.
134. Federman N, Fustiñana MS, Romano A. Histone acetylation is recruited in consolidation and reconsolidation as a molecular feature of stronger memories. *Learn Mem*. 2009;16: 600–606.
135. Misanin JR, Miller RR, Lewis DJ. Retrograde amnesia produced by electroconvulsive shock following reactivation of a consolidated memory trace. *Science*. 1968;16:554–555.
136. Nader K, Schafe GE, LeDoux JE. Fear memories require protein synthesis in the amygdala for reconsolidation after retrieval. *Nature*. 2000;406:722–726.
137. Sara SJ. Retrieval and reconsolidation: toward a neurobiology of remembering. *Learn Mem*. 2000;7:73–84.
138. Debiec J, LeDoux JE, Nader K. Cellular and systems reconsolidation in the hippocampus. *Neuron*. 2002;36:527–538.
139. Lee JLC, Everitt BJ, Thomas KL. Independent cellular processes for hippocampal memory consolidation and reconsolidation. *Science*. 2004;304:839–843.

140. Myers KM, Davis M. Behavioral and neural analysis of extinction. *Neuron*. 2002;36:567–584.
141. Salinska E, Bourne RC, Rose SP. Reminder effects: the molecular cascade following a reminder in young chicks does not recapitulate that following training on a passive avoidance task. *Eur J Neurosci*. 2004;19:3042–3047.
142. Merlo E, Maldonado H, Romano A. Activation of the transcription factor NF-κB by retrieval is required for long-term memory reconsolidation. *Learn Mem*. 2005;12:23–29.
143. Merlo E, Romano A. Memory extinction entails the inhibition of the transcription factor NF-κB. *PLoS ONE*. 2008;3:e3687.
144. Arenzana-Seisdedos F, Turpin P, Rodriguez M, et al. Nuclear localization of I kappa B alpha promotes active transport of NF-kappa B from the nucleus to the cytoplasm. *J Cell Sci*. 1997;110:369–378.
145. Dubnau J, Chiang AS, Tully T. Neural substrates of memory: from synapse to system. *J Neurobiol*. 2003;54:238–253.
146. Kandel ER. The molecular biology of memory storage: a dialog between genes and synapses. *Science*. 2001;294: 1030–1038.
147. Menzel R. Searching for the memory trace in a mini-brain, the honeybee. *Learn Mem*. 2001;8:53–62.
148. Roberts AC, Glanzman DL. Learning in *Aplysia*: looking at synaptic plasticity from both sides. *Trends Neurosci*. 2003;26: 662–670.
149. Hemmi J, Tomsic D. The neuroethology of escape in crabs: from sensory ecology to neurons and back. *Curr Opin Neurobiol*. 2012;22:194–200.
150. Perez-Cuesta LM, Hepp Y, Pedreira ME, Maldonado H. Memory is not extinguished along with CS presentation but within a few seconds after CS-offset. *Learn Mem*. 2007;14: 101–108.

4.4 Insects

4.4.1 Drosophila

CHAPTER

27

Drosophila Memory Research through Four Eras

Genetic, Molecular Biology, Neuroanatomy, and Systems Neuroscience

Seth M. Tomchik and Ronald L. Davis

The Scripps Research Institute Florida, Jupiter, Florida

INTRODUCTION

The study of learning and memory using *Drosophila melanogaster* has evolved across four broad approaches since learned behavior was initially demonstrated in 1974.[1] The genetic approach was first heralded by the isolation and description of the first learning mutant, *dunce* (*dnc*), by Dudai, Benzer, and colleagues in 1976.[2] This mutant was found to be impaired in its ability to learn about an odor paired with the aversive stimulus of electric shock, although its ability to detect and respond to the odors and electric shock was judged to be normal. This observation formed the pioneering bridge between the function of a single gene and learning.

The approaches of biochemistry and molecular biology came next, with the discovery by Byers, Davis, and colleagues that the *dnc* mutant was defective in one form of cyclic AMP (cAMP) phosphodiesterase (PDE) activity.[3] This connection was made through the coincidence of genetic mapping. A mutant isolated due to a defect in female sterility and in cAMP PDE activity was mapped to the distal region of the X chromosome, the same approximate region to which *dnc* had been mapped. Subsequent complementation experiments demonstrated that the mutants altered the function of the same gene. This biochemical discovery was followed by the cloning and characterization of the *dnc* gene by Davis, Davidson, and colleagues,[4–6] which demonstrated that the *dnc* gene is one of the most complex genes in the *Drosophila* genome, for reasons that remain unexplained, and that *dnc* encodes the structural gene for cAMP PDE as shown by sequence homology with other PDEs[7] and expression in heterologous systems.[8] The sum of these results indicated that cAMP signaling, at least as mediated by the product of the *dnc* gene, is essential for normal olfactory learning.

These discoveries were informative about the genetics, molecular biology, and biochemistry of learning, but the knowledge base about the brain structures that mediate learning remained empty. Elegant studies in the honeybee by Masuhr, Menzel, and colleagues provided the initial hints.[9] Cooling of the mushroom body (MB) neurons after the honeybee learned an odor/sucrose association produced amnesia, whereas cooling some other parts of the bee brain after learning was generally without effect. This initial hint that MBs might mediate learning catalyzed a genetic screen using *Drosophila* by Borst, Heisenberg, and colleagues for mutants that altered MB structure.[10] The mutants recovered were thus shown to alter olfactory learning ability. The bridge from MBs back to the gene and biochemistry/molecular biology was made through immunohistochemical experiments that showed that cAMP PDE is highly enriched in expression in the MBs.[11]

Systems neuroscience approaches for studying *Drosophila* learning and memory were introduced in the past decade. Two general systems approaches have been introduced that have offered tremendous power. First, transgenes have been developed that allow the experimenter to silence or activate specific sets of neurons during or after learning. For instance, Waddell, Quinn, and colleagues[12] cleverly employed a transgene that expresses *Shibire* (*Shi*^{ts})[13] to conditionally silence the neurons of interest after learning to probe the role for these neurons in learning processes. The second major systems approach introduced by Yu, Davis, and colleagues is functional cellular imaging of living

Invertebrate Learning and Memory.
DOI: http://dx.doi.org/10.1016/B978-0-12-415823-8.00027-7

flies.[14] This approach is based on the expression in specific neurons of GFP derivatives that offer a fluorescent readout of neural activity. This approach offers an optical assay for neural processes in the brain that may underlie learning and memory.

These four broad approaches have culminated in making the fly an ideal system for studying learning and memory using a combination of genetics, biochemistry, molecular biology, neuroanatomy, and systems neuroscience. The field is still young, and there exists much more to be discovered using these approaches.

THE GENETICS OF *DROSOPHILA* LEARNING

The discovery that *dnc* impairs cAMP PDE activity led to the direct testing of other learning-defective mutants for alterations in cAMP signaling. Remarkably, the *rutabaga* (*rut*) mutant was discovered to decrease Ca^{2+}/calmodulin-stimulated adenylyl cyclase (AC) activity.[15] Molecular biological studies then showed that the gene encoded this enzyme[16] and that remarkably, the AC, like the *dnc*-encoded PDE, exhibits enriched expression in the MBs.[17] This second case of mapping a gene product required for learning to the MBs sealed the importance of these structures for olfactory learning. Genetic screens conducted since in many laboratories have identified numerous genes and signaling pathways that are required for *Drosophila* learning and memory. Table 27.1 provides a listing of the genes implicated to date. These genes and gene products have been discussed in detail in prior reviews,[70–72] so only a few salient points are made here.

First, many of the genes involved in olfactory learning are highly expressed in the MBs (Table 27.1). The continued discovery of genes involved in olfactory learning with elevated expression in the MBs has repeatedly emphasized that these neurons are physiologically tuned for learning and memory processes and that disrupting this physiology with a mutation in an MB-expressed gene disrupts learning (see Chapters 2, 5, and 28). Nevertheless, the MBs, although often incorrectly referred to as 'the memory center' or some similar label, do not support olfactory learning in isolation from other brain neurons. Rather, many different cell types in the olfactory nervous system and other neural circuits support olfactory learning such that learning and memory emerges as a systems property.[73] In addition, although many of the identified genes exhibit enriched expression in the MBs, there are few cases in which the gene's function in learning and memory has been mapped specifically to the MBs. The notable exception is *rut*, whose wild-type and specific expression within the adult MBs has been shown to rescue the *rut* learning defect in an otherwise *rut* mutant animal.[74,75] Although we can reasonably accept from the accumulated data that the learning and memory impairments for many of the mutants are likely due to disrupting MB neuron physiology, how these physiological disruptions alter cellular properties such as synaptic plasticity, neuronal structure, or excitability that are important for learning remains relatively unexplored.

Second, some of the genes and their products can be assigned to discrete operations and/or temporal phases of memory. Numerous genes appear to be involved in the process of acquisition and/or short-term memory (STM). It is impossible to distinguish whether a given mutant alters acquisition or STM because there exists no assay for monitoring acquisition in real time and one must judge the level of acquisition based on behavioral memory performance after acquisition. However, acquisition curves that utilize short and discrete training trials offer an approach, albeit imperfect, toward this discrimination.[76] Such curves that suggest involvement in the acquisition process *per se* have been completed for *Nf1*, *rut*,[21] $G_o\alpha$,[20] and *fasII*.[18] Acquisition curves and memory time course experiments for the *amnesiac* (*amn*) mutant, in contrast, indicate predominant effects on memory stability over the first few hours after acquisition.[77] Thus, the role for *amn* appears to be in the temporal phase of intermediate-term memory (ITM), perhaps in the process of consolidation[78] that occurs within this time window. One novel protein, Radish, functions in what is termed anesthesia-resistant memory (ARM), a form of memory considered to be consolidated because it is resistant to anesthesia induced by cold shock. Numerous genes with diverse functions have been identified to participate rather specifically in the process of protein synthesis-dependent long-term memory (PSD-LTM).[62] ARM can also persist over the long term after massed conditioning but is independent of protein synthesis.[79] Overall, the genetic dissection experiments implicate more than 100 different molecules so far, and they indicate that the various temporal phases of memory are molecularly distinct.

Third, two different genes have recently been implicated in the process of active forgetting. There is good reason to believe that memory systems have evolved to both form and actively forget memories, although experimentation into the molecular and cellular biology of active forgetting has only recently taken hold.[45,46] Shuai, Zhong, and colleagues are credited with establishing the first observations on the molecular biology underlying active forgetting by showing that inhibiting the activity of the small G protein Rac1 in the MBs slows the decay of memory, whereas expressing a constitutively active Rac1 in the same neurons accelerates decay.[46] Berry, Davis, and colleagues[45] demonstrated that inhibiting dopamine

TABLE 27.1 Genes and Gene Products Required for *Drosophila* Olfactory Memory[a]

Process	Gene	Gene product	Expression	References
ACQ	*fasciclin II* (*fasII*)	Fasciclin II	High in MBs	18
	dDA1 (*DopR*)	Dopamine receptor	High in MBs	19
	$G_o\alpha$ (G-oα47A)	$G_o\alpha$ protein	—	20
	rutabaga (*rut*)	$Ca2^+$/calmodulin-stimulated AC	High in MBs	15
	neurofibromin (*nf1*)	Ras GAP	Widespread	21,22
ACQ/STM	*arouser* (*aru*)	EPS8 involved in actin dynamics		23
	tribbles (*trbl*)	Involved in protein degradation		24
	bruchpilot (*brp*)	Active zone protein		25
	Gilgamesh (*gish*)	Casein kinase 1γ	High in MBs	26
	Alk	Receptor tyrosine kinase interacting with *nf1*		27
	Nmdar1	Glutamate receptor subunit	Widespread	28
	Nmdar2	Glutamate receptor subunit	Widespread	28
	Tbh	Tyramine β-hydroxylase	Octopamine neurons	29
	dunce (*dnc*)	cAMP PDE	High in MBs	2
	DC0 (*Pka-C1*)	PKA catalytic subunit	High in MBs	30
	PKA-RI	PKA regulatory subunit	High in MBs	31
	Volado/scab (*Vol/scb*)	α-integrin	High in MBs	32
	leonardo (*leo*)	14-3-3ζ	High in MBs	33
	synapsin (*syn*)	Presynaptic vesicle protein	Widespread	25
	Shaker (*Sh*)	Voltage-gated K^+ channel	High in MBs	34
	ether-a-go-go (*eag*)	Voltage-gated K^+ channel	—	35
	DOPA decarboxylase (*Ddc*)	DOPA decarboxylase	—	36
	$G_s\alpha$ (*G-sα60A*)	Stimulatory G protein	—	37
	ignorant (*S6KII*)	Ribosomal S6 kinase	—	38
	nalyot (*Adf1*)	ADF1 transcription factor	Widespread	39
	nebula (*nla*)	Calcineurin inhibitor	—	40
	dPQBP1	Polyglutamine tract binding protein 1	—	41
	ent2	Equilibrative nucleoside transporter 2	High in MBs	42
	Nemy	Cytochrome B561	High in MBs	43
	DRK (*drk*)	Downstream receptor kinase adaptor protein	High in MBs	44
FGT	*damb* (*DopR2*)	D1/D5 type dopamine receptor	High in MBs	45
	rac	Small G protein		46
ITM	*amnesiac* (*amn*)	Neuropeptide(s) related to PACAP	High in DPMn	47
	tomosyn	Tomosyn	–	48
	neuralized (*neur*)	E3 ubiquitin ligase	High in MBs	49
	aPKC	Atypical PKC	—	50
ARM	*radish* (*rsh*)	Novel protein	High in MBs	51

(Continued)

TABLE 27.1 (Continued)

Process	Gene	Gene product	Expression	References
PSD-LTM	*Notch (N)*	Notch cell surface receptor	—	52,53
	GLD2 polyA polymerase	GLD2 polyA polymerase	—	54
	armitage (armi)	RISC helicase	—	55
	tequila (teq)	Neurotrypsin relative	MBs after training	56
	corkscrew (csw)	SHP2 tyrosine phosphatase	—	57
	Orb2	Cytoplasmic poly(A) element binding protein (CPEB)	—	58
	debra (dbr)	Novel protein possibly involved in ubiquitination		59
	dCREB2	cAMP-response element binding protein	—	60
	CaMKII	Ca^{2+}/calmodulin-dependent protein kinase II		55
	crammer (cre)	Cysteine protease inhibitor	In MBs and glia	61
	pumilio (pum)	RNA binding protein/translational repressor	High in MBs	62
	oskar (osk)	RNA binding protein	High in MBs	62
	eIF-5C	Translation initiation factor	High in MBs	62
	staufen (stau)	RNA binding protein	—	62
	dFabp	Fatty acid binding protein	—	63
	Su(H)	Suppressor of Hairless DNA binding protein	Widespread	64
	klingon (klg)	Homophilic cell adhesion molecule		65
	ben (be)	Novel protein	Widespread	66
	dube3a	Ubiquitin ligase E3	High in MBs	67
	yu (yu)	A-kinase anchoring protein	High in MBs	68
	chi (chi)	Protein tyrosine phosphatase	Widespread	69
	54 other LTM mutants	Various functions	Many high in MBs	62

[a]*Some genes known to influence olfactory learning and memory that are thought or known to have their effects from abnormal neural development are not included in the table, including* linotte/derailed, chico, *and* turnip. *It is also likely that some of the genes listed in the table have their effects on learning and memory by altering neural development, but the data delineating developmental versus physiological roles are not yet available.*

ACQ, acquisition; ARM, anesthesia-resistant memory; FGT, active forgetting; ITM, intermediate-term memory; PSD-LTM, protein synthesis-dependent long-term memory; STM, short-term memory.

signaling to the MBs after learning slows memory decay, whereas increasing this signaling erases memory. These observations represent a fruitful start in dissecting the molecular biology of active forgetting.

DROSOPHILA LEARNING

Numerous behavioral assays have been developed for testing the learning and memory capacity of *Drosophila*. These were described and discussed in detail by Davis[80] and are also discussed in Chapters 2 and 28; they include assays that (1) employ appetitive (sugar reward) and aversive (electric shock) stimuli, (2) challenge different sensory systems (olfaction, vision, etc.), and (3) measure different types of learning (classical conditioning, operant conditioning, habituation, sensitization, etc.). However, the majority of conceptual advances have come from employing olfactory classical conditioning using aversive or appetitive cues, and this is the focus here.

Classical aversive olfactory conditioning[81] is generally performed by presenting odors in succession, delivered with or without electric shock through a copper grid, to groups of flies sequestered in a small plastic tube (Figure 27.1A). The odor used as the conditioned stimulus (CS+) is presented for 60 sec along with shock pulses delivered every 5 sec such that electric shock serves as the unconditioned stimulus (US) (Figure 27.1B). The flies then receive 60 sec of a second odor, the CS−, without associated shock pulses. This conditioning, designated as 'long program training,' generates a memory of the CS+ odor associated with the aversive electric shock stimulus so that flies avoid the CS+ over the CS− when tested

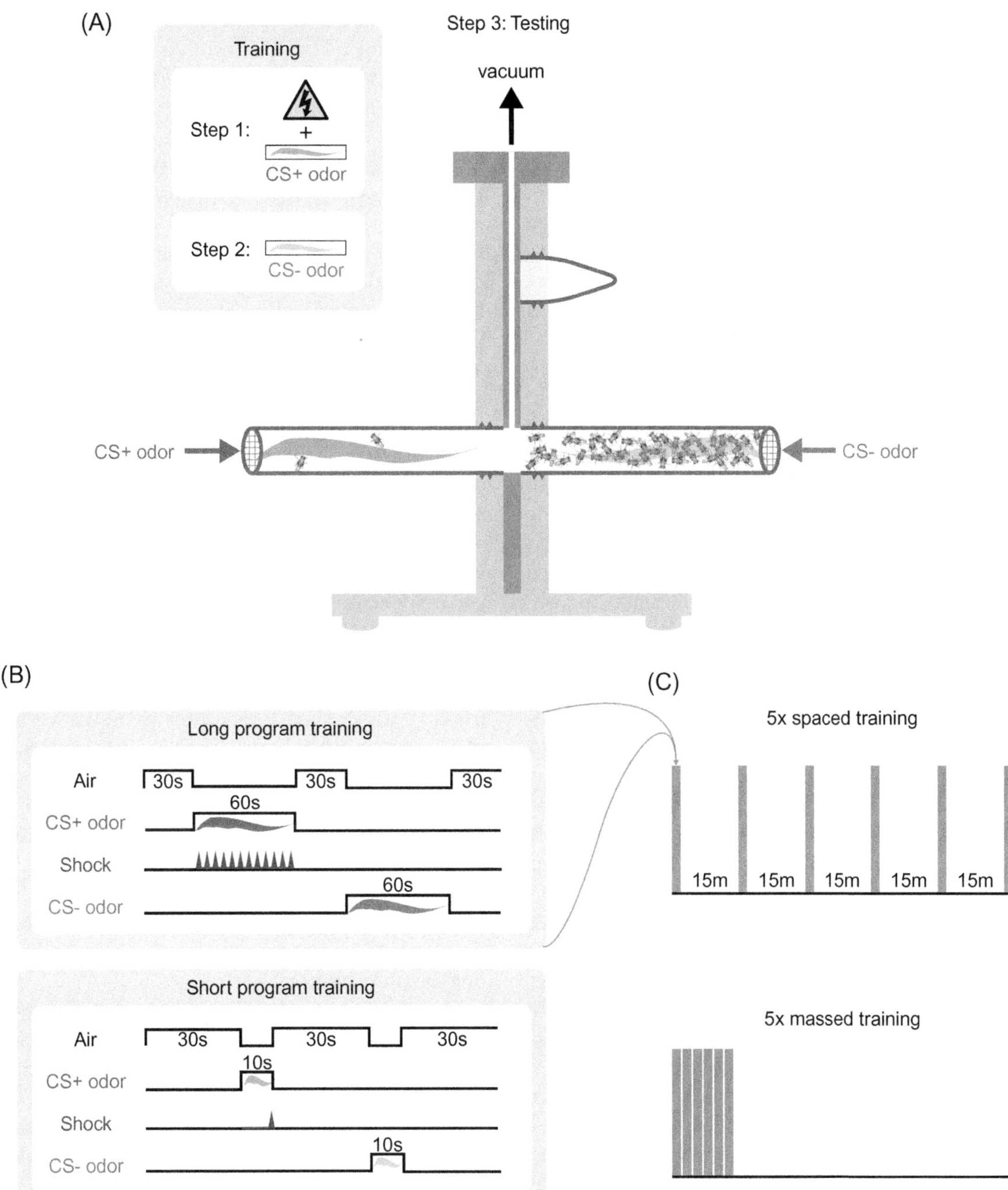

FIGURE 27.1 **Olfactory aversive classical conditioning assay in *Drosophila*.** (A) Flies are trained in a tube lined with a shock grid (not shown), where they receive a pairing of one odor (the CS+) with electric shock, followed by presentation of a second odor (the CS−) that is not paired with shock. Subsequently, the flies are tested in a T-maze, where each odor is pulled through one of two tubes. If flies remember the training, they move into the tube containing the CS− odor. (B) The training protocol involves either a long program with 60 sec of odor and 12 electric shocks or a short program with 10 sec of odor and 1 electric shock (or a variant of this). The long program is used in most experiments, whereas the short program is used to interrogate memory acquisition. (C) Long-term memory is generated using multiple (5–10) training trials, which are presented either back-to-back (with a 30-sec delay between trials; massed training) or spaced out over a longer period of time (with 15 min between each trial; spaced training).

subsequently in a T-maze (Figure 27.1A). *Drosophila* learning in this situation is robust and reproducible, with memory persisting for more than 24 hr. An alternative schedule for training referred to as 'short program training' employs short and discrete presentations of odor for 10 sec, each associated with only one shock pulse.[76] This alternative schedule produces modest performance gains with a single training trial, but multiple trials can be added back-to-back to evaluate the performance gains as a function of trial number. This provides the methodology for producing acquisition curves as discussed in the preceding section.

Olfactory memory can be extended by using 5–10 training trials performed back-to-back with no rest between trials (massed training) or allowing a rest between each training trial of usually 15 min (spaced training) (Figure 27.1C). Spaced conditioning produces PSD-LTM, whereas mass conditioning produces ARM or protein synthesis-independent LTM (PSI-LTM). There is debate in the literature regarding whether ARM and PSD-LTM coexist or whether one is formed at the expense of the other.[72,82,83]

The molecular basis for the spacing effect in generating PSD-LTM remained a mystery until Pagani, Zhong, and collaborators demonstrated that the phosphatase SHP2 is involved in the spacing effect.[57] Overexpression of the wild-type version of the SHP2 protein tyrosine phosphatase encoded by the gene *corkscrew* (*csw*) in the MB neurons shortened the optimal rest period required for LTM induction, whereas expression of a constitutively active phosphatase prolonged the optimal rest. The genetic effects on optimal spacing were interpreted molecularly as due to producing different kinetics on MAPK activity with each training trial, with repetitive waves of MAPK activation being necessary for PSD-LTM.

Olfactory classical conditioning with an appetitive US has recently become of greater widespread interest and allows interesting behavioral, molecular, and imaging contrasts to be made with aversive conditioning. For appetitive conditioning, a dry filter paper previously saturated with sugar replaces the electric shock grid used for aversive conditioning, but other aspects of the conditioning procedure remain essentially the same as those used for aversive training. Three interesting discoveries have been made about appetitive compared to aversive conditioning.

First, whereas dopamine (DA) signaling is essential for aversive conditioning and is believed to represent part of the US signal, it is not required for appetitive conditioning.[29] This was shown by blocking synaptic release from dopaminergic neurons (DANs) by expressing *Shibire*^{ts} in these neurons and training the flies at the restrictive temperature. *Shibire*^{ts} codes for a dominant-negative form of dynamin that blocks synaptic release at restrictive temperatures. Blocking synaptic release from DANs during aversive conditioning impaired memory formation, whereas the same block during appetitive conditioning did not. Rather, octopamine was shown to be required for appetitive conditioning instead of DA. This indicates that different neuromodulators are required for learning the valence of the same odor used as the CS+.

Second, whereas a single session of long program conditioning using an aversive US produces a memory that decays over approximately 24 hr, a single conditioning session with sucrose produces a nondecaying memory that persists for more than 3 days.[84,85] This long-lasting, appetitive memory formed from a single training session requires protein synthesis at the time of training and the activity of the transcription factor, dCreb2, in the α/β MB neurons. More recent studies have demonstrated that the essential ingredient for instilling a long-lasting memory from a single trial is not the hedonic character of the sugar reward but, rather, its nutrient value.[86,87] Sugar rewards that are sweet tasting but have no nutrient value, such as arabinose, instill only a decaying memory, similar to that observed after aversive conditioning. However, when supplemented with a nutrient carbon source such as sorbitol that is tasteless to flies, a long-lasting memory is achieved. Thus, the flies evaluate in some unknown way the nutrient value of the sugar used as the US for reward learning, and this calculation of nutrient versus non-nutrient somehow enters the memory processing circuit to determine the stability of memory formed.

Third, memory traces revealed by functional cellular imaging (discussed later) after aversive versus appetitive conditioning have distinct differences. Notably, the dorsal posterior medial (DPM) neuron memory trace forms in both neurite branches after appetitive conditioning and is longer lasting than the branch-specific trace that forms from aversive conditioning.[88] Thus, the qualitative and quantitative differences of the DPM neuron appetitive memory trace compared to the aversive memory trace seem likely to be related to long-lasting memory.

More complex behavioral variations of classical conditioning, such as extinction, second-order conditioning, trace conditioning, and latent inhibition, although well studied in other insects such as the honeybee, have been only infrequently queried in *Drosophila*. However, a few points are worth noting. Extinction of consolidated memory has been demonstrated by Qin and Dubnau.[89] Interestingly, mutants that impair the acquisition of olfactory aversive classical conditioning, such as *rut* and *NMDAr*1 (Table 27.1), leave extinction of PSD-LTM intact. This indicates that the molecular processes underlying initial learning are distinct from those underlying extinction of PSD-LTM. Flies have been shown to be capable of second-order conditioning,[90] but the genetics of this complex behavior have not been investigated. The molecular genetics of trace conditioning was recently explored.[91] Interestingly, the residual memory in the *rut* mutant forms even when trace intervals between the CS+ and US are extended to 60 sec, whereas the form of memory dependent on *rut* requires that the US follow the CS+ by less than 15 sec. Furthermore, the inhibition of Drac1 in the MBs lengthens the trace interval required for acquisition, presumably by stabilizing the *rut*-dependent form of memory.

THE OLFACTORY NERVOUS SYSTEM

The anatomy of the *Drosophila* olfactory nervous systems was reviewed in Davis,[92] so only a brief account is provided here. Flies have approximately 2600 olfactory receptor neurons (ORNs) in the antennae and maxillary palps that serve to detect the presence of odors in the environment. These neurons send axons to the antennal lobe (AL) to terminate in approximately 43 glomeruli that contain the synapses made between ORNs and neurons in the AL (Figure 27.2A). The ORNs form excitatory synapses with local interneurons, which are axonless and either inhibitory or excitatory, and the projection neurons (PNs). The PNs generally innervate a single glomerulus with their dendrites and transmit olfactory information to the MB neurons and neurons in the lateral horn. The MB neurons are categorized into three distinct classes based on morphology and positioning of their axons. Each MB neuron extends a dendrite into the calyx, the neuropil area in which it synapses with PNs to receive incoming information. The α/β MB neurons extend an axon in an anterior direction to a location just dorsal to the AL, at which point the axon divides to form a vertically oriented α branch and a horizontally oriented β branch. The neuropil regions that contain the α and β branches of the α/β MB neurons are referred to as the α′β′ neurons. The α′/β′ MB neurons are similar in structure to the α/β MB neurons, but their axon divides to form vertically and horizontally oriented branches in the α′ and β′ neuropil regions, respectively. These neuropil regions lie adjacent to the α and β neuropil. The third class of MB neurons, the γ MB neurons, does not have a branched axon. Their axons extend through the peduncle but turn medially to form the γ lobe.

Three classes of neurons that are extrinsic to the MBs have been extensively studied for their roles in learning and memory processes (Figure 27.2B). The two dorsal paired medial (DPM) neurons each have a large cell body that is located in the dorsal and medial aspect of each brain hemisphere. These neurons extend a single neurite that branches before ramifying the MB neuropil, with one branch innervating the vertical lobes of the MBs and the other ramifying the horizontal lobes (Figure 27.2B). The anterior paired lateral (APL) neuron provides GABAergic input to the MBs.[93] Its cell body resides lateral to the cell bodies of the MB neurons and, like the DPM neuron, appears to broadly innervate the MB neuropil through two main branches. There are eight clusters of DA neurons in the fly brain,

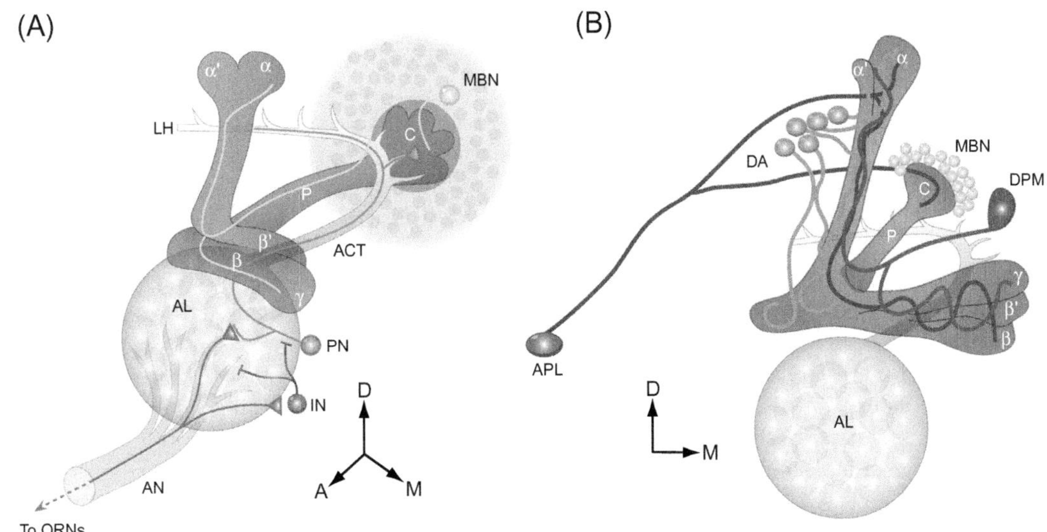

FIGURE 27.2 Structure of the *Drosophila* olfactory pathway. (A) Olfactory information is transduced by olfactory receptors neurons (ORNs) and enters the brain through the antennal nerve (AN). ORN axons enter the antennal lobe (AL), where they synapse with projection neurons (PN) in discrete glomeruli as well as interneurons (IN). Projection neurons carry this information to the mushroom body neurons (MBN), forming synapses with the MBN in the calyx (C) at the posterior edge of the brain, as well as the lateral horn (LH). Each MBN sends a single axon anterior through the puduncle (P). Near the anterior face of the brain, this neurite turns to innervate one or more lobes (vertical: α or α′; horizontal: β, β′, or γ) according to the MBN cell type (α/β, α′/β′, or γ). Directional abbreviations: A, anterior, D, dorsal, M, medial. (B) Mushroom body extrinsic neurons that are involved in learning and memory. APL, anterior paired lateral neuron; DA, dopaminergic neurons; DPM, dorsal paired medial neuron. DAL neurons are omitted in this schematic for the sake of clarity. Source: *Reprinted from Davis*[73] *with permission.*

but three of these are most important relative to MB neuron physiology and learning and memory. The protocerebral anterior medial neurons innervate a medial zone of the horizontal lobes, the protocerebral anterior lateral (PPL1) neurons innervate the vertical lobes, and the protocerebral posterior lateral 2ab (PPL2ab) cluster innervates the MB calyx. The PPL1 neurons are further divided into five distinct classes based on their targets: the tip of α lobe, the tip of α′ lobe, the upper stalk, the lower stalk/heel area, and the spur/distal peduncle (Figure 27.2B).

NEURAL CIRCUITS UNDERLYING LEARNING AND MEMORY

Drosophila research has been propelled to the forefront of systems-level analysis of brain function by the advent of multiple critical tools, including novel binary expression systems,[94–96] inducible expression systems,[75,97] genetically encoded optical imaging reporters,[98,99] neuronal activity modulators,[100–103] and electrophysiological approaches.[104] These techniques have pushed the *Drosophila* model system forward dramatically, enabling researchers to approach a range of previously intractable questions. In the remainder of the chapter, we highlight some conceptual areas that have seen recent significant progress.

The contribution of the different types of MB neurons to learning has been studied with several approaches. First, multiple studies have examined the temporal requirements for normal synaptic transmission from each set of MB neurons during learning, consolidation, and retrieval. These experiments used Shibire^{ts} to conditionally silence the synaptic output of a given subset of MB neurons with high temporal precision. Synaptic output from MB α/β neurons is required for memory retrieval but not for acquisition or consolidation.[105,106] In contrast, α′/β′ neurons are required to acquire and stabilize memories during consolidation, but they are dispensable during retrieval.[107] The DPM neuron is also involved in memory consolidation and is addressed later.

A major area of current investigation concerns the spatiotemporal encoding and storage of memories within the MBs focusing on two questions. First, are short- and long-term memories stored within the same brain regions or different regions (e.g., in two different subsets of the MB neurons)? Second, if the memories are stored in different brain regions, are they simultaneously encoded in both or initially encoded in one region and subsequently transferred to another? Experiments to address these questions use one of two approaches: blocking synaptic output with Shibire^{ts} or rescue with wild-type protein function in mutants with learning deficits (e.g., *rut*). The Shibire^{ts} experiments establish the necessity of a given area for learning and memory, whereas rescue experiments address sufficiency for memory performance.

Using *rut* rescue, Zars *et al.*[108] reported that restoration of Rut expression in γ neurons rescues STM performance in *rut* mutants. Conversely, the MB vertical lobes were implicated in LTM based on data from *alpha lobes absent* (*ala*) mutant flies that are missing the vertical lobes.[109] Blum *et al.*[110] reported that expression of Rut specifically in γ neurons was sufficient to rescue STM in *rut* mutants, whereas expression in α/β neurons was sufficient to rescue LTM. Trannoy *et al.*[111] reported similar results using appetitive olfactory training. Collectively, these experiments suggest that STM and LTM storage may be dissociated into the γ and α/β lobes, respectively. However, other experiments have painted a more complex picture. McGuire *et al.*[75] and Akalal *et al.*[112] reported that expression of wild-type Rut was required in both γ and α/β neurons for complete rescue of memory in *rut* mutants. In addition, Neurofibromin is necessary specifically in α/β neurons for memory acquisition and STM.[21] These experiments suggest a role for α/β neurons, in addition to γ neurons, in STM. The overall view that emerges from the literature seems to be that γ neurons are crucial for STM, whereas α/β neurons are particularly important for LTM but may play a role in STM as well. Necessity of each of these regions is clear, whereas sufficiency of each region is currently debatable.

If MB α/β neurons are required for LTM, but perhaps less important for STM storage, then when is the memory encoded in these neurons? Does the initial acquisition trigger a process in α/β neurons that results in later expression of LTM, or is the memory encoded first in the γ neurons and subsequently transferred to α/β during a systems consolidation process? Several sets of experiments have examined this question in various ways, with apparently conflicting results. First, Trannoy *et al.*[111] blocked synaptic transmission from γ neurons during training and consolidation and found that it did not impair LTM. This suggests that the memory is not transferred from γ to α/β neurons. Data from imaging experiments suggest that the putative machinery for CS/US coincidence detection may be in place in the vertical lobes (including both the α/β and α′/β′ neurons). Pairing of neuronal depolarization with DA application to the tip of the vertical lobes generates a synergistic elevation of cAMP in the α and α′ lobes that is *rut* dependent[113]; bath application produces a similar *rut*-dependent effect on PKA activity.[114] This suggests that CS/US coincidence detection through *rut* occurs at the time of training in both α/β and α′/β′ neurons. Alternatively,

Qin *et al.*[115] reported that providing DopR activity in γ neurons of DopR mutants is sufficient to rescue both STM and LTM, suggesting that memory is initially encoded in γ neurons and then transferred to α/β neurons.

The difficulty in reaching a consensus on the precise role of each MB neuron subset in short- versus long-term memory may lie in the number of experimental factors at play. None of the currently available enhancer trap lines drives GAL4 expression in all desired neurons without any undesired expression outside of the target region (i.e., we have no driver that expresses in all α/β neurons but no γ neurons and vice versa). Furthermore, multiple CS+/CS− odor pairs have been used in different experiments. Thus, these conflicting results could be explained, at least in part, by different experimental conditions, particularly the chosen GAL4 drivers and pairs of odors. Indeed, Akalal *et al.*[112] found that providing Rut activity in α/β neurons produced partial rescue with some CS+/CS− odor pairs but not others. Furthermore, blocking synaptic output with Shibire^{ts} blocked STM retrieval with some combinations of α/β-expressing GAL4 drivers and odor pairs but not others.

In addition to the MBs, other brain areas are involved in olfactory associative memory, and it may be difficult to assign precise roles to individual neuron types because of the systems interactions that may occur in the circuit. DPM neurons are included in the circuitry as previously mentioned and were postulated to offer reciprocal interactions with one or more classes of MB neurons.[77] Because DPM and MB α′/β′ neuron synaptic output is required at equivalent time points after training, it may be that the dominant reciprocal interactions occur between these two neuron types.[107] Complicating this is the claim that DPM neurons are also coupled to the GABAergic APL neuron via gap junctions, and that this coupling is necessary for the formation of anesthesia-sensitive memory.[116] Thus, the DPM neurons form connections with multiple cell types in the MB and appear to be involved in memory consolidation in some way.

The MB lobes and calyx are densely innervated by processes of the APL neuron.[93] Acting via the $GABA_A$ receptor RDL, GABA released from the APL neuron suppresses the responses of MB neurons to olfactory stimuli and inhibits learning in the fly.[93,117] In addition, a memory trace was observed in the APL neuron. Thus, GABAergic circuits vary the strength of olfactory memory via modulation of the CS pathway—that is, the MB responsiveness to odor.[118] The overall role of this modulation could be to provide an extra layer of control over the circuits responsible for learning, enabling the animal to concentrate on storing behaviorally relevant associations. Resolving how the APL neuron modulates CS strength and the DPM neuron modulates consolidation while being interconnected via gap junctions to each other presents a difficult conceptual problem.

Several studies have implicated components of the central complex in learning and memory. Heisenberg *et al.*[10] found that mutant fly lines with structural defects in the central complex exhibited defects in olfactory memory. Wu *et al.*[28] demonstrated that NMDA receptors are required specifically in the ellipsoid body for LTM. The ellipsoid body is a component of the central complex and is involved in motor control in insects. It remains unclear how the MB circuits interface with the ellipsoid body. However, given that memory retrieval after classical conditioning (avoidance of the CS+ and approaching the CS−) involves motor activity, it may be that the ellipsoid body neurons are involved during memory retrieval.

In addition to neurons, glia are required for normal LTM. Mutants in the *crammer* gene (*cer*) are deficient in LTM.[61] The Crammer gene product, a cathepsin inhibitor, is expressed in both the MBs and glia around the brain. Interestingly, expression of wild-type Cer in the glia but not the MB rescues memory performance in the *cer* mutant flies. Therefore, glia have a role in supporting LTM. These observations also highlight the fact that memory requires many functional links in the brain that reside outside the ORN−AL−MB circuit.

The dorsal anterior lateral (DAL) set of MB extrinsic neurons is involved in LTM, adding an additional anatomical link to the circuit.[119] Inhibiting protein synthesis in these neurons following training diminishes LTM, suggesting that this is a site for protein synthesis-dependent changes in neuronal circuit connectivity that underlie part of LTM stability. Multiple lines of evidence demonstrate the criticality of MBs—and specific gene products within the MBs—for LTM (see Table 27.1). However, protein synthesis-dependent events must occur in several different areas of the olfactory nervous system in support of PSD-LTM.[120] It would not be possible for DAL neurons by themselves to encode the range of odors that can be learned by flies. However, the DAL neurons seem to be part of an MB−DAL−MB circuit loop required for normal LTM, which may also include additional neuronal elements.[121]

Recently, our knowledge of the memory-encoding circuits in the *Drosophila* olfactory system progressed one synapse downstream of the MBs. The MB-V2 MB extrinsic neurons are a set of three cholinergic neurons, postsynaptic to the MB, that project to the lateral horn and parts of the protocerebrum.[122] Synaptic output of MB-V2 neurons is required for retrieval of memories, and a memory trace was found in these neurons following conditioning.[123] These experiments provide the first concrete step in following olfactory memory down the pathways in the brain beyond the MBs. Considering that some innate olfactory behaviors involve circuits consisting of as few as four neurons in the brain,[124] it is plausible that we will soon be able to delineate the entire olfactory classical conditioning pathway, from olfactory receptor neuron to motor neuron output.

The final major class of MB extrinsic neurons that influence learning and memory is the class consisting of DANs. DA plays a prominent role in learning and memory across vertebrate and invertebrate taxa. In *Drosophila*, DA is necessary for both aversive and appetitive conditioning. Mutants for the DA receptor DopR exhibit no memory at all in aversive conditioning paradigms, and appetitive memory is partially impaired.[19] This demonstrates how critical DA is for memory, because the DopR mutation is the only known mutation to completely abolish memory. The precise function of DA within the neuronal circuits that encode and store memories is not completely understood, although much progress has been made in the past decade. DANs that innervate the MB are activated during aversive events in *Drosophila*, such as when a fly receives electric shock during aversive olfactory classical conditioning.[125,126] Furthermore, DA is sufficient as an aversive reinforcer: stimulating DANs in lieu of providing electric shock generates aversion to the paired odor.[102,127] These observations have led to the hypothesis that DA conveys the aversive US to the MB, where it is integrated with information about the CS+ that is relayed to the MB via AL projection neurons.

The view that DA functions as an aversive reinforcer has been updated and further elaborated recently, driven in large part by the generation of new GAL4 drivers that allow manipulation of small subsets of DANs.[122] Aso *et al.*[128] reported that only two subsets of DANs, MB-MP1 and MB-M3 (innervating the heel and distal horizontal lobe regions, respectively), are sufficient for reinforcement in aversive learning paradigms. However, the MB-MP1 neurons are also involved in an apparently different aspect of learning—gating of memory retrieval following appetitive conditioning.[129] When starved flies are trained in an appetitive classical conditioning paradigm (odor + sucrose), the animals will approach the conditioned odor if they are kept in a starved state but will not if they are fed. This is classic state-dependent memory; the flies will only act on the memory if they are in an appropriate state, perhaps a state of motivation to acquire the sucrose reward. This state-dependent memory is dependent on NPF signaling and the DA MB-MP1 neurons. Studies have further demonstrated that stimulating a triad of DAN types—MB-V1, MB-MV1, and MB-MP1—induces forgetting of a previously acquired memory.[45] The contribution of each of these subsets of neurons is unclear because stimulating each set of DANs individually has no effect. Thus, the MB-MP1 neurons are involved in aversive reinforcement, gating of appetitive memory retrieval, and the modulation of memory stability. These findings point to a previously unrecognized level of complexity in the role of dopamine in learning and memory.

In addition to its explicit roles in learning and memory, DA directly modulates sensory perception. Two papers demonstrate that DA alters the sensitivity of neurons in the gustatory pathway to sucrose based on the hunger state of the animal.[130,131] Modulation of these neurons has the net effect of making sucrose potentially more rewarding to a hungry animal. This demonstrates that DA modulates multiple circuits involved in reward and punishment, ranging from taste receptors to the gustatory pathways in the subesophageal ganglion and higher order regions of memory processing (e.g., the MB). In addition, flies discriminate different sugars in appetitive learning paradigms, modulating their memory strength according to the nutritive value of the sugars.[86,87] This effect is dependent, at least in part, on postingestive effects of the sugars.

DA has been implicated in another aspect of rewarded behavior—the expression of conditioned preference for alcohol-paired odors.[132] When ethanol is used as a reinforcer in olfactory classical conditioning assays, flies initially exhibit a transient aversive behavioral response (avoiding the odor that was paired with ethanol). However, when tested 1 day later, the flies exhibit conditioned preference for the odor that was previously paired with ethanol, suggesting that the ethanol intoxication was positively reinforcing after a delay. Synaptic output from DA neurons was not required for the initial aversive response, but it was required later, at the time of retrieval, for the expression of conditioned preference. Thus, the role of DA in learning and memory appears to depend on the behavioral context.

The outlook for invertebrate models systems to significantly contribute to our understanding of the behavioral roles of DA in learning is excellent. DANs encode multiple aspects of stimuli in primates.[133,134] The *Drosophila* model system provides the advantage of being able to record from and manipulate individual, genetically identifiable neurons that have clearly delineated behavioral roles. In addition, there are approximately 282 DA neurons in the entire *Drosophila* brain; only 12 per hemisphere comprise the PPL1 cluster that is involved in olfactory aversive memory. Estimates of the number of DA neurons in the rat VTA alone are on the order of 18,000.[135] The ability to manipulate individual DA neurons in the fly, combined with the relatively small number of these neurons, has put big picture questions about the role of DA in learning squarely within the realm of experimental amenability. For further background in this area, readers are encouraged to consult Waddell.[136]

SUBCELLULAR SIGNALING DYNAMICS

The dynamics of cellular signaling phenomena underlying learning and memory has become accessible to *Drosophila* researchers due to the advent of genetically encoded fluorescent reporters. Multiple genes involved in cAMP signaling were identified in early genetic screens for learning and memory mutants (Table 27.1). One of these is the *rut*-encoded AC, a Ca^{2+}/calmodulin-sensitive cyclase that is similar to AC1 in mammals. Rut is sensitive to stimulation by G protein-coupled receptors via $G_{\alpha s}$ and to excitation state of the neuron via its Ca^{2+} sensitivity.[15,16] These two pathways would be activated in a subset of MB neurons during olfactory associative learning. The CS+ triggers action potentials in a subset of MB neurons that encode the odor, and the US (electric shock) triggers release of DA in the MB. On the basis of these results, *rut* was proposed to be a molecular coincidence detector for olfactory associative learning—a mechanism to trigger the plasticity underlying memory acquisition in the appropriate circumstances. However, this hypothesis remained untested until recently.

If the Rut AC indeed functions as a molecular coincidence detector, it would be expected to register the CS+ and US, generating more cAMP when they are paired than when either stimulus is presented individually. This integration could conceivably be either additive or synergistic; in either case, the level of cAMP must reach some critical threshold to produce the cellular plasticity that underlies memory. Studies in *Aplysia* suggested that the levels of cAMP generated by $Ca^{2+}/G_{\alpha s}$-sensitive ACs are in fact synergistic—that is, greater than the sum of the individual pathways stimulated in isolation. When serotonin, which activates $G_{\alpha s}$-coupled receptors, and Ca^{2+} were applied to cell membrane preparations, a synergistic elevation of cAMP was observed.[137,138]

Recent studies have provided support for the long-standing but previously untested hypothesis that Rut functions as a coincidence detector in *Drosophila*. Focal application of acetylcholine to the MB calyx to stimulate MB neurons followed by application of DA to the MB vertical lobes generates a synergistic elevation of cAMP (Figure 27.3).[113] The effect was spatially and temporally precise—reverse pairing, DA first followed by acetylcholine, or application of the two neurotransmitters to the calyx did not produce any synergy. Elevation of cAMP increased the ACh-evoked Ca^{2+} influx in the MBs, suggesting that the net effect of this elevation is to facilitate MB output. A follow-up study reinforced these conclusions by demonstrating a synergistic elevation of PKA when acetylcholine and DA were bath-applied to the brain.[114] This study reported activation of PKA specifically in the vertical and not the horizontal lobes of the MBs in response to DA.

MEMORY TRACES

At its core, a memory is a set of physiological changes in the brain that are often referred to as the engram or memory trace. The advent of *in vivo* functional imaging, its application to insects,[139] and its combination with genetically encoded fluorescent proteins in *Drosophila*[99] have enabled the direct observation of memory traces in the brain. The general principle behind these experiments is to use genetically encoded reporters to image neuronal activity in different brain regions, train the animal(s) using a classical conditioning paradigm, and then search for changes in neuronal activity following training. These changes in activity presumably reflect the physiological representation of memory in the brain. In olfactory classical conditioning, a fly must learn that an odor is paired with a reward or punishment and alter its behavior appropriately when it detects a subsequent presentation of that odor. Thus, to examine memory traces, researchers have examined responses to the conditioned odors, comparing the odor-evoked responses before and after training and measuring the onset and duration of the altered responses after conditioning (Figures 27.4 and 27.5). Direct observation of memory traces provides detailed information about how the properties of neurons and circuits are altered as memories are encoded, stored, and retrieved.

Several memory traces have been localized to the intrinsic components of the olfactory pathway, specifically the AL and MBs. The first memory trace was discovered in projection neurons of the AL.[14] This trace was manifested as a change in the pattern of synaptic vesicle release across the AL glomeruli (Figure 27.4). Specifically, an additional glomerulus, the D glomerulus, was recruited into the odor representation in the AL following training. This trace was specifically linked to STM because it was observed for only a few minutes following training (Figure 27.5). A change in the pattern of protein synthesis across AL glomeruli following training was found by Ashraf *et al.*,[55] linking plasticity in the AL to long-term PSD-LTM

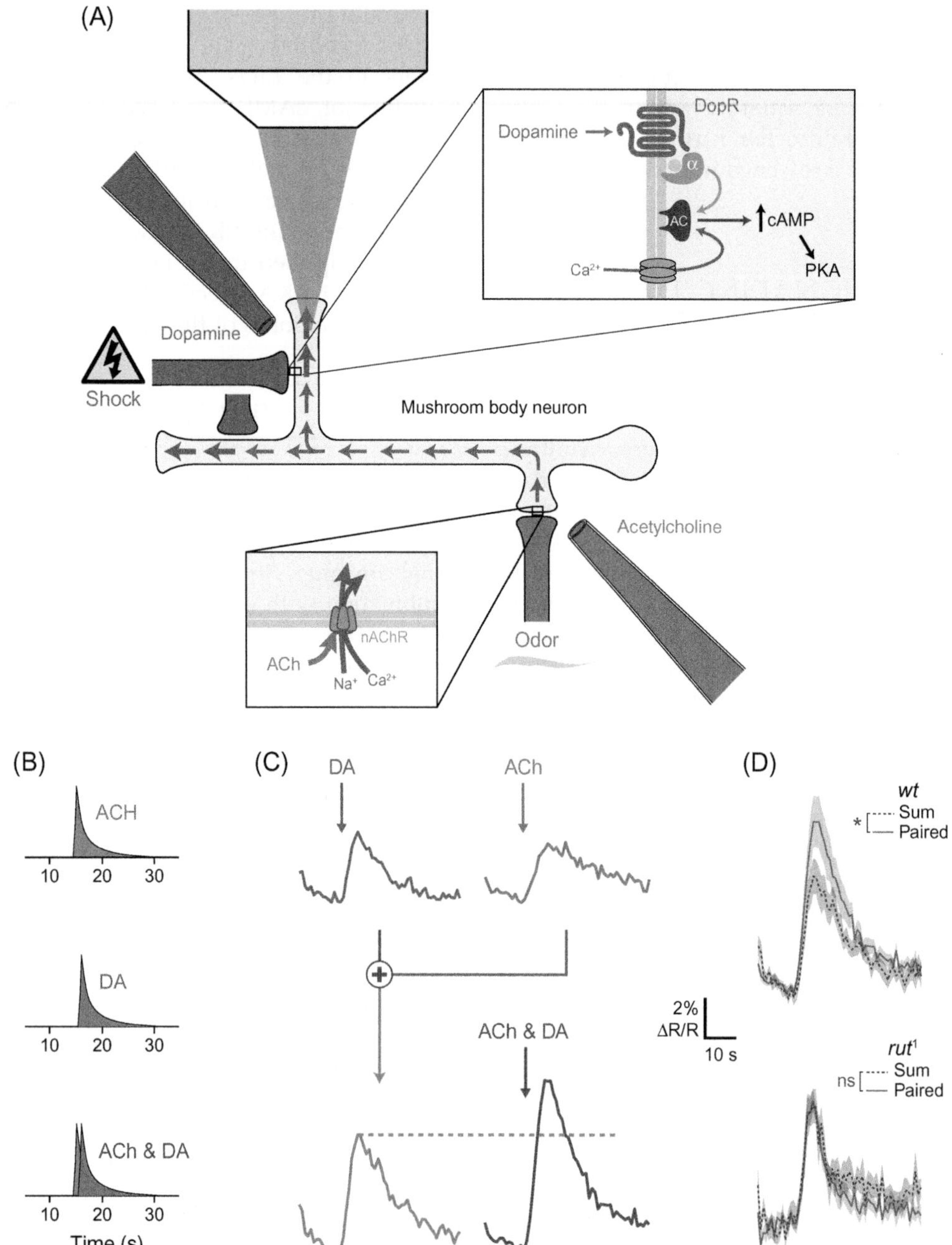

FIGURE 27.3 Schematic of the imaging preparation used to examine the dynamics of subcellular signal integration underlying learning. (A) Schematic of a single mushroom body neuron (MBN). The MBNs receive input from cholinergic projection neurons, which activate nicotinic acetylcholine receptors on the MBN dendrites. The cholinergic input is mimicked by focal application of acetylcholine to the mushroom body calyx. Dopaminergic neurons innervate the mushroom body lobes and activate DopR receptors on the MBNs, elevating cAMP via one or more adenylyl cyclases (AC). This pathway is activated with focal application of dopamine to the lobes. A genetically encoded cAMP reporter is expressed in MBNs, and the neurons are imaged with confocal microscopy. (B) Stimulation protocol. Three trials are performed: one application of acetylcholine (ACh), one of dopamine (DA), and one paired application of ACh and DA. (C) Synergistic elevation of cAMP when ACh and DA are applied to the calyx and lobes, respectively. The mean responses are plotted. The lower left trace (cyan) is the mean of the sum of the ACh and DA responses. (D) The synergistic elevation of cAMP is eliminated in *rut*[1] mutants. Responses are plotted as mean ± SEM. Source: *Data are modified from Tomchik and Davis*[113] *and reprinted with permission.*

(Figure 27.5). In the MBs, an STM trace was found in the α′/β′ neurons.[26,141] This α′/β′ memory trace was recorded as an increase in the ratio of Ca^{2+} responses to the CS + versus the CS− in both α′ and β′ lobes that lasted for approximately 1 hr after conditioning (Figures 27.4 and 27.5).

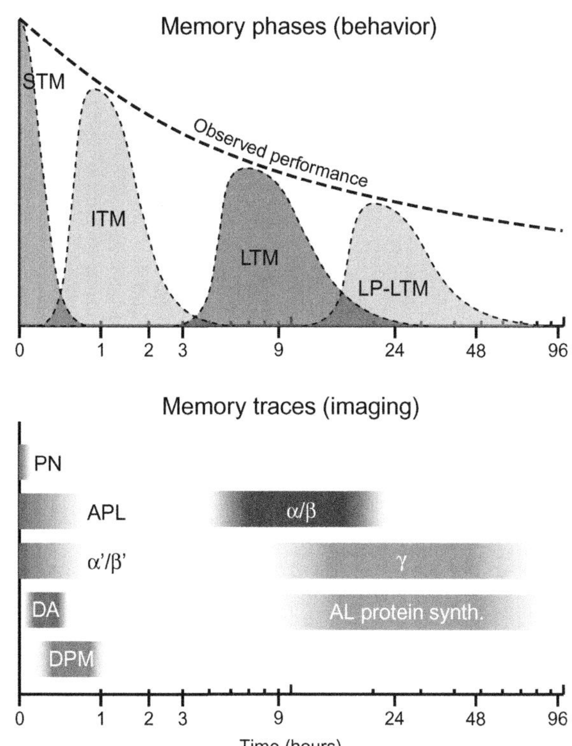

FIGURE 27.4 **Memory phases and memory traces.** Behavioral memory is a combination of separate memory phases: short-term memory (STM), intermediate-term memory (ITM), long-term memory (LTM), and late-phase long-term memory (LP-LTM). Memory traces have been observed in antennal lobe projection neurons (PN), anterior paired lateral (APL) neurons, α′/β′ MB neurons, the dorsal paired medial (DPM) neuron, α/β MB neurons, γ MB neurons, dopaminergic (DA) neurons, and in terms of protein synthesis in the antennal lobes. Source: *Modified from Davis*[73] *and reprinted with permission.*

An LTM trace has been localized in the α/β neurons of mushroom bodies (Figures 27.4 and 27.5). This trace was registered as an increase in Ca^{2+} responses to the CS+ 9–24 hr following conditioning (Figure 27.5).[140] Interestingly, the trace was branch specific, occurring in the vertical α lobe but not in the horizontal β lobe. In a subsequent study, the authors tested a large battery of LTM mutants—26 in total—and found that the trace was absent in all of them.[142] Therefore, the α/β memory trace appears to reflect a fundamental physiological process underlying LTM, which is disrupted in conditions that abolish memory. In addition to the α/β trace and the α′/β′ memory traces, a late-phase LTM trace was observed in γ neurons of the MB, which was registered as an increase in Ca^{2+} responses to the CS+ odor and was present across a time window of 18–48 hr after conditioning (Figures 27.4 and 27.5).[143]

The previously discussed memory traces were localized to the intrinsic components of the olfactory pathway, the AL and MBs. Additional memory traces have been found in neurons that have presumed neuromodulatory roles in the olfactory pathway and project to the MB neuropil. The DPM neuron, which may be required for memory consolidation and expresses the *amn* gene product, exhibits a memory trace 30–60 min post-training (Figures 27.4 and 27.5).[77] This time window is correlated with ITM, and the trace is abolished in *amn* mutants that are deficient in this temporal phase of memory. The DPM memory trace, like the α/β trace, is branch specific, occurring only in the branch of the DPM neuron that innervates the vertical lobes.

Another memory trace was reported in DA neurons that innervate the MB lobes.[126] These neurons, previously thought to encode only information about the aversive reinforcer, were found to respond to odors as well as electric shock. Following the pairing of odor with electric shock, the DA neurons exhibited a prolonged response to subsequent odor presentation. This punishment prediction is reminiscent (although inverse) of the response of DA neurons in mammals that encode a reward prediction error.[134] One caveat is that another group failed to find evidence for the memory trace in DA neurons.[125] Multiple review articles have covered the research on memory traces, and interested readers are encouraged to consult those articles for a detailed description of the research area.[70,73,144,145]

Two memory traces have been observed that are manifested as a decrease, rather than an increase, in responsiveness to the CS+. The first was found in the GABAergic APL neurons that suppress the responsiveness of the MB to olfactory stimuli.[93] Following conditioning, the Ca^{2+} responses of the APL neuron to the CS+ are reduced (Figure 27.5). Because the APL neurons are GABAergic, this reduction in their responses would have the net effect of reducing inhibition of the MB, increasing its output. Thus, this memory trace pushes the MB output up in response to the CS+, potentially in concert with the memory traces observed in the intrinsic MB neurons. Intriguingly, similar to the APL neuron memory trace, MB-V2 neurons exhibited a memory trace that was manifested as a reduction in response to the CS+ following training.[123] The MB-V2 neurons are MB output neurons that are involved in memory retrieval. Because our knowledge about the retrieval circuits is currently very sparse, it is not clear how the reduction in responses of MB-V2 is involved in modulating memory retrieval.

OUTLOOK

In computing systems, Moore's law refers to the rapid pace of increasing computer power—the number of transistors that can be placed inexpensively on an integrated circuit doubles approximately every 2 years.

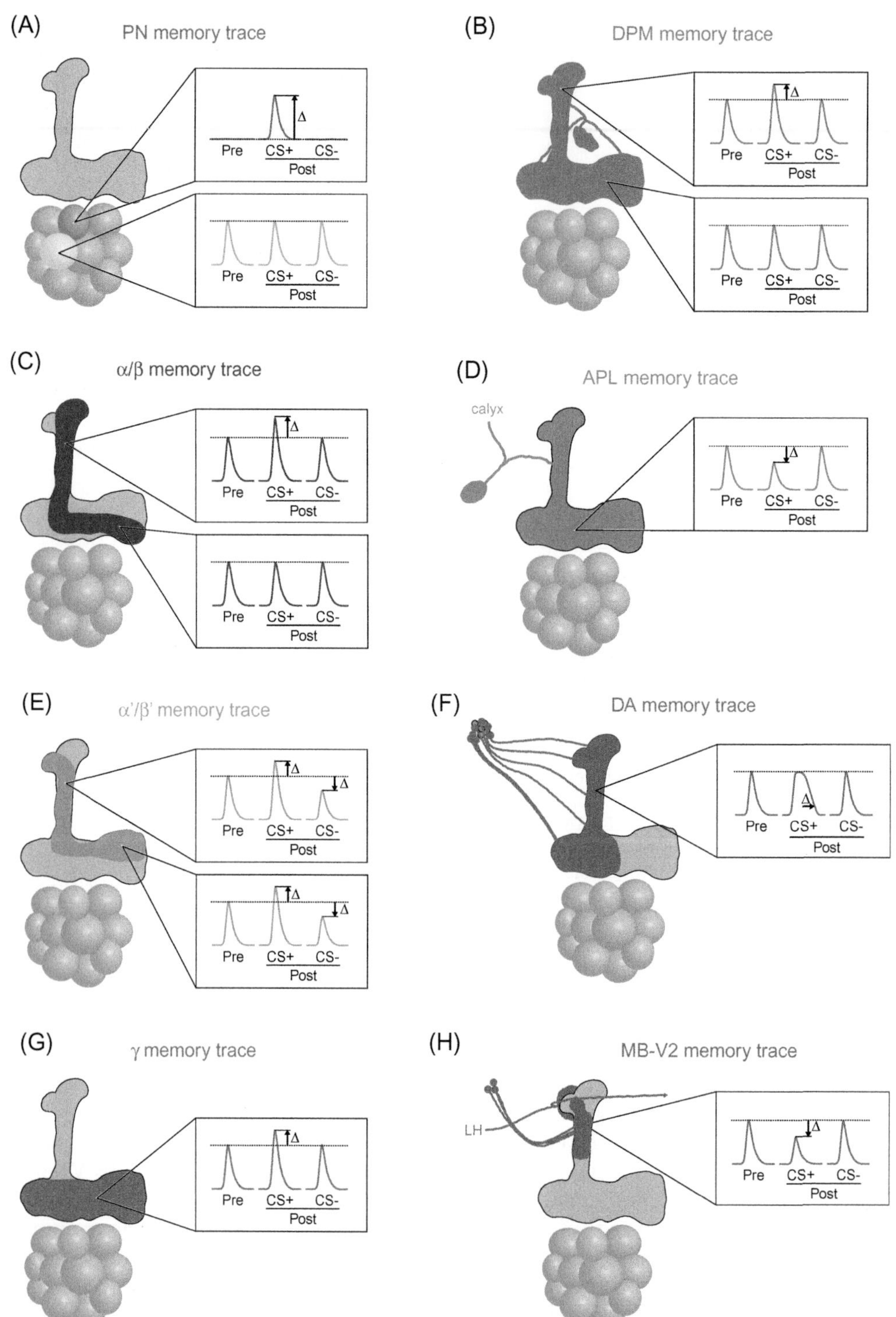

FIGURE 27.5 **Schematic summarizing the localization and features of memory traces.** The mushroom body lobes and antennal lobes are shown in gray, with relevant regions highlighted in color. The response to the CS+ and CS− before conditioning, assumed for simplicity to be equal, is marked as 'Pre.' Following conditioning, a change in the magnitude or duration of the responses to the CS+ and/or CS− occurred. (A) The projection neuron (PN) memory trace resulted in a recruitment of new antennal lobe glomeruli into the odor representation.[14] (B) The dorsal paired medial (DPM) neuron memory trace was observed as a vertical branch-specific increase in the response to the CS+.[77] (C) The α/β MB neuron memory trace was manifested as a vertical branch-specific increase in Ca^{2+} responses to the CS+, observed 9–24 hr following conditioning.[140] (D) The anterior paired lateral (APL) neuron memory trace is a decrease in the Ca^{2+} response to the CS+.[93] (E) The α′/β′ MB neuron memory trace consisted of an increase in the ratio of the responses to the CS+ versus the CS−, observed in both vertical and horizontal branches of the α′/β′ MB neurons.[26,141] (F) The DA neuron memory trace is a prolonged response to the CS+ following conditioning.[126] (G) The γ lobe memory trace was observed as an increase in the CS+ response.[142] (H) The MB-V2 memory trace, similar to the APL memory trace, was manifested as a decrease in the responses to the CS+.[123]

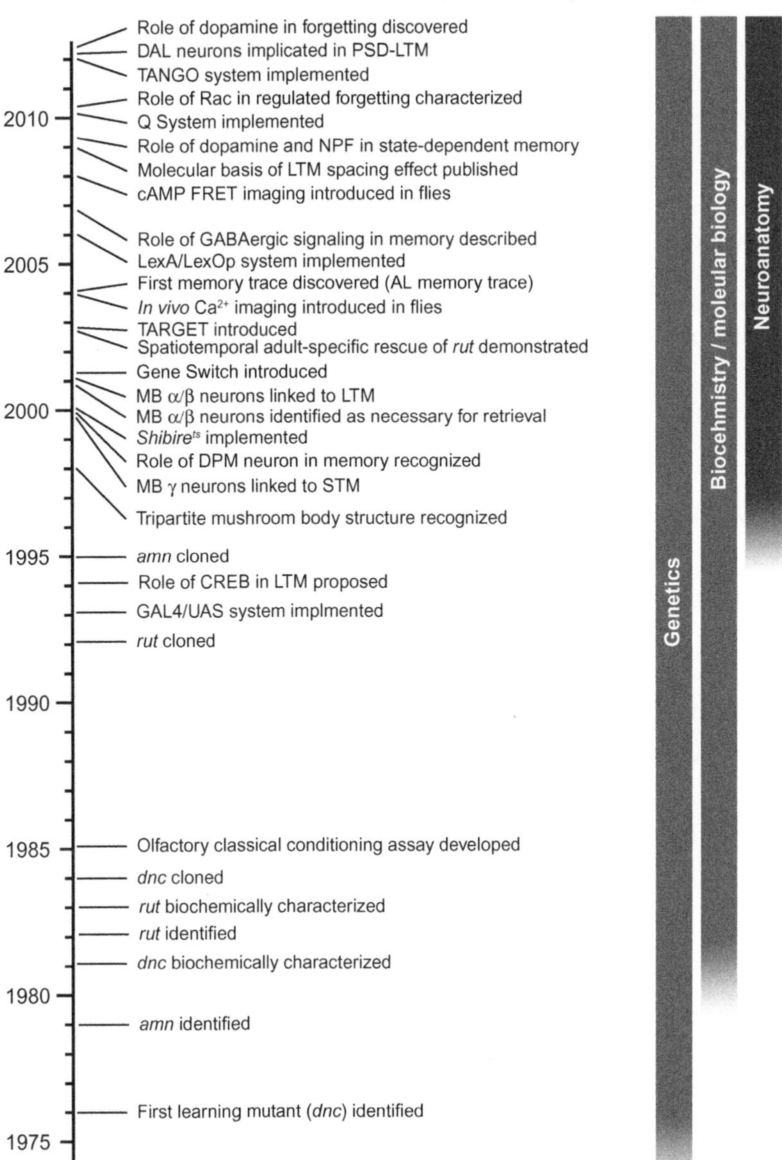

FIGURE 27.6 Selected milestones in the development of *Drosophila* as a model system for learning and memory. Emphasis here is given to technical and conceptual developments that are relevant to the four major eras identified in the text. Note the large increase in the number of experimental tools that have been invented or implemented in *Drosophila* since 2000.

Although it not nearly as simple to quantify scientific progress, we have clearly seen an analogous increase in the power of the *Drosophila* model system. This expansion has been fueled in large part by the development of new tools in the field. To get an idea of the increasing pace of progress, it is instructive to consider the evolution of the field in the recent past. Since 2000, we have seen the introduction of tools that allow precise spatiotemporal control of gene expression and synaptic transmission, optical imaging reporters for intracellular signaling cascades, two new binary transcriptional control systems (tripling the total number of options), and optogenetic/chemogenetic/thermogenetic tools to stimulate neurons, among many other advances (Figure 27.6).

In this chapter, we summarized research spanning four major epochs of *Drosophila* learning and memory: genetics, biochemistry/molecular biology, neuroanatomy, and systems neuroscience. These approaches highlight the confluence of experimental approaches that has enabled researchers to approach questions with a depth of analysis that was previously unimaginable. Perhaps the best example is the discovery and characterization of the role of cAMP signaling in memory formation, which spanned the levels of genetics (rut, dnc, DCO, etc.), biochemistry/molecular and cellular biology (Ca^{2+} sensitivity of Rut/synergistic response), neuroanatomy (cAMP signaling in mushroom body neurons), and systems neuroscience (roles of specific mushroom body neurons and dopaminergic neurons in memory circuits). This pathway for discovery, beginning with genetics and carrying through the systems levels, will likely continue to yield significant advances into the future. At the same time, new avenues are being explored in attempts to further increase the pace of discovery. For instance, tools of neuronal circuit analysis are being refined and applied in large-scale experiments in an attempt to 'reverse engineer' the *Drosophila* brain.[146] Although it is unclear where these new approaches will lead, *Drosophila* will undoubtedly remain a powerful model system to probe many aspects of brain function, including learning and memory.

References

1. Quinn WG, Harris WA, Benzer S. Conditioned behavior in *Drosophila* melanogaster. *Proc Natl Acad Sci USA.* 1974;71:708–712.
2. Dudai Y, Jan YN, Byers D, Quinn WG, Benzer S. *dunce*, a mutant of *Drosophila* deficient in learning. *Proc Natl Acad Sci USA.* 1976;73:1684–1688.
3. Byers D, Davis RL, Kiger JA. Defect in cyclic AMP phosphodiesterase due to the dunce mutation of learning in *Drosophila melanogaster. Nature.* 1981;289:79–81.
4. Chen CN, Malone T, Beckendorf S, Davis RL. At least two genes reside within a large intron of the *Drosophila dunce* gene. *Nature.* 1987;329:721–724.
5. Davis RL, Davidson N. Isolation of the *Drosophila melanogaster dunce+* chromosomal region and recombinational mapping of dunce sequences with restriction site polymorphisms as genetic markers. *Mol Cell Biol.* 1984;4:358–367.
6. Davis RL, Davidson N. The memory gene *dunce+* encodes a remarkable set of RNAs with internal heterogeneity. *Mol Cell Biol.* 1986;6:1464–1470.
7. Chen CN, Denome S, Davis RL. Molecular analysis of cDNA clones and the corresponding genomic coding sequences of the *Drosophila dunce+* gene, the structural gene for cAMP phosphodiesterase. *Proc Natl Acad Sci USA.* 1986;83:9313–9317.
8. Qiu Y, Chen CN, Malone T, Richter L, Beckendorf SK, Davis RL. Characterization of the memory gene *dunce* of *Drosophila melanogaster. J Mol Biol.* 1991;222:553–565.
9. Erber J, Masuhr T, Menzel R. Localization of short-term memory in the brain of the bee, *Apis mellifera. Physio Entomol.* 1980;5:343–358.
10. Heisenberg M, Borst A, Wagner S, Byers D. *Drosophila* mushroom body mutants are deficient in olfactory learning. *J Neurogenet.* 1985;2:1–30.
11. Nighorn A, Healy M, Davis RL. The cAMP phosphodiesterase encoded by the *Drosophila dunce* gene is concentrated in mushroom body neuropil. *Neuron.* 1991;6:455–467.
12. Waddell S, Armstrong JD, Kitamoto T, Kaiser K, Quinn WG. The *amnesiac* gene product is expressed in two neurons in the *Drosophila* brain. *Cell.* 2000;103:805–813.
13. Kitamoto T. Conditional modification of behavior in *Drosophila* by targeted expression of a temperature-sensitive shibire allele in defined neurons. *J Neurobiol.* 2001;47:81–92.
14. Yu D, Ponomarev A, Davis RL. Altered representation of the spatial code for odors after olfactory classical conditioning; memory trace formation by synaptic recruitment. *Neuron.* 2004;42:437–449.
15. Livingstone MS, Sziber PP, Quinn WG. Loss of calcium/calmodulin responsiveness in adenylate cyclase of *rutabaga*, a *Drosophila* learning mutant. *Cell.* 1984;37:205–215.
16. Levin LR, Han PL, Hwang PM, Feinstein PG, Davis RL. The *Drosophila* learning and memory gene *rutabaga* encodes a Ca^{2+}/calmodulin-responsive adenylyl cyclase. *Cell.* 1992;68:479–489.
17. Han PL, Levin LR, Reed RR, Davis RL. Preferential expression of the *Drosophila rutabaga* gene in mushroom bodies, neural centers for learning in insects. *Neuron.* 1992;9:619–627.
18. Cheng Y, Endo K, Wu K, Rodan AR, Heberlein U, Davis RL. *Drosophila fasciclinII* is required for the formation of odor memories and for normal sensitivity to alcohol. *Cell.* 2001;105:757–768.
19. Kim YC, Lee HG, Han KA. D1 dopamine receptor dDA1 is required in the mushroom body neurons for aversive and appetitive learning in *Drosophila. J Neurosci.* 2007;27:7640–7647.
20. Madalan A, Yang X, Ferris J, Zhang S, Roman GG. (o) activation is required for both appetitive and aversive memory acquisition in *Drosophila. Learn Mem.* 2011;19:26–34.
21. Buchanan ME, Davis RL. A distinct set of *Drosophila* brain neurons required for NF1-dependent learning and memory. *J Neurosci.* 2010;30:10135–10143.
22. Ho IS, Hannan F, Guo H-F, Hakker I, Zhong Y. Distinct functional domains of neurofibromatosis type 1 regulate immediate versus long-term memory formation. *J Neurosci.* 2007;27:6852–6857.
23. LaFerriere H, Ostrowski D, Guarnieri DJ, Zars T. The arouser EPS8L3 gene is critical for normal memory in *Drosophila. PLoS ONE.* 2011;6:e22867.
24. LaFerriere H, Guarnieri DJ, Sitaraman D, Diegelmann S, Heberlein U, Zars T. Genetic dissociation of ethanol sensitivity and memory formation in *Drosophila melanogaster. Genetics.* 2008;178:1895–1902.
25. Knapek S, Gerber B, Tanimoto H. Synapsin is selectively required for anesthesia-sensitive memory. *Learn Mem.* 2010;17:76–79.
26. Tan Y, Yu D, Pletting J, Davis RL. *Gilgamesh* is required for *rutabaga* independent olfactory learning in *Drosophila. Neuron.* 2010;67:810–820.
27. Gouzi JY, Moressis A, Walker JA, Apostolopoulou AA, Palmer RH, Bernards A, Skoulakis EM. The receptor tyrosine kinase Alk controls neurofibromin functions in *Drosophila* growth and learning. *PLoS Genet.* 2011;7:e1002281.
28. Wu CL, Xia S, Fu TF, Wang H, Chen YH, Leong D, Chiang AS, Tully T. Specific requirement of NMDA receptors for long-term memory consolidation in *Drosophila* ellipsoid body. *Nat Neurosci.* 2007;10:1578–1586.
29. Schwaerzel M, Monastirioti M, Scholz H, Friggi-Grelin F, Birman S, Heisenberg M. Dopamine and octopamine differentiate between aversive and appetitive olfactory memories in *Drosophila. J Neurosci.* 2003;23:10495–10502.
30. Skoulakis EM, Kalderson D, Davis RL. Preferential expression in mushroom bodies of the catalytic subunit of protein kinase A and its role in learning and memory. *Neuron.* 1993;11:1–14.
31. Goodwin SF, Del Vecchio MD, Velinzon K, Hogel C, Russell SRH, Tully T, Kaiser K. Defective learning in mutants of the *Drosophila* gene for a regulatory subunit of cAMP-dependent protein kinase. *J Neurosci.* 1997;17:8817–8827.
32. Grotewiel MS, Beck CDO, Wu K-H, Zhu X-R, Davis RL. Integrin-mediated short-term memory in *Drosophila. Nature.* 1998;391:455–460.
33. Skoulakis EMC, Davis RL. Olfactory learning deficits in mutants for Leonardo, a *Drosophila* gene encoding a 14-3-3 protein. *Neuron.* 1996;17:931–944.
34. Cowan TM, Siegel RW. *Drosophila* mutations that alter ionic conduction disrupt acquisition and retention of a conditioned odor avoidance response. *J Neurogenet.* 1986;3:187–201.

35. Griffith LC, Wang J, Zhong Y, Wu CF, Greenspan RJ. Calcium/calmodulin-dependent protein kinase II and potassium channel subunit Eag similarly affect plasticity in *Drosophila*. *Proc Natl Acad Sci USA*. 1994;91:10044–10048.
36. Tempel BL, Bonini N, Dawson DR, Quinn WG. Reward learning in normal and mutant *Drosophila*. *Proc Natl Acad Sci USA*. 1983;80:1482–1486.
37. Connolly JB, Roberts IJ, Armstrong JD, Kaiser K, Forte M, Tully T, O'Kane CJ. Associative learning disrupted by impaired Gs signaling in *Drosophila* mushroom bodies. *Science*. 1996;274: 2104–2107.
38. Putz G, Bertolucci F, Raabe T, Zars T, Heisenberg M. The S6KII (*rsk*) gene of *Drosophila melanogaster* differentially affects an operant and a classical learning task. *J Neurosci*. 2004;24:9745–9751.
39. DeZazzo J, Sandstrom D, deBelle S, Velinzon K, Smith P, Grady L, Del Vecchio M, Ramaswami M, Tully T. *Nalyot*, a mutation of the *Drosophila* myb-related *Adf1* transcription factor, disrupts synapse formation and olfactory memory. *Neuron*. 2000;27: 145–158.
40. Chang KT, Shi YJ, Min KT. The *Drosophila* homolog of Down's syndrome critical region 1 gene regulates learning: implications for mental retardation. *Proc Natl Acad Sci USA*. 2003;100:15794–15799.
41. Tamura T, Horiuchi D, Chen YC, Sone M, Miyashita T, Saitoe M, Yoshimura N, Chiang AS, Okazawa H. *Drosophila* PQBP1 regulates learning acquisition at projection neurons in aversive olfactory conditioning. *J Neurosci*. 2010;30:14091–14101.
42. Knight D, Harvey PJ, Iliadi KG, Klose MK, Iliadi N, Dolezelova E, Charlton MP, Zurovec M, Boulianne GL. Equilibrative nucleoside transporter 2 regulates associative learning and synaptic function in *Drosophila*. *J Neurosci*. 2010;30:5047–5057.
43. Iliadi KG, Avivi A, Iliadi NN, Knight D, Korol AB, Nevo E, Taylor P, Moran MF, Kamyshev NG, Boulianne GL. *nemy* encodes a cytochrome b561 that is required for *Drosophila* learning and memory. *Proc Natl Acad Sci USA*. 2008;105:19986–19991.
44. Moressis A, Friedrich AR, Pavlopoulos E, Davis RL, Skoulakis EM. A dual role for the adaptor protein DRK in *Drosophila* olfactory learning and memory. *J Neurosci*. 2009;29:2611–2625.
45. Berry JA, Cervantes-Sandoval I, Nicholas EP, Davis RL. Dopamine is required for learning and forgetting in *Drosophila*. *Neuron*. 2012;74:530–542.
46. Shuai Y, Lu B, Hu Y, Wang L, Sun K, Zhong Y. Forgetting is regulated through Rac activity in *Drosophila*. *Cell*. 2010;140:579–589.
47. Quinn WG, Sziber PP, Booker R. The *Drosophila* memory mutant *amnesiac*. *Nature*. 1979;277:212–214.
48. Chen K, Richlitzki A, Featherstone DE, Schwärzel M, Richmond JE. Tomosyn-dependent regulation of synaptic transmission is required for a late phase of associative odor memory. *Proc Natl Acad Sci USA*. 2011;108:18482–18487.
49. Pavlopoulos E, Anezaki M, Skoulakis EM. Neuralized is expressed in the α/β lobes of adult *Drosophila* mushroom bodies and facilitates olfactory long-term memory formation. *Proc Natl Acad Sci USA*. 2008;105:14674–14679.
50. Drier EA, Tello MK, Cowan M, Wu P, Blace N, Sacktor TC, Yin JC. Memory enhancement and formation by atypical PKM activity in *Drosophila melanogaster*. *Nat Neurosci*. 2002;5:316–324.
51. Folkers E, Drain P, Quinn WG. *Radish*, a *Drosophila* mutant deficient in consolidated memory. *Proc Natl Acad Sci USA*. 1993;90:8123–8127.
52. Presente A, Boyles RS, Serway CN, de Belle JS, Andres AJ. Notch is required for long-term memory in *Drosophila*. *Proc Natl Acad Sci USA*. 2004;101:1764–1768.
53. Ge X, Hannan F, Xie Z, Feng C, Tully T, Zhou H, Xie Z, Zhong Y. Notch signaling in *Drosophila* long-term memory formation. *Proc Natl Acad Sci USA*. 2004;101:10172–10176.
54. Kwak JE, Drier E, Barbee SA, Ramaswami M, Yin JC, Wickens M. GLD2 poly(A) polymerase is required for long-term memory. *Proc Natl Acad Sci USA*. 2008;105:14644–14649.
55. Ashraf SI, McLoon AL, Sclarsic SM, Kunes S. Synaptic protein synthesis associated with memory is regulated by the RISC pathway in *Drosophila*. *Cell*. 2006;124:191–205.
56. Didelot G, Molinari F, Tchénio P, Comas D, Milhiet E, Munnich A, Colleaux L, Preat T. Tequila, a neurotrypsin ortholog, regulates long-term memory formation in *Drosophila*. *Science*. 2006; 313:851–853.
57. Pagani MR, Oishi K, Gelb BD, Zhong Y. The phosphatase SHP2 regulates the spacing effect for long-term memory induction. *Cell*. 2009;139:186–198.
58. Majumdar A, Cesario WC, White-Grindley E, Jiang H, Ren F, Khan MR, Li L, Choi EM, Kannan K, Guo F, Unruh J, Slaughter B, Si K. Critical role of amyloid-like oligomers of *Drosophila* Orb2 in the persistence of memory. *Cell*. 2012;149: 515–529.
59. Kottler B, Lampin-Saint-Amaux A, Comas D, Preat T, Goguel V. Debra, a protein mediating lysosomal degradation, is required for long-term memory in *Drosophila*. *PLoS ONE*. 2011;6:e25902.
60. Yin JCP, Wallach JS, Del Vecchio M, Wilder EI, Zhou H, Quinn WG, Tully T. Induction of a dominant negative CREB transgene specifically blocks long-term memory in *Drosophila*. *Cell*. 1994;79:49–58.
61. Comas D, Petit F, Preat T. *Drosophila* long-term memory formation involves regulation of cathepsin activity. *Nature*. 2004;430: 460–463.
62. Dubnau J, Chiang AS, Grady L, Barditch J, Gossweiler S, McNeil J, Smith P, Buldoc F, Scott R, Certa U, Broger C, Tully T. The *staufen/pumilio* pathway is involved in *Drosophila* long-term memory. *Curr Biol*. 2003;13:286–296.
63. Gerstner JR, Vanderheyden WM, Shaw PJ, Landry CF, Yin JC. Fatty-acid binding proteins modulate sleep and enhance long-term memory consolidation in *Drosophila*. *PLoS Genet*. 2011;6: e15890.
64. Song Q, Sun K, Shuai Y, Lin R, You W, Zhong Y. Suppressor of Hairless is required for long-term memory formation in *Drosophila*. *J Neurogenet*. 2009;23:405–411.
65. Matsuno M, Horiuchi J, Tully T, Saitoe M. The *Drosophila* cell adhesion molecule Klingon is required for long-term memory formation and is regulated by Notch. *Proc Natl Acad Sci USA*. 2009;106:310–315.
66. Zhao H, Zheng X, Yuan X, Wang L, Wang X, Zhong Y, Xie Z, Tully T. *ben* Functions with *Scamp* during synaptic transmission and long-term memory formation in *Drosophila*. *J Neurosci*. 2009;29:414–424.
67. Wu Y, Bolduc FV, Bell K, Tully T, Fang Y, Sehgal A, Fischer JAA. *Drosophila* model for Angelman syndrome. *Proc Natl Acad Sci USA*. 2008;105:12399–12404.
68. Lu Y, Lu YS, Shuai Y, Feng C, Tully T, Xie Z, Zhong Y, Zhou HM. The AKAP Yu is required for olfactory long-term memory formation in *Drosophila*. *Proc Natl Acad Sci USA*. 2007;104: 13792–13797.
69. Qian M, Pan G, Sun L, Feng C, Xie Z, Tully T, Zhong Y. Receptor-like tyrosine phosphatase PTP10D is required for long-term memory in *Drosophila*. *J Neurosci*. 2007;27:4396–4402.
70. Davis RL. Olfactory memory formation in *Drosophila*: From molecular to systems neuroscience. *Annu Rev Neurosci*. 2005;28:275–302.
71. Keene AC, Waddell S. *Drosophila* olfactory memory: Single genes to complex neural circuits. *Nat Rev Neurosci*. 2007;8:341–354.
72. Margulies C, Tully T, Dubnau J. Deconstructing memory in *Drosophila*. *Curr Biol*. 2005;15:R700–R713.
73. Davis RL. Traces of *Drosophila* memory. *Neuron*. 2011;70:8–19.

74. Mao Z, Roman G, Zong L, Davis RL. Pharmacogenetic rescue in time and space of the *rutabaga* memory impairment using Gene-Switch. *Proc Natl Acad Sci USA*. 2004;101:198–203.
75. McGuire SE, Le PT, Osborn AJ, Matsumoto K, Davis RL. Spatio-temporal rescue of memory dysfunction in *Drosophila*. *Science*. 2003;302:1765–1768.
76. Beck CDO, Schroeder B, Davis RL. Learning performance of *Drosophila* after repeated conditioning trials with discrete stimuli. *J Neurosci*. 2000;20:2944–2953.
77. Yu D, Keene AC, Srivatsan A, Waddell S, Davis RL. *Drosophila* DPM neurons form a delayed and branch-specific memory trace after olfactory classical conditioning. *Cell*. 2005;123: 945–957.
78. Keene AC, Stratmann M, Keller A, Perrat PN, Vosshall LB, Waddell S. Diverse odor-conditioned memories require uniquely timed dorsal paired medial neuron output. *Neuron*. 2004;44:521–533.
79. Tully T, Preat T, Boynton SC, Del Vecchio M. Genetic dissection of consolidated memory in *Drosophila*. *Cell*. 1994;79:35–47.
80. Davis RL. Biochemistry and physiology of *Drosophila* learning mutants. *Physiol Rev*. 1996;76:299–317.
81. Tully T, Quinn WG. Classical conditioning and retention in normal and mutant *Drosophila* melanogaster. *J Comp Physiol*. 1985;157:263–277.
82. Isabel G, Pascual A, Preat T. Exclusive consolidated memory phases in *Drosophila*. *Science*. 2004;304:1024–1027.
83. Plaçais PY, Trannoy S, Isabel G, Aso Y, Siwanowicz I, Belliart-Guérin G, Verneir P, Birman S, Tanimoto H, Preat T. Slow oscillations in two pairs of dopaminergic neurons gate long-term memory formation in *Drosophila*. *Nat Neurosci*. 2012;15: 592–599.
84. Colomb J, Kaiser L, Chabaud MA, Preat T. Parametric and genetic analysis of *Drosophila* appetitive long-term memory and sugar motivation. *Genes Brain Behavior*. 2009;8:407–415.
85. Krashes MJ, Waddell S. Rapid consolidation to a *radish* and protein synthesis-dependent long-term memory after single-session appetitive olfactory conditioning in *Drosophila*. *J Neurosci*. 2008;28:3103–3113.
86. Burke CJ, Waddell S. Remembering nutrient quality of sugar in *Drosophila*. *Curr Biol*. 2011;21:746–750.
87. Fujita M, Tanimura T. *Drosophila* evaluates and learns the nutritional value of sugars. *Curr Biol*. 2011;21:751–755.
88. Cervantes-Sandoval I, Davis RL. Distinct traces for appetitive versus aversive olfactory memories in DPM neurons *Drosophila*. *Curr Biol*. 2012;22:1247–1252.
89. Qin H, Dubnau J. Genetic disruptions of *Drosophila* Pavlovian learning leave extinction learning intact. *Genes Brain Behavior*. 2010;9:203–212.
90. Tabone CJ, deBelle JS. Second-order conditioning in *Drosophila*. *Learn Mem*. 2011;18:250–253.
91. Shuai Y, Hu Y, Qin H, Campbell RA, Zhong Y. Distinct molecular underpinnings of *Drosophila* olfactory trace conditioning. *Proc Natl Acad Sci USA*. 2011;108:20201–20206.
92. Davis RL. Olfactory learning. *Neuron*. 2004;44:31–48.
93. Liu X, Davis RL. The GABAergic anterior paired lateral neuron of *Drosophila* suppresses and is suppressed by olfactory learning. *Nat Neurosci*. 2009;12:53–59.
94. Brand AH, Perrimon N. Targeted gene expression as a means of altering cell fates and generating dominant phenotypes. *Development*. 1993;118:401–415.
95. Lai A, Lee T. Genetic mosaic with dual binary transcriptional systems in *Drosophila*. *Nat Neurosci*. 2006;9:703–709.
96. Potter CJ, Tasic B, Russler EV, Liang L, Luo L. The Q system: A repressible binary system for transgene expression, lineage tracing, and mosaic analysis. *Cell*. 2010;141:536–548.
97. Roman G, Endo K, Zong L, Davis RL. P [Switch], a system for spatial and temporal control of gene expression in *Drosophila* melanogaster. *Proc Natl Acad Sci USA*. 2001;98:12602–12607.
98. Wang JW, Wong AM, Flores J, Vosshall LB, Axel R. Two-photon calcium imaging reveals an odor-evoked map of activity in the fly brain. *Cell*. 2003;112:271–282.
99. Yu D, Baird GS, Tsien RY, Davis RL. Detection of calcium transients in *Drosophila* mushroom body neurons with camgaroo reporters. *J Neurosci*. 2003;23:64–72.
100. Hamada FN, Rosenzweig M, Kang K, Pulver SR, Ghezzi A, Jegla TJ, Garrity PA. An internal thermal sensor controlling temperature preference in *Drosophila*. *Nature*. 2008;454:217–220.
101. Peabody NC, Pohl JB, Diao F, Vreede AP, Sandstrom DJ, Wang H, Zelensky PK, White BH. Characterization of the decision network for wing expansion in *Drosophila* using targeted expression of the TRPM8 channel. *J Neurosci*. 2009;29: 3343–3353.
102. Schroll C, Riemensperger T, Bucher D, Ehmer J, Voller T, Erbguth K, Gerber B, Hendel T, Nagei G, Buchner E, Fiala A. Light-induced activation of distinct modulatory neurons triggers appetitive or aversive learning in *Drosophila* larvae. *Curr Biol*. 2006;16:1741–1747.
103. White BH, Osterwalder TP, Yoon KS, Joiner WJ, Whim MD, Kaczmarek LK, Keshishian H. Targeted attenuation of electrical activity in *Drosophila* using a genetically modified K(+) channel. *Neuron*. 2001;31:699–711.
104. Wilson RI, Turner GC, Laurent G. Transformation of olfactory representations in the *Drosophila* antennal lobe. *Science*. 2004;303:366–370.
105. Dubnau J, Grady L, Kitamoto T, Tully T. Disruption of neurotransmission in *Drosophila* mushroom body blocks retrieval but not acquisition of memory. *Nature*. 2001;411:476–480.
106. McGuire SE, Le PT, Davis RL. The role of *Drosophila* mushroom body signaling in olfactory memory. *Science*. 2001;293: 1330–1333.
107. Krashes MJ, Keene AC, Leung B, Armstrong JD, Waddell S. Sequential use of mushroom body neuron subsets during *Drosophila* odor memory processing. *Neuron*. 2007;53:103–115.
108. Zars T, Fischer M, Schulz R, Heisenberg M. Localization of short-term memory in *Drosophila*. *Science*. 2000;288:672–675.
109. Pascual A, Preat T. Localization of long-term memory within the *Drosophila* mushroom body. *Science*. 2001;294:1115–1117.
110. Blum AL, Li W, Cressy M, Dubnau J. Short- and long-term memory in *Drosophila* require cAMP signaling in distinct neuron types. *Curr Biol*. 2009;19:1341–1350.
111. Trannoy S, Redt-Clouet C, Dura JM, Preat T. Parallel processing of appetitive short- and long-term memories in *Drosophila*. *Curr Biol*. 2011;21:1647–1653.
112. Akalal DBG, Wilson CF, Zong L, Davis RL. Roles for *Drosophila* mushroom body neurons in olfactory learning and memory. *Learn Mem*. 2006;13:659–668.
113. Tomchik S, Davis RL. Dynamics of learning-related cAMP signaling and stimulus integration in the *Drosophila* olfactory pathway. *Neuron*. 2009;64:510–521.
114. Gervasi N, Tchénio P, Preat T. PKA dynamics in a *Drosophila* learning center: Coincidence detection by rutabaga adenylyl cyclase and spatial regulation by dunce phosphodiesterase. *Neuron*. 2010;65:516–529.
115. Qin H, Cressy M, Li W, Coravos JS, Izzi SA, Dubnau J. Gamma neurons mediate dopaminergic input during aversive olfactory memory formation in *Drosophila*. *Curr Biol*. 2012;22:608–614.
116. Wu CL, Shih MF, Lai JS, Yang HT, Turner GC, Chen L, Chiang AS. Heterotypic gap junctions between two neurons in the *Drosophila* brain are critical for memory. *Curr Biol*. 2011;21: 848–854.

117. Liu X, Krause WC, Davis RL. GAB_{AA} receptor RDL inhibits *Drosophila* olfactory associative learning. *Neuron*. 2007;56: 1090–1102.
118. Liu X, Buchanan ME, Han KA, Davis RL. The GAB_{AA} receptor RDL suppresses the conditioned stimulus pathway for olfactory learning. *J Neurosci*. 2009;29:1573–1579.
119. Chen CC, Wu JK, Lin HW, Pai TP, Fu TF, Wu CL, Tully T, Chiang AS. Visualizing long-term memory formation in two neurons of the *Drosophila* brain. *Science*. 2012;335:678–685.
120. Davis RL, Giurfa M. Mushroom body memories: An obituary prematurely written? *Curr Biol*. 2012;22:R272–R275.
121. Dubnau J. Neuroscience: Ode to the mushroom bodies. *Science*. 2012;335:664–665.
122. Tanaka NK, Tanimoto H, Ito K. Neuronal assemblies of the *Drosophila* mushroom body. *J Comp Neurol*. 2008;508:711–755.
123. Séjourné J, Plaçais PY, Aso Y, Siwanowicz I, Trannoy S, Thoma V, Tedjakumala SR, Rubin GM, Tchénio P, Ito K, Isabel G, Tanimoto H, Preat T. Mushroom body efferent neurons responsible for aversive olfactory memory retrieval in *Drosophila*. *Nat Neurosci*. 2011;14:903–910.
124. Ruta V, Datta SR, Vasconcelos ML, Freeland J, Looger LL, Axel R. A dimorphic pheromone circuit in *Drosophila* from sensory input to descending output. *Nature*. 2010;468:686–690.
125. Mao Z, Davis RL. Eight different types of dopaminergic neurons innervate the *Drosophila* mushroom body neuropil: Anatomical and physiological heterogeneity. *Front. Neural Circuits*. 2009;3:5.
126. Riemensperger T, Völler T, Stock P, Buchner E, Fiala A. Punishment prediction by dopaminergic neurons in *Drosophila*. *Curr Biol*. 2005;15:1953–1960.
127. Claridge-Chang A, Roorda RD, Vrontou E, Sjulson L, Li H, Hirsh J, Miesenbock G. Writing memories with light-addressable reinforcement circuitry. *Cell*. 2009;139:405–415.
128. Aso Y, Siwanowicz I, Bräcker L, Ito K, Kitamoto T, Tanimoto H. Specific dopaminergic neurons for the formation of labile aversive memory. *Curr Biol*. 2010;20:1445–1451.
129. Krashes MJ, DasGupta S, Vreede A, White B, Armstrong JD, Waddell S. A neural circuit mechanism integrating motivational state with memory expression in *Drosophila*. *Cell*. 2009;139:416–427.
130. Inagaki HK, Ben-Tabou de-Leon S, Wong AM, Jagadish S, Ishimoto H, Barnea G, Kitamoto T, Axel R, Anderson DJ. Visualizing neuromodulation *in vivo*: TANGO-mapping of dopamine signaling reveals appetite control of sugar sensing. *Cell*. 2012;148:583–595.
131. Marella S, Mann K, Scott K. Dopaminergic modulation of sucrose acceptance behavior in *Drosophila*. *Neuron*. 2012;73:941–950.
132. Kaun KR, Azanchi R, Maung Z, Hirsh J, Heberlein UA. *Drosophila* model for alcohol reward. *Nat Neurosci*. 2011;14: 612–619.
133. Bromberg-Martin ES, Matsumoto M, Hikosaka O. Dopamine in motivational control: Rewarding, aversive, and alerting. *Neuron*. 2010;68:815–834.
134. Schultz W. Behavioral dopamine signals. *Trends Neurosci*. 2007;30:203–210.
135. Oades RD, Halliday GM. Ventral tegmental (A10) system: Neurobiology: 1. Anatomy and connectivity. *Brain. Res* 1987; 434:117–165.
136. Waddell S. Dopamine reveals neural circuit mechanisms of fly memory. *Trends Neurosci*. 2010;33:457–464.
137. Abrams TW, Karl KA, Kandel ER. Biochemical studies of stimulus convergence during classical conditioning in *Aplysia*: Dual regulation of adenylate cyclase by Ca^2/calmodulin and transmitter. *J Neurosci*. 1991;11:2655–2665.
138. Yovell Y, Abrams TW. Temporal asymmetry in activation of *Aplysia* adenylyl cyclase by calcium and transmitter may explain temporal requirements of conditioning. *Proc Natl Acad Sci USA*. 1992;89:6526–6530.
139. Joerges J, Küttner A, Galizia CG, Menzel R. Representation of odours and odour mixtures visualized in the honeybee brain. *Nature*. 1997;387:285–288.
140. Yu D, Akalal DB, Davis RL. *Drosophila* α/β mushroom body neurons form a branch-specific, long-term cellular memory trace after spaced olfactory conditioning. *Neuron*. 2006;52: 845–855.
141. Wang Y, Mamiya A, Chiang A-S, Zhong Y. Imaging of an early memory trace in the *Drosophila* mushroom body. *J Neurosci*. 2008;28:4368–4376.
142. Akalal DBG, Yu D, Davis RL. The long-term memory trace formed in the *Drosophila* α/β mushroom body neurons is abolished in long-term memory mutants. *J Neurosci*. 2011;31: 5643–5647.
143. Akalal DBG, Yu D, Davis RL. A late-phase, long-term memory trace forms in the γ neurons of *Drosophila* mushroom bodies after olfactory classical conditioning. *J Neurosci*. 2010;30: 16699–16708.
144. Berry J, Krause W, Davis RL. Olfactory memory traces in *Drosophila*. In: Sossin W, Lacaille JC, Castellucci VF, Belleville. S, eds. *Progress in Brain Research: The Essence of Memory*. Oxford: Elsevier; 2008:169.
145. Liu X, Davis RL. Insect olfactory memory in time and space. *Curr Opin Neurobiol*. 2006;16:679–685.
146. Moses K. Flies at the farm: *Drosophila* at Janelia. *Fly (Austin)*. 2007;1:139–141.

CHAPTER

28

Visual Learning and Decision Making in *Drosophila melanogaster*

Aike Guo[*,†], *Huimin Lu*[†], *Ke Zhang**, *Qingzhong Ren** and *Yah-Num Chiang Wong**

[*]Institute of Neuroscience, Shanghai Institutes for Biological Sciences, Chinese Academy of Sciences, Shanghai, China

[†]Institute of Biophysics, Chinese Academy of Sciences, Beijing, China

INTRODUCTION

The spontaneous responses to optic flow in walking and flying *Drosophila* are considered to initiate visual learning. Using their compound eyes as primary visual sensors, fruit flies transmit optic inputs to the deeper layers of the central nervous system, where the processed signals descend to the motor system to generate behavioral outputs. The neuropilar structures characterize the insects' visual systems adapted to serve functions such as color, motion, and polarization sensitivity, and they provide the necessary neural signals to help organize behaviors such as spatial learning, cross talk between sensory modalities, attention, and decision making. Many studies have been published since the canonical model of motion detection and behavioral control, termed the Hassenstein–Reichardt elementary motor detector (EMD), was presented decades ago. However, many issues related to visual integration and behavioral control are still enigmatic. In this chapter, we first discuss the structure and mechanism of the fly's visual neural machinery as related to visual learning, and then we present our current thoughts about the neural organization of the fly's visual system in the context of multimodal integration and visual decision making.

DROSOPHILA VISION

Neural Superposition

Visual information in high-level animals is transmitted to higher order processing centers along separate channels according to visual features such as shape, color, motion, and depth.[1] Insects are equipped with compound eyes that show diverse designs and organizations: apposition eye, superposition eye, and neural superposition eye.[2,3] The compound eye of the fruit fly belongs to the neural superposition type[4] and represents a mixture of the apposition and superposition types: At the retinal level, it is an optical apposition eye, whereas at the level of the lamina it is a neural superposition eye.[5] Both strategies improve signal-to-noise ratio in visual information processing and help to mediate several visual responses, such as fixation, escape, pursuit, flight stability, landing, optomotor course control, and collision avoidance.

Color Vision

Color vision of *Drosophila* has been analyzed to some extent but is still not very well understood. Each ommatidium of compound eye consists of eight photoreceptor cells, R1–R8, which may express the same or different visual pigments (rhodopsins) partly combined with antennal pigment. Rhodopsin is a light-sensitive pigment molecule composed of a genetically encoded opsin apoprotein that covalently attaches to the chromophore 3-hydroxyl-11-*cis*-retinal via Schiff's base linkage. The blue-sensitive rhodopsin Rh1 is a major rhodopsin in R1–R6 photoreceptor neurons that shares homolog of primary sequence with human and bovine rhodopsin.[6] Rh2 is a violet-sensitive pigment in simple eye ocelli on the vertex of the fly head.[7] Rh3 and Rh4 are ultraviolet-sensitive pigments in

Invertebrate Learning and Memory.
DOI: http://dx.doi.org/10.1016/B978-0-12-415823-8.00028-9

non-overlapping sets of R7 photoreceptor cells.[8,9] Rh3 is also expressed in R8 photoreceptor cells.[10] Rh5 is a blue-sensitive pigment expressed in R8 cells, strictly coordinated with Rh3 expression in the overlying R7 cells. The Rh6 pigment is green-sensitive and expressed in R8, paired with the expression of Rh4 in R7 cells of the respective ommatidium.[11] Thus, the expressive pattern of rhodopsin in different photoreceptor cell types shapes the color vision.[12,13] Phototransduction starts with capturing a photon by rhodopsin (or the antennal pigment), followed by the isomerization of the light-sensitive chromophore from the 11-*cis* to all-*trans* form. Then the conformation change of the protein ('opsin') transfers the inactivated rhodopsin to the activated metarhodopsin—a process that finally triggers the downstream biochemical signaling cascade.[14] In addition to the previously described phototransduction cascade, antennal pigment in R1–R6 photoreceptor neurons absorbs light quanta and transfers the energy to rhodopsins in R1–R6 photoreceptor.[15]

Elementary Motion Detector

A basic function of the visual system in a wide variety of species is the optomotor response to movement in their environment. The EMD model was the first network model for detecting motion direction[16] (Figure 28.1). It consists of two separate input elements, a nonlinear interaction between two crossover information channels and a time-averaging output.[17,18] The EMD senses the luminance changes induced by motion, but it is blind for the second-order motion.[19–21] Although the EMD model does not represent a biological wiring diagram in fly's brain, subsequent studies focused on exploring the neural substrate and a most optimized approximation of this model. The two most prominent pathways of the large monopolar cells L1 and L2 within the lamina were identified to be the respective inputs to the two branches of the EMD.[2,22,23]

Motion Vision

It is generally assumed that humans have two systems for motion perception: The first-order motion system extracts motion information based on coherent spatiotemporal correlations in luminance, whereas the second-order motion system gathers information about variations in second-order image characteristics such as contrast, flicker, texture, and local motion.[24,25] Both systems have been widely studied in humans and primates.[26,27] The second-order motion processing is thought to represent 'high-level' visual processing and therefore may require elaborate visual processing, for example, in the cortex. However, in 2000, second-order motion was found to be visible to zebra fish,[28] and a recent study showed that even fruit flies can see theta motion, a local motion-defined second-order motion.[29] Theta motion delivers conflicting information, in which the dots within a window or theta object move in the opposite direction to the window.[30] In contrast to the well-studied first-order motion, where and how the higher order motion signals are encoded and processed in the brain are poorly understood, even after intensive studies in mammals. By taking advantage of the rich genetic tools and quantitative behavioral analysis paradigm available in *Drosophila*, we have begun to explore the neural correlates of high-order motion vision.

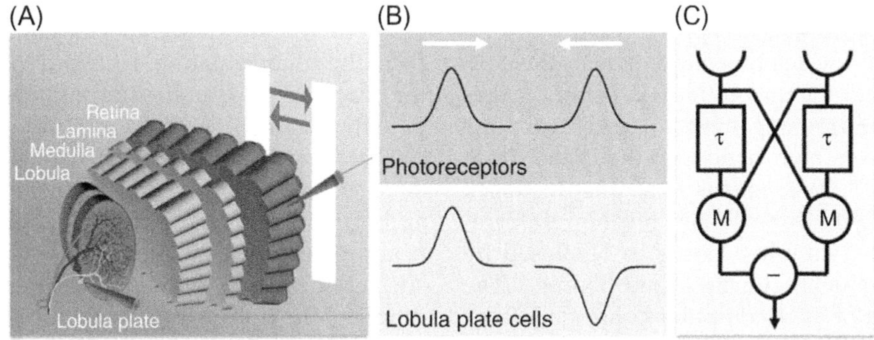

FIGURE 28.1 Direction sensitivity and Reichardt model. (A) Schematic diagram of the fly optic lobe. The retina (red) is followed by four retinotopically organized neuropile layers. In the lobula plate, three large tangential cells of the horizontal system (HS cells) are shown. (B) Schematic responses of photoreceptors and lobula plate tangential cells to bar motion. Whereas the photoreceptor responses are identical for right- and leftward motion, the responses of tangential cells are directionally selective. (C) The Reichardt detector calculates the direction of image motion by multiplying (M) the brightness values at two adjacent image points after one of them has been delayed by a low-pass filter with a time constant τ. This is done in two mirror-symmetrical subunits, the outputs of which are subtracted from one another. *Source: Reproduced with permission from Borst.*[2]

Motion Perception is Independent of Color Vision

The relationship between color and motion channels is highlighted by the following experimental findings.[31] The optomotor response in *Drosophila* to moving bars with two colors and high color contrast was examined. Flies have no optomotor responses to equiluminant color stripes. Such equiluminant colors were reached by adjusting the brightness ratio of the two colors. Interestingly, the point of equiluminance condition informs us about differences in the equipment of rhodopsins in R1–R6 cells compared to that in R7/R8. Specifically, flies lacking rhodopsin in R1–R6 do not respond to moving bars, whereas flies lacking rhodopsin in R7/R8 neither alter the strength of the optomotor response nor shift the point of equiluminance, suggesting that the color channel (R7/R8) does not contribute to optomotor response. Because photoreceptors R1–R6 have the same spectral sensitivity throughout the eye for motion vision and R7 and R8 exhibit heterogeneity of rhodopsins and are important for color vision, it is assumed that only the latter contribute to color vision, but this may not be the full story. Yamaguchi *et al.*[12] measured differential phototaxis by designing a 'color A versus color B' choice paradigm. In a 'blue versus UV' group, flies with only R1–R6 or only R7 and R8 photoreceptors preferred blue, suggesting a nonadditive interaction between the two subsystems.

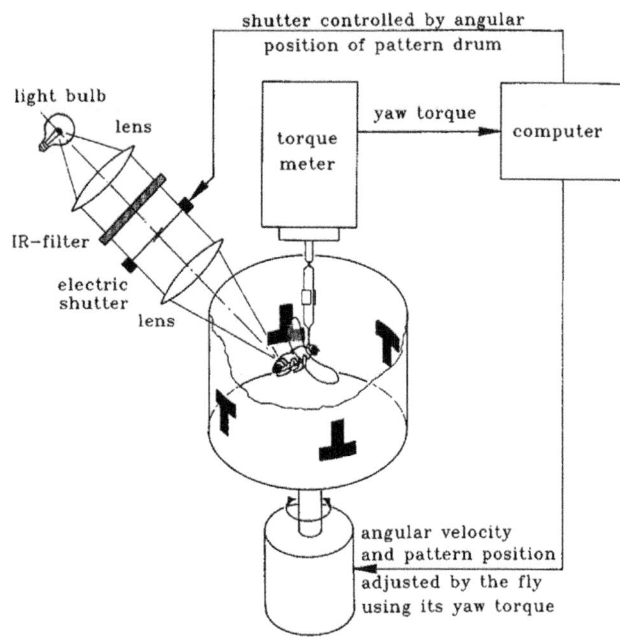

FIGURE 28.2 Simplified diagram of flight simulator. Yaw torque of the fixed flying fly is transduced continuously into DC voltage by the torque meter. From this signal, the angular image deviations (as seen from the fly) that would occur in the free flight from the same torque maneuvers are calculated online, and the fly's visual surround (arena) is turned to the calculated angular position by a stepping motor. The fly can be heated by an infrared light beam controlled by the same computer as the arena. In the conditioning experiment, heat is switched on whenever one of the two pattern types is in the frontal quadrant of the fly's visual field. *Source: Reproduced with permission from Wolf and Heisenberg.*[33]

CLASSICAL AND OPERANT CONDITIONING

Operant Conditioning in Flight Simulator

Since the early 1990s, *Drosophila* has been used as an animal model for the study of visual learning and memory by taking advantage of flight simulators.[32,33] In contrast to Pavlovian classical conditioning, in which animals learn the relationship between the conditioned stimulus and the unconditioned stimulus, animals in operant conditioning have to learn from the consequences of their own actions rather than from explicit teaching.[34,35] This means that the feedback loop between the animal's behavior and the reinforcement (unconditioned stimulus) needs to be closed by the animal's movement during conditioning. The visual flight simulator is a computer-controlled feedback system in which the fly is allowed to control the rotation of a panorama surrounding its yaw torque meter. It was designed by Götz[36] and modified by Heisenberg and Wolf.[37] It is interesting to note that Fuchs and Robinson[38] designed and applied an analog operant paradigm to eye movement studies in anesthetized monkeys. The basic principle of the flight simulator is to amplify the currents generated by the tethered flying fly in the torque compensator as body saccades happen to visual inputs. In a particular experiment, the individual test flies suspended at the visual flight simulator face a choice between two visual patterns and are punished by heat when they turn to one of them (Figure 28.2). In the closed-loop mode, the fly's flight information recorded by potentiometer is fed back to the simulator and then returns to the potentiometer to control the position of the cylinder.[39] The panorama surrounding the tethered flying fly can be set to contain four visual targets with one object in each quadrant, but the two different objects are in the near quadrants. The memory of the fly tested under such conditions is defined as the performance that conditioned fly tracks to a 'safe' visual object and turns away from the 'danger' one after the unconditioned stimulus is turned off.[33,40,41] The learning indices of individual test flies in the visual flight simulator are highly variable. Guo *et al.*[41] studied the causes of inter-individual differences in visual learning performance. They found that three potential variables—age, flight practice, and diet—account for the variability.

Different Visual Memory Phases

Using operant conditioning and pharmacological analyses, Xia *et al.*[42]established three distinct phases in *Drosophila* visual memory formation: (1) an anesthesia-sensitive memory (ASM) phase that lasts for approximately 20 min after training and can be disrupted by a potassium chloride (KC1) feeding regimen and cold anesthesia, (2) an anesthesia-resistant memory (ARM) phase that is preserved if cold anesthesia or cycloheximide (CXM)-treated animals are tested, and (3) a long-term memory (LTM) phase that depends on novel protein synthesis and can be disrupted by CXM. In addition, a very short-term memory phase (pre-STM) can also be identified by feeding flies KCl. Visual memory rapidly decays during the first 3 hr after conditioning but remains measurable 48 hr later. Gong *et al.*[43] further reported that the mutant *dunce* appears to show normal learning during training but impaired short-term memory (STM), ARM, as well as LTM, whereas the mutant *amnesiac* shows disrupted middle-term memory (MTM) but normal STM and LTM. The latter finding suggests mechanistic correspondence of visual and olfactory learning.[44]

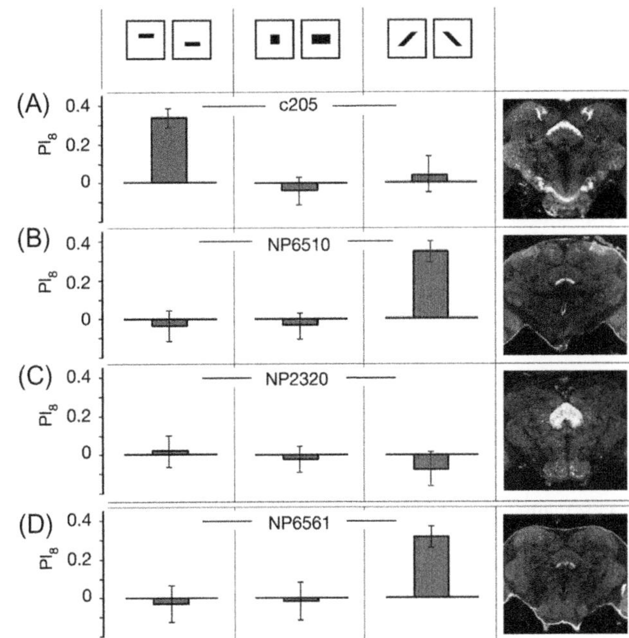

FIGURE 28.3 Distinct memory traces are spatially separated in the fan-shaped body. (A–C) Three GAL4 lines—c205 (A), NP6510 (B), and NP2320 (C)—driving *rut*$^+$ cDNA in a mutant *rut* background were tested for 3-min memory using patterns specifically testing for the single parameters 'elevation,', 'size,' or 'contour orientation.' Expression patterns in the fan-shaped body, using GFP as effector, are shown on the right. Note that in NP2320 (C), only columnar elements are stained. (D) Driver line NP6561 has a similar expression pattern as that of NP6510. In flies of the genotype *rut* 2080/Y;NP6561/UAS–*rut*$^+$, again only the memory for pattern parameter 'contour orientation' is rescued. *Source: Reproduced with permission from Liu* et al.*, 2006*[49]

INVARIANT RECOGNITION AND MEMORY TRACES

Invariant Recognition

Although flies have much less spatial resolution, they perceive the structured environment very well.[36,45] Flies can perceive five visual pattern parameters—size, color, elevation, vertical compactness, and contour orientation—as well as the spatial relationships between these characters.[46–50] Flies fail to remember the previous learned patterns when these patterns are shifted up or down by nine degrees or more in the arena of a flight simulator, a finding initially interpreted to support the retinotopic matching concept in visual encoding.[47] However, it was recently demonstrated that the memory retrieval in *Drosophila* in flight simulator is translation invariant and strict retinotopic matching is not indispensable.[50] Invariant visual pattern recognition is one of the important tasks in visual object recognition because the perceived signal of an object can undergo various changes (translation, rotation, scaling, and mirror symmetries) during the observation.[51]

Visual Memory

To explore the circuit mechanism underlying the translation invariance, it is necessary to know where the visual feature memory is stored in the fly brain. Olfactory memory has been assigned to the mushroom body (MB)[52] and the median bundle.[53] The finding of Chen *et al.*[54] suggests that CREB-dependent gene transcription and protein synthesis for LTM necessary for olfactory memory formation does not occur in MBs but, rather, in the two dorsal-anterior-lateral neurons. Thus, we may ask, where is the visual 'engram' stored in the *Drosophila* brain? By using the GAL4/UAS system to restore the Rutabaga (Rut) protein expression in *rutabaga* (*rut*) mutant flies,[55–57] Liu *et al.*[49] identified two groups of large-field neurons, F1 and F5 neurons, innervating the fan-shaped body (FB) responsible for visual memory formation (Figure 28.3). The F1 neurons in the first layer of the FB store the 'contour orientation,' whereas the F5 neurons in the fifth layer are used for 'elevation.' The morphological features of these two groups of neurons were characterized at the single neuron level.[58] The F1 and F5 neuron circuits have been suggested to be involved in invariant visual pattern recognition. Pan *et al.*[59] identified another subset of R neurons (R2/R4 m) in the ellipsoid body (EB) that appear to be involved in visual memory for all parameters in a parameter-independent manner. Thus, they concluded that FB and EB are involved in the fly's visual memory for invariant recognition. Quinn[60] noted that Liu and colleagues make very clear that "genetic trickery has converted *Drosophila* from one of the worst organisms for functional neuroanatomy to one of the better ones."

FEATURE EXTRACTION AND CONTEXT GENERALIZATION

Feature Extraction

Generalization means to deduce some 'general (abstract)' information from a certain set of stimuli and to make use of this information in further decision between new and unexpected sets of stimuli. —*Wehner*[61]

In abstract terms, generalization is the task of synthesizing a function that best represents the relationship between an input, X and an output, Y—by learning from a set of 'examples' x_i, y_i. —*Poggio and Bizzi*[62]

Do flies have some potential capability of generalization? Can flies be conditioned to recognize a particular visual feature from an object containing combinatorial conflicting features? Peng *et al.*[63] performed a series of learning experiments in the visual flight simulator to address these questions. Four different visual patterns (green upright and reverse T and blue upright and reverse T) were grouped by either shape feature or color feature (Figure 28.4). First, the flies were trained to extract the shape cue from these patterns containing combinatorial conflicting features. Both *wild-type Berlin* (*WTB*) and *Canton-S* (*CS*) flies failed to extract the shape feature from the four different objects with

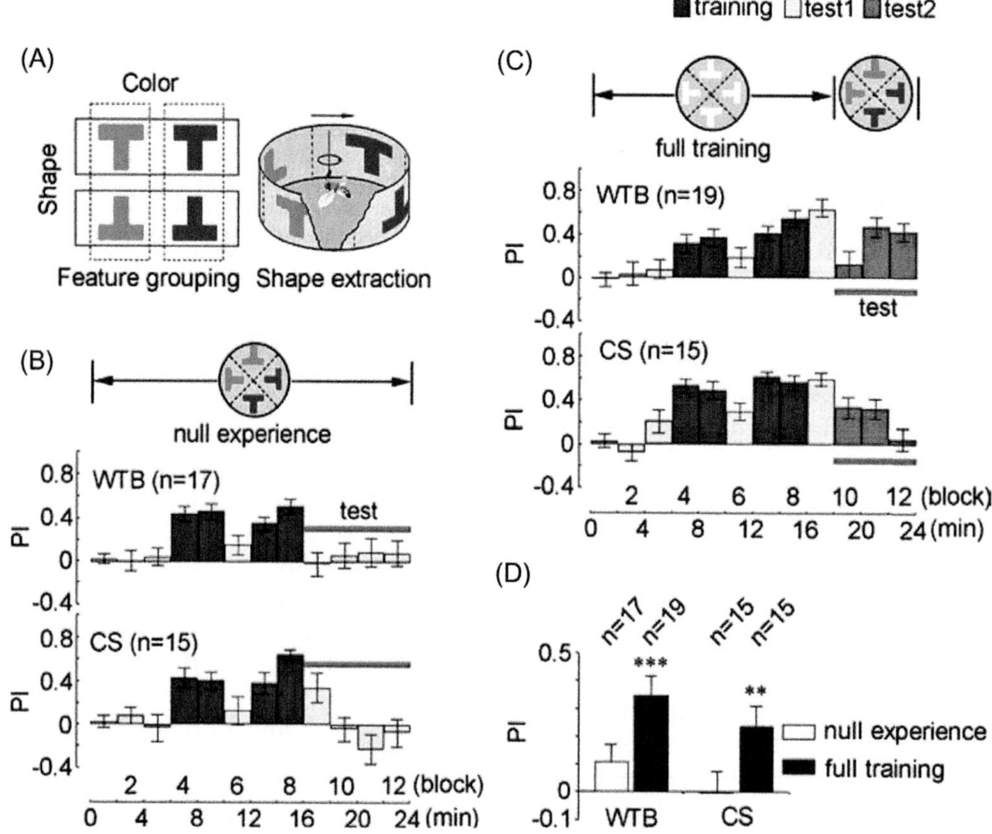

FIGURE 28.4 Experience-dependent shape feature extraction in *Drosophila*. (A) The paradigm of visual feature extraction. (Left) Feature grouping for four differently colored T patterns. The shape feature was the main component to distinguish the difference of the horizontal groups (solid line boxes), whereas the color feature was the main component to distinguish the vertical difference (dashed line boxes). (Right) Illustration of shape feature extraction in the arena of a flight stimulator, with the same shape but different color patterns in opposite quadrants. (B) The flies were directly trained to track the upright or inverted T's (shown in the top panel) with the 24-min protocol of pretest (yellow), training (brown), and test (yellow). Both WTB and CS wild-type flies exhibited poor ability (blocks 9–12) to extract the shape feature from objects with combinatorial shape–color features. (C) The flies were fully trained to learn the shape feature alone and then retested with combinatorial shape–color features (shown as the green histograms). Wild-type flies recognized the shape feature (blocks 10–12) after previous training. (D) Statistical analysis demonstrated that previous training significantly improves shape feature extraction in wild-type flies. *Source: Reproduced with permission from Peng* et al.[63]

combined shape and color features. Then, it was examined whether *Drosophila* can use prior experience to guide their feature extraction. Flies were initially trained to learn the shape cue alone and then were presented with conflicting features. Wild-type flies (*WTB*) were subsequently able to recognize the shape cue from the conflicting features consisting of color and shape in the testing period. Thus, prior experience with one particular feature can subsequently guide feature extraction from confused combinatorial features. In addition, flies recognized a shape feature from combinatorial shape–color features when the shape feature was reversely conditioned after prior training. These results suggest that the flies may extract the 'abstract' category of 'shape' feature in addition to a particular conditioned shape. The authors found that this capability was impaired in MB-ablated flies (mutant with miniature MB (*mbm*^{1}) and hydroxyurea ablation of MB (HU-treated flies)). Furthermore, they showed that MBs are also required for more complex feature extraction but dispensable for simple shape or color extractions. Thus, the ability to focus on a prior conditioned shape feature (or color) and to ignore the later-appearing distracters could be an important characteristic in concept formation. Poggio and Bizzi[62] noted, "One of the most obvious differences to machine learning is the ability of people and animals to learn from very few examples."

Context Generalization

The brain's ability to extract actual predictors of salient events from nonpredictive background stimuli as context is vital for visual encoding and recognition. Liu *et al.*[64] showed that visual memory retrieval in a visual flight simulator was partially context independent when the background color was changed between training and test. This so-called context generalization required MBs. Whether a stimulus plays as a predictor or context, it is not always fixed. Brembs and Wiener[65] found a shift in associative strength between the contextual color cues and the predictive punishment. Different memory templates can be evoked by the same color cues: They can be context-independent memory (color as context), context-dependent memory (color as occasion setters), or simple conditioned memory (color as conditioned stimulus). MBs are required for context-independent but dispensable for visual operant conditional discrimination. How does the fly brain accomplish this task? How do MBs decide between predictor and context? Brembs and Wiener proposed that the relative timing of stimuli is very important. In transgenic flies in which MB functions are impaired, their sensitivity to these subtle variations is altered, indicating that the MBs might stabilize visual memories against context changes.

In summary, feature and contest generalization can be considered as a kind of invariant pattern recognition, in which the local features in the perceived pattern have to be matched flexibly with their counterparts in the stored pattern, and both feature abstract and contest generalization may share common neural circuit mechanisms.[51] It is likely that flies are able to learn more complex patterns using their generalization abilities.

MULTISENSORY PERCEPTION AND CROSS-MODAL MEMORY

Multisensory Perception

Like most animals, fruit flies are equipped with multiple sensory channels through which they communicate with their environments. Multisensory integration should be especially advantageous for their high-speed flight activity.[66] For example, flies can easily track an attractive odor source hidden in the floor in an arena 1 m in diameter surrounded by a randomly textured, rather than a uniform white, visual background.[67] Interestingly, the flies could readily localize odor sources on the background with vertical stripes but failed to do so on a background with horizontal stripes. A further study showed that flies use large-field rather than small-field visual objects or landmarks to accurately locate their favorite food odor.[68] These results indicate that flies require specific visual cues for their olfactory-guided search activity. Optomotor flight in the presence of attractive odor cues can improve the fly's flight trajectory and benefit the optomotor control.[69] When the MBs were ablated by HU treatment, the olfactory enhancement of yaw optomotor response was attenuated, implying the MBs are indispensable in olfactory–vision interaction.[70] The integration of visual and mechanosensory cues in flight control has also been analyzed, and it has been found that flying flies can use mechanosensory cues produced by wind stimuli to maintain forward flight.[71]

Cross-Modal Memory

Gottfried and Dolan[72] reported cross-modal facilitation between vision and olfactory perception in humans. Little is currently known about cross-modal processing and learning in insects.[73] Honeybees can use an occasion-setting mechanism to solve uncertainty in bimodal (color–odor) conditioning, in which colors do not themselves elicit the operant learning but predict the odor–sucrose association.[74] Cross-modal spatial learning in fly has been studied with a heat maze paradigm, an analog of the Morris water maze for rodents.[75,76] It was found that flies can use either proximal or distal visual cues to improve their performance in safe-zone navigation.[75] This finding indicates that the fly's brain integrates multiple sensory modalities to improve its performance. Using a wing-beat processor that allows the tethered flying fly to choose the flight direction to the preferred visual object, Guo and Götz[77] demonstrated a cross-modal association between visual object and repellent odor under operant conditions. In order to further investigate the cross-modal learning between vision and olfaction in *Drosophila*, Guo and Guo[78] updated the flight simulator by adding an odor delivery system (Figure 28.5). By decreasing the vertical visual angle between two horizontal bars (upper and lower) in the arena, the authors identified the threshold of visual learning. In a similar way, olfactory threshold was also determined between the two odors, 3-octanol and 4-methylcyclohexanol. When individual flies were presented with both visual and olfactory thresholds, substantially superadditive memory scores were achieved. It is in line with the 'principle of inverse effectiveness,' which states that cross-modal enhancements can reach the maximum when the individual stimuli are minimally effective.[73] The integration of multiple sensory inputs can dramatically enhance the detection and discrimination of external stimuli and markedly speed responsiveness.[79–83] Using the 'sensory preconditioning' paradigm,[40,84] Guo and Guo[78] further found that training flies to discriminate between two horizontal bars and then presenting olfactory and visual stimuli without any reinforcements leads to an olfactory memory for the odor paired with the trained horizontal bars. What are the neural mechanisms underlying the cross-modal interaction? Multisensory integrative (MSI) neurons have been identified in cat cortex,[85,86] rats,[87] and monkeys.[88–90] In nonhuman primates, a region has been found to receive afferent inputs from both the primary olfactory cortex and the visual association areas.[91] As an alternative neural mechanism to MSI neurons, the apparent synthesis of information from different sensory modalities might be achieved through the synchronized activities of sensory-specific cortices.[92]

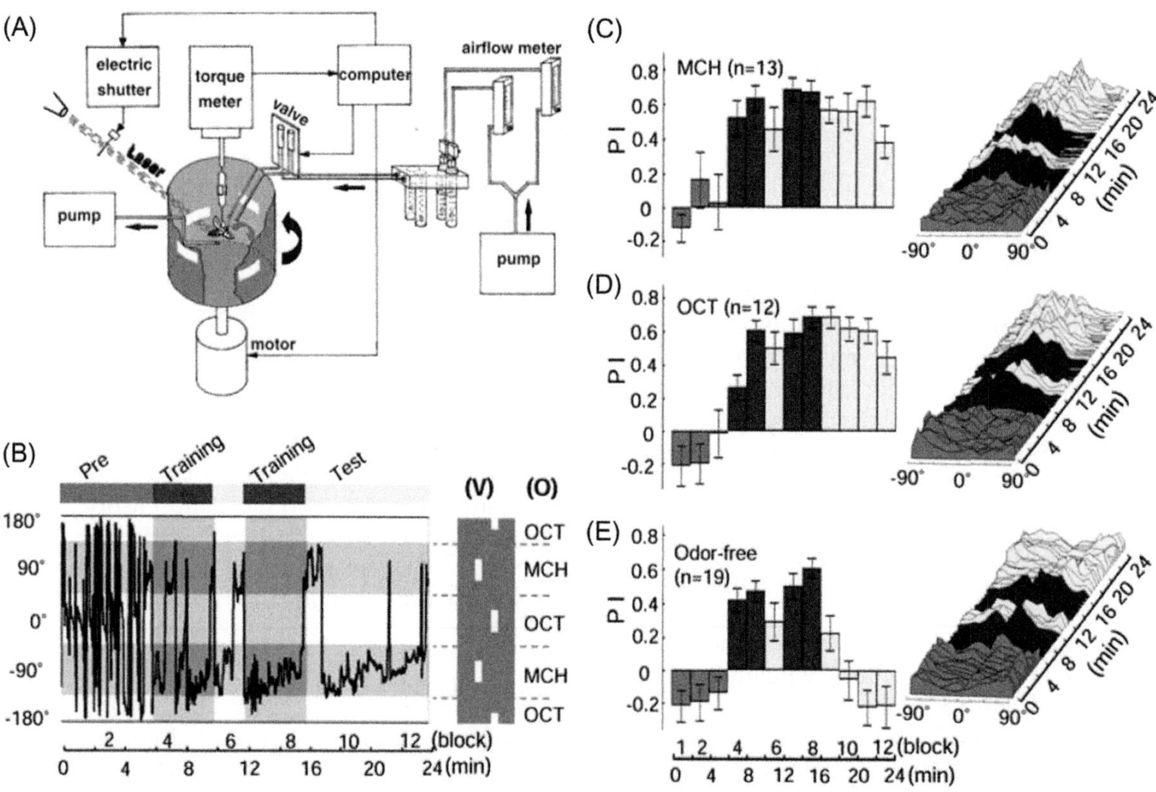

FIGURE 28.5 **Cross-modal interactions between olfactory and visual learning.** (A) The modifications for using the flight simulator in olfactory conditioning. (B) A representative trace of flight path of a single fly with time in arena position (−180° to 180°) during preconditioning ('pre'), training, and test period for unimodal olfactory conditioning in the dark. Shown on the right are corresponding visual (V) and olfactory (O) cue locations in the arena. (C and D) Summary of olfactory operant conditioning. Single CS flies were trained to prefer 3-octanol (OCT) to 4-methylcyclohexanol (MCH) and MCH to OCT, respectively. (E) The airflow was odor-free. *Source: Reproduced with permission from Guo and Guo.*[78]

Which mechanism is adopted by the fly brain? Regardless of extreme difficulties encountered, electrophysiological recordings from a fly's brain during ongoing cross-modal performance are required to address this question. In locusts, a local field potential (LFP) of 20–30 Hz oscillation has been recorded in the calyx of MB, induced by an odor puff to the antennae.[93–95] This kind of electrophysiological study can help us to understand the neural circuit mechanisms underlying cross-modal interaction.

SPATIAL LEARNING AND WORKING MEMORY

Spatial Working Memory

> The term working memory refers to a brain system that provides temporary storage and manipulation of the information necessary for such complex cognitive tasks as language comprehension, learning, and reasoning. —*Baddeley*[96]

Electrophysiological recordings in primates showed that sustained spiking of neurons represents the working memory encoding an object that has been presented to the subject a short time ago.[97] Do the flies have working memory? Interestingly, flies keep walking toward an object target even after the target disappears.[98] This persistence of orientation for objects is interpreted as a form of spatial working memory for objects. Neuser *et al.*[99] developed a detour paradigm to further explore the putative spatial working memory in flies. Individual flies put in a cylindrical virtual-reality arena were presented with two dark vertical stripes opposite each other (Figure 28.6). Wild-type flies spent much time walking toward each of the two targets. When the flies crossed over the invisible midline, they were confronted with a sudden disappearance of the stripes and a subsequent appearance of a novel target at a 90-degree angle. Flies were found to turn toward the new target if it was presented for more than 500 msec. If the new object to which flies oriented also disappeared and no other visual stimuli appeared on the cylinder wall, the flies returned to the original target, which had disappeared for a while. This result suggests that a spatial memory is formed in the fly brain storing position information of the target.

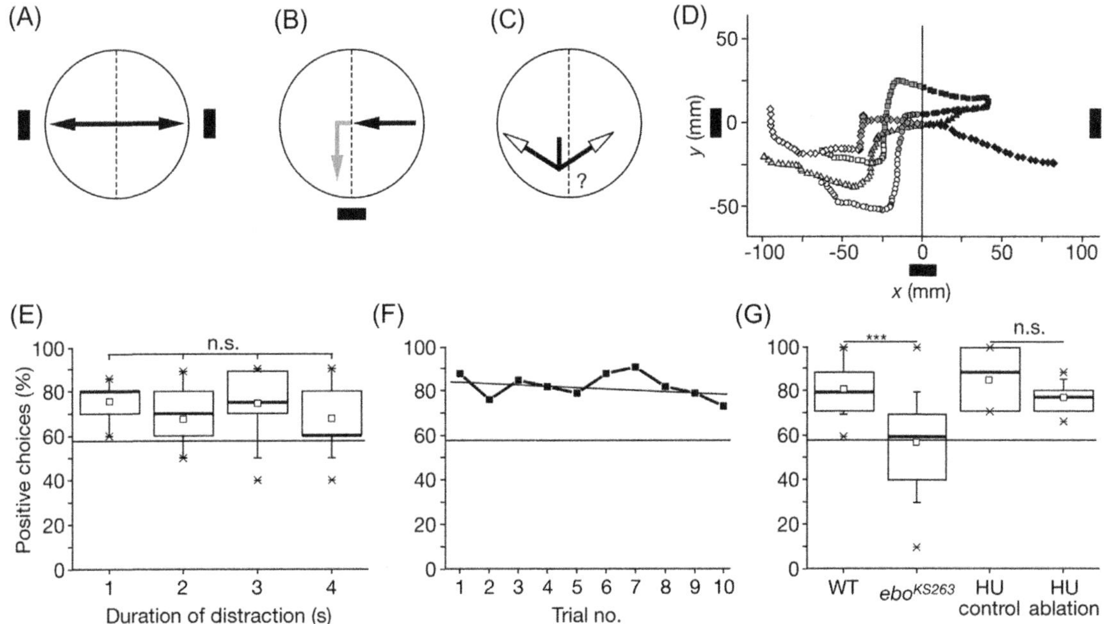

FIGURE 28.6 **Orientation memory in the detour paradigm.** (A) A fly patrols between two vertical stripes shown on a cylindrical screen. On crossing the midline, the stripes disappear and simultaneously another one appears laterally to the fly. (B) After the fly has turned toward the distracter for 1 sec, this stripe also disappears. (C) Subsequently, it is determined whether or not the fly turns back toward its original target. (D) Walking traces of four wild-type males. Black symbols, initial phase (a); gray symbols, distractor phase (b); open symbols, memory phase (c). The maximum duration of a trace is 20 sec. (E) Prolonged distraction does not change the orientation memory ($p = 0.22$). Bold horizontal lines represent the medians, squares represent the means, boxes represent the 25 and 75% quartiles, whiskers represent the 10 and 90% quantiles, and stars represent the extreme values. (F) Percentage of positive choices for each of 10 consecutive trials ($N = 73$; $r^2 = 0.08$). (G) Mushroom body ablation with hydroxyurea (HU) does not impair the memory in comparison with mock treatment ($p = 0.25$). In contrast, ebo^{KS263} mutants show a reduced performance compared to wild-type (WT) flies ($p < 10^{-3}$), as indicated by three asterisks. n.s., not significant. The horizontal line in panels E–G indicates the 58% chance level. *Source: Reproduced with permission from Neuser* et al.[99]

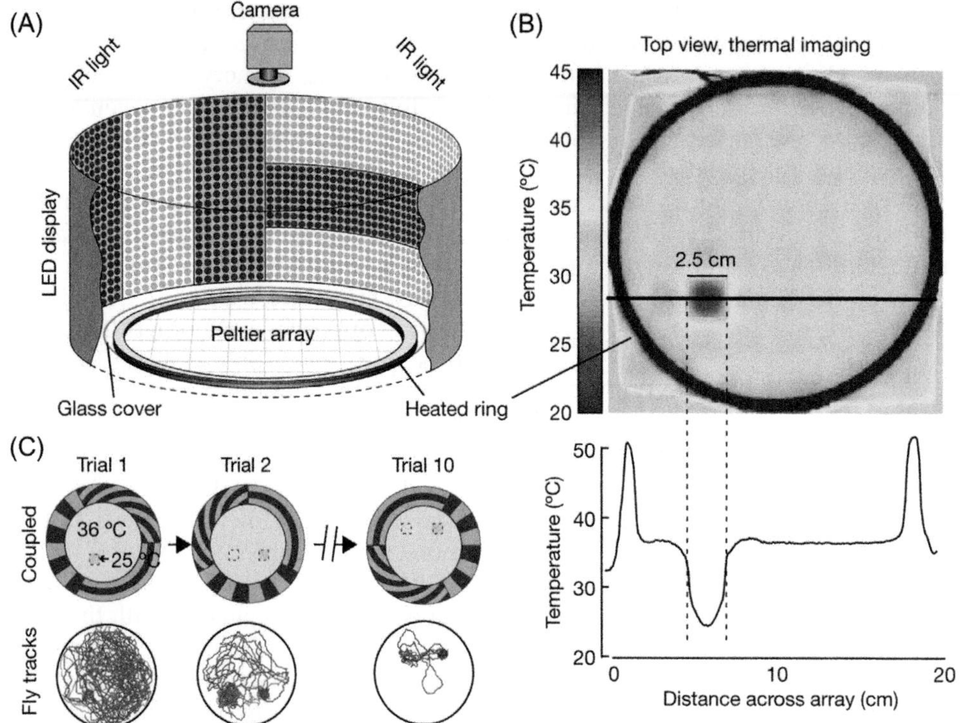

FIGURE 28.7 ***Drosophila*** **trained in thermal–visual arena show place learning.** (A) Illustration of the arena. The floor is composed of 64 thermoelectric modules (a Peltier array), the panorama is provided by a 24 × 192 light-emitting-diode (LED) display, and flies are recorded using a camera under infrared (IR) illumination. (B) (Top) Thermal imaging view of the arena's floor showing the uniformly warm surface with a single cool tile; also shown is the heated ring barrier. (Bottom) Temperature readings across the arena. (C) Trajectories of four representative flies from trials 1, 2, and 10 are shown below a diagrammatic representation of the visual panorama denoting the locations of the cool tile in the previous trial (dashed square) and in the current trial (blue square). In this coupled condition, the position of the cool tile relative to the visual panorama remains constant even as its absolute position changes between trials. *Source: Reproduced with permission from Ofstad* et al.[104]

Interestingly, flies with deficient γ-aminobutyric acid (GABA)ergic ring neurons in EB failed to track the original object. Protein kinase S6KII was also found to be indispensable for ring neurons to hold this working memory. The relevant ring neurons in EB express GABA and dDA1 dopamine receptors,[100] which seems to be functionally related to the GABA–dopamine system in the prefrontal cortex, a brain area responsible for working memory in mammals.[101]

Landmark Learning

Approximately 700,000 pyramidal cells are included in areas CA1 and CA3 of the rat hippocampus. Extracellular recordings in freely moving rats show that most pyramidal cells are 'place cells' that are context dependent.[102] The 'hippocampal cognitive map' in rats is thought to underlie spatial navigation and landmark learning in mammals.[103] It is unknown whether flies are able to identify and remember visual place in addition to the five pattern parameters mentioned previously. Ofstad *et al.*[104] designed a thermal–visual arena containing a hidden cool square (at 25°C) surrounded by a uniformly warm surface heated to an aversive temperature of 36°C. In this space learning paradigm, flies had to search for a hidden 'cool' target in an unappealing warm surrounding (Figure 28.7). Flies could find their favorite spot through a vision-guided navigation, indicating that the flies are capable of place learning and memory. Neurons in the EB are required for visual place learning. In contrast with dopamine (DA) as reward prediction error in reinforcement learning in nonhuman primates,[105] it has been proposed that DA is the only neuromodulator implicated in negatively reinforced learning in *Drosophila*.[106,107] Whether DA also plays a role in space learning needs further investigation. On the other hand, Sitaraman *et al.*[108,109] found that serotonin, but not octopamine, is necessary for forming and retrieving place memory in *Drosophila*. Further in-depth studies dissecting the cellular basis of place learning in *Drosophila* will be helpful to elucidate the neural and cellular processes involved in spatial learning and memory.

ATTENTION-LIKE FIXATION BEHAVIOR AND VISUAL SELECTIVE ATTENTION

Attention-Like Fixation

Attention was defined by William James[110] as "the taking possession of the mind in clear and vivid form of one out of what seem several simultaneous objects or trains of thought." The original studies on flight behaviors such as detection and tracking of moving objects or partners were carried out on housefly *Musca domestica*.[111] Do the flies use an attention mechanism to guide their flight orientation? It has been well-established that fruit flies exhibit spontaneous flight paths toward a black object either stationary or moving in the flight simulator.[112] A novel fixation paradigm has been designed by Xi *et al.*[113] to elucidate the neural circuits underlying attention-like processes in *Drosophila*. During the fixation and tracking of a visual stripe, visual and olfactory distracters have been applied to interfere with orientation. In another paradigm, visual choice among two or three stripes with the same or different contrast in the panorama was studied. Wild-type flies perform better fixation to the more salient target than do MB-deficient flies under conditions with visual background noise or some less salient distracters. These results suggest that flies restrict their visual attention to parts of the visual field during flight.[37] Thus, the flies display some attention-like elements, such as selection, suppression, inhibition, and fixation tracking.

Selective Attention

Visual selective attention is defined as "the ability to confine visual processing to a visual field region of choice".[114] Animals including humans are endowed with selective attention to focus primarily on the visual signals closely relating to their safety and to what they are searching for.[115–122] Because behavior observation alone in *Drosophila* is not sufficient to reveal the neural circuit mechanisms underlying high-order cognitive functions, such as salience-based selective attention, the electrophysiological recording approach has been tried in the central nervous system in *Drosophila*.

Recordings of LFPs were found to be related to phenomena such as visual novelty in attention-like processes.[123] LFPs appear more stereotypical in their temporal properties and are less modulated by *dunce* and *rutabaga* genes involved in visual short-term memory formation. These mutants show distinct optomotor control efforts from that in wild-type flies, and the attention-like performances are lacking in these mutants. It was further uncovered that the defects of selective attention-like performance were reflected in both vision- and olfactory-based behaviors. Given the fact that mutations affecting olfactory memory in *Drosophila* also produce distinct defects in visual attention-like behavior, van Swinderen *et al.*[124,125] proposed that attention and memory probably share one memory system in generating oscillatory brain activity required for attention-like process, and the MB seems to be a common structure of visual attention and olfactory memory in fly's brain.

DECISION MAKING

Perceptual Decision

There are two conceptual frameworks of decision making: perceptual and value-based.[126] A functional magnetic resonance imaging study by Heekeren *et al.*[127] suggests that the human brain computes perceptual decisions using a simple subtraction mechanism by the comparison of the outputs of different pools of selectively tuned neurons. Using multiunit array and classic electrophysiological approaches, Tang and Jousola[128] recorded the LFP from both left and right optic lobes in a tethered flying fly while it attempted to follow competing gratings and generated yaw torques, showing a synchronized switch-like fluctuation among torque and spiking pattern. Each switch-like attempt of a test fly to turn left or right was taken as its momentary choice of the stimulus. Both single neuron firing and LFPs were recorded before and during the decision making by two miniaturized electrodes implanted in the left and right optical lobes. Neural activity correlated with the visual stimuli and the fly's orientation choice, indicating that the fly's torque responses as behavioral choice output are initiated and modulated endogenously. When a fly chose one stimulus, the ipsilateral optic lobe was excited and the contralateral optic lobe was inhibited. The authors indicated that the coupled excitatory and inhibitory modulation seems unlikely to attribute to conventional binocular rivalry; rather, a fly may activate its attention-like circuit(s) to better discriminate relevant changes in the optic flow. This modulation is likely to arise from internal dynamics of neural circuits, and it may result in a gating-process that enhances/suppresses the overall output of the relevant/opposite optic lobe facing two sides of stimuli. The previously mentioned authors demonstrated that when a fly appears to decide to turn, intrinsic neural activity acts on the optic lobes to modulate visual input from the eyes.

Egg-Laying Site Choice

Yang *et al.*[129] reported a simple egg-laying site choice behavior in *Drosophila*. A sequence of egg-laying behaviors was observed in *Drosophila*, including the search program, the ovipositor motor program, and the clean and rest program. This sequence lasts from a few seconds to several minutes, during which the fly walks around and probes the media with its proboscis and ovipositor to evaluate egg-laying site quality. In addition to making a simple choice to accept or reject the site for egg-laying, the fly can make a context-dependent choice by valuing a given egg-laying site differently according to the availability of other laying options. The results show that flies can balance the 'desirability' of egg-laying sites versus the 'effort it takes to locate them' in making their decisions. Moreover, the insulin-like peptide 7 (ILP7)-expressing neurons were shown to be important for proper execution of egg-laying decisions, providing new anatomical and molecular evidence for egg-laying site selection.

Choice Between Goal-Directed and Habitual Response

A previous study found that classical learning and operant learning in the flight simulator could be differentiated at the molecular level.[40] Specifically, manipulation of the levels of *Rutabaga*, a type I adenylyl cyclase widely believed to function as a coincidence detector of the conditional and unconditional stimuli, was found to affect classical learning but not operant learning. Conversely, altering the function of protein kinase C impaired operant learning, whereas it left classical learning intact.

Interestingly, there is a hierarchical interaction between these two learning systems.[39] Flies preferentially learn the classical predictor over the operant predictor in composite learning situations in which both components are present. Thus, the results suggest that the classical learning system inhibits the operant learning system, allowing generalization of useful information and preventing premature habit formation. However, when trained more extensively in this task, flies fail to generalize the classical memory. Instead, the operant learning system dominates and the flies show habitual action-based choice behaviors accordingly. The formation of habits may enhance the efficiency in producing the required response during extended training scenarios. Nevertheless, habit formation is also accompanied by a loss of behavioral flexibility and, in extreme cases, may lead to irrational choice behavior. Intriguingly, the inhibition of the operant learning systems by the classical system is mediated by the MBs, especially the α/β lobe substructures.[130] Flies with silenced MB output recapitulated the performance of flies observed after more extended training; the generalization of the classical predictor was impaired and habitual responses governed behavior. These results provide a foundation for further studies on the molecular and circuit mechanisms of habit formation in *Drosophila*.

Salience-Based Decision

Decision making is essential for the survival of animals and is the origin of the rational aspects in our lives.[126,131–133] From a neuroeconomic standpoint, to make value-based decisions is to evaluate the costs and benefits of alternative options.[126,134] We asked whether value-based decision making is patent only for high-level animals or human beings. Does *Drosophila* make value-based decisions? If yes, what are the neural circuit mechanisms? To answer these questions, Guo and co-workers[135,136] designed both 'shape/color' and 'color/position' dilemmas in the flight simulator (Figure 28.8). In the 'color/position' dilemma, individual flies are conditioned to choose a flight path in accordance with the color (green and blue) and position (upper and lower) cues and then confronted with choice test with color and position cues reversed after training. Results revealed that wild-type flies *WTB* and *CS* are able to resolve this paradoxical situation by evaluating the 'value' of the current information and making a nonlinear 'winner-takes-all' decision. The decision curve exhibits a clear-cut and sigmoid-like function. Moreover, they have shown by sequential decision-making experiments that the flies can make the second consecutive choice after the first one. This is important for the balance between maintaining the pre-existing choice and switching to a new decision. Using different approaches to block the MBs (*mbm*[1] mutants lacking MB, hydroxyurea-treated flies, and genetic conditioning silencing MBs), they further found that flies lacking MBs intact function seem to have difficulty resolving 'conflicting' situations; the decision making becomes a linear process. Furthermore, they found that silencing dopaminergic transmission by genetic tools appears to selectively damage the decision-making process without affecting the fly's memory retrieval and sensory/motor control (Figure 28.9). Thus, DA and MBs are essential for flies to make a nonlinear discrete and firm decision when confronted with conflicting cues. The results suggest that the MB–DA circuit acts as the commander in fly's brain by providing a gating and a gain mechanism to suppress less important stimuli, enhance the relevant and salient stimuli, and thereby modulate the consequent behavioral outputs. The MB–DA circuit portrays an ideal model and a general principle to examine the neural process underlying cognitive

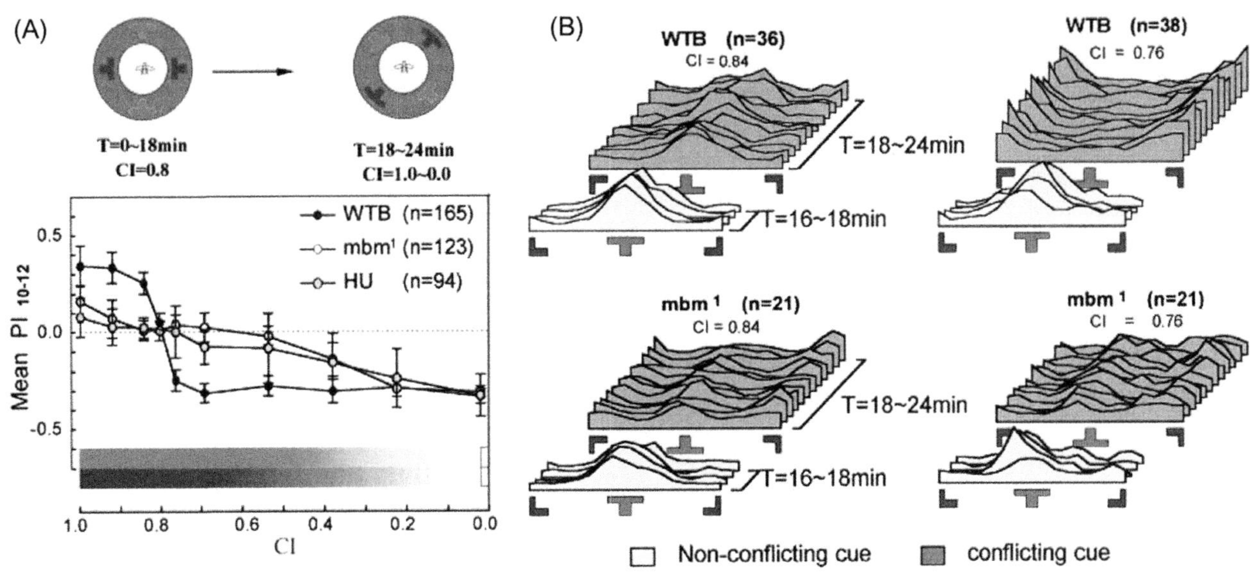

FIGURE 28.8 **Choice behaviors of WTB, *mbm*^{1}, and HU-treated flies facing the 'color/shape dilemma.'** (A) The flies were trained at CI = 0.8 to prefer a green upright T over a blue inverted T, but they were presented with conflicting cues (green inverted T and blue upright T) during the post-training session over the entire range of CI values (1.00 to 0.00). Data points represent mean PI_{10-12} values (SEM; n = 10–38 for each point). (B) Angular histograms during post-training sessions for WTB and *mbm*^{1} flies. The flies were presented with original nonconflicting cues during the first post-training test session (T = 16–18 min), followed by conflicting cues in subsequent test sessions (T = 18–24 min). *Source: Reproduced with permission from Tang and Guo.*[135]

behavior and even helps to promote the relevant research in high-level animals. It is worth noting that in primates, the DA neurons projecting from the ventral tegmental areas (DA system) to several cortical areas (prefrontal cortex, striatum, and basal ganglia) code properties such as error tracking that provide explanatory power for decision making.[105,126,137]

Computational Studies on Decision Making

Wu and Guo[138] adopted a computational modeling approach to investigate how different brain areas and the DA system work together to drive a fly to make salience-based decisions (Figure 28.10). In addition to a direct pathway from the feature binding area to the motor center, another pathway connecting two areas via MB, a target of DA release, is proposed to be critical for the decision-making behavior. Raised DA level is hypothesized to be induced by complex choice tasks and to enhance lateral inhibition and steepen the response gain of neurons in MBs. Computer simulation results showed that training before the choice task helps to assign values to formerly neutral features. The memorized values characterized by a linear choice curve were found to drive a circuit model in which the MBs are blocked to generate a simple choice. With respect to an intact circuit, enhanced lateral inhibition dependent on DA critically promotes competition between alternatives, turning the linear into nonlinear choice behavior. These results account well for the previous experimental data. In addition, several predictions can be tested experimentally based on these results. For instance, ambiguous/conflicting visual cues should evoke a phasic DA increase in the MB. More accurately, phasic DA release should begin when the animal perceives a difference in value between two visual targets. Another prediction is that activation of GABA receptors should be observed in the MB along with a raised DA level. In other words, it is predicted that flexible tasks in *Drosophila*, such as experience-dependent decision making, require the involvement of GABA receptors in the MB. In terms of computational modeling, it is interesting to note that network topology may affect behavioral output in decision making. By constructing a recurrent network model with four different topologies (regular, random, small-world, and scale-free), Lu *et al.*[139] demonstrated by computer simulation that the small-world network has the best performance in decision making for the low-noise condition, whereas the random network is most robust for high noise.

Collective Learning and Social Decision

Decision-making behaviors do not only emerge in individual animals but also can occur in animal groups. Decisions even independently made by individuals

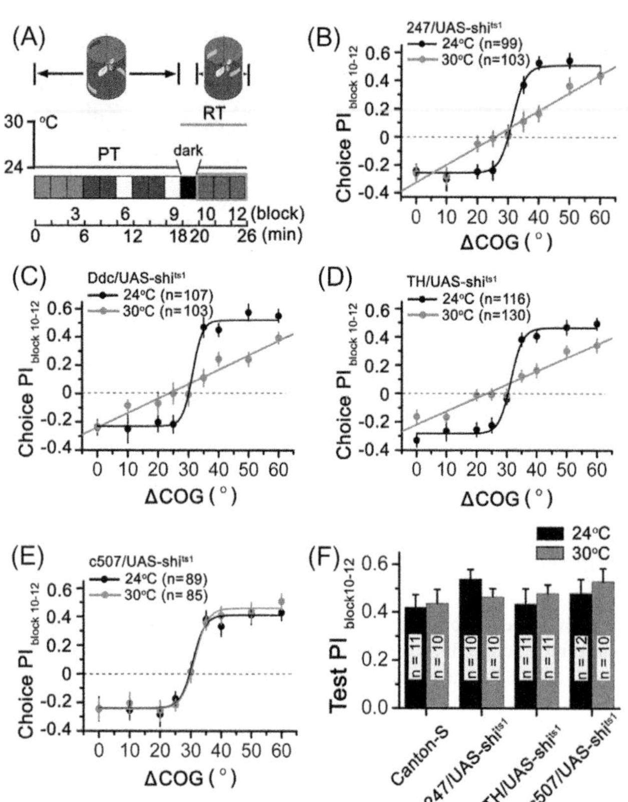

FIGURE 28.9 Choice behavior depends on dopamine and MBs. (A) Choice behavior–temperature shift paradigm involved training flies at PT (24°C) and testing at RT (30°C). (B) Choice PI in 247/UAS-*shi*^{ts1} flies fit a sigmoid curve at PT (Boltzmann fit, $r^2 = 0.99$). In contrast, RT resulted in genetic silencing of MB function and defective choice performance (linear fit, $r^2 = 0.97$). (C and D) Choice behaviors in Ddc-Gal4/UAS-*shi*^{ts1} flies (PT, Boltzmann fit, $r^2 = 0.99$; RT, linear fit, $r^2 = 0.95$) and TH-Gal4/UAS-*shi*^{ts1} flies (PT, Boltzmann fit, $r^2 = 0.99$; RT, linear fit, $r^2 = 0.95$). (E) Choice behavior in c507/UAS-*shi*^{ts1} flies (Boltzmann fit, $r^2 = 0.99$ at PT and RT). (F) Transgenic and control flies showed normal memory retrieval at both RT and PT. *Source: Reproduced with permission from Zhang* et al.[136]

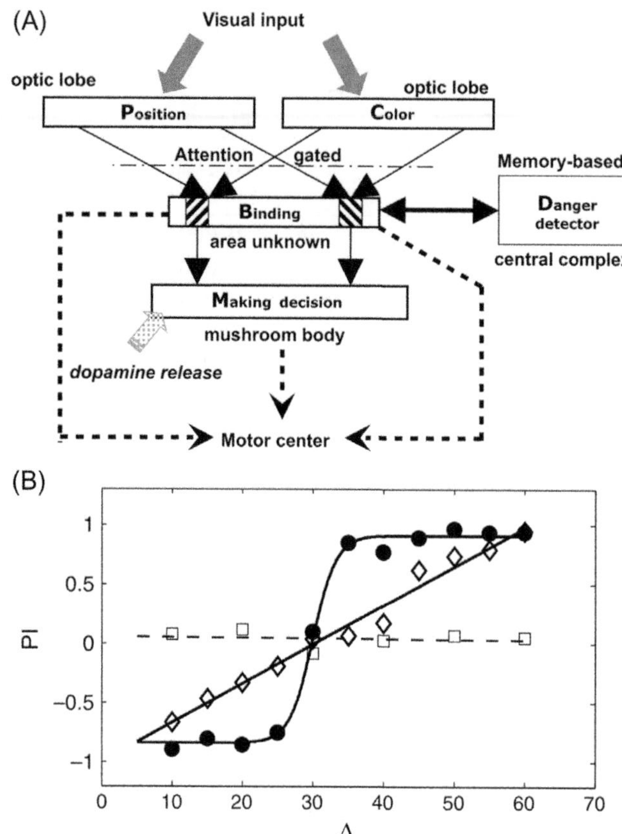

FIGURE 28.10 Network model for decision-making task in *Drosophila*. (A) Schematic model architecture. Input to binding module is gated by attention (dot–dashed line). Hatched boxes mark two recognized visual bars with different colors in the binding area. The MB is the only target of dopamine release (solid arrow). The transformation from a decision to an action (dashed lines) is not simulated. (B) Choice curves of computer simulation. Before training, the learning network shows no preference for two targets, which is evidenced by PIB near zero (dashed line with square markers). After training, a linear choice behavior emerges in the circuit with the M-module excluded (solid line with diamond markers). In an intact network, the small choice PI originating from the B-module is substantially increased with dopamine modulation in the M-module, resulting in a clear-cut decision behavior (solid sigmoid curve with solid circle markers). *Source: Reproduced with permission from Wu and Guo.*[138]

mostly show the trace of the group's tendency and are influenced by the ratio of experienced members or average level of attraction or repulsion to the stimuli within the group and so on. Failure to make a consensus decision on a critical choice may be costly for the colony, so grouping animals evolve to prevent it. In eusocial insect honeybee swarms, choosing a nesting site is a natural behavior produced in the spring. A cross-inhibition model, similar to that in the neural system, was proposed to address how this competing decision is made.[140] Each scout honeybee produces waggle dances advertising its choice of a particular nest site. The scout not only advertises for its own choice but also sends inhibitory stop signals to other scouts if its own goal differs from those indicated by the dance of other scouts. Thus, the decision of the swarm is made by dancers that receive the least stop signals and win the vote by majority. In contrast, in fish schools with conflicting interests among individuals, uninformed individuals can inhibit the choice process dictated by the strongly opinionated minority and help to achieve democratic consensus.[141] A variety of theoretical models have been proposed to explain these two modes controlling the collective decision making, but the underlying neural mechanisms are unknown in both cases.

Social decision is not exclusive to eusocial species. Solitary animals still influence the behavior of other animals of the same species—for example, mate copying behavior in *Drosophila*. Mery *et al.*[142] found increasing female preference for a male in poor condition after watching another female chose it as a partner, suggesting that fruit fly can generalize social information to help make difficult individual decisions or adapt to some sort of social criteria in mate-choice behavior. Analogously, flies can also share searching information in food exploration.[143] In another case, which is not as obvious as that previously discussed, flies in a T-maze were forced to make a choice between a conditioned odor and a secure odor in long-lasting memory retrieval tasks.[144] Flies tested in groups showed significantly better anesthesia-resistant memory (ARM) than those tested individually, independent of the context during learning (individually or in groups). Regardless of the mechanism involved, the group members facilitated ARM retrieval, demonstrating the positive effect of social circumstance on cognitive behavior. Interestingly, protein synthesis-dependent LTM, another form of long-lasting memory, being separated by a different training protocol from ARM, showed no social effects in group tests. These findings suggest that social effects on the decision of individuals may require a potential trigger that we do not yet understand.

PERSPECTIVES

Animal intelligence can be understood as "robust active maintenance of goals and other task-relevant information, and rapid updating of this information to keep pace with a changing environment or task".[145] Although it is difficult to build the bridge from fly learning to 'fly intelligence,' we are confident that "*Drosophila* is poised to help us decipher how the brain works"[146] because *Drosophila* implements even rather complex tasks, and it does so with a small brain.

We propose to explore the remote root of 'fly intelligence' along two threads in the future. The short-term goal is to precisely understand how the DA−GABA−MB circuit implements the dynamic gating gain inhibition process during decision making. We hypothesize that the DA−GABA−MB circuit suppresses noisy stimuli and emphasizes salience. In the long term, we need to find more evidence for the critical involvement of the MBs in a variety of high-order visual cognition, although MBs seem to be dispensable for basic visual learning. We hypothesize that there should exist functional connections between MBs and the central complex (CC) for regulating goal-directed behaviors. The MB−CC−DA−GABA circuit might provide integrated neural substrates underlying the fly intelligence.

Acknowledgments

We sincerely thank Zhihua Wu, Yan Zhu, Yueqin Peng, and Li Liu for critical reading of the manuscript and helpful comments. This work is supported by 973 Program (Major State Basic Research Program; grant 2011CBA00400 to A.G.) and the National Science Foundation of China (grants 30921064 and 31130027 to A.G.).

References

1. Livingstone M, Hubel D. Segregation of form, color, movement and depth: anatomy, physiology and perception. *Science*. 1988;240:740−749.
2. Borst A. *Drosophila*'s view on insect vision. *Curr Biol*. 2009;19: R36−R47.
3. Land MF, Fernald RD. The evolution of eyes. *Annu Rev Neurosci*. 1992;15:1−29.
4. Kirschfeld K. Das neural superpositionsauge. *Fortschr Zool*. 1973;21:229−257.
5. Trujillo-Cenoz O, Melamed J. Compound eye of dipterans: anatomical basis for integration—an electron microscope study. *J Ultrastruct Res*. 1966;16:395−397.
6. Zuker CS, Cowman AF, Rubin GM. Isolation and structure of a rhodopsin gene from *D. melanogaster*. *Cell*. 1985;40:851−858.
7. Cowman AF, Zuker CS, Rubin GM. An opsin gene expressed in only one photoreceptor cell type of the *Drosophila* eye. *Cell*. 1986;44:705−710.
8. Montell C, Jones K, Zuker CS, Rubin GM. A Second opsin gene expressed in the ultraviolet-sensitive R7 photoreceptor cells of *Drosophila melanogaster*. *J Neurosci*. 1987;7:1558−1566.
9. Zuker CS, Montell C, Jones K, Laverty T, Rubin GM. A rhodopsin gene expressed in photoreceptor cell R7 of the *Drosophila* eye: homologies with other signal-transducing molecules. *J Neurosci*. 1987;7:1550−1557.
10. Fortini ME, Rubin GM. Analysis of *cis*-acting requirements of the Rh3 and Rh4 genes reveals a bipartite organization to rhodopsin promoters in *Drosophila melanogaster*. *Gene Dev*. 1990;4: 444−463.
11. Chou W, Huber A, Bentrop J, et al. Patterning of the R7 and R8 photoreceptor cells of *Drosophila*: evidence for induced and default cell-fate specification. *Development*. 1999;126:607−616.
12. Yamaguchi S, Desplan C, Heisenberg M. Contribution of photoreceptor subtypes to spectral wavelength preference in *Drosophila*. *Proc Natl Acad Sci USA*. 2010;107:5634−5639.
13. Zuker CS. Biology of vision of *Drosophila*. *Proc Natl Acad Sci USA*. 1996;93:571−576.
14. Ranganathan R, Malicki DM, Zuker CS. Signal transduction in *Drosophila* photoreceptors. *Annu Rev Neurosci*. 1995;18:283−317.
15. Kirschfeld K, Franceschini N, Minke B. Evidence for a sensitising pigment in fly photoreceptors. *Nature*. 1977;269:387−389.
16. Hassenstein B, Reichardt W. Systemtheoretische analyse der Zeit-, Reihenfolgen- und vorzeichenauswertung bei der beweigungsperzeption des ruesselkaefeafers *Chroronphanus*. *Z Naturforsch B*. 1956;11b:513−524.
17. Buchner E. Elementary movement detectors in an insect visual system. *Biol Cybern*. 1976;24:85−101.
18. Reichardt WE. Autokorrelations-auswertung als funktionsprinzip des zentralnervensystems. *Z Naturforschung B*. 1957;12: 448−457.
19. Rau RMV, Vidyasagar TR. Apparent movement with subjective contours. *Vision Res*. 1973;13:1399−1401.
20. Reichardt WE. Movement perception in insects, processing of optical data by organisms and machines [Course XLIII] In: Fermi E, ed. *Proceedings of the International School of Physics*. New York: Academic Press; 1969.

21. Sperling G. Movement perception in computer-driven visual displays. *Behav Res Methods Instrum*. 1976;8:144–151.
22. Rister J, Pauls D, Schnell B, et al. Dissection of the peripheral motion channel in the visual system of *Drosophila melanogaster*. *Neuron*. 2007;56:155–170.
23. Takemura S, Karuppudurai T, Ting CY, Lu Z, Lee CH, Meinertzhagen IA. Cholinergic circuits integrate neighboring visual signals in a *Drosophila* motion detection pathway. *Curr Biol*. 2011;21:2077–2084.
24. Chubb C, Sperling G. Two motion perception mechanisms revealed through distance-driven reversal of apparent motion. *Proc Natl Acad Sci USA*. 1989;86:2985–2989.
25. Chubb C, Sperling G. Drift-balanced random stimuli: a general basis for studying non-fourier motion perception. *J Opt Soc Am A*. 1988;5:1986–2007.
26. Lu Z, Sperling G. Three-system theory of human visual motion perception: review and update. *J Opt Soc Am A*. 2001;18:2331–2370.
27. Mareschal I, Baker CL. Temporal and spatial response to second-order stimuli in cat area 18. *J Neurophysiol*. 1998;80:2811–2823.
28. Orger MB, Smear MC, Anstis SM, Baier H. Perception of fourier and non-fourier motion by larval zebra fish. *Nat Neurosci*. 2000;3:1128–1133.
29. Theobald JC, Duistermars BJ, Ringach DL, Frye MA. Flies see second-order motion. *Curr Biol*. 2008;18:R464–R465.
30. Zanker JM. Theta motion: a paradoxical stimulus to explore higher order motion extraction. *Vision Res*. 1993;33:553–569.
31. Yamaguchi S, Wolf R, Desplan C, Heisenberg M. Motion vision is independent of color in *Drosophila*. *Proc Natl Acad Sci USA*. 2008;105:4910–4915.
32. Wolf R, Heisenberg M. On the fine structure of yaw torque in visual flight orientation of *Drosophila melanogaster*. *J Comp Physiol*. 1980;140:69–80.
33. Wolf R, Heisenberg M. Basic organization of operant behavior as revealed in *Drosophila* flight orientation. *J Comp Physiol A*. 1991;169:699–705.
34. Skinner BF. *The Behavior of Organisms: An Experimental Analysis*. Oxford: Appleton-Century; 1938.
35. Sutton RS, Barto AG. *Reinforcement Learning: An Introduction*. Cambridge, MA: MIT Press; 1998.
36. Götz KG. Optomotorische untersuchung des visuellen systems einiger augenmutanten der fruchtfliege *Drosophila*. *Kybernetik*. 1964;2:77–92.
37. Heisenberg M, Wolf R. *Vision in* Drosophila. *Studies of Brain Function*. Berlin: Springer; 1984 [XII.]
38. Fuchs AF, Robinson DA. A method for measuring horizontal and vertical eye movement chronically in the monkey. *J Appl Physiol*. 1966;21:1068–1070.
39. Brembs B, Heisenberg M. The operant and the classical in conditioned orientation of *Drosophila melanogaster* at the flight simulator. *Learn Mem*. 2000;7:104–115.
40. Brembs B, Plendl W. Double dissociation of PKC and AC manipulations on operant and classical learning in *Drosophila*. *Curr Biol*. 2008;18:1168–1171.
41. Guo A, Liu L, Xia SZ, Feng CH, Wolf R, Heisenberg M. Conditioned visual flight orientation in *Drosophila*: dependence on age, practice, and diet. *Learn Mem*. 1996;3:49–59.
42. Xia S, Liu L, Feng C, Guo A. Memory consolidation in *Drosophila* operant visual learning. *Learn Mem*. 1997;4:205–218.
43. Gong Z, Xia S, Liu L, Feng C, Guo A. Operant visual learning and memory in *Drosophila* mutants dunce, amnesiac and radish. *J Insect Physiol*. 1998;44:1149–1158.
44. Folkers E. Visual learning and memory of *Drosophila melanogaster* wild type CS and the mutants, dance, amnesiac, turnip and rutabaga. *J Insect Physiol*. 1982;28:535–539.
45. Land MF. Visual acuity in insects. *Annu Rev Entomol*. 1997;42: 147–177.
46. Dill M, Heisenberg M. Visual pattern memory without shape recognition. *Philos Trans R Soc Lond B Biol Sci*. 1995;349: 143–152.
47. Dill M, Wolf R, Heisenberg M. Visual pattern recognition in *Drosophila* involves retinotopic matching. *Nature*. 1993;365:751–753.
48. Ernst R, Heisenberg M. The memory template in *Drosophila* pattern vision at the flight simulator. *Vision Res*. 1999;39:3920–3933.
49. Liu G, Seiler H, Wen A, et al. Distinct memory traces for two visual features in the *Drosophila* brain. *Nature*. 2006;439: 551–556.
50. Tang S, Wolf R, Xu S, Heisenberg M. Visual pattern recognition in *Drosophila* is invariant for retinal position. *Science*. 2004;305: 1020–1022.
51. Konen WK, Maurer T, von der Malsburg C. A fast dynamic link matching algorithm for invariant pattern recognition. *Neural Networks*. 1994;7:1019–1930.
52. Davis RL. Olfactory memory formation in *Drosophila*: from molecular to systems neuroscience. *Annu Rev Neurosci*. 2005;28: 275–302.
53. Zars T, Fischer M, Schulz R, Heisenberg M. Localization of a short-term memory in *Drosophila*. *Science*. 2000;288:672–675.
54. Chen CC, Wu JK, Lin HW, et al. Visualizing long-term memory formation in two neurons of the *Drosophila* brain. *Science*. 2012;335:78–685.
55. Mao Z, Roman G, Zong L, Davis RL. Pharmacogenetic rescue in time and space of the rutabaga memory impairment by using Gene-Switch. *Proc Natl Acad Sci USA*. 2004;101:198–203.
56. McGuire SE, Le PT, Osborn AJ, Matsumoto K, Davis RL. Spatiotemporal rescue of memory dysfunction in *Drosophila*. *Science*. 2003;302:1765–1768.
57. Zars T, Wolf R, Davis R, Heisenberg M. Tissue-specific expression of a type I adenylyl cyclase rescues the rutabaga mutant memory defect: in search of the engram. *Learn Mem*. 2000;7: 18–31.
58. Li W, Pan Y, Wang Z, Gong H, Gong Z, Liu L. Morphological characterization of single fan-shaped body neurons in *Drosophila melanogaster*. *Cell Tissue Res*. 2009;336:509–519.
59. Pan Y, Zhou Y, Guo C, Gong H, Gong Z, Liu L. Differential roles of the fan-shaped body and the ellipsoid body in *Drosophila* visual pattern memory. *Learn Mem*. 2009;16:289–295.
60. Quinn WG. Neurobiology: memories of a fruit fly. *Nature*. 2006;439:546–548.
61. Wehner R. Spatial vision in arthropods. In: Autrum H, ed. *Handbook of Sensory Physiology*. Vol. 7. Berlin: Springer; 1981:287–616.
62. Poggio T, Bizzi E. Generalization in vision and motor control. *Nature*. 2004;431:768–774.
63. Peng Y, Xi W, Zhang W, Zhang K, Guo A. Experience improves feature extraction in *Drosophila*. *J Neurosci*. 2007;27:5139–5145.
64. Liu L, Wolf R, Ernst R, Heisenberg M. Context generalization in *Drosophila* visual learning requires the mushroom bodies. *Nature*. 1999;400:753–756.
65. Brembs B, Wiener J. Context and occasion setting in *Drosophila* visual learning. *Learn Mem*. 2006;13:618–628.
66. Frye MA. Multisensory systems integration for high-performance motor control in flies. *Curr Opin Neurobiol*. 2010;20: 347–352.
67. Frye MA, Tarsitano M, Dickinson MH. Odor localization requires visual feedback during free flight in *Drosophila melanogaster*. *J Exp Biol*. 2003;206:843–855.
68. Duistermars BJ, Frye MA. Crossmodal visual input for odor tracking during fly flight. *Curr Biol*. 2008;18:270–275.

69. Chow DM, Frye MA. Context-dependent olfactory enhancement of optomotor flight control in *Drosophila*. *J Exp Biol*. 2008;211:2478–2485.
70. Chow DM, Theobald JC, Frye MA. An olfactory circuit increases the fidelity of visual behavior. *J Neurosci*. 2011;31: 15035–15047.
71. Budick SA, Reiser MB, Dickinson MH. The role of visual and mechanosensory cues in structuring forward flight in *Drosophila melanogaster*. *J Exp Biol*. 2007;210:4092–4103.
72. Gottfried JA, Dolan RJ. The nose smells what the eye sees: crossmodal visual facilitation of human olfactory perception. *Neuron*. 2003;9:375–386.
73. Stein BE, Meredith MA, Wallace MT. The visually responsive neuron and beyond: multisensory integration in cat and monkey. *Prog Brain Res*. 1993;95:79–90.
74. Mota T, Giurfa M, Sandoz JC. Color modulates olfactory learning in honeybees by an occasion-setting mechanism. *Learn Mem*. 2011;18:144–155.
75. Foucaud J, Burns JG, Mery F. Use of spatial information and search strategies in a water maze analog in *Drosophila melanogaster*. *PLoS ONE*. 2010;5:e15231.
76. Morris R. Developments of a water-maze procedure for studying spatial learning in the rat. *J Neurosci Method*. 1984;11:47–60.
77. Guo AK, Götz KG. Association of visual objects and olfactory cues in *Drosophila*. *Learn Mem*. 1997;4(2):192–205.
78. Guo J, Guo A. Crossmodal interactions between olfactory and visual learning in *Drosophila*. *Science*. 2005;309:307–310.
79. Frens MA, van Opstal AJ, van der Willigen RF. Spatial and temporal factors determine auditory-visual interactions in human saccadic eye movements. *Percept Psychophys*. 1995;57:802–816.
80. Hughes HC, Reuter-Lorenz PA, Nozawa G, Fendrich R. Visual-auditory interactions in sensorimotor processing: saccades versus manual responses. *J Exp Psychol Hum Percept Perform*. 1994;20:131–153.
81. Perrott DR, Saberi K, Brown K, Strybel TZ. Auditory psychomotor coordination and visual search performance. *Percept Psychophys*. 1990;48:214–226.
82. Stein BE, Meredith MA, Huneycutt WS, MacDade L. Behavioral indices of multisensory integration: orientation to visual cues is affected by auditory stimuli. *J Cogn Neurosci*. 1989;1:12–24.
83. Zahn JR, Abel LA, Dell'Osso LF. Audio-ocular response characteristics. *Sens Processes*. 1978;2:32–37.
84. Brembs B, Heisenberg M. Conditioning with compound stimuli in *Drosophila melanogaster* in the flight simulator. *J Exp Biol*. 2001;204:2849–2859.
85. Wallace MT, Meredith MA, Stein BE. Integration of multiple sensory modalities in cat cortex. *Exp Brain Res*. 1992;91:484–488.
86. Wilkinson LK, Meredith MA, Stein BE. The role of anterior ectosylvian cortex in cross-modality orientation and approach behavior. *Exp Brain Res*. 1996;112:1–10.
87. Barth DS, Goldberg N, Brett B, Di S. The spatiotemporal organization of auditory, visual and auditory-visual evoked potentials in rat cortex. *Brain Res*. 1995;678:177–190.
88. Duhamel J-R, Colby C, Goldberg M. Congruent representations of visual and somatosensory space in single neurons of monkey ventral intraparietal cortex (area VIP). In: Paillard J, ed. *Brain and Space*. New York: Oxford University Press; 1991:223–236.
89. Graziano MS, Gross CG. Spatial maps for the control of movement. *Curr Opin Neurobiol*. 1998;8:195–201.
90. Mistlin AJ, Perrett DI. Visual and somatosensory processing in the macaque temporal cortex: the role of 'expectation.' *Exp Brain Res*. 1990;82:437–450.
91. Carmichael ST, Price JL. Sensory and premotor connections of the orbital and medial prefrontal cortex of macaque monkeys. *J Comp Neurol*. 1995;363:642–664.
92. Ettlinger G, Wilson WA. Cross-modal performance: behavioral processes, phylogenetic considerations and neural mechanisms. *Behav Brain Res*. 1990;40:169–192.
93. Laurent G, Davidowitz H. Encoding of olfactory information with oscillating neural assemblies. *Science*. 1994;265:1872–1875.
94. Laurent G, Naraghi M. Odorant-induced oscillations in the mushroom bodies of the locust. *J Neurosci*. 1994;14:2993–3004.
95. Laurent G. Odor images and tunes. *Neuron*. 1996;16:473–476.
96. Baddeley A. Working memory. *Science*. 1992;255:556–559.
97. Chafee MV, Goldman-Rakic PS. Matching patterns of activity in primate prefrontal area 8a and parietal area 7ip neurons during a spatial working memory task. *J Neurophysiol*. 1998;79: 2919–2940.
98. Strauss R, Pichler J. Persistence of orientation toward a temporarily invisible landmark in *Drosophila melanogaster*. *J Comp Physiol A*. 1998;182:411–423.
99. Neuser K, Triphan T, Mronz M, Poeck B, Strauss R. Analysis of a spatial orientation memory in *Drosophila*. *Nature*. 2008;453: 1244–1248.
100. Kim YC, Lee HG, Seong CS, Han KA. Expression of a D1 dopamine receptor dDA1/DmDOP1 in the central nervous system of *Drosophila melanogaster*. *Gene Exp Patterns*. 2003;3: 237–245.
101. Williams GV, Castner SA. Prefrontal cortex and working memory processes. *Neuroscience*. 2006;139:251–261.
102. O'Keefe J. A review of the hippocampal place cells. *Prog Neurobiol*. 1979;13:419–439.
103. O'Keefe J, Nadel L. *The Hippocampus as a Cognitive Map*. Oxford: Oxford University Press; 1978.
104. Ofstad TA, Zuker CS, Reiser MB. Visual place learning in *Drosophila melanogaster*. *Nature*. 2011;474:204–207.
105. Schultz W. Behavioral theories and the neurophysiology of reward. *Annu Rev Psychol*. 2006;57:87–115.
106. Schroll C, Riemensperger T, Bucher D, et al. Light-induced activation of distinct modulatory neurons triggers appetitive or aversive learning in *Drosophila* larvae. *Curr Biol*. 2006;16: 1741–1747.
107. Schwaezel M, Monastirioti M, Scholz H, Friggi-Grelin F, Birman S, Heisenberg M. Dopamine and octopamine differentiate between aversive and appetitive olfactory memories in *Drosophila*. *J Neurosci*. 2003;23:10495–10502.
108. Sitaraman D, Zars M, LaFerriere H, et al. Serotonin is necessary for place memory in *Drosophila*. *Proc Natl Acad Sci USA*. 2008;105:5579–5584.
109. Sitaraman D, Zars M, Zars T. Place memory formation in *Drosophila* is independent of proper octopamine signaling. *J Comp Physiol A*. 2010;196:299–305.
110. James W. *The Principles of Psychology*. New York: Holt; 1890.
111. Virsik RP, Reichardt WE. Detection and tracking of moving objects by the fly *Musca domestica*. *Biol Cybern*. 1976; 23:83–98.
112. Wolf R, Heisenberg M. Visual control of straight flight in *Drosophila melanogaster*. *J Comp Physiol A*. 1990;167:269–283.
113. Xi W, Peng Y, Guo J, et al. Mushroom bodies modulate salience-based selective fixation behavior in *Drosophila*. *Eur J Neurosci*. 2008;27:1441–1451.
114. Posner MI, Snyder CR, Davidson BJ. Attention and the detection of signals. *J Exp Psychol*. 1980;109:160–174.
115. Chun MM, Marois R. The dark side of visual attention. *Curr Opin Neurobiol*. 2002;12:184–189.
116. Corbetta M, Shulman GL. Control of goal-directed and stimulus-driven attention in the brain. *Nat Rev Neurosci*. 2002;3:201–215.
117. Desimone R, Duncan J. Neural mechanisms of selective visual attention. *Annu Rev Neurosci*. 1995;18:193–222.

118. Kastner S, Ungerleider LG. Mechanisms of visual attention in the human cortex. *Annu Rev Neurosci.* 2000;23:315–341.
119. Posner RA. Rational choice, behavioral economics, and the law. *Stanford Law Review.* 1998;50:1551–1575.
120. Reynolds JH, Chelazzi L. Attentional modulation of visual processing. *Annu Rev Neurosci.* 2004;27:611–647.
121. Wu Z, Gong Z, Feng C, Guo A. An emergent mechanism of selective visual attention in *Drosophila. Biol Cybern.* 2000;82:61–68.
122. Yantis S, Serences JT. Cortical mechanisms of space-based and object-based attentional control. *Curr Opin Neurobiol.* 2003;13: 187–193.
123. van Swinderen B, Flores KA. Attention-like processes underlying optomotor performance in a *Drosophila* choice maze. *Dev Neurobiol.* 2007;67:129–145.
124. van Swinderen B. Attention-like processes in *Drosophila* require short-term memory genes. *Science.* 2007;315:1590–1593.
125. van Swinderen B, McCartney A, Kauffmann S, et al. Shared visual attention and memory systems in the *Drosophila* brain. *PLoS ONE.* 2009;4:e5989.
126. Sugru LP, Corrado GS, Newsome WT. Choosing the greater of two goods: neural currencies for valuation and decision making. *Nature Rev Neurosci.* 2005;6:363–375.
127. Heekeren HR, Marrrett S, Bandettini PA, Ungerleider LG. A general mechanism for perceptual decision-making in the human brain. *Nature.* 2004;431:859–862.
128. Tang S, Jousola M. Intrinsic activity in the fly brain gates visual information during behavioral choices. *PLoS ONE.* 2010;5: e14455.
129. Yang CH, Belawat P, Hafen E, Jan LY, Jan YN. *Drosophila* egg-laying site selection as a system to study simple decision making progress. *Science.* 2008;319:1679–1683.
130. Brembs B. Mushroom bodies regulate habit formation in *Drosophila. Curr Biol.* 2009;19:1351–1355.
131. Padoa-Schioppa C, Assad JA. Neurons in the orbitofrontal cortex encodes economic value. *Nature.* 2006;441:223–226.
132. Platt ML. Neural correlates of decisions. *Curr Opin Neurobiol.* 2002;12:141–148.
133. Yang T, Shadlen MN. Probabilistic reasoning by neurons. *Nature.* 2007;447:1075–1080.
134. Glimcher PW, Dorris MC, Bayer HM. Physiological utility theory and the neuroeconomics of choice. *Games Econ Behav.* 2005;52: 213–256.
135. Tang S, Guo A. Choice behavior of *Drosophila* facing contradictory visual cues. *Science.* 2001;294:1543–1547.
136. Zhang K, Guo J, Peng Y, Wang X, Guo A. Dopamine-mushroom body circuit regulates saliency-based decision-making in *Drosophila. Science.* 2007;316:1901–1904.
137. Gruber AJ, Dayan P, Gutkin BS, Solla SA. Dopamine modulation in the basal ganglia locks the gate to working memory. *J Comput Neurosci.* 2006;20:153–166.
138. Wu Z, Guo A. A model study on the circuit mechanism underlying decision-making in *Drosophila. Neural Networks.* 2011;24: 333–344.
139. Lu S, Fang J, Guo A, Peng Y. Impact of network topology on decision making. *Neural Networks.* 2009;22:39–40.
140. Seeley TD, Visscher PK, Schlegel T, Hogan PM, Franks NR, Marshall JAR. Stop signals provide cross inhibition in collective decision-making by honeybee swarms. *Science.* 2012;335: 108–111.
141. Couzin ID, Ioannou CC, Demirel G, et al. Uninformed individuals promote democratic consensus in animal groups. *Science.* 2011;334:1578–1580.
142. Mery F, Varela SAM, Danchin E, et al. Public versus personal information for mate copying in an invertebrate. *Curr Biol.* 2009;19:730–734.
143. Tinette S, Zhang L, Robichon A. Cooperation between *Drosophila* flies in searching behavior. *Genes Brain Behav.* 2004;3: 39–50.
144. Chabaud MA, Isabel G, Kaiser L, Preat T. Social facilitation of long-lasting memory retrieval in *Drosophila. Curr Biol.* 2009;19: 1654–1659.
145. O'Reilly RC. Biologically based computational models of high-level cognition. *Science.* 2006;314:91–94.
146. Desplan C. Time to pick the fly's brain. *Nature.* 2007;450:173.

4.4.2 Honeybees

CHAPTER

29

In Search of the Engram in the Honeybee Brain

Randolf Menzel

Freie Universität Berlin, Berlin, Germany

THE CONCEPT OF THE ENGRAM

Memories exist in multiple forms and have multiple functions. They may be categorized according to their cell-physiological substrates along a timescale defined as short-term, mid-term, and long-term memory (STM, MTM, and LTM, respectively). Ongoing neural activity serves as the storage device for STM; intracellular signaling cascades lead to MTM; and gene activation, protein synthesis, and new structures underlie LTM. The physiological substrates of these memory stages or phases can be sequential or in parallel, indicating that memory systems are highly dynamic. But what exactly is processed and stored in these different phases of memory formation? Memory is about something, namely objects, events, and relations between objects in the external world as well as internal body states such as hunger and satiety. Thus, memory stores information about the meaning of multiple signals, external and internal; in other words, it has content. Stimuli and actions are evaluated by the nervous system according to expected outcomes, and it is this loop into the future that defines the core of memory content. The ultimate goal of memory research is to uncover the neural mechanisms that allow the content of memory to be encoded, stored, processed, and retrieved. The content of memory is usually considered to be encoded as an engram or memory trace. Lashley[1] referred to the engram in the title of his famous paper and asked questions concerning the location of the engram(s) in different parts of the mammalian brain (cortex). Localization is indeed a major feature of memory, and in the mammalian brain memory localizations can be categorized according to the types of memory that are processed—for example, procedural memory (e.g., cerebellum), episodic memory (hippocampus and prefrontal cortex), and emotional memory (amygdala)—but the content of each of these memories involves many more parts of the brain. Another character of memory is content-sensitive processing.[2] Any retrieval from the memory store changes its content due to the updating process in working memory, a process referred to as 'reconsolidation.' It is this updating process that may reveal rules underlying generalization, categorization, and implicit (and explicit in humans) forms of abstraction. However, both localization and content-specific processing tell us little about the mechanisms of how content is encoded and stored in the nervous system, although knowledge of both aspects of the engram is requirements for hypothesis-driven research.

Cognitive psychology has struggled with the question of whether the engram or memory trace 'exists' if it is not retrieved. "Where is the memory trace when we are not remembering? ... The hunt for the engram (the physical manifestation of the memory trace that is independent of the operations needed to recover it) may prove to be fruitless as the hunting of the Snark".[3] Indeed, the engram is not yet a memory if it is not activated, but it is the necessary physical condition for memory to emerge through the readout process in the nervous system. In this sense, the engram, together with the neural processes of activating it and combining it with the information provided by the retrieval process, leads to memory. The informational content of the engram is therefore rather elusive,[4] and we characterize our efforts better by saying that we aim to uncover necessary physical components that we hope will define (at least to some extent) the informational content of the trace leading to the engram. These attempts will be very limited because in reading these physical components as separate entities, the emerging properties from the interaction of multiple components necessary to

Invertebrate Learning and Memory.
DOI: http://dx.doi.org/10.1016/B978-0-12-415823-8.00029-0

convert the isolated traces to the engram will be extremely difficult to discover.

Memory content requires ubiquitous molecules and chains of cellular reaction cascades but is certainly not encoded and stored in such elemental processes. Rather, memory content is expected to result from the spatial distribution of learning-dependent changes of neural activity and synaptic transmission resulting in reorganized neural nets. Such a view follows Ramon y Cajal's[5] view that the engram is expressed in novel brain structures. Addressing "Hebb's dream," as Nicolelis and co-workers[6] called it, the uncovering of the engram thus requires capturing the changing structures and the dynamic processes hidden in learning-dependent changes of whole neural nets. Ideally, one would like to elucidate all these components of neural nets with subcellular resolution, a demanding task for higher organisms but possibly not out of reach for invertebrate brains.

In this chapter, I discuss attempts to follow this line of argument by characterizing the olfactory engram in the honeybee brain that develops in the course of odor/reward learning. Early in my research career, I would have liked to address this question for visual (color) learning in bees because my major interest was studying behavior in bees. Because bees do not learn colors well (or do not reveal their visual learning of the proboscis response) under conditions that allow brain recordings, I needed to shift to olfaction, a perceptual modality that makes stimulus quantification much more difficult. Furthermore, at the time these efforts started (1985), very little was known about primary sensory coding of odors in the insect brain (and in other brains). In hindsight, this forced detour was favorable because it led us to think about methods that allow simultaneous measurements of neural excitation in neural nets under conditions in which an animal learns, encodes, stores, retrieves, and prepares for actions on the basis of a memory trace.

THE OLFACTORY LEARNING PARADIGM

Learning takes many forms and plays an essential role in honeybee behavior. Latent (observational) and associative learning (operant and classical conditioning) interact in natural behavior. On their first flights out of the colony, honeybees explore the environment and learn the spatial relations of landscape structures relative to their hive location within a reference system, the sun compass, by relating the sun azimuth time function to extended landmarks.[7] They attend to waggle dances, receive the information about distance and direction of the outbound flight toward the indicated location, and apply this information within the frame of the experienced landscape. Olfactory cues sensed during dance recruitment are learned and searched for when localizing the communicated place.[8] At a feeding site, they associate the signals (odors, colors, shape, manipulatory components, spatial location relative to nearby landmarks, and time of the day) with the quality and quantity of reward, nectar and pollen. Multiple visits to the feeding site allow them to extract features such as the change of reward quantity over time[9] and to the reliable components of the signals such as symmetry[10,11] or the consistent components within variable odor mixtures.[11]

The memory traces resulting from such rich forms of learning store not just the stimulus-response associations but also derived representations characterizing the where, when, and what components of evaluated experiences (see Chapter 3 for further arguments in favor of a cognitive interpretation of learned representations in honeybees). It is also important to recognize the dynamic structure of the memory trace. Four memory stages (in addition to a sensory memory in the seconds range) can be distinguished depending on the respective time courses controlling behavior, sensitivity to interference, and the molecular cascades involved,[12] resembling the general structure of memory dynamics in other invertebrates and mammals.[13] Consolidation from STM to MTM requires ongoing neural activity in the minutes range and activation of protein kinase M (PKM; mediated by the proteolytic cleavage of PKC in the hours range,[14] whereas consolidation of STM to LTM requires the activity of the cAMP-dependent PKA in the antennal lobes.[15] Early long-term memory (eLTM) depends on translation (1 to 2 days after conditioning) and late long-term memory (lLTM) on transcription and translation (>3 days after conditioning). Both forms of LTM develop in parallel.[16] LTM lasts for the lifetime of a bee, which may be more than 6 months in overwintering animals.[17] Formation of LTM requires multiple (three or more) learning trials, whereas STM and MTM can be formed after one learning trial.[18]

Control of ongoing behavior at any given moment is supervised by working memory, a "limited capacity system for maintaining and manipulating information ... allowing for complex and flexible cognition".[19] The span of working memory in freely behaving honeybees has been uncovered by several experimental procedures. Short-term working memory (in the seconds range) was observed in maze learning,[20] matching-to-sample tests,[21] and serial learning tasks.[22] Longer term working memory (in the range of several minutes) was reported in tests in which the quantitative reward conditions were made contingent on the animals' own behavior.[23] Very long working memory (in the range of hours to days) was found in tests that involved learning of incentive gradients.[9] In the latter two cases, working memory is characterized by the retrieval of context-specific memory that is used to evaluate and update current experience. Directed attention, a characteristic component of working memory, has yet to be addressed in honeybee research.

The search for the neural correlates of memory calls for experimental conditions in which a bee learns to associate a stimulus with reinforcement and forms a lasting memory trace. Olfactory reward conditioning of the proboscis extension response (PER) is a robust paradigm that allows combining behavioral with neural studies. Foraging bees are collected at the hive entrance, cooled, and harnessed in a tube so that the antennae and mouthparts are freely moving but the legs, wings, and abdomen are encased. The conditioned stimulus (CS+; odor, mechanical stimuli, CO_2, humidity, and temperature) is applied to the antennae and subsequently rewarded with sucrose reward (unconditioned stimulus (US). Hungry bees extend the proboscis when the sucrose receptors on the antennae are stimulated. The proboscis is then allowed to lick sucrose solution for a few seconds. The optimal time interval between onset of CS+ and US is 2–4 sec. Bees acquire the conditioned response to CS+ within a few trials and retain it after several training trials as long as they can be kept alive under these conditions (up to 1 week if fed to satiation every evening). Backward conditioning (first US and then CS+; optimal interval, 20 sec[24]) leads to inhibitory learning, as indicated by resistance to subsequent acquisition. Multiple conditioning procedures have been tested, including trial spacing effects, second-order conditioning, conditioned inhibition, extinction learning and spontaneous recovery from extinction, compound processing, and occasion setting (25; see also Chapter 33). Performance values are usually group-average learning curves; however, such group effects do not adequately represent the behavior of individual animals, an important finding because correlations between behavior and neural correlates need to be established on the basis of individuals. Individual behavior is characterized by a rapid and stable acquisition of the conditioned response (CR), as well as by a rapid and stable cessation of the CR following unrewarded stimuli (extinction). Two processes interact during classical conditioning—a gradual and an all-or-none learning process. Thus, individual behavior is a meaningful predictor for the internal state of a honeybee irrespective of the group-average behavioral performance.[26]

THE OLFACTORY PATHWAY IN THE BEE BRAIN AND POTENTIAL LOCATIONS OF THE ENGRAM

Sensory neurons expressing the same olfactory receptor converge on aggregates of glomerular neuropil structures in the antennal lobe (AL),where they synapse onto local interneurons and projection neurons (PNs) that connect the mushroom body (MB) and the lateral horn (LH) via two tracts, the median and later antennoprotocerebralis tracts (mAPT and lAPT, respectively) (Figure 29.1). In both vertebrates and invertebrates, the combinatorial pattern of neural activity in the glomeruli is odor specific. In insects, the PNs transmit this pattern to the next synapse in the calyces of the MB. Here, neural excitation diverges from approximately 800 PNs to a large number (>100,000) of MB intrinsic neurons, the Kenyon cells (KCs). Each of the more than 1 million presynaptic boutons of the PNs comprise a microcircuit composed of approximately 10 postsynaptic sites of KCs and both pre- and postsynaptic sites of putative inhibitory neurons of the protocerebral-calycal tract (PCT). KCs may collect their input either exclusively in one of the three calycal compartments (lip receiving input from olfactory PNs, collar receiving visual input, basal ring receiving mixed input from olfactory and mechanosensory input; Kenyon cell type I (KC I)) or across these calycal compartments and then project their axons to both lobes of

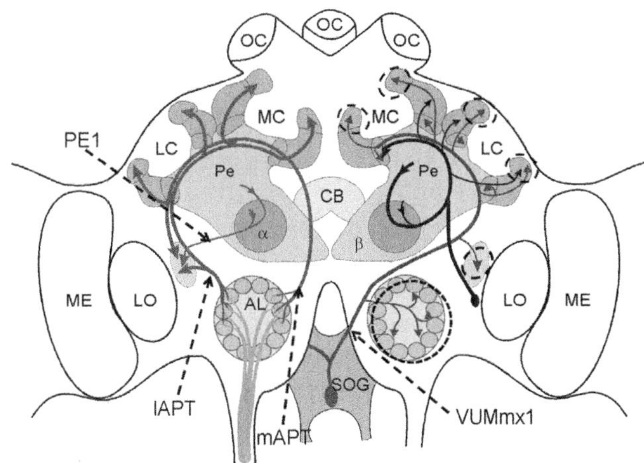

FIGURE 29.1 The olfactory pathway (left half of the brain) and reward pathway (right half of the brain) in the honeybee brain. The convergence sites between olfactory and reward pathways are marked by dotted circles in the right side of the brain. α and β, alpha and beta lobes of the mushroom body; AL, antennal lobe; CB, central body; LC and MC, lateral and median calyx of the mushroom body; LO, lobula (third visual ganglion); ME, medulla (second visual ganglion); OC, ocelli; Pe, peduncle of the mushroom body; SOG, subesophageal ganglion. The dashed arrows point to lAPT and mAPT (lateral and median antennoglomerular tract, respectively), the PE1 neuron, and the VUMmx1 neuron.

the MB (α and β lobes; Kenyon cell type II (KC II)). The axons of both types form collaterals halfway along the peduncle and project to the α and β lobes. KC I project one collateral to the dorsal α lobe and the other to the caudal part of the β lobe. KC II project one to the ventral part of the α lobe and the other to the proximal part of the β lobe.[27] The large number of KCs converge on a rather small number (~400) of MB extrinsic neurons (ENs), leaving the α lobe at three prominent exit points—the lateral, ventral, and ventromedian exit points.[28] These ENs are divided into seven subgroups (A1−A7) depending on the localization of their somata. Most of them are postsynaptic to KCs, as judged by their spiny-like structure and in some cases by electron microscopic evidence, but post- and presynaptic structures are known to occur in close vicinity. ENs project to many parts of the brain—some of them (e.g., the identified neurons PE1 and the A4 neurons) to different subregions of the LH, where they converge directly or indirectly with collaterals of mAPT and lAPT. Other ENs (A3 neurons) project via a recurrent pathway back to the calyx of the same MB along the GABA-immunoreactive (GABA-ir) PCT. Multiple ENs connect the ring neuropil around the α lobe (A1, A2, A4, and A7) with the MB on the ipsi- or contralateral side (A6 and A7) or with other protocerebral areas on the ipsi- or contralateral side (A4, A5, and A7). A single neuron has been identified that appears to project back to the ipsilateral AL (the AL1). Dendrites of ENs are often restricted within the α lobe to one of the horizontal bands, suggesting that they receive sensory modality-specific input via KCs. Others distribute their dendrites across the banded structure of the α and β lobes. The structural diversity of ENs reflects a multiplicity of functions concerning the readout from the MB and the information flow to other parts of the brain.

Neurons containing the neuromodulators octopamine (OA) and dopamine are related to the reinforcing functions during conditioning both in *Drosophila* (see Chapters 5 and 27) and in the bee. One OA immunoreactive neuron, the VUMmx1, was identified in a reward substitution experiment to be sufficient for the reward function of sucrose in olfactory conditioning in the bee.[29] VUMmx1 receives its input in the subesophageal ganglion and converges with the olfactory pathway at three pairs of symmetrical sites—the ALs, the LHs, and the lip regions of the MB calyces, respectively (Figure 29.1). Thus, it has been hypothesized that these convergence sites may constitute localizations of the olfactory engram as it develops in reward learning,[30] and therefore recordings of learning-related neural plasticity focused on the neurons and their synaptic connections so far on two of these three sites (AL and MB calyx). A functional MB was previously found to be required for the consolidation of olfactory STM into MTM.[31] The VUMmx1 neuron responds to sucrose and to many other stimuli. In the course of conditioning, it enhances its response to the forward paired conditioned olfactory stimulus (CS+) and reduces its response to the backward paired stimulus (CS−) (see Figure 29.10). Interestingly, regarding the notion of expectation and anticipation, the octopaminergic neuron VUMmx1 exhibits activity reflecting the animal's expectation: It responds to unexpected sucrose presentations but not to expected ones.[29] Although there is evidence in vertebrates of how this reduction in the error signal may be implemented biologically, such evidence is still lacking in invertebrates.

THE ANTENNAL LOBE

The AL of the honeybee is believed to constitute a component of a distributed network storing olfactory information, but evidence is controversial. As noted previously, convergence of the olfactory and reward pathway in the AL suggests a memory trace to be formed in the AL. Indeed, substituting reward in olfactory PER conditioning by local injection of octopamine (the putative transmitter of VUMmx1) into the AL leads to learning of the forward but not the backward paired odor.[32] Accordingly, blocking octopamine receptors in the AL with RNAi reduces olfactory learning.[33] Additional arguments in favor of a memory trace in the AL concern (1) the role of the AL in memory consolidation and (2) neural correlates of a memory trace. First, memory consolidation induced by a single learning trial was found to be blocked if the AL is cooled during the minute following the trial, whereas cooling even immediately after the last of multiple learning trials does not impair memory consolidation, suggesting that the AL possibly in connection with other brain parts stores a short-lasting memory trace necessary for consolidation.[31] Local uncaging of cAMP in the AL (cAMP promotes the transfer from STM to LTM in bees) shifts STM to LTM when it is uncaged soon after a single learning trial.[15] So far, it has not been possible to test directly whether a more permanent memory trace is stored in the AL because blocking neural activity in the AL during retrieval tests interferes with the processing of olfactory coding. Second, neural correlates of olfactory learning were collected with two methods—Ca^{2+} imaging of glomerular activity and extracellular recordings from PNs. In the first case, the Ca^{2+} signals came either predominantly from the presynaptic terminals of olfactory receptor neurons (and possibly also from glia cells) in the glomeruli or from the postsynaptic elements, the PNs. Presynaptic signals increased for the CS+ odor.[34] Controversial data exist for associative plasticity in PNs. Peele and co-workers[35] found no consistent changes for CS+ or CS− during differential PER conditioning, whereas Weidert and co-workers[36] did. These inconsistencies may be resolved on the basis of the data from multiunit recordings of PNs. In line with the interpretation that PNs undergo associative change is the finding by Fernandez and co-workers[37] that binary mixtures of odors are coded more differently for learned odors and this effect correlates with changes of neural responses as seen in Ca^{2+}-imaging. Rath and co-workers[38] reported that PNs undergo associative plasticity in differential PER conditioning depending on their response level to the respective CS+ and CS− before conditioning: Glomeruli responding to CS+ before conditioning enhanced their CS+ responses, those that responded to CS− did not change their responses, and those that responded before both CS+ and CS− either reduced or enhanced their respective odor responses depending on the strength of their responses (weak responses were enhanced, and strong responses were reduced). The model derived from these studies assumes two types of plastic synapses in the glomeruli: (1) synapses between olfactory receptor neurons and PNs and (2) synapses between olfactory receptor neurons and local interneurons. Taken together, these results indicate that odor learning improves spatial representations of the learned odors and facilitates their discrimination—forms of specified memory traces that contribute important components to corresponding memory traces stored somewhere else.

Multiunit extracellular recordings from PNs documented both increases and decreases in rate changes to the reinforced (CS+) and the specifically not reinforced (CS−) odor (Figure 29.2),[39] but it is unknown how these effects relate to the differential associative changes seen in Ca^{2+} imaging. If such associatively up- and downregulated PNs receive their inputs within the same glomerulus (e.g., for the CS+), it is not surprising that the overall associative changes seen by Ca^{2+} imaging of whole glomeruli may cancel each other out. A model implementing these findings and making specific assumptions about spike timing-dependent plasticity induced by activity of the reward neuron VUMmx1 on the connection between local interneurons and PN predicts asymmetric changes of PN responses to the rewarded neuron.[36]

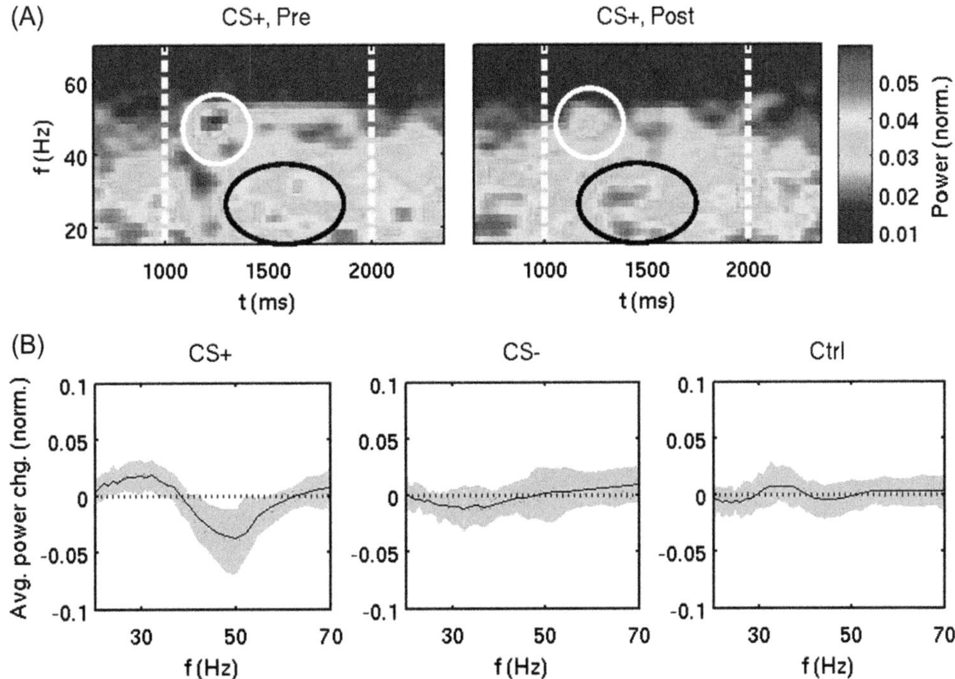

FIGURE 29.2 Changes in LFP power of projection neurons in the course of differential conditioning as recorded by multiple extracellular electrodes. (A) Time-resolved power spectra for CS+ tests before (left; Pre) and after (right; Post) differential conditioning averaged across three test trials for all animals. Dashed white lines indicate stimulus onset and offset; power is indicated in color scale. The white circles indicate a decrease of power in the high-frequency band, and the dark circles indicate an increase in the low-frequency band. (B) Average power change during the on-response resolved by individual frequency bands for CS+ (left), CS− (middle), and a control odor, Ctrl (right). Before averaging across animals, the differences between power before and after conditioning were calculated. Error bars (±2.5%) were obtained using 1000 bias-corrected standard bootstraps. Source: *After Denker* et al.[39]

The high temporal resolution of spike recordings allows for the analysis of the ensemble activity using odor-induced local field potentials (LFPs) and their relation to single-unit activity. The largest learning-related difference was found for CS+. LFP power increases for CS+ in the 15- to 40-Hz frequency band and decreases for frequencies higher than 45 Hz[39] (Figure 29.2). This learning-related power change correlates with the size of the neuronal ensemble that is phase-locked to the particular frequency: After learning, less units are entrained to the higher frequency band, and more units are entrained to the lower band. These results reflect associative plasticity in the AL resulting from a restructured odor coding network.

The memory trace in the AL as seen by opto- and electrophysiological recordings results from multiple training trials, suggesting that it represents a lasting trace. It optimizes primary odor coding both at the spatial and at the temporal domain. It is unknown whether other neuropils or neural tracks (e.g., feedback neurons from the MB) are required for its formation and readout and whether it contains information about the specifically learned odors. To test for this, it will be necessary to manipulate selectively the contribution of neural subsets within the AL separately for learning, consolidation, and retrieval.

INTRINSIC NEURONS OF THE MUSHROOM BODY: KENYON CELLS

MBs are expected to house the engram of insects. In 1896, Kenyon[40] stated the following:

> Ever since Dujardin[41] discovered the mushroom bodies and pointed out the relation between their size and the development of insect intelligence, nearly every writer on the subject of the hexapod brain who has referred to the matter of intelligence has recognized the fact. (p. 161)

However, even with the brilliant work in *Drosophila*, direct evidence is rather scarce. A first hint in favor of the idea that the MB is involved in memory storage came from the finding that the time course of retrograde amnesia induced by cooling the honeybee calyces matches the time course of cooling the whole animal.[31] Olfactory memory is expected to be located in the lips of the calyces because they comprise the second-order convergence sites of the olfactory pathway with the reward pathway, suggesting associative processing at the MB input. PNs are presynaptic to KCs in discrete microcircuits, the microglomeruli, composed of one large presynaptic bouton of PNs, 6–12 postsynaptic KC spines, and usually one GABA-ir profile, the presynaptic site of A3-v neurons of the PCT, and frequently a profile with dense core vesicles most likely from the OA-ir VUMmx1 neuron (Figure 29.3).

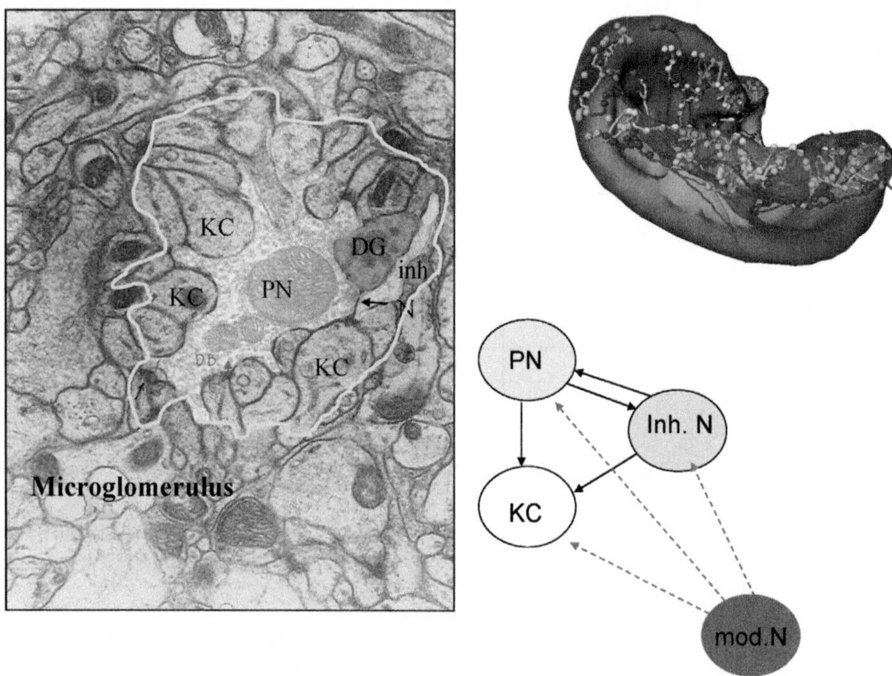

FIGURE 29.3 Synaptic organization of the microglomerulus in the lip region of the MB. (Top right) The terminals of two projection neurons (one in yellow and one in blue) in the lip region of the calyx with their multiple presynaptic swellings (boutons). (Left) Electron microscopic view: The large presynaptic bouton of a projection neuron (PN; surrounded by a yellow line) comprises the center of the microcircuit. It is presynaptic to multiple spines of Kenyon cells (KC) and postsynaptic to GABA-ir profiles (inh. N; blue) of the A3-v neurons. The bouton also receives input from profiles with dense core vesicles (DG; pink) interpreted to represent presynaptic sites of the reward pathway (VUMmx1). (Bottom right) The schematic representation of the microcircuit indicates directions of synaptic contacts and assumes modulatory input (mod. N) to all three partners of the circuit. Source: *After Ganeshina* et al.[42]

The density of these microglomeruli depends on the age and experience of the animals and increases during protein synthesis-dependent consolidation into olfactory LTM.[43] This latter finding was interpreted to document a structural correlate of the olfactory memory trace based on the growth of new synapses—an intriguing interpretation that will become even more convincing if it becomes possible to document stimulus-specific changes of microglomerulus patterns.

KCs feature a sparse odor code in a twofold manner: An odor activates a small proportion of highly odor-specific KCs, and in contrast to the presynaptic PNs, KCs respond with brief and phasic responses often combined with off-responses,[44] corroborating findings in the locust.[45] Stimuli of different modalities induce qualitatively similar responses, activating small subsets of KCs. Theoretically, temporal and population sparseness makes KCs potentially well suited as a memory store because the organization of the calyx can be conceptualized as an associative matrix comparable to the network of the hippocampus or cortex.[46] The memory trace as stored in such an associative matrix is characterized by features such as partial overlap between closely related traces, an optimal number of changes per trace (1–5% of the total number of synaptic contacts), and the ability to reconstruct the full pattern even if only part of the trace is activated.

Ca^{2+} imaging of the KC spines in the lip region of the calyx allowed the elucidation of learning-related plasticity of the matrix-like circuit of the MB.[47] For the first time, it was possible to reconstruct the spatial distribution of multiple changes in a neural net with a large range of partners (Figures 29.4A and 29.4B). Stimulus repetition leads to depressed responses in KCs, a form of nonassociative plasticity that is counteracted for the CS+ but not for the CS− in differential conditioning. This suggests that meaningless repetition of stimuli leads to depression, and meaningful repetition of stimuli (as indicated by the activation of the reward pathway) compensates depression possibly by selectively facilitating neural responses. Most important, KCs are either specifically recruited or eliminated from responding during odor learning, leading to a change of odor-induced activity pattern in KCs (Figure 29.4). Gain of activity (recruitment) was more often observed for the CS+ and loss of function (elimination) more frequent for the CS−, but both changes occur for both stimuli. Unchanged KC activity in the course of odor learning (shown in white in Figure 29.4A and in yellow in Figure 29.4B) are rare,

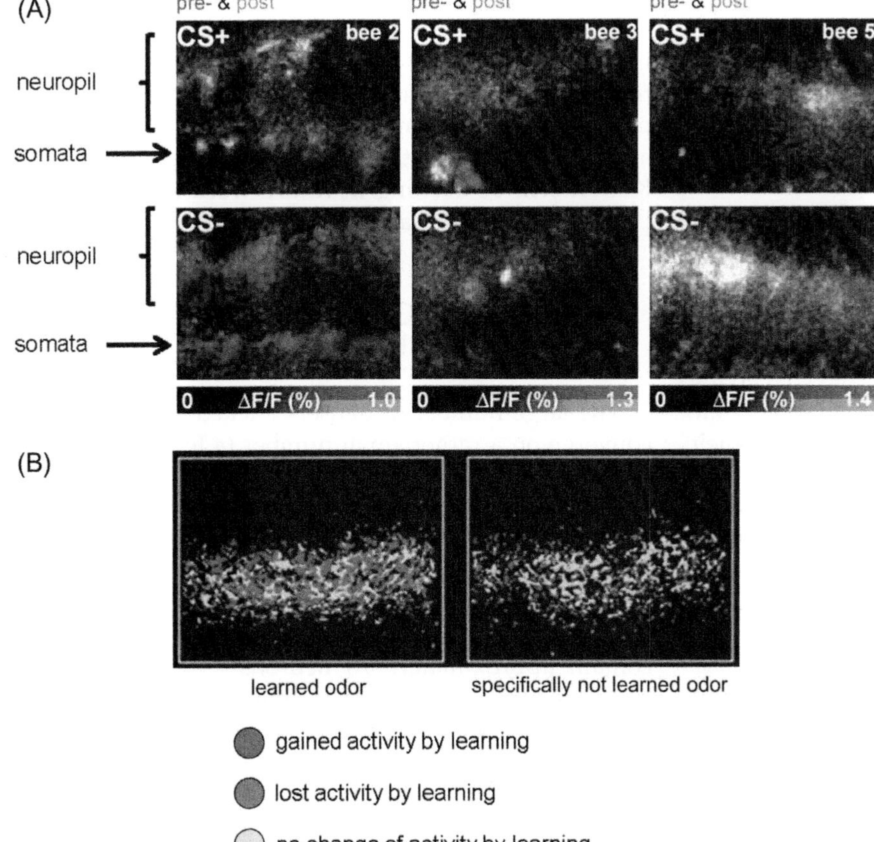

FIGURE 29.4 Two examples of color-coded changes of KC activation patterns during differential odor learning. (A) Response changes in three animals for CS+ and CS−. Ca^{2+} activity pattern imaged before differential PER conditioning (pre) is given in magenta, and that after conditioning (post) is shown in green separately for the CS+ (top) and the CS− (bottom). The imaged region of the MB lip shows both the somata of the clawed KCs and their synaptic neuropil. Somata and neuropil whose activity does not change during conditioning appear in white, those active only before conditioning in magenta, and those active only after conditioning in green. Odor learning leads to recruitment (green), loss of activity (magenta), and no changes (white). Bee 2 (left) is representative for the population of tested animals because it shows recruitment of activity predominantly for the CS+ and loss of activity for the CS−. (B) Example of a fourth animal in which gained activity in the course of differential conditioning is expressed in red, lost activity in blue, and no changes of activity in yellow. *Sources: Panel A after Szyszka* et al.[47]*; panel B courtesy of P. Szyszka.*

indicating that learning leads to a drastic rearrangement of odor representation in the MB input. Because odors activate primarily non-overlapping KC ensembles, the parallel representations of multiple odor traces allow for an effective and robust memory trace. In the future, it will be necessary to compare patterns of changes for odors generalized more or less. Furthermore, it will be necessary to show that multimodal stimuli lead to a more precise KC activity pattern rather than to a higher number of activated KCs.

In addition to these changes in activity patterns, KCs also undergo dynamic changes. Before conditioning, their odor responses are short, even during long-lasting odor stimulation.[43] During odor–sucrose pairing, the odor-activated KCs become reactivated, leading to coincident activity in odor coding and reward coding neurons, a possible mechanism for delayed and trace conditioning.[48] The picture emerging from Ca^{2+} imaging studies assumes an intracellular trace for the CS+, possibly in the form of a lasting increase in Ca^{2+}[49] that is associatively paired with a delayed OA input from VUMmx1 leading to an enhancement of KC activity. A reduced response to CS+ in KCs may result from a similar mechanism in inhibitory inputs to KCs—for example, via A3 neurons of the PCT, a mechanism suggested by the close apposition of OA-ir profiles and GABA-ir profiles in microglomeruli of the calyx lip (Figure 29.3). Taken together, it is conceivable that the olfactory engram in the MB lip comprises a combinatorial pattern of predominantly enhanced synaptic transmission to KCs but also reduced transmission, leading to the conclusion that KCs store a memory trace in stimulus-specific sparse activation patterns.

As noted previously, KC I receive their input selectively via small dendritic fields from lip, collar, or basal ring and project one axon collateral to the dorsal half of the α lobe and the other to the caudal part of the β lobe. KC II, to which the imaged clawed KCs belong, collect input across the calyx, thus receiving input from multiple sensory modalities via elaborate and clawed dendritic fields, and project one of their axons to the ventral part of the α lobe and the other to the proximal part of the β lobe. KC II converge on a small number of MB ENs, whereas KC I serve more ENs. One would expect the two types of KCs to process high-order sensory input and value signals differently, possibly leading to two parallel coding, storing, and retrieval schemes within the MB—a concept of high relevance in *Drosophila* MB function[50] (see Chapter 27 for further discussion of controversial data). Unfortunately, electrophysiological recordings from KCs of either type have been unsuccessful so far despite intensive efforts, and imaging experiments have not been performed in KC I. It will be an important task for the future to unravel the specific coding schemes in the two KC types as well as elucidate the potentially different roles of the median and lateral calyces with their KC projections to the inner and outer part of the lobes, respectively.

Although the MB in bees is large in comparison to the whole brain, its total volume ($25 \times 10^6\ \mu m^3$[51]) is small given the high number of densely packed KCs. Witthöft's[52] estimate of 170,000 KCs needs revision to a lower number, but even 100,000–150,000 KCs as derived from a comparison between the volume of single KCs and total MB volume (see Chapter 4) is a very high number. Obviously, MB intrinsic circuitry is designed to take advantage of small neuron size (less material, lower energy consumption) and particularly effective interneuron cross talk via short connections. It thus can be concluded that the miniaturized MB circuitry pushes information processing capacity (IPC) to an upper limit with the lowest possible volume of neural tissue, highest efficiency of use of material and metabolic energy, and shortest interneuron connections. However, axon diameters well below 0.5 μm cause problems of reliable transfer of action potentials because reduced numbers of ion channels in such small membrane areas lead to a decline in signal-to-noise ratio.[53] One important component in optimization of IPC in the MB may be related to the low spiking activity in KCs, a phenomenon well documented in the MB of locusts[45] but only indirectly assumed in the honeybee MB. IPC, rather than absolute or relative brain volumes, is considered to be the major determining factor in brain–intelligence relations.[54] The MB appears to optimize this factor by its dense packing of KCs. It will be exciting in future work to unravel the physiological and anatomical measures of IPC in KCs.

EXTRINSIC NEURONS OF THE MUSHROOM BODY

The large number of densely packed KCs in the MB converge on a rather small number (a few hundred) of MB ENs in the three output regions of the MB—the peduncle and the α and β lobes. Rybak and Menzel[28] characterized eight different groups of ENs (counting A3-v and A3-d as different groups, pooling A1 and A2 in one group, and counting PE1 as a neuron different from all other ENs) (Figure 29.5). Most of these groups contain approximately 70 neurons as judged by the number of somata (one exception: A5 comprises only 4 neurons). Four main dendritic target areas were found: (1) the ring neuropil of the α lobe to which all α lobe ENs project at least with parts of their dendrites, (2) the LH (A4 and Pe1) and optical tubercle (A5 and A7), (3) the contralateral protocerebrum

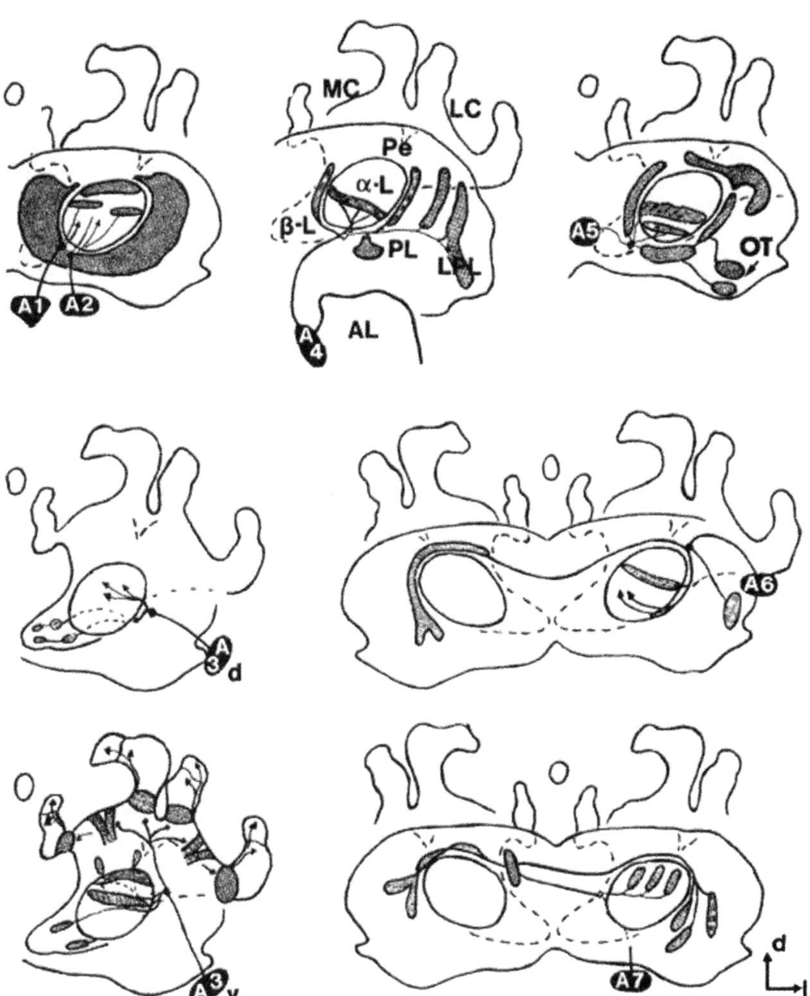

FIGURE 29.5 Schematic depiction of the α lobe ENs showing their respective clusters of somata (black circles with numbers A1–A7) and their dendritic branching areas. Source: *After Rybak and Menzel.*[28]

(A6 and A7), and (4) the feedback neurons to the calyx (A3). The multiplicity of connections established by these ENs makes it very likely that each group serves a different function. To date, these differences have not been able to be interpreted because we are ignorant about the functional characteristics of many of the target areas. In particular, we do not know the functional properties of the ring neuropil around the α lobe and the various subregions of the unstructured lateral protocerebrum. In any case, this structural multiplicity between EN groups and the number of neurons per group suggests forms of combinatorial coding of neural processing categories that are defined by the respective input and output regions. What are these categories?

Because many of these ENs receive input across the modality-specific regions of the MB, it is not surprising that they respond to a large range of sensory stimuli, indicating a different coding scheme than the highly specific combinatorial sensory code at the input of the MB. One large EN, the PE1, offers the unique possibility to repeatedly record from the same identified neuron during olfactory PER conditioning. The PE1 was found to reduce its responses to the learned odor[55] (Figure 29.6). This unique neuron receives excitatory input across the whole peduncle of the MB from KCs, also indicated by its multimodal sensitivity,[55] and inhibitory input presumably from GABA-ir A3 neurons of the PCT.[56] These latter neurons develop associative plasticity, and therefore it was tentatively concluded that enhanced inhibition via A3 neurons constitutes learned response reduction in PE1. However, PE1 also possesses intrinsic associative plasticity because KC excitation paired with PE1 depolarization induces long-term potentiation (LTP) in PE1 (Figure 29.6D).[57] It is possible that as yet unknown modulatory processes regulate transitions between LTP and long-term depression (LTD) in PE1 similar to what is known from principal cells in the mammalian cortex and hippocampus,[58] leading to associative response reduction via LTD under conditions of behavioral learning and to response

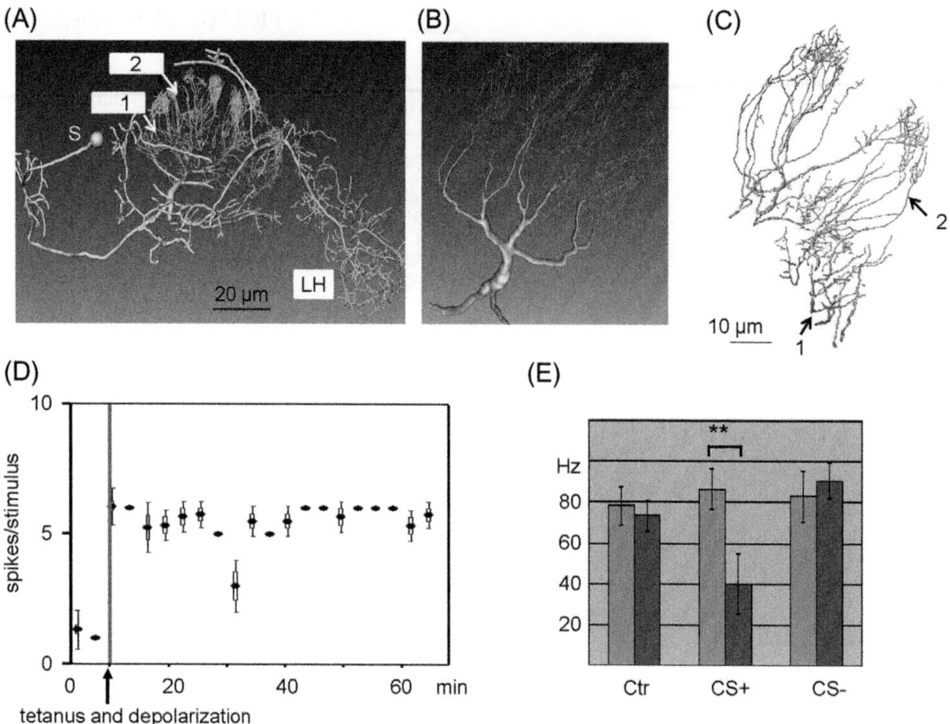

FIGURE 29.6 **Structure and plasticity of the identified MB extrinsic neuron PE1.** (A) The whole neuron with its soma (S), the dense dendrites in the MB peduncle, and the two axons—one (the neurite) leading to the soma and the other projecting to the lateral horn (LH). 1 and 2 mark two domains of the dendritic tree. (B) A higher resolution view of the dendritic tree, with the thick integrating segment and the branch point of the two axons. (C) Part of the dendritic tree (green) together with close attachments of GABA-ir profiles (red) from A3 neurons. These close attachments are found predominantly in domain 1. (D) Pairing of tetanic stimulation of KCs (arrow, 1 sec of 100 Hz) together with intracellular depolarization of PE1 leads to long-lasting enhancement of synaptic transmission (associative LTP) as seen in the number of spikes induced by each test stimulus. (E) Differential odor conditioning leads to selective reduction of CS+-induced activity. Blue bars indicate responses before and red bars indicate responses after conditioning. Ctr is a control odor. **Significant difference between the responses before and after conditioning. Source: *Panels C and E after Okada* et al.[56]; *panel D after Menzel and Manz.*[57]

enhancement via LTP under conditions of tetanic KC stimulation as used in the study by Menzel and Manz.[57] In such a scenario, associative response reduction to the CS+ would not reflect enhanced inhibition via A3 neurons but would represent an additional PE1-specific associative mechanism. Associative LTP and LTD in PE1 could also reflect spike timing-dependent plasticity (STDP), leading to either enhancement or reduction of synaptic efficiency, depending on the precise timing of spikes from KCs—a mechanism modeled for associative plasticity in the antennal lobe and reported for ENs in the locust.[59] However, STDP has yet to be proven to be related to behavioral learning in an insect.

The recorded properties of the MB from stimulus specificity to value-based responses are also demonstrated by other ENs recorded at the α exit close to the PE1.[60] Most of these neurons respond to odors with a broad chemo-profile and multiple other stimuli (visual and mechanosensory) including the sucrose reward. The sucrose responsiveness is likely to result from input of sucrose-responsive KCs that receive gustatory input via a specific ascending tract.[61] The kind of odor learning-induced plasticity in ENs varies considerably. Figure 29.7 shows four examples. Most ENs develop a response enhancement to CS+. Some of these neurons responded initially only to the US and after conditioning only to the CS+ (Figure 29.7A). Other ENs changed from excitatory CS+ responses to transient CS+ inhibitory responses (Figure 29.7C). Approximately half of the ENs recorded by Strube-Bloss *et al.*[60] changed their responses to the reinforced stimuli. Interestingly, most changed their responses not during the acquisition process but, rather, after a consolidation phase of a few hours. Two kinds of changes were observed—qualitative changes (referred to as switching; Figures 29.7A and 29.7C) and quantitative changes (referred to as modulation; Figures 29.7B and 29.7D). Switching neurons dropped responses and/or they developed new responses to one or several odors. All switches observed with respect to the CS+ odor were recruitments; those to the CS− could be recruitment or loss of response. Modulating neurons increased and/or decreased their response rates to different

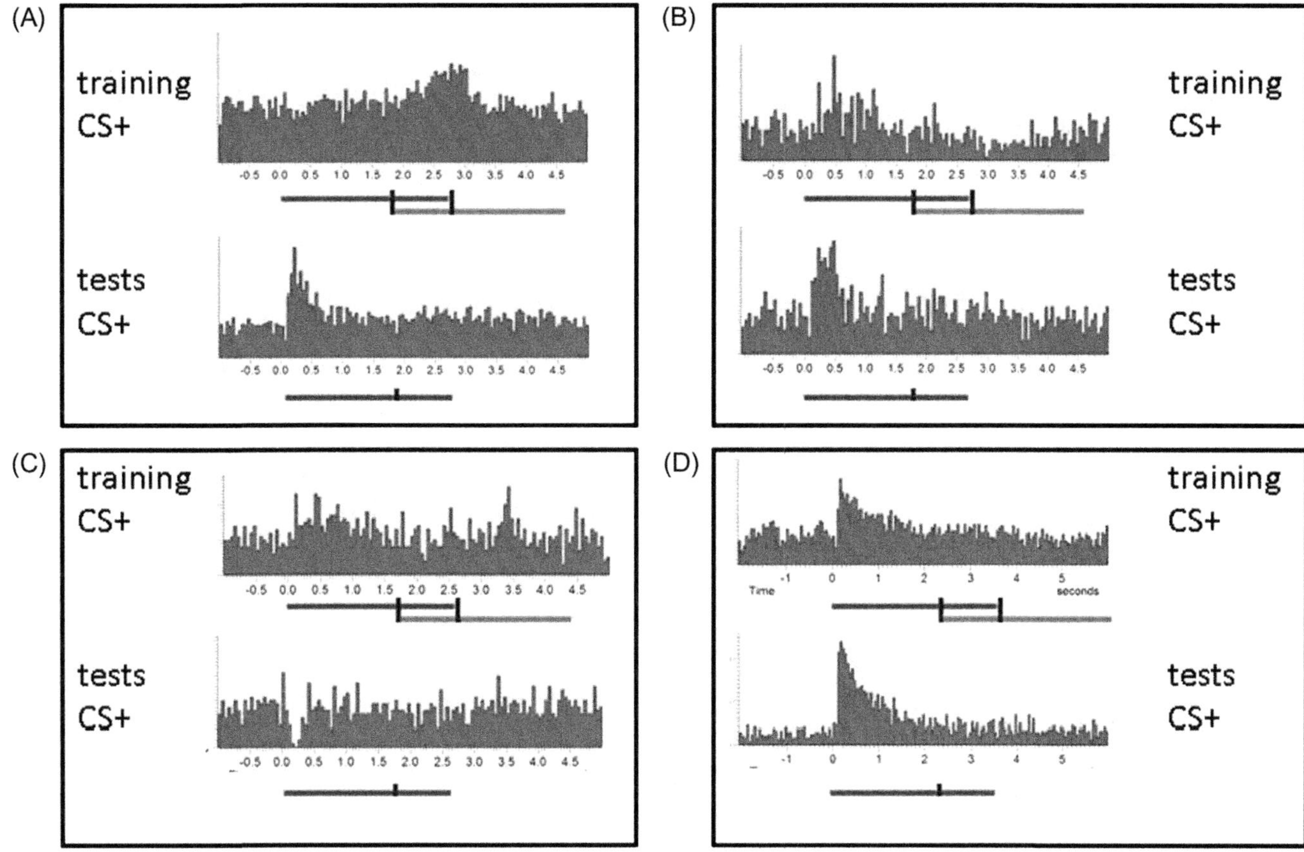

FIGURE 29.7 Four types of associative plasticity in MB extrinsic neurons as they develop during olfactory reward conditioning. The graphs give the sum of spikes during 10 repetitions of CS+/US pairing (training) or CS+ only (tests) for time bins of 50 msec. The top bar during conditioning and the bar during tests mark the CS+, and the lower bar marks the US during conditioning. The tick on the odor bar during tests marks the onset of the US during conditioning. (A) Before and during conditioning, the neuron responds only to the US; after conditioning, it responds to the CS+. This neuron is categorized as a switching EN. (B) The neuron responds before, during, and after conditioning to the CS+ and develops rather small quantitative changes of CS+ responding. This neuron is categorized as a modulating EN. (C) The neuron responds before and during conditioning with excitation to CS+; after conditioning, the CS+ induces an ON inhibition. This neuron is categorized as a switching EN. (D) The neuron responds before and during conditioning with phasic/tonic excitation to the CS+. After conditioning, the response to CS+ becomes more phasic. The time interval between conditioning and tests was 3 hr. This neuron is categorized as a modulating EN. Source: *After Strube-Bloss* et al.[60]

odors: CS+ always provoked increased responses, whereas CS− and control odors (the latter were used to test generalization phenomena) decreased or increased responses in approximately equal proportions. It was argued that the dichotomy of 'switching' and 'modulating' neurons may result from morphologically distinct ENs because switched and modulated neurons were rarely observed in the same recording. The delayed expression of associative plasticity in the switched and modulated ENs could reflect memory consolidation that depends on prolonged neural activities because consolidation in the MB can be blocked by cooling.

ENs appear to reflect both KC-related and own endogenous plasticity. The multiplicity of response changes during and after associative learning in several to many ENs could indicate a coding dimension at the MB output according to the meaning of the stimulus. Such meaning may be related to the prediction of the appetitive value, but it could also reflect different indicative categories of stimuli, such as a differentiation according to context and cue. Free-flying honeybees are known to learn the context in order to solve a discrimination task. They also learn contexts (light, colors, and temperature) quickly in the olfactory PER paradigm and use them for better discrimination.[62] They even master a trans-switching task in which the cue/context is reversed. ENs of the ventral α lobe close to the PE1 were found to reduce their responses to the cue (odor), whereas they increased their responses to the context (Figure 29.8). Therefore, these ENs do not simply code the appetitive value of a set of stimuli but,

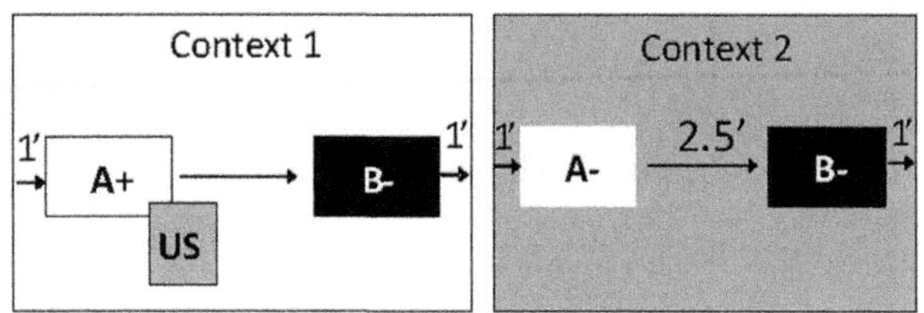

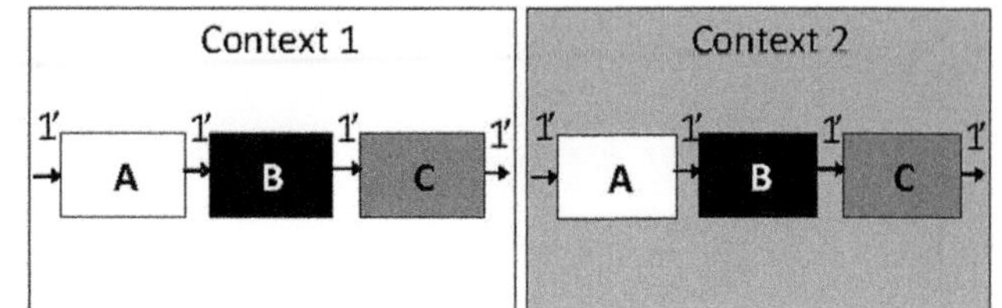

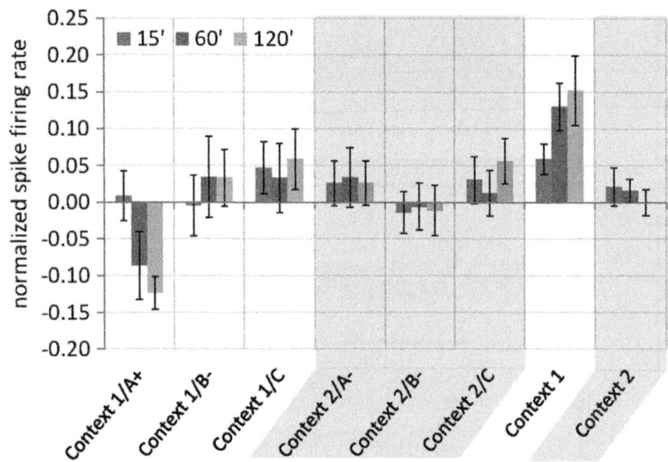

FIGURE 29.8 **Neural activity of ENs during context-dependent learning.** (Top) The procedure during conditioning and testing (post-conditioning). During conditioning, bees were presented with context 1 (bright light) and odor A was presented together with the sugar reward (US), whereas odor B was presented without any US. Then, context 1 was turned off and context 2 (dark) was presented, after which the two odors A and B were presented without any US. The five conditioning trials ended when context 2 was turned off. During testing (post-conditioning), each trial was presented once at 15, 60, and 120 min after conditioning. A trial consisted of context odor combinations or only context presentations. (Bottom) Normalized spike firing rate for different odor/context combinations as indicated. Spike firing rate toward odor A in context 1 was significantly reduced at 60 min ($p<0.05$) and 120 min ($p<0.01$) post-conditioning compared to the other two odors B and C. There was no difference in firing rate between odors A, B, and C in context 2. Firing rate toward context 1 increased at 15, 60, and 120 min ($p<0.01$) after conditioning, whereas that toward context 2 remained unchanged. Plots show average normalized spike firing rate for each group, and error bars represent standard error of mean. Source: *After Hussaini and Menzel.*[62]

rather, differentiate according to a different dimension, cue and context.

Cue and context learning is represented differently in A3 neurons—those neurons in the GABA-ir PCT that serve inhibitory feedback locally in the α lobe and in a recurrent loop that projects back into the calyx (Figure 29.1). Ca^{2+} imaging experiments revealed increased neuronal responses to CS+ after training, decreased response to repeated odor stimulation attenuating specifically for the CS+, and decreased response that was strongest for the CS−. These neuronal changes were linked to the behavioral changes as seen in retention tests on the first[63] or the second[64] day. Multiunit extracellular recordings of both subgroups of A3 neurons (A3-v and A3-d) revealed qualitatively similar associative plasticity for the cue (odor) and the context (color) in a double discrimination task in which a particular color indicated a particular odor–reward association and another color an odor–nonreward association (Figure 29.9).[65] These A3 neurons develop their learning-related plasticity during acquisition or during the course of days—some on the first day, and others on the second or third day.

Two response profiles with respect to learning-related plasticity have been identified. Neurons either increased (as shown in Figure 29.9) or decreased (not shown) their rate responses to the reinforced cue and context, and they also expressed the inverse rate changes to the non-reinforced cue and context. The first group was tentatively related to A3-d neurons and the second group to A3-v neurons. These antagonistic rate changes were stronger when the animals were able to discriminate between the conditioned odors behaviorally, and they peaked at discrete time windows for different neurons over a recording period of up to 3 days. With their output within the lobes, A3-d neurons are expected to enhance inhibition locally on other MB lobe ENs (e.g., PE1) and thus may act as the source of learning-related CS+ response reduction documented for the PE1 neuron.[56] Reduced learning-related inhibition via A3-v neurons feeding back into the calyx might enhance synaptic strength in

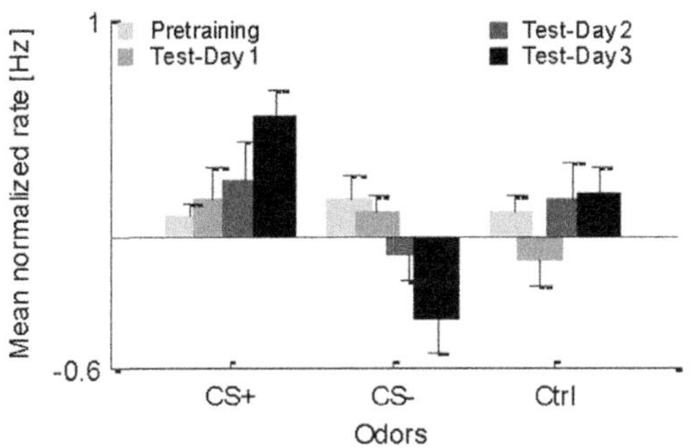

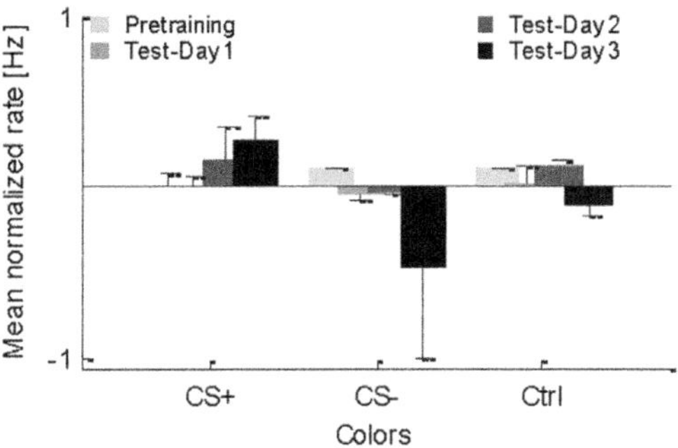

FIGURE 29.9 **Response changes of A3 neurons following context-cue training.** The graphs show responses separately for the cue (odors) and the context (colors) before training and during retention tests on three consecutive days. During retention tests, the specifically reinforced (CS+) odor (left) and color (right), the non-reinforced odor/color, and a control odor/color not presented during training were applied during extinction trials (tests without rewards). The graphs give the normalized spike rates during extinction tests separately for odors and colors of those neurons that developed their highest associative plasticity (the strongest rate difference between the conditioned stimuli) on the third day. Cues and contexts are represented in A3 neurons with qualitatively similar spike rate changes. Source: *After Filla and Menzel.*[65]

specific neurons in the MB main input microcircuits (e.g., that of the PN boutons[66]) through reduced inhibition. Decreased GABAergic input to KCs would consequently favor the induction of synaptic plasticity for reinforced stimuli. Evidence in favor of this hypothesis arises from the GABAergic anterior paired lateral (APL) neuron of *Drosophila*, which has striking morphological similarities with the A3-v cluster of the PCT.[67] The APL neuron suppresses and is suppressed by olfactory learning, suggesting that reduced inhibition promotes learning.[67] Reduced recurrent inhibition may also be related to an attention mechanism. In any case, the information stored in the recurrent pathway appears to modulate associative plasticity at a strategically important site, namely that part of the MB where precise coding of stimulus conditions (within and across sensory modalities) dominates.

THE LATERAL HORN

Olfactory and reward pathways also converge at the LH (Figure 29.1), suggesting that this structure may also form associative memory traces. The LH is part of the lateral protocerebrum, a rather unstructured neuropil with multiple subregions as indicated by different projection areas of PNs, ENs, and other protocerebral neurons.[28] The LH receives olfactory input directly from the AL via collaterals of the median and lateral tracts of PNs and indirectly via ENs of the MB (Figure 29.1). Some of the neurons extrinsic to the LH may contact descending neurons either directly or indirectly (e.g., in the case of the cockroach[68]; see Chapter 5), possibly allowing for rather direct sensorimotor loops. Such loops may underlie innate and fast odor-controlled responses such as pheromone-driven behavior,[69] but other views relate the lateral protocerebrum more closely to the MB and other high-order integrating centers in the insect brain.[70] The direct sensory premotor connections bypass and shortcut the olfactory pathways via the MB, and because MB ENs (e.g., PE1) also project directly to the LH, it has been suggested that the LH output is controlled by the experience-dependent signals from the MB. Thus, in addition to the possibly stereotypical sensorimotor connections, three sources of learning-related plasticity may control premotor output (and across brain connections)—plasticity in the AL, MB, and that intrinsic to the LH. Unfortunately, no data exist on LH function and plasticity in the honeybee. In *Drosophila*, the activity induced by fruit odors and pheromones in the LH is well segregated, suggesting a spatial organization with respect to odor classes,[71] but again learning-related plasticity needs to be demonstrated.

MEMORY TRACES IN THE REWARD PATHWAY

The neuron sufficient for the reinforcing function of sucrose in olfactory conditioning, the VUMmx1,[29] responds to a large range of stimuli before conditioning, but its response to sucrose stimulation at both the antennae and the proboscis is particularly strong and long-lasting (Figure 29.10A). The dendritic arbors suggested that it may be involved in the processing of olfactory information (Figure 29.1), and indeed it carries the reward signal in olfactory conditioning. Furthermore, during differential conditioning, the response to CS+ is enhanced, and that to CS− is reduced until no response is seen for the CS−. This finding indicates an excitatory memory for CS+ and

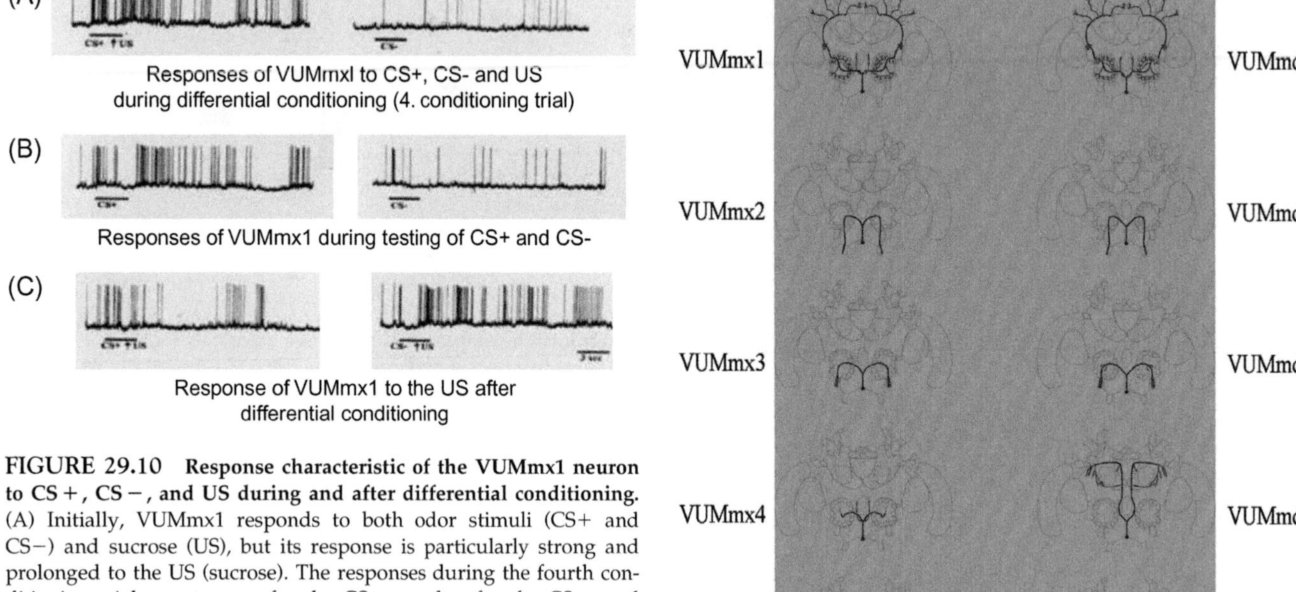

FIGURE 29.10 **Response characteristic of the VUMmx1 neuron to CS+, CS−, and US during and after differential conditioning.** (A) Initially, VUMmx1 responds to both odor stimuli (CS+ and CS−) and sucrose (US), but its response is particularly strong and prolonged to the US (sucrose). The responses during the fourth conditioning trial are stronger for the CS+, weaker for the CS−, and prolonged for the US. (B) In an extinction test, VUMmx1 responds strongly to CS+ and not to CS−. Note the similar response pattern to the CS+ and that to the US during conditioning. (C) If a US is presented after CS+, the sustained response is reduced, indicating that an expected US leads to inhibition of the sustained response, whereas an unexpected US as after CS− excites VUMmx1. Note the delayed burst of spikes occurring after both CS+/US and CS−/US stimulations.[72]

FIGURE 29.11 Five pairs of ventral unpaired median neurons immunoreactive to octopamin. Source: *After Schröter* et al.[73]

an inhibitory memory for CS− (Figure 29.10B). In addition, VUMmx1 develops a memory trace for the US because after conditioning and some consolidation period, an expected US induces inhibition in the sustained response. However, an unexpected US, such as after CS− application, excites VUMmx1 (Figure 29.10C). Interestingly, a delayed burst of spikes characteristic for the US long after its offset is seen in CS+-only stimulations and in CS+/US as well as CS−/US (unexpected reward) stimulations (Figures 29.10B and 29.10C).

VUMmx1 is one of 10 ventral unpaired OA-ir neurons of the subesophageal ganglion.[73] Five VUM neurons are localized in the maxillary and mandibulary neuromere and express pairwise corresponding dendritic structures (Figure 29.11). VUMmx1 and VUMmd1 match each other in dendritic arbors and response physiology; VUMmx2, VUMmd2, VUMmx3, and VUMmd3 send their dendrites along the antennal and mandibular nerves; VUMmx5 and VUMmd5 are likely to arbor in the antennal lobe and the ring neuropil around the α lobe (Figure 29.11 shows that they could not fully be reconstructed); and VUMmx5 and VUMmd5 innervate neuropils of the subesophageal ganglion. The responses to sucrose in most of these neurons make it likely that they carry information about the arousal and reward function of appetitive stimuli possibly under control of the level of satiation. Five additional corresponding VUM neurons have been reported for the labial neuromere. Because 6–8 OA-ir ventral cell bodies were seen in each of the three neuromeres of the subesophageal ganglion,[74] it is likely that there is a total of at least 18 VUM neurons in the bee brain. Assuming that all these VUM neurons are related to transmitting various components of appetitive stimuli and that as yet undetected VUM neurons reach all sensory (and possibly premotor) processing areas, it is likely that the positive value-based neural system is highly multifaceted. Additional aminergic neurons may participate in such an appetitively modulatory and reinforcing system, including particular dopamine neurons in the *Drosophila* brain. Each of these neurons or neuron pairs may store its own particular reward-related nonassociative and associative memory, and in combination with its respective target areas may constitute particular subsets of stimulus evaluation during learning, retrieval, and evaluation of stimulus compounds. It is thus not surprising that stimulation of octopamin receptors enhances feeding behavior, reward-seeking behavior, arousal, sensitivity to multiple sensory stimuli,[75] and social

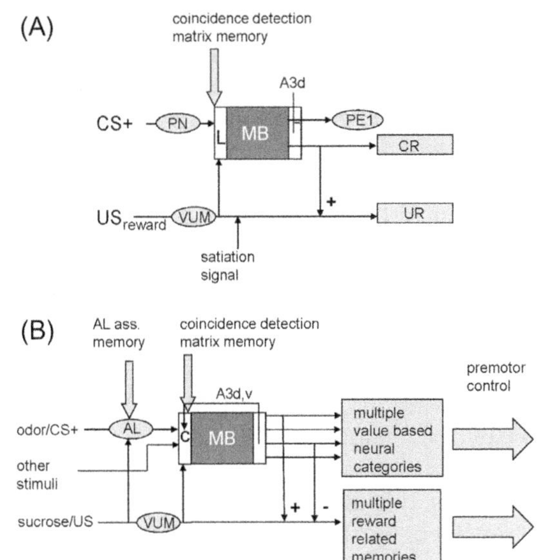

FIGURE 29.12 Two models of memory traces in the honeybee brain. (A) A radically simplified model. (B) A more adequate model.

interactions,[76,77] whereas blocking octopamin receptors reduces appetitive arousal, learning, and retrieval.[33,78]

THE DISTRIBUTED NATURE OF THE ENGRAM

Olfactory PER conditioning comprises a simple form of associative learning, but multiple processes at multiple sites are involved in forming the respective engram. The components observed so far are likely to comprise only a fraction of all associative changes comprising the full engram. It is also likely that these multiple sites are differently contributing to the sequential engrams during system's consolidation. Despite this complexity, it will be helpful for guiding future experiments to formulate a working hypothesis that attempts to integrate, at least partially, the current knowledge. Such a radically simplified model of the engram in the bee brain assumes a single, stable memory trace in the MB (Figure 29.12A). Spatial/temporal odor coding in the AL and the lip region of the MB calyx is thought to lead to an odor representation at the level of KCs by sparse and specific population activities. These network activities are associated with reward via convergence with the reward-encoding neurons (VUMmx1 and VUMmd1). Subpopulations of KCs differ with respect to their input from evaluating signals as indicated by specific gene activation patterns for octopamine and dopamine receptors[79]; thus, they are assumed to differ with respect to their coding scheme of different sensory modalities. It is therefore likely that the olfactory input needs to be distributed over the whole calyx, ensuring that the olfactory stimuli can be evaluated differently and combined with other stimuli in multiple combinations. Associatively enhanced activity in KCs driven by the CS+ will lead to stronger CS+ responses in ENs, as found in some ENs exciting the α lobe in its ventral and lateral aspect. The ENs expressing an inhibitory transmitter such as A3 neurons could inhibit other ENs such as PE1 within the lobes, reducing their CS+ responses. A subpopulation of ENs may transmit their enhanced CS+ excitation onto VUMmx1, causing the reward neuron to respond to CS+ and not to respond anymore to the US due to an intrinsic form of postexcitatory inhibition after it had responded to the CS+.

This radically simplified model could include various forms of nonassociative plasticity. For example, modulation of odor coding in the AL via VUMmx1 could lead to transient facilitation that could mimic enhanced odor responses following sucrose stimulation. Similarly, facilitation via VUMmx1 activity could counteract depression after stimulus repetition in KCs. Both modulatory phenomena could account at least partially for motivational and/or attentional effects induced by the appetitive stimulus in hungry bees. The associative effects observed in pre- and postsynaptic elements of the AL glomeruli would result in part from modulation via VUM neurons.

Such a radically simplified model of the memory trace offers a concept for explaining the recoding phenomenon in the MB from stimulus specificity to stimulus value based on the high divergence from PNs to KCs at the input and the extreme convergence of more than 100,000 KCs onto a few hundred ENs at the output. In addition, it assigns stable odor encoding to the AL, allowing the PNs to transmit an experience-independent odor code to the lateral protocerebrum and to the MB calyx. However, this model does not incorporate a whole range of findings both at the level of the AL and at the level of the MB. Furthermore, it is unconceivable that the enormously rich forms of learning in honeybees as seen under natural conditions and even partially in PER conditioning (context dependence and reward expectation) could be adequately conceptualized by such a simple model. A more realistic model of the distributed engram in the bee brain will have to incorporate the AL and the subtypes of KCs (and in further studies, the premotor and motor centers), and it needs to search for processing categories as they are read out from the MB. In this context, it will be helpful to consider the large range of connectivity patterns established by the ENs between the MB lobes and different brain parts.[28]

Evidence is strong for an independent memory trace in the AL. Although the stronger and more synchronized responses of PNs to the learned odor could result from feed-forward loops to the AL carrying information about the learned odor—for example, via the reward pathway (VUMmx1) or/and via inputs from the MB—additional assumptions are necessary to include the different forms of associative plasticity in glomeruli as seen in both Ca^{2+} imagining and multi-electrode recordings. PNs are either up- or downregulated for the CS + depending on whether or not they responded to the CS + before learning. It is unlikely that such plasticity could result from VUMmx1 modulatory signals or from a single forward loop from the MB. The stable enhancement of odor mixture coding after learning requires plasticity of the odor coding network in the AL and cannot be provided by general modulatory signals. Additional arguments in favor of an independent olfactory memory trace in the AL come from US substitution experiments by local injection of octopamine into the AL and the finding that blocking of octopamin receptors in the AL interferes with olfactory learning. In addition, it was shown that the transition from short- to long-term memory can be facilitated by activating cAMP-dependent PKC in the AL. Taken together, both the MB calyx and the LH appear to receive odor signals from the AL that encode experience with the particular odors and counteract the concept that the AL codes odors in a stable and experience-independent way. However, what exactly is transmitted about experience is unclear. It may well be that it is limited to enhanced attentional effects rather than to indexing a specific odor.

The simplified model assigns the stimulus-specific memory trace to the associative matrix of divergent PNs onto KCs in the lip of the MB calyx and, more generally, to all neurons reaching the calyx and feeding the more than 100,000 KCs of the MB. The matrix memory could indeed store the rich content of the memory trace, including all relevant combinations of external (cues and contexts) and internal stimuli, because formally it could possess the necessary intrinsic properties—divergent and convergent connectivity, sparse population and temporal coding, and high thresholds in KCs—making them respond only to convergent input. Neuroanatomical evidence supports the assumption that all sensory inputs more or less processed converge in the MB calyx onto KCs. A tiny glimpse into such a storage device is given in Figure 29.4 for a specific subtype of KCs (clawed KCs of KC II) receiving input across the modality-specific regions (lip, collar, and basal ring) of the calyx. As mentioned previously, the different types of KCs combine different subsets of inputs, some of which keep the sensory modalities apart and others combine them. Modulating and evaluating pathways reach the calyx (VUMmx1 and VUMmd1) or the peduncle (neurons immunoreactive to dopamine and serotonin). Their pattern of convergence with the different KC types is unknown except for the fact that the two VUM neurons reach only the olfactory input (lip region of the calyx) and not other sensory inputs. It will be necessary to demonstrate how visual and other modalities are evaluated by reward and whether such an associative matrix-based model also applies to these sensory modalities. It is also not known how aversive stimuli are evaluated by the MB, except for the hint that dopamine neurons are likely to be involved,[80] which is also corroborated by findings in *Drosophila*.[81] Because the expression of aminergic receptor genes differs between groups of KCs, this indicates that subsets may be selectively involved in coding appetitive and aversive forms of learning.[79] These authors also provide evidence in favor of different subpopulations of KCs to store short- and long-term olfactory memory.

Important neural components of the calycal matrix memory also include the presynaptic sites—for example, the boutons of the PNs for olfactory memory. Because these microcircuits can be easily quantified histologically, their structural plasticity in the course of natural life history and olfactory learning is well documented, indicating a structural substrate of the lasting memory trace at the MB input site. Surprisingly, Ca^{2+} imaging during learning reveals rather small associative effects in the presynaptic boutons of PNs (Yamagata, personal communication). Hypothetically, protein synthesis-dependent restructuring of PN boutons as seen after olfactory conditioning[43] may be orchestrated by postsynaptic effects of KCs, and short-term associative effects may therefore not be seen. If this interpretation is correct, different patterns of change could correspond to short- and long-term memory traces storing the same content—a concept supported for the MB of *Drosophila*.[50]

The most deficient aspect of the radically simplified model of memory trace in the MB is the assumption that the readout of the matrix memory is limited to very few types of ENs establishing direct connection to premotor centers in the bee brain (e.g., the LH). An alternative view interprets the outputs as processing circuits that represent acquired and value-based categories of stimulus combinations and assumes multiple forms of reward-related memories stored in the VUM neurons (Figure 29.10B). Although the number of ENs is small compared to that of KCs, their structures and response properties are enormously rich. Both their connectivity patterns and their response changes during learning and memory formation indicate that ENs are involved differently in the readout of the MB. As noted previously (Figure 29.5), eight different groups of ENs have been characterized on the basis of their somata loci and their arborizations patterns. The rich structural variability makes it likely that ENs of different groups serve different functions. Although these differences cannot be interpreted yet, the structural multiplicity suggests forms of combinatorial coding of neural processing categories that are defined by the respective input and output regions. What are these categories?

A common property of these categories could be that they represent acquired and value-based information. The following value-based categories (VBCs) come to mind:

1. Detection of novel versus already learned stimulus conditions
2. Distinction between appetitively and aversively learned stimulus conditions
3. Separation between cues and contexts
4. Separation between self-generated stimuli as experienced during active exploration and passively experienced stimuli as during classical conditioning
5. Storing stimulus traces for later learning, particularly under latent learning conditions as in navigation
6. Recognition and learning of symmetrical inputs to paired sense organs across sensory modality
7. Activation of specific memory traces for consolidation, such as during sleep or other forms of neural self-organization

These and other VBCs may define higher order neural processes brought about by the cooperation and combination of the lower level neural processes occurring in other parts of the brain, including (1) defining global context conditions (within the social context vs. acting individually; foraging for food vs. foraging for information during exploration), (2) working memory as a neural platform for the evaluation of expected outcomes, and (3) wakefulness versus sleep.

Although the assumption of defined VBCs is speculative, it may help in future work to relate some of the groups of ENs to these or other VBCs on the basis of their morphology. Consider several examples. First, A3-v projecting back from the lobes to the calyx could be involved in the distinction between novel and learned stimulus conditions and may be involved in facilitating the respective stimuli according to the response strength of the KCs. (2) ENs of groups A1–A4 are characterized by their branches in the ring neuropil of the same MB from which they receive input. Although we know nothing about the function of the ring neuropil, it could be that it houses long-term memory outside the active circuits of the MB, and neurons communicating between lobes and ring neuropil may be involved in memory consolidation and memory retrieval. Third, ENs connecting the two MBs in the two hemispheres of the brain (A6 and A7) may help to detect symmetrical stimuli across sensory modalities, transfer memory content from one MB to the other as proposed by Sandoz and Menzel,[82] and/or coordinate consolidation processes between the two MBs.

CONCLUSION

The engram of olfactory stimuli in the bee brain is characterized by its distributed nature with different prevailing processing categories at different sites. I conclude from the limited existing data that the trace in the AL relates predominantly to attention-generating properties, the matrix trace in the calyx to high-order combinatorics of all sensory inputs, the system of VUM neurons to appetitive internal states of the animal controlling nonassociative and associative traces, and the ENs of the MB to multiple processing categories that represent the acquired values and provide neural commands for goal-directed behavior and decision making. Although speculative, this framework offers a structure for experimental and modeling approaches and prevents us from believing that the properties of the memory trace can be captured by simply assuming flexible and experience-dependent sensory–interneuron–motor connections. Rather, we have to search for the coding/recoding, evaluating, and predicting processes involved in storing the contents of memory, the engram.

Gerber and co-workers[83] asked whether it is possible to localize a memory trace to a subset of cells in the brain. According to them, it needs to be shown that (1) neuronal plasticity occurs in the respective cells, (2) neuronal plasticity in these cells is sufficient for memory recall, (3) neuronal plasticity in these cells is necessary for memory formation, (4) memory content is lost if these cells do not function during retrieval tests, (5) and memory formation is abolished if these cells do not receive input during learning. This list of requirements, although difficult to meet experimentally (possibly only in *Drosophila* so far), is not complete and suffers from the focus on processes involved in neural plasticity rather than asking where and how the content of memory, the engram, is stored. The engram will not be found in a single type of neuron. It results from distributed network properties that add their respective contents when memory is formed, processed (consolidated), and retrieved. In other words, the engram is not a property of particular neurons but, rather, that of highly interacting networks of neurons. This form of interaction is different during memory formation, consolidation, and retrieval, meaning that different engrams (for the same content) exist depending on what happens to them and for what they are used. In this way, the engram does not 'exist' but develops over time and in relation to actions of the brain as mirrored in incorporating new contents into existing ones, in consulting different contents during decision making and planning, and during execution of behavioral acts.

References

1. Lashley KS. In search of the engram. *Symp Soc exp Biol.* 1950;4:454–482.
2. Dudai Y. The neurobiology of consolidations, or, how stable is the engram? *Annu Rev Psychol.* 2004;55:51–86.
3. Craik FI. Levels of processing: past, present. and future? *Memory.* 2002;10(5-6):305–318.
4. Moscovitch M. Memory: why the engram is elusive. In: Roediger HL, Dudai Y, Fristzpatrick SM, eds. *Science of Memory: Concepts.* Oxford: Oxford University Press; 2007:17–21.
5. Ramon Y, Cajal S. Einige hypothesen über den anatomischen mechanismus der ideenbildung, der association und der aufmerksamkeit. *Archiv für Anatomie und Physiologie.* 1895;25: 367–378.
6. Nicolelis MAL, Fanselow EE, Ghazanfar AA. Hebb's dream: the resurgence of cell assemblies. *Neuron.* 1997;19:219–221.
7. von Frisch K. *The Dance Language and Orientation of Bees.* Cambridge, MA: Harvard University Press; 1967.
8. Farina WM, Gruter C, Acosta L, Mc CS. Honeybees learn floral odors while receiving nectar from foragers within the hive. *Naturwissenschaften.* 2007;94:55–60.
9. Gil M, De Marco RJ, Menzel R. Learning reward expectations in honeybees. *Learn Mem.* 2007;14(491):496.
10. Giurfa M, Eichmann B, Menzel R. Symmetry as a perceptual category in honeybee vision. In: Elsner N, Menzel R, eds. *Learning and Memory. Proceedings of the 23rd Göttingen Neurobiology Conference.* Stuttgart: G. Thieme Verlag; 1995:423.
11. Wright GA, Skinner BD, Smith BH. Ability of honeybee, *Apis mellifera*, to detect and discriminate odors of varieties of canola (*Brassica rapa* and *Brassica napus*) and snapdragon flowers (*Antirrhinum majus*). *J Chem Ecol.* 2002;28(4):721–740.
12. Menzel R. Memory dynamics in the honeybee. *J Comp Physiol A.* 1999;185:323–340.
13. Davis RL. Traces of *Drosophila* memory. *Neuron.* 2011;70(1):8–19.
14. Grünbaum L, Müller U. Induction of a specific olfactory memory leads to a long-lasting activation of protein kinase C in the antennal lobe of the honeybee. *J Neurosci.* 1998;18:4384–4392.
15. Müller U. Prolonged activation of cAMP-dependent protein kinase during conditioning induces long-term memory in honeybees. *Neuron.* 2000;27:159–168.
16. Friedrich A, Thomas U, Müller U. Learning at different satiation levels reveals parallel functions for the cAMP-protein kinase a cascade in formation of long-term memory. *J Neurosci.* 2004;24 (18):4460–4468.
17. Lindauer M. Allgemeine sinnesphysiologie. Orientierung im raum. *Fortschr Zool.* 1963;16:58–140.
18. Menzel R. Das gedächtnis der honigbiene für spektralfarben. I. Kurzzeitiges und langzeitiges behalten. *Z vergl Physiol.* 1968;60:82–102.
19. Baddeley AD. *Working Memory, Thought and Action.* Oxford: Oxford University Press; 2007.
20. Zhang SW, Bartsch K, Srinivasan MV. Maze learning by honeybees. *Neurobiol Learn Mem.* 1996;66:267–282.
21. Dacke M, Srinivasan MV. Evidence for counting in insects. *Anim Cogn.* 2008;11(4):683–689.
22. Menzel R. Serial position learning in honeybees. *PLoS ONE.* 2009;4(3):e4694–e4701.
23. Greggers U, Menzel R. Memory dynamics and foraging strategies of honeybees. *Behav Ecol Sociobiol.* 1993;32:17–29.
24. Hellstern F, Malaka R, Hammer M. Backward inhibitory learning in honeybees: a behavioral analysis of reinforcement processing. *Learn Mem.* 1998;4:429–444.
25. Menzel R, Giurfa M. Dimensions of cognition in an insect, the honeybee. *Behav Cogn Neurosci Rev.* 2006;5:24–40.
26. Pamir E, Chakroborty NK, Stollhoff N, Gehring KB, Antemann V, Morgenstern L, et al. Average group behavior does not represent individual behavior in classical conditioning of the honeybee. *Learn Mem.* 2011;18(11):733–741.
27. Mobbs PG. The brain of the honeybee *Apis mellifera*: I. The connections and spatial organization of the mushroom bodies. *Phil Trans R Soc Lond B.* 1982;298:309–354.
28. Rybak J, Menzel R. Anatomy of the mushroom bodies in the honey bee brain: the neuronal connections of the alpha-lobe. *J Comp Neurol.* 1993;334(3):444–465.
29. Hammer M. An identified neuron mediates the unconditioned stimulus in associative olfactory learning in honeybees. *Nature.* 1993;366:59–63.
30. Hammer M, Menzel R. Learning and memory in the honeybee. *J Neurosci.* 1995;15(3):1617–1630.
31. Menzel R, Erber J, Masuhr T. Learning and memory in the honeybee. In: Barton-Browne L, ed. *Experimental Analysis of Insect Behaviour.* Berlin: Springer; 1974:195–217.
32. Hammer M, Menzel R. Multiple sites of associative odor learning as revealed by local brain microinjections of octopamine in honeybees. *Learn Mem.* 1998;5:146–156.
33. Farooqui T, Vaessin H, Smith BH. Octopamine receptors in the honeybee (*Apis mellifera*) brain and their disruption by RNA-mediated interference. *J Insect Physiol.* 2004;50(8):701–713.
34. Faber T, Joerges J, Menzel R. Associative learning modifies neural representations of odors in the insect brain. *Nature Neuroscience.* 1999;2(1):74–78.
35. Peele P, Ditzen M, Menzel R, Galizia CG. Appetitive odor learning does not change olfactory coding in a subpopulation of honeybee antennal lobe neurons. *J Comp Physiol A Neuroethol Sensory Neural Behav Physiol.* 2006;192(10):1083–1103.
36. Schmuker M, Weidert M, Menzel R. A network model for learning-induced changes in odor representation in the antennal lobe. In: Laurent UP, Emmanuel D, eds. *Proceedings of the Second French Conference on Computational Neuroscience.* Marseille, France; 2008.
37. Fernandez PC, Locatelli FF, Person-Rennell N, Deleo G, Smith BH. Associative conditioning tunes transient dynamics of early olfactory processing. *J Neurosci.* 2009;29(33):10191–10202.
38. Rath L, Giovanni GC, Szyszka P. Multiple memory traces after associative learning in the honey bee antennal lobe. *Eur J Neurosci.* 2011;34(2):352–360.
39. Denker M, Finke R, Schaupp F, Grun S, Menzel R. Neural correlates of odor learning in the honeybee antennal lobe. *Eur J Neurosci.* 2010;31(1):119–133.
40. Kenyon FC. The brain of the bee: a preliminary contribution to the morphology of the nervous system of the arthropoda. *J Comp Neurol.* 1896;6:134–210.
41. Dujardin J. Memoire sur le systeme nerveux des insects. *Ann Sci Nat Zool.* 1850;14:196–206.
42. Ganeshina OT, Vorobyev MV, Menzel R. Synaptogenesis in the mushroom body calyx during metamorphosis in the honeybee

Apis mellifera: an electron microscopic study. *J Comp Neurol.* 2006;497(6):876–897.
43. Hourcade B, Muenz TS, Sandoz JC, Rossler W, Devaud JM. Long-term memory leads to synaptic reorganization in the mushroom bodies: a memory trace in the insect brain. *J Neurosci.* 2010;30(18):6461–6465.
44. Szyszka P, Ditzen M, Galkin A, Galizia CG, Menzel R. Sparsening and temporal sharpening of olfactory representations in the honeybee mushroom bodies. *J Neurophysiol.* 2005;94 (5):3303–3313.
45. Perez-Orive J, Mazor O, Turner GC, Cassenaer S, Wilson RI, Laurent G. Oscillations and sparsening of odor representations in the mushroom body. *Science.* 2002;297(5580):359–365.
46. Rolls ET. An attractor network in the hippocampus: theory and neurophysiology. *Learn Mem.* 2007;14(11):714–731.
47. Szyszka P, Galkin A, Menzel R. Associative and non-associative plasticity in Kenyon cells of the honeybee mushroom body. *Front Syst Neurosci.* 2008;2:1–10.
48. Szyszka P, Demmler C, Oemisch M, Sommer L, Biergans S, Birnbach B, et al. Mind the gap: olfactory trace conditioning in honeybees. *J Neurosci.* 2011;31(20):7229–7239.
49. Grünewald B, Wersing A, Wüstenberg D. Learning channels: cellular physiology of odor processing neurons within the honeybee brain. *Acta Biol Hungarica.* 2004;55(1-4):53–63.
50. Pascual A, Preat T. Localization of long-term memory within the *Drosophila* mushroom body. *Science.* 2001;294(5544):1115–1117.
51. Brandt R, Rohlfing T, Rybak J, Krofczik S, Maye A, Westerhoff M, et al. A three-dimensional average-shape atlas of the honeybee brain and its applications. *J Comp Neurol.* 2005;492 (1):1–19.
52. Witthöft W. Absolute anzahl und verteilung der zellen im hirn der honigbiene. *Z Morph Tiere.* 1967;61:160–184.
53. White JA, Rubinstein JT, Kay AR. Channel noise in neurons. *Trends Neurosci.* 2000;23(3):131–137.
54. Roth G, Dicke U. Evolution of the brain and intelligence. *Trends Cogn Sci.* 2005;9(5):250–257.
55. Mauelshagen J. Neural correlates of olfactory learning in an identified neuron in the honey bee brain. *J Neurophysiol.* 1993;69:609–625.
56. Okada R, Rybak J, Manz G, Menzel R. Learning-related plasticity in PE1 and other mushroom body-extrinsic neurons in the honeybee brain. *J Neurosci.* 2007;27(43):11736–11747.
57. Menzel R, Manz G. Neural plasticity of mushroom body-extrinsic neurons in the honeybee brain. *J Exp Biol.* 2005;208(Pt 22):4317–4332.
58. Bear MF, Malenka RC. Synaptic plasticity: LTP and LTD. *Curr Opin Neurobiol.* 1994;4(3):389–399.
59. Cassenaer S, Laurent G. Hebbian STDP in mushroom bodies facilitates the synchronous flow of olfactory information in locusts. *Nature.* 2007;448(7154):709–713.
60. Strube-Bloss MF, Nawrot MP, Menzel R. Mushroom body output neurons encode odor reward associations. *J Neurosci.* 2011;31 (8):3129–3140.
61. Schröter U, Menzel RA. New ascending sensory tract to the calyces of the honeybee mushroom body, the subesophageal-calycal tract. *J Comp Neurol.* 2003;465:168–178.
62. Hussaini SA, Menzel R. Mushroom body extrinsic neurons in the honeybee brain encode cues and contexts differently. *J Neurosci.* 2012, in press.
63. Haehnel M, Menzel R. Sensory representation and learning-related plasticity in mushroom body extrinsic feedback neurons of the protocerebral tract. *Front Neurosci.* 2010;4:1–16.
64. Haehnel M, Menzel R. Long-term memory and response generalization in mushroom body extrinsic neurons in the honeybee *Apis mellifera*. *J Exp Biol.* 2012;215(559):565.
65. Filla I, Menzel R. Visual and olfactory associative plasticity in an inhibitory local and recurrent pathway in the honeybee. 2012 [In revision.]
66. Ganeshina OT, Menzel R. GABA-immunoreactive neurons in the mushroom bodies of the honeybee: an electron microscopic study. *J Comp Neurol.* 2001;437(3):335–349.
67. Liu X, Davis RL. The GABAergic anterior paired lateral neuron suppresses and is suppressed by olfactory learning. *Nat Neurosci.* 2009;12(1):53–59.
68. Okada R, Sakura M, Mizunami M. Distribution of dendrites of descending neurons and its implications for the basic organization of the cockroach brain. *J Comp Neurol.* 2003;459(3):158–174.
69. Gerber B, Stocker RF. The *Drosophila* larva as a model for studying chemosensation and chemosensory learning: a review. *Chem Senses.* 2006;32(1):65–89.
70. Li Y, Strausfeld NJ. Multimodal efferent and recurrent neurons in the medial lobes of cockroach mushroom bodies. *J Comp Neurol.* 1999;409(4):647–663.
71. Jefferis GS, Potter CJ, Chan AM, Marin EC, Rohlfing T, Maurer Jr. CR, et al. Comprehensive maps of *Drosophila* higher olfactory centers: spatially segregated fruit and pheromone representation. *Cell.* 2007;128(6):1187–1203.
72. Hammer M. The neural basis of associative reward learning in honeybees. *Trends Neurosci.* 1997;20(6):245–252.
73. Schröter U, Malun D, Menzel R. Innervation pattern of suboesophageal VUM neurons in the honeybee brain. *Cell Tissue Res.* 2006;326(3):647–667.
74. Kreissl S, Eichmüller S, Bicker G, Rapus J, Eckert M. Octopamine-like immunoreactivity in the brain and suboesophageal ganglion of the honeybee. *J Comp Neurol.* 1994;348:583–595.
75. Bicker G, Menzel R. Chemical codes for the control of behaviour in arthropods. *Nature.* 1989;337:33–39.
76. Barron AB, Maleszka R, Vander Meer RK, Robinson GE. Octopamine modulates honey bee dance behavior. *Proc Natl Acad Sci USA.* 2007;104(5):1703–1707.
77. Johnson RN, Oldroyd BP, Barron AB, Crozier RH. Genetic control of the honey bee (*Apis mellifera*) dance language: segregating dance forms in a backcrossed colony. *J Hered.* 2002;93 (3):170–173.
78. Farooqui T, Robinson K, Vaessin H, Smith BH. Modulation of early olfactory processing by an octopaminergic reinforcement pathway in the honeybee. *J Neurosci.* 2003;23(12):5370–5380.
79. McQuillan HJ, Nakagawa S, Mercer AR. Mushroom bodies of the honeybee brain show cell population-specific plasticity in expression of amine-receptor genes. *Learn Mem.* 2012;19 (4):151–158.
80. Vergoz V, Roussel E, Sandoz JC, Giurfa M. Aversive learning in honeybees revealed by the olfactory conditioning of the sting extension reflex. *PLoS ONE.* 2007;2:e288.
81. Schwaerzel M, Monastirioti M, Scholz H, Friggi-Grelin F, Birman S, Heisenberg M. Dopamine and octopamine differentiate between aversive and appetitive olfactory memories in *Drosophila*. *J Neurosci.* 2003;23(33):10495–10502.
82. Sandoz J-C, Menzel R. Side-specificity of olfactory learning in the honeybee: generalization between odors and sides. *Learn Mem.* 2001;8:286–294.
83. Gerber B, Tanimoto H, Heisenberg M. An engram found? Evaluating the evidence from fruit flies. *Curr Opin Neurobiol.* 2004;14(6):737–744.

CHAPTER

30

Neural Correlates of Olfactory Learning in the Primary Olfactory Center of the Honeybee Brain

The Antennal Lobe

Jean-Christophe Sandoz

CNRS, Gif-sur-Yvette, France

INTRODUCTION

A general question in the study of associative learning and memory is how stimulus-specific and outcome-related information is acquired, stored, and retrieved by the nervous system. To answer such questions, appropriate learning paradigms, a good description of the nervous system, and the availability of neurophysiological techniques allowing to record the activity of neural networks are all necessary conditions. In this context, the honeybee *Apis mellifera* L. is an interesting model for the study of the mechanisms of learning and memory because (1) it shows robust learning abilities in nature, which can be reproduced in controlled laboratory conditions; (2) its brain has been extensively described by previous neuroanatomical work; and (3) its robustness allows for its routine use in neurophysiological experiments.[1–4]

In nature, honeybee foragers optimize their search for food by using a combination of remarkable individual learning skills and elaborate communication with nestmates.[5–7] At the individual level, foraging bees learn and memorize the features of a given floral species and exploit its resources as long as this remains profitable.[8] Such floral cues include its color, odor, shape, and texture. However, among these, odors play a prominent role because they are most readily associated with a reward of nectar or pollen.[5,6] Due to its particular implication in ecological conditions and its amenability to laboratory experiments, olfactory learning has become the most studied form of learning in honeybees (for reviews, see[1,3,4]). One of the central questions that has been pursued is how the bee brain, comprising only approximately 1 million neurons, learns, memorizes, and retrieves odor-reward associations.

Learning and memory formation rely on changes in the intrinsic excitability of neurons, in the strength of synaptic transmission, and/or on the creation of new connections.[9,10] Such processes have been extensively studied at the cellular and *in vitro* levels in a variety of species, among which rodents and mollusks are prominent.[9,10] At a macroscopic level, involving the whole animal brain, neural plasticity can be observed as structural and/or functional changes within brain networks. Functional changes usually correspond to changes in the intensity or in the temporal pattern of responses of particular neurons to sensory stimuli, whereas structural changes are usually observed as changes in the conformation, shape, or volume of brain structures or particular neuronal processes. An incentive for studying such phenomena in an insect such as the honeybee is the unique possibility of observing such plasticity in a living animal while it is learning, memorizing, or retrieving olfactory information. In addition, the relative simplicity and modularity of its brain made the search for plasticity within individual neurons and/or recognizable structures easier.[11] In particular, the remarkable structure of its first olfactory center, the antennal lobe, with its clearly identifiable

Invertebrate Learning and Memory.
DOI: http://dx.doi.org/10.1016/B978-0-12-415823-8.00030-7

glomerular units, has been a choice model system for studying olfactory learning-induced plasticity.

This chapter reviews the search for functional and structural plasticity in the bee brain as a result of olfactory conditioning, focusing on the antennal lobe. After detailing the different forms of olfactory learning studied in laboratory conditions, the chapter describes the olfactory system of honeybees with emphasis on the antennal lobe, detailing the reasons why plasticity may be expected in this structure. Then, the chapter reviews the results procured by studies coupling different olfactory conditioning protocols with neurophysiological and neuroanatomical quantifications of neural activity or structure. Finally, the chapter discusses where research should focus next in order to understand how stimulus-specific and outcome-related memories are formed and stored in the bee brain.

STUDIED FORMS OF OLFACTORY LEARNING IN HONEYBEES

Although bees' learning abilities can be effectively studied using free-flying individuals visiting artificial feeders or Y-mazes, the search for neural plasticity demands learning protocols that are amenable to the laboratory.[1–3] One of the major advantages of the honeybee as an experimental model is related to the wide range of such protocols, the most important of which are detailed here.

Pavlovian (classical) conditioning is by far the most studied associative learning form in bees.[1–3] In most cases, experiments have involved appetitive associations between odors and a sucrose reward, but recently an aversive learning protocol has also been proposed. The most influential protocol allowing the study of olfactory learning on restrained individuals is the appetitive conditioning of the proboscis extension response (PER; Figure 30.1A). When the antennae, mouthparts, or tarsi of a hungry bee are touched with sucrose solution, the animal reflexively extends its proboscis to suck the sucrose. This response was initially conditioned by Kuwabara[13] and Takeda[14] by associating visual and olfactory stimuli, respectively, with a sucrose reward. Perfecting the olfactory version of this protocol, Bitterman *et al.*[15] also showed that it is a case of Pavlovian conditioning. Odors to the antennae do not normally release a PER in naive animals (Figure 30.1A, left). If an odor is presented immediately before the sucrose solution (forward pairing, usually with a 3-sec interstimulus interval), an association is formed (Figure 30.1A, middle). During conditioning, in the interval of odor presentation before sucrose is applied or in a later test, the presentation of the odor alone will trigger the PER (Figure 30.1A, right). Thus, the odor can be viewed as the conditioned stimulus (CS) and sucrose solution as the reinforcing unconditioned stimulus (US). This association is thought to recapitulate the final phase of the foraging behavior, when bees drink nectar from an odorous flower.

More recently, an aversive Pavlovian conditioning protocol has been proposed on restrained individuals, the conditioning of the sting extension response (SER) (see Chapter 36). The SER is a defensive response of bees to potentially noxious stimuli, which can be experimentally elicited by delivering a mild electric shock to the thorax of the bees. During conditioning, harnessed bees learn to associate an initially neutral odor (CS) with the electric shock (US).[16] Whereas PER conditioning is appetitive and induces attraction toward the CS in a choice test, SER conditioning is aversive and bees will accordingly avoid the CS.[17] Hence, the effect of associative olfactory learning can be studied and compared with respect to different reinforcement modalities. However, until now, most studies of learning-induced plasticity have concentrated on appetitive conditioning.

Different types of conditioning protocols have been used. In absolute conditioning (A+), bees learn to associate an odor A with the US in a procedure with multiple conditioning trials (Figure 30.1B). In the course of training, an increasing proportion of bees respond to odor A. In a later test, bees still respond to A, and if presented with a different (novel) odor B, they will tend to generalize their response to B, especially if A and B are chemically similar.[18,19] Neural signals to odor A in this group of animals (usually called the 'paired group' because CS and US are forward paired) are compared to those recorded in a group of control animals that received the same number of CS and US presentations, but not in temporal association, so that no learning can take place (called the 'unpaired group').[15] The problem with this strategy is that it necessitates a high number of animals to counterbalance interindividual variability in neurophysiological signals. Another strategy, which has been most extensively used for studying plasticity, is based on obtaining a within-animal control for the learned association. In differential conditioning (A + B −), a bee has to learn that an odorant A (CS+) is associated with the reinforcement but that another odorant B (CS −) is not (Figure 30.1C). Usually, bees quickly learn to respond to the CS + and not to the CS −, but successful differentiation can take more time if the two odorants are chemically similar. In this protocol, each bee acts as its own control because different neural responses could be expected to the two odorants differing in their outcome. In such differential conditioning, explicitly not pairing an odor B with reward within a series of forward-paired experiences of A with reward induces

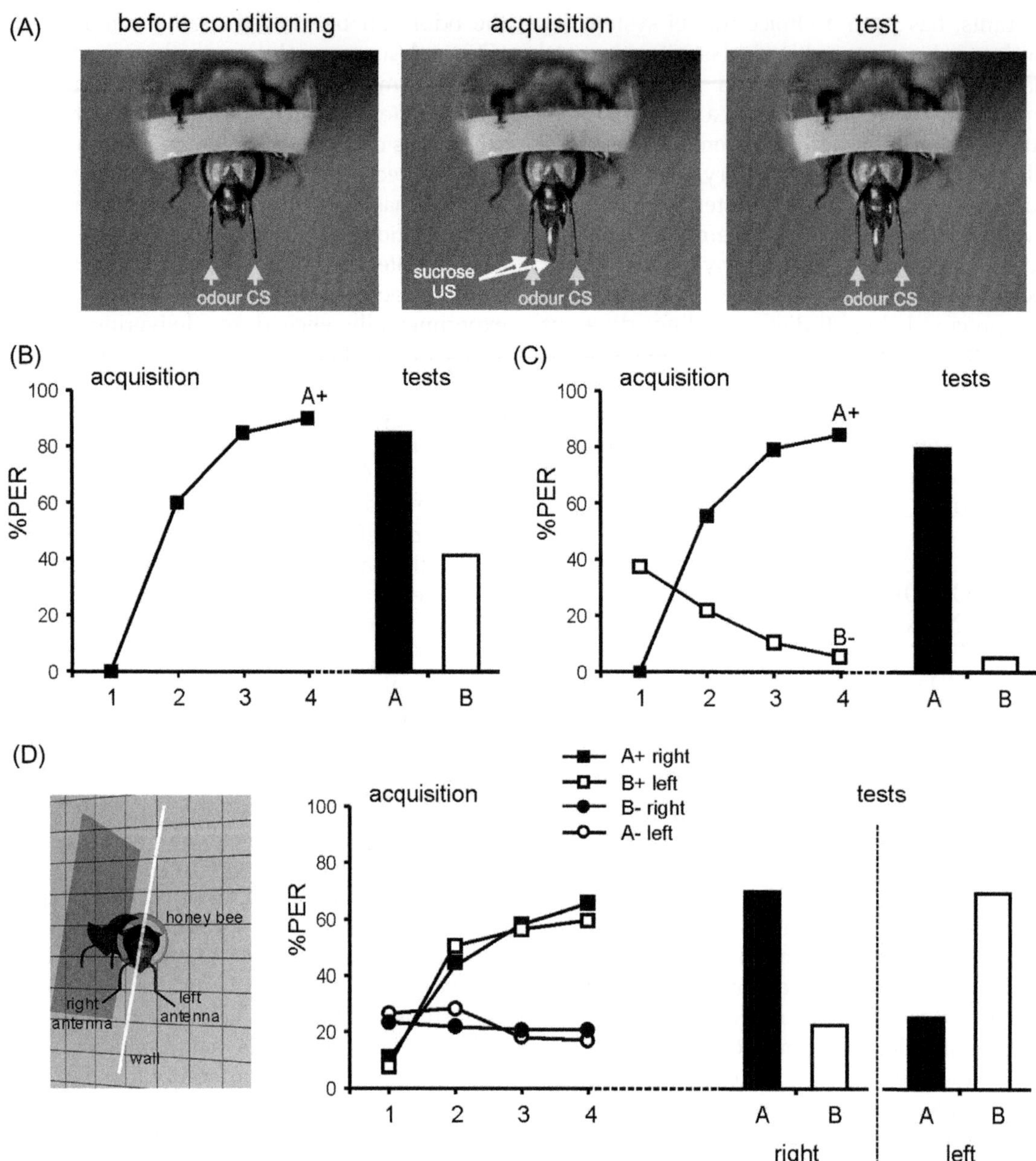

FIGURE 30.1 Different olfactory learning protocols available using the conditioning of the proboscis extension response (PER) on restrained bees. (A) Principle of PER conditioning. Before conditioning (left), odors usually do not trigger the PER in experimentally naive bees. However, a response (extension of the proboscis) is triggered in hungry bees when their antennae, tarsi, or mouthparts are contacted with sucrose solution. During conditioning (middle), an odor (conditioned stimulus; CS) is presented in close temporal association with sucrose solution to the antennae and to the proboscis (unconditioned stimulus; US). If conditioning is successful, presentation of the odor alone during a test trial (right) triggers the PER. (B) Absolute conditioning (A +). Bees are subjected to multiple conditioning trials with one odorant A associated with sucrose reward. In the course of conditioning, an increasing proportion of bees respond to the odor (which starts 3 sec before the US) with a PER. In a test performed after conditioning, bees respond to the learned odor A but also tend to generalize their responses to other odorants (see the intermediate level of response to odor B). (C) Differential conditioning (A + B −). During acquisition, bees receive in a pseudo-randomized order trials with an odor A associated with sucrose reward (CS +) and with an odor B without reward (CS −). During training, an increasing proportion of bees learn to respond to odor A but not to odor B. In a later test, bees maintain a clear differentiation between the two stimuli, responding to odor A but not to B. (D) Side-specific conditioning (A + B −/B + A −). In such experiments, two airflows allow stimulating separately the right and the left antennae separated by a plastic wall placed on the bee head. During acquisition, bees have to learn contrary information presented on the two antennae. While odor A is rewarded when presented on the right side (and B is not), the reversed contingency is given on the left side. Bees learn the task and respond specifically to each odor on the side on which it was rewarded. (Right) This pattern of responses is maintained in a later test. Source: *Adapted from unpublished data and Sandoz* et al.[12]

an additional form of learning of B, namely that it is not rewarded (CS–no US association; see Chapter 33). Many variants of differential conditioning protocols exist with varying degrees of complexity.[1] Only one such protocol was used in the search for plasticity. In the side-specific discrimination (A + B −/B + A −), bees learn opposite discrimination patterns on the two brain sides.[20] Thus, bees have to learn that odorant A is reinforced when presented on the right side (and odorant B is not), and that B is reinforced on the left side (and odorant A is not; Figure 30.1D). Bees learn this task slowly but efficiently and respond specifically to each odor on the side on which it was rewarded. This protocol not only provides a within-animal control for the CS–US association, as for differential conditioning, but also provides a control for the possible interaction between odor quality and reinforcement, as each odorant is associated or not with the US.[12]

PER conditioning has permitted dissecting some of the molecular and cellular aspects of olfactory memory formation in bees. Performance after associative conditioning is supported by a number of sequential or parallel memory phases, which may occur at different times within different brain structures[3,11] (see Chapter 29). Therefore, the understanding of learning-induced plasticity in the bee brain is also closely linked to the understanding of these memory phases. At least five types of olfactory memory phases have been identified. After a single conditioning trial, responses to the CS are high soon after conditioning (1 or 2 min) and then decrease, showing a 'dip' at approximately 3 min, and increase again after 7 min until approximately 1 day, when performance definitively decays. Two different memory phases are thought to underlie this performance: In the first minutes after conditioning, performance depends on a strongly nonassociative short-term memory (STM), due to sensitization from the US. Whereas STM decays after 2 or 3 min, a consolidation process leads to a highly odor-specific midterm memory (MTM) lasting approximately 1 day. This consolidation process is characterized by a prolonged activity of the cAMP-dependent protein kinase (PKA)[21] (see Chapter 31). Multiple-trial conditioning, on the other hand, triggers the formation of different memory phases, which rely on different cellular actors. After such conditioning, performance remains very high for several days or even several weeks. After an initial STM phase (which may include two forms, early and late STM), consolidation leads to a different MTM phase (multiple-trial MTM) characterized by a selective increase in Ca^{2+}/calmodulin-dependent protein kinase C activity.[22] As for single-trial MTM, multiple-trial MTM decays within approximately 1 day. In the day's range, performance is controlled by two long-term memory (LTM) phases formed in parallel. The early phase (e-LTM) depends on translation of already existing mRNA and is predominantly induced by massed training with short intertrial intervals (usually 1 min). The late phase (l-LTM) is critically dependent on transcription and is only formed after spaced training with long intertrial intervals (usually 10 min). Therefore, depending on the number and the distribution of associative events, multiple memory forms are formed. As the cellular mechanisms differ for these different phases, so also may differ the forms of functional and structural plasticity found at these time points.

THE OLFACTORY SYSTEM

Odor Detection

Peripheral odor detection starts at the level of olfactory receptor neurons (ORNs), which are located below cuticular structures on the antennae, called sensillae. In honeybees, poreplate sensilla (sensilla placodea) are the main olfactory sensilla and are innervated by 5–35 ORNs.[23] Odorant molecules reach the dendrites of ORNs by diffusing through a receptor hemolymph located in the sensillum cavity. Odorant binding proteins may help transport the odorants through the hemolymph; however, their role in bees has not been confirmed *in vivo*. When reaching the ORN membrane, odorant molecules interact with olfactory receptor proteins (ORs). The functional receptor is a heteromeric complex of an OR and the co-receptor *Orco* expressed in all ORNs (previously called AmOr2).[24]

Primary Processing: The Antennal Lobe

ORN axons form the antennal nerve to reach the antennal lobe (AL), the primary olfactory center of the insect brain (Figure 30.2). The antennal nerve splits into six sensory tracts upon entrance into the AL. Four of these tracts (T1–T4) innervate distinct portions of the AL, whereas the two remaining tracts (T5 and T6) bypass the AL. The bee AL is compartmentalized in 165 anatomical and functional units, the glomeruli. Glomeruli can be recognized based on their position, size, and shape.[25] In *Drosophila*, axons of ORNs expressing the same OR converge onto the same glomerulus.[26] Thus, the array of AL glomeruli would correspond to the array of approximately 163 OR types found in the genome.[24] Two main neuron types comprise the AL.

Local neurons (LNs) have branching patterns restricted to the AL. They are especially numerous in the honeybee (~4000 LNs) and can be classified in two main types.[27,28] Homogeneous LNs innervate most, if not all, glomeruli in a uniform manner; heterogeneous LNs innervate one dominant glomerulus with

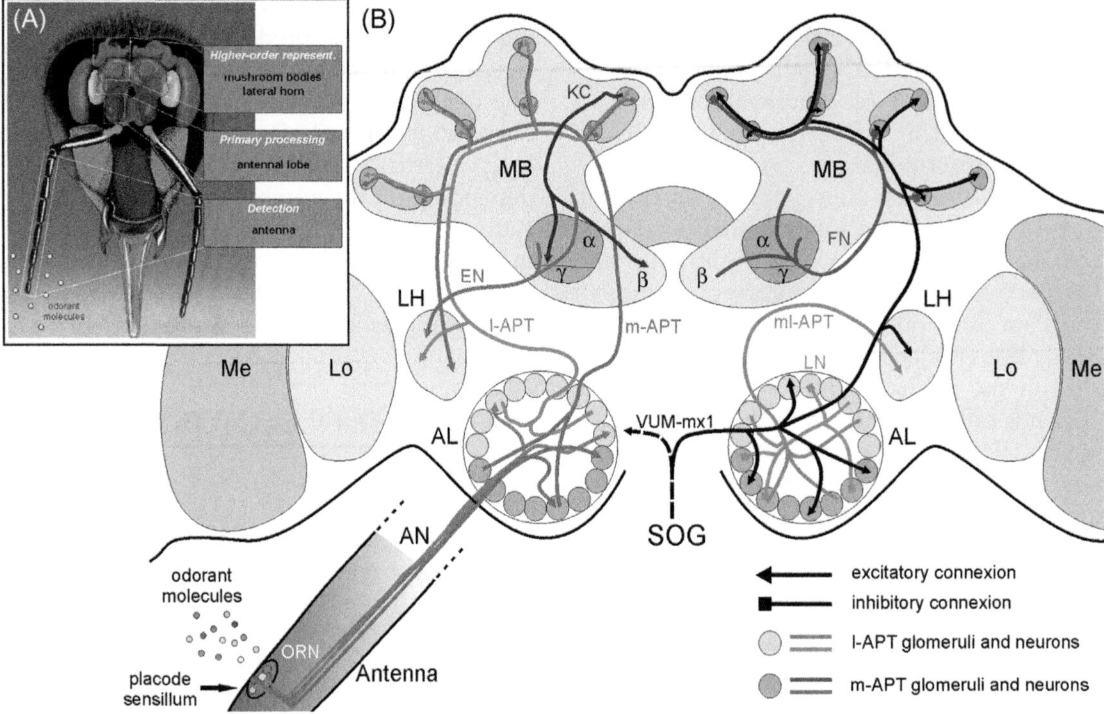

FIGURE 30.2 The honeybee brain and the olfactory pathway. For clarity, different neuron types are represented separately in the two brain hemispheres. On the left, major excitatory pathways involved in the transmission of olfactory information in the brain are shown. On the right, mostly inhibitory connections and modulatory neurons are presented. The antennal lobe (AL), first-order olfactory neuropil, receives input from ~60,000 olfactory receptor neurons (ORNs), which detect odorants within placode sensilla on the antenna. Within the AL's anatomical and functional units, the 160 glomeruli, ORNs contact ~4000 inhibitory local neurons (LNs), which carry out local computations, and ~800 projection neurons, which further convey processed information via different tracts. The lateral antenno-protocerebral tract (l-APT) projects first to the lateral horn (LH) and then to the mushroom body (MB) calyces (lips and basal ring), whereas the medial tract (m-APT) projects to the same structures but in the reverse order. Both tracts are uniglomerular, with each neuron taking information within a single glomerulus. They form two parallel, mostly independent olfactory subsystems (green and magenta) from the periphery until higher order centers, where they project in mostly non-overlapping regions. Multiglomerular projection neurons form three mediolateral tracts (ml-APT), which convey information directly to the medial protocerebrum and to the LH. The dendrites of the Kenyon cells (KC), the mushroom bodies' 170,000 intrinsic neurons, form the calyces, whereas their axons form the peduncle. The output regions of the MB are the α and β lobes, formed by two collaterals of each KC axon. Within the MBs, feedback neurons (FN) project from the peduncles and lobes back to the calyces, providing inhibitory feedback to the MB input regions. Extrinsic neurons (EN) take information from the peduncle and the lobes and project to different parts of the protocerebrum and most conspicuously to the LH. It is thought that descending neurons from these areas are then involved in the control of olfactory behavior. The scheme also presents a single identified octopaminergic neuron, VUMmx1, which was shown to represent reinforcement during appetitive conditioning. This neuron projects from the subesophageal ganglion (SOG), where it gets gustatory input from sucrose receptors, to the brain and converges with the olfactory pathway in three areas—the AL, the MB calyces, and the LH. Source: *Reproduced with permission from Sandoz.*[4]

very dense innervation and a few other glomeruli with sparse processes. Approximately 750 LNs are GABAergic, but glutamate, histamine, and several peptides have also been identified in the AL. To date, there is no evidence of excitatory LNs in honeybees.

Projection neurons (PNs) connect the AL with higher order brain areas, the mushroom bodies (MBs) and the lateral horn (LH), following five different pathways, called antenno-protocerebral tracts (APTs).[29,30] PNs can be classified as two types. Uniglomerular projection neurons (uPNs) branch in a single glomerulus and project to the MBs and to the LH using the two major APT tracts. Multiglomerular projection neurons (mPNs) branch in most glomeruli. Their axons form three lesser tracts, the mediolateral APTs, leading not to the MBs but to different regions of the medial protocerebrum and the LH.[29,31]

The more numerous uPNs (~800) form two roughly equal tracts toward higher-order brain centers—the lateral (l-APT) and the medial (m-APT) tract. The l-APT runs on the lateral side of the protocerebrum, forming collaterals in the LH, and continues to the MB calyces. The m-APT runs along the brain midline first toward the MBs, where collaterals enter into the calyces, and then travels laterally to the LH.[29,31] l-APT neurons take information from glomeruli receiving input from the

T1 tract of ORNs, whereas m-APT neurons receive input from T2, T3, and T4 glomeruli.[29] This corresponds to non-overlapping groups of 84 and 77 glomeruli, respectively,[31] so that each tract conveys information about two different subparts of the AL.

Higher Order Integration: Mushroom Bodies and Lateral Horn

The MBs are prominent structures in the honeybee brain involved in olfactory learning and memory, as well as in the integration of olfactory information with other sensory modalities.[1,3,11] By contrast, the LH belongs to the lateral protocerebrum and is not clearly structurally isolated from the rest of the brain. Although its exact function is unknown, it may be a premotor center involved in triggering innate behaviors in response to pheromones but also possibly for learned responses.[32]

MB intrinsic neurons are the Kenyon cells (KCs), which form two cup-shaped regions called calyces in each hemisphere. MB calyces are anatomically and functionally subdivided into the lip, the collar, and the basal ring.[30] The lip region and the inner half of the basal ring receive olfactory input, whereas the collar and outer half of the basal ring receive visual input,[30] in addition to mechanosensory and gustatory pathways. The projections of individual PNs extend in most parts of each calyx, but l- and m-APT project to different subregions.[31] PN boutons form multisynaptic microcircuits in the MB lips, with GABAergic input and KC output connections arranged to form particular structures termed microglomeruli. KC axons project in bundles into the central brain, forming the pedunculus and the α and β lobes. The calyx is topologically represented in the lobes.[30] Approximately 55 GABAergic feedback neurons from the MB output lobes (the protocerebral calycal tract (PCT); see Chapter 29) project back to the calyces. KCs mostly provide bifurcating axons to both α and β lobes. In bees, approximately 800 PNs diverge onto a major proportion of the 170,000 KCs of each MB (olfactory KCs). Each PN contacts many KCs, and each KC receives input from many PNs. This organization appears ideal for a combinatorial readout across PNs. The second major target area of all PNs is the LH. The LH shows relative PN tract-specific compartmentalization.[31] Processing within this structure as well as the connectivity between PNs and other neurons are still mostly unknown.

Several neuron populations and single neurons project from the MBs toward other brain areas (Figure 30.2), with major output regions in the α and β lobes.[30] Approximately 400 extrinsic neurons (ENs) from the α lobe have been studied in details.[33] Some are unilateral neurons with projection fields restricted to the ipsilateral protocerebrum, whereas others are bilateral neurons connecting both α lobes or projecting from one lobe to the contralateral protocerebrum around the α lobe.[33] A single neuron in each MB, called Pe-1, forms a major output pathway from the MB peduncle[34] and projects to the LH, where it synapses directly or via interneurons onto descending neurons. Some centrifugal neurons project back from the MBs toward the AL.[31,33]

Convergence between Olfactory and Reinforcement Pathways

In theory, associative learning is possible because at one or several locations in the brain, the pathways of the CS and of the US converge and plasticity originates at these locations. It is therefore particularly important to observe where in the brain pathways mediating appetitive and aversive reinforcement (US) are located and where they converge with the olfactory pathway (CS).

The olfactory pathway receives input from different modulatory systems using biogenic amines as neurotransmitters. As in other insects, appetitive reinforcement in bees is known to depend on octopamine,[35,36] whereas aversive reinforcement is thought to depend on dopamine.[16] A single, putatively octopaminergic neuron in the bee brain, VUMmx1 (Figure 30.2), was shown to represent a neural substrate of the sucrose US pathway because the forward (but not backward) pairing of an odor CS with an artificial depolarization of VUMmx1 produces an associative memory trace.[37] VUMmx1, which has its cell body in the subesophageal ganglion (SOG), converges with the olfactory pathway bilaterally at three sites: the AL, the MB calyces, and the LH. Another neuron (VUMmd1[38]), with its cell body in another neuromere of the SOG, shows a similar projection pattern as that of VUMmx1 and may serve a similar purpose (see Chapter 29). On the other hand, dopaminergic neurons are widely present in the bee brain.[39,40] Dopamine receptors are found in the AL and in the MBs,[41,42] suggesting that, as for appetitive learning, these areas may harbor aversive learning-induced plasticity. However, until now, there has not been any description of a dopaminergic neuron that may be sufficient for providing aversive reinforcement.

Demonstrations of the Involvement of ALs and MBs in Olfactory Learning

The demonstration that the AL and the MBs are involved in olfactory conditioning in bees was obtained early on in appetitive conditioning experiments studying the resistance of olfactory memory to amnestic treatments.[43,44] Olfactory memory after a single trial is transiently sensitive to a range of amnestic treatments, such as CO_2 or N_2 narcosis, cooling, or electroconvulsive shocks.[45] Cooling provided the interesting opportunity to interfere locally with brain function. Thus, local cooling of the ALs within 1 or 2 min after single-trial conditioning was shown to disrupt olfactory memory formation because conditioned responses were strongly reduced in a test performed 20 min after conditioning.[43,44] Likewise, local cooling of the MB calyces or of the α lobes disrupted memory formation if applied in the first 5 min after conditioning. Thus, normal function in the ALs and in the MBs is needed for normal memory formation, although at slightly different time scales. Later, it was shown that in both structures, local injection of octopamine (representing the US) in association with an odor presentation is sufficient for producing a significant level of olfactory memory, which was not the case when the neurotransmitter was injected into the lateral protocerebrum, including the LH.[36] Interestingly, injection in the ALs and in the MBs did not have the same effects on bees' performances because only injection in the ALs produced performance during acquisition, whereas injections at both sites supported significant performance in retention tests 30 min later. Thus, octopamine signaling within the AL may be important for performance increase during acquisition, whereas both structures would be involved in memory formation. The involvement of the AL in memory formation is also supported by experiments interfering with the early molecular cascades leading to the formation of olfactory LTM. As mentioned previously, multiple-trial conditioning induces a prolonged activation of PKA in the ALs, leading to LTM. Local imitation of such prolonged PKA activation by photorelease of cAMP in the AL is sufficient for inducing significant LTM after single-trial conditioning[21] (see Chapter 31). All these data indicate that olfactory acquisition and memory consolidation imply at least two main brain structures, with a significant role played by the primary olfactory center, the AL. It is thus a prominent structure in which learning-induced plasticity could appear.

THE SEARCH FOR EXPERIENCE-INDUCED PLASTICITY IN THE ANTENNAL LOBE

Neural plasticity related to olfactory learning has been searched for in different parts of the olfactory pathway, from the periphery to MB extrinsic neurons. Most of these studies, however, concentrated on the AL because its modular structure is more tractable for studying functional or structural changes. Tables 30.1 and 30.2 provide a list of the most important studies on neural plasticity induced by olfactory learning in the AL of bees. Although the main focus of this chapter is on the AL, it also briefly discusses plasticity found in other brain structures.

Peripheral Plasticity: ORNs

Some of the first studies searching for neural plasticity related to olfactory conditioning used the electroantennogram technique (EAG), a global electrophysiological recording of whole-antenna activity. Such recordings sum the depolarization from all ORNs within the antenna upon odor presentation and may thus allow measuring learning-induced plasticity at a peripheral level. Although initial studies found increases in EAG responses to odors after PER conditioning,[55,56] later accounts could not confirm these findings.[20,57] One possible reason for this discrepancy is that the earlier studies used odor mixtures as CS (essential oils and a synthetic six-component mixture, respectively), whereas the later ones used pure odorants. However, it must be noted that the studies that found significant modifications of peripheral sensitivity with mixtures did not include any controls for possible nonassociative effects (e.g., US only, CS only, or unpaired training).[55,56] This question remains unsolved because in recent years the advent of more efficient recording methods led to neglect of this approach.

Functional Plasticity in the Antennal Lobe: ORNs, LNs, and PNs

The search for plasticity in the AL has greatly benefited from the development of *in vivo* optical imaging techniques. Such techniques allow measuring the activity of whole neuronal populations simultaneously and may thus be used to follow changes in neural network activity due to experience. The first demonstration of combinatorial activity patterns in AL glomeruli was indeed obtained in the honeybee.[58] Since then, calcium imaging has been successfully applied to record neural activity from several olfactory structures, although most studies have focused on the AL (for review, see Sandoz[4]). In the AL, odors elicit glomerular response patterns corresponding to the subset of ORN types that were activated by this odor at the periphery.[58] Accordingly, such patterns are highly conserved between individuals.[59] These glomerular activity patterns actually correspond to a perceptual representation of odors because the olfactory behavior of honeybees in a generalization experiment

TABLE 30.1 Experimental Studies of Functional Plasticity Related to Olfactory Learning in the Antennal Lobe

Reference	Recording Technique	Neuronal Population	Conditioning Protocol	Conditioning ITI	Reinforced Trials	Retention Delay[a]	Main Result	
							Intensity	Similarity
Faber *et al.*, 1999[46]	Calcium imaging Bath-applied dye	Global (ORNs)	A + B −	1 min	5	10–30 min MTM	Increased to A + (and novel odor)	Decorrelation A to B
Sandoz *et al.*, 2003[12]	Calcium imaging Bath-applied dye	Global (ORNs)	A + B − / B + A −	5 min (10 min/side)	8 (4/side)	24 hr e-LTM	No effect	Decorrelation between sides
Peele *et al.*, 2006[47]	Calcium imaging Retrograde staining	l-PNs	A + and A + B −	1–2 min	1, 3, 5	5–15 min MTM	No effect	No effect
Fernandez *et al.*, 2009[48]	Calcium imaging Retrograde staining	l-PNs	A + B −	6 min	8	24 h e-LTM	No effect	Decorrelation A to B
Hourcade *et al.*, 2009[49]	Calcium imaging Bath-applied dye	Global (ORNs)	A +	10 min	5	3 days l-LTM	No effect	No effect
Arenas *et al.*, 2009[50]	Calcium imaging Bath-applied dye	Global (ORNs)	A + ; 4 days *ad libitum*	Unknown	Unknown	9 days l-LTM	Increased to odors	Modified
Denker *et al.*, 2010[51]	Multiunit recording	All PNs	A + B −	1 min	5	5–13 min MTM	A: both increase and decrease local field potential (LFP) increase in 15- to 40-Hz band	Not tested
Roussel *et al.*, 2010[52]	Calcium imaging Bath-applied dye	Global (ORNs)	Aversive conditioning; A + B −	5 min	4	During conditioning	No effect	No effect
Rath *et al.*, 2011[53]	Calcium imaging Retrograde staining	l- PNs	A + B −	10 min	6	2–5 hr MTM	No effect	Decorrelation A to B

[a]*Delays considered: STM, <3 min; MTM, 3 min–8 hr; e-LTM, 8–48 hr; l-LTM, 48 + hr.*
e-LTM, early long-term memory; ITI, intertrial interval; LFP, local field potential; l-LTM, late long-term memory; MTM, midterm memory; ORNs, olfactory receptor neurons; l-PNs, l-APT projection neurons.

TABLE 30.2 Experimental Studies of Structural Plasticity Related to Olfactory Learning in the Antennal Lobe

Reference	Recording Technique	Neuronal Population	Conditioning Protocol	Conditioning ITI	Reinforced Trials	Retention Delay	Main Result
Hourcade *et al.*, 2009[49]	Glomerular staining 3D reconstruction	All	A +	10 min	5	3 days l-LTM	Increased glomerular volume
Arenas *et al.*, 2012[54]	Glomerular staining 3D reconstruction	All	A + ; 4 days *ad libitum*	Unknown	Unknown	9 days l-LTM	Increased glomerular volume

ITI, intertrial interval; l-LTM, late long-term memory.

(absolute conditioning followed by tests with different odorants; see Figure 30.1B) can be predicted on the basis of this code.[18] Two complementary strategies based on two staining protocols have been used:

Global staining (e.g.,[58]): A calcium-sensitive fluorescent dye (e.g., Calcium Green-2, AM) is bath-applied onto the brain, potentially staining all cells in the imaged structure. This strategy is interesting in a first approach aiming to couple optical imaging with learning because it reveals overall activity from the studied structure and keeps the brain intact. In the AL, bath application has been shown to emphasize activity from ORN presynaptic terminals.[60,61]

Selective staining of known neuronal populations (e.g.[62]): In this case, a dextran-coupled calcium dye (e.g., fura-2 dextran) is placed on a particular neuronal tract and migrates within the neurons back to the structure one wants to image. This provides precise knowledge about the activity of a particular population of neurons. In the AL, it allowed recording specifically the activity from the PNs.[62]

Other neural activity-dependent signals may be recorded using optical imaging. For instance, voltage-sensitive dyes have been successfully applied to record depolarizations at the glomerular level upon odor presentation in bees.[63] However, their low signal-to-noise ratio did not allow their use for plasticity studies.

Plasticity in ORNs

Whereas EAG recordings attempted to measure ORN activity within the antenna, possible plasticity at the level of AL glomeruli was initially studied using calcium imaging and bath-applied dyes. The first such study compared odor-evoked signals before and after a differential conditioning procedure (Figure 30.1C).[46] At each time point, the CS+, the CS−, and a third, novel, odor were tested. At the time, glomerular identification was only beginning, so signals were analyzed globally over the entire imaged surface. This study showed that within 30 min after conditioning, the intensity of the calcium response was increased for the CS+ and the novel odor but not for the CS−. Moreover, this study showed that training decorrelated the activity maps of the CS+ and CS− so that the representations of the two stimuli were more differentiated after training than before training.

A later study using the same staining method made use of the within-animal controls available in the side-specific conditioning (Figure 30.1D).[12] Bees were thus trained to solve the A+B− discrimination when odors were presented on the right antenna and the opposite B+A− rule when odors were presented on the left antenna. Thus, A and B were both CS+ and CS− in the same individual depending on the recorded side. Twenty-four hours after training, bilateral AL recordings were performed to compare odor-evoked activity in the two ALs between side-specifically conditioned and naive individuals (Figure 30.3). Due to a glomerulus identification step, responses could be compared on a glomerular level. In naive bees, odor response patterns were highly symmetrical; that is, before conditioning, the same odorant elicited the same activation pattern in both antennal lobes (Figure 30.3A). In conditioned bees, however, small differences between sides were found in the activity of some glomeruli. To analyze the effect of subtle changes appearing in many glomeruli simultaneously on the global representation of odors, Euclidian distances were calculated in the putative olfactory space represented by 23 dimensions corresponding to the 23 imaged glomeruli. Such a measure of dissimilarity between different glomerular response maps takes into account both changes in signal intensity and changes in the relative activation of the different glomeruli. After side-specific conditioning, the left and right representations of the same odorant became different so that Euclidian distances between sides were increased for odors A and B but not for several control odorants tested in the same experiment (Figures 30.3B and 30.3C). In other words, here again, a discriminative conditioning task induced a decorrelation of glomerular response patterns. Interestingly, in the same task (A+B−/B+A−), the patterns of A and B were not decorrelated within each side, meaning that this side-specific task probably involved for the bees a stronger effort for differentiating the same odorant between sides than the two, clearly different, odorants within each side (Figure 30.3D).[12]

Not all tested learning tasks induced such functional plasticity in the calcium responses recording using bath-applied dyes. In a study using five-trial absolute conditioning, no significant change was found in calcium responses to the CS or to other odors.[49] Following the line of reasoning presented previously, one may explain this finding by the fact that an absolute conditioning task does not imply to differentiate the CS from any other specific stimulus. However, in this study, significant glomerular volume changes were found (see below).

Plasticity in AL responses to odors was also studied after precocious olfactory learning. In such experiments, odors are mixed with sucrose solution and provided to bees during early adulthood.[64] Treated bees usually associate the odor with the sucrose reward and show long-term memory performance in a PER test at a later stage (17 days). There seems to be a sensitive period because the formed odor−sucrose association is

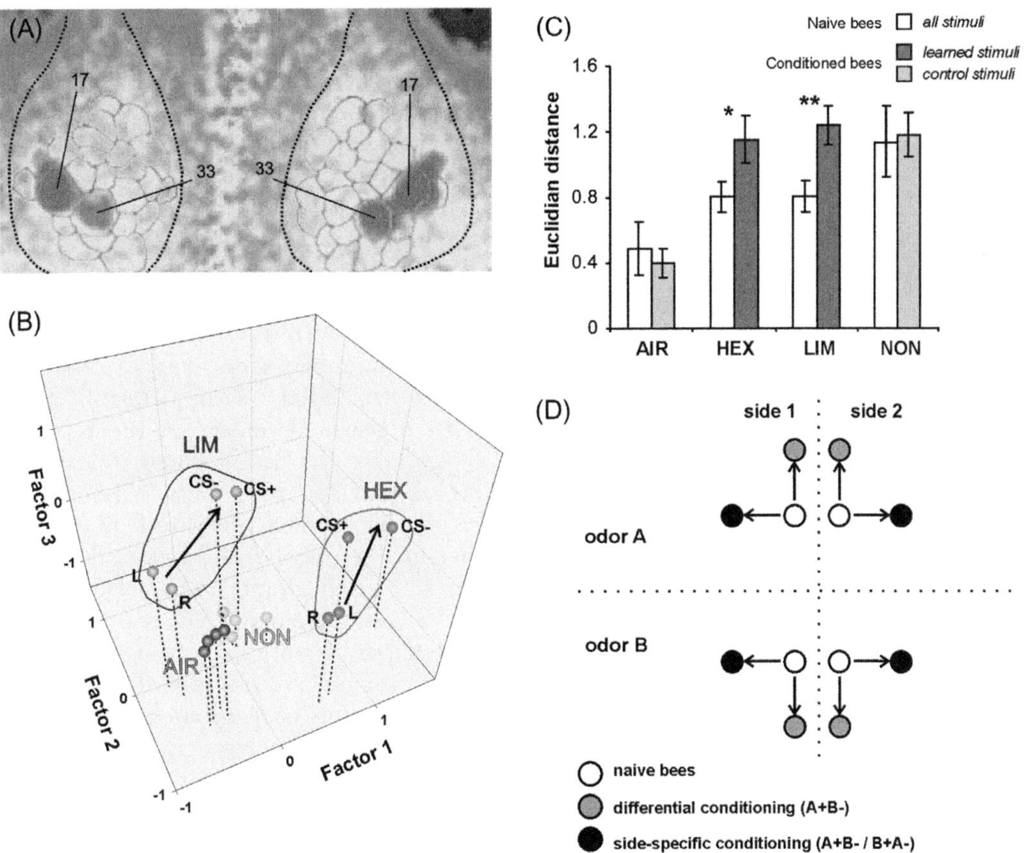

FIGURE 30.3 **Examples of functional plasticity: Side-specific decorrelation of odor activity patterns in the bee antennal lobe.**[12] (A) Simultaneous calcium imaging recording of both antennal lobes (delimited by the dashed lines). The response of naive (untrained) bees to 1-nonanol (NON) is shown. Glomeruli 17 and 33 are activated by this odor, and this activation is symmetrical between sides. (B) Effect of side-specific conditioning (A + B − /B + A −) on odor representation in conditioned and naive bees. The graph shows the left and right representations of each odorant according to the first three factors extracted from the activation of 23 glomeruli accessible to optical imaging by a principal component analysis. The arrows show the migration of the representations of learned odors between naive (L, left side; R, right side) and conditioned bees (CS + , side on which the odor was reinforced; CS − , side on which the odor was not reinforced). Air controls and NON (novel odorant) did not show any change related to conditioning (see grouped representations). (C) Euclidian distance between left and right representations of the same odorant in the original 23-dimension space. For 1-hexanol (HEX) and limonene (LIM), the two odors used in the side-specific conditioning, the distance between left and right representations increased significantly as a consequence of training (red bars), thus showing that left and right representations of the same odorant became different. For NON and the air control, the responses on the right and left antennal lobes were similar. * = $p < 0.05$, ** = $p < 0.01$. (D) A model of changes in odor representations after different types of conditioning. In naive bees (white circles), the representations on each brain side for the two odors are very similar (close together), whereas the representations of two different odors A and B are farther apart. As shown previously,[46,48,53] simple differential conditioning A + B − decorrelates the representations of A and B so that A and B are farther apart than in naive bees. Because differential conditioning is carried out with bilateral olfactory stimulations, the same process takes place on both sides of the brain. During side-specific conditioning, honeybees have to give each odor a different value on each side. As shown here, the distance of the representations of each odor A and B increases between sides, but the distance between the two training odors within sides does not change (black circles).

stronger when bees are exposed between 5 and 8 days of age than when the same exposure is performed before (1–4 days old) or, surprisingly, after this critical period (9–12 days old).[64] *In vivo* calcium imaging experiments with a bath-applied dye showed that such precocious olfactory experience increases general odor-induced activity as well as the number of glomeruli activated by the learned odor in the adult AL. It also affected qualitative odor representations, which appeared shifted in the neural space of treated animals relative to controls.[65] Such effects were not limited to the experienced odor but were in part generalized to other perceptually similar odors. Thus, early olfactory experiences made within the hive at a time when the olfactory system is still maturing[66] may have long-lasting effects on odor representation and affect bees' behavioral responses to odorants and concomitant neural activity throughout their adult life. Supporting the idea that these effects are related to developmental plasticity, treated bees were found to have a better ability to memorize novel odor–sucrose associations than control bees.[50]

Plasticity in PNs

The development of a technique for specifically staining particular neuronal populations with calcium-sensitive dyes[62] led to the progressive neglect of bath application because signal-to-noise ratio was much higher with this new technique and the neural structures producing the calcium signals were better defined. The first study to use this novel technique for measuring plasticity in the PNs concluded that PN activity was stable, independent of the conditioning procedure used (absolute or differential conditioning with different numbers of trials; see Figures 30.1B and 30.1C).[47] This conclusion was at odds with the previous observations of plasticity in the AL in similar training conditions and at approximately the same time after learning[46] because if changes appeared at the ORN level, the result of these changes should be observed in downstream neurons. A few years later, however, another study using differential conditioning with two odorants that were highly similar (two binary mixtures containing the same constituents in different proportions) demonstrated a significant increase of Euclidian distances between CS+ and CS− 24 hr after training.[48] Thus, here again, the constraint for the olfactory system of differentiating between similar sensory stimuli could have led to measurable response changes at the PN level. As in the study by Sandoz *et al.*,[12] this study did not find any clear changes in the responses of individual glomeruli but, rather, observed distributed changes that led to a global differentiation of the patterns for the two odorants (i.e., an increase in the Euclidian distance between CS+ and CS− representations). The problem of identifying the role of individual glomeruli in learning-induced plasticity was addressed in a study in which the learning performance of each imaged bee was monitored by direct observation of PER.[53] Managing such direct observation had long been a major hurdle because PER movements induced brain movements during recordings. Based on their behavior, bees were classified as discriminators (they learned to differentiate between CS+ and CS−) or nondiscriminators (see also Roussel *et al.*[52]). First, this study showed that only discriminators displayed an increased distance between the representations of CS+ and CS−, supporting the idea that only successful learning induced the observed plasticity. Second, even in discriminators, mapping response changes to identified glomeruli failed to show any systematic rule. Rather, response changes appeared to affect different glomeruli in different animals, such that the type of change observed depended on how strongly each glomerulus responded to the CS+ and CS− before conditioning. In other words, glomeruli that responded to the CS+, but not to the CS−, before conditioning had their response increased, whereas glomeruli responsive to both CS+ and the CS− had their response decreased. Glomeruli responding specifically to the CS− before training did not change, and glomeruli that responded to neither odor slightly increased their responses.[53] These results also demonstrated that if a glomerulus' response changed to the CS+, it changed in the same way to other odorants, even novel ones. Based on these results, the authors proposed a plasticity model of the AL with two levels of plasticity: One level would modify the strength of ORN-to-PN synapses only when reinforcement is present (i.e., when VUMmx1 liberates octopamine), whereas the second level would affect LN-to-ORN synapses in all glomeruli in a Hebbian way irrespective of reinforcement (i.e., by coincident pre- and postsynaptic activity).[53]

The previous studies recorded the activity of multiple PNs within each glomerulus. The following is a recurrent question: How do individual PNs change with learning? The advent of extracellular recordings using silicon multiprobes allowed the study of the responses of multiple PNs recorded at the same time before, during, and after olfactory conditioning.[51] This electrophysiological approach demonstrated that response changes of individual neurons can be extremely diverse after training. Thus, both learning-induced increases and learning-induced decreases in PN spike rates were found in response to odors after conditioning, with the strongest effects for the CS+.[51] Although this study could not localize the glomeruli to which each recorded PN belonged, it suggested that plasticity at the network level should be considered as the product of multiple complex changes within individual neurons, emphasizing the future need for computational efforts to understand the different phenomena at work within the AL. In particular, if learning-related activity increases and decreases should happen in PNs belonging to the same glomerulus, they could level out, precluding the recording of any learning-induced changes in optical imaging experiments (e.g.,[47]).

Most previous work has addressed the plasticity of ORNs or PNs because they are the most accessible neurons in the AL. However, many results point to possible plasticity within LNs, especially because recent data suggest that some LNs possess octopamine receptors.[67] Until now, LN activity was recorded using intracellular electrophysiology,[27,28,68] but such difficult recordings could not be coupled with learning experiments.

Structural Plasticity in the Antennal Lobe

Researchers have also attempted to find structural changes in the bee brain in relation to development and olfactory experience in general,[69,70] and only in rare cases have such studies aimed to narrow the observed effect to olfactory conditioning.[49,71] Because the glomeruli of the antennal lobe can be easily stained using histological dyes, the first strategy was to stain the whole brains of bees with different olfactory experiences and to obtain optical slices using a confocal microscope to measure the volume of individual glomeruli. In recent years, the development of three-dimensional (3D) reconstruction software has allowed precise volumetric reconstruction and volume assessment. Olfactory experience gained during foraging was shown to induce glomerular volume and structure changes.[72,73] It was unclear for quite some time whether the observed changes were actually due to olfactory learning *per se* because during foraging bees receive both associative olfactory experience and non-associative exposure to many odors. However, this question was addressed by comparing the volumes of 17 identified glomeruli in groups of bees subjected to five-trial absolute conditioning with an odor (paired group) and bees receiving the same number of stimulations with CS and US but without any temporal pairing (unpaired group).[49] This study found a specific glomerular volume increase in a subset of the assessed glomeruli as a result of the formation of a long-term appetitive olfactory memory (i.e., after 72 hr; Figures 30.4A and 30.4B). Moreover, this study showed that structural changes related to distinct odorant-specific long-term memories are restricted to different subsets of glomeruli when learning different odors. Because this study also carried out optical imaging recordings on the same subset of glomeruli, it could directly address the relationship between the spatial representation of an odor CS in the AL and the localization of structural plasticity.

A direct explanation for this glomerular specificity would be that the glomeruli responding to the CS would be those that are most plastic during conditioning (Figures 30.4C and 30.4D). The AL glomeruli are primary convergence sites between the CS and the US pathways.[37] Thus, CS–US pairing would induce concomitant activation of ORNs responding to the odor and of the pathway signaling sucrose reinforcement (VUMmx1), triggering synaptic plasticity mostly in the glomeruli activated by the CS. However, no correlation was found between glomerular volume changes and CS-induced activity. Rather, the study pointed to an alternative model based on a significant role of LN networks (Figures 30.4C and 30.4E). As a result of previous optical imaging and computational work,[59,74] a putative matrix of interglomerular inhibition was available for the glomeruli under study. When calculating how much inhibition each glomerulus would receive when the CS is presented, Hourcade *et al.*[49] showed that this variable correlated negatively with glomerulus volume change. In other words, inhibition between glomeruli during CS presentations would hinder plasticity so that the glomeruli that are inhibited the least would show the strongest volume increases. Hetero-LNs with a heterogeneous glomerular distribution of synaptic weights[28] could introduce nonlinear changes that would decorrelate activity and plasticity maps (see also[12,48]).

Significant structural plasticity also appears in the AL after precocious olfactory learning. Using the protocol mentioned previously, in which bees receive scented sugar water between 5 and 8 days after emergence, Arenas *et al.*[54] showed that precocious olfactory learning induces significant changes in the relative size of AL glomeruli, when measured on Day 17, compared to controls receiving unscented solution. Interestingly, comparison of the identity of plastic glomeruli found in the functional study (calcium imaging,[65]) and in the structural study[54] showed a significant correlation: Glomeruli that showed stronger calcium responses after precocious learning were also those that showed a volume increase. In other words, in these studies, a direct link between functional and structural plasticity was found at the glomerular level, suggesting that the location of these effects depends on the quality of the learned odorant, possibly increasing its detectability/discriminability from other odorants in the bees' environment. Compared to the results of the previously discussed experiments using PER conditioning, these results also support the idea that the rules underlying olfactory plasticity during the maturation of the olfactory system and during learning and memory formation at the adult age are different.

Which cellular mechanisms underlie such structural plasticity observed as a glomerular volume increase? Studies on honeybee glomeruli found that experience-dependent changes in glomerular volume can be correlated with variations in synapse numbers,[75] although this is not always the case and other processes may be at work.[72] In particular, glomeruli are wrapped by glial processes that play a crucial role, for instance, during AL development.[76] It is thus conceivable that glial cells may play a role in experience-dependent glomerular volume changes, particularly during development.

Plasticity in Other Brain Areas

Whereas many studies of honeybees have searched for neural plasticity in the AL, several influential

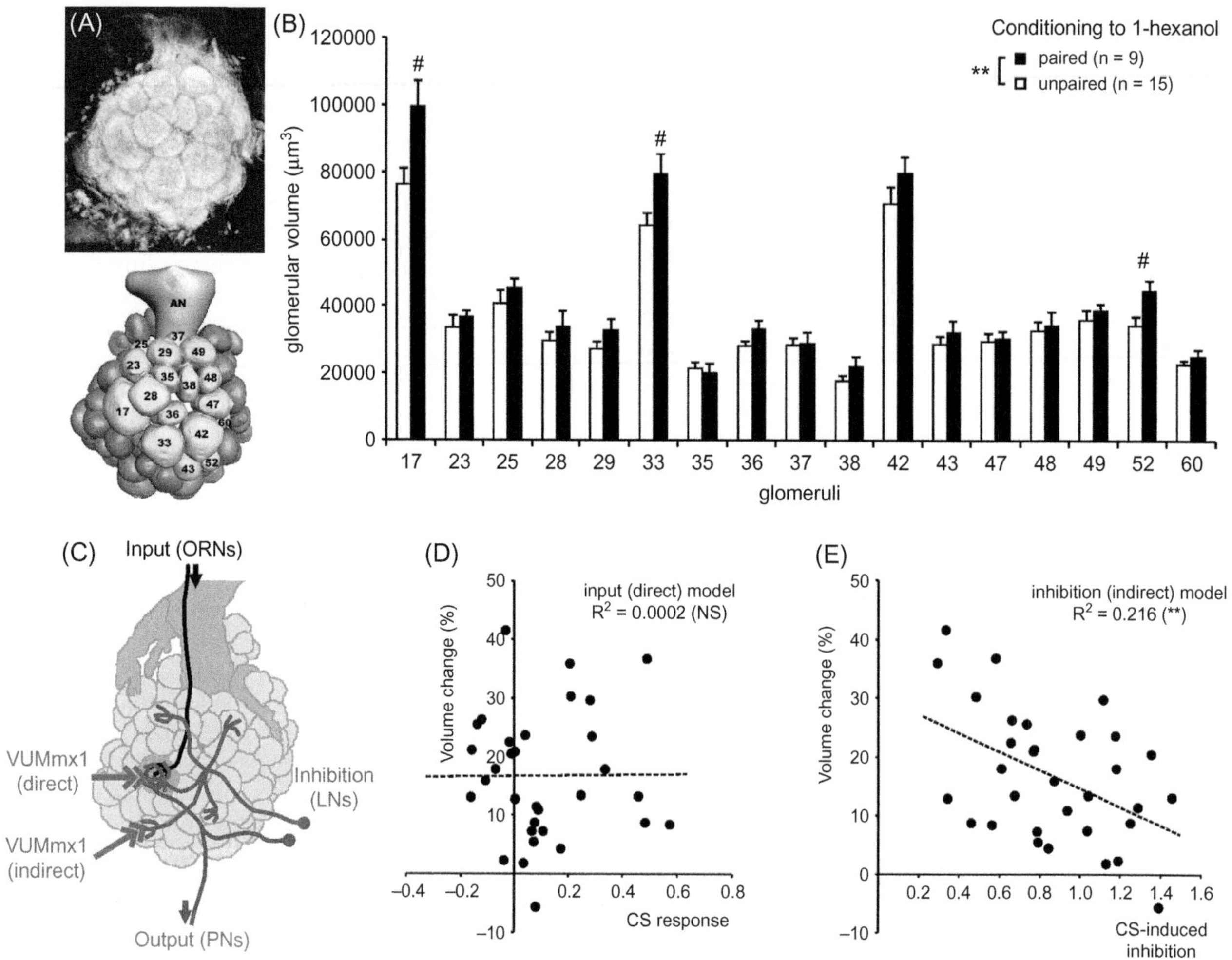

FIGURE 30.4 **Examples of structural plasticity: Increased glomerular volume related to long-term appetitive memory.**[49] (A) (Top) Confocal stack of an antennal lobe stained with neutral red, in which the glomeruli are clearly visible. (Bottom) The same antennal lobe after 3D reconstruction for volume analysis of 17 identified glomeruli. (B) Comparison of mean glomerular volumes measured 3 days after learning in bees from the paired group trained with 1-hexanol (showing LTM) and controls from the unpaired group (showing no LTM). For some of the glomeruli, the volume is significantly higher in bees from the paired group. # = $p < 0.05$ (posthoc comparisons), ** = $p < 0.01$. (C) Two different models of glomerular-based plasticity can be tested based on the convergence of the olfactory and the reward pathways. In the direct model, reinforcement mediated by the VUMmx1 neuron acts on ORN–PN synapses and increases the volume of the glomeruli activated by the CS (see panel D). In the indirect model, reinforcement acts on synapses between local interneurons (LNs) and PNs and therefore plasticity can be found in glomeruli that are not directly activated by the CS (see panel E). (D and E) Correlation between the relative volume change of each glomerulus after conditioning to 1-hexanol or 1-nonanol and the response of the glomerulus to the CS (D) or the amount of inhibition received by this glomerulus from other glomeruli (E). Only the latter is significant and suggests a negative effect of inhibition on glomerular volume increase.

studies have demonstrated changes downstream of this structure. Modified odor-evoked responses to a learned odor were found in the MB calyces soon after conditioning (10–30 min).[77,78] In particular, specific imaging of KC activity showed that repeated presentation of an odor induces a reduction of the evoked response (interpreted as habituation), whereas appetitive training induced a recovery from this decrease.[78] On a structural level, a long-term olfactory memory trace 72 hr after training was revealed as an increase in the density of microglomeruli in the MB lips.[71] MB output neurons are also subject to changes through associative learning, as exemplified by the Pe-1 neuron,[79] by recurrent PCT neurons,[80] or by other ENs.[81] Also, in some cases, specific changes are found for the CS+, and response differences between CS+ and CS– are increased.[81] Currently, it is still difficult to relate changes observed in the AL with those observed in other brain areas. It seems logical to think that changes taking place in the AL will affect odor-evoked signals downstream of this structure. The opposite may also be true. Because some neurons project back from the MBs to the AL (e.g., see[31,33]), plasticity in the MBs may also affect responses in the AL. To date, however, no data substantiate this possibility.

WHERE DO WE GO FROM HERE? THE MULTIFACTORIAL QUALITY OF THE SEARCH FOR NEURAL PLASTICITY

As demonstrated by the examples cited previously (see Tables 30.1 and 30.2), the search for learning-induced plasticity in the bee brain has been applied to different structures (AL and MBs) and neuron types (ORNs, PNs, KCs, and ENs), using different recording techniques (electrophysiology, optical imaging with different staining methods, and histological stainings), after various conditioning protocols (absolute, differential, and side-specific conditioning), and at widely differing times after conditioning (from 1 min to 72 hr) corresponding to different memory phases (STM, MTM, e-LTM, and l-LTM). This diversity makes it difficult to draw a comprehensive picture of how the bee brain may learn about odorants. The work performed to date demonstrates that plasticity is distributed in the bee brain and may affect at the same time different neuron types in different brain structures. Moreover, these changes may take different forms with different types of functional changes (increases but also decreases in CS-evoked responses and changes also in responses to non-reinforced odors) and structural changes (volumetric increases reflecting possibly modified synapse or altered microcircuits).

In addition to the multifactorial quality of the search for plasticity, another problem resides in the strong bias for publishing only studies in which response changes were observed after conditioning. It is likely that many labs worldwide have acquired data showing that particular neural responses in the bee brain are not modified by conditioning. Because it is very difficult to publish such negative results, only a few of these data sets are available, usually presented together with other data showing some neural response changes.[47,49,51,52,78,81]

Finding learning-induced plasticity in the brain does not mean that we understand the cause of this plasticity. It is thus often difficult to relate the observed neural plasticity to its function. Are the observed changes related to modifications of odor processing, modulating the neural representation of the learned odors so that it can be better distinguished from environmental background? Or are they related to an 'engram,' revealing the storage of odor–reinforcement associations in the brain? (See Chapter 29.). The current hypothesis is that plasticity at the primary processing center, the AL, is mostly responsible for the former, whereas plasticity in the MBs, the downstream multimodal integration center, would be crucial for the latter. However, considerable work is still needed to confirm this hypothesis. In particular, future neurobiological studies need to manipulate this plasticity so that we can understand whether the observed circuits (and their plasticity) are necessary and/or sufficient for the expression of memory at the behavioral level and for which part they are responsible—improved sensory processing or storing an engram.[82]

Thus, although our understanding of plasticity related to olfactory experience has greatly improved in the past several years due to well-controlled olfactory conditioning protocols and state-of-the-art recording techniques, entire brain regions and neuron types have yet to be explored. Concerning the AL, several studies have indicated a substantial role of LNs in learning-induced plasticity. Currently, experimental data on potential LN plasticity is completely lacking. It would therefore be critical to invest in studying LN responses specifically in relation to olfactory conditioning procedures. Another prominent question relates to m-APT-dependent parts of the AL. The functional dichotomy of PNs in the bee AL has recently been addressed in more detail, attempting to understand the possible differences in the rules underlying olfactory coding in the l-APT and m-APT subsystems.[83–85] Although both systems appear mostly redundant in the range of odors to which they respond, some differences have been observed, such as finer coding of chain length in the l-APT system[83] or stronger inhibitory interactions in the l-APT system when processing mixture information.[84] The interesting hypothesis that a subsystem may be plastic through olfactory learning, whereas another subsystem may remain stable, informing higher order centers in a learning-independent manner (e.g.,[47]), has not been tested. The development of imaging techniques allowing direct recording of m-APT glomeruli may be instrumental for answering this question.[83]

Even if we can observe plasticity in different brain areas at different moments after learning, how the brain learns probably depends on complex interactions between different brain structures, involving different neuron types, so that the use of computational approaches based on experimental data may help us to understand this process. Plasticity appears in multiple regions of the olfactory pathway, but their respective implications for tuning the olfactory system or for storing outcome-related memories is still unknown. It shall be the goal of future research to answer these questions so that comprehensive models of olfactory learning can be devised—models in which the role of the AL can be adequately evaluated. As mentioned previously, a bias toward the study of olfactory plasticity at the primary processing level should not cause us to forget that remarkable and probably more crucial plasticity with respect to an olfactory memory trace resides at later stages in the olfactory pathway, such as in the MB and other protocerebral structures.[78,79,81]

Acknowledgments

I thank the members of the EVOLBEE team at CNRS Gif-sur-Yvette for their intellectual contribution to the ideas developed here. This research is financed by the CNRS and by grants from the National Research Agency (ANR-BLANC-2010) as well as from NERF and R2DS Ile-de-France Programs.

References

1. Giurfa M. Behavioral and neural analysis of associative learning in the honeybee: a taste from the magic well. *J Comp Physiol A*. 2007;193(8):801–824.
2. Giurfa M, Sandoz JC. Invertebrate learning and memory: fifty years of olfactory conditioning of the proboscis extension response in honeybees. *Learn Mem*. 2012;19(2):54–66.
3. Menzel R. Memory dynamics in the honeybee. *J Comp Physiol A*. 1999;185:323–340.
4. Sandoz JC. Behavioral and neurophysiological study of olfactory perception and learning in honeybees. *Front Syst Neurosci*. 2011;5:98.
5. Menzel R, Greggers U, Hammer M. Functional organization of appetitive learning and memory in a generalist pollinator, the honey bee. In: Lewis AC, ed. *Insect Learning*. New York: Chapman & Hall; 1993:79–125.
6. von Frisch K. *The Dance Language and Orientation of Bees*. Cambridge, MA: Harvard University Press; 1967.
7. Winston ML. *The Biology of the Honey Bee*. Cambridge, MA: Harvard University Press; 1987.
8. Grant V. The flower constancy of bees. *Bot Rev*. 1950;16:379–398.
9. De Roo M, Klauser P, Garcia PM, Poglia L, Muller D. Spine dynamics and synapse remodeling during LTP and memory processes. *Prog Brain Res*. 2008;169:199–207.
10. Kandel ER. The molecular biology of memory storage: a dialogue between genes and synapses. *Science*. 2001;294 (5544):1030–1038.
11. Menzel R, Giurfa M. Cognitive architecture of a mini-brain: the honeybee. *Trends Cogn Sci*. 2001;5(2):62–71.
12. Sandoz JC, Galizia CG, Menzel R. Side-specific olfactory conditioning leads to more specific odor representation between sides but not within sides in the honeybee antennal lobes. *Neuroscience*. 2003;120:1137–1148.
13. Kuwabara M. Bildung des bedingten Reflexes von Pavlovs Typus bei der Honigbiene, Apis mellifica. *J Fac Hokkaido Univ Ser VI Zool*. 1957;13:458–464.
14. Takeda K. Classical conditioned response in the honey bee. *J Insect Physiol*. 1961;6:168–179.
15. Bitterman ME, Menzel R, Fietz A, Schafer S. Classical conditioning of proboscis extension in honeybees. *J Comp Psychol*. 1983; 97(2):107–119.
16. Vergoz V, Roussel E, Sandoz JC, Giurfa M. Aversive learning in honeybees revealed by the olfactory conditioning of the sting extension reflex. *PLoS ONE*. 2007;2(3):e288.
17. Carcaud J, Roussel E, Giurfa M, Sandoz JC. Odour aversion after olfactory conditioning of the sting extension reflex in honeybees. *J Exp Biol*. 2009;212(Pt 5):620–626.
18. Guerrieri F, Schubert M, Sandoz JC, Giurfa M. Perceptual and neural olfactory similarity in honeybees. *PLoS Biol*. 2005;3(4):e60.
19. Smith BH, Menzel R. The use of electromyogram recordings to quantify odourant discrimination in the honey bee, *Apis mellifera*. *J Insect Physiol*. 1989;35(5):369–375.
20. Sandoz JC, Menzel R. Side-specificity of olfactory learning in the honeybee: generalization between odors and sides. *Learn Mem*. 2001;8:286–294.
21. Müller U. Prolonged activation of cAMP-dependent protein kinase during conditioning induces long-term memory in honeybees. *Neuron*. 2000;27:159–168.
22. Grünbaum L, Müller U. Induction of a specific olfactory memory leads to a long-lasting activation of protein kinase C in the antennal lobe of the honeybee. *J Neurosci*. 1998;18:4384–4392.
23. Esslen J, Kaissling KE. Zahl und Verteilung antennaler Sensillen bei der Honigbiene (Apis mellifera L.). *Zoomorphology*. 1976;83:227–251.
24. Robertson HM, Wanner KW. The chemoreceptor superfamily in the honey bee, *Apis mellifera*: expansion of the odorant, but not gustatory, receptor family. *Genome Res*. 2006;16(11): 1395–1403.
25. Galizia CG, McIlwrath SL, Menzel R. A digital three-dimensional atlas of the honeybee antennal lobe based on optical sections acquired using confocal microscopy. *Cell Tissue Res*. 1999;295:383–394.
26. Vosshall LB, Wong AM, Axel R. An olfactory sensory map in the fly brain. *Cell*. 2000;102:147–159.
27. Flanagan D, Mercer AR. Morphology and response characteristics of neurones in the deutocerebrum of the brain in the honeybee Apis mellifera. *J Comp Physiol A*. 1989;164:483–494.
28. Fonta C, Sun XJ, Masson C. Morphology and spatial distribution of bee antennal lobe interneurones responsive to odours. *Chem Senses*. 1993;18(2):101–119.
29. Abel R, Rybak J, Menzel R. Structure and response patterns of olfactory interneurons in the honeybee, *Apis mellifera*. *J Comp Neurol*. 2001;437:363–383.
30. Mobbs PG. The brain of the honeybee *Apis mellifera*: I. The connections and spatial organization of the mushroom bodies. *Philos Trans R Soc Lond B*. 1982;298:309–354.
31. Kirschner S, Kleineidam CJ, Zube C, Rybak J, Grünewald B, Rossler W. Dual olfactory pathway in the honeybee, *Apis mellifera*. *J Comp Neurol*. 2006;499(6):933–952.
32. Hammer M. The neural basis of associative reward learning in honeybees. *Trends Neurosci*. 1997;20(6):245–251.
33. Rybak J, Menzel R. Anatomy of the mushroom bodies in the honey bee brain: the neuronal connections of the alpha-lobe. *J Comp Neurol*. 1993;334:444–465.
34. Mauelshagen J. Neural correlates of olfactory learning paradigms in an identified neuron in the honeybee brain. *J Neurophysiol*. 1993;69(2):609–625.
35. Farooqui T, Robinson K, Vaessin H, Smith BH. Modulation of early olfactory processing by an octopaminergic reinforcement pathway in the honeybee. *J Neurosci*. 2003;23(12):5370–5380.
36. Hammer M, Menzel R. Multiple sites of associative odor learning as revealed by local brain microinjections of octopamine in honeybees. *Learn Mem*. 1998;5:146–156.
37. Hammer M. An identified neuron mediates the unconditioned stimulus in associative olfactory learning in honeybees. *Nature*. 1993;366:59–63.
38. Schröter U, Malun D, Menzel R. Innervation pattern of suboesophageal ventral unpaired median neurones in the honeybee brain. *Cell Tissue Res*. 2007;327(3):647–667.
39. Schäfer S, Rehder V. Dopamine-like immunoreactivity in the brain and suboesophageal ganglion of the honeybee. *J Comp Neurol*. 1989;280:43–58.
40. Schürmann FW, Elekes K, Geffard M. Dopamine-like immunoreactivity in the bee brain. *Cell Tissue Res*. 1989;256:399–410.

41. Beggs KT, Hamilton IS, Kurshan PT, Mustard JA, Mercer AR. Characterization of a D2-like dopamine receptor (AmDOP3) in honey bee, *Apis mellifera*. *Insect Biochem Mol Biol*. 2005;35(8):873–882.
42. Kokay IC, Ebert PR, Kirchhof BS, Mercer AR. Distribution of dopamine receptors and dopamine receptors homologs in the brain of the honey bee, *Apis mellifera* L. *Microsc Res Tech*. 1999;44:179–189.
43. Erber J, Masuhr TH, Menzel R. Localization of short-term memory in the brain of the bee, *Apis mellifera*. *Physiol Entomol*. 1980;5:343–358.
44. Menzel R, Erber J, Masuhr T. Learning and memory in the honeybee. In: Browne LB, ed. *Experimental Analysis of Insect Behaviour*. Berlin: Springer Verlag; 1974:195–217.
45. Erber J. The dynamics of learning in the honey bee (Apis mellifica carnica): II. Principles of information processing. *J Comp Physiol*. 1975;99:243–255.
46. Faber T, Joerges J, Menzel R. Associative learning modifies neural representations of odors in the insect brain. *Nat Neurosci*. 1999;2(1):74–78.
47. Peele P, Ditzen M, Menzel R, Galizia CG. Appetitive odor learning does not change olfactory coding in a subpopulation of honeybee antennal lobe neurons. *J Comp Physiol A Neuroethol Sens Neural Behav Physiol*. 2006;192(10):1083–1103.
48. Fernandez PC, Locatelli FF, Person-Rennell N, Deleo G, Smith BH. Associative conditioning tunes transient dynamics of early olfactory processing. *J Neurosci*. 2009;29(33): 10191–10202.
49. Hourcade B, Perisse E, Devaud JM, Sandoz JC. Long-term memory shapes the primary olfactory center of an insect brain. *Learn Mem*. 2009;16(10):607–615.
50. Arenas A, Fernandez VM, Farina WM. Associative learning during early adulthood enhances later memory retention in honeybees. *PLoS ONE*. 2009;4(12):e8046.
51. Denker M, Finke R, Schaupp F, Grün S, Menzel R. Neural correlates of odor learning in the honeybee antennal lobe. *Eur J Neurosci*. 2010;31(1):119–133.
52. Roussel E, Sandoz JC, Giurfa M. Searching for learning-dependent changes in the antennal lobe: simultaneous recording of neural activity and aversive olfactory learning in honeybees. *Front Behav Neurosci*. 2010;4:155.
53. Rath L, Giovanni Galizia C, Szyszka P. Multiple memory traces after associative learning in the honey bee antennal lobe. *Eur J Neurosci*. 2011;34(2):352–360.
54. Arenas A, Giurfa M, Sandoz JC, Hourcade B, Devaud JM, Farina WM. Early olfactory experience induces structural changes in the primary olfactory center of an insect brain. *Eur J Neurosci*. 2012;35(5):682–690.
55. De Jong R, Pham-Delègue MH. Electroantennogram responses related to olfactory conditioning in the honey bee (*Apis mellifera ligustica*). *J Insect Physiol*. 1991;:37.
56. Wadhams LJ, Blight MM, Kerguelen V, et al. Discrimination of oilseed rape volatiles by honey bee: novel combined gas chromatographic-electrophysiological behavioral assay. *J Chem Ecol*. 1994;20(12):3221–3231.
57. Bhagavan S, Smith BH. Olfactory conditioning in the honey bee *Apis mellifera* : effects of odor intensity. *Physiol Behav*. 1997; 61(1):107–117.
58. Joerges J, Küttner A, Galizia CG, Menzel R. Representations of odours and odour mixtures visualized in the honeybee brain. *Nature*. 1997;387:285–288.
59. Galizia CG, Sachse S, Rappert A, Menzel R. The glomerular code for odor representation is species specific in the honeybee *Apis mellifera*. *Nat Neurosci*. 1999;2(5):473–478.
60. Galizia CG, Nägler K, Hölldobler B, Menzel R. Odour coding is bilaterally symmetrical in the antennal lobe of honeybees (*Apis mellifera*). *Eur J Neurosci*. 1998;10:2964–2974.
61. Sachse S, Galizia CG. The coding of odour-intensity in the honeybee antennal lobe: local computation optimizes odour representation. *Eur J Neurosci*. 2003;18(8):2119–2132.
62. Sachse S, Galizia CG. The role of inhibition for temporal and spatial odor representation in olfactory output neurons: a calcium imaging study. *J Neurophysiol*. 2002;87:1106–1117.
63. Galizia CG, Küttner A, Joerges J, Menzel R. Odour representation in honeybee olfactory glomeruli shows slow temporal dynamics: an optical recording study using a voltage-sensitive dye. *J Insect Physiol*. 2000;46(6):877–886.
64. Arenas A, Farina WM. Age and rearing environment interact in the retention of early olfactory memories in honeybees. *J Comp Physiol A Neuroethol Sens Neural Behav Physiol*. 2008;194 (7):629–640.
65. Arenas A, Giurfa M, Farina WM, Sandoz JC. Early olfactory experience modifies neural activity in the antennal lobe of a social insect at the adult stage. *Eur J Neurosci*. 2009;30(8):1498–1508.
66. Masson C, Pham-Delègue MH, Fonta C, et al. Recent advances in the concepts of adaptation to natural odour signals in the honeybee, *Apis mellifera* L. *Apidologie*. 1993;24: 169–194.
67. Sinakevitch I, Mustard JA, Smith BH. Distribution of the octopamine receptor AmOA1 in the honey bee brain. *PLoS ONE*. 2011;6 (1):e14536.
68. Meyer A, Galizia CG. Elemental and configural olfactory coding by antennal lobe neurons of the honeybee (*Apis mellifera*). *J Comp Physiol A Neuroethol Sens Neural Behav Physiol*. 2012;198 (2):159–171.
69. Durst C, Eichenmüller S, Menzel R. Development and experience lead to increased volume of subcompartments of the honeybee mushroom body. *Behav Neural Biol*. 1994;62(3):259–263.
70. Withers GS, Fahrbach SE, Robinson GE. Selective neuroanatomical plasticity and division of labour in the honeybee. *Nature*. 1993;364(6434):238–240.
71. Hourcade B, Muenz TS, Sandoz JC, Rossler W, Devaud JM. Long-term memory leads to synaptic reorganization in the mushroom bodies: a memory trace in the insect brain? *J Neurosci*. 2010;30(18):6461–6465.
72. Brown SM, Napper RM, Thompson CM, Mercer AR. Stereological analysis reveals striking differences in the structural plasticity of two readily identifiable glomeruli in the antennal lobes of the adult worker honeybee. *J Neurosci*. 2002;22 (19):8514–8522.
73. Sigg D, Thompson CM, Mercer AR. Activity-dependent changes to the brain and behavior of the honey bee, *Apis mellifera* (L.). *J Neurosci*. 1997;17(18):7148–7156.
74. Linster C, Sachse S, Galizia CG. Computational modeling suggests that response properties rather than spatial position determine connectivity between olfactory glomeruli. *J. Neurophysiol*. 2005;93(6):3410–3417.
75. Brown SM, Napper RM, Mercer AR. Foraging experience, glomerulus volume, and synapse number: a stereological study of the honey bee antennal lobe. *J Neurobiol*. 2004;60(1):40–50.
76. Boeckh J, Tolbert LP. Synaptic organization and development of the antennal lobe in insects. *Microsc Res Tech*. 1993;24 (3):260–280.
77. Faber T, Menzel R. Visualizing a mushroom body response to a conditioned odor in honeybees. *Naturwissenschaften*. 2001;88 (10):472–476.
78. Szyszka P, Galkin A, Menzel R. Associative and non-associative plasticity in Kenyon cells of the honeybee mushroom body. *Front Syst Neurosci*. 2008;2:3.
79. Okada R, Rybak J, Manz G, Menzel R. Learning-related plasticity in PE1 and other mushroom body-extrinsic neurons in the honeybee brain. *J Neurosci*. 2007;27(43):11736–11747.

80. Haehnel M, Menzel R. Sensory representation and learning-related plasticity in mushroom body extrinsic feedback neurons of the protocerebral tract. *Front Syst Neurosci.* 2010;4:161.
81. Strube-Bloss MF, Nawrot MP, Menzel R. Mushroom body output neurons encode odor-reward associations. *J Neurosci.* 2011; 31(8):3129–3140.
82. Gerber B, Tanimoto H, Heisenberg M. An engram found? Evaluating the evidence from fruit flies. *Curr Opin Neurobiol.* 2004;14(6):737–744.
83. Carcaud J, Hill T, Giurfa M, Sandoz JC. Differential coding by two olfactory subsystems in the honey bee brain. *J Neurophysiol.* 2012;108:1106–1121.
84. Krofczik S, Menzel R, Nawrot MP. Rapid odor processing in the honeybee antennal lobe network. *Front Comput Neurosci.* 2009;2:9.
85. Yamagata N, Schmuker M, Szyszka P, Mizunami M, Menzel R. Differential odor processing in two olfactory pathways in the honeybee. *Front Syst Neurosci.* 2009;3:16.

CHAPTER

31

Memory Phases and Signaling Cascades in Honeybees

Uli Müller

Saarland University, Saarbrücken, Germany

APPETITIVE OLFACTORY LEARNING IN HONEYBEES: BEHAVIOR AND NEURONAL CIRCUITRY

During their lifetime, honeybees perform a multitude of different behaviors according to their age and caste. For example, workers are engaged in in-hive social communication, in-hive labor, foraging including navigation, and subsequent recruitment of other honeybees to new food sources. The essential task of foraging is enabled by their capability to associate features such as colors and odors with food quality[1] and to memorize these associations. The transfer of field observations to the controlled laboratory situation was a breakthrough for scientific approaches to bee behavior, especially to learning. A major field of interest in this respect is appetitive olfactory learning.[2,3] The characterization of appetitive conditioning of the proboscis extension response (PER) not only demonstrated that the cellular mechanisms of associative learning in honeybees are quite similar to those of associative learning in other species but also provided and still provides the possibility to study the cellular and molecular processes under well-controlled conditions.

An associative learning trial comprises the temporal pairing of two stimuli: the conditioned stimulus (CS) followed by the unconditioned stimulus (US). The correct sequence and temporal relation of CS and US determine the success of learning. In olfactory appetitive conditioning of the PER, the odor stimulus (CS) is immediately followed by the sucrose reward. Sucrose is first applied to the antennae, inducing the extension of the proboscis, and then to the proboscis, allowing the honeybee to lick the sucrose. A single CS/US pairing that lasts a few seconds induces a rather short-lived associative memory for the learned odor. Although the memory performance is at a high level in the range of hours, it decays over days. To induce a memory that remains at a high level for days requires repeated CS/US pairings that are separated by intervals in the range of several minutes. In contrast to a single-trial conditioning, the stable long-lasting memory induced by three subsequent conditioning trials (intertrial interval, 2 min) shows all properties of a long-term memory (LTM)[4]; that is, it requires translation and transcription processes.[5,6] In the early days of these laboratory assessments of behavior, cooling experiments provided the first evidence that the processes induced by a single-trial conditioning differ from those induced by repeated trial conditioning. Cooling the brain after conditioning impairs early memory triggered by single-trial conditioning, whereas memory after three-trial conditioning is insensitive to immediate cooling.[7–9] Moreover, because cooling the antennal lobes (ALs) and the mushroom bodies (MBs) led to differences in the acquisition phase, these experiments provided the first functional evidence that the ALs and the MBs contribute differently to olfactory learning.

In addition to these behavioral experiments, a series of studies using anatomy, immunohistochemistry, electrophysiology, and optical recordings characterized the neuronal pathways that mediate the odor (CS) and the sucrose (US) information in the honeybee brain (Figure 31.1). During olfactory conditioning, the odor information activates chemoreceptor neurons on the antennae (~60,000 per antenna) that project to the ALs. In the ALs, the chemosensory neurons terminate in the glomeruli (~160 per AL), where they relay their information to local interneurons (~4000) and

Invertebrate Learning and Memory.
DOI: http://dx.doi.org/10.1016/B978-0-12-415823-8.00031-9

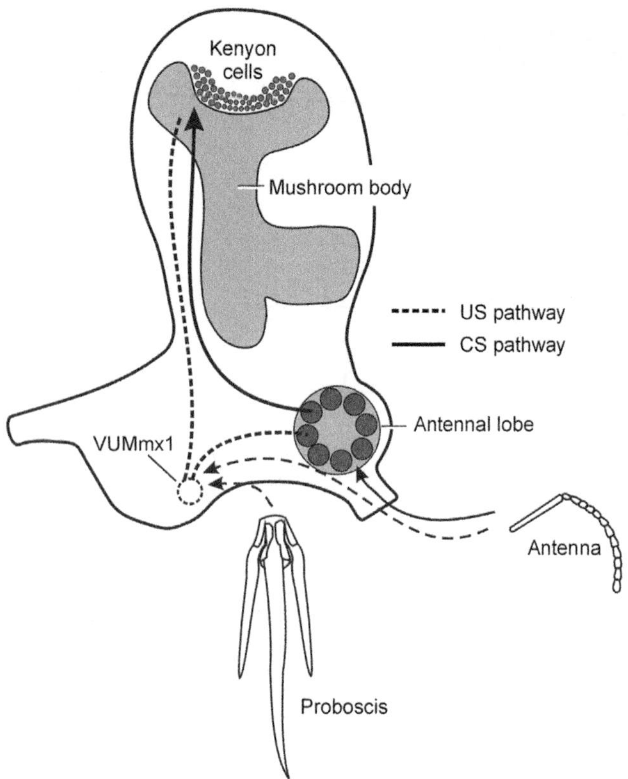

FIGURE 31.1 Neuronal circuits involved in appetitive olfactory learning. The scheme (sagittal section) illustrates the neuronal circuits that are involved in the processing of olfactory and appetitive stimuli used in appetitive olfactory conditioning. The olfactory information from the sensory neurons on the antenna is processed in the antennal lobes (ALs), the primary olfactory centers in insects. Projection neurons that leave the ALs transmit the information to the calyces of the mushroom bodies (MBs) and the lateral horn (not shown). The latter brain areas are of special interest because they are innervated by the VUMmx1 neuron, which has been shown to substitute the US function in associative conditioning. During conditioning, the appetitive stimulus is sensed by the antenna and the proboscis. Sensory neurons of both connect to VUMmx1, which projects to the ALs and the MBs. Thus, the ALs and the MBs are sites of CS/US convergence in olfactory conditioning. Anatomically, they are well accessible for analyzing the function of learning-induced signaling cascades.

projection neurons (PNs; ~800).[10,11] The local interneurons connect between different glomeruli and build inhibitory networks that modulate the overall activity but also enhance odor discrimination of the PNs.[12] The PNs transmit the odor information via two different pathways to the MBs and the lateral horn in the lateral protocerebrum. In the median antennocerebral tract, the PNs response is delayed (odor-specific), whereas PNs in the lateral antennocerebral tract react very fast (odor-unspecific).[13] Not much is known regarding the target network of the PNs in the lateral horn, whereas the MBs as the other target of the PNs are well-characterized structures in insects[14,15] that play a critical role in insect learning.[16–19] The PNs use acetylcholine as transmitter and exclusively project to the lip region of the MBs.[20] In summary, these studies show that the ALs, the lateral horn, and the lip areas receive odor information during associative olfactory conditioning (Figure 31.1).

During conditioning, the sucrose stimulus applied to the antennae is sensed by taste hairs that transmit the information to the dorsal lobes, and connections between the antennal nerve and the subesophageal ganglion can transmit the information close to the input area of the ventral unpaired median (VUM) neurons.[21,22] How this antennal circuitry connects to the yet unidentified neuronal circuitry of the proboscis is unknown. However, it is known that the ingestion of sucrose is necessary to form an appetitive LTM.[23] This suggests that in addition to sensory neurons on antennae and proboscis, intrinsic sensors evaluate the nutritional intake. Neuronal circuits in the subesophageal ganglion are involved in the processing of the proboscis response, and especially the VUM neurons play a critical role in the processing of the reward (US) information.[24,25] Recording and activation of the VUM neuron 1 of the maxillary neuromere 1 (VUMmx1) during olfactory conditioning demonstrated that VUMmx1 can substitute for the US function.[24] The VUMmx1 neuron innervates the lip region of the MBs, the lateral horn, and the ALs and thus overlaps with the areas implicated in olfactory processing. This makes the VUMmx1 neuron and its neurotransmitter octopamine[26] essential components of reward processing (Figure 31.1).

REWARD AND ODOR STIMULI INDUCE FAST AND TRANSIENT ACTIVATION OF THE cAMP- AND CA^{2+}-DEPENDENT SIGNALING CASCADES IN THE ANTENNAL LOBES

To address the molecular signaling cascades underlying associative learning in the honeybee, two factors were of great advantage: the knowledge of the brain areas involved in CS/US processing and the reliable PER paradigm in associative conditioning. On this basis, techniques that allow the monitoring of *in vivo*-induced changes in defined signaling cascades at distinct time points could be established (Box 31.1). The short duration of an associative conditioning trial (a few seconds) and the short time window required for repeated training sessions that induce LTM (a few minutes) guarantee a clear separation between acquisition and consolidation of memory. In this way, both the molecular processes involved in memory formation and the processes that maintain memory can be

BOX 31.1

MONITORING IN VIVO-INDUCED CHANGES IN SIGNALING CASCADES

The scheme depicts the critical steps in monitoring stimulus-induced changes in signaling cascades in defined brain areas of the honeybee.[27,28] At desired times after stimulation (sucrose, CS, conditioning, etc.), the whole animal is quick-frozen in liquid nitrogen in order to stop and thus conserve all enzyme activities. The shortest possible handling time and thus temporal resolution from end of the stimuli to quick-freezing is approximately 0.5 sec. It is extremely important that the tissue remains frozen in all subsequent steps. After freeze-drying (−15°C), the brain areas of interest (e.g., antennal lobes and mushroom bodies) are dissected under liquid nitrogen. The dissected tissue is transferred into microcapillaries containing frozen buffer, grinded on the surface of the frozen buffer (2 sec), and immediately stored in liquid nitrogen again. To measure the enzyme activity of interest, the microcapillary is warmed up (~1 sec), and the content with the sample is pushed into the reaction mixture. The reaction is terminated after 10–15 sec and processed according to the desired assay.

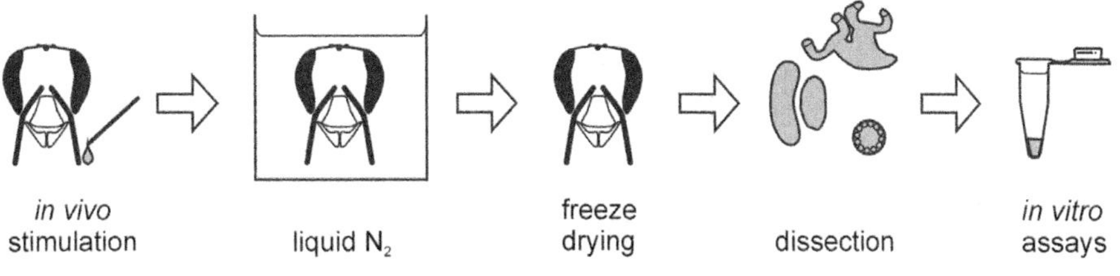

studied. Due to the overlap of odor-mediating circuits and the reward-processing VUMmx1 neuron, the analysis focused on the ALs and the MBs. These two neuropils are anatomically well-defined and accessible for dissection.

Octopamine released by VUMmx1 can in principle activate receptors that are coupled to cAMP- or Ca^{2+}-regulated intracellular signaling cascades.[29,30] A short stimulation of an antenna with sucrose at a concentration used in olfactory conditioning induces the activation of the cAMP-dependent protein kinase A (PKA) in the ipsilateral AL.[27] This sucrose-induced PKA activation in the AL is fast and transient: The maximal activation is reached in less than 1 sec after the stimulus and then decays within 3 sec to background level. Application of mechanical or odor stimuli to the antennae does not affect PKA activity in the ALs.[27,28] A sucrose stimulus applied to the proboscis elevates PKA activity in both ALs, mediated by octopamine.[27,28] This suggests that the VUMmx1 neuron mediates the sucrose-induced PKA activation. Immunostaining shows that PKA is mainly localized in the local interneurons[31] that connect the glomeruli within the ALs. Elevation of PKA activity upon a sucrose stimulus is observed in all parts of the AL. This observation suggests that US-induced modulation of PKA-dependent processes may occur throughout the ALs. The US stimulus also activates Ca^{2+}-dependent signaling cascades in the ALs, similar to an odor stimulus.[5] The neuronal circuitries and the transmitter systems mediating the US- and CS-induced Ca^{2+}-dependent processes have not been identified.

Although the lip areas of the MBs also receive input from the octopaminergic VUMmx1 neuron, a sucrose stimulus does not change PKA activity in this input area of Kenyon cells. Given the fact that octopamine elevates PKA activity in cultured Kenyon cells,[31] it is most likely that VUMmx1 in the lip area of the MBs acts on octopamine receptors that are coupled to Ca^{2+}-regulated pathways.[29,30] Because Kenyon cells differ in their amine receptor spectrum,[32] VUMmx1 may regulate different signaling pathways depending on the target neurons. The broad projection of VUMmx1 within the ALs, or the lip area of the MBs, points to a global modulation of neuronal circuitries by the US stimulus (sucrose). This rather general modulation by the US (sucrose) clearly differs from the specific activation triggered by a CS (odor). Optical recordings of Ca^{2+} signals in the ALs and the MBs show that each odor induces highly characteristic, locally defined patterns of activity in specific neuronal circuits.[11,33,34] Thus, a key feature of appetitive olfactory conditioning is the temporal processing of US-triggered molecular events that occur in larger neuronal networks, possibly including all glomeruli in the ALs and CS-triggered molecular events in highly specific neuronal circuitries (distinct glomeruli in the ALs).

THE LINK BETWEEN TRAINING PARAMETERS AND MEMORY FORMATION: THE SPECIFIC ROLE OF SECOND MESSENGER-REGULATED SIGNALING CASCADES

Training parameters critically influence the properties of the induced memories.[4] In the honeybee, three successive conditioning trials given within 4 min induce an LTM, which can be divided into a translation-dependent early phase (eLTM, 1 or 2 days) and a transcription-dependent late phase (lLTM, ≥3 days).[5,6,35,36] In this respect, it is important to mention that the tested blockers of translation and transcription reduce memory only to a level that is observed after a single-trial conditioning. The memory induced by a single-trial conditioning shows obvious similarities to the amnesia-resistant memory observed in *Drosophila*[37] and cannot be affected by any pharmacological tool tested so far.

The cAMP/PKA cascade, identified as indispensable for learning in *Aplysia* and *Drosophila*,[38] also plays a critical role in nonassociative and associative learning in honeybees. Inhibition of cAMP-dependent processes during three-trial conditioning specifically impairs the formation of LTM without affecting acquisition or memory up to 1 day.[39,40] Blocking PKA does not affect any aspect of single-trial-induced memory.[40]

PKA activity is only required for a few minutes during the short period of a three-trial conditioning. This limitation of the PKA requirement permitted the direct measurement of learning-induced dynamics of PKA activation in distinct brain areas using the stop-freeze technique (Box 31.1). Although the MBs and the ALs are potential circuits mediating cAMP-dependent processes induced by olfactory conditioning, the used method could detect learning-induced changes in PKA activity only in the ALs. Whereas a US stimulus triggers a PKA activation that lasts only 3 sec,[27,28] a single-trial conditioning (CS/US pairing) induces a PKA activity in the ALs that stays elevated for 1 min. After three-trial conditioning that induces LTM, PKA activity in the ALs is elevated for more than 3 min.[40] The amplitude of PKA activation does not differ from that after a single-trial conditioning. The relevance of this prolonged PKA activation in LTM formation was tested by photolytic release of caged cAMP in the ALs (Box 31.2). Extending the PKA activation by photolytic release of caged cAMP in the ALs after a single-trial conditioning is sufficient to induce a long-lasting memory (Figure 31.2), supporting the importance of the three-trial-induced prolongation of PKA activity in the ALs for the induction of LTM.

Prolonged PKA activation triggered by repeated conditioning trials is mediated by nitric oxide (NO) (Figure 31.2). NO is an unconventional signaling molecule that diffuses from its site of production through membranes to act on targets as the soluble guanylyl cyclase (sGC) within the same or in neighboring cells.[41] Blocking of NO synthase, which is abundant in the ALs and the MB calyces, impairs LTM and erases the prolonged PKA activation in the ALs.[9,40] Inhibitors of cGC cause the same impairments, suggesting that NO and cGMP are critical components required for the conditioning-induced prolonged PKA activation. Especially cGMP is of major interest. It can act at two sites: It activates the cGMP-dependent protein kinase and interacts with the cAMP pathway via cGMP-regulated phosphodiesterases or cyclic nucleotide-gated channels. In honeybees, cGMP has a third target: At low cAMP levels, cGMP can directly activate PKA (Figure 31.2).[42]

The photorelease technique *in vivo* verified the participation of cGMP in the formation of a long-lasting memory. A single-trial conditioning followed by photorelease of cGMP leads to formation of a long-lasting memory comparable to that observed after release of caged cAMP.[40] Photorelease of NO, however, leads to a total impairment of all memory phases,[40] pointing to the interference of the released NO with other important processes in the ALs. In this context, two findings with regard to the NO/cGMP system in the ALs of honeybees are important: The NO/cGMP system in the ALs contributes to the processing of appetitive chemosensory information during habituation[9,36,41,43,44] and to odor discrimination.[45] This makes it very likely that artificial release of NO during conditioning interferes with one of these processes in the neuronal circuitry of the ALs causing a general learning impairment.

Given the important role of the NO/cGMP system in sensory processing and learning, it is very likely that the NO/cGMP-mediated processes are directly linked with the cGMP-dependent protein kinase, also known as the product of the *foraging* gene.[46] This is supported by the observation that the *foraging* gene is implicated in the transition from hive bee to forager. In the course of this transition, drastic changes in the sensory system and in behavior provide the basis to cope with the very different tasks in the different environments.[47]

Taken together, the analysis of the learning-induced modulation of second messenger cascades

BOX 31.2

IN VIVO MANIPULATION OF SIGNALING CASCADES BY PHOTORELEASE OF CAGED COMPOUNDS

To prove the functional relevance of learning-induced changes of signaling cascades in learning (see Box 31.1), the dynamics of the measured activation were mimicked in the particular brain area during conditioning.[40] For the photolytic illumination of the brain, a small window is cut into the head capsule above the antennae. The caged compound (here, caged cAMP) is injected into the hemolymph 20 min prior to the photolysing experiment. The illumination is provided by a UV flash using an appropriate binocular and an aperture to ensure the exact illumination of the ALs (e.g., two holes). The intensity and interval of the UV flashes are adjusted by measuring the PKA activation in the antennal lobes by the rapid freezing technique (see Box 31.1). The scheme shows the photolytic release of cAMP from its inactive form (caged cAMP) and its action on the intrinsic cAMP/PKA cascade. The released cAMP, like the intrinsic cAMP, binds to the regulatory subunit (R) of the cAMP-dependent PKA (R_2C_2), resulting in dissociation and release of the active catalytic subunit (C). Thus, using caged compounds enables a locally and temporally defined activation of distinct signaling processes to manipulate the function of the cascades in question *in vivo*.

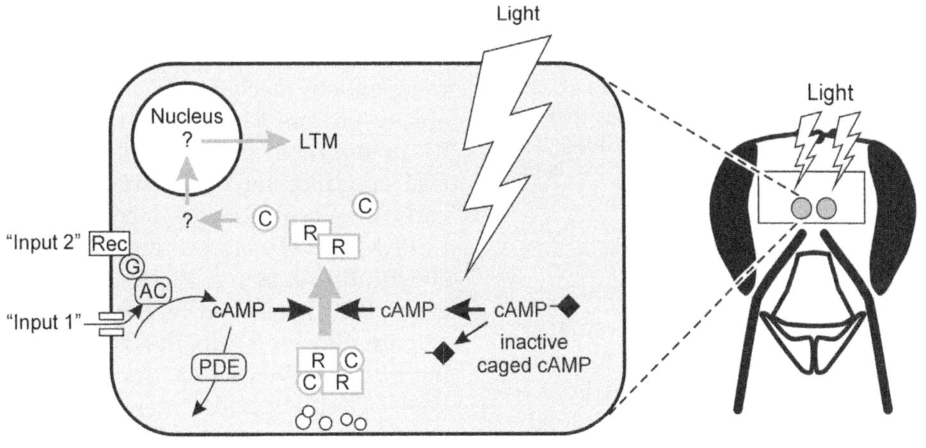

demonstrates that PKA-dependent processes in the range of minutes during conditioning trigger molecular processes that become evident days later (Figure 31.2). Due to their important role in olfactory learning, as demonstrated in *Drosophila*,[17] neuronal circuitries in the MBs are expected targets of the short-lasting molecular events in the ALs.

SATIATION AFFECTS FORMATION OF APPETITIVE MEMORY VIA MOLECULAR PROCESSES DURING CONDITIONING

In addition to parameters such as the number of conditioning trials, the intertrial interval, and the strength of the reward, the satiation level of the animal determines the performance in appetitive learning. A reliable induction of LTM requires conditioning of hungry animals, whereas feeding before appetitive conditioning or memory retrieval results in a suppressed behavioral performance, pointing to a motivational influence on the underlying signaling processes.[35,48–50] Although honeybees learn to associate an odor with a sucrose stimulus to the antenna,[2] the formation of LTM requires ingestion of the rewarding sucrose in addition.[23]

Whereas conditioning of hungry animals leads to an obvious acquisition and memory formation, feeding before conditioning (4 hr) impairs acquisition and memory formation independent of the training strength.[35] Studies in this regard revealed a connection to the cAMP/PKA cascade. Honeybees fed 4 hr before measurement show a lower basal brain PKA activity than that of hungry animals.[35] Elevating the low basic PKA activity in fed honeybees during three-trial conditioning led to a rescue of the transcription-dependent

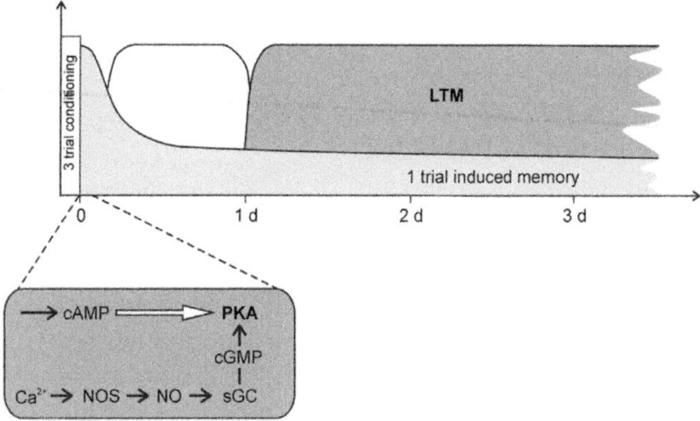

FIGURE 31.2 The cAMP/PKA signaling in the antennal lobes is essential for the induction of long-term memory (LTM). A single conditioning trial induces a memory that decays over time and triggers a very short-lasting activation of the cAMP/PKA cascade in the ALs. In contrast, three-trial conditioning that reliably induces LTM leads to a prolonged (~3 min) activation of PKA in the ALs. This prolonged PKA activation is mediated via the production of nitric oxide (NO) by the NO synthase (NOS). NO activates the soluble guanylyl cyclase (sGC). The latter produces cGMP, which acts synergistically on PKA and prolongs its activation. Whereas inhibition of NOS, cGC, or PKA leads to a specific loss of LTM, the photorelease of cGMP or cAMP in the ALs in combination with a single-trial conditioning is able to induce LTM. Thus, a very short activation of the NO/cGMP and cAMP/PKA signaling cascades in the ALs is essential for the induction of LTM.

lLTM (≥3 days), whereas acquisition and memory up to 2 days remained at a low level. This is remarkable because both eLTM and lLTM require PKA for their induction, which implies the existence of at least two different cAMP/PKA-pathways that are triggered by strong training and are implicated in LTM formation. One of these learning-induced cAMP/PKA pathways is obviously influenced by the satiation level.

Single-trial-induced memory is also impaired in satiated animals. This is of particular interest because only amnestic treatment by cooling immediately after learning is known to interfere with memory induced by a single conditioning trial.[7] Thus, the satiation-dependent impairment of memory induced by a single-trial conditioning may serve as a basis to analyze the underlying molecular mechanisms.

MIDTERM MEMORY REQUIRES THE INTERACTION OF A CA^{2+}-REGULATED PROTEASE AND PROTEIN KINASE C

In the ALs, odors induce transient changes in Ca^{2+} levels in subsets of glomeruli.[11,33] Each odor induces its own specific activation pattern of glomeruli. Monitoring the stimulus-induced activation of Ca^{2+}-dependent signaling cascades in the ALs (Box 31.1) revealed a transient activation of the Ca^{2+}-phospholipid-dependent protein kinase C (PKC) by both US and CS stimulation.[5] A similar PKC activation is induced by pairing of CS–US, US–CS, and multiple pairings of CS and US. In any case, the transient activation of PKC in the ALs lasts for a few minutes. In the MBs, this stimulus-induced activation of the Ca^{2+}-dependent PKC could not be detected. Inhibition of this immediate PKC activation has no effect on learning and memory formation, indicating that the PKC activation triggered by CS or US stimuli is not essential for associative learning or memory formation.

Interestingly, determination of PKC activity in the ALs at later time points after training uncovered a highly specific role of PKC-mediated processes in memory formation. Approximately 1 hr after three-trial conditioning, PKC activity in the ALs rises again. This increase in PKC activity in the ALs lasts up to 3 days and is not observed after single-trial training.[5] Further investigation of this long-lasting PKC activation revealed two mechanistically distinct and also independent mechanisms.

The process responsible for elevation of PKC activity in the time window from 1 h to 16 hr after three-trial conditioning was characterized in detail. In this early time window, the increased PKC activity is due to PKM, a cleavage product of PKC (Figure 31.3). The constitutively active PKM is formed when the activated PKC is cleaved by the Ca^{2+}-dependent protease calpain. Blocking calpain activity during the conditioning phase prevents PKM formation and thus decreases PKM activity and also memory in the time window of 1–16 hr. Interestingly, acquisition, memory up to 30 min, and memory after 16 hr are not affected at all. Preventing PKM formation, which occurs approximately 30 min to −2 hr after conditioning, impairs memory between 1 and 16 hr, suggesting that PKM is required for processes that maintain memory within this time window (Figure 31.3). Blocking calpain affects neither increased PKC activity nor memory in the time window of 1–3 days after three-trial conditioning.

Injection of translation and transcription blockers during and immediately after conditioning erases the elevated PKC activity and memory from 1 to 3 days after three-trial conditioning.[5] Currently, the contribution of the translation/transcription-dependent elevation of PKC activity (1–3 days) to LTM is unclear. PKM production and memory up to 1 day are not affected by translation and transcription blockers, demonstrating the independency of the early and late phase of conditioning-induced elevation of PKC activity. Studies in *Aplysia*, *Drosophila*, and mammals point to a well-conserved function of PKM in memory formation.[51–53]

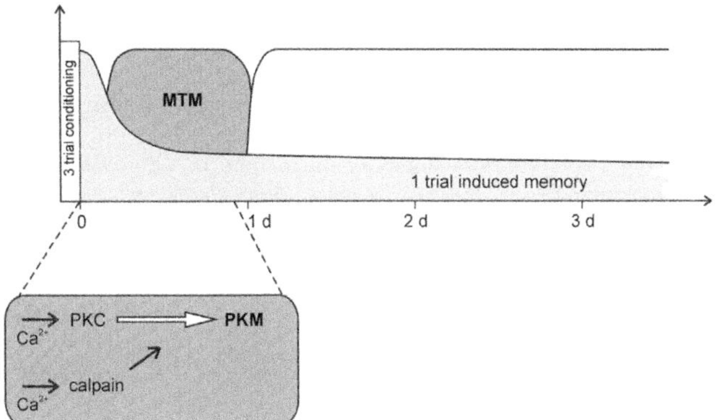

FIGURE 31.3 Calcium-dependent PKC/PKM signaling in the antennal lobes is critical for the induction and maintenance of midterm memory (MTM). In contrast to a single trial, three-trial conditioning activates the Ca^{2+}-dependent PKC and the Ca^{2+}-dependent protease calpain in the ALs. Calpain cleaves the activated PKC, and the constitutive active PKM is formed. Inhibition of calpain within a short time window after conditioning results in a specific loss of PKM and MTM, demonstrating that PKM is required for maintaining MTM. The blocking of calpain does not affect acquisition, memory up to 30 min, and memory greater than 1 day. Thus, MTM is formed in parallel to but independently of LTM.

MUSHROOM BODY GLUTAMATE TRANSMISSION IS IMPLICATED IN MEMORY FORMATION

Compared to the detailed knowledge about glutamate function in the mammalian brain,[54] the contribution of the neurotransmitter glutamate in insect learning and memory formation is not well understood. Critical components of glutamate transmission, such as glutamate receptors, glutamate transporters, as well as glutamate-induced currents, have been identified in the honeybee brain.[55–58] Pharmacological interference with these receptors or transporters causes deficits in learning and memory.[59–61] Because the specificity and efficiency of the used drugs are not characterized in honeybees, the interpretation of the pharmacological data is not without problems. The specific manipulation of the NMDA-type glutamate receptors, however, impairs olfactory learning and long-term memory in *Drosophila*[62] and in honeybees,[63] arguing for a functional role of glutamate in insect learning and memory formation (see Chapter 34).

The photolytic uncaging of glutamate offered the possibility to gain information about the brain area and the temporal requirement of glutamate during associative conditioning[64] (Box 31.2). These studies showed that only the release of glutamate in the MBs, but not the ALs, influences olfactory memory. Moreover, only release immediately (~3 sec) after a single-trial conditioning, but not after three-trial conditioning, improves memory 2 days after training to a level usually observed after three-trial conditioning. Uncaging glutamate 1 min before single-trial conditioning does not affect acquisition or memory formation, demonstrating that the function of glutamate is restricted to a very narrow time window after single-trial conditioning. Enhanced glutamate release after conditioning either imitates a situation similar to that after three-trial conditioning and facilitates acquisition processes or contributes to processes of memory formation. In both cases, however, the enhanced glutamate release in the MBs specifically facilitates processes that contribute to the formation of transcription-dependent lLTM. These findings argue in favor of a contribution of glutamatergic neurotransmission in the honeybee MBs comparable to the induction of LTM known from mammals.[54]

PARALLEL SIGNALING PROCESSES IN THE ALS AND THE MBS CONTRIBUTE TO MEMORY FORMATION

Identification and characterization of the signaling cascades underlying the induction and maintenance of olfactory appetitive memory in the honeybee revealed a network of processes in the ALs and the MBs, acting in parallel.[36,65] The learning-induced, prolonged activation of the cAMP/PKA cascade, mediated by the NO/cGMP cascade in the ALs, is critical for the induction of LTM (eLTM and lLTM) (Figure 31.2). During approximately the same time window, a glutamate-mediated process localized in the MBs also contributes to the formation of LTM.[64] This demonstrates that during the short time course of associative conditioning, distinct signaling cascades in different brain areas are active. The interactions between these processes and the molecular targets (e.g., transcription factors) that finally contribute to the maintenance of LTM have not been identified. The neuronal circuitry of the ALs also houses the molecular signaling cascades that contribute to the induction and maintenance of the MTM. These Ca^{2+}-dependent signaling processes act independently of the cAMP/PKA signaling processes in the ALs (Figure 31.3).[5]

Thus, studies in the honeybee uncovered different molecular processes located in the antennal lobes that contribute to distinct aspects of memory formation. Investigations in *Drosophila* focused on the role of the MBs in olfactory learning[16,18,19,66] and elegantly demonstrated that small subsets of the MB intrinsic Kenyon cells mediate distinct features of olfactory learning and memory formation.[17] Taken together, the findings in honeybees and *Drosophila* provide a consistent picture. Signaling cascades located in the ALs contribute to memory induction and the maintenance of MTM, whereas signaling cascades located in the MBs are critical for establishing memories during all phases. It is not known whether different, content-related memories are stored in the ALs and MBs in the honeybee or in *Drosophila*. It is feasible that ALs and MBs are specialized regarding their contribution to distinct aspects such as attention, motivation, generalization, or specificity of the memory content. In such a scenario, memory retrieval would require the interaction between ALs and MBs, and probably other brain areas as well (see Chapter 29).

Acknowledgments

I thank Dr. S. Meuser and A. Gardezi for their help in preparing the manuscript.

References

1. Menzel R, Müller U. Learning and memory in honeybees: from behavior to neural substrates. *Annu Rev Neurosci.* 1996;19:379–404.
2. Bitterman ME, Menzel R, Fietz A, Schäfer S. Classical olfactory conditioning of proboscis extension in honeybees (*Apis mellifera*). *J Comp Physiol.* 1983;97:107–119.
3. Kuwabara M. Bildung des bedingten Reflexes von Pavlovs Typus bei der Honigbiene, *Apis mellifica*. *J Fac Sci Hokkaido Univ Ser VI Zool.* 1957;13:458–464.
4. Davis HP, Squire LR. Protein synthesis and memory: a review. *Psychol Bull.* 1984;96:518–559.
5. Grünbaum L, Müller U. Induction of a specific olfactory memory leads to a long-lasting activation of protein kinase C in the antennal lobe of the honeybee. *J Neurosci.* 1998;18:4384–4392.
6. Wüstenberg D, Gerber B, Menzel R. Short communication: long- but not medium-term retention of olfactory memories in honeybees is impaired by actinomycin D and anisomycin. *Eur J Neurosci.* 1998;10:2742–2745.
7. Erber J, Masuhr T, Menzel R. Localization of short-term memory in the brain of the bee, *Apis mellifera*. *Physiol Entomol.* 1980;5:343–358.
8. Menzel R, Erber J, Masuhr T. Learning and memory in the honeybee. In: Browne LB, ed. *Experimental Analysis of Insect Behaviour*. Berlin: Springer Verlag; 1974:195–217.
9. Müller U. Inhibition of nitric oxide synthase impairs a distinct form of long-term memory in the honeybee, *Apis mellifera*. *Neuron.* 1996;16:541–549.
10. Flanagan D, Mercer AR. An atlas and 3-D reconstruction of the antennal lobes in the worker honeybee, *Apis mellifera* L. *Int J Insect Morphol Embryol.* 1989;18:145–159.
11. Galizia CG, Menzel R. Odour perception in honeybees: coding information in glomerular patterns. *Curr Opin Neurobiol.* 2000;10:504–510.
12. Sachse S, Galizia CG. The role of inhibition for temporal and spatial odor representation in olfactory output neurons: a calcium imaging study. *J Neurophysiol.* 2002;87:1106–1117.
13. Müller D, Abel R, Brandt R, Zockler M, Menzel R. Differential parallel processing of olfactory information in the honeybee, *Apis mellifera* L. *J Comp Physiol A Neuroethol Sens Neural Behav Physiol.* 2002;188:359–370.
14. Mobbs PG. The brain of the honeybee *Apis mellifera* I: the connections and spatial organization of the mushroom bodies. *Philos Trans R Soc Lond Biol.* 1982;298:309–354.
15. Witthöft W. Absolute Anzahl und Verteilung der Zellen im Hirn der Honigbiene. *Z Morphol Oekol Tiere.* 1967;61:160–184.
16. Davis RL. Olfactory memory formation in *Drosophila*: from molecular to systems neuroscience. *Annu Rev Neurosci.* 2005;28:275–302.
17. Davis RL. Traces of *Drosophila* memory. *Neuron.* 2011;70:8–19.
18. Heisenberg M, Borst A, Wagner S, Byers D. *Drosophila* mushroom body mutants are deficient in olfactory learning. *J Neurogenet.* 1985;2:1–30.
19. Zars T, Fischer M, Schulz R, Heisenberg M. Localization of a short-term memory in *Drosophila*. *Science.* 2000;288:672–675.
20. Kreissl S, Bicker G. Histochemistry of acetylcholinesterase and immunocytochemistry of an acetylcholine receptor-like antigen in the brain of the honeybee. *J Comp Neurol.* 1989;286:71–84.
21. Haupt SS. Antennal sucrose perception in the honey bee (*Apis mellifera* L.): behaviour and electrophysiology. *J Comp Physiol A Neuroethol Sens Neural Behav Physiol.* 2004;190:735–745.
22. Haupt SS. Central gustatory projections and side-specificity of operant antennal muscle conditioning in the honeybee. *J Comp Physiol A Neuroethol Sens Neural Behav Physiol.* 2007;193: 523–535.
23. Wright GA, Mustard JA, Kottcamp SM, Smith BH. Olfactory memory formation and the influence of reward pathway during appetitive learning in honey bees. *J Exp Biol.* 2007;210:4024–4033.
24. Hammer M. An identified neuron mediates the unconditioned stimulus in associative olfactory learning in honeybees. *Nature.* 1993;366:59–63.
25. Rehder V. Sensory pathways and motoneurons of the proboscis reflex in the suboesophageal ganglion of the honeybee. *J Comp Neurol.* 1989;279:499–513.
26. Kreissl S, Eichmüller S, Bicker G, Rapus J, Eckert M. Octopamine-like immunoreactivity in the brain and suboesophageal ganglion of the honeybee. *J Comp Neurol.* 1994;348:583–595.
27. Hildebrandt H, Müller U. PKA activity in the antennal lobe of honeybees is regulated by chemosensory stimulation *in vivo*. *Brain Res.* 1995;679:281–288.
28. Hildebrandt H, Müller U. Octopamine mediates rapid stimulation of protein kinase A in the antennal lobe of honeybees. *J Neurobiol.* 1995;27:44–50.
29. Balfanz S, Strunker T, Frings S, Baumann A. A family of octopamine receptors that specifically induce cyclic AMP production or Ca^{2+} release in *Drosophila melanogaster*. *J Neurochem.* 2005;93:440–451.
30. Grohmann L, Blenau W, Erber J, Ebert PR, Strunker T, Baumann A. Molecular and functional characterization of an octopamine receptor from honeybee (*Apis mellifera*) brain. *J Neurochem.* 2003;86:725–735.
31. Müller U. Neuronal cAMP-dependent protein kinase type II is concentrated in mushroom bodies of *Drosophila melanogaster* and the honeybee *Apis mellifera*. *J Neurobiol.* 1997;33:33–44.
32. McQuillan HJ, Nakagawa S, Mercer AR. Mushroom bodies of the honeybee brain show cell population-specific plasticity in expression of amine-receptor genes. *Learn Mem.* 2012;19:151–158.
33. Joerges J, Küttner A, Galizia G, Menzel R. Representations of odours and odour mixtures visualized in the honeybee brain. *Nature.* 1997;387:285–288.
34. Szyszka P, Ditzen M, Galkin A, Galizia CG, Menzel R. Sparsening and temporal sharpening of olfactory representations in the honeybee mushroom bodies. *J Neurophysiol.* 2005;94:3303–3313.

35. Friedrich A, Thomas U, Müller U. Learning at different satiation levels reveals parallel functions for the cAMP-protein kinase A cascade in formation of long-term memory. *J Neurosci.* 2004;24:4460–4468.
36. Müller U. Learning in honeybees: from molecules to behaviour. *Zoology.* 2002;105:313–320.
37. Isabel G, Pascual A, Preat T. Exclusive consolidated memory phases in *Drosophila. Science.* 2004;304:1024–1027.
38. Kandel ER. The molecular biology of memory storage: a dialogue between genes and synapses. *Science.* 2001;294:1030–1038.
39. Fiala A, Müller U, Menzel R. Reversible downregulation of protein kinase A during olfactory learning using antisense technique impairs long-term memory formation in the honeybee, *Apis mellifera. J Neurosci.* 1999;19:10125–10134.
40. Müller U. Prolonged activation of cAMP-dependent protein kinase during conditioning induces long-term memory in honeybees. *Neuron.* 2000;27:159–168.
41. Müller U. The nitric oxide system in insects. *Prog Neurobiol.* 1997;51:363–381.
42. Leboulle G, Müller U. Synergistic activation of insect cAMP-dependent protein kinase A (type II) by cyclicAMP and cyclicGMP. *FEBS Lett.* 2004;576:216–220.
43. Müller U, Hildebrandt H. The nitric oxide/cGMP system in the antennal lobe of *Apis mellifera* is implicated in integrative processing of chemosensory stimuli. *Eur J Neurosci.* 1995;7: 2240–2248.
44. Müller U, Hildebrandt H. Nitric oxide/cGMP-mediated protein kinase A activation in the antennal lobes plays an important role in appetitive reflex habituation in the honeybee. *J Neurosci.* 2002;22:8739–8747.
45. Hosler JS, Buxton KL, Smith BH. Impairment of olfactory discrimination by blockade of GABA and nitric oxide activity in the honey bee antennal lobes. *Behav Neurosci.* 2000;114: 514–525.
46. Ben-Shahar Y, Robichon A, Sokolowski MB, Robinson GE. Influence of gene action across different time scales on behavior. *Science.* 2002;296:741–744.
47. Ben-Shahar Y. The foraging gene, behavioral plasticity, and honeybee division of labor. *J Comp Physiol A Neuroethol Sens Neural Behav Physiol.* 2005;191:987–994.
48. Ben-Shahar Y, Robinson GE. Satiation differentially affects performance in a learning assay by nurse and forager honey bees. *J Comp Physiol A Neuroethol Sens Neural Behav Physiol.* 2001;187:891–899.
49. Chabaud MA, Devaud JM, Pham-Delègue MH, Preat T, Kaiser L. Olfactory conditioning of proboscis activity in *Drosophila melanogaster. J Comp Physiol A Neuroethol Sens Neural Behav Physiol.* 2006;192:1335–1348.
50. Krashes MJ, Waddell S. Rapid consolidation to a radish and protein synthesis-dependent long-term memory after single-session appetitive olfactory conditioning in *Drosophila. J Neurosci.* 2008;28:3103–3113.
51. Drier EA, Tello MK, Cowan M, et al. Memory enhancement and formation by atypical PKM activity in *Drosophila melanogaster. Nat Neurosci.* 2002;5:316–324.
52. Sacktor TC. PKMzeta, LTP maintenance, and the dynamic molecular biology of memory storage. *Prog Brain Res.* 2008;69:27–40.
53. Sutton MA, Bagnall MW, Sharma SK, Shobe J, Carew TJ. Intermediate-term memory for site-specific sensitization in *Aplysia* is maintained by persistent activation of protein kinase C. *J Neurosci.* 2004;24:3600–3609.
54. Riedel G, Platt B, Micheau J. Glutamate receptor function in learning and memory. *Behav Brain Res.* 2003;140:1–47.
55. Barbara GS, Zube C, Rybak J, Gauthier M, Grünewald B. Acetylcholine, GABA and glutamate induce ionic currents in cultured antennal lobe neurons of the honeybee, *Apis mellifera. J Comp Physiol A Neuroethol Sens Neural Behav Physiol.* 2005;191:823–836.
56. Funada M, Yasuo S, Yoshimura T, et al. Characterization of the two distinct subtypes of metabotropic glutamate receptors from honeybee, *Apis mellifera. Neurosci Lett.* 2004;359:190–194.
57. Kucharski R, Ball E, Hayward D, Maleszka R. Molecular cloning and expression analysis of a cDNA encoding a glutamate transporter in the honeybee brain. *Gene.* 2000;244:399–405.
58. Zannat MT, Locatelli F, Rybak J, Menzel R, Leboulle G. Identification and localisation of the NR1 sub-unit homologue of the NMDA glutamate receptor in the honeybee brain. *Neurosci Lett.* 2006;398:274–279.
59. Lopatina N, Ryzhova I, Chesnokova E. The role of non-NMDA-receptors in the process of associative learning in the honeybee *Apis mellifera. J Evol Biochem Physiol.* 2002;38:211–217.
60. Maleszka R, Helliwell P, Kucharski R. Pharmacological interference with glutamate re-uptake impairs long-term memory in the honeybee, *Apis mellifera. Behav Brain Res.* 2000;115:49–53.
61. Si A, Helliwell P, Maleszka R. Effects of NMDA receptor antagonists on olfactory learning and memory in the honeybee (*Apis mellifera*). *Pharmacol Biochem Behav.* 2004;77:191–197.
62. Xia S, Miyashita T, Fu TF, et al. NMDA receptors mediate olfactory learning and memory in *Drosophila. Curr Biol.* 2005;15: 603–615.
63. Müssig L, Richlitzki A, Rössler R, Eisenhardt D, Menzel R, Leboulle G. Acute disruption of the NMDA receptor subunit NR1 in the honeybee brain selectively impairs memory formation. *J Neurosci.* 2010;30:7817–7825.
64. Locatelli F, Bundrock G, Müller U. Focal and temporal release of glutamate in the mushroom bodies improves olfactory memory in *Apis mellifera. J Neurosci.* 2005;25:11614–11618.
65. Schwärzel M, Müller U. Dynamic memory networks: dissecting molecular mechanisms underlying associative memory in the temporal domain. *Cell Mol Life Sci.* 2006;63:989–998.
66. deBelle JS, Heisenberg M. Associative odor learning in *Drosophila* abolished by chemical ablation of mushroom bodies. *Science.* 1994;263:692–695.

CHAPTER

32

Pheromones Acting as Social Signals Modulate Learning in Honeybees

*Elodie Urlacher**,†, *Jean-Marc Devaud*† *and Alison R. Mercer**

*University of Otago, Dunedin, New Zealand †National Center for Scientific Research, University Paul Sabatier, Toulouse, France

INTRODUCTION

Foraging honeybees quickly learn which flowers offer the best rewards, and their ability to communicate this information to potential recruits in the colony is central to the success of this highly social insect.[1–4] However, not all foraging experiences are rewarding. If a bee enters an alfalfa (*Medicago sativa* L.) flower, for example, it is likely to trigger the release of spring-loaded stamens and anthers that burst forth and hit the bee firmly on its head.[5] Bumblebees apparently tolerate this punishment,[5,6] but honeybees learn to avoid alfalfa flowers or to solve the problem in some other way. For example, nectar foragers discover that if they approach alfalfa flowers from the side and push their proboscis between the petals, they can steal nectar from the nectary without tripping the flower.[5] Bees may once have been thought of as robots, programmed by their DNA to survive, but evidence that bees learn and that learning contributes to the development of new skills in this insect is now overwhelming.[3,4,7,8]

Although it is generally accepted that the ability to learn from experience is critical for survival, recent studies suggest that in certain situations, learning behavior is better suppressed.[9,10] Honeybees, like other social insects, have evolved a highly sophisticated communication system that involves the production, release, and recognition of chemical signals, called pheromones.[11–13] These important social signals have been found to modulate learning behavior in bees. For example, queen bees produce a pheromone that can block aversive learning in young bees,[10] and guards confronted by an intruder release chemical signals that not only elicit defensive behavior but also suppress appetitive learning in bees.[9] In this chapter, we examine what is known about the mechanisms that support pheromone modulation of learning behavior, and we consider the possible adaptive value to the colony as a whole of such examples of modulation.

PHEROMONES AND THEIR ROLES

The term *pheromone* was introduced by Peter Karlson and Martin Lüscher more than 50 years ago.[14] The word is derived from the Greek *pherein* (to transfer) and *hormon* (to excite), and it is used to describe a substance released by an animal that elicits a specific behavioral or physiological response in individuals of the same species.[15] Queen bees produce more than 50 different compounds that act, either alone or in combination, to influence the behavior and physiology of workers and drones.[13] Workers (sterile females) and drones (males) also produce pheromones, and chemical communication between colony members in a variety of different contexts helps maintain the colony as a functional unit.[12,13] Of particular interest to us here are two complex pheromones: queen mandibular pheromone (QMP) and sting alarm pheromone (SAP)—pheromones produced by honeybee queens and workers, respectively. Of the many chemical signals produced by honeybees, QMP and SAP are arguably the best understood. Both are complex mixtures, both prompt a range of responses, and both affect learning behavior in the bee. A detailed description of the many functions of these two multicomponent pheromones lies well beyond the scope of this chapter, but there are many excellent reviews available that

Invertebrate Learning and Memory.
DOI: http://dx.doi.org/10.1016/B978-0-12-415823-8.00032-0

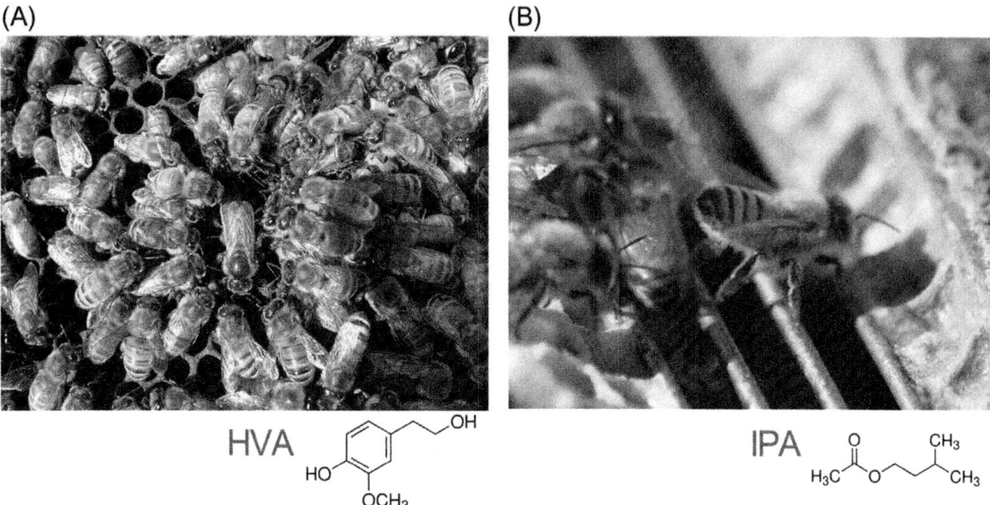

FIGURE 32.1 Responses of worker bees to queen mandibular pheromone (QMP) and sting alarm pheromone (SAP). (A) Retinue behavior exhibited by young worker bees in response to the queen's pheromones. Homovanillyl alcohol (HVA), which is a key component of QMP, has been shown to modulate aversive learning behavior in young worker bees.[10] (B) Guard bee releasing SAP, a key component of which is isopentylacetate (IPA). IPA has been shown to modulate appetitive learning behavior in bees.[9] Photographs by Fanny Mondet (A) and Elodie Urlacher (B).

describe the chemistry of these pheromones and their functions.[12,13,16–19]

QMP contains five key components,[19,20] one of which is an aromatic compound commonly referred to as homovanillyl alcohol (HVA; Figure 32.1). HVA suppresses aversive learning in young bees,[10] an effect that appears to be enhanced by other components of QMP. One critical function of QMP that is supported by other pheromones is to attract young workers to feed and groom the queen (Figure 32.1A).[13,18,21] As young attendants groom the queen's body, they pick up samples of her pheromones, and through antennation and food exchange—two hallmark behaviors of this highly social insect—the pheromones are rapidly distributed throughout the colony.[22–25] Widespread distribution of queen pheromone is important not only for the queen but also for the colony as a whole. Among its many other functions, QMP assists in inhibiting the rearing of new queens[26–28] and suppressing ovary development in workers,[29] two outcomes that typically result in a single queen serving as the sole reproductive female in the colony. Because the presence of a healthy egg-laying queen is central to the survival of the colony as a whole, it is not surprising that evolution has provided honeybees with a plethora of chemicals that enable colony members to identify the queen and that facilitate queen survival.

SAP also plays a critical role in the survival of the colony. Produced by guard bees when they perceive a threat, SAP serves to sound the alarm and to provoke responses required for colony defense.[17] SAP is produced in the Koschenikov gland, which is located close to the sting chamber of the bee. Of the many (<40) chemicals detected in SAP,[30,31] isopentylacetate (IPA; Figure 32.1B) appears to be the principal component. Presentation of IPA alone is sufficient to trigger many of the responses elicited by the whole blend.[11,32,33] Typical responses to this pheromone include not only wing flickering, raising of the abdomen, and release of SAP (Figure 32.1B) but also defensive responses that involve attack and stinging behavior (reviewed in Hunt[17]). Interestingly, these behaviors are more pronounced among groups of bees than in individual animals, and they are associated with physiological changes such as an increased respiratory rate.[34] Like QMP, SAP has recently been shown to affect learning behavior in bees, but evidence suggests that these two pheromones operate in markedly different ways.

PHEROMONE MODULATION OF LEARNING BEHAVIOR

Associative olfactory learning in harnessed bees can be demonstrated readily by taking advantage of highly predictable responses, including sting extension in response to an aversive stimulus such as a brief electric shock (Figure 32.2A) or proboscis extension in response to sucrose stimulation of the antennae (Figure 32.2D). A bee that is presented several times with an odor paired with electric shock, for example, begins extending its sting in response to the odor alone in expectation of punishment to follow (Figure 32.2B).[36,37] Appetitive learning in bees is also simple to

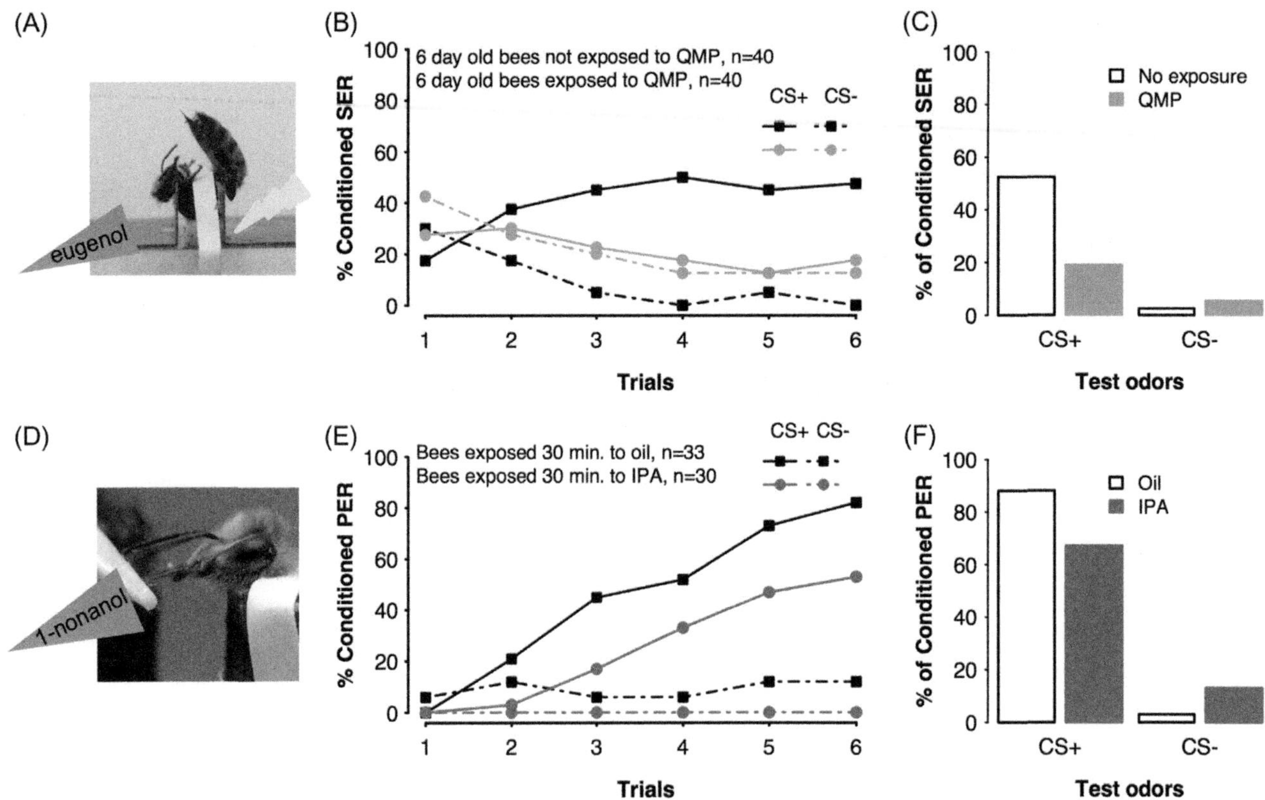

FIGURE 32.2 **(Top) QMP exposure inhibits aversive learning in young bees.** (A) Bee harnessed for differential aversive conditioning. The bee receives six presentations of the conditioned stimulus (CS + eugenol) paired with electric shock (the unconditioned stimulus), which elicits reflexive sting extension. When 1-hexanol (CS −) is presented, no shock is given. (B) Percentage of bees displaying a conditioned sting extension response (SER) to CS + (continuous line) and CS − (dashed line) across trials. (C) Memory test: The percentage of bees displaying a conditioned response to CS + and CS − 1 hr after the end of conditioning, in the absence of electric shock. (Bottom) IPA exposure decreases appetitive learning in forager/guard bees. (D) Bee harnessed for differential appetitive conditioning. The bee receives six presentations of the conditioned stimulus (CS + 1-nonanol) paired with sucrose (the unconditioned stimulus), which elicits reflexive proboscis extension. When 1-hexanol (CS −) is presented, no reward is given. (E) Percentage of conditioned proboscis extension responses (PER) to CS + (continuous line) and CS − (dashed line) across trials. (F) Memory test: Percentage of bees displaying a conditioned PER to CS + and CS − 1 hr after the end of conditioning, in the absence of sucrose. Source: *Data used to create Figures 32.2B and 32.2C from Vergoz* et al.[10] *Data used to create Figures 32.2E and 32.2F from Urlacher*[35] *and Urlacher* et al.[9]

demonstrate. A bee that is presented several times with an odor paired with a sucrose reward begins extending its proboscis in response to the odor alone, in expectation of a food reward (Figure 32.2E).[38–40]

Using the sting extension paradigm, Vergoz and colleagues[10] found that young bees exposed to QMP from the time of their emergence as adults showed a severe deficit in their ability to associate an odor with punishment (Figures 32.2B and 32.2C). The modulatory effects of QMP on aversive learning were found to be age dependent; whereas learning in young (4- to 6-day-old) workers was suppressed by QMP, it remained unaffected in older bees. Appetitive learning in the young animals was not affected by the pheromone.

Intriguingly, SAP also modulates learning in a specific manner, but in this case appetitive learning rather than aversive learning is impaired in bees exposed to SAP (Figures 32.2E and 32.2F). Urlacher and colleagues[9] showed that when exposed to the natural alarm pheromones produced by conspecifics, bees performed less well in an appetitive olfactory conditioning assay than controls. The same effect was observed when bees were exposed to IPA, the main component of SAP. Importantly, the effects they observed were the same irrespective of the odor used or the type of (appetitive) conditioning protocol adopted (absolute or differential conditioning). Their results also showed that the effects of SAP on appetitive learning were not due to impaired perception of either the conditioned stimulus (CS; the odorant) or the unconditioned stimulus (sucrose solution). Indeed, learning was reduced without any impairment of odor discrimination and independently of any reduction in responsiveness to sucrose after IPA exposure.

Although there is no direct evidence that SAP reduces appetitive learning in a natural context, some experiments report decreased foraging when the hive is under attack[41] or when the bees are exposed to one key component of alarm pheromone at the foraging site.[42]

COINCIDENTAL OR ADAPTIVE?

These effects of QMP and SAP on aversive learning and appetitive learning, respectively, suggest that there may be a selective advantage to bees in being able to modulate learning behavior via these social signals. If so, what might the biological significance of such 'metaplasticity' be? Modulating the efficiency of learning in a specific (aversive or appetitive) context might be of importance to colony survival. It is possible, for example, that impeding the formation of aversive associations in young bees might help the queen secure continuous care from workers, particularly if some of the chemicals she produces, or the effects they induce, are unpleasant to workers.[26,43,44] Consistent with this possibility, older bees are not strongly attracted to QMP, and bees of foraging age have been reported to avoid contact with this pheromone.[44,45]

Whereas aversive learning is not affected by SAP,[35] IPA exposure is able to modulate a bee's sensitivity to a noxious stimulus: Bees exposed to high doses of the main component of alarm pheromone (IPA) show reduced sensitivity to electric shocks.[46] This effect is reminiscent of the stress-induced analgesia state reported in mammals,[47] which might serve to promote defensive behavior in the face of injury. Thus, the reduced sensitivity to painful stimuli shown by IPA-exposed bees could potentially explain why bees fight rather than flee. However, effects of IPA are dose dependent: In low doses, IPA increases responsiveness to aversive stimuli, an effect that might contribute to recruitment for defense upon detection of alarm pheromone.[48]

Searching for adaptive value in the effects of pheromones on learning behavior assumes that there may be a selective advantage in social modulation of learning behavior in bees. However, an alternative explanation is that the impact of pheromones on learning is one manifestation, among others, of a change in balance between behavioral activities required to achieve different goals. Decreased appetitive learning resulting from exposure to alarm pheromone, for example, might simply occur as a consequence of a shift away from foraging activities toward activities involved in colony defense. The goals of a forager bee are markedly different from those of a nurse bee or guard. Depending on the needs of the colony, pheromones might provide an efficient way to promote certain suites of behaviors (or behavioral 'syndromes'[49–51]) in order to enhance the performance of particular tasks. Reducing the drive to forage and perform tasks associated with foraging, for example, may be critical to redirecting foragers to the task of defending the colony. Similarly, reduction of responses to pheromones that elicit defensive behavior might be important for ensuring that even during times of threat, some bees continue to care for the queen. Indeed, it is known that individuals differ in their probability to engage in defensive behavior,[52] and the level of 'priming' may depend on experience of previous pheromone exposure, for example. Individual bees perform different behavioral tasks during their normal behavioral maturation, and importantly they can respond in an adaptive way to colony needs.[53] As a key component of the foraging syndrome, appetitive learning might be reduced as a consequence of an increased probability of transition from foraging to defense. This probabilistic action of SAP would explain the recruitment of defenders at the hive entrance[54] and the reduced foraging activity observed at the collective level in the presence of SAP.[11,41,42] Similarly, one might consider a retinue (or nursing) syndrome, from which transition to defense would be blocked in the presence of QMP. In such a case, impaired aversive learning might reflect the repression of the defensive syndrome. It is interesting to note, however, that appetitive learning remains strong in young bees exposed to QMP, suggesting that learning behavior plays an important role not only in the context of foraging but also within the hive.

Like many effects of SAP and QMP,[13,55–57] the impact of these pheromones on learning performance is age dependent[10] (Urlacher, personal observation), and in this regard, the actions of SAP and QMP are intriguingly complementary: Young bees are more strongly attracted to QMP than older bees, but older bees are more responsive to SAP than young workers. Age-related shifts in responsiveness to these pheromones provide an important reference point for studies exploring the mechanisms that underlie pheromone regulation of learning behavior.

MODES OF ACTION

As is evident from the chapters of this book, there is intense current interest in cellular and molecular mechanisms that underlie the acquisition, storage, and recall of information in the brain. Understanding how QMP and SAP modulate learning behavior should advance our knowledge and understanding of learning and memory mechanisms.

In insects, the negative reinforcing properties of an aversive stimulus are conveyed in the brain by nerve cells that release the biogenic amine, dopamine,[58–60] and dopamine released at the level of the mushroom bodies of the brain is known to play a central role in aversive learning (reviewed by[61,62]). Consistent with these findings, QMP, which impairs aversive learning in young worker bees, has been found to target dopamine-signaling pathways of the brain. Exposing young bees to QMP reduces brain dopamine levels and alters levels of dopamine receptor gene expression not only in the brain[63,64] but also in the antennae of the bee.[44] It is not yet clear which of QMP's effects are primarily responsible for suppressing aversive learning in young bees, but evidence suggests that HVA, a key component of QMP, contributes significantly to these effects. Interestingly, a strong correlation has been identified between the survival of introduced queens and levels of HVA produced by the queen.[65]

Vergoz and colleagues[10] found that a young bee's ability to associate an odor with punishment can be suppressed by exposing the bee to HVA. Structural similarities between HVA and dopamine led to the hypothesis that HVA might target dopamine receptors directly. In *Drosophila*, two distinct types of dopamine receptor have been implicated in aversive learning—dDA1 and DAMB.[66,67] The honeybee orthologs of these receptors (*Am*DOP1 and *Am*DOP2, respectively) have been cloned and characterized.[68,69] Their functional properties are very similar to those of their counterparts in the fly [reviewed by70,61]. Activation of either *Am*DOP1 or *Am*DOP2 receptors leads to an increase in intracellular levels of cAMP, but *Am*DOP2, like its *Drosophila* ortholog,[71,72] also generates a calcium signal when activated.[73] In bees, as in flies, both receptor types are expressed in the mushroom bodies of the brain.[74]

Using heterologous expression techniques, Beggs and Mercer[75] examined the effects of HVA on honeybee dopamine receptors expressed *in vitro* in cell lines normally devoid of dopamine receptors. Although they could find no evidence that HVA activated or inhibited *Am*DOP1 or *Am*DOP2 receptors, HVA was found to selectively activate the honeybee 'D2-like' dopamine receptor, *Am*DOP3.[75] *Am*DOP3 is referred to as a D2-like receptor because vertebrate D2 dopamine receptors downregulate intracellular levels of cAMP, a feature shared by *Am*DOP3,[76] and its *Drosophila* ortholog, DD2R.[77] *Am*DOP3 receptors are also expressed in the mushroom bodies of the brain,[76,78] but their role, if any, in associative olfactory learning remains unknown. It is intriguing therefore that HVA blocks aversive learning in young worker bees, but where does HVA act?

It is known that HVA is detected by olfactory receptor neurons housed in the antennae of the bee,[79] and there is evidence that QMP's actions may be mediated, at least in part, through HVA acting at this level.[44] Young bees that show strong attraction to QMP have higher levels of expression of *Am*DOP3 (and the octopamine receptor *Am*OA1) in their antennae than bees of the same age that show little or no attraction to this pheromone.[44] Therefore, it is possible that the dramatic decline in *Am*dop3 expression levels that occurs in the antennae of worker bees during the first days of adult life[44] contributes to the age-related decline in responsiveness to QMP. However, young bees that attend the queen not only feed her, and touch her body with their antennae, but also groom her.[13] As they lick her body, they are likely to ingest queen pheromones, including HVA. Whether HVA crosses the blood–brain barrier is unclear, but HVA in the hemolymph could potentially target receptors located peripherally, for example, in the antennae, in endocrine organs such as the corpora allata, or in the ovaries. Because changes in hormone titers are known to have significant effects on biogenic amine titers in insects (reviewed by[80]), QMP's ability to reduce the rate of synthesis of juvenile hormone (JH)[13,81,82] is likely to contribute also to QMP's effects on learning behavior in young bees.[83]

Similarly, modulation of appetitive learning by IPA may be the result of hormonal action. In their investigations of the effects of SAP on learning, Urlacher and colleagues were drawn to the finding that the opioid receptor agonist, morphine, was able to mimic the effects of IPA (the principal component of SAP). This suggested to them that an opioid-like signaling system might be involved in the modulation of learning behavior in the bee, as previously proposed for the analgesic effect of IPA.[46] Indeed, these authors have shown that not only can IPA-induced 'stress-like analgesia' be abolished by treatment with the opioid antagonist, naloxone, but also the canonical agonist of mammalian opioid receptors, morphine, can at least in part mimic this effect. Similarly, morphine, like IPA, has been found to decrease appetitive learning in bees, whereas naloxone restores normal learning abilities in IPA-exposed bees.[35] By taking advantage of the sequenced genome of the honeybee,[84] Urlacher and colleagues identified an opioid-like receptor in the bee. They showed that when expressed in a heterologous system, the honeybee receptor, like its *Drosophila* ortholog,[85] could be activated by allatostatin C.[35] Allatostatins are neuropeptides known to act also as neurohormones.[86,87] Consistent with this, bees injected with allatostatin C show learning deficits that are similar to those observed following IPA exposure. Interestingly, allatostatin C also leads to an 'analgesia-like' state, and both these effects are counteracted by the opioid antagonist naloxone.[35] Taken together, these results suggest that various effects of IPA, including modulation of appetitive learning, are mediated by allatostatin C acting on as yet unidentified cellular targets that apparently trigger molecular cascades that share similarities with opioid signaling in mammals.

A FOCUS FOR FUTURE STUDIES

What are the targets of allatostatins in the bee? Allatostatins were identified in insects as inhibitors of the corpora allata, endocrine glands that produce JH.[87] Although allatostatin C has not been tested for its allatostatic activity in the bee, we can hypothesize that it might decrease JH titers and possibly octopamine levels as well. Consistent with this hypothesis, JH has been shown to have a stimulatory effect on octopamine release,[88,89] and as described elsewhere in this book, octopamine has been strongly implicated in appetitive learning in the bee[90–92] as well as in division of labor.[93–95] There is evidence for interplay between biogenic amines and signaling systems involving hormones such as JH and ecdysone,[80,89] and QMP itself inhibits JH synthesis.[57,81,82] One challenge for the future will be to identify how pheromones such as QMP and SAP interact with these critically important signaling systems.

Evolution has provided us with a unique set of tools to examine learning and memory mechanisms in the honeybee. Understanding how QMP and SAP modulate learning behavior promises important insights into the cellular and molecular mechanisms that underpin learning behavior in this highly social insect.

References

1. von Frisch K. *Bees, Their Vision, Chemical Senses and Language.* New York: Cornell University Press; 1950.
2. von Frisch K. *The Dance Language and Orientation of Bees.* Cambridge, MA: Harvard University Press; 1967.
3. Menzel R. Learning, memory and 'cognition' in honey bees. In: Kesner RP, Olten DS, eds. *Neurobiology of Comparative Cognition.* New York: Erlbaum; 1990:237–292.
4. Menzel R. Memory dynamics in the honeybee. *J Comp Physiol A.* 1999;185:323–340.
5. Reinhardt JF. Some responses of honey bees to alfalfa flowers. *Am Nat.* 1952;830:257–275.
6. Brunet J, Stewart CM. Impact of bee species and plant density on alfalfa pollination and potential for gene flow. *Psyche.* 2010;201010.1155/2010/201858.
7. Menzel R, Müller U. Learning and memory in honeybees: from behavior to neural substrates. *Annu Rev Neurosci.* 1996;19:379–404.
8. Menzel R, Erber J, Masuhr T. Learning and memory in the honeybee. In: Browne LB, ed. *Experimental Analysis of Insect Behaviour.* Berlin: Springer; 1974:195–217.
9. Urlacher E, Francés B, Giurfa M, Devaud J-M. An alarm pheromone modulates appetitive olfactory learning in the honeybee (*Apis mellifera*). *Front Behav Neurosci.* 2010;4:157.
10. Vergoz V, Schreurs HA, Mercer AR. Queen pheromone blocks aversive learning in young worker bees. *Science.* 2007;317: 384–386.
11. Free JB. *Pheromones of Social Bees.* Ithaca, NY: Comstock; 1987.
12. Le Conte Y, Hefetz A. Primer pheromones in social hymenoptera. *Annu Rev Entomol.* 2008;53:523–542.
13. Slessor KN, Winston ML, Le Conte Y. Pheromone communication in the honeybee (*Apis mellifera* L.). *J Chem Ecol.* 2005;31: 2731–2745.
14. Karlson P, Lüscher M. 'Pheromones': a new term for a class of biologically active substances. *Nature.* 1959;183:55–56.
15. Wyatt TD. Pheromones and signature mixtures: defining species-wide signals and variable cues for identity in both invertebrates and vertebrates. *J Comp Physiol A.* 2010;196: 685–700.
16. Alaux C, Maisonnasse A, Le Conte Y, Gerald L. Pheromones in a superorganism: from gene to social regulation. *Vit Horm.* 2010;83:401–423.
17. Hunt GJ. Flight and fight: a comparative view of the neurophysiology and genetics of honey bee defensive behavior. *J Insect Physiol.* 2007;53:399–410.
18. Maisonnasse A, Alaux C, Beslay D, et al. New insights into honey bee (*Apis mellifera*) pheromone communication: is the queen mandibular pheromone alone in colony regulation? *Front Zool.* 2010;7:18.
19. Slessor KN, Kaminski LA, King GGS, Winston ML. Semiochemicals of the honeybee queen mandibular glands. *J Chem Ecol.* 1990;16:851–860.
20. Slessor KN, Kaminski LA, King G, Borden JH, Winston ML. Semiochemical basis of the retinue response to queen honey bees. *Nature.* 1988;332:354–356.
21. Keeling CI, Slessor KN, Higo HA, Winston ML. New components of the honey bee (*Apis mellifera* L.) queen retinue pheromone. *Proc Natl Acad Sci USA.* 2003;100:4486–4491.
22. Ferguson AW, Free JB. Queen pheromone transfer within honeybee colonies. *Physiol Entomol.* 1980;5(4):359–366.
23. Naumann K, Winston ML, Slessor KN, Prestwich GD, Webster FX. Production and transmission of honey-bee queen (*Apis mellifera* L.) mandibular gland pheromone. *Behav Ecol Sociobiol.* 1991;29:321–332.
24. Seeley TD. Queen substance dispersal by messenger workers in honeybee colonies. *Behav Ecol Sociobiol.* 1979;5(4):391–415.
25. Velthuis HHW. Observations on the transmission of queen substances in the honey bee colony by the attendants of the queen. *Behaviour.* 1972;41:105–129.
26. Pettis JS, Winston ML, Collins AM. Suppression of queen rearing in European and Africanized honey bees *Apis mellifera* L. by synthetic queen mandibular gland pheromone. *Insect Soc.* 1995;42(2):113–121.
27. Winston ML, Higo HA, Slessor KN. Effect of various dosages of queen mandibular gland pheromone on the inhibition of queen rearing in the honey bee (Hymenoptera: Apidae). *Ann Ent Soc Am.* 1990;83(2):234–238.
28. Winston ML, Slessor KN, Willis LG, et al. The influence of queen mandibular pheromones on worker attraction to swarm clusters and inhibition of queen rearing in the honey bee (*Apis mellifera* L.). *Insect Soc.* 1989;36(1):15–27.
29. Hoover SR, Keeling C, Winston M, Slessor K. The effect of queen pheromones on worker honey bee ovary development. *Naturwiss.* 2003;90:477–480.
30. Blum MS, Fales HM. Chemical releasers of alarm behavior in the honey bee: informational 'plethora' of the sting apparatus signal. In: Needham GR, Page RE, Delfinado-Baker M, Bowman CE, eds. *Africanized Honey Bees and Bee Mites.* New York: Wiley; 1988:141–148.

31. Lensky Y, Cassier P. The alarm pheromones of queen and worker honey bees. *Bee World*. 1995;76:119–129.
32. Collins AM, Blum MS. Bioassay of compounds derived from the honeybee sting. *J Chem Ecol*. 1982;8:463–470.
33. Collins AM, Blum MS. Alarm responses caused by newly identified compounds derived from the honeybee sting. *J Chem Ecol*. 1983;8:57–65.
34. Moritz RFA, Southwick EE, Breh M. A metabolic test for the quantitative-analysis of alarm behavior of honeybees (*Apis mellifera* L.). *J Exp Zool*. 1985;235:1–5.
35. Urlacher E. *A Novel Peptidergic Pathway Modulating Learning in the Honeybee*. PhD thesis, Toulouse, France: Université Paul Sabatier; 2011.
36. Giurfa M, Fabre E, Flaven-Pouchon J, et al. Olfactory conditioning of the sting extension reflex in honeybees: memory dependence on trial number, interstimulus interval, intertrial interval, and protein synthesis. *Learn Mem*. 2009;16:761–765.
37. Vergoz V, Roussel E, Sandoz JC, Giurfa M. Aversive learning in honeybees revealed by the olfactory conditioning of the sting extension reflex. *PLoS ONE*. 2007;2:e288.
38. Bitterman ME, Menzel R, Fietz A, Schafer S. Classical conditioning of proboscis extension in honeybees (*Apis mellifera*). *J Comp Psychol*. 1983;97:107–119.
39. Giurfa M, Sandoz J-C. Invertebrate learning and memory: fifty years of olfactory conditioning of the proboscis extension response in honeybees. *Learn Mem*. 2012;19(2):54–66.
40. Takeda K. Classical conditioned response in the honeybee. *J Insect Physiol*. 1961;6:168–179.
41. Ken T, Hepburn HR, Radloff SE, et al. Heat-balling wasps by honeybees. *Naturwiss*. 2005;92(10):492–495.
42. Nieh JC. A negative feedback signal that is triggered by peril curbs honey bee recruitment. *Curr Biol*. 2010;20(4):310–315.
43. Moritz RFA, Crewe RM, Hepburn HR. Attraction and repellence of workers by the honeybee queen (*Apis mellifera* L.). *Ethology*. 2001;107(6):465–477.
44. Vergoz V, McQuillan HJ, Geddes LH, et al. Peripheral modulation of worker bee responses to queen mandibular pheromone. *Proc Natl Acad Sci USA*. 2009;106:20930–20935.
45. Fan YL, Richard FJ, Rouf N, Grozinger CM. Effects of queen mandibular pheromone on nestmate recognition in worker honeybees, *Apis mellifera*. *Anim Behav*. 2010;79:649–656.
46. Núñez J, Almeida L, Balderrama N, Giurfa M. Alarm pheromone induces stress analgesia via an opioid system in the honeybee. *Physiol Behav*. 1998;63:75–80.
47. Fanselow MS. Conditioned fear-induced opiate analgesia: a competing motivational state theory of stress analgesia. *Ann N Y Acad Sci*. 1986;467(1):40–54.
48. Balderrama N, Núñez J, Guerrieri F, Giurfa M. Different functions of two alarm substances in the honeybee. *J Comp Physiol A Neuroethol Sens Neural Behav Physiol*. 2002;188:485–491.
49. Pankiw T. The honey bee foraging behavior syndrome: quantifying the response threshold model of division of labor. *Proceedings of the 2005 IEEE Swarm Intelligence Symposium*; 2005;5:1–6.
50. Roussel E, Carcaud J, Sandoz JC, Giurfa M. Reappraising social insect behavior through aversive responsiveness and learning. *PLoS ONE*. 2009;4:e4197.
51. Sih A, Bell A, Johnson JC. Behavioral syndromes: an ecological and evolutionary overview. *TREE*. 2004;19(7):372–378.
52. Breed MD, Guzmán-Novoa E, Hunt GJ. Defensive behavior of honey bees: organization, genetics, and comparisons with other bees. *Annu Rev Entomol*. 2004;49:271–298.
53. Robinson GE. Regulation of division of labor in insect societies. *Annu Rev Entomol*. 1992;37(1):637–665.
54. Alaux C, Robinson GE. Alarm pheromone induces immediate-early gene expression and slow behavioral response in honey bees. *J Chem Ecol*. 2007;33:1346–1350.
55. Peters L, Zhu-Salzman K, Pankiw T. Effect of primer pheromones and pollen diet on the food producing glands of worker honey bees (*Apis mellifera* L.). *J Insect Physiol*. 2010;56(2):132–137.
56. Robinson GE. Modulation of alarm pheromone perception in the honey bee: evidence for division of labor based on hormonally regulated response thresholds. *J Comp Physiol A*. 1987;160(5):613–619.
57. Robinson GE, Winston ML, Huang Z-Y, Pankiw T. Queen mandibular gland pheromone influences worker honey bee (*Apis mellifera* L.) foraging ontogeny and juvenile hormone titers. *J Insect Physiol*. 1998;44(7–8):685–692.
58. Riemensperger T, Voller T, Stock P, Buchner E, Fiala A. Punishment prediction by dopaminergic neurons in *Drosophila*. *Curr Biol*. 2005;15:1953–1960.
59. Schroll C, Riemensperger T, Bucher D, et al. Light-induced activation of distinct modulatory neurons triggers appetitive or aversive learning in *Drosophila* larvae. *Curr Biol*. 2006;16:1741–1747.
60. Schwärzel M, Monastirioti M, Scholz H, Friggi-Grelin D, Birman S, Heisenberg M. Dopamine and octopamine differentiate between aversive and appetitive olfactory memories in *Drosophila*. *J Neurosci*. 2003;23:10495–10502.
61. Mustard JA, Vergoz V, Mesce KA, et al. Dopamine signaling in the bee. In: Eisenhardt D, Galizia CG, Giurfa M, eds. *Honeybee Neurobiology and Behavior: A Tribute to Randolf Menzel*. New York: Springer Verlag; 2012:199–209.
62. Waddell S. Dopamine reveals neural circuit mechanisms of fly memory. *Trends Neurosci*. 2010;33:457–464.
63. Beggs KT, Glendining KA, Marechal NM, et al. Queen pheromone modulates brain dopamine function in worker honey bees. *Proc Natl Acad Sci USA*. 2007;104:2460–2464.
64. Grozinger CM, Sharabash NM, Whitfield CW, Robinson GE. Pheromone-mediated gene expression in the honeybee brain. *Proc Natl Acad Sci USA*. 2003;100:14519–14525.
65. Rhodes J, Somerville D. *Introduction and Early Performance of Queen Bees: Some Factors Affecting Success*. Kingston, Australia: Rural Industries Research and Development Corporation, Publication No. 03/049.
66. Kim Y-C, Lee G-H, Han K-A. D_1 dopamine receptor dDA1 is required in the mushroom body neurons for aversive and appetitive learning in *Drosophila*. *J Neurosci*. 2007;27:7640–7647.
67. Selcho M, Pauls D, Han K-A, Stocker FR, Thum AS. The role of dopamine in *Drosophila* larval classical olfactory conditioning. *PLoS ONE*. 2009;4(6):e5897.
68. Blenau W, Erber J, Baumann A. Characterization of a dopamine D1 receptor from *Apis mellifera*: cloning, functional expression, pharmacology, and mRNA localization in the brain. *J Neurochem*. 1998;70:15–23.
69. Humphries M, Mustard JA, Hunter SJ, Mercer A, Ward V, Ebert PR. Invertebrate D2 type dopamine receptor exhibits age-based plasticity of expression in the mushroom bodies of the honeybee brain. *J Neurobiol*. 2003;55:315–330.
70. Mustard JA, Beggs KT, Mercer AR. Molecular biology of the invertebrate dopamine receptors. *Arch Insect Biochem Physiol*. 2005;59:103–117.
71. Feng G, Hannan F, Reale V, et al. Cloning and functional characterization of a novel dopamine receptor from *Drosophila melanogaster*. *J Neurosci*. 1996;16:3925–3933.
72. Han KA, Millar NS, Grotewiel MS, Davis RL. DAMB, a novel dopamine receptor expressed specifically in *Drosophila* mushroom bodies. *Neuron*. 1996;16:1127–1135.

73. Beggs KT, Tyndall JDA, Mercer AR. Honey bee dopamine and octopamine receptors linked to intracellular calcium signaling have a close phylogenetic and pharmacological relationship. *PLoS ONE*. 2011;6:e26809.

74. Kurshan PT, Hamilton IS, Mustard J, Mercer AR. Developmental changes in expression patterns of two dopamine receptor genes in mushroom bodies of the honeybee, *Apis mellifera*. *J Comp Neurol*. 2003;466:91–103.

75. Beggs KT, Mercer AR. Dopamine receptor activation by honey bee queen pheromone. *Curr Biol*. 2009;19:1206–1209.

76. Beggs KT, Hamilton IS, Kurshan PT, Mustard JA, Mercer AR. Characterization of a D2-like dopamine receptor (*Am*DOP3) in honey bee, *Apis mellifera*. *Insect Biochem Molec Biol*. 2005;35: 873–882.

77. Hearn MG, Ren Y, McBride EW, Reveillaud I, Beinborn M, Kopin AS. A *Drosophila* dopamine 2-like receptor: molecular characterization and identification of multiple alternatively spliced variants. *Proc Natl Acad Sci USA*. 2002;99:14554–14559.

78. McQuillan HJ, Nakagawa S, Mercer AR. Mushroom bodies of the honey bee brain show cell-population-specific plasticity in expression of amine-receptor genes. *Learn Mem*. 2012;19: 151–158.

79. Sandoz J-C. Odour-evoked responses to queen pheromone components and to plant odours using optical imaging in the antennal lobe of the honey bee drone *Apis mellifera* L. *J Exp Biol*. 2006;209:3587–3598.

80. Gruntenko NE, Rauschenbach IY. Interplay of JH, 20E and biogenic amines under normal and stress conditions and its effect on reproduction. *J Insect Physiol*. 2008;54:902–908.

81. Kaatz H-H, Hildebrandt H, Engels W. Primer effect of queen pheromone on juvenile hormone biosynthesis in adult worker honey bees. *J Comp Physiol B*. 1992;162(7):588–592.

82. Pankiw T, Huang ZY, Winston ML, Robinson GE. Queen mandibular gland pheromone influences worker honey bee (*Apis mellifera* L.) foraging ontogeny and juvenile hormone titers. *J Insect Physiol*. 1998;44:685–692.

83. Jarriault D, Mercer AR. Queen mandibular pheromone: questions that remain to be resolved. *Apidologie*. 2012;43(3):292–307.

84. Honey Bee Genome Sequencing Consortium. Insights into social insects from the genome of the honeybee *Apis mellifera*. *Nature*. 2006;443:931–949.

85. Kreienkamp HJ, Larusson HJ, Witte I, et al. Functional annotation of two orphan G-protein-coupled receptors, Drostar1 and -2, from *Drosophila melanogaster* and their ligands by reverse pharmacology. *J Biol Chem*. 2002;277:39937–39943.

86. Audsley N, Matthews J, Weaver RJ. Neuropeptides associated with the frontal ganglion of larval Lepidoptera. *Peptides*. 2005;26:11–21.

87. Stay B, Tobe SS. The role of allatostatins in juvenile hormone synthesis in insects and crustaceans. *Annu Rev Entomol*. 2007;52: 277–299.

88. Schulz DJ, Barron AB, Robinson GE. A role for octopamine in honey bee division of labor. *Brain Behav Evol*. 2002;60: 350–359.

89. Schulz DJ, Sullivan JP, Robinson GE. Juvenile hormone and octopamine in the regulation of division of labor in honey bee colonies. *Horm Behav*. 2002;42:222–231.

90. Hammer M. An identified neuron mediates the unconditioned stimulus in associative olfactory learning in honeybees. *Nature*. 1993;366:59–63.

91. Hammer M, Menzel R. Multiple sites of associative odor learning as revealed by local brain microinjections of octopamine in honeybees. *Learn Mem*. 1998;5(1–2):146–156.

92. Scheiner R, Baumann A, Blenau W. Aminergic control and modulation of honeybee behaviour. *Curr Neuropharmacol*. 2006;4 (4):259–276.

93. Schulz DJ, Robinson GE. Biogenic amines and division of labor in honey bee colonies: behaviorally related changes in the antennal lobes and age-related changes in the mushroom bodies. *J Comp Physiol A*. 1999;184(5):481–488.

94. Schulz DJ, Robinson GE. Octopamine influences division of labor in honey bee colonies. *J Comp Physiol A*. 2001;187: 53–61.

95. Wagener-Hulme C, Kuehn JC, Schulz DJ, Robinson GE. Biogenic amines and division of labor in honey bee colonies. *J Comp Physiol A*. 1999;184:471–479.

CHAPTER

33

Extinction Learning and Memory Formation in the Honeybee

Dorothea Eisenhardt

Freie Universität Berlin, Berlin, Germany

EXTINCTION RESEMBLES AN ANIMAL'S ADAPTATION TO A FLUCTUATING ENVIRONMENT

The availability of food as well as the occurrence of potential threats are signaled to animals by naturally occurring stimuli. Animals learn the association of these stimuli with an appetitive stimulus (i.e., a food reward) or an aversive stimulus (i.e., a potential punishment). Accordingly, these stimuli predict the occurrence of food and punishment and elicit an animal's behavior toward or away from it. However, an animal's environment is seldom stable, and the previously learned association between the predicting stimulus and the appetitive or the aversive stimulus might not always hold true. Thus, in order to adapt to their environment, animals need to learn that a previously learned association is not relevant anymore.

Indeed, animals learn about the failure of a stimulus association. This has been demonstrated in vertebrates and invertebrates using different classical conditioning paradigms. In these learning paradigms, the decrease of an animal's behavior is observed once the predicting stimulus is not reinforced anymore. This decrease of behavior is termed *extinction*.[1,2]

Interestingly, the memory about the initial stimulus association is not erased after extinction. Rather, a time- and context-dependent reappearance of the reinforced behavior is observed.[2,3] Therefore, several authors argue that extinction does not comprise the destruction of the conditioned stimulus (CS)–unconditioned stimulus (US) association but, rather, a new form of learning about the failure of the reinforcement. Accordingly, extinction learning results in an extinction memory (i.e., a memory about the CS–no US association) that is stored in parallel with the previously formed associative memory. As a consequence, the predicting stimulus acquires a second meaning during extinction learning so that it becomes an ambiguous stimulus for the animal.[4]

Thus, studies on extinction learning suggest that animals do not update their previously formed memory about a certain stimulus association by 'overwriting' this previously formed memory with the most recently experienced association. Rather, a new memory is formed about every stimulus association an animal experiences. By this, the animal gains as much information about its environment as possible, enabling an optimal adaptation to a fluctuating environment.

However, the formation of several memories about one stimulus is only useful if each memory is retrieved in the appropriate situation. Retrieval of a maladaptive memory might have severe consequences. In humans, for example, retrieval of a maladaptive memory might underlie anxiety disorders and drug addiction.[5,6] Therefore, the question remains how contrasting memories are organized and in which circumstances each memory can be retrieved.

CLASSICAL CONDITIONING IN HARNESSED HONEYBEES

The organization of contrasting memories formed about the same stimulus is studied in extinction of honeybees. Honeybees are particularly well-suited for studies on memory formation because of their pronounced learning ability that is rooted in their social behavior and their biology as a pollinator.[7,8] The

Invertebrate Learning and Memory.
DOI: http://dx.doi.org/10.1016/B978-0-12-415823-8.00033-2

adaptation of free-flying honeybees to failing rewards resembles extinction, suggesting that extinction plays a role in the foraging context and hence in their natural environment.[9] In addition, classical conditioning and extinction can be studied in individually harnessed honeybees in the lab utilizing a well-established conditioning paradigm, the olfactory conditioning of the proboscis extension response (PER).[10,11] A honeybee extends its proboscis when the antennae are touched with sucrose solution. This response to sucrose solution is called the PER. During the acquisition phase, the presentation of a neutral odor (the CS) precedes the presentation of the sucrose reward (the US). Honeybees learn to associate the odor with the sucrose reward. The odor by itself elicits the PER once the association of odor and reward has been formed. This reaction to the odor is called the conditioned response (CR). Exposure to the learned odor can elicit the CR immediately after acquisition and up to several days later. This indicates the formation of short- as well as long-term memories.[8,12]

SPONTANEOUS RECOVERY FROM EXTINCTION DEMONSTRATES THE EXISTENCE OF TWO MEMORIES

When the CS is presented multiple times after successful conditioning, a successive decrease of the CR is observed. This decrease in the CR is termed extinction and can be interpreted as an extinction learning curve.[13–15]

Bitterman *et al.*[13] demonstrated extinction after a CS was presented five times soon after olfactory conditioning. A subsequent memory test 30 min after extinction revealed the reappearance of the CR. Thus, the decreased CR recovered spontaneously.[13] Therefore, the reappearance of the CR after extinction is termed *spontaneous recovery*[3] (Figure 33.1A).

The occurrence of spontaneous recovery in honeybee olfactory conditioning suggests that after extinction, two memories about the odor (the CS) are formed: one memory about the association of the odor with the reward (the US), here referred to as the reward memory, and one memory about the absence of the reward, the extinction memory. Sandoz and Pham-Delègue[14] found that the occurrence and the extent of spontaneous recovery depend on the learning parameters of initial reward learning and extinction learning, namely the number of CS–US pairings, the number of extinction trials, and the time interval between acquisition and extinction. How can these findings be conceptualized? In honeybees, several memory phases are formed after reward learning that are defined by the time interval after learning during which they control behavior and by biochemical processes that are necessary for their formation. The formation of these different memory phases depends on the number of learning trials and the intertrial interval applied during acquisition.[8,16,17] Accordingly, spontaneous recovery depends on the parameters of reward learning and extinction learning and also on the interval between these two forms of learning, indicating that the reward memory phase during which extinction takes place is crucial for the occurrence and the extent of spontaneous recovery.

Furthermore, the stability of a memory phase depends on the number of conditioning trials applied during acquisition.[14,18] Accordingly, the finding that spontaneous recovery depends on both the number of conditioning trials during reward learning and the number of extinction trials during extinction learning suggests that the stability of reward memory and the stability of the extinction memory are crucial for the occurrence and the extent of spontaneous recovery. Taken together, both reward memory and extinction memory are formed and control conditioned responding after extinction.

(A) Spontaneous Recovery

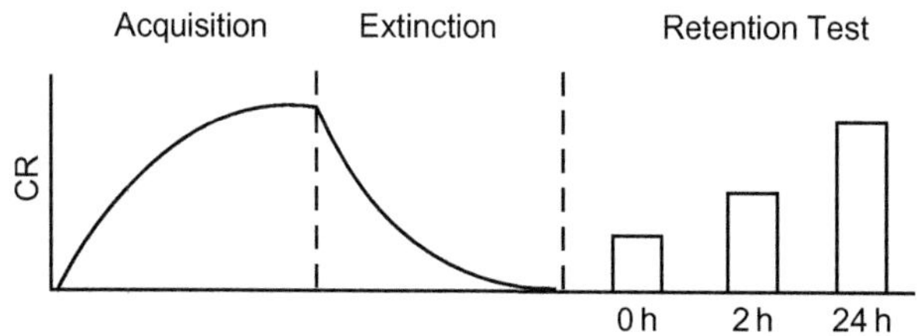

(B) Reinstatement

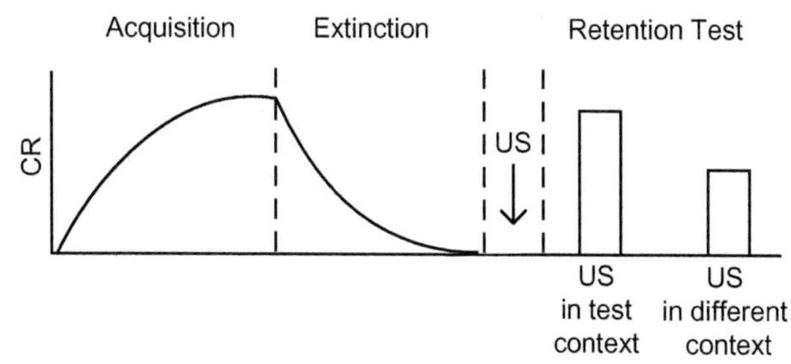

Modified from Myers & Davis, 2002

FIGURE 33.1 Spontaneous recovery and reinstatement. Schematic diagram of two behavioral phenomena related to extinction. (A) Spontaneous recovery is defined as the reappearance of the CR after extinction learning. (B) Reinstatement is defined as the context-dependent reappearance of the CR when the US is presented during the interval between extinction learning and a subsequent retention test. Source: *Modified from Myers and Davis.*[3]

REINSTATEMENT OF THE EXTINGUISHED MEMORY IS CONTEXT DEPENDENT

In addition to spontaneous recovery, reinstatement is a second phenomenon that relates to extinction. In reinstatement, the CR reappears when the US is presented during the interval between the extinction session and a subsequent retention test, but only when US presentation and retention test take place in the same context[3] (Figure 33.1B). Thus, the occurrence of reinstatement again demonstrates that the extinguished reward memory is not destructed after extinction and that, more important, the context is critical for memory retrieval after extinction.

Reinstatement has only recently been demonstrated in honeybees[19] using the PER conditioning paradigm. Honeybees were trained with one CS–US conditioning trial and extinguished 3 hr later with three extinction trials. Ten minutes later, the PER was elicited with the US (i.e., a sucrose solution), and memory retention was tested 2 hr later. Reappearance of the CR in the final retention test was observed in significantly more animals that received the US than in those that did not receive the US. However, the reappearance of the CR was observed only when the US was presented in the context of acquisition, extinction, and the final memory test. When the US was presented in a different context, the reappearance of the CR was not observed.[19] Thus, in honeybees, the US induces the recovery of the CR in a context-dependent manner. Because the US presentation induces the reappearance of the CR, reinstatement is another indication that the initial acquisition memory still exists after extinction in honeybees. Because the initial reward memory is not visible directly after extinction, it is most likely suppressed by the extinction memory. Most important, reinstatement is context dependent in honeybees,[19] as it has been demonstrated in vertebrates.[2,20] Accordingly, also in honeybees, the context is learned and plays a role in determining to what extent the initial reward memory and the extinction memory contribute to control behavior after extinction.

EXTINCTION OF A CONSOLIDATED LONG-TERM MEMORY

Previously, we discussed memory formation during extinction within the first hour(s) after acquisition.[13,14,19] However, extinction in honeybees occurs independently of the time interval between acquisition and retrieval. Stollhoff *et al.*[15] demonstrated extinction 24 hr after acquisition. In honeybees, the formation of a reward memory that controls behavior 24 hr after learning depends on translation taking place within the first hours after reward learning.[15,18,21] Translation is thought to stabilize long-term memories by the synthesis of new proteins that contribute to structural changes within the nervous system. This stabilization process is termed *memory consolidation* and the respective memories are defined as *consolidated memories*.[22] Thus, in honeybees, a reward memory that controls behavior 24 hr after acquisition underwent a consolidation process and is regarded as being particularly stable. Because of this stability and longevity, one might assume that consolidated memories are not suppressed by an extinction memory. However, Stollhoff *et al.*[15] demonstrated the opposite in honeybees, namely that extinction of a consolidated memory can be observed even after rather weak extinction learning by two extinction trials.

CONSOLIDATING EXTINCTION MEMORY

One day after extinction of a consolidated reward memory with two extinction trials, retention of a long-term extinction memory can be observed.[15] However, the percentage of animals that do not show the CR after extinction is rather low, although significantly different, compared to the percentage of animals of the nonextinguished control group. Accordingly, after extinction with two trials, the acquisition memory is only slightly suppressed by the long-term extinction memory. A recovery from this mild suppression does not occur.

Nevertheless, the extinction memory formed after extinction with two extinction trials is protein synthesis dependent.[15] Furthermore, shaking honeybees for 15 hr after two-trial extinction blocks the formation of a long-term extinction memory.[23] Shaking honeybees during their sleeping phase results in a sleep rebound during the subsequent night, and it has therefore been concluded that shaking results in sleep deprivation. Accordingly, sleep deprivation might be the reason for the disturbed formation of a long-term extinction memory.[23] Both the susceptibility to protein synthesis inhibition and the sensitivity to sleep deprivation indicate that the extinction memory undergoes a stabilization process after the last extinction trial during which it is sensitive to amnestic treatments. Thus, an extinction memory in honeybees is formed that undergoes a consolidation process.

EXTINCTION MEMORY FORMATION DEPENDS ON REWARD LEARNING

Consolidation of the extinction memory depends on the parameters of reward learning. Consolidation is only observed when the duration of the US presentation (i.e., the reward duration) during reward learning exceeds 2 sec.[24] When the reward duration is only 2 sec, the long-term extinction memory is not susceptible to protein synthesis inhibition. How can these results be reconciled? In their model for classical conditioning, Rescorla and Wagner[25] proposed that the mismatch between the US presentation during acquisition and the absence of the US during extinction is the cause for extinction and extinction learning. The extent of this mismatch is smaller when animals are rewarded with a short US than with a long US. It seems that the extent of this mismatch has to exceed a certain threshold to trigger extinction memory consolidation. Accordingly, honeybees only form a stable and long-lasting memory about the reward's failure if a substantial mismatch between the previously memorized reward and its failure is detected.[24] This might be a mechanism to ensure that only meaningful changes in the reward magnitude are memorized after extinction.

RECONSOLIDATION OF REWARD MEMORY

As discussed previously, extinction memories undergo a consolidation process. Interestingly, spontaneous recovery is also protein synthesis dependent.[15] Spontaneous recovery is observed 1, 2, 4, 24, and 48 hr after extinction when a consolidated reward memory is extinguished with five extinction trials but not when two extinction trials are applied. Spontaneous recovery observed 24 hr after extinction is inhibited when a protein synthesis inhibitor is applied 30 min before extinction. Accordingly, the spontaneous recovery of the reward memory depends on protein synthesis.[15] The occurrence of spontaneous recovery is interpreted as the recovery of the initial memory that is transiently suppressed by an extinction memory. Thus, when the animals experience extinction, the reward memory becomes susceptible to protein synthesis inhibition and undergoes a consolidation process. This is surprising because 1 day after acquisition, the reward memory can no longer be disturbed by amnesic treatment (i.e., inhibition of protein synthesis).[15] Thus, by the application of extinction trials, the consolidated reward memory seems to become labile again and undergoes a second round of consolidation. This interpretation resembles reconsolidation. Reconsolidation is the process of memory stabilization after a consolidated memory has been reactivated and a state of plasticity has been induced.[26,27] Accordingly, extinction in honeybees seems to activate the consolidated reward memory, which then undergoes a reconsolidation process that underlies spontaneous recovery. Thus, the effect of extinction in honeybees seems to be twofold: On the one hand, the application of extinction trials triggers a learning event about the failure of the previously experienced reward; on the other hand, it activates the previously formed reward memory, which undergoes a second round of protein synthesis-dependent processes.

PROTEIN DEGRADATION CONSTRAINS THE REWARD MEMORY

The hypothesis that extinction reactivates the initial reward memory is supported by experiments on the role of protein degradation in reward and extinction learning. With these experiments, Felsenberg *et al.*[28] aimed toward elucidating the role of protein degradation in long-term memory formation. The authors blocked protein degradation with MG132, an inhibitor of the ubiquitin–proteasome system that was injected 1 hr after reward learning. MG132 enhanced memory retention 1–3 days after acquisition depending on the number of acquisition trials and the intertrial interval: An enhancement of memory retention was observed when the animals were conditioned with three CS–US trials with an intertrial interval of 2 or 10 min or with two CS–US trials with an intertrial interval of 10 min, but not after training with two CS–US trials with an intertrial interval of 2 min or with one CS–US trial.[28] This enhancement of memory retention by an inhibitor of protein degradation indicates that protein degradation constrains the formation of long-term memories.

Furthermore, Felsenberg *et al.*[28] tested the impact of MG132 on extinction memory formation. They extinguished long-term memory 24 hr after acquisition with two extinction trials and injected MG132 1 hr after extinction. One day later, they tested the extinction memory. Because in reward learning MG132 enhanced the reward memory, the authors proposed an enhancement of extinction memory after extinction learning. However, the opposite was observed. When the application of MG132 was combined with extinction, an enhancement of the reward memory was observed such that the reward memory controlled behavior although extinction occurred. Accordingly, protein degradation constrains the reward memory after extinction, enabling the behavioral expression of the extinction memory.

Next, Felsenberg *et al.*[28] asked whether the enhancement of reward memory after initial learning and extinction learning is linked. They varied the parameters of reward learning—that is, the intertrial interval and the number of CS–US trials—but kept the number of extinction trials (two trials) and the intertrial interval (10 min) constant. It was found that the inhibition of extinction memory retention by MG132 depends in the same way on the parameters of reward learning as the enhancement of reward memory: An enhancement of memory retention was observed when the animals were conditioned with three CS–US trials with an intertrial interval of 10 min or with two CS–US trials with an intertrial interval of 10 min, but not after training with two CS–US trials with an intertrial interval of 2 min or with one CS–US trial. This held true only when the application of MG132 was combined with two extinction trials. When MG132 was injected 24 hr later without the presentation of two extinction trials, an enhancement of memory retention was not observed another 24 hr later in the MG132-treated group. Thus, the learning parameters of reward memory are crucial for the MG132 effect after reward learning and after extinction learning. Accordingly, protein degradation occurring after reward learning seems to be reactivated by extinction learning. This effect resembles reconsolidation because extinction learning seems to reactivate consolidation of the reward memory. Interestingly, these results parallel findings on the role of the reward duration on extinction memory formation[24] demonstrating that the molecular mechanisms triggered by the application of extinction trials depend on the parameters of reward learning. Given that the molecular mechanisms of memory formation are indicative for a memory's stability, the underlying processes might ensure that only meaningful changes of the reward experience are memorized after extinction.

EPIGENETIC MECHANISMS IMPACT ON MEMORY FORMATION AND THE RESISTANCE TO EXTINCTION

The formation of consolidated memories is based on long-lasting neuronal changes. During their formation, consolidated memories undergo a labile phase of transcription and translation, indicating that learning triggers several signaling cascades that converge onto the genome, finally inducing changes in gene expression. Induction of gene expression is twofold. On the one hand, transcription factors such as the cAMP responsive element binding protein (CREB) are modified during learning, activating, or repressing the expression of their target genes. On the other hand, the accessibility of DNA is directly modified by active chromatin remodeling processes. These epigenetic processes, such as histone acetylation, histone methylation, and DNA methylation, result in a long-term change of gene expression.[29]

In honeybees, epigenetic mechanisms play a role in long-term memory formation.[30,31] Reward learning induces an enhancement of histone acetylation. When a learning protocol is used that leads to consolidated memory, a long-lasting enhancement of histone acetylation is observed that is necessary for memory consolidation. In contrast, when a learning protocol is used that does not lead to a consolidation process, the enhancement of histone acetylation is only transient.[30] Furthermore, upregulation of the gene that is encoding the DNA methylating protein, DNA methyltransferase 3, 30 min after reward learning has been demonstrated in honeybees,[32] and the ability of honeybees to differentiate between the learned odor and a novel odor 1 and 3 days after conditioning is affected by the inhibition of DNA methylation.[31]

Experiments by Lockett *et al.*[32] indicate that DNA methylation is also important for the retention of extinction memory. The authors demonstrated that the application of a DNA methyltransferase 3 inhibitor 24 hr before and immediately after conditioning enhances extinction when extinction trials are applied 24 hr later. These data suggest that ongoing methylation before and during acquisition prevents rapid extinction 24 hr later; that is, it increases the resistance to extinction. Thus, gene methylation during conditioning might be the prerequisite for the formation of a stable reward memory counteracting the control of behavior by extinction memories.

In contrast, when the same inhibitor was applied 22 hr after acquisition (i.e., 2 hr before extinction), an inhibition of extinction was observed.[32] This result indicates that methylation 22 hr after acquisition enhances extinction 2 hr later. Thus, ongoing methylation at a later time point after acquisition might decrease the resistance to extinction. Whether this decreased resistance to extinction is based on an enhancement of extinction or an inhibition of the reward memory remains to be determined.

Taken together, data by Biergans *et al.*[33] and Lockett *et al.*[32] demonstrate that methylation plays a crucial role in reward memory formation. Furthermore, data by Lockett *et al.* indicate that DNA methylation participates in the balance between reward memory and extinction memory, thus controlling retention of these two memories. To date, it is unclear how DNA methylation impacts the balance of these two memories. To understand the impact of methylation, it will be especially important to determine which of the honeybee genes that get methylated[33] play a role in learning and memory formation.

Interestingly, epigenetic mechanisms also play a role in extinction in vertebrates.[34] It has been hypothesized that stable modifications of chromatin might be the substrate of a long-lasting extinction memory trace rather than changes of neuronal structures and their connectivity.[34] However, it remains to be shown whether this holds true.

EXTINCTION IN VERTEBRATES AND HONEYBEES: CONSERVED BEHAVIOR, CONSERVED MOLECULAR MECHANISMS, BUT DIFFERENT BRAINS?

Several findings on extinction learning in honeybees resemble findings in vertebrates, namely the observation of two behavioral phenomena—spontaneous recovery and reinstatement—both indicating the existence of two contrasting memories after extinction (honeybee, see references[14,19]; vertebrates, reviewed in reference[3]), context dependence of reinstatement indicating context dependency of extinction memory formation (honeybee, see reference[19]; vertebrates, see references[2,4]), and the reactivation and reconsolidation of the initially formed acquisition memory by the application of extinction trials (honeybee, see references[15,35]; vertebrates, see references[36,37]).

These similarities at the behavioral level suggest that extinction learning and memory formation in honeybees and in vertebrates are based on closely related molecular mechanisms. Indeed, in honeybees, extinction learning and memory formation depend on protein synthesis, protein degradation, and epigenetic mechanisms, resembling findings from studies in vertebrates (honeybee, see references[15,24,28,32]; vertebrates, see references[36,38–40]). What might be the cause of these similarities between extinction in honeybees and vertebrates? Two possibilities can be envisioned. First, extinction and its underlying mechanisms might have evolved independently in insects and vertebrates due to the need to adapt to similar conditions of an ever changing environment. Second, extinction learning might be an ancient behavior that was already part of the behavioral repertoire of a common ancestor of insects and vertebrates. In this case, extinction might be rooted in conserved neuronal networks and might even hint toward a homology of honeybee and vertebrate brains. At first glance, this possibility seems to be rather unlikely because honeybee brains do not appear to be particularly similar to vertebrate brains. Honeybee brains consist of far fewer neurons (950,000 neurons)[41] than vertebrate brains (e.g., mouse, 75 million neurons),[42] and their internal organization is fundamentally different, as holds true for insect brains in general. Nevertheless, a homology in head and brain development between insects and vertebrates has been suggested based on the pattern of gene activation during development.[43,44] Furthermore, Strausfeld *et al.*[45] proposed that the insect mushroom body and the cerebral cortex of vertebrates share some structural similarities. Findings by Tomer *et al.*[46] support this hypothesis by demonstrating striking similarities between gene expression patterns during development of the vertebrate cerebral cortex and the Annelid mushroom body. Therefore, Tomer *et al.* proposed that the vertebrate cerebral cortex and the invertebrate mushroom body evolved from the same structure in the last common ancestor of these organisms. Given that cerebral cortex and mushroom bodies share some developmental programs, it is interesting to note that both structures, the cerebral cortex and the mushroom body, are also implicated in extinction. In vertebrates, the prefrontal cortex plays a pivotal role in regulating the expression of appetitive and aversive memories after extinction (for review, see references[6,47,48]). In honeybees, the repeated presentation of a CS after initial acquisition results in a decrease of Ca^{2+} signals in mushroom body intrinsic feedback neurons of the protocerebral–calycal tract[49] (see Chapter 29). In addition, in the fruit fly *Drosophila melanogaster*, the neuronal mechanisms that underlie extinction are confined to the intrinsic neurons of the mushroom body, the Kenyon cells.[50] Thus, the finding that cerebral cortex and mushroom bodies might derive from the same structure of one common ancestor and that both brain structures are involved in extinction suggests that extinction might be a phylogenetically very old behavior that may have been part of the behavioral repertoire of a common ancestor of insects and vertebrates.

CONCLUSION

Extinction learning and the molecular mechanisms underlying extinction memory formation are surprisingly similar between honeybees and vertebrates. In honeybees and in vertebrates, a context dependency of extinction has been shown. Furthermore, a reactivation and reconsolidation of the initially formed acquisition memory and the existence of two contrasting memories after extinction have been demonstrated. Recent results on extinction in honeybees extend these findings by showing that extinction memory formation (i.e., its underlying molecular mechanisms) is dependent on the learning parameters of reward learning. This might ensure that only meaningful changes of the reinforcing stimulus are memorized.

Extinction resembles a basic learning phenomenon that enables animals to adequately react to a fluctuating environment by providing them with the ability to learn about environmental changes and memorize this information. The similarity of extinction learning and its underlying molecular mechanisms between honeybees and vertebrates suggests that extinction is a conserved, phylogenetically old mechanism. Given such conserved functions, findings on the mechanisms of extinction in honeybees will be most important to further elucidate the basic mechanisms of extinction learning in both invertebrates and vertebrates.

References

1. Pavlov IP. *Conditioned Reflexes: An Investigation of the Physiological Activity of the Cerebral Cortex.* Oxford: Oxford University Press; 1927.
2. Bouton ME, Moody EW. Memory processes in classical conditioning. *Neurosci Biobehav Rev.* 2004;28(7):663–674.
3. Myers KM, Davis M. Behavioral and neural analysis of extinction. *Neuron.* 2002;36(4):567–584.
4. Bouton ME. Context, ambiguity, and unlearning: sources of relapse after behavioral extinction. *Biol Psychiatry.* 2002;52(10):976–986.
5. Milton AL, Everitt BJ. The persistence of maladaptive memory: addiction, drug memories and anti-relapse treatments. *Neurosci Biobehav Rev.* 2012;36(4):1119–1139.
6. Milad MR, Quirk GJ. Fear extinction as a model for translational neuroscience: ten years of progress. *Annu Rev Psychol.* 2012;63:129–151.
7. Menzel R, Leboulle G, Eisenhardt D. Small brains, bright minds. *Cell.* 2006;124(2):237–239.
8. Menzel R. Memory dynamics in the honeybee. *J Comp Physiol A Neuroethol Sens Neural Behav Physiol.* 1999;185(4):323–340.
9. Eisenhardt D. Extinction learning in honeybees. In: Eisenhardt D, Galizia G, Giurfa M, eds. *Honeybee Neurobiology and Behaviour.* Dordrecht: Springer-Verlag; 2012.
10. Menzel RE, Erber J, Masuhr T. Learning and memory in the honeybee. In: Browne LB, ed. *Experimental Analysis of Insect Behaviour.* New York: Springer-Verlag; 1974:195–217.
11. Giurfa M, Sandoz JC. Invertebrate learning and memory: fifty years of olfactory conditioning of the proboscis extension response in honeybees. *Learn Mem.* 2012;19(2):54–66.
12. Menzel R. Learning, memory, and "cognition" in honey bees. In: Kesner RPO, Olton DS, eds. *Neurobiology of Comparative Cognition.* Hillsdale, NJ: Erlbaum; 1990:237–292.
13. Bitterman ME, Menzel R, Fietz A, Schafer S. Classical conditioning of proboscis extension in honeybees (*Apis mellifera*). *J Comp Psychol.* 1983;97(2):107–119.
14. Sandoz JC, Pham-Delegue MH. Spontaneous recovery after extinction of the conditioned proboscis extension response in the honeybee. *Learn Mem.* 2004;11(5):586–597.
15. Stollhoff N, Menzel R, Eisenhardt D. Spontaneous recovery from extinction depends on the reconsolidation of the acquisition memory in an appetitive learning paradigm in the honeybee (*Apis mellifera*). *J Neurosci.* 2005;25(18):4485–4492.
16. Muller U. Learning in honeybees: from molecules to behaviour. *Zoology (Jena).* 2002;105(4):313–320.
17. Hourcade B, Muenz TS, Sandoz JC, Rossler W, Devaud JM. Long-term memory leads to synaptic reorganization in the mushroom bodies: a memory trace in the insect brain? *J Neurosci.* 2010;30(18):6461–6465.
18. Friedrich A, Thomas U, Muller U. Learning at different satiation levels reveals parallel functions for the cAMP-protein kinase A cascade in formation of long-term memory. *J Neurosci.* 2004;24(18):4460–4468.
19. Plath JA, Felsenberg J, Eisenhardt D. Reinstatement in honeybees is context-dependent. *Learn Mem.* 2012;19:543–549.
20. Bouton ME. Context and behavioral processes in extinction. *Learn Mem.* 2004;11(5):485–494.
21. Wustenberg D, Gerber B, Menzel R. Short communication: long- but not medium-term retention of olfactory memories in honeybees is impaired by actinomycin D and anisomycin. *Eur J Neurosci.* 1998;10(8):2742–2745.
22. Dudai Y. The neurobiology of consolidations, or, how stable is the engram? *Annu Rev Psychol.* 2004;55:51–86.
23. Hussaini SA, Bogusch L, Landgraf T, Menzel R. Sleep deprivation affects extinction but not acquisition memory in honeybees. *Learn Mem.* 2009;16(11):698–705.
24. Stollhoff N, Eisenhardt D. Consolidation of an extinction memory depends on the unconditioned stimulus magnitude previously experienced during training. *J Neurosci.* 2009;29(30):9644–9650.
25. Rescorla RA, Wagner AW. A theory of Pavlovian conditioning: variations in the effectiveness of reinforcement and nonreinforcement. In: Prokasy AH, Black WF, eds. *Classical Conditioning II. Current Theory and Research.* New York: Appleton-Century-Crofts; 1972:64–99.
26. Nader K. Memory traces unbound. *Trends Neurosci.* 2003;26(2):65–72.
27. Sara SJ. Retrieval and reconsolidation: toward a neurobiology of remembering. *Learn Mem.* 2000;7(2):73–84.
28. Felsenberg J, Dombrowski V, Eisenhardt D. A role of protein degradation in memory consolidation after initial learning and extinction learning in the honeybee (*Apis mellifera*). *Learn Mem.* 2012;19(10):470–477.
29. Puckett RE, Lubin FD. Epigenetic mechanisms in experience-driven memory formation and behavior. *Epigenomics.* 2011;3(5):649–664.
30. Merschbaecher K, Haettig J, Mueller U. Acetylation-mediated suppression of transcription-independent memory: bidirectional modulation of memory by acetylation. *PLoS ONE.* 2012;7(9):e45131.
31. Biergans SD, Jones JC, Treiber N, Galizia CG, Szyszka P. DNA methylation mediates the discriminatory power of associative long-term memory in honeybees. *PLoS ONE.* 2012;7(6):e39349.
32. Lockett GA, Helliwell P, Maleszka R. Involvement of DNA methylation in memory processing in the honey bee. *Neuroreport.* 2010;21(12):812–816.
33. Foret S, Kucharski R, Pellegrini M, et al. DNA methylation dynamics, metabolic fluxes, gene splicing, and alternative phenotypes in honey bees. *Proc Natl Acad Sci USA.* 2012;109(13):4968–4973.
34. Stafford JM, Lattal KM. Is an epigenetic switch the key to persistent extinction? *Neurobiol Learn Mem.* 2011;96(1):35–40.
35. Stollhoff N, Menzel R, Eisenhardt D. One retrieval trial induces reconsolidation in an appetitive learning paradigm in honeybees (*Apis mellifera*). *Neurobiol Learn Mem.* 2008;89(4):419–425.
36. Eisenberg M, Kobilo T, Berman DE, Dudai Y. Stability of retrieved memory: inverse correlation with trace dominance. *Science.* 2003;301(5636):1102–1104.
37. Duvarci S, Mamou CB, Nader K. Extinction is not a sufficient condition to prevent fear memories from undergoing reconsolidation in the basolateral amygdala. *Eur J Neurosci.* 2006;24(1):249–260.
38. Lattal KM, Radulovic J, Lukowiak K. Extinction: [corrected] does it or doesn't it? The requirement of altered gene activity and new protein synthesis. *Biol Psychiatry.* 2006;60(4):344–351.
39. Lee SH, Choi JH, Lee N, et al. Synaptic protein degradation underlies destabilization of retrieved fear memory. *Science.* 2008;319(5867):1253–1256.
40. Maddox SA, Schafe GE. Epigenetic alterations in the lateral amygdala are required for reconsolidation of a Pavlovian fear memory. *Learn Mem.* 2011;18(9):579–593.
41. Witthöft W. Absolute Anzahl und Verteilung der Zellen im Hirn der Honigbiene. *Z Morophol Tiere.* 1967;61.

42. Williams RW. Mapping genes that modulate mouse brain development: a quantitative genetic approach. *Results Probl Cell Differ.* 2000;30:21–49.
43. Posnien N, Koniszewski ND, Hein HJ, Bucher G. Candidate gene screen in the red flour beetle *Tribolium* reveals *six3* as ancient regulator of anterior median head and central complex development. *PLoS Genet.* 2011;7(12):e1002416.
44. Demilly A, Simionato E, Ohayon D, Kerner P, Garces A, Vervoort M. Coe genes are expressed in differentiating neurons in the central nervous system of protostomes. *PLoS ONE.* 2011;6 (6):e21213.
45. Strausfeld NJ, Hansen L, Li Y, Gomez RS, Ito K. Evolution, discovery, and interpretations of arthropod mushroom bodies. *Learn Mem.* 1998;5(1–2):11–37.
46. Tomer R, Denes AS, Tessmar-Raible K, Arendt D. Profiling by image registration reveals common origin of annelid mushroom bodies and vertebrate pallium. *Cell.* 2010;142(5):800–809.
47. Peters J, Kalivas PW, Quirk GJ. Extinction circuits for fear and addiction overlap in prefrontal cortex. *Learn Mem.* 2009;16(5):279–288.
48. Pape HC, Pare D. Plastic synaptic networks of the amygdala for the acquisition, expression, and extinction of conditioned fear. *Physiol Rev.* 2010;90(2):419–463.
49. Haehnel M, Menzel R. Sensory representation and learning-related plasticity in mushroom body extrinsic feedback neurons of the protocerebral tract. *Front Syst Neurosci.* 2010;4:161.
50. Schwaerzel M, Heisenberg M, Zars T. Extinction antagonizes olfactory memory at the subcellular level. *Neuron.* 2002;35 (5):951–960.

CHAPTER

34

Glutamate Neurotransmission and Appetitive Olfactory Conditioning in the Honeybee

Gérard Leboulle

Freie Universität Berlin, Berlin, Germany

INTRODUCTION

The honeybee is a valuable model for the study of learning and memory. It can be investigated under laboratory conditions using the proboscis extension reflex (PER) conditioning, a classical conditioning procedure.[1] In one version of PER conditioning, the animal learns to associate a neutral odor, the conditioned stimulus (CS), with a sucrose reward, the unconditioned stimulus (US). Acetylcholine is recognized as the main excitatory neurotransmitter in the central nervous system (CNS) of insects.[2] The olfactory and the reward pathways are well described in the honeybee.[3] Olfactory sensory neurons projecting to the glomeruli of the antennal lobe (AL), the first relay station of the olfactory system, are cholinergic. Projection neurons connecting the AL with the lateral horn and the calyces of the mushroom body (MB) are also cholinergic. The calyces are formed by the dendrites of the Kenyon cells (KCs), whose neurotransmitters are not firmly identified. The calyces are divided in different regions receiving inputs from different sensory modalities—the lip, the collar, and the basal ring. Olfactory projection neurons make synaptic connections in the lip and the basal ring. The US pathway in appetitive conditioning is represented by an octopaminergic neuron of the subesophageal ganglion, VUMmx1, which projects bilaterally to the AL, the lateral horn, and the calyces. It is expected that the simultaneous activation of neurons of the CS and of the US, during conditioning, induces a modification of the synaptic connections between these pathways. Indeed, it was shown that the AL and the MB, but not the lateral horn, are important sites for memory, the cholinergic and the octopaminergic neurotransmissions are implicated in this process, and VUMmx1 activity is increased in response to CS stimulation after conditioning.[3,4]

The presentation of sucrose induces the release of octopamine and the activation of cAMP-dependent protein kinase A (PKA) in the AL but not in the MB.[5,6] If the CS and the US are presented in association, a prolonged PKA activation is measured in the AL of the honeybee[7] (see Chapter 31). Studies in *Drosophila* and *Aplysia* suggest that octopamine receptors are coupled to adenylyl cyclase, which produces cAMP, the principal PKA activator. The simultaneous binding of octopamine to the receptor and the calcium influx induced by cholinergic CS neurons probably enhances adenylyl cyclase activity and cAMP production at presynaptic sites of CS neurons. Repeated CS–US presentations induce a longer PKA activation in the AL that is crucial for the consolidation of long-term memory (LTM).[7] These molecular events are conserved in invertebrates, and they play a central role in synaptic plasticity related to learning and memory.

Glutamate is the main excitatory neurotransmitter in mammals,[8] and it is widely accepted that postsynaptic mechanisms dominate synaptic plasticity. The related phenomena are best described by long-term potentiation (LTP), a cellular model of memory. Glutamate mediates its effects by acting on different kind of receptors: *N*-methyl-D-aspartate (NMDA), non-NMDA (amino-3-hydroxy-5-methylisoxazole-4-propionic acid (AMPA) and kainate), and metabotropic receptors. Glutamate induces exclusively excitatory currents that depend principally on the activation of non-NMDA receptors (AMPA and kainate), which are mainly permeable to Na^+ and K^+ ions. The NMDA receptor, which is principally permeable to Ca^{2+}, contributes only to a small extent to the glutamate-induced current. The NMDA

Invertebrate Learning and Memory.
DOI: http://dx.doi.org/10.1016/B978-0-12-415823-8.00034-4

receptor plays a central role in the induction of LTP due to its molecular properties. Its activation requires two simultaneous stimuli: a strong depolarization of the postsynaptic membrane and the release of glutamate from the presynaptic neuron. This convergent activation of the receptor allows calcium entry into the postsynaptic neuron and the activation of calcium-dependent signals that ultimately facilitate synaptic transmission by modifying the conformation and the composition of non-NMDA receptors expressed at the membrane.[9]

There is ample evidence that glutamate is also a neurotransmitter in the CNS of the honeybee. Glutamate-induced currents have been characterized, and receptors homologous to their mammalian counterparts as well as particular class of glutamate receptors mediating inhibitory currents have been identified. Several studies indicate that they are implicated in memory.

GLUTAMATE AND COMPONENTS OF THE GLUTAMATE NEUROTRANSMISSION IN THE HONEYBEE NERVOUS SYSTEM

Glutamate is the neurotransmitter at the neuromuscular junction. Glutamate-like immunoreactivity (Glu-ir) is found in motor neurons,[10] and glutamate-induced currents are characterized in honeybee muscles.[11] There is also evidence that glutamate is a neurotransmitter in the CNS: Glu-ir, components of the biosynthesis and the recycling of glutamate, and specific receptors have been partially characterized at different levels in interneurons of the brain.

Glutamate is transported into synaptic vesicles by vesicular glutamate transporters called VGLUT. A *vglut* gene was identified in the honeybee genome,[12] and the corresponding mRNA was detected in brain and MB calyx extracts (Gérard Leboulle, unpublished observation). A putative excitatory amino acid transporter, AmEAAT, has been identified.[13] This membrane protein is involved in the recycling of the neurotransmitter by removing it from the synaptic cleft. AmEAAT shares the highest level of identity to the human EAAT2 glutamate transporter. It is predominantly localized in the brain in comparison with other body parts, and expression levels are higher in late pupae than in adults. Several transcripts are probably produced from this gene, and sequence analysis indicates that the 10 transmembrane domains, characteristic for these proteins, are conserved. Glia cells are implicated in the recycling of glutamate. In these cells, glutamate is transformed into glutamine by the enzyme glutamine synthase.[14] Glutamine is transported back into neurons, where it is recycled in glutamate that is packed in synaptic vesicles. Glutamine synthase was also detected in brain and MB calyx extracts (Gérard Leboulle, unpublished observation).

Several receptors homologous to their vertebrate counterparts have been identified. Several genes potentially encoding non-NMDA glutamate receptors are found in the genome.[12,15] Cultured KCs express excitatory currents induced by a non-NMDA glutamate receptor.[3] These receptors are sensitive to agonists and an antagonist of the AMPA receptor but not to an agonist and an antagonist of the NMDA receptor. Note that the pharmacology of insects and mammalian receptors is often different and does not allow a uniform classification between the two classes. Three genes (*nmdar1*, *nmdar2*, and *nmdar3*) encode NMDA receptor subunits in the honeybee.[12] The AmNR1 subunit is encoded by *nmdar1*, and several variants of the mRNA have been identified.[16,17] One of them, AmNR1-1, encodes the complete subunit, whereas the others encode truncated isoforms. Sequence analysis reveals that key regions of the NR1 subunit are conserved in AmNR1-1.[17] The highest homology level is found in the membrane domains. In particular, an asparagine that influences the divalent cation affinity of the receptor is conserved as well as PKA, PKC, and PKG phosphorylation sites.

In contrast to ionotropic glutamate receptors that mediate fast synaptic neurotransmission, metabotropic glutamate receptors are implicated in modulatory synaptic actions. In the honeybee, two metabotropic glutamate receptors—AmGluRA and AmGluRB—have been described.[18,19] AmGluRA mRNA is detected in the brain, the abdomen, and the thorax, whereas AmGluRB mRNA is found exclusively in the brain. Phylogenetic analysis suggests that AmGluRA belongs to type II, whereas AmGluRB is an orphan receptor. There are three types of identified glutamate metabotropic receptors in mammals. Type II receptors are negatively coupled to adenylyl cyclase and diminish cAMP levels. The expression of both receptors in insect cells shows that AmGluRA has a higher affinity for glutamate than AmGluRB,[18] and it has been proposed that the natural ligand of AmGluRB is not glutamate.[19] The expression of *amglura* is developmentally regulated; the mRNA levels are more important in adults than in pupae. Sequence analysis shows that the characteristic domains of metabotropic glutamate receptors are present. Functional AmGluRA receptors expressed in HEK (human embryonic kidney) cells can be modulated by agonists and an antagonist of type II glutamate metabotropic receptors but not by type I or type III agonists.[19] Thus, AmGluRA is probably a type II glutamate metabotropic glutamate receptor.

In invertebrates, glutamate-gated chloride (GluCl) channels constitute a particular class of glutamate receptors that mediate inhibitory currents.[20] Thus, in addition to GABA, glutamate can be an inhibitory neurotransmitter in invertebrates. Glutamate-induced chloride currents are expressed on the vast majority of cultured AL neurons of pupae and adult honeybees.[21,22] In addition to these currents, most of the recorded cells also express currents induced by GABA and acetylcholine and might thus be connected to cholinergic neurons and to local GABAergic neurons.[21] GluCl channels belong to the *cys*-loop ligand-gated ion channel superfamily, and the annotation of the honeybee genome reveals only one gene, *am_glucl*, encoding two alternatively spliced variants.[23] Phylogenetic analysis suggests that it belongs to the α subtype.[24] Nothing is known about the conformation of the functional receptor in the honeybee. Its physiological properties suggest that different subtypes are co-expressed, but the GluCl subunit can probably not co-assemble with GABA receptor subunits, as observed in *Drosophila*.[21]

ARCHITECTURE OF THE GLUTAMATERGIC NEUROTRANSMISSION

The localization of Glu-ir,[10] AmNR1,[17] AmGluRA,[18] AmGluClα,[24] and Am-EAAT[13] suggests that the glutamatergic neurotransmission is widespread in the brain.

Glu-ir and AmNR1 subunit signals are most intensive in the protocerebral lobe, the central complex, the subesophageal ganglion, and the optic lobe (Figures 34.1 and 34.2). In the optic lobe, the most prominent Glu-ir signals are detected in the monopolar cells in the lamina and the medulla, whereas the retinula cells are devoid of staining. Signals are also detected in the lobula and the optic tubercle. The AmNR1 subunit shows a similar expression pattern in specific layers of the optic lobe. The mRNAs of AmNR1, AmGluRA, AmGluClα, and Am-EAAT are found in the somata regions of these neuropiles. Am-EAAT mRNA levels are higher in the optic lobe, and the detection of AmGluClα in different brain extracts by Western blot shows also higher levels in this neuropile.

In the AL, Glu-ir and the AmNR1 subunit are localized in the glomeruli, where the signals are in general more intense at the periphery that is enriched with the terminals of olfactory sensory neurons (Figures 34.1 and 34.2). Only a few somata of AL neurons are positive for Glu-ir. It is not known if these neurons provide input to AL neurons or if they project to other brain regions. Projection neurons and local interneurons probably receive many glutamatergic inputs because most AL neurons express glutamate-induced inhibitory currents[21] and the mRNAs of AmGluClα, AmNR1, AmGluRA, and Am-EAAT are detected in most AL somata.

The calyces, the input site of the MB, show very low levels of Glu-ir and AmNR1 subunit compared with the surrounding neuropiles. The signals are restricted to the lip and the basal ring regions. Glu-ir probably originates from unidentified projection neurons and AmNR1 from the dendrites of KCs because the NMDA receptor is mainly expressed postsynaptically.[25] Only weak Glu-ir signals are detected in the somata of the innermost type I KCs, located within the calycal cup and innervating the basal ring, and in type II KCs located outside the calycal cup (Figure 34.1). Interestingly, the Am-EAAT mRNA levels are predominant in the same regions. Therefore, glutamate might be the neurotransmitter of only a small subset of KCs. Studies in *Drosophila* support this assumption.[26,27]

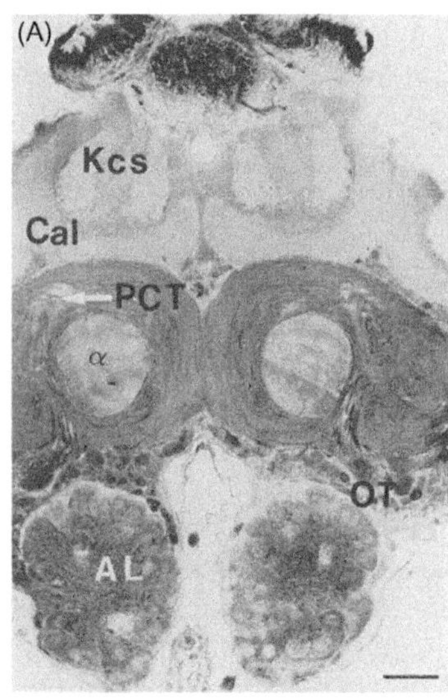

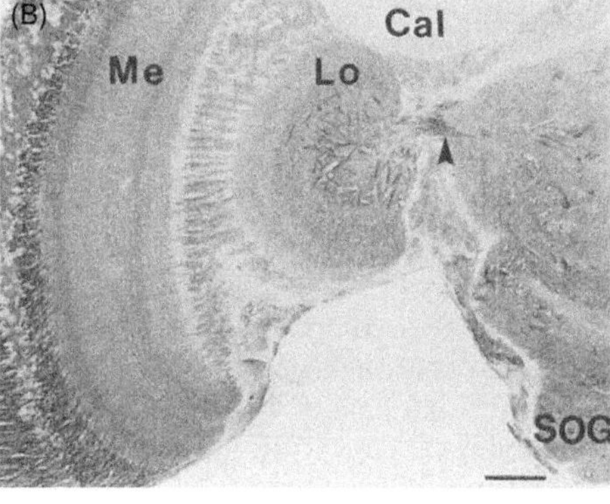

FIGURE 34.1 Survey of Glu-ir in neuropiles of the central brain and optic lobe of the honeybee. (A) The optic tubercles (OT) contain Glu-ir fibers. The α lobe (also called vertical lobe, α) contains rather low levels of staining. The calycal neuropil (Cal) shows only faint labeling except for the lip region, which is similar in staining intensity to that of the innermost type I Kenyon cells (Kcs). No labeling is found in the protocerebrocalycal tract (PCT). (B) Glu-ir in the medulla (Me), lobula (Lo), protocerebral lobe, and subesophageal ganglion (SOG). Glu-ir fibers (arrowhead) extend from the lobula into the posterior protocerebrum. Scale bars = 100 μm. Source: *Reproduced with permission from Bicker et al.,*[10] *Figure 6.*

Highly variable AmNR1mRNA levels are found in KCs. In some sections of the calyces, the mRNA is restricted to the somata of KCs innervating the lip and the basal ring; as in other sections, it is detected in almost all somata. The mRNAs of AmGluRA and AmGluClα are localized in almost all somata. The axons of KCs project in the pedunculus and terminate in the lobes, the output sites of the MB where they make up layers. Glu-ir and AmNR1 are detected in several layers of the α lobe, probably representing the lip, collar, and basal ring, and in its ventral part (also called the vertical γ lobe) (Figures 34.1 and 34.2). However, only weak Glu-ir signals are detected in the somata of type I KCs of the basal ring and of type II KCs. Thus, the structure of the neurotransmission has to be clarified in the MB. As in the AL, there are a few glutamatergic KCs and most of the cells receive glutamatergic inputs.

GLUTAMATERGIC NEUROTRANSMISSION IS IMPORTANT FOR LEARNING AND MEMORY

The role of glutamatergic neurotransmission in learning and memory was evaluated with PER conditioning.[1] This procedure is extremely robust; one CS–US pairing leads to a conditioned response lasting several hours in more than 50% of animals. In more than 80% of animals, multiple conditioning trials

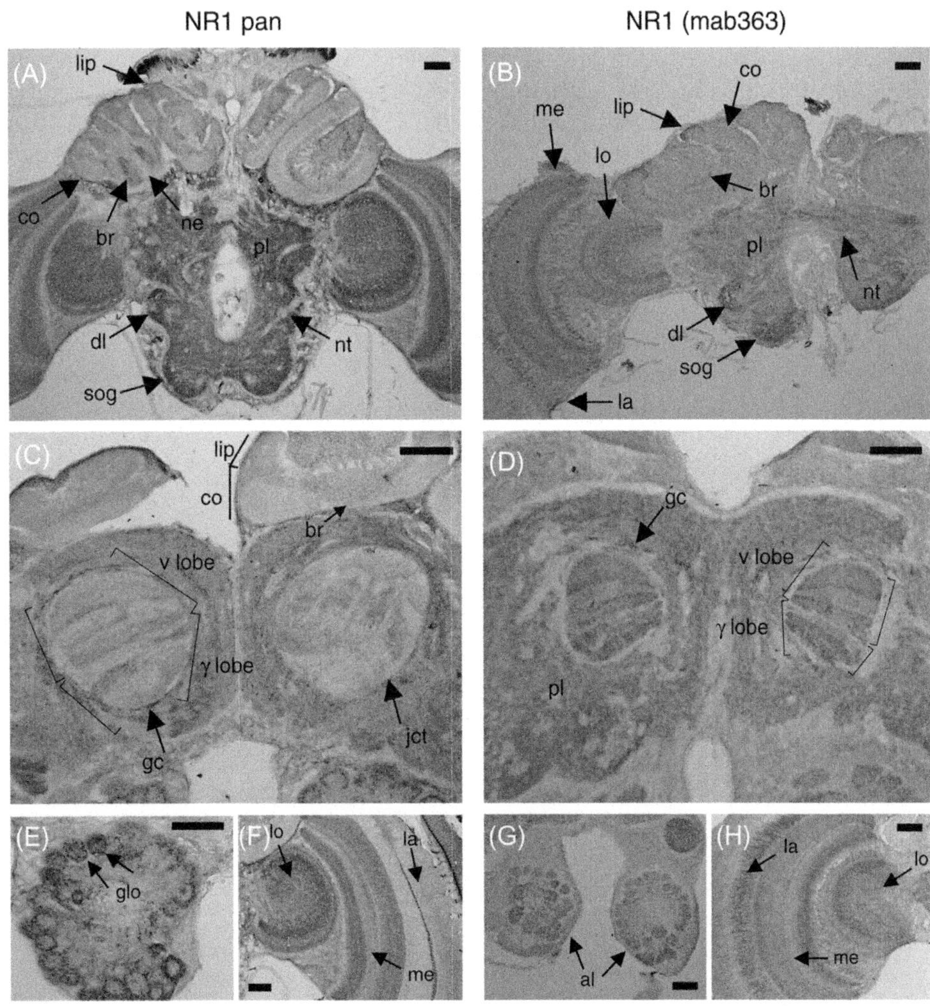

FIGURE 34.2 **Detection of AmNR1 in the honeybee brain.** The NR1 subunit was detected with two antibodies directed against the NR1 subunit (NR1 pan (A, C, E, F) and mab363 (B, D, G, H)). Signals were found in the protocerebral lobe (pl), antennal mechanosensory and motor center (dl), subesophageal ganglion (sog), optic lobe, lobula (lo), medulla (me), and lamina (la) (A, B, F, H). In the mushroom bodies, signals were detected: in the lip (lip), basal ring (br), and neck (ne), and only limited signals were found in the collar (co) (A–C). The output region of the mushroom bodies revealed specific patterns in the α (v lobe and γ lobe), at the pedunculus divide (jct) between the vertical and the medial lobe, and in glial cells (gc) (C, D). In the antennal lobe (al), signals were found in the glomeruli (glo) (E, G). Scale bars = 100 μm. Source: *Reproduced with permission from Zannat* et al.,[17] *Figure 3.*

induce the formation of a midterm memory (MTM) immediately after learning and a consolidated LTM lasting days. LTM can be subdivided into early LTM (eLTM; 1 and 2 days after conditioning) and late LTM (lLTM; from day 3 after conditioning).[28–31] eLTM depends on translation and lLTM on transcription and translation. These particular physiological processes are not unique to olfactory conditioning in the honeybee. In mammals, LTP, a cellular correlate of learning and memory, can be decomposed in similar phases. Early LTP (eLTP or LTP 1) depends on covalent modification at synapses and is of short duration.[32] Intermediate LTP (iLTP or LTP 2) is of intermediate duration and depends on translation,[33] and late LTP (lLTP or LTP 3) depends on translation and transcription and is of longer duration or even permanent.[34] In *Aplysia*, different phases of facilitation, a cellular model of sensitization, have similar properties. Intermediate-term facilitation depends on protein translation for its induction,[35] and long-term facilitation is translation and transcription dependent.[36,37] Corresponding memory phases of sensitization have also been described.[38] Olfactory reward learning in the honeybee appears to be particularly well suited to study these different phases at the behavioral level. Several studies indicate that the glutamatergic neurotransmission plays a role in different aspects of olfactory memory in the honeybee. The results of the most relevant studies are summarized in Figure 34.3.

The release of a caged-glutamate complex on the MB calyces of honeybees just after a single CS–US pairing consolidates a stable eLTM and has no effect on memory evaluated 2 hr after conditioning.[39] The release of glutamate just before CS–US pairing does not affect memory performance. In the same manner, L-*trans*-2,4-pyrrolidine dicarboxylate, an inhibitor of glutamate reuptake acting on EAAT in vertebrates, induces a strong impairment of eLTM without affecting the acquisition and MTM when injected shortly before a multiple training procedure.[40]

The agonists and the antagonist of type II metabotropic receptors were administered before multiple-trials conditioning to evaluate the role of AmGluRA in learning and memory.[19] Surprisingly, the agonists as well as the antagonist impair eLTM but have no effect during the acquisition and during the MTM test. Applying AmGluRA modulators 1 hr after conditioning does not affect memory performance or memory

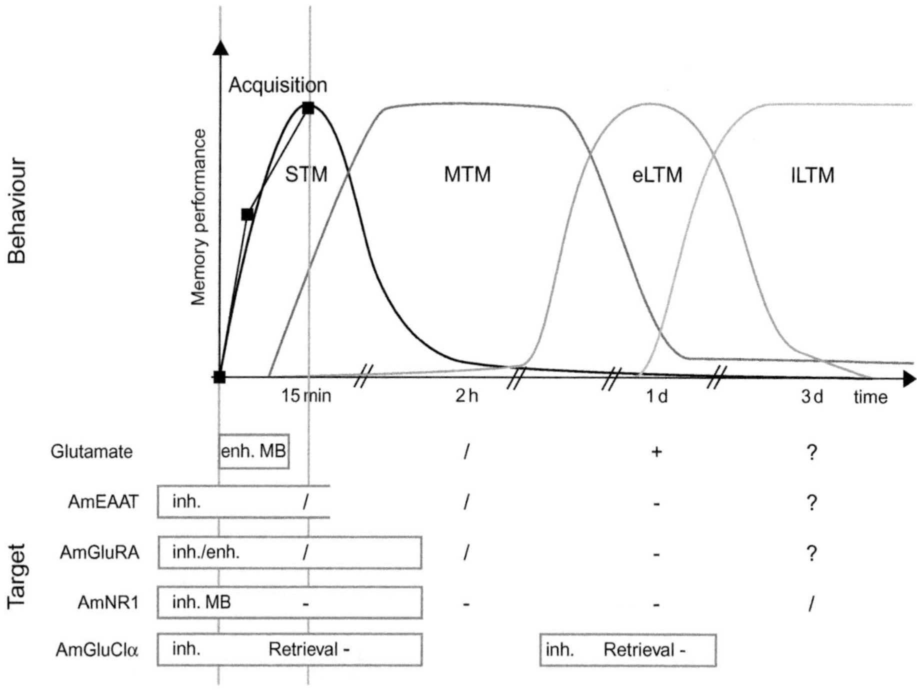

FIGURE 34.3 Summary of the most relevant studies on the role of the glutamate neurotransmission in appetitive olfactory conditioning. The top panel (behavior) shows the temporal memory phases: A short-term memory (STM) phase develops during the acquisition phase. It is followed by midterm memory (MTM), early long-term memory (eLTM), and late long-term memory (lLTM). Several components of the glutamatergic neurotransmission were targeted: glutamate,[39] AmEAAT,[40] AmGluRA,[19] AmNR1,[41] and AmGluClα.[24] The kind of manipulation (inhibition (inh.) or enhancement (enh.)) and the duration are indicated by the boxes. In the study on glutamate, honeybees received one training trial. In the case of AmEAAT, the box is open because the duration of the effect is not known. Retrieval was also evaluated in the study on AmGluRA and AmNR1 and is not affected. The effect of the manipulation on the different memory phases is indicated: improvement (+), impairment (−), no effect (/), not tested (?), or impairment of retrieval (Retrieval −).

retrieval.[19] These results indicate that glutamate metabotropic receptors are implicated in the consolidation of the translation-dependent eLTM. Studies on LTP in mammals have reported that metabotropic glutamate receptors, probably of type I, are implicated in local protein synthesis and the priming of LTP,[42] suggesting a similarity between the honeybee and mammals. This should be considered with caution because AmGluRA has some properties of type II receptors. However, the second messenger system associated with this receptor has not been firmly identified. It might be that AmGluRA is characterized by type II receptor pharmacology but that it has a type I receptor physiology associated with the inositol-3-phosphate and diacylglycerol systems.

Pharmacological studies using NMDA receptor antagonists suggested a role in memory processes.[16,43] We used RNA interference (RNAi) against the AmNR1 subunit, which is probably the obligatory subunit of the honeybee NMDA receptor.[41] The treatment led to a 30% reduction of protein levels in the MB region. Interestingly, 2 hr after conditioning, the protein levels returned to normal. This acute reduction of the NR1 subunit affected the acquisition phase, MTM, and eLTM formation but left lLTM intact (Figure 34.4). The fact that lLTM was not affected by the treatment supports the previous finding that lLTM develops independently from earlier memory phases.[28] The inhibition of AmNR1 during retrieval does not affect the performance of the animal.

It is surprising that in our study, lLTM was not affected. Studies in other model systems show that NMDA receptors are required for the induction of memory,[44–46] and our results suggest a role in memory formation. On the one hand, NMDA receptors might be important for appetitive learning. Their activation during conditioning in the MB region might be required for the CS–US association leading to MTM and eLTM, and activation in another region (e.g., the central complex) might be required for CS–US association underlying lLTM. This hypothesis is supported by a study in *Drosophila* showing that NMDARs of the MB are important for MTM and those of the ellipsoid body for LTM.[47] NMDARs of the AL are probably not important for lLTM because the release of glutamate[39] or the inhibition of the AmNR1 subunit (Gérard Leboulle, unpublished observation) in the AL at the moment of conditioning do not modulate memory. Thus, different signaling pathways in the AL and the MB contribute to the formation of the olfactory memory trace in appetitive learning.[48,49] Glutamatergic neurotransmission plays an important role in the MB calyces, whereas the PKA cascade is implicated in memory formation in the AL. On the other hand, NMDA receptors might have a role in memory formation. In our work, inhibition of the NR1 subunit did not last longer than 2 hr. It is thus conceivable that a longer activation or a reactivation of the receptor is required for the consolidation of lLTM. Reactivation of NMDARs is required for the consolidation of lLTP and of late memory phases in addition to their role during acquisition.[50,51] Finally, it might be that stronger inhibition of the NR1 subunit is required to inhibit lLTM. NMDAR might play a predominant role in early memories. In contrast, lLTM formation could depend on several signaling cascades converging in the nucleus to induce translation and transcription. It is conceivable that the NMDAR signaling cascade is not the predominant one and that only a massive inhibition of NMDAR activation would impair lLTM formation. In a study of mammals, eLTP, iLTP, and lLTP were induced by increasing theta burst stimulations of hippocampal neurons.[52] Interestingly, these authors showed that eLTP and iLTP depend completely on

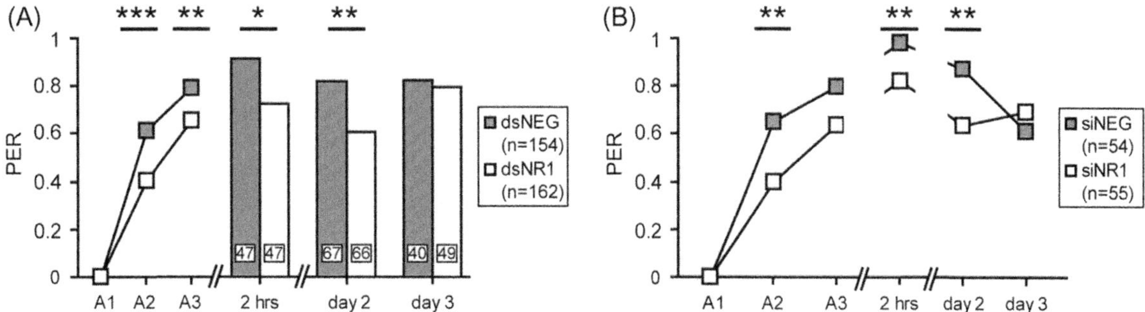

FIGURE 34.4 The inhibition of the NR1 subunit selectively affects memory formation. One day after the injection of dsRNA (A; dsNEG, gray; dsNR1, white) or siRNA (B; siNEG, gray; siNR1, white), animals were subjected to three CS–US pairings (A1–A3). Memory was retrieved 2 hr and on day 2 and day 3 after conditioning. In the dsRNA experiment, the animals were tested only once. Data from the acquisition phase were pooled for all subgroups (n). The numbers on the bars represent the number of animals tested for each time point. In the siRNA experiment, n animals received multiple post-training tests. Asterisks indicate significant differences between groups: $^{*}p < 0.05$, $^{**}p < 0.01$, and $^{***}p < 0.001$; χ^2 test. PER represents the percentage of animals that showed a PER during the CS presentation. Source: *Reproduced with permission from Müßig* et al.,[41] *Figure 4.*

NMDAR, whereas lLTP depends only partially on the receptor's activation. L-type voltage-gated calcium channels appear to be more important for the expression of lLTP. These receptors induce calcium signals mainly in the cell bodies and proximal dendrites, whereas NMDA receptors are more expressed in distal dendrites.[52,53] These different types of LTP coexist in the same neuron, but their induction depends on the spatial distribution of different calcium sources.

Several pharmacological studies targeting GluCl channels suggested that they control other aspects of memory processes.[22,54,55] To characterize the functional role of GluCl channels in the honeybee brain, we studied their implication in olfactory learning and memory by means of RNAi against AmGluClα.[24] AmGluClα is necessary for the retrieval of olfactory memories; specifically, injection of dsRNA or siRNA results in a decrease of retention performance approximately 24 hr after injection. Knock-down of AmGluClα subunit impairs neither olfaction nor sucrose sensitivity, and the capacity to associate odor and sucrose is not affected. These data provide the first evidence for the involvement of glutamate-gated chloride channels in olfactory memory in an invertebrate. Cholinergic transmission from the MB lobe is required during memory retrieval in the honeybee and in *Drosophila*.[56–58] Therefore, it was proposed that the α lobe of the honeybee MB could be a potential site of cholinergic and glutamatergic neuron interactions.[24] This proposal emphasizes a difference with mammals, in which retrieval relies on non-NMDA receptors.[59] Interestingly, a form of LTP, involving interaction between astrocytes and cholinergic and glutamatergic neurons, was recently described in mammals.[60] Cholinergic activity increases calcium in hippocampal astrocytes, which stimulate astrocyte glutamate release that activates metabotropic glutamate receptors and induces LTP of CA3–CA1 synapses.

CONCLUSION

Components of the biosynthesis, recycling, and synaptic release of glutamate in the honeybee brain are rather well described. Receptors homologous to their vertebrate counterparts and unique to invertebrates characterize the glutamatergic neurotransmission of the honeybee brain. However, we have an incomplete understanding of the neurotransmission's architecture. Localization and physiological studies suggest that it is widespread within the brain. In the MB, it does not prevail in the calyces, whereas it is important in several layers of the α lobe. The AL and MB neuropiles probably comprise a few glutamatergic neurons, but most of their intrinsic neurons express glutamate receptors. The inhibitory glutamatergic neurotransmission is better characterized than the excitatory one, and it is not known whether one of them is dominant anywhere in the brain.

Several studies have shown that glutamate neurotransmission within the MB plays a role in appetitive olfactory memory. Although there is good evidence of glutamatergic neurotransmission in the AL, its modulation during or just after conditioning does not affect memory. It is worth noting that in some pharmacological studies, the specificity of the drugs used and the interpretation of their effects are questionable. In this regard, molecular tools such as RNAi allow a more precise manipulation of the brain physiology, although they also have drawbacks (e.g., limited amplitude of the inhibition) and need to be improved.

It is well-established that the synapses between cholinergic neurons of the olfactory pathway and octopaminergic neurons of the reward pathway constitute the fundamental basis of appetitive olfactory conditioning in the honeybee. It involves the cAMP second messenger cascade probably by the activation of adenylyl cyclase at presynaptic sites of olfactory neurons. These mechanisms are conserved and also described in *Aplysia* and *Drosophila*. In mammals, NMDAR-dependent LTP is one of the main forms of associative plasticity underlying learning and memory. It occurs principally at postsynaptic sites and involves NMDA, non-NMDA, and metabotropic glutamate receptors. These conditions are considered to be a major difference between vertebrates and invertebrates. However, several studies in the honeybee show similar functions for the glutamatergic neurotransmission in appetitive olfactory conditioning and LTP. The neurotransmission is required during or soon after conditioning for the formation of specific memory phases. Metabotropic glutamate receptors are required for eLTM consolidation, whereas NMDARs are important for MTM and eLTM formation but not for lLTM. It is not known if these mechanisms occur at postsynaptic sites in the honeybee brain comparable to the postsynaptic mechanisms involving calcium signaling and NMDA receptors described in *Aplysia*. Thus, synaptic plasticity might occur at pre- and postsynaptic sites in invertebrates. Interestingly, β-adrenergic receptors expressed by hippocampus neurons facilitate the induction of translation-dependent LTP, possibly through the PKA system.[61] The conserved role of the neurotransmission between insects and mammals is remarkable because glutamate is the main excitatory neurotransmitter in mammals and at the sensory-to-motor synapse in *Aplysia*, whereas acetylcholine fulfills the latter role in insects. Taken together, these findings suggest that glutamatergic networks dedicated to synaptic plasticity are evolutionary conserved independently of the principal excitatory neurotransmitter of the CNS.

A particularity of invertebrates is the inhibitory function of glutamate transmission through the activation of GluCl channels. It has been shown that these channels are necessary for memory retrieval, probably in MB neurons. Interestingly, cholinergic neurons are also important for memory retrieval in the MB lobe. The two neurotransmissions might be interconnected at these sites, as observed in the AL, or they might run in parallel. These mechanisms differ in mammals, in which retrieval depends on non-NMDAR. It is not known whether non-NMDAR receptors play a role in learning and memory in insects.

LTP was successfully induced in the PE1 neuron, an extrinsic output neuron of the MB.[62] Other studies showed that PE1 reduces its response to the CS after conditioning.[63,64] It cannot be excluded that glutamatergic neurotransmission is implicated in these plasticity processes.

References

1. Bitterman ME, Menzel R, Fietz A, Schafer S. Classical conditioning of proboscis extension in honeybees (*Apis mellifera*). *J Comp Psychol.* 1983;97(2):107–119.
2. Breer H, Sattelle DB. Molecular properties and functions of insect acetylcholine receptors. *J Insect Physiol.* 1987;33(11):771–790.
3. Galizia CG, Eisenhardt D, Giurfa M. *Honeybee Neurobiology and Behavior. A Tribute to Randolf Menzel.* New York: Springer; 2012.
4. Hammer M. An identified neuron mediates the unconditioned stimulus in associative olfactory learning in honeybees. *Nature.* 1993;366(6450):59–63.
5. Hildebrandt H, Müller U. Octopamine mediates rapid stimulation of protein kinase A in the antennal lobe of honeybees. *J Neurobiol.* 1995;27(1):44–50.
6. Hildebrandt H, Müller U. PKA activity in the antennal lobe of honeybees is regulated by chemosensory stimulation *in vivo*. *Brain Res.* 1995;679(2):281–288.
7. Müller U. Prolonged activation of cAMP-dependent protein kinase during conditioning induces long-term memory in honeybees. *Neuron.* 2000;27(1):159–168.
8. Weinberg RJ. Glutamate: an excitatory neurotransmitter in the mammalian CNS. *Brain Res Bull.* 1999;50(5–6):353–354.
9. Malinow R, Malenka RC. AMPA receptor trafficking and synaptic plasticity. *Annu Rev Neurosci.* 2002;25:103–126.
10. Bicker G, Schafer S, Ottersen OP, Storm-Mathisen J. Glutamate-like immunoreactivity in identified neuronal populations of insect nervous systems. *J Neurosci.* 1988;8(6):2108–2122.
11. Collet C, Belzunces L. Excitable properties of adult skeletal muscle fibres from the honeybee *Apis mellifera*. *J Exp Biol.* 2007;210 (Pt 3):454–464.
12. The Honeybee Genome Sequencing Consortium. Insights into social insects from the genome of the honeybee *Apis mellifera*. *Nature.* 2006;443(7114):931–949.
13. Kucharski R, Ball EE, Hayward DC, Maleszka R. Molecular cloning and expression analysis of a cDNA encoding a glutamate transporter in the honeybee brain. *Gene.* 2000;242 (1–2):399–405.
14. Norenberg MD, Martinez-Hernandez A. Fine structural localization of glutamine synthetase in astrocytes of rat brain. *Brain Res.* 1979;161(2):303–310.
15. Croset V, Rytz R, Cummins SF, Budd A, Brawand D, Kaessmann H, et al. Ancient protostome origin of chemosensory ionotropic glutamate receptors and the evolution of insect taste and olfaction. *PLoS Genet.* 2010;6(8):e1001064.
16. Si A, Helliwell P, Maleszka R. Effects of NMDA receptor antagonists on olfactory learning and memory in the honeybee (*Apis mellifera*). *Pharmacol Biochem Behav.* 2004;77(2):191–197.
17. Zannat T, Locatelli F, Rybak J, Menzel R, Leboulle G. Identification and localisation of the NR1 sub-unit homologue of the NMDA glutamate receptor in the honeybee brain. *Neurosci Lett.* 2006;398(3):274–279.
18. Funada M, Yasuo S, Yoshimura T, Ebihara S, Sasagawa H, Kitagawa Y, et al. Characterization of the two distinct subtypes of metabotropic glutamate receptors from honeybee, *Apis mellifera*. *Neurosci Lett.* 2004;359(3):190–194.
19. Kucharski R, Mitri C, Grau Y, Maleszka R. Characterization of a metabotropic glutamate receptor in the honeybee (*Apis mellifera*): implications for memory formation. *Invert Neurosci.* 2007;7 (2):99–108.
20. Raymond V, Sattelle DB. Novel animal-health drug targets from ligand-gated chloride channels. *Nat Rev Drug Discov.* 2002;1 (6):427–436.
21. Barbara GS, Zube C, Rybak J, Gauthier M, Grunewald B. Acetylcholine, GABA and glutamate induce ionic currents in cultured antennal lobe neurons of the honeybee, *Apis mellifera*. *J Comp Physiol A Neuroethol Sens Neural Behav Physiol.* 2005;191 (9):823–836.
22. El Hassani AK, Dupuis JP, Gauthier M, Armengaud C. Glutamatergic and GABAergic effects of fipronil on olfactory learning and memory in the honeybee. *Invert Neurosci.* 2009;9 (2):91–100.
23. Jones AK, Sattelle DB. The cys-loop ligand-gated ion channel superfamily of the honeybee, *Apis mellifera*. *Invert Neurosci.* 2006;6(3):123–132.
24. El Hassani AK, Schuster S, Dyck Y, Demares F, Leboulle G, Armengaud C. Identification, localization and function of glutamate-gated chloride channel receptors in the honeybee brain. *Eur J Neurosci.* 2012;36:2409–2420.
25. Dingledine R, Borges K, Bowie D, Traynelis SF. The glutamate receptor ion channels. *Pharmacol Rev.* 1999;51(1):7–61.
26. Daniels RW, Gelfand MV, Collins CA, DiAntonio A. Visualizing glutamatergic cell bodies and synapses in *Drosophila* larval and adult CNS. *J Comp Neurol.* 2008;508(1):131–152.
27. Sinakevitch I, Grau Y, Strausfeld NJ, Birman S. Dynamics of glutamatergic signaling in the mushroom body of young adult *Drosophila*. *Neural Dev.* 2010;5:10.
28. Friedrich A, Thomas U, Müller U. Learning at different satiation levels reveals parallel functions for the cAMP-protein kinase A cascade in formation of long-term memory. *J Neurosci.* 2004;24 (18):4460–4468.
29. Hourcade B, Muenz TS, Sandoz JC, Rossler W, Devaud JM. Long-term memory leads to synaptic reorganization in the mushroom bodies: a memory trace in the insect brain? *J Neurosci.* 2010;30(18):6461–6465.
30. Menzel R, Manz G, Menzel R, Greggers U. Massed and spaced learning in honeybees: the role of CS, US, the intertrial interval, and the test interval. *Learn Mem.* 2001;8(4): 198–208.
31. Wüstenberg D, Gerber B, Menzel R. Short communication: long- but not medium-term retention of olfactory memories in honeybees is impaired by actinomycin D and anisomycin. *Eur J Neurosci.* 1998;10(8):2742–2745.

32. Malinow R, Madison DV, Tsien RW. Persistent protein kinase activity underlying long-term potentiation. *Nature*. 1988;335 (6193):820–824.
33. Kelleher III RJ, Govindarajan A, Tonegawa S. Translational regulatory mechanisms in persistent forms of synaptic plasticity. *Neuron*. 2004;44(1):59–73.
34. Nguyen PV, Abel T, Kandel ER. Requirement of a critical period of transcription for induction of a late phase of LTP. *Science*. 1994;265(5175):1104–1107.
35. Sutton MA, Carew TJ. Parallel molecular pathways mediate expression of distinct forms of intermediate-term facilitation at tail sensory-motor synapses in *Aplysia*. *Neuron*. 2000;26(1):219–231.
36. Casadio A, Martin KC, Giustetto M, Zhu H, Chen M, Bartsch D, et al. A transient, neuron-wide form of CREB-mediated long-term facilitation can be stabilized at specific synapses by local protein synthesis. *Cell*. 1999;99(2):221–237.
37. Martin KC, Casadio A, Zhu H, Yaping E, Rose JC, Chen M, et al. Synapse-specific, long-term facilitation of *Aplysia* sensory to motor synapses: a function for local protein synthesis in memory storage. *Cell*. 1997;91(7):927–938.
38. Sutton MA, Masters SE, Bagnall MW, Carew TJ. Molecular mechanisms underlying a unique intermediate phase of memory in *Aplysia*. *Neuron*. 2001;31(1):143–154.
39. Locatelli F, Bundrock G, Muller U. Focal and temporal release of glutamate in the mushroom bodies improves olfactory memory in *Apis mellifera*. *J Neurosci*. 2005;25(50):11614–11618.
40. Maleszka R, Helliwell P, Kucharski R. Pharmacological interference with glutamate re-uptake impairs long-term memory in the honeybee, *Apis mellifera*. *Behav Brain Res*. 2000;115(1):49–53.
41. Müßig L, Richlitzki A, Rossler R, Eisenhardt D, Menzel R, Leboulle G. Acute disruption of the NMDA receptor subunit NR1 in the honeybee brain selectively impairs memory formation. *J Neurosci*. 2010;30(23):7817–7825.
42. Raymond CR, Thompson VL, Tate WP, Abraham WC. Metabotropic glutamate receptors trigger homosynaptic protein synthesis to prolong long-term potentiation. *J Neurosci*. 2000;20 (3):969–976.
43. Lopatina NG, Ryzhova IV, Chesnokova EG, Dmitrieva LA. *N*-methyl-D-aspartate receptors in the short-term memory development in the honey bee *Apis mellifera*. *Zh Evol Biokhim Fiziol*. 2000;36(3):223–228.
44. Martin SJ, Grimwood PD, Morris RG. Synaptic plasticity and memory: an evaluation of the hypothesis. *Annu Rev Neurosci*. 2000;23:649–711.
45. Roberts AC, Glanzman DL. Learning in *Aplysia*: looking at synaptic plasticity from both sides. *Trends Neurosci*. 2003;26 (12):662–670.
46. Xia S, Miyashita T, Fu TF, Lin WY, Wu CL, Pyzocha L, et al. NMDA receptors mediate olfactory learning and memory in *Drosophila*. *Curr Biol*. 2005;15(7):603–615.
47. Wu CL, Xia S, Fu TF, Wang H, Chen YH, Leong D, et al. Specific requirement of NMDA receptors for long-term memory consolidation in *Drosophila* ellipsoid body. *Nat Neurosci*. 2007;10 (12):1578–1586.
48. Hammer M, Menzel R. Multiple sites of associative odor learning as revealed by local brain microinjections of octopamine in honeybees. *Learn Mem*. 1998;5(1–2):146–156.
49. Menzel R, Muller U. Learning and memory in honeybees: from behavior to neural substrates. *Annu Rev Neurosci*. 1996;19:379–404.
50. Gong LQ, He LJ, Dong ZY, Lu XH, Poo MM, Zhang XH. Postinduction requirement of NMDA receptor activation for late-phase long-term potentiation of developing retinotectal synapses *in vivo*. *J Neurosci*. 2011;31(9):3328–3335.
51. Shimizu E, Tang YP, Rampon C, Tsien JZ. NMDA receptor-dependent synaptic reinforcement as a crucial process for memory consolidation. *Science*. 2000;290(5494):1170–1174.
52. Raymond CR, Redman SJ. Spatial segregation of neuronal calcium signals encodes different forms of LTP in rat hippocampus. *J Physiol*. 2006;570(Pt 1):97–111.
53. Westenbroek RE, Ahlijanian MK, Catterall WA. Clustering of L-type Ca^{2+} channels at the base of major dendrites in hippocampal pyramidal neurons. *Nature*. 1990;347(6290):281–284.
54. El Hassani AK, Dacher M, Gauthier M, Armengaud C. Effects of sublethal doses of fipronil on the behavior of the honeybee (*Apis mellifera*). *Pharmacol Biochem Behav*. 2005;82(1):30–39.
55. El Hassani AK, Giurfa M, Gauthier M, Armengaud C. Inhibitory neurotransmission and olfactory memory in honeybees. *Neurobiol Learn Mem*. 2008;90(4):589–595.
56. Lozano VC, Armengaud C, Gauthier M. Memory impairment induced by cholinergic antagonists injected into the mushroom bodies of the honeybee. *J Comp Physiol A*. 2001;187(4):249–254.
57. Krashes MJ, Keene AC, Leung B, Armstrong JD, Waddell S. Sequential use of mushroom body neuron subsets during *Drosophila* odor memory processing. *Neuron*. 2007;53(1):103–115.
58. Sejourne J, Placais PY, Aso Y, Siwanowicz I, Trannoy S, Thoma V, et al. Mushroom body efferent neurons responsible for aversive olfactory memory retrieval in *Drosophila*. *Nat Neurosci*. 2011;14(7):903–910.
59. Bast T, da Silva BM, Morris RG. Distinct contributions of hippocampal NMDA and AMPA receptors to encoding and retrieval of one-trial place memory. *J Neurosci*. 2005;25(25):5845–5856.
60. Navarrete M, Perea G, Fernandez DS, Gomez-Gonzalo M, Nunez A, Martin ED, et al. Astrocytes mediate *in vivo* cholinergic-induced synaptic plasticity. *PLoS Biol*. 2012;10(2): e1001259.
61. Gelinas JN, Nguyen PV. Beta-adrenergic receptor activation facilitates induction of a protein synthesis-dependent late phase of long-term potentiation. *J Neurosci*. 2005;25(13):3294–3303.
62. Menzel R, Manz G. Neural plasticity of mushroom body-extrinsic neurons in the honeybee brain. *J Exp Biol*. 2005;208(Pt 22):4317–4332.
63. Mauelshagen J. Neural correlates of olfactory learning paradigms in an identified neuron in the honeybee brain. *J Neurophysiol*. 1993;69(2):609–625.
64. Okada R, Rybak J, Manz G, Menzel R. Learning-related plasticity in PE1 and other mushroom body-extrinsic neurons in the honeybee brain. *J Neurosci*. 2007;27(43):11736–11747.

CHAPTER

35

Cellular Mechanisms of Neuronal Plasticity in the Honeybee Brain

Bernd Grünewald

Goethe-Universität Frankfurt am Main, Germany

INTRODUCTION

Learning often correlates with changes in the spike activity of single neurons or the synaptic transmission between neurons. Such learning-induced neuronal modulations are manifold and may include covalent modifications of voltage-sensitive currents, altered transmitter release, or postsynaptic modulations. Honeybees are valuable models for research on the cell physiological mechanisms of learning-related plasticity (for review, see[1]), which comprises at least three levels of analyses: (1) the identification and characterization of those ionic currents that may be involved in cellular plasticity, (2) the physiological analysis of ionic current modulation, and (3) the study of the roles of the various currents and receptors during behavioral learning. This chapter summarizes what is known about these different levels with regard to the honeybee. It provides an overview of the characterized ionic currents and transmitter receptors expressed by neurons within the honeybee brain. Second, it summarizes the evidence regarding the modulation of these currents. Finally, it discusses how the identified transmitters and currents may relate to behavioral plasticity. These points are discussed with respect to olfactory learning because the physiology of the neuronal elements of this behavior is well described in honeybees.

CELLULAR PHYSIOLOGY OF MEMBRANE EXCITABILITY

Several neurons that are involved in olfactory information processing have been physiologically described (Figure 35.1). Intracellular and extracellular recordings *in vivo* were performed from antennal lobe neurons,[3–9] from Kenyon cells of the mushroom body,[10] and from mushroom body extrinsic neurons.[11–16] Odor stimuli induce a complex spatiotemporal pattern within the antennal lobe neurons[10,17–19] (for review, see[20]). Axons of olfactory receptor cells synapse onto projection neurons (PNs) and onto local interneurons (LNs) within the antennal lobe (AL). The spike activity within these neuron classes is determined by a balanced excitatory (cholinergic) and inhibitory (γ-aminobutyric acid (GABA)ergic and glutamatergic) input. Thus, particular neurons may exhibit significant spontaneous spike activity, whereas other neurons rarely spike in the absence of odor stimuli. Stimulation with an odor or an odor mixture may induce an increase of spike frequency in some AL neurons and a decrease in others.[3,9] In addition, odor specificity appears to be encoded by response latencies in some PN classes.[8,9] These response patterns are mainly determined by the AL synaptic network. The neurons' spike shapes and their frequencies, propagations, and thresholds depend on the composition of voltage-sensitive ionic currents. These currents differ between AL neurons and mushroom body (MB) intrinsic Kenyon cells.[21] Consequently, the spiking activities of these two neuron classes differ as well. The sparse responses of Kenyon cells during odor stimulation[22] are probably determined by strong inhibitory input onto these neurons. In addition, Kenyon cells also express a high density of rapidly activating, voltage-sensitive K^+ currents that hinder the generation of an action potential.[23,24]

Next we ask which voltage-sensitive ionic currents have been characterized in honeybee AL neurons and MB Kenyon cells. To compare the ionic currents of

Invertebrate Learning and Memory.
DOI: http://dx.doi.org/10.1016/B978-0-12-415823-8.00035-6

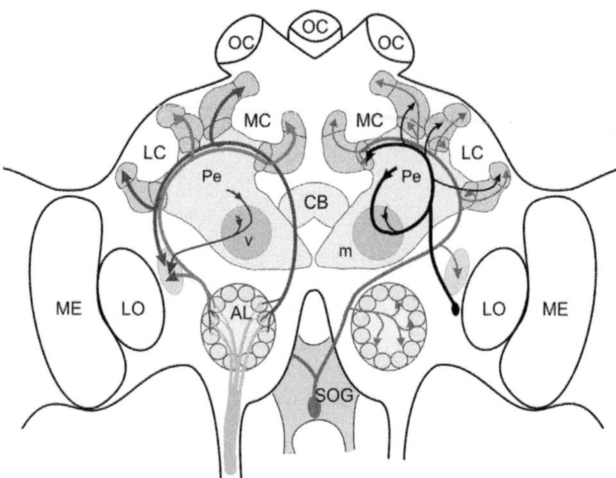

FIGURE 35.1 Anatomy of olfactory and reward information processing neural pathways within the honeybee brain. Axons of the ORN project via the antennal nerve and synapse onto postsynaptic PN and local interneurons within the glomeruli of the AL. Olfactory information enters the brain via the antennal nerve. From there, projection neurons (blue) innervate the lateral protocerebral lobe, the calyces of the mushroom bodies, and various other regions of the protocerebrum. Neurons of the mAPT (dark blue) show strong acetylcholine esterase activity and are probably cholinergic. By contrast, the transmitter(s) of PN projecting via the lAPT (light blue) is unknown. Feedback neurons (black) connect the lobes and peduncle with the calyces of the mushroom bodies (green). Output neurons such as the PE1 neuron (violet) transmit Kenyon cell information from the lobes and peduncle to various regions of the ipsi- and contralateral protocerebral lobe. The VUMmx1 neuron (red) encodes reward information and converges with odor information processing neurons in the antennal lobes, the calyces, and the lateral protocerebral lobes. AL, antennal lobe; CB, central body; LC, lateral calyx; Lo, lobula; m, medial lobe; MC, median calyx; Me, medulla; OC, ocelli; PE, pedunculus; SOG, suboesophageal ganglion; v, ventral lobe. Source: *Reproduced with permission from Barbara* et al.[2]

these neurons, their somata were taken into primary cell culture.[25] In some cases, the neurons were identified prior to recording by retrograde labeling using dextran-rhodamine injections.[21] The somata express voltage-sensitive Na^+ (I_{Na}), K^+ (I_K), and Ca^{2+} (I_{Ca}) currents (Figure 35.2; for review, see[26]). Kenyon cells and projection neurons express similar voltage-sensitive Na^+ currents.[24] A fast transient, TTX-sensitive I_{Na} that activates at voltages more positive than -40 mV is probably mediating the depolarizing phase of the action potential. In addition, Kenyon cells express a small sustained I_{Na} (<1% of the total I_{Na}), which is voltage sensitive and TTX sensitive but shows little or no inactivation during prolonged voltage command pulses.[24,27]

All neurons studied so far possess a delayed rectifier type K^+ current ($I_{K,V}$),[21,27–29] which does not inactivate during prolonged depolarizing voltage commands. The *shaker*-like K^+ current (also called A-current, $I_{K,A}$) of the honeybee is a fast-activating, transient current that is sensitive to the blocker 4-aminopyridine.[23,29] Whereas Kenyon cells express pronounced *shaker*-like K^+ currents, such inactivating outward currents are much smaller in PNs.[21] Interestingly, this finding resembles the situation in the *Drosophila* brain, in which *shaker* channel proteins are highly expressed in the MB neuropil but not in the AL.[30] A Ca^{2+}-dependent K^+ current ($I_{K,Ca}$) is expressed by PNs but has rarely been found in Kenyon cells.[21,24,29] The Ca^{2+} inflow at negative clamp potentials activates this outward K^+ current. Finally, Kenyon cells express a slow transient K^+ current ($I_{K,ST}$),[24] which is not sensitive to 4-aminopyridine and activates more slowly than A-type currents. Slowly inactivating potassium currents have been described in many cells. Interestingly, Wright and Zhong[31] described two transient outward currents in cultured *Drosophila* Kenyon cells, one of which was insensitive to 4-AP and might therefore correspond to the newly identified component in honeybee Kenyon cells.

Voltage-sensitive Ca^{2+} currents may contribute to action potential generation, synaptic transmission, or neuromodulation (for reviews, see[32,33]). The voltage-sensitive Ca^{2+} currents of honeybee neurons activate rapidly and show a slow inactivation.[21,24,27] Their steady-state activation, Cd^{2+} sensitivity, and inactivation are very similar to those described in other insect preparations.

Integration of the various voltage-sensitive ionic currents into a computer simulation using Hodgkin–Huxley-type equations yields the following scenario (Figure 35.2): The fast-inactivating I_{Na} mediates the membrane depolarizing during the action potential. The $I_{K,V}$ is largely responsible for the repolarization of the membrane potential during a single spike.[24] Thus, I_{Na} and $I_{K,V}$ are the primary currents of the action potential. In Kenyon cells, $I_{K,A}$ plays only a minor role during spike repolarization. Rather, $I_{K,A}$ modulates the spike duration and spike threshold.[24] The Ca^{2+}-dependent K^+ currents play a major role in the control of neuronal excitability and, for example, mediate after-hyperpolarization or spike repolarization. Computer modeling indicates that $I_{K,ST}$ is the primary determinant of the delayed spiking responses of Kenyon cells during constant current stimuli, and $I_{K,ST}$ prevents the model from responding to oscillatory stimuli.[24] These findings suggest that the spiking characteristics of Kenyon cells *in vivo* could be profoundly altered by the modulation of $I_{K,ST}$. Modulation of the action potential duration can be mediated by $I_{K,A}$ and $I_{K,Ca}$. The high expression level of $I_{K,A}$ in Kenyon cells may partially explain the very low spike rate and high spike threshold of these cells. Therefore, analyzing the modulation of $I_{K,Ca}$ or $I_{K,A}$ is most promising for

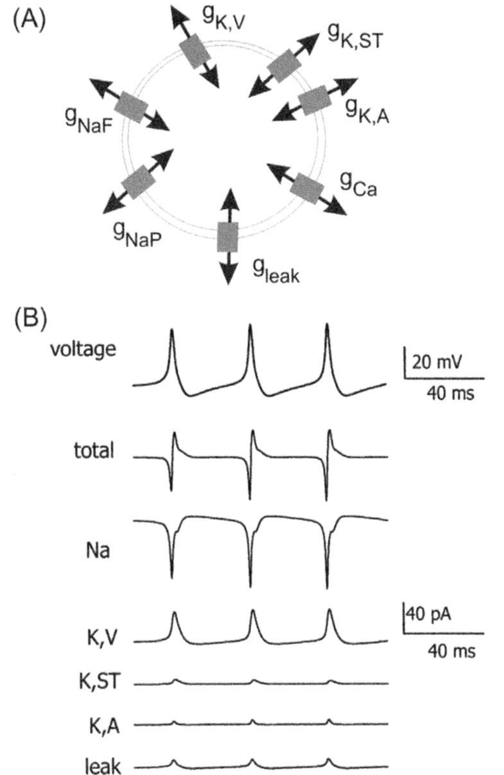

FIGURE 35.2 Ionic currents contributing to Kenyon cell action potentials. (A) Kenyon cells express six different voltage-sensitive ionic currents and a leak current. (B) Integration of the physiologically characterized voltage-sensitive currents into a realistic Hodgkin–Huxley-type mathematical model revealed the individual membrane currents flowing during action potentials in the simulated Kenyon cell. The primary currents are I_{Na} and $I_{K,V}$, whereas the two other outward currents are relatively small. The voltage-sensitive I_{Ca} is not integrated because its steady state and kinetic properties are not yet characterized in sufficient detail. Source: *Panel B reproduced from Grünewald.*[26]

understanding the cellular plasticity of intrinsic spiking behavior of honeybee central neurons.

In summary, the characterization of the voltage-sensitive currents of cultured Kenyon cells explains the spike behavior of these cells *in vivo* quite well.[155] Nevertheless, the identification and the physiology of voltage-gated Ca^{2+} currents and Ca^{2+}-dependent ionic currents require further experimental analyzes. It would also be very interesting to determine whether the various neuron types of the AL (various projection neurons of the different tracts and local interneurons) express differential sets of ionic currents.

SYNAPTIC TRANSMITTERS AND THEIR RECEPTORS

Several transmitter receptors engaged in honeybee olfactory information processing and olfactory learning and memory have been characterized (for review, see[26]). Excitatory synaptic transmission is mediated via nicotinic cholinergic receptors and probably via ionotropic glutamate receptors. Inhibition within the AL and the MB networks depends on ionotropic GABA and chloride-permeable glutamate receptors. The existence of metabotropic receptors for glutamate, GABA, or acetylcholine in the honeybee brain has been made plausible (for review, see[34]). However, these receptors are not included in this review because they are not yet physiologically characterized. Honeybee neurons express a variety of receptors for the biogenic amines, octopamine, serotonin, dopamine, and tyramine,[35] which provide a modulatory network upon fast synaptic transmission.

Acetylcholine Receptors

Acetylcholine (ACh) is the most abundant excitatory transmitter within the olfactory pathway of insects (for reviews, see[36]). The axons of the olfactory receptor neurons (ORNs) probably release ACh onto postsynaptic neurons within the ALs (Figure 35.3), and a subpopulation of projection neurons from the ALs form cholinergic synapses with Kenyon cells within the MB lip regions.[37] Honeybee AL neurons as well as Kenyon cells stain against nicotinic AChR (nAChR) antibodies.[37] Sequence analyses identified 11 different nAChR subunits in the bee genome—9 α subunits, Amelα1–9, and 2 β subunits, Amelβ1–2—with an as yet unidentified stoichiometry.[38] The expression pattern of 4 nAChR subunits in the bee brain was described by *in situ* hybridization. Amelα8 is closest to the vertebrate α_3 subunit and is found in MB Kenyon cells and in AL neurons.[39] Amelα5 and Amelα7 subunits are expressed in MB type II and clawed Kenyon cells and AL neurons.[40] Amelα2 subunits are found in type I and type II Kenyon cells but not in the AL. Amelα7 is also expressed in type I Kenyon cells.[39]

The honeybee nAChR is a pentameric receptor of the *cys*-loop receptor family with an as yet unknown stoichiometry.[41] Pressure application of ACh induces rapidly activating inward currents in cultured bee neurons.[42] The native honeybee nAChRs in Kenyon cells and AL neurons are cation-selective channels with a pharmacological profile indicating a neuronal nAChR.[2,24,155] They are blocked by the nicotinergic blockers curare, methyllycaconitine, dihydroxy-β-erythroidine, and mecamylamine. ACh as well as carbamylcholine are full agonists, whereas nicotine, epibatidine, cytisine, and the neonicotinoid imidacloprid are partial agonists.[43] The nAChR of Kenyon cells has a high Ca^{2+} permeability,[42] and calcium imaging experiments *in vitro* revealed a strong intracellular Ca^{2+} signal after application of nicotinic agonists[44] (Himmelreich and

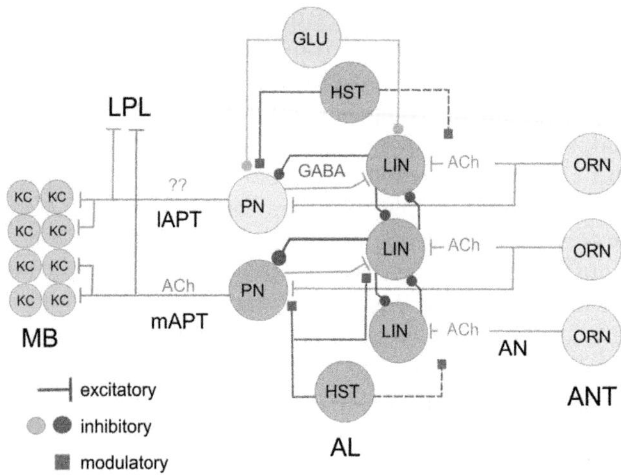

FIGURE 35.3 Schematic synaptic wiring diagram of the honeybee olfactory pathways based on the physiologically identified ligand-gated ionic currents and on ultrastructural findings from other insect species. Olfactory receptor neurons (ORN) provide excitatory cholinergic input onto local inhibitory neurons (LIN) and projection neurons (PN). Most, if not all, LIN are GABAergic and presynaptic to PN and other LIN. They receive cholinergic input from PN, which are connecting the antennal lobe with the mushroom body (MB) and the lateral protocerebral lobe (LPL). Glutamatergic neurons (GLU) provide a second inhibitory network and putatively synapse onto both PN and LIN. Finally, a group of histaminergic neurons (HST) may modulate the neural activity within the honeybee AL and may regulate the input into the AL by presynaptically inhibiting sensory afferences from the antenna (dashed lines). Output from the AL to the LPL and the MB is provided by PN of the lAPT, whose transmitter is unknown, and by cholinergic PN of the mAPT. The several hundred PN diverge onto several thousand Kenyon cells (KC).ik Source: *Modified from Barbara* et al.[2]

Grünewald, manuscript in preparation). Therefore, the nAChR mediates influx of Ca^{2+} and a membrane depolarization upon activation. The unique expression pattern of the different AChR subtypes between Kenyon cells and AL neurons may indicate different physiological roles of cholinergic synaptic transmission in the honeybee brain.[45] For instance, it may explain why highly potent blockers such as dihydroxy-β-erythroidine (DHE) or methyllylcaconitine (MLA) do not block odor perception completely but, rather, affect certain aspects of olfactory learning[46] (for review, see[34]).

Behavioral pharmacological studies have shown that nAChRs are involved in various phases of classical conditioning, memory formation, and retrieval (for review, see[34]). However, the effects caused by nicotinic antagonists are complex and often contradictory. Injections of the nACh antagonists mecamylamine, α-bungarotoxin (BGT), or MLA into the honeybee brain impaired acquisition (mecamylamine[47]) or long-term memory (BGT and MLA[46]). Given that various blockers target different nAChR subtypes, it was assumed that at least two nAChRs (one BGT-sensitive and one BGT-insensitive nAChR) are differentially involved during olfactory learning and memory formation in bees.[46,48] However, the precise identity of these receptors awaits future studies.

The nAChRs of insects are target molecules of neonicotinoid insecticides, which are commonly used in agriculture.[49] They act as agonists also on honeybee receptors.[43,45,50,51] Sublethal doses of neonicotinoids such as imidacloprid, clothianidin, thiacloprid, or thiamethoxam affect olfactory learning,[52–57] habituation,[56,57] or foraging behavior[58–60] (for review, see[61]). Because bees are coming in contact with these substances during foraging flights, it will be increasingly important to investigate the effects of agrochemicals on managed honeybee colonies[62,63] but also on wild bee populations that are under threat as well.[64,65]

GABA Receptors

The inhibitory transmitter GABA was detected throughout the honeybee brain.[66–68] GABA immunoreactivity is highly abundant in the neuropils involved in olfactory learning in bees, the ALs and the MBs. Most local interneurons of the AL are probably GABAergic, and inhibition within these neuropils appears crucial for proper odor perception, discrimination, or learning.[10,12,19] Similarly, a large group of GABA immunoreactive feedback neurons that connect the output and the input sites of the MB is a common feature of insect MBs.[11,69–71] The activation of honeybee ionotropic GABA receptors induces fast Cl^- currents in cultured neurons.[2,72,73] Accordingly, applications of GABAergic inhibitors in the AL increase neuronal activity *in vivo*[19] and abolished odor-induced oscillatory activity.[10] Based on sequence analyses, the honeybee genome contains genes coding for three different $GABA_A$ receptor subunits (Amel_GRD, Amel_RDL, Amel_LCCH3). These GABA receptors are pentameric structures like vertebrate ionotropic GABA receptors.[41] In bee central neurons, GABA receptors are probably composed of RDL and LCCH3 receptor subunits with an as yet unknown stoichiometry.[72,73] The native receptor shows a typical insect GABA receptor pharmacology. Muscimol and *cis*-4-aminocotronic acid act as agonists; the receptor is sensitive to picrotoxin (PTX) but insensitive to bicuculline. The insecticide fipronil blocks GABA-induced currents in honeybee neurons.[2] GABA-induced currents are modulated by intracellular Ca^{2+}.[73] This modulation may be mediated via Ca^{2+}-dependent phosphorylation, although the signaling cascades have not yet been identified.

The role of inhibitory synaptic transmission during olfactory learning is not very well understood. Whereas blocking chloride currents with PTX strongly affects odor processing within the AL,[10,19] GABAergic modulations on learning and memory are rather subtle. Injections of PTX lead to an increased generalization rate after training.[10] Because the insecticide fipronil acts as a chloride channel blocker, it affects olfactory learning and memory in honeybees.[74,75]

Glutamate Receptors

The following different glutamate receptors exist in the honeybee brain: (1) a cation-selective current that is induced by glutamate or AMPA ($GluR_{AMPA}$; Grünewald, unpublished data), (2) a $GluR_{NMDA}$ whose subunits were found in AL neurons and Kenyon cells[76,77] (see Chapter 34), and (3) a chloride current that is activated by glutamate applications onto AL neurons ($GluR_{Cl}$).[2] The $GluR_{Cl}$ currents comprise a rapidly activating, a desensitizing, and a sustained component. This honeybee $GluR_{Cl}$ is partially sensitive to PTX and bicuculline and is blocked by fipronil. Therefore, it shares several properties with the $GluR_{Cl}$ of other insects[78–83] (for review, see[84]). The honeybee genome contains a gene encoding for a putative glutamate-sensitive chloride channel subunit (Amel_GluCl[41]). Immunocytochemistry using peptide antibodies indicates expression of Amel_GluCl in the glomeruli of the AL.[85] Thus, two independent inhibitory systems within the AL may exist—a glutamatergic inhibitory network and the GABAergic network.[2]

A putative NMDA receptor subunit (NR1) has been detected in the bee brain by Western blot analyses.[86] Although the physiology of the honeybee $GluR_{NMDA}$ has not yet been described, it may function like the *Drosophila* NMDAR, which shares several pharmacological similarities with its mammalian counterpart (heterologous expression of NR1[87] and native receptor[88]). It is obvious that the $GluR_{NMDA}$ is a key cellular component of the neural pathway of olfactory learning in insects: (1) Photorelease of caged glutamate within the MBs immediately after a single-trial conditioning induces a stable long-term memory[89]; (2) injections of NMDA receptor blockers impaired 24-hr long-term memory, leaving short-term memory intact[90]; and (3) acutely inhibiting NR1 expression by RNA interference affected acquisition as well as midterm and early long-term memory.[86] Likewise, in *Drosophila*, NMDA receptor functioning is essential during olfactory learning and memory formation[87,88,91] (for reviews, see[92,93]).

Histamine Receptors

Live calcium imaging of the honeybee AL indicated the existence of histamine-dependent neural inhibition.[94] Two potential histamine-gated Cl^- channel gene sequences (Amel_HisCl1 and Amel_HisCl2) were found within the bee's genome.[41] Bornhauser and Meyer[95] reported the presence of histamine-immunoreactive neurons in the honeybee AL. Histamine is found in many neurons throughout the insect brain and is probably the transmitter released by insect photoreceptors (for review, see[96]). Although no histamine-immunoreactive neurons have been detected in the MB, the insect AL contains histamine-immunoreactive fibers (cockroaches,[97,98] moths,[97,99] and crickets[96]). Histamine fails to elicit any current in cultured honeybee AL neurons or in Kenyon cells,[2] but histamine blocked odor-induced calcium signals in the AL *in vivo* at high concentrations (10–50 mM).[94]

G Protein-Coupled Biogenic Amine Receptors

Honeybees express G protein-coupled receptors for the biogenic amines octopamine, serotonin, dopamine, and tyramine (for comprehensive reviews, see[35,100]). They are differentially involved in controlling honeybee behavior.

Octopamine plays a key role as the transmitter of the reward processing neuron, VUMmx1 (a ventral unpaired median neuron of the maxillary neuropil of the subesophageal ganglion), during classical conditioning of the proboscis extension reflex in honeybees.[101] The VUMmx1 neuron projects to the antennal lobes, the lateral protocerebral lobes, and the lip and basal ring regions of the mushroom body calyces[101,102] (see Chapter 29). Neurons within these areas presumably receive octopaminergic input after unpredicted sucrose reward stimulation, and the function of octopamine during appetitive learning has been repeatedly shown: (1) Substituting the unconditioned stimulus with experimental depolarization of VUMmx1 during appetitive conditioning of the PER induced behavioral learning[101]; (2) pairing intraneuropilar octopamine injections into the ALs or the MBs with odor stimulations mimicked the actual sucrose reward during learning[103]; and (3) injections of octopamine into reserpinized bees, whose biogenic amine stores were depleted, rescued acquisition of the conditioned odor during classical conditioning.[104] Probably multiple octopamine receptors (OARs) are expressed by neurons of the honeybee brain. Based on findings of OAR subtypes in *Drosophila*,[105] five metabotropic G protein-coupled OAR candidates have been annotated from the honeybee genome (for review, see[100]). One octopamine receptor, AmOA1R, has a widespread distribution in the honeybee brain, including the AL, MB, central complex, optic lobes, and subesophageal ganglion; it is also expressed in Kenyon cells, GABA-immunoreactive interneurons of the AL, and feedback neurons of the MB.[106–108] Activation of heterologously expressed AmOA1R induces intracellular Ca^{2+} oscillations by applying nanomolar concentrations of OA. In addition, small increases in the concentration of cAMP were observed after applying OA at micromolar concentrations,[106,109] which also led to the activation of PKA.[110]

Dopaminergic neuromodulation is commonly associated with odor avoidance learning in flies (for reviews, see[111,112]). Immunohistochemistry of the honeybee brain demonstrated several dopamine-immunoreactive cell clusters containing approximately 330 somata and dense innervation by dopamine-immunoreactive fibers around and inside the MBs and within the ALs.[114,115] Three honeybee dopamine receptor genes have been cloned, and the receptors were characterized using heterologous expression and pharmacological techniques[114,116,117] (for review, see[118]). One receptor, AmDOP2, is similar to the honeybee octopamine receptor, AmOA1R. Both are coupled to intracellular Ca^{2+} signals.[109] Dopamine perfusion reversibly reduces the amplitude of the $I_{K,Ca}$ in cultured honeybee AL neurons.[29] This effect may explain the impairment of conditioned responses after dopamine injections into the AL,[119] but the underlying dopamine receptor has not yet been identified. The behavioral functions of dopamine for bees are less well understood. Injections of dopamine into reserpinized bees restored the impaired motor component of the PER, indicating a role of dopamine in motor control.[104] Dopamine is also crucial during conditioning of the sting reflex in bees because injections of dopaminergic antagonists impaired this form of olfactory learning.[120] Finally, dopaminergic neuromodulation is involved in the control of worker behavior by the queen mandibular pheromone. One of the major components of this pheromone, homovanillyl alcohol (HVA), activates honeybee AmDOP3 receptors.[121] HVA also blocks aversive learning[122] and enhances attraction to queen mandibular pheromone in young worker bees and reduces the expression of AmDOP1 in their antennae[123] (see Chapter 32).

Approximately 75 serotonin-immunoreactive neurons send their fibers into all neuropils of the honeybee brain.[124] Early studies indicated that serotonin may act antagonistically to octopamine because it reduced the learning performance during appetitive olfactory conditioning of the PER.[104,125–127] However, this view appears to be too simplified because recent experiments have indicated a more complex picture. Sitaraman *et al.*[128] showed that serotonin is necessary for proper appetitive odor learning in *Drosophila*. Wright *et al.*[129] found that serotonergic antagonists impair the ability of bees to learn to avoid odors that signal toxic food. The serotonin receptors and serotonergic neurons involved in food avoidance learning have not been identified in bees. Two receptors have been cloned in bees. Am5-HT_{1A} is negatively coupled to cAMP production and involved in phototactic behavior.[130] Because it reduces the cAMP signal, it may be responsible for the inhibition of olfactory learning. By contrast, heterologously expressed Am5-HT_7 receptors increase cAMP levels upon activation.[131]

CELL PHYSIOLOGICAL EVENTS UNDERLYING OLFACTORY LEARNING

The ALs and MBs are differentially involved during classical conditioning of the PER, and learning-dependent neural plasticity has been observed in both neuropils[4,12,13,15,17,132–137] (see Chapters 29 and 30). Because the neural pathways of the conditioned stimuli converge with the reward-processing VUMmx1 neuron in the ALs and MBs (see Chapter 29), the temporal coincidence of CS–US pairings should be detected or leave physiological traces in both neuropils.[138] How can the described receptors and ionic currents be integrated in a model of cellular physiology underlying reward learning in bees?

Axons of the olfactory receptor cells probably release ACh onto postsynaptic PNs and local interneurons within the glomeruli of the ALs. Binding of ACh to its receptors depolarizes the postsynaptic cell and increases the intracellular calcium concentration.[2,6,139] PNs projecting to the MB via the median or lateral antennal-protocerebral tract (mAPT and lAPT, respectively)[140,141] (for review, see[20]) transfer CS information into the lip region of the MB. PNs of the mAPT are probably cholinergic,[37] and Kenyon cells express functional nAChR. Assuming coexpression of nAChR and OAR in AL and MB neurons (Figure 35.4), we hypothesize that the coincident activation of both receptors during classical appetitive conditioning alters the postsynaptic cell physiology specifically.[138] Two potential intracellular pathways may be involved: (1) a calcium-dependent pathway comprising the α-adrenergic-like AmOA1R[106] coupled to a Ca^{2+} release from internal stores and/or (2) a cAMP/PKA-dependent pathway via an unidentified OAR (possibly a β-adrenergic-like OAR) that is coupled to an adenylyl cyclase.[142–145] These two pathways may modulate cholinergic or GABAergic synaptic transmission,[73,138] induce CREB phosphorylation,[146] or activate a Ca^{2+}-dependent protein kinase C.[147] Inhibitory input via GABAergic (and in ALs glutamatergic) fibers is abundant within the bee brain. Not surprisingly, blocking inhibition also affects learning performance. However, the specific role of inhibition during the various phases

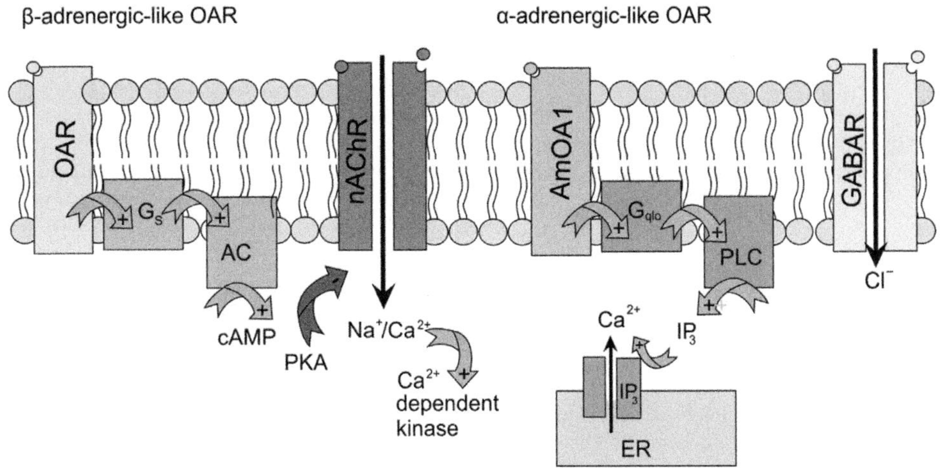

FIGURE 35.4 Cellular model of the signaling cascades underlying olfactory classical conditioning. Odor stimulation activates nAChR, which in turn depolarizes the postsynaptic membrane (the model may apply to both AL and MB neurons) representing the olfactory CS pathway. The reward pathway activates either α-adrenergic-like AmOA1, leading to Ca^{2+} release out of the endoplasmic reticulum (ER), or β-adrenergic-like OAR, which are positively coupled to an adenylyl cyclase (AC). Both pathways may modulate the ACh-induced currents. The elevated cytosolic Ca^{2+} level may also activate as yet unknown Ca^{2+}-dependent kinases, which may phosphorylate the nAChR and thus modulate ACh-induced currents. The coincidence between the odor and reward may be detected by two independent Ca^{2+} signals that may converge onto Ca^{2+}-dependent kinases. This pathway would require AmOA1 receptors. Alternatively, CS/reward coincidence could be detected by a Ca^{2+}-dependent AC, which requires β-adrenergic-like OARs and a Ca^{2+} signal through nAChR activation. Inhibitory input is provided by GABA-induced Cl^- influx, which balances the ACh-dependent membrane depolarization. The other ion channels and transmitter receptors mentioned in the text were omitted because either physiological data are missing (NMDAR and histamine receptors) or their contribution to learning-dependent plasticity is unclear (voltage-sensitive currents and receptors for 5-HT or dopamine). Source: *Modified from Himmelreich and Grünewald.*[138]

of odor learning has not yet been unraveled (but compare studies on *Drosophila*[148,149]). In the honeybee, MB feedback neurons undergo learning-dependent plasticity[12,150,151] (see Chapter 29), and the GABAergic synapse is modulated by intracellular Ca^{2+}.[73]

The physiological consequences of excitatory glutamatergic synaptic transmission via Glu_{NMDA} are not yet understood in bees. The CS pathway may comprise glutamatergic neurons (e.g., projection neurons), or Kenyon cells or MB extrinsic neurons may release glutamate. Co-release of GABA and glutamate from local interneurons of the *Drosophila* antennal lobes has been proposed.[152] It would be interesting to learn whether this also occurs in honeybee MB feedback neurons or AL local interneurons. NMDA receptors are involved in honeybee olfactory learning[153] (see Chapter 34), and LTP-like processes have been observed at the synapse between Kenyon cells and MB output neurons, which may be explained by the activation of NMDA receptors at this synapse.[14] The cellular consequences of serotonin receptor activation[154] for appetitive learning have also not yet been revealed in bees. Serotonin is required for olfactory reward learning in flies[128] and odor avoidance learning in bees,[129] but the biochemical nature of the underlying receptors has not yet been identified.

CONCLUSIONS

Olfactory conditioning of the honeybee proboscis extension proved to be a particularly useful behavioral paradigm to identify the neural bases of learning and memory formation. The obvious advantage of this paradigm is that one can combine it with genetic, molecular, or cellular approaches targeted to network properties of the neuronal pathways that are already quite well-known—the pathways of the conditioned and unconditioned stimuli. In addition, hypotheses derived from subcellular levels of analyses can be tested rather directly on the behavioral level. As shown in this review, cell physiological results help to explain mechanisms and processes involved in learning-dependent plasticity of the bee brain. The excitatory transmitters, acetylcholine and glutamate, are crucial for acquisition and memory formation, consolidation, or recall. Octopamine is an essential component of the rewarding property of the unconditioned stimulus during classical olfactory conditioning. Coincident activation of the conditioned and unconditioned stimuli leads to experience-dependent odor response modulations of antennal lobe and mushroom body neurons. These forms of plasticity may depend on the modulation of voltage-sensitive and/or ligand-gated ionic currents (I_K, I_{Ca}, nAChR, and GABAR), on the activation of glutamate receptors (NMDAR), and on the activation of Ca^{2+}-dependent signaling cascades via nAChR and/or OAR stimulation. Finally, these short-term modulations of cell physiology translate into lasting changes of neural activity and altered behavior.

References

1. Menzel R. The honeybee as a model for understanding the basis of cognition. *Nat Rev Neurosci.* 2012;13:758–768.
2. Barbara GS, Zube C, Rybak J, Gauthier M, Grünewald B. Acetylcholine, GABA and glutamate induce ionic currents in cultured antennal lobe neurons of the honeybee, *Apis mellifera. J Comp Physiol A Neuroethol Sens Neural Behav Physiol.* 2005;19:823–836.
3. Abel R, Rybak J, Menzel R. Structure and response patterns of olfactory interneurons in the honeybee, *Apis mellifera. J Comp Neurol.* 2001;437:363–383.
4. Denker M, Finke R, Schaupp F, Grün S. Neural correlates of odor learning in the honeybee antennal lobe. *Eur J Neurosci.* 2010;31:119–133.
5. Flanagan D, Mercer AR. Morphology and response characteristics of neurones in the deutocerebrum of the brain in the honeybee *Apis mellifera. J Comp Phys A.* 1989;164:483–494.
6. Galizia CG, Kimmerle B. Physiological and morphological characterization of honeybee olfactory neurons combining electrophysiology, calcium imaging and confocal microscopy. *J Comp Physiol A Neuroethol Sens Neural Behav Physiol.* 2004;190(1):21–38.
7. Homberg U. Processing of antennal information in extrinsic mushroom body neurons of the bee brain. *J Comp Phys A.* 1984;154:825–836.
8. Krofczik S, Menzel R, Nawrot MP. Rapid odor processing in the honeybee antennal lobe network. *Front Comput Neurosci.* 2009;2:1–13.
9. Müller D, Abel R, Brandt R, Zöckler M. Differential parallel processing of olfactory information in the honeybee, *Apis mellifera* L. *J Comp Phys A.* 2002;188:359–370.
10. Stopfer M, Bhagavan S, Smith BH, Laurent G. Impaired odour discrimination on desynchronization of odour-encoding neural assemblies. *Nature.* 1997;390:70–74.
11. Gronenberg W. Anatomical and physiological properties of feedback neurons of the mushroom bodies in the bee brain. *Exp Biol.* 1987;46:115–125.
12. Grünewald B. Physiological properties and response modulations of mushroom body feedback neurons during olfactory learning in the honeybee, *Apis mellifera. J Comp Phys A.* 1999;185:565–576.
13. Mauelshagen J. Neural correlates of olfactory learning paradigms in an identified neuron in the honeybee brain. *J Neurophysiol.* 1993;69(2):609–625.
14. Menzel R, Manz G. Neural plasticity of mushroom body-extrinsic neurons in the honeybee brain. *J Exp Biol.* 2005;208:4317–4332.
15. Okada R, Rybak J, Manz G, Menzel R. Learning-related plasticity in PE1 and other mushroom body-extrinsic neurons in the honeybee brain. *J Neurosci.* 2007;27(43):11736–11747.
16. Rybak J, Menzel R. Integrative properties of the Pe1 neuron, a unique mushroom body output neuron. *Learn Mem.* 1998;5:133–145.
17. Faber T, Joerges J, Menzel R. Associative learning modifies neural representations of odors in the insect brain. *Nat Neurosci.* 1999;2(1):74–78.
18. Joerges J, Küttner A, Galizia CG, Menzel R. Representations of odours and odour mixtures visualized in the honeybee brain. *Nature.* 1997;387:285–288.
19. Sachse S, Galizia CG. Role of inhibition for temporal and spatial odor representation in olfactory output neurons: a calcium imaging study. *J Neurophysiol.* 2002;87(2):1106–1117.
20. Galizia CG, Rössler W. Parallel olfactory systems in insects: anatomy and function. *Annu Rev Entomol.* 2010;55:399–420.
21. Grünewald B. Differential expression of voltage-sensitive K^+ and Ca^{2+} currents in neurons of the honeybee olfactory pathway. *J Exp Biol.* 2003;206:117–129.
22. Szyszka P, Ditzen M, Galkin A, Galizia CG, Menzel R. Sparsening and temporal sharpening of olfactory representations in the honeybee mushroom bodies. *J Neurophysiol.* 2005;94(5):3303–3313.
23. Pelz C, Jander J, Rosenboom H, Hammer M, Menzel R. I_A in Kenyon cells of the mushroom body of honeybees resembles shaker currents: kinetics, modulation by K^+, and simulation. *J Neurophysiol.* 1999;81:1749–1759.
24. Wüstenberg D, Boytcheva M, Grünewald B, Byrne JH, Menzel R, Baxter DA. Current- and voltage-clamp recordings and computer simulations of Kenyon cells in honeybee. *J Neurophysiol.* 2004;92:2589–2603.
25. Kreissl S, Bicker G. Dissociated neurons of the pupal honeybee brain in cell culture. *J Neurocytol.* 1992;21:545–556.
26. Grünewald B. Cellular physiology of the honey bee brain. In: Galizia CG, Eisenhardt D, Giurfa M, eds. *Honeybee Neurobiology and Behaviour.* Dordrecht: Springer; 2012:185–198.
27. Schäfer S, Rosenboom H, Menzel R. Ionic currents of Kenyon cells from the mushroom body of the honeybee. *J Neurosci.* 1994;14(8):4600–4612.
28. Kloppenburg P, Kirchhof BS, Mercer AR. Voltage-activated currents from adult honeybee (*Apis mellifera*) antennal motor neurons recorded *in vitro* and *in situ*. *J Neurophysiol.* 1999;81:39–48.
29. Perk CG, Mercer AR. Dopamine modulation of honey bee (*Apis mellifera*) antennal-lobe neurons. *J Neurophysiol.* 2006;95:1147–1157.
30. Rogero O, Hämmerle B, Tejedor FJ. Diverse expression and distribution of Shaker potassium channels during the development of the *Drosophila* nervous system. *J Neurosci.* 1997;17(13):5108–5118.
31. Wright NJD, Zhong Y. Characterization of K^+ currents and the cAMP-dependent modulation in cultured *Drosophila* mushroom body neurons identified by lacZ expression. *J Neurosci.* 1995;15(2):1025–1034.
32. Jeziorski MC, Greenberg RM, Anderson PA. The molecular biology of invertebrate voltage-gated Ca(2+) channels. *J Exp Biol.* 2000;203(Pt 5):841–856.
33. Wicher D, Walther C, Wicher C. Non-synaptic ion channels in insects: basic properties of currents and their modulation in neurons and skeletal muscles. *Prog Neurobiol.* 2001;64:431–525.
34. Gauthier M, Grünewald B. Neurotransmitter systems in the honeybee brain: functions in learning and memory. In: Galizia CG, Eisenhardt D, Giurfa M, eds. *Honeybee Neurobiology and Behavior.* Dordrecht: Springer; 2012:155–169.
35. Scheiner R, Baumann A, Blenau W. Aminergic control and modulation of honeybee behaviour. *Curr Neuropharmacol.* 2006;4:259–276.
36. Thany SH. Insect nicotinic acetylcholine receptors. *Adv Exp Med Biol.* 2010;683:1–118.

37. Kreissl S, Bicker G. Histochemistry of acetylcholinesterase and immunocytochemistry of an acetylcholine receptor-like antigen in the brain of the honeybee. *J Comp Neurol.* 1989;286:71–84.
38. Jones AK, Sattelle DB. Diversity of insect nicotinic acetylcholine receptor subunits. *Adv Exp Med Biol.* 2010;683:25–43.
39. Thany SH, Lenaers G, Crozatier M, Armengaud C, Gauthier M. Identification and localization of the nicotinic acetylcholine receptor alpha3 mRNA in the brain of the honeybee, *Apis mellifera. Insect Mol Biol.* 2003;12(3):255–262.
40. Thany SH, Crozatier M, Raymond-Delpech V, Gauthier M, Lenaers G. Apisalpha2, Apisalpha7-1 and Apisalpha7-2: three new neuronal nicotinic acetylcholine receptor alpha-subunits in the honeybee brain. *Gene.* 2005;344:125–132.
41. Jones AK, Sattelle DB. The *cys*-loop ligand-gated ion channel superfamily of the honeybee, *Apis mellifera. Invert Neurosci.* 2006;6(3):123–132.
42. Goldberg F, Grünewald B, Rosenboom H, Menzel R. Nicotinic acetylcholine currents of cultured Kenyon cells from the mushroom bodies of the honey bee *Apis mellifera. J Physiol.* 1999;514:759–768.
43. Deglise P, Grünewald B, Gauthier M. The insecticide imidacloprid is a partial agonist of the nicotinic receptor of honeybee Kenyon cells. *Neurosci Lett.* 2002;321(1–2):13–16.
44. Bicker G. Transmitter-induced calcium signalling in cultured neurons of the insect brain. *J Neurosci Methods.* 1996;69:33–41.
45. Dupuis JP, Gauthier M, Raymond-Delpech V. Expression patterns of nicotinic subunits alpha2, alpha7, alpha8, and beta1 affect the kinetics and pharmacology of ACh-induced currents in adult bee olfactory neuropiles. *J Neurophysiol.* 2011;106 (4):1604–1613.
46. Gauthier M, Dacher M, Thany SH, et al. Involvement of alpha-bungarotoxin-sensitive nicotinic receptors in long-term memory formation in the honeybee (*Apis mellifera*). *Neurobiol Learn Mem.* 2006;86(2):164–174.
47. Cano Lozano V, Bonnard E, Gauthier M, Richard D. Mecamylamine-induced impairment of acquisition and retrieval of olfactory conditioning in the honeybee. *Behav Brain Res.* 1996;81:215–222.
48. Gauthier M. State of the art on insect nicotinic acetylcholine receptor function in learning and memory. *Adv Exp Med Biol.* 2010;683:97–115.
49. Jeschke P, Nauen R. Neonicotinoids—From zero to hero in insecticide chemistry. *Pest Manag Sci.* 2008;64:1084–1098.
50. Nauen R, Ebbinghaus-Kintscher U, Schmuck R. Toxicity and nicotinic acetylcholine receptor interaction of imidacloprid and its metabolites in *Apis mellifera* (Hymenoptera: Apidae). *Pest Manag Sci.* 2001;57(7):577–586.
51. Tomizawa M, Casida JE. Neonicotionoid insecticide toxicology: mechanisms of selective action. *Annu Rev Pharmacol Toxicol.* 2005;45:247–268.
52. Aliouane Y, Hassani Vincent Gary AK, Armengaud C, Lambin M, Gauthier M. Subchronic exposure of honeybee to sublethal doses of pesticides: effects on behaviour. *Environ Toxicol Chem.* 2008;28:17–26.
53. Decourtye A, Armengaud C, Renou M, et al. Imidacloprid impairs memory and brain metabolism in the honeybee (*Apis mellifera* L.). *Pest Biochem Physiol.* 2004;78:83–92.
54. Decourtye A, Devillers J, Genecque E, et al. Comparative sublethal toxicity of nine pesticides on olfactory learning performances of honeybee *Apis mellifera. Arch Environ Contam Toxicol.* 2005;48:242–250.
55. Decourtye A, Lacassie E, Delegue M-HP. Learning performances of honeybees (*Apis mellifera* L) are differentially affected by imidacloprid according to the season. *Pest Manag Sci.* 2003;59: 269–278.
56. Guez D, Belzunces LP, Maleszka R. Effects of imidacloprid metabolites on habituation in honeybees suggest the existence of two subtypes of nicotinic receptors differentially expressed during adult development. *Pharmacol Biochem Behav.* 2003;75(1):217–222.
57. Guez D, Suchail S, Gauthier M, Maleszka R, Belzunces LP. Contrasting effects of imidacloprid on habituation. *Neurobiol Learn Mem.* 2001;76(2):183–191.
58. Decourte A, Devillers J, Aupinel P, Brun F, Bagnis C. Honeybee tracking with microchips: a new methodology to measure the effects of pesticides. *Ecotoxicology.* 2011;20:429–437.
59. Henry M, Beguin M, Requier F, Rollin O. A common pesticide decrease foraging success and survival in honey bees. *Science.* 2012;336:348–350.
60. Schneider CW, Tautz J, Grünewald B, Fuchs S. RFID tracking of sublethal effects of two neonicotinoid insecticides on the foraging behaviour of *Apis mellifera. PLoS ONE.* 2012;7:1–10.
61. Belzunces LP, Tchamitchan S, Brunet J-L. Neural effects of insecticides in the honey bee. *Apidologie.* 2012;43(3):348–370.
62. Grünewald B. Is pollination at risk? Current threats to and conservation of bees. *GAIA.* 2010;19:61–67.
63. Thompson HM. Behavioural effects of pesticides in bees-their potential for use in risk assessment. *Ecotoxicology.* 2003;12:317–330.
64. Gill RJ, Ramos-Rodriguez O, Raine NE. Combined pesticide exposure severely affects individual- and colony-level traits in bees. *Nature.* 2012;491:105–108.
65. Whitehorn PR, O'Connor S, Wackers FL, Goulson D. Neonicotinoid pesticide reduces bumble bee colony growth and queen production. *Science.* 2012;336:351–352.
66. Kiya T, Kubo T. Analysis of GABAergic and non-GABAergic neuron activity in the optic lobes of the forager and re-orienting worker honeybee (*Apis mellifera* L.). *PLoS ONE.* 2010;5:1–8.
67. Schäfer S, Bicker G. Common projection areas of 5-HT- and GABA-like immunoreactive fibers in the visual system of the honeybee. *Brain Res.* 1986;380:368–370.
68. Schäfer S, Bicker G. Distribution of GABA-like immunoreactivity in the brain of the honeybee. *J Comp Neurol.* 1986;246:287–300.
69. Bicker G, Schäfer S, Kingan TG. Mushroom body feedback interneurons in the honeybee show GABA- like immunoreactivity. *Brain Res.* 1985;360:394–397.
70. Ganeshina O, Menzel R. GABA-immunoreactive neurons in the mushroom bodies of the honeybee: an electron microscopic study. *J Comp Neurol.* 2001;437(3):335–349.
71. Grünewald B. Morphology of feedback neurons in the mushroom body of the honeybee, *Apis mellifera. J Comp Neurol.* 1999;404:114–126.
72. Dupuis JP, Bazelot M, Barbara GS, Paute S, Gauthier M, Raymond-Delpech V. Homomeric RDL and heteromeric RDL/LCCH3 GABA receptors in the honeybee antennal lobes: Two candidates for inhibitory transmission in olfactory processing. *J Neurophysiol.* 2010;103:458–468.
73. Grünewald B, Wersing A. An ionotropic GABA receptor in cultured mushroom body Kenyon cells of the honeybee and its modulation by intracellular calcium. *J Comp Physiol A.* 2008;194: 329–340.
74. El Hassani AK, Dacher M, Gauthier M, Armengaud C. Effects of sublethal doses of fipronil on the behavior of the honeybee (*Apis mellifera*). *Pharmacol Biochem Behav.* 2005;82:30–39.
75. El Hassani AK, Dupuis JP, Gauthier M, Armengaud C. Glutamatic and GABAergic effects of fipronil on olfactory learning and memory in the honeybee. *Invert Neurosci.* 2009;9:91–100.
76. Zachepilo TG, Il'inykh YF, Lopatina NG, et al. Comparative analysis of the locations of the NR1 and NR2 NMDA receptor subunits in honeybee (*Apis mellifera*) and fruit fly (*Drosophila melanogaster*, Canton-S wild-type) cerebral ganglia. *Neurosci Behav Physiol.* 2008;38(4):369–372.

77. Zannat MT, Locatelli F, Rybak E, Menzel R, Leboulle G. Identification and localisation of the NR1 sub-unit homologue of the NMDA glutamate receptor in the honeybee brain. *Neurosci Lett.* 2006;398(3):274–279.
78. Cully DF, Paress PS, Liu KK, Schaeffer JM, Arena JP. Identification of a *Drosophila melanogaster* glutamate-gated chloride channel sensitive to the antiparasitic agent avermectin. *J Biol Chem.* 1996;271(33):20187–20191.
79. Dubas F. Actions of putative amino acid neurotransmitters on the neuropile arborizations of locust flight motoneurons. *J Exp Biol.* 1991;155:337–356.
80. Horoszok L, Raymond V, Sattelle DB, Wolstenholme AJ. GLC-3: a novel fipronil and BIDN-sensitive, but picrotoxinin-insensitive, L-glutamate-gated chloride channel subunit from *Caenorhabditis elegans. Br J Pharmacol.* 2001;132(6):1247–1254.
81. Raymond V, Sattelle DB, Lapied B. Co-existence in DUM neurones of two GluCl channels that differ in their picrotoxin sensitivity. *NeuroReport.* 2000;11(12):2695–2701.
82. Usherwood PNR. Insect glutamate receptors. *Adv Insect Physiol.* 1994;24:309–341.
83. Wafford KA, Sattelle DB. A novel kainate receptor in the insect nervous system. *Neurosci Lett.* 1992;141:273–276.
84. Cleland TA. Inhibitory glutamate receptor channels. *Mol Neurobiol.* 1996;13:97–136.
85. Demares F, Raymond V, Armengaud C. Expression and localization of glutamate-gated chloride channel variants in honeybee brain (*Apis mellifera*). *Insect Biochem Mol Biol.* 2012; [Epub ahead of print]
86. Müßig L, Richlitzki A, Rößler R, Eisenhardt D, Menzel R, Leboulle G. Acute disruption of the NMDA receptor subunit NR1 in the honeybee brain selectively impairs memory formation. *J Neurosci.* 2010;30:7817–7825.
87. Xia S, Miyashita T, Fu TF, et al. NMDA receptors mediate olfactory learning and memory in *Drosophila. Curr Biol.* 2005;15(7): 603–615.
88. Miyashita T, Oda Y, Horiuchi J, Yin JCP. Mg^{2+} block of *Drosophila* NMDA receptors is required for long-term memory formation and CREB-dependent gene expression. *Neuron.* 2012;74:887–898.
89. Locatelli F, Bundrock G, Muller U. Focal and temporal release of glutamate in the mushroom bodies improves olfactory memory in *Apis mellifera. J Neurosci.* 2005;25(50):11614–11618.
90. Si A, Helliwell P, Maleszka R. Effects of NMDA receptor antagonists on olfactory learning and memory in the honeybee (*Apis mellifera*). *Pharmacol Biochem Behav.* 2004;77(2): 191–197.
91. Wu CL, Xia S, Fu TF, et al. Specific requirement of NMDA receptors for long-term memory consolidation in *Drosophila* ellipsoid body. *Nat Neurosci.* 2007;10:1578–1586.
92. Tabone CJ, Ramaswami M. Is NMDA receptor-coincidence detection required for learning and memory? *Neuron.* 2012;74:767–769.
93. Xia S, Chiang A-S. NMDA receptors in *Drosophila*. In: Van Dongen AM, ed. *Biology of the NMDA Receptor.* Boca Roca, FL: CRC Press; 2009.
94. Sachse S, Peele P, Silbering AF, Guhmann M, Galizia CG. Role of histamine as a putative inhibitory transmitter in the honeybee antennal lobe. *Front Zool.* 2006;3:22.
95. Bornhauser BC, Meyer EP. Histamine-like immunoreactivity in the visual system and brain of an orthopteran and a hymenopteran insect. *Cell Tissue Res.* 1997;287(1):211–221.
96. Nässel DR. Histamine in the brain of insects: a review. *Microsc Res Tech.* 1999;44:121–136.
97. Loesel R, Homberg U. Histamine-immunoreactive neurons in the brain of the cockroach *Leucophaea maderae. Brain Res.* 1999;842(2):408–418.
98. Pirvola U, Tuomisto L, Yamatodani A, Panula P. Distribution of histamine in the cockroach brain and visual system: an immunocytochemical and biochemical study. *J Comp Neurol.* 1988;276:514–526.
99. Homberg U, Hildebrand JG. Histamine-immunoreactive neurons in the midbrain and suboesophageal ganglion of sphinx moth *Manduca sexta. J Comp Neurol.* 1991;307(4):647–657.
100. Hauser F, Cazzamali G, Williamson M, Blenau W. A review of neurohormone GPCRs present in the fruit fly *Drosophila melanogaster* and the honey bee *Apis mellifera. Prog Neurobiol.* 2006;80:1–19.
101. Hammer M. An identified neuron mediates the unconditioned stimulus in associative olfactory learning in honeybees. *Nature.* 1993;366:59–63.
102. Kreissl S, Eichmüller S, Bicker G, Rapus J, Eckert M. Octopamine-like immunoreactivity in the brain and subesophageal ganglion of the honeybee. *J Comp Neurol.* 1994;348: 583–595.
103. Hammer M, Menzel R. Multiple sites of associative odor learning as revealed by local brain microinjections of octopamine in honeybees. *Learn Mem.* 1998;5:146–156.
104. Menzel R, Heyne A, Kinzel C, Gerber B, Fiala A. Pharmacological dissociation between the reinforcing, sensitizing, and response-releasing functions of reward in honeybee classical conditioning. *Behav Neurosci.* 1999;113:744–754.
105. Evans PD, Maqueira B. Insect octopamine receptors: a new classification scheme based on studies of cloned *Drosophila* G-protein coupled receptors. *Invert Neurosci.* 2005;5(3–4):111–118.
106. Grohmann L, Blenau W, Erber J, Ebert PR, Strünkes T, Baumann A. Molecular and functional characterization of an octopamine receptor from honeybee (*Apis mellifera*) brain. *J Neurochem.* 2003;86:725–735.
107. Sinakevitch I, Mustard JA, Smith BH. Distribution of the octopamine receptor AmOA1 in the honey bee brain. *PLoS ONE.* 2011;6:1–16.
108. Sinakevitch I, Niwa M, Strausfeld NJ. Octopamine-like immunoreactivity in the honey bee and cockroach: comparable organization in the brain and subesophageal ganglion. *J Comp Neurol.* 2005;488(3):233–254.
109. Beggs KT, Tyndall JDA, Mercer AR. Honey bee dopamine and octopamine receptors linked to intracellular calcium signaling have a close phylogenetic and pharmacological relationship. *PLOS ONE.* 2011;6:1–10.
110. Müller U. Neuronal cAMP-dependent protein kinase type II is concentrated in mushroom bodies of *Drosophila melanogaster* and the honeybee *Apis mellifera. J Neurobiol.* 1997;33:33–44.
111. Davis RL. Traces of *Drosophila* memory. *Neuron.* 2011;70:8–19.
112. Heisenberg M. Mushroom body memoir: from maps to models. *Nat Rev Neurosci.* 2003;4:266–275.
113. Perry CJ, Barron AB. Neural mechanisms of reward in insects. *Annu Rev Entomol.* 2012;58:543–562.
114. Blenau W, Erber J, Baumann A. Characterization of a dopamine D1 receptor from *Apis mellifera*: cloning, functional expression, pharmacology, and mRNA localization in the brain. *J Neurochem.* 1998;70(1):15–23.
115. Schäfer S, Rehder V. Dopamine-like immunoreactivity in the brain and suboesophageal ganglion of the honeybee. *J Comp Neurol.* 1989;280(1):43–58.
116. Beggs KT, Hamilton IS, Kurshan PT, Mustard JA, Mercer AR. Characterization of a D2-like dopamine receptor (AmDOP3) in honey bee, *Apis mellifera. Insect Biochem Mol Biol.* 2005; 35(8):873–882.
117. Humphries MA, Mustard JA, Hunter SJ, Mercer A, Ward P, Ebert PR. Invertebrate D2 type dopamine receptor exhibits age-based plasticity if expression in the mushroom bodies of the honeybee brain. *J Neurobiol.* 2003;55:315–330.

118. Mustard JA, Beggs KTMAR. Molecular biology of the invertebrate dopamine receptors. *Arch Insect Biochem Physiol.* 2005;59:103–117.
119. Macmillan CS, Mercer AR. An investigation of the role of dopamine in the antennal lobes of the honeybee, *Apis mellifera. J Comp Phys A.* 1987;160:359–366.
120. Vergoz V, Roussel E, Sandoz JC, Giurfa M. Aversive learning in honeybees revealed by the olfactory conditioning of the sting extension reflex. *PLoS ONE.* 2007;2(3):e288.
121. Beggs KT, Mercer AR. Dopamine receptor activation by honey bee queen pheromone. *Current Biology.* 2009;19:1206–1209.
122. Vergoz V, Schreurs HA, Mercer AR. Queen pheromone blocks aversive learning in young worker bees. *Science.* 2007;317 (5836):384–386.
123. Vergoz V, McQuillan HJ, Geddes LH, Pullar K, Nicholoson BJ. Peripheral modulation of worker bee responses to queen mandibular pheromone. *Proc Natl Acad Sci USA.* 2009;106: 20930–20935.
124. Schürmann F-W, Klemm N. Serotonin-immunoreactive neurons in the brain of the honeybee. *J Comp Neurol.* 1984;225:570–580.
125. Menzel R, Durst C, Erber J, et al. The mushroom bodies in the honeybee: from molecules to behaviour. In: Schildberger K, Elsner N, eds. *Fortschritte der Zoologie, Vol. 39: Neural Basis of Behavioural Adaptations.* Stuttgart: Gustav Fischer; 1994:81–102.
126. Menzel R, Wittstock S, Sugawa M. Chemical codes of learning and memory in honey bees. In: Squire L, Lindenlaub K, eds. *The Biology of Memory, Medica Hoechst Volume 23.* Stuttgart: Schattauer Verlagsgesellschaft; 1990:335–355.
127. Mercer AR, Menzel R. The effects of biogenic amines on conditioned and unconditioned responses to olfactory stimuli in the honeybee *Apis mellifera. J Comp Phys A.* 1982;145:363–368.
128. Sitaraman D, LaFerriere, Birman S, Zars T. Serotonin is critical for rewarded olfactory short-term memory in *Drosophila. J Neurogenet.* 2012;26(2):238–244.
129. Wright GA, Mustard JA, Simcock NK, et al. Parallel reinforcement pathways for conditioned food aversions in the honeybee. *Curr Biol.* 2010;21:2234–2240.
130. Thamm M, Balfanz S, Scheiner R, Baumann A, Blenau W. Characterization of the 5 HT 1A receptor of the honeybee (*Apis mellifera*) and involvement of serotonin in phototactic behaviour. *Cell Mol Life Sci.* 2010;67:2467–2479.
131. Schlenstedt J, Balfanz S, Baumann A, Blenau W. Am5-HT7: molecular and pharmacological characterization of the first serotonin receptor of the honeybee (*Apis mellifera*). *J Neurochem.* 2006;98:1985–1998.
132. Erber J, Masuhr T, Menzel R. Localization of short-term memory in the brain of the bee, *Apis mellifera. Physl Entomol.* 1980;5:343–358.
133. Giurfa M, Fabre E, Flaven-Pouchon J, et al. Olfactory conditioning of the sting extension reflex in honeybees: memory dependence on trial number, interstimulus interval, intertrial interval, and protein synthesis. *Learn Mem.* 2011;16:761–765.
134. Rath L, Galizia CG, Szyszka P. Multiple memory traces after associative learning in the honey bee antennal lobe. *Eur J Neurosci.* 2011;34:352–360.
135. Roussel E, Sandoz J-C, Giurfa M. Searching for learning-dependent changes in the antennal lobe: simultaneous recording of neural activity and aversive olfactory learning in honeybees. *Front Syst Neurosci.* 2010;4:1–12.
136. Strube-Bloss M, Nawrot MP, Menzel R. Mushroom body output neurons encode odor–reward associations. *J Neurosci.* 2011;23:3129–3140.
137. Szyszka P, Galkin A, Menzel R. Associative and non-associative plasticity in Kenyon cells of the honeybee mushroom body. *Front Syst Neurosci.* 2008;2:1–10.
138. Himmelreich S, Grünewald B. Cellular physiology of olfactory learning in the honeybee brain. *Apidologie.* 2012;43(3): 308–321.
139. Barbara GS, Grünewald B, Paute S, Gauthier M, Raymond-Delpech V. Study of nicotinic acetylcholine receptors on cultured antennal lobe neurones from adult honeybee brains. *Invert Neurosci.* 2008;8:19–29.
140. Kirschner S, Kleineidam CJ, Zube C, Rybak J, Grünewald B, Rossler W. Dual olfactory pathway in the honeybee, *Apis mellifera. J Comp Neurol.* 2006;499(6):933–952.
141. Mobbs PG. The brain of the honeybee *Apis mellifera*: I. The connections and spatial organization of the mushroom bodies. *Philos Trans R Soc Lond B.* 1982;298:309–354.
142. Eisenhardt D. Learning and memory formation in the honeybee (*Apis mellifera*) and its dependency on the cAMP-protein kinase A pathway. *Anim Biol.* 2006;56(2):259–278.
143. Fiala A, Müller U, Menzel R. Reversible downregulation of protein kinase A during olfactory learning using antisense technique impairs long-term memory formation in the honeybee, *Apis mellifera. J Neurosci.* 1999;19(22):10125–10134.
144. Hildebrandt H, Müller U. Octopamine mediates rapid stimulation of protein kinase a in the antennal lobe of honeybees. *J Neurobiol.* 1995;27(1):44–50.
145. Maqueira B, Chatwin H, Evans PD. Identification and characterization of a novel family of *Drosophila* beta-adrenergic-like octopamine G-protein coupled receptors. *J Neurochem.* 2005; 94(2):547–560.
146. Eisenhardt D, Kühn C, Leboulle G. The PKA–CREB system encoded in the honeybee genome. *Insect.* 2006;15(5):551–561.
147. Grünbaum L, Müller U. Induction of a specific olfactory memory leads to a long-lasting activation of protein kinase C in the antennal lobe of the honeybee. *J Neurosci.* 1998;18(11): 4384–4392.
148. Liu X, Davis RL. The GABAergic anterior paired lateral neuron suppresses and is suppressed by olfactory learning. *Nat Neurosci.* 2009;12:53–57.
149. Liu X, Krause WC, Davis RL. GABA(A) receptor RDL inhibits *Drosophila* olfactory associative learning. *Neuron.* 2007;20: 1090–1102.
150. Haehnel M, Menzel R. Sensory representation and learning-related plasticity in mushroom body extrinsic feedback neurons of the protocerebral tract. *Front Syst Neurosci.* 2010; 4:161.
151. Haehnel M, Menzel R. Long-term memory and response generalization in mushroom body extrinsic neurons in the honeybee *Apis mellifera. J Exp Biol.* 2012;215(Pt 3):559–565.
152. Das S, Sadanandappa MK, Dervan A, Larkin A, Lee JA, Sudhakaran JP. Plasticity of local GABAergic interneurons drives olfactory habituation. *Proc Natl Acad Sci USA.* 2011;108: E646–E654.
153. Leboulle G. Glutamate neurotransmission in the honey bee central nervous system. In: Galizia CG, Eisenhardt D, Giurfa M, eds. *Honeybee Neurobiology and Behavior.* Dordrecht: Springer; 2012:171–184.
154. Ellen CW, Mercer AR. Modulatory actions of dopamine and serotonin on insect antennal lobe neurons: insights from studies in vitro. *J Mol Hist.* 2012;43:401–404.
155. Wüstenberg DG, Grünewald B. Pharmacology of the neuronal nicotinic acetylcholine receptor of cultured Kenyon cells of the honeybee, Apis mellifera.. *J Comp Physiol A.* 2004;190 (10):807–821.

CHAPTER

36

Behavioral and Neural Analyses of Punishment Learning in Honeybees

Stevanus Rio Tedjakumala[*,†] *and Martin Giurfa*[*,†]

[*]Université de Toulouse, Centre de Recherches sur la Cognition Animale, Toulouse, France [†]Centre National de la Recherche Scientifique, Centre de Recherches sur la Cognition Animale, Toulouse, France

INTRODUCTION

Associative learning allows extracting the logical structure of the world because it enables making predictions about stimuli and their potential outcomes. Honeybees (*Apis mellifera*) constitute a traditional invertebrate model for the study of associative learning at the behavioral, cellular, and molecular levels.[1–5] For almost a century, research on honeybee learning and memory has focused almost exclusively on appetitive learning, exploiting the fact that bees can learn about a variety of sensory stimuli or to perform certain behaviors if these are rewarded with sucrose solution, the equivalent of nectar collected in flowers.[6] Since the discovery of the immense potential of this appetitive behavior by Karl von Frisch,[7] researchers interested in bee learning have concentrated on appetitive learning. Indeed, a single Pavlovian protocol, the olfactory conditioning of the proboscis extension reflex (PER),[8–10] has been used for approximately 50 years as the unique tool to access the neural and molecular bases of learning and memory in honeybees.[1,10] This protocol relies on PER, the appetitive reflex exhibited by a harnessed honeybee to sugar reward (the unconditioned stimulus (US)) delivered to its antennae and mouthparts.[9] After pairing a neutral odorant (the conditioned stimulus (CS)) and sucrose, the bee learns the association between odorant and food and therefore extends its proboscis in response to the odorant alone.[8–10]

In contrast, less was known about the capacity of honeybees to learn about aversive events in their environment. Can bees learn to avoid specific stimuli that have been associated with undesirable consequences? In the fruit fly *Drosophila melanogaster*, the other insect model that has emerged as a powerful model for the study of learning and memory,[11–14] aversive learning has been the dominant framework. The typical procedure consists of training groups of flies alternatively presented with two different odors, one paired with an electric shock (CS+) and another nonpaired with the shock (CS−).[15] Retention is measured in a T-maze, in which conditioned flies must choose between the CS+ and the CS− and in which they avoid the CS+ in case of successful learning and retention. Due to obvious differences in behavioral and motivational contexts, and to the impossibility to equate US nature and strength, caution is needed when comparing appetitive and aversive learning in bees and flies, respectively.

To determine if and how bees learn about punishment in their environment, and to fill the gap existing between bee and fruit fly research on learning and memory, we studied punishment learning in honeybees and established a new conditioning protocol based on the sting extension reflex (SER), which is a defensive response to potentially noxious stimuli.[16] This unconditioned response can be elicited by means of electric-shock delivery to a harnessed bee.[17,18] Because no appetitive responses are involved in this experimental context, true punishment learning could be studied for the first time in harnessed honeybees.

OLFACTORY CONDITIONING OF THE STING EXTENSION REFLEX

Using odorants forward-paired with electric shocks, we conditioned the SER so that bees learned to extend their sting in response to odorants previously punished.[19]

Invertebrate Learning and Memory.
DOI: http://dx.doi.org/10.1016/B978-0-12-415823-8.00036-8

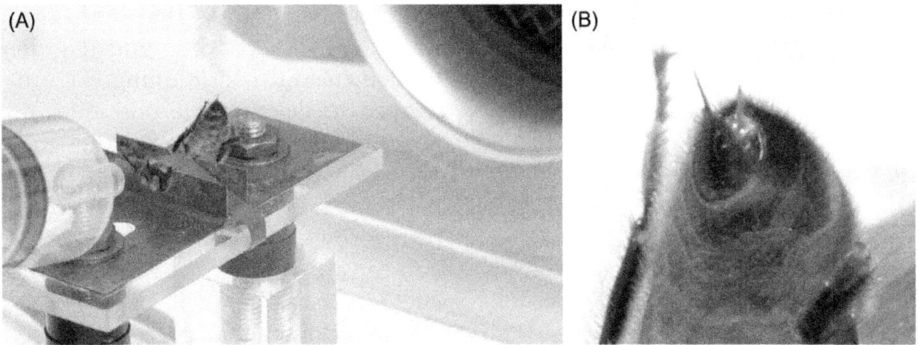

FIGURE 36.1 (A) View of a honeybee in the experimental setup. The bee is fixed between two brass plates set on a plexiglass plate, with EEG cream smeared on the two notches to ensure good contact between the plates and the bee and also a girdle that clamped the thorax to restrain mobility. The bee closes a circuit and receives a mild electric shock (7.5 V) that induces the sting extension reflex. An originally neutral odorant is delivered through a 20-mL syringe placed 1 cm from the antennae. Odorant stimulation lasted 5 sec. The electric shock started 3 sec after odorant onset and lasted 2 sec so that it ended with odorant offset. Contamination with remains of odorants used for conditioning or pheromones is avoided via an air extractor that is on continuously. (B) The sting extension response elicited upon stimulation with an electric shock of 7.5 V. Source: *Adapted from Vergoz* et al.[19]

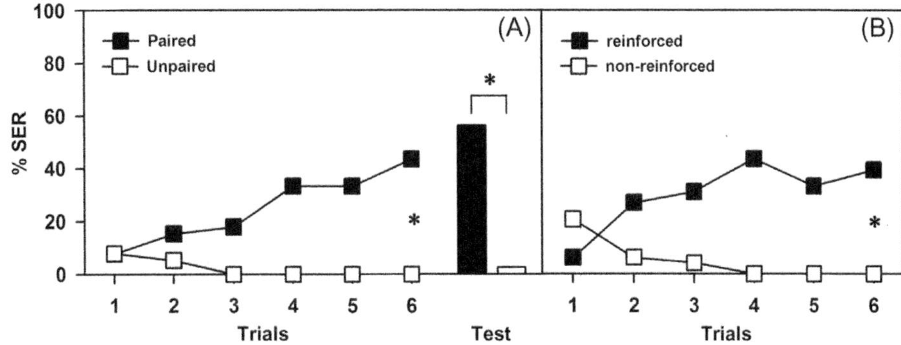

FIGURE 36.2 (A) Responses (SER) of two groups of bees, one trained with an odorant explicitly paired with an electric shock (black squares; $n = 38$) and the other with the same odorant and an unpaired electric shock (white squares; $n = 39$), during 6 trials. Only the bees in the paired group learned the association and extended their sting as a response to the odorant. One hour after conditioning, an olfactory aversive memory was present in the paired (black bar) but not in the unpaired group (white bar). (B) Responses (SER) of a group of bees ($N = 48$) trained in a differential conditioning procedure to discriminate an odorant reinforced with an electric shock (black squares) and a non-reinforced odorant (white squares) during 12 trials (6 reinforced and 6 non-reinforced). Bees learned to discriminate between odorants as a result of conditioning. $^*p < 0.0001$. Source: *Adapted from Vergoz* et al.[19]

Honeybees were fixed individually on a metallic holder so that they built a bridge between two brass plates through which a 2-sec mild electric shock (7.5 V) was delivered by a stimulator (60-Hz AC current) (Figure 36.1A). Bees treated in this way extend their sting in response to the electric shock (Figure 36.1B).[17,18] Bees were trained with explicitly paired presentations of an odorant and the electric shock. In parallel, bees of an unpaired group were presented with explicitly unpaired presentations of odorant and shock.[19]Figure 36.2 shows that bees from the paired group significantly increased conditioned SER to the odorant that preceded the electric shock, whereas bees in the unpaired group showed no significant change in responsiveness to the unpaired odorant during trials. Thus, the increase in SER observed in the paired group was due to associative learning and not to the simple experience with the odorant and the shock. One hour after conditioning, bees of the paired group still remembered the conditioned odorant, whereas bees of the unpaired did not respond to the odorant (Figure 36.2A, black and white bars). An aversive memory retrievable 1 hr after learning was therefore established in the paired but not in the unpaired group.[19]

Bees could also learn to extend their sting to an odorant paired with an electric shock and not to respond to a non-reinforced odorant (differential conditioning). This procedure is a typical within-subject control in studies of associative learning. Bees were conditioned during six reinforced and six non-reinforced trials, presented in a pseudorandom sequence. The resulting learning curves (Figure 36.2B) show that bees learned to discriminate between odorants as a result of conditioning. Thus, olfactory conditioning of SER is truly associative and does not depend on the simple exposure to the training stimuli, independent of their outcome.[19]

OLFACTORY CONDITIONING OF SER IS A TRUE CASE OF AVERSIVE LEARNING

Pairing an odor with the electric shock results in the odor gradually gaining control over the SER. Because the animals are restrained in individual holders, their eventual avoidance of the punished odor cannot be assessed. In *Drosophila*, the aversive nature of conditioning is clear because after successful conditioning the animals clearly avoid the CS + in choice tests. In the case of olfactory SER conditioning, the term *aversive* could be considered inappropriate given that no response inhibition or avoidance is observed during conditioning and that the orientation behavior of bees toward the CS has not been evaluated. Would conditioned bees explicitly avoid the CS, showing that the odor acquired an aversive value?

We studied whether freely moving bees avoid the CS after successful SER conditioning. To provide a comparison framework with appetitive conditioning, we analyzed the orientation behavior of freely walking honeybees presented with odors in a mini Y-maze under red light (i.e., in the dark for bees) after successful olfactory PER conditioning or SER conditioning.[20] We explicitly asked whether SER-conditioned bees avoid the CS in accordance with the aversive punishment associated with it, whereas PER-conditioned bees approach it in accordance with the appetitive nature of sucrose reward used in this conditioning.[20]

Figure 36.3 shows that both groups of bees (those trained in PER conditioning (Figure 36.3A) and those trained in SER conditioning (Figure 36.3B)) efficiently learned to discriminate odorants. Appetitively trained bees significantly increased PER to the rewarded odor and decreased it to the non-rewarded odor. Bees that performed correctly in the last two blocks of trials, responding only to the rewarded and not to the non-rewarded odor, were tested in the Y-maze 1 hr later. Aversively trained bees also learned to differentiate the odorant that preceded the electric shock from the non-reinforced odorant in the course of training. They significantly increased their sting extension responses to the punished odor and decreased their responses to the nonpunished odor. Bees that responded correctly in the two last blocks of trials were tested in the Y-maze 1 hr later. Once in the maze, bees that learned the appetitive discrimination preferred the odor previously paired with sucrose in the Y-maze (Figure 36.3C). On the contrary, bees that learned the aversive discrimination avoided the odor previously paired with shock in the Y-maze, thus preferring the previously non-reinforced odor (Figure 36.3C). Therefore, the inhibitory, aversive nature of SER conditioning was revealed by this avoidance behavior expressed when the bees had the opportunity to freely choose between CS+ and CS −.[20]

The bees used in our experiments were mostly foragers captured at the hive entrance when departing from the hive. Therefore, we cannot exclude that repeating the same experiments with guards yields a different result (i.e., bees orienting toward the odor paired with shock and exhibiting SER) or that providing contextual stimuli such as odors from the hive or social pheromones within the Y-maze also changes the

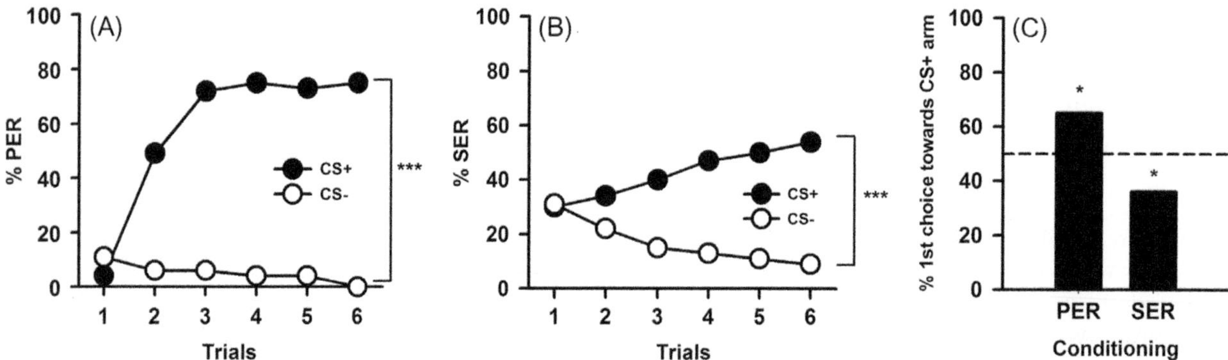

FIGURE 36.3 (A) Appetitive conditioning. Percentage of PER in bees trained with an odorant explicitly reinforced with sucrose solution (CS + , black circles, $n = 142$) and with a non-reinforced odorant (CS − , white circles). Bees learned to differentiate between CS + and CS − in the course of training (***$p < 0.001$). (B) Aversive conditioning. Percentage of SER in bees trained with an odorant explicitly reinforced with electric shock (black circles, $n = 238$) and with an odorant explicitly non-reinforced (white circles). Bees learned to differentiate between CS + and CS − in the course of training (***$p < 0.001$). (C) Orientation of honeybees in the Y-maze 1 hr after associative olfactory conditioning. The graphs show the first choice toward the arm containing the CS + , after PER conditioning ($n = 79$) and SER conditioning ($n = 72$). The dashed line at 50% indicates random choice between CS + and CS − arms. After PER conditioning, honeybees significantly chose the CS + . On the contrary, after SER conditioning, honeybees significantly avoided the CS + (*$p < 0.05$). Source: *Adapted from Carcaud* et al.[20]

response of the bees toward the odor previously punished. Despite these particularities, the results obtained so far demonstrate that SER conditioning in honeybees is a true case of aversive conditioning.[20]

OLFACTORY CONDITIONING OF SER LEADS TO THE FORMATION OF LONG-TERM MEMORIES

In restraining conditions, retention tests showed the presence of aversive memories 1 hr after conditioning.[19] This period corresponds, in appetitive learning, to mid-term memory, which is independent of protein synthesis and thus relatively labile.[1] Does aversive SER conditioning induce the formation of long-term memories retrievable several days after learning, as is the case in appetitive learning[1]? In the honeybee, one pairing of an odorant with sucrose (i.e., one conditioning trial) leads to an early long-term memory (eLTM) that can be retrieved 24–48 hr after conditioning. This eLTM depends on translation but not on gene transcription and thus is not affected by transcription inhibitors such as actinomycin D. On the other hand, three conditioning trials lead to a stable late long-term memory (lLTM) that can be retrieved 72 hr or more after conditioning. Contrary to eLTM, lLTM requires gene transcription and thus can be inhibited by actinomycin D.[1,2,10,21]

We thus asked whether olfactory SER conditioning leads to the formation of memories retrievable 1, 24, 48, and 72 hr after differential conditioning, the latter period corresponding to lLTM as characterized in olfactory PER conditioning.[1,2,10,21] An independent group of bees was used for each retention time. All groups learned to discriminate the CS+ from the CS− and reached comparable levels of discrimination at the end of training. After conditioning, bees exhibited significant retention and responded more to the CS+ than to the CS− (Figure 36.4A). Therefore, these results show that SER conditioning leads to robust long-term memories that are retrievable even 3 days after training.[22]

We then asked whether the 3-day LTM depends on *de novo* protein synthesis. The conditioning procedure was identical to that of the previous experiment. In the 2 hr following conditioning, bees were injected with PBS (control group), 10^{-2} M anisomycin (a translation inhibitor), or 1.5×10^{-3} M actinomycin D (a transcription inhibitor). Seventy-two hours after conditioning, retention performances varied depending on treatment (Figure 36.4B). Retention was significant in control bees injected with PBS but not in bees injected with either anisomycin or actinomycin D. Thus, both translation and transcription are essential for LTM formation and retrieval of the odor–shock association 72 hr after training.[22]

Therefore, these results show that aversive learning can induce a robust and stable lLTM that relies on protein synthesis because it depends both on translation and on transcription. Bees thus have the capacity to remember aversive experiences long after the experiences have taken place. The biological contexts in which such capacity could be applied are multiple. On the one hand, foragers could avoid in this way returning to

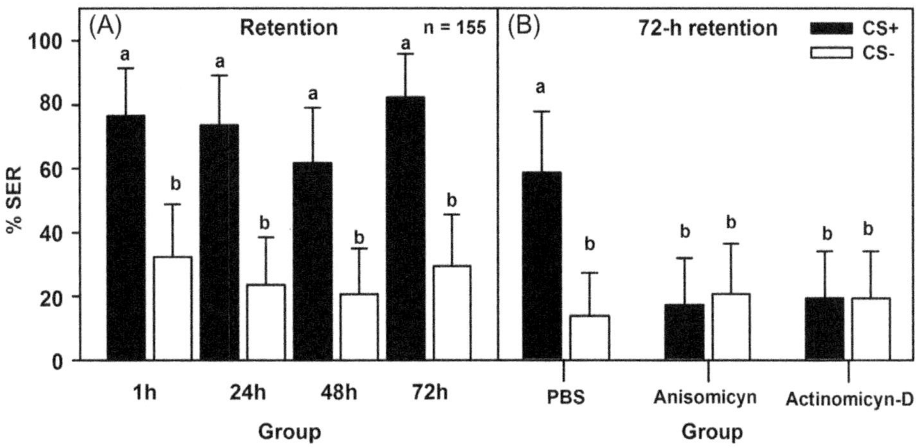

FIGURE 36.4 (A) Memory retention after SER differential conditioning. Percentage of SER (+ 95% confidence interval) to the CS+ (black bars) and to the CS − (white bars). Four groups of bees were trained in parallel (acquisition) and then tested after different retention intervals (1, 24, 48, and 72 hr postconditioning). Each group was tested once. Letters indicate significant differences. All groups remembered the discrimination learned during training. (B) Dependency of l-LTM (72-hr retention) on translation and transcription. Three groups of bees were trained in parallel (acquisition) and tested 72 hr after the last acquisition trial and after injection of PBS, anisomycin, or actinomycin D. Each group was tested once. Letters indicate significant differences. Only the group injected with PBS (control) remembered the discrimination learned during training; inhibition of transcription (actinomycin D) or translation (anisomycin) resulted in the absence of l-LTM. Source: *Adapted from Giurfa* et al.[22]

food places in which negative experiences, or eventually unfulfilled expectations, occurred, thus enhancing foraging efficiency. On the other hand, aversive memories could help organizing defensive responses against enemies whose odors have been previously experienced. It may thus be adaptive to memorize and remember during long periods the smell of predators in order to exhibit appropriate defensive responses.

THE NEURAL BASIS OF AVERSIVE LEARNING

CS Signaling

Odorants are processed at different stages in the bee brain (Figure 36.5). Olfactory detection starts at the level of the antennae, where olfactory receptor neurons are located within specialized hairs called sensilla. Sensory neurons endowed with molecular olfactory receptors convey information about odorants to the antennal lobe. Each antennal lobe is composed of 160 globular structures called glomeruli. Glomeruli are synaptic interaction sites between olfactory receptor neurons, local inhibitory interneurons interconnecting glomeruli, and projection neurons conveying processed olfactory information to higher order centers such as the lateral horn and the mushroom bodies. The latter are considered to be higher order integration centers because they receive input from visual, gustatory, and mechanosensory pathways in addition to the olfactory pathway.

Neural activity has been measured at the level of the antennal lobe, and these measurements were

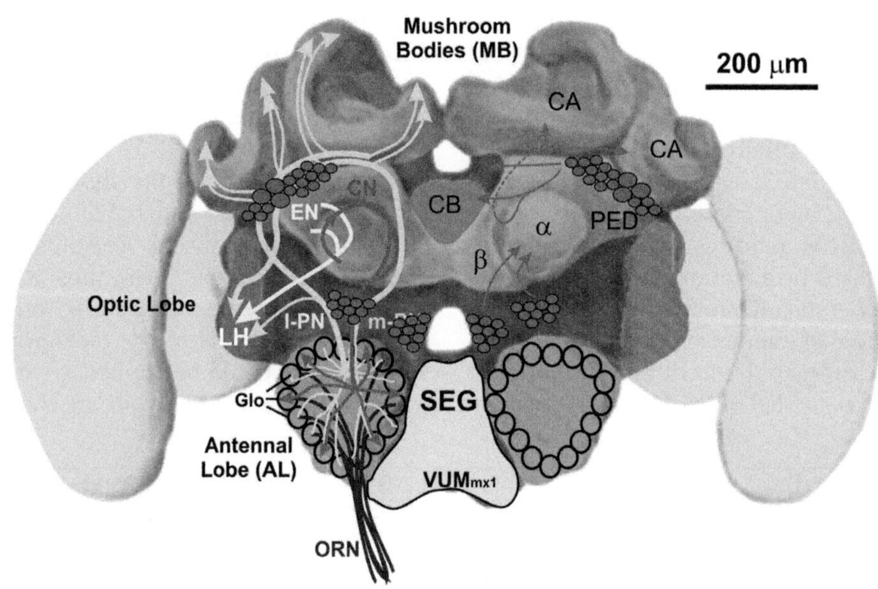

FIGURE 36.5 **Neural substrates for CS and aversive-US information in the honeybee brain.** The CS pathway is shown in more detail on the left side. The antennal lobe (AL), first-order olfactory neuropil, receives input from approximately 60,000 olfactory receptor neurons (ORNs), which detect odorants within sensilla on the antenna. Within the AL's anatomical and functional units, the 160 glomeruli (Glo), ORNs contact approximately 4000 inhibitory local neurons (not shown), which carry out local computations, and approximately 800 projection neurons (PN), which convey processed information to higher brain centers via different tracts. The lateral antenno-protocerebralis tract (l-PN) projects first to the lateral horn (LH) and then to the mushroom body calyces (CA), within the lips and the basal ring. The medial tract of projection neurons (m-PN) projects to the same structures but in the reverse order. The dendrites of the Kenyon cells, the mushroom bodies' 170,000 intrinsic neurons, form the CA, whereas their axons form the pedunculus (PED) constituted by two output lobes—the vertical (or α) lobe and the horizontal (or β) lobe, formed by two collaterals of each KC axon. Within the MBs, feedback neurons (not shown) project from the PED and lobes back to the CA, providing inhibitory feedback to the MB input regions. Extrinsic neurons (ENs) take information from the pedunculus and the lobes and project to different parts of the protocerebrum, but most conspicuously to the LH. Moreover, centrifugal neurons (CN) are thought to be involved in a retrograde modulation of AL circuits. Dopaminergic neuron clusters C1–C3, whose activity may mediate aversive US reinforcement, are shown in red. Red arrows indicate possible dendritic arborizations/axonal projections as estimated in Schäfer S, Rehder.[33] C1 clusters are located in the inferior medial protocerebrum. The almost adjacent C2 clusters are found inferiorly to the α lobe (α). Expanding themselves from the most anterior to the most posterior part of the cerebrum, the C3 clusters are observed at the superior border of protocerebrum, below the CA of the MBs. The C1 and C2 clusters, each consisting of approximately 60–70 cell bodies, send their processes ventromedially into the α lobes. Three main processes emanate from the C3 clusters, which consist of approximately 140 cell bodies. The first goes to a small, most anterior region of the superior medial protocerebrum; the second goes to the central body (CB); and the third goes along the dorsal border of the α lobe, makes a turn at the border of the CB, and innervates straight both calyces homogeneously. Various dopaminergic cell bodies (1–10) are observed sporadically in the cerebrum as well as in the subesophageal ganglion (SEG).

coupled with SER conditioning to determine whether punishment learning induces changes in the neural representation of odorants learned.[23] Studies on olfactory coding at the level of the antennal lobe showed that odorants are encoded as odor-specific spatiotemporal patterns of glomerular activity.[24] Simultaneous recordings of conditioned SER upon differential conditioning and calcium activity at the level of the antennal lobe revealed no differences between glomerular responses to the CS+ and those to the CS− in bees that learned the discrimination. Thus, the olfactory memory traces generated by SER conditioning could be located downstream of the antennal lobe, for instance, in the mushroom bodies.[25]

US Signaling

A significant advantage of olfactory SER conditioning is that bees exhibit learning and retention performances while being harnessed. Immobilized animals offer the possibility of accessing the neural mechanism of learning and memory through different disruptive/invasive procedures.[5] Such studies have been initiated to determine the neural pathways and brain structures mediating aversive olfactory learning and memory.

We used neuropharmacological blocking to determine which neurotransmitter system mediates punishment signals in the honeybee brain. Previous work had shown that octopamine is crucial for representing sucrose reward in the bee brain.[26–28] Indeed, pairing an odor with injections of octopamine in the bee brain, which substitute for sucrose reward, leads to olfactory learning in harnessed bees.[27] On the other hand, dopamine mediates aversive reinforcement in fruit flies[29–31] and crickets.[32] We thus studied the role and implication of these two biogenic amines in punishment learning in honeybees.

Ringer solution (control), octopaminergic receptor antagonists (mianserine and epinastine), or dopaminergic receptor antagonists (fluphenazine and flupentixol) were injected into the bee brain 30 min before differential conditioning with two odors, one paired with the shock and the other nonpaired with shock. Ringer-injected bees learned to discriminate the reinforced from the non-reinforced odorant (Figure 36.6A). One hour later, they remembered the aversive association and extended their sting in response to the previously punished odorant. Octopaminergic antagonists (mianserine or epinastine) did not affect performance at any of the concentrations used in these experiments (Figure 36.6B): Mianserine- and epinastine-injected bees learned to discriminate the two odorants and responded with SER only to the odorant paired with the electric shock. Retention tests also showed significant discrimination. Thus, octopaminergic antagonists did not impair aversive olfactory learning in honeybees. On the contrary, dopaminergic antagonists (fluphenazine and flupentixol) had a dramatic effect on aversive olfactory learning. Flupentixol-injected bees did not learn to discriminate between odorants. Consequently, they did not show discrimination in the tests performed 1 hr later (Figure 36.6C). Fluphenazine had a similar effect although with less effectiveness. These results therefore show that dopamine, but not octopamine, signaling is necessary for aversive olfactory learning in honeybees.[19]

DOPAMINERGIC NEURONS IN THE BEE BRAIN

Precise morphological characterization of dopaminergic neurons in the honeybee brain is still lacking. Immunocytochemistry studies using an antiserum against dopamine were performed in 1989,[33] but the technique used to stain candidate dopaminergic neurons did not allow determining whether labeled neurons were neurons producing dopamine (true dopaminergic neurons) or neurons incorporating dopamine.

Dopamine-like immunoreactive neurons were identified in most parts of the brain and in the subesophageal ganglion[33] (Figure 36.5). Only the optic lobes are devoid of label. Approximately 330 dopamine immunoreactive cell bodies were found in each brain hemisphere plus respective subesophageal hemiganglion. Most of the labeled cell bodies were situated within three clusters: One mostly below the lateral calyx (C3) and two (C1 and C2) below the α lobe in the inferior medial protocerebrum. Other labeled cell bodies lied dispersed or in small groups around the protocerebral bridge, below the optic tubercles, proximal to the inferior rim of the lobula, and in the lateral and inferior somatal rind of the subesophageal ganglion. However, due to limitations of the staining technique, not all of the dendritic arborizations and axons of these neurons could be visualized so that where and how dopaminergic circuits contact the olfactory pathway remain to be determined.[33]

MODULARITY OF REWARD AND PUNISHMENT SYSTEMS IN HONEYBEES

Reward and punishment systems are mediated by different sets of aminergic neurons, with reward being signaled by octopaminergic neurons and punishment being signaled by dopaminergic neurons. These two systems could thus constitute insulated modules, each acting as a separate value system, allowing strict separation of appetitive and aversive experiences. As a

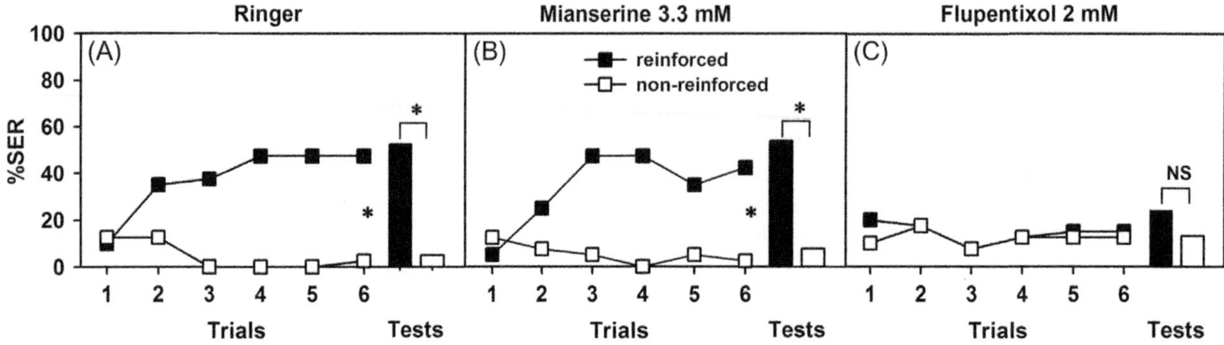

FIGURE 36.6 The effect of octopaminergic and dopaminergic receptor antagonists on olfactory conditioning of the SER. Responses (SER) of bees trained to discriminate an odorant reinforced with an electric shock (black squares) and a non-reinforced odorant (white squares) during 12 acquisition trials (6 reinforced and 6 non-reinforced). A retention test was conducted 1 hr after the last acquisition trial (black bar: odorant previously reinforced; white bar: odorant previously non-reinforced). (A) Responses (SER) of control bees injected with Ringer into the brain ($n = 40$); (B) responses (SER) of bees injected with the octopaminergic antagonist mianserine 3.3 mM into the brain ($n = 40$); (C) responses (SER) of bees injected with the dopaminergic antagonist flupentixol 2 mM into the brain ($n = 40$). Ringer- and mianserine-injected bees learned to discriminate the reinforced from the non-reinforced odorant and remembered the difference 1 hr later. Flupentixol-injected bees did not learn to discriminate the reinforced from the non-reinforced odorant, nor did they respond appropriately in the retention tests. Similar results were obtained with other concentrations of octopaminergic and dopaminergic antagonists. These results show that dopamine but not octopamine receptors are required for aversive olfactory learning in honeybees. $*p < 0.0001$. Source: *Adapted from Vergoz* et al.[19]

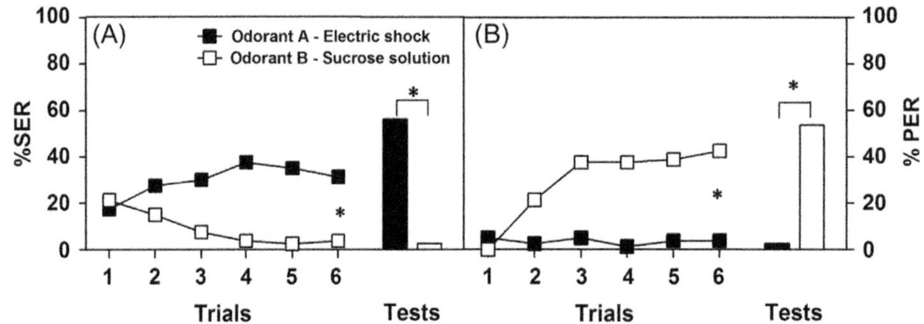

FIGURE 36.7 Simultaneous aversive and appetitive learning in honeybees. A single group of bees (SER–PER group; $n = 80$) was trained in a double discrimination task with an odorant ('A') paired with an electric shock that elicited SER and with another odorant ('B') paired with sucrose solution delivered to the antennae and proboscis that elicited PER. (A) Bees responded significantly with a SER to the odorant associated with the electric shock (black squares and bar) but not to that associated with sucrose (white squares and bar). (B) The same bees responded significantly with a PER to the odorant associated with sucrose (white squares and bar) but not to that associated with electric shock (black squares and bar). One hour after the last conditioning trial, bees still responded correctly to the odorants (bars) in the absence of their respective USs. Appetitive and aversive learning can thus be mastered simultaneously. $*p < 0.0001$. Source: *Adapted from Vergoz* et al.[19]

consequence, bees should be able to master efficiently appetitive and aversive associations simultaneously.

We tested this idea by training honeybees in a double-association task in which one odor was paired with electric shock and the other was paired with sucrose solution.[19] Their performance is presented in Figures 36.7A and 36.7B. Bees responded significantly with a SER to the odorant associated with the shock but not to that associated with sucrose, whereas they responded significantly with a PER to the odorant associated with sucrose but not to that associated with shock. One hour after the last conditioning trial, bees still responded correctly to the odorants even in the absence of punishment or reward. Control experiments (not shown) suggested that whereas the formation of the appetitive association did not interfere with aversive conditioning, the formation of the aversive odorant–shock association (and not the shock alone) induced a performance decrease during appetitive conditioning. However, the fact that some bees manage to learn both associations simultaneously supports the notion that appetitive and aversive olfactory learning are mediated by relatively independent neural systems dedicated to the processing of appetitive and aversive associations.

CONCLUSION

Punishment learning in honeybees became experimentally accessible due to the development of a conditioning protocol in which harnessed bees learn to associate odorants with an electric shock. In addition to behavioral quantifications, this protocol has enabled accessing the bee brain while the insect learns and memorizes, thus uncovering neural principles of associative aversive learning in a framework that is distinct from any appetitive behavior.

Previous studies focused on avoidance learning in foraging bees and therefore preserved an appetitive framework. In such studies, free-flying bees foraging for food learned to avoid flower patches infested with crab spiders[34,35] or artificial flowers penalized either with quinine[36,37] or with a puff of compressed air.[38] They also learned to avoid landing on five out of six petals of a mechanical flower that flicked forward and hit them upon landing.[39] Electric shock has also been used, although seldom, to generate avoidance of visited food sources in free-flying honeybees.[40,41] All these studies have in common the impossibility of accessing the nervous system in parallel to behavioral recording because they used free-flying bees. Furthermore, they all maintain an appetitive framework as they aim to inhibit the appetitive response of food search. The appetitive framework is also present in a variant of olfactory PER conditioning in which after pairing an odorant and sucrose, an electric shock is delivered to the proboscis so that bees learn to retract it in response to the odorant.[42] Our assay, in contrast, precludes confounding appetitive responses and converges on experimental conditioning procedures traditionally used for *Drosophila* in which odorants are directly paired with electric shock[15] without involving an appetitive context. In the fruit fly, this assay led to significant progress in the study of learning and memory.[11,12,14,43–45] However, olfactory SER conditioning has a significant advantage with respect to olfactory conditioning in *Drosophila* because it does not contain orienting or locomotion components, which could be interpreted as an operant component in an otherwise classical conditioning paradigm. When these components are allowed, by putting trained bees in a Y-maze in which they can freely express their choice behavior, bees, like flies, avoid the odorant previously punished, thus demonstrating the true aversive nature of the learning acquired. Therefore, we expect that our assays will facilitate new research and comparative studies on the neurobiology of aversive learning and memory, which have not been feasible until now.

That this is indeed the case is shown by the discovery that dopaminergic signaling mediates the reinforcing properties of the electric shock in bees. More generally, we posit that most forms of aversive reinforcement would be signaled by dopaminergic neurons. In this way, appetitive and aversive learning would be supported by independent modules corresponding to separate neural systems dedicated to the processing of the different unconditioned stimuli. Octopamine has been strongly implicated in appetitive olfactory learning in bees, and octopamine injections in the brain can substitute for sucrose reward and induce olfactory learning.[27] We found that dopamine, but not octopamine, underlies aversive olfactory learning, thus suggesting that dopamine is linked to aversive learning across insect species.[29–32]

Our results suggest that dopamine plays an instructive function in aversive learning, possibly conveying information about punitive stimuli. Dopaminergic neurons capable of mediating and predicting aversive stimuli have been found in the *Drosophila* brain.[31,46] Specific subsets of these neurons are required for the formation of different types of aversive memories.[30] Two types of dopaminergic neurons, MB-M3 and MB-MP1, are required for the formation of labile olfactory memories, but they target distinct subdomains of the mushroom bodies. It has been thus proposed that these reinforcement circuits might induce different forms of aversive memory in spatially segregated synapses in the mushroom bodies.[30]

However, dopaminergic signaling also seems to be involved in appetitive learning in fruit flies[47] and in appetitive control of sugar sensing.[48] Therefore, we suggest that an important function of modulatory, aminergic neurons in insects is to act as value systems in associative learning phenomena—that is, as a system allowing ordering, prioritizing, and assigning appropriate labels to learned stimuli.[49] Alternatively, they may play a fundamental role in perceptual suppression, thus directing the insect's attention toward relevant cues that need to be learned or that require appropriate responding.[50]

In our opinion, SER conditioning constitutes a significant contribution in the study of different aspects of honeybee behavior. We expect it to complement the monofaceted view offered during approximately the past 100 years by the equivalent protocol in the appetitive domain, the olfactory conditioning of PER. The results presented in this chapter show that this expectation is justified and that new research pathways have been opened through this new protocol.

References

1. Menzel R. Memory dynamics in the honeybee. *J Comp Physiol A*. 1999;185:323–340.
2. Menzel R. Searching for the memory trace in a mini-brain, the honeybee. *Learn Mem*. 2001;8:53–62.
3. Menzel R, Giurfa M. Cognitive architecture of a mini-brain: the honeybee. *Trends Cognit Sci*. 2001;5:62–71.
4. Giurfa M. Cognitive neuroethology: dissecting non-elemental learning in a honeybee brain. *Curr Opin Neurobiol*. 2003;13:726–735.

5. Giurfa M. Behavioral and neural analysis of associative learning in the honeybee: a taste from the magic well. *J Comp Physiol A*. 2007;193:801–824.
6. von Frisch K. *The Dance Language and Orientation of Bees*. Cambridge, MA: Harvard University Press; 1967.
7. von Frisch K. Der farbensinn und formensinn der biene. *Zool Jahrb Abt Allg Zool Physiol Tiere*. 1914;37:1–238.
8. Takeda K. Classical conditioned response in the honey bee. *J Insect Physiol*. 1961;6:168–179.
9. Bitterman ME, Menzel R, Fietz A, Schäfer S. Classical conditioning of proboscis extension in honeybees (*Apis mellifera*). *J Comp Psychol*. 1983;97:107–119.
10. Giurfa M, Sandoz JC. Invertebrate learning and memory: fifty years of olfactory conditioning of the proboscis extension response in honeybees. *Learn Mem*. 2012;19:54–66.
11. Davis RL. Olfactory memory formation in *Drosophila*: from molecular to systems neuroscience. *Annu Rev Neurosci*. 2005;28:275–302.
12. Keene AC, Waddell S. *Drosophila* olfactory memory: single genes to complex neural circuits. *Nat Rev Neurosci*. 2007;8:341–354.
13. Heisenberg M. Mushroom body memoir: from maps to models. *Nat Rev Neurosci*. 2003;4:266–275.
14. Margulies C, Tully T, Dubnau J. Deconstructing memory in *Drosophila*. *Curr Biol*. 2005;15:R700–713.
15. Tully T, Quinn WG. Classical conditioning and retention in normal and mutant *Drosophila melanogaster*. *J Comp Physiol Psychol*. 1985;156:263–277.
16. Breed MD, Guzman-Novoa E, Hunt GJ. Defensive behavior of honey bees: organization, genetics, and comparisons with other bees. *Annu Rev Entomol*. 2004;49:271–298.
17. Núñez J, Almeida L, Balderrama N, Giurfa M. Alarm pheromone induces stress analgesia via an opioid system in the honeybee. *Physiol Behav*. 1997;63:75–80.
18. Burrell BD, Smith BH. Age- but not case-related regulation of abdominal mechanisms underlying the sting reflex of the honey bee, *Apis mellifera*. *J Comp Physiol A*. 1994;174:581–592.
19. Vergoz V, Roussel E, Sandoz JC, Giurfa M. Aversive learning in honeybees revealed by the olfactory conditioning of the sting extension reflex. *PLoS ONE*. 2007;2:e288.
20. Carcaud J, Roussel E, Giurfa M, Sandoz JC. Odour aversion after olfactory conditioning of the sting extension reflex in honeybees. *J Exp Biol*. 2009;212:620–626.
21. Schwärzel M, Müller U. Dynamic memory networks: dissecting molecular mechanisms underlying associative memory in the temporal domain. *Cell Mol Life Sci*. 2006;63:989–998.
22. Giurfa M, Fabre E, Flaven-Pouchon J, et al. Olfactory conditioning of the sting extension reflex in honeybees: memory dependence on trial number, interstimulus interval, intertrial interval, and protein synthesis. *Learn Mem*. 2009;16:761–765.
23. Roussel E, Sandoz JC, Giurfa M. Searching for learning-dependent changes in the antennal lobe: simultaneous recording of neural activity and aversive olfactory learning in honeybees. *Front Behav Neurosci*. 2010;4:155.
24. Joerges J, Küttner A, Galizia CG, Menzel R. Representation of odours and odour mixtures visualized in the honeybee brain. *Nature*. 1997;387:285–288.
25. Gerber B, Tanimoto H, Heisenberg M. An engram found? Evaluating the evidence from fruit flies. *Curr Opin Neurobiol*. 2004;14:737–744.
26. Hammer M. An identified neuron mediates the unconditioned stimulus in associative olfactory learning in honeybees. *Nature*. 1993;366:59–63.
27. Hammer M, Menzel R. Multiple sites of associative odor learning as revealed by local brain microinjections of octopamine in honeybees. *Learn Mem*. 1998;5:146–156.
28. Farooqui T, Robinson K, Vaessin H, Smith BH. Modulation of early olfactory processing by an octopaminergic reinforcement pathway in the honeybee. *J Neurosci*. 2003;23:5370–5380.
29. Schwaerzel M, Monastirioti M, Scholz H, Friggi-Grelin F, Birman S, Heisenberg M. Dopamine and octopamine differentiate between aversive and appetitive olfactory memories in *Drosophila*. *J Neurosci*. 2003;23:10495–10502.
30. Aso Y, Siwanowicz I, Bracker L, Ito K, Kitamoto T, Tanimoto H. Specific dopaminergic neurons for the formation of labile aversive memory. *Curr Biol*. 2010;20:1445–1451.
31. Claridge-Chang A, Roorda RD, Vrontou E, et al. Writing memories with light-addressable reinforcement circuitry. *Cell*. 2009;139:405–415.
32. Unoki S, Matsumoto Y, Mizunami M. Participation of octopaminergic reward system and dopaminergic punishment system in insect olfactory learning revealed by pharmacological study. *Eur J Neurosci*. 2005;22:1409–1416.
33. Schäfer S, Rehder V. Dopamine-like immunoreactivity in the brain and suboesophageal ganglion of the honey bee. *J Comp Neurol*. 1989;280:43–58.
34. Dukas R. Effects of perceived danger on flower choice by bees. *Ecol Lett*. 2001;4:327–333.
35. Dukas R, Morse DH. Crab spiders affect flower visitation by bees. *Oikos*. 2003;101:157–163.
36. Chittka L, Dyer AG, Bock F, Dornhaus A. Psychophysics: bees trade off foraging speed for accuracy. *Nature*. 2003;424:388-388
37. Avarguès-Weber A, de Brito Sanchez MG, Giurfa M, Dyer AG. Aversive reinforcement improves visual discrimination learning in free-flying honeybees. *PLoS ONE*. 2010;5:e15370.
38. Gould JL. Pattern learning by honey bees. *Anim Behav*. 1986;34:990–997.
39. Gould JL. Honey bees store learned flower landing behaviour according to time of day. *Anim Behav*. 1987;35:1579–1581.
40. Núñez JA, Denti A. Respuesta de abejas recolectoras a un estímulo nociceptivo. *Acta Physiol Latinoam*. 1970;20:140–146.
41. Abramson CI. Aversive conditioning in honeybees (*Apis mellifera*). *J Comp Psychol*. 1986;100:108–116.
42. Smith BH, Abramson CI, Tobin TR. Conditional withholding of proboscis extension in honeybees (*Apis mellifera*) during discriminative punishment. *J Comp Psychol*. 1991;105:345–356.
43. Fiala A. Olfaction and olfactory learning in *Drosophila*: recent progress. *Curr Opin Neurobiol*. 2007;17:720–726.
44. Busto GU, Cervantes-Sandoval I, Davis RL. Olfactory learning in *Drosophila*. *Physiology*. 2010;25:338–346.
45. Davis RL. Traces of *Drosophila* memory. *Neuron*. 2011;70:8–19.
46. Riemensperger T, Voller T, Stock P, Buchner E, Fiala A. Punishment prediction by dopaminergic neurons in *Drosophila*. *Curr Biol*. 2005;15:1953–1960.
47. Kim YC, Lee HG, Han KA. D1 dopamine receptor dDA1 is required in the mushroom body neurons for aversive and appetitive learning in *Drosophila*. *J Neurosci*. 2007;27:7640–7647.
48. Inagaki HK, Ben-Tabou de-Leon S, Wong AM, et al. Visualizing neuromodulation *in vivo*: TANGO-mapping of dopamine signaling reveals appetite control of sugar sensing. *Cell*. 2012;148:583–595.
49. Giurfa M. Associative learning: the instructive function of biogenic amines. *Curr Biol*. 2006;16:R892–895.
50. van Swinderen B, Andretic R. Dopamine in *Drosophila*: setting arousal thresholds in a miniature brain. *Proc R Soc B*. 2011;278:906–913.

CHAPTER

37

Brain Aging and Performance Plasticity in Honeybees

Daniel Münch and Gro V. Amdam*,†*

*Norwegian University of Life Sciences, Ås, Norway †Arizona State University, Tempe, Arizona

SOCIAL CASTE, SOCIAL ENVIRONMENT, AND FLEXIBLE LIFE HISTORIES IN THE HONEYBEE

The honeybee is domesticated as a source of honey, as an important pollinator, and it is also a recognized research system for understanding the differentiation of eusocial individuals into reproductive and sterile forms.[1–3] In addition, the bee is a model in systems theory,[4,5] behavioral ecology,[6] neurobiology,[7] and aging.[8,9] The increasing attention it receives in molecular research is fueled by an emerging availability of functional genomic tools,[10–12] by an annotated genome sequence,[13] and by the presence of a DNA methylation machinery with parallels to mammalian epigenetic regulation.[14,15]

A key characteristic of eusocial insect societies (termites, ants, and some bees and wasps) is the reproductive division of labor within the colony.[16] In honeybees, this is achieved via a diphenism between females: Highly fecund queens perform the reproductive function of the colony, and essentially sterile workers (helpers) carry out the vast majority of other tasks. Life span differences among female castes are extreme: The queen can survive for several years, whereas helpers live from a few weeks to almost 1 year.[17,18] That the queen is both highly reproductive and long-lived challenges views that suggest a universal tradeoff between reproductive efforts and survival.[19,20] To explain this extreme life span, a model has linked the queen's high nutritional status with reduced insulin signaling—an unconventional relation that might make further regulatory feedback interactions with hormonal and antioxidant pathways.[21] The queen's life history is determined by metabolic responses to a queen-specific diet during early development.[22,23] In contrast, the life histories of the different female helper castes—the subjects of this review—are plastic and shaped by colony demography as well as by caste-specific traits, including behavior and resource allocation.[9,24,25]

Adult female workers go through an age-related task schedule, which requires them to change between nurse, forager, and winter (diutinus) stages (Figure 37.1).[26] Within the first 3 or 4 days of adulthood, the majority of workers mature into nurse bees that perform tasks within the nest.[27] Nurse bees are characterized by enlarged jelly-synthesizing hypopharyngeal glands and hypertrophied abdominal fat bodies (analogous to the mammalian liver and adipose tissue). Among worker bees, nurse bees reach intermediate life spans and can perform sib care tasks until they are at least 130 days old (Figure 37.1; for review, see Amdam (2)). However, at the age of 2–4 weeks, nurse bees typically develop into foragers.[28] The differentiation into foragers is driven by social signals that modulate a regulatory feedback loop between the *vitellogenin* gene, which is expressed in the fat body, and the systemic juvenile hormone secreted by the brain-associated corpora allata complex.[29,30] Foragers work in the nest's periphery, collecting nectar, pollen, water, and propolis in the field. A forager is characterized by atrophy and apoptosis of the hypopharyngeal glands[31] and the abdominal fat body.[32] Foragers are shorter lived than nurse bees and typically survive only 7–10 days once foraging commences (Figure 37.1).[33]

Winter bees—the most enduring, long-lived phenotype—develop during unfavorable periods without opportunities for brood rearing (Figure 37.1).[26,34,35] They differentiate from nurse bees that no longer take care for brood.[24,35,36] Winter bees are characterized by hypertrophied hypopharyngeal glands and the

Invertebrate Learning and Memory.
DOI: http://dx.doi.org/10.1016/B978-0-12-415823-8.00037-X

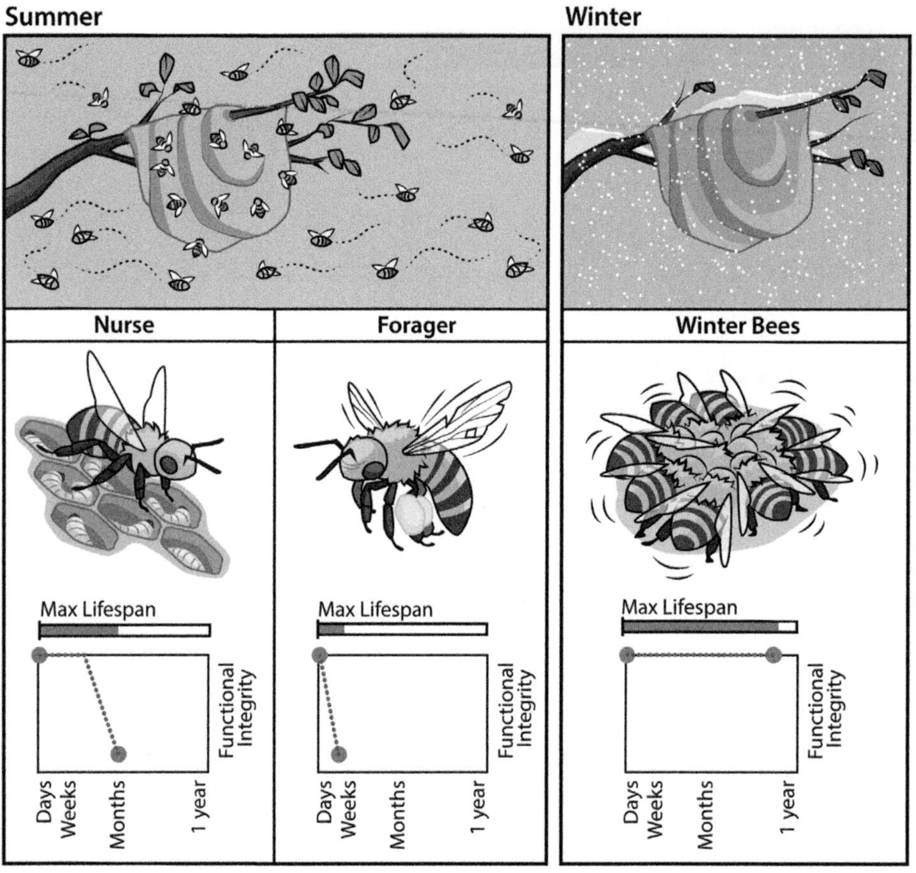

FIGURE 37.1 Flexible patterns of longevity and functional senescence in honeybee workers. Worker types with very short to intermediate life spans develop during favorable summer conditions. Young adults (nurses) first engage in caregiving tasks and after approximately 2 or 3 weeks transform into food-collecting foragers. Foragers typically perish within the next 1 or 2 weeks ('max life span'). Nurse bees, however, can survive more than 2 months by delaying the transition to foraging. An age-related functional decline is documented for both nurse and forager bees, but a decline is much more rapidly detected in foragers ('functional integrity'). The longest lived winter (diutinus) bees develop in autumn, when brood load is ceasing. In contrast to summer worker types, behavioral decline was not detected at the population level, even in groups of bees with individuals that are more than 6 months old.

abdominal fat body.[32,37,38] At the onset of the next favorable season, the queen begins to lay eggs and winter workers eventually develop physiological and behavioral attributes of nurse bees and foragers.[32,39]

The temporal differentiation of honeybee workers typically follows a trajectory in which the forager stage follows the nurse stage, as described previously. However, worker bees also have the unique plasticity to reverse this normal ontogeny.[40,41] Specifically, differentiated foragers can go through a physiological reversion characterized by reduced juvenile hormone signaling, elevated vitellogenin protein activity (see Aging Interventions), and the reactivation of the hypopharyngeal glands.[42,43] Collectively, the survival capacity of worker bees is flexible and linked to specific social caste behaviors, with foragers being the shortest and winter bees representing the longest lived workers (Figure 37.1).

BEHAVIORAL SENESCENCE IN HONEYBEES

In classical olfactory conditioning, bees can learn to associate a neutral odor (conditioned stimulus (CS)) with a sucrose reward (unconditioned stimulus).[44] In this paradigm, individual performance is measured by quantifying how readily an acquired response, monitored as the proboscis extension response to the CS, is detected during consecutive learning trials. The first study that used olfactory conditioning in restrained bees to investigate functional aging failed to detect age-related deficits.[45] However, soon after, another study documented significantly reduced learning performance during honeybee aging.[46] Both studies focused on worker bees. However, whereas Rueppell *et al.*[45] considered effects of chronological age in bees of unknown caste identity, Behrends *et al.*[46] contrasted age-matched groups of nurses with foragers. In agreement with Rueppell *et al.*, Behrends *et al.* did not detect signs of reduced learning function for older groups of nurse bees. The foragers, on the other hand, showed a significant decline of acquisition performance starting from approximately 2 weeks after foraging began.[46] Scheiner and Amdam[47] later found that similar to olfactory learning performance, foragers also show an age-related decline in tactile learning ability.

Since then, several studies have confirmed that behavioral senescence, measured as reduced acquisition performance, is accelerated in forager bees.[9,48] However, rapid decline is documented not only for learning functions. For example, after initial improvements of flight abilities in young foragers, flight performance declines as foragers age,[49] as does the efficiency of food collection from artificial flowers.[50] A sharp increase in mortality after approximately 10 days of foraging, with practically no survivors after 18 days,[33] corroborates that critical survival functions are indeed compromised by aging. These data are in line with previous studies, which already had suggested that social task performance (i.e., nursing vs. foraging) affects worker survival.[17,51,52] This is not to say that aging is exclusive to forager bees. Nest workers that continue nursing for more than 40 days also develop symptoms of senescence, including increased mortality and reduced stress resilience.[45,53] Such 'pre-foraging senescence'[54] is also indicated by the fact that when older nest bees change to foraging, their remaining foraging period (foraging age) will be shortened compared to that of bees that shift tasks earlier in life.

A common observation in vertebrate models is that aging rarely affects only one function; instead, decline can be detected across multiple, often unrelated functions (for cognitive disorders in humans, see Hedden and Gabrieli[55]). This prompted us to design an experiment that screens for associated decline patterns in a heterogeneous population of fully mature to old individuals by using a sequence of free-flight and in-lab behavioral testing (Figures 37.2A–37.2C).[56] We found that screening for an established symptom of old age (poor olfactory learning) could predict spatial orientation toward a previously abandoned home location—a hive box that bees had been trained to avoid. In contrast, bees that performed better in olfactory learning did prefer alternative hive boxes over the previously abandoned location (Figure 37.2D). This suggests that age-related decline can be associated across very different functions—that is, poor spatial extinction of an abandoned home location and poor olfactory learning. Of similar importance, however, our experiment (Figure 37.2) established how the olfactory learning tool can be utilized to identify behavioral decline patterns that so far are unknown.

HETEROGENEITY OF BEHAVIORAL AGING

Olfactory learning tests in honeybees are often used to assess how acquisition and memory performance are influenced by different stimulation protocols or to evaluate effects of invasive or noninvasive treatments. All such studies aim to reduce disturbances or variation that can cause experimental noise. For example, to reduce disturbing influences from satiation and caste-specific behavioral differences, experiments often make use of starved foragers as a 'standard bee.' A problem is that performance variability is increased in old foragers compared to younger individuals, even when the satiation level is controlled (Figure 37.3, top panel).[56] Such increase in variability, or heterogeneity, is a common finding in diverse model systems of aging and was demonstrated for behavioral as well as cellular degeneration.[57,58]

The mechanistic basis of the increased performance heterogeneity between old animals is not well understood. This increased heterogeneity is also observed in old individuals of isogenic populations, and even within a single tissue,[59] suggesting that stochastic (i.e., low-frequency, random) events may cause large differences in aging symptoms between individuals and can add to documented effects of genetic influences, for example.[60] In combination, these factors can create considerable variation and may drastically reduce the resolution power of studies when old cohorts—comprising age-afflicted but also nonsenesced individuals—are included. What is the relevance of these considerations for research on honeybee learning and memory? If starved foragers are desirable 'standard bees,' then the identification of 'forager age' and the exclusive use of young mature bees can reduce undesirable noise caused by functional senescence and by performance heterogeneity between individuals.

Age-related heterogeneity is also observed between functions (Figure 37.3, middle panel). Although aged individuals are most often afflicted with losses across several and often unrelated behavioral functions, some behavioral expressions may be less sensitive to deterioration. For example, in groups of forager bees that display a significant decline for learning function, no decline was detected for gustatory responsiveness, memory retention, or olfactory memory extinction.[46,47,56] This might be partially explained by uneven mortality in long-term memory tests that—in contrast to common acquisition tests—can last for several days. In such long-lasting test paradigms, a higher mortality in old groups compared to mature controls can bias the old test groups toward more resilient and healthy individuals; consequently, declined performance may not be detected in the surviving subgroup of old foragers.[56] In contrast, tests on gustatory sensitivity and acquisition performance are not biased by unequal mortality because they are measured almost simultaneously. Consequently, the lack of detectable age-related changes in gustatory responsiveness clearly indicates that gustatory sensitivity is less afflicted by aging than olfactory learning.[46,47,56] Similarly, heterogeneity across different behaviors is also documented in the *Drosophila* model: Geotaxis and learning decline progressively, different forms of consolidated memory show different decline patterns, but electrical shock

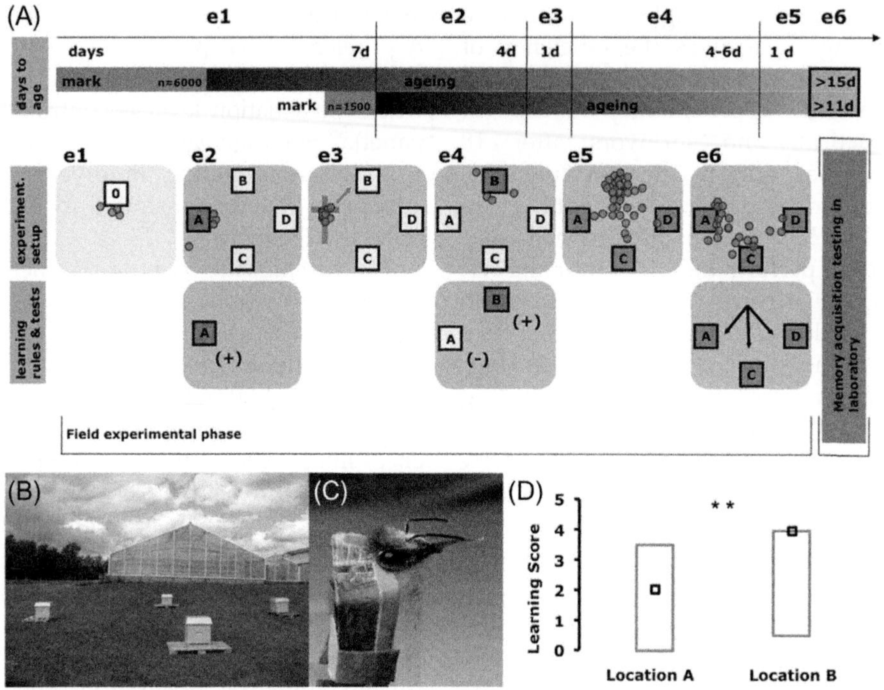

FIGURE 37.2 Age-related learning deficits that are detected in the lab can predict free-flight performance in a spatial extinction task. (A) Experimental design. The sequence of events e1–e6 served to test individual behavior in a free-flight memory task and is followed by in-lab testing of a symptom of old age—that is, olfactory memory acquisition. The top row ('days to age') depicts the duration of the single events as well as the minimum foraging age for two tested groups of foragers. At the time foragers were finally tested for free-flight orientation and olfactory learning in the lab, they constituted a heterogeneous age cohort with fully mature to old individuals (minimum foraging age of more than 11 and 15 days). Such an age group is typically characterized by increased performance heterogeneity between individuals. The middle row ('experiment. setup') depicts the sequence of events used to extinct the spatial memory of nest site A and to form a new memory of nest site B. The bottom row ('learning rules and tests') schematically depicts the learning rules. In e1, bees returning from foraging flights were marked to later confirm foraging age. They were subsequently moved to the test arena with four hive boxes (e2; see also panel B). Initially, all hive boxes, except for A, were empty ('dummies'), and bees were given 4 days for learning to orient toward the location A. In e3, an artificial swarm was produced and moved to location B (e4). The previous home site A now was closed off. Bees were given 4–6 days to learn the spatial setting of the new home site B while learning to extinguish the memory of the previous, now defunct home site A. Finally, the entire worker population of the colony was shaken to the ground (e5). The hive box at the recent location B was removed to force foragers to orient toward the previous location A or alternative locations C and D (e6). At the time point of the forced-choice orientation task, all locations A, C, and D were similarly equipped with unrelated queens, young workers (<48 hr old), and combs to resemble equally functioning hives. Marked foragers were collected at the different hives. These individuals with a known orientation toward the different hive boxes could then be subjected to olfactory memory acquisition in the laboratory. (B) An example of the arenas in which spatial extinction performance was examined. (C) To screen for a symptom of aging, differences in olfactory learning performance were quantified in the lab. (D) Poor olfactory learning ability, typical for old foragers, predicts orientation behavior in mixed-age forager groups. Foragers, which oriented toward a novel location C, had a higher median learning ability (boxes indicate interquartile ranges, and asterisks denote statistical significance level). In contrast, foragers that oriented toward location A, the location bees were trained to extinguish, also performed more poorly in the olfactory learning task. Source: *Modified from Munch et al.*[56]

avoidance and the response to free fall have been found to be intact, even in older flies.[61–63]

In summary, aging in honeybees can cause rapid loss of unrelated functions, such as olfactory and tactile learning but also flight performance. That the decline is not detectable for some other functions may be explained by more failure-proof computational circuitries; for example, in receptor systems with a high degree of redundant elements, neuronal loss may be more easily compensated. Alternatively, certain neuronal circuitries and brain regions may be better protected from age-related damage (see the following section).

AGING INTERVENTIONS

Honeybee colonies are accessible for manipulations that invoke transitions from one worker type to another. Such approaches might help identify how the progression of aging can be altered. Established paradigms include the removal of foragers to induce precocious foraging behavior in nurse bees[40] or, conversely, the removal of nurse bees from colonies to induce the reversal of foragers back to nursing tasks.[41] The latter method is probably the most fascinating because a phenotype with progressive aging (forager)

Performance Heterogeneity Among Individuals

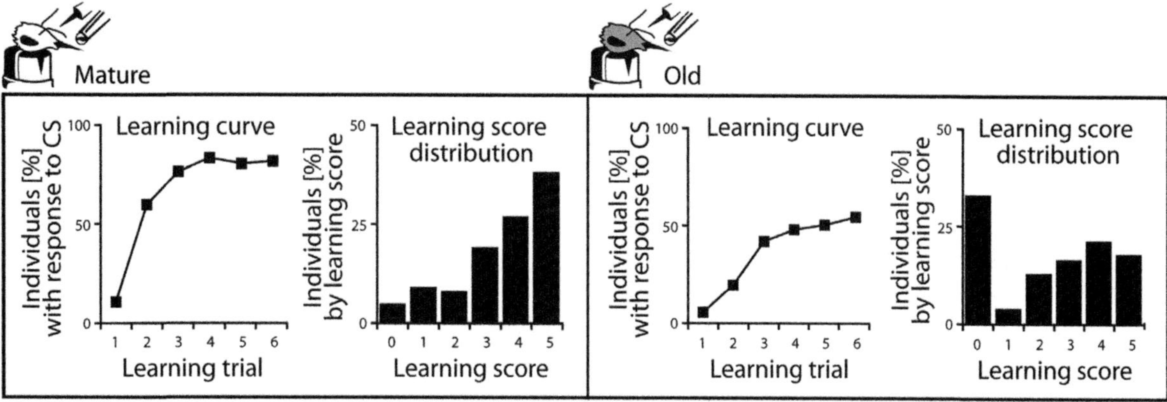

Heterogeneity Across Functions

Age Related Changes Detected	Age Related Changes Not Detected
Olfactory/Tactile Learning	Gustatory Resposiveness
Flight	Long-Term Memory
Spatial Extinction	
Survival	

Heterogeneity Across Tissues

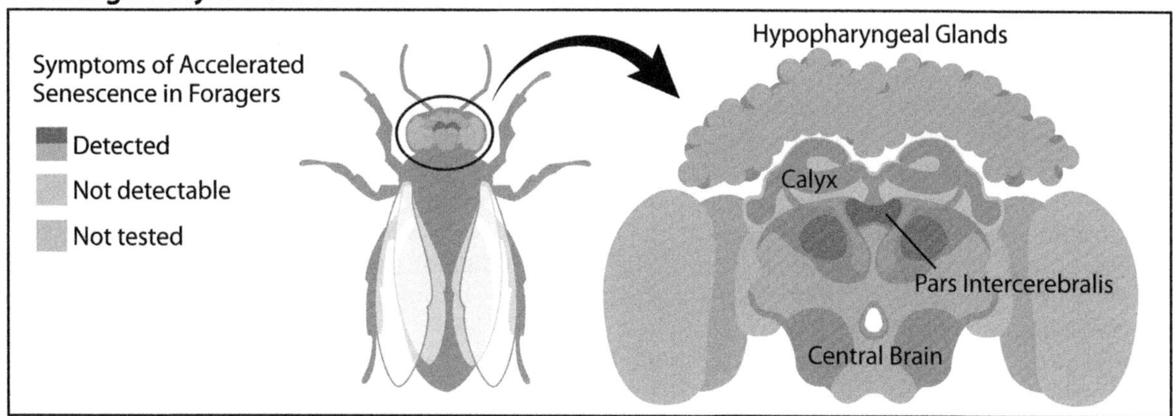

FIGURE 37.3 **Heterogeneity and aging.** (Top) Olfactory learning studies in foragers indicate that old groups are characterized not only by reduced average performance but also by increased performance heterogeneity. Representative learning curves show that in mature controls, more individuals show a learned behavior at the end of a typical six-trial association test. The histograms ('learning score distribution') depict a more variable distribution of performance scores among old individuals: Some old individuals still perform excellent (higher learning scores at $\sim$5), whereas others show moderate to poor performance (learning scores $\leq$3). Such increased performance variability in old groups is often attributed to the stochastic nature of aging. This can give rise to more unpredictable health outcomes in old compared to younger groups (see Heterogeneity of Behavioral Aging). (Middle) Functional aging studies on foragers could identify a rapid decline for several, sometimes unrelated, behaviors and functions (left; see also Behavioral Senescence in Honeybees; Heterogeneity of Behavioral Aging). For other functions, including gustatory responsiveness, such a decline could not be detected (right). (Bottom) Likewise, symptoms of cellular aging are heterogeneously expressed across different tissues. In the mushroom bodies' calyx regions, for example, decline dynamics may be slowed compared to those of other regions within the central brain and in the head's hypoharyngeal glands (see Underpinnings of Brain Aging).

is forced to revert to a life stage with typically slowed senescence (nurse). Perhaps supporting this proposition, Behrends *et al.*[46] found that reverted workers performed better in learning tasks than aged forager bees. However, this study could not exclude that only nonsenesced foragers reverted to nursing tasks, whereas their senesced sisters continued to forage. To address this challenge, we performed a study that tested worker groups more rigorously, both before and after the reversion. This experiment demonstrated that reverted workers initially display declined performance, but their learning capacity improves over several days after reversal to nursing.[64] How this and similar studies can provide more information about the reversal of aging symptoms[8,42,43] is discussed later (see Immune Defenses and Aging; Proteome, Aging, and the Reversal of Aging Symptoms).

NEGLIGIBLE SENESCENCE

The lack of detectable senescence over very long periods is often based on assumptions derived from survival data of long-lived individuals but can only be confirmed by testing of functional senescence. However, many species that are considered to show negligible senescence do not lend themselves readily to laboratory research, including some deep-dwelling, cold-water fish (e.g., rockfish), tortoise species, bivalves, and the bowhead whale.[65] In contrast, honeybees and naked mole rats represent well-established systems that are suitable to model elementary aspects of negligible or slowed senescence. Both species show large phenotype-dependent variation in longevity, providing the opportunity to address how slowed senescence can emerge among closely related individuals. Most important, honeybees and naked mole rats can reach extraordinary healthy life spans (without apparent functional decline), which are followed by very short periods of increased mortality and senescence ('sudden death'[66]). As a proposed hallmark feature of species with negligible senescence, sudden death contrasts to the more commonly observed situation of gradual decline.[67]

In honeybees, the longest lived winter worker type can survive for nearly 1 year (Figure 37.1). Winter bees show active behavior during their entire life, which implies that longer survival is not aided by dormancy states that are typical for long-lived phenotypes in other insects.[68] Winter bee physiology and extended longevity emerge in response to the removal of brood (eggs, larvae, and pupae) from the colony, and the lack of brood pheromones appears to be a key signal for this transition.[34,69,70]

In accord with the assumptions on negligible senescence, winter bees do not develop detectable performance decline, as measured with standard tests for olfactory and tactile conditioning (Figure 37.1).[71] In a separate study, the lack of senescence symptoms was confirmed to last at least from October to May, a vast difference to the rapid aging in summer foragers, which develops over only 2 weeks (D. Münch, C. Kreibich, and G. V. Amdam, unpublished results; see also Behavioral Senescence in Honeybees). To better understand the conditions under which such extreme longevity and retained performance can emerge, we tested paradigms that interrupt or extend the winter state. When winter bee colonies were transferred to indoor flight rooms, typical summer traits such as the transition to foraging and increased brood load were readily induced. Former winter bees that now engaged in foraging developed a detectable drop in learning performance within only 2 weeks after onset of foraging. In contrast, no decline was detected in non-senesced controls, as well as in outside winter bees. This rapid senescence following a long period without detectable senescence ('sudden death') was also observed under natural conditions in spring (D. Münch, C. Kreibich, and G. V. Amdam; unpublished results).

It has been suggested that harmful metabolic effects of intense flight activity are a cause of senescence in insects.[72–74] In support of this proposition, we found that brain function could be retained in old summer foragers that were constrained by *ad libitum* foraging.[48] In addition, we found that the removal of brood could retard the onset of detectable senescence after long-lived winter bees had naturally changed to summer tasks, including foraging. When brood was absent in these postwinter colonies, learning performance remained intact even in over-aged foragers—that is, more than 70 days after they had begun foraging (D. Münch, C. Kreibich, and G. V. Amdam, unpublished results). Because the absence of brood is known to also reduce foraging activity,[75,76] our results collectively suggest that negligible senescence in honeybees emerges through the joined action of social signals and activity levels.

UNDERPINNINGS OF PLASTIC BRAIN AGING

Aging can be caused by a multitude of factors, and no consensus has been reached on the relative contributions of the many mechanisms proposed to cause aging.[77] Holliday[78] therefore suggests that "if we accept the fact that there are multiple causes of aging, then it follows that many of the important theories of aging have some truth."

Free Radicals

One such framework—the free radical theory of aging—posits that reactive oxygen species (ROS) damage biomolecules and represent a major cause of aging.[79,80] However, recent research draws a more complex picture. Although mitochondria are a main source of harmful ROS, they also function as the principal antioxidant of the cell.[81–83] Even ROS, traditionally considered as harmful, can positively affect health,[84,85] whereas antioxidants, traditionally considered as beneficial, may also shorten life expectancy.[86]

Such ambivalences of ROS action are mirrored by findings in bees. The classical view of harmful ROS effects is supported by the highest carbonylation levels—a marker for oxidative damage—found in the brains of short-lived summer foragers. In contrast, significantly lower levels were detected for the much longer lived winter bees.[87] Similarly, one would expect that slowed behavioral senescence is associated with lower levels of oxidative damage. However, when behavioral senescence is slowed by restricting flight, old foragers with still intact learning function did accrue higher levels of oxidative stress markers in their brains (malondialdehyde protein adducts (MDA)) than a senesced control with unrestricted foraging activity and poor learning activity.[48] Hence, detectable functional decline in honeybees may not be a simple function of ROS-related damage accumulation.

Also contrary to expectations based on the free radical theory of aging, antioxidant enzymes were found to be downregulated in the heads of the longest lived social caste—queens.[88,89] Similarly, levels of the antioxidant enzyme superoxide dismutase were lowest in long-lived castes of ants, another highly social insect.[90] Therefore, the overexpression of antioxidant enzymes might not be required for extreme longevity[90]; alternatively, other antioxidant mechanisms might be in place to fight excessive ROS action.

Such alternative mechanisms could involve the phosphoglycolipoprotein vitellogenin, a common egg yolk precursor that in non-reproducing honeybee castes has an acquired role in social feeding.[9] Vitellogenin is most abundant in longer lived social castes of the honeybee,[91] whereas low-protein titers and low *vitellogenin* gene expression cause shortened survival under oxidative stress. That vitellogenin is strongly carbonylated by paraquat, a pro-oxidative agent, suggests that this protein can scavenge highly reactive oxygen.[52] This assumption is supported by similar findings in other systems. such as *Caenorhabditis elegans*[92] and fish (*Anguilla japonica*).[93] Higher vitellogenin levels that can compensate for the apparent lack of other antioxidant systems in long-lived casts is a tempting scenario, but it remains to be shown if and how vitellogenin can interact with mitochondrial respiration in the various vital organs of honeybees.

Mitochondrial Aging and Lipofuscin

Although complex network interactions of mitochondrial ROS homeostasis will remain a source of scientific controversy, mitochondrial degeneration and eventual dysfunction is without doubt central to survival and aging. Compared to the nuclear genome, mitochondrial DNA is more prone to damage, likely due to the lack of a compact chromatin structure and an inferior replication and repair system.[94,95] In several animal systems, mitochondrial DNA damage and copy number can change during aging.[96,97] In contrast, none of the age-related changes we and co-workers identified in mitochondrial DNA of bee brains point to degenerative effects (E. M. Hystad, G. V. Amdam, and L. Eide, unpublished results). Specifically, the copy number was similar in young and old foragers. Old foragers had a higher abundance of intact mitochondrial DNA and also showed increased gene transcription for Nd1, a mitochondrial complex 1 enzyme. Age-related degeneration of mitochondrial DNA is not yet indicated to be a strong player in bee brain aging.

Lipofuscin is composed of nondegradable lipid–protein aggregates and is the most abundant fluorescent 'pigment' in the aged human brain.[98] Its abundance and specific optical properties have made it a well-established biomarker of cellular aging and organismal integrity (for insects and other arthropods, see Fonseca *et al.*[99]). Lipofuscin accumulation results from incomplete cellular, including mitochondrial, turnover.[98,100] It was therefore suggested that lipofuscin can deplete the autophagic capacity of a cell,[101,102] providing a link between aging, impaired autophagy, and mitochondrial dysfunction in old individuals.[101–103] Recently, an association between lipofuscin accumulation and chronological age in honeybees was demonstrated by a study of abdominal fat body cells.[104]

Because studies on learning performance have repeatedly shown a strong link between aging and social caste but less with chronological age, we asked how chronological age and social caste transitions affect lipofuscin accumulation in long-lived worker bees (see Negligible Senescence). In contrast to the lack of behavioral senescence in winter bees, we found that old winter workers accrued higher lipofuscin levels than chronologically much younger controls (D. Münch, C. Kreibich, and G. V. Amdam, unpublished results). However, similar to behavioral decline, increased levels of the cellular senescence marker became rapidly detectable after disrupting the winter state and transition to another social task (forager).

Tissue-Dependent Heterogeneity of Cellular Senescence

Accumulation of lipofuscin is not uniform across tissues: The transition to foraging in postwinter bees is associated with higher lipofuscin levels in hypopharyngeal glands and in the brain's pars intercerebralis, a major neurosecretory area (D. Münch, C. Kreibich, and G. V. Amdam, unpublished results; compare also Fonseca *et al.*[99]). Within the same individuals, however, we found that lipofuscin levels remained unchanged in Kenyon cells of the mushroom bodies' calyx region, a higher brain center with multimodal sensory input and a role in memory formation. Our data on lipofuscin therefore suggest a spatial heterogeneity of cellular senescence patterns and, perhaps, slower aging dynamics after social task transition for the mushroom bodies' calyx regions (Figure 37.3, bottom panel).

In addition, using mass spectrometry and quantitative anatomy, we failed to detect changes in the calyx' protein and synapse matrix during rapid aging in summer foragers.[105] This is despite (1) significant protein level changes in the honeybee's entire central brain,[105] (2) the repeated confirmation of decline in learning function,[46,47,56] and (3) a steep increase in mortality during aging in foragers.[33]

In summary, evidence is mounting to suggest that a major center for memory and sensory integration might be spared from highly accelerated senescence patterns found for other structures and functions. This is not to say the calyx region is less modulated by life-history transitions than other regions. On the contrary, a number of studies have capitalized on features that are linked to a prominent plasticity in this brain area.[106] This includes persistent volumetric growth and neuronal branching during maturation,[107–109] as well as experience-dependent changes of synaptic organization.[106,110,111] How enduring plasticity may intersect with region-dependent aging patterns is a rewarding theme, and it holds promise to identify how tissue-specific adaptations may confer resilience against factors that cause aging.

IMMUNE DEFENSES AND AGING

Loss of immune function is an important cause of higher frailty and increased mortality that is typical for old cohorts. Hemocytes are circulating blood cells that perform important functions in the insect innate immune system, including phagocytosis, nodulation, and encapsulation.[112] Reduced rates of tactile interactions between nurse bees and foragers cause hemocytes of nurse bees to self-terminate through apoptosis.[42] This response is one component of the workers' transition from nest tasks to foraging activities. Loss of hemocytes is associated with deterioration of the nodulation immune response,[113] a predominant defense reaction to infection in insects.[114] Infections are more frequently observed in foragers than in nurse bees,[42] and in general, increased susceptibility to pathogens and toxins is characteristic of the forager stage.[115]

The increased frailty in foragers appears nonintuitive from the perspective that foragers, to a much larger extent than other workers, encounter a variety of pathogens as they engage in foraging trips outside the protected nest. Flowers have rich faunas of bacteria and fungi, and some pollens and nectars are poisonous to bees.[32,34] Conversely, having a robust forager caste may not be advantageous for the society. If resistant foragers repeatedly brought virulent strains of bacteria and fungi in addition to loads of poisonous nectars to the nest, the situation could soon become disastrous for the superorganism. Therefore, the increased frailty of honeybee foragers can be interpreted to reflect a colony-level selection pressure for disease control. In line with these arguments, a wide range of physiological stressors, such as cooling, oxygen deprivation, mechanical wounding, and parasite exposure, all cause nurse bees to differentiate into foragers.[116–119] Apart from triggering task transitions, stressors can also directly affect forager caste behaviors, with afflicted foragers removing themselves from the colony ('altruistic suicide').[120] Specifically, health-compromised foragers treated with CO_2 or hydroxyurea (cytostatic drug) leave nests quicker and more often than controls, display shortened stays within the nest, are less inclined to forage, and drastically reduce food transfers to nest mates. Together, cumulative evidence suggests that individuals in poor condition can be driven into a physiological state in which they rapidly perish.[121]

In vitro, programmed cell death is triggered if hemocytes are cultured in a zinc-deprived medium.[121] The zinc concentration of worker bee hemolymph is a function of the bee's social role in that foragers have much lower levels (0–4 ppm) than nurse bees and winter workers (up to 28 ppm). The vitellogenin protein appears to be the major circulating zinc ligand in the bee,[121] and thus the dramatic decline in the hemolymph zinc concentration of new foragers can be linked to the feedback suppression of vitellogenin synthesis that occurs with workers changing from nest to foraging activities. This scenario suggests that hemocyte apoptosis, and possibly also patterns of apoptotic and necrotic senescence of the hypopharyngeal glands[31] and abdominal fat body (S. C. Seehuus and G. V. Amdam, unpublished data), is downstream of a causal chain of forager aging driven by reduced *vitellogenin* gene activity.[8]

PROTEOME, AGING, AND THE REVERSAL OF AGING SYMPTOMS

During normal maturation, aging, and social reversal, honeybees undergo considerable changes in gene expression and protein matrix. We have studied changes in the proteomic composition of worker brains to identify proteins that correlate with recovery of brain function. We previously contrasted senesced foragers that had reverted to nurse activity.[64] Reverted workers were grouped based on 'good' versus 'poor' learning ability, assuming that bees had succeeded or failed to express a recovery-related performance improvement. Among the eight proteins that after reversion were associated with unequal performance were structural proteins (α-tubulins), proteins involved in stress response and cellular maintenance (heat shock proteins and the antioxidant peroxiredoxin), as well as proteins essential for neuronal function and signaling (fatty acid binding protein and glutamate transporter). Five of these proteins passed the additional criterion of a twofold or greater abundance difference between the two groups: tubulin α-3, heat shock protein 8, peroxiredoxin, fatty acid binding protein, and glutamate transporter. The heat shock proteins, peroxiredoxin, and fatty acid binding protein were more abundant in individuals that performed well, whereas the tubulins and glutamate transporter were more abundant in the individuals that performed poorly. This study shows that the central brains of previously senesced workers with 'good' learning ability differ for a subset of proteins compared to brains from bees that failed to express recovery-related plasticity and thus performed poorly in learning after behavioral reversal. In particular, our finding of a positive link between heat shock proteins, peroxiredoxin, and brain function after reversal may indicate that recovery-related brain plasticity in honeybees may rely on the classical roles of these proteins in cellular stress resilience, maintenance, and repair.[64]

Studies with a focus on the abdominal fat body found that proteins involved in lipid and cholesterol metabolism tend to be associated with worker phenotype and not with worker age.[122] This correlation may reflect the different metabolic requirements and strategies for nurse bees and foragers (reviewed by Wang *et al.*[123]). In contrast, worker age-related changes are seen in odorant binding proteins (OBPs) that occur at higher levels in younger bees. OBPs are small, soluble proteins that bind other more hydrophobic substances often associated with odor recognition,[124] but this is a complex group of proteins and much more work is required to understand the roles of OBPs in honeybees.

APPLICATION-ORIENTED RESEARCH: SCREENING FOR TREATMENTS THAT MAY EXTEND LIFE SPAN AND IMPROVE HEALTH

Insects as Models to Study Neurodegeneration and Cognitive Disability

Insects are recognized models to study the function of the intact nervous system and to assess relevant concepts on aging. However, can they be equally valuable for applied research that aims at understanding the etiology of age-related disease and the discovery of drugs to retard these conditions? A reflex caveat that is often phrased as a major obstacle is the long evolutionary distance between insects and mammals. However, genomic data point to an astounding degree of possible transferability. For example, the *Drosophila* genome contains homologs for more than 85% of genes linked to cognitive disability.[125] A pioneering example for applied studies with clear medical relevance is given by studies on the *Drosophila fragile X mental retardation 1* gene (*dfmr1*). Mutations in *dfmr1* and its human counterpart give rise to the fragile X syndrome, probably the most common inherited cause of intellectual disability with typical symptoms such as behavioral malfunction, abnormal neuronal shape, and—in *Drosophila*—aberrant mushroom body organization.[126] Work in *Drosophila* was the first to establish a successful treatment—a pharmacological rescue for *dfmr1*-related impairment of memory functions through application of mGluR antagonists.

Screenings for drugs and other treatments that can impede the onslaught of age-related conditions demand approaches that are economical and time efficient. It is here where insect systems can probably excel the most. In insects, moderate life spans and ease of use allow for whole life treatment studies, and well-established learning tests can be readily applied in high-throughput screenings to test for behavioral dysfunction. Although these benefits are not yet fully recognized, they may become invaluable to the honeybee model as funding policies increasingly call for applied research. This section therefore reviews attempts that illustrate how the virtues of insect models, particularly the honeybee, can be utilized toward the identification of treatments that may slow aging and control diseases.

Identifying Compounds with Life-Extending and Life-Shortening Effects

A long-standing challenge to medical research is posed by the sheer number of bioactive compounds that are considered but not yet tested for their benefits on health and brain aging.[127] Among the few compounds with documented effects on life span in insects are turmeric and cocoa, as well as extracts from *Rosa damascena* (damask rose) and *Vitis vinifera* (grapevine).[128] However, apart from its benefits, uninformed treatment with 'natural' herbal extracts can be risky: Several compounds seem to be life extending at low concentrations but show adverse effects at higher concentrations. To explain this, it was suggested that plant adaptogens at lower doses could act pro-oxidative and induce mild stress, which in turn triggers beneficial cellular defense mechanisms (hormesis).[129] To better understand these aspects, research in *Drosophila* will gain from its leading role as a genetic tool set, whereas the honeybee can complement for its more sophisticated behavioral repertoire, its larger body size, and flexible, identifiable aging patterns. The latter may enable comparing potential health risks in groups of more frail, aged individuals with effects on more resilient, young phenotypes.

Testing bees in their natural context can be problematic when a constant environment and controllable treatment administration of a specific diet, for example, are desirable. Here, cage studies that house bees in small social groups (<100 individuals), often in the controlled environment of an incubator, can be the method of choice and have been sporadically used for several decades.[130–133] Such studies successfully established, for example, how nutrition and social environment affect head gland and ovarian development—two cornerstones of honeybee physiology.[132,133] Experiments in our lab indicate that survival curves for caged bees mirror the common mortality distributions for aging in environments with few external hazards[67,134]: A very low initial mortality suggests negligible acute effects of environmental stressors (e.g., social deprivation), followed by a period with rapidly increasing mortality.

Among the plant compounds that are currently studied in the bee are caffeine and *Rhodiola* (roseroot). Caffeine has a well-acknowledged role in behavioral improvement, such as in learning and alertness, but it is also considered to mediate life span effects. In the honeybee, acute caffeine exposure reduces the age at which young bees successfully perform olfactory association tasks.[135] In mature bees, caffeine was found to positively affect learning in more complex, free-flight tasks.[135] This behavioral improvement may be explained by caffeine-mediated upregulation of genes involved in synaptic and calcium signaling, among others.[136] Another study that used olfactory learning in restrained bees concluded that high caffeine doses impaired the expression of learned responses during training trials but left the formation of early long-term memory intact.[137] Therefore, high caffeine doses may specifically affect responsiveness during learning rather than inhibiting the circuits necessary for memory formation.

In contrast to these studies, our lab recently set out to address survival effects of continued caffeine exposure during early to late adult life. Because this demands a constant environment and controlled diets, the aforementioned caging approaches were employed. For honeybees that were caged at the age of 5 days and from then on received continued oral treatment with caffeine, we found life span shortening at high and life span extension at low concentrations.[138] For roseroot, another plant extract with a potential health benefit, we found significant life span shortening in caged, mature honeybees (L. Rojahn, B. Egelandsdal, and D. Münch; unpublished results), whereas an often-proposed positive effect on survival at lower doses was not clear. To better understand the risks associated with the application of this top-selling plant extract, we are investigating whether life span shortening at higher doses is linked to the pro-oxidant action of roseroot and its effects on mitochondrial activity.

Applied Approaches to Study Harmful Influences of Ecotoxic Compounds and Parasites on Honeybee Health

Apart from the bee's use as a model system, the understanding of behavioral decline and disease etiology also has a direct applied relevance due to the bee's role as a main pollinator and the recent threat of colony collapse disorder (CCD)—a still not well-understood syndrome.[139] Studies on bee pests and ecotoxicology utilize the behavioral tool sets developed for the bee and aim to identify factors that cause functional loss, which in turn may weaken colony health. It was demonstrated that learning and other brain functions deteriorate when bees are afflicted with viral infections[140] or when exposed to pesticides and genetically modified plants that produce such substances.[141–144] Compared to survival assays on acute toxicity, the advantage of behavioral test tools can be a higher sensitivity when identifying deleterious effects of common agricultural practices. This was convincingly demonstrated by behavioral studies that established harmful effects of pesticides at sublethal doses, suggesting that common pesticide use in agriculture indeed can put colonies at risk.[143,145,146]

Although ecotoxicity and bee pests are not necessarily associated with aging and its plasticity, we lastly touch on an argument that may link these aspects. In bees, stressors such as disease and frailty can drive social task transitions, with afflicted animals changing precociously to foraging tasks.[116,118,147] Under normal conditions, this would not critically affect the balanced demographical equilibrium of a hive; on the contrary, colony survival may benefit from a lowered risk of spreading diseases by health-compromised foragers (see Immune Defenses and Aging). However, a higher pesticide or parasite load can cause disproportionate numbers of afflicted workers to transform into the rapidly aging phenotype, which also is most exposed to external hazards. Under such conditions, nest workers will rapidly vanish, with queen and brood being abandoned—a common symptom of CCD.[148] If this scenario holds true, then the flexible division of colony workforce and the associated plasticity of aging may represent the Achilles' heel of an evolutionary success model that is social organization.

SYNTHESIS

Despite the honeybee's long history as a model in neurobiology and social organization, it was not until the past decade that research took advantage of these assets to study the plasticity of brain aging. The honeybee is now a prolific system to examine the various aspects of plastic aging, including transformations between states of accelerated and negligible behavioral senescence, as well as the reversion of aging symptoms. The fact that changes in aging dynamics can be closely linked to worker caste and demography makes the honeybee an excellent system to identify how social context influences aging. Here, the adapted functions of vitellogenin—including roles in social feeding, nutrient storage, endocrine signaling, and free radical defense—may hold the key to linking social factors with plastic aging. Although not yet widely recognized, we argue that the age-related increase in heterogeneity between individual performance scores, among different functions and across different tissues, can be relevant to other fields of neurobiological research. In conclusion, we believe that the unique features of honeybee aging will not only inform concept-oriented biogerontology but also represent a rich resource for applied research.

Acknowledgments

We are very grateful to Sabine Deviche and James Baxter (SOLS Visualization lab) for their work on the figures. We thank Olav Rueppell, Marla Spivak, and Nicholas Baker for valuable comments on the manuscript. This work was supported by the Research Council of Norway (grants 180504, 185306, and 191699), the National Institute on Aging (grant NIA P01 AG22500), and Pew Charitable Trusts.

References

1. Amdam GV, Norberg K, Fondrk MK, Page RE. Reproductive ground plan may mediate colony-level selection effects on individual foraging behavior in honey bees. *Proc Natl Acad Sci USA*. 2004;101:11350–11355.
2. Bloch G, Wheeler D, Robinson GE. Endocrine influences on the organization of insect societies. In: Pfaff D, Arnold A, Etgen A, Fahrbach S, Moss R, Rubin R, eds. *Hormones, Brain and Behavior*. Vol 3. San Diego: Academic Press; 2002:195–235.
3. Robinson GE, Grozinger CM, Whitfield CW. Sociogenomics: social life in molecular terms. *Nat Rev Genet*. 2005;6(4):257–270.
4. Mitchell SD. *Biological Complexity and Integrative Pluralism*. Cambridge, MA: Cambridge University Press; 2003.
5. Page RE, Erber J. Levels of behavioral organization and the evolution of division of labor. *Naturwissenschaften*. 2002;89:91–106.
6. Seeley TD. *The Wisdom of the Hive*. Cambridge, MA: Harvard University Press; 1995.
7. Giurfa M, Sandoz JC. Invertebrate learning and memory: fifty years of olfactory conditioning of the proboscis extension response in honeybees. *Learn Mem*. 2012;19(2):54–66.
8. Amdam GV. Social context, stress, and plasticity of aging. *Aging Cell*. 2011;10(1):18–27.
9. Munch D, Amdam GV. The curious case of aging plasticity in honey bees. *FEBS Lett*. 2010;584(12):2496–2503.
10. Amdam GV, Simões ZLP, Guidugli KR, Norberg K, Omholt SW. Disruption of vitellogenin gene function in adult honeybees by intra-abdominal injection of double-stranded RNA. *BMC Biotechnol*. 2003;3:1–8.
11. Jarosch A, Moritz RF. Systemic RNA-interference in the honeybee *Apis mellifera*: tissue dependent uptake of fluorescent siRNA after intra-abdominal application observed by laser-scanning microscopy. *J Insect Physiol*. 2011;57(7):851–857.
12. Patel A, Fondrk MK, Kaftanoglu O, Emore C, Hunt G, Amdam GV. The making of a queen: TOR pathway governs diphenic caste development. *PLoS ONE*. 2007;6:e509.
13. The honey bee genome consortium. Insights into social insects from the genome of the honey bee *Apis mellifera*. *Nature*. 2006;443:931–949.
14. Kucharski R, Maleszka J, Foret S, Maleszka R. Nutritional control of reproductive status in honeybees via DNA methylation. *Science*. 2008;319:1827–1830.
15. Wang Y, Jorda M, Jones PL, et al. Functional CpG methylation system in a social insect. *Science*. 2006;314:645–647.
16. Wilson EO. *Sociobiology: The New Synthesis*. Cambridge, MA: Belknap; 1975.
17. Page Jr. RE, Peng CY. Aging and development in social insects with emphasis on the honey bee, *Apis mellifera* L. *Exp Gerontol*. 2001;36(4–6):695–711.
18. Seeley TD. Life-history strategy of honey bee, *Apis mellifera*. *Oecologia*. 1978;32(1):109–118.
19. De Loof A. Longevity and aging in insects: is reproduction costly; cheap; beneficial or irrelevant? A critical evaluation of the 'trade-off' concept. *J Insect Physiol*. 2011;57(1):1–11.

20. Kirkwood TBL. Evolution of ageing. *Nature*. 1977;270:301–304.
21. Corona M, Velarde RA, Remolina S, et al. Vitellogenin, juvenile hormone, insulin signaling, and queen honey bee longevity. *Proc Natl Acad Sci USA*. 2007;104(17):7128–7133.
22. Kamakura M. Royalactin induces queen differentiation in honeybees. *Nature*. 2011;473(7348):478–483.
23. Patel A, Fondrk MK, Kaftanoglu O, et al. The making of a queen: TOR pathway is a key player in diphenic caste development. *PLoS ONE*. 2007;2(6):e509.
24. Amdam GV, Omholt SW. The regulatory anatomy of honeybee lifespan. *J Theor Biol*. 2002;216(2):209–228.
25. Lee RD. Rethinking the evolutionary theory of aging: transfers, not births, shape senescence in social species. *Proc Natl Acad Sci USA*. 2003;100(16):9637–9642.
26. Amdam GV, Page RE. Intergenerational transfers may have decoupled physiological and chronological age in a eusocial insect. *Aging Res Rev*. 2005;4:398–408.
27. Naiem E-S, Hrassnigg N, Crailsheim K. Nurse bees support the physiological development of young bees (*Apis mellifera* L.). *J Comp Physiol B*. 1999;169:271–279.
28. Seeley TD. Adaptive significance of the age polyethism schedule in honeybee colonies. *Behav Ecol Sociobiol*. 1982;11:287–293.
29. Amdam GV, Omholt SW. The hive bee to forager transition in honeybee colonies: the double repressor hypothesis. *J Theor Biol*. 2003;223:451–464.
30. Guidugli KR, Nascimento AM, Amdam GV, et al. Vitellogenin regulates hormonal dynamics in the worker caste of a eusocial insect. *FEBS Lett*. 2005;579:4961–4965.
31. De Moraes R, Bowen ID. Modes of cell death in the hypopharyngeal gland of the honey bee (*Apis mellifera* L). *Cell Biol Int*. 2000;24(10):737–743.
32. Maurizio A. Pollenernahrung und Lebensvorgange bei der Honigbiene (*Apis mellifera* L.). *Landwirtsch Jahrb Schweiz*. 1954;245:115–182.
33. Dukas R. Mortality rates of honey bees in the wild. *Insect Soc*. 2008;55:252–255.
34. Maurizio A. The influence of pollen feeding and brood rearing on the length of life and physiological condition of the honeybee preliminary report. *Bee World*. 1950;31:9–12.
35. Omholt SW, Amdam GV. Epigenic regulation of aging in honeybee workers. *Sci Aging Knowl Environ*. 2004;26:pe28.
36. Huang Z-Y, Robinson GE. Seasonal changes in juvenile hormone titers and rates of biosynthesis in honey bees. *J Comp Physiol B*. 1995;165:18–28.
37. Deseyn J, Billen J. Age-dependent morphology and ultrastructure of the hypopharyngeal gland of *Apis mellifera* workers (Hymenoptera, Apidae). *Apidologie*. 2005;36(1):49–57.
38. Koehler A. Beobachtungen über Veranderungen am Fettkörper der Biene. *Schweiz Bienen-Zeitung*. 1921;44:424–428.
39. Sekiguchi K, Sakagami SF. Structure of foraging population and related problems in the honeybee, with considerations on division of labour in bee colonies. *Hakkaido Natl Agric Exp Stat*. 1966; [Report No. 69:1–58]
40. Huang Z-Y, Robinson GE. Regulation of honey bee division of labor by colony age demography. *Behav Ecol Sociobiol*. 1996;39:147–158.
41. Robinson GE, Page RE, Strambi C, Strambi A. Colony integration in honey bees: mechanisms of behavioral reversion. *Ethology*. 1992;90:336–348.
42. Amdam GV, Aase AL, Seehuus SC, Kim Fondrk M, Norberg K, Hartfelder K. Social reversal of immunosenescence in honey bee workers. *Exp Gerontol*. 2005;40(12):939–947.
43. Huang ZY, Robinson GE. Regulation of honey bee division of labor by colony age demography. *Behav Ecol Sociobiol*. 1996;39 (3):147–158.
44. Bitterman ME, Menzel R, Fietz A, Schafer S. Classical conditioning of proboscis extension in honeybees (*Apis mellifera*). *J Comp Psychol*. 1983;97(2):107–119.
45. Rueppell O, Christine S, Mulcrone C, Groves L. Aging without functional senescence in honey bee workers. *Curr Biol*. 2007;17: R274–R275.
46. Behrends A, Scheiner R, Baker N, Amdam GV. Cognitive aging is linked to social role in honey bees (*Apis mellifera*). *Exp Gerontol*. 2007;42(12):1146–1153.
47. Scheiner R, Amdam GV. Impaired tactile learning is related to social role in honeybees. *J Exp Biol*. 2009;212(Pt 7):994–1002.
48. Tolfsen CC, Baker N, Kreibich C, Amdam GV. Flight restriction prevents associative learning deficits but not changes in brain protein-adduct formation during honeybee ageing. *J Exp Biol*. 2011;214(Pt 8):1322–1332.
49. Vance JT, Williams JB, Elekonich MM, Roberts SP. The effects of age and behavioral development on honey bee (*Apis mellifera*) flight performance. *J Exp Biol*. 2009;212(Pt 16):2604–2611.
50. Tofilski A. Senescence and learning in honeybee (*Apis mellifera*) workers. *Acta Neurobiol Exp*. 2000;60:35–39.
51. Amdam GV. Social control of aging and frailty in bees. In: Carey JR, Robine J-M, Michel J-P, Christen Y, eds. *Longevity and Frailty*. Berlin: Springer-Verlag; 2005:17–26.
52. Seehuus SC, Norberg K, Gimsa U, Krekling T, Amdam GV. Reproductive protein protects functionally sterile honey bee workers from oxidative stress. *Proc Natl Acad Sci USA*. 2006;103 (4):962–967.
53. Remolina SC, Hafez DM, Robinson GE, Hughes KA. Senescence in the worker honey bee *Apis mellifera*. *J Insect Physiol*. 2007;53 (10):1027–1033.
54. Rueppell O, Bachelier C, Fondrk MK, Page Jr. RE. Regulation of life history determines lifespan of worker honey bees (*Apis mellifera* L.). *Exp Gerontol*. 2007;42(10):1020–1032.
55. Hedden T, Gabrieli JD. Insights into the ageing mind: a view from cognitive neuroscience. *Nat Rev Neurosci*. 2004;5(2): 87–96.
56. Munch D, Baker N, Kreibich CD, Braten AT, Amdam GV. In the laboratory and during free-flight: old honey bees reveal learning and extinction deficits that mirror mammalian functional decline. *PLoS ONE*. 2010;5(10):e13504.
57. Rapp PR, Amaral DG. Individual differences in the cognitive and neurobiological consequences of normal aging. *Trends Neurosci*. 1992;15(9):340–345.
58. Ylikoski R, Ylikoski A, Keskivaara P, Tilvis R, Sulkava R, Erkinjuntti T. Heterogeneity of cognitive profiles in aging: successful aging, normal aging, and individuals at risk for cognitive decline. *Eur J Neurol*. 1999;6(6):645–652.
59. Herndon LA, Schmeissner PJ, Dudaronek JM, et al. Stochastic and genetic factors influence tissue-specific decline in ageing *C. elegans*. *Nature*. 2002;419(6909):808–814.
60. Kirkwood TB, Feder M, Finch CE, et al. What accounts for the wide variation in life span of genetically identical organisms reared in a constant environment? *Mech Ageing Dev*. 2005;126 (3):439–443.
61. Cook-Wiens E, Grotewiel MS. Dissociation between functional senescence and oxidative stress resistance in *Drosophila*. *Exp Gerontol*. 2002;37(12):1347–1357.
62. Mery F. Aging and its differential effects on consolidated memory forms in *Drosophila*. *Exp Gerontol*. 2007;42(1–2):99–101.
63. Simon AF, Liang DT, Krantz DE. Differential decline in behavioral performance of *Drosophila melanogaster* with age. *Mech Ageing Dev*. 2006;127(7):647–651.
64. Baker N, Wolschin F, Amdam GV. Age-related learning deficits can be reversible in honeybees *Apis mellifera*. *Exp Gerontol*. 2012;47:764–772.

65. Finch CE, Austad SN. History and prospects: symposium on organisms with slow aging. *Exp Gerontol.* 2001;36 (4–6):593–597.
66. Buffenstein R. Negligible senescence in the longest living rodent, the naked mole-rat: insights from a successfully aging species. *J Comp Physiol B.* 2008;178(4):439–445.
67. Finch CE. *Longevity, Senescence, and the Genome.* Chicago: University of Chicago Press; 1990.
68. Tatar M, Yin C. Slow aging during insect reproductive diapause: why butterflies, grasshoppers and flies are like worms. *Exp Gerontol.* 2001;36(4–6):723–738.
69. Fluri P, Lüscher M, Wille H, Gerig L. Changes in weight of the pharyngeal gland and haemolymph titres of juvenile hormone, protein and vitellogenin in worker honey bees. *J Insect Physiol.* 1982;28:61–68.
70. Smedal B, Brynem M, Kreibich CD, Amdam GV. Brood pheromone suppresses physiology of extreme longevity in honeybees (*Apis mellifera*). *J Exp Biol.* 2009;212(Pt 23):3795–3801.
71. Behrends A, Scheiner R. Learning at old age: a study on winter bees. *Front Behav Neurosci.* 2010;4:15.
72. Magwere T, Pamplona R, Miwa S, et al. Flight activity, mortality rates, and lipoxidative damage in *Drosophila*. *J Gerontol A Biol Sci Med Sci.* 2006;61(2):136–145.
73. Sohal RS. Aging changes in insect flight muscle. *Gerontology.* 1976;22(4):317–333.
74. Williams JB, Roberts SP, Elekonich MM. Age and natural metabolically-intensive behavior affect oxidative stress and antioxidant mechanisms. *Exp Gerontol.* 2008;43(6):538–549.
75. Free JB. Factors determining collection of pollen by honeybee foragers. *Anim Behav.* 1967;15(1):134.
76. Huang ZY, Otis GW. Factors determining hypopharyngeal gland activity of worker honey bees (*Apis mellifera* L). *Insectes Soc.* 1989;36(4):264–276.
77. Hughes KA, Reynolds RM. Evolutionary and mechanistic theories of aging. *Annu Rev Entomol.* 2005;50:421–445.
78. Holliday R. Aging is no longer an unsolved problem in biology. *Ann N Y Acad Sci.* 2006;1067:1–9.
79. Harman D. Aging: a theory based on free radical and radiation chemistry. *J Gerontol.* 1956;11(3):298–300.
80. Muller FL, Lustgarten MS, Jang Y, Richardson A, Van Remmen H. Trends in oxidative aging theories. *Free Radic Biol Med.* 2007;43(4):477–503.
81. Harman D. The biologic clock: the mitochondria? *J Am Geriatr Soc.* 1972;20(4):145–147.
82. Miquel J, Economos AC, Fleming J, Johnson JE. Mitochondrial role in cell aging. *Exp Gerontol.* 1980;15(6):575–591.
83. Schulz TJ, Westermann D, Isken F, et al. Activation of mitochondrial energy metabolism protects against cardiac failure. *Aging.* 2010;2(11):843–853.
84. Lee J, Giordano S, Zhang J. Autophagy, mitochondria and oxidative stress: cross-talk and redox signalling. *Biochem J.* 2012;441(2):523–540.
85. Richter C, Gogvadze V, Laffranchi R, et al. Oxidants in mitochondria: from physiology to diseases. *Biochim Biophys Acta.* 1995;1271(1):67–74.
86. Schulz TJ, Zarse K, Voigt A, Urban N, Birringer M, Ristow M. Glucose restriction extends *Caenorhabditis elegans* life span by inducing mitochondrial respiration and increasing oxidative stress. *Cell Metab.* 2007;6(4):280–293.
87. Seehuus SC, Krekling T, Amdam GV. Cellular senescence in honey bee brain is largely independent of chronological age. *Exp Gerontol.* 2006;41(11):1117–1125.
88. Corona M, Hughes KA, Weaver DB, Robinson GE. Gene expression patterns associated with queen honey bee longevity. *Mech Ageing Dev.* 2005;126(11):1230–1238.
89. Corona M, Robinson GE. Genes of the antioxidant system of the honey bee: annotation and phylogeny. *Insect Mol Biol.* 2006;15(5):687–701.
90. Parker JD, Parker KM, Sohal BH, Sohal RS, Keller L. Decreased expression of Cu–Zn superoxide dismutase 1 in ants with extreme lifespan. *Proc Natl Acad Sci USA.* 2004;101(10):3486–3489.
91. Fluri P, Wille H, Gerig L, Lüscher M. Juvenile hormone, vitellogenin and haemocyte composition in winter worker honeybees (*Apis mellifera*). *Experientia.* 1977;33:1240–1241.
92. Nakamura A, Yasuda K, Adachi H, Sakurai Y, Ishii N, Goto S. Vitellogenin-6 is a major carbonylated protein in aged nematode, *Caenorhabditis elegans*. *Biochem Biophys Res Comm.* 1999;264 (2):580–583.
93. Ando S, Yanagida K. Susceptibility to oxidation of copper-induced plasma lipoproteins from Japanese eel: protective effect of vitellogenin on the oxidation of very low density lipoprotein. *Comp Biochem Physiol C Pharmacol Toxicol Endocrinol.* 1999;123(1):1–7.
94. Jeppesen DK, Bohr VA, Stevnsner T. DNA repair deficiency in neurodegeneration. *Prog Neurobiol.* 2011;94(2):166–200.
95. Yakes FM, Van Houten B. Mitochondrial DNA damage is more extensive and persists longer than nuclear DNA damage in human cells following oxidative stress. *Proc Natl Acad Sci USA.* 1997;94(2):514–519.
96. Barazzoni R, Short KR, Nair KS. Effects of aging on mitochondrial DNA copy number and cytochrome c oxidase gene expression in rat skeletal muscle, liver, and heart. *J Biol Chem.* 2000;275(5):3343–3347.
97. Tyynismaa H, Suomalainen A. Mouse models of mitochondrial DNA defects and their relevance for human disease. *EMBO Rep.* 2009;10(2):137–143.
98. Double KL, Dedov VN, Fedorow H, et al. The comparative biology of neuromelanin and lipofuscin in the human brain. *Cell Mol Life Sci.* 2008;65(11):1669–1682.
99. Fonseca DB, Brancato CL, Prior AE, Shelton PM, Sheehy MR. Death rates reflect accumulating brain damage in arthropods. *Proc Biol Sci.* 2005;272(1575):1941–1947.
100. Riga D, Riga S, Halalau F, Schneider F. Brain lipopigment accumulation in normal and pathological aging. *Ann N Y Acad Sci.* 2006;1067:158–163.
101. Brunk UT, Terman A. The mitochondrial-lysosomal axis theory of aging: accumulation of damaged mitochondria as a result of imperfect autophagocytosis. *Eur J Biochem.* 2002;269(8):1996–2002.
102. Terman A, Dalen H, Brunk UT. Ceroid/lipofuscin-loaded human fibroblasts show decreased survival time and diminished autophagocytosis during amino acid starvation. *Exp Gerontol.* 1999;34(8):943–957.
103. Terman A, Kurz T, Navratil M, Arriaga EA, Brunk UT. Mitochondrial turnover and aging of long-lived postmitotic cells: the mitochondrial-lysosomal axis theory of aging. *Antioxid Redox Signal.* 2010;12(4):503–535.
104. Hsieh YS, Hsu CY. Honeybee trophocytes and fat cells as target cells for cellular senescence studies. *Exp Gerontol.* 2011;46 (4):233–240.
105. Wolschin F, Munch D, Amdam GV. Structural and proteomic analyses reveal regional brain differences during honeybee aging. *J Exp Biol.* 2009;212(Pt 24):4027–4032.
106. Groh C, Meinertzhagen IA. Brain plasticity in Diptera and Hymenoptera. *Front Biosci.* 2010;2:268–288.
107. Durst C, Eichmuller S, Menzel R. Development and experience lead to increased volume of subcompartments of the honeybee mushroom body. *Behav Neural Biol.* 1994;62(3):259–263.
108. Fahrbach SE, Farris SM, Sullivan JP, Robinson GE. Limits on volume changes in the mushroom bodies of the honey bee brain. *J Neurobiol.* 2003;57:141–151.

109. Farris SM, Robinson GE, Fahrbach SE. Experience- and age-related outgrowth of intrinsic neurons in the mushroom bodies of the adult worker honeybee. *J Neurosci.* 2001;21(16):6395–6404.
110. Hourcade B, Muenz TS, Sandoz JC, Rossler W, Devaud JM. Long-term memory leads to synaptic reorganization in the mushroom bodies: a memory trace in the insect brain? *J Neurosci.* 2010;30(18):6461–6465.
111. Krofczik S, Khojasteh U, de Ibarra NH, Menzel R. Adaptation of microglomerular complexes in the honeybee mushroom body lip to manipulations of behavioral maturation and sensory experience. *Dev Neurobiol.* 2008;68(8):1007–1017.
112. Ribeiro C, Brehelin M. Insect haemocytes: what type of cell is that? *J Insect Physiol.* 2006;52(5):417–429.
113. Bedick JC, Tunaz H, Aliza ARN, Putnam SM, Ellis MD, Stanley DW. Eicosanoids act in nodulation reactions to bacterial infections in newly emerged adult honey bees, *Apis mellifera*, but not in older foragers. *Comp Biochem Physiol.* 2001;130:107–117.
114. Franssens V, Simonet G, Bronckaers A, Claeys I, De Loof A, Broeck JV. Eicosanoids mediate the laminarin-induced nodulation response in larvae of the flesh fly, *Neobellieria bullata*. *Arch Insect Biochem Physiol.* 2005;59:32–41.
115. Meled M, Thrasyvoulou A, Belzunces LP. Seasonal variations in susceptibility of *Apis mellifera* to the synergistic action of prochloraz and deltamethrin. *Environ Toxicol Chem.* 1998;17:2517–2520.
116. Bühler A, Lanzrein B, Wille H. Influence of temperature and carbon dioxide concentration on juvenile hormone titre and dependent parameters of adult worker honey bees (*Apis mellifera* L.). *J Insect Physiol.* 1983;29:885–893.
117. Ebadi R, Gary NE, Lorenzen K. Effects of carbon dioxide and low temperature narcosis on honey bees, *Apis mellifera*. *Environ Entomol.* 1980;9:144–147.
118. Kovac H, Crailsheim K. Lifespan of *Apis mellifera carnica* Pollm. Infested by *Varroa jacobsoni* Oud. in relation to season and extent of infestation. *J Apic Res.* 1988;27:230–238.
119. Tustain RCR, Faulke J. Effect of carbon dioxide anesthesia on the longevity of honey bee in the laboratory. *N Z J Exp Agric.* 1979;7:327–329.
120. Rueppell O, Hayworth MK, Ross NP. Altruistic self-removal of health-compromised honey bee workers from their hive. *J Evol Biol.* 2010;23(7):1538–1546.
121. Amdam GV, Simões ZLP, Hagen A, et al. Hormonal control of the yolk precursor vitellogenin regulates immune function and longevity in honeybees. *Exp Gerontol.* 2004;39:767–773.
122. Wolschin F, Amdam GV. Comparative proteomics reveal characteristics of life-history transitions in a social insect. *Proteome Sci.* 2007;5(1):10.
123. Wang Y, Brent C, Fennern E, Amdam GV. Gustatory perception and fat body energy metabolism are jointly affected by vitellogenin and juvenile hormone in honey bees. *PLoS Genet.* 2012;8:e1002779.
124. Pelosi P, Zhou JJ, Ban LP, Calvello M. Soluble proteins in insect chemical communication. *Cell Mol Life Sci.* 2006;63(14):1658–1676.
125. Inlow JK, Restifo LL. Molecular and comparative genetics of mental retardation. *Genetics.* 2004;166(2):835–881.
126. Restifo LL. Mental retardation genes in *Drosophila*: new approaches to understanding and treating developmental brain disorders. *Ment Retard Dev D R.* 2005;11(4):286–294.
127. Wink M. Functions and biotechnology of plant secondary metabolites. *Annu Plant Rev.* 2009;39:1–20.
128. Kim SI, Jung JW, Ahn YJ, Restifo LL, Kwon HW. *Drosophila* as a model system for studying lifespan and neuroprotective activities of plant-derived compounds. *J Asia-Pac Entomol.* 2011;14(4):509–517.
129. Wiegant FA, Surinova S, Ytsma E, Langelaar-Makkinje M, Wikman G, Post JA. Plant adaptogens increase lifespan and stress resistance in *C. elegans*. *Biogerontology.* 2009;10(1):27–42.
130. Hess G. Über den Einfluss der Weisellosigkeit und des Fruchtbarkeitsvitamins E auf die Ovarien der Bienenarbeiterin. *Beih Schweiz Bienenzeitung.* 1942;1:929–939.
131. Lass A, Crailsheim K. Influence of age and caging upon protein metabolism, hypopharyngeal glands and trophallactic behavior in the honey bee (*Apis mellifera* L.). *Insectes Soc.* 1996;43:347–358.
132. Lin H, Winston ML. The role of nutrition and temperature in the ovarian development of the worker honey bee (*Apis mellifera*). *Can Entomologist.* 1998;130:883–891.
133. Wegener J, Huang ZY, Lorenz MW, Bienefeld K. Regulation of hypopharyngeal gland activity and oogenesis in honey bee (*Apis mellifera*) workers. *J Insect Physiol.* 2009;55(8):716–725.
134. Gompertz B. On the nature of the function expressive of the law of human mortality, and a new mode of determining the value of life contingencies. *Philos Trans R Soc Lond.* 1825;115:513–585.
135. Si A, Zhang SW, Maleszka R. Effects of caffeine on olfactory and visual learning in the honey bee (*Apis mellifera*). *Pharmacol Biochem Behav.* 2005;82(4):664–672.
136. Kucharski R, Maleszka R. Microarray and real-time PCR analyses of gene expression in the honeybee brain following caffeine treatment. *J Mol Neurosci.* 2005;27(3):269–276.
137. Mustard JA, Dews L, Brugato A, Dey K, Wright GA. Consumption of an acute dose of caffeine reduces acquisition but not memory in the honey bee. *Behav Brain Res.* 2012;232 (1):217–224.
138. Yusaf M, Münch D, Amdam GV. *Long Term Effects of Caffeine on Honeybee (Apis mellifera) Lifespan and Learning Ability in Old Age.* Aas, Norway: MS thesis, University of Life Sciences; 2012.
139. Ratnieks FL, Carreck NL. Clarity on honey bee collapse? *Science.* 2010;327(5962):152–153.
140. Iqbal J, Mueller U. Virus infection causes specific learning deficits in honeybee foragers. *Proc Biol Sci.* 2007;274(1617):1517–1521.
141. Bernadou A, Demares F, Couret-Fauvel T, Sandoz JC, Gauthier M. Effect of fipronil on side-specific antennal tactile learning in the honeybee. *J Insect Physiol.* 2009;55(12):1099–1106.
142. Decourtye A, Lacassie E, Pham-Delegue MH. Learning performances of honeybees (*Apis mellifera* L) are differentially affected by imidacloprid according to the season. *Pest Manag Sci.* 2003;59(3):269–278.
143. Han P, Niu CY, Lei CL, Cui JJ, Desneux N. Use of an innovative T-tube maze assay and the proboscis extension response assay to assess sublethal effects of GM products and pesticides on learning capacity of the honey bee *Apis mellifera* L. *Ecotoxicology.* 2010;19(8):1612–1619.
144. Ramirez-Romero R, Chaufaux J, Pham-Delegue MH. Effects of Cry1Ab protoxin, deltamethrin and imidacloprid on the foraging activity and the learning performances of the honeybee *Apis mellifera*, a comparative approach. *Apidologie.* 2005;36 (4):601–611.
145. Henry M, Beguin M, Requier F, et al. A common pesticide decreases foraging success and survival in honey bees. *Science.* 2012;336(6079):348–350.
146. Stokstad E. Agriculture: field research on bees raises concern about low-dose pesticides. *Science.* 2012;335(6076):1555.
147. Tofilski A. Shorter-lived workers start foraging earlier. *Insectes Soc.* 2009;56(4):359–366.
148. Khoury DS, Myerscough MR, Barron AB. A quantitative model of honey bee colony population dynamics. *PLoS ONE.* 2011;6 (4):e18491.

4.4.3 Ants

CHAPTER

38

Learning and Recognition of Identity in Ants

Patrizia d'Ettorre

University of Paris 13, Sorbonne Paris Cité, Villetaneuse, France

The tremendous ecological success of social insects is due to the remarkable organization of their colonies based on a multilevel division of labor, first between reproductive and nonreproductive individuals (typically queens and workers) and then among workers, with individuals specializing in different tasks, such as brood care, colony maintenance, and foraging.[1] Efficient communication systems are required to coordinate social interactions and to ensure that altruistic acts are directed toward colony members.[2] Because colonies are full of resources (stored food and brood), a crucial task for social insect workers is to defend their colony against competitors and social parasites by keeping the colony as closed as a fortress. This requires the ability to discriminate between colony members and strangers, the so-called nestmate recognition, which is typically expressed by rejecting non-nestmate intruders (reviewed in[3,4]). Social insect colonies are often composed of highly related individuals, resulting in the functional equivalence between nestmate and kin recognition, but a clear distinction should be made between the two processes: Kin recognition is the assessment of the degree of relatedness toward another individual, whereas nestmate recognition is the mere binary recognition of group membership.[5,6] The large majority of social insect literature on communication of group identity addresses nestmate and not kin recognition.

What are the mechanisms underlying nestmate recognition in social insects? It is believed that nestmate recognition involves matching the label perceived on another individual with an internal template of the discriminating individual. The label is chemical, the so-called colony odor, and the template is thought to be a neural representation of the colony odor possibly stored in the memory of the individual animal.[6–8] Depending on the similarity between label and template, an individual will be accepted or rejected; if the dissimilarity exceeds a certain threshold, the individual will be rejected.[4] A great deal of research has been devoted to the identification and characterization of the chemical label underlying nestmate recognition in ants and other social insects; the chemical label is a blend of long-chain hydrocarbons that is present on the insect cuticle. Cuticular hydrocarbons are complex mixtures of linear and branched alkanes and alkenes, usually ranging between 20 and 40 carbon atoms (Figure 38.1). There are up to 60–70 different hydrocarbons on the cuticle of an ant.[3] The original role of these compounds was likely preventing desiccation, and they were later co-opted to serve a communication role acting as 'signature mixtures,' which are defined as a variable set of molecules allowing to distinguish individuals or colonies, as opposed to pheromones, which are relatively invariable species-wide signals.[9]

Are all the different compounds present on the cuticle necessary to convey information about colonial identity, or is there a particular class of compounds that is actually encoding the recognition signal? It is difficult to identify a general rule because different results have been found in different ant species. For instance, in *Camponotus herculeanus*, dimethyl alkanes appear to play a more important role than linear alkanes in nestmate recognition,[10] whereas *Formica exsecta* alkenes, which are abundant on the cuticle together with linear alkanes, are likely to be essential.[11] In *Formica japonica*, both linear alkanes and alkenes are needed to discriminate nestmates from aliens.[12] Similarly, in *Linepithema humile* and *Aphaenogaster cockerelli*, a combination of at least two classes of hydrocarbons is necessary for nestmate recognition.[13] Deciphering the nestmate recognition code and identifying the nature of the colony signal are challenging because most of the hydrocarbons present on the

Invertebrate Learning and Memory.
DOI: http://dx.doi.org/10.1016/B978-0-12-415823-8.00038-1

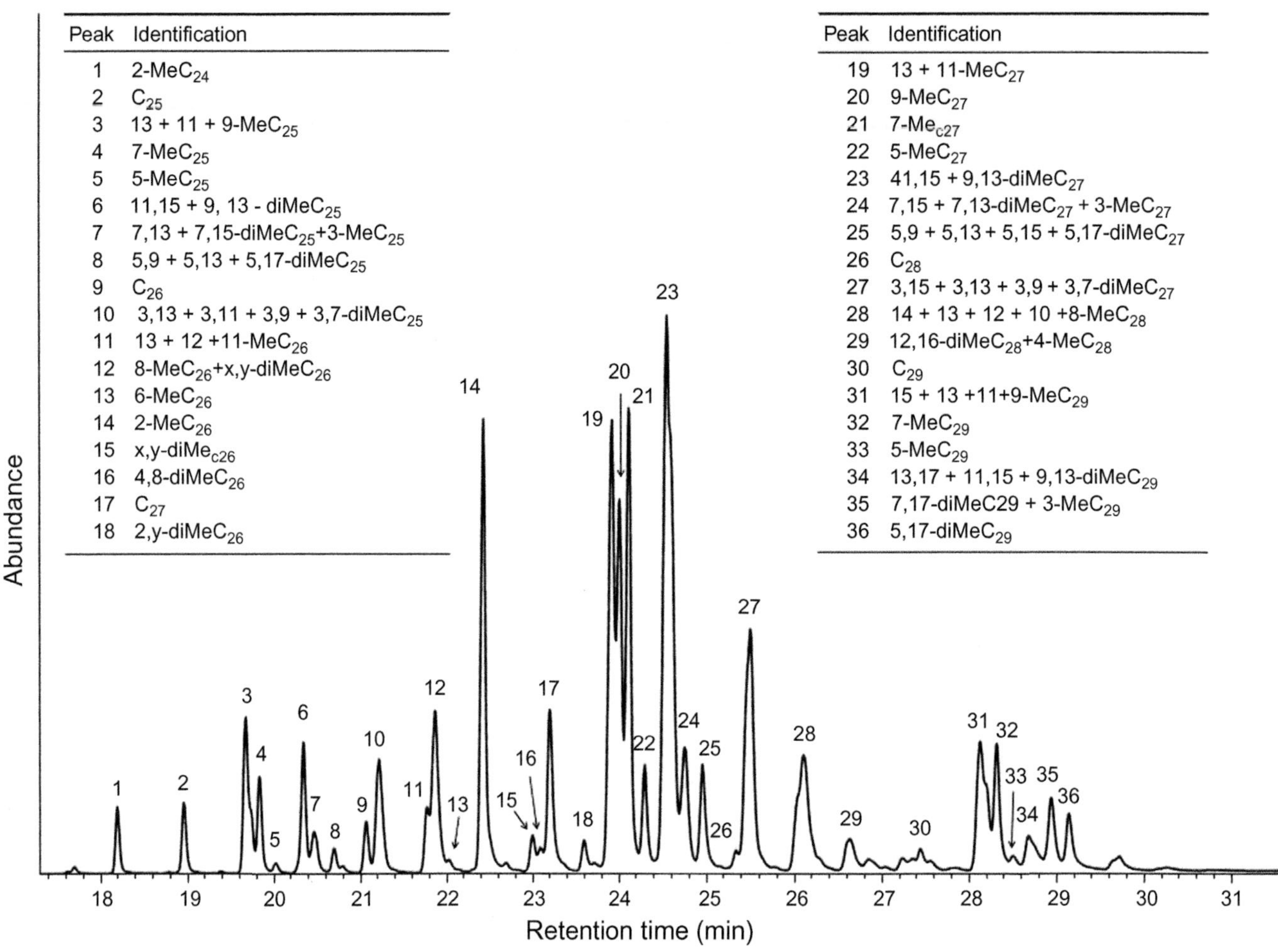

Peak Identification
1 2-MeC_{24}
2 C_{25}
3 13 + 11 + 9-MeC_{25}
4 7-MeC_{25}
5 5-MeC_{25}
6 11,15 + 9, 13 - diMeC_{25}
7 7,13 + 7,15-diMeC_{25}+3-MeC_{25}
8 5,9 + 5,13 + 5,17-diMeC_{25}
9 C_{26}
10 3,13 + 3,11 + 3,9 + 3,7-diMeC_{25}
11 13 + 12 +11-MeC_{26}
12 8-MeC_{26}+x,y-diMeC_{26}
13 6-MeC_{26}
14 2-MeC_{26}
15 x,y-diMe$_{c26}$
16 4,8-diMeC_{26}
17 C_{27}
18 2,y-diMeC_{26}
Peak Identification
19 13 + 11-MeC_{27}
20 9-MeC_{27}
21 7-Me$_{c27}$
22 5-MeC_{27}
23 41,15 + 9,13-diMeC_{27}
24 7,15 + 7,13-diMeC_{27} + 3-MeC_{27}
25 5,9 + 5,13 + 5,15 + 5,17-diMeC_{27}
26 C_{28}
27 3,15 + 3,13 + 3,9 + 3,7-diMeC_{27}
28 14 + 13 + 12 + 10 +8-MeC_{28}
29 12,16-diMeC_{28}+4-MeC_{28}
30 C_{29}
31 15 + 13 +11+9-MeC_{29}
32 7-MeC_{29}
33 5-MeC_{29}
34 13,17 + 11,15 + 9,13-diMeC_{29}
35 7,17-diMeC29 + 3-MeC_{29}
36 5,17-diMeC_{29}
Abundance
18 19 20 21 22 23 24 25 26 27 28 29 30 31
Retention time (min)

FIGURE 38.1 Typical gas chromatogram of the cuticular hydrocarbon profile of an ant. This is from a worker of *Camponotus aethiops*, characterized by linear, monomethyl, and dimethyl branched alkanes. Source: *Modified from van Zweden et al.*[29]

cuticle of ants and other social insects are not commercially available and need to be synthesized ad hoc for use in bioassays. Moreover, it is currently not possible to remove cuticular hydrocarbons from a live ant; therefore, any experimental application of synthetic hydrocarbon will result in a supplementation. The use of dummies is possible, and it has produced good results,[11] but dummies do not always elicit a response that is as clear as that elicited by live insects.

Nestmate recognition cues have both genetic and environmental origin; they can be influenced, for instance, by food and nest material.[3] To function as colony-specific visa, the cuticular hydrocarbon profile needs to be homogenized among nestmates in order to achieve an integrated label, the so-called Gestalt odor,[6,14] that will narrow the cue distribution within a given colony, thus facilitating discrimination between colony members and aliens by reducing cue overlapping between colonies.[4] Ants have evolved a number of mechanisms to enhance cue mixing, such as allogrooming (mutual licking and cleaning of the body surface) and trophallaxis (mouth-to-mouth exchange of liquid food). Hydrocarbons are synthesized by enocytes, either associated with epidermal tissue or present in the peripheral fat body, taken up by lipophorin and transported via the hemolymph to the cuticle and to a storage organ situated in the ant head, the post-pharyngeal gland, which opens to the mouth cavity.[15] The composition of the hydrocarbons on the cuticle and in the post-pharyngeal gland is qualitatively and often quantitatively similar,[16,17] suggesting that this organ plays an active role in homogenizing the nestmate recognition cues. Indeed, the mixture of hydrocarbons can be easily spread from the post-pharyngeal gland to the cuticle via self-grooming and can be transferred among individuals via allogrooming and trophallaxis. This was confirmed by using *in vivo* radiochemical assays, which allowed the description of the dynamics of hydrocarbon transfer between nestmates.[17] Interestingly, a cross-fostering experiment using different colonies of *Formica rufibarbis* showed that some classes of hydrocarbons, namely methyl branched alkanes, are heritably synthesized but at the same time extensively transferred among nestmates, whereas linear hydrocarbons are less heritable and are transferred to a modest extent.[18] The mechanisms allowing this differential transfer are unknown, although the post-pharyngeal gland may play a major role in this process.

Although the general characteristics of the recognition label, such as the identification, production, and distribution of cuticular hydrocarbons, have been investigated in many ant species, there are still numerous open questions about the process of template formation.

IS LEARNING INVOLVED IN THE FORMATION OF THE NESTMATE RECOGNITION TEMPLATE?

Preimaginal Learning

Brood care is an essential component of the social repertoire, and workers are predicted to care exclusively for highly related brood. Indeed, when newly eclosed workers of the Formicine ant *Cataglyphis cursor* are reared in isolation or in groups composed of only young workers, without contact with older nestmates and brood, and then tested for preference between nestmate larvae and alien larvae, they clearly prefer caring for nestmate larvae. However, if eggs are transferred from the mother colony to a different, adoptive colony and are allowed to develop until the stage of pupae and then transferred back to the mother colony, the young workers tested 5 days after emergence from the pupal stage prefer caring for larvae of the adoptive colony rather than for their sister larvae.[19] This suggests that some kind of learning of the characteristics of the adoptive colony, probably the colony odor, has been taking place at the larval stage and could be retained through metamorphosis. Similar results were found in another Formicine species, *Camponotus floridanus*, in which not eggs but first-instar larvae were transferred to an adoptive colony until emergence,[20] supporting the finding that preimaginal experience might play an important role in nestmate brood recognition, at least in Formicine ants. There is evidence that queen recognition might also be influenced by larval experience because *C. cursor* workers that are adopted by an alien queen soon after emergence and that stay with this queen for up to 2 months are still able to recognize their mother queen.[21] To date, no studies have tested whether preimaginal experience might influence nestmate recognition of adult workers in the context of colony defense, although the role of experience gained by young imagos soon after eclosion has been investigated in this respect.

Early Learning

Artificially mixing different ant species in the laboratory has proven to be a useful tool to address several fundamental questions underlying nestmate recognition systems. Some ant species can adopt heterospecific larvae, as in the case of carpenter ants. If larvae of a *Camponotus* species are reared by the queen of a different species, once they reach adulthood, the adopted ants attack and are attacked by their unfamiliar genetic sisters, suggesting that they have incorporated heterospecific cues in their label (eliciting aggression from their sisters) and have formed a template based on those heterospecific cues (causing them to attack their sisters). This suggests that the template is not inherent and it is possibly learned soon after eclosion.[22]

The fact that newly eclosed (callow) ants lack nestmate recognition cues on their cuticle soon after emergence, or at least have very low amounts of cuticular hydrocarbons,[6] allows mixing callow ants from different species without eliciting mutual aggression. In this way, experimentally mixed species groups can be created containing ants of the same age that develop without contact with older nestmates. Although this is an artificial condition, it resembles in some aspects the natural situation of slave-making ants and other social parasites that use a different species as host.[23,24] Slave-making ants conduct raids and pillage brood (especially pupae) from neighboring host nests. These pupae are brought into the slave-making nest and allowed to eclose. The young host workers will then integrate into the mixed nest and start working as if they were in their original colony. The occurrence of this natural phenomenon led to the hypothesis that newly eclosed ant workers do not have an inherent knowledge of their colony odor but, rather, learn it from their nestmates and/or from the nest environment soon after emergence via an imprinting-like phenomenon.[25] Parasites and hosts are typically phylogenetically close, and this could facilitate the integration process if we assume that closely related species have similar recognition cues and thus the mismatch between the cues produced by the host and the parasite is minimal. However, in the laboratory, ants from different subfamilies, such as Myrmicinae and Formicinae, can be mixed experimentally.

In one behavioral experiment, newly eclosed callow workers of *Manica rubida* (Myrmicinae) and *Formica selysi* (Formicinae) were kept together for 15 days after their emergence and then separated into homospecific groups. The ants were subsequently reunited again after 8, 15, 30, or 90 days. Aggression was very low up to 30 days from the separation and increased only after 90 days, although it was still significantly lower than aggression observed in unfamiliar heterospecific groups.[26] This suggests the occurrence of early olfactory learning of recognition cues, which is long-lasting. A similar experiment[27] was then conducted in which ants reared in mixed species groups were separated in homospecific groups and tested for aggression toward heterospecific individuals in dyadic encounters after the same periods as mentioned previously and also after 6 months and 1 year. The results showed that even after 1 year of separation, ants could recognize familiar individuals of the other species. Chemical analyses of the cuticular hydrocarbon profile revealed that during the cohabitation period, the two ant species exchanged hydrocarbons. Indeed, the cuticular profiles of ants reared in heterospecific groups were intermediate between those of *F. selysi* and *M. rubida* reared in single-species groups, with each species carrying on its cuticle some of the cues of the other species. During the separation period, the ants progressively lost the heterospecific hydrocarbons more or less quickly, depending on the species. However, even after 1 year of separation, traces of heterospecific hydrocarbons were still present on the cuticle of ants reared in mixed-species groups.

These results can thus be interpreted according to two different hypotheses. One postulates the occurrence of template learning early in adult life, when workers of one species were cohabiting with workers of the other species and could form a mixed template including both homospecific and heterospecific chemical cues; this template would serve as a reference later in life. The second hypothesis does not require early learning of the template but is based on the fact that the presence of heterospecific cues on the ant cuticle, even if in low amount, would allow recognition because the encountered individual does not bear cues that are different from self. Further studies are needed to better understand the role of early learning in nestmate recognition, although recent data appear to support the second hypothesis.

Habituation and Sensory Adaptation

It is possible to incorporate a novel hydrocarbon into the cuticule of ants by feeding them with a synthetic hydrocarbon mixed with liquid food. Two groups of sister workers were created from a colony of *Camponotus herculeanus*; one group was fed honey mixed with a synthetic hydrocarbon not present in the natural chemical profile of this species, and the other group was fed honey mixed with solvent only. In this way, the only difference in the cuticular chemical profile of the two groups was the presence of one additional hydrocarbon on the cuticle of the first group. When ants from the two groups confronted each other, aggression was asymmetrical: The ants with the additional compound on their cuticle did not aggress the ants lacking the compound, whereas the ants lacking the compound were aggressive toward the ants bearing the novel compound, particularly when this was a dimethyl branched alkane, which is supposed to be a class of hydrocarbons important for recognition. Thus, it is the presence and not the absence of an additional hydrocarbon on the cuticle that elicits aggression.[10] These results do not support the model based on a matching between a perceived label and a template stored in the long-term memory for recognition to occur, because also in the case of the absence of a compound, there would be a mismatch between label and template. Alternatively, the mechanism could be based on habituation, a nonassociative form of elemental learning. Ants are habituated to their own cuticular chemical profile, to the colony odor to which they are constantly exposed, and thus respond only to what can be perceived as different. Multicomponent odor profiles likely activate more than one family of olfactory receptors that will elicit a given pattern of activity in the glomeruli (the processing units of the antennal lobes, the first-order integration centers of olfactory information in the brain; see relevant chapters in this book). Under the habituation model, the pattern of glomeruli elicited by cuticular odors of nestmates does not cause aggression, as well as a pattern of active glomeruli that overlaps (i.e., one compound missing), because this is included in the nestmate pattern. To reject an encountered individual, additional activity in the glomeruli would be needed, and this is elicited by compounds that are not part of the colony odor or that are present in much higher amount in the label of the encountered individual.

Imaging of the ant antennal lobe has been proven to be problematic because ants possess more than 400 glomeruli, a very high number compared to the 160 glomeruli in the honeybee. However, a recent study succeeded in performing calcium imaging of the antennal lobe following presentation of cuticular hydrocarbon profiles in the ant *Camponotus floridanus*. The results suggest that the spatial activity patterns of glomeruli upon repeated stimulation with the same colony odor are as variable as activity patterns elicited by different colony odors, thus increasing our challenge in understanding what kind of information is used to classify nestmates and non-nestmates.[28]

The habituation model can parsimoniously explain how ants might cope with the fact that the colony odor is dynamic and changes over time, due also to influences of the external environment.[29,30] This requires a continuous updating of the template throughout the lifetime of an individual. Indeed, adult worker ants can be familiarized with the odor of an alien colony. If a group of *Camponotus aethiops* workers from one colony is exposed to cuticular chemical extracts only from another colony for 24 hr, they significantly reduce aggression toward individuals of this other colony in subsequent encounters between ants, suggesting that a template updating occurred during the familiarization period by simple exposure to the odor.[31] To further investigate the information processing of nestmate recognition cues, tethered ants were exposed to the odor of an alien colony of one antenna only by inserting the antenna in a glass capillary coated with the cuticular extract of the alien colony (the other antenna was inserted in a glass capillary coated with the solvent only). By selectively excising the cuticular hydrocarbon or the sham-exposed antennae before aggression tests involving the treated ant and an ant from the alien colony (donor of the odor), it was possible to show that information processing of recognition cues is side specific.[31] When the antenna that was left was the sham exposed, the treated ant was aggressive against the alien ant, but when the antenna that was left was the one exposed to the cuticular hydrocarbons of the alien colony, aggression toward the alien ant significantly decreased (Figure 38.2). This suggests that decision making with respect to nestmate recognition cues does not require information transfer between brain hemispheres and that the process might be decentralized.

There is evidence that discrimination among nestmate and non-nestmate odors may occur even at a more peripheral level. A chemosensory sensillum sensitive to cuticular hydrocarbons was discovered on the antennae of the ant *Camponotus japonicus*. The cuticle wall of this sensillum has numerous tiny olfactory pores, and a single sensillum contains approximately 130 receptor neurons, making it suitable to respond to the multicomponent cuticular hydrocarbon mixtures encoding recognition cues. In electrophysiological experiments, it was shown that this sensillum selectively responds to non-nestmate cuticular hydrocarbons only. Indeed, surprisingly, it does not discharge impulses when stimulated with cuticular hydrocarbons extracts of nestmates.[32,33] This suggests that sensory adaptation to the constantly present nestmate colony odor might occur. The sensillum would act as a sensory filter so that only alien odors (sufficiently different from nestmate odors) are relayed to the central nervous system and elicit a behavioral reaction such as aggression.

However, in *Camponotus floridanus*, reduction of aggression after exposure of the antennae to non-nestmate odors takes between 2 and 15 hr, suggesting a different process than sensory adaptation.[34] More important, electroantennography performed on the whole antenna showed that both nestmate and non-nestmate odors elicit voltage responses of similar signal amplitude, and calcium imaging detected different and variable activity patterns in the antennal lobes upon stimulation with both nestmate and non-nestmate hydrocarbons.[28] Thus, it appears that ants are not anosmic to nestmate colony odors, as it is expected under the sensory adaptation model.

These results in the two *Camponotus* species are not completely incompatible if we postulate the existence of two kinds of sensilla on the ant antennae: the highly efficient 'nestmate recognition sensilla' specifically tuned to nestmate cuticular hydrocarbons and a second type of sensilla that respond to hydrocarbons but are not tuned to specific ones and can thus transfer a coarse image to the antennal lobes.[35] Additional studies are needed to understand the relative importance of early learning, habituation, and sensory adaptation in the formation and updating of the recognition template and the exact mechanisms underlying nestmate discrimination in ants and other social insects.

WHEN LEARNING AND MEMORY ARE INDISPENSABLE

In a colony containing hundreds or even thousands of individuals, it is important to know who belongs to the group (nestmate recognition) and whether or not a fertile queen is present (via the detection of queen pheromones[36]). It may also be advantageous to recognize classes of individuals that perform a certain task (e.g., patrollers eliciting the onset of foraging activity[37]) or even selfish workers that try to reproduce in order to police them by aggression or by destroying their eggs.[38] However, recognizing the exact identity of each single individual appears to be extremely challenging and not necessarily beneficial for the organization of a populous colony. Nevertheless, in certain circumstances, individual recognition might be advantageous, and it has indeed been demonstrated in some social wasps and ants.

Ant Queens

Queens of the ponerine tropical ant *Pachycondyla villosa* usually associate with unrelated queens to start a new colony. They establish a dominance hierarchy leading to a division of labor: Subordinate queens take up the risky task of foraging, whereas dominant queens stay in the nest and care for the brood.

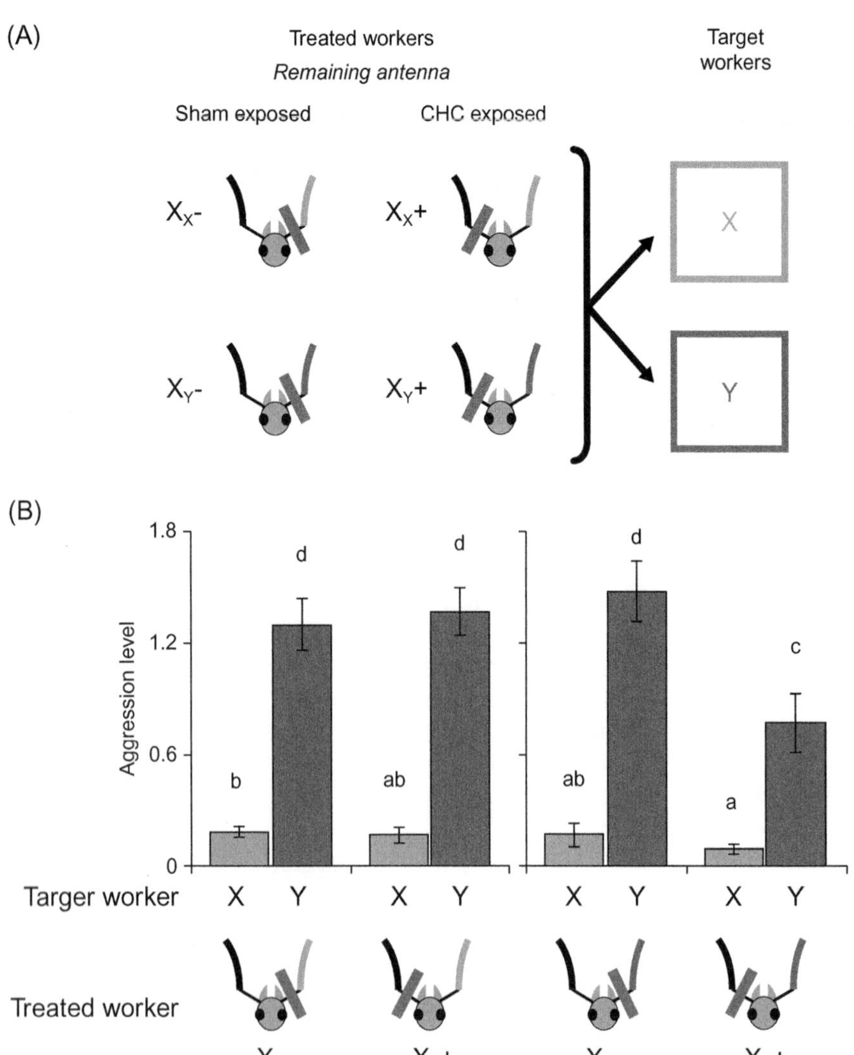

FIGURE 38.2 Side-specific perceptual discrimination: Effect of unilateral antennal exposure to alien colony odor. (A) Experimental design. Antennae of workers from colony X were inserted into two capillaries, one of which was treated with solvent (sham exposed, −) and the other was coated with cuticular hydrocarbons (CHC-exposed, +) from either nestmate workers (control X_X) or non-nestmates from colony Y (test X_Y). After an 18-hr exposure, one antenna was selectively excised. Aggression tests between treated workers and anaesthetized target workers from colonies X or Y were performed immediately after excision, as indicated by the arrows. (B) Aggression level of treated workers toward targets from colonies X (yellow bars) and Y (blue bars). Columns and error bars indicate mean and standard error of aggression indices, respectively. Different letters indicate significant differences between categories (mixed-effects model with least square means post hoc comparisons, $p < 0.05$). When their remaining antenna had been exposed to non-nestmate CHCs (X_Y^+), workers were significantly less aggressive toward non-nestmates from colony Y than when their remaining antenna had been sham exposed. X_Y^+ workers were also less aggressive toward Y individuals than control workers, which had been exposed to nestmate odor. The aggressiveness of control workers (X_X) toward Y individuals was always high and did not depend on which antenna was excised. Source: *Modified from Stroeymeyt et al.*[31]

Pleometrosis (cooperative colony founding) is relatively common in ants, but usually only one queen survives the founding stage.[1] In *Pachycondyla*, however, the queens stay together after the emergence of workers and thus the original hierarchy set up during the colony foundation might influence the social structure of established colonies, which usually do not grow very large (up to 100 individuals, but frequently a few dozens).

Do founding queens recognize the individual identity of their associates? In a first experiment, it was tested whether founding queens recognize individual identity or merely social status (dominant/subordinate) of other queens by using two rounds of binary interactions. A series of two unfamiliar, likely unrelated founding queens were housed together for 24 hr, during which a clear dominance relationship is usually established. Then, each subordinate queen was housed in a new nest either with the familiar dominant or with an unfamiliar dominant. Subordinates were highly aggressive toward the unfamiliar but not toward the familiar dominant, thus ruling out status recognition. A Y-maze choice experiment showed that chemical cues are important in this recognition process, and chemical analyses revealed that the cuticular hydrocarbon profiles of queens are highly variable and do not cluster according to social status, thus suggesting that individual recognition may be based on mutual learning of the idiosyncratic cuticular chemical profile characterizing each queen in an association.[39]

In another study, founding queens of two sister species of *Pachycondyla* with a very similar ecology and life history, *P. villosa* and *P. inversa*, were tested for individual recognition with a modified experimental design. Each experimental session involved four queens confronted in dyadic encounters four consecutive times. In the first round, pairs of unfamiliar queens were housed together for 24 hr and then switched so that each queen met a second unfamiliar queen belonging to the other pair for another 24 hr.

Aggression was high in both rounds of encounters for both species, as expected for unfamiliar queens that have to establish a dominance hierarchy. Afterwards, each queen met the familiar queen encountered in the first round. Aggression was low in both species, demonstrating that queens are able to recognize each other individually after more than 24 hr of separation. The decrease in aggression in the third round was not due to fatigue or habituation to the experimental conditions because when queens encountered an additional unfamiliar queen in the fourth round of interactions, they again showed significantly higher aggression.[40]

These results indicate that recognition is based on long-term memory of individual identity, which is very robust because it is not erased by interaction with different individuals (occurring in the second round). In natural conditions, co-founding queens of *Pachycondyla* may be separated for a relatively long time because subordinates leave the nest to engage in foraging trips. At the same time, unfamiliar queens may try to join established associations given that new queens in these species are produced year-round. These new queens should be rejected. Long-term memory of individual identity is therefore an efficient mechanism to minimize the cost of repeated assessment, thus decreasing aggressive contests and risks of injuries, and to ensure that a partner with whom a stable relationship has already been established is recognized when it comes back to the nest.

Assuming that the distinctive cuticular hydrocarbon signature of each queen allows individual recognition, the question arises as to whether cue variability evolves to signal identity only when queens are selected to be easily identifiable. Is it the fact that individual recognition is advantageous in systems characterized by stable dominance hierarchies that favors the evolution of an individually distinctive chemical profile? In the black garden ant, *Lasius niger*, queens may also associate to found a new colony, but they remain in a nest chamber throughout the founding period and rely on their body reserves to raise the first brood instead of engaging in foraging trips. In contrast to that of *Pachycondyla*, division of labor is not required in the *L. niger* social system. Moreover, at worker emergence, only one queen survives. It was experimentally shown that co-founding queens of *L. niger* do not recognize each other individually, although the variation in their cuticular chemical profile is comparable to that of *Pachycondyla* queens,[41] ruling out informational constraints hampering recognition of individual identity. Thus, high variability in the cuticular hydrocarbon profile of queens might be present for other reasons and depending on different social contexts and environmental constraints has evolved or not to signal individual identity. It remains to be tested whether memory capacities are also selected depending on the social context.

Ant Workers

Using a classical habituation/discrimination paradigm, it was possible to show that adult workers of the ant *Cataglyphis niger* can learn the individual characteristics of non-nestmates, both homospecific and heterospecific (*C. cursor*). A worker ant encountered the same non-nestmate ant 10 times for 3 min, with an intertrial interval of 10 min, and was then immediately confronted with a novel ant. Aggression decreased over the repeated encounters with the same ant but was high again when the focal ant encountered a novel non-nestmate (belonging to the same colony as the ant encountered 10 times), suggesting that the focal ant had learned the individual odor of a particular non-nestmate ant during the habituation period.[42] This implies that there is enough variability in the chemical profile of different worker ants belonging to the same colony allowing the observed discrimination; however, chemical analyses were not performed in this study.

Another study used a similar procedure to investigate memory retention of learned individual characteristics in the ant *C. cursor*. An adult *C. cursor* worker ant encountered the same anesthetized *Camponotus aethiops* worker ant in four successive encounters of 3 min each (habituation trials), with a 10-min intertrial interval. Similar to the previous study, aggression decreased over the repeated encounters. Afterwards, the *C. cursor* ant was tested for discrimination between the familiar and an unfamiliar (from the same colony) *Ca. aethiops* ant, with an interval of 10, 30, or 60 min from the end of the habituation period. The results showed that *C. cursor* workers could remember the identity of the ant encountered during the habituation period for up to 30 min (measured by the lack of aggression), but this memory of individual identity disappeared after 60 min of separation.[43] Although retention scores were low, possibly indicating loss of memory, a lasting memory cannot be totally excluded, and other factors, such as motivation and context, may influence retrieval from memory.

Although the adaptive function of individual recognition in founding queens establishing a dominance hierarchy is apparent, the fitness advantage of developing such a sophisticated recognition ability in ant workers is not clear. The ants tested in the two studies discussed previously were mostly foraging workers; in this context, it might be beneficial to learn the identity of foragers from other colonies that might be repeatedly encountered during the same foraging trip in order to minimize the possible cost of aggressive interactions. A phenomenon known as 'dear enemy' effect—that is, responding less aggressively to familiar neighbors than to complete strangers—is relatively widespread in the animal kingdom, and it has also been reported in some ant species.[44]

TOOLS TO STUDY OLFACTORY LEARNING AND MEMORY IN ANTS

During the past several decades, the honeybee and the fruit fly have become model systems for the study of insect learning and memory (see relevant chapters in this book). Ants' learning abilities have been extensively studied in the context of navigation[45–47] and foraging,[48,49] and only recently has attention been focused on individual olfactory learning under controlled conditions.

The first study showing that individual ants can perceive and learn odors in controlled laboratory conditions used a Y-maze in which ants were individually trained to forage.[50] Training followed a differential conditioning procedure in which one odor was positively reinforced with sucrose solution and another odor was negatively reinforced with quinine solution (a bitter substance). After each training trial in the Y-maze, the ant was returned to its own colony, and it was picked up for the next trial when it climbed on a vertical stick placed in the foraging arena of the nest, from which it was then transferred again to the Y-maze. A motivated forager would typically come back to the stick within 4 or 5 min. After 24 training trials, the ants were immediately tested with the two odors but no reinforcer. Workers of two different *Camponotus* species (*Ca. mus* and *Ca. fellah*) successfully learned to associate an odor with the reward and discriminated both between structurally dissimilar compounds (limonene and octanal) and between structurally similar compounds (heptanal and 2-heptanone). This pioneering study set the stage for novel investigations.

By using a similar procedure and the same experimental apparatus, another study tested whether olfactory learning leads to stable long-term memory in *Ca. fellah* worker ants.[51] Individual ants were differentially conditioned during 16 training trials using two odors (limonene and octanal) and then subjected to five consecutive retention tests (with no reinforcer) performed either 24 or 72 hr after training. Ants were able to retrieve the learned information in both cases, thus showing that they have the ability to store olfactory information in their long-term memory. This could be advantageous in the foraging context if ants feed on the same sources over several days, especially in the case of *Ca. fellah*, which is a species active mostly at night and should thus rely more on olfactory than on possible visual cues.

Using free-walking ants in a Y-maze has proven to be effective for investigating individual olfactory learning and memory; however, this procedure does not allow perfect control of some important parameters, such as stimulus sequence, stimulation time, and intertrial interval. Indeed, a free-walking ant could detect both odor stimuli before encountering the unconditioned stimulus; it could spend more or less time in contact with an odor stimulus; and, once back in its own colony, it could take a variable amount of time to come back to the stick and start the next training trial. A procedure using harnessed insects, restrained in an experimental apparatus, allows controlling precisely all the key parameters during conditioning. This kind of technique has been widely used in the honeybee to quantify response to chemical stimulation via conditioning of the proboscis extension response (PER).[52] When an odor is forward-paired with a sucrose reward, the bee develops a conditioned PER response after a few learning trials.[53] A similar procedure has been developed in ants, using *Ca. aethiops* workers individually restrained in a harness. Upon stimulation on the antennae with sucrose, the ant extends its maxilla–labium; this response was thus called the maxilla–labium extension response (MaLER).[54] Ant workers could be trained in both absolute and differential conditioning using octan-1-ol and hexanal as odor stimuli and sucrose solution or quinine solution as positive and negative reinforcement, respectively (Figure 38.3). Approximately 60% of the ants could be successfully conditioned in six learning trials, and memory lasted for at least 1 hr.[54] The development of this novel conditioning protocol represents an important step forward in the study of chemoreception, learning, and memory formation in ants.

Indeed, the MaLER protocol could be used to answer the question of whether memories formed upon olfactory learning are retrievable several days after training and whether they are based on *de novo* protein synthesis. Individually harnessed *Ca. fellah* ant workers were differentially conditioned to discriminate between two long-chain hydrocarbons (docosane (n-C_{22}) and octacosane (n-C_{28})), one paired with sucrose and the other with quinine. The memory retention tests were performed 1, 12, and 72 hr after conditioning, which correspond to distinct memory phases in the honeybee[55]: early (1 hr) and late (12 hr) midterm memory and long-term memory (72 hr). Half of the ants were treated with a protein synthesis blocker (cycloheximide) mixed with sucrose solution before training (control ants received only sucrose). The acquisition phase, as well as memory retention, was

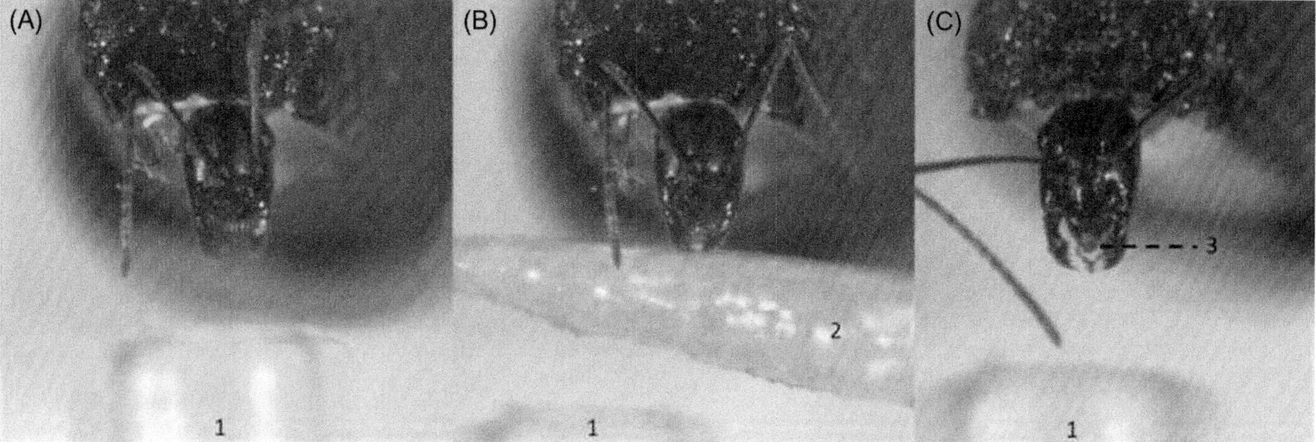

FIGURE 38.3 The *maxilla–labium* extension response. (A) The response of an ant to a stimulus before conditioning. (B) Conditioning: The stimulus is presented together with a sugar reward. (C) The response to the stimulus after conditioning: Upon presentation of the conditioned stimulus, the ant extends her maxilla–labium in expectance of the reward. 1, Syringe containing the conditioned stimulus (odor); 2, toothpick containing the unconditioned stimulus (reward); 3, extended maxilla–labium. Source: *Photos courtesy of Fernando Guerrieri and Nick Bos.*

not affected by cycloheximide 1 and 12 hr after conditioning. However, control ants could retrieve the olfactory memory 72 hr after conditioning, whereas ants treated with the protein synthesis blocker could not.[56] This demonstrates that olfactory memories retrievable 3 days after training in *Ca. fellah* fulfill both temporal and molecular requirements of long-term memory and that earlier memories in ants do not require protein synthesis, similar to the honeybee.[55] These results are important not only for the characterization of insect olfactory memories in general but also to better understand the evolutionary constraints that might have shaped the foraging dynamics of ant species such as *Ca. fellah*, which rely on individual foragers collecting nectar from predictable resources represented, for instance, by extrafloral nectaries.

MaLER conditioning was used successfully in the Argentine ant *Linepithema humile* to study learning and discrimination of synthetic hydrocarbons that are present on the cuticle of these ants. According to the hypothesis that some classes of hydrocarbons may play a more salient role in nestmate recognition than others, ant workers learned more easily trimethyl alkanes than monomethyl or linear alkanes. In addition, Argentine ants could distinguish between hydrocarbons with methyl branches in different positions but not necessarily between hydrocarbons with methyl branches in the same position but with different chain length.[57]

Another study used free-walking *Ca. aethiops* ant workers that could associate a synthetic hydrocarbon to sugar reward to study learning and generalization among different classes of hydrocarbons that are typically present on the ants' cuticle.[58] Ants learned both linear and monomethyl branched alkanes with the same efficiency. However, they generalized among linear alkanes when the novel molecule was smaller than the conditioned one but not the other way around, suggesting a perceptual strategy based on an 'inclusion criterion' (when the novel molecule is shorter, it is included in the conditioned one). Generalization among methyl branched alkanes was usually high, independent of the position of the branch; however, when hydrocarbons differed in the presence or absence of the methyl group, the ants could easily discriminate between two compounds. This suggests that chain length and functional group might be coded independently in the olfactory system of this ant species.

These results indicate that different ant species might have evolved different perceptual and coding strategies; consequently, deciphering the nestmate recognition process in one species would not necessarily permit understanding the generalities of this process. Ants are a very diverse taxon characterized by a plethora of life histories, and they have evolved under very different ecological constrains. Nevertheless, we now have a set of tools that will allow the exploration of several aspects of the developmental, cellular, and molecular bases of olfactory learning and memory in ants, which are among the most advanced social insects.

Acknowledgments

I thank the editors for inviting me to contribute to this volume. I thank my collaborators Nick Bos, Martin Giurfa, Fernando Guerrieri, Volker Nehring, and Jelle van Zweden for insightful discussions.

References

1. Hölldobler B, Wilson EO. *The Ants*. Cambridge, MA: Belknap; 1990.
2. d'Ettorre P, Moore AJ. Chemical communication and the coordination of social interactions in insects. In: d'Ettorre P, Hughes DP, eds. *Sociobiology of Communication*. Oxford: Oxford University Press; 2008:81–96.
3. d'Ettorre P, Lenoir A. Nestmate recognition. In: Lach L, Parr C, Abbott K, eds. *Ant Ecology*. Oxford: Oxford University Press; 2010:194–209.
4. van Zweden JS, d'Ettorre P. The role of hydrocarbons in nestmate recognition. In: Blomquist GC, Bagnères AG, eds. *Insect Hydrocarbons: Biology, Biochemistry and Chemical Ecology*. Cambridge, UK: Cambridge University Press; 2010:222–243.
5. Arnold G, Quenet B, Cornuet J-M, et al. Kin recognition in honeybees. *Nature*. 1996;379:498.
6. Lenoir A, Fresneau D, Errard C, Hefetz A. The individuality and the colonial identity in ants: the emergence of the social representation concept. In: Detrain C, Deneubourg JL, Pasteels J, eds. *Information Processing in Social Insects*. Basel: Birkhäuser Verlag; 1999:219–237.
7. Hölldobler B, Michener CD. Mechanisms of identification and discrimination in social Hymenoptera. In: Markl H, ed. *Evolution of Social Behavior: Hypotheses and Empirical Tests*. Weinheim: Chemie Verlag; 1980:35–58.
8. Vander Meer RK, Morel L. Nestmate recognition in ants. In: Vander Meer RK, Breed MD, Espelie K, Winston ML, eds. *Pheromone Communication in Social Insects: Ants, Wasps, Bees and Termites*. Boulder, CO: Westview; 1998:79–103.
9. Wyatt TD. Pheromones and signature mixtures: defining species-wide signals and variable cues for identity in both invertebrates and vertebrates. *J Comp Physiol A*. 2010;196: 685–700.
10. Guerrieri FJ, Nehring V, Jørgensen CG, Nielsen J, Galizia CG, d'Ettorre P. Ants recognize foes and not friends. *Proc R Soc Lond B*. 2009;276:2461–2468.
11. Martin SJ, Vitikainen E, Helanterä H, Drijfhout FP. Chemical basis of nestmate discrimination in the ant *Formica exsecta*. *Proc R Soc Lond, B*. 2008;275:1271–1278.
12. Akino T, Yamamura N, Wakamura S, Yamaoka R. Direct behavioral evidence for hydrocarbons as nestmate recognition cues in *Formica japonica* (Hymenoptera: Formicidae). *Appl Entomol Zool*. 2004;39:381–387.
13. Greene MJ, Gordon DM. Structural complexity of chemical recognition cues affects the perception of group membership in the ants *Linepithema humile* and *Aphaenogaster cockerelli*. *J Exp Biol*. 2007;210:897–905.
14. Crozier RH, Dix MW. Analysis of two genetic models for the innate components of colony odor in social Hymenoptera. *Behav Ecol Sociobiol*. 1979;4:217–224.
15. Bagnères AG, Blomquist GJ. Site of synthesis, mechanisms of transport and selective deposition of hydrocarbons. In: Blomquist GC, Bagnères AG, eds. *Insect Hydrocarbons: Biology, Biochemistry and Chemical Ecology*. Cambridge, UK: Cambridge University Press; 2010:75–99.
16. Bagnères AG, Morgan ED. The post-pharyngeal glands and the cuticle of Formicidae contain the same hydrocarbons. *Experientia*. 1991;47:106–111.
17. Soroker V, Vienne C, Hefetz A. Hydrocarbon dynamics within and between nestmates in *Cataglyphis niger* (Hymenoptera: Formicidae). *J Chem Ecol*. 1995;21:365–378.
18. van Zweden JS, Brask JB, Christensen JH, Boomsma JJ, Linksvayer TA, d'Ettorre P. Blending of heritable recognition cues among ant nestmates creates distinct colony gestalt odours but prevents within-colony nepotism. *J Evol Biol*. 2010;23: 1498–1508.
19. Isingrini M, Lenoir A, Jaisson P. Preimaginal learning as a basis of colony-brood recognition in the ant *Cataglyphis cursor*. *Proc Natl Acad Sci USA*. 1985;82:8545–8547.
20. Carlin NF, Schwartz PH. Pre-imaginal experience and nestmate brood recognition in the carpenter ant, *Camponotus floridanus*. *Anim Behav*. 1989;38:89–95.
21. Berton F, Lenoir A, Nowbahari E, Barreau S. Ontogeny of queen attraction to workers in the ant *Cataglyphis cursor* (Hymenoptera, Formicidae). *Insectes Soc*. 1991;38:293–305.
22. Carlin NF, Hölldobler B. Nestmate and kin recognition in interspecific mixed colonies of ants. *Science*. 1983;222:1027–1029.
23. d'Ettorre P, Heinze J. Sociobiology of slave-making ant. *Acta Ethologica*. 2001;3:67–82.
24. Lenoir A, d'Ettorre P, Errard C, Hefetz A. Chemical ecology and social parasitism in ants. *Annu Rev Entomol*. 2001;46:573–599.
25. Jaisson P. Kinship and fellowship in ants and social wasps. In: Hepper PG, ed. *Kin Recognition*. Cambridge, UK: Cambridge University Press; 1991:60–93.
26. Errard C. Role of early experience in mixed-colony odor recognition in the ants *Manica rubida* and *Formica selysi*. *Ethology*. 1986;72:243–249.
27. Errard C. Long-term memory involved in nestmate recognition in ants. *Anim Behav*. 1994;48:263–271.
28. Brandstaetter AS, Rössler W, Kleineidam CJ. Friends and foes from an ant brain's point of view: neuronal correlates of colony odors in a social insect. *PLoS ONE*. 2011;6:e21383.
29. van Zweden JS, Dreier S, d'Ettorre P. Disentangling environmental and heritable nestmate recognition cues in a carpenter ant. *J Insect Physiol*. 2009;55:158–163.
30. Vander Meer RK, Saliwanchik D, Lavine B. Temporal changes in colony cuticular hydrocarbon patterns of *Solenopsis invicta*: implications for nestmate recognition. *J Chem Ecol*. 1989;15: 2115–2125.
31. Stroeymeyt N, Guerrieri FJ, van Zweden JS, d'Ettorre P. Rapid decision-making with side-specific perceptual discrimination in ants. *PLoS ONE*. 2010;5:e12377.
32. Ozaki M, Wada-Katsumata A, Fujikawa K, et al. Ant nestmate and non-nestmate discrimination by a chemosensory sensillum. *Science*. 2005;309:311–314.
33. Ozaki M, Wada-Katsumata A. Perception and olfaction of cuticular compounds. In: Blomquist GC, Bagnères AG, eds. *Insect Hydrocarbons: Biology, Biochemistry and Chemical Ecology*. Cambridge, UK: Cambridge University Press; 2010:207–221.
34. Leonhardt SD, Brandstaetter AS, Kleineidam CJ. Reformation process of the neuronal template for nestmate-recognition cues in the carpenter ant *Camponotus floridanus*. *J Comp Physiol A*. 2007;193:993–1000.
35. Bos N, d'Ettorre P. Recognition of social identity in ants. *Front Comp Psychol*. 2012;3:83.
36. Holman L, Jørgensen CG, Nielsen J, d'Ettorre P. Identification of an ant queen pheromone regulating worker sterility. *Proc R Soc Lond B*. 2010;277:3793–3800.
37. Greene MJ, Gordon DM. Cuticular hydrocarbons inform task decision. *Nature*. 2003;423:32.
38. van Zweden JS, Fürst MA, Heinze J, d'Ettorre P. Specialization in policing behaviour among workers of the ant *Pachycondyla inversa*. *Proc R Soc Lond B*. 2007;274:1421–1428.
39. d'Ettorre P, Heinze J. Individual recognition in ant queens. *Curr Biol*. 2005;15:2170–2174.
40. Dreier S, van Zweden JS, d'Ettorre P. Long-term memory of individual identity in ant queens. *Biol Lett*. 2007;3:459–462.
41. Dreier S, d'Ettorre P. Social context predicts recognition systems in ant queens. *J Evol Biol*. 2009;22:644–649.

42. Nowbahari E. Learning of colonial odor in the ant *Cataglyphis niger* (Hymenoptera: Formicidae). *Learn Behav.* 2007;35:87–94.
43. Foubert E, Nowbahari E. Memory span for heterospecific individuals' odors in an ant, *Cataglyphis cursor. Learn Behav.* 2008;36:319–326.
44. Dimarco RD, Farji-Brener AG, Premoli AC. Dear enemy phenomenon in the leaf-cutting ant *Acromyrmex lobicornis*: behavioural and genetic evidence. *Behav Ecol.* 2010;21:304–310.
45. Chameron S, Schatz B, Pastergue Ruiz I, Beugnon G, Collett TS. The learning of a sequence of visual patterns by the ant *Cataglyphis cursor. Proc R Soc Lond B.* 1998;265:2309–2313.
46. Collet TS, Graham P, Harris RA, Hempel de Ibarra N. Navigational memories in ants and bees: memory retrieval when selecting and following routes. *Adv Study Behav.* 2006;36:123–172.
47. Riabinina O, Hempel de Ibarra N, Howard L, Collet TS. Do wood ants learn sequences of visual stimuli? *J Exp Biol.* 2011;214:2739–2748.
48. Saverschek N, Herz H, Wagner M, Roces F. Avoiding plants unsuitable for the symbiotic fungus: learning and long-term memory in leaf-cutting ants. *Anim Behav.* 2010;79: 689–698.
49. Schatz B, Lachaud J-P, Beugnon G. Spatio-temporal learning by the ant *Ectatomma ruidum. J Exp Biol.* 1999;2010:1897–1907.
50. Dupuy F, Sandoz J-C, Giurfa M, Josens R. Individual olfactory learning in *Camponotus* ants. *Anim Behav.* 2006;72:1081–1091.
51. Josens R, Eschbach C, Giurfa M. Differential conditioning and long-term olfactory memory in individual *Camponotus fellah* ants. *J Exp Biol.* 2009;212:1904–1911.
52. Takeda K. Classical conditioned response in the honey bee. *J Insect Physiol.* 1961;6:168–179.
53. Bitterman ME, Menzel R, Fietz A, Schäfer S. Classical conditioning of proboscis extension in honeybees (*Apis mellifera*). *J Comp Psychol.* 1983;97:107–119.
54. Guerrieri FJ, d'Ettorre P. Associative learning in ants: conditioning of the maxilla-labium extension response in *Camponotus aethiops. J Insect Physiol.* 2010;56:88–92.
55. Menzel R. Searching for the memory trace in a mini-brain, the honeybee. *Learn Mem.* 1999;8:53–62.
56. Guerrieri FJ, d'Ettorre P, Devaud J-M, Giurfa M. Long-term olfactory memories are stabilised via protein synthesis in *Camponotus fellah* ants. *J Exp Biol.* 2011;214:3300–3304.
57. van Wilgenburg E, Felden A, Choe D-H, et al. Learning and discrimination of cuticular hydrocarbons in a social insect. *Biol Lett.* 2012;8:17–20.
58. Bos N, Dreier S, Jørgensen CG, Nielsen J, Guerrieri FJ, d'Ettorre P. Learning and perceptual similarity among cuticular hydrocarbons in ants. *J Insect Physiol.* 2012;58:138–146.

CHAPTER

39

Bounded Plasticity in the Desert Ant's Navigational Tool Kit

Rüdiger Wehner and Wolfgang Rössler†*

*University of Zürich, Zürich, Switzerland †University of Würzburg, Würzburg, Germany

Insect behavior is often characterized as being governed by rigidly preordained routines merely operating on the basis of strict stimulus–response associations. This is a rather inappropriate view, if one considers the rich behavioral repertoires that social insects employ in navigating through their (in insect terms) vast foraging territories under varying environmental conditions. Although the navigational tool box of bees and ants contains quite a number of task-specific routines, or modules, recent studies have unraveled an unexpectedly large potential of experience-dependent flexibility in the interaction of these modules and, to a certain extent, within the modules themselves.[1,2] The evolutionary biologist Theodosius Dobzhansky remarked that it is "the norm of reaction of the organism to the environment" that changes in evolution,[3] and we can add that what subsequently changes during the individual lifetime of an organism is the expression and molding of this species-specific norm under particular environmental conditions, and hence its transformation into the actual individual-specific behavior. Such molding occurs during the early ontogenetic development of the organism and continues, to various degrees, during the life span of the adult. The plasticity of this molding process and thus the flexibility of the resulting behavior might help one to disentangle what has been built into the nervous system by the evolutionary past and what is gained by present-day individual experience.

Of course, the view that insects behave in a stereotyped, automaton-like way is not totally fictitious. An insect's short lifetime and small brain size might well limit the amount of information that could be taken up and the ease with which information is exchanged among various modules. A desert ant (*Cataglyphis fortis*) is endowed with a 0.1-mg brain[4] and has an outdoor life expectancy of approximately 6 days.[5] Nevertheless, it is able to acquire, store, retrieve, handle, and use amazing amounts of spatial information. One solution to this obvious conundrum might lie in the fact that the acquisition of information occurs in a rather prestructured way within the navigation network (predictive learning). Dobzhansky's species-specific norm implanted in the nervous system by natural selection defines what current information is taken up, when and how quickly this happens, where in the network information transfer occurs, and where it does not. This 'bounded plasticity' might guarantee that insects accomplish even complex tasks in a fast, frugal, robust, and relatively error-free way.

Saharan desert ants of the genus *Cataglyphis* as well as their ecological analogs *Ocymyrmex* and *Melophorus* in the southern African and central Australian deserts, respectively, have become model organisms in the study of animal navigation. These thermophilic ants are visually guided, exclusively diurnal foragers that search individually for arthropod corpses during the hottest time periods of day and year.[6] Furthermore, due to the rather low prey densities prevailing in their desert habitats, they have to cover large foraging distances. These ecological constraints have set the stage for sophisticated means of navigation to evolve.[1,2,7,8]

THE MAJOR TRANSITION IN THE ANT'S ADULT LIFETIME

The workers of thermophilic desert ants, as well as those of most other ant species,[9,10] exhibit a marked temporal, age-dependent caste polyethism.[5,11–15] They start their foraging lives as callows (i.e., newly hatched

Invertebrate Learning and Memory.
DOI: http://dx.doi.org/10.1016/B978-0-12-415823-8.00039-3

workers), which are usually not older than 24 hr and which can be easily recognized by their pale cuticle. Next comes the replete (interior I) stage, which is characterized by the workers' swollen gasters and stretched abdominal intersegmental membranes. In contrast to these repletes, which mostly sit still inside the nest chambers and, if disturbed, move around only sluggishly, the workers of the subsequent interior II stage are actively moving within the colony, where they accomplish various brood-caring and nest-maintaining tasks. Toward the end of this stage, they can often be observed carrying soil particles out of the nest and dumping refuse. Finally, after an entire indoor lifetime of approximately 28 days, the workers start their relatively short outdoor activities as foragers (Figure 39.1).

During the entire sequence of worker life stages, several inner organs, such as ovarioles, fat bodies, and labial glands, gradually decrease in size and reach their most reduced state in the foragers. Concomitantly, the sensory world changes dramatically during the transition from indoor to outdoor life. Whereas olfactory cues used, for example, in the context of intranidal communication systems are of paramount importance during the first stages of life, the foragers later depend heavily on visual (celestial and terrestrial) cues. This transition is correlated with a marked change in visual behavior. For example, in laboratory experiments, indoor workers

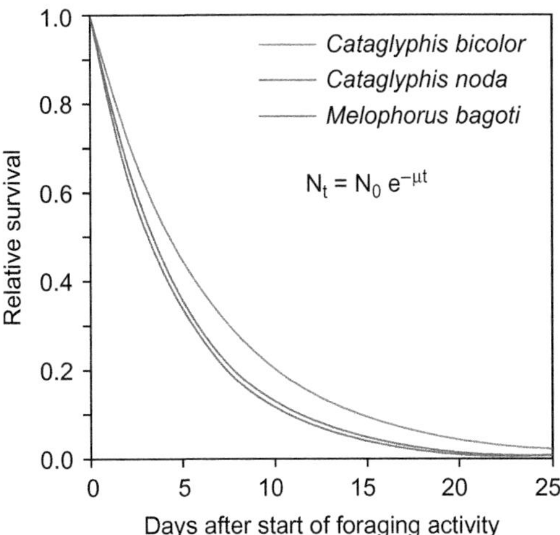

FIGURE 39.1 Forager survival frequencies in three species of thermophilic desert ant as recorded in their natural habitats: *Cataglyphis bicolor* of the lowland steppes of Tunisia (34°32′ N, 10°29′ E), *C. noda* of the Thessalian plain in Greece (40°00′ N, 22°22′ E), and *Melophorus bagoti* of the arid-zone low scrubland of central Australia (23°45′ S, 133°53′ E). In all species, the survival frequencies follow an exponential decay function with life expectancies of $\mu^{-1} = 6.3$ days (*C. bicolor*), 4.7 days (*C. noda*), and 4.9 days (*M. bagoti*); half-life times of $t_{0.5} = 4.4$ days (*C. bicolor*), 3.2 days (*C. noda*), and 3.4 days (*M. bagoti*); and relative daily losses of $(1 - e^{-\mu}) = 0.147$ (*C. bicolor*), 0.193 (*C. noda*), and 0.186 (*M. bagoti*). Source: *Data from references.*[5,11–13,15]

spontaneously move toward the black sections of the wall of a cylindrical arena, whereas outdoor workers prefer the white sectors. If certain illumination programs are applied, this behavior can be flexibly adapted and even reversed relative to what occurs in the normal sequence of life stages. Indoor and outdoor workers kept for an increasing number of days under light and dark conditions increasingly prefer white and black wall sections, respectively, with the latter (reversed) process occurring much slower than the former one.[14]

As outdoor life proceeds, the workers gradually increase the duration and spatial extent of their foraging trips as well as their foraging efficiency—that is, the number of food items detected and retrieved per unit distance traveled.[5,12,15] These distinct changes in foraging behavior are certainly correlated with improvements in navigational performances and food detection skills, especially because the parameter that increases most remarkably during a forager's lifetime is sector fidelity—that is, the tendency to remain faithful to a particular spatial foraging sector,[15] within which an increasing amount of navigational information can be gained.[16] In the beginning, when the ants start their foraging careers, the only navigational strategy available to them for guidance is an inherited one: path integration. It later serves as a scaffold for acquiring learned landmark information. During their small-scale explorative 'learning walks,' the ants often stop, read out their path integrator, and orient themselves in the direction of the starting point of their learning walks, the nest entrance (best studied in *Ocymyrmex robustior*[17]; for *Cataglyphis* species, see references.[15,18] It is most likely during these short (~150 msec) stopping phases that the foragers acquire their first information about the visual scenery of the nest environs. From then on, an intricate set of flexibly intertwined landmark-based navigational strategies comes into play. The adaptive plasticity within the organization of the ant's navigational tool kit containing quite a number of interacting routines is the topic of the following sections.

PATH INTEGRATION

Central place foragers such as ants and other hymenopteran social insects employ a path integration (PI) routine when they venture out along devious paths into unfamiliar or featureless terrain. This PI strategy provides them with a continually updated 'home vector,' an online estimate of their current direction and distance from the starting point. Hence, PI can be considered as some kind of vector navigation.[19–23] Once the ants have returned to the origin of their foraging journey, the home vector is reset to zero, but a copy of it is retained in memory. With its compass polarity reversed, this stored PI vector (reference vector) can later be used as a 'food vector' to steer the animal to a previously visited profitable food site. This path integration behavior is an inborn navigational routine that works already during an ant's first foraging excursion. With increasing foraging experience, the ants acquire substantial amounts of information about terrestrial signposts, but whatever form of landmark guidance routine they later employ, their path integrator keeps running uninterruptedly all the time while they are outside the nest. It provides them with an ever available safety line firmly fixed at the origin.

Calibrating the Path Integrator

As the desert ants continually explore their foraging area for new food items, their path integrator must be recalibrated repeatedly. In this calibration process, the condition that the outbound PI vector (food vector) is the 180° reversal of the inbound PI vector (home vector) invariably holds. This can be demonstrated best by subjecting the ants to an open-jaw round-trip paradigm—that is, by displacing them sideways from their food site just before they start their home run.[24] As shown in Figure 39.2A, the ants released at what they still consider to be their food site (hence, the fictive position of the feeder, F*) first run off their home vector and then start a systematic search around the fictive position of the nest (N*). If the original displacement has not been too large, their search routine will enable them to arrive at the real position of the nest (N) quite soon. There, they experience a mismatch between the current home vector pertaining to the post-displacement inbound run (F* → N) and their reference vector (F → N). This mismatch causes them to recalibrate their stored food vector (N → F). When they set out for their next foraging journey, they take an intermediate course between the inverse of the most recently computed home vector, N → F*, and the previous food vector, N → F, but the deviation from the latter is rather small (δ_{out} in Figure 39.2B). If the ants having finally arrived at F are again subjected to the same open-jaw displacement paradigm as before, they deviate from their previous homeward course F* → N* by exactly the same angular amount by which they have deviated from N → F on their outbound run. This $\delta_{in} = \delta_{out}$ relation—that is, the relation PI vector$_{in}$ = PI vector$_{out}$ ± 180°—holds irrespective of how often this training-around-a-circuit procedure is repeated.

The calibration angle δ maintains a constant value even if the displacement paradigm is repeated up to 50 times. This constant value is reached very quickly, often after the first displacement, but at the latest after three or four such experimental treatments, and it then remains constant. In absolute terms, it is rather small (in the displacement paradigm depicted in Figure 39.2,

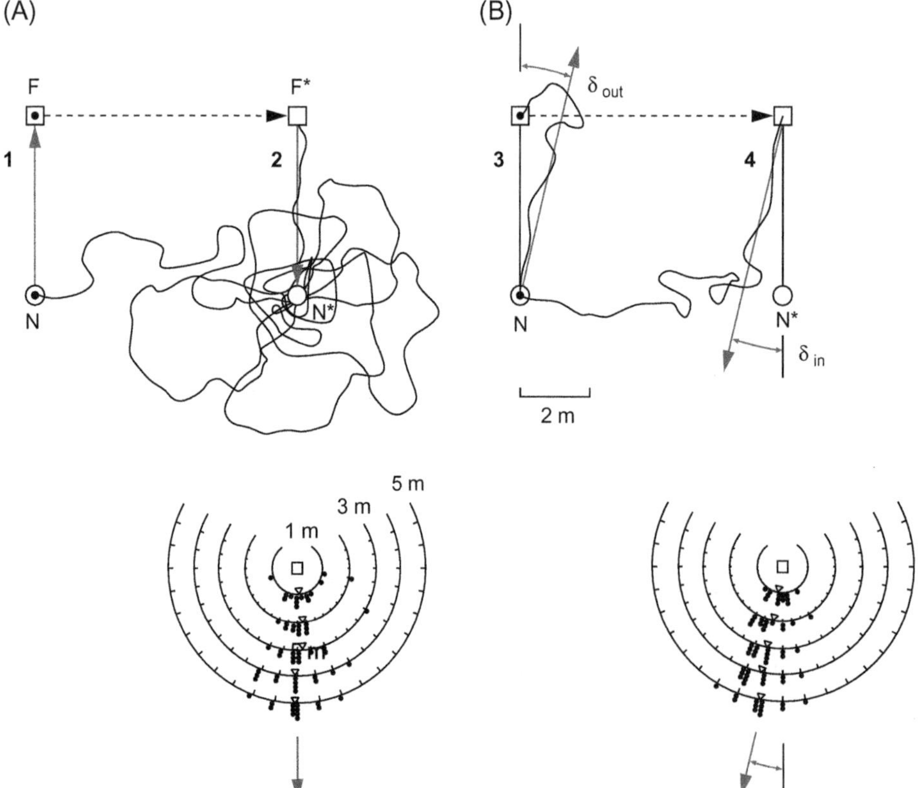

FIGURE 39.2 Recalibration of path-integration vectors in *Cataglyphis fortis*. (A) In an open-jaw round-trip test paradigm ants (**1**) trained from the nest (N) to a feeder (F) were captured there and displaced (dashed arrow) 7.5 m sideways. Hence, the point of release corresponded to the fictive site of the feeder (F*). (**2**) The inbound ants, having run off their home vector, searched around the fictive position of the nest (N*). During this area-concentrated search, they finally happened to arrive at N. (B) (**3**) On their subsequent outbound runs (N → F) and (**4**) post-displacement inbound runs (F* → N), the ants deviated by δ_{out} and δ_{in} from N → F and F* → N*, respectively. The angle δ ($\delta_{in} = \delta_{out}$) represents the experimentally induced recalibration of the ant's path-integration system. The circular diagrams (bottom) depict the angular distributions of the inbound courses of 13 ants, which have been tested under the experimental conditions illustrated in the top of the figure for one of them. Source: *Adapted from Wehner* et al.[24]

$\delta = 9.6°$) and significantly different from the bisector of the angle between N→F and N→F* (28°) and the direction of the mean vector (36°). Several factors might account for the size of the calibration angle δ. For example, the ant's previous familiarity with the food site—that is, the number of pre-displacement round trips N→F→N—could play a role, and so could the time that the ants spent searching during their post-displacement inbound run from F* to N. The influence of such factors might be responsible for the idiosyncratic behavior of particular ants. For example, some ants did not recalibrate their previous PI vector at all ($\delta = 0°$), whereas in others, the calibration angle amounted up to $\delta = 20°$. The factors mentioned previously might add another level of plasticity to the ants' vector navigation system. However, this plasticity is bounded at least insofar as the inbound and outbound vectors are always 180° reversals of each other, as it applies to the natural foraging situation. In the experimental open-jaw paradigm, one never succeeds in causing the ants to deviate from this rule—that is, in training them to take a straight inbound course F*→N and at the same time to remain faithful to a previously learned outbound course N→F (see the attempt to do so in Wehner and Flatt[25]).

Under natural conditions, the ants' reference vector is constantly updated. Because *Cataglyphis* is a single-prey loader collecting individual and widely scattered food items (arthropod corpses), it must recalibrate its reference vector whenever prey is found at a new location. Usually, the ants first return to the site where they have found food most recently, and then they continue to search in the same direction from the nest nest[13,26] (sector fidelity[15,27]). However, they can also be trained to repeatedly visit two food sites located in different directions and thus are able to simultaneously acquire and store two reference vectors pertaining to two food sources, A and B. For example, by retrieving reference vector B after food source A has been found to be depleted, they can travel along a novel, shortcut route from A to B by employing their PI system.[28] Hence, they do not have to first return to the origin, the nest, before setting out for another familiar food site. Seen in this light, path integration alone can provide the ants with some flexible and efficient means of navigating between known places along novel routes.

Estimating Path Integration Errors

All PI systems are prone to cumulative errors. Because the ant's path integrator is reset only after the animal has returned to the starting point,[29,30] all errors occurring within the PI processing networks will accumulate until the ant has arrived at this point. It has been argued that honeybees prevent this excessive accumulation of errors by resetting their path integrator whenever prominent landmarks are encountered.[31] However, the improved accuracy observed in bees that had passed a familiar landmark might well be due to the use of a local vector associated with the landmark and employed in addition to the PI system, which nevertheless keeps running uninterruptedly in the background.

To investigate the effect of foraging distance on error accumulation, *Cataglyphis* ants were trained in landmark-free terrain to feeders located at different distances from the nest. Having been experimentally displaced from the feeder, they performed their homeward runs in an unfamiliar test field. As expected, the end points of their straight homeward runs—the tips of the home vectors as estimated by the ants—scattered: The more widely around the fictive position of the goal (home), the larger the ants' foraging distances had been.[32] This end point distribution defines the 'target probability function' of the ant's path integrator. Furthermore, it is not only this probability function that widens with increasing foraging distance but also the 'search density profile' (i.e., the search distribution exhibited by the ants when looping around the end points of their home runs in search of the goal[33,34]) that broadens as foraging distance increases.[21,32] In short, the farther the ants have ventured out on their foraging journeys, the less accurate they become in locating the starting point right away, and the wider are their subsequent search distributions from the beginning. This means that the ants' search density profile is adapted to the target probability function and thus reflects the ants' uncertainty about the position of the goal.

This is a vivid demonstration of how flexibly the ants adapt their search behavior to the errors to be expected (i.e., to the uncertainty of their path integration system). However, because uncertainty is an inherent property of the path integrator and its constituent parts such as the compass and the odometer, the ants cannot assess it directly. Indirect measures could be the foraging distance, d, or the overall length of the foraging path, l. That the ants use the former rather than the latter follows from experiments in which two groups of ants were compared. In either group, the ants covered the same foraging (nest–feeder) distance, d, but differed in the length of their paths, l, needed to cover this distance.[35] The two groups differed in their end point distributions (target probability functions), with the long path group exhibiting a significantly wider distribution than the short path group, but they coincided in their search density profiles. At first glance, the latter correspondence might be astounding because it is the tortuousness of the outbound path reflected, for example, in the number of steps made by the ants that correlates with the extent of error accumulation and, as mentioned previously, determines the target probability function. Why, then, do the inbound ants adjust their search density distributions on the basis of the outbound distance, even if outbound path length were a more powerful predictor of uncertainty? It may be that foraging distance (i.e., the length of the home vector) is the only parameter finally accessible to the ant's navigational tool kit (but see the two-odometer hypothesis proposed for honeybees by Dacke and Srinivasan[36]). It may also be that given the geometry of the ants' foraging paths,[13,26,27] the length of such a path is correlated sufficiently well with maximum foraging distance so that the latter can be used as a reasonably good estimate of PI uncertainty. Further consider that the search density profiles are much wider than the target probability distributions so that in terms of minimizing search time, not much might be gained by fine-tuning the search distribution to the structural details of a foraging path.

As shown in Figure 39.3A1, the end points of the ants' straight homeward runs are centered on the goal, the fictive position of the nest, only for relatively short nest–feeder distances. For larger distances ($d_{nf} > 20$ m), the ants start searching before they have run off their full home vector so that their search distribution is shifted toward the starting point of the homeward run.[38]

Under natural conditions, ants of all *Cataglyphis* species investigated to date concentrate their outdoor activities to particular, individual-specific sectors of their foraging terrain. With increasing foraging age, their foraging distances get larger and their sector fidelities stronger[15] so that they repeatedly travel in their near-nest area along particular routes. Landmark-based information acquired along such frequently traveled routes reduces search time substantially (compared to landmark-free situations in which the ants have to exclusively rely on path integration[41]). Hence, it is a likely hypothesis that the ants shift the search distribution of their path integration system into that part of their nest environs in which they can rely on additional guides such as habitual visual landmarks and where they thus can compensate for PI uncertainty most efficiently (Figure 39.3A). The same argument applies to the finding, perplexing at first, that after an L-shaped outbound detour, the inbound ants do not head directly toward the starting point but, rather, exhibit systematic inward errors[37] (Figure 39.3A_2). Again, these inward trajectories place the ants' search distribution directly into familiar-route terrain.

Flexible Use of Path Integration Vectors

Previously, we described a situation in which the ants had to recalibrate their stored PI vector. Here, we discuss a case in which the ants maintain this reference vector but in paying it out on the way to a habitual food site 'deliberately' deviate from it. The functional significance of this angular deviation might be the same as that of the linear deviation described in the previous section: shifting the target probability distribution provided by the path integration system into that section of the goal environs in which additional information is available to finally pinpoint the goal (Figure 39.3B).

What is this additional information? Even though *Cataglyphis* ants are visually guided animals to the extreme, food items are detected by smell and hence always approached against the wind. This means that the ants combine olfactory and anemotactic (wind-based) orientation strategies,[39] making use of their well-developed olfactory systems.[42] Because in their desert habitats wind usually blows steadily from a rather constant direction, at least during foraging times, the ants increase their probability of locating a familiar food site, F, by steering a downwind course toward the leeward side of F rather than directly toward F. After having picked up the odor plume that emanates from the food source, they move upwind until F is reached. By adopting this indirect detour-approach strategy, they transform their food vector into a currently traveled two-leg vector course consisting of an initial long downwind path toward the odor plume and a final short path upwind directly toward F. It is only under windless or tailwind conditions that they steer directly toward F. In the latter case, they usually overshoot the food site slightly, and after having traveled a short overshoot distance of 0.1 or 0.2 m they perform a U-turn and start their final approach toward F.

In dissecting the direct PI course into an indirect two-leg detour course and thus performing vector subtraction while moving toward the food site, the ants exhibit a high degree of flexibility in employing their PI system. This flexibility raises the question of what factors and environmental cues ants use in adjusting their two-leg courses most efficiently? First, the ants could consider F and its leeward odor plume as an expanded target, the size and shape of which would be affected by wind speed and direction. However, the ants' downwind approach distance d_a does not depend on either of these parameters. Instead, it varies systematically, in fact linearly, with the nest–feeder distance d_{nf}[40]: The farther the food site, the larger the ants' final downwind approach distance becomes (Figure 39.3B_2). The linear relationship between d_a and d_{nf} means that the ants, when leaving the nest, deviate by a rather constant downwind angle α from their food vector course ($\alpha = 3{-}6°$). According to the goal expansion hypothesis mentioned previously, the angle α should

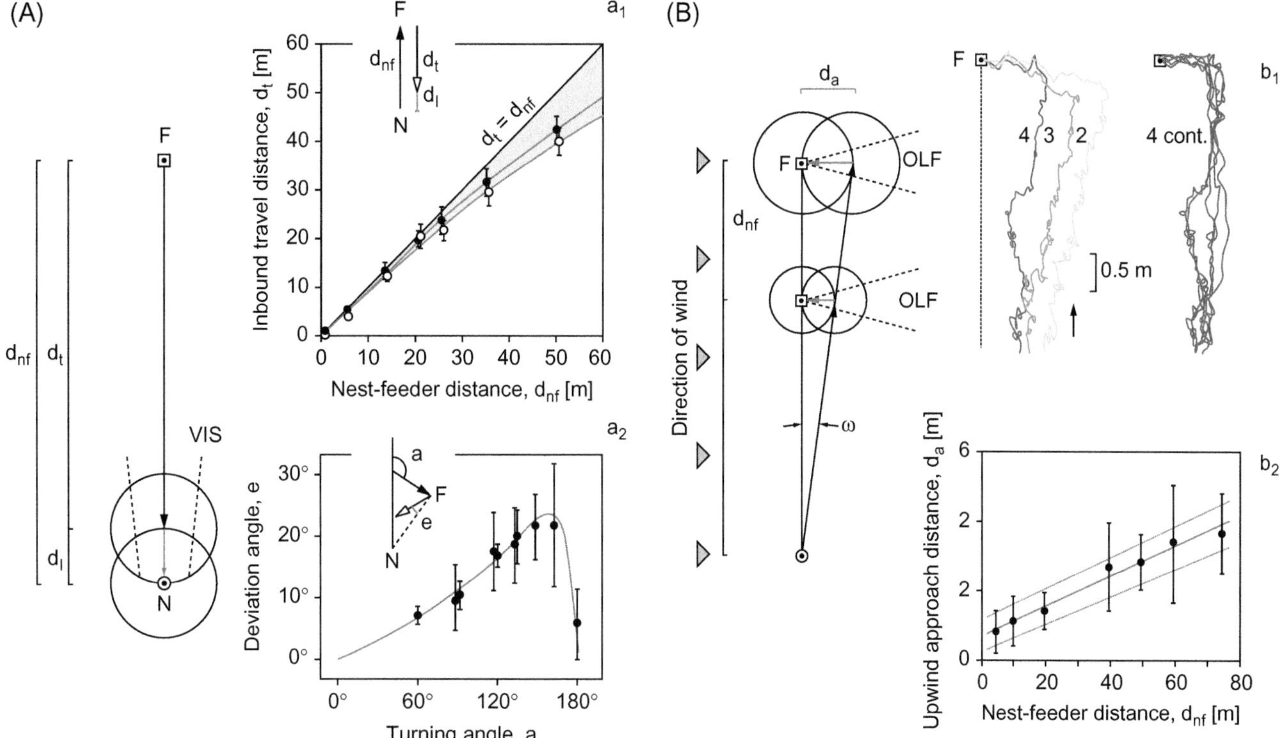

FIGURE 39.3 **Systematic linear and angular deviations from the direct path integration (PI) vector courses.** The deviations cause the target probability distributions defined by the PI vectors (and symbolized by the circular area around the tip of the PI vector) to be shifted away from the goal into areas that are characterized by familiar (A) visual and (B) olfactory cues. (A) Underestimating the linear components of inward courses (after straight outward journeys; A_1) and the angular components of one-sided turns (after L-shaped outward journeys; A_2). In panel A_1, the solid and open diamonds represent the ants' first turns and the centers of search, respectively. (B) Steering downwind courses (heavy black arrows) to a familiar food source. The downwind courses are followed by short courses upwind once the odor plume emanating from the food source is reached. (B_1) Walking trajectories of an ant approaching a familiar food site for the second, third, fourth, and the following times (4 cont.). (B_2) Final approach distances upwind toward the food source (runs 4 cont.; standard deviations and 95% confidence intervals). F, feeder; N, nest; d_a, upwind approach distance; d_l, distance by which the ants underestimate the full PI vector length: $d_l = d_{nf} - d_t$; d_{nf}, nest–feeder distance; d_t, inbound travel distance; α, turning angle in L-shaped detour experiment; ε, inward error angle; ω, angle by which the ants deviate downwind from the direct PI vector course. Source: *Combined and adapted from references.*[37–40]

decrease with increasing nest–feeder distance, but it does not.

The detour angle α by which an ant steers downwind from the reference vector course can be considered as a measure of the target probability function—that is, the ant's estimate of the uncertainty inherent in its PI system. By steering downwind courses, the ants compensate for this uncertainty. If they centered their searches directly on the goal (the feeder in the present case and the nest in the former one), there would be a 50% chance that they would end up in an area that lacked familiar (olfactory or visual) cues so that walking distance and time would substantially increase (Figure 39.3B). Furthermore, depending on experiences made during previous visits to the feeder, the ants flexibly adjust their initial downwind courses α and final approach distances d_a. In successive visits, they shorten and straighten their approach paths considerably, until from the third to the fifth visit onward, α and d_a remain rather constant (Figure 39.3B_1).[39,40]

INTERPLAY BETWEEN PATH INTEGRATION AND LANDMARK GUIDANCE ROUTINES

Location Memories, Site-Based Steering Commands, and Route Memories

In navigating toward familiar places and along habitual routes, desert ants use visual landmarks in a variety of ways. In the past few decades, an impressive amount of literature has accumulated that deals with various landmark-based navigational routines, and various models have been designed to account for the observed behavior[43,44] (reviewed in[28]). What seems common to most of these experiment cum modeling approaches is the notion that the location memories rely on view-based information. For example, in using landmarks for pinpointing a goal location, the ants acquire panoramic two-dimensional retinotopic images, so-called 'snapshots,' from the visual environment seen from the goal. 'View-based' means that it is the change in the retinal images between stored and current views rather than the absolute metric information about the distances and size of the surrounding landmarks that is used in goal navigation. Desert ants acquire this information in well-structured 'learning walks' consisting of a sequence of loops around the goal, rotations on the spot ('pirouettes'), and stopping phases.[17,18] How they combine the visual information gained during as many as 30 pirouettes performed during a single learning walk—how they process, store, and later use this information when subsequently returning to the goal—is one of the most intriguing questions open to future research.

View-based landmark memories are used not only to guide the ants to particular sites (i.e., cause them to stay there once image matching has occurred) but also to guide the animals in a particular direction away from a site. In the latter case, the familiar view is used as a signpost associated with a site-based steering command, a local vector (with its direction derived, e.g., from a skylight compass). Such site-based vectors could be used in route navigation by sequentially linking familiar views, like pearls on a string, and thus leading the animal from one waypoint to the next, from one familiar view to the catchment area of the subsequent view. However, rather than dissecting a route into a sequence of path segments, recent research favors a more holistic view of route navigation[44–46] and even provides a model that completely abandons the storage of views.[47]

Indeed, as many field experiments—especially in *Melophorus bagoti*—show, route memories are quite impressive.[16,48–50] The ants can remember several idiosyncratic routes, faithfully travel along them even with their path integrator at zero state, and retain these various route memories for their entire several-day foraging lifetimes. In *Cataglyphis bicolor* and *C. fortis*, location memories defining the nest site have been found to persist even for periods of weeks when the animals are experimentally kept in isolation for extended periods of time, far beyond their natural life spans.[51,52] On the other hand, PI vectors decay much more rapidly. The theoretical approaches currently promoted for route navigation by Paul Graham and colleagues[45,47,53] offer fascinating new ways of understanding how desert ants are able to navigate through cluttered environments over large spatial scales.

Functional Adaptability within a Modular System of Navigation

Under natural conditions, the various modes of navigation considered in the previous section operate in concert and can be studied individually only by careful experimental dissection of the ant's navigational tool kit. Some paradigms applied in this context are sketched out in Figure 39.4. In all test situations described in this figure, the layout of landmarks or detour devices used in the training situations (left column) was altered in a way that the global path integration vector (PI-V) and the local site-based vectors (SB-Vs) pointed in different directions. The experimental animals tested under these conditions in novel territory were either taken from the feeder (i.e., before they had started their home runs; middle column) or after they had finished their home runs and arrived at the nest (right column). Hence, their PI processor was still filled (full PI state) or already emptied (zero PI state), respectively. In the test situation depicted in Figure 39.4A, an alley of cylindrical landmarks was rotated by 45° relative to the training situation. The full vector ants either followed their PI-V course directly from the start or deviated from it by entering the alley and, when leaving the alley at various positions, again heading for the fictive position of the nest. This means that the path integrator kept running in the background while the ants had followed the landmark-defined rather than their PI-defined course, and that it had come to the fore again after the ants had reached the end of the alley or had broken out of it earlier. As expected, all the zero vector ants entered the alley and upon leaving it played out the SB-Vs they had previously associated with the landmarks.

In general, in the kind of cue competition experiment described previously, how consistently the full vector ants were guided by SB-Vs rather than their global PI-V depended on the conspicuousness of the landmark configuration. The ambiguity in relying on one or another of these vector programs is also borne out by the experimental paradigm depicted in Figure 39.4B, in which a linear channel was used to force the ants on an L-shaped detour route (left side of figure). When the inbound ants exited the east-pointing channel, they had to move southwards over open ground to finally reach the nest. In this case, they associated the distal end of the channel with an SB-V oriented at right angles to the channel and pointing to the south. In the test situation, in which the channel was shortened, the majority of the full vector ants left the channel by first following their SB-V course—that is, going south. However, after they had covered some distance in this direction, they turned and followed their updated PI-V course—that is, aimed at the fictive position of the nest (middle of figure). The zero vector ants, of course, could only play out their SB-V program (right side of figure). The fact that their southward courses prevailed even after the channel had been rotated by 45° is clear evidence that their SB-V courses

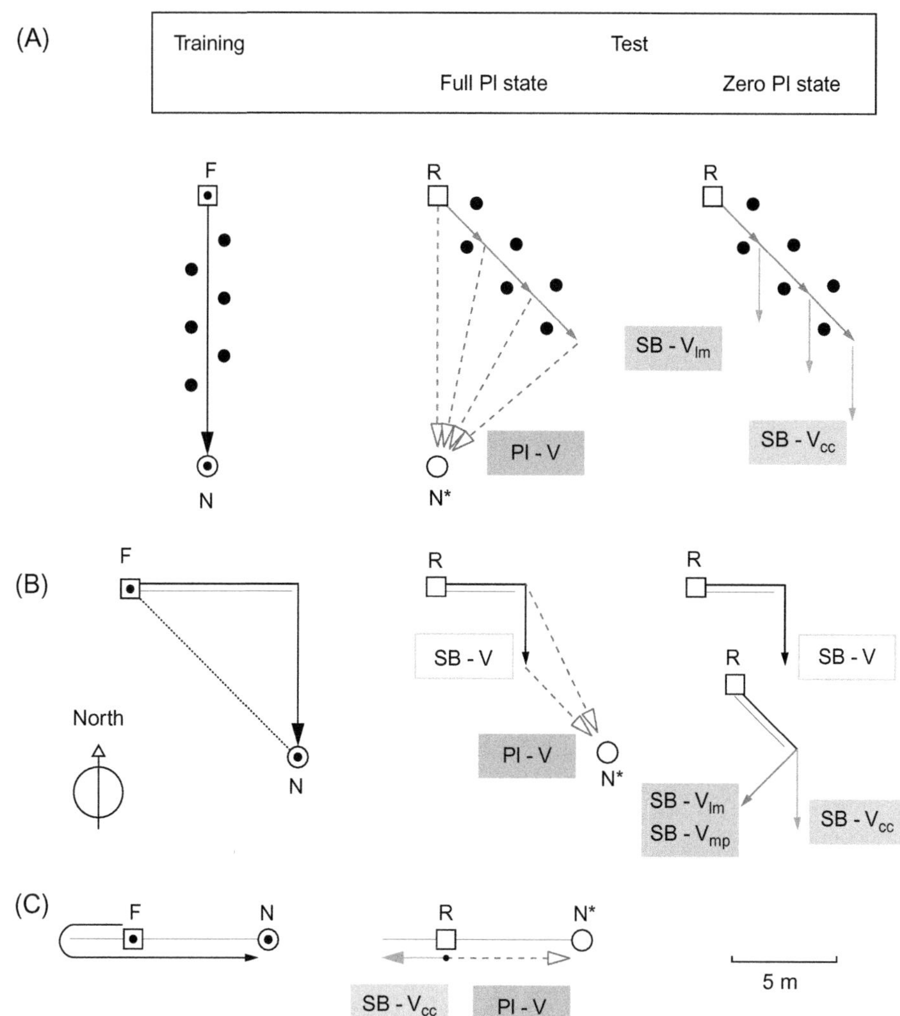

FIGURE 39.4 The flexible use of local site-based vectors (SB-Vs; solid lines and solid arrowheads) and the global path integration vector (PI-V; dotted lines and open arrowheads). The figure sketches out the design principles of experiments performed by various authors in *Cataglyphis fortis*: left, training setup; middle and right, experimental situations, in which either full-vector ants (ants taken from the feeder: full PI state) or zero-vector ants (ants taken after arrival at the nest: zero PI state) have been tested. (A) SB-Vs associated with visual landmarks. Training along a landmark route (alley of cylindrical landmarks; solid circles), which in the test is rotated by 45°. (B) SB-Vs linked to the end of a forced detour (the end of an open-topped channel sunk into the ground, or a landmark barrier; for the latter; see also Figure 39.5). (C) SB-V associated with the feeder and pointing in the opposite direction of the PI-V. Ants are trained along a U-turn route and tested within a straight channel. Whereas in all cases (A–C) the zero-vector ants (right) always displayed their SB-Vs, the full-vector ants (middle) followed, to various degrees, either the PI-V or a SB-V or first a SB-V and subsequently the PI-V. F, feeder; N, nest; N*, fictive position of nest; R, point of release; SB-V, site-based vector based on celestial compass information (cc), landmark views (lm), or motor programs (mp). Source: *Data from references*[54,55] *(A)*,[56–58] *(B) and (C)*.[59]

were based on global (celestial) compass information (top of figure; SB-V$_{cc}$, the canonical 'local vector'). Moreover, the rotation of a detour obstacle—in the following two cases a barrier—further showed that in setting SB-V courses, terrestrial visual scenes as provided, for example, by extended landmarks can override celestial compass information (dashed arrow in bottom of figure; SB-V$_{lm}$). Even if the extended landmark was made to vanish from the ant's field of view once the ant had reached the end of it, the zero vector ants could still behave in a compass-independent way by keeping a constant angle relative to their previous walking trajectories (Figure 39.5). They might have learned a motor program: Turn around a particular angle once the end of an obstacle has been reached (dashed arrow; SB-V$_{mp}$). Keeping a constant angle relative to a previous path segment is also apparent from the channel-detour experiments described previously (Figure 39.4B, lower right). When the rotated test channel had the same length as the training channel, half of the ants behaved according to the SB-V$_{cc}$ hypothesis, whereas the other half exhibited SB-V$_{mp}$ responses. Only when the rotated channel had been shortened substantially did the SB-V$_{cc}$ response dominate the ants' behavior. Obviously, in the latter case, the path segment preceding the turn had been too short to trigger the SB-V$_{mp}$ response. This experimental result demonstrates that at any one time, which navigational routine will contribute to what extent to the ant's overall behavior will largely depend on context. It again reflects the large amount of plasticity inherent in the insect's navigational tool kit.

Finally, it was in an extreme case of a two-leg detour route, in which homebound ants were forced to perform a U-turn before heading for the nest (Figure 39.4C), that the significance of site-based vector memories could be demonstrated most strikingly. With the SB-V (linked to the feeder) and the PI-V pointing in opposite directions, the vast majority of full vector ants tested in a straight channel decided in favor of the former—that is, headed in the counterdirection of the nest. Thereafter, and in contrast to what

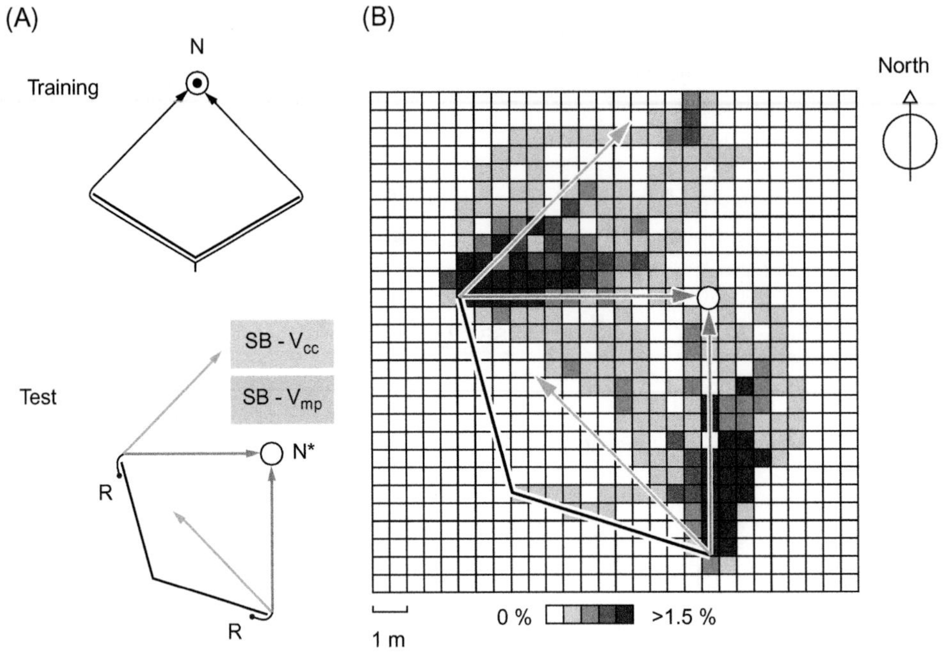

FIGURE 39.5 Local, site-based vectors associated with the end of an obstacle, which caused the ants to walk along a devious route. (A) Experimental paradigm: During training, the ants had to detour around either the left or the right end of a V-shaped barrier. In the test, zero-vector ants were presented with the barrier rotated by 45°. The fictive position of the nest is marked by an open circle. Hypotheses: Site-based vector based on either celestial compass information (SB-V_{cc}; solid arrows) or matching the rotation of the barrier (dashed arrows). (B) Result: As the path density plot shows, the ants behaved according to the latter hypothesis. They did not keep a constant compass course (SB-V_{cc} according to Figure 39.4B, right), nor did they determine their course on the basis of a learned heading direction relative to a landmark (SB-V_{lm}) because the barrier became invisible once the ants had detoured around it. Instead, they played out a learned motor program (SB-V_{mp}). Source: *Adapted from Bisch-Knaden and Wehner*[56]

had happened in the other two test paradigms (Figures 39.5A and 39.5B, middle), they did not reactivate their PI-V program.

The interplay between path integration vectors, site-based vectors, and other navigational routines not mentioned here provides the ant's system of navigation with the flexibility necessary to cope with the animal's foraging demands, for example, in exploiting new resource areas and dealing with unforeseen changes in its landmark environment. At any one time, various navigational routines will be active, operating in unison and supplementing each other. For example, while detouring around an obstacle, the ant can deduce the direction of the next segment of its path allothetically from either celestial compass cues[54,57] or the retinal position of landmarks.[58] In addition, the ant can determine it idiothetically in terms of a particular turn to be made relative to the previous segment of the path[59] (for honeybees, see[60]). As discussed previously (Figure 39.4), only by proper experimental interference can the coherent cooperation of these various routines be interrupted and the network of navigational strategies disentangled. Under natural conditions, which routine will become more effective in governing the ant's behavior might largely depend on context, the salience of cues, and the state of the path integrator. Depending on such contextual cues site-based vectors can dominate the output of the path integrator, and in determining the direction of a local, site-based vector, terrestrial cues can supersede celestial ones.

Furthermore, temporal aspects of acquiring and maintaining navigational information via one or another routine might influence the functional relevance of a given routine at any one time. Snapshot memories and local vector memories are formed while the animal navigates by path integration. In this case, one routine acquires its navigational information while the ant's behavior is currently governed by another one, but later it can operate without the commands provided by the latter. Moreover, the stability of a given vector memory can depend on the ongoing operation of others.[61] It can even be difficult, and may afford long disruptive training procedures, to experimentally separate two naturally combined navigational strategies.[62] There may be several explanations for the observation that in nearly all of the experimental paradigms outlined in Figure 39.4, some animals relied more on path integration vectors than on site-based vectors, and some more on one kind of site-based vector than on another kind, whereas other animals exhibited just the opposite behavior. It could have resulted from differences in experiences made in the immediate past (e.g., in different numbers of training runs preceding the test procedures), or it could have been the consequence of differences in the foraging ages of the animals employed in the experiments. The latter aspect, which has rarely been considered until now, would add another level of bounded plasticity to the operation of the ant's navigational tool kit—and would bring us back to the ant's life history pattern.

PLASTICITY OF THE ADULT ANT'S BRAIN

The various navigational routines flexibly used by *Cataglyphis* desert ants reflect the impressive sophistication of the underlying neuronal machinery in the ants' navigational systems. The drastic changes in sensory input and motor performances at the ants' major transition from the dark nest interior to outside foraging predict that the underlying neuronal substrates must have high levels of plasticity. Hence, the next major challenge is to (1) unravel the neuronal mechanisms underlying these behaviors and their interactions and (2) to understand their plasticity in the course of life history transitions and daily adjustments. The navigational tool kit is housed in a mini-brain built from less than 1 million neurons. In recent years, combined behavioral manipulations (both in the field and in the laboratory) and analyses of synaptic microcircuits in the mushroom bodies have resulted in important progress in beginning to understand the plasticity and neuronal changes of neuronal circuits involved in sensory integration and adult neuronal plasticity of *Cataglyphis* ants.[18,42,63–65]

Plasticity in Mushroom Bodies—Sensory Association Centers of the Insect Brain

The insect mushroom bodies (MBs) were shown to be centers in the insect brain associated with sensory integration, learning, memory, and orientation.[66–72] In Hymenoptera, particularly in ants,[73,74] bees,[75,76] and social wasps,[77] they are very prominent structures in relation to other brain regions. A volumetric study of MBs in various Hymenoptera promoted the hypothesis that the elaboration of MBs in parasitoid wasps may represent a preadaptation to social lifestyle.[78] This certainly requires further investigations and comparison among closely related species. In the neuroanatomically well-investigated honeybee, the MB intrinsic neurons—the Kenyon cells (KCs)—comprise approximately 45% of the total number of brain neurons,[79,80] highlighting their importance in sensory integration in order to cope with the challenges of orientation associated with central place foraging and collective brood care in social insects.[76] Based on anatomical observation, these numbers may be even higher in *Cataglyphis* (W. Rössler, personal observation; Figure 39.6). For comparison, in the fruit fly, only approximately 4% of all brain neurons are KCs.[82,83] In bees and ants, the MB calyces comprise three sensory input regions: the olfactorily innervated lip, the visually innervated collar, and the basal ring receiving input from both sensory modalities (Figure 39.6).[18,65,84–87]

The role of the MBs has been under debate for many years.[88] It is interesting to note in this context that a recent study on developmental gene expression in an annelid worm suggests that the invertebrate MBs and vertebrate pallium (including the cerebral cortex) may have a common origin in deep phylogeny.[89] Striking structural similarities in the neuronal circuitry of the hippocampal dentate gyrus and the MB microcircuits further add to this aspect.[88] In the same line, based on structural and functional considerations,[90] Frambach *et al.*[91] and Farris *et al.*[92] promoted the idea that microcircuits within the vertebrate cerebellum and the MBs may share functional similarities. Under functional aspects, the two views may not be contradictory because both the hippocampus and the cerebellum are substrate for associative multimodal learning in the context of spatial orientation and during the control of motor programs.

Previous studies have shown that both age and sensory experience lead to a remarkable volumetric expansion of the adult MBs. Volumetric changes associated with the transition from interior workers to foragers were reported for bees,[92–94] wasps,[95] and ants.[63,73] However, the neuronal mechanisms and causes remained largely unknown for a long time. A Golgi study[92] showed that the branches of KC dendrites in the visual collar expand in foragers compared to nurse bees. Selective immunolabeling of individual pre- and postsynaptic compartments of microglomeruli (MG), large synaptic complexes in the MB calyx, allowed the investigation of the cellular and subcellular bases underlying this structural synaptic plasticity in the MB calyx (Figure 39.6).[18,65,75,81,91,96–98]

Anatomical details of MG were pioneered by electron microscopic (EM) studies in ants[99] and followed by immuno-EM studies in the honeybee[100] showing synaptic microcircuits formed between olfactory projection neuron boutons, KC dendrites, and recurrent GABAergic extrinsic neurons (Figures 39.6C and 39.6D). Further EM studies by Yasuyama *et al.*[90] revealed a similar organization in the MB calyx of *Drosophila*, indicating that MG may represent evolutionary conserved functional units. Finally, several studies provided increasing evidence that the behavioral transition from inside the nest to

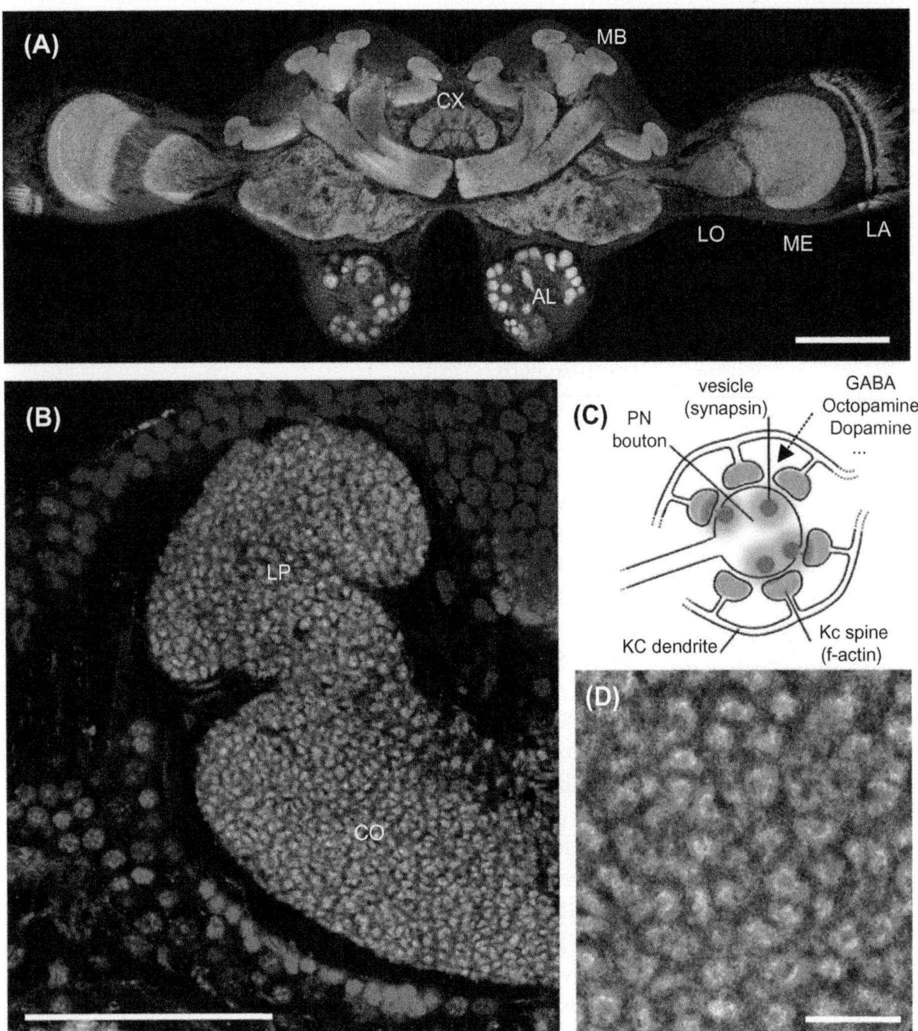

FIGURE 39.6 Immunolabeling and quantitative analyses of synaptic structures in the brain of *Cataglyphis fortis*. (A) Frontal overview of the brain with the large mushroom bodies (MB), central complex (CX), antennal lobe (AL), and optic lobes with lamina (LA), medulla (ME), and lobula (LO). Triple staining with anti-synapsin, F-actin phalloidin, and cell nuclei (Hoechst). (B) Detail showing the inner branch of the medial MB calyx on the right side in panel A. Subdivision of synaptic complexes (microglomeruli) in the lip (LP), which receives olfactory input, and the collar (CO), which receives visual input. (C) Scheme depicting the organization of one microglomerulus. (D) High magnification of synaptic complexes in the collar region double labeled with anti-synapsin antibody (magenta) and F-actin phallodin (green). Scale bars = 200 μm in panel A, 50 μm in panel B, and 5 μm in panel D. Source: *Modified from Groh* et al.[81] *and Stieb* et al.[18]

outdoor foraging is associated with presynaptic pruning and massive dendritic growth in KCs, which represents the predominant cause for the robust volume increase in the MB calyces.[18,65,75,78,97,101]

Light-Triggered Rewiring of Mushroom Body Microcircuits

Age cohort experiments in the honeybee[98] indicated that both sensory experience and age are involved in structural plasticity in olfactory subregions of the MB calyx. However, the precise trigger for these changes remained unclear. Investigations of age cohorts and experimentally treated ants in *C. fortis*[18,65] clearly demonstrated that rewiring of synaptic complexes within the visual subregions of the MB calyx is triggered by the exposure to light during the first exploration (learning) walks of the ants. The first exposure may start even earlier during the late interior II stage, when the ants can often be observed carrying soil particles out of the nest and dumping refuse. This phase is followed by approximately 2 days of exploration walks.[15,18] The effects were partly reversible when foragers were returned into darkness,[18] which may explain differences that had been observed in the behavior.[14]

The experiments by Stieb *et al.* clearly separated the effects of age and sensory experience. In contrast to the light-triggered pruning and dendritic expansion observed in age cohorts in natural colonies and under artificial light stimulation, ants reared in complete darkness did not show a similar effect at the age of natural behavioral transition, ruling out an internal developmental program controlling this plasticity. Instead, dark-reared ants showed a very slow progressive age-dependent increase in MB calyx volume and in the resulting number of synaptic complexes (Figure 39.7).[63,65] Most interestingly, a standard light-pulse program was able to trigger synaptic reorganization in the MB microcircuits not only in 1-day-old

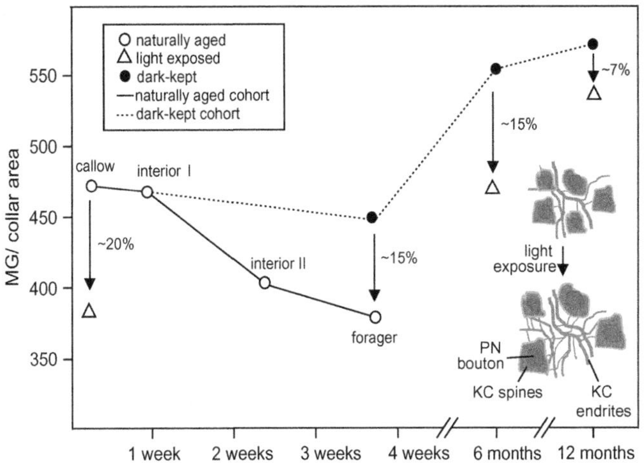

FIGURE 39.7 Summary of age-related and light-triggered plasticity in synaptic complexes (microglomeruli (MG)) in the visual subregion (collar) of the mushroom body calyces. For each group, averaged MG numbers are extrapolated to the area of the collar and plotted versus the ant's age. Naturally aged ants show a decrease in MG density (decrease in MG is highlighted by the arrows and numbers are given in %) initiated by the first light input at the transition from interior workers to foragers (interior II). In dark-kept ants, the MG density remains unchanged for the first weeks of life, but MG numbers increase very slowly at older age (6 and 12 months). In all tested age groups (1 day, 6 months, and 12 months), light exposure results in a decrease in MG density comparable to what occurs during the natural transition in behavior. The schematic drawing on the right side illustrates the expansion of MG caused by massive dendritic growth and, at the same time, pruning of presynaptic boutons. Source: *Modified from Stieb* et al.[65]

callows but also in ants reared in total darkness for 6–12 months.[65] This precocious or delayed synaptic reorganization triggered by exposure to light pulses shows that the experience-dependent changes are independent of an internal developmental program and age (Figure 39.7). Interestingly, synaptic rewiring in the MBs was accompanied by significant changes in the ants' behavior.[18] After exposure to light pulses for at least 2 days, the ants showed a significant increase in general locomotor activities in comparison to the dark-kept controls.

Sensory Experience, Age, and the Formation of Long-Term Memory

What could be the functional consequences of rewiring MB calyx microcircuits following first visual exposure? Dendritic labeling of cytoskeletal proteins (F-actin and tubulin) together with ultrastructural investigations show that the massive expansion of KC dendrites during adult maturation occurs simultaneously with changes in the ratio between multiple active zones in the presynaptic projection neuron boutons and their divergence on postsynaptic partners.[64,65,97] In the honeybee, serial quantitative EM studies have revealed that individual active zones within the large synaptic boutons contact a higher number of postsynaptic profiles (KC dendritic spines) in a forager bee compared to a young bee.[97] For projection neuron boutons, this means a greater divergence of information, whereas for the postsynaptic KCs, this leads to a stronger convergence and, consequently, higher associative power. Furthermore, an increase in the number of ribbon versus nonribbon synapses was observed between the two stages, indicating that active zones, at the same time, increase their synaptic efficiency (ribbon synapses are known to be more efficient in synaptic transmission).[97] These drastic changes in synaptic wiring may represent the neuronal calibration from a 'default state' to a 'functional state' to prepare the system for the massive change in sensory input occurring at this major behavioral transition (Figure 39.7, inset).

Are structural synaptic changes in the MB calyx associated with the formation of long-term memory? In mammals, the consolidation of stable memories is accompanied by structural plasticity involving variations in synapse number and/or size, such as in the hippocampus.[102] In fact, a recent study was able to demonstrate that protein synthesis-dependent olfactory long-term memory in adult, age-controlled honeybees leads to structural changes in synaptic complexes in the olfactory lip of the MB calyx.[69] MG numbers increased after differential associative conditioning, suggesting that new MG are embedded into the dendritic matrix that had preformed after sensory exposure alone (Figure 39.7). Both long-term memory after 3 days and the increase in MG were abolished by a protein synthesis inhibitor. Taking effects of long-term memory formation into consideration, it appears reasonable to speculate that one major function of exploration (learning) walks in *Cataglyphis* could be to trigger the transformation from a default to a functional state in order to prepare neuronal microcircuits necessary for long-term memory storage.

What is the function of a slow age-related increase in MG numbers? The study by Stieb *et al.*[65] revealed a slow addition of new MG running in the background and leading to a slow volume increase (Figure 39.7). Whereas the MG increase associated with long-term memory resulted in an increase in the MG density at a constant volume of the MB calyx lip,[69] the addition of MG with age is associated with an increase in the overall volume of the MB calyx.[63,65] It is interesting to speculate whether this age-dependent change results in larger storage capacities and/or a higher degree of plasticity of the MB calyx. The presynaptic boutons of projection neurons within MG contain large mitochondria and have approximately 50–70 active zones connected to numerous (>100) postsynaptic profiles (mostly KC dendritic spines).[64,97] Thus, the MB calyx neuropil represents very 'expensive' neuronal substrate in terms of energy consumption and maintenance. It is therefore reasonable that it reflects the biological role of the age-dependent addition of new MG, which appears very different from aging processes normally associated with degeneration. During the summer season, *Cataglyphis* workers remain in the nest for approximately 4 weeks and then go out and forage for 6 or 7 days before they die (Figure 39.1). Therefore, during the normal foraging period, the age-dependent increase of MG might not play a significant role (Figure 39.7). The situation may be different during overwintering, particularly in those ants that had already gained some foraging experience. In any case, the maintenance of plasticity in the expanded system is in accord with the fact that light exposure after 6 or even 12 months was still able to trigger synaptic reorganization (Figure 39.7).[65]

Outlook: Neuronal Basis of Behavioral Plasticity

The synaptic changes observed in the MB calyx with age, sensory exposure, or in association with long-term memory show that the processes underlying these modes of structural synaptic plasticity clearly differ. Most important, they can be experimentally separated and narrowed down by using defined experiments and/or stimulation programs. This is important for future experiments because we still do not understand how these different processes interfere with each other. With the brief and clear transition between the indoor and foraging periods, *Cataglyphis* ants are excellent models to investigate these fundamental features of neuronal and behavioral plasticity.

The major challenge is to determine how these changes in the neuronal circuitry relate to navigation-relevant information and behavioral performances during orientation. What are the functional consequences of the reformation of neuronal microcircuits in the MBs? How are synaptic changes in neuronal connectivity functionally organized, and how do they relate to navigationally important information? An earlier study[71] demonstrated that cockroaches had deficits in spatial orientation after surgical ablation of the MBs, indicating their potential role in navigation. Considering the enormous size and multimodal input of the MBs in *Cataglyphis* (Figure 39.6A), it appears reasonable to hypothesize that the MBs in *Cataglyphis* may play a role in landmark orientation. One hypothesis is that MBs may function like a filter network that compares stored information with current information (e.g., similarity vs. novelty)[103] comparable to the comparisons of stored and current landmark views.[28] If this is true, it will be critical to determine how the changes in the synaptic circuitry relate to orientation-relevant information that must be integrated and transferred to a limited number of output neurons.

As behavioral analyses have shown, landmark guidance and path integration routines can operate quite independently of each other. This may indicate that both types of information are, to a large extent, processed in different neuronal substrates, most likely the MBs and the central complex, respectively. Because no direct connection between the polarization pathway in the central complex and the MBs has been found,[88,104] it is pertinent to ask how and where in the brain interaction between landmark information and path integration takes place. Another question is whether parts of the polarization pathway show a similar degree of plasticity at the transition to foraging or, on a daily basis, to accommodate the solar ephemeris function. Furthermore, how is the hierarchical organization of the two types of information implemented in the nervous system?

Finally, increasing knowledge about the molecular mechanisms that link sensory experience with structural synaptic plasticity and the changes in behavior will enable us to manipulate the neuronal substrate. Recent studies on proteins and molecules (CaMKII and rhoGTPase) involved in structural synaptic plasticity[105,106] as well as studies on gene expression[107,108] have opened up new avenues for functional analyses. In any case, the large repertoires of behavioral as well as functional anatomical and even molecular techniques currently available provide fascinating possibilities in integrative research on insect navigation.

Acknowledgments

We thank Stefan Sommer for computing the exponential decay functions in Figure 39.1. We received financial support from the Deutsche Forschungsgemeinschaft (grant 3675/1-1 to R. W. and grants SFB 554-A8 and SPP 1392 to W. R.).

References

1. Collett TS, Graham P, Harris RA, Hempel-de-Ibara N. Navigational memories in ants and bees: memory retrieval when selecting and following routes. *Adv Stud Behav*. 2006;36:123–172.
2. Wehner R. The desert ant's navigational tool kit: procedural rather than positional knowledge. *J Navigation*. 2008;55:101–114.
3. Dobzhansky TG. *Genetics and the Origin of Species*. New York: Columbia University Press; 1937.
4. Wehner R, Fukushi T, Isler K. On being small: ant brain allometry. *Brain Behav Evol*. 2007;69:220–228.
5. Schmid-Hempel P, Schmid-Hempel R. Life duration and turnover of foragers in the ant *Cataglyphis bicolor* (Hymenoptera, Formicidae). *Insectes Soc*. 1984;31:345–360.

6. Wehner R, Wehner S. Parallel evolution of thermophilia: daily and seasonal foraging patterns of heat-adapted desert ants, *Cataglyphis* and *Ocymyrmex* species. *Physiol Entomol.* 2011;36:271–281.
7. Cheng K, Narendra A, Sommer S, Wehner R. Traveling in clutter: navigation in the central Australian desert ant *Melophorus bagoti*. *Behav Proc.* 2009;80:261–268.
8. Wehner R. Desert ant navigation: how miniature brains solve complex tasks. Karl von Frisch Lecture. *J Comp Physiol A.* 2003;189:579–588.
9. Hölldobler B, Wilson EO. *The Ants.* Cambridge, MA: Harvard University Press; 1990.
10. Oster GF, Wilson EO. Caste and ecology in the social insects. *Monogr Popul Biol.* 1978;12:1–352.
11. Harkness RD. The carrying of ants (*Cataglyphis bicolor*) by others of the same nest. *J Zool.* 1977;183:419–430.
12. Muser B, Sommer S, Wolf H, Wehner R. Foraging ecology of the thermophilic Australian desert ant, *Melophorus bagoti*. *Aust J Zool.* 2005;53:301–311.
13. Wehner R, Harkness RD, Schmid-Hempel P. *Foraging Strategies in Individually Searching Ants, Cataglyphis bicolor (Hymenoptera: Formicidae).* Stuttgart: G. Fischer; 1983.
14. Wehner R, Herrling PL, Brunnert A, Klein R. Periphere Adaptation und zentralnervöse Umstimmung im optischen System von Cataglyphis bicolor (Hymenoptera: Formicidae). *Rev Suisse Zool.* 1972;77:239–255.
15. Wehner R, Meier C, Zollikofer C. The ontogeny of foraging behaviour in desert ants, *Cataglyphis bicolor*. *Ecol Entomol.* 2004;29:240–250.
16. Sommer S, von Beeren C, Wehner R. Multiroute memories in desert ants. *Proc Natl Acad Sci USA.* 2008;105:317–322.
17. Müller M, Wehner R. Path integration provides a scaffold for landmark learning in desert ants. *Curr Biol.* 2010;20:1368–1371.
18. Stieb SM, Hellwig A, Wehner R, Rössler W. Visual experience affects both behavioral and neuronal aspects in the individual life history of the desert ant *Cataglyphis fortis*. *Dev Neurobiol.* 2012;72:729–742.
19. Collett M, Collett TS. How do insects use path integration for their navigation? *Biol Cybern.* 2000;83:245–259.
20. Wehner R. Himmelsnavigation bei Insekten. Neurophysiologie und Verhalten. *Neujahrsbl Naturf Ges Zürich.* 1982;184:1–132.
21. Wehner R. Arthropods. In: Papi F, ed. *Animal Homing.* London: Chapman & Hall; 1992.
22. Wehner R, Boyer M, Loertscher F, Sommer S, Menzi U. Desert ant navigation: one-way routes rather than maps. *Curr Biol.* 2006;16:75–79.
23. Wehner R, Srinivasan MV. Path integration in insects. In: Jefferey KJ, ed. *The Neurobiology of Spatial Behaviour.* Oxford: Oxford University Press; 2003:9–30.
24. Wehner R, Gallizzi K, Frei C, Vesely M. Calibration processes in desert ant navigation: vector courses and systematic search. *J Comp Physiol A.* 2002;188:683–693.
25. Wehner R, Flatt I. The visual orientation of desert ants, *Cataglyphis bicolor*, by means of terrestrial cues. In: Wehner R, ed. *Information Processing in the Visual Systems of Arthropods.* New York: Springer; 1972:295–302.
26. Schmid-Hempel P. Foraging characteristics of the desert ant *Cataglyphis*. *Experientia Suppl.* 1987;54:43–61.
27. Wehner R. Spatial organization of foraging behaviour in individually searching desert ants, *Cataglyphis* (Sahara Desert) and *Ocymyrmex* (Namib Desert). *Experientia Suppl.* 1987;54:15–42.
28. Wehner R, Cheng K, Cruse H. Visual navigation strategies in insects: lessons from deserts. In: Werner J, Chalupa L, eds. *The New Visual Neurosciences.* Cambridge, MA: MIT Press.
29. Knaden M, Wehner R. Nest mark orientation in desert ants *Cataglyphis*: what does it do to the path integrator? *Anim Behav.* 2005;70:1349–1354.
30. Knaden M, Wehner R. Ant navigation: resetting the path integrator. *J Exp Biol.* 2006;209:26–31.
31. Srinivasan MV, Zhang SW, Bidwell NJ. Visually mediated odometry in honeybees. *J Exp Biol.* 1997;200:2513–2522.
32. Merkle T, Knaden M, Wehner R. Uncertainty about nest position influences systematic search strategies in desert ants. *J Exp Biol.* 2006;209:3545–3549.
33. Müller M, Wehner R. The hidden spiral: systematic search and path integration in desert ants, *Cataglyphis fortis*. *J Comp Physiol A.* 1994;175:525–530.
34. Wehner R, Srinivasan MV. Searching behaviour of desert ants, *Cataglyphis bicolor* (Formicidae, Hymenoptera). *J Comp Physiol.* 1981;142:325–338.
35. Merkle T, Wehner R. Desert ants use foraging distance to adept the nest search to the uncertainty of the path integrator. *Behav Ecol.* 2010;21:349–355.
36. Dacke M, Srinivasan MV. Two odometers in honeybees? *J Exp Biol.* 2008;211:3281–3286.
37. Müller M, Wehner R. Path integration in desert ants, *Cataglyphis fortis*. *Proc Natl Acad Sci USA.* 1988;85:5287–5290.
38. Sommer S, Wehner R. The ant's estimation of distance travelled: experiments with desert ants, *Cataglyphis fortis*. *J Comp Physiol A.* 2004;190:1–6.
39. Wolf H, Wehner R. Pinpointing food sources: olfactory and anemotactic orientation in desert ants, *Cataglyphis fortis*. *J Exp Biol.* 2000;203:857–868.
40. Wolf H, Wehner R. Desert ants, *Cataglyphis fortis*, compensate navigation uncertainty. *J Exp Biol.* 2005;208:4223–4230.
41. Wehner R, Müller M. Piloting in desert ants: pinpointing the goal by discrete landmarks. *J Exp Biol.* 2010;213:4174–4179.
42. Stieb SM, Kelber C, Wehner R, Rössler W. Antennal-lobe organization in desert ants of the genus *Cataglyphis*. *Brain Behav Evol.* 2011;77:136–146.
43. Möller R, Vardy A. Local visual homing by matched-filter descent in image distance. *Biol Cybern.* 2006;95:413–430.
44. Zeil J, Hoffmann MI, Chahl JS. The catchment areas of panoramic snapshots in outdoor scenes. *J Opt Soc Am A.* 2003;20:450–469.
45. Baddeley B, Graham P, Philippides A, Husbands P. Holistic visual encoding of ant-like routes: navigation without waypoints. *Adapt Behav.* 2011;19:3–15.
46. Stürzl W, Zeil J. Depth, contrast and view-based homing in outdoor scenes. *Biol Cybern.* 2007;96:519–531.
47. Baddeley B, Graham P, Husbands P, Philippides A. Model of ant route navigation driven by scene familiarity. *PLoS Comput Biol.* 2012;8(1):e1002336.
48. Kohler M, Wehner R. Idiosyncratic route-based memories in desert ants, *Melophorus bagoti*: how do they interact with path-integration vectors? *Neurobiol Learn Mem.* 2005;83:1–12.
49. Wystrach A, Beugnon G, Cheng K. Landmarks or panoramas: what do navigating ants attend to for guidance? *Frontiers Zool.* 2011;8:21.
50. Wystrach A, Beugnon G, Cheng K. Ants might use different view-matching strategies on and off the route. *J Exp Biol.* 2012;215:44–55.
51. Wehner R. Spatial vision in arthropods. In: Autrum J, ed. *Handbook of Sensory Physiology.* Vol VII/6C. New York: Springer; 1981:287–616.
52. Ziegler PE, Wehner R. Time-courses of memory decay in vector-based and landmark-based systems of navigation in desert ants, *Cataglyphis fortis*. *J Comp Physiol A.* 1997;181:13–20.
53. Graham P. Insect navigation. In: Breed MD, Moore J, eds. *Encyclopedia of Animal Behavior.* Vol 2. Oxford: Academic Press; 2010:167–175.

54. Bisch S, Wehner R. Visual navigation in ants: evidence for site-based vectors. *Proc Neurobiol Conf Göttingen*. 1998;26:417.
55. Sassi S, Wehner R. Dead reckoning in desert ants, *Cataglyphis fortis*: can homeward-bound vectors be reactivated by familiar landmark configurations? *Proc Neurobiol Conf Göttingen*. 1997;25:484.
56. Bisch-Knaden S, Wehner R. Egocentric information helps desert ants to navigate around familiar obstacles. *J Exp Biol*. 2001;204:4177–4184.
57. Collett M, Collett TS, Bisch S, Wehner R. Local and global vectors in desert ant navigation. *Nature*. 1998;394:269–272.
58. Collett TS, Collett M, Wehner R. The guidance of desert ants by extended landmarks. *J Exp Biol*. 2001;204:1635–1639.
59. Knaden M, Lange C, Wehner R. The importance of procedural knowledge in desert-ant navigation. *Curr Biol*. 2006;16:916–917.
60. Collett TS, Fry SN, Wehner R. Sequence learning by honeybees. *J Comp Physiol A*. 1993;172:693–706.
61. Collett M, Collett TS. The learning and maintenance of local vectors in desert ant navigation. *J Exp Biol*. 2009;212:895–900.
62. Wehner R. Die Konkurrenz von Sonnenkompass- und Horizontmarken-Orientierung bei der Wüstenameise Cataglyphis bicolor (Hymenoptera: Formicidae). *Verh Dtsch Zool Ges*. 1970;64:238–242.
63. Kühn-Bühlmann S, Wehner R. Age-dependent and task-related volume changes in the mushroom bodies of visually guided desert ants, *Cataglyphis bicolor*. *Dev Neurobiol*. 2006;66:511–521.
64. Seid MA, Wehner R. Delayed axonal pruning in the ant brain: a study of developmental trajectories. *Dev Neurol*. 2009;69:1–15.
65. Stieb SM, Muenz TS, Wehner R, Rössler W. Visual experience and age affect synaptic organization in the mushroom bodies of the desert ant *Cataglyphis fortis*. *Dev Neurobiol*. 2010;70:408–423.
66. Davis RL. Olfactory memory formation in *Drosophila*: from molecular to systems neuroscience. *Ann Review Neurosci*. 2005;28:275–302.
67. Gerber B, Tanimoto R, Heisenberg M. An engram found? Evaluating the evidence from fruit flies. *Curr Opin Neurobiol*. 2004;14:737–744.
68. Giurfa M. Behavioral and neural analysis of associative learning in the honeybee: a taste from the magic well. *J Comp Physiol A*. 2007;193:801–824.
69. Hourcade B, Muenz TS, Sandoz JC, Rössler W, Devaud JM. Long-term memory leads to synaptic reorganization in the mushroom bodies: a memory trace in the insect brain? *J Neurosci*. 2010;30:6461–6465.
70. Menzel R. Memory dynamics in the honeybee. *J Comp Physiol*. 1999;185:323–340.
71. Mizunami M, Weibrecht JM, Strausfeld NJ. Mushroom bodies of the cockroach: their participation in place memory. *J Comp Neurol*. 1998;402:520–537.
72. Strausfeld NJ, Hansen L, Li Y, Gomez RS, Ito K. Evolution, discovery, and interpretations of arthropod mushroom bodies. *Learn Mem*. 1998;5:11–37.
73. Gronenberg W, Heeren S, Hölldobler B. Age-dependent and task-related morphological changes in the brain and the mushroom bodies of the ant *Camponotus floridanus*. *J Exp Biol*. 1996;199:2011–2019.
74. Seid MA, Harris KM, Traniello JFA. Age-related changes in the number and structure of synapses in the lip region of the mushroom bodies in the ant *Pheidole dentata*. *J Comp Neurol*. 2005;488:269–277.
75. Groh C, Rössler W. Comparison of microglomerular structures in the mushroom-body calyx of neopteran insects. *Arthopod Struct Dev*. 2011;40:358–367.
76. Rössler W, Groh C. Plasticity of synaptic microcircuits in the mushroom-body calyx of the honeybee. In: Eisenhardt D, Girufa M, Galizia CG, eds. *Honeybee Neurobiology and Behavior: A Tribute to Randolf Menzel*. New York: Springer; 2012:141–153.
77. Ehmer B, Hoy R. Mushroom bodies of vespid wasps. *J Comp Neurol*. 2000;416:93–100.
78. Farris SM, Schulmeister S. Parasitoism, not sociality, is associated with the evolution of insects elaborate mushroom bodies in the brains of hymenopteran insects. *Proc Roy Soc B*. 2011;278:940–951.
79. Strausfeld NJ. Organization of the honey bee mushroom body: representation of the calyx within the vertical and gamma lobes. *J Comp Neurol*. 2002;450:4–33.
80. Witthöft W. Absolute Anzahl und Verteilung der Zellen im Gehirn der Honigbiene. *Zeitschrift für Morphologie der Tiere*. 1967;61:160–184.
81. Groh C, Ahrens D, Rössler W. Environment- and age-dependent plasticity of synaptic complexes in the mushroom bodies of honeybee queens. *Brain Behav Evol*. 2006;68:1–14.
82. Armstrong JD, Kaiser K, Müller A, Fischbach KF, Merchant N, Strausfeld NJ. Flybrain, an on-line atlas and database of the *Drosophila* nervous system. *Neuron*. 1995;15:17–20.
83. Aso Y, Grübel K, Busch S, Friedrich AB, Siwakowicz I, Tanimoto H. The mushroom body of adult *Drosophila* characterized by GAL4 drivers. *J Neurogenet*. 2009;23:156–172.
84. Farris SM, Sinakevitch I. Development and evolution of the insect mushroom bodies: towards the understanding of conserved developmental mechanisms in a higher brain center. *Arthropod Struct Dev*. 2003;32:79–101.
85. Gronenberg W. Subdivisions of hymenopteran mushroom body calyces by their afferent supply. *J Comp Neurol*. 2001;435:474–489.
86. Mobbs PG. The brain of the honeybee *Apis mellifera*: the connections and spatial organization of the mushroom bodies. *Phil Trans R Soc Lond B*. 1982;298:309–354.
87. Paulk AC, Gronenberg W. Higher order visual input to the mushroom bodies in the bee, *Bombus impatiens*. *Arthropod Struct Dev*. 2008;37:443–458.
88. Strausfeld NJ. *Arthropod Brains: Evolution, Functional Elegance, and Historical Significance*. Cambridge, MA: Harvard University Press; 2012.
89. Tomer R, Denes AS, Tessmar-Raible K, Arendt D. Profiling by image registration reveals common origin of annelid mushroom bodies and vertebrate pallium. *Cell*. 2010;142:800–809.
90. Yasuyama K, Meinertzhagen IA, Schürmann FW. Synaptic organization of the mushroom body calyx in *Drosophila melanogaster*. *J Comp Neurol*. 2001;445:211–226.
91. Frambach I, Rössler W, Winkler M, Schürmann FW. F-actin at identified synapses in the mushroom body neuropil of the insect brain. *J Comp Neurol*. 2004;475:303–314.
92. Farris SM, Robinson GE, Fahrbach SE. Experience- and age-related outgrowth of intrinsic neurons in the mushroom bodies of the adult worker honeybee. *J Neurosci*. 2001;21:6395–6404.
93. Durst C, Eichmüller S, Menzel R. Development and experience lead to increased volume of subcompartments of the honeybee mushroom body. *Behav Neural Biol*. 1994;62:259–263.
94. Withers GS, Fahrbach SE, Robinson GE. Selective neuroanatomical plasticity and division of labour in the honeybee. *Nature*. 1993;364:238–240.
95. O'Donnell S, Donlan NA, Jones TA. Mushroom body structural change is associated with division of labor in eusocial wasp workers (*Polybia aequatorialis*, Hymenoptera: Vespidae). *Neurosci Lett*. 2004;356:159–162.
96. Groh C, Tautz J, Rössler W. Synaptic organization in the adult honey bee brain is influenced by brood-temperature control during pupal development. *Proc Natl Acad Sci USA*. 2004;101:4268–4273.
97. Groh C, Lu Z, Meinertzhagen IA, Rössler W. Age-related plasticity in the synaptic ultrastructure of neurons in the mushroom body calyx of the adult honeybee *Apis mellifera*. *J Comp Neurol*. 2012;520:3509–3527.

98. Krofczik S, Khojasteh U, Hempel de Ibarra N, Menzel R. Adaptation of microglomerular complexes in the honeybee mushroom body lip to manipulations of behavioral maturation and sensory experience. *Dev Neurobiol.* 2008;68:1007–1017.
99. Steiger U. Über den Feinbau des Neuropils im Corpus pedunculatum der Waldameise. *Z Zellforsch.* 1967;81:511–536.
100. Ganeshina O, Menzel R. GABA-immunoreactive neurons in the mushroom bodies of the honeybee: an electron microscopic study. *J Comp Neurol.* 2001;437:335–349.
101. Dobrin SE, Herlihy JD, Robinson GE, Fahrbach SE. Muscarinic regulation of Kenyon cell dendritic arborizations in adult worker honey bees. *Arthropod Struct Dev.* 2011;40:409–419.
102. Bramham CR. Local protein synthesis, actin dynamics, and LTP consolidation. *Curr Opin Neurobiol.* 2008;18:524–531.
103. Farris SM. Are mushroom bodies cerebellum-like structures? *Arthropod Struct Dev.* 2011;40:368–379.
104. Homberg U, Heinze S, Pfeiffer K, Kinoshita M, el Jundi B. Central neural coding of sky polarization in insects. *Phil Trans R Soc B.* 2011;366:680–687.
105. Dobrin SE, Fahrbach SE. Rho GTPase activity in the honey bee mushroom bodies is correlated with age and foraging experience. *J Insect Physiol.* 2012;58:228–234.
106. Pasch E, Muenz TS, Rössler W. CaMKII protein is differentially localized in dendritic compartments of mushroom body intrinsic neurons in the honeybee brain. *J Comp Neurol.* 2011;519:3700–3712.
107. Kiya T, Kubo T. Dance type and flight parameters are associated with different mushroom body neural activities in worker bee brains. *PLoS ONE.* 2011;6:e19301.
108. Sarma MS, Rodriguez-Zas SL, Gernat T, Nguyen T, Newman T, Robinson GE. Distance-responsive genes found in dancing honey bees. *Genes Brain Behav.* 2010;9:825–830.

CHAPTER

40

Learning and Decision Making in a Social Context

Nigel R. Franks and Ana B. Sendova-Franks†*

*University of Bristol, Bristol, United Kingdom †University of the West of England, Bristol, United Kingdom

INTRODUCTION

Our goal in this chapter is to show that insects are ideal models for studies of learning in a social context. Colonies of social insects such as ants, bees, wasps, and termites represent the last great evolutionary transition[1]—in this case, from solitary organisms to highly social ones. Insect societies are unrivalled systems for testing hypotheses through manipulative experimentation because they have a modular structure (e.g., the workers are one type of module) and so can be taken apart and put together again in different configurations. Hence, they represent an ideal opportunity to understand learning in a social context. Moreover, studies of insects in general and social insects in particular provide an antidote to anthropomorphic biases that have entered the field of animal behavior from psychology. Consider, for example, the important textbook on cognition, evolution, and behavior by Shettleworth.[2] One entire chapter is devoted to social learning, which Shettleworth defines admirably as follows: "Social learning refers to any learning resulting from the behavior of other animals" (p. 467). However, should we not be shocked that throughout that chapter almost all the examples are from studies of vertebrates? After all, a moment's thought would indicate that the waggle dance of the honeybee fits this definition of social learning to a 'T.'

Note that our aim in this chapter is not to review social learning in insects in general. Indeed, such a review has been made recently.[3] Instead, in this chapter, we discuss two special examples of learning in a social context that highlight some of the key issues. The first is the controversial case of teaching in ants, and the second is the almost equally controversial case of colony-level learning. We consider both cases in the light of the definition of social learning provided by Shettleworth[2] (p. 467) as mentioned previously. We much admire this definition. It is extraordinarily concise yet it is almost audacious in its boldness—"*any learning* resulting from the behavior of other animals" (emphasis added). Thus, for example, this definition would encompass the case of one ant following the trail of another, and learning landmarks as it does so, to help it find food and avoid becoming lost and, as mentioned previously, the waggle dance of the honeybee. These examples contrast with mostly from studies of vertebrates in that they are based on signals and not cues. One vertebrate eavesdropping on another may do so to pick up inadvertent cues that will help it, purely selfishly, to learn to solve particular problems. By contrast, social insects actively provide signals that have been tuned by natural selection via influences on inclusive fitness: They are likely directly to benefit a nestmate in the short term and the mutual inclusive fitness of both parties in the long term.

SOCIAL LEARNING THROUGH TEACHING

Of course, biases to vertebrate studies are not new. For example, when Caro and Hauser[4] defined and reviewed teaching in nonhuman animals, they seemed to ignore the possibility that any invertebrates at all might exhibit such behavior. This is all the more surprising given that Caro and Hauser were adamant that teaching in nonhuman animals should be considered

Invertebrate Learning and Memory.
DOI: http://dx.doi.org/10.1016/B978-0-12-415823-8.00040-X

in ways that did not necessarily require those animals to have a theory of mind—that is, thoughts about the thoughts of others.

At this point, it might well be helpful to strike a cautionary note. As members of human society, and especially as academics, we may well be practicing at least some teaching and social learning every day of our working lives. Moreover, we suspect that all human beings would believe that they know teaching when they encounter it. So why define the obvious? Yet further consideration will recognize the ubiquity of teaching in human society and the vast plethora of diverse activities that belong in this particular special case of social learning. So clearly a definition of teaching is needed.

Consider teaching as defined by the *Oxford English Dictionary*—"to show by way of information or instruction." In many ways, this is an admirable definition because it is both short and simple. Furthermore, it makes no mention of a theory of mind, but neither does it exclude this possibility. For example, a human teacher might think, "I know what I know and I know what you do not. So I will teach you to rectify this discrepancy." But to reiterate, neither the *Oxford English Dictionary* nor Caro and Hauser[4] have sought to invoke theories of mind as a *necessary* component for teaching. Indeed, we believe that it is perfectly reasonable to think of some forms of teaching as simply rather mechanical and algorithmic. Thus, we could imagine one robot teaching another or indeed one computer providing a learning opportunity for another.

Thus, given this stance, it is all the more remarkable that teaching and social learning were neglected for so long in studies of the most highly social of all nonhuman animals, namely the eusocial insects.

Caro and Hauser[4] defined teaching as follows:

> An individual actor A can be said to teach if it modifies its behavior only in the presence of a naive observer, B, at some cost or at least without obtaining an immediate benefit for itself. A's behavior thereby provides B with experience, or sets an example for B. As a result, B acquires knowledge or learns a skill earlier in life or more rapidly or efficiently than it might otherwise do, or that it would not learn at all. (p. 153)

One tentative example of teaching that Caro and Hauser put forward in 1992 was the possibility that mother cheetahs provide their cubs with living antelope fawns so that they can practice their hunting skills. Although this example might fulfill all of the clauses of the Caro and Hauser definition, rigorous evidence that cheetah cubs learn more quickly than they otherwise do by being provided with mobile but easy prey was missing. Indeed, the only way to demonstrate teaching according to Caro and Hauser's definition is through rigorous experimentation.

Nevertheless, the Caro and Hauser[4] definition and review, published as it was in the extremely respected *Quarterly Review of Biology*, remains a benchmark for attempts to demonstrate teaching. The first published case of teaching in a nonhuman animal that rigorously met the Caro and Hauser criteria was that of tandem running in the rock ant *Temnothorax albipennis*. Tandem running is a recruitment behavior in which one ant leads a single nestmate to a food source or a new nest site. Franks and Richardson[5] showed that such tandem running not only met all the Caro and Hauser criteria but also met an additional one, namely that there was feedback in both directions between the teacher and the pupil. We discuss the massive importance of this additional criterion in due course.

Franks and Richardson[5] used the following criteria based on the Caro and Hauser[4] definition: An individual is a teacher if

1. it modifies its behavior in the presence of a naive observer,
2. at some initial cost to itself,
3. in order to set an example,
4. so that the other individual can learn more quickly.

Tandem running meets criterion 1 because the tandem leader clearly modifies its behavior in the presence of a naive observer. It produces an attractive pheromone, moves very slowly, and will only proceed when touched on its hind legs and gaster by a follower's antennae. We must emphasize a critical point here. In this case, it is not necessary for the tandem leader to identify another ant as naive. All that needs to occur is for an ant to decide to follow the tandem leader. In a sense, when such an ant elects to follow a tandem leader, it is declaring itself as naive. So in this very simple way, there is obviously absolutely no need to invoke any theory of mind in this particular case. The leader does not need to 'think' about what the follower may or may not know.

With regard to the second criterion, there is clearly some cost to the leader because she moves so slowly when accompanied by a follower that her return to a food source or a new nest site takes three times longer than if she was unencumbered by a pupil. Of course, such a cost is proximate, and one ant teaching its nestmate about the location of valuable resources should ultimately boost the potential inclusive fitness of both individuals and their society.

Regarding the third criterion, the tandem leader clearly sets an example of how to proceed from the old nest site to a new one or to a valuable source of forage. Franks and Richardson[5] were also able to show that worker ants are typically much quicker in learning the location of a valuable resource if they follow a tandem run to it than if they simply have to search for it for themselves. In so doing, they showed that the fourth criterion for teaching also applied to tandem running.

Intriguingly, Franks and Richardson[5] raised the stakes beyond the Caro and Hauser[4] definition by adding a fifth criterion, namely (5) active teaching may involve feedback both between the teacher and the pupil and between the pupil and the teacher. They called this bidirectional feedback between the teacher and the pupil. In the case of the ants, such feedback was clearly indicated by the movements of the tandem leader and the tandem follower. If the leader moves too quickly so that it is no longer frequently contacted by the follower, the leader decelerates so that contact is re-established. The behavior of the follower mirrors this pattern of acceleration and deceleration. When the leader is just too far away, the follower will accelerate to re-establish contact, and when they are too close together the follower will decelerate. Only when the leader and follower are separated by exactly the length of the outstretched antennae of the follower do both parties move at a constant velocity. The reason tandem runs are so slow is that neither party can proceed at full speed because each is responding with a lag to the movements and contact, or lack of contact, of the other. In addition, tandem runs are very slow because contact between the leader and the follower is often broken by the follower producing a rather convoluted path in the wake of her leader. Indeed, followers often seem to be looking around and may even be learning landmarks. In fact, followers will often loop round completely (i.e., through 360°) before re-establishing contact with the leader. In these brief periods in which contact is broken, the leader will remain almost perfectly still until the follower once again touches, with her antennae, the leader's hind legs and gaster.

What first alerted us to the possibility that tandem running in ants might be a form of teaching? It was the observation that *T. albipennis* workers have two distinctly different ways to take a nestmate from one place to another: Tandem running is one; the other is social carrying, in which one individual simply picks up another and runs off with it. Both of these behaviors occur during nest emigrations. However, social carrying is approximately three times quicker than tandem running. So why do the ants engage in tandem running when it is so slow? They do so because an ant that is carried has its head orientated backwards and upside down and its feet do not touch the ground. Hence, a carried ant cannot learn the route to the new nest site either by learning landmarks or by path integration. Indeed, carried ants typically stay where they are put, at least for quite a while. Probably this is because they do not know where the new nest is with respect to the old one. This contrasts with an ant that was led in a tandem run. Such followers often become leaders of subsequent tandem runs—teaching others in turn. In this way, tandem running, although slow, creates a chain reaction. One knowledgeable ant transfers information through tandem running to another, and both can then teach; two becomes four and then eight and so on. This cascade of information thus snowballs. Indeed, Möglich[6] rightly described tandem running as the recruitment of recruiters. Intriguingly, *T. albipennis* may use tandem running to recruit either to new nest sites or to food. However, they have not yet been observed to carry another ant to food even though that is much quicker.[7] Again, this might be because carried ants are disorientated. Thus, it seems that tandem running, although slow, 'adds value,' just as other forms of teaching add value. Therefore, it was careful observation of the behavior of *T. albipennis* that caused us to think about the possibility that they taught—rather than any convoluted philosophical debates about the deep nature of teaching.

Nevertheless, the response to the claim that tandem running is a form of teaching was strong. Some authors even seemed so aghast that certain ants met all of the criteria of the Caro and Hauser[4] definition that they immediately queried the value of that definition. This was a truly remarkable response, given that people working on vertebrates had been comfortable with the Caro and Hauser definition for the 14 years between its publication and the report by Franks and Richardson.[5] Others claimed that if tandem running qualified as teaching, then the waggle dance of honeybees might also do so,[8] or perhaps even pheromone trail laying in ants. Such might be the case regarding the Caro and Hauser definition—but the waggle dance and pheromone trails are, or at least seem to be, excluded by the fifth criterion that Franks and Richardson introduced. To the best of our knowledge, there is no feedback in either direction between a waggle dancing bee and the bees that usually follow its dancing. Indeed, occasionally a dancing bee might dance on an empty dance floor (Tom Seeley, personal communication). Recently, it has been shown that honeybees may inhibit dancing that advertises either foraging sites[9] or nest sites[10] through the deployment of stop signals. However, this is a kind of remote social feedback and not the intimate feedback between a specific teacher and her pupil that occurs in tandem running ants. Similarly, in most cases, ants that lay recruitment pheromone trails may seemingly do so without immediately or directly involving others. Franks and Richardson highlighted this distinction by referring to certain communication in social insects as broadcasting rather than teaching. In broadcasting, information is sent out with the expectation that it might be received, but exactly who will receive it is unknown. A simple parable should make this clearer.

The term broadcasting was not invented for the mass media. An old-fashioned way of sowing seeds was to broadcast them—scattering them rather than carefully planting them one by one. Indeed, some broadcast seeds might fall on stony ground (and fail to germinate), just as broadcast information may fall on deaf ears or no ears at all. Thus, a pheromone trail might evaporate before anyone can respond to it or a waggle dance may find no one to follow and learn from it. Of course, broadcasting works in large populations because statistically someone is likely to be available and respond to it. Indeed, in a large society, it may be statistically quite predictable that a sizeable portion of the population may be available and fall under the influence of the broadcast. Similarly, a peasant farmer broadcasting seeds should be able to judge from experience, and be confident, that a sufficient number of seeds will fall on receptive and fertile ground. Hence, Franks and Richardson suggested that broadcast communication might be the norm in large, well-integrated societies, whereas teaching (with feedback) was more likely in small societies in which information might be both easily lost and costly to recover.

Another response to the demonstration that tandem running is teaching was from the Premack and Premack[11] school of thought that true teaching involves evaluation by the teacher—that is, "a combination of observation, judgment, and intervention." Indeed, Marc Hauser and David Premack (2006, personal communication) wrote to Franks and Richardson, stating their view that

> evaluation presupposes an image or mental representation of a standard, a preferred action or product, to which the teacher compares the action or product of the novice. There is no evidence for evaluation—standards and correction of a deficient novice—in nonhumans. Teaching still looks to be unique to humans.

The value of this new viewpoint was that it encouraged further experiments that actually showed three different forms of evaluation in tandem-running ants. All three forms of evaluation focus on the amount of time a leader will remain stationary, waiting for new contact from her follower, before giving up and walking away. Our method[12] was simply to remove the follower and determine how long the leader remained stationary, and indeed we interrupted no fewer than 738 tandem runs to examine these teacher giving-up times in different circumstances. We showed first that the longer the tandem run had proceeded up until the interruption point, the longer the leader was prepared to wait for her missing follower. Thus, for example, if the tandem run had proceeded for 20 sec, leaders, on average, were prepared to wait for 26 sec, whereas if

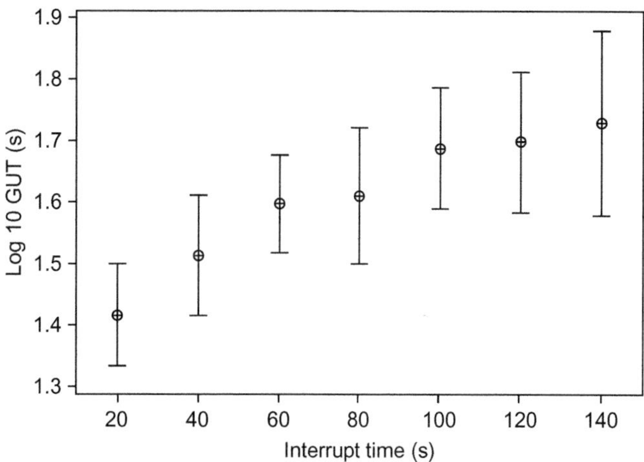

FIGURE 40.1 Leader giving-up times (GUT) vary according to the length of time the tandem has proceeded. Source: *Adapted from* Current Biology, *17, T. O. Richardson, P. A. Sleeman, J. M. McNamara, A. I. Houston, and N. R. Franks, Teaching with evaluation in ants, 1520–1526, Fig. 1, Copyright 2007, with permission from Elsevier.*

the tandem run had proceeded for 140 sec, leaders were prepared to wait, on average, 55 sec (Figure 40.1). In simple terms, this seems to show that the leader evaluates how much investment she has put into a tandem run. Second, we were able to show that teacher giving-up times depend on the quality of the resource to which she was heading and the direction of the tandem she was leading. This might suggest that leaders evaluate the worth of the lesson that they are teaching. Third, we were able to create a class of slow pupils by amputating one of their antennae. Such partially disabled followers make for slow tandem runs (Figure 40.2), and we were able to show that tandem leaders have longer giving-up times when the tandem run had been proceeding quickly up to that point rather than if it had been proceeding slowly (Figure 40.3). One might imagine that the results of the first and third experiments are just reflecting the same phenomena. However, such is not the case. In the third experiment, the follower was taken away at a set point, not at a set time. Pupils with both antennae would get to that set point faster than pupils with one antenna. Yet teachers are more 'patient' with pupils with two antennae, even though they will have invested less time in teaching them. Thus, this strongly suggests that here teachers are evaluating the quality of the pupil or at least the rate at which the lesson had been proceeding (rather than the time invested). Therefore, there really do seem to be three different forms of evaluation that rock ants may use in teaching by tandem running.

Of course, even though we have shown three distinct forms of evaluation, one might reasonably argue that none of these presupposes an image or mental

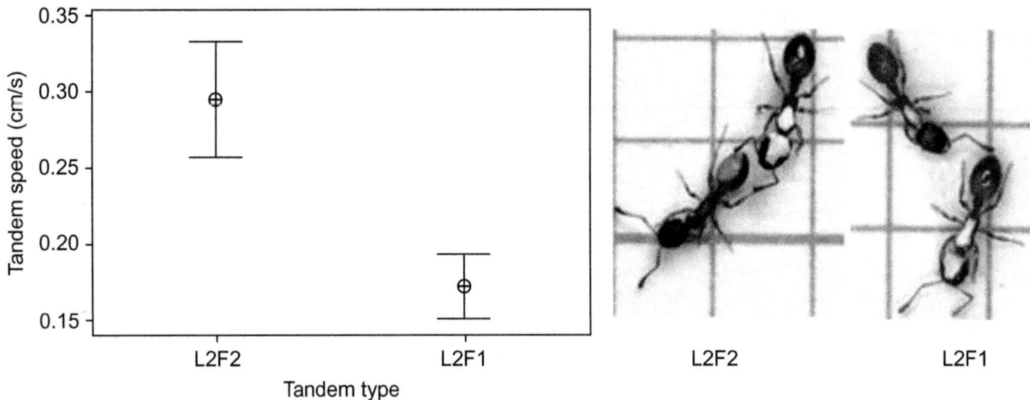

FIGURE 40.2 Comparison of overall speed of tandem pairs. (Left) Followers with two antennae (L2F2). (Right) Followers with one antenna (L2F1). Error bars represent 95% confidence intervals for the means. Source: *Adapted from* Current Biology, *17, T. O. Richardson, P. A. Sleeman, J. M. McNamara, A. I. Houston, and N. R. Franks, Teaching with evaluation in ants, 1520–1526, Fig. 3, Copyright 2007, with permission from Elsevier.*

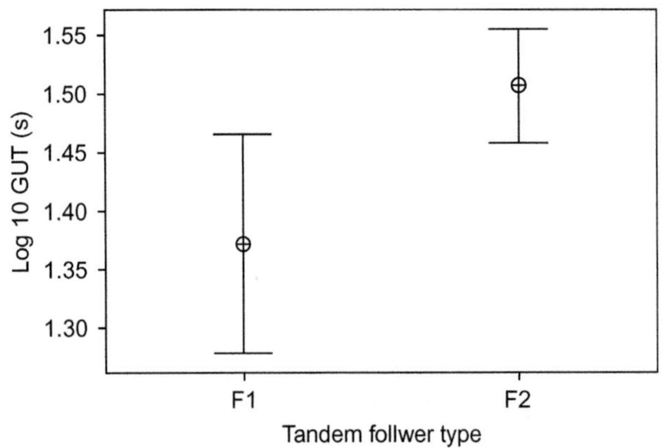

FIGURE 40.3 Leader giving-up times (GUT) vary according to follower antennal number. F1, one antenna; F2, two antennae. Source: *Adapted from* Current Biology, *17, T. O. Richardson, P. A. Sleeman, J. M. McNamara, A. I. Houston, and N. R. Franks, Teaching with evaluation in ants, 1520–1526, Fig. 4, Copyright 2007, with permission from Elsevier.*

representation of a standard, a preferred action or product, to which the teacher compares the action or product of the novice. Of course, all of this hinges on what one imagines "an image or a mental representation of a standard" to be. However, we would reason that such a 'standard' could also be essentially algorithmic. Nevertheless, when Richardson, Franks, *et al.*[12] published their evidence of three forms of evaluation in tandem-running ants, David Premack (e-mail, June 13, 2007) graciously wrote to Tom Richardson,

> Marvelous article of great importance. [What] I've written on the topic in the last 20 or 30 years seems to have worked—provoked you, irritated you enough to do the work I'd not been able to get biologists to do despite years of trying.

Nevertheless, controversy about tandem running as an example of teaching was expressed by Hölldobler and Wilson[13]:

> Recently, this already well-known behavior, informing the naive nestmates by tandem running about a route to a new target area, has been referred to as a case of 'teaching' in ants—the leader ant 'the teacher' and the follower ant 'the pupil.' This might be a charming metaphor, but it adds little, if anything, to our understanding of this fascinating recruitment behavior. (p. 494)

Both Wilson and Hölldobler are experts on tandem running in ants. Indeed, Wilson had named the phenomenon 'tandem running' in 1959,[14] although the behavior had been first described in 1896 by Adlerz.[15] Hölldobler had done absolutely exquisite experimental work, in the 1970s and 1980s,[16,17] showing, for example, which glands were used in a variety of tandem-running ants. Perhaps one of the reasons Hölldobler and Wilson[13] were so dismissive is that they appear not to have seen the follow-up paper to the one published in *Nature*,[5] showing the evaluation that occurs during tandem running.[12] That study and others that have followed it,[18–20] at the very least, reveal that tandem running is a much richer behavior than anyone had previously expected.

By contrast, when Thornton and McAuliffe[21] showed the second rigorous example in a nonhuman animal, they fully accepted the rigor of the experimental demonstration of teaching by Franks and Richardson.[5] Indeed, they stated, "To date, only one study provides firm evidence for teaching (Franks and Richardson 2006)." Thornton and McAuliffe also endorsed the view that teaching does not require a theory of mind. They stated this very clearly as follows:

> It is often assumed that teaching requires awareness of the ignorance of pupils and a deliberate attempt to correct that

ignorance ..., but viewed from the functional perspective ..., teaching could be based on simple mechanisms without the need of intentionality and attribution of mental states. (p. 229)

What Thornton and McAuliffe had shown was that experienced meerkats provide structured learning opportunities for young inexperienced meerkats. These fascinating animals often need to handle dangerous prey such as scorpions, and those meerkats looking after very inexperienced youngsters provide them with disabled scorpions, whereas older pups are provided with less disabled prey. In this way, meerkat pups can safely develop skills to handle dangerous prey. Thornton and McAuliffe showed that teaching in wild meerkats is the provision of a learning opportunity. The format of the opportunity changes in response to pup calls, which change with pup age. However, that meerkats base their provisioning on pup calls and could therefore be fooled by playback experiments suggests that meerkats use a 'rule of thumb' rather than direct evaluation of the prey-handling abilities of the youngsters. Meerkats seem to base the teaching opportunity they provide on perceived pup age but not on actual skill. Nevertheless, this rule of thumb is probably highly effective. Rarely, if ever, would meerkats hear pup calls that were misleading as to the actual age and likely associated skills of the youngsters. Nevertheless, the evaluation shown in tandem-running ants, underpinned by the bidirectional feedback between the teacher and the pupil in this case, seems to show a more sensitive and quantitative form of evaluation than that so far shown in the meerkats.

Our work on evaluation during tandem runs used the protocol of removing tandem followers to determine what leaders would do. We have also performed experiments in which we took leaders away from followers to determine what the latter would do.[20] In these new experiments, we removed tandem leaders at set points during tandem runs. We found that former followers often first engage in a 'Brownian search' for almost exactly the time that their former leader should have waited for them (Figure 40.4). We use the term 'Brownian search' because the movement of the former follower is initially like that of a slowly defusing particle executing a random walk (i.e., like Brownian motion). As Berg[22] stated, in a different context, such a "particle tends to explore a given region of space rather thoroughly." That former followers do so for almost exactly the appropriate time—that is, for almost exactly as long as the former leader should have waited for them—shows how well the algorithms of followers and leaders are tuned to one another. After such a bout of local search, former followers typically switch to a superdiffusive search (Figure 40.4), becoming ever more wide ranging so that in the absence of

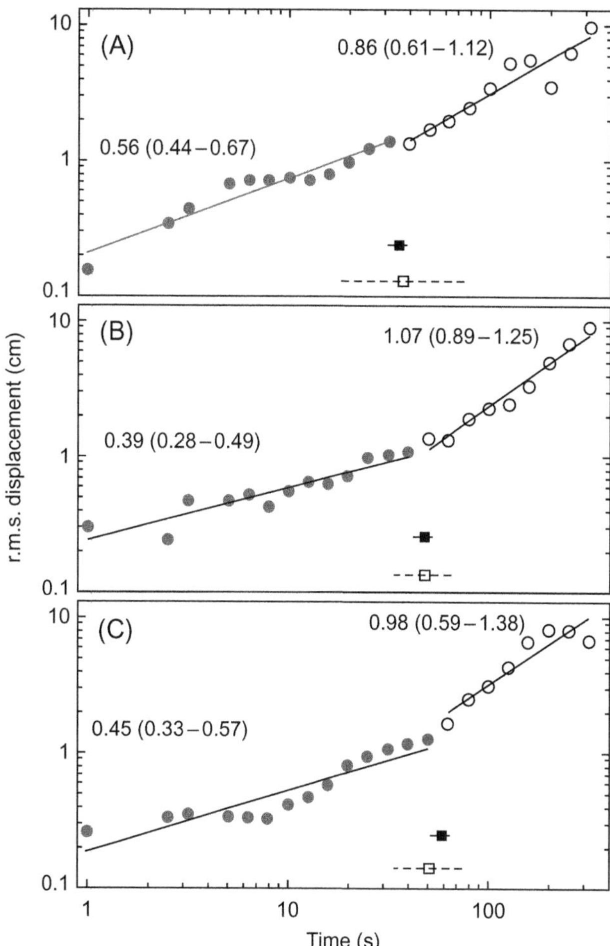

FIGURE 40.4 Log-binned root mean square (r.m.s.) displacement for the search process of former followers as a function of the log time since their leader was removed. The diffusive movement of a Brownian walker will be represented by a slope of 0.5, slower or subdiffusive spreading will have a slope <0.5, whereas a slope >0.5 will indicate faster, superdiffusive motion. The data for all former followers have been pooled for each of three interruption points—(A) 1, (B) 2, and (C) 3—which were at 1/4, 1/2, and 3/4 of the way to the new nest, respectively. Former followers switch from Brownian search to a superdiffusive search at the time when their leader is likely to have left. The expected leader giving-up times[12] are represented by solid squares (mean) and solid lines (±95% CI) and those estimated from the breakpoints of the r.m.s. curves by open squares (mean) and broken lines (±95% CI). The slopes of the six regression lines and their 95% CI are shown. Source: *Adapted from Franks* et al.,[20] *Fig. 6, p. 1703, with permission from The Company of Biologists.*

their former leader they can often find the new nest, re-encounter the old one, or meet a new leader. Because tandem followers often later become tandem leaders, it is not surprising that they should be equipped with similar algorithms that presumably enable them to estimate how long their lost leader should be waiting for them nearby. Nevertheless, the observation that former followers first search

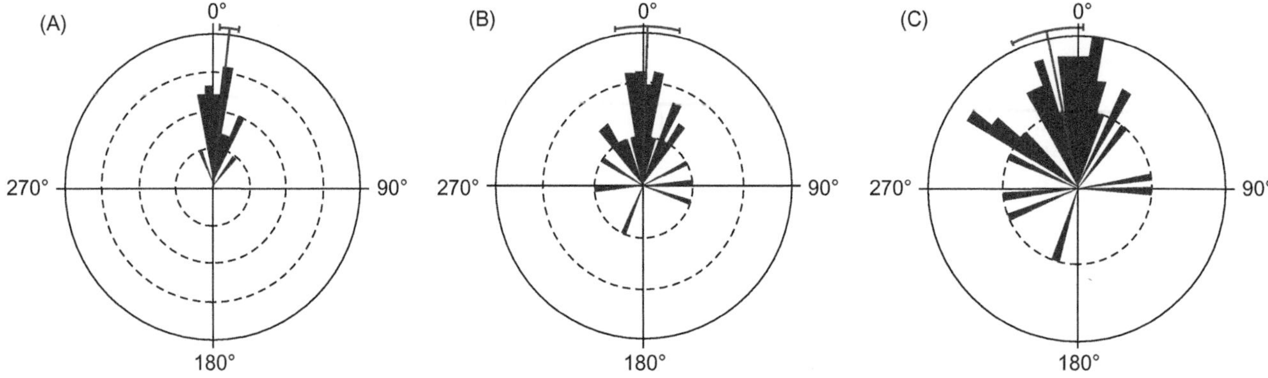

FIGURE 40.5 **Path vectors (compass headings) for (A) all tandems and (B) former followers.** Each path was reduced to a single mean compass heading. Wedge area is proportional to frequency. The new nest was at 0°, and the old nest was at 180°. (C) Difference between paired compass headings for former followers and their tandems. We subtracted the mean compass heading for each former follower from that for its tandem. Then we converted all differences to angles of up to 180°. Finally, we placed all differences where the former follower had the greater mean compass heading to the left of 0° (360° − difference) and all differences where the tandem had the greater mean compass heading to the right of 0° (0° + difference). The paired compass headings do not differ significantly. All calculations were carried out in Oriana, version 1.06, 1994; Warren L. Kovach, Kovach Computing Services, Pentraeth, Wales, UK. The T-bar shows the mean heading with the crossbar displaying the 95% CI. Source: *Adapted from Franks* et al.,[20] *Fig. 5, p. 1702, with permission from The Company of Biologists.*

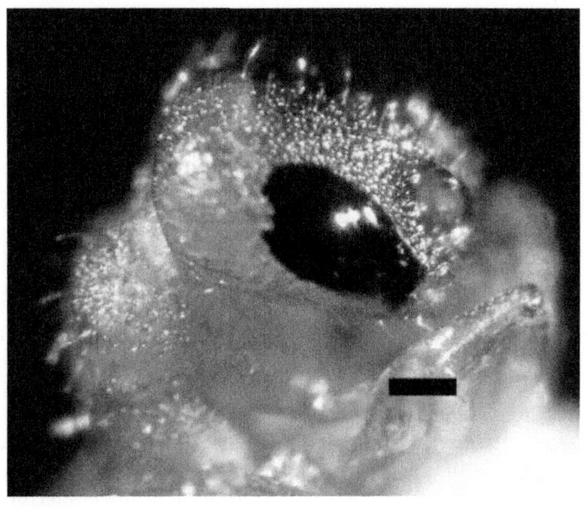

FIGURE 40.6 **A dead *Temnothorax albipennis* worker that had participated in a tandem run, with black paint over the whole of its right eye and beyond.** The scale bar is approximately 0.1 mm long. (The right antenna has been removed for clarity; photograph by Saki Okuda). Source: *From* Behavioral Ecology and Sociobiology *65, 2011, p. 571, Blinkered teaching: Tandem running by visually impaired ants, E. L. Franklin, T. O. Richardson, A. B. Sendova-Franks, E. J. H. Robinson, and N. R. Franks, Fig. 1, with kind permission from Springer Science + Business Media.*

intensively and then extensively seems highly appropriate. Once a leader should have abandoned its lost follower (to proceed by itself to the new nest), the chances that the original follower and leader will reunite become almost vanishingly small. Hence, this explains why former followers switch to a much more extensive superdiffusive search for other static targets such as the new or the old nest. We also showed that followers gain useful information even from incomplete tandem runs. In a substantial minority of cases, followers that have lost their leader after the tandem has progressed for some distance seem to extrapolate from the overall bearing of the partial tandem run, after they have terminated their local search, in the direction to which the tandem had been heading, and they successfully find the new nest (Figure 40.5). These observations point to the important principle that sophisticated communication behaviors may have evolved as anytime algorithms—that is, procedures that are beneficial even if they do not run to completion.[20]

We then continued our work on tandem running in *Temnothorax* by examining the behavior of visually impaired ants.[18] We investigated how dependent tandem leaders and followers are on visual cues by painting over their compound eyes to impair their vision (Figure 40.6). There are two ways in which *T. albipennis* might use vision during tandem running. First, the follower might track the movements of the leader by keeping it in sight. Our results suggest that the ants do not use vision in this way. For example, in all four classes of tandem run (those with either leader or follower, both, or neither of its participants with visual impairments), progress was most smooth at approximately 3 mm/sec. This suggests that communication between leaders and followers during tandem runs does not rely on vision and is mainly tactile and pheromonal. Work on landmark navigation in *T. albipennis* had estimated that their sight can resolve an object that subtends an angle of approximately 7°.[23] Given that each of the compound eyes of a *T. albipennis* worker has only 84 ommatidia (personal observation) compared to 4500 ommatidia in each of a honeybee's compound eyes, and such a bee's ability to resolve approximately 2.8°,[24] the vision of rock ant workers seems surprisingly rather acute. So why tandem followers do not rely on vision to track their leader is a fascinating mystery.

Second, the leader and the follower might be using vision to navigate, and our results support this possibility but also suggest that these ants have other methods of navigation. Ants with visual impairments were more likely to follow than to lead, but they could occupy either role, even though they had many fully sighted nestmates. This might help to explain why the ants in these experiments[18] did not focus grooming on their most visually impaired nestmates. 'Wild-type' tandem runs, with both participants fully sighted and presumably taking time to learn landmarks, were overall significantly slower, smoother, and slightly less tortuous (Figure 40.7) than the other treatments. All four classes of tandem run significantly increased mean instantaneous speeds and mean absolute changes in instantaneous acceleration over their journeys. Moreover, tandems with sighted followers increased their speed with time more than the other treatments. In general, these findings suggest that eyesight is used for navigation during tandem running but that these ants also probably use other orientation systems during such recruitment and to learn how to get to new nest sites. Again, these results suggest that the ants' methods of teaching and learning are very robust and flexible.[18]

We then examined tandem-running abilities in callow workers compared to older, more experienced workers.[19] Callow workers are recognizable because their cuticle is a much lighter shade than those of their darker brown, well-tanned, and older nestmates. In *T. albibennis*, workers can remain recognizably callow for approximately 2 months (Franks and Sendova-Franks, personal observations). At least in the lab, *T. albipennis* workers can survive for 5 or more years (Sendova-Franks, personal observations). The division of labor in *T. albipennis* colonies seems not to be strongly driven by age,[25] but nevertheless one might expect callow workers mostly to remain within the safety of the nest and therefore to have little knowledge of the outside world and rarely to engage in tandem running. Nevertheless, our goal was to determine experimentally to what extent participation in, and efficient execution of, tandem running depends on either the age or the experience of worker ants. To investigate these issues, we constructed colonies of the ant *T. albipennis* with different compositions of inexperienced and experienced workers from different age cohorts and then examined which ants participated in tandem runs when they emigrated. Our results revealed that the ability to participate actively in recruitment by tandem running is present in all worker age groups, but the propensity to participate varies with experience but not age *per se*. Experienced individuals were more likely to engage in tandem runs, either as leaders or as followers, than young inexperienced individuals, and older experienced ants were more likely to lead tandems than older inexperienced ants. Young inexperienced ants led faster (Figure 40.8), more rapidly dispersing, and less accurately orientated (Figure 40.9) tandem runs than the older experienced ants. This study suggests that experience (rather than age per se) coupled to stimulus threshold responses might interact to promote a division of labor so that a suitable number of workers actively participate in tandem runs.

To some extent, the link we have discovered between experience and the propensity to participate in tandem runs suggests that worker ants might learn to lower their thresholds for taking up this task. Moreover, our observations that callows produce

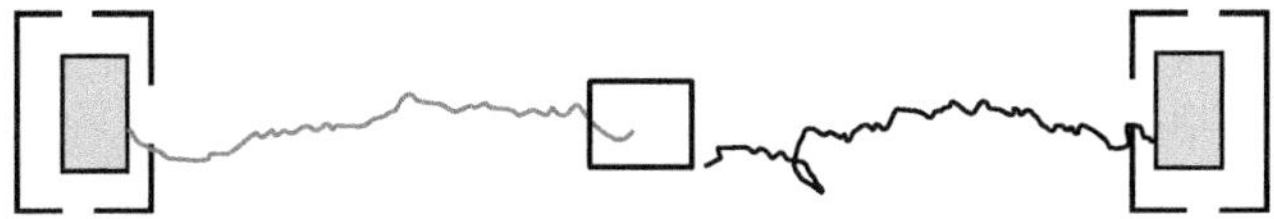

FIGURE 40.7 Experimental setup: The destroyed old nest in the center and the two alternative new intact nest sites at *each* end of the experimental arena. The light gray path is the course of a leader in a tandem run in which neither the leader nor the follower is blinkered (LN/FN). The black path is the course of a leader in a tandem run in which both the leader and the follower are blinkered (LB/FB). Source: *Adapted from* Behavioral Ecology and Sociobiology *65, 2011, p. 571, Blinkered teaching: Tandem running by visually impaired ants, E. L. Franklin, T. O. Richardson, A. B. Sendova-Franks, E. J. H. Robinson, and N. R. Franks, Fig. 2, with kind permission from Springer Science + Business Media.*

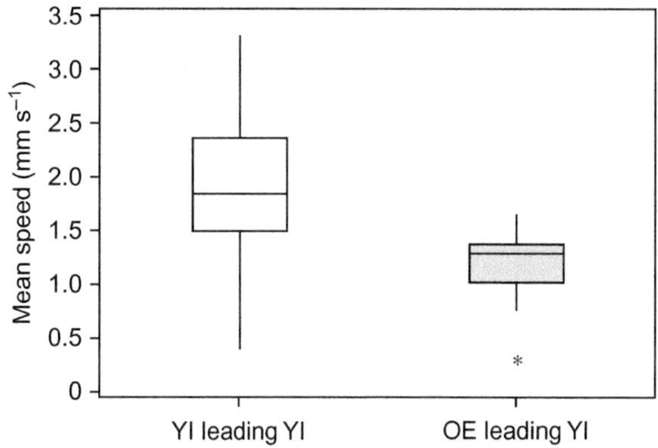

FIGURE 40.8 Box plot showing the mean speed of tandems, comparing young inexperienced (YI) ants leading young inexperienced ants and old experienced (OE) ants leading young inexperienced ants. The central line within the box is the median, the box encloses the interquartile range, whiskers show the 1.5 times interquartile range, and outliers are marked with an asterisk. Source: *Adapted from Franklin* et al.,[19] *Fig. 2, p. 1290, with permission from The Company of Biologists.*

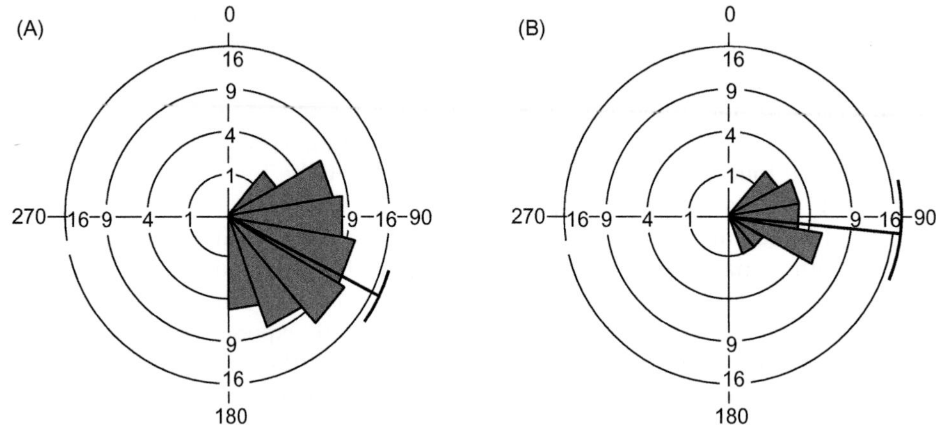

FIGURE 40.9 **Histograms of the mean angle of orientation of young inexperienced ants leading young inexperienced ants (A) or old experienced ants leading young inexperienced ants in tandems (B).** The 360° data are folded into 180° to combine paths to both nests and facilitate analysis. The new nests would be located at 90°. Mean heading and 95% CI drawn as in Figure 40.5. Source: *Adapted from Franklin* et al.,[19] *Fig. 4, p. 1291, with permission from The Company of Biologists.*

tandem runs that may be a little too fast and rather poorly orientated may suggest that with experience and learning, workers can become more skilled in this very intriguing form of recruitment.

In summary, all of our work on tandem running so far suggests that it is a remarkably sophisticated and robust behavior. It seems to be based on a large number of relatively simple rules of thumb that in concert produce some very sophisticated and effective algorithms. Moreover, learning seems to play an important part in the fine-tuning of these rules of thumb and especially in the ability of tandem leaders and followers to work well together.

Tandem running is typically a crucial behavior in *T. albipennis* new nest site selection and colony emigrations. Thus, this provides a very natural link to the second major theme of this chapter. We can best express this as two linked questions. Is there such a thing as colony-level learning? Can colonies learn to emigrate faster?

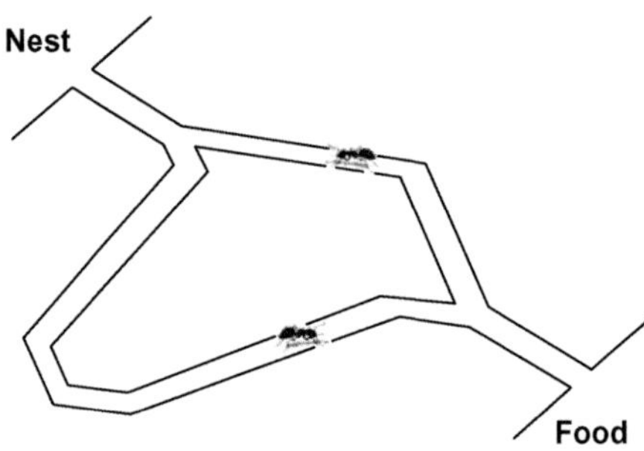

FIGURE 40.10 **An asymmetrical bridge between an ant nest and a food source such that there are two paths, one much longer than the other.** Source: *Adapted from Goss* et al.,[27] *Fig. 1a, p. 579.*

COLONY-LEVEL LEARNING

The main theme of the second part of this chapter is the possibility of colony-level learning in superorganisms. However, to put this in a broader context, we first consider colony-level cognition.[26] Let's begin with a classical example of colony-level decision making. Consider the case of shortcut selection during foraging by an ant colony. Figure 40.10 shows an asymmetrical bridge between an ant nest and a food source such that there are two paths, one much longer than the other. It can be shown, both theoretically and experimentally, that a colony can choose the shorter path even though individual ants may have no awareness whatsoever of the alternative path lengths.[27,28] The system simply works as follows. If the first ant happens to choose the longer path, it will take so long to get to the food and return by that path that it can be beaten by a later ant that simply happened to take the shorter path. If such ants lay, choose, and reinforce pheromone recruitment trails, the attractiveness of the shorter path will typically grow much more rapidly than that of the longer path. In this way, the colony is able actively to choose the shorter path even though no individual ant has made the comparison. Intriguingly, this does not actually mean that individual ants would be incapable of comparing the two path lengths, and this early work in self-organization of ant foraging might now benefit from investigations to determine whether these systems actually combine individual-level decision making and colony-level decision making.

Similarly, in the model systems we focus on in this chapter—new nest site selection by rock ants and their relatives—there is considerable controversy regarding whether decision making involves colony-level comparisons between nest sites or individual-level comparisons of nest sites or some combination of the two.[29–31] Such issues are even more intriguing when we begin to consider the possibility of colony-level learning.

Before we describe the natural history of nest choice and emigration in rock ants, let us first consider a useful working definition of 'cognition.' The definition of 'cognition' that we now favor is one kindly brought to our attention by one of the editors of this book, Prof. Menzel: "Cognition refers to all processes by which the sensory input is transformed, reduced, elaborated, stored, recovered, and used"[32] (p. 4). Given this, is it reasonable to consider that a superorganism might exhibit collective cognition—that is, cognition that is more than the sum of the separate cognitive processes of the organisms that make up the superorganism? Recall the example of the colony choosing a shorter path for foraging rather than a longer one (Figure 40.10). Individuals receive and provide sensory input associated with the strength of the pheromone trail on a particular path. However, it is the population of workers, or as shorthand the colony, that is able to choose the shorter path as a result of the pheromones accumulating more quickly on the shorter path. Thus, indeed, in albeit a rather strange and unfamiliar way, sensory input has been transformed, reduced, elaborated, stored, recovered, and used. Of course, one might balk at the issue of the form of the sensory input. Conventionally, this would mean that the results of certain sensory activities have entered the organism and presumably some memory-like processes have been involved in the transformation, reduction, elaboration, storage, recovery, and use. However, we would reason that pheromone trails, etc. can act as memory-like processes. We do not doubt that some might find this controversial. For example, one might argue that memories encoded only in neural circuits are implicit in cognition. We disagree. A written treatise as an 'aide memoire' or, more prosaically, a Post-it® note or indeed this book chapter are all examples of information stored for future use. Thus, we reason that many memory-like processes can be external to organisms or indeed superorganisms and that such processes are integral to thinking about the extended phenotypes of organisms.[33] Of course, these external memory-like processes are often coupled to, and act in concert with, neuronally based memories. Moreover, science proceeds by rigorously testing important concepts to the brink of destruction and beyond. So we advocate that thinking about colony cognition is potentially very valuable precisely because it may be highly controversial.

Returning to the issue of nest site choice in rock ants, we suggest that one of the pivotal phenomena in this process, that is at the heart of collective cognition during nest site selection, is the use of quorum thresholds.[34,35] In essence, *T. albipennis* colonies typically use the abundance of workers within alternative nest sites to help choose between them. These quorum thresholds in effect allow separate sources of information to be combined into more reliable collective information. The way in which the quorum thresholds work is as follows. Individual ants discover potential nest sites and if they find them acceptable will start to recruit to them through tandem running. Not only does the follower of a tandem run learn where a new nest site is but also it seems likely that each ant makes its own separate evaluation of the nest site to which it is taken. In this way, a tandem follower can also assess for itself whether a nest site is of sufficient quality. Tandem followers often become leaders of new tandem runs. So the process of tandem running, albeit a slow one, can also build a geometrical increase in the number of ants visiting a particular nest site. Visitors to nest sites not only assess their intrinsic qualities but also will respond to the abundance of their nestmates in a new nest site, if they exceed a quorum threshold, by switching from slow tandem running to the recruitment of nestmates by simply carrying them from the old nest site to the new one. The key reason why quorum thresholds can represent a crucial step in colony-level cognition is that typically quorum thresholds are set at a high level. Hence, this would imply that many ants independently have deemed a particular nest site as being of a sufficiently high quality to attract the full commitment of a colony to it. Indeed, quorum thresholds, when met, trigger such rapid recruitment by social carrying that an entire colony of rock ants of up to 400 workers, their single queen, plus several hundred brood can complete their emigration to a new nest within 60 min.

Before we conclude the topic of quorum thresholds and their use by house-hunting rock ants, we note that they seem to be flexible in a way that is likely to be adaptive. For example, if a colony is emigrating under particularly harsh conditions, the ants use low quorum thresholds,[36] whereas if the colony is emigrating under benign conditions, the ants typically use relatively high thresholds.[37] Thus, in harsh conditions, the ants minimize their exposure by rapidly deciding on a new nest site. By contrast, in benign conditions, such as when their old nest is intact but they are offered a much superior one, colonies deploy many individuals in slow and careful nest site evaluation. Furthermore, many ants would have been taught the route to the new nest site so they would be able to emigrate to it very quickly indeed. This work on variable quorum thresholds has demonstrated speed–accuracy tradeoffs in these decision-making systems.[36,37]

We hope the above account of nest choice by rock ants will have begun to establish that not only individuals but also colonies are very sophisticated in their behavior. We now turn to the issue of whether there might be colony-level learning in these superorganisms.

It has long been known that individual ants can learn. A classic example is the ability of ant workers to solve mazes with their nest as the 'reward'.[38] Although ants take longer than rats to learn the solution and there is no evidence they can transfer what they have learned to a reversed maze as rats can,[38] their individual learning abilities fare well when compared to those of rats considering ants have two or three orders of magnitude fewer neurons than rats. In addition, individuals in colonies of neotropical nectivorous ants are able to learn to forage at particular places at particular times.[39,40] Individual learning also plays an important role in the division of labor in ant colonies because individuals can adapt their task performance according to the positive or negative reinforcement they receive from the outcome of their labor.[41] In addition to such evidence at the behavioral level, recent studies of individual learning in ants at the neurophysiological level have demonstrated the existence of short-term and long-term memory in individual workers,[42] thus paving the way for linking learning at the collective level to the neuronal substrate of memory at the individual level (for recent results suggesting the existence of colony-wide memory in ants, see Gill *et al.*[43]).

Given that individual learning is well-established in ants, the new issue is this: Is it reasonable to hypothesize colony-level learning? Simply stated, what we mean by colony-level learning is that an entire society may change its behavior adaptively in response to its experience. Indeed, although other authors, besides Shettleworth,[2] define *social learning* broadly as "learning that is influenced by observation of, or interaction with, a conspecific, or its products",[44,45] the study of social learning has been dominated by an emphasis on imitation and little attention has been paid to the possibility that it can have effects other than the production of an observer (or learner) and a demonstrator (or observed animal).[44] Our aim in this section is to draw attention to learning at the collective level as a new, neglected outcome of social learning that results from interactions with conspecifics. The improvement of collective performance with experience[46] has already been described in the literature as social learning or a process akin to social learning.[47] Here, we will make this point explicitly.

To highlight some of the issues, recall G. C. Williams' distinction between a herd of fleet deer and a fleet herd of deer[48] (p. 16). What Williams was driving at is whether we should think of the rate of progress of a group being nothing more than the average speed of its individuals, on the one hand, or an attribute of a group as a whole, on the other hand. To demystify this, consider the herd of deer as a relatively small one composed of related individuals. Imagine that they are running away from a pack of wolves. In one case, each deer might try to run away as quickly as it possibly can, but in so doing the resulting scramble might mean that they block one another's progress so that the speed of the group is actually restricted. Contrast this with deer that are sensitive to one another's movements so that they do not impede one another. Finally, consider a group of deer that take turns to beat a path through thick snow or underbrush so that by working as a group effectively, they are much better at escaping the wolves that are snapping at their heels. In this final case, the progress of the group could be more than the sum of its parts—and here one might really be able to talk of a fleet herd of deer rather than a herd of fleet deer.

To further highlight some of the conceptual issues, consider the case of possible improvement in a football team. Is it reasonable to hypothesize that a team might learn and crucially that there might be collective processes in this learning? Thus, for example, during the course of a season, the individual players on a team might become more skillful. However, we would only wish to think of this as the *team* learning if the individuals were learning to interact with one another more skillfully. Undoubtedly, individual players can learn to hone their personal skills, but it seems likely that the very best players also learn about the qualities and abilities of their teammates. Thus, for example, in soccer or football, the best players also learn to move more effectively 'off the ball'—in other words, when they are not actually in possession of the football. In this way, they can 'create space' so that they can receive passes and the team can make its next move more successfully. Or indeed by running off the ball they can drag defenders along with them to create more space for other teammates to pass and receive the ball. So what intrigues us here is the ability of the members of a team to become increasingly more effective at the group level. To return to biology, could the members of the fleet herd of deer learn to interact more effectively to enhance the performance of the whole group?

Some key aspects of what we call colony-level learning can be emphasized by comparing our studies with that of Johnson.[49] Superficially, Johnson's study and our own may appear somewhat similar; for example, Johnson refers to 'colony memory' when examining seed harvesting efficiency in two species of desert ant. However, Johnson focused only on learning behavior of foragers as they became more efficient at handling and retrieving new seed types. By contrast, we focused on a mass action, which involves the whole colony, such as an emigration to a new nest site.[46,50,51] Johnson showed that foragers in study colonies became more efficient in terms of reduced handling times, retrieval times, and carrying more seeds per trip. He also showed that study colonies lost such skills if they could not practice them. A separate investigation of one or two individuals per colony marked with fluorescent pigment to allow identification over 8–15 consecutive trips revealed that individual learning mostly paralleled the average learning performance of foragers that were not individually marked. However, because the investigation of individuals was a separate one, Johnson could not distinguish between changes in the behavior of individual foragers and changes associated with differences in the number of foragers. Thus, it is not clear if individual foragers forgot their learned skills or if colonies lost such skilled workers over time (i.e., thorough mortalities, which Johnson states as unlikely given the time frames involved) or lost such skilled foragers to other roles within colonies. Furthermore, Johnson did not determine if workers learned to interact with one another more skillfully, or even if the whole colony learned to process new seed types better. For example, it could be that workers that stay in the nest also learn to be more efficient in processing the new seed types that forgers bring into the nest. Thus, even though Johnson's study is important and pioneering, his use of the term 'colony memory' is misleading in much the same way as the confusion of a herd of fleet deer with a fleet herd of deer[48] (p. 16). Johnson inferred learning by individuals within specific colonies. He did not show that whole colonies can learn.

Given such thoughts, we wished to investigate whether emigrating rock ant colonies could learn to emigrate more effectively and whether this could be attributed simply to changes in individuals or to the way in which individuals interact with one another.

Langridge *et al.*[46] showed that *T. albipennis* colonies can improve their collective performance progressively when they have to repeat the same process, namely colony emigration. Typically, colonies get quicker at relocating themselves to a new nest site over repeated emigrations (Figure 40.11). Crucially, Langridge *et al.* showed that such improvement was experience-based and involved some memory-like process rather than a concomitant change in development. A key demonstration in this study was that the benefits of experience can be lost (i.e., forgotten) if the interval between successive emigrations is too long (Figure 40.12). Why are these last two points so important? They hinge on the following logic. To really demonstrate learning, one has to demonstrate forgetting. Why? Because demonstrating forgetting helps to exclude the possibility that an improvement through time occurred simply because of some non-memory-based concurrent change over the same time period. For example, older children typically run faster than younger ones, and one might attribute this to their larger size and strength rather than them learning to run more quickly. So a child's improvement in this task would be attributable to a developmental process rather than a memory-based process.

Langridge *et al.*[46] also showed that the benefits of experience in *T. albipennis* emigrations are more likely to be retained over a longer period if the collective performance has been repeated several times (Figure 40.12). The next question was whether this collective improvement in performance was based only on the summation of more effective activity of individuals or whether it also involved improvements in the ways in which the ants interacted with one another. In other words, has the colony as a whole learned to be more fleet or, on the basis of experience, are colonies composed simply of more fleet individuals. In yet other words, has learning helped the whole to become more than the sum of its parts?

Langridge *et al.*[50] next focused on how *T. albipennis* colonies improved their emigration speeds in terms of progressive changes in the behavior of individual workers. This is a major task because all of the workers in several colonies have to be identified with multiple, multicolored paint marks, and changes in their behavior over many tasks need to be quantified over repeated emigrations. Langridge *et al.* showed that (1) those transporters of nestmates that also transported in the preceding emigration began to transport earlier in the current emigration and, in the majority of emigrations, transported more items than those transporters that had not transported in the preceding emigration; (2) the time that elapsed before the very first item was transported into the new nest as well as the intervals between the first items transported by individual transporters were reduced over successive emigrations (Figure 40.13), and the very first item was, in the majority of emigrations, carried by a transporter that had also transported in the preceding emigration; and (3) the number of adults that were transported was reduced over successive emigrations (Figure 40.14). These results strongly suggested that the behavior of transporters was greatly influenced by their recent prior experience if they took an active role in an earlier emigration.

It is perhaps not surprising that reduced emigration times were associated with changes in the behavior of those ants that transport nestmates during emigrations. Only small time savings are likely to be made in the discovery and assessment of a new nest site because thoroughness is required for the ants to make accurate decisions and they take many variables into account. By contrast, *T. albipennis* colonies frequently have many more brood items than adult workers, and all of these items, plus the queen and the rather passive

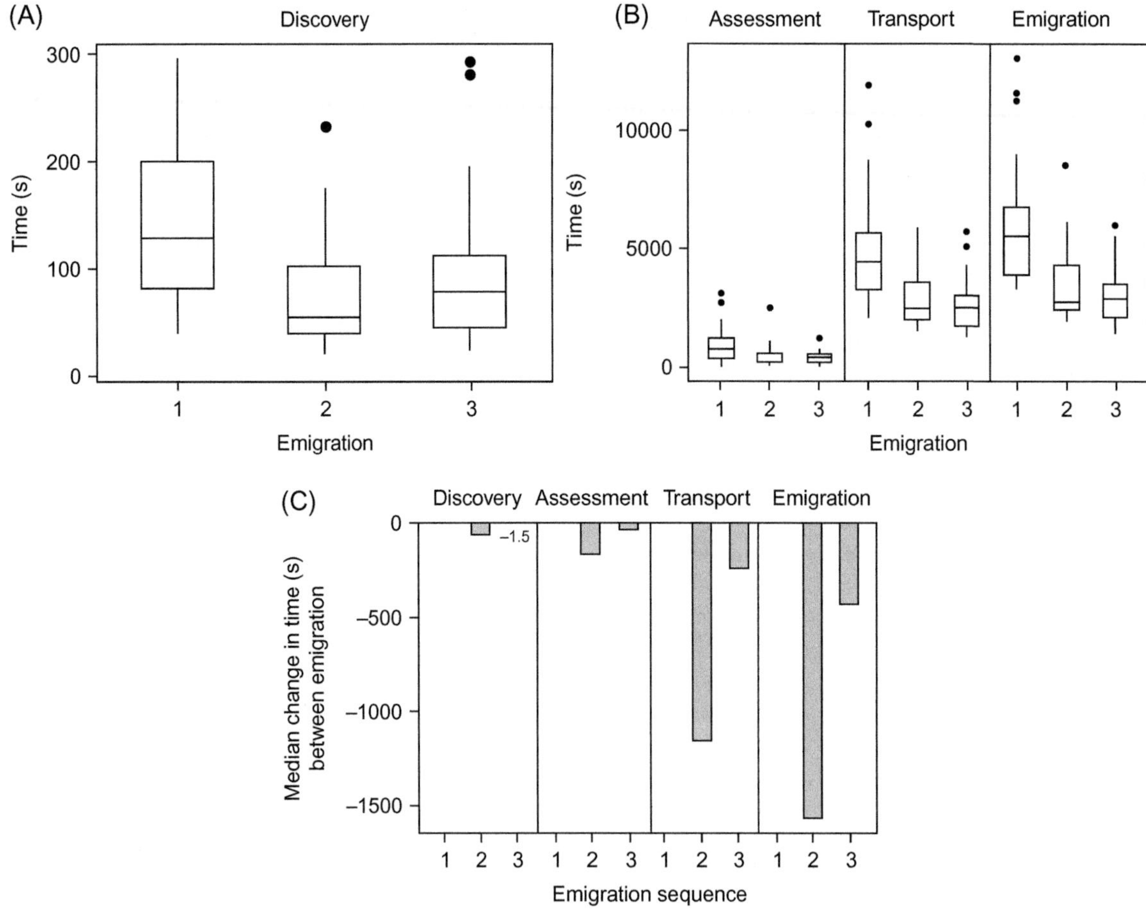

FIGURE 40.11 Results for three successive emigrations of 30 colonies. (A) Discovery time (from when the old nest was opened until the first worker entered the new nest). (B) Assessment time (from when the first worker entered the new nest until the first adult or brood item was transported into the new nest), transport time (from when the first adult or brood item was transported into the new nest until the last brood item was transported into the new nest), and emigration time (from when the old nest was opened until the last brood item was transported into the new nest). The numbers 1, 2, and 3 refer to the first, second, and third emigration, respectively. Box plots are drawn as in Figure 40.8, except that outliers are marked with a black circle. (C) Median change in discovery, assessment, transport, and emigration time between emigrations. The value of the median change in time for discovery between emigrations 2 and 3 is shown as a number due to its small size. Source: *Adapted from* Behavioral Ecology and Sociobiology *56, 2004, p. 526, Improvement in collective performance with experience in ants, E. A. Langridge, N. R. Franks, and A. B. Sendova-Franks, Fig. 2, with kind permission from Springer Science + Business Media.*

brood workers, are carried to a new nest during an emigration. Thus, the relatively small minority of workers, that discover the new nest and determine that it is suitable, may have to move almost 1000 items (brood, passive workers, and the queen) to the new nest. When workers are transporting brood or adults to a new nest site, they typically run at a high and constant speed, and they tend to converge on fairly direct routes. Therefore, improvements in transport performance are probably largely associated with how many journeys an ant makes per unit time, and this may be mostly influenced by their behavior in the old and new nest sites. Langridge *et al.*[51] indeed showed that transporters spent less time at the old and new nests in successive emigrations.

Transporters expedited choosing and picking up brood items at the old nest and depositing them in the new nest. Overall, such improvements were usually associated with brood transport. Brood transporters visited several locations in the new nest before depositing their load but did not interact with other adults. Therefore, reductions in depositing times for brood transported into the new nest represented the sum of time savings made by individual transporters. By contrast, transporters at the old nest spent most of their time interacting with other adults before picking up a brood item. The frequency of such interactions did not decline with successive emigrations. Therefore, we suggest that the overall reduction in brood picking-up times at the old nest occurred because transporters

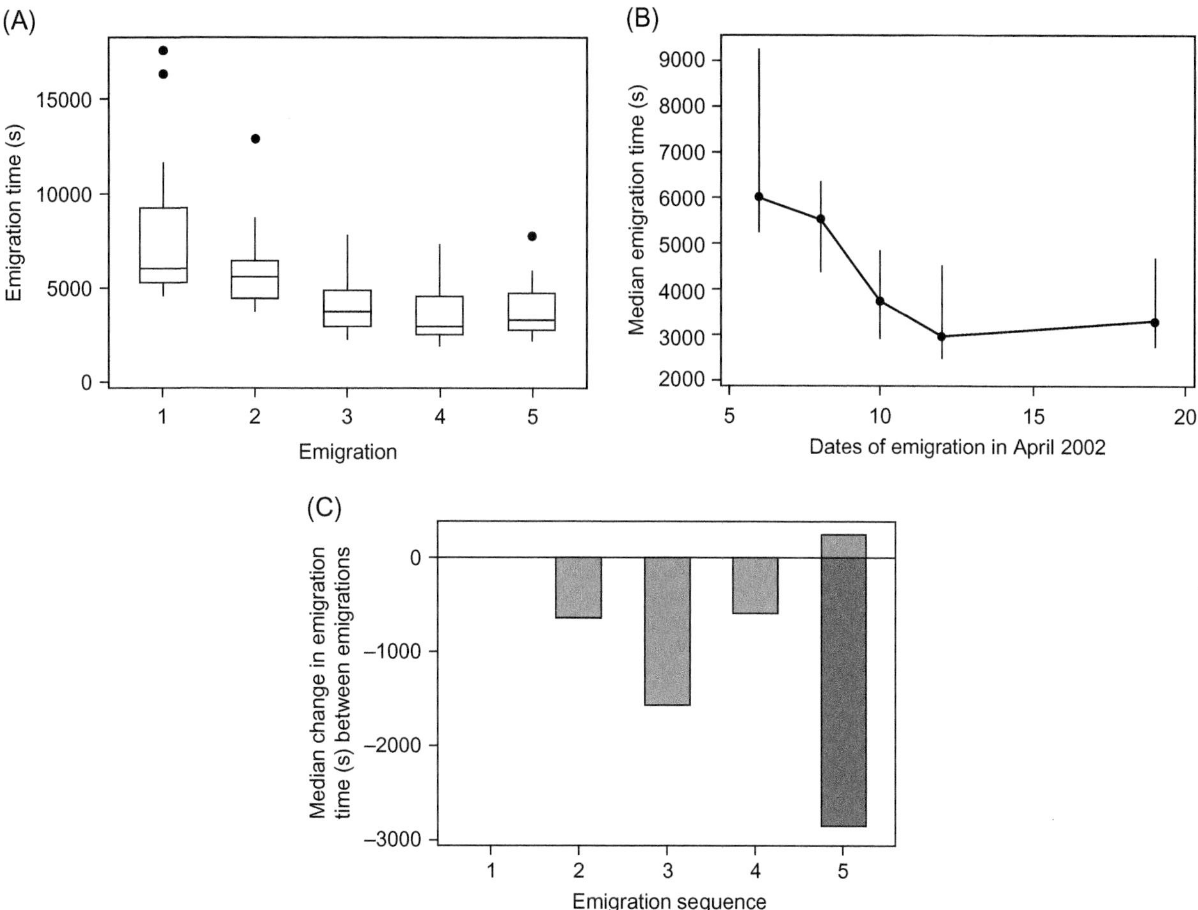

FIGURE 40.12 **Results for five successive emigrations of 13 colonies.** (A) Box plots of emigration times. Colonies were emigrated four successive times with 1-day intertrial intervals and then for a fifth time after a 6-day interval. The box plots are drawn as in Figure 40.11. (B) Median emigration time plotted against date of emigration to illustrate the relationship between the intertrial interval and the subsequent emigration time. Lines denote the interquartile range. (C) Median change in emigration time between emigrations. Median changes between emigrations 2 and 1, 3 and 2, 4 and 3, and 5 and 4 are shown as solid bars; median change between emigrations 5 and 1 is shown as a darker bar. Emigration 4 was significantly faster than emigration 3, which was significantly faster than emigration 2, which was significantly faster than emigration 1. By contrast, emigration 5 was not significantly faster than emigration 4 but was still significantly faster than emigration 1, showing that 6 days is sufficient for colonies to forget a single earlier experience but that memories could be reinforced with repeated experience. Source: *Adapted from* Behavioral Ecology and Sociobiology *56, 2004, p. 528, Improvement in collective performance with experience in ants, E. A. Langridge, N. R. Franks, and A. B. Sendova-Franks, Fig. 3, with kind permission from Springer Science + Business Media.*

modified their behaviors so that time at the old nest was saved as a result of adults reducing the duration of interactions with nestmates.[51]

One distinct possibility is that over successive emigrations, transporters of brood can pick up brood more quickly at the old nest because the brood care workers learn to relinquish it more quickly. In all of the nest emigration experiments of Langridge *et al.*,[46,50,51] the old nest was destroyed and hence the brood in that site was vulnerable. In the field, such brood can be captured by other ants and carried away. One role of brood care workers in a destroyed nest, therefore, is to guard the brood, and they often do so by literally straddling the largest and most valuable brood items (i.e., fully grown larvae and pupae). Strong evidence for the role of nest workers in guarding defenseless nestmates during emigrations comes from our discovery that the colony's single queen is typically moved in the middle of the emigration.[52] Thus, she is moved when the population of workers in the new nest is roughly equal to the population of workers in the old nest. The queen is the colony's only 'vital organ' (i.e., irreplaceable component), so moving her in the middle of an emigration means that her protection through a safety in numbers is maximized. Brood items are actively guarded in a destroyed nest by brood workers, and it seems that such workers learn, over successive emigrations, to release them to transporting workers more rapidly, thus making the task of brood transporters much quicker. In this way,

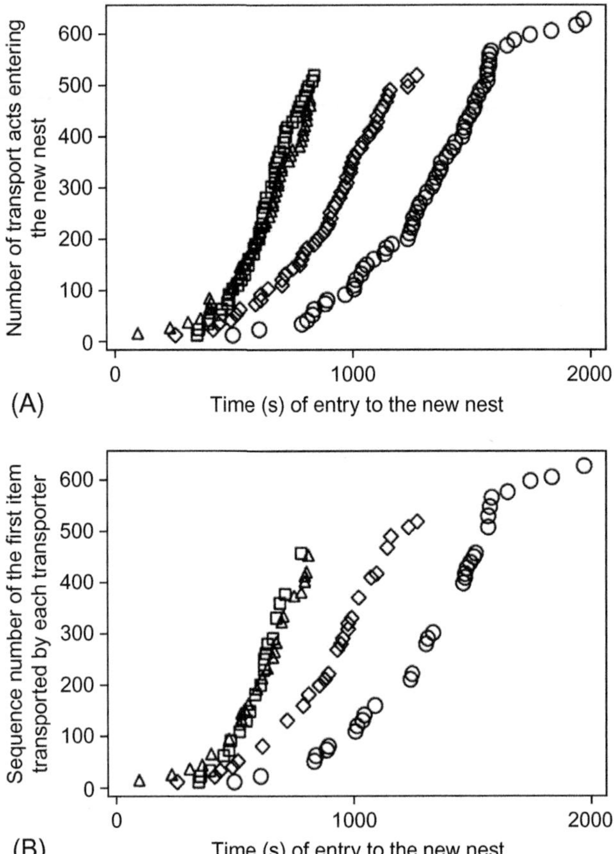

FIGURE 40.13 The four successive emigrations of one colony illustrating a behavior common to all five studied colonies, namely that in any emigration, individual transporters can begin to transport at any point during the course of that emigration. (A) Number of transport acts (brood and adults) recorded when entering the new nest in relation to the time elapsed since the old nest was opened at $t = 0$. (B) Sequence number of the first item transported into the new nest by each transporter active in each emigration (note that the intervals between first items reduce over successive emigrations). The last point in each emigration represents the last worker to begin transporting. Circles, first emigration; diamonds, second emigration; squares, third emigration; triangles, fourth emigration. Source: *Adapted from* Behavioral Ecology and Sociobiology *62, 2008, p. 452, How experienced individuals contribute to an improvement in collective performance in ants, E. A. Langridge, A. B. Sendova-Franks, and N. R. Franks, Fig. 5, with kind permission from Springer Science + Business Media.*

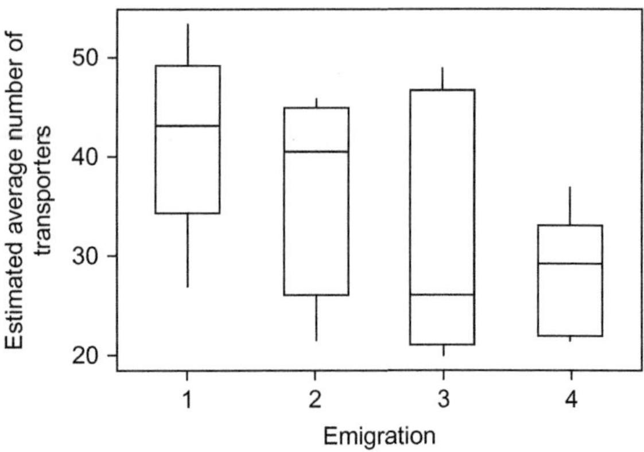

FIGURE 40.14 Estimated average number of transporters for successive emigrations of the same five colonies mentioned in the legend to Figure 40.13. Source: *Adapted from* Behavioral Ecology and Sociobiology *62, 2008, p. 450, How experienced individuals contribute to an improvement in collective performance in ants, E. A. Langridge, A. B. Sendova-Franks, and N. R. Franks, Fig. 2a, with kind permission from Springer Science + Business Media.*

workers with different roles seem to learn to interact with one another more quickly and effectively, and it is these enhanced interactions that help the colony to learn to emigrate more quickly.

To return to our football analogy, it is as if the players on a team are learning, over successive games, to trust one another and hence to pass the ball to one another more readily—and as every football fan might acknowledge, such behavior can increase the tempo of a team's performance massively.

CONCLUSION

Our goal in this chapter was to examine in detail two possibly controversial examples of social learning in ant colonies. One was the first formal demonstration of teaching in any nonhuman animal; the other was an analysis of colony-level learning. The two examples we chose were taken from our own laboratory. We have done this not only for egotistical reasons but also because a chapter such as this provides an unrivalled opportunity to explain our motivation and approach to exploring certain aspects of ant biology. Much more importantly, however, we believe that these examples illustrate in general that certain social insect colonies are model systems for understanding how relatively simple problem-solving algorithms coupled to learning and plasticity in behavior can generate great sophistication. This emphasizes our overarching point that the comparative neglect of social learning in insects in general and social insects in particular has held back progress in this fascinating field within biology as a whole. We are convinced that a new focus on social learning in diverse insect societies will be important not only in contributing to sociobiology but also in pointing to general biological principles.

References

1. Maynard-Smith J, Szathmary E. *The Major Transitions in Evolution.* Oxford: Oxford University Press; 1997.
2. Shettleworth SJ. *Cognition, Evolution, and Behavior.* 2nd ed. Oxford: Oxford University Press; 2010.
3. Leadbeater E, Chittka L. Social learning in insects: from miniature brains to consensus building. *Curr Biol.* 2007;17:R703–R713.

4. Caro TM, Hauser MD. Is there teaching in nonhuman animals? *Q Rev Biol.* 1992;67:151–174.
5. Franks NR, Richardson T. Teaching in tandem-running ants. *Nature.* 2006;439:153.
6. Möglich M. Social organization of nest emigration in *Leptothorax* (Hym., Form.). *Insectes Soc.* 1978;25:205–225.
7. Guénard B, Silverman J. Tandem carrying, a new foraging strategy in ants: description, function, and adaptive significance relative to other described foraging strategies. *Naturwissenschaften.* 2011;98:651–659.
8. Leadbeater E, Raine NE, Chittka L. Social learning: ants and the meaning of teaching. *Curr Biol.* 2006;16:R323–R325.
9. Nieh JC. A negative feedback signal that is triggered by peril curbs honey bee recruitment. *Curr Biol.* 2010;20:310–315.
10. Seeley TD, Visscher PK, Schlegel T, Hogan PM, Franks NR, Marshall JAR. Stop signals provide cross inhibition in collective decision-making by honeybee swarms. *Science.* 2012;335:108–111.
11. Premack D, Premack AJ. Why animals lack pedagogy and some cultures have more of it than others. In: Olson DR, Torrance N, eds. *The Handbook of Human Development and Education.* Oxford: Blackwell; 1996:302–344.
12. Richardson TO, Sleeman PA, McNamara JM, Houston AI, Franks NR. Teaching with evaluation in ants. *Curr Biol.* 2007; 17:1520–1526.
13. Hölldobler B, Wilson EO. *The Superorganism: The Beauty, Elegance and Strangeness of Insect Societies.* New York: Norton; 2009.
14. Wilson EO. Communication by tandem running in the ant genus *Cardiocondyla. Psyche.* 1959;66:29–34.
15. Stuart JR. An early record of tandem running in Leptothoracine ants: Gottfrid Adlerz, 1896. *Psyche.* 1986;93:103–106.
16. Hölldobler B, Traniello JFA. Tandem running pheromone in ponerine ants. *Naturwissenshaften.* 1980;67:360.
17. Hölldobler B, Möglich M, Maschwitz U. Communication by tandem running in the ant *Camponotus sericeus. J Comp Physiol.* 1974;90:105–127.
18. Franklin EL, Richardson TO, Sendova-Franks AB, Robinson EJH, Franks NR. Blinkered teaching: tandem running by visually impaired ants. *Behav Ecol Sociobiol.* 2011;65:569–579.
19. Franklin EL, Robinson EJH, Marshall JAR, Sendova-Franks AB, Franks NR. Do ants need to be old and experienced to teach? *J Exp Biol.* 2012;215:1287–1292.
20. Franks NR, Richardson TO, Keir S, Inge SJ, Bartumeus F, Sendova-Franks AB. Ant search strategies after interrupted tandem runs. *J Exp Biol.* 2010;213:1697–1708.
21. Thornton A, McAuliffe K. Teaching in wild meerkats. *Science.* 2006;313:227–229.
22. Berg HC. *Random Walks in Biology.* Princeton, NJ: Princeton University Press; 1983.
23. Pratt SC, Brooks SE, Franks NR. The use of edges in visual navigation by the ant *Leptothorax albipennis. Ethol.* 2001;107:1125–1136.
24. Srinivasan MV. Honey bees as a model for vision, perception, and cognition. *Annu Rev Entomol.* 2010;55:267–284.
25. Sendova-Franks AB, Franks NR. Social resilience in individual worker ants and its role in division of labour. *Proc R Soc B.* 1994;256:305–309.
26. Marshall JAR, Franks NR. Colony-level cognition. *Curr Biol.* 2009;19:R395–R396.
27. Goss S, Aron S, Deneubourg JL, Pasteels JM. Self-organized shortcuts in the Argentine ant. *Naturwissenschaften.* 1989;76:579–581.
28. Deneubourg JL, Goss S. Collective patterns and decision-making. *Ethol Ecol Evol.* 1989;1:295–311.
29. Edwards SC, Pratt SC. Rationality in collective decision-making by ants. *Proc R Soc B.* 2009;276:3655–3661.
30. Robinson EJH, Smith FD, Sullivan KME, Franks NR. Do ants make direct comparisons? Proc R Soc *B.* 2009;276:2635–2641.
31. Sasaki T, Pratt SC. Emergence of group rationality from irrational individuals. *Behav Ecol.* 2011;22:276–281.
32. Neisser U. *Cognitive Psychology.* New York: Appleton-Century-Crofts; 1967.
33. Dawkins R. *The Extended Phenotype.* Oxford: Oxford University Press; 1982.
34. Pratt SC. Quorum sensing by encounter rates in the ant *Temnothorax albipennis. Behav Ecol.* 2005;16:488–496.
35. Pratt SC, Mallon EB, Sumpter DJT, Franks NR. Quorum sensing, recruitment, and collective decision-making during colony emigration by the ant *Leptothorax albipennis. Behav Ecol Sociobiol.* 2002;52:117–127.
36. Franks NR, Dornhaus A, Fitzsimmons JP, Stevens M. Speed versus accuracy in collective decision-making. *Proc R Soc B.* 2003; 270:2457–2463.
37. Dornhaus A, Franks NR, Hawkins RM, Shere HNS. Ants move to improve: colonies of *Leptothorax albipennis* emigrate whenever they find a superior nest site. *Anim Behav.* 2004;67: 959–963.
38. Schneirla TC. Modifiability in insect behavior. In: Roeder KD, ed. *Insect Physiology.* New York: Wiley; 1953:723–747.
39. Harrison JM, Breed MD. Temporal learning in the giant tropical ant, *Paraponera clavata. Physiol Ent.* 1987;12:317–320.
40. Schatz B, Lachaud J-P, Beugnon G. Spatio-temporal learning by the ant *Ectatomma ruidum. J Exp Biol.* 1999;202: 1897–1907.
41. Ravary F, Lecoutey E, Kaminski GI, Châline N, Jaisson P. Individual experience alone can generate lasting division of labor in ants. *Curr Biol.* 2007;17:1–6.
42. Guerrieri FJ, d'Ettorre P, Devaud J-M, Giurfa M. Long-term olfactory memories are stabilised via protein synthesis in *Camponotus fellah* ants. *J Exp Biol.* 2011;214:3300–3304.
43. Gill KP, van Wilgenburg E, Taylor P, Elgar MA. Collective retention and transmission of chemical signals in a social insect. *Naturwissenschaften.* 2012;99:245–248.
44. Heyes CM. Social learning in animals: categories and mechanisms. *Biol Rev.* 1994;69:207–231.
45. Hoppitt W, Laland KN. Social processes influencing learning in animals: a review of the evidence. *Adv Stud Behav.* 2008;38: 105–165.
46. Langridge EA, Franks NR, Sendova-Franks AB. Improvement in collective performance with experience in ants. *Behav Ecol Sociobiol.* 2004;56:523–529.
47. Coolen I, Dangles O, Casas J. Social learning in noncolonial insects? *Curr Biol.* 2005;15:1931–1935.
48. Williams GC. *Adaptation and Natural Selection.* Princeton, NJ: Princeton University Press; 1966.
49. Johnson RA. Learning, memory, and foraging efficiency in two species of desert seed-harvester ants. *Ecology.* 1991;72: 1408–1419.
50. Langridge EA, Sendova-Franks AB, Franks NR. How experienced individuals contribute to an improvement in collective performance in ants. *Behav Ecol Sociobiol.* 2008;62: 447–456.
51. Langridge EA, Sendova-Franks AB, Franks NR. The behaviour of ant transporters at the old and new nests during successive colony emigrations. *Behav Ecol Sociobiol.* 2008; 62:1851–1861.
52. Franks NR, Sendova-Franks AB. Queen transport during ant colony emigration: a group-level adaptive behaviour. *Behav Ecol.* 2000;11:315–318.

4.4.4 Other Insect Species

CHAPTER

41

Olfactory and Visual Learning in Cockroaches and Crickets

Makoto Mizunami, Yukihisa Matsumoto*, Hidehiro Watanabe† and Hiroshi Nishino**

***Hokkaido University, Sapporo, Japan †Fukuoka University, Fukuoka, Japan**

INTRODUCTION

Insects, like other animals, live in constantly changing environments, in which the availability of food sources varies over time. Therefore, the ability to learn to associate a certain cue with an abundant food source must be of great significance for all species of insects. In addition, many species of insects possess territory or a constant foraging area, and they have developed the capability to perceive and learn conspicuous landmarks in the environment. Previous studies on learning and memory in insects have focused on only two species—the fruit fly *Drosophila melanogaster*[1,2] and the honeybee *Apis mellifera*.[3–6] Therefore, studies on other insect species are needed to understand the general principles of learning and memory in insects and their species-specific adaptations. Indeed, studies on crickets and cockroaches, as well as other species of insects such as ants,[7] bumblebees,[8] moths,[9] and parasitoid wasps,[10] have helped researchers to appreciate the richness of learning and memory capabilities of insects and the diversity of their neural mechanisms. In this chapter, we review recent progress in the study of olfactory and visual learning in crickets and cockroaches, focusing on our own research.

OLFACTORY AND VISUAL LEARNING IN CRICKETS

In an earlier stage of the study of learning in crickets, a Y-maze learning procedure was used to examine the effects of injections of amino acids and opioids into the hemolymph on memory acquisition and consolidation.[11,12] Subsequent establishment of an operant conditioning procedure[13] and a classical conditioning procedure[14] facilitated quantitative analysis of learning and memory. In this section, we focus on findings obtained using the classical conditioning procedure.

Experimental Procedure

We developed a 'classical conditioning and operant testing procedure' for crickets that is based on the transfer of memory formed during classical conditioning to an operant testing situation.[14,15] For conditioning, crickets were individually placed in a beaker. A filter paper soaked with an odor (conditioned stimulus (CS)) was placed within 2 cm of the head of the cricket for 2 sec, and then a drop (4.5–5 μL) of water or 20% sodium chloride solution (appetitive or aversive unconditioned stimulus (US)) was applied to the mouth (Figure 41.1A). In the operant odor preference test, crickets were individually placed in a test chamber and allowed to freely visit two odor sources on the floor (Figure 41.1B). The amount of time that the crickets explored each odor source with the mouth or palpi was measured to evaluate relative odor preference of the crickets.

In most experiments, crickets were subjected to an absolute appetitive or aversive conditioning procedure, in which an odor (CS+ or CS−) was paired with appetitive US or aversive US, and their preferences between the CS and a control odor were tested before and after conditioning. In other experiments, we used

Invertebrate Learning and Memory.
DOI: http://dx.doi.org/10.1016/B978-0-12-415823-8.00041-1

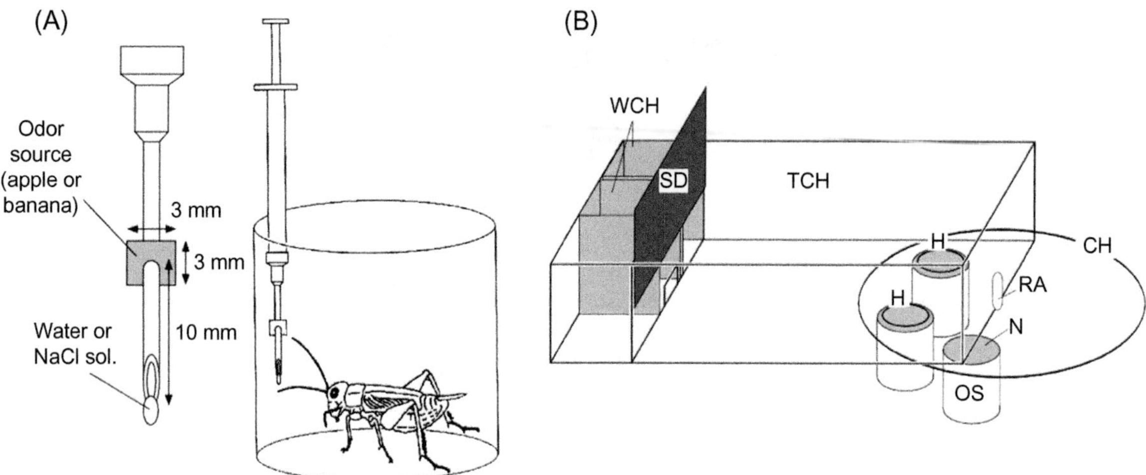

FIGURE 41.1 Procedures for olfactory conditioning in crickets. (A) For appetitive or aversive conditioning, an odor (e.g., apple or banana odor) was paired with water (appetitive US) or 20% sodium chloride solution (aversive US). A syringe containing water or sodium chloride solution was used for US delivery. A filter paper soaked with apple or banana essence was attached to the needle of the syringe. The filter paper was placed near the cricket's antennae, and then water or sodium chloride solution was presented to the mouth. (B) Apparatus for the odor preference test. Two holes (H) connecting the chamber with odor sources (OS) were inserted in the floor of the test chamber (TCH). Each odor source consisted of a container with a filter paper soaked with apple or banana essence, covered with fine gauze net (N). Three containers were mounted on a rotating container holder (CH), and two of three odor sources could be presented at the same time. A cricket was placed in the waiting chamber (WCH) for 4 min for acclimation and then allowed to enter the test chamber to visit odor sources by opening a sliding door (SD). Two minutes later, the relative positions of the apple and banana sources were changed. The preference test lasted for 4 min. RA, rotating axle. Source: *Modified from Matsumoto and Mizunami.*[14]

a differential conditioning procedure, in which an odor (CS+) was paired with appetitive US and another odor (CS−) was paired with aversive US.

For visual pattern conditioning, either a white-center and black-surround pattern or a black-center and white-surround pattern was presented for 2 sec and then water or sodium chloride solution was presented to the mouth.[15] In the visual pattern preference test, two visual patterns were simultaneously presented on the wall of the test chamber, and the amount of time that the crickets touched each of the patterns was measured to evaluate relative preference between the two patterns.[15]

Olfactory Learning in Crickets

Crickets exhibited excellent olfactory learning. In appetitive olfactory conditioning, for example, one conditioning trial was sufficient to establish a memory lasting for several hours (midterm memory; Figure 41.2A).[16] Two[15] or four (Figure 41.2A[16]) conditioning trials (with a 5-min intertrial interval (ITI)) induced memory that lasted for at least 1 day, which reflects a protein synthesis-dependent long-term memory (LTM).[17] After aversive olfactory conditioning, one trial was sufficient to establish 30-min retention, and six trials with a 5-min ITI were needed to establish 1-day retention.[15] Our subsequent study demonstrated that the time course of memory retention after aversive conditioning and that after appetitive conditioning are fundamentally different in crickets.[18]

Next, we examined olfactory learning of crickets with respect to (1) durability of memory retention[19]; (2) capacity of memory storage[20]; and (3) higher order learning, namely context-dependent discrimination learning[21] and second-order conditioning.[22]

First, we showed that crickets retain memory for life.[19] Previously, convincing reports of lifetime memory retention in insects were limited to adults of honeybees.[23] Third- or fourth-instar nymphal crickets were trained to associate one odor (CS+) with water and another odor (CS−) with sodium chloride solution. Six and 10 weeks after training, adult crickets exhibited higher preferences for CS+ compared to CS−. The learned preference was altered when they were given reversal training 6 weeks after training.

Next, we investigated whether crickets simultaneously memorize seven odor pairs.[20] Fourteen odors were grouped into seven A/B pairs, and crickets in one group were subjected to differential conditioning to associate A odors with water (appetitive US) and B odors with sodium chloride solution (aversive US) for all seven pairs. Crickets in another group were trained with the opposite stimulus arrangement. Crickets in both groups exhibited significantly greater preference

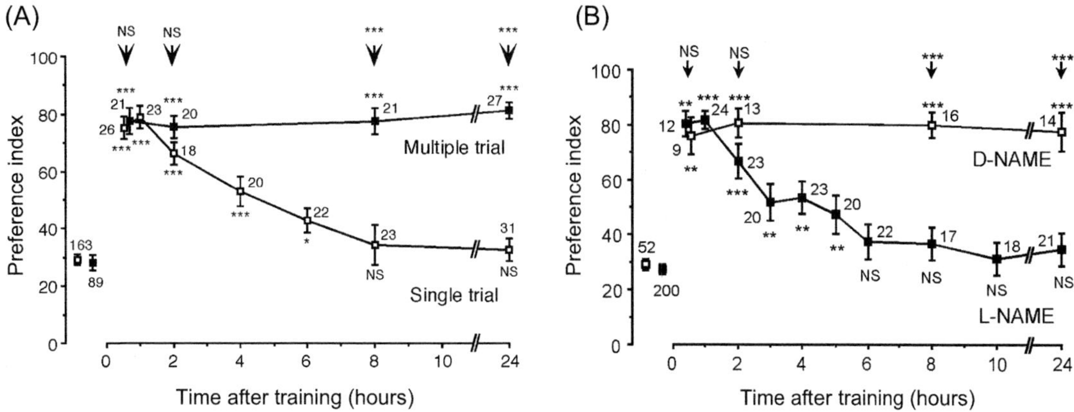

FIGURE 41.2 (A) Retention scores after single- and multiple-trial appetitive olfactory conditioning. Seven groups of animals were subjected to single-trial conditioning (open squares) and another 4 groups were subjected to four-trial conditioning, with an ITI of 5 min (solid squares). (B) Effects of L-NAME, an inhibitor of NO synthase, or D-NAME, a noneffective isomer, on LTM formation. Twenty minutes prior to the four-trial conditioning, 10 groups were each injected with 3 μL saline containing 400 μM L-NAME (solid squares) and animals in another 4 control groups were each injected with 3 μL saline containing 400 μM D-NAME (open squares). Odor preference tests were given to all animals before and at various times after conditioning. Preference indexes for the rewarded odor are shown as means $\pm$ SEM. To simplify the figure, the preference indexes (PIs) before conditioning are shown as pooled data for each category of animal groups. Statistical comparisons of odor preferences were made before and after conditioning for each group (Wilcoxon's test) and between single- and multiple-trial groups at each time after conditioning (Mann–Whitney test), and the results are shown at each data point and above the arrows, respectively: *$p<0.05$; **$p<0.01$; ***$p<0.001$; NS, $p>0.05$. The number of animals is shown at each data point. The preferences for rewarded odor remained unchanged from 30 min to 24 hr after conditioning in the multiple-trial group ($p>0.05$, Mann–Whitney test). Source: *Modified from Matsumoto* et al.[16]

for the odors associated with appetitive US compared to the odors associated with aversive US for all seven odor pairs, demonstrating that crickets can retain memory of seven pairs of odors at the same time.

We then studied whether crickets select one of a pair of odors and avoid the other in one context and the opposite pairing in another context.[21] One group of crickets received differential conditioning to associate one of a pair of odors (CS1) with water and another odor (CS2) with sodium chloride solution under illumination and to the reversed pair in the dark (CS1 with aversive US and CS2 with appetitive US). Another group of crickets received training of the opposite stimulus arrangement. One day after completion of the 3-day training, the former group preferred CS1 over CS2 under illumination but preferred CS2 over CS1 in the dark, and the latter group exhibited the opposite odor preference. Results of control experiments showed that background light conditions had no significant effects on memory formation or retrieval unless the background light was explicitly associated with US during training. We concluded that crickets use visual context stimuli to disambiguate the meaning of CSs and to predict USs.

Second-order conditioning is a procedure for testing whether a stimulus (CS1) can acquire the reinforcing properties of a US by conditioning with the US and whether the stimulus can support a new conditioning thereafter. In our experiment, an olfactory stimulus (CS1) was paired with water or sodium chloride solution (US), and then a visual stimulus (CS2) was paired with the CS1. Crickets exhibited significantly changed preference for the CS2 that had never been paired with the US, indicating that second-order conditioning was successful.[22]

Whether and to what extent insects perform social or observational learning are interesting topics in neurobiology and behavioral biology (see Chapters 29 and 40). Coolen *et al.*[24] showed that wood crickets, *Nemobius sylvestris*, exhibit predator-avoidance behavior after having observed other crickets being attacked by a spider. This finding, as well as findings in other species of insects, including honeybee foragers transferring information about the location of a profitable food source to nestmates by waggle dances[25] and foraging bumblebees copying flower choice of other individuals,[8] raise the possibility that social learning is more widespread in insects than previously considered.

Role of NO–cGMP Signaling in the Formation of Long-Term Memory

Studies in honeybees, *A. mellifera*, suggested that the nitric oxide (NO)–cyclic GMP signaling pathway and

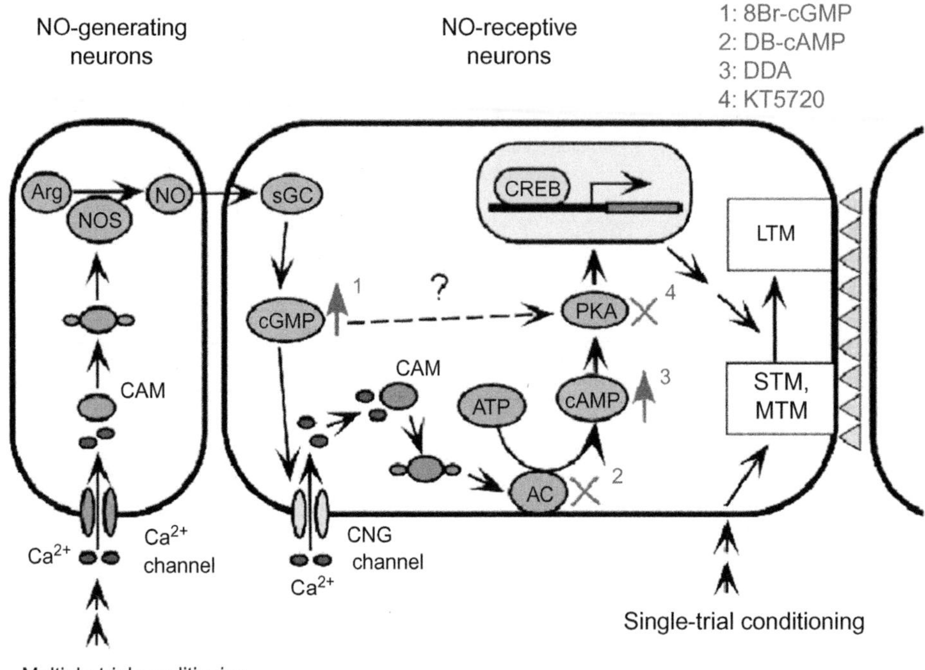

FIGURE 41.3 **A scheme of the experimental design for determining parallel or serial arrangement of the NO–cGMP system and the cAMP system for stimulating PKA for LTM formation in crickets.** The points of pharmacological manipulation (1–4) are indicated for experiments aiming to study whether cGMP activates PKA (dotted line with a question mark). Single-trial conditioning induces only short-term synaptic plasticity that underlies amnesic treatment-sensitive short-term memory (STM) and amnesic treatment-resistant midterm memory (MTM).[17] Multiple-trial conditioning activates the NO–cGMP system, and this in turn activates adenylyl cyclase (AC) and then PKA via the cyclic nucleotide-gated (CNG) channel and calcium/calmodulin (CAM) system. Activation of PKA is assumed to activate a transcription factor, cAMP responsive element binding protein (CREB), which leads to protein synthesis that is necessary to achieve long-term plasticity of synaptic connection (gray triangles) in other neurons assumed to be necessary for LTM. The drugs (1–4) shown at the top right were used to inhibit (marked by X) or activate (marked by up arrows) the elements of the biochemical systems. Arg, arginine; NOS, NO synthase; sGC, soluble guanylyl cyclase. Source: *Modified from Matsumoto* et al.[29]

the cAMP pathway act in parallel and are complementary for the formation of LTM[26,27] (see Chapter 31). We studied whether this is also the case in crickets.[16] NO is a membrane-permeable molecule that functions in intercellular signaling. It is produced by NO synthase (NOS), diffuses into neighboring cells, and stimulates soluble guanylyl cyclase to produce cGMP.[28] In crickets, multiple (two or more) appetitive olfactory conditioning trials lead to LTM that lasts for at least 1 day, whereas memory induced by single-trial conditioning decays within several hours, as mentioned previously (Figure 41.2A). Injection of inhibitors of the enzyme catalyzing the formation of NO, cGMP, or cAMP into hemolymph prior to multiple-trial conditioning blocked LTM (e.g., Figure 41.2B), whereas injection of an NO donor, a cGMP analog, or a cAMP analog prior to single-trial conditioning induced LTM, suggesting participation of the NO–cGMP pathway and the cAMP pathway in LTM formation. Induction of LTM by injection of an NO donor or a cGMP analog paired with single-trial conditioning was blocked by inhibition of the cAMP pathway, but induction of LTM by a cAMP analog was unaffected by inhibition of the NO–cGMP pathway, suggesting the cAMP pathway is a downstream target of the NO–cGMP pathway for LTM formation. Inhibitors of the cyclic nucleotide-gated (CNG) channel or calmodulin blocked induction of LTM by the cGMP analog paired with single-trial conditioning, but they did not affect induction of LTM by the cAMP analog. The results suggest that the CNG channel and calcium/calmodulin are downstream targets of the NO–cGMP pathway and are upstream of the cAMP pathway (Figure 41.3).

In honeybees, the NO–cGMP system and the cGMP system have been suggested to converge on cAMP-dependent protein kinase A (PKA), based on the *in vitro* finding that cGMP activates PKA when a suboptimal dose of cAMP is present[27] (see Chapter 31). Following this line of argument, we performed experiments to confirm that the NO–cGMP system and the cGMP system are arranged in series in crickets (Figure 41.3). First, we compared the effect of multiple-trial conditioning with the effect of an externally applied cGMP analog prior to single-trial conditioning on LTM formation, in the presence of a suboptimal dose of a cAMP analog and under the condition in which adenylyl cyclase was inhibited (Figure 41.3).[29] The results suggest that an externally applied cGMP analog activates PKA when a suboptimal dose of a cAMP analog is present, as has been suggested in honeybees, but cGMP produced by multiple-trial conditioning could not activate PKA when adenylyl cyclase was inhibited, even when a suboptimal dose of a cAMP analog was present. The results indicate that cGMP produced by multiple-trial conditioning is not accessible to PKA. We conclude that the NO–cGMP system stimulates the cAMP system for LTM formation in crickets.

We also demonstrated that RNA interference (RNAi) of the *NOS* gene impairs LTM formation in crickets.[30] *In situ* hybridization demonstrated a high level of expression of *NOS* mRNA in a subset of Kenyon cells of the mushroom body, in addition to some neurons around the antenna lobe and the base of the optic lobe. The mushroom body has been suggested to play roles in olfactory learning.[1,2,5,6] Crickets injected with double-stranded RNA (dsRNA) into the hemolymph 2 days before conditioning exhibited complete impairment of 1-day memory retention, although 30-min retention was intact. This impairment was fully rescued by injection of an NO donor, suggesting that *NOS* dsRNA acts through inhibition of NOS. We thus conclude that silencing of *NOS* expression by systemic RNAi specifically impairs LTM formation. RNAi is a promising method for study of the molecular mechanisms of learning and memory in crickets.

Roles of Dopaminergic and Octopaminergic Neurons in Acquisition and Memory Retrieval

Roles of Aminergic Neurons in Olfactory Memory Formation

We studied the effects of octopamine (OA) and dopamine (DA) receptor antagonists on appetitive and aversive olfactory conditioning in crickets.[15] Previous studies on honeybees[4] and fruit flies[31] suggested that OA- and DA-ergic neurons play roles in appetitive and aversive olfactory conditioning, respectively. Thus, we examined whether this is the case in crickets. Crickets injected with an OA receptor antagonist (epinastine or mianserin) into the hemolymph 30 min before conditioning exhibited a complete impairment of appetitive conditioning of an odor with water reward (Figure 41.4A). On the other hand, these animals exhibited no impairment of aversive conditioning with sodium chloride punishment, indicating that OA receptor antagonists do not impair sensory function, motor function, or motivation necessary for learning. We thus concluded that OA-ergic neurons are specifically involved in conveying water reward. We also found that injection of a DA receptor antagonist (fluphenazine, chlorpromazine, or spiperone) completely impaired aversive learning with sodium chloride punishment (Figure 41.4B). In contrast, it did not impair appetitive learning with water reward, indicating that

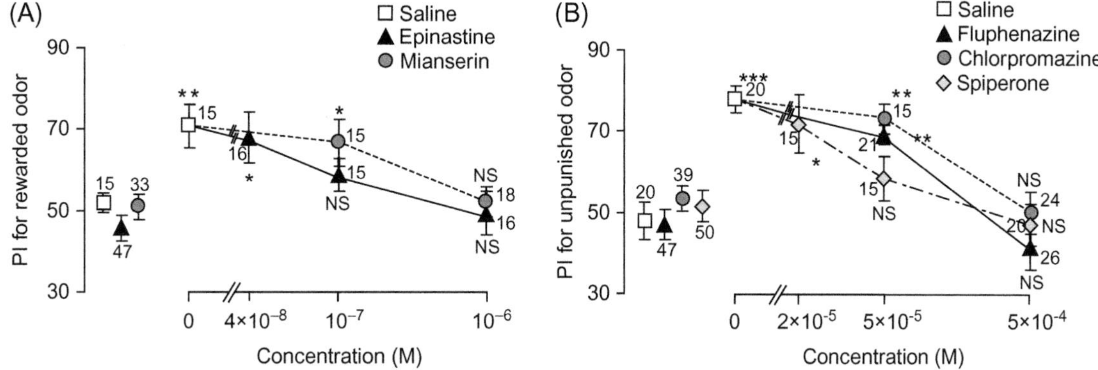

FIGURE 41.4 Effects of OA or DA receptor antagonists on appetitive and aversive olfactory conditioning. (A) Dose-dependent effects of OA receptor antagonists on appetitive olfactory conditioning. Six groups of crickets were injected with 3 μL saline (white squares) or saline containing 0.04, 0.1, or 1 μM epinastine (black triangles) or 0.1 or 1 μM mianserin (gray circles). (B) Dose-dependent effects of DA receptor antagonists on aversive olfactory conditioning. Eight groups of crickets were injected with 3 μL saline (white squares) or saline containing 50 or 500 μM fluphenazine (black triangles), 50 or 500 μM chlorpromazine (gray circles), or 20, 50, or 500 μM spiperone (white diamonds). Relative odor preferences were measured as preference indexes for rewarded odor (A) or unpunished control odor (B) before (data points on the left) and 30 min after conditioning (data points on the right) and are shown with means ± SEM. The number of animals is shown at each data point. The results of statistical comparison before and after conditioning are shown as asterisks (Wilcoxon's test; *$p<0.05$; **$p<0.01$; ***$p<0.001$; NS, $p>0.05$). Source: *Modified from Unoki* et al.[15]

DA receptor antagonists do not impair the sensory function, motor function, or motivation necessary for learning. These results suggest that DA-ergic neurons are specifically involved in conveying sodium chloride punishment. We concluded that OA- and DA-ergic neurons convey information about appetitive and aversive US, respectively, for olfactory conditioning in crickets.

Roles of Aminergic Neurons in the Formation of Memory for Visual Patterns and Colors

We next studied the effects of OA and DA receptor antagonists on appetitive and aversive conditioning of a visual pattern[32] and a color cue.[18] For conditioning of a visual pattern, crickets injected with an OA receptor antagonist (epinastine or mianserin) into the hemolymph exhibited an impairment of appetitive learning, whereas aversive learning of a visual pattern was unaffected.[32] In contrast, a DA receptor antagonist (fluphenazine, chlorpromazine, or spiperone) impaired aversive learning but not appetitive learning.[32] In color conditioning, injection of an OA receptor antagonist impaired appetitive color learning without affecting aversive color learning.[18] In contrast, injection of a DA receptor antagonist impaired aversive color learning without affecting appetitive color learning.[18] These results indicate that the roles of OA-ergic and DA-ergic neurons in conveying information about appetitive and aversive US, respectively, are ubiquitous in learning of odor, visual pattern, and color stimuli. Thus, we propose that OA- and DA-ergic neurons serve as general reward and punishment systems for learning in insects.[33] This is in contrast to learning in mammals, in which DA-ergic neurons are thought to convey both reward and punishment signals.[34]

Participation of OA-ergic Neurons and DA-ergic Neurons in Appetitive and Aversive Memory Retrieval

We then studied the effects of OA and DA receptor antagonists on appetitive and aversive memory retrieval.[33] Crickets were subjected to appetitive or aversive olfactory conditioning and were injected with an OA or a DA receptor antagonist before a retention test. Injection of an OA receptor antagonist (epinastine or mianserin) completely impaired appetitive olfactory memory retrieval but had no effect on aversive olfactory memory retrieval (Figure 41.5A). On the other hand, injection of a DA receptor antagonist (fluphenazine, chlorpromazine, or spiperone) completely impaired aversive memory retrieval but had no effect on appetitive memory retrieval (Figure 41.5B). This is in accordance with the finding in honeybees that disruption of OA-ergic transmission in the antennal lobe, the primary olfactory center, by an OA receptor antagonist (mianserin) or by RNAi of the OA receptor gene disrupted appetitive olfactory memory retrieval.[35] Moreover, we observed that injection of OA and DA

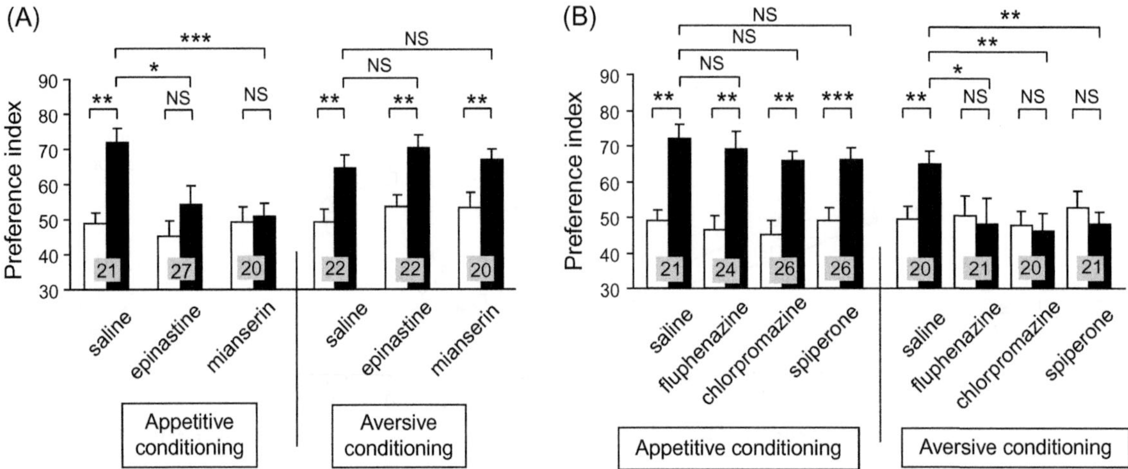

FIGURE 41.5 OA and DA receptor antagonists impair appetitive and aversive olfactory memory retrieval, respectively. Effects of OA (A) and DA (B) receptor antagonists on olfactory memory retrieval. Twelve groups of crickets were each subjected to two-trial appetitive (left) or six-trial aversive (right) olfactory conditioning trials. On the next day, each group was injected with 3 μL of saline or saline containing 1 μM epinastine, 1 μM mianserin, 500 μM fluphenazine, 500 μM chlorpromazine, or 500 μM spiperone at 30 min before the final test. Preference indexes for the rewarded odor (in the case of appetitive conditioning) or unpunished control odor (in the case of aversive conditioning) before (white bars) and 1 day after (black bars) conditioning are shown with means + SEM. The number of crickets is shown at each data point. The results of statistical comparison before and after conditioning (Wilcoxon's test) and between experimental and saline-injected control groups (Mann–Whitney test) are shown as asterisks (*$p < 0.05$; **$p < 0.01$; ***$p < 0.001$; NS, $p > 0.05$). Source: *Modified from Mizunami* et al.[22]

receptor antagonists impaired appetitive and aversive memory retrieval, respectively, in visual pattern conditioning.[22] Therefore, we concluded that OA- and DA-ergic neurons participate in the retrieval of appetitive memory and aversive memory, respectively, in both olfactory learning and visual pattern learning.[33]

Our findings are not consistent with those of conventional neural models of insect classical conditioning. Figure 41.6A depicts a model proposed by Schwaerzel *et al.*[31] to account for the roles of extrinsic and intrinsic neurons of the mushroom body in appetitive or aversive olfactory conditioning with sucrose reward or electric shock punishment in *Drosophila*. In this model, it is assumed that (1) 'CS' neurons (intrinsic neurons of the mushroom body called Kenyon cells) that convey signals about a CS make synaptic connections with dendrites of 'CR' neurons (efferent (output) neurons of the mushroom body lobe), the activation of which leads to a conditioned response (CR) that mimics an unconditioned response (UR), but these synaptic connections are silent or very weak before conditioning; (2) OA- and DA-ergic efferent neurons projecting to the lobes ('OA/DA' neurons), which convey signals for appetitive and aversive US, respectively, make synaptic connections with axon terminals of 'CS' neurons; and (3) the efficacy of synaptic transmission from 'CS' neurons to 'CR' neurons that induces a conditioned response (CS–CR connection) is strengthened by coincident activation of 'CS' neurons and 'OA/DA' neurons during conditioning.

We have proposed a new model (Figure 41.6B) with minimal modifications of the model proposed by Schwaerzel *et al.*[31] In our model, we assumed that (1) activation of 'OA/DA' neurons and the resulting release of OA or DA are needed to 'gate' the synaptic pathway from 'CS' neurons to 'CR' neurons after conditioning, and (2) synaptic connection from 'CS' neurons to 'OA/DA' neurons representing US is strengthened by coincident activation of 'CS' neurons and 'OA/DA' neurons by pairing of a CS with a US. Results of our pharmacological analysis using the second-order conditioning procedure supported this model.[33,36] Further confirmation of this model will be one of the major goals of our future studies.

VISUAL AND OLFACTORY LEARNING IN COCKROACHES

Leg position learning of headless cockroaches was demonstrated by Horridge in 1962,[37] and this paradigm was widely used for the next three decades[38,39]. Recent studies on learning and memory in cockroaches have focused on the visual place learning procedure,[36] visual–olfactory associative learning,[40–43] operant conditioning,[44–46] and classical conditioning.[47] In this section, we focus on our studies of visual place learning and classical conditioning.

Visual Place Learning

The capability of visual place learning has been well documented in some species of insects, including solitary wasps,[48] honeybees,[25] and ants.[7] We found evidence that cockroaches exhibit visual place learning and that the mushroom body, an olfactory and multisensory center in the insect brain,[2,5,6] plays a crucial role in place learning.[36] Behavioral tests, based on paradigms similar to the Morris water maze used for testing place learning in rodents,[49] demonstrated the

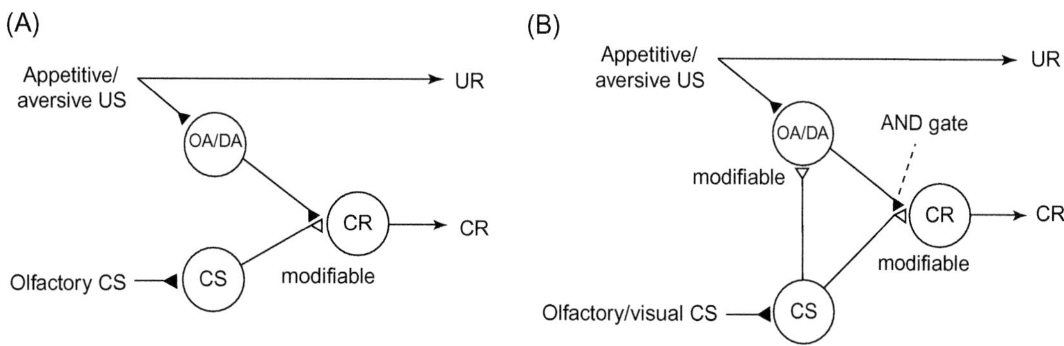

FIGURE 41.6 Conventional and new models of classical conditioning in insects. (A) A model proposed to account for the roles of intrinsic and extrinsic neurons of the mushroom body in olfactory conditioning in fruit-flies.[31] OA-ergic and DA-ergic neurons ('OA/DA' neurons) convey signals for appetitive and aversive US, respectively. 'CS' neurons, which convey signals for CS, make synaptic connections with 'CR' neurons that induce a conditioned response (CR), the efficacy of the connection being strengthened by conditioning. 'OA/DA' neurons make synaptic connections with axon terminals of 'CS' neurons. (B) A new model of classical conditioning. The model assumes that efficacy of synaptic transmission from 'CS' neurons to 'OA/DA' neurons is strengthened by conditioning and that coincident activation of 'OA/DA' neurons and 'CS' neurons is needed to activate 'CR' neurons to lead to a CR (AND gate). Source: *Modified from Mizunami* et al.[22]

capability of cockroaches to locate a hidden cool spot (goal) on a floor heated to an aversive temperature by using distant visual cues placed on the wall of an arena. Bilateral lesions of the mushroom bodies abolished this ability but left unimpaired the ability to locate a visible goal.

Wessnitzer *et al.*[50] also documented place learning in crickets (*Gryllus bimaculatus*) by using a similar experimental setup. Adult female crickets were released into an arena with a floor heated and with one hidden cool spot. Over 10 trials, the time taken to find the cool spot decreased significantly. When the scene on the wall was rotated, the crickets preferentially approached the fictive target position corresponding to the rotation.

Visual place learning has been reported in *Drosophila*, and the central complex, but not the mushroom body, has been implicated in this learning.[51] Interestingly, the cockroach mushroom body receives visual input,[52–55] but no evidence of visual input has been found in the mushroom body of *Drosophila*. Further studies are needed to determine whether different brain areas participate in visual place memory in different insect species.

Neural correlates of spatial learning have remained elusive. We have chronically recorded activities of extrinsic neurons of the mushroom body from cockroaches freely walking in an arena,[56,57] but no evidence of neural activities associated with spatial learning has been obtained, except that some neurons exhibited activities prior to the onset of some specific locomotory actions and thus may participate in the preparation of locomotory behavior. More studies on this subject are needed. Accumulated knowledge on the neural organization of the cockroach mushroom body[58–65] and physiology of its neurons[52,53,66] should facilitate such studies.

Olfactory Conditioning in Cockroaches

We established a differential conditioning procedure for the cockroach in which peppermint odor was used as CS+ and vanilla odor was used as CS−. Sucrose solution was used as appetitive US, and sodium chloride solution was used as aversive US.[47] Relative preference of cockroaches was tested by allowing them to choose between CS+ and CS−. Cockroaches that had undergone one set of differential conditioning trials exhibited a significantly greater preference for CS+ than did untrained cockroaches. Memory formed by three sets of differential conditioning trials, with 5-min ITI, was retained for at least 4 days after conditioning.

Cockroaches exhibited, as did crickets, context-dependent discriminatory olfactory learning.[67] One group of cockroaches received differential conditioning to associate peppermint odor (CS1) with sucrose solution (appetitive US) and vanilla odor (CS2) with sodium chloride solution (aversive US) under illumination and to associate CS1 with aversive US and CS2 with appetitive US in the dark. Another group received training with the opposite stimulus arrangement. Before training, both groups exhibited preference for CS2 over CS1. After training, the former group preferred CS1 over CS2 under illumination but preferred CS2 over CS1 in the dark, and the latter group exhibited the opposite odor preference.

Salivary Conditioning

Conditioning of Salivation and of Activities of Salivary Neurons

Secretion of saliva to aid swallowing and digestion is a basic physiological function found in many vertebrates and invertebrates. In dogs, classical conditioning of salivation was reported by Pavlov in the 1900s.[68] However, to our knowledge, until very recently, reports on conditioning of salivation have been restricted to dogs and humans.[69] In cockroaches, salivary neurons of the subesophageal ganglion control secretion of saliva from salivary glands.[70] We found that salivary neurons of the cockroach exhibited a prominent response to sucrose solution applied to the mouth and a weak response to odors applied to an antenna, and we studied the effect of conditioning of an odor with sucrose reward on responses of salivary neurons to the odor.[70] After three sets of differential conditioning trials in which an odor was presented just before the presentation of sucrose solution and the other odor was presented alone, the response of salivary neurons to sucrose-associated odor increased (Figure 41.7A), but the response to the odor presented alone was unchanged. Backward pairing trials in which an odor was presented after the presentation of sucrose solution were not effective in achieving conditioning (Figure 41.7B).[70] Our study on change in the level of saliva secretion in response to electrical stimulation of salivary neurons suggested that the magnitude of increase in odor response of salivary neurons by conditioning is sufficient to lead to an increased level of salivation.

We also found that the amount of saliva secretion in response to an odor increases after pairing of the odor with sucrose solution.[69] Untrained cockroaches exhibited salivary responses to sucrose solution applied to the mouth but not to peppermint or vanilla odor applied to an antenna. After differential conditioning trials in which an odor was paired with sucrose solution and another odor was presented without pairing with sucrose solution, sucrose-associated odor induced

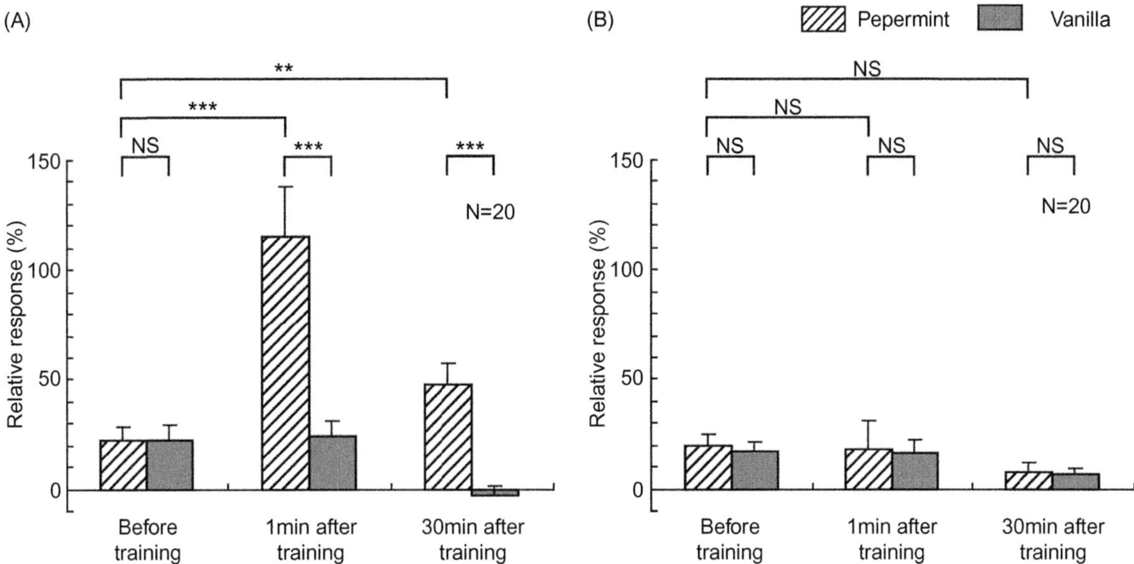

FIGURE 41.7 **Effects of forward and backward odor–sucrose pairing trials on odor responses of salivary neurons (SNs).** (A) Summed responses of SN1 and SN2 to peppermint (hatched bars) or vanilla (shaded bars) odor before and 1 and 30 min after five sets of forward-pairing trials, in which peppermint odor was presented before presentation of sucrose solution and then vanilla odor was presented alone. (B) Summed responses of SN1 and SN2 to odors before and 1 and 30 min after five sets of backward-pairing trials, in which peppermint odor was presented after presentation of sucrose solution and then vanilla was presented alone. The responses are shown as means + SEM. The results of statistical comparison are shown as asterisks (t test; NS, $p > 0.05$; $^{**}p < 0.01$; $^{***}p < 0.001$). Source: *Modified from Watanabe and Mizunami.*[70]

an increase in the level of salivation, but the odor presented alone did not.

We also found that a gustatory stimulus presented to an antenna could serve as an effective US for producing salivary conditioning.[71] Presentation of sucrose or sodium chloride solution to an antenna induced salivation and also increased activities of salivary neurons. After one set of differential conditioning trials in which an odor (CS+) was paired with antennal presentation of sucrose solution (US) and another odor (CS_0) was presented alone without pairing with the US, CS+ induced an increase in salivation or of activities of salivary neurons, but CS_0 did not. Five sets of conditioning trials led to a conditioning effect that lasted for 1 day. Water or tactile stimulus presented to an antenna was not effective for producing conditioning. The results demonstrate that gustatory US presented to an antenna is as effective as that presented to the mouth for producing salivary conditioning. This procedure is suitable for electrophysiological study of changes in activities of brain neurons during conditioning because the application of sucrose to the mouth triggers feeding movement of the jaw muscles, which often accompanies movement of the brain and prevents stable intracellular recordings from brain neurons, but sucrose stimulation to the antenna does not induce feeding movement.

Context-Dependent Discrimination Learning as Monitored by the Activity of Salivary Neurons

Cockroaches exhibited context-dependent discriminatory olfactory learning of salivation, which was monitored as activity changes of salivary neurons.[72] A group of cockroaches was differentially conditioned to associate peppermint odor (CS1) with sucrose solution while vanilla odor (CS2) was presented alone under a flickering light condition and also trained to associate CS2 with sucrose reward while CS1 was presented alone under a steady light condition. After training, the responses of salivary neurons to CS1 were greater than those to CS2 under steady illumination, and the responses to CS2 were greater than those to CS1 in the presence of flickering light. This conditioning procedure is ideal for the study of activity changes of brain neurons associated with this form of higher order associative learning because it can be performed in strictly immobilized cockroaches.

Participation of the Mushroom Body in Salivary Conditioning

Finally, we showed in a study using the effect of local injection of an acetylcholine receptor antagonist in several regions of the central olfactory pathway in cockroaches that the mushroom body participates in salivary conditioning.[73] In insects, cholinergic neurons are suggested to transmit olfactory CS to the sites for associating the CS with US.[74] In cockroaches, a type of nicotinic acetylcholine receptor specifically antagonized by mecamylamine has been characterized.[75] We investigated the possible roles of neurons that possess mecamylamine-sensitive acetylcholine receptors and are inhibited by mecamylamine, which we call 'mecamylamine-sensitive neurons,' in olfactory conditioning of salivation, monitored by changes in activities of salivary neurons.[73] Local and bilateral microinjections of mecamylamine into each of the three olfactory centers—antennal lobes, calyces of the mushroom body, and lateral protocerebra—impaired olfactory responses of salivary neurons, indicating that mecamylamine-sensitive neurons in all olfactory centers participate in pathways mediating olfactory responses of salivary neurons. Conditioning of olfactory CS with sucrose US was impaired by injection of mecamylamine into the antennal lobes or calyces; that is, conditioned responses were absent even after recovery from mecamylamine injection, suggesting that the CS–US association occurs in mecamylamine-sensitive neurons in calyces (most likely Kenyon cells) or in neurons in downstream pathways. In contrast, conditioned responses appeared after recovery from mecamylamine injection into the lateral protocerebra, suggesting that mecamylamine-sensitive neurons in the lateral protocerebra are downstream of the association sites. Because the lateral protocerebra are major termination areas of efferent neurons of the mushroom body,[52,53] we suggested that input synapses of mecamylamine-sensitive Kenyon cells, or their output synapses on efferent neurons of the mushroom body, are the sites for CS–US association for conditioning of salivation. Further study is needed to clarify whether CS–US association occurs in the calyces or the lobes of the mushroom body or both.

CONCLUSION AND FUTURE PERSPECTIVE

Insects are useful models for the study of the neural basis of learning and memory,[2,5,6] and studies on crickets and cockroaches have provided many insights into the neural mechanisms underlying olfactory and visual learning. New technologies such as RNAi (in crickets[30,76] and in cockroaches[77]) and transgenesis (in crickets[78]) are now available for the study of learning and memory in these insects. Thus, further studies on crickets and cockroaches should contribute to our knowledge of learning and memory and their neural basis.

References

1. Davis RL. Traces of *Drosophila* memory. *Neuron*. 2011;70:8–19.
2. Heisenberg M. Mushroom body memoir: from maps to models. *Nat Rev Neurosci*. 2003;4:266–275.
3. Giurfa M. Behavioral and neural analysis of associative learning in the honeybee: a taste from the magic well. *J Comp Physiol A*. 2007;193:801–824.
4. Hammer MR, Menzel R. Multiple sites of associative odor learning as revealed by local brain microinjections of octopamine in honeybees. *Learn Mem*. 1998;5:146–156.
5. Menzel R, Giurfa M. Cognitive architecture of a mini-brain: the honeybee. *Trends Cogn Sci*. 2001;5:62–71.
6. Menzel R, Giurfa M. Dimensions of cognition in an insect, the honeybee. *Behav Cogn Neurosci Rev*. 2006;5:24–40.
7. Wehner R, Michel B, Antonsen P. Visual navigation in insects: coupling of egocentric and geocentric information. *J Exp Biol*. 1996;199:129–140.
8. Worden BD, Papaj DR. Flower choice copying in bumblebees. *Biol Lett*. 2005;22:504–507.
9. Ito I, Ong RC, Raman B, Stopfer M. Sparse odor representation and olfactory learning. *Nat Neurosci*. 2008;11:1177–1184.
10. van Nouhuys S, Kaartinen R. A parasitoid wasp uses landmarks while monitoring potential resources. *Proc Biol Sci*. 2008;275: 377–385.
11. Jaffe K, Zabala NA, de Bellard ME, Granier M, Aragort W, Tablante A. Amino acid and memory consolidation in the cricket II: effect of injected amino acids and opioids on memory. *Pharmacol Biochem Behav*. 1990;35:133–136.
12. Jaffe K, Blanco ME. Involvement of amino acids, opioids, nitric oxide, and NMDA receptors in learning and memory consolidation in crickets. *Pharmacol Biochem Behav*. 1994;47:493–496.
13. Matsumoto Y, Mizunami M. Olfactory learning in the cricket *Gryllus bimaculatus*. *J Exp Biol*. 2000;203:2581–2588.
14. Matsumoto Y, Mizunami M. Temporal determinants of olfactory long-term retention in the cricket *Gryllus bimaculatus*. *J Exp Biol*. 2002;205:1429–1437.
15. Unoki S, Matsumoto Y, Mizunami M. Participation of octopaminergic reward system and dopaminergic punishment system in insect olfactory learning revealed by pharmacological study. *Eur J Neurosci*. 2005;22:1409–1416.
16. Matsumoto Y, Unoki S, Aonuma H, Mizunami M. Critical role of nitric oxide-cGMP cascade in the formation of cAMP-dependent long-term memory. *Learn Mem*. 2006;13:35–44.
17. Matsumoto Y, Noji S, Mizunami M. Time course of protein synthesis-dependent phase of olfactory memory in the cricket *Gryllus bimaculatus*. *Zool Sci*. 2003;20:409–416.
18. Nakatani Y, Matsumoto Y, Mori Y, et al. Why the carrot is more effective than the stick: different dynamics of punishment memory and reward memory and its possible biological basis. *Neurobiol Learn Mem*. 2009;92:370–380.
19. Matsumoto Y, Mizunami M. Lifetime olfactory memory in the cricket *Gryllus bimaculatus*. *J Comp Physiol A*. 2002;188: 295–299.
20. Matsumoto Y, Mizunami M. Olfactory memory capacity of the cricket *Gryllus bimaculatus*. *Biol Letters*. 2006;2:608–610.
21. Matsumoto Y, Mizunami M. Context-dependent olfactory learning in an insect. *Learn Mem*. 2004;11:288–293.
22. Mizunami M, Unoki S, Mori Y, Hirashima D, Hatano A, Matsumoto Y. Roles of octopaminergic and dopaminergic neurons in appetitive and aversive memory recall in an insect. *BMC Biol*. 2009;7:46.
23. Lindauer M. Allgemeine sinnesphysiologie. Orientierung im raum. *Fortschr Zool*. 1963;16:58–140.
24. Coolen I, Dangles O, Casas J. Social learning in noncolonial insects? *Curr Biol*. 2005;15:1931–1935.

25. Frisch KV. The Dance Language and Orientation of Bees. Cambridge, MA: Harvard University Press; 1967.
26. Müller U. Inhibition of nitric oxide synthase impairs a distinct form of long-term memory in the honeybee, *Apis mellifera*. *Neuron*. 1996;16:541–549.
27. Müller U. Prolonged activation of cAMP-dependent protein kinase during conditioning induces long-term memory in honeybees. *Neuron*. 2000;27:159–168.
28. Garthwaite J, Charles SL, Chess-Williams R. Endothelium-derived relaxing factor release on activation of NMDA receptors suggests a role as intracellular messenger in the brain. *Nature*. 1988;336:385–388.
29. Matsumoto Y, Hatano A, Unoki S, Mizunami M. Stimulation of the cAMP system by the nitric oxide-cGMP system underlying the formation of long-term memory in an insect. *Neurosci Lett*. 2009;467:81–85.
30. Takahashi T, Hamada A, Miyawaki K, et al. Systemic RNA interference for the study of learning and memory in an insect. *J Neurosci Methods*. 2009;179:9–15.
31. Schwaerzel M, Monastirioti M, Scholz H, Friggi-Grelin F, Birman S, Heisenberg M. Dopamine and octopamine differentiate between aversive and appetitive olfactory memories in *Drosophila*. *J Neurosci*. 2003;23:10495–10502.
32. Unoki S, Matsumoto Y, Mizunami M. Roles of octopaminergic and dopaminergic neurons in mediating reward and punishment signals in insect visual learning. *Eur J Neurosci*. 2006;24:2031–2038.
33. Mizunami M, Matsumoto Y. Roles of aminergic neurons in formation and recall of associative memory in crickets. *Frontiers Behav Neurosci*. 2010;4:172.
34. Schultz W. Behavioral theories and the neurophysiology of reward. *Annu Rev Psychol*. 2006;57:87–115.
35. Farooqui T, Robinson K, Vaessin H, Smith BH. Modulation of early olfactory processing by an octopaminergic reinforcement pathway in the honeybee. *J Neurosci*. 2003;23:5370–5380.
36. Mizunami M, Weibrecht JM, Strausfeld NJ. Mushroom bodies of the cockroach: their participation in place memory. *J Comp Neurol*. 1998;402:520–537.
37. Horridge GA. Learning of leg position by the ventral nerve cord in headless insects. *Proc R Soc Lond B*. 1962;157:33–52.
38. Eisenstein EM, Carlson AD. Leg position learning in the cockroach nerve cord using an analog technique. *Physiol Behav*. 1994;56:687–691.
39. Harris CL. An improved Horridge procedure for studying leg-position learning in cockroaches. *Physiol Behav*. 1991;49:543–548.
40. Brown S, Strausfeld N. The effect of age on a visual learning task in the American cockroach. *Learn Mem*. 2009;16:210–223.
41. Kwon HW, Lent DD, Strausfeld NJ. Spatial learning in the restrained American cockroach *Periplaneta americana*. *J Exp Biol*. 2004;207:377–383.
42. Lent DD, Pintér M, Strausfeld NJ. Learning with half a brain. *Dev Neurobiol*. 2007;67:740–751.
43. Pintér M, Lent DD, Strausfeld NJ. Memory consolidation and gene expression in *Periplaneta americana*. *Learn Mem*. 2005;12:30–38.
44. Balderrama N. One trial learning in the American cockroach, *Periplaneta americana*. *J Insect Physiol*. 1980;26:499–504.
45. Sakura M, Mizunami M. Olfactory learning and memory in the cockroach *Periplaneta americana*. *Zool Sci*. 2001;18:21–28.
46. Sakura M, Okada R, Mizunami M. Olfactory discrimination of structurally similar alcohols by cockroaches. *J Comp Physiol A*. 2002;188:787–797.
47. Watanabe H, Kobayashi Y, Sakura M, Matsumoto Y, Mizunami M. Classical olfactory conditioning in the cockroach *Periplaneta americana*. *Zool Sci*. 2003;20:1447–1454.
48. Tinbergen N. *The Study of Instinct*. New York: Oxford University Press; 1951.
49. Morris RGM, Garrud P, Rawlins JNP, O'Keefe J. Place navigation impaired in rats with hippocampal lesions. *Nature*. 1982;297:681–683.
50. Wessnitzer J, Mangan M, Webb B. Place memory in crickets. *Proc R Sci B*. 2008;275:915–921.
51. Ofstad TA, Zuker CS, Reiser MB. Visual place learning in *Drosophila melanogaster*. *Nature*. 2011;474:204–207.
52. Li Y, Strausfeld NJ. Morphology and sensory modality of mushroom body extrinsic neurons in the brain of the cockroach, *Periplaneta americana*. *J Comp Neurol*. 1997;387:631–650.
53. Li Y, Strausfeld NJ. Multimodal efferent and recurrent neurons in the medial lobes of cockroach mushroom bodies. *J Comp Neurol*. 1999;409:647–663.
54. Nishikawa M, Nishino H, Mizunami M, Yokohari F. Function-specific distribution patterns of axon terminals of input neurons in the calyces of the mushroom body of the cockroach, *Periplaneta americana*. *Neurosci Lett*. 1998;245:33–36.
55. Nishino H, Iwasaki M, Yasuyama K, Hongo H, Watanabe H, Mizunami M. Visual and olfactory input segregation in the mushroom body calyces in a basal neopteran, the American cockroach. *Arthropod Struct Dev*. 2012;41:3–16.
56. Mizunami M, Okada R, Li Y, Strausfeld NJ. Mushroom bodies of the cockroach: the activity and identities of neurons recorded in freely moving animals. *J Comp Neurol*. 1998;402:501–519.
57. Okada R, Ikeda J, Mizunami M. Sensory responses and movement-related activities in extrinsic neurons of the cockroach mushroom bodies. *J Comp Physiol A*. 1999;185:115–129.
58. Farris SM, Strausfeld NJ. A unique mushroom body substructure common to basal cockroaches and to termites. *J Comp Neurol*. 2003;456:305–320.
59. Iwasaki M, Mizunami M, Nishikawa M, Itoh T, Tominaga Y. Ultrastructural analysis of modular subunits in the mushroom bodies of the cockroach. *J Electron Microsc*. 1999;48:5–62.
60. Mizunami M, Iwasaki M, Nishikawa M, Okada R. Modular structures in the mushroom body of the cockroach. *Neurosci Lett*. 1997;229:153–156.
61. Mizunami M, Iwasaki M, Okada R, Nishikawa M. Topography of modular subunits in the mushroom bodies of the cockroach. *J Comp Neurol*. 1998;399:153–161.
62. Mizunami M, Iwasaki M, Okada R, Nishikawa M. Topography of four classes of Kenyon cells in the mushroom bodies of the cockroach. *J Comp Neurol*. 1998;399:162–175.
63. Sinakevitch I, Niwa M, Strausfeld NJ. Octopamine-like immunoreactivity in the honey bee and cockroach: comparable organization in the brain and subesophageal ganglion. *J Comp Neurol*. 2005;488:233–254.
64. Strausfeld NJ, Li Y. Organization of olfactory and multimodal afferent neurons supplying the calyx and pedunculus of the cockroach mushroom bodies. *J Comp Neurol*. 1999;409:603–625.
65. Strausfeld NJ, Li Y. Representation of the calyces in the medial and vertical lobes of cockroach mushroom bodies. *J Comp Neurol*. 1999;409:626–646.
66. Nishino H, Mizunami M. Giant input neurons of the mushroom body: intracellular recording and staining in the cockroach. *Neurosci Lett*. 1998;246:57–60.
67. Sato C, Matsumoto Y, Sakura M, Mizunami M. Contextual olfactory learning in cockroaches. *NeuroReport*. 2006;17:553–557.
68. Pavlov IP. In: Anrep GV, (Trans.). *Conditioned Reflexes: An Investigation of the Physiological Activity of the Cerebral Cortex*. Oxford: Oxford University Press; 1927.
69. Watanabe H, Mizunami M. Pavlov's cockroach: classical conditioning of salivation in an insect. *PLoS ONE*. 2007;6:e529.

70. Watanabe H, Mizunami M. Classical conditioning of activities of salivary neurons in an insect. *J Exp Biol.* 2006;209:766–779.
71. Watanabe H, Sato C, Kuramochi T, Nishino H, Mizunami M. Salivary conditioning with antennal gustatory unconditioned stimulus in an insect. *Neurobiol Learn Mem.* 2008;90:245–254.
72. Matsumoto CS, Matsumoto Y, Watanabe H, Nishino H, Mizunami M. Context-dependent olfactory learning monitored by activities of salivary neurons in cockroaches. *Neurobiol Learn Mem.* 2012;97:30–36.
73. Watanabe H, Matsumoto SC, Nishino H, Mizunami M. Critical roles of mecamylamine-sensitive mushroom body neurons in insect olfactory learning. *Neurobiol Learn Mem.* 2011;95:1–13.
74. Kreissl S, Bicker G. Histochemistry of acetylcholinesterase and immunocytochemistry of an acetylcholine receptor-like antigen in the brain of the honeybee. *J Comp Neurol.* 1989;286:71–84.
75. Thany SH, Lenaers G, Raymond-Delpech V, Sattelle DB, Lapied B. Exploring the pharmacological properties of insect nicotinic acetylcholine receptors. *Trends Pharmacol Sci.* 2007;28: 14–22.
76. Nakamura T, Mito T, Miyawaki K, Ohuchi H, Noji S. EGFR signaling is required for re-establishing the proximodistal axis during distal leg regeneration in the cricket *Gryllus bimaculatus* nymph. *Dev Biol.* 2008;319:46–55.
77. Marie B, Blagburn JM. Differential roles of engrailed paralogs in determining sensory axon guidance and synaptic target recognition. *J Neurosci.* 2003;23:7854–7862.
78. Nakamura T, Yoshizaki M, Ogawa S, et al. Imaging of transgenic cricket embryos reveals cell movements consistent with a syncytial patterning mechanism. *Curr Biol.* 2010;20: 1641–1647.

CHAPTER

42

Individual Recognition and the Evolution of Learning and Memory in *Polistes* Paper Wasps

Elizabeth A. Tibbetts and Michael J. Sheehan

University of Michigan, Ann Arbor, Michigan

INTRODUCTION

There has been extensive interest in how social insects coordinate large societies with complex social interactions despite having relatively small brains.[1,2] This question has attracted particular attention because vertebrates that live in large groups must manage many individual social relationships, a demanding task that is hypothesized to select for increased cognitive capacity and/or brain size.[3,4] As a result, vertebrates that live in large social groups often have larger brains and more sophisticated learning and memory abilities than vertebrates that live in smaller social groups.[5–7] The relationship between large group size and high social complexity in vertebrates contrasts sharply with the pattern in social insects. Insects are thought to avoid some of the cognitive demands of large societies because they rely on simplified social networks rather than the individually differentiated social relationships of vertebrate societies.[8,9] Instead of using detailed information about individual conspecifics to coordinate social interactions, social insects are typically thought to rely on simpler decision rules or forms of category recognition that do not require learning or memory (e.g., nestmate recognition and queen vs. worker recognition).[8,10] In fact, social insects were long thought to lack individual recognition,[11,12] although the occurrence of individual recognition was not specifically tested.

Recent work demonstrates that some social insects rely on individual recognition to coordinate behavior within their societies (*Polistes* paper wasps[13] and *Pachycondyla* ants[13,14]). Individual recognition has the potential to dramatically increase the cognitive demands of social life because it allows insects to develop differentiated social relationships with other colony members. Furthermore, individual recognition is often considered the most cognitively challenging form of recognition. After all, individual recognition depends on learning and memory.[15–17] During individual recognition, the receiver must learn the unique features of another individual, associate the features with information about the individual, and then recall that information during subsequent interactions.

In this review, we use *Polistes* paper wasps as a model to explore how individual recognition influences the evolution of learning and memory. *Polistes* paper wasps provide a good model for comparative analyses of learning and memory because at least one species of paper wasp, *Polistes fuscatus*, has variable facial patterns that are used for individual recognition (Figure 42.1).[13] Other *Polistes* species lack individual recognition.[18,19] This review addresses (1) the evolution of individual recognition in paper wasps, (2) memory of specific individuals during social interactions, (3) a new negative reinforcement training method for wasps and other invertebrates, and (4) how individual recognition has shaped the evolution of visual learning in paper wasps.

EVOLUTION OF INDIVIDUAL RECOGNITION

Polistes paper wasps are a large, cosmopolitan genus of primitively eusocial insects that nest in relatively small colonies (typically less than 100 individuals).[20] There is extensive variation in nest founding behavior within the genus. In some species, nests are started by a single queen. In other species, nests are started by

Invertebrate Learning and Memory.
DOI: http://dx.doi.org/10.1016/B978-0-12-415823-8.00042-3

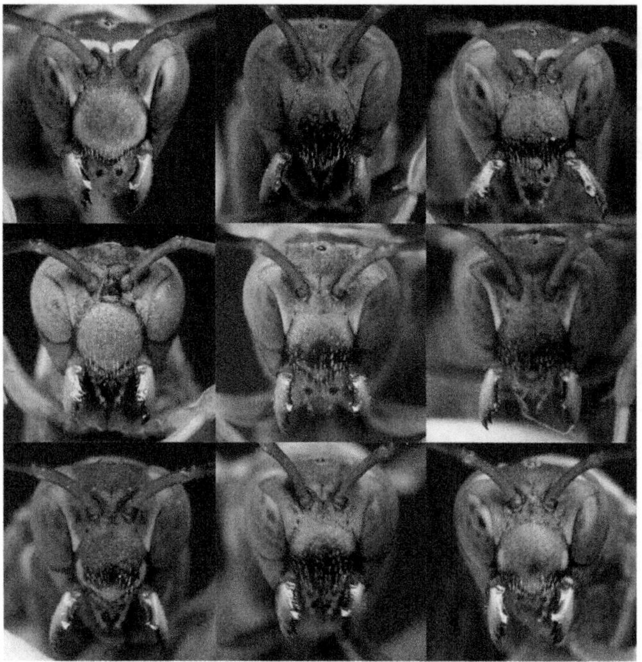

FIGURE 42.1 Portraits of nine female *P. fuscatus* paper wasps, illustrating the variable facial patterns that are used for individual recognition.

multiple queens. Still other species have flexible nest founding strategies because nests can either be started by a single queen or by a group of cooperating queens.[21,22] Flexible nesting species form linear dominance hierarchies among cooperating queens that determine the division of reproduction and work.[20,22] Because rank is an important determinant of fitness, foundresses have intense battles with many rivals before settling down to start a nest.[21] After nests are started, aggression declines and dominance behavior becomes more ritualized. *Polistes* are primitively eusocial insects that lack discrete castes, so workers are potential reproductives. As workers emerge, they incorporate into the colony dominance hierarchy.[23]

Multiple lines of evidence suggest that individual recognition has evolved in the paper wasps to mediate dominance interactions among nest founding females. First, experimental work shows that individual recognition provides social benefits by reducing aggression and stabilizing social interactions among *P. fuscatus* foundresses. Foundresses that have been experimentally manipulated to prevent individual recognition receive more aggression from conspecifics and have difficulty establishing stable dominance hierarchies.[24] Second, altering the facial patterns of foundresses on wild nests to make known individuals unrecognizable increases aggression and disrupts colony stability.[13]

Comparative analysis of the evolution of individual recognition also suggests that individual recognition is particularly beneficial in species with linear dominance hierarchies among nest founding queens. The type of variable facial patterns used for individual recognition has likely evolved multiple times in *Polistes* species with flexible nest-founding strategies (i.e., nests are started by a single queen or multiple queens). Flexible nesting species have linear dominance hierarchies and the type of variable markings used for individual recognition. In contrast, species without flexible nest-founding strategies have low marking variability and are thought to lack linear dominance hierarchies.[18] The significant association between social behavior and marking variability suggests that individual recognition has evolved because it is beneficial in species with dominance hierarchies among nesting queens. This comparative analysis also suggests that worker interactions are unlikely to be the selective pressure that has driven the evolution of individual recognition[18] because all *Polistes* species, with and without individual recognition, have workers. Although individual recognition appears to be primarily selected in foundresses, it is not surprising that workers are capable of individual recognition[13] because *Polistes* are a primitively eusocial group that lacks preimaginal castes.[23] As a result, workers and queens have broadly similar phenotypes and behavior. Overall, experimental and comparative work indicates that individual recognition provides social benefits for paper wasp foundresses.

There is one challenge associated with the comparative analysis exploring the evolution of individual recognition. Individual recognition is not the only visual signaling system in the *Polistes*. Some *Polistes* species have variable facial patterns that are visual signals of agonistic ability (also known as quality signals; *P. dominulus*[25–27] (but see[28]), *P. satan*,[29] and *P. exclamans*[30]). Recent work indicates that the type of variation in species with individual recognition and quality signals is significantly different.[19,31] However, both types of variation were included in the original comparative analysis that analyzed the evolution of facial pattern variation.[18] This review primarily focuses on *P. fuscatus*, the *Polistes* species with individual recognition. Although *P. fuscatus* is currently the only *Polistes* species in which individual recognition has been identified, the patterns of social behavior and phenotypic variation across species suggest that individual recognition is likely to be found in other paper wasp species.[18]

Both quality and identity signals are important in paper wasps, although these two signaling systems are quite different. First, the signals have different cognitive requirements. Quality signaling does not require learning and memory to function effectively. Receivers need only compare the sender's quality signal to a template; the template may be innate or learned.[32] In contrast, individual recognition requires extensive learning and memory; receivers must learn the unique characteristics of every sender.[17] The patterns of variation in quality and identity signals are also different.[33,34] Quality signals are typically continuous, unimodal traits with condition-dependent development.[34,35] Identity signals are much more variable, composed of multiple traits that vary independently.[34] In addition, identity signals are not condition dependent.[31] Thus far, the empirical evidence suggests that paper wasps have either quality or identity signals but not both.[19]

It is not clear why flexible nesting strategies favor the evolution of individual identity signals in some species and quality signals in other species. Typically, individual recognition and quality signals are used in different social contexts.[17,26] Agonistic signals are particularly useful in species with numerous aggressive interactions among unfamiliar conspecifics,[36,37] and they are often ignored during interactions with familiar individuals.[38] In contrast, identity signals are used in stable social groups and are useless during interactions with unknown rivals.[17] Both social contexts occur in *Polistes* wasps with flexible nesting strategies. Foundresses compete with numerous unfamiliar foundresses during the first few weeks of the nesting cycle. At the same time, foundresses form stable cooperative groups in which individual recognition is likely to be more useful than agonistic signaling.

A more comprehensive comparative analysis will be important to establish why some species evolved agonistic quality signals, whereas other species evolved identity signals. One interesting idea is that reproductive transactions may be the key selective pressure favoring individual recognition. Transactional skew theory explores the rules used to divide reproduction in social groups.[39] Aspects of skew theory are controversial,[40] but there is good experimental evidence that *P. fuscatus* use complex transaction-based decision rules to divide reproduction. Precise reproductive shares in *P. fuscatus* depend on factors such as dominance rank, relatedness, independent breeding opportunities, and contribution to nest productivity.[39] Perhaps reproductive transactions are involved in the evolution of individual recognition in *P. fuscatus*. Alternatively, signal evolution may not be deterministic. Instead, historical contingency may influence which communication systems evolve in which lineage.[19]

INDIVIDUAL RECOGNITION AND SOCIAL MEMORY

Individual recognition requires flexible learning and memory. Individuals must learn the unique features of conspecifics and then recall that information during subsequent social interactions. As a result, individual recognition is often considered the most complex form of recognition. A few studies in vertebrates have examined the extent of memory for specific individuals, finding that some species have remarkable social memory. For example, northern fur seals remember the call of their pup after a 4-year separation.[41] Migratory hooded warblers retain the memory of individual neighbors' songs across breeding seasons, even though they spend the 8 months between breeding seasons in Central America and do not sing during this period.[42]

Invertebrates may also have robust memories of individuals, although most work on social memory in invertebrates has been done in controlled situations that lack the interference common in the wild. Hermit crabs maintain memory of familiar versus unfamiliar conspecifics for 4 days if kept in isolation.[43] Similarly, lobsters may have a 1-week memory for familiar versus unfamiliar conspecifics when kept in isolation during the intervening week.[44] Ant individual memory has also been tested, and it was found that in some *Pachycondyla* species (which have multiple queen colonies), queens are able to remember a familiar queen after being separated for 24 hr.[45] The ants maintained this memory despite interacting with one other ant during the 24-hr separation, suggesting that their individual memory is somewhat robust to interference. *Polistes fuscatus* have even more impressive individual memory.[46] Nest founding wasp queens can recall particular individuals after being separated for 1 week, even if they live in a complex social environment in which they interact with approximately 10 other individuals during the intervening week.

Polistes fuscatus memory was tested following the method used by Dreier *et al.*[45] Aggression was measured during staged interactions among wasp foundresses. Initially, wasps encountered a new social partner from a distant collection location (Day 0). Then, we separated the pair and housed them in different communal cages, each of which contained 10 other wasps. One week later (Day 7), the same wasps interacted again. The start date of trials was staggered to ensure that results were due to treatment effects rather than environmental variation. Previous work demonstrated that *P. fuscatus* are more aggressive to individuals with unknown appearances.[13] Therefore, if wasps have robust social memories, they should be less aggressive toward individuals with which they have interacted previously than toward individuals they have not encountered before. Indeed, familiar pairs were less aggressive and had more nonaggressive contacts than unfamiliar pairs (Figure 42.2).[42]

The individual memory in *P. fuscatus* is notable because it is robust to social interference. Storing and retaining one type of information in

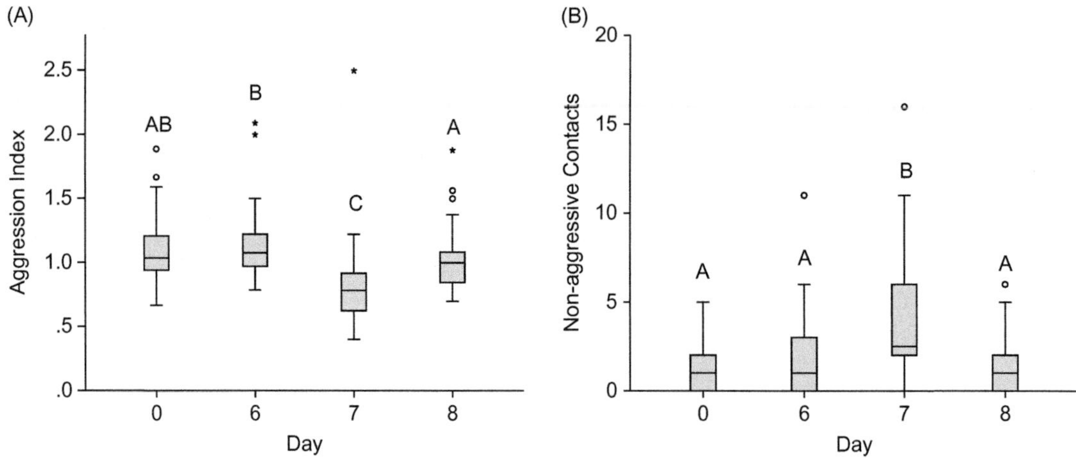

FIGURE 42.2 Long-term social memory in a paper wasp. Wasps were (A) less aggressive and (B) engaged in more nonaggressive interactions with individuals they interacted with previously than with unfamiliar individuals. Aggression indices for the 4 days of dyadic encounters between foundresses are shown. The aggression index weights interactions based on the intensity of aggression, with lower scores indicating less intense aggression (nonaggressive physical contacts (0 points) to grappling (4 points)). On Days 0, 6, and 8, wasps interacted with a new social partner for the first time. On Day 7, wasps re-encountered the same partner from Day 0. Different letters indicate significant differences between days at $p = 0.05$. Box plots show the medians and quartiles. Source: *Reproduced with permission from Sheehan and Tibbetts.*[46]

isolation is likely to be more straightforward than organizing and correctly retrieving relevant memories across a range of contexts.[47,48] As a result, the ability to maintain one memory in isolation is a very different type of task than correctly using memory in a natural social context.

Given the strong context dependence of memory, additional work on memory across a range of species and contexts will be important to learn more about the diversity and flexibility of invertebrate memory. Additional experiments will be particularly useful in *P. fuscatus* to establish the limits of their individual memory. Previous work suggests that individuals learn and remember the face of every individual on their nest, and large nests contain approximate 20 individuals.[13] Is 20 individuals the limit for *P. fuscatus* memory, or can they learn more individuals? For example, foundresses typically spend a few weeks prior to nest foundation flying around and competing with numerous other foundresses.[21] How many of those individuals can *P. fuscatus* learn and recall? How long do they retain these memories? Similarly, how much social interaction is required to form a lasting individual memory? Can wasps form lasting memories after brief interactions or are hours of social interactions required for stable memory formation?

Individual memory was also tested in *P. dominulus* and *P. metricus*, two paper wasp species that lack the type of facial pattern variation associated with individual recognition. These species were tested using a method similar to the one used for *P. fuscatus*, although the task was simplified.[19] Foundresses were only separated for 1 day prior to their next interaction and were kept in isolation during the separation. Nevertheless, there was no evidence of individual memory in either species. Familiar and unfamiliar foundresses received similar amounts of aggressive and nonaggressive interactions. The lack of individual memory is not entirely surprising because both species lack the type of variable facial patterns used for individual recognition. Nevertheless, the results indicate that *P. metricus* and *P. dominulus* do not perform individual recognition in another sensory modality (e.g., chemical).

Interspecific variation in individual recognition provides an excellent opportunity for comparative analyses of learning and memory. The requirements of visual individual recognition may shape a species' ability to learn and remember visual information. The challenge is that it is difficult to compare learning across species and stimuli without a standardized training method. As a result, we developed a negative reinforcement training method that allows rigorous comparative analyses of visual learning.

DETAILED METHODS FOR TRAINING WASPS

Most studies on learning in social insects have used positive reinforcement to train workers. Typically, foraging workers are given a glucose solution for correct choices.[49] Occasionally, positive and negative reinforcement are combined, with workers receiving glucose for correct choices and quinine or another distasteful substance for incorrect choices.[50] Learning is rarely tested in social insect queens or foundresses.

For our comparative studies on paper wasps, we needed to study visual learning in foundresses rather than workers. As previously described , individual recognition is most important during the founding stage of the nesting cycle. Unfortunately, we were unable to train free-flying foundresses to differentiate among images using positive reinforcement. This is partially because foundresses are not very motivated by food. Unlike the workers in honeybee and bumblebee colonies, wasp foundresses spend relatively little time foraging for nectar[51] (E. A. Tibbetts, unpublished data). Furthermore, wasps have relatively low food requirements compared to worker bees. Whereas bees will die of starvation in a matter days,[52] wasp foundresses can typically survive more than 1 month without food and workers can last upwards of 1 week[53] (E. A. Tibbetts, unpublished data). As a result, it is difficult to train wasp foundresses using foraging-like tasks. Instead, we developed a paradigm for training wasps using electric shock, similar to negative reinforcement. The negative reinforcement training is logistically straightforward and provides a closer parallel with individual recognition behavior than positive reinforcement. During individual recognition, learning occurs through negative reinforcement (aggression) rather than positive reinforcement (food).

We developed a simple and effective apparatus for training wasps that is based on previous studies of negative reinforcement in insects.[54] Wasps were trained to differentiate between two images in a T-shaped maze with a small electric shock providing negative reinforcement (Figure 42.3). The correct image was associated with a nonelectrified safety zone. The location of the safety zone was changed between trials using standard randomization methods. Although both sides of the maze chamber appear identical (other than the images with which they are associated), the chamber on the incorrect side will shock the wasp, whereas the chamber on the correct side will not. Therefore, the wasp must learn the image associated with safety to avoid being shocked.

Mazes were constructed by gluing walls of white foam board to a clear acrylic sheet (Optix brand by Plaskolite) that acted as a ceiling. The maze was shallow enough to ensure that wasps walked through the maze and maintained contact with the charged floor at all times. For *P. fuscatus*, we used a wall depth of 7 mm, although the depth can be readily altered to accommodate larger or smaller species. The floor of the maze was anti-static conductive foam. This commercially available foam is impregnated with graphite, which allows it to conduct an electric charge. We electrified the foam by running two copper wires through the length of the foam piece at the top and bottom of the maze. When electrified, the foam pad has a consistent, low-level electrical charge. The charge experienced by the wasps can be regulated using a Variac transformer to modulate the electric current. In addition, the charge can be altered by changing the size of the foam pad (i.e., the distance between the two copper wires). Larger pads provide less shock to the

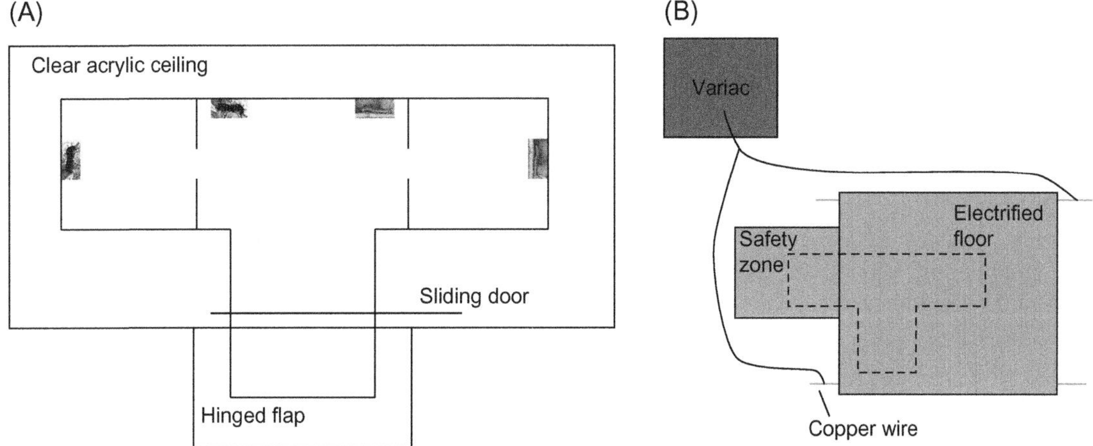

FIGURE 42.3 (A) Design of the T-shaped maze use to train wasps. The ceiling of the maze was made from a large piece of clear acrylic and a smaller piece of hinged, clear acrylic that opened so wasps could be placed in the maze. A sliding door was created from a long, thin strip of card stock that was opened manually. Pictures were placed both in the central section of the maze and at the ends of the 'T.' (B) A schematic of the apparatus for delivering the electric shock used to train wasps. Only the large piece of anti-static foam with parallel copper wires carries an electric charge. The smaller piece of foam does not carry a charge and can be moved from the right to left side of the electrified foam as needed for training. The outline shows the approximate location of the T-maze during training.

wasps. In our experiments, we set the current so that wasps received a shock of approximately 2–4 V. This level of shock elicits a reaction from the wasps but does not inhibit movement or harm the wasps. The safety zone was created by placing a separate piece of nonelectrified foam against the electrified foam. The maze was not attached to the floor, so the unelectrified foam could be moved as the safety zone changed from the right to the left side of the maze in a pseudorandom manner. The maze was illuminated by directing two full-spectrum incandescent lights at the maze.

Each trial began by manually placing the wasp in the antechamber and allowing it a few seconds to acclimate to the maze. Once the wasp acclimated, we turned on the electricity and opened a sliding door, which allowed the wasp access to the rest of the maze. We recorded the wasp's choice as whichever arm of the maze the wasp entered first. During the trials, wasps often appeared to assess the images on the walls of the maze before making a choice. Frequently, wasps would examine both images more than once before making a decision. The trials ended 2 min after the wasp was released into the maze. At the end of the trial, the electricity was turned off and the wasp was removed and placed in a container with sugar and water while the maze was reset. In rare cases, wasps had not made a choice by the end of the 2-min trial, so we let the trial continue until the wasp had chosen. We chose 2-min trials because they seemed to provide wasps with ample time to experience the reward of the safety zone. In some preliminary studies, we successfully trained wasps using shorter trial lengths (e.g., 1 min). Typically, wasps chose a side within the first 30 sec of the trials. During the remaining time, the wasps were free to move about the maze. In the first few trials, when wasps were new to the maze, they often explored the entire maze after making a choice, even if they had located the safety zone. In later trials, wasps typically remained in the safety zone once they located it.

This T-maze training paradigm is quite versatile. We used this method to successfully train 11 species of paper wasps to differentiate among a range of images[55] (M. J. Sheehan and E. A. Tibbetts, unpublished data). Our negative reinforcement paradigm provides a number of advantages for cognition research. First, subjects do not need to be pretrained on the task before they are tested. Some other training techniques require pretraining on the apparatus; only the individuals that learn the apparatus are used for further experiments.[56,57] All our data reflect the wasps' first experience in our maze setup. Second, we can train individuals that are not sufficiently food motivated to train using typical sucrose rewards. We used the paradigm most extensively to examine visual cognition in wasp foundresses, although we also successfully trained workers, males, and gynes. Our ability to train diverse groups is exciting because visual cognition in nonworkers has been largely ignored. This is presumably because males and queens are difficult to train using the foraging-based tasks commonly used to examine visual cognition. Third, the paradigm is reasonably flexible and can be adapted to study a variety of questions. In preliminary studies, we adapted the basic paradigm to examine memory, visual acuity, and delayed match-to-sample tasks. Although we have not yet attempted to train wasps using non-visual stimuli, the paradigm could presumably be used with odors or vibrations as well.

There are some limitations to the negative reinforcement training method. First, there is a limit to the number of trials that a subject can feasibly undergo in a single training session. As the number of trials increases, the wasps appear to become sluggish and less responsive, even though their performance accuracy is maintained or improves over time. We successfully trained wasps for more than 60 trials in a single day, although we believe that this is likely the maximum number that can be performed without decreasing accuracy. If more trials are needed, wasps can be trained over successive days. Second, the depth of the walls limits the size of the stimuli. In effect, this means that the images used for training must be relatively small and that the subject must be able to approach the stimuli at close range to inspect them. In our maze design, images were posted outside each choice chamber so that wasps could assess the stimuli as closely as they wanted (Figure 42.3A). Close interactions with stimuli make sense in the ecological context of social interactions, although they may be less realistic in other contexts such as navigation. Third, the possible designs of the maze are somewhat limited because the anti-static foam pad used as the charged floor must be rectangular. The copper wires that provide the charge need to run parallel for the entire length of the foam to generate an even charge. Despite these limitations, we believe that this negative reinforcement paradigm can be used to train diverse taxa and can be used to provide a valuable comparative perspective on invertebrate learning and memory.

SPECIALIZED VERSUS GENERALIZED VISUAL LEARNING

Although animals exhibit a wide array of specially adapted physiologies, morphologies, and behaviors, the occurrence of specially adapted learning has been a subject of intense debate.[58–61] The adaptive specialization view of cognitive evolution proposes that cognitive processes such as learning are adapted to solve particular ecological problems.[61,62] As a result, cognitive processes may be specialized for specific tasks and vary based on a species' behavioral requirements. A classic example of specialization occurs in caching birds; species that collect and store food often have better spatial memory than relatives that do not store food.[58,63] Although there are a few notable examples of specialized cognition, it is difficult to understand how all the flexible, complex behavior exhibited by animals could be mediated by specialized cognitive modules. Therefore, much cognition is thought to be generalized, involving phylogenetically widespread processes that are applicable to a variety of learning problems. For example, there are remarkable similarities in learning mechanisms across animals as diverse as humans, birds, bees, and mollusks.[64,65] Because most cognitive processes appear generalized, there has been extensive interest in the few examples of specialization. In particular, how frequently and in what circumstances does cognitive specialization evolve?

Some of the best evidence for fine-grain specialization in visual learning comes from research on individual face recognition. Humans and monkeys use conspecific faces to recognize individuals. In both groups, individuals are also better at learning and remembering conspecific faces than any other visual stimuli.[66,67] The specialization for learning and remembering one particular type of visual pattern is somewhat surprising.[68,69] After all, there is no dramatic difference in the light bouncing off faces versus nonface images. Why, then, does the brain treat face images differently from other images?

Paper wasps are a good system to study the evolution of face specialization because they appear to be the only invertebrate with individual face recognition.[13] Furthermore, closely related *Polistes* species differ in their ability to individually recognize conspecific faces. We tested the adaptive evolution of specialized face learning by comparing face and pattern learning in *P. fuscatus*, a species with individual face recognition, and *P. metricus*, a species that lacks individual face recognition.[19]

If wasps are specialized for face learning, they should learn faces faster and/or more accurately than any other image. Specialization was tested by comparing the ability of *P. fuscatus* to learn to discriminate pairs of faces, nonface images, and manipulated faces (Figure 42.4). Wasps were trained to discriminate two different types of nonface images. First, wasps were trained to discriminate simple black-and-white images originally developed to train honeybees. These images were chosen because they are known to be straightforward for hymenopteran eyes to discriminate.[70,71]

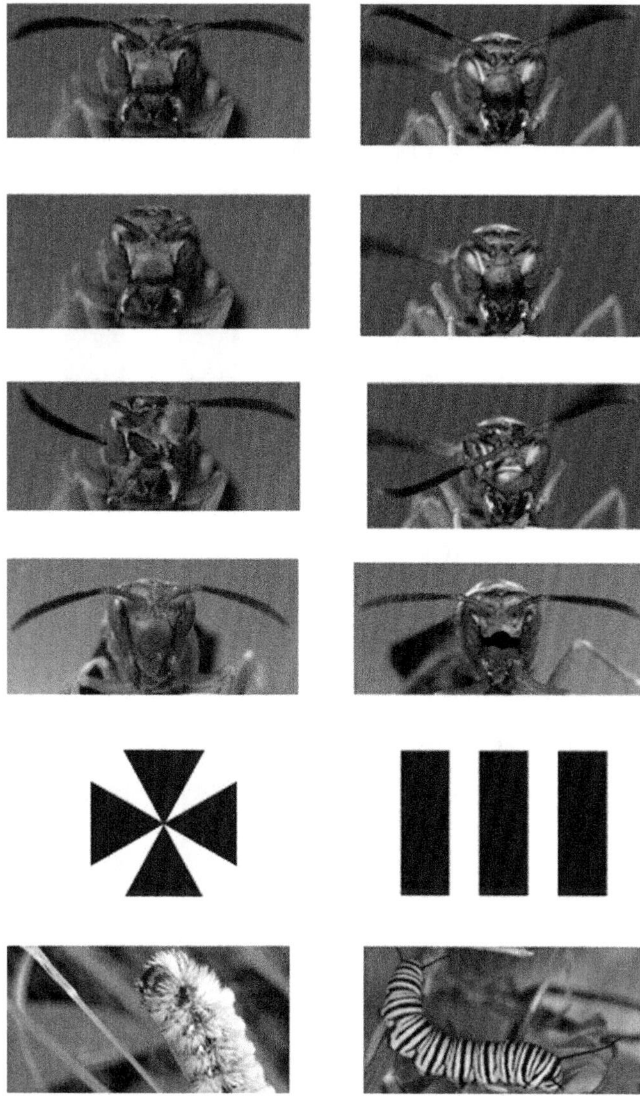

FIGURE 42.4 Example pairs of stimuli used to study specialized learning. Top to bottom: *P. fuscatus* faces, antennae-less *P. fuscatus* faces, jumbled *P. fuscatus* faces, *P. metricus* faces, black-and-white patterns, and caterpillars.

Second, wasps were trained to discriminate pairs of colored caterpillar images. *Polistes* are visual predators of caterpillars, so the caterpillar pictures provide complex, colored stimuli that *Polistes* commonly encounter in the wild. In addition to nonface images, *P. fuscatus* were trained to discriminate pairs of manipulated face images. Comparing learning abilities for manipulated versus nonmanipulated faces provides a particularly good test for face specialization because a manipulated face is composed of the same colors and patterns as the nonmanipulated face. However, the rearrangement prevents the perceptual system from identifying it as a face.[50,66,72,73] Two different forms of manipulation were performed in Adobe Photoshop: jumbling and

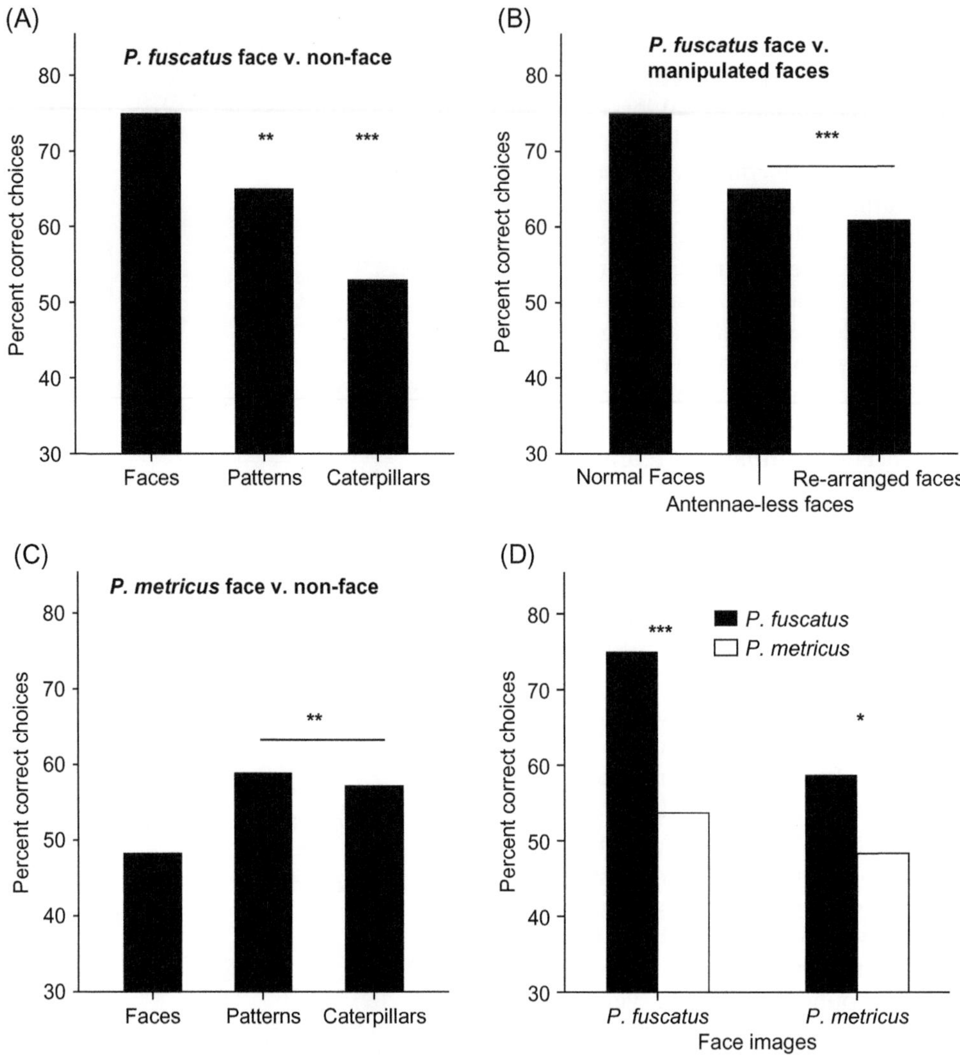

FIGURE 42.5 (A) *Polistes fuscatus* made more correct choices when trained to discriminate between pairs of conspecific face images than patterns or caterpillars. (B) *Polistes fuscatus* made more correct choices when trained to discriminate between pairs of conspecific face images than manipulated faces. (C) *Polistes metricus* made fewer correct choices when trained to discriminate between pairs of conspecific face images than non-face images. (D) *Polistes fuscatus* learned to discriminate *P. fuscatus* and *P. metricus* face images better than *P. metricus* learned to discriminate the same images. Bars show the sum of correct choices across all wasps in a treatment as a percentage of trials. Random choice is 50%; $n = 12$ wasps for each treatment. Asterisks denote the statistical significance level for comparisons to normal conspecific faces: $*p < 0.01$; $**p < 0.001$; $***p < 0.0001$. Source: *Adapted from Sheehan and Tibbetts.*[55]

removing antennae. In jumbled faces, facial features were rearranged and placed in abnormal configurations while maintaining the overall shape of the face. In the antennaeless faces, the antennae were removed while the rest of the face remained unaltered.

The training results provide striking evidence that *P. fuscatus* are specialized for learning conspecific faces (Figure 42.5).[55] *Polistes fuscatus* learned normal faces faster and more accurately than either patterns or manipulated faces. Although the fact that *P. fuscatus* learned facial images more rapidly than patterns or caterpillars is interesting, the question of specialization is most directly addressed by data on manipulated face images. The two image manipulations suggest that there is something special about normal, intact faces that allows wasps to distinguish them more rapidly and accurately than slightly altered faces. If wasps simply treated the faces like any other type of image, the manipulation would have had minimal effect on learning because the normal and manipulated images were composed of the same colors, shapes, and patterns. Similar sensitivity to altered facial images is a hallmark of specialized face learning in humans.[74] The marked decrease in learning in the absence of antennae is particularly notable because this is a minor image manipulation. Furthermore, the antennae are not variable among wasps, so they are unlikely to be good cues for distinguishing among individuals. Nevertheless, all wasps have antennae, so they are good features to cue wasps into faces. Wasps may have an 'antennae detector' similar to the eye detector in human face recognition.

We tested the evolution of face specialization by analyzing face and pattern learning in *P. metricus*, a close relative of *P. fuscatus* that lacks individual face recognition. If specialized face learning in *P. fuscatus* is the result of adaptive evolution to facilitate individual recognition, *P. metricus* is predicted to lack specialization for face learning. As predicted, *P. metricus* had difficulty learning conspecific faces (Figure 42.5).[55] In fact, *P. metricus* foundresses learned to discriminate pairs of black-and-white patterns and pairs of caterpillars faster and more accurately than face images. These results suggest that face specialization is evolutionarily labile. Species without individual face recognition have difficulty learning to discriminate pairs of face images.

An alternative explanation for the difficulty associated with *P. metricus* face learning is that the *P. metricus* face images used during training were more difficult for wasps to learn than *P. fuscatus* face images. *Polistes metricus* naturally lack face variation, so we created variable face images for training by adding variation to natural *P. metricus* faces. The altered *P. metricus* face images had variation similar to that found among *Polistes* species with variable color patterning, but it is possible that the stimuli were still more difficult to learn than *P. fuscatus* faces.

We tested whether the specific face images used during training influenced learning by training each species to discriminate conspecific and heterospecific face images. *Polistes fuscatus* learned both *P. metricus* and *P. fuscatus* face images faster and more accurately than *P. metricus* learned the same images (Figure 42.5).[55] Therefore, *P. fuscatus* are generally better at face learning than *P. metricus*. Our results also indicate that face learning may be somewhat species specific because *P. fuscatus* learn conspecific face images better than heterospecific face images. However, *P. metricus* also learned *P. fuscatus* face images better than *P. metricus* face images. Therefore, *P. fuscatus* faces may simply be particularly easy for wasps to learn. Previous experimental work indicates that *P. fuscatus* foundress benefit by being easily recognizable,[24] suggesting that distinctive foundresses with faces that are easy to learn and remember may be favored. Therefore, both sender evolution and receiver evolution are likely to be involved in the evolution of specialized face learning in *P. fuscatus*. Senders are favored to have easily identifiable faces,[17,24] and receivers are selected to rapidly and efficiently learn conspecific faces.[55]

The differences in face learning across species are not caused by general differences in visual learning in *P. fuscatus* and *P. metricus*. Both species were trained to discriminate the same nonface images, and they learned caterpillars and black-and-white patterns similarly.[60] *Polistes* rely on visual learning for a wide range of tasks, so it is not surprising that both species are good visual learners. *Polistes* are visual caterpillar predators and must learn to find, identify, and process palatable prey items.[75,76] Furthermore, *Polistes* are long-distance foragers and use visual information to navigate.[77] Much research on hymenopteran learning and memory has focused on how challenges of homing and navigation have shaped honeybee cognition.[49,70] *Polistes* paper wasps face similar navigational challenges, although there has been no research on the role of navigation in shaping paper wasp cognition.

Overall, the training results indicate that face specialization is surprisingly evolutionarily labile and covaries with individual recognition. *Polistes fuscatus* are specialized for learning faces, whereas a close relative that lacks individual recognition, *P. metricus*, has trouble learning faces. Although the precise divergence time for *P. fuscatus* and *P. metricus* is not clear, the two species are very closely related. Both are in the subgenus fuscopolistes, and they are typically closely grouped within the subgenus.[78,79] There have been no previous comparative analyses exploring the evolution of face specialization in vertebrates or invertebrates. Instead, previous work has focused on a few primates, all of which have both face specialization and individual face recognition.[66,80] Without additional analyses of species that lack face recognition, it is difficult to analyze how frequently and in what circumstances face specialization evolves. Paper wasps are a large genus with extensive variation in recognition and social behavior.[18,20] As a result, they have potential to be an important model system for rigorous comparative analyses of the evolution of specialization

Given the general importance of visual learning in the *Polistes*, it is interesting that *P. fuscatus* makes use of specialized learning mechanisms to discriminate faces instead of using general pattern learning mechanisms for face learning. The evolution of specialization indicates that the generalized learning process may be less efficient than specialized processes for learning cognitively challenging tasks. Our data suggest that specialized learning of faces likely increases the speed at which wasps can discriminate individuals compared to other visual stimuli (Figure 42.5). We speculate that specialization may also reduce the cognitive costs associated with individual recognition in wasps. Specialized face learning is thought to streamline recognition in humans.[81] Therefore, specialized learning may be a way for animals to reduce the processing costs associated with complex visual tasks that are encountered frequently. Previous research on learning specialization has primarily focused on vertebrates,[61,82] perhaps because invertebrate learning is sometimes assessed narrowly, focusing on a few training techniques or contexts. However, given the potential cognitive limitations of the 'mini-brain',[48,83] specialization for complex behavioral tasks may be even more widespread in invertebrates than vertebrates.

References

1. Gronenberg W, Riveros AJ. Social brains and behavior—Past and present. In: Gadau J, Fewell J, eds. *Organization of Insect Societies from Genome to Sociocomplexity*. Cambridge, MA: Harvard University Press; 2009:377–401.
2. Chittka L, Niven J. Are bigger brains better? *Curr Biol*. 2009;19 (21):R995–R1008.
3. Dunbar RIM. The social brain hypothesis. *Evol Anthropol*. 1998;6 (5):178–190.
4. Adolphs R. The neurobiology of social cognition. *Curr Opin Neurobiol*. 2001;11(2):231–239.
5. Dunbar RIM, Shultz S. Evolution in the social brain. *Science*. 2007;317(5843):1344–1347.
6. Perez-Barberia FJ, Shultz S, Dunbar RIM. Evidence for coevolution of sociality and relative brain size in three orders of mammals. *Evolution*. 2007;61(12):2811–2821.
7. Healy SD, Rowe C. A critique of comparative studies of brain size. *Proc R Soc, Ser B*. 2007;274:453–464.
8. Bonabeau E, Theraulaz G, Deneubourg JL, Aron S, Camazine S. Self-organization in social insects. *Trends Ecol Evol*. 1997;12 (5):188–193.
9. Beshers SN, Fewell JH. Models of division of labor in social insects. *Annu Rev Entomol*. 2001;46:413–440.
10. Anderson C, McShea DW. Individual versus social complexity, with particular reference to ant colonies. *Biol Rev*. 2001;76(2): 211–237.
11. Wilson EO. *The Insect Societies*. New York: Belknap; 1974.
12. Maynard-Smith J. Conflict and cooperation in human societies. In: Keller L, ed. *Levels of Selection in Evolution*. New York: Princeton University Press; 1999:197–208.
13. Tibbetts EA. Visual signals of individual identity in the wasp *Polistes fuscatus*. *Proc R Soc Biol Sci Ser B*. 2002;269(1499): 1423–1428.
14. D'Ettorre P, Heinze J. Individual recognition in ant queens. *Curr Biol*. 2005;15(23):2170–2174.
15. Proops L, McComb K, Reby D. Cross-modal individual recognition in domestic horses (*Equus caballus*). *Proc Natl Acad Sci USA*. 2009;106(3):947–951.
16. Brennan PA, Kendrick KM. Mammalian social odours: attraction and individual recognition. *Philos Trans R Soc B-Biol Sci*. 2006;361:2061–2078.
17. Tibbetts EA, Dale J. Individual recognition: it is good to be different. *Trends Ecol Evol*. 2007;22:529–537.
18. Tibbetts EA. Complex social behaviour can select for variability in visual features: a case study in *Polistes* wasps. *Proc R Soc Biol Sci Ser B*. 2004;271(1551):1955–1960.
19. Sheehan MJ, Tibbetts EA. Selection for individual recognition and the evolution of polymorphic identity signals in *Polistes* paper wasps. *J Evol Biol*. 2010;23(3):570–577.
20. Reeve HK. Polistes. In: Ross KG, Matthews RW, eds. *The Social Biology of Wasps*. Ithaca, NY: Comstock; 1991:99–148.
21. Roseler PF. Reproductive competition during colony establishment. In: Ross KG, Matthews RW, eds. *The Social Biology of Wasps*. London: Comstock; 1991:309–335.
22. West-Eberhard M. The social biology of polistine wasps. *Misc Publ Mus Zool Univ Mich*. 1969;140:1–101.
23. O'Donnell S. Reproductive caste determination in eusocial wasps (Hymenoptera : Vespidae). *Annu Rev Entomol*. 1998;43: 323–346.
24. Sheehan MJ, Tibbetts EA. Evolution of identity signals: frequency-dependent benefits of distinctive phenotypes used for individual recognition. *Evolution*. 2009;63:3106–3113.
25. Tibbetts EA, Dale J. A socially enforced signal of quality in a paper wasp. *Nature*. 2004;432(7014):218–222.
26. Tibbetts EA, Lindsay R. Visual signals of status and rival assessment in *Polistes dominulus* paper wasps. *Biol Lett*. 2008;4: 237–239.
27. Zanette L, Field J. Cues, concessions and inheritance: dominance hierarchies in the paper wasp *Polistes dominulus*. *Behav Ecol*. 2009;20:773–780.
28. Cervo R, Dapporto L, Beani L, Strassmann JE, Turillazzi S. On status badges and quality signals in the paper wasp *Polistes dominulus*: body size, facial colour patterns and hierarchical rank. *Proc R Soc B-Biol Sci*. 2008;275(1639):1189–1196.
29. Tannure-Nascimento IC, Nascimento FS, Zucchi R. The look of royalty: visual and odour signals of reproductive status in a paper wasp. *Proc R Soc B-Biol Sci*. 2008;275:2555–2561.
30. Tibbetts EA, Sheehan MJ. Facial patterns are a conventional signal of agonistic ability in *Polistes exclamans* paper wasps. *Ethology*. 2011;117:1138–1146.
31. Tibbetts EA, Curtis TR. Rearing conditions influence quality signals but not individual identity signals in *Polistes* wasps. *Behav Ecol*. 2007;18(3):602–607.
32. Bradbury J, Vehrencamp S. *Principles of Animal Communication*. Sunderland, MA: Sinauer; 1998.
33. Dale J, Lank DB, Reeve HK. Signaling individual identity versus quality: a model and case studies with ruffs, queleas, and house finches. *Am Nat*. 2001;158(1):75–86.
34. Dale J. Intraspecific variation in coloration. In: Hill GE, McGraw KJ, eds. *Bird Coloration, Vol. 2: Function and Evolution*. Cambridge, MA: Harvard University Press; 2006:36–86.
35. Tibbetts EA. The condition-dependence and heritability of signaling and non-signaling color traits in paper wasps. *Am Nat*. 2010;175:495–503.
36. Tibbetts EA, Safran RJ. Co-evolution of plumage characteristics and winter sociality in New and Old World sparrows. *J Evol Biol*. 2009;22(12):2376–2386.
37. Maynard-Smith JM, Harper DGC. The evolution of aggression: can selection generate variability? *Philos Trans R Soc Lond Ser B-Biol Sci*. 1988;319(1196):557–570.
38. Vedder O, Schut E, Magrath MJL, Komdeur J. Ultraviolet crown colouration affects contest outcomes among male blue tits, but only in the absence of prior encounters. *Funct Ecol*. 2010;24(2): 417–425.
39. Reeve HK, Keller L. Tests of reproductive-skew models in social insects. *Annu Rev Entomol*. 2001;46:347–385.
40. Nonacs P, Hager R. The past, present and future of reproductive skew theory and experiments. *Biol Rev*. 2011;86(2):271–298.
41. Insley SJ. Long-term vocal recognition in the northern fur seal. *Nature*. 2000;406(6794):404–405.
42. Godard R. Long-term memory of individual neighbours in a migratory songbird. *Nature*. 1991;350(6315):228–229.
43. Gherardi F, Atema J. Memory of social partners in hermit crab dominance. *Ethology*. 2005;111(3):271–285.
44. Karavanich C, Atema J. Individual recognition and memory in lobster dominance. *Anim Behav*. 1998;56:1553–1560.

45. Dreier S, van Zweden JS, D'Ettorre P. Long-term memory of individual identity in ant queens. *Biol Lett.* 2007;3(5):459–462.
46. Sheehan MJ, Tibbetts EA. Robust long-term social memories in a paper wasp. *Curr Biol.* 2008;18:R851–R852.
47. Chittka L. Sensorimotor learning in bumblebees: long-term retention and reversal training. *J Exp Biol.* 1998;201(4):515–524.
48. Burns JG, Foucaud J, Mery F. Costs of memory: lessons from 'mini' brains. *Proc R Soc B-Biol Sci.* 2011;278(1707):923–929.
49. Menzel R, Giurfa M. Dimensions of cognition in an insect, the honeybee. *Behav Cogn Neurosci Rev.* 2006;5:24–40.
50. Avargues-Weber A, de Brito Sanchez MG, Giurfa M, Dyer AG. Aversive reinforcement improves visual discrimination learning in free-flying honeybees. *PLoS ONE.* 2010;5:10.
51. de Souza AR, Rodrigues IL, Rocha IVA, Reis WAA, Lopes JFS, Prezoto F. Foraging behavior and dominance hierarchy in colonies of the neotropical social wasp *Polistes ferreri* (Hymenoptera, Vespidae) in different stages of development. *Sociobiology.* 2008;52(2):293–303.
52. Moret Y, Schmid-Hempel P. Survival for immunity: the price of immune system activation for bumblebee workers. *Science.* 2000;290(5494):1166–1168.
53. Tibbetts EA, Banan M. Advertised quality, caste and food availability influence the survival cost of juvenile hormone in paper wasps. *Proc R Soc B-Biol Sci.* 2010;277(1699):3461–3467.
54. Abramson CI, Morris AW, Michaluk LM, Squire J. Antistatic foam as a shocking surface for behavioral studies with honey bees (Hymenoptera: Apidae) and American cockroaches (Orthoptera: Blattelidae). *J Entomol Sci.* 2004;39(4):562–566.
55. Sheehan MJ, Tibbetts EA. Specialized face learning is associated with individual recognition in paper wasps. *Science.* 2011;334 (6060):1272–1275.
56. Cheng K, Wignall AE. Honeybees (*Apis mellifera*) holding on to memories: response competition causes retroactive interference effects. *Anim Cogn.* 2006;9:141–150.
57. Zhang SW, Schwarz S, Pahl M, Zhu H, Tautz J. Honeybee memory: a honeybee knows what to do and when. *J Exp Biol.* 2006;209(22):4420–4428.
58. Healy SD, de Kort SR, Clayton NS. The hippocampus, spatial memory and food hoarding: a puzzle revisited. *Trends Ecol Evol.* 2005;20(1):17–22.
59. Macphail EM, Bolhuis JJ. The evolution of intelligence: adaptive specializations versus general process. *Biol Rev.* 2001;76(3): 341–364.
60. Bolhuis JJ, Macphail EM. A critique of the neuroecology of learning and memory. *Trends Cogn Sci.* 2001;5(10):426–433.
61. Shettleworth SJ. Modularity and the evolution of cognition. In: Heyes C, Huber L, eds. *The Evolution of Cognition.* London: Bradford; 2000:42–60.
62. Shettleworth SJ. Clever animals and killjoy explanations in comparative psychology. *Trends Cogn Sci.* 2010;14(11):477–481.
63. Sherry DF, Jacobs LF, Gaulin SJC. Spatial memory and adaptive specialization of the hippocampus. *Trends Neurosci.* 1992;15 (8):298–303.
64. Papini MR. Pattern and process in the evolution of learning. *Psychol Rev.* 2002;109(1):186–201.
65. Menzel R, Fischer J. *Animal Thinking: Contemporary Issues in Comparative Cognition.* Cambridge, MA: MIT Press; 2011.
66. Tsao DY, Freiwald WA, Knutsen TA, Mandeville JB, Tootell RBH. Faces and objects in macaque cerebral cortex. *Nat Neurosci.* 2003;6(9):989–995.
67. Kanwisher N. Domain specificity in face perception. *Nat Neurosci.* 2000;3(8):759–763.
68. Gauthier I, Bukach C. Should we reject the expertise hypothesis? *Cognition.* 2007;103(2):322–330.
69. McKone E, Kanwisher N, Duchaine BC. Can generic expertise explain special processing for faces? *Trends Cogn Sci.* 2007;11 (1):8–15.
70. Srinivasan MV. Honey bees as a model for vision, perception, and cognition. *Annu Rev Entomol.* 2010;55:267–284.
71. Horridge A. What the honeybee sees: a review of the recognition system of *Apis mellifera*. *Physiol Entomol.* 2005;30(1):2–13.
72. Brown SD, Dooling RJ. Perception of conspecific faces by Budgerigars (*Melopsittacus undulatus*): 2. Synthetic models. *J Comp Psychol.* 1993;107(1):48–60.
73. Tanaka JW, Farah MJ. Parts and wholes in face recognition. *Q J Exp Psychol Sect A-Hum Exp Psychol.* 1993;46(2):225–245.
74. Kanwisher N, Tong F, Nakayama K. The effect of face inversion on the human fusiform face area. *Cognition.* 1998;68(1):B1–B11.
75. Stamp NE. Effects of prey quantity and quality on predatory wasps. *Ecol Entomol.* 2001;26(3):292–301.
76. Weiss MR, Wilson EE, Castellanos I. Predatory wasps learn to overcome the shelter defences of their larval prey. *Anim Behav.* 2004;68:45–54.
77. Ugolini A, Cannicci S. Homing in paper-wasps. In: Turillazzi S, West-Eberhard MJ, eds. *Natural History and Evolution of Paper-Wasps.* New York: Oxford University Press; 1996:126–143.
78. Arevalo E, Zhu Y, Carpenter JM, Strassmann JE. The phylogeny of the social wasp subfamily Polistinae: evidence from microsatellite flanking sequences, mitochondrial COI sequence, and morphological characters. *Bmc Evol Biol.* 2004; 4:8.
79. Pickett KM, Carpenter JM, Wheeler WC. Systematics of *Polistes* (Hymenoptera : Vespidae), with a phylogenetic consideration of Hamilton's haplodiploidy hypothesis. *Ann Zoologici Fennici.* 2006;43(5–6):390–406.
80. Dufour V, Pascalis O, Petit O. Face processing limitation to own species in primates: a comparative study in brown capuchins, Tonkean macaques and humans. *Behav Processes.* 2006;73 (1):107–113.
81. Tsao DY, Livingstone MS. Mechanisms of face perception. *Annu Rev Neurosci.* 2008;31:411–437.
82. Shettleworth SJ. *Cognition, Evolution, and Behavior.* New York: Oxford University Press; 1998.
83. Menzel R, Giurfa M. Cognitive architecture of a mini-brain: the honeybee. *Trends Cogn Sci.* 2001;5(2):62–71.

Index

Note: Page numbers followed by "*f*" and "*t*" refers to figures and tables, respectively.

B

C

M

Printed and bound by CPI Group (UK) Ltd, Croydon, CR0 4YY
08/05/2025
01865035-0001